Essential
Cell Biology

Essential
Cell Biology

An Introduction to the Molecular Biology of the Cell

Bruce Alberts
Dennis Bray
Alexander Johnson
Julian Lewis
Martin Raff
Keith Roberts
Peter Walter

Garland Publishing, Inc.
New York & London

Editorial Director: Miranda Robertson
Text Editors: Eleanor Lawrence and Valerie Neal
Managing Editor: Anne Vinnicombe
Illustrator: Nigel Orme
Molecular Model Drawings: Kate Hesketh-Moore
Director of Electronic Publishing: John M-Roblin
Computer Specialists: John Shea and Chuck Bartelt
Copy Editor: Douglas Goertzen
Project Coordinator: Emma Hunt
Editorial Assistants: Nasreen Arain and Matthew Day
Production Coordinator: Perry Bessas
Indexer: Maija Hinkle

Bruce Alberts received his Ph.D. from Harvard University and is currently
President of the National Academy of Sciences in Washington, D.C., and
Professor of Biochemistry and Biophysics at the University of California,
San Francisco. *Dennis Bray* received his Ph.D. from the Massachusetts
Institute of Technology and is currently a Medical Research Council Fellow
in the Department of Zoology, University of Cambridge. *Alexander Johnson*
received his Ph.D. from Harvard University and is currently Professor of
Microbiology and Immunology at the University of California, San
Francisco. *Julian Lewis* received his D.Phil. from the University of Oxford
and is currently a Senior Scientist in the Imperial Cancer Research Fund.
Martin Raff received his M.D. from McGill University and is currently a
Professor in the MRC Laboratory for Molecular Cell Biology and the
Biology Department, University College London. *Keith Roberts* received his
Ph.D. from the University of Cambridge and is currently Head of the Cell
Biology Department at the John Innes Centre, Norwich. *Peter Walter*
received his Ph.D. from The Rockefeller University and is currently
Professor of Biochemistry and Biophysics and Director of the Cell Biology
Program at the University of California, San Francisco.

Library of Congress Cataloging-in-Publication Data
Essential cell biology : an introduction to the molecular
       biology of the cell / [Bruce] Alberts . . . [et al.].
       p.   cm.
    Includes bibliographical references and index.
    ISBN 0-8153-2045-0.—ISBN 0-8153-2971-7 (pbk.)
    1. Cytology. 2. Molecular biology. 3. Biochemistry.
    I. Alberts, Bruce.
QH581.2.E78   1997
571.6—DC21   97-17039
                                     CIP

Published by Garland Publishing, Inc.
717 Fifth Avenue, New York, NY 10022

Printed in the United States of America
15 14 13 12 11 10 9 8 7 6 5 4 3 2 1

Front cover: The photograph shows
mitosis in the early syncytial
Drosophila embryo. The metaphase
spindles are stained for microtubules
(*red*) and DNA (*blue*). (Courtesy of
Douglas R. Daily and William Sullivan,
University of California at Santa Cruz.)

Back cover: The authors, Bruce Alberts,
Dennis Bray, Alexander Johnson, Julian
Lewis, Peter Walter, Keith Roberts, and
Martin Raff (*clockwise from upper left*).
(Photograph by Craig Dawson, Media
Solutions, San Francisco.)

Preface

What does it take to be educated? The question of what core knowledge needs to be passed on to students has provoked strong arguments as long as educational institutions have existed. As we approach the twenty-first century, a basic understanding of the cell as the unit of living matter must surely become an integral part of that core. The revolutionary advances over the past 50 years in our understanding of how cells work are among the great triumphs of human discovery.

We can now explain the chemistry that makes life possible and allows us to move, think, talk, and experience the world around us. We have learned how to trace back the ancestry of each of the large molecules in our cells through other organisms that share them. We have gained a dramatically improved appreciation of who we are in relation to other living things. The new knowledge has also had many practical benefits, leading to biological discoveries of importance for our future health and prosperity. Genetic testing for health screening, genetic engineering of foods, the invention of new medicines, the use of DNA fingerprinting in court cases, and the balancing of environmental risks with benefits are but a few of the biology-based issues all of us now need to grapple with. The successful application of the new wealth of knowledge in the next century will require many difficult decisions by local citizens, who will need a basic understanding of cell biology to make them.

Our purpose in writing this book, then, is to provide a straightforward explanation of the workings of a living cell. By "workings," we mean principally the way in which the molecules of the cell—especially the proteins, the DNA, and the RNA—cooperate to create a system that feeds, moves, responds to stimuli, grows, and divides—one, in short, that is alive. By "straightforward," we mean an account that can be easily understood by first- or second-year undergraduates with little background in biology. The need for a short, clear account of the essentials of cell biology became apparent to us while we were writing *Molecular Biology of the Cell* (*MBoC*). *MBoC*, which is now in its third edition, is aimed at advanced undergraduates specializing in the life sciences or medicine, and it is clear that many students requiring a general account of cell biology find it too specialized and too heavy for their needs.

An initial attempt to write an abbreviated version of *MBoC* by simple pruning proved futile. We painfully learned that writing an introductory text requires a new approach and that the clay must be thrown again. The present book, then, is freshly written. We have retained the same stylistic and graphical features and the same emphasis on central concepts over facts as in *MBoC*. But the scope and level are very different. Here we focus on the properties that are common to most eucaryotic cells and that are necessary to an understanding of how any individual cell lives and reproduces itself. The organs and systems of multicellular organisms, the process of development, the myriad disorders that affect humans and which increasingly can be understood in cell biological terms have all been subordinated to our central theme.

The book has been designed for clarity. The text is as short and simple as we can make it, and we have reduced technical vocabulary to a minimum. In order to present the central ideas without distractions we have omitted the names of the scientists involved, as well as the personal stories that lie behind each scientific conclusion. The diagrams have been drawn to emphasize concepts and are stripped of unnecessary details. Key terms introduced in each chapter are highlighted when they first appear and are collected together at the end of the book in a large, illustrated glossary. We have not listed references for further reading; in a textbook at this level, we feel that this choice is best left to the individual teacher. Readers wishing to explore a subject in greater depth are encouraged to consult the extensive reading lists in *MBoC*.

A central feature of the book is the many questions that are presented in the text margins and at the end of each chapter. These are designed to provoke students to think about what they have read and to encourage them to pause to test their understanding. Many questions challenge the student to place the newly acquired information in a broader biological context, and some have more than one valid answer. Others invite speculation. Answers to all the questions are offered at the end of the book; in many cases these give a commentary or an alternative perspective on material in the main text.

As with *MBoC*, each chapter of this book is the product of a communal effort, with individual drafts circulating from one author to another. In addition we were helped by many people. A number of scientists gave advice on specific areas, including Raoul Andino, Elizabeth Blackburn, Christine Guthrie, Tim Hunt, Joachim Li, and Norman Pace. We received extensive comments from university teachers and students, which were collected and collated by Valerie Neal. The teachers were Jerry Brand, University of Texas at Austin; Heinz Gert de Couet, University of Hawaii; Michael A. Goldman, San Francisco State University; W. Michael Gray, Bob Jones University; Michael Lewitt, Anglia University; Herbert Lin, National Research Council; Linda Matsuuchi, University of British Columbia; Sheldon S. Shen, Iowa State University; and Jim Shinkle, Trinity University. The students were from Bob Jones University and San Francisco State University. Eleanor Lawrence edited each chapter, often several times, and taught us to write at what we hope is the appropriate level. We are enormously grateful to all of these individuals for their invaluable help.

The staff at Garland Publishing were ever helpful and encouraging. Ruth Adams saw long before anyone else the potential value of such a book. The illustrations, which were created by Keith Roberts, were constructed on the computer with skill and flair by Nigel Orme. The book was designed by John M-Roblin, copyedited by Douglas Goertzen, and produced by Perry Bessas. Donna Scholes led an enthusiastic team of market researchers who collected opinions and suggestions from many teachers. Anne Vinnicombe, the Production Chief, was a recent but vital addition to the Garland team, taking on our hectic schedules at short notice.

Most of the book was written in St. John's Wood, London, where we were supported by the staff at Garland's London office. We are grateful to Emma Hunt, Nasreen Arain, Sheila Archibald, and Matthew Day for their unflagging help. Miranda Robertson led this team and supervised every detail of the writing process—from the editing of text, to the marshaling of secretaries, stationery supplies, and food. As in the past, we were nourished in style by Emily Preece Foden. Perhaps some day she will put down her Sabatier and write the "The *MBoC* Cookbook." But not yet, Emily, not yet.

Finally, we are indebted to Libby Borden, President of Garland Publishing. Six years after the mantle of authority was thrust upon her by the untimely death of Gavin Borden, Libby runs the company with humanity and style. Her encouragement and support have never faltered. Without her there would be no book.

Despite our best efforts, it is inevitable that there will be errors in the book. We encourage readers who find them to let us know, so that we can correct them in the next printing (e-mail: ecb@garland.com; fax: 212-308-9399).

Contents in Brief

Special Features

List of Topics

Chapter 3   Energy, Catalysis, and Biosynthesis

Chapter 4   How Cells Obtain Energy from Food

Chapter 5   Protein Structure and Function

Chapter 6   DNA

Chapter 7   From DNA to Protein

Chapter 8   Chromosomes and Gene Regulation

Chapter 9   Genetic Variation

Chapter 10 DNA Technology

Chapter 11 Membrane Structure

Chapter 12 Membrane Transport

Chapter 13 Energy Generation in Mitochondria and Chloroplasts

Chapter 14 Intracellular Compartments and Transport

Chapter 15 Cell Communication

Chapter 16 Cytoskeleton

Chapter 17 Cell Division

Chapter 18 Cell-Cycle Control and Cell Death

Chapter 19 **Tissues**

Essential
Cell Biology

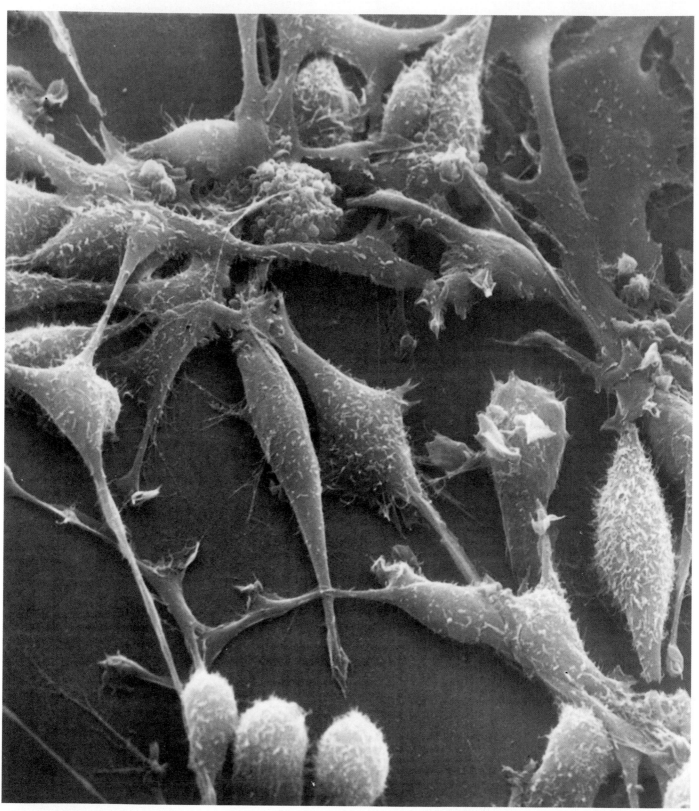

As long as they are supplied with an appropriate mixture of nutrients and growth factors, cells taken from multicellular organisms can often be induced to survive and proliferate in culture. This scanning electron micrograph shows a group of rat cells growing on the plastic surface of a culture dish. The rounded cells at the bottom are dividing. The ability to grow cells in culture in this way makes cells accessible for study in ways that are not possible in intact tissues. (Courtesy of Guenter Albrecht-Buehler.)

1 Introduction to Cells

All living creatures are made of **cells**—small membrane-bounded units filled with a concentrated aqueous solution of chemicals and endowed with the extraordinary ability to create copies of themselves by growing and dividing in two. The simplest forms of life are solitary cells. Higher organisms, such as ourselves, are communities of cells derived by growth and division from a single founder cell—vast cellular cities whose individual inhabitants perform specialized functions and are coordinated by intricate communication systems.

While cells can be components of larger living things, nothing less than a cell can truly be called living. Viruses, for example, contain some of the same types of molecules as cells but have no ability to reproduce themselves by their own efforts; they get themselves copied only by parasitizing the reproductive machinery of cells that they invade. Cells, therefore, are the fundamental units of life, and it is to *cell biology* that we must turn for an answer to the question of what life is and how it works.

In this chapter we consider first how cells can be viewed and what we see when we peer inside them. We shall then look at the great variety of forms that cells can show, and also take a first glimpse at the chemical machinery that all cells have in common.

Cells Under the Microscope

Cells are small, and the first practical problem in cell biology is how to see them. The invention of the **microscope** in the seventeenth century made cells visible for the first time, and for hundreds of years afterward all that was known about cells was discovered using this simple device. *Light microscopes* are still central pieces of equipment for cell biologists. But although these instruments now incorporate many sophisticated improvements, the properties of light itself set a limit to the fineness of detail they can reveal. *Electron microscopes*, invented in the 1930s, go beyond this limit by using beams of electrons instead of beams of light as the source of illumination, greatly extending our ability to see the fine details of cells and even making some of the larger molecules visible. A survey of the principal types of microscopy used to examine cells is given in Panel 1–1 (pp. 4–5).

The Invention of the Light Microscope Led to the Discovery of Cells

The development of the light microscope depended on advances in the production of glass lenses. In the seventeenth century, lenses were refined to the point that they could be used to make simple microscopes, and in 1665 Robert Hooke reported to the Royal Society of London that he had examined a piece of cork and found it to be composed of a mass of minute chambers, which he called "cells." The name "cell" stuck, even though the structures Hooke described were only the *cell walls* that remained after the living cells inside them had died. Later, Hooke and some of his contemporaries were able to see living cells.

For almost 200 years, the light microscope remained an exotic instrument, available only to a few wealthy individuals. It was not until the nineteenth century that it began to be widely used to look at living cells. The emergence of cell biology as a distinct science was a gradual process to which many individuals contributed, but its official birth is generally said to be signaled by two publications: one by the botanist Matthias Schleiden in 1838 and the other by the zoologist Theodor Schwann in 1839. In these papers, the authors documented the results of a systematic investigation of plant and animal tissues with the light microscope, showing that cells were the universal building blocks of all living tissues. Their work and that of other nineteenth-century microscopists, as in the example illustrated in Figure 1–1, slowly led to the realization that all living cells are formed by the division of existing

Figure 1–1 Early images of cells. (A) Drawings of a living plant cell (a hair cell from a *Tradescantia* flower), observed dividing into two daughter cells over a period of 2.5 hours by Eduard Strasburger in 1880. (B) A comparable living cell photographed recently through a modern light microscope. (B, courtesy of Peter Hepler.)

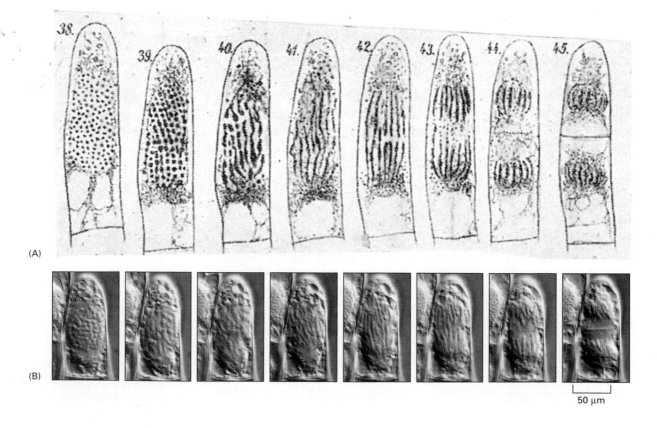

(A)

(B)

50 µm

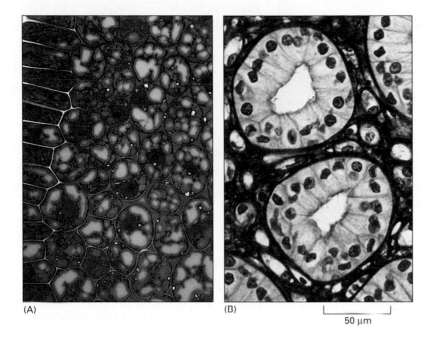

(A)　(B)

50 μm

Figure 1–2 **Cells in plant and animal tissues.** (A) Cells in a plant root tip, each surrounded by an extracellular matrix (cell wall), stained *red*. (B) Cells in the urine-collecting ducts of the kidney. Each duct is made of closely packed cells (with nuclei stained *red*) that form a ring. The ring is surrounded by extracellular matrix, stained *purple*. (B, from P.R. Wheater et al., Functional Histology, 2nd ed., 1987, Churchill Livingstone.)

cells—a doctrine sometimes referred to as the *cell theory*. The implication that living organisms do not arise spontaneously but can be generated only from existing organisms was hotly contested, but it was finally confirmed by experiments performed in the 1860s by Louis Pasteur.

The principle that cells are generated only from preexisting cells and inherit their characteristics from them underlies all of biology and gives the subject a unique flavor: in this branch of science, questions about the present are inescapably linked to questions about the past. To understand why present-day cells and organisms behave as they do, we need to understand their history, all the way back through the generations to the misty origins of the first cells on earth. Darwin's theory of evolution, published in 1859, provided the key insight that makes this history comprehensible, by showing how random variation and natural selection can drive the production of organisms with novel features. As we shall discuss later in the chapter, the theory of evolution explains how diversity has arisen among organisms that share a common ancestry. When combined with the cell theory, it leads us to a view of all life, from its beginnings to the present day, as one vast family tree of individual cells. Although this book is primarily about how cells work now, we shall encounter the theme of evolution repeatedly.

Cells, Organelles, and Even Molecules Can Be Seen Under the Microscope

If you cut a very thin slice of a suitable plant or animal tissue and place it under a light microscope, you will see that the tissue is divided into thousands of small cells, which may be either closely packed or separated from one another by material known as *extracellular matrix* (Figure 1–2). Each cell is typically about 5–20 **micrometers,** or μm, in diameter (Figure 1–3). If you have taken care to keep your specimen under the right conditions, you will see that the cells show signs of life: particles move around inside them, and if you watch patiently, you may see a cell slowly change shape and divide into two (see Figure 1–1).

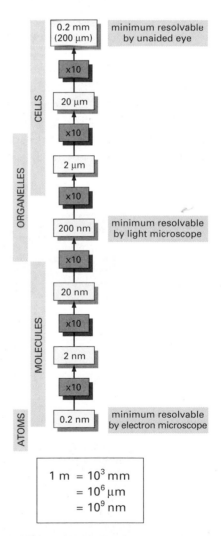

0.2 mm (200 μm)	minimum resolvable by unaided eye
x10	
20 μm	
x10	
2 μm	
x10	
200 nm	minimum resolvable by light microscope
x10	
20 nm	
x10	
2 nm	
x10	
0.2 nm	minimum resolvable by electron microscope

CELLS
ORGANELLES
MOLECULES
ATOMS

$$1 \text{ m} = 10^3 \text{ mm}$$
$$= 10^6 \text{ μm}$$
$$= 10^9 \text{ nm}$$

Figure 1–3 **The sizes of cells and of their component parts, and the units in which they are measured.**

THE LIGHT MICROSCOPE

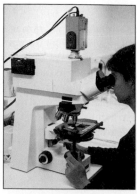

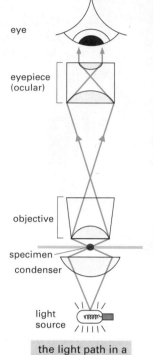

eye

eyepiece (ocular)

objective

specimen

condenser

light source

the light path in a light microscope

The light microscope allows us to magnify cells up to a thousand times, and to resolve details as small as 0.2 μm (a limitation imposed by the wavelike nature of light, not by the quality of the lenses). Three things are required for viewing cells in a light microscope. First, a bright light must be focused onto the specimen by lenses in the condenser. Second, the specimen must be carefully prepared to allow light to pass through it. Third, an appropriate set of lenses (objective and eyepiece) must be arranged to focus an image of the specimen in the eye.

FLUORESCENCE MICROSCOPY

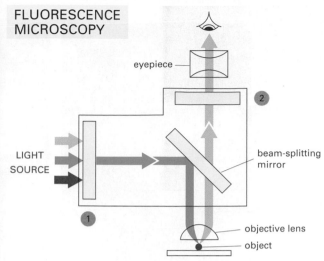

eyepiece

2

LIGHT SOURCE

beam-splitting mirror

objective lens

object

1

Fluorescent dyes used for staining cells are detected with the aid of a *fluorescence microscope*. This is similar to an ordinary light microscope except that the illuminating light is passed through two sets of filters. The first (1) filters the light before it reaches the specimen, passing only those wavelengths that excite the particular fluorescent dye. The second (2) blocks out this light and passes only those wavelengths emitted when the dye fluoresces. Dyed objects show up in bright color on a dark background.

THREE TYPES OF LIGHT MICROSCOPY

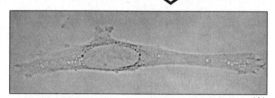

(A)

(B)

(C)

50 μm

Three images of the same unstained animal cell in culture. (A) is viewed with straightforward (bright-field) optics. More structure can be seen using more complex optical systems, such as *phase-contrast optics* (B) and *differential-interference-contrast optics* (C). These systems exploit differences in the way light travels through regions of the cell with differing refractive indexes. All three images can be obtained readily on the same microscope simply by interchanging optical components.

THIN SECTIONS

Most tissues are neither small enough nor transparent enough to examine directly in the microscope. Typically, therefore, they are chemically fixed and cut into very thin slices, or *sections*, that can be mounted on a glass microscope slide and subsequently stained to reveal different components of the cells. A stained section of a plant root tip is shown here.

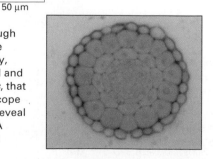

FLUORESCENT PROBES

Dividing cells seen with a fluorescence microscope after staining with specific fluorescent dyes.

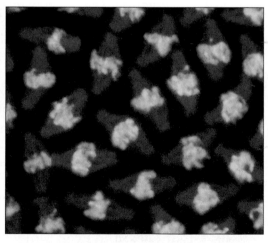

Fluorescent dyes absorb light at one wavelength and emit it at another, longer wavelength. Some such dyes bind specifically to particular molecules in cells and can reveal their location when examined with a fluorescence microscope. An example is the stain for DNA shown here (*green*). Other dyes can be coupled to antibody molecules, which then serve as highly specific and versatile staining reagents that bind selectively to particular macromolecules, allowing us to see their distribution in the cell. In the example shown, a microtubule protein in the mitotic spindle is stained *red* with a fluorescent antibody. (Courtesy of William Sullivan.)

SCANNING ELECTRON MICROSCOPY

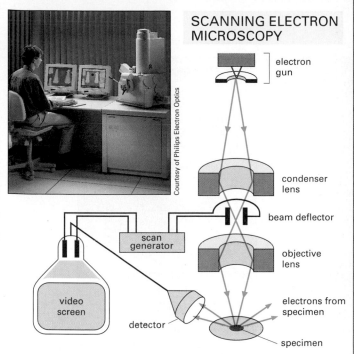

Courtesy of Philips Electron Optics

electron gun

condenser lens

beam deflector

scan generator

objective lens

video screen

detector

electrons from specimen

specimen

In the scanning electron microscope (SEM) the specimen, which has been coated with a very thin film of a heavy metal, is scanned by a beam of electrons brought to a focus on the specimen by the electromagnetic coils that, in electron microscopes, act as lenses. The quantity of electrons scattered or emitted as the beam bombards each successive point on the surface of the specimen is measured by the detector, and is used to control the intensity of successive points in an image built up on a video screen. The microscope creates striking images of three-dimensional objects with great depth of focus and a resolution between 3 nm and 20 nm, depending on the instrument.

5 µm

Scanning electron micrograph of the stereocilia projecting from a hair cell in the inner ear (*left*). For comparison, the same structure is shown by light microscopy, at the limit of its resolution (*above*). (Courtesy of Richard Jacobs and James Hudspeth.)

1 µm

TRANSMISSION ELECTRON MICROSCOPY

Courtesy of Philips Electron Optics

electron gun

condenser lens

specimen

objective lens

projector lens

viewing screen or photographic film

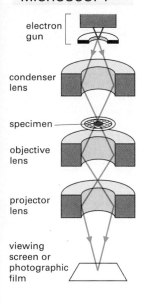

The electron micrograph below shows a small region of a cell in a piece of testis. The tissue has been chemically fixed, embedded in plastic, and cut into very thin sections that have then been stained with salts of uranium and lead, before being put into the vacuum of the microscope. (Courtesy of Daniel S. Friend.)

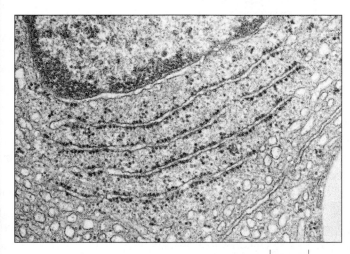

0.5 µm

The transmission electron microscope (TEM) is in principle similar to an inverted light microscope, but it uses a beam of electrons instead of a beam of light and magnetic coils to focus the beam instead of glass lenses. The specimen, which is placed in a vacuum, must be very thin. Contrast is usually introduced by electron-dense heavy-metal stains that locally absorb or scatter electrons, removing them from the beam as it passes through the specimen. The TEM has a useful magnification of up to a million-fold and a resolution, with biological specimens, of about 2 nm.

GOLD-LABELED PROBES

In the same way that fluorescent dyes are coupled to molecular probes to detect specific molecules in the light microscope, minute gold particles can be coupled to molecular probes to reveal the location of specific molecules in the electron microscope. Typically, antibodies of known specificity are attached to electron-dense colloidal gold spheres (5–20 nm in diameter) and used to label thin sections of cells or tissues, in a process called immunogold electron microscopy. The gold spheres appear in the electron microscope as tiny black dots. The example shown is a thin section of an insulin-secreting cell in the pancreas, in which a gold-labeled anti-insulin antibody has revealed the subcellular location of the insulin. (From L. Orci, *Diabetologia* 28:528–546, 1985.)

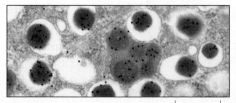

0.5 µm

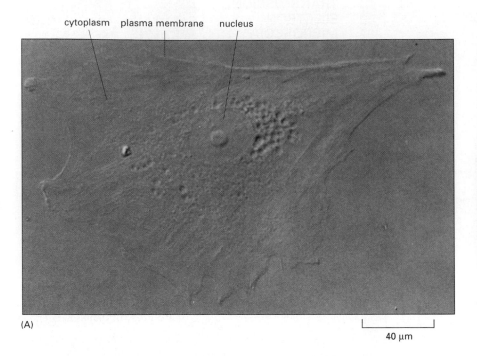

cytoplasm   plasma membrane   nucleus

(A)

⊢―――――⊣
40 μm

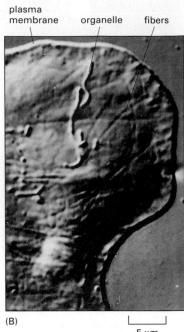

plasma
membrane   organelle   fibers

(B)

⊢―――⊣
5 μm

To see the internal structure of a cell is more difficult, not only because the parts are small but also because they are transparent and mostly colorless. One approach is to stain cells with dyes that color particular components differently (see Figure 1–2). Alternatively, one can exploit the fact that cell components differ slightly from one another in refractive index (just as glass differs from water), and these small differences can be made visible by optical tricks (see Panel 1–1, pp. 4–5). Today, the microscope image can be further enhanced by electronic processing.

The cell thus revealed has a distinct anatomy (Figure 1–4). It has a sharply defined boundary, suggesting the presence of an enclosing *membrane*. In the middle, a large round body, the *nucleus*, is prominent. Around the nucleus lies a transparent substance that fills the rest of the cell's interior, the **cytoplasm,** crammed with what seems at first to be a jumble of tiny miscellaneous objects. With a good light microscope, one can begin to distinguish specific components in the cytoplasm (Figure 1–4B) and to classify them, but details smaller than about 0.2 μm—about half the wavelength of visible light—cannot be resolved.

For higher magnification one must turn to an electron microscope, which can reveal details down to a few **nanometers,** or nm (see Figure 1–3). This requires careful preparation. Even for light microscopy, tissue usually has to be *fixed* (that is, preserved by pickling in a reactive chemical solution), supported by *embedding* in a solid wax or resin, *sectioned* into thin slices, and *stained* before it is viewed. For electron microscopy, similar procedures are required, but the sections have to be much thinner and there is no possibility of looking at living, wet cells. But when the sections are cut, stained, and placed in the electron microscope, much of the jumble of cell components becomes resolved into distinct **organelles**—which can now be seen in far greater detail than with a light microscope. A delicate membrane, about 5 nm thick, is visible enclosing the cell, and similar membranes form the boundary of many of the organelles inside (Figure 1–5A and B). The external membrane is called

Figure 1–4 A living cell seen under a light microscope, showing nucleus and cytoplasm. (A) A cell taken from human skin and growing in tissue culture was photographed through a light microscope. Fibers and organelles can be distinguished in the cytoplasm. (B) Detail of part of a newt cell growing in culture. The video image, at high magnification, has been computer enhanced, and numerous organelles and fibers can be seen. (A, courtesy of Casey Cunningham; B, courtesy of Lynne Cassimeris.)

Figure 1–5 (*opposite page*) **Fine structure of a cell seen in a transmission electron microscope.** (A) Thin section showing the enormous amount of detail that is visible. Some of the components to be discussed later in the chapter are labeled, identified by their size and shape. (B) A small region of the cell cytoplasm at somewhat higher magnification. The smallest structures that are clearly visible are the ribosomes, each of which is made of 80–90 or so individual large molecules. (C) A preparation of DNA isolated from a cell, showing that individual large molecules can be seen with the electron microscope. (A and B, courtesy of Daniel S. Friend; C, courtesy of Mei Lie Wong.)

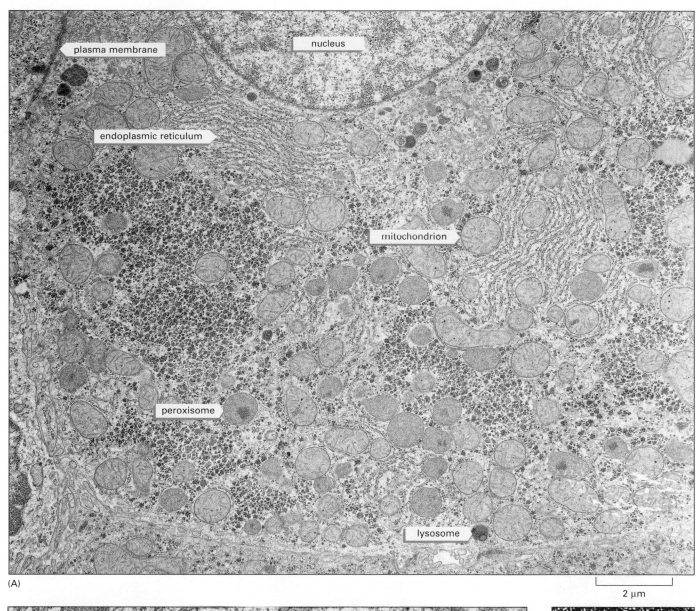

(A)

2 μm

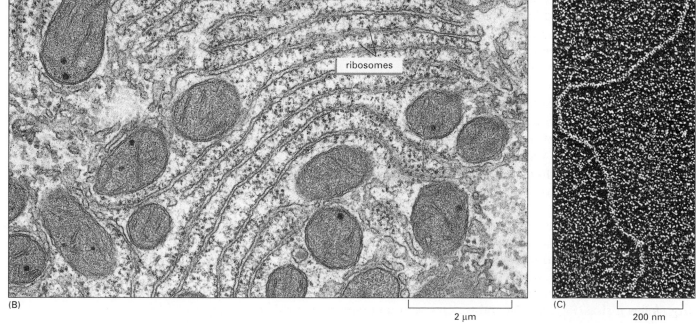

(B)

ribosomes

2 μm

(C)

200 nm

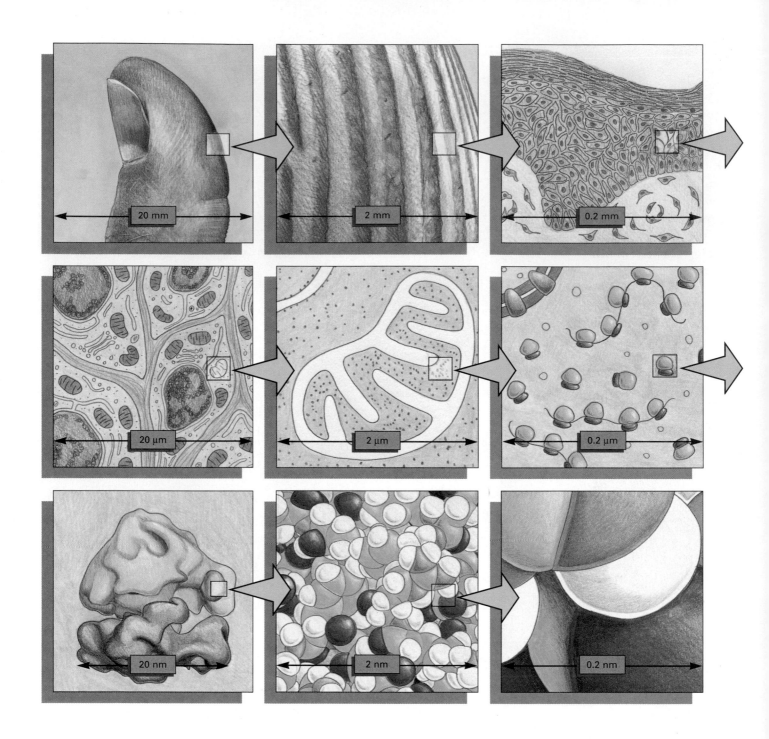

the *plasma membrane*, while the membranes surrounding organelles are called *internal membranes*. With an electron microscope, even some of the individual large molecules in a cell can be seen (Figure 1–5C). Electron microscopes that are used to look at thin sections of tissue are known as *transmission electron microscopes*. Another type of electron microscopy—*scanning electron microscopy*—is used to look at the surface detail of cells and other structures (see Panel 1–1, pp. 4–5). Even with the most powerful electron microscopes, however, one cannot routinely see the individual atoms that make up molecules (Figure 1–6).

Figure 1–6 A sense of scale between living cells and atoms. Each diagram shows an image magnified by a factor of ten in an imaginary progression from a thumb, through skin cells, to a ribosome, to a cluster of atoms forming part of one of the many protein molecules in our bodies. Details of molecular structure, as shown in the last two panels, are beyond the power of the electron microscope.

The Eucaryotic Cell

Cells do not all contain the same organelles in the same proportions. Indeed, one vast and evolutionarily ancient class of cells, the *bacteria*, contain essentially no organelles—not even a nucleus. The presence or absence of a nucleus is used as the basis for a simple but fundamental classification of all living organisms. Those whose cells have a nucleus are called **eucaryotes** (from the Greek words *eu*, meaning "well" or "truly," and *karyon*, a "kernel" or "nucleus"). Those whose cells do not have a nucleus are called **procaryotes** (from *pro*, meaning "before"). The terms "bacteria" and "procaryotes" are generally used interchangeably. Most procaryotes live as single-celled organisms, although some form chains, clusters, or other organized multicellular structures. All the more complex multicellular organisms, including plants, animals, and fungi, are formed from eucaryotic cells; many single-celled organisms, ranging from yeasts to amoebas, are also eucaryotic (Figure 1–7). Possession of a nucleus goes hand-in-hand with possession of a variety of other organelles, and we shall now take a look at the main organelles found in eucaryotic cells.

The Nucleus Is the Information Store of the Cell

The **nucleus** is usually the most prominent organelle in a eucaryotic cell (Figure 1–8). It is enclosed within two concentric membranes that form

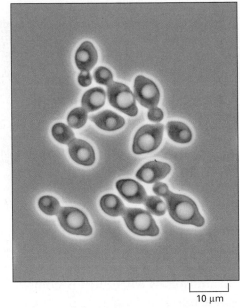

10 μm

Figure 1–7 Yeast cells, as seen with the light microscope. The species shown here are the cells that make dough rise and turn grape juice into wine. They reproduce by forming a bud and then dividing asymmetrically into a large and a small daughter cell.

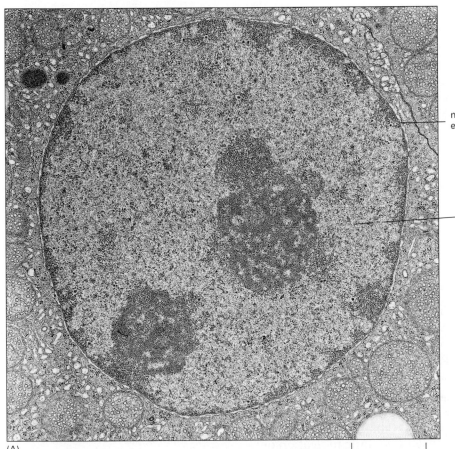

nuclear envelope

(B)

nucleus

(A)

2 μm

Figure 1–8 The nucleus. (A) This organelle, containing most of the DNA of the eucaryotic cell, is seen here in a thin section of a mammalian cell examined in the electron microscope. Individual chromosomes are not visible because the DNA is dispersed as fine threads throughout the nucleus at this stage of the cell's growth. (B) In this schematic diagram of a typical animal cell, illustrating its extensive system of membrane-bounded organelles, the nucleus is colored *brown*, the nuclear envelope is *green*, and the cytoplasm (the interior of the cell outside the nucleus) is *white*. (A, courtesy of Daniel S. Friend.)

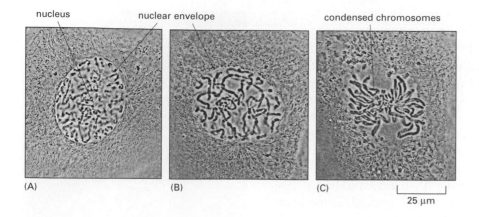

nucleus nuclear envelope condensed chromosomes

(A) (B) (C)

25 μm

Figure 1–9 Chromosomes in a cell that is about to divide. As a cell prepares to divide, its DNA condenses into threadlike chromosomes that can be distinguished in the light microscope. The photographs show three successive steps in this process in a cultured cell from a newt's lung. (Courtesy of Conly L. Rieder.)

the *nuclear envelope,* and it contains molecules of *DNA*—extremely long polymers that encode the genetic specification of the organism. These giant molecules become individually visible in the light microscope as **chromosomes** when they become more compact as a cell prepares to divide into two daughter cells (Figure 1–9). DNA also acts as the genetic information store in procaryotic cells; these cells, however, do not keep the DNA enclosed in a nuclear envelope, and so no distinct nucleus is seen in them.

Long before DNA itself was identified as the genetic material, careful observation using the light microscope had already suggested that the chromosomes were the carriers of genetic information. All the cells in all the individuals of the same sex of a given species were seen to have identical sets of chromosomes—8 chromosomes in a fly, 46 in a human, 104 in a carp, for example—which differed visibly in their shapes and sizes from those of other species. Moreover, when a cell divided into two daughter cells, or when an egg and a sperm cell were generated and fused to form a new individual by sexual reproduction, it was seen that the allocation of chromosomes corresponded exactly to the rules discovered in the nineteenth century by Gregor Mendel for the transmission of inherited characteristics. For example, the egg and the sperm each contribute half of the chromosomes to the new organism, reflecting the equal contributions of mother and father to the inheritance of traits by their offspring.

Mitochondria Generate Energy from Food to Power the Cell

Among the most conspicuous organelles in the cytoplasm, **mitochondrion** are present in essentially all eucaryotic cells (Figure 1–10). They have a very distinctive structure when seen in the electron microscope: each mitochondrion is sausage- or worm-shaped; it is from one to many micrometers long; and it is enclosed in two separate membranes, the inner one being thrown into folds that project into the interior of the mitochondrion (Figure 1–11). Mitochondria contain their own DNA and reproduce by dividing in two, and because of their many resemblances to bacteria they are thought to derive from bacteria that were engulfed by some ancestor of present-day eucaryotic cells (Figure 1–12).

Observation under the microscope by itself gives little indication of what mitochondria do. Their function was discovered by breaking up cells and collecting mitochondria by spinning the soup of cell fragments

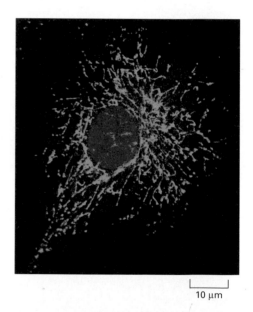

10 μm

Figure 1–10 Mitochondria, as seen with a light microscope. These organelles are power generators that oxidize food molecules to produce useful chemical energy in almost all eucaryotic cells. Mitochondria are quite variable in shape; in this cultured mammalian cell they are stained *green* with a fluorescent dye and appear wormlike. The nucleus is stained *blue.* (Courtesy of Lan Bo Chen.)

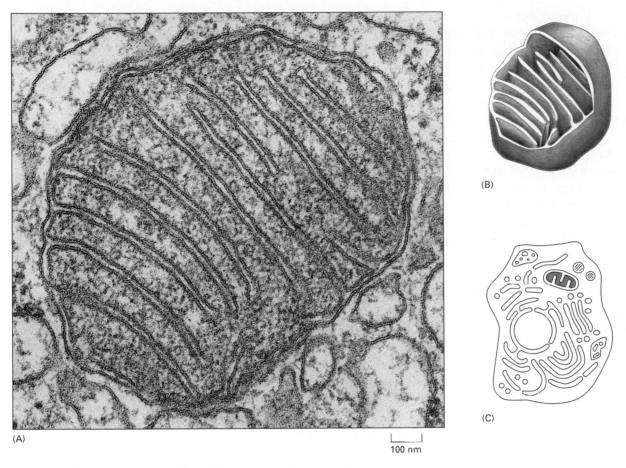

(B)

(C)

(A)

100 nm

Figure 1–11 A mitochondrion, as seen with an electron microscope.
(A) Cross section. (B) Three-dimensional representation of the
arrangement of the mitochondrial membranes. Note the smooth outer
membrane and the highly convoluted inner membrane. This inner
membrane contains most of the proteins responsible for cellular respiration, and it is highly folded to provide a large surface area for this
activity. (C) In the schematic cell, the interior space of the mitochondrion is colored. (A, courtesy of Daniel S. Friend.)

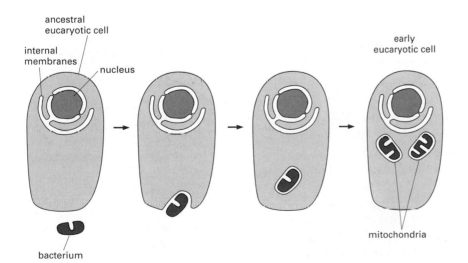

ancestral
eucaryotic cell

internal
membranes

nucleus

early
eucaryotic cell

bacterium

mitochondria

Figure 1–12 The origin of mitochondria.
It has been suggested that these organelles
originated as bacteria that were engulfed
by an ancestral eucaryotic cell and survived inside it, living in symbiosis with
their host.

in a centrifuge. Purified mitochondria were then tested to see what chemical processes they could perform. This revealed that mitochondria are generators of chemical energy for the cell. They harness the energy from the oxidation of food molecules, such as sugars, to produce the basic chemical fuel *adenosine triphosphate,* or *ATP*—the gasoline, so to speak, that powers most of the cell's activities. Because the mitochondrion consumes oxygen and releases carbon dioxide in the course of this activity, the entire process is called *cellular respiration,* from its similarity to breathing. Without mitochondria, animals, fungi, and plants would be unable to use oxygen to extract the maximum amount of energy from the food molecules that nourish them. Organisms that can use oxygen in this way, including all plants and animals, are said to be *aerobic.* A few eucaryotic organisms (as well as many procaryotes) are unable to live in environments containing oxygen; they are said to be *anaerobic,* and they lack mitochondria.

Chloroplasts Capture Energy from Sunlight

Found only in cells of plants and algae, not in cells of animals or fungi, **chloroplasts** are large green organelles that have an even more complex structure than mitochondria: in addition to the two surrounding membranes, there are stacks of membrane containing the green pigment *chlorophyll* inside the organelle (Figure 1–13). When a plant is kept in the dark, its greenness fades; when put back in the light, its greenness returns. This suggests that the chlorophyll and the chloroplasts that contain it are crucial to the special relationship that plants and algae have with light. But what is that relationship?

Animals and plants both need energy to live, grow, and reproduce. Animals can use only chemical energy obtained by feeding on the products of other living things. But plants can get their energy directly from sunlight, and chloroplasts are the organelles that enable them to do so. From the standpoint of life on earth, chloroplasts carry out an even more essential task than mitochondria: they perform *photosynthesis*—that is,

Figure 1–13 Chloroplasts. These organelles capture the energy of sunlight in plant cells. (A) Leaf cells in a moss seen in a light microscope showing the green chloroplasts. (B) Electron micrograph of a chloroplast in a grass leaf showing its extensive system of internal membranes. The flattened sacs of membrane contain chlorophyll and are arranged in stacks. (C) Interpretative drawing of (B). (B, courtesy of Eldon Newcomb.)

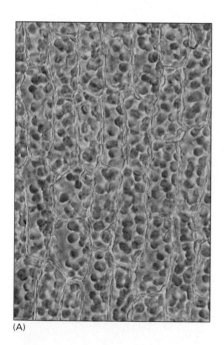

(A)

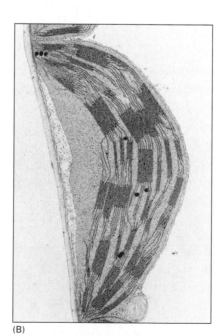

(B)

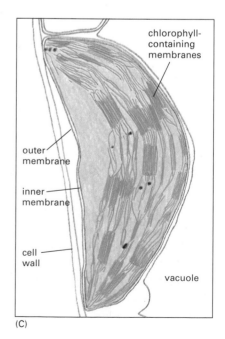

(C)

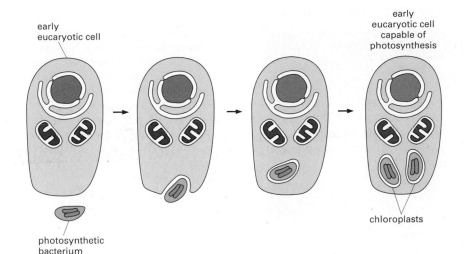

early
eucaryotic cell

early
eucaryotic cell
capable of
photosynthesis

chloroplasts

photosynthetic
bacterium

Figure 1–14 The origin of chloroplasts. Chloroplasts are thought to have originated as symbiotic photosynthetic bacteria, which were taken up by early eucaryotic cells that already contained mitochondria.

they trap the energy of sunlight in chlorophyll molecules and use this energy to drive the manufacture of energy-rich sugar molecules. In the process they release oxygen as a waste product. Plant cells, like animal cells, can then extract this stored chemical energy when required, by oxidizing these sugars using their mitochondria. The chloroplasts generate both the food molecules and the oxygen that all mitochondria use.

Like mitochondria, chloroplasts contain their own DNA, reproduce themselves by dividing in two, and are thought to have evolved from bacteria—in this case from photosynthetic bacteria that were somehow engulfed by an early eucaryotic cell (Figure 1–14). This evidently created a *symbiotic* relationship—one in which the host eucaryote and the engulfed bacterium helped one another to survive and reproduce together.

Internal Membranes Create Intracellular Compartments with Different Functions

Nuclei, mitochondria, and chloroplasts are not the only membrane-bounded organelles inside eucaryotic cells. The cytoplasm contains a profusion of other organelles, most of them enclosed by a single membrane, that perform many distinct functions. Most of these are concerned with the cell's need to import raw materials and to export manufactured substances and waste products. Some organelles are enormously enlarged in cells that are specialized for secretion of proteins; others are especially plentiful in cells specialized for digestion of foreign bodies; and so on.

An irregular maze of spaces enclosed by membrane, called the *endoplasmic reticulum (ER)* (Figure 1–15), is the site at which most cell membrane components, as well as materials destined for export from the cell, are made. Stacks of flattened membrane-bounded sacs constitute the *Golgi apparatus* (Figure 1–16), which receives and often modifies chemically the molecules made in the endoplasmic reticulum, and then directs them to the exterior of the cell or to various other locations. *Lysosomes* are small, irregularly shaped organelles in which intracellular digestion occurs, releasing nutrients from food particles and breaking down unwanted molecules for recycling or excretion. *Peroxisomes* are small, membrane-bounded vesicles that provide a contained environ-

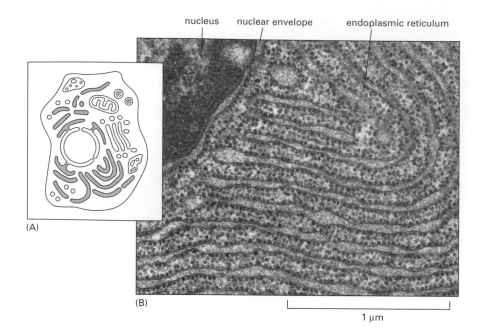

nucleus   nuclear envelope   endoplasmic reticulum

(A)

(B)

1 μm

Figure 1–15 The endoplasmic reticulum (ER). (A) Schematic diagram of an animal cell showing the endoplasmic reticulum in *green*. (B) Electron micrograph of a thin section of a mammalian cell showing the endoplasmic reticulum. Note that it is continuous with the membrane of the nuclear envelope. The black particles studding the particular region of the ER shown here are ribosomes—the molecular assemblies that perform protein synthesis. (B, courtesy of Lelio Orci.)

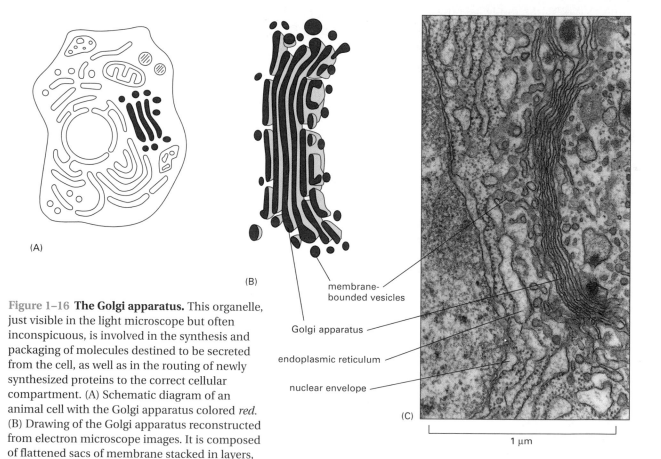

(A)

(B)

membrane-bounded vesicles

Golgi apparatus

endoplasmic reticulum

nuclear envelope

(C)

1 μm

Figure 1–16 The Golgi apparatus. This organelle, just visible in the light microscope but often inconspicuous, is involved in the synthesis and packaging of molecules destined to be secreted from the cell, as well as in the routing of newly synthesized proteins to the correct cellular compartment. (A) Schematic diagram of an animal cell with the Golgi apparatus colored *red*. (B) Drawing of the Golgi apparatus reconstructed from electron microscope images. It is composed of flattened sacs of membrane stacked in layers, from which small vesicles of membrane pinch off. Only one stack is shown here, but several can be present in each cell. (C) Electron micrograph of the Golgi apparatus from a typical animal cell. (C, courtesy of Brij J. Gupta.)

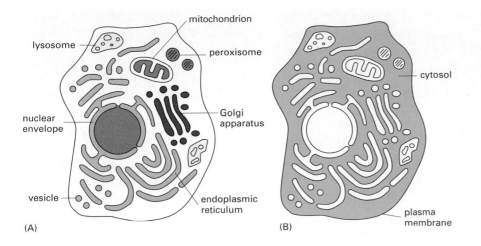

Figure 1–17 Internal membranes and the cytosol. (A) A variety of membrane-bounded compartments exist within eucaryotic cells, each specialized to perform a different function. (B) The rest of the cell, excluding all these organelles, is called the cytosol, and is the site of many vital cellular activities.

ment for reactions where a dangerously reactive chemical, hydrogen peroxide, is generated and degraded. Membranes also form many different types of small *vesicles* involved in the transport of materials between one membrane-bounded organelle and another. This whole system of related organelles is shown in Figure 1–17A.

A continual exchange of materials takes place between the endoplasmic reticulum, the Golgi apparatus, the lysosomes, and the outside of the cell. The exchange is mediated by small membrane-bounded vesicles that pinch off from the membrane of one organelle and fuse with another. At the surface of the cell, for example, portions of the plasma membrane invaginate (that is, tuck inward) and pinch off to form vesicles that carry into the cell material captured from the external medium (Figure 1–18). These generally fuse ultimately with lysosomes, where the captured material is digested. Very large particles, or even entire foreign cells, can be engulfed by animal cells in this way. The reverse process, whereby vesicles from inside the cell fuse with the plasma membrane and release their contents into the external medium, is also a common activity of cells (see Figure 1–18).

Question 1–2   Suggest a reason why it would be advantageous for eucaryotic cells to evolve elaborate internal membrane systems that allow them to import substances from the outside, as shown in Figure 1–18.

The Cytosol Is a Concentrated Aqueous Gel of Large and Small Molecules

If we were to strip the plasma membrane from a eucaryotic cell and then remove all of its membrane-enclosed organelles, including mitochondria and chloroplasts, we would be left with the **cytosol** (Figure 1–17B). In most cells this is by far the largest single compartment, and in bacteria it is the only intracellular compartment. It contains a host of large and small molecules, crowded together so closely that it behaves more like a water-based gel than a liquid solution (Figure 1–19). The cytosol is

Figure 1–18 Import and export. Cells can import materials from the external medium by capturing them in vesicles that pinch off from the plasma membrane. The vesicles ultimately fuse with lysosomes, where intracellular digestion occurs. By a converse process, cells export materials that they have synthesized in intracellular compartments: the materials are stored in the intracellular vesicles and released to the exterior when these vesicles fuse with the plasma membrane.

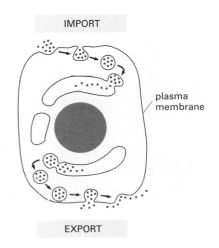

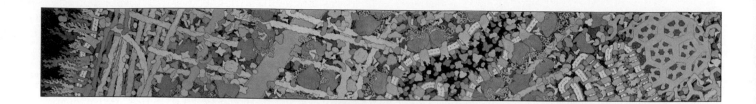

the site of many chemical reactions that are fundamental to the cell's existence. The early steps in the breakdown of nutrient molecules take place in the cytosol, for example, and it is here too that the cell performs one of its key synthetic processes—the manufacture of *proteins*. The tiny molecular machines that make protein molecules are visible with the electron microscope as small particles in the cytosol: they are called *ribosomes* (see Figures 1–5B and 1–15B).

The Cytoskeleton Is Responsible for Cell Movements

The cytosol is not just a structureless soup of chemicals, however. Under the electron microscope one can see that in eucaryotic cells (but not in bacteria) it is crisscrossed by long, fine filaments of protein. Frequently the filaments can be seen to be anchored at one end to the plasma membrane or to radiate out from a central site adjacent to the nucleus. This system of filaments is called the **cytoskeleton** (Figure 1–20). The thinnest of the filaments are *actin filaments*, which are present in all eucaryotic cells but occur in especially large numbers inside muscle cells, where they serve as part of the machinery that generates contractile forces. The thickest filaments are called *microtubules*, because they have the form of minute hollow tubes. They become reorganized into spectacular arrays in dividing cells, where they help pull the duplicated chromosomes in opposite directions and distribute them equally to the two daughter cells (Figure 1–21). Intermediate in thickness between actin filaments and microtubules are the *intermediate filaments*, which serve to strengthen the cell mechanically. These three types of filaments,

Figure 1–19 The cytosol. This schematic drawing, based on the known sizes and concentrations of molecules in the cytosol, shows how crowded it is. The panorama begins on the far left (*this page*) at the cell surface; moves through the endoplasmic reticulum, Golgi apparatus, and a mitochondrion; and ends on the far right (*next page*) in the nucleus. (Courtesy of D. Goodsell.)

Figure 1–20 The cytoskeleton. Filaments made of protein provide all eucaryotic cells with an internal framework that helps organize the internal activities of the cell and underlies its movements and changes of shape. Shown here are (A) actin filaments, (B) microtubules, and (C) intermediate filaments, in cultured cells. The different types of filaments are revealed with different fluorescent stains. (A, courtesy of Simon Barry and Chris D'Lacey; B, courtesy of Nancy Kedersha; C, courtesy of Clive Lloyd.)

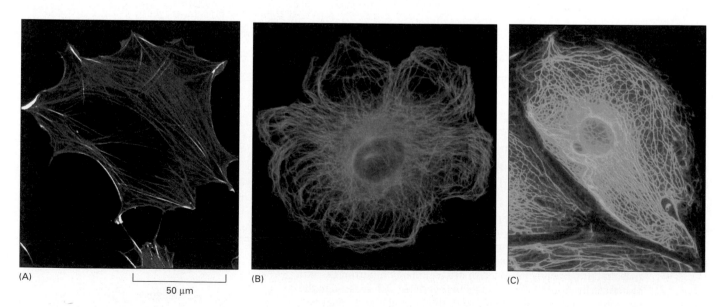

(A)

50 μm

(B)

(C)

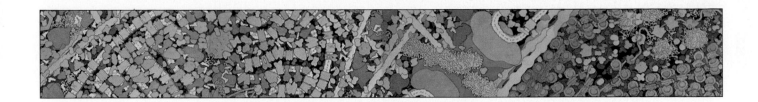

together with other proteins that attach to them, form a system of girders, ropes, and motors that gives the cell its mechanical strength, controls its shape, and drives and guides its movements. Because this apparatus governs the internal organization of the cell as well as its external features, it is as necessary to a plant cell, boxed in by a rigid wall of extracellular matrix, as it is to an animal cell that freely bends, stretches, swims, or crawls. In a plant cell, for example, organelles such as mitochondria are driven in a constant stream around the cell interior along cytoskeletal tracks, and animal cells and plant cells alike depend on the cytoskeleton to separate their internal components into two daughter sets during cell division.

Panel 1–2, on the next page, summarizes the differences between animal, plant, and bacterial cells, and Table 1–1, on page 19, lists some important dates in the discovery of cells and their organelles.

Question 1–3 Discuss the relative advantages and disadvantages of light and electron microscopy. How could you best visualize (a) a living skin cell, (b) a yeast mitochondrion, (c) a bacterium, and (d) a microtubule?

Unity and Diversity of Cells

Cell biologists often speak of "the cell" without specifying any particular cell, as though all cells were alike. But cells, as we have already noted, are not all alike; in fact, they can be wildly different. It is estimated that there are at least ten million distinct species of living things in the world—perhaps a hundred million, according to some estimates. Before delving deeper into cell biology, we must take stock: how much do the cells of these species have in common—the bacterium and the butterfly, the rose and the dolphin? In what ways do they vary?

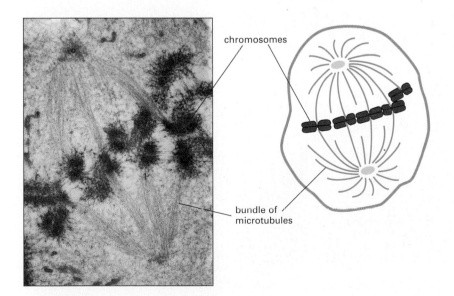

chromosomes

bundle of microtubules

Figure 1–21 Microtubules in a dividing cell. When a cell divides, its nuclear envelope breaks down and its DNA condenses into pairs of visible chromosomes, which are pulled apart into separate cells by microtubules. The microtubules radiate from foci at opposite ends of the dividing cell. (Photograph courtesy of Conly L. Rieder.)

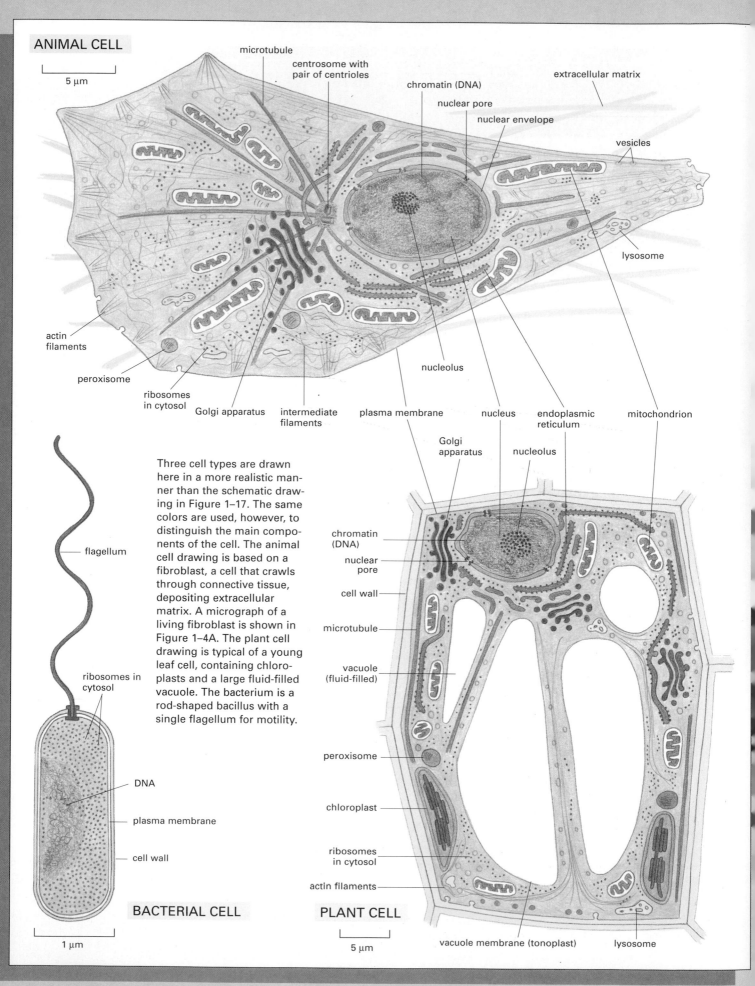

ANIMAL CELL

5 μm

microtubule

centrosome with
pair of centrioles

chromatin (DNA)

nuclear pore

nuclear envelope

extracellular matrix

vesicles

lysosome

actin
filaments

peroxisome

ribosomes
in cytosol

Golgi apparatus

intermediate
filaments

nucleolus

plasma membrane

nucleus

endoplasmic
reticulum

mitochondrion

Three cell types are drawn
here in a more realistic manner than the schematic drawing in Figure 1–17. The same
colors are used, however, to
distinguish the main components of the cell. The animal
cell drawing is based on a
fibroblast, a cell that crawls
through connective tissue,
depositing extracellular
matrix. A micrograph of a
living fibroblast is shown in
Figure 1–4A. The plant cell
drawing is typical of a young
leaf cell, containing chloroplasts and a large fluid-filled
vacuole. The bacterium is a
rod-shaped bacillus with a
single flagellum for motility.

flagellum

ribosomes in
cytosol

DNA

plasma membrane

cell wall

BACTERIAL CELL

1 μm

Golgi
apparatus

nucleolus

chromatin
(DNA)

nuclear
pore

cell wall

microtubule

vacuole
(fluid-filled)

peroxisome

chloroplast

ribosomes
in cytosol

actin filaments

PLANT CELL

5 μm

vacuole membrane (tonoplast)

lysosome

Panel 1–2 **Cells: the principal features of animal, plant, and bacterial cells.**

Table 1–1 Historical Landmarks in the Microscopy of Cells

1655 **Hooke** used a primitive microscope to describe small pores in sections of cork that he called "cells."

1674 **Leeuwenhoek** reported his discovery of protozoa. He saw bacteria for the first time nine years later.

1833 **Brown** published his microscopic observations of orchids, clearly describing the cell nucleus.

1838 **Schleiden** and **Schwann** proposed the cell theory, stating that the nucleated cell is the universal building block of plant and animal tissues.

1857 **Kölliker** described mitochondria in muscle cells.

1879 **Flemming** described with great clarity chromosome behavior during mitosis in animal cells.

1881 **Cajal** and other histologists developed staining methods that revealed the structure of nerve cells and the organization of neural tissue.

1898 **Golgi** first saw and described the Golgi apparatus by staining cells with silver nitrate.

1902 **Boveri** related chromosomes to heredity by observing their behavior during sexual reproduction.

1952 **Palade**, **Porter**, and **Sjöstrand** developed methods of electron microscopy that enabled many intracellular structures to be seen for the first time. In one of the first applications of these techniques, **Huxley** showed that muscle contains arrays of protein filaments—the first evidence of a cytoskeleton.

1957 **Robertson** described the bilayer structure of the cell membrane, seen for the first time in the electron microscope.

Cells Vary Enormously in Appearance and Function

Let us begin with size. A bacterial cell—say a *Lactobacillus* in a piece of cheese—is a few micrometers long. A frog's egg—which is also a single cell—has a diameter of about 1 millimeter. If we scaled them both up to make the *Lactobacillus* the size of a person, the frog's egg would be half a mile high.

Cells vary no less widely in their shape and in their movements. Consider the gallery of cells displayed in Figure 1–22. A typical nerve cell in your brain is enormously extended; it sends out its electrical signals along a fine cytoplasmic protrusion that is 10,000 times longer than it is thick, dividing into hundreds of branches at its far end. A *Paramecium* in a drop of pond water is shaped like a submarine and is covered with tens of thousands of *cilia*—hairlike extensions whose sinuous beating sweeps the cell forward, rotating as it goes. A cell in the surface layer of a plant is a squat immobile prism that forms around itself a rigid box of cellulose, with an outer waterproof coating of wax. A *Bdellovibrio* bacterium is a sausage-shaped torpedo shape driven forward by a rotating flagellum attached to its stern, which acts as a propeller. A neutrophil or a macrophage in the body of an animal crawls through tissues like an amoeba, constantly pouring itself into new shapes and engulfing debris, foreign microorganisms, and dead or dying cells.

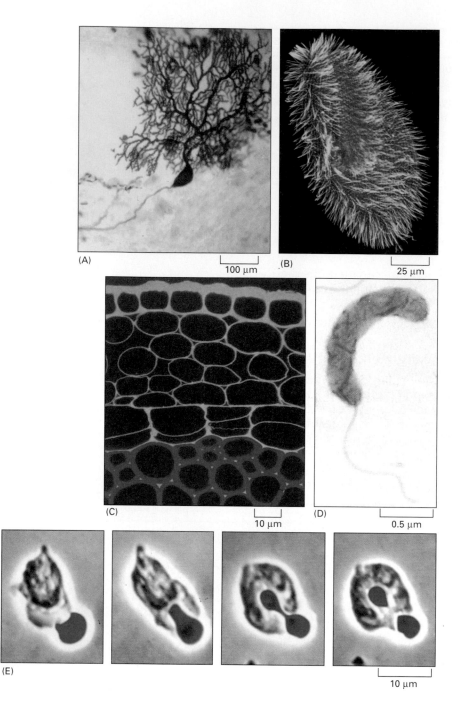

Figure 1–22 Some examples of the variety of cell shapes and sizes. (A) A nerve cell from the cerebellum (a part of the brain that controls movement). This cell has a huge branching tree of processes, through which it receives signals from as many as 100,000 other nerve cells. (B) *Paramecium*. This protozoan swims by means of the beating cilia that cover its surface. (C) A section of a young plant stem in which cellulose is stained *orange* and lignin *red.* The outermost layer of cells is at the top of the photo. (D) A tiny bacterium, *Bdellovibrio bacteriovorus*, with a single terminal flagellum. This bacterium attacks, kills, and feeds on other larger bacteria. (E) A human white blood cell (a neutrophil) approaching and engulfing a red blood cell. (A, courtesy of Constantino Sotelo; B, courtesy of Anne Fleury, Michel Laurent, and André Adoutte; D, courtesy of M.A. Stein and D.L. Diedrich; E, courtesy Stephen E. Malawista and Anne de Boisfleury Chevance.)

Some cells are clad only in a flimsy plasma membrane; others make themselves an outer coat of slime, build around themselves rigid cell walls, or surround themselves with a hard mineralized extracellular matrix such as that of bone.

Cells are also enormously diverse in their chemical requirements and activities. Some require oxygen to live; for others it is a deadly poison. Some consume little more than air, sunlight, and water as their raw materials; others need a complex mixture of molecules produced by other cells. Some appear to be specialized factories for the production of particular substances, such as hormones, starch, fat, latex, or pigments. Some are engines, like muscle, burning fuel to do mechanical work, or electricity generators, like the modified muscle cells in the electric eel.

Some specializations modify a cell so much that they spoil its chances of leaving any progeny. Such specialization would be senseless for a species of cell that lived a solitary life. In a multicellular organism, however, there is a division of labor between cells, allowing some cells to become specialized to an extreme degree for particular tasks and leaving them dependent on their fellow cells for many basic requirements. Even the most basic need of all, that of passing on the genetic instructions to the next generation, is delegated to specialists—the egg and the sperm.

Living Cells All Have a Similar Basic Chemistry

In spite of the extraordinary diversity of plants and animals, people have recognized from time immemorial that these organisms have something in common, something that entitles them all to be called living things. With the invention of the microscope, it became clear that plants and animals are assemblies of cells, that cells can also exist as independent organisms, and that cells individually are living. But while it was easy enough to recognize life, it remained remarkably difficult to say in what sense all living things were alike. Textbooks struggled to define life in abstract general terms.

The discoveries of biochemistry and molecular biology have made this problem evaporate in a most spectacular way. We now know that all living things on our planet are similar not just in the abstract sense that they all grow, reproduce, convert energy from one form into another, and so on, but to an astonishing degree in the details of their chemistry. In all of them, genetic instructions—*genes*—are stored in DNA molecules, written in the same chemical code, constructed out of the same chemical building blocks, interpreted by essentially the same chemical machinery, and duplicated in the same way to allow the organism to reproduce. The DNA directs the production of a huge variety of large protein molecules that dominate the behavior of the cell, serving as structural materials, chemical catalysts, molecular motors, and so on. In every living thing, the same 20 different types of chemical subunits are linked to make proteins. But the subunits are linked in different sequences, conferring different chemical properties on the protein molecules, just as different sequences of letters spell different words. In this way the same basic biochemical machinery has served to generate the whole gamut of living things (Figure 1–23).

All Present-Day Cells Have Apparently Evolved from the Same Ancestor

A cell reproduces by duplicating its DNA and then dividing in two, passing a copy of the genetic instructions encoded in the DNA to each of its daughter cells. That is why the daughters resemble the parent. The copying is not always perfect and the instructions are occasionally corrupted. That is why the daughters do not always match the parent exactly. *Mutations*—that is, changes in the DNA—can result in a daughter that is changed for the worse, in the sense that it is less able to survive and reproduce; or changed for the better, in that it is better able to survive and reproduce; or changed neutrally, being different, but equally viable. The struggle for survival eliminates the first, favors the second, and tol-

Figure 1–23 **Bacteria, a butterfly, a rose, and a dolphin.** All these living organisms are constructed from cells that have a fundamentally similar chemistry and operate according to the same basic principles. (A, courtesy of Tony Brain and Science Photo Library; B, courtesy of J.S. and E.J. Woolmer © Oxford Scientific Films; C, courtesy of the John Innes Foundation; D, courtesy of Jonathan Gordon, IFAW.)

(A)

(B)

(C)

(D)

erates the third. Whichever descendants survive in the next generation will inherit the altered instructions that brought them into being. In addition, the genetic cards can be shuffled and redealt through *sexual reproduction,* where two cells of the same species fuse, pooling their DNA and allowing the genetic instructions to be shared out in new combinations in the next generation.

These simple principles of change and selection, applied repeatedly over thousands of millions of cell generations, are the basis of **evolution,** the process in which living species become gradually modified and adapted to their environment in more and more sophisticated ways. Evolution offers a startling but compelling explanation of why present-day cells are so similar in their fundamentals: they have all inherited their genetic instructions from the same ultimate common ancestor. It is estimated that this ancestral cell existed between 3.5 billion and 3.8 billion years ago, and we must suppose that it contained a prototype of the universal machinery of all life on earth today. Through mutation, its descendants have gradually diverged to fill every corner of the earth with living things, exploiting the potential of the machinery in an endless variety of ways.

Question 1–4 Mutations are mistakes in the DNA that change the genetic plan from the previous generation. Envision a shoe factory. Would you expect mistakes (i.e., unintentional changes) in copying the shoe design to lead to improvements in the shoes produced? Explain your answer.

Bacteria Are the Smallest and Simplest Cells

Among present-day cells, **bacteria** have the simplest structure and come closest to showing us life stripped down to its essentials. They are typically spherical, rod-shaped, or spiral cells, a few micrometers long

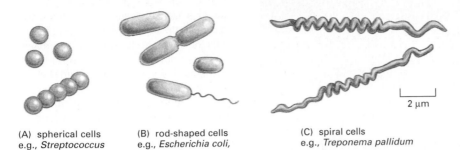

Figure 1–24 Bacterial shapes and sizes. Drawing showing typical spherical, rodlike, and spiral-shaped bacteria drawn to scale. The spiral cells shown are the causal agent of syphilis.

(A) spherical cells
e.g., *Streptococcus*

(B) rod-shaped cells
e.g., *Escherichia coli, Salmonella*

(C) spiral cells
e.g., *Treponema pallidum*

2 μm

(Figure 1–24). They often have a tough protective coat, called a cell wall, beneath which a plasma membrane encloses a single compartment containing the cytoplasm, including the DNA. In the electron microscope this cell interior typically appears as a matrix of varying texture without any obvious organized internal structure (Figure 1–25).

Bacteria typically are small and reproduce quickly by dividing in two. Under optimum conditions, when food is plentiful, a bacterium can duplicate itself in this way in as little as 20 minutes; in less than 11 hours, by repeated divisions, it can give rise to 5 billion progeny (approximately equal to the present total number of humans on earth). Because of their large numbers and rapid growth rates, populations of bacteria can evolve fast, rapidly acquiring the ability to use a new food source or to resist being killed by a new antibiotic.

In shape and structure bacteria may seem simple and limited, but in their chemistry they are the most diverse and inventive class of cells. Bacteria exploit an enormous range of habitats, from hot puddles of volcanic mud to the interiors of other living cells, and they vastly outnumber other living organisms. Some are aerobic, using oxygen to oxidize food molecules; some are anaerobic, shunning the oxygen-rich environments that would poison them. Since eucaryotic mitochondria are thought to have evolved from aerobic bacteria that took to living inside the anaerobic ancestors of today's eucaryotic cells (see Figure 1–12), our own aerobic metabolism can be seen as based ultimately on the activities of bacterial cells.

Virtually any organic material, from wood to petroleum, can be utilized as food by one sort of bacterium or another. Still more remarkable, there are bacteria that can live entirely on inorganic substances, getting their carbon from CO_2 in the atmosphere, their nitrogen from atmospheric N_2, and their oxygen, hydrogen, sulfur, and phosphorus from air, water, and inorganic minerals. Some of these bacterial cells, like plant

Note on biological names: Species of living organisms are officially identified by a pair of Latin words, usually printed in italics, analogous to a person's given name and family name. The genus (*Escherichia*, corresponding to a family name) is stated first; the second term (*coli*) qualifies this, identifying a particular species belonging to that genus. For short, the genus name may be abbreviated (*E. coli*), or the species label may be dropped (so that we often speak of the fly *Drosophila*, meaning *Drosophila melanogaster*).

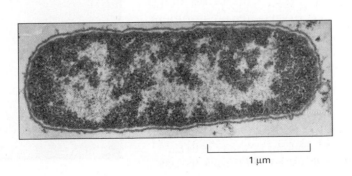

Figure 1–25 The bacterium *Escherichia coli* (*E. coli*). An electron micrograph of a longitudinal section is shown here; the cell's DNA is concentrated in the lightly stained region. (Courtesy of E. Kellenberger.)

1 μm

Unity and Diversity of Cells

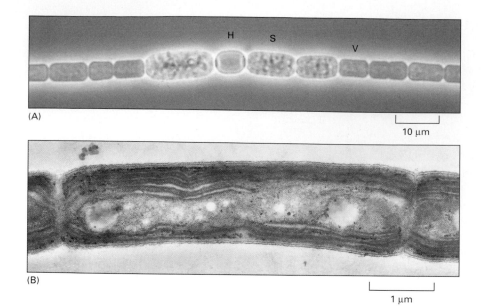

(A)

10 µm

(B)

1 µm

Figure 1–26 Two types of photosynthetic bacteria. (A) *Anabaena cylindrica* viewed in the light microscope. These procaryotic cells form long multicellular filaments, in which specialized cells (labeled *H*) fix nitrogen (that is, capture N_2 from the atmosphere and incorporate it into organic compounds), while the other cells are photosynthetic and fix CO_2 (*V*), or become resistant spores (*S*). (B) Electron micrograph of *Phormidium laminosum*, showing its intracellular membranes where photosynthesis occurs. (A, courtesy of Dave G. Adams; B, courtesy of D.P. Hill and C.J. Howe.)

cells, are photosynthetic, getting the energy they need for biosynthesis from sunlight (Figure 1–26); others derive it from the chemical reactivity of the inorganic substances in the environment (Figure 1–27). In either case, such bacteria play a unique and fundamental part in the economy of life on earth: other living things depend on the organic compounds that these cells generate from inorganic materials.

Plants are a partial exception: they can capture energy from sunlight and carbon from atmospheric CO_2. But plants unaided by bacteria cannot capture N_2 from the atmosphere, and in a sense even plants depend on bacteria for photosynthesis. As we noted earlier, it is almost certain that the organelles in the plant cell that perform photosynthesis—the chloroplasts—have evolved from photosynthetic bacteria that found a home inside the plant cell's cytoplasm (see Figure 1–14).

Traditionally, all bacteria have been regarded as belonging to the same large group. But molecular studies reveal that there is a gulf within the class of procaryotes, dividing it into two distinct kingdoms, called the *eubacteria* and the *archaebacteria*. Remarkably, at a molecular level, the members of these two kingdoms differ as much from one another as either does from the eucaryotes. Most of the bacteria familiar in everyday life—the species that live in the soil or make us ill—are eubacteria. Archaebacteria are typically found in environments hostile to most other cells: there are species that live in concentrated brine, in hot acid volcanic springs, in the airless depths of marine sediments, in the sludge of sewage treatment plants, and in the acidic, oxygen-free environment of a cow's stomach, where they break down cellulose and generate methane gas. Many of these environments resemble the harsh conditions that must have existed on the primitive earth, where living things first evolved, before the atmosphere became rich in oxygen.

Question 1–5 A bacterium weighs about 10^{-12} g and can divide every 20 minutes. If a single bacterial cell carried on dividing at this rate, how long would it take before the mass of bacteria would equal that of the earth (6×10^{24} kg)? Contrast your result with the fact that bacteria originated at least 3.5 billion years ago and have been dividing ever since. Explain the apparent paradox. (The number of cells N in a culture at time t is described by the equation $N = N_0 \times 2^{t/G}$, where N_0 is the number of cells at zero time, and G is the generation time.)

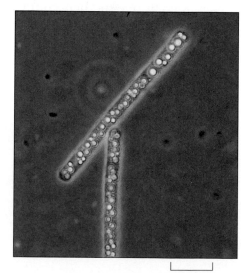

6 µm

Figure 1–27 A gliding sulfur bacterium. *Beggiatoa*, which lives in sulfurous environments, gets its energy by oxidizing H_2S and can fix carbon even in the dark. Note the yellow deposits of sulfur inside the cells. (Courtesy of Ralph W. Wolfe.)

Molecular Biologists Have Focused on *E. coli*

Because living organisms are so complex, the more we learn about any particular species, the more attractive it becomes as an object of study: each discovery raises new questions and provides new tools to tackle those questions in the context of the chosen organism. In this way, large communities of biologists become dedicated to studying the same organism. In the whole world of bacteria, the spotlight of molecular biology has fallen chiefly on just one species: *Escherichia coli*, or *E. coli* for short (see Figure 1–25). This small, rod-shaped eubacterial cell normally lives in the gut of humans and other vertebrates, but it can be grown easily in a simple nutrient broth in a culture bottle. *E. coli* can cope with variable chemical conditions in its environment, and it reproduces rapidly. Its genetic instructions are contained in a single molecule of DNA, and it makes about 4000 different kinds of proteins. A human cell, for comparison, contains 600 times as much DNA, coding for 50,000–100,000 types of proteins.

In molecular terms, we understand the workings of *E. coli* more thoroughly than those of any other living organism. Most of our knowledge of the fundamental mechanisms of life, such as the replication of DNA, or the decoding of the genetic instructions, has come from studies of *E. coli*. Subsequent research has confirmed that these basic processes occur in essentially the same way in our own cells as they do in *E. coli*.

Giardia May Represent an Intermediate Stage in the Evolution of Eucaryotic Cells

By definition, eucaryotic cells, such as our own bodies are made of, have their DNA contained in a separate compartment, the nucleus, enclosed by a double layer of membrane. As we have seen, they possess in addition a whole collection of other features—a cytoskeleton, mitochondria, and other organelles—that set them apart from bacteria.

Figure 1–28 The single-celled parasitic microorganism *Giardia*. (A) Drawing, as seen in the light microscope. (B) Electron micrograph of a cross section through the broad, flattened body of the cell. *Giardia* is thought to be one of the most primitive types of eucaryotic cells. It is nucleated (in fact, it has, strangely, two identical nuclei) and possesses a cytoskeleton, but it has no mitochondria or chloroplasts and no detectable endoplasmic reticulum or Golgi apparatus. (A, after G.D. Schmidt and L.S. Roberts, Foundations of Parasitology, 4th ed. St Louis: Times Mirror/Mosby, 1989; B, courtesy of Dennis Feely.)

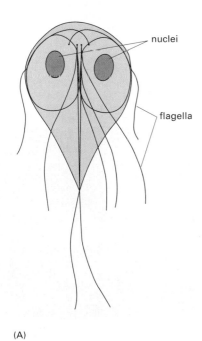

(A)

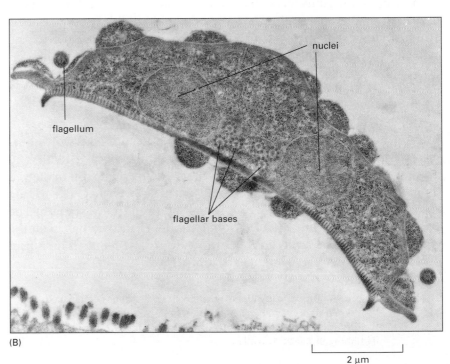

(B)

2 μm

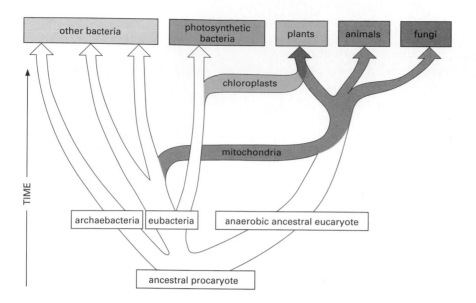

Figure 1–29 **The evolutionary origins of present-day eucaryotes.** The eucaryotic, eubacterial, and archaebacterial lineages diverged from one another very early in the evolution of life on earth. Much later, the eucaryotes acquired mitochondria; later still, a subset of eucaryotes acquired chloroplasts. Mitochondria are essentially the same in plants, animals, and fungi, and therefore are thought to have been acquired before these lines diverged.

Eucaryotes (also known as Eucarya), eubacteria (Bacteria), and archaebacteria (Archaea) must have diverged from one another very early in the history of life on earth (discussed in Chapter 13). Not surprisingly, the eucaryotes did not acquire all of their distinctive features at the same time. Some of the steps in eucaryote evolution can be deduced by comparing modern eucaryotic species with one another and with bacteria. An intriguing organism from this point of view is *Giardia*. This odd-looking single-celled creature (Figure 1–28) lives anaerobically as a parasite in the gut and can cause disease in humans. It appears to be a sort of living fossil, representing a halfway house in the evolution of eucaryotic cells. Although it has a nucleus (in fact it has two) and a cytoskeleton, and is therefore classed as a eucaryote, analysis of its DNA indicates that *Giardia* is almost as closely related to bacteria as it is to other eucaryotes, and it lacks several of the other standard organelles of eucaryotic cells. In particular, it has no mitochondria or chloroplasts.

Eucaryotic organisms such as *Giardia* that can live without oxygen suggest that the ancestral eucaryotic cell originated before the earth's atmosphere became rich in oxygen, and also indicate that a nucleus and cytoskeleton had evolved before the eucaryotic cell acquired mitochondria or chloroplasts (Figure 1–29).

Question 1–6 *Giardia* has a nucleus but no mitochondria, and it is considered a halfway house in the evolution of eucaryotic cells. But one could turn this argument around and conclude that *Giardia* may have evolved from eucaryotic cells that have lost their mitochondria because, living in an anaerobic environment, they have no use for them. What additional arguments or evidence would strengthen the former interpretation?

Brewer's Yeast Is a Simple Eucaryotic Cell

We are preoccupied with eucaryotes because we are eucaryotes ourselves. But human cells are complicated and difficult to work with, and if we want to understand the fundamentals of eucaryotic cell biology, it is often more effective to concentrate on a species that, like *E. coli* among the bacteria, is simple and robust and reproduces rapidly. The popular choice for this role of minimal model eucaryote has been the yeast *Saccharomyces cerevisiae* (Figure 1–30)—the same microorganism that is used by brewers, winemakers, and bakers.

Figure 1–30 **Scanning electron micrograph of cells of the yeast *Saccharomyces cerevisiae*.** A light micrograph of the same species of cells is shown in Figure 1–7. (Courtesy of Ira Herskowitz and Eric Schabatach.)

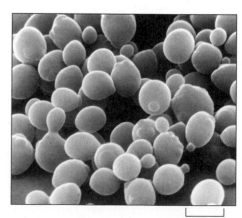

10 μm

S. cerevisiae is a small, single-celled fungus and thus, according to modern views, is at least as closely related to animals as it is to plants. Like other fungi, it has a rigid cell wall, is relatively immobile, and possesses mitochondria but not chloroplasts. When nutrients are plentiful, it reproduces itself almost as rapidly as a bacterium. As its nucleus contains only about 2.5 times as much DNA as *E. coli,* yeast is also a good subject for genetic analysis. Genetic and biochemical studies in yeast have been crucial to understanding many basic mechanisms in eucaryotic cells, including the cell-division cycle—the chain of events by which the nucleus and all the other components of a cell are duplicated and parceled out to create two daughter cells.

Question 1–7 Your next-door neighbor has donated $20 in support of cancer research and is horrified to learn that her money is being spent on studying brewer's yeast. How could you put her mind at ease?

Single-celled Organisms Can Be Large, Complex, and Fierce: The Protozoans

Yeasts are small, simple, and innocuous cells, feeding on sugar, secreting alcohol, and burping carbon dioxide. But there is a huge diversity of other free-living single-cell eucaryotes, and many of these have very different characteristics. The single-celled organisms known as **protozoans** include some of the most complex and (in their own tiny way) ferocious cells known. Figure 1–31 conveys something of the variety of forms of protozoans, and their behavior is just as varied: they can be photosyn-

Figure 1–31 An assortment of protozoans, illustrating the enormous variety within this class of single-celled microorganisms. These drawings are done to different scales, but in each case the scale bar represents 10 μm. The organisms in (A), (B), (E), (F), and (I) are ciliates; (C) is a euglenoid; (D) is an amoeba; (G) is a dinoflagellate; and (H) is a heliozoan. (From M.A. Sleigh, The Biology of Protozoa. London: Edward Arnold, 1973.)

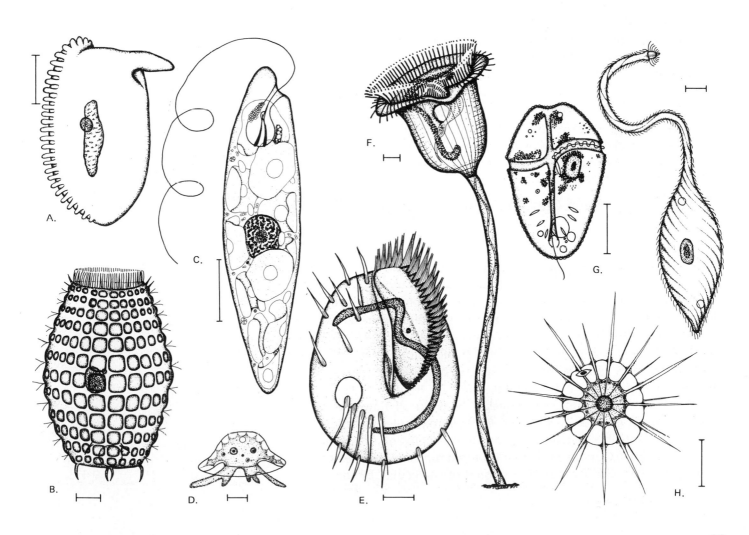

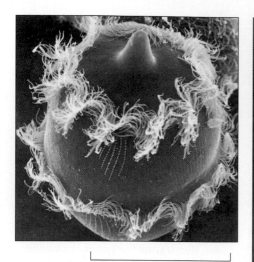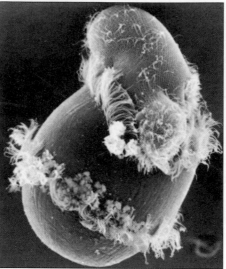

Figure 1–32 **One protozoan eating another.** The micrograph on the left shows *Didinium* on its own, with its circumferential rings of beating cilia and its "snout" at the top; in the picture on the right, it is seen ingesting another ciliated protozoan, *Paramecium*. (Courtesy of D. Barlow.)

100 µm

thetic or carnivorous, motile or sedentary. Their cellular anatomy is often elaborate and includes such structures as sensory bristles, photoreceptors, beating cilia, stalklike appendages, mouth parts, stinging darts, and musclelike contractile bundles. Although they are single cells, protozoans can be as intricate and versatile as many multicellular organisms, as the following example shows.

Didinium is a large carnivorous protozoan, with a diameter of about 150 µm—perhaps 10 times that of an average human cell. It has a globular body encircled by two fringes of cilia; its front end is flattened except for a single protrusion rather like a snout (Figure 1–32). *Didinium* swims at high speed by means of its beating cilia. When it encounters a suitable prey, usually another type of protozoan, it releases numerous small paralyzing darts from its snout region. Then *Didinium* attaches to and devours the other cell, inverting like a hollow ball to engulf its victim, which is almost as large as itself.

Arabidopsis Has Been Chosen Out of 300,000 Species as a Model Plant

The large multicellular organisms that we see around us—the flowers and trees and animals—seem fantastically varied, but they are much closer to one another in their evolutionary origins, and more similar in their basic cell biology, than the great host of microscopic single-cell organisms. While bacteria and eucaryotes separated from each other more than 3000 million years ago, plants, animals, and fungi are separated by only about 1500 million years, fish and mammals by only about 400 million years, and the different species of flowering plants by less than 200 million years.

The close evolutionary relationship between all flowering plants means that we can get insight into the cell and molecular biology of flowering plants by focusing on just a few convenient species for detailed analysis. Out of the several hundred thousand species of flowering plants on earth today, the choice of molecular biologists has

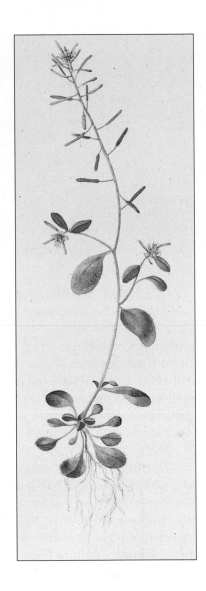

Figure 1–33 *Arabidopsis thaliana,* **the common wall cress.** This small weed has become the favorite organism of plant molecular and developmental biologists. (Courtesy of Toni Hayden and the John Innes Centre.)

recently fallen on a small weed, the common wall cress *Arabidopsis thaliana* (Figure 1–33), which can be grown indoors in large numbers and produces thousands of offspring per plant within 8 to 10 weeks. *Arabidopsis* has only three or four times as much DNA per cell as yeast, and we are beginning to understand how the genetic instructions that it carries enable its cells to collaborate in forming a typical flowering plant.

The World of Animals Is Represented by a Fly, a Worm, a Mouse, and *Homo Sapiens*

Multicellular animals account for the majority of all named species of living organisms, and the majority of animal species are insects. It is fitting, therefore, that an insect, the small fly *Drosophila melanogaster* (Figure 1–34), should occupy a central place in biological research. A concentrated attack has been made on the genetics of *Drosophila,* and especially on the genetic mechanisms underlying its embryonic and larval development. Through this work on *Drosophila,* we are at last beginning to understand in detail how living cells achieve their most spectacular feat: how a single fertilized egg cell (or *zygote*) develops into a multicellular organism comprising vast numbers of cells of differing types, organized in an exactly predictable way. The pattern of the adult fly, with its body and head, its legs and wings and eyes, is determined by the genes carried in the zygote—which are copied and passed on to every cell in the body. These genes define how each cell will behave in its social interactions with its sisters and cousins, and in that way they control the structures that the cells create. *Drosophila,* more than any other organism, has shown us how to trace the chain of cause and effect from the genetic instructions encoded in the DNA to the structure of the adult multicellular organism.

Figure 1–34 *Drosophila melanogaster.* Molecular genetic studies on this small fly have provided a key to the understanding of how all animals develop. (Courtesy of E.B. Lewis.)

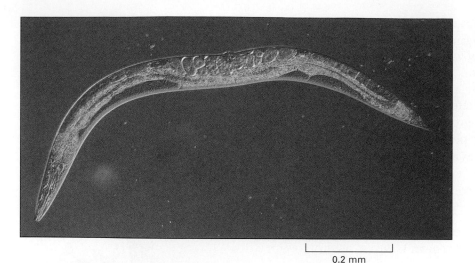

0.2 mm

Figure 1–35 *Caenorhabditis elegans.* This small nematode worm lives in the soil. Its development, from the fertilized egg cell to the 959 cells of the adult body, has been traced in extraordinary detail, and a great deal is known about the underlying genetic mechanisms. Most individuals are hermaphrodites, producing both eggs and sperm. (Courtesy of Ian Hope.)

Another widely studied organism, smaller and simpler than *Drosophila*, is the nematode worm *Caenorhabditis elegans* (Figure 1–35), a harmless relative of the eelworms that attack crops. This creature develops with clockwork precision from a fertilized egg cell into an adult with exactly 959 body cells (plus a variable number of egg and sperm cells), and we now have a minutely detailed description of the sequence of events by which this occurs—as the cells divide, move, and become specialized, according to strict and predictable rules.

At the other extreme, mammals are among the most complex of animals, with 10 times as many genes as *Drosophila*, and millions of times more cells in their adult bodies. The mouse has long been used as the model organism in which to study mammalian genetics, development, immunology, and cell biology. New techniques have given it even greater importance. It is now possible to breed mice with deliberately engineered mutations in any specific gene, or with artificially constructed genes introduced into them. In this way, one can test what a given gene is required for and how it functions.

Last but not least in the list of model organisms are human beings themselves. Research in cell biology has been largely driven by medical interests, and a great deal of what we know has come from studies of human cells. A human being cannot be studied the way one can study a fly, a worm, or a mouse, but a human cell can be studied in a culture dish. The medical database on human cells is enormous, and although naturally occurring mutations in any given gene are rare, the consequences of mutations in thousands of different genes are known without resort to genetic engineering.

Nevertheless, the extent of our ignorance is still daunting. Given the complexity of the mammalian body, one might despair of ever understanding how the DNA in a fertilized mouse egg cell makes it generate a mouse, or how the DNA in a human egg cell directs the development of a human. But the discoveries of molecular biology have given us hope, revealing that mammalian genes have close counterparts in *Caenorhabditis* and in *Drosophila*, apparently serving similar functions. We all have a common evolutionary origin, and under the surface it seems that we share the same molecular mechanisms (Figure 1–36). Flies, worms, mice, and humans thus provide a key to understanding how animals in general are made.

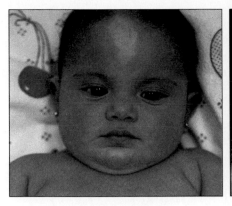 

Figure 1–36 **Different living species share the same molecular mechanisms.** The human baby and the mouse shown here have similar white patches on their foreheads because they both have defects in the same gene (called *kit*), required for the development and maintenance of pigment cells. (Courtesy of R.A. Fleischman, from *Proc. Natl. Acad. Sci. USA* 88:10885–10889, 1991. © 1991 National Academy of Sciences.)

Cells in the Same Multicellular Organism Can Be Spectacularly Different

The cells in an individual animal or plant are extraordinarily varied (Panel 1–3, pp. 32–33). Fat cells, skin cells, bone cells, and nerve cells seem as dissimilar as any cells could be. Yet all these *differentiated cell types* are generated during embryonic development from a single fertilized egg cell, and all contain identical copies of the DNA of the species. Once again, cells looked at from one angle are amazingly different; looked at from another angle, they are amazingly the same. In a multicellular organism, the explanation lies in the way that individual cells use their genetic instructions. Different cells express different genes, depending on the cues that they and their ancestor cells have received from their surroundings.

We shall see that the DNA is not just a shopping list specifying the molecules that every cell must have, and a cell is not just an assembly of all the items on the list. Rather, each cell behaves as a multipurpose machine, possessing sensors to receive environmental signals, built-in computing power to control a range of possible activities, and many megabytes of genetic information available for access in the DNA. Even a simple solitary cell such as *E. coli* or yeast uses its genetic information in a similarly controlled way. The living, growing, self-reproducing cell is truly a wonderful device. In the rest of this book, we shall try to explain how it works.

CELL TYPES

There are over 200 types of cells in the human body. These are assembled into a variety of types of tissue such as

epithelia

connective tissue

muscle

nervous tissue

Most tissues contain a mixture of cell types.

EPITHELIA

Epithelial cells form coherent cell sheets called epithelia, which line the inner and outer surfaces of the body. There are many specialized types of epithelia.

Absorptive cells have numerous hairlike projections called microvilli on their free surface to increase the area for absorption.

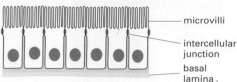

microvilli

intercellular junction

basal lamina

Adjacent epithelial cells are bound together by cell junctions that give the sheet mechanical strength and also make it impermeable to small molecules. The sheet rests on a basal lamina.

Ciliated cells have cilia on their free surface that beat in synchrony to move substances (such as mucus) over the epithelial sheet.

Secretory cells are found in most epithelial layers. These specialized cells secrete substances onto the surface of the cell sheet.

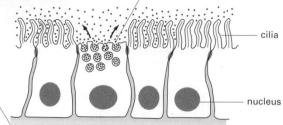

cilia

nucleus

CONNECTIVE TISSUE

The spaces between organs and tissues in the body are filled with connective tissue made principally of a network of tough protein fibers embedded in a polysaccharide gel. This extracellular matrix is secreted mainly by fibroblasts.

fibroblasts in loose connective tissue

Two main types of extracellular protein fiber are collagen and elastin.

Bone is made by cells called osteoblasts. These secrete an extracellular matrix in which crystals of calcium phosphate are later deposited.

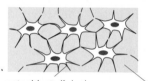

osteoblasts linked together by cell processes

extracellular matrix

Calcium salts are deposited in the extracellular matrix.

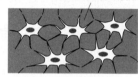

Fat cells (or adipose cells), among the largest cells in the body, are responsible for the production and storage of fat. The nucleus and cytoplasm are squeezed by a large lipid droplet.

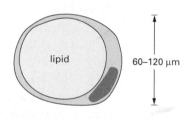

lipid

 60–120 μm

NERVOUS TISSUE

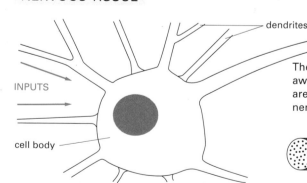

dendrites

OUTPUT

axon

INPUTS

cell body

The axon conducts electrical signals away from the cell body. These signals are produced by a flux of ions across the nerve cell plasma membrane.

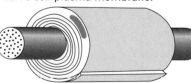

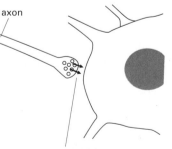

Nerve cells, or neurons, are specialized for communication. The brain and spinal cord, for example, are composed of a network of neurons among supporting glial cells.

Specialized glial cells wrap around an axon to form a multilayered membrane sheath.

A synapse is where a neuron forms a specialized junction with another neuron (or with a muscle cell). At synapses, signals pass from one neuron to another (or from a neuron to a muscle cell).

Secretory epithelial cells are often collected together to form a gland that specializes in the secretion of a particular substance. As illustrated, exocrine glands secrete their products (such as tears, mucus, and gastric juices) into ducts. Endocrine glands secrete hormones into the blood.

secreted material

duct of gland

secretory cells of gland

BLOOD

Erythrocytes (red blood cells) are very small cells, and in mammals have no nucleus or internal membranes. When mature they are stuffed full of the oxygen-binding protein hemoglobin.

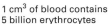

1 cm³ of blood contains 5 billion erythrocytes

their normal shape is a biconcave disc

Leucocytes (white blood cells) protect against infections. Blood contains about one leucocyte for every 100 red blood cells. Although leucocytes travel in the circulation, they can pass through the walls of blood vessels to do their work in the surrounding tissues. There are several different kinds, including

lymphocytes—responsible for immune responses such as the production of antibodies.

macrophages and neutrophils—move to sites of infection, where they ingest bacteria and debris.

wall of small blood vessel

bacterial infection in connective tissue

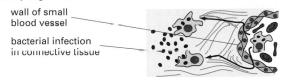

GERM CELLS

Both sperm and egg are *haploid,* that is, they carry only one set of chromosomes. A sperm from the male fuses with an egg from the female, which then forms a new diploid organism by successive cell divisions.

egg with sperm drawn to scale

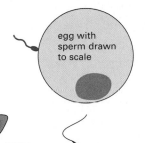

sperm

MUSCLE

Muscle cells produce mechanical force by their contraction. In vertebrates there are three main types:

skeletal muscle—this moves joints by its strong and rapid contraction. Each muscle is a bundle of muscle fibers, each of which is an enormous multinucleated cell.

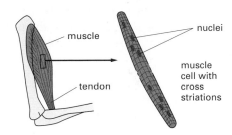

muscle

nuclei

tendon

muscle cell with cross striations

smooth muscle—present in digestive tract, bladder, arteries, and veins. It is composed of thin elongated cells (not striated), each of which has one nucleus.

cardiac muscle—intermediate in character between skeletal and smooth muscle. It produces the heart beat. Adjacent cells are linked by electrically conducting junctions that cause the cells to contract in synchrony.

SENSORY CELLS

Among the most strikingly specialized cells in the vertebrate body are those that detect external stimuli. Hair cells of the inner ear are primary detectors of sound. They are modified epithelial cells that carry special microvilli (stereocilia) on their surface. The movement of these in response to sound vibrations causes an electrical signal to pass to the brain.

stereocilia are very rigid because they are packed with actin filaments

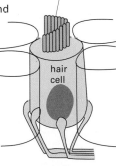

hair cell

Rod cells in the retina of the eye are specialized to respond to light. The photosensitive region contains many membranous discs (*red*) in whose membranes the light-sensitive pigment rhodopsin is embedded. Light evokes an electrical signal (*green arrow*), which is transmitted to nerve cells in the eye, which relay the signal to the brain.

Essential Concepts

- Cells are the fundamental units of life.

- Cells of animal and plant tissue are typically 5–20 μm in diameter and can be seen using a light microscope, which also reveals some of their internal components, or organelles.

- The electron microscope permits the smaller organelles and even individual molecules to be seen, but specimens require elaborate preparation and cannot be viewed alive.

- The most prominent organelle in most plant and animal cells is the nucleus. This contains the genetic information of the organism, stored in the structure of DNA molecules. The rest of the cell's contents, excluding the nucleus, constitutes the cytoplasm.

- The cytoplasm of plant and animal cells contains a variety of other internal membrane-bounded organelles with specialized chemical functions. Such organelles include mitochondria, which carry out the oxidation of food molecules, and, in plant cells, chloroplasts, which perform photosynthesis.

- The remaining intracellular compartment, excluding the membrane-bounded organelles, is the cytosol. This contains a concentrated mixture of large and small molecules that carry out many essential biochemical processes.

- A system of protein filaments called the cytoskeleton extends throughout the cytosol. This controls the shape and movement of the cell and enables organelles and molecules to be transported from one location to another in the cytoplasm.

- All present-day cells are believed to have evolved from the same ancestral cell that existed more than three billion years ago.

- All cells contain DNA as a store of genetic information and use it to guide the synthesis of proteins, and all cells make their DNA and their proteins from the same two small sets of building blocks.

- Bacteria, the simplest of present-day living cells, are procaryotes: although they contain DNA, they lack a nucleus and other organelles and probably resemble most closely the ancestral cell.

- Different species of bacteria are diverse in their chemical capabilities and inhabit an amazingly wide range of habitats. Two fundamental evolutionary subdivisions are recognized: eubacteria and archaebacteria.

- Eucaryotic cells possess a nucleus. They probably evolved in a series of stages from cells more similar to bacteria. An important step appears to have been the acquisition of mitochondria, originating as engulfed bacteria living in symbiosis with larger anaerobic cells.

- Free-living single-celled eucaryotic microorganisms include some of the most complex eucaryotic cells known, and they are able to swim, mate, hunt, and devour food.

- Other types of eucaryotic cells cooperate to form large, complex multicellular organisms such as ourselves, composed of thousands of billions of cells.

- Biologists have chosen a small number of organisms as a focus for intense investigation. These include the bacterium *E. coli*, brewer's yeast, a nematode worm, a fly, a small plant, a mouse, and the human species itself.

- Cells in a multicellular organism, though they all contain the same DNA, can be very different. They use different parts of their genetic information selectively, according to cues accumulated from their environment.

Key Terms

bacterium
cell
chloroplast

chromosome
cytoplasm
cytoskeleton
cytosol
eucaryote

evolution
microscope
micrometer
mitochondrion
nanometer

nucleus
organelle
procaryote
protozoan

Questions

Question 1–8   By now you should be familiar with the following cellular components. Briefly define what they are and what function they provide for cells.

A. cytosol

B. cytoplasm

C. mitochondria

D. nucleus

E. chloroplasts

F. lysosomes

G. chromosomes

H. Golgi apparatus

I. peroxisomes

J. plasma membrane

K. endoplasmic reticulum

L. cytoskeleton

Question 1–9   Which of the following statements are correct? Explain your answers.

A. The hereditary information of a cell is passed on by its proteins.

B. Bacterial DNA is found in the cytosol.

C. Plants are composed of procaryotic cells.

D. All cells of the same organism have the same number of chromosomes (with the exception of egg and sperm cells).

E. The cytosol contains membrane-bounded organelles, such as lysosomes.

F. Nuclei and mitochondria are surrounded by a double membrane.

G. Protozoans are complex organisms with a set of specialized cells that form tissues, such as flagella, mouth parts, stinging darts, and leglike appendages.

H. Lysosomes and peroxisomes are the site of degradation of unwanted materials.

Question 1–10   To get a feeling for the size of cells (and to practice the use of the metric system) consider the following: the human brain weighs about 1 kg and contains about 10^{11} cells. Calculate the average size of a brain cell (although we know that their sizes vary widely), assuming that each cell is entirely filled with water (1 cm³ of water weighs 1 g). What would be the length of one side of this average-sized brain cell if it were a simple cube? If the cells were spread out as a thin layer that is only a single cell thick, how many pages of this book would this layer cover?

Question 1–11   Identify the different organelles indicated with letters in the electron micrograph shown in Figure Q1–11. Estimate the length of the scale bar in the figure.

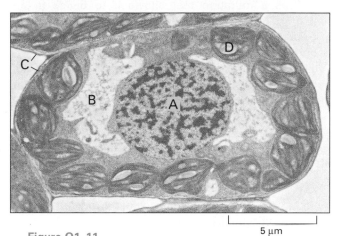

Figure Q1–11

Question 1–12   There are three major classes of filaments that make up the cytoskeleton. What are they and what are the differences in their functions?

Question 1–13   Natural selection is such a powerful force in evolution because cells with even a small growth advantage quickly outgrow their competitors. To illustrate this process, consider a cell culture that contains one million bacterial cells that double every 20 minutes. A single cell in this culture acquires a mutation that allows it to divide faster with a generation time of only 15 minutes. Assuming that there is an unlimited food supply and no cell death, how long would it take before the progeny of the mutated cell became predominant in the culture? (Before you go through the calculation, make a guess: do you think it would take about a day, a week, a month, or a year?) How many cells of either species are present in the culture at this time? (The number of cells N in the culture at time t is described by the equation $N = N_0 \times 2^{t/G}$, where N_0 is the number of cells at zero time and G is the generation time.)

Question 1–14   When bacteria are grown under adverse conditions, i.e., in the presence of a poison such as an antibiotic, most cells grow slowly. But it is not uncommon that the growth rate of a bacterial culture kept in the presence of the poison is restored after a few days to that observed in its absence. Suggest why this may be the case.

Question 1–15   Apply the principle of exponential growth as described in Question 1–13 to the cells in a multicellular organism, such as yourself. There are about 10^{13} cells in your body. Assume that one cell acquires a mutation that allows it to divide in an uncontrolled manner (that is, it becomes a cancer cell). Some cancer cells can grow with a generation time of about 24 hours. How long would it take before 10^{13} cells in your body would be cancer cells? (Use the equation $N = N_0 \times 2^{t/G}$, with t, the time, and G, the length of each generation. Hint: $10^{13} \cong 2^{43}$.)

Question 1–16   Discuss the following statement: "The structure and function of a living cell are dictated by the laws of physics and chemistry."

Question 1–17   What, if any, are the advantages in being multicellular?

Question 1–18   Draw to scale the outline of two spherical cells, one a bacterium with a diameter of 1 μm, the other an animal cell with a diameter of 15 μm. Calculate the volume, surface area, and surface-to-volume ratio for each cell. How would this latter value change if you included the internal membranes of the cell in the calculation of surface area (assume internal membranes have 15 times the area of the plasma membrane)? (The volume of a sphere is given by $^{4}/_{3}\pi R^{3}$ and its surface by $4\pi R^{2}$, where R is its radius.) Discuss the following hypothesis: "Internal membranes allowed bigger cells to evolve."

Question 1–19   What are the arguments that all living cells evolved from a common ancestor cell? Envision the very early days of evolution of life on earth. Would you assume that the primordial ancestor cell was the first and only cell to form?

2 Chemical Components of Cells

It is at first sight difficult to accept that living creatures are merely a chemical system. Their incredible diversity of form, their seemingly purposeful behavior, and their ability to grow and reproduce all seem to set them apart from the world of solids, liquids, and gases that chemistry normally describes. Indeed, until the nineteenth century it was widely accepted that animals contained a Vital Force—an "animus"—that was responsible for their distinctive properties.

We now know there is nothing in living organisms that disobeys chemical and physical laws. However, the chemistry of life is indeed of a special kind. First, it is based overwhelmingly on carbon compounds, whose study is therefore known as *organic chemistry.* Second, it depends almost exclusively on chemical reactions that take place in aqueous solution and in the relatively narrow range of temperatures experienced on earth. Third, it is enormously complex: even the simplest cell is vastly more complicated in its chemistry than any other chemical system known. Finally, it is dominated and coordinated by enormous *polymeric molecules*—chains of chemical subunits linked end-to-end—whose unique properties enable cells and organisms to grow and reproduce and to do all the other things that are characteristic of life.

Chemical Bonds

Matter is made of combinations of *elements*—substances such as hydrogen or carbon that cannot be broken down or converted into other substances by chemical means. The smallest particle of an element that still retains its distinctive chemical properties is an *atom.* However, the characteristics of substances other than pure elements—including the materials from which living cells are made—depend on the way their atoms are linked together in groups to form *molecules.* In order to understand how living organisms are built from inanimate matter, therefore, it is crucial to know how the chemical bonds that hold atoms together in molecules are formed.

Cells Are Made of Relatively Few Types of Atoms

Each **atom** has at its center a positively charged massive nucleus, which is surrounded at some distance by a cloud of negatively charged **electrons** (Figure 2–1), held in orbit by electrostatic attraction to the nucleus. The nucleus consists of two kinds of subatomic particles: **protons,** which are positively charged, and *neutrons,* which are electrically neutral. The number of protons in the atomic nucleus gives the *atomic number.* An atom of hydrogen has a nucleus composed of a single proton; so hydrogen, with an atomic number of 1, is the lightest element. An atom of carbon has six protons in its nucleus and an atomic number of 6 (Figure 2–2). The electric charge carried by each proton is exactly equal and opposite to the charge carried by a single electron. Since an atom as a whole is electrically neutral, the number of negatively charged electrons surrounding the nucleus is equal to the number of positively charged protons that the nucleus contains; thus the number of electrons in an atom equals the atomic number. The atomic number is the same for all atoms of a given element, and we shall shortly see that it determines the chemical behavior of the element.

Neutrons are uncharged subatomic particles of essentially the same mass as protons. They contribute to the structural stability of the nucleus—if there are too many or too few, the nucleus may disintegrate by radioactive decay—but they do not alter the chemical properties of the atom. Thus an element can exist in several physically distinguishable but chemically identical forms, called *isotopes,* each isotope having a different number of neutrons but the same number of protons. Multiple isotopes of almost all the elements occur naturally, including some that are unstable. For example, while most carbon on earth exists as the stable isotope carbon 12, with six protons and six neutrons, there are also small amounts of an unstable isotope, the radioactive carbon 14, whose atoms have six protons and eight neutrons. Carbon 14 undergoes radioactive decay at a slow but steady rate, which is the basis of the technique of carbon 14 dating of organic material in archaeology.

The **atomic weight** of an atom, or the **molecular weight** of a molecule, is its mass relative to that of a hydrogen atom. This is essentially equal to the number of protons plus neutrons that the atom or molecule contains, since the electrons are much lighter and contribute almost nothing to the total. Thus the major isotope of carbon has an atomic weight of 12 and is symbolized as ^{12}C, whereas the unstable isotope just

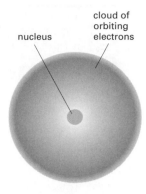

Figure 2–1 Schematic view of an atom. The dense, positively charged nucleus contains most of the atom's mass. The much lighter and negatively charged electrons move rapidly around the nucleus, governed by the laws of quantum mechanics. The electrons are depicted as a continuous cloud, as there is no way of predicting exactly where an electron is at any given instant of time. The density of shading of the cloud is an indication of the probability that electrons will be found there. The diameter of the electron cloud ranges from about 0.1 nm (for hydrogen) to about 0.4 nm (for atoms of high atomic number). The nucleus is very much smaller: about 2×10^{-5} nm for carbon, for example.

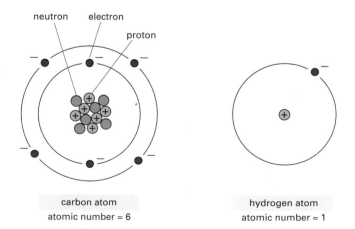

carbon atom
atomic number = 6

hydrogen atom
atomic number = 1

Figure 2–2 Schematic representations of an atom of carbon and an atom of hydrogen. In contrast to Figure 2–1, the electrons are shown here as individual particles. The nucleus of every atom except hydrogen consists of both positively charged protons and electrically neutral neutrons. The number of electrons in an atom is equal to its number of protons (the atomic number), so that the atom has no net charge. The concentric *black circles* represent in a highly schematic form the orbits of the electrons. The neutrons, protons, and electrons are in reality minute in relation to the atom as a whole; their size is greatly exaggerated here.

discussed has an atomic weight of approximately 14 and is written as ^{14}C. The mass of an atom or a molecule is often specified in *daltons,* one dalton being an atomic mass unit approximately equal to the mass of a hydrogen atom.

Atoms are so small that it is hard to imagine their size. An individual carbon atom is roughly 0.2 nm in diameter, so that it would take about 5 million of them, laid out in a straight line, to span a millimeter. One proton or neutron weighs approximately $1/(6 \times 10^{23})$ gram, so one gram of hydrogen contains 6×10^{23} atoms. This huge number (6×10^{23}, called **Avogadro's number**) is the key scale factor describing the relationship between everyday quantities and quantities measured in terms of individual atoms or molecules. If a substance has a molecular weight of M, 6×10^{23} molecules of it will have a mass of M grams. This quantity is called one *mole* of the substance (Figure 2–3).

There are 92 naturally occurring elements, each differing from the others in the number of protons and electrons in its atoms. Living organisms, however, are made of only a small selection of these elements, four of which—carbon (C), hydrogen (H), nitrogen (N), and oxygen (O)—make up 96.5% of an organism's weight. This composition differs markedly from that of the nonliving inorganic environment (Figure 2–4) and is evidence of a distinctive type of chemistry. The most common elements in living organisms are listed in Table 2–1 with some of their atomic characteristics.

The Outermost Electrons Determine How Atoms Interact

To understand how atoms bond together to form the molecules that make up living organisms, we have to pay special attention to their electrons. Protons and neutrons are welded tightly to one another in the nucleus and change partners only under extreme conditions—during radioactive decay, for example, or in the interior of the sun or of a nuclear reactor. In living tissues, it is only the electrons of an atom that

A mole is M grams of a substance, where M is its relative molecular mass (molecular weight). It will contain 6×10^{23} molecules of the substance.

1 mole of carbon weighs 12 g
1 mole of glucose weighs 180 g
1 mole of sodium chloride weighs 58 g

Molar solutions have a concentration of 1 mole of the substance in 1 liter of solution. A molar solution (1 M) of glucose, for example, has 180 g/l, while a millimolar solution (1 mM) has 180 mg/l.

Figure 2–3 **Moles and molar solutions.**

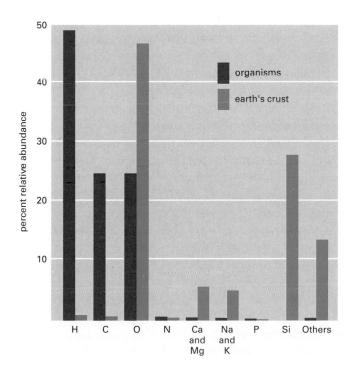

Figure 2–4 **The abundance of some chemical elements in the nonliving world (the earth's crust) compared to their abundance in the tissues of an animal.** The abundance of each element is expressed as a percentage of the total number of atoms present in the sample. Thus, for example, nearly 50% of the atoms in a living organism are hydrogen atoms. The survey here excludes mineralized tissues such as bone and teeth, as they contain large amounts of inorganic salts of calcium and phosphorus. The relative abundance of elements is similar in all living organisms.

common elements in living organisms

element		protons	neutrons	electrons	atomic number	atomic weight
Hydrogen	H	1	0	1	1	1
Carbon	C	6	6	6	6	12
Nitrogen	N	7	7	7	7	14
Oxygen	O	8	8	8	8	16

less common elements

element		protons	neutrons	electrons	atomic number	atomic weight
Sodium	Na	11	12	11	11	23
Magnesium	Mg	12	12	12	12	24
Phosphorus	P	15	16	15	15	31
Sulfur	S	16	16	16	16	32
Chlorine	Cl	17	18	17	17	35
Potassium	K	19	20	19	19	39
Calcium	Ca	20	20	20	20	40

Table 2–1 Atomic characteristics of the most abundant elements in living tissues. The four elements shown in *blue* account for 96.5% of human body weight; those shown in *green* each account for between 0.1% and 1.5%. The atomic number, equal to the number of protons in each nucleus, defines each element uniquely. The atomic weight, given by the sum of protons and neutrons in the nucleus, can vary with the particular isotope of the element, as emphasized in the text. The atomic weights given in this table are those of the most common isotope of each element.

undergo rearrangements. They form the exterior of an atom and specify the rules of chemistry by which atoms combine to form molecules.

Electrons are in continuous motion around the nucleus, but motions on this submicroscopic scale obey different laws from those we are familiar with in everyday life. These laws dictate that electrons in an atom can exist only in certain discrete states of movement—roughly speaking, discrete orbits—and that there is a strict limit to the number of electrons that can be accommodated in an orbit of a given type—a so-called *electron shell*. The electrons closest on average to the positive nucleus are attracted most strongly to it and occupy the innermost, most tightly bound shell. This shell can hold a maximum of two electrons. The second shell is farther away from the nucleus, and its electrons are less tightly bound. This second shell can hold up to eight electrons. The third shell contains electrons that are even less tightly bound; it can also hold up to eight electrons. The fourth and fifth shells can hold 18 electrons each. Atoms with more than four shells are very rare in biological molecules.

The electron arrangement of an atom is most stable when all the electrons are in the most tightly bound states that are possible for them—that is, when they occupy the innermost shells. Therefore, with certain exceptions in the larger atoms, the electrons of an atom fill the shells in order—the first before the second, the second before the third, and so on. An atom whose outermost shell is entirely filled with electrons is especially stable and therefore chemically unreactive. Examples are helium with 2 electrons, neon with 2 + 8, and argon with 2 + 8 + 8; these are all inert gases. Hydrogen, by contrast, with only one electron, and therefore only a half-filled shell, is highly reactive. Likewise, the other atoms found in living tissues all have incomplete outer electron shells and are therefore able to react with one another to form molecules (Figure 2–5).

Because an unfilled electron shell is less stable than a filled one, atoms with incomplete outer shells have a strong tendency to interact with other atoms so as to either gain or lose enough electrons to achieve a completed outermost shell. This electron exchange can be achieved

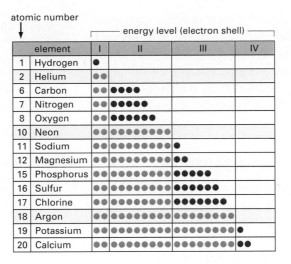

atomic number	element	I	II	III	IV
1	Hydrogen	●			
2	Helium	●●			
6	Carbon	●●	●●●●		
7	Nitrogen	●●	●●●●●		
8	Oxygen	●●	●●●●●●		
10	Neon	●●	●●●●●●●●		
11	Sodium	●●	●●●●●●●●	●	
12	Magnesium	●●	●●●●●●●●	●●	
15	Phosphorus	●●	●●●●●●●●	●●●●●	
16	Sulfur	●●	●●●●●●●●	●●●●●●	
17	Chlorine	●●	●●●●●●●●	●●●●●●●	
18	Argon	●●	●●●●●●●●	●●●●●●●●	
19	Potassium	●●	●●●●●●●●	●●●●●●●●	●
20	Calcium	●●	●●●●●●●●	●●●●●●●●	●●

energy level (electron shell)

Figure 2–5 Filled and unfilled electron shells in some common elements. All the elements commonly found in living organisms have unfilled outermost shells (*red*) and can thus participate in chemical reactions with other atoms. For comparison, some elements that have only filled shells (*yellow*) are shown; these are chemically unreactive.

either by transferring electrons from one atom to another or by sharing electrons between two atoms. These two strategies generate two types of **chemical bonds** between atoms: an *ionic bond* is formed when electrons are donated by one atom to another, whereas a *covalent bond* is formed when two atoms share a pair of electrons (Figure 2–6). Often, the pair of electrons is shared unequally, with partial transfer between the atoms; this intermediate strategy results in a *polar covalent bond*, as we shall discuss later.

An H atom, which needs only one more electron to fill its shell, generally acquires it by sharing, forming one covalent bond with another atom; in many cases this bond is polar. The other most common elements in living cells—C, N, and O, with an incomplete second shell, and P and S, with an incomplete third shell (see Figure 2–5)—generally share electrons and achieve a filled outer shell of eight electrons by forming several covalent bonds. The number of electrons an atom must acquire or lose (either by sharing or by transfer) to attain a filled outer shell is known as its *valence*.

The crucial role of the outer electron shell in determining the chemical properties of an element means that when the elements are listed in

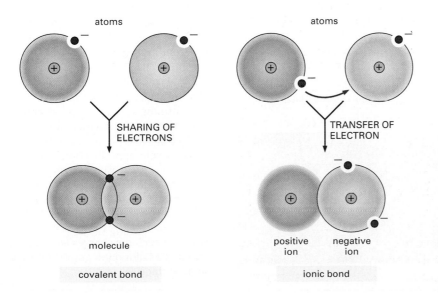

Figure 2–6 Comparison of covalent and ionic bonds. Atoms can attain a more stable arrangement of electrons in their outermost shell by interacting with one another. An ionic bond is formed when electrons are transferred from one atom to the other. A covalent bond is formed when electrons are shared between atoms. The two cases shown represent extremes; often, covalent bonds form with a partial transfer (unequal sharing of electrons), resulting in a polar covalent bond (see Figure 3–11).

order of their atomic number there is a periodic recurrence of elements with similar properties: an element with, say, an incomplete second shell containing one electron will behave in much the same way as an element that has filled its second shell and has an incomplete third shell containing one electron. The metals, for example, all have incomplete outer shells with just one or a few electrons, whereas, as we have just seen, the inert gases have full outer shells.

Ionic Bonds Form by the Gain and Loss of Electrons

Ionic bonds are most likely to be formed by atoms that have just one or two electrons in addition to a filled outer shell or are just one or two electrons short of acquiring a filled outer shell. They can often more easily attain a completely filled outer electron shell by transferring electrons to or from another atom than by sharing them. For example, from Figure 2–5 we see that a sodium (Na) atom, with atomic number 11, can strip itself down to a filled shell by giving up the single electron external to its second shell. By contrast, a chlorine (Cl) atom, with atomic number 17, can complete its outer shell by gaining just one electron. Consequently, if a Na atom encounters a Cl atom, an electron can jump from the Na to the Cl, leaving both atoms with filled outer shells. The offspring of this marriage between sodium, a soft and intensely reactive metal, and chlorine, a toxic green gas, is table salt (NaCl).

When an electron jumps from Na to Cl, both atoms become electrically charged **ions.** The Na atom that lost an electron now has one less electron than it has protons in its nucleus; it therefore has a single positive charge (Na^+). The Cl atom that gained an electron now has one more electron than it has protons and has a single negative charge (Cl^-). Positive ions are called *cations*, and negative ions, *anions*. Ions can be further classified according to how many electrons are lost or gained. Thus sodium and potassium (K) have one electron to lose and form cations with a single positive charge (Na^+ and K^+), whereas magnesium

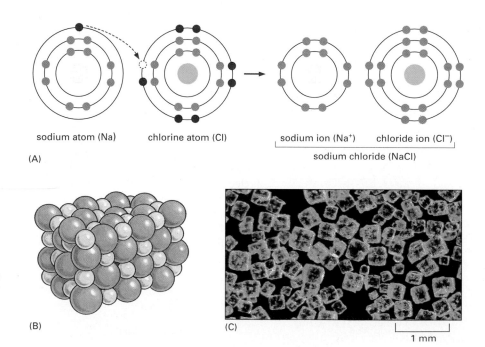

(A) sodium atom (Na)   chlorine atom (Cl)   sodium ion (Na⁺)   chloride ion (Cl⁻)

sodium chloride (NaCl)

(B)

(C)

1 mm

Figure 2–7 Sodium chloride: an example of ionic bond formation. (A) An atom of sodium (Na) reacts with an atom of chlorine (Cl). Electrons of each atom are shown in their different energy levels; electrons in the chemically reactive (incompletely filled) shells are shown in *red*. The reaction takes place with transfer of a single electron from sodium to chlorine, forming two electrically charged atoms, or ions, each with complete sets of electrons in their outermost levels. The two ions with opposite charge are held together by electrostatic attraction. (B) The product of the reaction between sodium and chlorine, crystalline sodium chloride, consists of sodium and chloride ions packed closely together in a regular array in which the charges are exactly balanced. (C) Color photograph of crystals of sodium chloride.

bond type		length (nm)	strength (kcal/mole)	
			in vacuum	in water
Covalent		0.15	90	90
Noncovalent:	ionic	0.25	80	3
	hydrogen	0.30	4	1
	van der Waals	0.35	0.1	0.1
	attraction (per atom)			

Table 2–2 Covalent and noncovalent chemical bonds. Bond strengths are measured by the energy required to break them, in kilocalories per mole (see Glossary for definitions of these units). The length of a hydrogen bond X . . . H . . . X is the distance between the two nonhydrogen atoms (X). The bond strengths and lengths listed are approximate, because the exact values will depend on the atoms involved. The different types of noncovalent bonds are described later in the chapter (see Panel 2–7, pp. 70–71).

and calcium have two electrons to lose and form cations with two positive charges (Mg^{2+} and Ca^{2+}).

Because of their opposite charges, Na^+ and Cl^- are attracted to each other and are thereby held together in an **ionic bond.** A salt crystal contains astronomical numbers of Na^+ and Cl^- (about 2×10^{19} ions of each type in a crystal 1 mm across) packed together in a precise three-dimensional array with their opposite charges exactly balanced (Figure 2–7). Substances such as NaCl, which are held together solely by ionic bonds, are generally called *salts* rather than molecules. Ionic bonds are just one of several types of *noncovalent bonds* that can exist between atoms. We shall meet another example of a noncovalent bond, the hydrogen bond, later in this chapter. Because of a favorable interaction between water molecules and ions, many salts (including NaCl) are highly soluble in water—dissociating into individual ions (such as Na^+ and Cl^-), each surrounded by a group of water molecules. For the same reason, the strength of a hydrogen bond between two molecules is significantly reduced in water. In contrast, covalent bond strengths are not affected in this way (Table 2–2).

Covalent Bonds Form by the Sharing of Electrons

All the characteristics of a cell depend on the molecules it contains. A **molecule** is a cluster of atoms held together by **covalent bonds,** in which electrons are shared rather than transferred between atoms. The shared electrons complete the outer shells of both atoms. In the simplest possible molecule—a molecule of hydrogen (H_2)—two H atoms, each with a single electron, share two electrons, which is the number required to fill the first shell. The shared electrons form a cloud of negative charge that is densest between the two positively charged nuclei and helps to hold them together, in opposition to the mutual repulsion between like charges that would otherwise force them apart. The attractive and repulsive forces are in balance when the nuclei are separated by a characteristic distance, called the *bond length* (Figure 2–8).

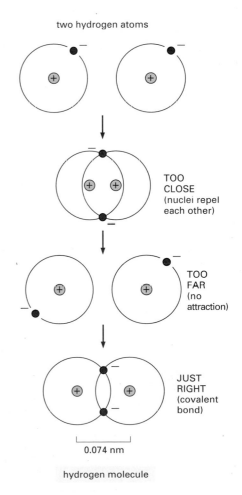

two hydrogen atoms

TOO CLOSE (nuclei repel each other)

TOO FAR (no attraction)

JUST RIGHT (covalent bond)

0.074 nm

hydrogen molecule

Figure 2–8 The hydrogen molecule: a simple example of covalent bond formation. Each hydrogen atom in isolation has a single electron so that its first electron shell is incompletely filled. By coming together the two atoms are able to share the two electrons, and each obtains a completely filled first shell, with the shared electrons adopting modified orbits around the two nuclei. The covalent bond between the two atoms has a definite length. If the atoms were closer together, the positive nuclei would repel each other; if they were farther apart than this distance, they would not be able to share electrons effectively.

Chemical Bonds

A further crucial property of any bond—covalent or noncovalent—is its strength. *Bond strength* is measured by the amount of energy that must be supplied to break that bond, usually expressed in units of kilocalories per mole (kcal/mole), where a kilocalorie is the amount of energy needed to raise the temperature of one liter of water by one degree centigrade. Thus if 1 kilocalorie must be supplied to break 6×10^{23} bonds of a specific type (that is, 1 mole of these bonds), then the strength of that bond is 1 kcal/mole. Typical strengths and lengths of the main classes of chemical bonds are given in Table 2–2; an equivalent, widely used measure of energy is the kilojoule, which is equal to 0.239 kilocalories.

To get an idea of what bond strengths mean, it is helpful to compare them with the average energies of the impacts that molecules are constantly undergoing from collisions with other molecules in their environment (their thermal, or heat, energy). Typical covalent bonds are stronger than these thermal energies by a factor of 100, so they are resistant to being pulled apart by thermal motions and are normally broken only during specific chemical reactions with other atoms and molecules. The making and breaking of covalent bonds are violent events, and in living cells they are carefully controlled by highly specific catalysts, called *enzymes*. Noncovalent bonds as a rule are much weaker; we shall see later that they are important in the cell in the many situations where molecules have to associate and dissociate readily to carry out their functions.

Whereas an H atom can form only a single covalent bond, the other common atoms that form covalent bonds in cells—O, N, S, and P, as well as the all-important C atom—can form more than one. The outermost shell of these atoms, as we have seen, can accommodate up to eight electrons, and they form covalent bonds with as many other atoms as necessary to reach this number. Oxygen, with six electrons in its outer

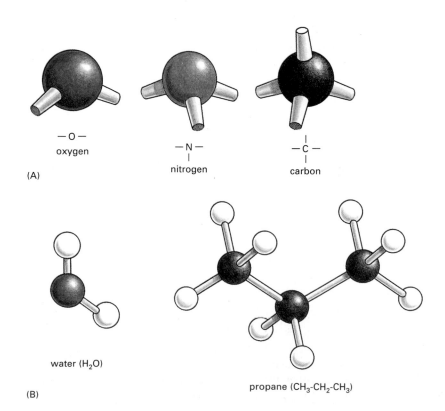

— O —
oxygen

— N —
|
nitrogen

|
— C —
|
carbon

(A)

water (H₂O)

propane (CH₃-CH₂-CH₃)

(B)

Figure 2–9 **Geometry of covalent bonds.** (A) The spatial arrangement of the covalent bonds that can be formed by oxygen, nitrogen, and carbon. (B) Molecules formed from these atoms therefore have a precise three-dimensional structure, as shown here for water and propane, defined by the bond angles and bond lengths for each covalent linkage. A water molecule, for example, forms a "V" shape with an angle close to 109°.

shell, is most stable when it acquires an extra two electrons by sharing with other atoms and therefore forms up to two covalent bonds. Nitrogen, with five outer electrons, forms a maximum of three covalent bonds, while carbon, with four outer electrons, forms up to four covalent bonds—thus sharing four pairs of electrons (see Figure 2–5).

When one atom forms covalent bonds with several others, these multiple bonds have definite orientations in space relative to one another, reflecting the orientations of the orbits of the shared electrons. Covalent bonds between multiple atoms are therefore characterized by specific bond angles as well as bond lengths and bond energies (Figure 2–9). The four covalent bonds that can form around a carbon atom, for example, are arranged as if pointing to the four corners of a regular tetrahedron. The precise orientation of covalent bonds is the basis for the three-dimensional geometry of organic molecules.

There Are Different Types of Covalent Bonds

Most covalent bonds involve the sharing of two electrons, one donated by each participating atom; these are called *single bonds*. Some covalent bonds, however, involve the sharing of more than one pair of electrons. Four electrons can be shared, for example, two coming from each participating atom; such a bond is called a *double bond*. Double bonds are shorter and stronger than single bonds and have a characteristic effect on the three-dimensional geometry of molecules containing them. A single covalent bond between two atoms generally allows the rotation of one part of a molecule relative to the other around the bond axis. A double bond prevents such rotation, producing a more rigid and less flexible arrangement of atoms (Figure 2–10 and Panel 2–1, pp. 46–47).

Some molecules contain sets of bonds that share electrons and consequently have a hybrid character intermediate between single and double bonds. The highly stable benzene molecule, for example, comprises a ring of six carbon atoms in which the bonding electrons are evenly distributed (although usually depicted as an alternating sequence of single and double bonds, as shown in Panel 2–1.

When the atoms joined by a single covalent bond belong to different elements, the two atoms usually attract the shared electrons to different degrees. Compared with a C atom, for example, O and N atoms attract electrons relatively strongly, whereas an H atom attracts electrons relatively weakly. By definition, a **polar** structure (in the electrical sense) is one with positive charge concentrated toward one end (the positive pole) and negative charge concentrated toward the other (the negative pole). Covalent bonds in which the electrons are shared unequally in this way are therefore known as polar bonds, as we mentioned earlier (Figure 2–11). For example, the covalent bond between oxygen and hydrogen, –O–H, or between nitrogen and hydrogen, –N–H, is polar, whereas that between carbon and hydrogen, –C–H, has the electrons attracted much more equally by both atoms and is relatively nonpolar.

Polar covalent bonds are extremely important in biology because they allow molecules to interact through electrical forces. Any large molecule with many polar groups will have a pattern of partial positive and negative charges on its surface. When such a molecule encounters a second molecule with a complementary set of charges, the two molecules will be attracted to each other by weak noncovalent ionic bonds that

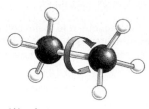

(A) ethane

(B) ethene

Figure 2–10 Carbon-carbon double bonds and single bonds compared. (A) The ethane molecule, with a single covalent bond between the two carbon atoms, illustrates the tetrahedral arrangement of single covalent bonds formed by carbon. One of the CH_3 groups joined by the covalent bond can rotate relative to the other around the bond axis. (B) The double bond between the two carbon atoms in a molecule of ethene (ethylene) alters the bond geometry of the carbon atoms and brings all the atoms into the same plane; the double bond prevents the rotation of one CH_2 group relative to the other.

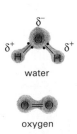

water

oxygen

Figure 2–11 Polar covalent bonds. The electron distributions in the polar water molecule (H_2O) and the nonpolar oxygen molecule (O_2) are compared (δ^+, partial positive charge; δ^-, partial negative charge).

CARBON SKELETONS

Carbon has a unique role in the cell because of its ability to form strong covalent bonds with other carbon atoms. Thus carbon atoms can join to form chains.

or branched trees

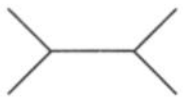

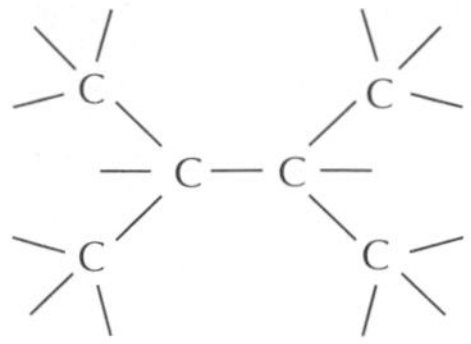

or rings

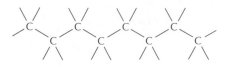

also written as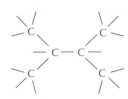

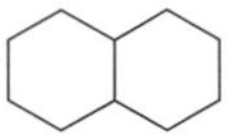

also written as

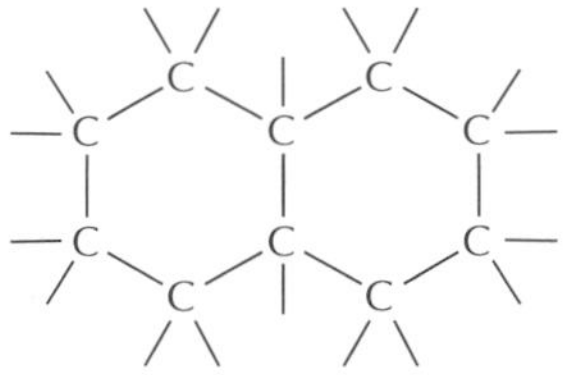

also written as

COVALENT BONDS

A covalent bond forms when two atoms come very close together and share one or more of their electrons. In a single bond one electron from each of the two atoms is shared; in a double bond a total of four electrons are shared.

Each atom forms a fixed number of covalent bonds in a defined spatial arrangement. For example, carbon forms four single bonds arranged tetrahedrally, whereas nitrogen forms three single bonds and oxygen forms two single bonds arranged as shown below.

Double bonds exist and have a different spatial arrangement.

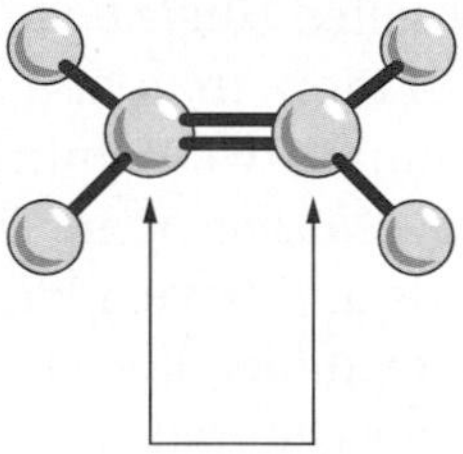

Atoms joined by two or more covalent bonds cannot rotate freely around the bond axis. This restriction is a major influence on the three-dimensional shape of many macromolecules.

C–H COMPOUNDS

Carbon and hydrogen together make stable compounds (or groups) called hydrocarbons. These are nonpolar, do not form hydrogen bonds, and are generally insoluble in water.

methane

methyl group

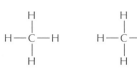

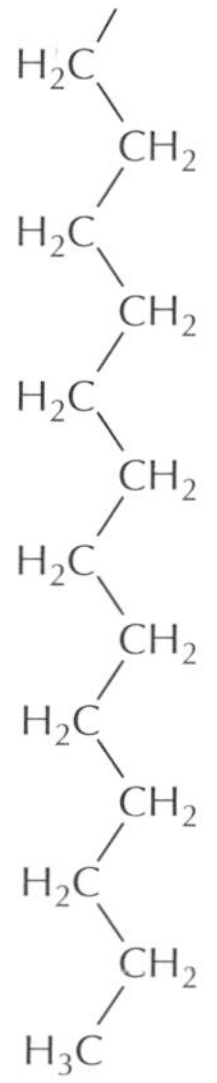

part of the hydrocarbon "tail" of a fatty acid molecule

ALTERNATING DOUBLE BONDS

The carbon chain can include double bonds. If these are on alternate carbon atoms, the bonding electrons move within the molecule, stabilizing the structure by a phenomenon called resonance.

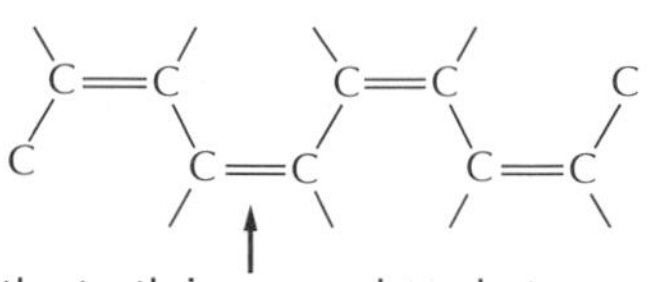

Alternating double bonds in a ring can generate a very stable structure.

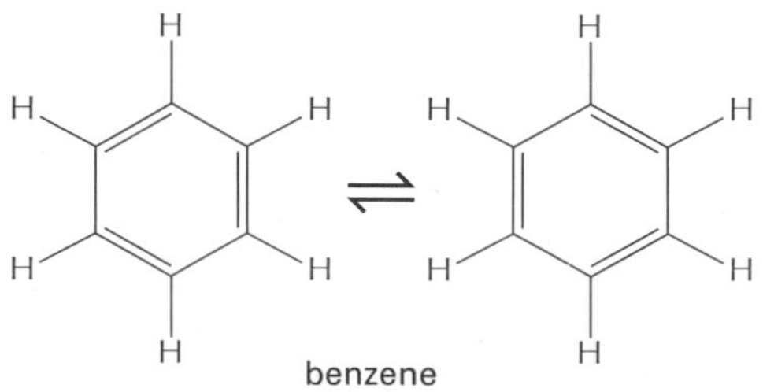

the truth is somewhere between these two structures

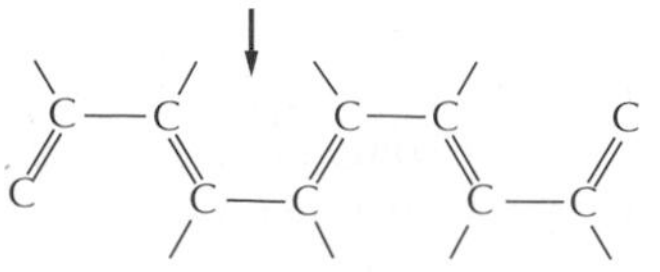

benzene

often written as

Panel 2–1 Chemical bonds and groups commonly encountered in biological molecules.

C–O COMPOUNDS

Many biological compounds contain a carbon bonded to an oxygen. For example,

alcohol

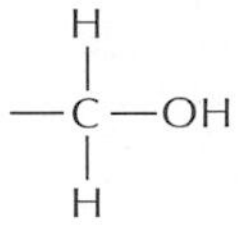

The –OH is called a hydroxyl group.

aldehyde

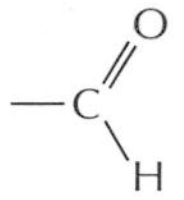

ketone

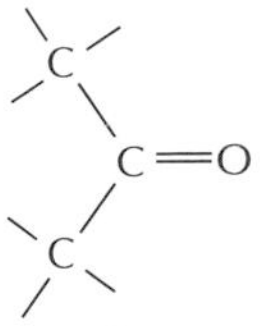

The C=O is called a carbonyl group.

carboxylic acid

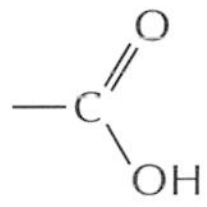

The –COOH is called a carboxyl group. In water this loses an H^+ ion to become –COO⁻.

esters   Esters are formed by combining an acid and an alcohol.

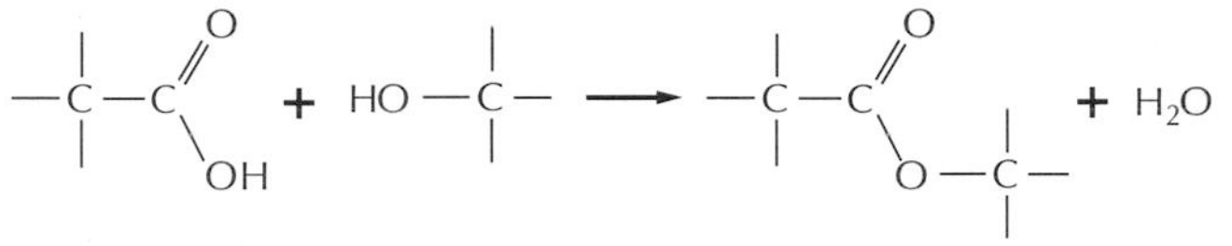

acid   alcohol   ester

C–N COMPOUNDS

Amines and amides are two important examples of compounds containing a carbon linked to a nitrogen.

Amines in water combine with an H^+ ion to become positively charged.

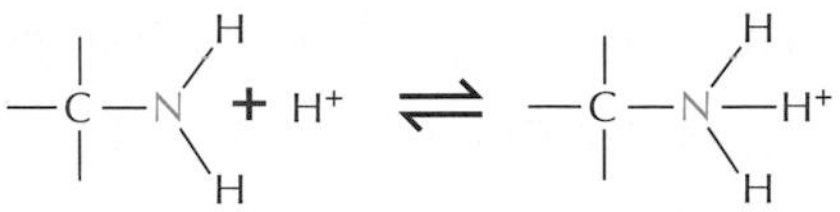

Amides are formed by combining an acid and an amine. Unlike amines, amides are uncharged in water. An example is the peptide bond that joins amino acids in a protein.

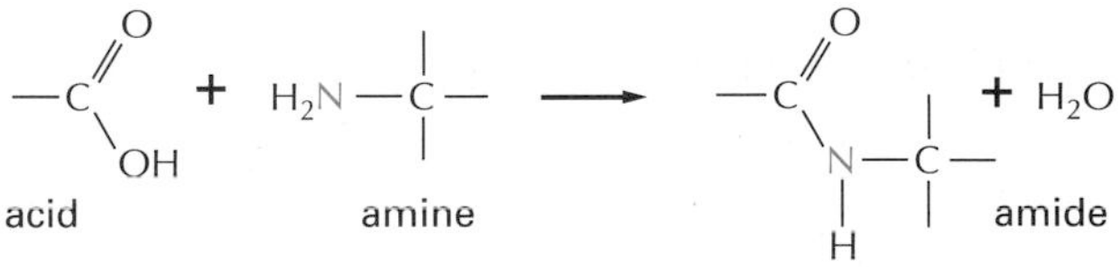

acid   amine   amide

Nitrogen also occurs in several ring compounds, including important constituents of nucleic acids: purines and pyrimidines.

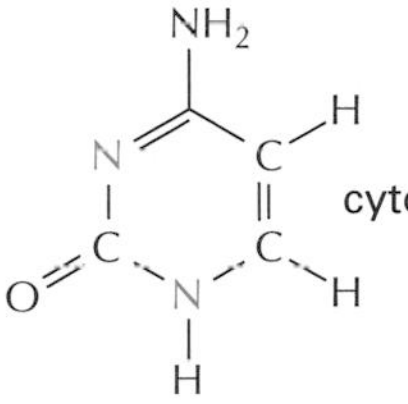

cytosine (a pyrimidine)

PHOSPHATES

Inorganic phosphate is a stable ion formed from phosphoric acid, H_3PO_4. It is often written as P_i.

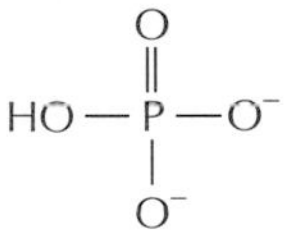

Phosphate esters can form between a phosphate and a free hydroxyl group. Phosphate groups are often attached to proteins in this way.

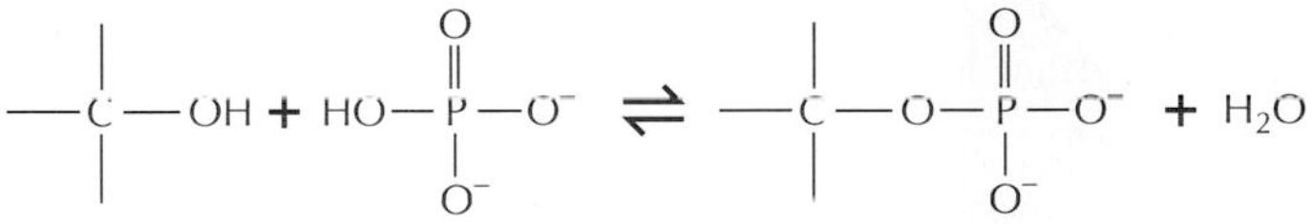

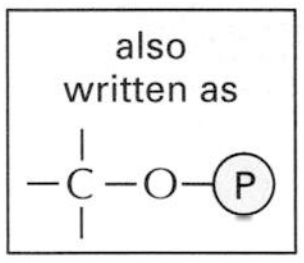

also written as

The combination of a phosphate and a carboxyl group, or two or more phosphate groups, gives an acid anhydride.

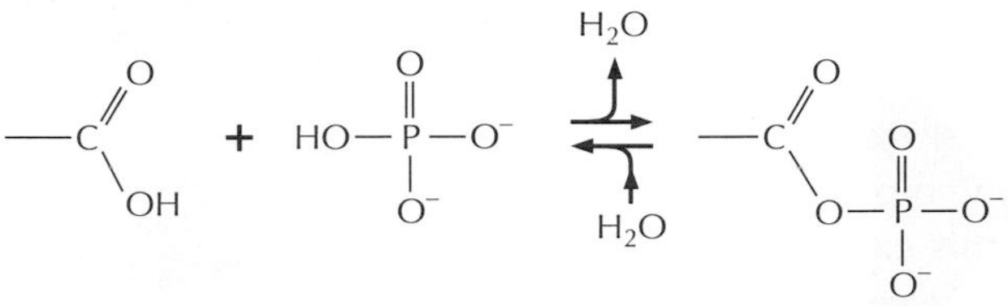

high-energy acyl phosphate bond (carboxylic–phosphoric acid anhydride) found in some metabolites

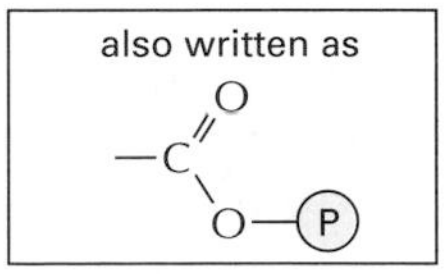

also written as

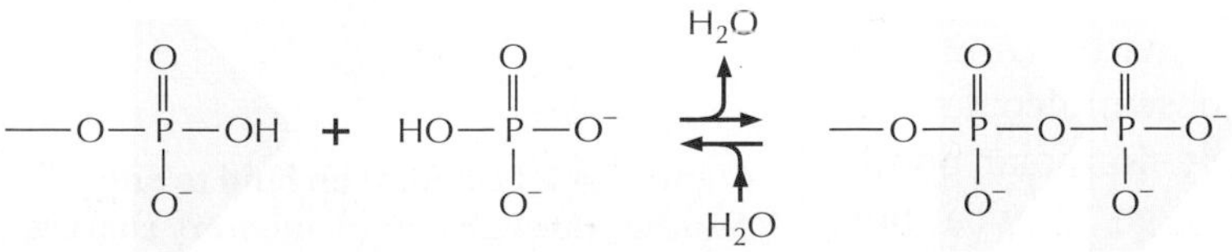

phosphoanhydride—a high-energy bond found in molecules such as ATP

also written as

resemble (but are weaker than) those discussed previously for NaCl (Figure 2–12). When enough weak noncovalent bonds are formed in this manner between two surfaces, the surfaces will stick specifically to each other, as described later in the chapter.

Water Is the Most Abundant Substance in Cells

Water accounts for about 70% of a cell's weight, and most intracellular reactions occur in an aqueous environment. Life on earth began in the ocean, and the conditions in that primeval environment put a permanent stamp on the chemistry of living things. Life therefore hinges on the properties of water (Panel 2–2, pp. 50–51).

In each water molecule (H_2O) the two H atoms are linked to the O atom by covalent bonds. The two bonds are highly polar because the O is strongly attractive for electrons, whereas the H is only weakly attractive. Consequently, there is an unequal distribution of electrons in a water molecule, with a preponderance of positive charge on the two H atoms and of negative charge on the O (see Figure 2–11 and Panel 2–2). When a positively charged region of one water molecule (that is, one of its H atoms) comes close to a negatively charged region (that is, the O) of a second water molecule, the electrical attraction between them can result in a weak bond called a **hydrogen bond.** These bonds are much weaker than covalent bonds and are easily broken by the random thermal motions due to the heat energy of the molecules, so each bond lasts only an exceedingly short time. But the combined effect of many weak bonds is far from trivial. Each water molecule can form hydrogen bonds through its two H atoms to two other water molecules, producing a network in which hydrogen bonds are being continually broken and formed (Panel 2–2). It is only because of the hydrogen bonds that link water molecules together that water is a liquid at room temperature, with a high boiling point and high surface tension, and not a gas.

In general, a hydrogen bond can form whenever a positively charged H held in one molecule by a polar covalent linkage comes close to a negatively charged atom—typically an oxygen or a nitrogen—belonging to another molecule. Hydrogen bonds can also occur between different parts of a single large molecule, where they often help maintain structural stability. They are one member of a family of weak noncovalent bonds that are crucial for allowing large molecules to fold up in unique ways and bind selectively to other molecules, as we shall discuss later in this chapter.

Molecules, such as alcohols, that contain polar bonds and that can form hydrogen bonds with water dissolve readily in water. As mentioned previously, molecules carrying plus or minus charges (ions) likewise interact favorably with water. Such molecules are termed *hydrophilic*, meaning that they are water-loving. A large proportion of the molecules in the aqueous environment of a cell necessarily fall into this category, including sugars, DNA, RNA, and a majority of proteins. *Hydrophobic* (water-hating) molecules, by contrast, are unchanged and form few or no hydrogen bonds, and so do not dissolve in water. Hydrocarbons are an important example (see Panel 2–1, pp. 46–47). In these molecules the H atoms are covalently linked to C atoms by a largely nonpolar bond. Because the H atoms have almost no net positive charge, they cannot participate in hydrogen-bonding to other molecules. This makes the

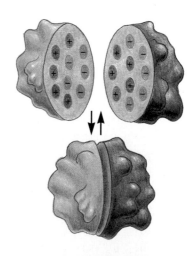

Figure 2–12 Proteins can bind to one another through complementary charges on their surfaces.

hydrocarbon as a whole hydrophobic—a property that is exploited in cells, whose membranes are constructed from molecules that have long hydrocarbon tails, as we shall see in Chapter 11.

Some Polar Molecules Form Acids and Bases in Water

One of the simplest kinds of chemical reaction, and one that has profound significance in cells, takes place when a molecule possessing a highly polar covalent bond between a hydrogen and a second atom dissolves in water. The hydrogen atom in such a molecule has largely given up its electron to the companion atom and so exists as an almost naked positively charged hydrogen nucleus—in other words, a *proton (H+)*. When the polar molecule becomes surrounded by water molecules, the proton is attracted to the partial negative charge on the O atom of an adjacent water molecule and can dissociate from its original partner to associate instead with the oxygen atoms of the water molecule to generate a **hydronium ion (H_3O^+)** (Figure 2–13A). The reverse reaction also takes place very readily, so one has to imagine an equilibrium state in which billions of protons are constantly flitting to and fro from one molecule in the solution to another.

Substances that release protons to form H_3O^+ when they dissolve in water are termed **acids**. The higher the concentration of H_3O^+, the more acidic the solution. H_3O^+ is present even in pure water, at a concentration of 10^{-7} M, as a result of the movement of protons from one water molecule to another (Figure 2–13B). By tradition, the H_3O^+ concentration is usually referred to as the H^+ concentration, even though most H^+ in an aqueous solution is present as H_3O^+. To avoid the use of unwieldy numbers, the concentration of H^+ is expressed using a logarithmic scale called the **pH scale**, as illustrated in Panel 2–2 (pp. 50–51). Pure water has a pH of 7.0.

Because the proton of a hydronium ion can be passed readily to many types of molecules in cells, altering their character, the concentration of H_3O^+ inside a cell (the acidity) must be closely regulated. Molecules that can give up protons will do so more readily if the concentration of H_3O^+ in solution is low and will tend to receive them back if the concentration in solution is high.

The opposite of an acid is a **base.** Just as the defining property of an acid is that it donates protons so as to raise the concentration of H_3O^+ ions, formed by addition of a proton to a water molecule, so the defining

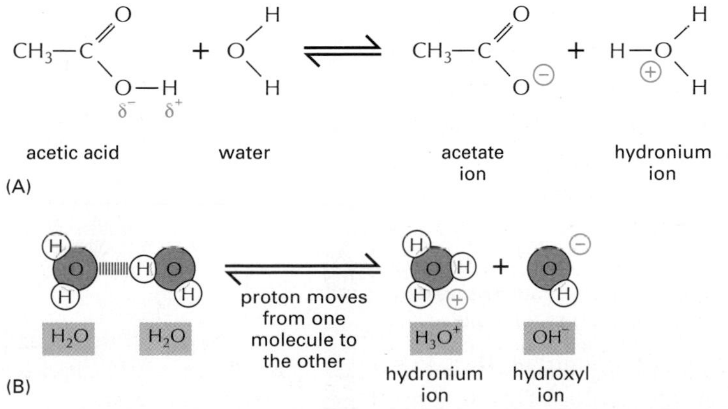

Question 2–3 What, if anything, is wrong with the following statement: "When NaCl is dissolved in water, the water molecules closest to the ions will tend to preferentially orient themselves such that their oxygen atoms face the sodium ions and face away from the chloride ions." Explain your answer.

(A)

(B)

Figure 2–13 Acids in water. (A) The reaction that takes place when a molecule of acetic acid dissolves in water. (B) Water molecules are continuously exchanging protons with each other to form hydronium and hydroxyl ions. These ions in turn rapidly recombine to form water molecules.

HYDROGEN BONDS

Because they are polarized, two adjacent H_2O molecules can form a linkage known as a hydrogen bond. Hydrogen bonds have only about 1/20 the strength of a covalent bond.

Hydrogen bonds are strongest when the three atoms lie in a straight line.

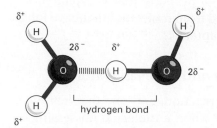

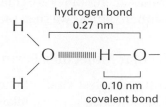

WATER

Two atoms, connected by a covalent bond, may exert different attractions for the electrons of the bond. In such cases the bond is polar, with one end slightly negatively charged (δ^-) and the other slightly positively charged (δ^+).

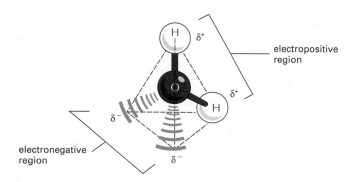

Although a water molecule has an overall neutral charge (having the same number of electrons and protons), the electrons are asymmetrically distributed, which makes the molecule polar. The oxygen nucleus draws electrons away from the hydrogen nuclei, leaving these nuclei with a small net positive charge. The excess of electron density on the oxygen atom creates weakly negative regions at the other two corners of an imaginary tetrahedron.

WATER STRUCTURE

Molecules of water join together transiently in a hydrogen-bonded lattice. Even at 37°C, 15% of the water molecules are joined to four others in a short-lived assembly known as a "flickering cluster."

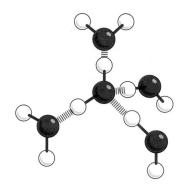

The cohesive nature of water is responsible for many of its unusual properties, such as high surface tension, specific heat, and heat of vaporization.

HYDROPHILIC MOLECULES

Substances that dissolve readily in water are termed hydrophilic. They are composed of ions or polar molecules that attract water molecules through electrical charge effects. Water molecules surround each ion or polar molecule on the surface of a solid substance and carry it into solution.

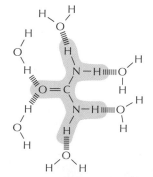

Ionic substances such as sodium chloride dissolve because water molecules are attracted to the positive (Na^+) or negative (Cl^-) charge of each ion.

Polar substances such as urea dissolve because their molecules form hydrogen bonds with the surrounding water molecules

HYDROPHOBIC MOLECULES

Molecules that contain a preponderance of non-polar bonds are usually insoluble in water and are termed hydrophobic. This is true, especially, of hydrocarbons, which contain many C–H bonds. Water molecules are not attracted to such molecules and so have little tendency to surround them and carry them into solution.

WATER AS A SOLVENT

Many substances, such as household sugar, dissolve in water. That is, their molecules separate from each other, each becoming surrounded by water molecules.

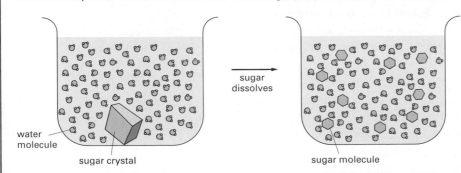

When a substance dissolves in a liquid, the mixture is termed a solution. The dissolved substance (in this case sugar) is the solute, and the liquid that does the dissolving (in this case water) is the solvent. Water is an excellent solvent for many substances because of its polar bonds.

ACIDS

Substances that release hydrogen ions into solution are called acids

HCl $\longrightarrow$ H$^+$ + Cl$^-$
hydrochloric acid   hydrogen ion   chloride ion
(strong acid)

Many of the acids important in the cell are only partially dissociated, and they are therefore weak acids—for example, the carboxyl group (−COOH), which dissociates to give a hydrogen ion in solution

(weak acid)

Note that this is a reversible reaction.

HYDROGEN ION EXCHANGE

Positively charged hydrogen ions (H$^+$) can spontaneously move from one water molecule to another, thereby creating two ionic species.

hydronium ion   hydroxyl ion
(water acting as   (water acting as
a weak base)   a weak acid)

often written as: H$_2$O $\rightleftharpoons$ H$^+$ + OH$^-$
hydrogen   hydroxyl
ion   ion

Since the process is rapidly reversible, hydrogen ions are continually shuttling between water molecules. Pure water contains a steady state concentration of hydrogen ions and hydroxyl ions (both 10^{-7} M).

pH

The acidity of a solution is defined by the concentration of H$^+$ ions it possesses. For convenience we use the pH scale, where

$$pH = -\log_{10}[H^+]$$

For pure water

$$[H^+] = 10^{-7} \text{ moles/liter}$$

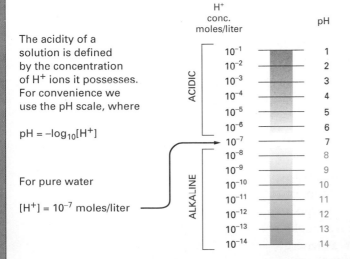

H$^+$ conc. moles/liter		pH
	10^{-1}	1
	10^{-2}	2
	10^{-3}	3
ACIDIC	10^{-4}	4
	10^{-5}	5
	10^{-6}	6
	10^{-7}	7
	10^{-8}	8
	10^{-9}	9
	10^{-10}	10
ALKALINE	10^{-11}	11
	10^{-12}	12
	10^{-13}	13
	10^{-14}	14

BASES

Substances that reduce the number of hydrogen ions in solution are called bases. Some bases, such as ammonia, combine directly with hydrogen ions.

NH$_3$ + H$^+$ $\longrightarrow$ NH$_4^+$
ammonia   hydrogen ion   ammonium ion

Other bases, such as sodium hydroxide, reduce the number of H$^+$ ions indirectly, by making OH$^-$ ions that then combine directly with H$^+$ ions to make H$_2$O.

NaOH $\longrightarrow$ Na$^+$ + OH$^-$
sodium hydroxide   sodium   hydroxyl
(strong base)   ion   ion

Many bases found in cells are partially dissociated and are termed weak bases. This is true of compounds that contain an amino group (−NH$_2$), which has a weak tendency to reversibly accept an H$^+$ ion from water, increasing the quantity of free OH$^-$ ions.

−NH$_2$ + H$^+$ $\rightleftharpoons$ −NH$_3^+$

property of a base is that it raises the concentration of hydroxyl (OH^-) ions, formed by removal of a proton from a water molecule. Thus sodium hydroxide (NaOH) is basic (the term *alkaline* is also used) because it dissociates in aqueous solution to form Na^+ ions and OH^- ions. Another class of bases, especially important in living cells, are those that contain NH_2 groups. These groups can generate OH^- by taking a proton from water: $-NH_2 + H_2O \rightarrow -NH_3^+ + OH^-$.

Because an OH^- ion combines with a H_3O^+ ion to form two water molecules, an increase in the OH^- concentration forces a decrease in the concentration of H_3O^+, and vice versa. A pure solution of water contains an equally low concentration (10^{-7} M) of both ions; it is neither acidic nor basic and is therefore said to be *neutral,* having a pH of 7.0. The inside of cells is kept close to neutrality.

Question 2–4

A. Are there any H_3O^+ ions present in pure water at neutral pH (i.e., at pH = 7.0)? If so, how are they formed?

B. If they exist, what is the ratio of H_3O^+ ions to H_2O molecules at neutral pH? (Hint: the molecular weight of water is 18, and 1 liter of water weighs 1 kg.)

Molecules in Cells

Having looked at the ways atoms combine into small molecules and how these molecules behave in an aqueous environment, we now examine the main classes of small molecules found in cells and their biological roles. We shall see that a few basic categories of molecules, formed from a handful of different elements, give rise to all the extraordinary richness of form and behavior shown by living things.

A Cell Is Formed from Carbon Compounds

If we disregard water, nearly all the molecules in a cell are based on carbon. Carbon is outstanding among all the elements in its ability to form large molecules; silicon is a poor second. Because it is small and has four electrons and four vacancies in its outermost shell, a carbon atom can form four covalent bonds with other atoms. Most important, one carbon atom can join to other carbon atoms through highly stable covalent C–C bonds to form chains and rings and hence generate large and complex molecules with no obvious upper limit to their size (see Panel 2–1, pp. 46–47). The small and large carbon compounds made by cells are called *organic molecules.*

Certain combinations of atoms, such as the methyl ($-CH_3$), hydroxyl ($-OH$), carboxyl ($-COOH$), carbonyl ($-C{=}O$), phosphate ($-PO_3^{2-}$), and amino ($-NH_2$) groups, occur repeatedly in organic molecules. Each such group has distinct chemical and physical properties that influence the behavior of the molecule in which the group occurs. The most common **chemical groups** and some of their properties are summarized in Panel 2–1.

Cells Contain Four Major Families of Small Organic Molecules

The small organic molecules of the cell are carbon compounds with molecular weights in the range 100 to 1000 and containing up to 30 or so carbon atoms. They are usually found free in solution in the cytoplasm and have many different fates. Some are used as *monomer* subunits to construct the giant polymeric *macromolecules*—the proteins, nucleic acids, and large polysaccharides—of the cell. Others act as energy sources and are broken down and transformed into other small mole-

Table 2–3 The Approximate Chemical Composition of a Bacterial Cell

	Percent of Total Cell Weight	Number of Types of Each Molecule
Water	70	1
Inorganic ions	1	20
Sugars and precursors	1	250
Amino acids and precursors	0.4	100
Nucleotides and precursors	0.4	100
Fatty acids and precursors	1	50
Other small molecules	0.2	~300
Macromolecules (proteins, nucleic acids, and polysaccharides)	26	~3000

cules in a maze of intracellular metabolic pathways. Many small molecules have more than one role in the cell—for example, acting both as a potential subunit for a macromolecule and as an energy source. Small organic molecules are much less abundant than organic macromolecules, accounting for only about one-tenth of the total mass of organic matter in a cell (Table 2–3). As a rough guess, there may be a thousand different kinds of these small molecules in a typical cell.

All organic molecules are synthesized from and are broken down into the same set of simple compounds. Both their synthesis and their breakdown occur through sequences of chemical changes that are limited in scope and follow definite rules. As a consequence, the compounds in a cell are chemically related and most can be classified into a small number of distinct families. Broadly speaking, cells contain four major families of small organic molecules: the *sugars*, the *fatty acids*, the *amino acids*, and the *nucleotides* (Figure 2–14). Although many compounds present in cells do not fit into these categories, these four families of small organic molecules, together with the macromolecules made by linking them into long chains, account for a large fraction of cell mass (see Table 2–3).

Question 2–5 Have a close look at the ball-and-stick and the space-filling representations of the glucose molecule shown in Figure 2–15C and D. Note that in both illustrations there are hydrogen atoms of two different sizes. Do we need to apologize because the artist made a mistake? Explain your answer.

Sugars Are Energy Sources for Cells and Subunits of Polysaccharides

The simplest **sugars**—the *monosaccharides*—are compounds with the general formula $(CH_2O)_n$, where n is usually 3, 4, 5, 6, or 7. Sugars, and the molecules made from them, are also called *carbohydrates* because of this simple formula. Glucose, for example, has the formula $C_6H_{12}O_6$ (Figure 2–15 and Panel 2–3, pp. 56–57). The formula, however, does not fully define the molecule: the same set of carbons, hydrogens, and oxygens can be joined together by covalent bonds in a variety of ways, cre-

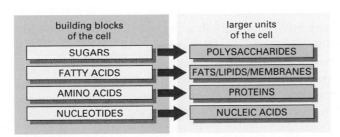

Figure 2–14 The four main families of small organic molecules in cells. They form the monomeric building blocks, or subunits, for most of the macromolecules and other assemblies of the cell. Some, like the sugars and the fatty acids, are also energy sources.

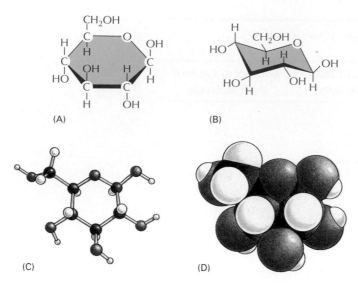

(A)

(B)

(C)

(D)

Figure 2–15 **The structure of glucose, a simple sugar.** The molecule can be represented in several ways. In the structural formulas shown in (A) and (B), the atoms are shown as chemical symbols linked together by *solid lines* representing the covalent bonds. The *thickened lines* are used to indicate the plane of the sugar ring and to show that the –H and –OH groups are not in the same plane as the ring. (B) Schematic representation of the three-dimensional structure of glucose and other similar sugars. (C) A ball-and-stick model in which the three-dimensional arrangement of the atoms in space is indicated. The different *colored balls* are the atoms, and the *sticks* are the covalent bonds. (D) A space-filling model, which, as well as depicting the three-dimensional arrangement of the atoms, also gives some idea of their relative sizes and the surface contours of the molecule. The atoms in (C) and (D) are colored as follows: H, *white*; C, *black*; O, *red*. This is the conventional color coding for these atoms and will be used throughout this book.

ating structures with different shapes. As shown in Panel 2–3 (pp. 56–57), for example, glucose can be converted into a different sugar—mannose or galactose—simply by switching the orientations of specific OH groups relative to the rest of the molecule. Each of these sugars, moreover, can exist in either of two forms, called the D-form and the L-form, which are mirror images of each other. Sets of molecules with the same chemical formula but different structures are called *isomers,* and mirror-image pairs of molecules are called *optical isomers.* Isomers are widespread among organic molecules in general, and they play a major part in generating the enormous variety of sugars.

Monosaccharides can be linked together by covalent bonds to form larger carbohydrates. Two monosaccharides linked together form a disaccharide, such as sucrose, which is composed of a glucose and a fructose unit. Larger sugar polymers range from the *oligosaccharides* (trisaccharides, tetrasaccharides, and so on) up to giant *polysaccharides,* which can contain thousands of monosaccharide units. An outline of sugar structures and chemistry is given in Panel 2–3.

The way that sugars are linked together illustrates some common features of biochemical bond formation. A bond is formed between an –OH group on one sugar and an –OH group on another by a **condensation** reaction, in which a molecule of water is expelled as the bond is formed (Figure 2–16). Subunits in other biological polymers, such as nucleic acids and proteins, are also linked by condensation reactions in which water is expelled. The bonds created by all of these condensation reactions can be broken by the reverse process of **hydrolysis,** in which a molecule of water is consumed (see Figure 2–16).

Because each monosaccharide has several free hydroxyl groups that can form a link to another monosaccharide (or to some other compound), sugar polymers can be branched, and the number of possible polysaccharide structures is extremely large. For this reason it is a much more complex task to determine the arrangement of sugars in a polysaccharide than to determine the nucleotide sequence of a DNA molecule, where each unit is joined to the next in exactly the same way.

The monosaccharide *glucose* has a central role as an energy source for cells. In a series of reactions, it is broken down to smaller molecules, releasing energy that the cell can harness to do useful work, as we shall

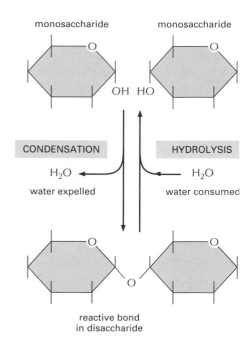

monosaccharide   monosaccharide

OH   HO

CONDENSATION   HYDROLYSIS

H_2O ←   ← H_2O

water expelled   water consumed

reactive bond
in disaccharide

Figure 2–16 **The reaction of two monosaccharides to form a disaccharide.** This reaction belongs to a general category of reactions termed *condensation reactions,* in which two molecules join together due to the loss of a water molecule. The reverse reaction (in which water is added) is termed *hydrolysis.*

explain in Chapter 4. Cells use simple polysaccharides composed only of glucose units—principally *glycogen* (see Panel 2–3, pp. 56–57) in animals and *starch* in plants—as long-term stores of energy.

Sugars do not function exclusively in the production and storage of energy. They can be used, for example, to make mechanical supports. Thus, the most abundant organic chemical on earth—the *cellulose* of plant cell walls—is a polysaccharide of glucose. Another extraordinarily abundant organic substance, the *chitin* of insect exoskeletons and fungal cell walls, is also a polysaccharide—in this case a linear polymer of a sugar derivative called *N*-acetylglucosamine (see Panel 2–3). Polysaccharides of various other sorts are the main components of slime, mucus, and gristle.

Smaller oligosaccharides can be covalently linked to proteins to form glycoproteins and to lipids to form *glycolipids* (Panel 2–4, pp. 58–59), which are found in cell membranes. The surfaces of most cells are clothed and decorated with sugar polymers belonging to glycoproteins and glycolipids in the cell membrane. These sugar side chains are often recognized selectively by other cells. Differences between people in the details of their cell-surface sugars are the molecular basis for the major different human blood groups, for example.

Fatty Acids Are Components of Cell Membranes

A **fatty acid** molecule, such as *palmitic acid* (Figure 2–17 and Panel 2–4), has two chemically distinct regions. One is a long hydrocarbon chain, which is hydrophobic and not very reactive chemically. The other is a carboxyl (–COOH) group, which behaves as an acid (carboxylic acid): it is ionized in solution (–COO⁻), extremely hydrophilic, and chemically reactive. Almost all the fatty acid molecules in a cell are covalently linked to other molecules by their carboxylic acid group (see Panel 2–4).

The hydrocarbon tail of palmitic acid is *saturated:* it has no double bonds between carbon atoms and contains the maximum possible number of hydrogens. Stearic acid, another one of the common fatty acids in animal fat, is also saturated. Some other fatty acids, such as oleic acid, have *unsaturated* tails, with one or more double bonds along their length. The double bonds create kinks in the molecules, interfering with their ability to pack together in a solid mass. It is this that accounts for the difference between hard (saturated) and soft (polyunsaturated) margarine. The many different fatty acids found in cells differ only in the length of their hydrocarbon chains and the number and position of the carbon-carbon double bonds (see Panel 2–4).

Fatty acids serve as a concentrated food reserve in cells, as they can be broken down to produce about six times as much usable energy, weight for weight, as glucose. They are stored in the cytoplasm of many cells in the form of droplets of *triacylglycerol* molecules, which consist of three fatty acid chains joined to a glycerol molecule (see Panel 2–4); these molecules are the animal fats found in meat, butter, and cream, and the plant oils like corn oil and olive oil. When required to provide energy, the fatty acid chains are released from triacylglycerols and broken down into two-carbon units. These two-carbon units are identical to those derived from the breakdown of glucose and enter the same energy-yielding reaction pathways, as will be described in Chapter 4.

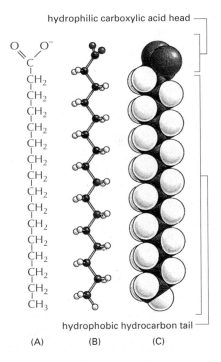

hydrophilic carboxylic acid head

hydrophobic hydrocarbon tail

(A)   (B)   (C)

Figure 2–17 A fatty acid. A fatty acid is composed of a hydrophobic hydrocarbon chain to which is attached a hydrophilic carboxylic acid group. Palmitic acid is shown here. Different fatty acids have different hydrocarbon tails. (A) Structural formula. The carboxylic acid head group is shown in its ionized form. (B) Ball-and-stick model. (C) Space-filling model.

MONOSACCHARIDES

Monosaccharides usually have the general formula $(CH_2O)_n$, where n can be 3, 4, 5, 6, 7, or 8, and have two or more hydroxyl groups. They either contain an aldehyde group ($-C{\overset{O}{\underset{H}{<}}}$) and are called aldoses or a ketone group ($>C=O$) and are called ketoses.

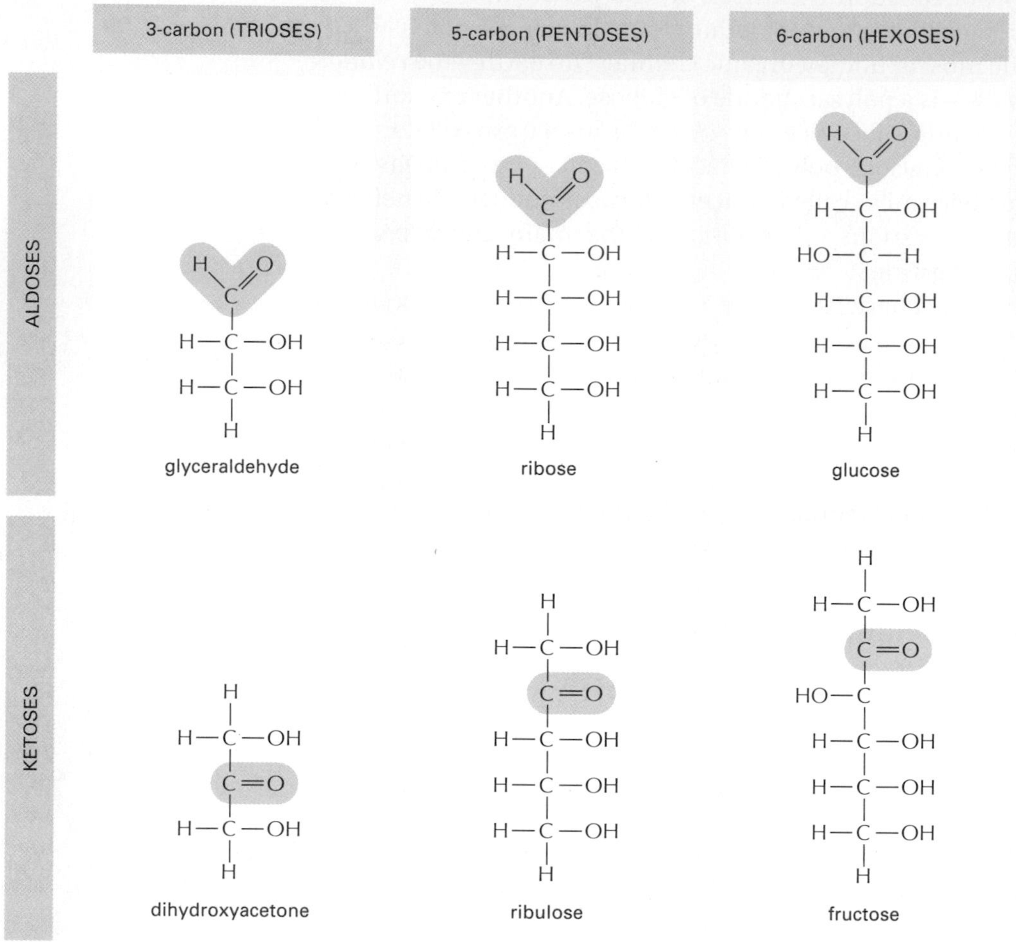

3-carbon (TRIOSES)	5-carbon (PENTOSES)	6-carbon (HEXOSES)

ALDOSES

glyceraldehyde

ribose

glucose

KETOSES

dihydroxyacetone

ribulose

fructose

RING FORMATION

In aqueous solution, the aldehyde or ketone group of a sugar molecule tends to react with a hydroxyl group of the same molecule, thereby closing the molecule into a ring.

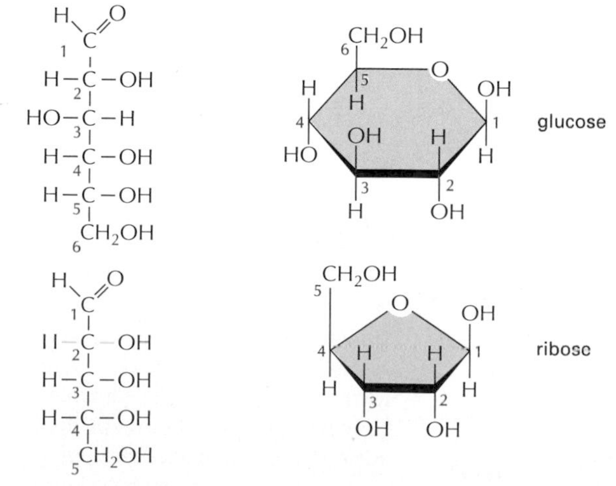

glucose

ribose

Note that each carbon atom has a number.

ISOMERS

Many monosaccharides differ only in the spatial arrangement of atoms—that is, they are isomers. For example, glucose, galactose, and mannose have the same formula $(C_6H_{12}O_6)$ but differ in the arrangement of groups around one or two carbon atoms.

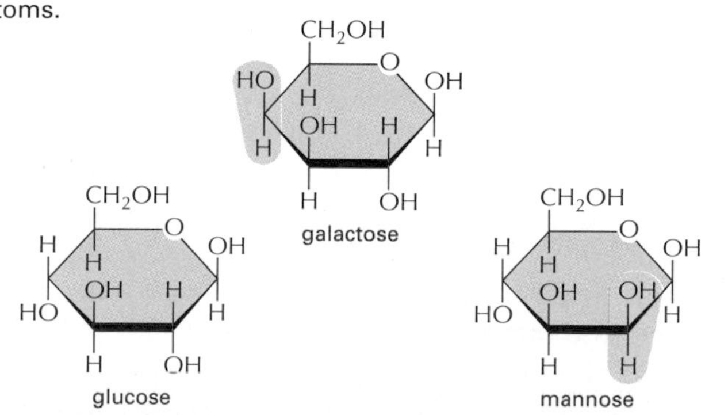

galactose

glucose

mannose

These small differences make only minor changes in the chemical properties of the sugars. But they are recognized by enzymes and other proteins and therefore can have important biological effects.

Panel 2–3 **An outline of some of the types of sugars commonly found in cells.**

α AND β LINKS

The hydroxyl group on the carbon that carries the aldehyde or ketone can rapidly change from one position to the other. These two positions are called α and β.

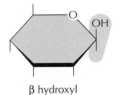

β hydroxyl

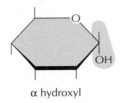

α hydroxyl

As soon as one sugar is linked to another, the α or β form is frozen.

SUGAR DERIVATIVES

The hydroxyl groups of a simple monosaccharide can be replaced by other groups. For example,

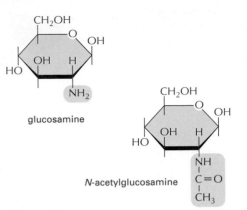

glucosamine

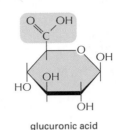

glucuronic acid

N-acetylglucosamine

DISACCHARIDES

The carbon that carries the aldehyde or the ketone can react with any hydroxyl group on a second sugar molecule to form a disaccharide Three common disaccharides are

   maltose (glucose + glucose)
   lactose (galactose + glucose)
   sucrose (glucose + fructose)

The reaction forming sucrose is shown here.

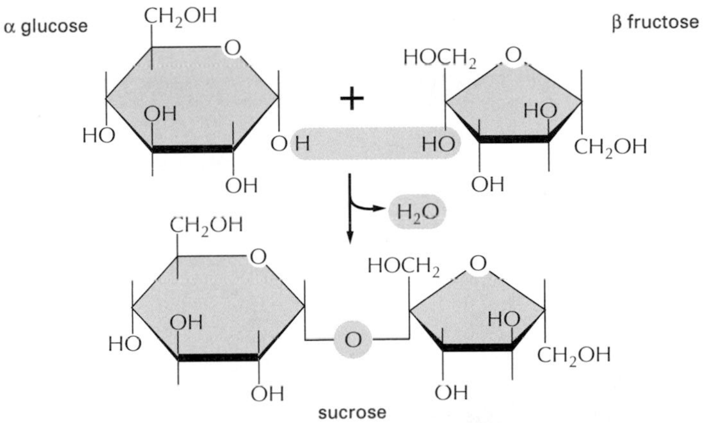

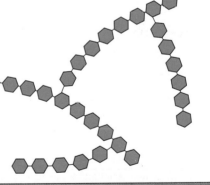

α glucose   β fructose

H_2O

sucrose

OLIGOSACCHARIDES AND POLYSACCHARIDES

Large linear and branched molecules can be made from simple repeating units. Short chains are called oligosaccharides, while long chains are called polysaccharides. Glycogen, for example, is a polysaccharide made entirely of glucose units joined together.

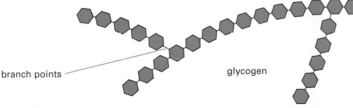

branch points

glycogen

COMPLEX OLIGOSACCHARIDES

In many cases a sugar sequence is nonrepetitive. Many different molecules are possible. Such complex oligosaccharides are usually linked to proteins or to lipids, as is this oligosaccharide, which is part of a cell-surface molecule that defines a particular blood group.

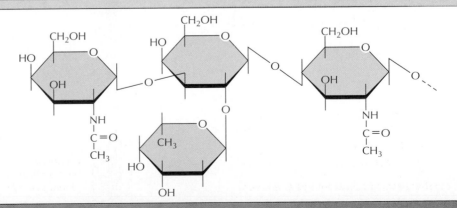

COMMON FATTY ACIDS

These are carboxylic acids with long hydrocarbon tails.

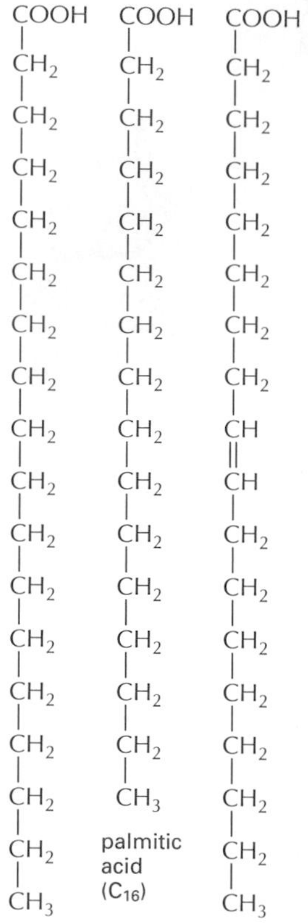

Hundreds of different kinds of fatty acids exist. Some have one or more double bonds in their hydrocarbon tail and are said to be unsaturated. Fatty acids with no double bonds are saturated.

COOH	COOH	COOH
CH₂	CH₂	CH₂
CH₂	CH₂	CH₂
CH₂	CH₂	CH₂
CH₂	CH₂	CH₂
CH₂	CH₂	CH₂
CH₂	CH₂	CH₂
CH₂	CH₂	CH₂
CH₂	CH₂	CH₂
CH₂	CH₂	CH
CH₂	CH₂	CH
CH₂	CH₂	CH₂
CH₂	CH₂	CH₂
CH₂	CH₂	CH₂
CH₂	CH₂	CH₂
CH₂	CH₃	CH₂
CH₂	palmitic acid (C₁₆)	CH₂
CH₃		CH₃
stearic acid (C₁₈)		oleic acid (C₁₈)

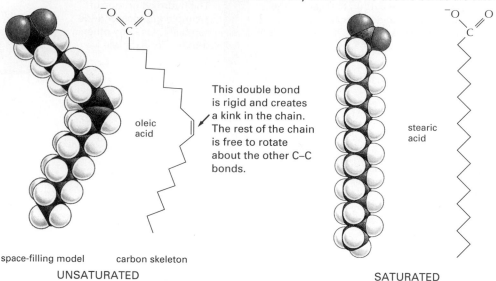

oleic acid

This double bond is rigid and creates a kink in the chain. The rest of the chain is free to rotate about the other C–C bonds.

stearic acid

space-filling model   carbon skeleton

UNSATURATED   **SATURATED**

TRIACYLGLYCEROLS

Fatty acids are stored as an energy reserve (fats and oils) through an ester linkage to glycerol to form triacylglycerols.

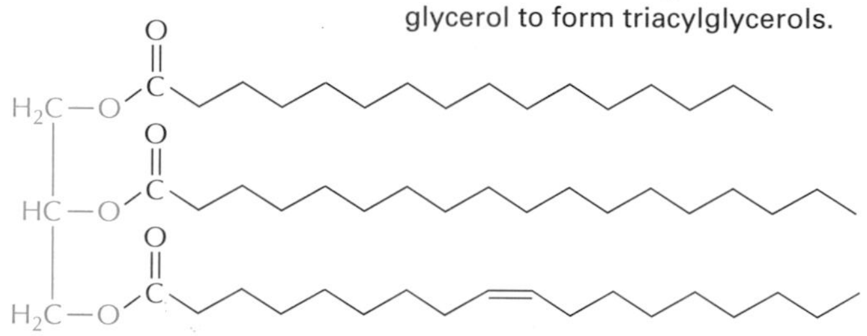

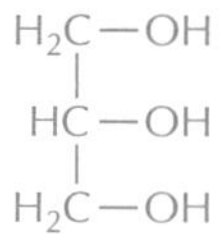

H₂C—OH
HC—OH
H₂C—OH

glycerol

CARBOXYL GROUP

If free, the carboxyl group of a fatty acid will be ionized.

But more usually it is linked to other groups to form either esters

or amides.

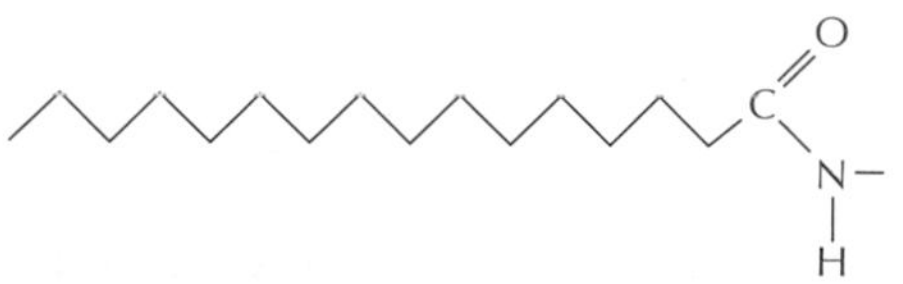

PHOSPHOLIPIDS

Phospholipids are the major constituents of cell membranes.

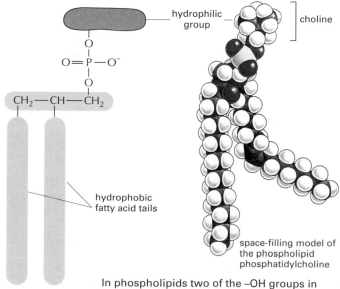

hydrophilic group

choline

$O=P-O^-$

CH₂—CH—CH₂

hydrophobic fatty acid tails

space-filling model of the phospholipid phosphatidylcholine

general structure of a phospholipid

In phospholipids two of the –OH groups in glycerol are linked to fatty acids, while the third –OH group is linked to phosphoric acid. The phosphate is further linked to one of a variety of small polar groups (alcohols).

LIPID AGGREGATES

Fatty acids have a hydrophilic head and a hydrophobic tail.

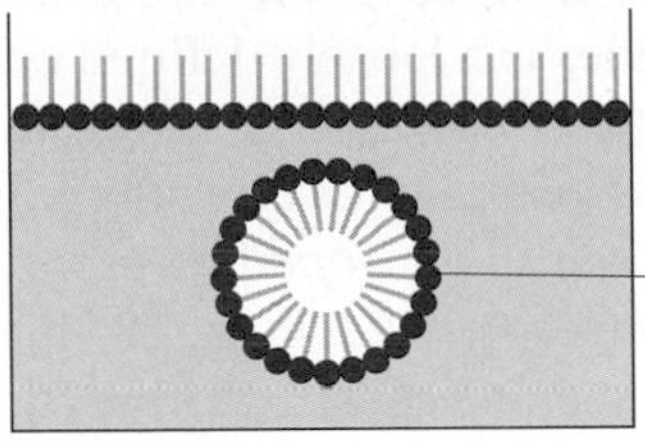

— micelle

In water they can form a surface film or form small micelles.

Their derivatives can form larger aggregates held together by hydrophobic forces:

Triglycerides form large spherical fat droplets in the cell cytoplasm.

200 nm or more

Phospholipids and glycolipids form self-sealing lipid bilayers that are the basis for all cellular membranes.

|← 4 nm →|

OTHER LIPIDS

Lipids are defined as the water-insoluble molecules in cells that are soluble in organic solvents. Two other common types of lipids are steroids and polyisoprenoids. Both are made from isoprene units.

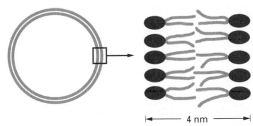

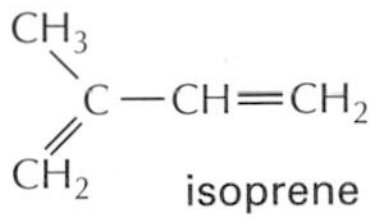

isoprene

STEROIDS

Steroids have a common multiple-ring structure.

cholesterol—found in many membranes

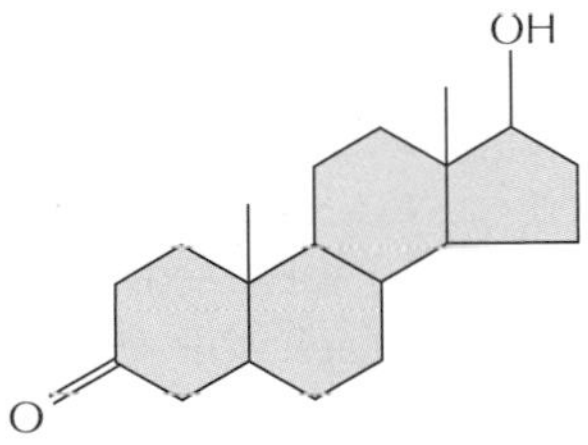

testosterone—male steroid hormone

GLYCOLIPIDS

Like phospholipids, these compounds are composed of a hydrophobic region, containing two long hydrocarbon tails, and a polar region, which, however, contains one or more sugar residues and no phosphate.

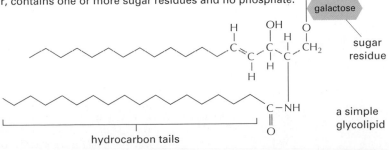

galactose

sugar residue

a simple glycolipid

hydrocarbon tails

POLYISOPRENOIDS

long chain polymers of isoprene

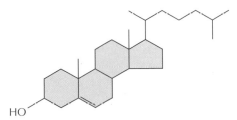

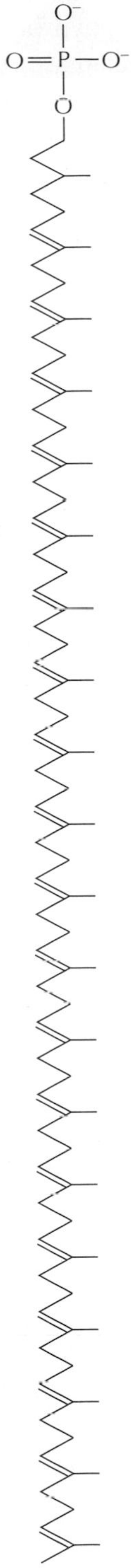

dolichol phosphate—used to carry activated sugars in the membrane-associated synthesis of glycoproteins and some polysaccharides

Fatty acids and their derivatives such as triacylglycerols are examples of *lipids*. This class of biological molecules is a loosely defined collection with the common feature that they are insoluble in water and soluble in fat and organic solvents such as benzene. Lipids typically contain long hydrocarbon chains, as in the fatty acids and *isoprenes*, or multiple linked aromatic rings, as in the *steroids* (see Panel 2–4, pp. 58–59).

The most important function of fatty acids in cells is in the construction of cell membranes. These thin sheets enclose all cells and surround their internal organelles. They are composed largely of *phospholipids*, which are small molecules that, like triacylglycerols, are constructed mainly from fatty acids and glycerol. In phospholipids the glycerol is joined to two fatty acid chains, rather than to three, as in triacylglycerols. The "third" site on the glycerol is linked to a hydrophilic phosphate group, which is in turn attached to a small hydrophilic compound such as choline (see Panel 2–4). Each phospholipid molecule, therefore, has a hydrophobic tail composed of the two fatty acid chains and a hydrophilic head, where the phosphate is located. This gives them different physical and chemical properties from triacylglycerols, which are predominantly hydrophobic. Molecules like phospholipids, with both hydrophobic and hydrophilic regions, are termed *amphipathic*.

The membrane-forming property of phospholipids results from their amphipathic nature. Phospholipids will spread over the surface of water to form a monolayer of phospholipid molecules, with the hydrophobic tails facing the air and the hydrophilic heads in contact with the water. Two such molecular layers can combine tail-to-tail in water to make a phospholipid sandwich, or *lipid bilayer*, which is the structural basis of all cell membranes (Figure 2–18) (discussed in Chapter 11).

Amino Acids Are the Subunits of Proteins

Amino acids are a varied class of molecules with one defining property: they all possess a carboxylic acid group and an amino group, both linked to a single carbon atom called the α-carbon (Figure 2–19). The chemical variety comes from the side chain also attached to the α-carbon. The importance of amino acids to the cell comes from their role in making

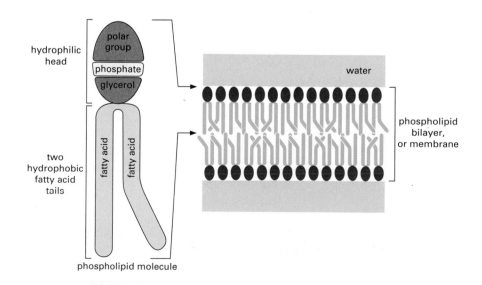

phospholipid molecule

Figure 2–18 Phospholipid structure and the orientation of phospholipids in membranes. In an aqueous environment, the hydrophobic tails of phospholipids pack together to exclude water, forming a bilayer with the hydrophilic head of each phospholipid facing the water.

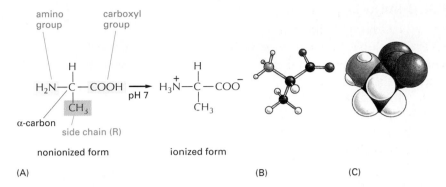

amino group

carboxyl group

$$H_2N-\overset{\overset{\displaystyle H}{|}}{\underset{\underset{\displaystyle CH_3}{|}}{C}}-COOH \xrightarrow{pH\ 7} H_3\overset{+}{N}-\overset{\overset{\displaystyle H}{|}}{\underset{\underset{\displaystyle CH_3}{|}}{C}}-COO^-$$

α-carbon

side chain (R)

nonionized form

ionized form

(A)   (B)   (C)

Figure 2–19 **The amino acid alanine.** (A) In the cell, where the pH is close to 7, the free amino acid exists in its ionized form; but when it is incorporated into a polypeptide chain, the charges on the amino and carboxyl groups disappear. (B) A ball-and-stick model and (C) a space-filling model of alanine (H, *white;* C, *black;* O, *red;* N, *blue*).

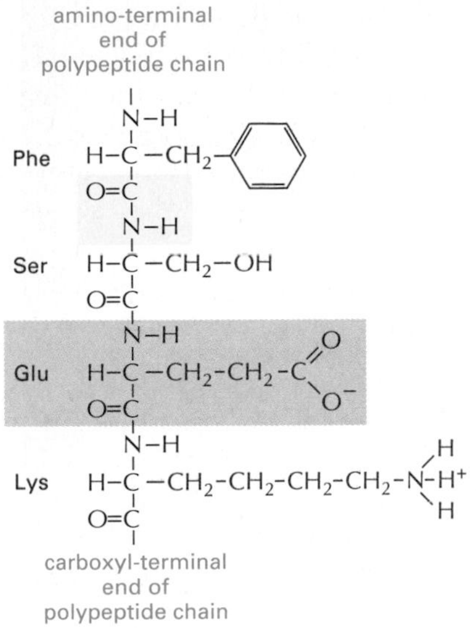

amino-terminal end of polypeptide chain

Phe

Ser

Glu

Lys

carboxyl-terminal end of polypeptide chain

Figure 2–20 **A small part of a protein molecule.** The four amino acid residues shown are linked together by three peptide bonds, one of which is highlighted in *yellow.* One of the amino acids is shaded in *gray.* The amino acid side chains are shown in *red.* The two ends of a polypeptide chain are chemically distinct. One end, the N-terminus, terminates in an amino group, and the other, the C-terminus, in a carboxyl group.

proteins, which are polymers of amino acids joined head-to-tail in a long chain that is then folded into a three-dimensional structure unique to each type of protein. The covalent linkage between two adjacent amino acids in a protein chain is called a *peptide bond;* the chain of amino acids is also known as a *polypeptide* (Figure 2–20). Regardless of the specific amino acids from which it is made, the polypeptide has an amino (NH_2) group at one end (its *N-terminus*) and a carboxyl (COOH) group at its other end (its *C-terminus*). This gives it a definite directionality—a structural (as opposed to an electrical) polarity.

Twenty types of amino acids are found commonly in proteins, each with a different side chain attached to the α-carbon atom (Panel 2–5, pp. 62–63). The same 20 amino acids occur over and over again in all proteins, whether from bacteria, plants, or animals. How this precise set of 20 amino acids came to be chosen is one of the mysteries surrounding the evolution of life; there is no obvious chemical reason why other amino acids could not have served just as well. But once the choice was established, it could not be changed; too much depended on it.

Like sugars, all amino acids, except glycine, exist as optical isomers in D- and L-forms (see Panel 2–5). But only L-forms are ever found in proteins (although D-amino acids occur as part of bacterial cell walls and in some antibiotics). The origin of this exclusive use of L-amino acids to make proteins is another evolutionary mystery.

The chemical versatility that the 20 standard amino acids provide is vitally important to the function of proteins. Five of the 20 amino acids have side chains that can form ions in solution and thereby can carry a charge (as in the case of alanine, shown in Figure 2–19). The others are uncharged; some are polar and hydrophilic, and some are nonpolar and hydrophobic (see Panel 2–5). As we shall discuss in Chapter 5, the collective properties of the amino acid side chains underlie all the diverse and sophisticated functions of proteins.

Nucleotides Are the Subunits of DNA and RNA

A **nucleotide** is a molecule made up of a nitrogen-containing ring compound linked to a five-carbon sugar. This sugar can be either ribose or

Question 2–6 Why do you suppose only L-amino acids and not a random mixture of the L- and D-forms of each amino acid are used to make proteins?

FAMILIES OF AMINO ACIDS

The common amino acids are grouped according to whether their side chains are

    acidic
    basic
    uncharged polar
    nonpolar

These 20 amino acids are given both three-letter and one-letter abbreviations.

Thus: alanine = Ala = A

BASIC SIDE CHAINS

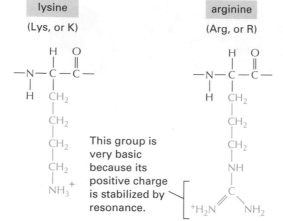

lysine (Lys, or K)

arginine (Arg, or R)

histidine (His, or H)

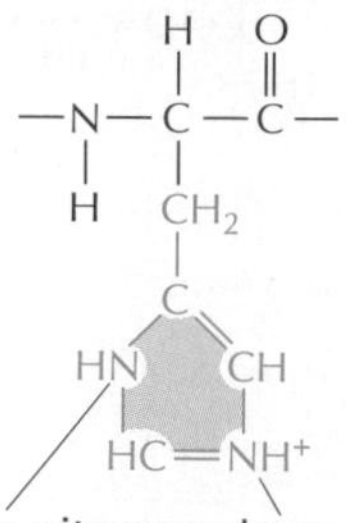

This group is very basic because its positive charge is stabilized by resonance.

These nitrogens have a relatively weak affinity for an H^+ and are only partly positive at neutral pH.

THE AMINO ACID

The general formula of an amino acid is

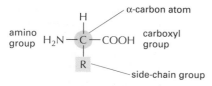

amino group — H_2N
α-carbon atom
carboxyl group — COOH
side-chain group — R

R is commonly one of 20 different side chains. At pH 7 both the amino and carboxyl groups are ionized.

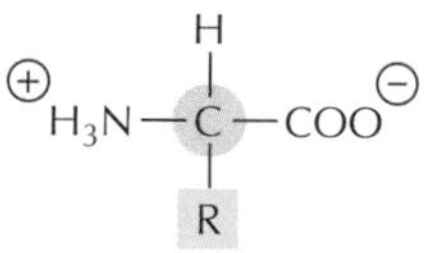

$^{(+)}H_3N - C - COO^{(-)}$

OPTICAL ISOMERS

The α-carbon atom is asymmetric, which allows for two mirror image (or stereo-) isomers, L and D.

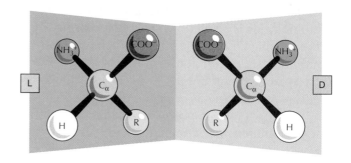

L   D

Proteins consist exclusively of L-amino acids.

PEPTIDE BONDS

Amino acids are commonly joined together by an amide linkage, called a peptide bond.

H_2O

Peptide bond: The four atoms in each *gray box* form a rigid planar unit. There is no rotation around the C–N bond.

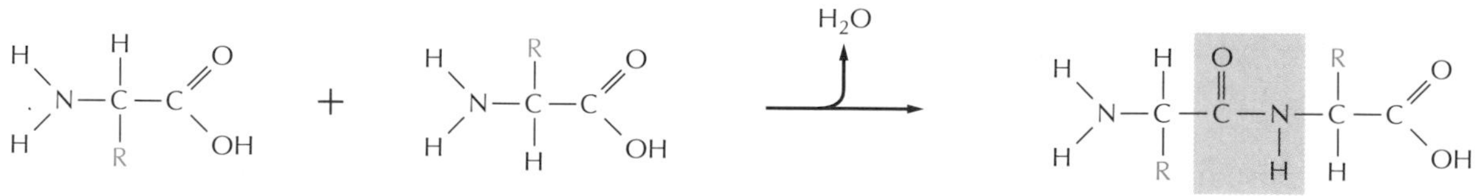

Proteins are long polymers of amino acids linked by peptide bonds, and they are always written with the N-terminus toward the left. The sequence of this tripeptide is histidine-cysteine-valine.

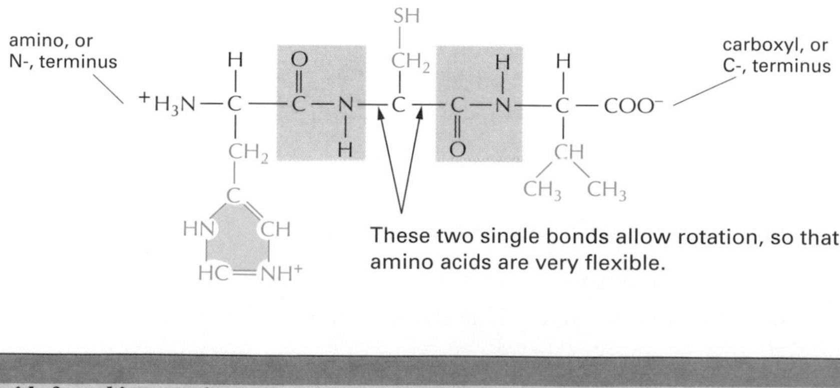

amino, or N-, terminus

carboxyl, or C-, terminus

These two single bonds allow rotation, so that long chains of amino acids are very flexible.

ACIDIC SIDE CHAINS

aspartic acid
(Asp, or D)

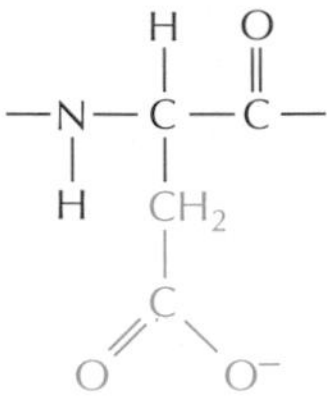

glutamic acid
(Glu, or E)

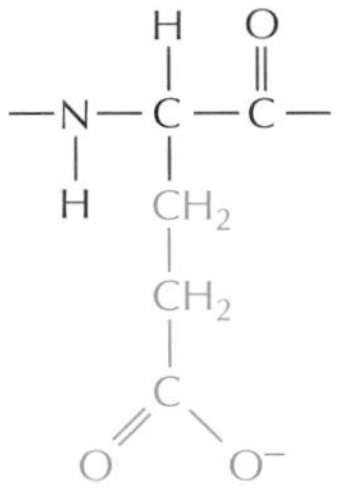

UNCHARGED POLAR SIDE CHAINS

asparagine
(Asn, or N)

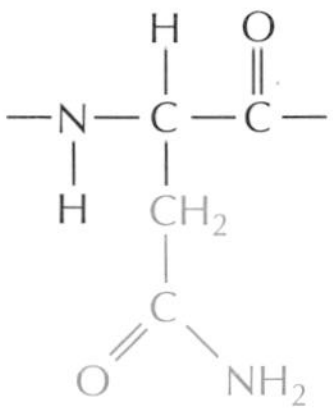

glutamine
(Gln, or Q)

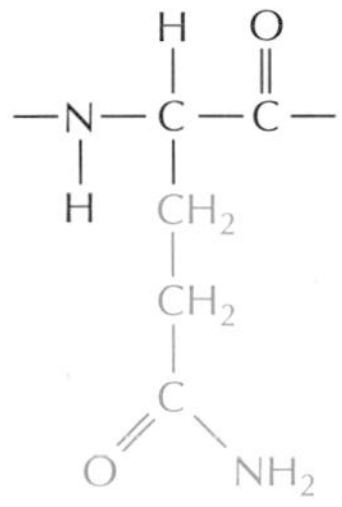

Although the amide N is not charged at neutral pH, it is polar.

serine
(Ser, or S)

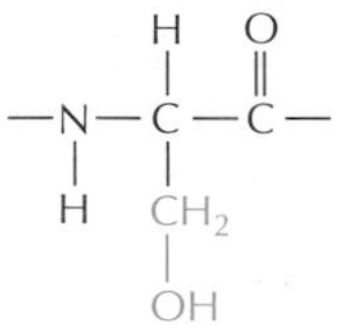

threonine
(Thr, or T)

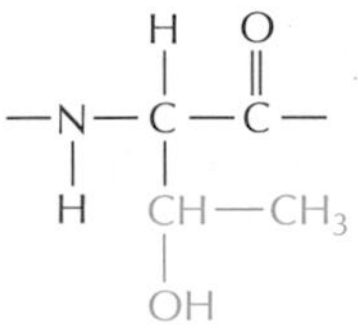

tyrosine
(Tyr, or Y)

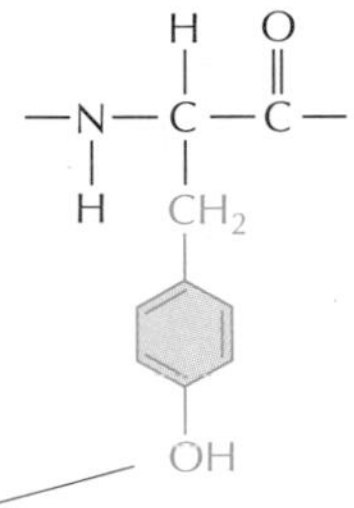

The –OH group is polar.

NONPOLAR SIDE CHAINS

alanine
(Ala, or A)

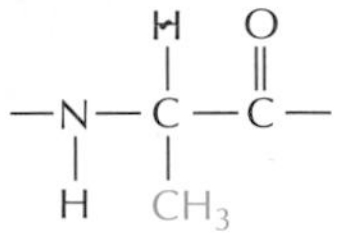

valine
(Val, or V)

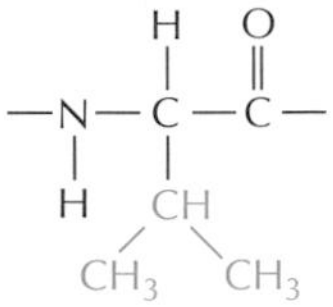

leucine
(Leu, or L)

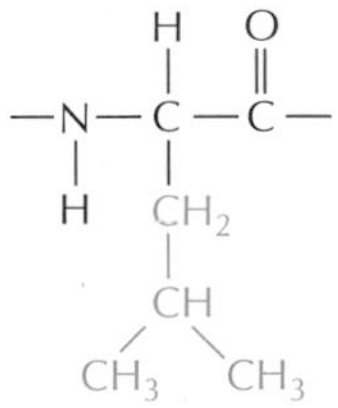

isoleucine
(Ile, or I)

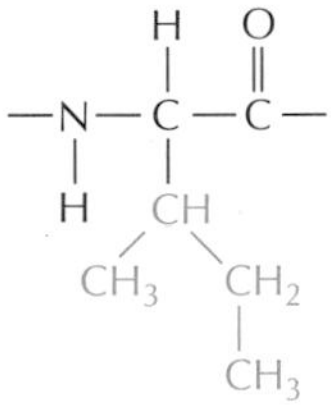

proline
(Pro, or P)

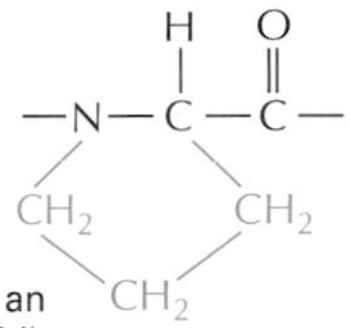

(actually an imino acid)

phenylalanine
(Phe, or F)

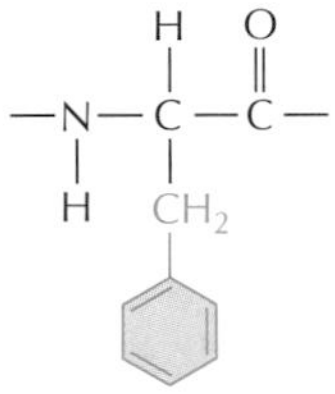

methionine
(Met, or M)

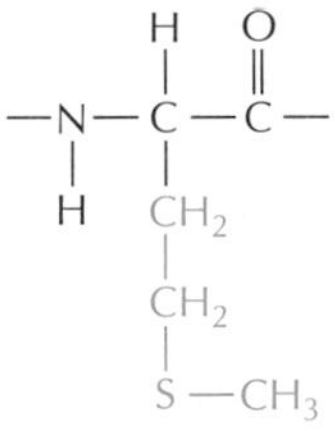

tryptophan
(Trp, or W)

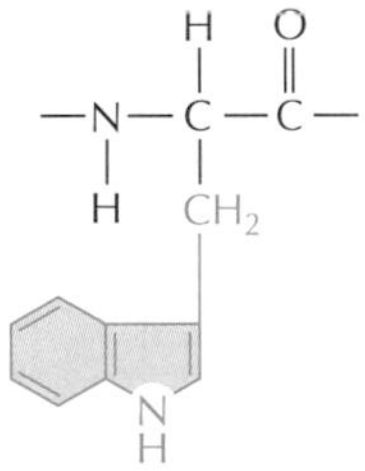

glycine
(Gly, or G)

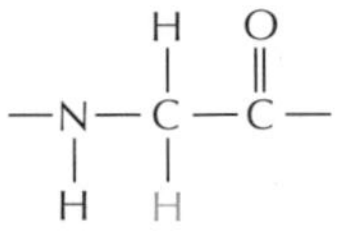

cysteine
(Cys, or C)

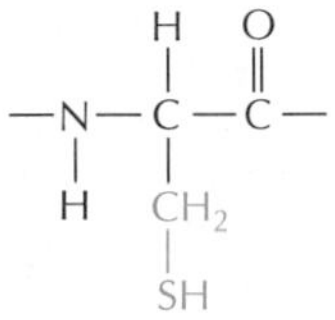

Disulfide bonds can form between two cysteine side chains in proteins.

$- -CH_2 - S - S - CH_2 - -$

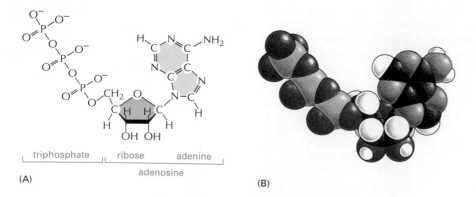

Figure 2–21 Chemical structure of adenosine triphosphate (ATP). (A) Structural formula. (B) Space-filling model. In (B) the colors of the atoms are C, *black;* N, *blue;* H, *white;* O, *red;* and P, *green.*

(A)

triphosphate | ribose | adenine

adenosine

(B)

deoxyribose, and it carries one or more phosphate groups (Panel 2–6, pp. 66–67). Nucleotides containing ribose are known as ribonucleotides, and those containing deoxyribose as deoxyribonucleotides. The nitrogen-containing rings are generally referred to as *bases* for historical reasons: under acidic conditions they can each bind an H⁺ (proton) and thereby increase the concentration of OH⁻ ions in aqueous solution. There is a strong family resemblance between the different bases. *Cytosine (C), thymine (T),* and *uracil (U)* are called *pyrimidines* because they all derive from a six-membered pyrimidine ring; *guanine (G)* and *adenine (A)* are *purine* compounds, with a second, five-membered ring fused to the six-membered ring. Each nucleotide is named from the base it contains (see Panel 2–6).

Nucleotides can act as short-term carriers of chemical energy. Above all others, the ribonucleotide **adenosine triphosphate,** or **ATP** (Figure 2–21), participates in the transfer of energy in hundreds of cellular reactions. ATP is formed through reactions that are driven by the energy released by the oxidative breakdown of foodstuffs. Its three phosphates are linked in series by two *phosphoanhydride bonds* (see Panel 2–6), whose rupture releases large amounts of useful energy. The terminal phosphate group in particular is frequently split off by hydrolysis, often transferring a phosphate to other molecules and releasing energy that drives energy-requiring biosynthetic reactions elsewhere in the cell

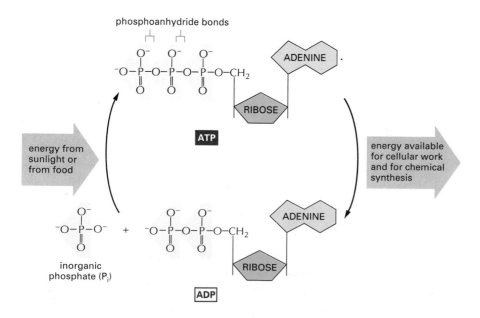

Figure 2–22 The ATP molecule serves as an energy carrier in cells. The energy-requiring formation of ATP from ADP and inorganic phosphate is coupled to the energy-yielding oxidation of foodstuffs (in animal cells, fungi, and some bacteria) or to the capture of light (in plant cells and some bacteria). The hydrolysis of this ATP back to ADP and inorganic phosphate in turn provides the energy to drive many cellular reactions.

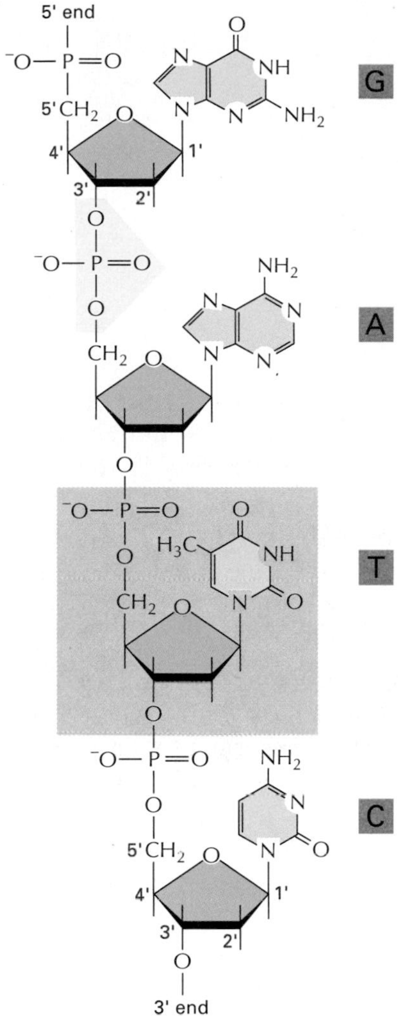

Figure 2–23 **A short length of one chain of a deoxyribonucleic acid (DNA) molecule, showing four nucleotide residues.** One of the phosphodiester bonds that links adjacent nucleotide residues is highlighted in *yellow,* and one of the nucleotides is enclosed in a *gray box.* Nucleotides are linked together by a phosphodiester linkage between specific carbon atoms of the ribose, known as the 5′ and 3′ atoms. For this reason, one end of a polynucleotide chain, the 5′ end, will have a free phosphate group and the other, the 3′ end, a free hydroxyl group. The linear sequence of nucleotide residues in a polynucleotide chain is commonly abbreviated by a one-letter code, and the sequence is always read from the 5′ end. In the example illustrated the sequence is G-A-T-C.

(Figure 2–22). Other nucleotide derivatives serve as carriers for the transfer of other chemical groups, as will be described in Chapter 3.

The most fundamental role of nucleotides in the cell, however, is in the storage and retrieval of biological information. Nucleotides serve as building blocks for the construction of *nucleic acids*—long polymers in which nucleotide subunits are covalently linked by the formation of a *phosphodiester bond* between the phosphate group attached to the sugar of one nucleotide and a hydroxyl group on the sugar of the next nucleotide. Nucleic acid chains are synthesized from energy-rich nucleoside triphosphates by a condensation reaction that releases inorganic pyrophosphate during phosphodiester bond formation.

There are two main types of nucleic acids, differing in the type of sugar in their sugar-phosphate backbone. Those based on the sugar *ribose* are known as **ribonucleic acids,** or **RNA,** and contain the bases A, G, C, and U. Those based on *deoxyribose* (in which the hydroxyl at the 2′ position of the ribose carbon ring is replaced by a hydrogen (see Panel 2–6) are known as **deoxyribonucleic acids,** or **DNA,** and contain the bases A, G, C, and T (T is chemically similar to the U in RNA) (Figure 2–23). RNA usually occurs in cells in the form of a single polynucleotide chain, but DNA is virtually always in the form of a double-stranded molecule composed of two polynucleotide chains running antiparallel to each other and held together by hydrogen-bonding between the bases of the two chains.

The linear sequence of nucleotides in a DNA or an RNA encodes the genetic information of the cell. The ability of the bases in different nucleic acid molecules to recognize and pair with each other by hydrogen-bonding (called *base-pairing*)—G with C, and A with either T or U—underlies all of heredity and evolution and is explained in Chapters 7 and 8.

Macromolecules Contain a Specific Sequence of Subunits

On a weight basis, macromolecules are by far the most abundant of the carbon-containing molecules in a living cell (Figure 2–24). They are the principal building blocks from which a cell is constructed and also the components that confer the most distinctive properties of living things. The macromolecules in cells are polymers that are constructed simply by covalently linking small organic molecules (called *monomers,* or *subunits*) into long chains (Figure 2–25). Yet they have many unexpected properties that could not have been predicted from their simple constituents. For example, DNA and RNA molecules (the nucleic acids) store and transmit hereditary information.

Question 2–7 What is meant by "polarity" of a polypeptide chain and by "polarity" of a chemical bond? How do the meanings differ?

BASES

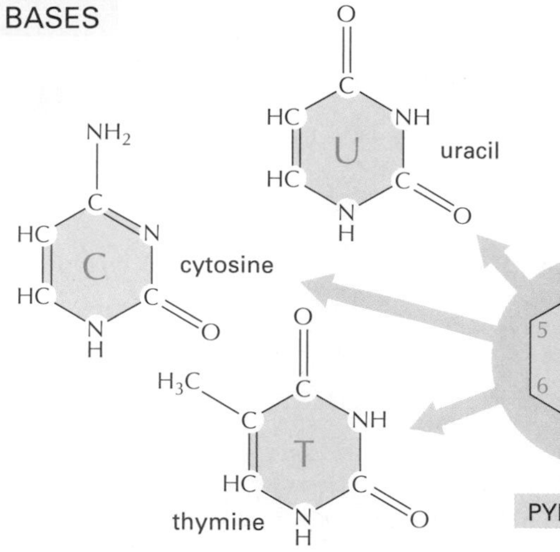

NH₂ is part of the structures shown.

cytosine

uracil

thymine

The bases are nitrogen-containing ring compounds, either pyrimidines or purines.

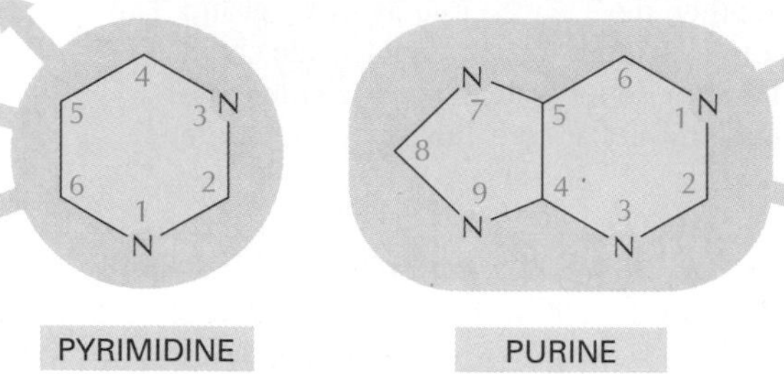

PYRIMIDINE

PURINE

adenine

guanine

PHOSPHATES

The phosphates are normally joined to the C5 hydroxyl of the ribose or deoxyribose sugar (designated 5'). Mono-, di-, and triphosphates are common.

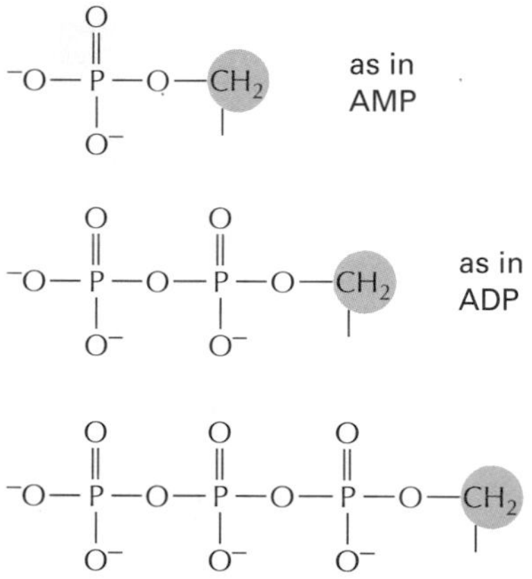

as in AMP

as in ADP

as in ATP

The phosphate makes a nucleotide negatively charged.

NUCLEOTIDES

A nucleotide consists of a nitrogen-containing base, a five-carbon sugar, and one or more phosphate groups.

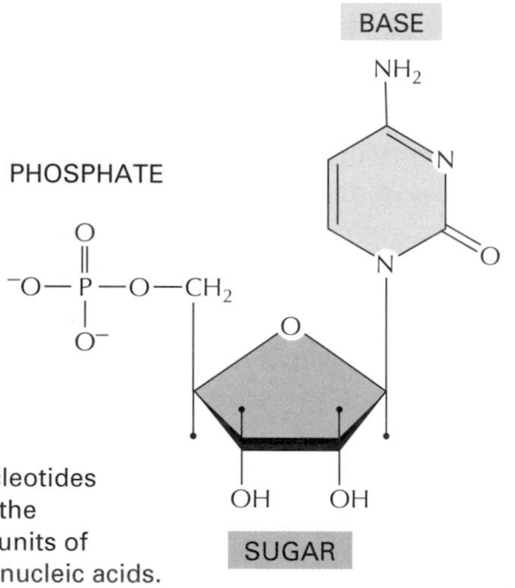

BASE

PHOSPHATE

SUGAR

Nucleotides are the subunits of the nucleic acids.

BASIC SUGAR LINKAGE

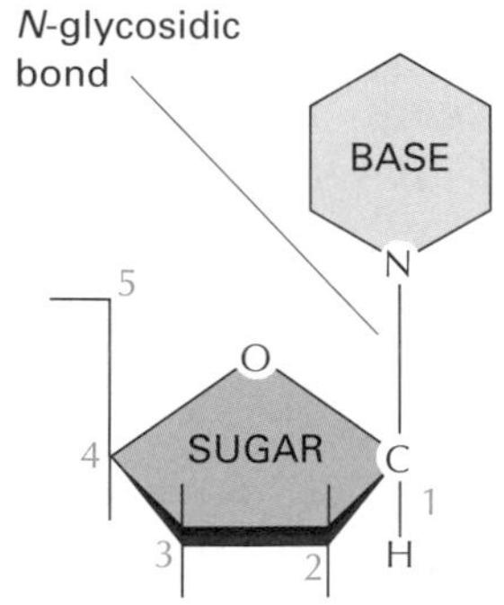

N-glycosidic bond

BASE

SUGAR

The base is linked to the same carbon (C1) used in sugar-sugar bonds.

SUGARS

PENTOSE

a five-carbon sugar

two kinds are used

Each numbered carbon on the sugar of a nucleotide is followed by a prime mark; therefore, one speaks of the "5-prime carbon," etc.

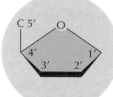

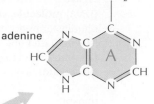

β-D-ribose
used in ribonucleic acid

β-D-2-deoxyribose
used in deoxyribonucleic acid

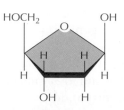

NOMENCLATURE   .The names can be confusing, but the abbreviations are clear.

BASE	NUCLEOSIDE	ABBR.
adenine	adenosine	A
guanine	guanosine	G
cytosine	cytidine	C
uracil	uridine	U
thymine	thymidine	T

Nucleotides are abbreviated by three capital letters. Some examples follow:

AMP   = adenosine monophosphate
dAMP = deoxyadenosine monophosphate
UDP   = uridine diphosphate
ATP   = adenosine triphosphate

BASE + SUGAR = NUCLEO<u>S</u>IDE

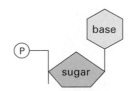

BASE + SUGAR + PHOSPHATE = NUCLEO<u>T</u>IDE

NUCLEIC ACIDS

Nucleotides are joined together by a phosphodiester linkage between 5' and 3' carbon atoms to form nucleic acids. The linear sequence of nucleotides in a nucleic acid chain is commonly abbreviated by a one-letter code, A—G—C—T—T—A—C—A, with the 5' end of the chain at the left.

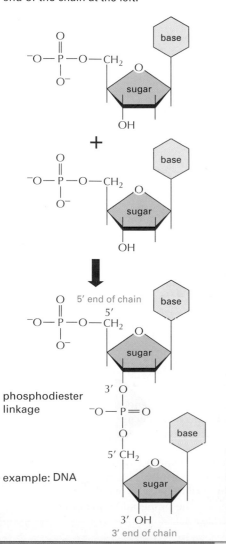

example: DNA

NUCLEOTIDES HAVE MANY OTHER FUNCTIONS

1 They carry chemical energy in their easily hydrolyzed phosphoanhydride bonds.

phosphoanhydride bonds

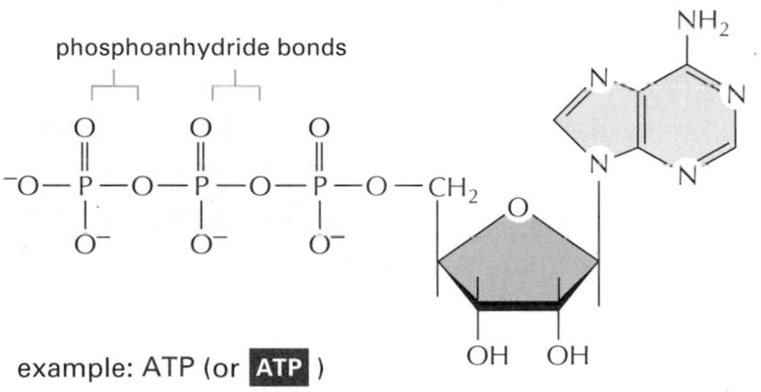

example: ATP (or ATP)

2 They combine with other groups to form coenzymes.

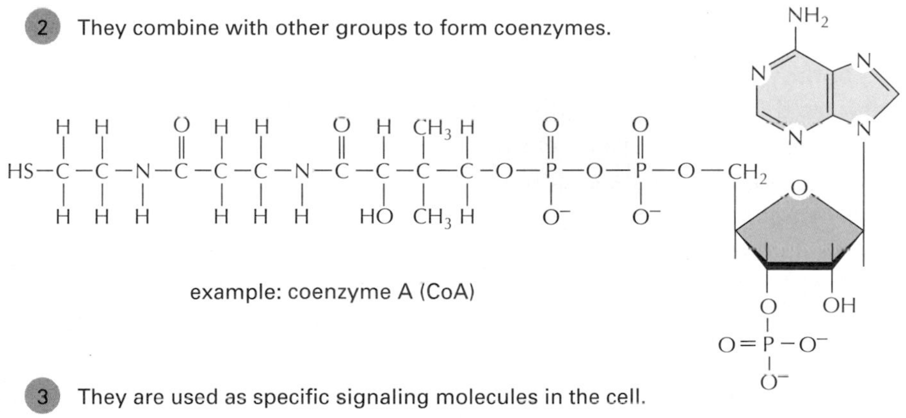

example: coenzyme A (CoA)

3 They are used as specific signaling molecules in the cell.

example: cyclic AMP (cAMP)

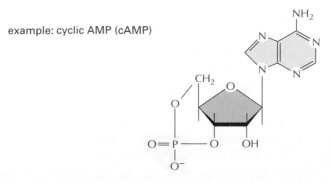

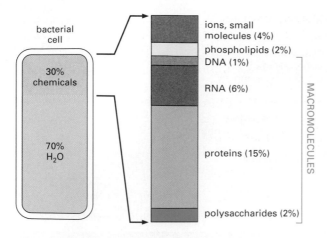

Figure 2–24 **Macromolecules are abundant in cells.** The approximate composition of a bacterial cell is shown. The composition of an animal cell is similar (see also Table 2–3, p. 53).

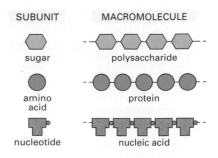

Figure 2–25 **Three families of macro-molecules.** Each is a polymer formed from small molecules (called monomers, or subunits) linked together by covalent bonds.

Proteins are especially versatile and perform thousands of distinct functions in cells. Many proteins serve as enzymes to catalyze reactions that the cell needs: all of the reactions whereby cells extract energy from food molecules are catalyzed by proteins serving as enzymes, for example, and an enzyme called ribulose bisphosphate carboxylase resides in chloroplasts and converts CO_2 to sugars in plants. Other proteins are used to build structural components, such as tubulin, a protein that self-assembles to make the cell's microtubules (see Figure 1–20), or histones, proteins that compact the DNA in chromosomes. Yet other proteins act as molecular motors to produce force and movement, as in the case of myosin in muscle. Proteins can also have a wide variety of other functions, and we shall examine the molecular basis of many of them later in this book. Here we consider some general principles of macromolecular chemistry that make such functions possible.

Although the chemical reactions for adding subunits to each polymer are different in detail for proteins, nucleic acids, and polysaccharides, they share important features. Each polymer grows by the addition of a monomer onto the end of a growing polymer chain in a *condensation reaction,* in which a molecule of water is lost with each subunit added (Figure 2–26; see also Figure 2–16). In all cases the reactions are catalyzed by enzymes, which ensure that only monomers of the appropriate type are incorporated. The stepwise polymerization of monomers into a long chain is a simple way to manufacture a large, complex molecule, since the subunits are added by the same reaction performed over and over again by the same set of enzymes. In a sense, the process resembles the repetitive operation of a machine in a factory—except in one crucial respect. Apart from some of the polysaccharides, most macromolecules are made from a set of monomers that are slightly different from one another, for example, the 20 different amino acids from which proteins are made (see Panel 2–5, pp. 62–63). Most important, the polymer chain is not assembled at random from these subunits; instead they are added in a particular order, or **sequence.**

The mechanisms that specify polymer sequence in the cell are discussed in Chapters 6 and 7. These mechanisms are central to biology because the biological function of proteins, nucleic acids, and many

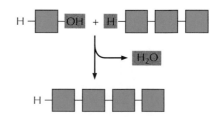

Figure 2–26 **The general reaction by which a macromolecule is made.** In a condensation reaction, a molecule of water is lost with the addition of each monomer to one end of the growing chain. The reverse reaction—the break-down of all three types of polymers—occurs by the simple addition of water (hydrolysis).

polysaccharides is absolutely dependent on the particular sequence of subunits in the linear chain. The possibility of varying the sequence of subunits creates enormous diversity in the polymeric molecules that can be produced. Thus, for a protein chain 200 amino acids long there are 20^{200} possible combinations ($20 \times 20 \times 20 \times 20 \ldots$ multiplied 200 times), while for a DNA molecule 10,000 nucleotides long (small by DNA standards) with its four different nucleotides there are $4^{10,000}$ different possibilities, an almost unimaginably large number. Thus the machinery of polymerization must be subject to a sensitive control that allows it to specify exactly which subunit should be added next to the growing polymer.

Question 2–8 In principle, there are many different, chemically diverse ways in which small molecules can be linked to form polymers. For example, the small molecule ethene ($CH_2{=}CH_2$) is used commercially to make the plastic polyethylene ($\ldots$ $-CH_2-CH_2-CH_2-CH_2-CH_2-$ $\ldots$). The individual subunits of the three major classes of biological macromolecules, however, are all linked by similar reaction mechanisms, i.e., by condensation reactions that eliminate water. Can you think of any benefits that this chemistry offers and why it might have been selected in evolution?

Noncovalent Bonds Specify the Precise Shape of a Macromolecule

Most of the single covalent bonds in a macromolecule allow rotation of the atoms they join, so that the polymer chain has great flexibility. In principle, this allows a macromolecule to adopt an almost unlimited number of shapes, or **conformations,** as the polymer chain writhes and rotates under the influence of random thermal energy. However, the shapes of most biological macromolecules are highly constrained because of weaker **noncovalent bonds** that form between different parts of the molecule. If these weaker bonds are formed in sufficient numbers, they will prevent the random movements and the polymer chain may then prefer one particular conformation, determined by the linear sequence of monomers in its chain. Virtually all protein molecules and many of the small RNA molecules found in cells fold tightly into one highly preferred conformation in this way (Figure 2–27).

The noncovalent bonds important in biological molecules include two types described earlier in this chapter—ionic bonds and hydrogen bonds. A third type of weak bond results from *van der Waals attractions,* which are a form of electrical attraction caused by fluctuating electric charges, arising whenever two atoms come within a very short distance of each other. All these noncovalent forces are described in Panel 2–7 (pp. 70–71). Another important noncovalent force is created by the three-dimensional structure of water, which forces hydrophobic groups together in order to minimize their disruptive effect on the hydrogen-bonded network of water molecules (see Panel 2–7, and Panel 2–2, pp.

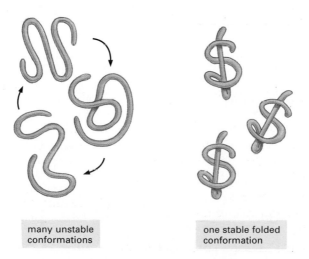

many unstable conformations

one stable folded conformation

Figure 2–27 Most proteins and many RNA molecules fold into only one stable conformation. If the weak bonds maintaining this stable conformation are disrupted, the molecule becomes a flexible chain that usually has no biological value.

VAN DER WAALS FORCES

At very short distances any two atoms show a weak bonding interaction due to their fluctuating electrical charges. If the two atoms are too close together, however, they repel each other very strongly.

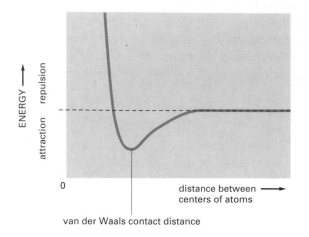

van der Waals contact distance

Each atom has a characteristic "size," or van der Waals radius: the contact distance between any two atoms is the sum of their van der Waals radii.

H	C	N	O
0.12 nm	0.2 nm	0.15 nm	0.14 nm

Two atoms will be attracted to each other by van der Waals forces until the distance between them equals the sum of their van der Waals radii. Although they are individually very weak, van der Waals attractions can become important when two macromolecular surfaces fit very close together.

WEAK CHEMICAL BONDS

Organic molecules can interact with other molecules through short-range noncovalent forces.

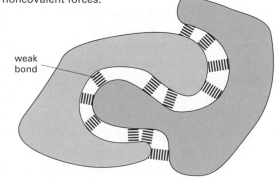

weak bond

Weak chemical bonds have less than 1/20 the strength of a strong covalent bond. They are strong enough to provide tight binding only when many of them are formed simultaneously.

HYDROGEN BONDS

As already described for water (see Panel 2–2, pp. 50–51) hydrogen bonds form when a hydrogen atom is "sandwiched" between two electron-attracting atoms (usually oxygen or nitrogen).

Hydrogen bonds are strongest when the three atoms are in a straight line:

Examples in macromolecules:

Amino acids in polypeptide chains hydrogen-bonded together.

Two bases, G and C, hydrogen-bonded in DNA or RNA.

HYDROGEN BONDS IN WATER

Any molecules that can form hydrogen bonds to each other can alternatively form hydrogen bonds to water molecules. Because of this competition with water molecules, the hydrogen bonds formed between two molecules dissolved in water are relatively weak.

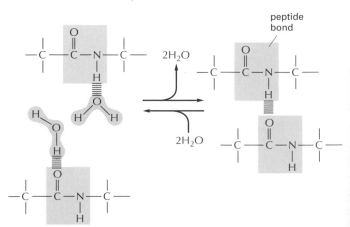

peptide bond

HYDROPHOBIC FORCES

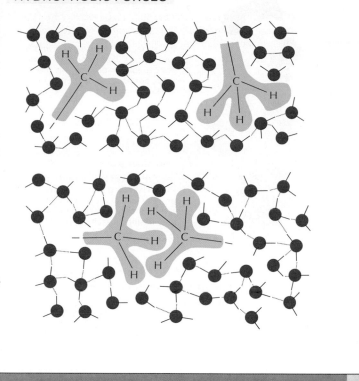

Water forces hydrophobic groups together in order to minimize their disruptive effects on the hydrogen-bonded water network. Hydrophobic groups held together in this way are sometimes said to be held together by "hydrophobic bonds," even though the attraction is actually caused by a repulsion from the water.

IONIC BONDS IN AQUEOUS SOLUTIONS

Charged groups are shielded by their interactions with water molecules. Ionic bonds are therefore quite weak in water.

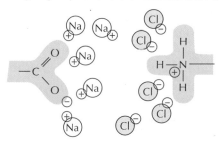

Similarly, other ions in solution can cluster around charged groups and further weaken ionic bonds.

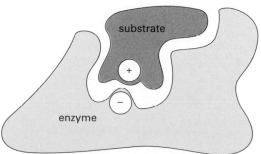

Despite being weakened by water and salt, ionic bonds are very important in biological systems; an enzyme that binds a positively charged substrate will often have a negatively charged amino acid side chain at the appropriate place.

IONIC BONDS

Ionic interactions occur either between fully charged groups (ionic bond) or between partially charged groups.

The force of attraction between the two charges, δ^+ and δ^-, falls off rapidly as the distance between the charges increases.

In the absence of water, ionic forces are very strong. They are responsible for the strength of such minerals as marble and agate.

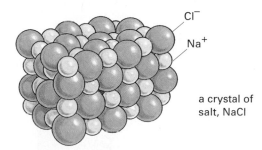

Cl^-
Na^+

a crystal of salt, NaCl

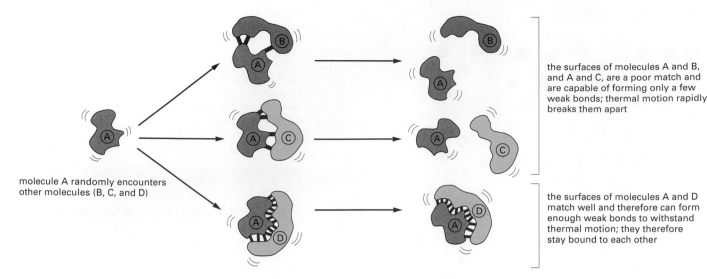

the surfaces of molecules A and B, and A and C, are a poor match and are capable of forming only a few weak bonds; thermal motion rapidly breaks them apart

molecule A randomly encounters other molecules (B, C, and D)

the surfaces of molecules A and D match well and therefore can form enough weak bonds to withstand thermal motion; they therefore stay bound to each other

50–51). This expulsion from the aqueous solution generates what is sometimes thought of as a fourth kind of weak noncovalent bond, called a *hydrophobic interaction*. This interaction forces phospholipid molecules together in cell membranes, and it also gives most protein molecules a compact, globular shape.

Figure 2–28 How noncovalent bonds mediate interactions between macromolecules.

Noncovalent Bonds Allow a Macromolecule to Bind Other Selected Molecules

Although individually very weak (see Table 2–2, p. 43), noncovalent bonds can add up to create a strong attraction between two molecules when these molecules fit together very closely, like a hand in a glove, so that many noncovalent bonds can form between them (see Panel 2–7, pp. 70–71). This form of molecular interaction provides for great specificity in the binding of macromolecules to other molecules, since the multipoint contacts required for strong binding make it possible for a macromolecule to select out—through binding—just one of the many thousands of different molecules present inside a cell. Moreover, because the strength of the binding depends on the number of noncovalent bonds that are formed, interactions of almost any strength are possible.

Binding of this type underlies all biological catalysis, making it possible for proteins to function as enzymes. Noncovalent bonds can also stabilize associations between two different macromolecules if their surfaces match closely (Figure 2–28). These bonds thereby allow macromolecules to be used as building blocks for the formation of larger struc-

Figure 2–29 Small molecules, proteins, and a ribosome drawn to scale. Ribosomes are part of the machinery the cell uses to make proteins: each ribosome is composed of about 90 macromolecules (protein and RNA molecules).

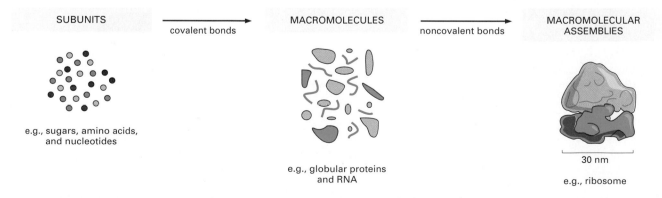

SUBUNITS

covalent bonds

MACROMOLECULES

noncovalent bonds

MACROMOLECULAR ASSEMBLIES

e.g., sugars, amino acids, and nucleotides

e.g., globular proteins and RNA

30 nm

e.g., ribosome

tures. For example, proteins often bind together into multiprotein complexes, thereby forming intricate machines with multiple moving parts that perform such complex tasks as DNA replication and protein synthesis (Figure 2–29).

Question 2–9 Why could covalent bonds not be used in place of noncovalent bonds to mediate most of the interactions of macromolecules?

Essential Concepts

- Living cells obey the same chemical and physical laws as nonliving things. Like all other forms of matter, they are composed of atoms, which are the smallest units of chemical elements.

- Atoms are made up of smaller particles. The nucleus of an atom contains protons, which are positively charged, and uncharged neutrons. The nucleus is surrounded by a cloud of negatively charged electrons.

- The number of electrons in an atom is equal to the number of protons in its nucleus. The nuclei of different isotopes of the same element contain the same number of protons but different numbers of neutrons.

- Living cells are made up of a limited number of elements, six of which—CHNOPS—make up 99% of their mass.

- The chemical properties of an atom are determined by the number and arrangement of its electrons. An atom is most stable when all of its electrons are at their lowest possible energy level and when each electron shell is completely filled with electrons.

- Chemical bonds form between atoms as electrons move to reach a more stable arrangement. Clusters of two or more atoms held together by chemical bonds are known as molecules.

- When an electron jumps from one atom to another, two ions of opposite charge are formed; ionic bonds are formed by the mutual attraction of these charged atoms.

- A covalent bond is formed when a pair of electrons is shared between adjacent atoms. If two pairs of electrons are shared, a double bond is formed.

- Living organisms contain a distinctive and restricted set of small carbon-based molecules that are essentially the same for every living species. The main categories are sugars, fatty acids, amino acids, and nucleotides.

- Sugars are a primary source of chemical energy for cells and can be incorporated into polysaccharides for energy storage.

- Fatty acids are also important for energy storage, but their most essential function is in the formation of cell membranes.

- Polymers consisting of amino acids constitute the remarkably diverse and versatile macromolecules known as proteins.

- Nucleotides play a central part in energy transfer and are the subunits from which the informational macromolecules, RNA and DNA, are made.

- Macromolecules are intermediate both in size and complexity between small molecules and cell organelles. They have many remarkable properties that are not easily deduced from the subunits from which they are made.

- Macromolecules are made as polymers of subunits by repetitive condensation reactions. Their remarkable diversity arises from the fact that each macromolecule has a unique sequence of subunits.

- Weak noncovalent bonds form between different regions of a macromolecule. These can cause the macromolecule to fold up into a unique three-dimensional shape, as seen most conspicuously in proteins.

Key Terms

acid	chemical bond	hydrolysis	polar
amino acid	chemical group	hydronium ion	protein
atom	condensation	ion	proton
atomic weight	conformation	ionic bond	RNA
ATP	covalent bond	molecular weight	sequence
Avogadro's number	DNA	molecule	sugar
base	electron	noncovalent bond	
	fatty acid	nucleotide	
	hydrogen bond	pH scale	

Questions

Question 2–10 Which of the following statements are correct? Explain your answers.

A. An atomic nucleus contains protons and neutrons.

B. An atom has more electrons than protons.

C. The nucleus is surrounded by a double membrane.

D. All atoms of the same element have the same number of neutrons.

E. The number of neutrons determines whether the nucleus of an atom is stable or radioactive.

F. Both fatty acids and polysaccharides can be important energy stores in the cell.

G. Hydrogen bonds are weak and can be broken by thermal energy, yet they contribute significantly to the specificity of interactions between macromolecules.

Question 2–11 To gain a better feeling for atomic dimensions, assume that the page on which this question is printed is made entirely of the polysaccharide cellulose, whose molecules are described by the formula $(C_nH_{2n}O_n)$, where n can be a quite large number and is variable from one molecule to another. The atomic weights of carbon, hydrogen, and oxygen are 12, 1, and 16, respectively, and this page weighs 5 g.

A. How many carbon atoms are there in this page?

B. In cellulose, how many carbon atoms would be stacked on top of each other to span the thickness of this page (the size of the page is 21 cm × 27.5 cm, and it is 0.07 mm thick).

C. Now consider the problem from a different angle. Assume that the page is composed only of carbon atoms. A carbon atom has a diameter of 2×10^{-10} m (0.2 nm); how many carbon atoms of 0.2 nm diameter would it take to span the thickness of the page?

D. Compare your answers from parts B and C and explain any differences.

Question 2–12

A. How many electrons can be accommodated in the first, second, and third electron shells of an atom?

B. How many electrons would atoms of the elements listed below preferentially gain or lose in order to obtain completely filled sets of energy levels?

hydrogen	gain __	lose __
helium	gain __	lose __
oxygen	gain __	lose __
carbon	gain __	lose __
sodium	gain __	lose __
chlorine	gain __	lose __

C. What do the answers tell you about the chemical properties of the elements and/or the bonds that can form between the elements (e.g., between sodium and chlorine, between oxygen and hydrogen, between carbon and oxygen, and between carbon and hydrogen)?

Question 2–13 Oxygen and sulfur have similar chemical properties because both elements have six electrons in their outermost electron shells. Indeed, both elements form molecules with two hydrogen atoms, water (H_2O) and hydrogen sulfide (H_2S). Surprisingly, water is a liquid, yet H_2S is a gas, despite the fact that sulfur is much larger and heavier than oxygen. Explain why this might be the case.

Question 2–14 Write the chemical formula for a condensation reaction of two amino acids to form a peptide bond. Write the formula for its hydrolysis.

Question 2–15 Which of the following statements are correct? Explain your answers.

A. Proteins are so remarkably diverse because each is made from a unique mixture of amino acids that are linked in random order.

B. Lipid bilayers are macromolecules that are made up mostly of phospholipid subunits.

C. Nucleic acids contain sugar groups.

D. Many amino acids have hydrophobic side chains.

E. The hydrophobic tails of phospholipid molecules are repelled from water.

F. DNA contains the four different bases A, G, U, and C.

Question 2–16

A. How many different molecules composed of (a) two, (b) three, and (c) four amino acids, linked together by peptide bonds, can be made from the set of 20 naturally occurring amino acids?

B. Assume you were given a mixture consisting of one molecule each of all possible sequences of a smallish protein of molecular weight 4800. How big a container would you need to hold this sample? Assume that the average molecular weight of an amino acid is 120.

C. What does this calculation tell you about the fraction of possible proteins that are currently in use by living organisms (the average molecular weight of proteins is about 30,000)?

Question 2–17
This is a biology textbook. Explain why the chemical principles that are described in this chapter are important in the context of modern cell biology.

Question 2–18

A. Describe the similarities and differences between van der Waals attractions and hydrogen bonds.

B. Which of the two bonds would form (a) between two hydrogens bound to carbon atoms, (b) between a nitrogen atom and a hydrogen bound to a carbon atom, and (c) between a nitrogen atom and a hydrogen bound to an oxygen atom?

Question 2–19
What are the forces that determine the folding of a macromolecule into a unique shape?

Question 2–20
Fatty acids are said to be "amphipathic." What is meant by this term, and how does an amphipathic molecule behave in water? Draw a diagram to illustrate your answer.

Question 2–21
Are the following formulas correct or incorrect? Explain your answer in each case.

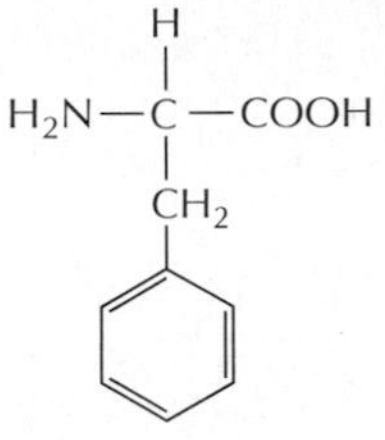

(A)

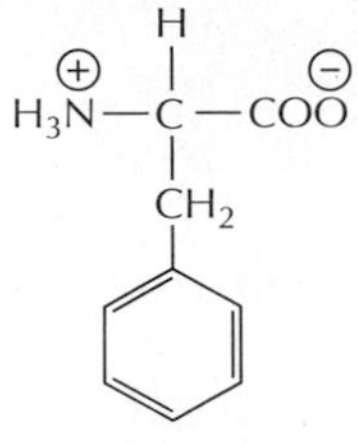

(B)

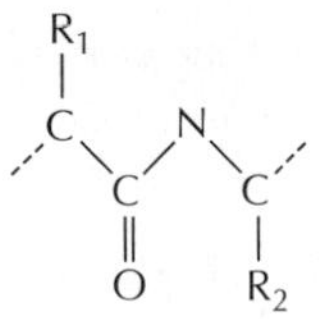

(C)

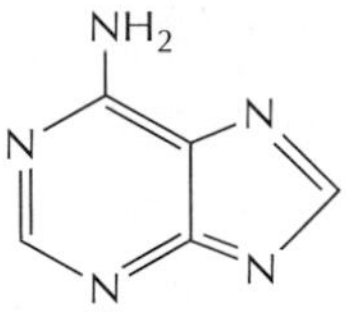

(D)

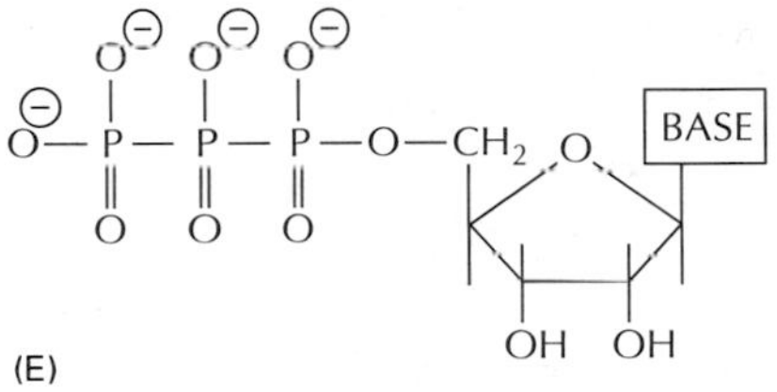

(E)

$$CH_3-CH_2-OH$$

(F)

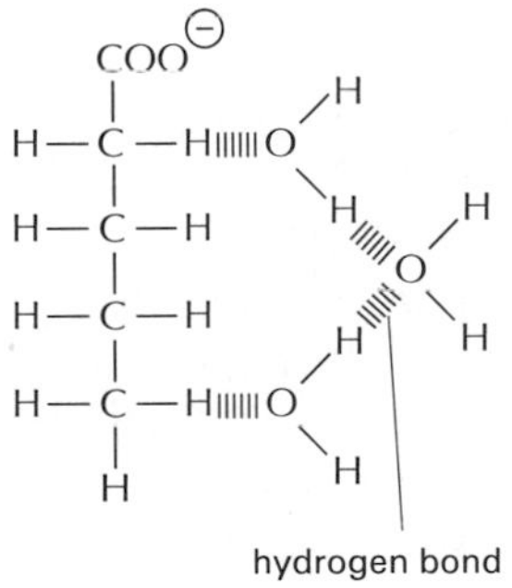

(G)

hydrogen bond

$$Na-Cl$$

(H)

$$\overset{\delta^+}{O}=\overset{\delta^-}{C}=\overset{\delta^+}{O}$$

(I)

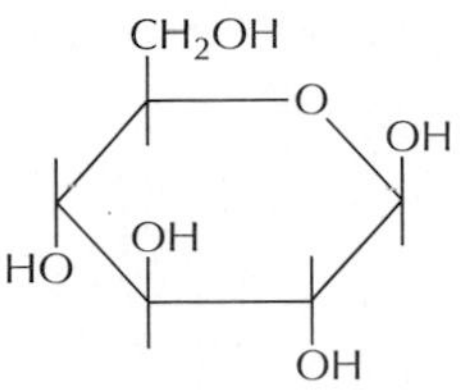

(J)

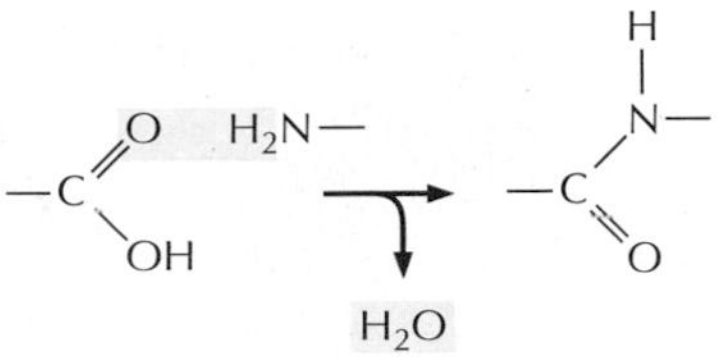

(K)

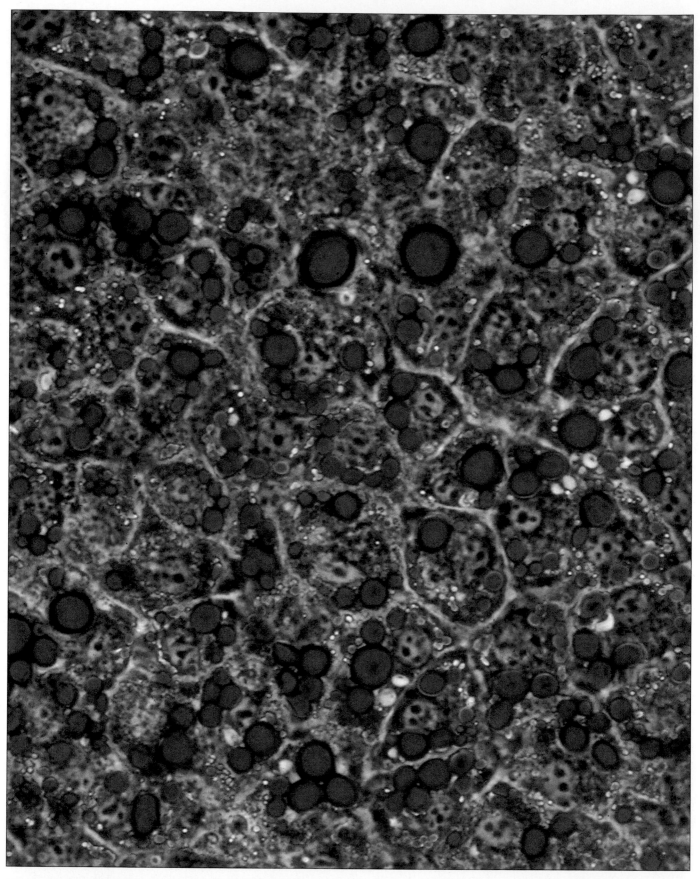

All cells are tiny chemical factories that use food molecules as a source of both building blocks and energy. Fat is an important concentrated store of energy for all cells, and in animals it is deposited in fat cells. Fat droplets are seen here (*red*) beginning to accumulate in developing fat cells. (Courtesy of Ronald M. Evans.)

3 Energy, Catalysis, and Biosynthesis

One property of living things above all makes them seem almost miraculously different from nonliving matter: they create and maintain order, in a universe that is tending always to greater disorder. To create this order, the cells in a living organism must perform a never-ending stream of chemical reactions. In some of these reactions, small organic molecules—amino acids, sugars, nucleotides, and lipids—are being taken apart or modified to supply the many other small molecules that the cell requires. In other reactions, these small molecules are being used to construct an enormously diverse range of proteins, nucleic acids, and other macromolecules that endow living systems with all of their most distinctive properties. Each cell can be viewed as a tiny chemical factory, performing many thousands of reactions every second.

To carry out the many chemical reactions needed to sustain it, a living organism requires not only a source of atoms in the form of food molecules, but also a source of energy. Both atoms and energy must come, ultimately, from the nonliving environment. In this chapter we discuss why cells require energy, and how they use this energy and the atoms from their environment to create the order at the molecular level that makes life possible.

The chemical reactions that a cell carries out would normally occur only at temperatures that are much higher than those existing inside cells. For this reason, each reaction requires a specific boost in chemical reactivity. This requirement is crucial, because it allows each reaction to be controlled by the cell. The control is exerted through specialized proteins called *enzymes*, each of which accelerates, or *catalyzes*, just one of the many possible kinds of reactions that a particular molecule might undergo. Enzyme-catalyzed reactions are usually connected in series, so that the product of one reaction becomes the starting material, or *substrate*, for the next (Figure 3–1). These long linear reaction pathways are in turn linked to one another, forming a maze of interconnected reactions that enable the cell to survive, grow, and reproduce (Figure 3–2).

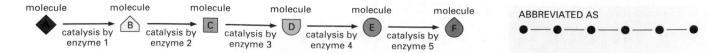

Two opposing streams of chemical reactions occur in cells. (1) The *catabolic* pathways break down foodstuffs into smaller molecules, thereby generating both a useful form of energy for the cell and some of the small molecules that the cell needs as building blocks, and (2) the *anabolic*, or *biosynthetic*, pathways use the energy harnessed by catabolism to drive the synthesis of the many other molecules that form the cell. Together these two sets of reactions constitute the **metabolism** of the cell (Figure 3–3).

Many of the details of cell metabolism form the subject of *biochemistry* and need not concern us here. But the general principles by which cells obtain energy from their environment and use it to create order are central to cell biology. These principles are outlined in this chapter, starting with a discussion of why a constant input of energy is needed to sustain living organisms.

Figure 3–1 How a set of enzyme-catalyzed reactions generates a metabolic pathway. Each enzyme catalyzes a particular chemical reaction, leaving the enzyme unchanged. In this example, a set of enzymes acting in series converts molecule A to molecule F, forming a metabolic pathway.

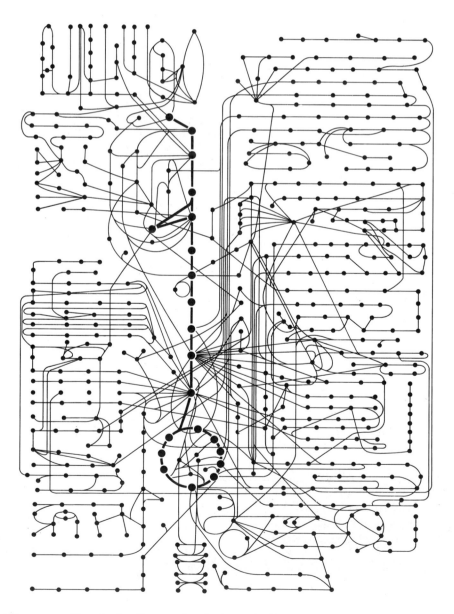

Figure 3–2 Some of the metabolic pathways and their interconnections in a typical cell. About 500 common metabolic reactions are shown diagrammatically, with each molecule in a metabolic pathway represented by a filled circle, as in the *yellow box* in Figure 3–1.

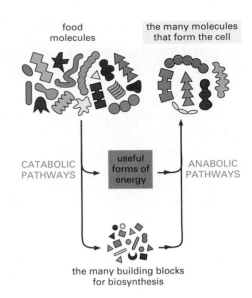

Figure 3–3 Schematic representation of the relationship between catabolic and anabolic pathways in metabolism. As suggested here, since a major portion of the energy stored in the chemical bonds of food molecules is dissipated as heat, the mass of food required by any organism that derives all of its energy from catabolism is much greater than the mass of the molecules that can be produced by anabolism.

Catalysis and the Use of Energy by Cells

Nonliving things left to themselves eventually become disordered: buildings crumble and dead organisms decay. Living cells, by contrast, not only maintain but actually generate order at every level, from the large-scale structure of a butterfly or a flower to the organization of the atoms in the molecules from which these organisms are made (Figure 3–4). This property of life is made possible by elaborate cellular mechanisms that extract energy from the environment and convert it into forms that can be used by the cell to drive the constant generation of biological order.

Biological Order Is Made Possible by the Release of Heat Energy from Cells

The universal tendency of things to become disordered is expressed in a fundamental law of physics the *second law of thermodynamics*—which states that in the universe, or in any isolated system (a collection of matter that is completely isolated from the rest of the universe), the degree of disorder can only increase. This law has such profound implications for all living things that it is worth restating in several ways.

For example, we can present the second law in terms of probability and state that systems will change spontaneously toward those arrangements that have the greatest probability. If we consider, for example, a box of 100 coins all lying heads up, a series of accidents that disturbs the box will tend to move the arrangement toward a mixture of 50 heads and 50 tails. The reason is simple: there is a huge number of possible arrangements of the individual coins in the mixture that can achieve the 50-50 result, but only one possible arrangement that keeps all of the coins oriented heads up. Because the 50-50 mixture is therefore the most probable, we say that it is more "disordered." For the same reason, it is a common experience that one's living space will become increasingly disordered without intentional effort: the movement toward disorder is a *spontaneous process*, requiring a periodic effort to reverse it (Figure 3–5).

Figure 3–4 Order in biological structures. Well-defined, ornate, and beautiful spatial patterns can be found at every level of organization in living organisms. In order of increasing size: (A) protein molecules in the coat of a virus; (B) the regular array of microtubules seen in a cross section of a sperm tail; (C) surface contours of a pollen grain (a single cell); (D) close-up of the wing of a butterfly showing the pattern created by scales, each scale being the product of a single cell; (E) spiral array of seeds, made of millions of cells, in the head of a sunflower. (A, courtesy of Robert Grant, Stéphane Crainic, and James M. Hogle; B, courtesy of Lewis Tilney; C, courtesy of Colin MacFarlane and Chris Jeffree; D and E, courtesy of Kjell B. Sandved.)

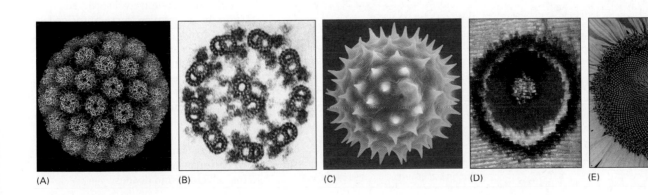

(A) (B) (C) (D) (E)

"SPONTANEOUS" REACTION
as time elapses

ORGANIZED EFFORT REQUIRING ENERGY INPUT

Figure 3–5 An everyday illustration of the spontaneous drive toward disorder. Reversing this tendency toward disorder requires an intentional effort and an input of energy: it is not spontaneous. In fact, from the second law of thermodynamics, we can be certain that the human intervention required will release enough heat to the environment to more than compensate for the reordering of the items in this room.

The amount of disorder in a system can be quantified. The quantity that we use to measure this disorder is called the **entropy** of the system: the greater the disorder, the greater the entropy. Thus, a third way to express the second law of thermodynamics is to say that systems will change spontaneously toward arrangements with greater entropy.

Living cells—by surviving, growing, and forming complex organisms—are generating order and thus might appear to defy the second law of thermodynamics. How is this possible? The answer is that a cell is not an isolated system: it takes in energy from its environment in the form of food, or as photons from the sun (or even, as in some chemosynthetic bacteria, from inorganic molecules alone), and it then uses this energy to generate order within itself. In the course of the chemical reactions that generate order, part of the energy that the cell uses is converted into heat. The heat is discharged into the cell's environment and disorders it, so that the total entropy—that of the cell plus its surroundings—increases, as demanded by the laws of physics.

To understand the principles governing these energy conversions, think of a cell as sitting in a sea of matter representing the rest of the universe. As the cell lives and grows, it creates internal order at the expense

Figure 3–6 A simple thermodynamic analysis of a living cell. In the diagram on the left the molecules of both the cell and the rest of the universe (the sea of matter) are depicted in a relatively disordered state. In the diagram on the right the cell has taken in energy from food molecules and released heat by a reaction that orders the molecules the cell contains. Because the heat increases the disorder in the environment around the cell (depicted by the *jagged arrows* and distorted molecules, indicating increased molecular motions), the second law of thermodynamics, which states that the amount of disorder in the universe must always increase, is satisfied as the cell grows and divides.

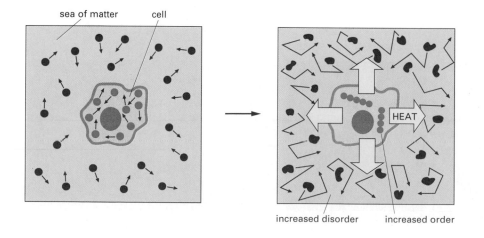

sea of matter   cell

HEAT

increased disorder   increased order

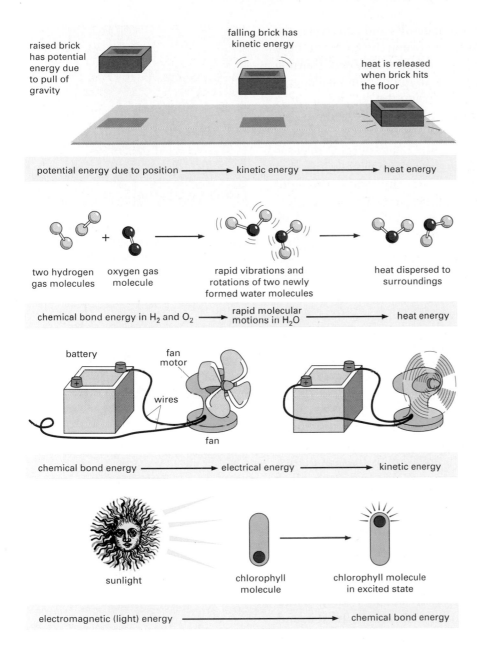

potential energy due to position ⟶ kinetic energy ⟶ heat energy

two hydrogen gas molecules   oxygen gas molecule → rapid vibrations and rotations of two newly formed water molecules → heat dispersed to surroundings

chemical bond energy in H_2 and O_2 ⟶ rapid molecular motions in H_2O ⟶ heat energy

battery   fan motor
wires   fan

chemical bond energy ⟶ electrical energy ⟶ kinetic energy

sunlight   chlorophyll molecule   chlorophyll molecule in excited state

electromagnetic (light) energy ⟶ chemical bond energy

Figure 3–7 Some interconversions between different forms of energy. All energy forms are, in principle, interconvertible. In all these processes the total amount of energy is conserved; thus, for example, from the height and weight of the brick in the first example, we can predict exactly how much heat will be released when it hits the floor. In the second example, note that the large amount of chemical bond energy released when water is formed is initially converted to very rapid thermal motions in the two new water molecules; but collisions with other molecules almost instantaneously spread this kinetic energy evenly throughout the surroundings (heat transfer), making the new molecules indistinguishable from all the rest.

of the sea, because it releases heat energy as it synthesizes molecules and assembles them into cell structures. Heat is energy in its most disordered form—the random jostling of molecules. When the cell releases heat to the sea, it increases the intensity of molecular motions there (thermal motion)—thereby increasing the randomness, or disorder, of the sea. The second law of thermodynamics is satisfied because the increase in the amount of order inside the cell is more than compensated by a greater decrease in order (increase in entropy) in the surrounding sea of matter (Figure 3–6).

Where does the heat that the cell releases come from? Here we encounter another important law of thermodynamics. The *first law of thermodynamics* states that energy can be converted from one form to another, but that it cannot be created or destroyed. Some forms of energy are illustrated in Figure 3–7. The amount of energy in different forms will change as a result of the chemical reactions inside the cell, but the first law tells us that the total amount of energy must always be the same. For example, an animal cell takes in foodstuffs and converts some of the energy present in the chemical bonds between the atoms of these food molecules (chemical bond energy) into the random thermal motion of molecules (heat energy). This conversion of chemical energy into heat energy is essential if the reactions inside the cell are to cause the universe as a whole to become more disordered—as required by the second law.

The cell cannot derive any benefit from the heat energy it releases unless the heat-generating reactions inside the cell are directly linked to the processes that generate molecular order. It is the tight coupling of heat production to an increase in order that distinguishes the metabolism of a cell from the wasteful burning of fuel in a fire. Later in this chapter, we shall illustrate how this coupling occurs. For the moment, it is sufficient to recognize that a direct linkage of the "burning" of food molecules to the generation of biological order is required if cells are to be able to create and maintain an island of order in a universe tending toward chaos.

Radiant energy of sunlight. Trapped in plants and some microorganisms by photosynthesis, light from the sun is the ultimate source of all energy for humans and other animals. (Courtesy of Museum Folkwang, Essen.)

Photosynthetic Organisms Use Sunlight to Synthesize Organic Molecules

All animals live on energy stored in the chemical bonds of organic molecules made by other organisms, which they take in as food. The molecules in food also provide the atoms that animals need to construct new living matter. Some animals obtain their food by eating other animals. But at the bottom of the animal food chain are animals that eat plants. The plants, in turn, trap energy directly from sunlight. As a result, all of the energy used by animal cells is derived ultimately from the sun (Figure 3–8).

Solar energy enters the living world through **photosynthesis** in plants and photosynthetic bacteria. Photosynthesis allows the electromagnetic energy in sunlight to be converted into chemical bond energy in the cell. Plants are able to obtain all the atoms they need from inorganic sources: carbon from atmospheric carbon dioxide, hydrogen and oxygen from water, nitrogen from ammonia and nitrates in the soil, and other elements needed in smaller amounts from inorganic salts in the soil. They use the energy they derive from sunlight to build these atoms into sugars, amino acids, nucleotides, and fatty acids. These small molecules in turn are converted into the proteins, nucleic acids, polysaccharides, and lipids that form the plant. All of these substances serve as food molecules for animals, if the plants are later eaten.

The reactions of photosynthesis take place in two stages (Figure 3–9). In the first stage, energy from sunlight is captured and transiently stored as chemical bond energy in specialized small molecules that act as carriers of energy and reactive chemical groups. (We discuss these

Figure 3–9 **Photosynthesis.** The two stages of photosynthesis. The energy carriers created in the first stage are two molecules that we discuss shortly—ATP and NADPH.

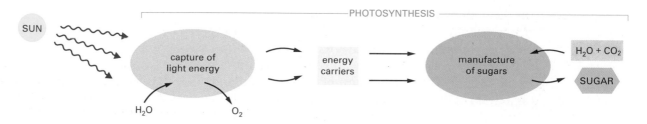

activated carrier molecules later.) Molecular oxygen (O_2 gas) derived from the splitting of water by light is released as a waste product of this first stage.

In the second stage, the molecules that serve as energy carriers are used to help drive a *carbon fixation* process in which sugars are manufactured from carbon dioxide gas (CO_2) and water (H_2O), thereby providing a useful source of stored chemical bond energy and materials—both for the plant itself and for any animals that eat it. We describe the elegant mechanisms that underlie these two stages of photosynthesis in Chapter 13.

The net result of the entire process of photosynthesis, so far as the green plant is concerned, can be summarized simply in the equation

$$\text{light energy} + CO_2 + H_2O \rightarrow \text{sugars} + O_2 + \text{heat energy}$$

The sugars produced are then used both as a source of chemical bond energy and as a source of materials to make the many other small and large organic molecules that are essential to the plant cell.

Cells Obtain Energy by the Oxidation of Organic Molecules

All animal and plant cells are powered by chemical energy stored in the chemical bonds of organic molecules, whether these be sugars that a plant has photosynthesized as food for itself or the mixture of large and small molecules that an animal has eaten. In order to use this energy to live, grow, and reproduce, organisms must extract it in a usable form. In both plants and animals, energy is extracted from food molecules by a process of gradual oxidation, or controlled burning.

The earth's atmosphere contains a great deal of oxygen, and in the presence of oxygen the most energetically stable form of carbon is as CO_2 and that of hydrogen is as H_2O. A cell is therefore able to obtain energy from sugars or other organic molecules by allowing their carbon and hydrogen atoms to combine with oxygen to produce CO_2 and H_2O, respectively—a process called **respiration.**

Photosynthesis and respiration are complementary processes (Figure 3–10). This means that the transactions between plants and animals are not all one way. Plants, animals, and microorganisms have existed together on this planet for so long that many of them have become an essential part of the others' environments. The oxygen released by photosynthesis is consumed in the combustion of organic molecules by nearly all organisms. And some of the CO_2 molecules that are fixed today into organic molecules by photosynthesis in a green leaf were yesterday released into the atmosphere by the respiration of an ani-

Question 3–1 Consider the equation

light energy + CO_2 + $H_2O \rightarrow$ sugars + O_2

Would you expect this reaction to be carried out by a single enzyme? Why is heat also generated in the reaction? Explain your answers.

Figure 3–10 Photosynthesis and respiration as complementary processes in the living world. Photosynthesis uses the energy of sunlight to produce sugars and other organic molecules. These molecules in turn serve as food for other organisms. Many of these organisms carry out respiration, a process that uses O_2 to form CO_2 from the same carbon atoms that had been taken up as CO_2 and converted into sugars by photosynthesis. In the process, the organisms that respire obtain the chemical bond energy that they need to survive. The first cells on the earth are thought to have been capable of neither photosynthesis nor respiration (see Chapter 13). However, photosynthesis must have preceded respiration on the earth, since there is strong evidence that billions of years of photosynthesis were required before O_2 had been released in sufficient quantity to create an atmosphere rich in this gas. (The earth's atmosphere presently contains 20% O_2.)

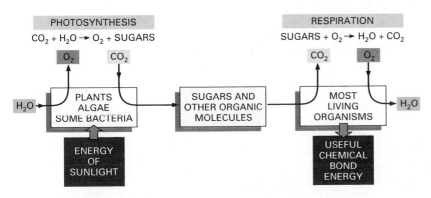

PHOTOSYNTHESIS
$CO_2 + H_2O \rightarrow O_2$ + SUGARS

RESPIRATION
SUGARS + $O_2 \rightarrow H_2O + CO_2$

O_2 CO_2 CO_2 O_2

$H_2O \rightarrow$ PLANTS ALGAE SOME BACTERIA → SUGARS AND OTHER ORGANIC MOLECULES → MOST LIVING ORGANISMS $\rightarrow H_2O$

ENERGY OF SUNLIGHT

USEFUL CHEMICAL BOND ENERGY

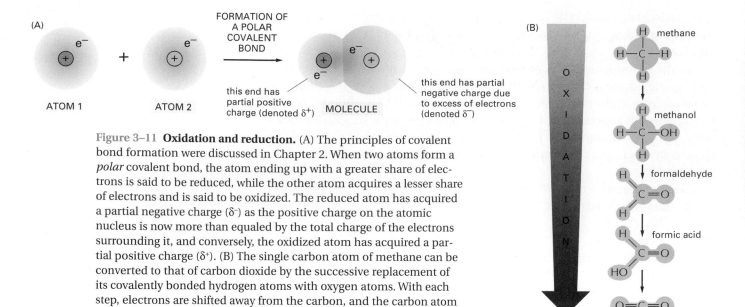

Figure 3–11 **Oxidation and reduction.** (A) The principles of covalent bond formation were discussed in Chapter 2. When two atoms form a *polar* covalent bond, the atom ending up with a greater share of electrons is said to be reduced, while the other atom acquires a lesser share of electrons and is said to be oxidized. The reduced atom has acquired a partial negative charge (δ^-) as the positive charge on the atomic nucleus is now more than equaled by the total charge of the electrons surrounding it, and conversely, the oxidized atom has acquired a partial positive charge (δ^+). (B) The single carbon atom of methane can be converted to that of carbon dioxide by the successive replacement of its covalently bonded hydrogen atoms with oxygen atoms. With each step, electrons are shifted away from the carbon, and the carbon atom becomes progressively more oxidized. Each of these steps is energetically favorable under the conditions present inside a cell.

mal—or by that of a fungus or bacterium decomposing dead organic matter. We therefore see that carbon utilization forms a huge cycle that involves the *biosphere* (all of the living organisms on earth) as a whole, crossing boundaries between individual organisms. Similarly, atoms of nitrogen, phosphorus, and sulfur move between the living and nonliving worlds in cycles that involve plants, animals, fungi, and bacteria.

Oxidation and Reduction Involve Electron Transfers

The cell does not oxidize organic molecules in one step, as occurs when organic material is burned in a fire. Through the use of enzyme catalysts, metabolism takes the molecules through a large number of reactions that only rarely involve the direct addition of oxygen. Before we consider some of these reactions and the purpose behind them, we need to discuss what is meant by the process of oxidation.

Oxidation, in the sense used above, does not mean only the addition of oxygen atoms; rather, it applies more generally to any reaction in which electrons are transferred from one atom to another. Oxidation in this sense refers to the removal of electrons, and **reduction**—the converse of oxidation—means the addition of electrons. Thus, Fe^{2+} is oxidized if it loses an electron to become Fe^{3+}, and a chlorine atom is reduced if it gains an electron to become Cl^-. Since the number of electrons is conserved (no loss or gain) in a chemical reaction, oxidation and reduction always occur simultaneously: that is, if one molecule gains an electron in a reaction (reduction), a second molecule loses the electron (oxidation). When a sugar molecule is oxidized to CO_2 and H_2O, for example, the O_2 molecules involved in forming H_2O gain electrons and thus are said to have been reduced.

The terms "oxidation" and "reduction" apply even when there is only a partial shift of electrons between atoms linked by a covalent bond (Figure 3–11A). When a carbon atom becomes covalently bonded to an atom with a strong affinity for electrons, such as oxygen, chlorine, or sulfur, for example, it gives up more than its equal share of electrons and forms a *polar* covalent bond: the positive charge of the carbon nucleus is now somewhat greater than the negative charge of its electrons, and the atom therefore acquires a partial positive charge and is said to be oxidized. Conversely, a carbon atom in a C–H linkage has somewhat more than its share of electrons, and so it is said to be reduced (Figure 3–11B).

When a molecule in a cell picks up an electron (e^-), it often picks up a proton (H^+) at the same time (protons being freely available in water). The net effect in this case is to add a hydrogen atom to the molecule

$$A + e^- + H^+ \rightarrow AH$$

Even though a proton plus an electron is involved (instead of just an electron), such *hydrogenation* reactions are reductions, and the reverse, *dehydrogenation* reactions, are oxidations. It is especially easy to tell whether an organic molecule is being oxidized or reduced: reduction is occurring if its number of C–H bonds increases, whereas oxidation is occurring if its number of C–H bonds decreases.

Cells use enzymes to catalyze the oxidation of organic molecules in small steps, through a sequence of reactions that allows useful energy to be harvested. We now need to explain how enzymes work and some of the constraints under which they operate.

Enzymes Lower the Barriers That Block Chemical Reactions

Consider the reaction

$$\text{paper} + O_2 \rightarrow \text{smoke} + \text{ashes} + \text{heat} + CO_2 + H_2O$$

The paper burns readily, releasing to the atmosphere both energy as heat and water and carbon dioxide as gases, but the smoke and ashes never spontaneously retrieve these entities from the heated atmosphere and reconstitute themselves into paper. When the paper burns, its chemical energy is dissipated as heat—not lost from the universe, since energy can never be created or destroyed, but irretrievably dispersed in the chaotic random thermal motions of molecules. At the same time, the atoms and molecules of the paper become dispersed and disordered. In the language of thermodynamics, there has been a loss of *free energy*, that is, of energy that can be harnessed to do work or drive chemical reactions. This loss reflects a loss of orderliness in the way the energy and molecules were stored in the paper. We shall discuss free energy in more detail shortly, but the general principle is clear enough intuitively: chemical reactions proceed only in the direction that leads to a loss of free energy; in other words, the spontaneous direction for any reaction is the direction that goes "downhill." A "downhill" reaction in this sense is often said to be *energetically favorable*.

Although the most energetically favorable form of carbon under ordinary conditions is as CO_2, and that of hydrogen is as H_2O, a living organism does not disappear in a puff of smoke, and the book in your hands does not burst into flames. This is because the molecules both in the living organism and in the book are in a relatively stable state, and they cannot be changed to a state of lower energy without an input of energy: in other words, a molecule requires **activation energy**—a kick over an energy barrier—before it can undergo a chemical reaction that leaves it in a more stable state (Figure 3–12). In the case of a burning

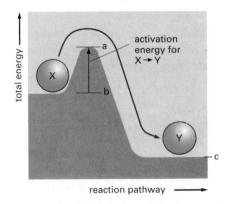

Figure 3–12 The principle of activation energy. Compound X is in a stable state, and energy is required to convert it to compound Y, even though Y is at a lower overall energy level than X. This conversion will not take place, therefore, unless compound X can acquire enough activation energy (*energy a minus energy b*) from its surroundings to undergo the reaction that converts it into compound Y. This energy may be provided by means of an unusually energetic collision with other molecules. For the reverse reaction, Y → X, the activation energy will be much larger (*energy a minus energy c*); this reaction will therefore occur much more rarely. Activation energies are always positive; note, however, that the total energy change for the energetically favorable reaction X → Y is *energy c minus energy b,* a negative number.

Catalysis and the Use of Energy by Cells   **85**

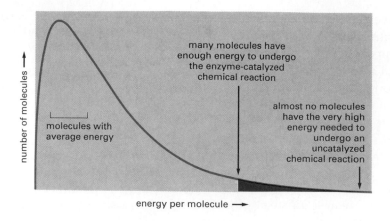

number of molecules ↑

molecules with average energy

many molecules have enough energy to undergo the enzyme-catalyzed chemical reaction

almost no molecules have the very high energy needed to undergo an uncatalyzed chemical reaction

energy per molecule →

Figure 3–13 Lowering the activation energy greatly increases the probability of reaction. A population of identical substrate molecules will have a range of energies that is distributed as shown on the graph at any one instant. The varying energies come from collisions with surrounding molecules, which make the substrate molecules jiggle, vibrate, and spin. For a molecule to undergo a chemical reaction, the energy of the molecule must exceed the activation energy for that reaction; for most biological reactions, this almost never happens without enzyme catalysis. Even with enzyme catalysis, the substrate molecules must experience a particularly energetic collision to react.

book, the activation energy is provided by the heat of a lighted match. For the molecules in the watery solution inside a cell, the kick is delivered by an unusually energetic random collision with surrounding molecules—collisions that become more violent as the temperature is raised.

In a living cell, the kick over the energy barrier is greatly aided by a specialized class of proteins—the **enzymes.** Each enzyme binds tightly to one or two molecules, called **substrates,** and holds them in a way that greatly reduces the activation energy of a particular chemical reaction that the bound substrates can undergo. A substance that can lower the activation energy of a reaction is termed a **catalyst;** catalysts increase the rate of chemical reactions because they allow a much larger proportion of the random collisions with surrounding molecules to kick the substrates over the energy barrier, as illustrated in Figure 3–13. Enzymes are among the most effective catalysts known, often speeding up reactions by a factor of as much as 10^{14}, and they thereby allow reactions that would not otherwise occur to proceed rapidly at normal temperatures.

Enzymes are also highly selective. Each enzyme usually catalyzes only one particular reaction: in other words, it selectively lowers the activation energy of only one of the several possible chemical reactions that its bound substrate molecules could undergo. In this way enzymes direct each of the many different molecules in a cell along specific reaction pathways (Figure 3–14).

The success of living organisms is attributable to a cell's ability to make enzymes of many types, each with precisely specified properties. Each enzyme has a unique shape containing an *active site,* a pocket or groove in the enzyme into which only particular substrates will fit (Figure 3–15). Like all other catalysts, enzyme molecules themselves remain unchanged after participating in a reaction and therefore can function over and over again. In Chapter 5, we discuss further how enzymes work, after we have looked in detail at the molecular structure of proteins.

How Enzymes Find Their Substrates: The Importance of Rapid Diffusion

A typical enzyme will catalyze the reaction of about a thousand substrate molecules every second. This means that it must be able to bind a new substrate molecule in a fraction of a millisecond. But both enzymes and their substrates are present in relatively small numbers in a cell. How do

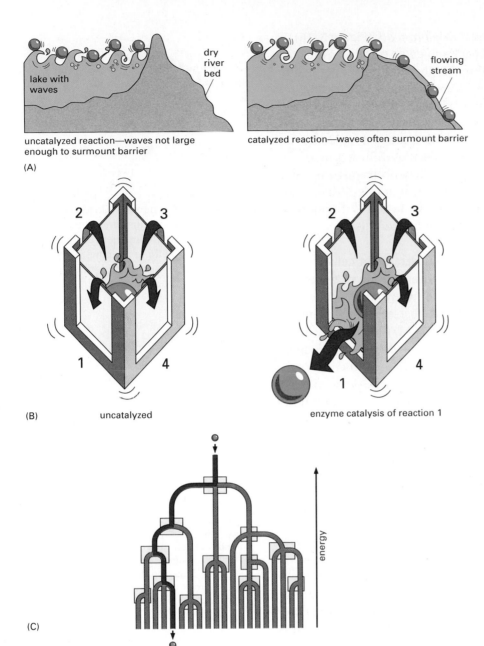

(A)

uncatalyzed reaction—waves not large enough to surmount barrier

catalyzed reaction—waves often surmount barrier

(B) uncatalyzed   enzyme catalysis of reaction 1

(C)

Figure 3–14 Floating ball analogies for enzyme catalysis. (A) A barrier dam is lowered to represent enzyme catalysis. The *green ball* represents a potential enzyme substrate (compound X) that is bouncing up and down in energy level due to constant encounters with waves (an analogy for the thermal bombardment of the substrate with the surrounding water molecules). When the barrier (activation energy) is lowered significantly, it allows the energetically favorable movement of the ball (the substrate) downhill. (B) The four walls of the box represent the activation energy barriers for four different chemical reactions that are all energetically favorable, in the sense that the products are at lower energy levels than the substrates. In the left-hand box, none of these reactions occurs because even the largest waves are not large enough to surmount any of the energy barriers. In the right-hand box, enzyme catalysis lowers the activation energy for reaction number 1 only; now the jostling of the waves allows passage of the molecule over this energy barrier only, inducing reaction 1. (C) A branching river with a set of barrier dams (*yellow boxes*) serves to illustrate how a series of enzyme-catalyzed reactions determines the exact reaction pathway followed by each molecule inside the cell.

they find each other so fast? Rapid binding is possible because motions are enormously fast at the molecular level. Due to heat energy, molecules are in constant motion and consequently will explore the space inside the cell very efficiently by wandering through it—a process called **diffusion.** In this way, every molecule in a cell collides with a huge number of other molecules each second. As the molecules in a liquid collide and bounce off one another, an individual molecule moves first one way

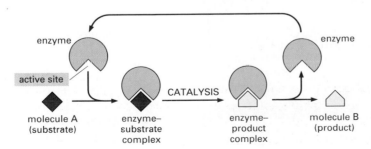

Figure 3–15 How enzymes work. Each enzyme has an active site to which one or two substrate molecules bind, forming an enzyme–substrate complex. A reaction occurs at the active site, producing an enzyme–product complex. The product is then released, allowing the enzyme to bind additional substrate molecules.

and then another, its path constituting a *random walk* (Figure 3–16). In such a walk, the average distance that each molecule travels (as the crow flies) from its starting point is proportional to the square root of the time involved: that is, if it takes a molecule 1 second on average to travel 1 μm, it takes 4 seconds to travel 2 μm, 100 seconds to travel 10 μm, and so on.

The inside of a cell is very crowded (Figure 3–17). Nevertheless, experiments in which fluorescent dyes and other labeled molecules are injected into cells show that small organic molecules diffuse through the watery gel of the cytosol nearly as rapidly as they do through water. A small organic molecule, for example, takes only about one-fifth of a second on average to diffuse a distance of 10 μm. Diffusion is therefore an efficient way for small molecules to move the limited distances in the cell.

Since enzymes move more slowly than substrates in cells, we can think of them as sitting still. The rate of encounter of each enzyme molecule with its substrate will depend on the concentration of the substrate molecule. For example, some abundant substrates are present at a concentration of 0.5 mM. Since pure water is 55 M, there is only about one such substrate molecule in the cell for every 10^5 water molecules. Nevertheless, the active site on an enzyme molecule that binds this substrate will be bombarded by about 500,000 random collisions with the substrate molecule per second. (For a substrate concentration tenfold lower, the number of collisions drops to 50,000 per second, and so on.) A random encounter between the surface of an enzyme and the matching surface of its substrate molecule often leads immediately to the formation of an enzyme–substrate complex that is ready to react. A reaction in which a covalent bond is broken or formed can now occur extremely rapidly. When one appreciates how quickly molecules move and react, the observed rates of enzymatic catalysis do not seem so amazing.

Once an enzyme and substrate have collided and snuggled together properly at the active site, they form multiple weak bonds with each other that persist until random thermal motion causes the molecules to dissociate again. In general, the stronger the binding of the enzyme and substrate, the slower their rate of dissociation. However, when two colliding molecules have poorly matching surfaces, few noncovalent bonds are formed and their total energy is negligible compared with that of thermal motion. In this case the two molecules dissociate as rapidly as they come together (see Figure 2–28). This is what prevents incorrect and unwanted associations from forming between mismatched molecules, such as between an enzyme and the wrong substrate.

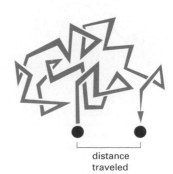

Figure 3–16 A random walk. Molecules in solution move in a random fashion due to the continual buffeting they receive in collisions with other molecules. This movement allows small molecules to diffuse rapidly from one part of the cell to another.

Question 3–3 The enzyme carbonic anhydrase is one of the speediest enzymes known. It catalyzes the hydration of CO_2 to HCO_3^- ($CO_2 + H_2O \rightleftharpoons HCO_3^- + H^+$). The rapid conversion of CO_2 gas into the much more soluble bicarbonate ion (HCO_3^-) is very important for the efficient transport of CO_2 in the bloodstream—from tissue, where CO_2 is produced by respiration, to the lungs, from where it is exhaled. Carbonic anhydrase accelerates the reaction by 10^7-fold, hydrating 10^5 CO_2 molecules per second at its maximal speed. What do you suppose limits the speed of the enzyme? Sketch a diagram analogous to the one shown in Figure 3–13 and indicate which portion of your diagram has been designed to display the 10^7-fold acceleration.

100 nm

Figure 3–17 The structure of the cytoplasm. The drawing is approximately to scale and emphasizes the crowding in the cytoplasm. Only the macromolecules are shown: RNAs are shown in *blue*, ribosomes in *green*, and proteins in *red*. Enzymes and other macromolecules diffuse relatively slowly in the cytoplasm, in part because they interact with many other macromolecules; small molecules, by contrast, diffuse nearly as rapidly as they do in water. (Adapted from D.S. Goodsell, *Trends in Biochem. Sci.* 16:203–206, 1991.)

The Free-Energy Change for a Reaction Determines Whether It Can Occur

We must now digress briefly to introduce some fundamental chemistry. Cells are chemical systems that must obey all chemical and physical laws. Although enzymes speed up reactions, they cannot by themselves force energetically unfavorable reactions to occur. In terms of a water analogy, enzymes by themselves cannot make water run uphill. Cells, however, must do just that in order to grow and divide: they must build highly ordered and energy-rich molecules from small and simple ones. We shall see that this is done through enzymes that directly *couple* energetically favorable reactions, which release energy and produce heat, to energetically unfavorable reactions, which produce biological order.

Before examining how such coupling is achieved, we must consider more carefully the term "energetically favorable." According to the second law of thermodynamics, a chemical reaction can proceed spontaneously only if it results in a net increase in the disorder of the universe (see Figure 3–6). Disorder increases when useful energy that could be harnessed (to do work) is dissipated as heat. The criterion for an increase of disorder can be expressed most conveniently in terms of a quantity called the **free energy, G,** of a system. The value of G is of interest only when a system undergoes a change, so that the *change* in G, denoted **ΔG** (delta G), can be specified. Suppose that the system being considered is a collection of molecules. Because of the way free energy is defined, ΔG measures the amount of disorder created in the universe when a reaction takes place that involves these molecules. *Energetically favorable reactions*, by definition, are those that decrease free energy, or, in other words, have a *negative* ΔG and create disorder (Figure 3–18).

A familiar example of an energetically favorable reaction on a macroscopic scale is the "reaction" by which a compressed spring relaxes to an expanded state, releasing its stored elastic energy as heat to its surroundings; an example on a microscopic scale is the dissolving of salt in water. Conversely, *energetically unfavorable reactions*, with a *positive* ΔG—such as those in which two amino acids are joined together to form a peptide bond—by themselves create order in the universe. Therefore, these reactions can take place only if they are coupled to a second reaction with a negative ΔG so large that the ΔG of the entire process is negative (Figure 3–19). These concepts are summarized, with examples, in Panel 3–1 (pp. 90–91).

The Concentration of Reactants Influences ΔG

As we have just described, a reaction A $\rightleftharpoons$ B will go in the direction A $\rightarrow$ B when the associated free-energy change, ΔG, is negative, just as a tensed spring left to itself will relax and lose its stored energy to its surroundings as heat. For a chemical reaction, however, ΔG depends not only on the energy stored in each individual molecule, but also on the concentrations of the molecules in the reaction mixture. Remember that ΔG reflects the degree to which a reaction creates a more disordered—in other words, a more probable—state of the universe. Recalling our coin analogy, it is very likely that a coin will flip from a head to a tail orientation if a jiggling box contains 90 heads and 10 tails, but this is a less probable event if the box contains 10 heads and 90 tails. For exactly the same

ENERGETICALLY FAVORABLE REACTION

The free energy of Y is greater than the free energy of X. Therefore $\Delta G < 0$, and the disorder of the universe increases during the reaction.

this reaction can occur spontaneously

ENERGETICALLY UNFAVORABLE REACTION

If the reaction X→Y occurred, ΔG would be > 0, and the universe would become more ordered.

this reaction can occur only if it is coupled to a second, energetically favorable reaction

Figure 3–18 The distinction between energetically favorable and energetically unfavorable reactions.

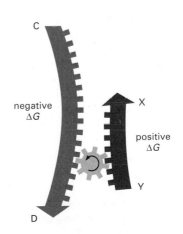

The energetically unfavorable reaction Y → X is driven by the energetically favorable reaction C → D, because the free-energy change for the pair of coupled reactions is less than zero.

Figure 3–19 How reaction coupling can drive an energetically unfavorable reaction.

FREE ENERGY

The molecules of a living cell possess energy because of their vibrations, rotations, and translations, and because of the energy that is stored in the bonds between individual atoms.

The free energy (*G*) measures the energy of a molecule that could in principle be used to do useful work at constant temperature (as in a living cell). *G* is measured in units of kilocalories per mole (kcal/mole) or kilojoules per mole (kJ/mole).

1 kJ = 0.239 kcal/mole
1 kcal = 4.18 kJ/mole

REACTIONS CAUSE DISORDER

Think of a chemical reaction occurring in an isolated cell at constant temperature and without change of volume. This reaction can produce disorder in two ways.

1 Changes of bond energy of the reacting molecules can cause heat to be released, which disorders the environment.

2 The reaction can decrease the amount of order in the reacting molecules—for example, by breaking apart a long chain of molecules, or by disrupting an interaction that prevents bond rotations.

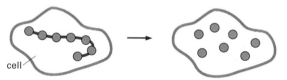

REACTION RATES

When we talk of a spontaneous reaction, we do not consider time or rate: a reaction with a negative free-energy change (Δ*G*) will not necessarily occur rapidly by itself. Thus, for the combustion of glucose in oxygen:

$$+\ 6O_2 \longrightarrow 6CO_2 + 6H_2O$$

$$\Delta G° = -686 \text{ kcal/mole}$$

But even this highly favorable reaction may not occur for centuries unless there are enzymes to speed up the process. Conversely, enzymes are able to catalyze reactions and speed up their rate, but they do not change the Δ*G°* of the reaction.

Δ*G* ("DELTA G")

Changes in free energy occurring in a reaction are denoted by Δ*G*, where "Δ" indicates a difference. Thus for the reaction:

$$A + B \longrightarrow C + D$$

Δ*G* = free energy (C + D) minus free energy (A + B)

Δ*G* is defined in such a way that it measures the amount of disorder caused by a reaction: the change in order inside the cell caused by the reaction, plus the change in order of the surroundings caused by the heat released.

Δ*G* is useful to biochemists because it measures how far away from equilibrium a reaction is. Thus the reaction

has a large negative Δ*G* because cells keep it a long way from equilibrium by continually making fresh ATP. However, if the cell dies, then most of its ATP becomes hydrolyzed, until equilibrium is reached and Δ*G* = 0. At this point the forward and backward reactions occur at equal rates, so that no further change takes place.

SPONTANEOUS REACTIONS

From the second law of thermodynamics, we know that the disorder of the universe can only increase (see page 79). The change in free energy, Δ*G*, measures the change in order produced by a reaction and is defined in such a way that Δ*G* is *negative* if the disorder of the universe (reaction plus surroundings) *increases*.

In other words, a chemical reaction that occurs spontaneously must have a negative Δ*G*

$$G_{products} - G_{reactants} = \Delta G < 0$$

EXAMPLE: the difference in free energy of 100 ml of 10 mM glucose (common sugar) and 100 ml of 10 mM sucrose plus 10 mM fructose is about −5.5 calories Therefore, the hydrolysis reaction (sucrose ⟶ glucose + fructose) can proceed spontaneously.

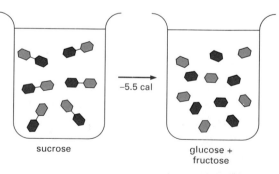

sucrose

glucose + fructose

In contrast, the reverse reaction (glucose + fructose ⟶ sucrose), which has a Δ*G* of +5.5 calories, could not occur without an input of energy from a coupled reaction (see page 89).

Panel 3–1 **Free energy and biological reactions.**

PREDICTING REACTIONS

In order to predict the outcome of a reaction (does it proceed to the right or to the left? At what point will it stop?), we must measure its standard free energy change ($\Delta G°$).

This quantity represents the gain or loss of free energy as one mole of reactant is converted to one mole of product under "standard conditions" (all molecules present at a concentration of 1 M and pH 7.0).

$\Delta G°$ for some reactions

glucose-1-P $\longrightarrow$ glucose-6-P	–1.7 kcal/mole
sucrose $\longrightarrow$ glucose + fructose	–5.5 kcal/mole
ATP $\longrightarrow$ ADP + P_i	–7.3 kcal/mole
glucose + $6O_2 \longrightarrow 6CO_2 + 6H_2O$	–686 kcal/mole

driving force

CHEMICAL EQUILIBRIA

There is a fixed relationship between the standard free energy change of a reaction, $\Delta G°$, and its equilibrium constant K. For example, the reversible reaction

$$A \rightleftharpoons B$$

will proceed until the ratio of concentrations [B]/[A] is equal to K (note: we use square brackets [ ] to indicate concentration). At this point the free energy of the system will have its lowest value.

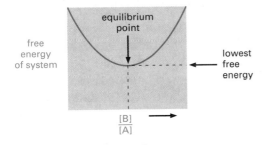

$$\Delta G° = -1.36 \log_{10} K$$
$$K = 10^{-\Delta G°/1.36}$$

For example, the reaction

glucose-1-P → glucose-6-P

has $\Delta G° = -1.74$ kcal/mole. Therefore, its equilibrium constant

$$K = 10^{(1.74/1.36)} = 10^{(1.28)} = 19$$

So the reaction will reach steady state when
[glucose-6-P]/[glucose-1-P] = 19

COUPLED REACTIONS

Reactions can be "coupled" together if they share one or more intermediates. In this case, the overall free energy change is simply the sum of the individual $\Delta G°$ values. A reaction that is unfavorable (has a positive $\Delta G°$) can for this reason be driven by a second highly favorable reaction.

SINGLE REACTION

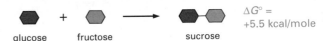

glucose + fructose → sucrose $\Delta G° = +5.5$ kcal/mole

NET RESULT: will not occur!

COUPLED REACTION

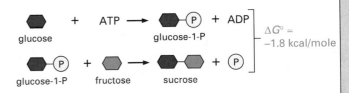

glucose + ATP → glucose-1-P + ADP
glucose-1-P + fructose → sucrose + P $\Delta G° = -1.8$ kcal/mole

NET RESULT: sucrose is made in a reaction driven by the hydrolysis of ATP.

HIGH-ENERGY BONDS

One of the most common reactions in the cell is hydrolysis, in which a covalent bond is split by adding water.

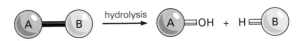

The $\Delta G°$ for this reaction is sometimes loosely termed the "bond energy." Compounds such as acetyl phosphate and ATP that have a large negative $\Delta G°$ of hydrolysis are said to have "high-energy" bonds.

	$\Delta G°$ (kcal/mole)
acetyl P $\longrightarrow$ acetate + P_i	–10.3
ATP $\longrightarrow$ ADP + P_i	–7.3
glucose-6-P $\longrightarrow$ glucose + P_i	–3.3

(Note that, for simplicity, water is omitted from the above equations.)

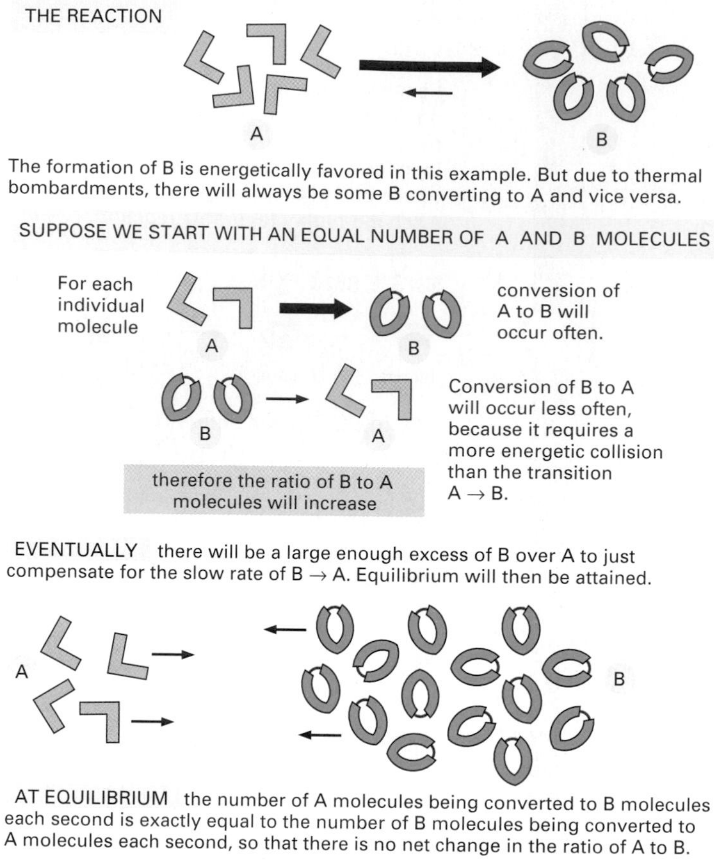

THE REACTION

The formation of B is energetically favored in this example. But due to thermal bombardments, there will always be some B converting to A and vice versa.

SUPPOSE WE START WITH AN EQUAL NUMBER OF A AND B MOLECULES

For each individual molecule

conversion of A to B will occur often.

Conversion of B to A will occur less often, because it requires a more energetic collision than the transition A → B.

therefore the ratio of B to A molecules will increase

EVENTUALLY there will be a large enough excess of B over A to just compensate for the slow rate of B → A. Equilibrium will then be attained.

AT EQUILIBRIUM the number of A molecules being converted to B molecules each second is exactly equal to the number of B molecules being converted to A molecules each second, so that there is no net change in the ratio of A to B.

Figure 3–20 **Chemical equilibrium.** When a reaction reaches equilibrium, the forward and backward flux of reacting molecules are equal and opposite.

reason, for a reversible reaction A ⇌ B, a large excess of A over B will tend to drive the reaction in the direction A → B; that is, there will be a tendency for there to be more molecules making the transition A → B than there are molecules making the transition B → A. Therefore, the ΔG becomes more negative for the transition A → B (and more positive for the transition B → A) as the ratio of A to B increases.

How much of a concentration difference is needed to compensate for a given decrease in chemical bond energy (and accompanying heat release)? The answer is not intuitively obvious, but it can be determined from a thermodynamic analysis that makes it possible to separate the concentration-dependent and the concentration-independent parts of the free-energy change. The ΔG for a given reaction can thereby be written as the sum of two parts: the first, called the *standard free-energy change*, $\Delta G°$, depends on the intrinsic characters of the reacting molecules; the second depends on their concentrations. For the simple reaction A → B at 37°C,

$$\Delta G = \Delta G° + 0.616 \ln \frac{[B]}{[A]}$$

where ΔG is in kilocalories per mole, [A] and [B] denote the concentrations of A and B, 0.616 is a constant, and ln is the natural logarithm.

Note that ΔG equals the value of $\Delta G°$ when the molar concentrations of A and B are equal (ln 1 = 0). As expected, ΔG becomes more negative as the ratio of B to A decreases (the ln of a number < 1 is negative).

Chemical **equilibrium** is reached when the concentration effect just balances the push given to the reaction by $\Delta G°$, so that there is no net

Table 3–1 **Relationship Between Free-Energy Change and Equilibrium Constant**

equilibrium constant $\dfrac{[B]}{[A]} = K$ (liters/mole)	free energy of B minus free energy of A (kcal/mole)
10^5	−7.1
10^4	−5.7
10^3	−4.3
10^2	−2.8
10	−1.4
1	0
10^{-1}	1.4
10^{-2}	2.8
10^{-3}	4.3
10^{-4}	5.7
10^{-5}	7.1

Values of the equilibrium constant were calculated for the simple chemical reaction A ⇌ B using the equation given in the text.

The $\Delta G°$ given here is in kilocalories per mole at 37°C (1 kilocalorie is equal to 4.184 kilojoules). As explained in the text, $\Delta G°$ represents the free-energy difference under standard conditions (where all components are present at a concentration of 1.0 mole/liter).

We see from this table that if there is a favorable free-energy change of −4.3 kcal/mole for the transition A → B, there will be 1000 times more molecules in state B than in state A.

change of free energy to drive the reaction in either direction (Figure 3–20). Here $\Delta G = 0$, and so the concentrations of A and B are such that

$$-0.616 \ln \frac{[B]}{[A]} = \Delta G°$$

which means that there is chemical equilibrium at 37°C when

$$\frac{[B]}{[A]} = e^{-\Delta G°/0.616}$$

Table 3–1 shows how the equilibrium ratio of A to B (expressed as an equilibrium constant, K) depends on the value of $\Delta G°$.

It is important to recognize that when an enzyme (or any catalyst) lowers the activation energy for the reaction A → B, it also lowers the activation energy for the reaction B → A by exactly the same amount (see Figure 3–12). The forward and backward reactions will therefore be accelerated by the same factor by an enzyme, and the equilibrium point for the reaction (and $\Delta G°$) remains unchanged (Figure 3–21).

For Sequential Reactions, $\Delta G°$ Values Are Additive

The course of most reactions can be predicted quantitatively. A large body of thermodynamic data has been collected that makes it possible to calculate the standard change in free energy, $\Delta G°$, for most of the important metabolic reactions of the cell. The overall free-energy change for a metabolic pathway is then simply the sum of the free-energy changes in each of its component steps. Consider, for example, two sequential reactions

$$X \rightarrow Y \quad \text{and} \quad Y \rightarrow Z$$

where the $\Delta G°$ values are +5 and –13 kcal/mole, respectively. (Recall that a mole is 6×10^{23} molecules of a substance.) If these two reactions occur sequentially, the $\Delta G°$ for the coupled reaction will be –8 kcal/mole. Thus, the unfavorable reaction X → Y, which will not occur spontaneously, can be driven by the favorable reaction Y → Z, provided that the second reaction follows the first.

Cells can therefore cause the energetically unfavorable transition, X → Y, to occur if an enzyme catalyzing the X → Y reaction is supplemented by a second enzyme that catalyzes the energetically *favorable* reaction, Y → Z. In effect, the reaction Y → Z will then act as a "siphon" to drive the conversion of all of molecule X to molecule Y, and thence to molecule Z (Figure 3–22). For example, several of the reactions in the long pathway that converts sugars into CO_2 and H_2O are energetically unfavorable. But the pathway nevertheless proceeds rapidly to comple-

Question 3–4 Consider again the analogy of the jiggling box containing coins that was described in the text. The reaction, the flipping of coins that either face heads up (H) or tails up (T), is described by the equation H ⇌ T.

A. What are ΔG and $\Delta G°$ in this analogy?

B. What corresponds to the temperature at which the reaction proceeds? What corresponds to the activation energy of the reaction? Assume you have an "enzyme," called jigglase, that catalyzes this reaction. What would the effect of jigglase be and what, mechanically, might jigglase do in this analogy?

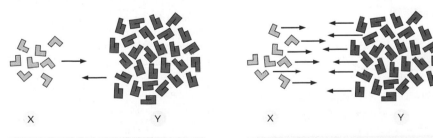

(A) UNCATALYZED REACTION   (B) ENZYME-CATALYZED REACTION

Figure 3–21 **Enzymes cannot change the equilibrium point for reactions.** Enzymes, like all catalysts, speed up the forward and backward rates of a reaction by the same factor. Therefore, for both the catalyzed and the uncatalyzed reactions shown here, the number of molecules undergoing the transition X → Y is equal to the number of molecules undergoing the transition Y → X when the ratio of Y molecules to X molecules is 3.5 to 1. In other words, the two reactions reach equilibrium at exactly the same point.

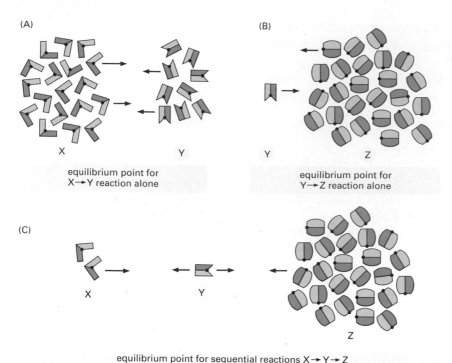

(A)

X   Y

equilibrium point for
X→Y reaction alone

(B)

Y   Z

equilibrium point for
Y→Z reaction alone

(C)

X   Y

Z

equilibrium point for sequential reactions X→Y→Z

Figure 3–22 How an energetically un-favorable reaction can be driven by a second, following reaction. (A) At equilibrium, there are twice as many X molecules as Y molecules, because X is of lower energy than Y. (B) At equilibrium, there are 25 times more Z molecules than Y molecules, because Z is of much lower energy than Y. (C) If the reactions in (A) and (B) are coupled, nearly all of the X molecules will be converted to Z molecules, as shown.

tion because the total $\Delta G°$ for the series of sequential reactions has a large negative value.

But forming a sequential pathway is not adequate for many purposes. Often the desired pathway is simply X → Y, without further conversion of Y to some other product. Fortunately, there are other more general ways of using enzymes to couple reactions together. How these work are the topic we discuss next.

Question 3–5   Look carefully at Figure 3–22. Sketch an energy diagram similar to that in Figure 3–12 for the two reactions alone and for the combined reactions. Indicate the standard free-energy changes for the reactions X → Y, Y → Z, and X → Z in the graph. Indicate how enzymes that catalyze these reactions would change the energy diagram.

Activated Carrier Molecules and Biosynthesis

The energy released by the oxidation of food molecules must be stored temporarily before it can be channeled into the construction of other small organic molecules and of the larger and more complex molecules needed by the cell. In most cases, the energy is stored as chemical bond energy in a small set of activated "carrier molecules," which contain one or more energy-rich covalent bonds. These molecules diffuse rapidly

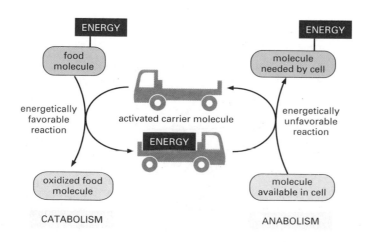

ENERGY

food
molecule

energetically
favorable
reaction

activated carrier molecule

ENERGY

oxidized food
molecule

CATABOLISM

ENERGY

molecule
needed by cell

energetically
unfavorable
reaction

molecule
available in cell

ANABOLISM

Figure 3–23 Energy transfer and the role of activated carriers in metabolism. By serving as energy shuttles, activated carrier molecules perform their function as go-betweens that link the breakdown of food molecules and the release of energy (*catabolism*) to the energy-requiring biosynthesis of small and large organic molecules (*anabolism*).

throughout the cell and thereby carry their bond energy from sites of energy generation to the sites where energy is used for biosynthesis and other needed cell activities (Figure 3–23).

The **activated carriers** store energy in an easily exchangeable form, either as a readily transferable chemical group or as high-energy electrons, and they can serve a dual role as a source of both energy and chemical groups in biosynthetic reactions. For historical reasons, these molecules are also sometimes referred to as *coenzymes*. The most important of the activated carrier molecules are ATP and two molecules that are closely related to each other, NADH and NADPH—as we discuss in detail shortly. We shall see that cells use activated carrier molecules like money to pay for reactions that otherwise could not take place.

The Formation of an Activated Carrier Is Coupled to an Energetically Favorable Reaction

When a fuel molecule such as glucose is oxidized in a cell, enzyme-catalyzed reactions ensure that a large part of the free energy that is released by oxidation is captured in a chemically useful form, rather than being released wastefully as heat. This is achieved by means of a **coupled reaction,** in which an energetically favorable reaction is used to drive an energetically unfavorable one that produces an activated carrier molecule or some other useful energy store. Coupling mechanisms require enzymes and are fundamental to all the energy transactions of the cell.

The nature of a coupled reaction is illustrated by a mechanical analogy in Figure 3–24, in which an energetically favorable chemical reaction is represented by rocks falling from a cliff. The energy of falling rocks would normally be entirely wasted in the form of heat generated by friction when the rocks hit the ground (see the falling brick diagram in Figure 3–7). By careful design, however, part of this energy could be used to drive a paddle wheel that lifts a bucket of water (Figure 3–24B). Because the rocks can now reach the ground only after moving the paddle wheel, we say that the energetically favorable reaction of rock

Figure 3–24 A mechanical model illustrating the principle of coupled chemical reactions. The spontaneous reaction shown in (A) could serve as an analogy for the direct oxidation of glucose to CO_2 and H_2O, which produces heat only. In (B) the same reaction is coupled to a second reaction; this second reaction could serve as an analogy for the synthesis of activated carrier molecules. The energy produced in (B) is in a more useful form than in (A) and can be used to drive a variety of otherwise energetically unfavorable reactions (C).

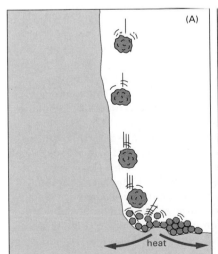

(A)

kinetic energy transformed into heat energy only

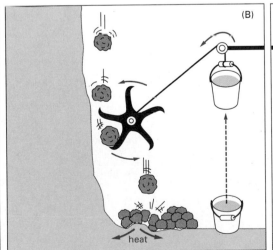

(B)

part of the kinetic energy is used to lift a bucket of water, and a correspondingly smaller amount is transformed into heat

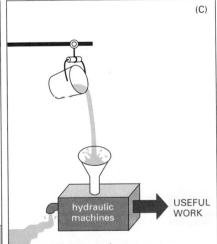

(C)

the potential kinetic energy stored in the raised bucket of water can be used to drive hydraulic machines that carry out a variety of useful tasks

falling has been directly *coupled* to the energetically unfavorable reaction of lifting the bucket of water. Note that because part of the energy is used to do work in (B), the rocks hit the ground with less velocity than in (A), and correspondingly less energy is wasted as heat.

Exactly analogous processes occur in cells, where enzymes play the role of the paddle wheel in our analogy. By mechanisms that will be discussed in Chapter 4, they couple an energetically favorable reaction, such as the oxidation of foodstuffs, to an energetically unfavorable reaction, such as the generation of an activated carrier molecule. As a result, the amount of heat released by the oxidation reaction is reduced by exactly the amount of energy that is stored in the energy-rich covalent bonds of the activated carrier molecule. The activated carrier molecule in turn picks up a packet of energy of a size sufficient to power a chemical reaction elsewhere in the cell.

ATP Is the Most Widely Used Activated Carrier Molecule

The most important and versatile of the activated carriers in cells is **ATP** (adenosine 5′-triphosphate). Just as the energy stored in the raised bucket of water in Figure 3–24B can be used to drive a wide variety of hydraulic machines, ATP serves as a convenient and versatile store, or currency, of energy to drive a variety of chemical reactions in cells. As shown in Figure 3–25, ATP is synthesized in an energetically unfavorable phosphorylation reaction in which a phosphate group is added to **ADP** (adenosine 5′-diphosphate). When required, ATP gives up its energy packet, represented by a molecule of inorganic phosphate that can be released, in an energetically favorable hydrolysis to ADP. The regenerated ADP is then available to be used for another round of the phosphorylation reaction that forms ATP.

The energetically favorable reaction of ATP hydrolysis is coupled to many otherwise unfavorable reactions through which other molecules are synthesized. We shall encounter several of these reactions later in this chapter. Many of them involve the transfer of the terminal phosphate in ATP to another molecule, as illustrated by the phosphorylation reaction in Figure 3–26.

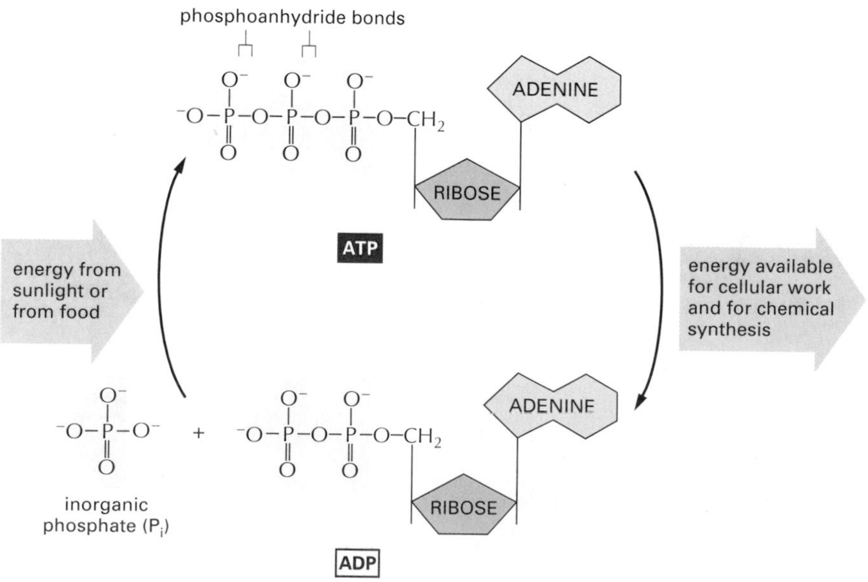

Question 3–6   Use Figure 3–24B to illustrate the following reaction driven by hydrolysis of ATP.

$$X + ATP \rightarrow Y + ADP + ⓟ$$

A. In this case, which molecule or molecules would be analogous to (a) rocks at top of cliff, (b) broken debris at bottom of cliff, (c) bucket at its highest point, (d) bucket on the ground?

B. What would be analogous to (a) the rocks hitting the ground in the absence of the paddle wheel in Figure 3–24A and (b) the hydraulic machine in Figure 3–24C?

Figure 3–25 The interconversion of ATP and ADP. The two outermost phosphates in ATP are held to the rest of the molecule by high-energy phosphoanhydride bonds and are readily transferred. Water can be added to ATP to form ADP and inorganic phosphate (P_i). This hydrolysis of the terminal phosphate of ATP yields between 11 and 13 kcal/mole of usable energy, depending on the intracellular conditions. The large negative ΔG of this reaction arises from a number of factors. Release of the terminal phosphate group removes an unfavorable repulsion between adjacent negative charges; in addition, the inorganic phosphate ion (P_i) released is stabilized by resonance and by favorable hydrogen-bond formation with water. The formation of ATP from ADP and P_i reverses the hydrolysis reaction, and it requires a coupling of this energetically unfavorable reaction to an even more favorable one.

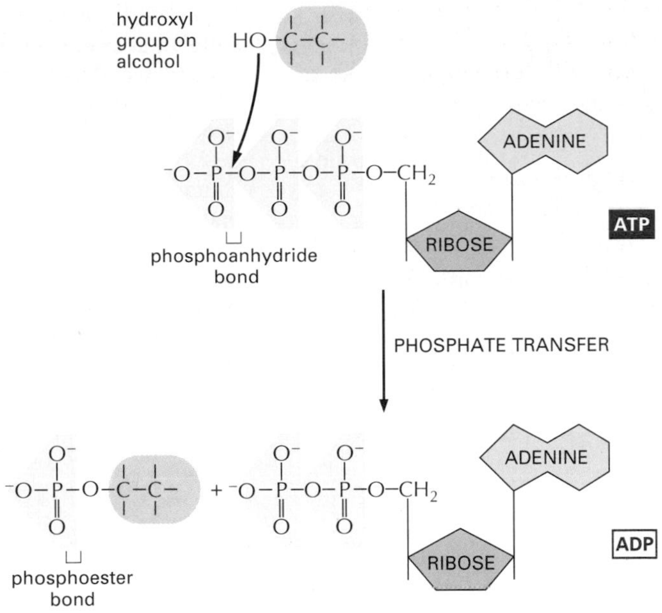

Figure 3–26 **An example of a phosphate transfer reaction.** Because an energy-rich phosphoanhydride bond in ATP is converted to a phosphoester bond, this reaction is energetically favorable, having a large negative ΔG. Reactions of this type are involved in the synthesis of phospholipids and in the initial steps of reactions that catabolize sugars.

ATP is the most abundant active carrier in cells. As one example, it is used to supply energy for many of the pumps that transport substances into and out of the cell (Chapter 12); it also powers the molecular motors that enable muscle cells to contract and nerve cells to transport materials from one end of their long axons to another (Chapter 16).

Energy Stored in ATP Is Often Harnessed to Join Two Molecules Together

We have previously discussed one way in which an energetically favorable reaction can be coupled to an energetically unfavorable reaction, $X \to Y$, so as to enable it to occur. In that scheme a second enzyme catalyzes the energetically favorable reaction $Y \to Z$, pulling all of the X to Y in the process (see Figure 3–22). But when the required product is Y and not Z, this mechanism is not useful.

A frequent type of reaction that is needed for biosynthesis is one in which two molecules, A and B, are joined together to produce A–B in the energetically unfavorable condensation reaction

$$A–H + B–OH \to A–B + H_2O$$

There is an indirect pathway that allows A–H and B–OH to form A–B, in which a coupling to ATP hydrolysis makes the reaction go. Here energy from ATP hydrolysis is first used to convert B–OH to a higher-energy intermediate compound, which then reacts directly with A–H to give A–B. The simplest possible mechanism involves the transfer of a phosphate from ATP to B–OH to make $B–OPO_3$, in which case the reaction pathway contains only two steps:

    1. $B–OH + ATP \to B–O–PO_3 + ADP$

    2. $A–H + B–O–PO_3 \to A–B + P_i$

    Net result: $B–OH + ATP + A–H \to A–B + ADP + P_i$

The condensation reaction, which by itself is energetically unfavorable, is forced to occur by being directly coupled to ATP hydrolysis in an enzyme-catalyzed reaction pathway (Figure 3–27A).

Question 3–7 The phosphoanhydride bond that links two phosphate groups in ATP in a high-energy linkage has a $\Delta G°$ of –7.3 kcal/mole. Hydrolysis of this bond liberates from 11 to 13 kcal/mole of usable energy. How can this be? Why do you think a range of energies is given, rather than a precise number as for $\Delta G°$?

(A)

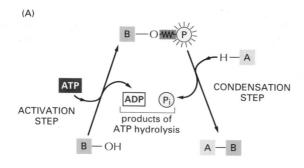

Figure 3–27 An example of an energetically unfavorable biosynthetic reaction driven by ATP hydrolysis. (A) Schematic illustration of the formation of A–B in the condensation reaction described in the text. (B) The biosynthesis of glutamine. Glutamic acid is first converted to a high-energy phosphorylated intermediate (corresponding to the compound B–O–PO₃ described in the text), which then reacts with ammonia (corresponding to A–H) to form glutamine. In this example both steps occur on the surface of the same enzyme, *glutamine synthase*. Note that, for clarity, the amino acids are shown in their uncharged form.

(B)

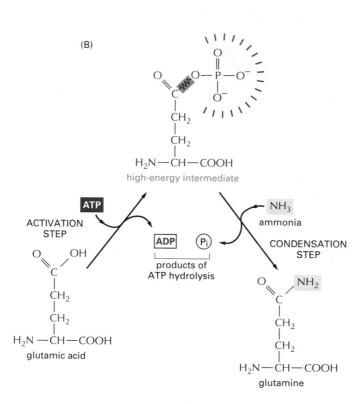

A biosynthetic reaction of exactly this type is employed to synthesize the amino acid glutamine, as illustrated in Figure 3–27B. We will see shortly that very similar (but more complex) mechanisms are also used to produce nearly all of the large molecules of the cell.

NADH and NADPH Are Important Electron Carriers

Other important activated carrier molecules participate in oxidation-reduction reactions and are commonly part of coupled reactions in cells. These activated carriers are specialized to carry high-energy electrons and hydrogen atoms. The most important of these electron carriers are **NAD⁺** (nicotinamide adenine dinucleotide) and the closely related molecule **NADP⁺** (nicotinamide adenine dinucleotide phosphate). Later, we examine some of the reactions in which they participate. NAD⁺ and NADP⁺ each pick up a "packet of energy" corresponding to two high-energy electrons plus a proton (H⁺), becoming **NADH** (*reduced* nicotinamide adenine dinucleotide) and **NADPH** (*reduced* nicotinamide adenine dinucleotide phosphate), respectively. These molecules can therefore also be regarded as carriers of hydride ions (the H⁺ plus two electrons, or H⁻).

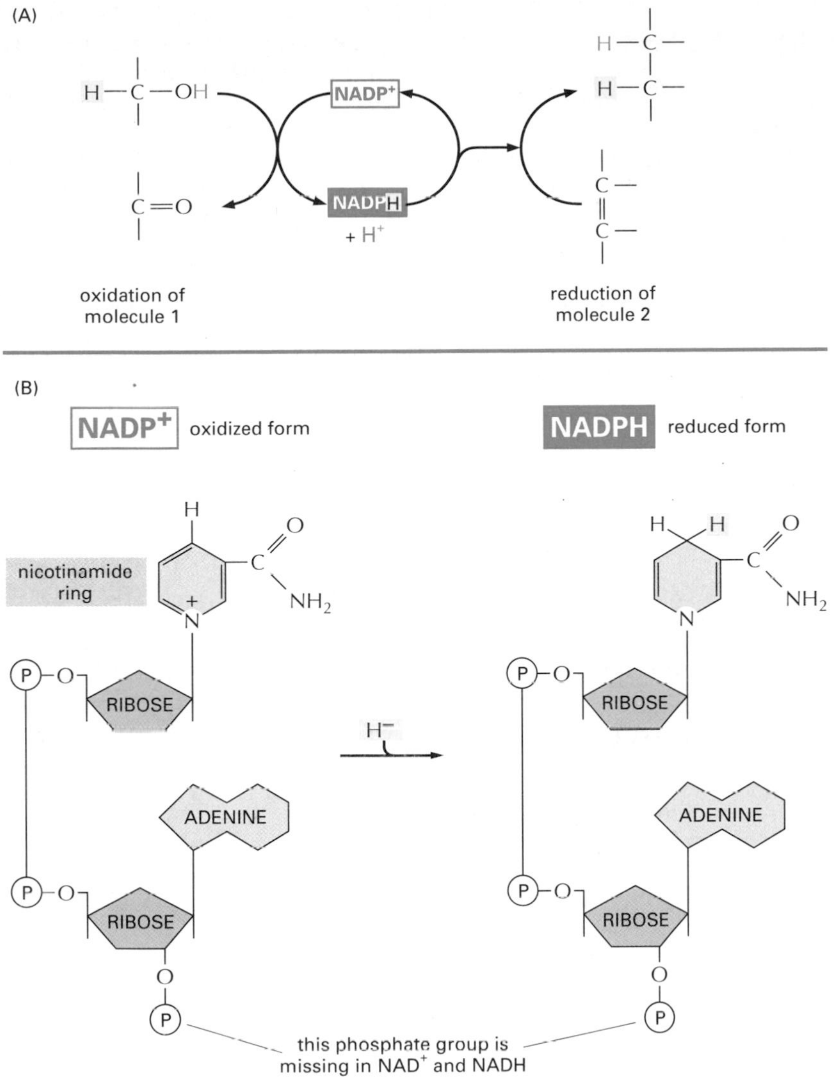

(A)

oxidation of
molecule 1

reduction of
molecule 2

(B)

NADP⁺ oxidized form

NADPH reduced form

nicotinamide
ring

H^-

this phosphate group is
missing in NAD⁺ and NADH

Figure 3–28 NADPH, an important carrier of electrons. (A) NADPH is produced in reactions of the general type shown on the left, in which two hydrogen atoms are removed from a substrate. The oxidized form of the carrier molecule, NADP⁺, receives one hydrogen atom plus an electron (a hydride ion), and the proton (H⁺) from the other H atom is released into solution. Because NADPH holds its hydride ion in a high-energy linkage, the added hydride ion can easily be transferred to other molecules, as shown on the right.

(B) The structure of NADP⁺ and NADPH. The part of the NADP⁺ molecule known as the nicotinamide ring accepts two electrons together with a proton (the equivalent of a hydride ion, H⁻), forming NADPH. The molecules NAD⁺ and NADH are identical in structure to NADP⁺ and NADPH, respectively, except that the indicated phosphate group is absent from both.

Like ATP, NADPH is an activated carrier that participates in many important biosynthetic reactions that would otherwise be energetically unfavorable. The NADPH is produced according to the general scheme shown in Figure 3–28. During a special set of energy-yielding catabolic reactions, a hydrogen atom plus two electrons are removed from the substrate molecule and added to the nicotinamide ring of NADP⁺ to form NADPH. This is a typical oxidation-reduction reaction; the substrate is oxidized and NADP⁺ is reduced.

The hydride ion carried by NADPH is given up readily in a subsequent oxidation-reduction reaction, because the ring can achieve a more stable arrangement of electrons without it. In this subsequent reaction, which regenerates NADP⁺, it is the NADPH that becomes oxidized and the substrate that becomes reduced. The NADPH is an effective donor of its hydride ion to other molecules for the same reason that ATP readily transfers a phosphate: in both cases the transfer is accom-

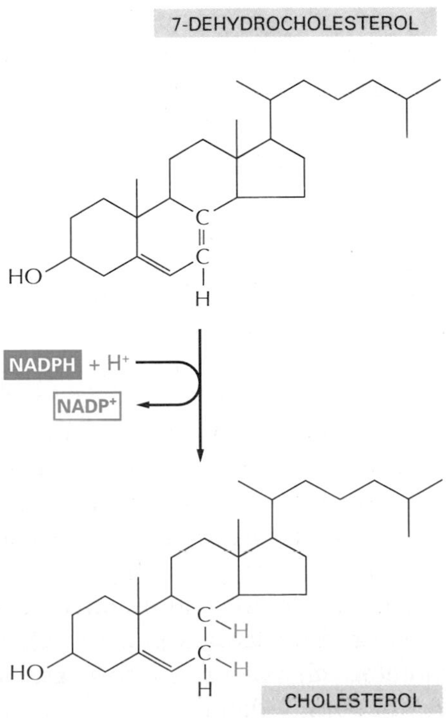

7-DEHYDROCHOLESTEROL

NADPH + H⁺

NADP⁺

CHOLESTEROL

Figure 3–29 The final stage in one of the biosynthetic routes leading to cholesterol. As in many other biosynthetic reactions, the reduction of the C=C bond is achieved by the transfer of a hydride ion from the carrier molecule NADPH, plus a proton (H⁺) from the solution.

Table 3–2 Some Activated Carrier Molecules Widely Used in Metabolism

Activated Carrier	Group Carried in High-Energy Linkage
ATP	phosphate
NADH, NADPH, FADH$_2$	electrons and hydrogens
Acetyl CoA	acetyl group
Carboxylated biotin	carboxyl group
S-Adenosylmethionine	methyl group
Uridine diphosphate glucose	glucose

panied by a large negative free-energy change. One example of the use of NADPH in biosynthesis is shown in Figure 3–29.

The difference of a single phosphate group has no effect on the electron-transfer properties of NADPH compared with NADH, but it is crucial for their distinctive roles. The extra phosphate group on NADPH is far from the region involved in electron transfer (see Figure 3–28B) and is of no importance to the transfer reaction. It does, however, give a molecule of NADPH a slightly different shape from that of NADH, and so NADPH and NADH bind as substrates to different sets of enzymes. Thus the two types of carriers are used to deliver electrons (or hydride ions) to different destinations.

Why should there be this division of labor? The answer lies in the need to regulate two sets of electron-transfer reactions independently. NADPH operates chiefly with enzymes that catalyze anabolic reactions, supplying the high-energy electrons needed to synthesize energy-rich biological molecules. NADH, by contrast, has a special role as an intermediate in the catabolic system of reactions that generate ATP through the oxidation of food molecules, as we discuss in Chapter 4. The genesis of NADH from NAD$^+$ and that of NADPH from NADP$^+$ occur by different pathways and are independently regulated, so that the cell can independently adjust the supply of electrons for these two contrasting purposes. Inside the cell the ratio of NAD$^+$ to NADH is kept high, whereas the ratio of NADP$^+$ to NADPH is kept low. This provides plenty of NAD$^+$ to act as an oxidizing agent and plenty of NADPH to act as a reducing agent—as required for their special roles in catabolism and anabolism, respectively.

There Are Many Other Activated Carrier Molecules in Cells

Other activated carriers also pick up and carry a chemical group in an easily transferred, high-energy linkage (Table 3–2). For example, coenzyme A carries an acetyl group in a readily transferable linkage, and in this activated form is known as **acetyl CoA** (acetyl coenzyme A). The structure of acetyl CoA is illustrated in Figure 3–30; it is used to add two carbon units in the biosynthesis of larger molecules.

In acetyl CoA and the other carrier molecules in Table 3–2, the transferable group makes up only a small part of the molecule. The rest consists of a large organic portion that serves as a convenient "handle," facil-

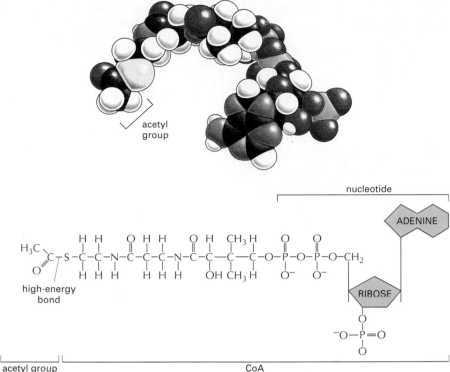

Figure 3–30 The structure of the important activated carrier molecule acetyl CoA. A space-filling model is shown above the structure. The sulfur atom (*yellow*) forms a thioester bond to acetate. Because this is a high-energy linkage, releasing a large amount of free energy when it is hydrolyzed, the acetate molecule can be readily transferred to other molecules.

itating the recognition of the carrier molecule by specific enzymes. As with acetyl CoA, this handle portion very often contains a nucleotide, a curious fact that may be a relic from an early stage of evolution. It is thought that the main catalysts for early life-forms on the earth were RNA molecules (or their close relatives) and that proteins were a later evolutionary addition (see Chapter 7). It is tempting to speculate that many of the carrier molecules that we find today originated in an earlier RNA world, where their nucleotide portions could have been useful for binding them to RNA enzymes.

Figure 3–31 A carboxyl group transfer reaction using an activated carrier molecule. Carboxylated biotin is used by the enzyme *pyruvate carboxylase* to transfer a carboxyl group in the production of oxaloacetate, a molecule needed for the citric acid cycle. The acceptor molecule for this group transfer reaction is pyruvate; other enzymes use biotin to transfer carboxyl groups to other acceptor molecules. Note that synthesis of carboxylated biotin requires energy derived from ATP—a general feature of many activated carriers.

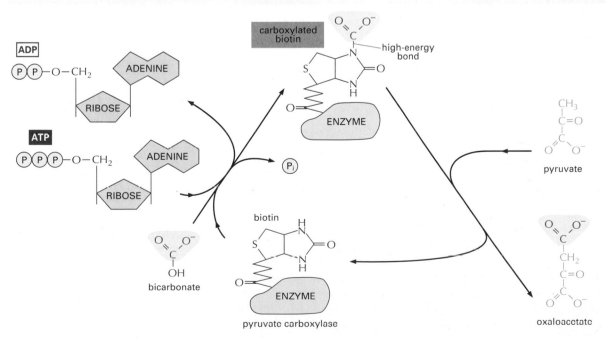

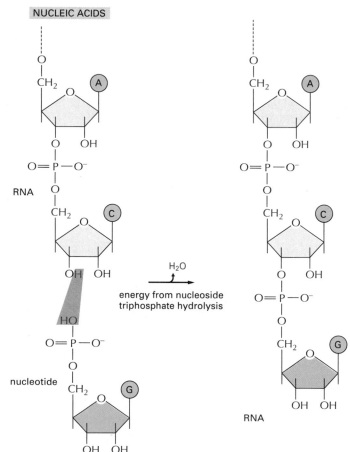

Figure 3–32 **Condensation and hydrolysis as opposite reactions.** The macromolecules of the cell are polymers that are formed from subunits (or monomers) by a condensation reaction and broken down by hydrolysis. The condensation reactions are all energetically unfavorable.

Examples of the type of transfer reactions catalyzed by the activated carrier molecules ATP (transfer of phosphate) and NADPH (transfer of electrons and hydrogen) have been presented in Figures 3–26 and 3–29, respectively. Other reactions might involve the transfers of methyl, carboxyl, and glucose groups from activated carrier molecules, for the purpose of biosynthesis. The activated carriers are usually generated in reactions coupled to ATP hydrolysis (Figure 3–31). Therefore, the energy that enables their groups to be used for biosynthesis ultimately comes from the catabolic reactions that generate ATP. Similar processes occur in the synthesis of the very large molecules of the cell—the nucleic acids, proteins, and polysaccharides—that we discuss next.

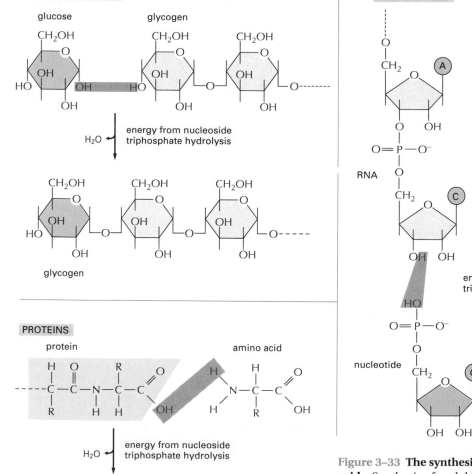

Figure 3–33 **The synthesis of polysaccharides, proteins, and nucleic acids.** Synthesis of each kind of biological polymer involves the loss of water in a condensation reaction. Not shown is the consumption of high-energy nucleoside triphosphates that is required to activate each monomer prior to its addition. In contrast, the reverse reaction—the breakdown of all three types of polymers—occurs by the simple addition of water (hydrolysis).

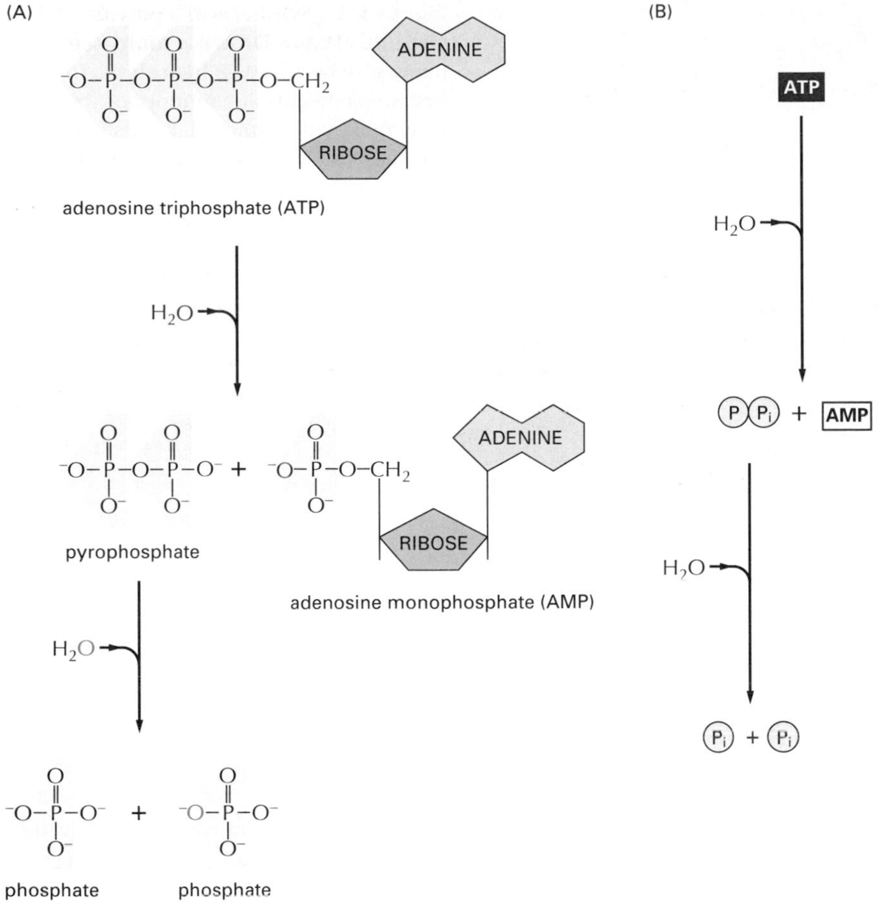

(A)

adenosine triphosphate (ATP)

$H_2O \rightarrow$

pyrophosphate + adenosine monophosphate (AMP)

$H_2O \rightarrow$

phosphate + phosphate

(B)

ATP

$H_2O \rightarrow$

P P_i + AMP

$H_2O \rightarrow$

P_i + P_i

Figure 3–34 **An alternative route for the hydrolysis of ATP, in which pyrophosphate is first formed and then hydrolyzed.** This route releases about twice as much free energy as the reaction shown earlier in Figure 3–25. (A) In the two successive hydrolysis reactions, oxygen atoms from the participating water molecules are retained in the products but the hydrogen atoms dissociate to form free hydrogen ions, H+. (B) Overall reaction shown in summary form.

The Synthesis of Biological Polymers Requires an Energy Input

The macromolecules of the cell constitute the vast majority of its dry mass—that is, of the mass not due to water. These molecules are made from subunits (or monomers) that are linked together in a *condensation* reaction, in which the constituents of a water molecule (OH plus H) are removed from the two reactants. Consequently, the reverse reaction—the breakdown of all three types of polymers—occurs by the enzyme-catalyzed addition of water (*hydrolysis*). This hydrolysis reaction is energetically favorable, whereas the biosynthetic reactions require an energy input and are more complex (Figure 3–32).

The nucleic acids (DNA and RNA), proteins, and polysaccharides are all polymers that are produced by the repeated addition of a *subunit* (also called a monomer) onto one end of a growing chain. The mode of synthesis of each of these types of macromolecules is outlined in Figure 3–33. As indicated, the condensation step in each case depends on energy from nucleoside triphosphate hydrolysis. And yet, except for the nucleic acids, there are no phosphate groups left in the final product molecules. How then are reactions that release the energy of ATP hydrolysis coupled to polymer synthesis?

For each type of macromolecule, an enzyme-catalyzed pathway exists which resembles that discussed previously for the synthesis of the amino acid glutamine (see Figure 3–27). The principle is exactly the

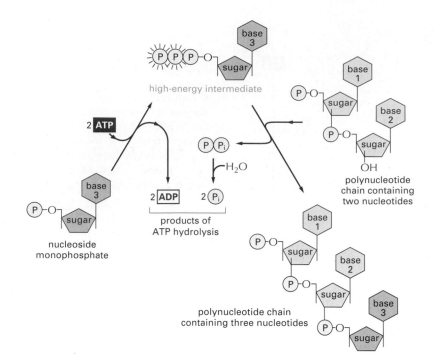

Figure 3–35 Synthesis of a polynucleotide, RNA or DNA, is a multistep process driven by ATP hydrolysis. In the first step a nucleoside monophosphate is activated by the sequential transfer of the terminal phosphate groups from two ATP molecules. The high-energy intermediate formed—a nucleoside triphosphate—exists free in solution until it reacts with the growing end of an RNA or a DNA chain with release of pyrophosphate. Hydrolysis of the latter to inorganic phosphate is highly favorable and helps to drive the overall reaction in the direction of polynucleotide synthesis.

same, in that the OH group that will be removed in the condensation reaction is first activated by becoming involved in a high-energy linkage to a second molecule. However, the actual mechanisms used to link ATP hydrolysis to the synthesis of proteins and polysaccharides are more complex than that used for glutamine synthesis, since a series of high-energy intermediates generates the final high-energy bond that is broken during the condensation step (as discussed in Chapter 7 for protein synthesis).

There are limits to what each activated carrier can do in driving biosynthesis. The ΔG for the hydrolysis of ATP to ADP and inorganic phosphate (P_i) depends on the concentrations of all of the reactants, but under the usual conditions in a cell it is between –11 and –13 kcal/mole. In principle, this hydrolysis reaction can be used to drive an unfavorable reaction with a ΔG of, perhaps, +10 kcal/mole, provided that a suitable reaction path is available. For some biosynthetic reactions, however, even –13 kcal/mole may not be enough. In these cases the path of ATP hydrolysis can be altered so that it initially produces AMP and pyrophosphate (PP_i), which is itself then hydrolyzed in a subsequent step (Figure 3–34). The whole process makes available a total free-energy change of about –26 kcal/mole. An important biosynthetic reaction that is driven in this way, nucleic acid (polynucleotide) synthesis, is illustrated in Figure 3–35.

Essential Concepts

- Living organisms are able to exist because of a continual input of energy. Part of this energy is used to carry out essential functions—such as maintenance, growth, and reproduction—and the remainder is lost in the form of heat.

- The primary source of energy for most living organisms is the sun. Plants and photosynthetic bacteria use solar energy to produce organic molecules from carbon dioxide. Animals obtain food by eating plants or by eating animals that feed on plants.

- Each of the many hundreds of chemical reactions that occur in a cell is specifically catalyzed by an enzyme. Large numbers of different enzymes work in sequence to form chains of reactions, called metabolic pathways, each performing a particular function in the cell.

- Catabolic reactions break down food molecules through oxidative pathways and release energy. Anabolic reactions build up the many complex molecules needed by the cell and require an energy input. Both the building blocks and the energy required for anabolic reactions are obtained by catabolism in animal cells.

- Enzymes catalyze reactions by binding to particular substrate molecules in a way that lowers the activation energy required for making and breaking specific covalent bonds.

- The only chemical reactions possible are those that increase the total amount of disorder in the universe. The free-energy change for a reaction, ΔG, measures this disorder, and it must be less than zero for a reaction to proceed.

- The free-energy change for a chemical reaction, ΔG, depends on the concentrations of the reacting molecules and may be calculated from these concentrations if the equilibrium constant (K) of the reaction (or the standard free-energy change $\Delta G°$ for the reactants) is known.

- By creating a reaction pathway that couples an energetically favorable reaction to an energetically unfavorable one, enzymes cause otherwise impossible reactions to occur.

- A small set of activated carrier molecules, in particular, ATP, NADH, and NADPH, plays a central part in these coupling events. ATP carries high-energy phosphate groups, whereas NADH and NADPH carry high-energy electrons.

- Food molecules provide the carbon skeletons for the formation of larger molecules. The covalent bonds of these larger molecules are typically produced in reactions coupled to energetically favorable bond changes in activated carrier molecules such as ATP and NADPH.

Key Terms

acetyl CoA	coupled reaction	$\Delta G, \Delta G°$	reduction
activated carrier	diffusion	metabolism	respiration
activation energy	entropy	NAD$^+$, NADH	substrate
ADP, ATP	enzyme	NADP$^+$, NADPH	
catalyst	equilibrium	oxidation	
	free energy	photosynthesis	

Questions

Question 3–9 Which of the following statements are correct? Explain your answers.

A. Some enzyme-catalyzed reactions cease completely if their enzyme is absent.

B. High-energy electrons (such as those found in the activated carriers NADH and NADPH) move faster around the atomic nucleus.

C. ATP hydrolysis to form AMP can provide almost twice the energy of ATP hydrolysis to form ADP.

D. A partially oxidized carbon atom has a smaller diameter than a more reduced one.

E. Some activated carrier molecules can transfer energy and chemical groups.

F. The rule that oxidations release energy, whereas reductions require energy input, applies to all chemical reactions, not just those that occur in living cells.

G. Cold-blooded animals have an energetic disadvantage because they give less heat to the environment than warm-blooded animals do. This slows their ability to make ordered macromolecules.

H. Linking the reaction X → Y to a second, energetically favorable reaction Y → Z will shift the equilibrium constant of the first reaction.

Question 3–10   Consider the transition of A → B in Figure 3–20. The molecules B differ from A by the presence of noncovalent bonds. Assume that the only difference between A and B is the presence of three hydrogen bonds in B that are absent in A. What is the ratio of A to B when the reaction is in equilibrium? Approximate your answer by using Table 3–1 (p. 92), with 1 kcal/mole as the energy of each hydrogen bond. Assume B has three additional hydrogen bonds, i.e., a total of six, that distinguish it from A. How would that change the ratio?

Question 3–11   Discuss the statement: "Whether the ΔG for a reaction is larger, smaller, or the same as $\Delta G°$ depends on the concentration of the compounds that participate in the reaction."

Question 3–12

A. How many ATP molecules could maximally be generated from one molecule of glucose, if the complete oxidation of 1 mole of glucose to CO_2 and H_2O yields 686 kcal, and the useful chemical energy available in the high-energy phosphate bond of 1 mole of ATP is 12 kcal?

B. Respiration produces 30 moles of ATP from 1 mole of glucose. Compare this number with your answer above. What is the overall efficiency of ATP production from glucose?

C. If the cells of your body oxidize 1 mole of glucose, by how much would the temperature of your body (assume that your body consists of 75 kg of water) increase if the heat were not dissipated into the environment? (Recall that a kilocalorie [kcal] is defined as that amount of energy that heats 1 kg of water by 1°C.)

D. What would the consequences be if the cells of your body could convert the energy in food substances with only 20% efficiency? Would your body—as it is presently constructed—(a) work just fine, (b) overheat, or (c) freeze?

E. A resting human hydrolyzes about 40 kg of ATP every 24 hours. The oxidation of how much glucose would produce this amount of energy? (Hint: look up the structure of ATP in Figure 2–21 to cal-

culate its molecular weight; the atomic weights of H, C, N, O, and P are 1, 12, 14, 16, and 31, respectively.)

Question 3–13   A prominent scientist claims to have isolated mutant cells that can convert 1 molecule of glucose into 57 molecules of ATP. Should this discovery be celebrated, or do you suppose that something might be wrong with it? Explain your answer.

Question 3–14   A reaction in a single-step biosynthetic pathway that converts a metabolite into a particularly vicious poison is energetically highly unfavorable (metabolite ⇌ poison). The reaction is normally driven by ATP hydrolysis. Assume that a mutation in the enzyme that catalyzes the reaction prevents it from utilizing ATP, but still allows it to catalyze the reaction.

A. Do you suppose it might be safe for you to eat the mutant organism? Base your answer on an estimation of how much less poison the organism would produce, assuming the reaction is in equilibrium and most of the energy stored in ATP is used to drive the unfavorable reaction.

B. Would your answer be different for another mutant enzyme that couples the reaction to ATP hydrolysis, but works 100 times slower?

Question 3–15   In a simple reaction A ⇌ A* a molecule is interconvertible between two forms that differ in standard free energy by 4.3 kcal/mole, with A* having the higher $G°$. (A) Use Table 3–1 (p. 92) to find how many more molecules will be in state A* compared to state A at equilibrium. (B) If an enzyme lowered the activation energy of the reaction by 2.8 kcal/mole, how would the ratio of A to A* change?

Question 3–16   Consider the effects of two enzymes. Enzyme A catalyzes the reaction

$$ATP + GDP \rightleftharpoons ADP + GTP$$

whereas enzyme B catalyzes the reaction

$$NADH + NADP^+ \rightleftharpoons NAD^+ + NADPH$$

Discuss whether the enzymes would be beneficial or detrimental to cells.

4 How Cells Obtain Energy from Food

Cells require a constant supply of energy to generate and maintain the biological order that keeps them alive. This energy is derived from the chemical bond energy in food molecules, which thereby serve as fuel for cells.

Sugars are particularly important fuel molecules. Plants make their own sugars by photosynthesis, whereas animal cells obtain sugars, and other molecules such as starch that are converted to sugars, by eating other organisms. However, the process whereby sugars are oxidized to generate energy is very similar in both animals and plants. The general principle that guides this process is illustrated in Figure 4–1. Useful energy is derived from the chemical bond energy stored in the sugar, as a sugar molecule is broken down and oxidized to CO_2 and H_2O. This energy is saved as high-energy chemical bonds in activated carrier molecules, such as ATP and NADPH, which in turn serve as portable sources of the chemical groups and electrons needed for biosynthesis (see Chapter 3).

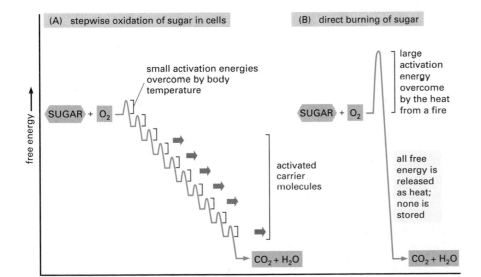

Figure 4–1 **Schematic representation of the controlled stepwise oxidation of sugar in a cell, compared with ordinary burning.** In the cell, enzymes catalyze oxidation via a series of small steps in which free energy is transferred in conveniently sized packets to carrier molecules—most often ATP and NADH. At each step, an enzyme controls the reaction by reducing the activation energy barrier that has to be surmounted before the specific reaction can occur. The total free energy released is exactly the same in (A) and (B).

Activated carrier molecules carry a packet of energy or a transferable chemical group of an appropriate size to be used in a single enzyme-catalyzed step in a biosynthetic pathway. If a fuel molecule such as glucose is oxidized to CO_2 and H_2O in a single step, it releases an amount of energy many times larger than any carrier molecule could capture. Instead, cells use enzymes to carry out the oxidation in a highly controlled series of reactions: the glucose molecule is degraded step-by-step, paying out energy in small packets to activated carrier molecules by means of coupled reactions (see Figure 3–23). In this way, much of the energy released by oxidizing glucose is saved and made available to do useful work for the cell.

In this chapter we trace the major steps in the breakdown, or catabolism, of sugars and show how they produce ATP, NADH, and other activated carrier molecules in animal cells. We concentrate on glucose breakdown, since it dominates energy production in most animal cells. In addition, a very similar pathway operates in plants, fungi, and many bacteria. Other molecules, such as fatty acids and proteins, can also serve as energy sources if they are funneled through appropriate enzymatic pathways.

The Breakdown of Sugars and Fats

Animal cells make ATP in two ways. One is by a series of enzyme-catalyzed reactions in the cytosol that ends in the partial oxidation of the food molecules. The second takes place in mitochondria and uses the energy from activated carrier molecules to drive ATP production. The mechanism of this second process is described in detail in Chapter 13. Here we focus on the sequence of reactions by which food molecules are partially oxidized, thereby producing both ATP and the activated carrier molecules needed to drive energy production in mitochondria.

Food Molecules Are Broken Down in Three Stages to Produce ATP

The proteins, lipids, and polysaccharides that make up most of the food we eat must be broken down into smaller molecules before our cells can use them—either as a source of energy or as building blocks for other molecules. The breakdown processes must act on food taken in from outside, but not on the macromolecules inside our own cells. Stage 1 in the enzymatic breakdown of food molecules is therefore *digestion*, which occurs either in our intestine outside cells, or in a specialized organelle within cells, the lysosome. (A membrane that surrounds the lysosome keeps its digestive enzymes separated from the cytosol; see Chapter 14.) In either case, the large polymeric molecules in food are broken down during digestion into their monomer subunits—proteins into amino acids, polysaccharides into sugars, and fats into fatty acids and glycerol—through the action of enzymes. After digestion, the small organic molecules derived from food enter the cytosol of the cell, where their gradual oxidation begins. As illustrated in Figure 4–2, oxidation occurs in two further stages of cellular catabolism: stage 2 starts in the cytosol and ends in mitochondria, whereas stage 3 is confined to the mitochondria.

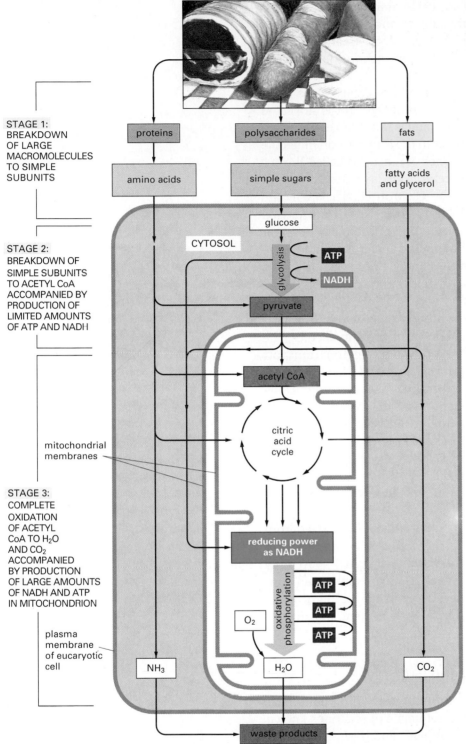

Figure 4–2 Simplified diagram of the three stages of cellular metabolism that lead from food to waste products in animal cells. This series of reactions produces ATP, which is then used to drive biosynthetic reactions and other energy-requiring processes in the cell. Stage 1 occurs outside cells. Stage 2 occurs mainly in the cytosol, except for the final step of conversion of pyruvate to acetyl groups on acetyl CoA, which occurs in mitochondria. Stage 3 occurs in mitochondria.

In stage 2 a chain of reactions called *glycolysis* converts each molecule of glucose into two smaller molecules of pyruvate. Sugars other than glucose are similarly converted to pyruvate after their conversion to one of the sugar intermediates in this glycolytic pathway. During pyruvate formation, two types of activated carrier molecules are produced—ATP and NADH. The pyruvate then passes from the cytosol into mitochondria. There, each pyruvate molecule is converted into CO_2 plus a two-carbon acetyl group—which becomes attached to coenzyme A (CoA), forming acetyl CoA, another of the activated carrier molecules discussed in Chapter 3 (see Figure 3–30). Large amounts of acetyl CoA are also produced by the stepwise breakdown and oxidation of fatty acids derived from fats, which are carried in the bloodstream, imported into cells as fatty acids, and then moved into mitochondria for acetyl CoA production.

Stage 3 of the oxidative breakdown of food molecules takes place entirely in mitochondria. The acetyl group in acetyl CoA is linked to coenzyme A through a high-energy linkage, and is therefore easily transferable to other molecules. After its transfer to the four-carbon molecule oxaloacetate, the acetyl group enters a series of reactions called the *citric acid cycle*. As we discuss shortly, the acetyl group is oxidized to CO_2 in these reactions, and large amounts of the electron carrier NADH are generated. Finally, the high-energy electrons from NADH are passed along an electron-transport chain within the mitochondrial inner membrane, where the energy released by their transfer is used to drive a process that produces ATP and consumes molecular oxygen (O_2). It is in these final steps that most of the energy released by oxidation is harnessed to produce most of the cell's ATP.

Because the energy to drive ATP synthesis in mitochondria ultimately derives from the oxidative breakdown of food molecules, the phosphorylation of ADP to form ATP that is driven by electron transport in the mitochondrion is known as *oxidative phosphorylation*. The fascinating events that occur within the mitochondrial inner membrane during oxidative phosphorylation are the major focus of Chapter 13.

Through the production of ATP, the energy derived from the breakdown of sugars and fats is redistributed as packets of chemical energy in a form convenient for use elsewhere in the cell. Roughly 10^9 molecules of ATP are in solution in a typical cell at any instant, and in many cells, all this ATP is turned over (that is, used up and replaced) every 1–2 minutes.

In all, nearly half of the energy that could in theory be derived from the oxidation of glucose or fatty acids to H_2O and CO_2 is captured and used to drive the energetically unfavorable reaction $P_i + ADP \rightarrow ATP$. (By contrast, a modern combustion engine, such as a car engine, can convert no more than 20% of the available energy in its fuel into useful work.) The rest of the energy is released by the cell as heat, making our bodies warm.

Glycolysis Is a Central ATP-producing Pathway

The most important process in stage 2 of the breakdown of food molecules is the degradation of glucose in the sequence of reactions known as **glycolysis**—from the Greek *glykos*, "sugar," and *lysis*, "splitting." Glycolysis produces ATP without the involvement of molecular oxygen (O_2 gas). It occurs in the cytosol of most cells, including many anaerobic microorganisms (those that can live without utilizing molecular oxygen). Glycolysis probably evolved early in the history of life, before the activities of photosynthetic organisms introduced oxygen into the atmosphere. During glycolysis, a glucose molecule with six carbon atoms is converted into two molecules of *pyruvate*, each of which contains three carbon atoms. For each molecule of glucose, two molecules of ATP are hydrolyzed to provide energy to drive the early steps, but four molecules of ATP are produced in the later steps. At the end of glycoly-

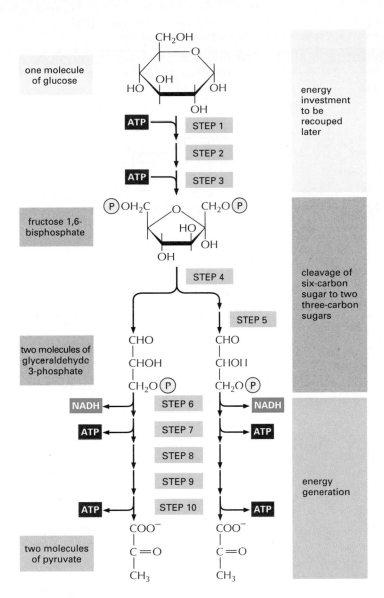

one molecule
of glucose

ATP STEP 1

STEP 2

ATP STEP 3

fructose 1,6-
bisphosphate

STEP 4

STEP 5

two molecules of
glyceraldehyde
3-phosphate

NADH STEP 6 NADH

ATP STEP 7 ATP

STEP 8

STEP 9

ATP STEP 10 ATP

two molecules
of pyruvate

energy
investment
to be
recouped
later

cleavage of
six-carbon
sugar to two
three-carbon
sugars

energy
generation

Figure 4–3 An outline of glycolysis. Each of the 10 steps shown is catalyzed by a different enzyme. Note that step 4 cleaves a six-carbon sugar into two three-carbon sugars, so that the number of molecules at every stage after this doubles. As indicated, step 6 begins the energy generation phase of glycolysis, which causes the net synthesis of ATP and NADH molecules (see also Panel 4–1, pp. 112–113).

sis, there is consequently a net gain of two molecules of ATP for each glucose molecule broken down.

The glycolytic pathway is presented in outline in Figure 4–3, and in more detail in Panel 4–1 (pp. 112–113). Glycolysis involves a sequence of 10 separate reactions, each producing a different sugar *intermediate* and each catalyzed by a different enzyme. Like most enzymes, these enzymes all have names ending in *ase*—like isomer*ase* and dehydrogen*ase*—which indicate the type of reaction they catalyze.

Although no molecular oxygen is involved in glycolysis, oxidation occurs, in that electrons are removed by NAD⁺ (producing NADH) from some of the carbons derived from the glucose molecule. The stepwise nature of the process allows the energy of oxidation to be released in small packets, so that much of it can be stored in carrier molecules rather than all of it being released as heat (see Figure 4–1). Thus, some of the energy released by oxidation drives the direct synthesis of ATP molecules from ADP and P_i, and some remains with the electrons in the high-energy electron carrier NADH.

Two molecules of NADH are formed per molecule of glucose in the course of glycolysis. In aerobic organisms (those that require molecular

The Breakdown of Sugars and Fats

For each step, the part of the molecule that undergoes a change is shadowed in blue, and the name of the enzyme that catalyzes the reaction is in a yellow box.

Step 1 Glucose is phosphorylated by ATP to form a sugar phosphate. The negative charge of the phosphate prevents passage of the sugar phosphate through the plasma membrane, trapping glucose inside the cell.

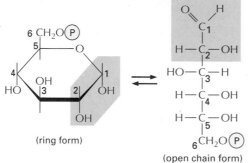

glucose + ATP → glucose 6-phosphate + ADP + H^+

(hexokinase)

Step 2 A readily reversible rearrangement of the chemical structure (isomerization) moves the carbonyl oxygen from carbon 1 to carbon 2, forming a ketose from an aldose sugar. (See Panel 2–3, pp. 56–57.)

glucose 6-phosphate (ring form) ⇌ (open chain form) — phosphoglucose isomerase — fructose 6-phosphate (open chain form) ⇌ (ring form)

Step 3 The new hydroxyl group on carbon 1 is phosphorylated by ATP, in preparation for the formation of two three-carbon sugar phosphates. The entry of sugars into glycolysis is controlled at this step, through regulation of the enzyme *phosphofructokinase*.

fructose 6-phosphate + ATP → phosphofructokinase → fructose 1,6-bisphosphate + ADP + H^+

Step 4 The six-carbon sugar is cleaved to produce two three-carbon molecules. Only the glyceraldehyde 3-phosphate can proceed immediately through glycolysis.

fructose 1,6-bisphosphate (ring form) ⇌ (open chain form) → aldolase → dihydroxyacetone phosphate + glyceraldehyde 3-phosphate

Step 5 The other product of step 4, dihydroxyacetone phosphate, is isomerized to form glyceraldehyde 3-phosphate.

dihydroxyacetone phosphate ⇌ triose phosphate isomerase ⇌ glyceraldehyde 3-phosphate

Panel 4–1 Details of the 10 steps of glycolysis.

Step 6 The two molecules of glyceraldehyde 3-phosphate are oxidized. The energy generation phase of glycolysis begins, as NADH and a new high-energy anhydride linkage to phosphate are formed (see Figure 4–5).

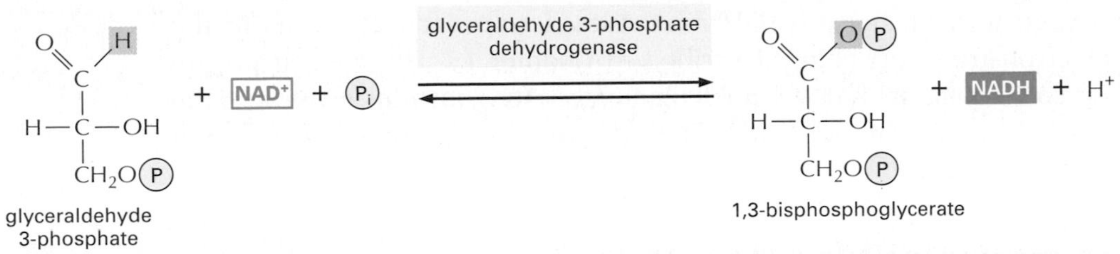

$$\text{glyceraldehyde 3-phosphate} + \boxed{\text{NAD}^+} + \textcircled{P}_i \xrightleftharpoons[]{\text{glyceraldehyde 3-phosphate dehydrogenase}} \text{1,3-bisphosphoglycerate} + \boxed{\text{NADH}} + \text{H}^+$$

glyceraldehyde 3-phosphate

1,3-bisphosphoglycerate

Step 7 The transfer to ADP of the high-energy phosphate group that was generated in step 6 forms ATP.

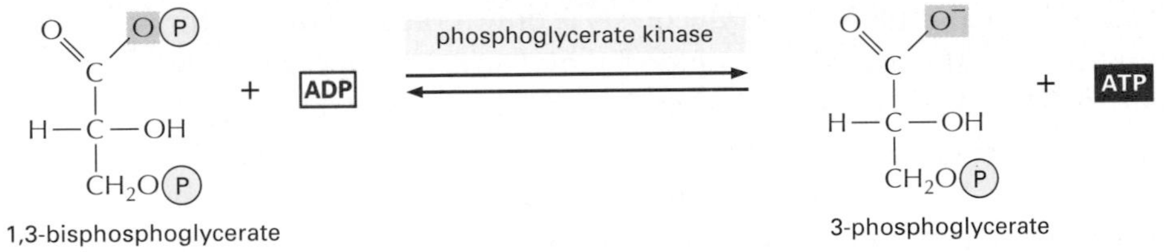

$$\text{1,3-bisphosphoglycerate} + \boxed{\text{ADP}} \xrightleftharpoons[]{\text{phosphoglycerate kinase}} \text{3-phosphoglycerate} + \boxed{\text{ATP}}$$

1,3-bisphosphoglycerate

3-phosphoglycerate

Step 8 The remaining phosphate ester linkage in 3-phosphoglycerate, which has a relatively low free energy of hydrolysis, is moved from carbon 3 to carbon 2 to form 2-phosphoglycerate.

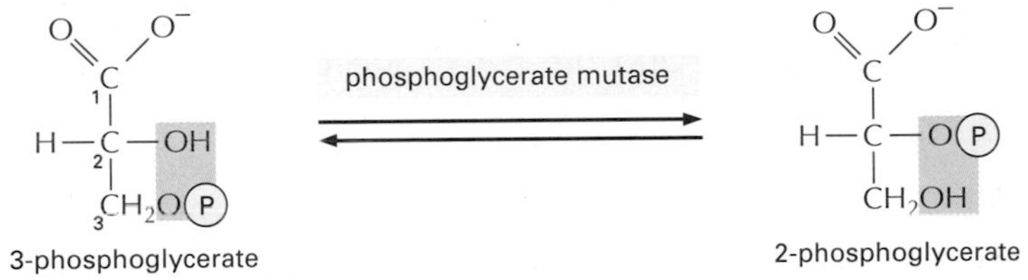

$$\text{3-phosphoglycerate} \xrightleftharpoons[]{\text{phosphoglycerate mutase}} \text{2-phosphoglycerate}$$

3-phosphoglycerate

2-phosphoglycerate

Step 9 The removal of water from 2-phosphoglycerate creates a high-energy enol phosphate linkage.

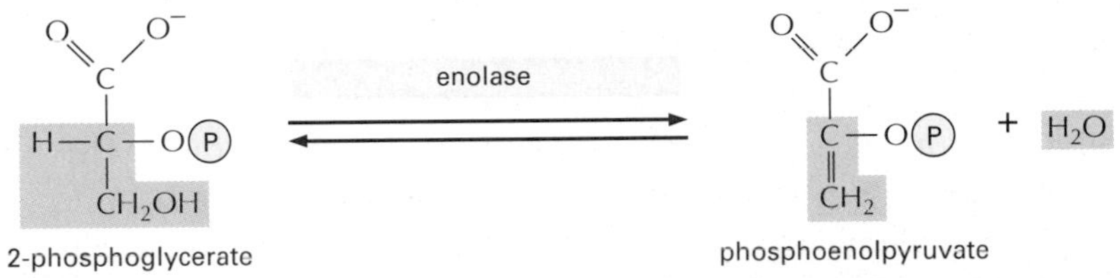

$$\text{2-phosphoglycerate} \xrightleftharpoons[]{\text{enolase}} \text{phosphoenolpyruvate} + \text{H}_2\text{O}$$

2-phosphoglycerate

phosphoenolpyruvate

Step 10 The transfer to ADP of the high-energy phosphate group that was generated in step 9 forms ATP, completing glycolysis.

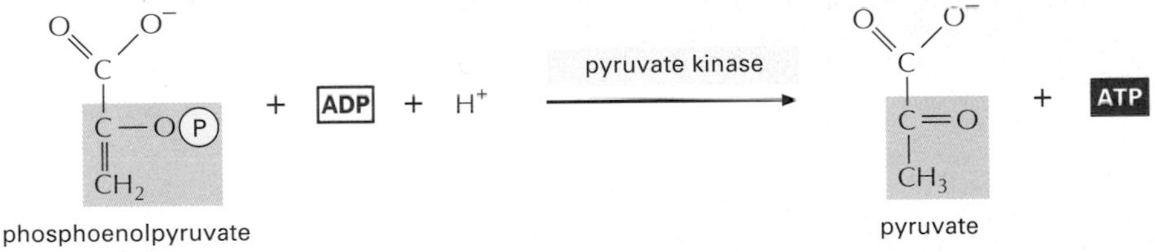

$$\text{phosphoenolpyruvate} + \boxed{\text{ADP}} + \text{H}^+ \xrightarrow{\text{pyruvate kinase}} \text{pyruvate} + \boxed{\text{ATP}}$$

phosphoenolpyruvate

pyruvate

NET RESULT OF GLYCOLYSIS

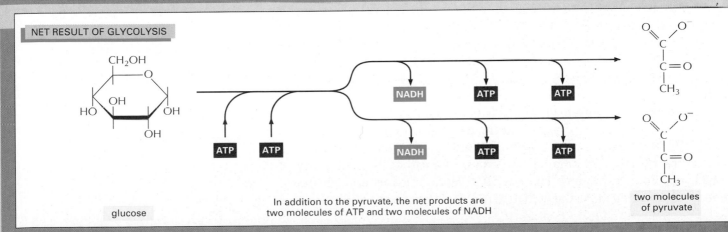

glucose

In addition to the pyruvate, the net products are two molecules of ATP and two molecules of NADH

two molecules of pyruvate

113

oxygen to live), these NADH molecules donate their electrons to the electron-transport chain described in Chapter 13. These electrons are passed along this chain to molecular oxygen (O_2), forming water, and the NAD^+ formed from the NADH is used again for glycolysis (see step 6 in Panel 4–1, p. 113).

Fermentations Allow ATP to Be Produced in the Absence of Oxygen

For most animal and plant cells, glycolysis is only a prelude to the third and final stage of the breakdown of food molecules. In this case the pyruvate formed at the last step of stage 2 is rapidly transported into the mitochondria, where it is converted into CO_2 plus acetyl CoA, which is then completely oxidized to CO_2 and H_2O. But for many anaerobic organisms, which do not utilize molecular oxygen and can grow and divide without it, glycolysis is the principal source of the cell's ATP. This is also true for certain animal tissues, such as skeletal muscle, that can continue to function when molecular oxygen is limiting. In these anaerobic conditions, the pyruvate and the NADH electrons stay in the cytosol. The pyruvate is converted into products excreted from the cell: for example, into ethanol and CO_2 in the yeasts used in brewing and breadmaking, or into lactate in muscle. In this process, the NADH gives up its electrons and is converted back into NAD^+. This regeneration of NAD^+ is required to maintain the reactions of glycolysis (Figure 4–4).

Anaerobic energy-yielding pathways like these are called **fermentations.** Studies of the commercially important fermentations carried out by yeasts inspired much of early biochemistry. Work in the nineteenth century led in 1896 to the then startling recognition that these processes could be studied outside living organisms, in cell extracts. This revolutionary discovery eventually made it possible to dissect out and study each of the individual reactions in the fermentation process. The piecing together of the complete glycolytic pathway in the 1930s was a major triumph of biochemistry, and it was quickly followed by the recognition of the central role of ATP in cellular processes. Therefore, most of the fundamental concepts discussed in this chapter have been understood for more than 50 years.

Question 4–1 At first glance, fermentation appears to be an optional add-on reaction to glycolysis. Explain why cells growing in the absence of oxygen could not simply discard pyruvate as a waste product. Which products derived from glucose would accumulate in cells unable to generate either lactate or ethanol by fermentation?

Glycolysis Illustrates How Enzymes Couple Oxidation to Energy Storage

We have previously used a "paddle wheel" analogy to explain how cells harvest useful energy from the oxidation of organic molecules by using enzymes to couple an energetically unfavorable reaction to an energetically favorable one (see Figure 3–24). Enzymes play the part of the paddle wheel in our analogy, and we now return to a step in glycolysis that we have previously discussed, in order to illustrate exactly how coupled reactions occur.

Two central reactions in glycolysis (steps 6 and 7 in Panel 4–1, p. 113) convert the three-carbon sugar intermediate glyceraldehyde 3-phosphate (an aldehyde) into 3-phosphoglycerate (a carboxylic acid). This entails the oxidation of an aldehyde group to a carboxylic acid group, which occurs in two steps. The overall reaction releases enough free energy to convert a molecule of ADP to ATP and to transfer two electrons

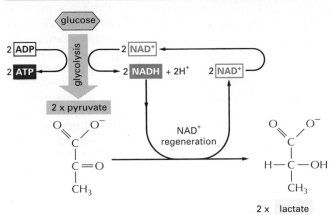

(A) FERMENTATION LEADING TO EXCRETION OF LACTATE

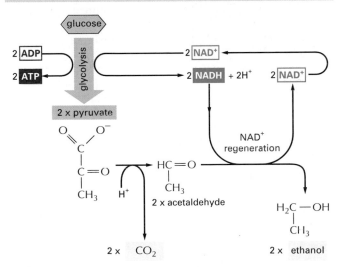

(B) FERMENTATION LEADING TO EXCRETION OF ALCOHOL AND CO_2

Figure 4–4 Anaerobic breakdown of pyruvate. (A) When inadequate oxygen is present, for example, in a muscle cell undergoing vigorous contraction, the pyruvate produced by glycolysis is converted to lactate as shown. This reaction restores the NAD^+ consumed in step 6 of glycolysis, but the whole pathway yields much less energy overall than complete oxidation. (B) In some organisms that can grow anaerobically, such as yeasts, pyruvate is converted via acetaldehyde into carbon dioxide and ethanol. Again, this pathway regenerates NAD^+ from NADH, as required to enable glycolysis to continue. Both (A) and (B) are examples of fermentations.

from the aldehyde to NAD^+ to form NADH, while still releasing enough heat to the environment to make the overall reaction energetically favorable ($\Delta G°$ for the overall reaction is –3.0 kcal/mole).

The means by which this remarkable feat is accomplished is outlined in Figure 4–5. The indicated chemical reactions are precisely guided by two enzymes to which the sugar intermediates are tightly bound. In fact, as detailed in Figure 4–5, the first enzyme (*glyceraldehyde 3-phosphate dehydrogenase*) forms a short-lived covalent bond to the aldehyde through a reactive –SH group on the enzyme, and catalyzes its oxidation while still in the attached state. The reactive enzyme-substrate bond is then displaced by an inorganic phosphate ion to produce a high-energy phosphate intermediate, which is released from the enzyme. This intermediate binds to the second enzyme (*phophoglycerate kinase*), which catalyzes the energetically favorable transfer of the high-energy phosphate just created to ADP, forming ATP and completing the process of oxidizing an aldehyde to a carboxylic acid (see Figure 4–5).

We have shown this particular oxidation process in some detail because it provides a clear example of enzyme-mediated energy storage through coupled reactions (Figure 4–6). These reactions (steps 6 and 7) are the only ones in glycolysis that create a high-energy phosphate

The Breakdown of Sugars and Fats

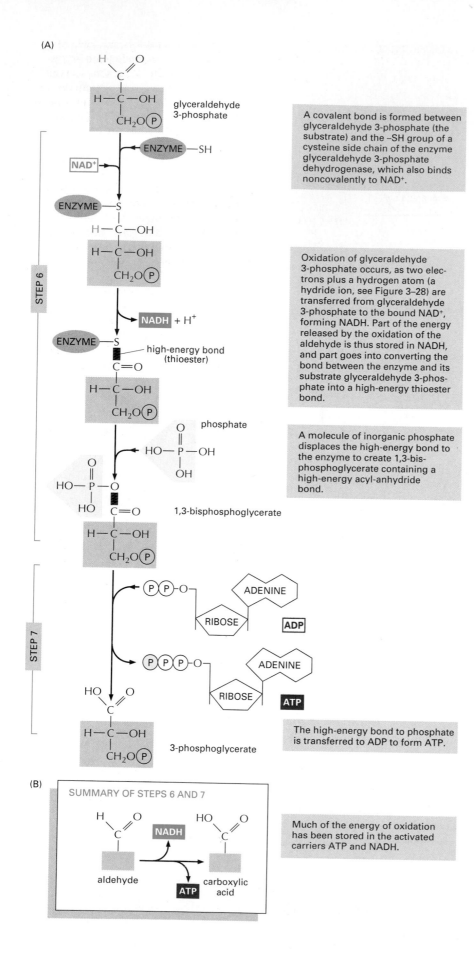

(A)

glyceraldehyde 3-phosphate

STEP 6

ENZYME—SH

NAD⁺

ENZYME—S

NADH + H⁺

ENZYME—S

high-energy bond (thioester)

phosphate

1,3-bisphosphoglycerate

STEP 7

ADENINE
RIBOSE
ADP

ADENINE
RIBOSE
ATP

3-phosphoglycerate

A covalent bond is formed between glyceraldehyde 3-phosphate (the substrate) and the –SH group of a cysteine side chain of the enzyme glyceraldehyde 3-phosphate dehydrogenase, which also binds noncovalently to NAD⁺.

Oxidation of glyceraldehyde 3-phosphate occurs, as two electrons plus a hydrogen atom (a hydride ion, see Figure 3–28) are transferred from glyceraldehyde 3-phosphate to the bound NAD⁺, forming NADH. Part of the energy released by the oxidation of the aldehyde is thus stored in NADH, and part goes into converting the bond between the enzyme and its substrate glyceraldehyde 3-phosphate into a high-energy thioester bond.

A molecule of inorganic phosphate displaces the high-energy bond to the enzyme to create 1,3-bis-phosphoglycerate containing a high-energy acyl-anhydride bond.

The high-energy bond to phosphate is transferred to ADP to form ATP.

(B)

SUMMARY OF STEPS 6 AND 7

aldehyde

NADH

ATP

carboxylic acid

Much of the energy of oxidation has been stored in the activated carriers ATP and NADH.

Figure 4–5 Energy storage in steps 6 and 7 of glycolysis. In these steps the oxidation of an aldehyde to a carboxylic acid is coupled to the formation of ATP and NADH. (A) Step 6 begins with the formation of a covalent bond between the substrate (glyceraldehyde 3-phosphate) and an –SH group exposed on the surface of the enzyme (glyceraldehyde 3-phosphate dehydrogenase). The enzyme then catalyzes transfer of hydrogen (as a hydride ion—a proton plus two electrons) from the bound glyceraldehyde 3-phosphate to a molecule of NAD⁺. Part of the energy released in this oxidation is used to form a molecule of NADH and part is used to convert the original linkage between the enzyme and its substrate to a high-energy thioester bond (shown in *red*). A molecule of inorganic phosphate then displaces this high-energy bond on the enzyme, creating a high-energy sugar-phosphate bond instead (*red*). At this point the enzyme has not only stored energy (in NADH), but also coupled the energetically favorable oxidation of an aldehyde to the energetically unfavorable formation of a high-energy phosphate bond. The second reaction has been driven by the first, thereby acting like the "paddle wheel" coupler in Figure 3–24.

In reaction step 7, the high-energy molecule just made, 1,3-bisphosphoglycerate, binds to a second enzyme, phosphoglycerate kinase. The reactive phosphate is transferred to ADP, forming a molecule of ATP and leaving a free carboxylic acid group on the oxidized sugar. (B) Summary of the overall chemical change produced by reactions 6 and 7.

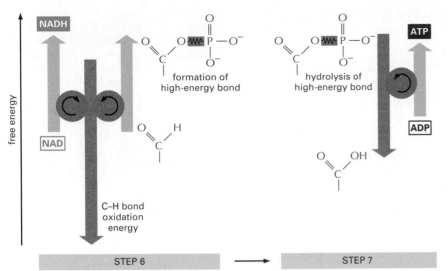

Figure 4–6 Schematic view of the coupled reactions that form NADH and ATP in steps 6 and 7 of glycolysis. The C–H bond oxidation energy drives the formation of both NADH and a high-energy phosphate bond. The breakage of the high-energy bond then drives ATP formation.

Figure 4–7 (*below*) **Some phosphate bond energies.** The transfer of a phosphate group from any molecule 1 to any molecule 2 is energetically favorable if the standard free-energy change ($\Delta G°$) for the hydrolysis of the phosphate bond in molecule 1 is more negative than that for hydrolysis of the phosphate bond in molecule 2. Thus, for example, a phosphate group is readily transferred from 1,3-bisphosphoglycerate to ADP, forming ATP. Note that the hydrolysis reaction can be viewed as the transfer of the phosphate group to water.

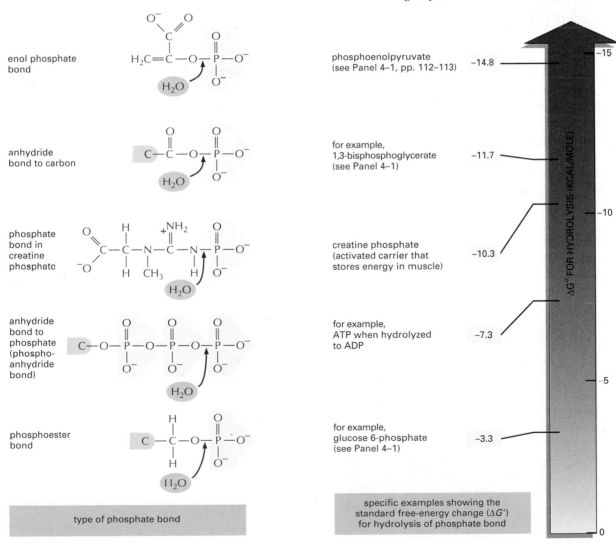

linkage directly from inorganic phosphate. As such, they account for the net yield of two ATP molecules and two NADH molecules per molecule of glucose (see Panel 4–1, pp. 112–113).

As we have just seen, ATP can be formed readily from ADP when reaction intermediates are formed with higher-energy phosphate bonds than those in ATP. Phosphate bonds can be ordered in energy by comparing the standard free-energy change ($\Delta G°$) for the breakage of each bond by hydrolysis, and Figure 4–7 compares the high-energy phosphoanhydride bonds in ATP with some other phosphate bonds generated during glycolysis.

Question 4–2 Arsenate (AsO_4^{3-}) is chemically very similar to phosphate (PO_4^{3-}) and is used as an alternative substrate by many phosphate-requiring enzymes. In contrast to phosphate, however, an anhydride bond between arsenate and carbon is very quickly hydrolyzed in water. Knowing this, suggest why arsenate is a compound of choice for murderers but not for cells. Formulate your explanation in the context of Figure 4–6.

Sugars and Fats Are Both Degraded to Acetyl CoA in Mitochondria

We now move on to consider stage 3 of catabolism, a process that requires abundant molecular oxygen (O_2 gas). Since the earth is thought to have developed an atmosphere containing O_2 gas between one and two billion years ago, whereas abundant life-forms are known to have existed on the earth for 3.5 billion years, the use of O_2 in the reactions that we discuss next is thought to be of relatively recent origin. In contrast, the mechanism used to produce ATP in Figure 4–5 does not require oxygen, and relatives of this elegant pair of coupled reactions could have arisen very early in the history of life on earth.

In aerobic metabolism, the pyruvate produced by glycolysis is rapidly decarboxylated by a giant complex of three enzymes, called the *pyruvate dehydrogenase complex.* The products of pyruvate decarboxylation are a molecule of CO_2 (a waste product), a molecule of NADH, and acetyl CoA. The three-enzyme complex is located in the mitochondria of eucaryotic cells; its structure and mode of action are outlined in Figure 4–8.

The enzymes that degrade the fatty acids derived from fats likewise produce acetyl CoA in mitochondria. Each molecule of fatty acid (as the activated molecule *fatty acyl CoA*) is broken down completely by a cycle

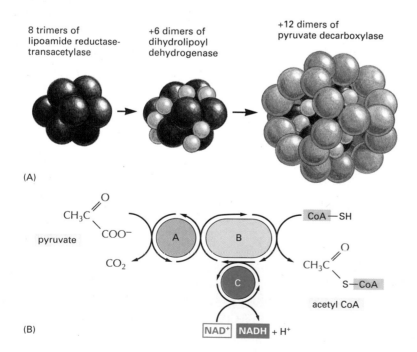

(A)

(B)

Figure 4–8 The oxidation of pyruvate to acetyl CoA and CO_2. (A) The structure of the pyruvate dehydrogenase complex, which contains 60 polypeptide chains. This is an example of a large multienzyme complex in which reaction intermediates are passed directly from one enzyme to another. In eucaryotic cells it is located in the mitochondrion. (B) The reactions carried out by the pyruvate dehydrogenase complex. The complex converts pyruvate to acetyl CoA in the mitochondrial matrix; NADH is also produced in this reaction. A, B, and C are the three enzymes *pyruvate decarboxylase, lipoamide reductase-transacetylase,* and *dihydrolipoyl dehydrogenase,* respectively, which were illustrated in (A); their activities are linked as shown.

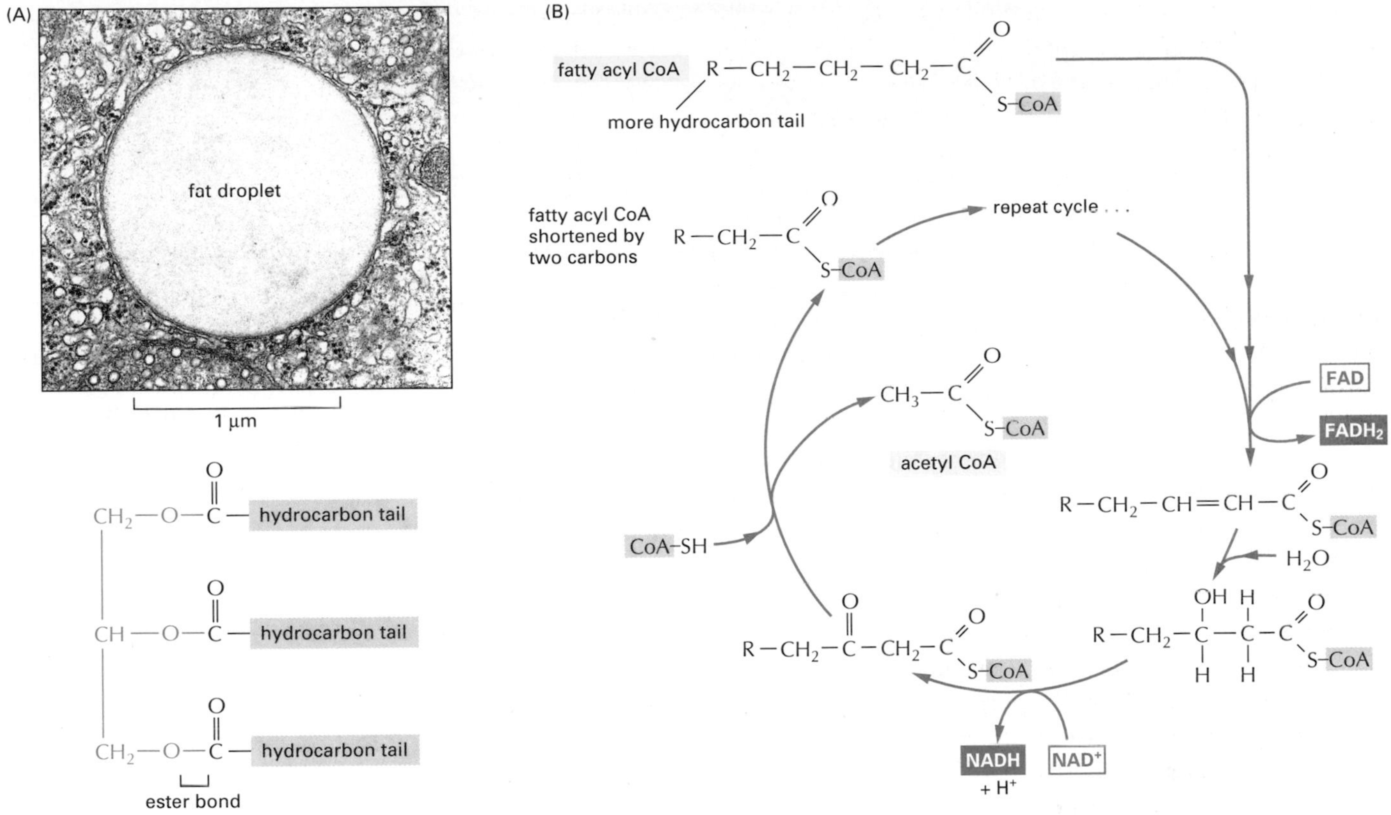

(A)

fat droplet

1 μm

$CH_2-O-\overset{\overset{\displaystyle O}{\|}}{C}-$ hydrocarbon tail

$CH-O-\overset{\overset{\displaystyle O}{\|}}{C}-$ hydrocarbon tail

$CH_2-O-\overset{\overset{\displaystyle O}{\|}}{C}-$ hydrocarbon tail

ester bond

(B)

fatty acyl CoA $R-CH_2-CH_2-CH_2-\overset{\overset{\displaystyle O}{\|}}{C}$
more hydrocarbon tail   S-CoA

fatty acyl CoA shortened by two carbons $R-CH_2-\overset{\overset{\displaystyle O}{\|}}{C}$
S-CoA

repeat cycle . . .

$CH_3-\overset{\overset{\displaystyle O}{\|}}{C}$
S-CoA
acetyl CoA

FAD
FADH$_2$

$R-CH_2-CH=CH-\overset{\overset{\displaystyle O}{\|}}{C}$
S-CoA
H_2O

CoA-SH

$R-CH_2-\overset{\overset{\displaystyle O}{\|}}{C}-CH_2-\overset{\overset{\displaystyle O}{\|}}{C}$
S-CoA

$R-CH_2-\overset{\overset{\displaystyle OH\ \ \ H}{|\ \ \ \ |}}{\underset{\underset{\displaystyle H\ \ \ \ H}{|\ \ \ \ |}}{C-C}}-\overset{\overset{\displaystyle O}{\|}}{C}$
S-CoA

NADH + H$^+$ NAD$^+$

of reactions that trims two carbons at a time from its carboxyl end, generating one molecule of acetyl CoA for each turn of the cycle. A molecule of NADH and a molecule of FADH$_2$ are also produced in this process (Figure 4–9).

Sugars and fats provide the major energy sources for most nonphotosynthetic organisms, including humans. However, the majority of the useful energy that can be extracted from the oxidation of both types of foodstuffs remains stored in the acetyl CoA molecules that are produced by the two types of reactions just described. The citric acid cycle of reactions, in which the acetyl group in acetyl CoA is oxidized to CO_2 and H_2O, is therefore central to the energy metabolism of aerobic organisms. In eucaryotes these reactions all take place in mitochondria, the organelle to which pyruvate and fatty acids are directed for acetyl CoA production (Figure 4–10). We should therefore not be surprised to discover that the mitochondrion is the place where most of the ATP is produced in animal cells. In contrast, aerobic bacteria carry out all of their reactions in a single compartment, the cytosol, and it is here that the citric acid cycle takes place in these cells.

The Citric Acid Cycle Generates NADH by Oxidizing Acetyl Groups to CO_2

In the nineteenth century, biologists noticed that in the absence of air (anaerobic conditions) cells produce lactic acid (for example, in muscle) or ethanol (for example, in yeast), while in its presence (aerobic conditions) they consume O_2 and produce CO_2 and H_2O. Intensive efforts to define the pathways of aerobic metabolism eventually focused on the oxidation of pyruvate and led in 1937 to the discovery of the **citric acid**

Figure 4–9 The oxidation of fatty acids to acetyl CoA. (A) Electron micrograph of a lipid droplet in the cytoplasm (top), and the structure of fats (bottom). Fats are triacylglycerols. The glycerol portion, to which three fatty acids are linked through ester bonds, is shown here in *green*. Fats are insoluble in water and form large lipid droplets in the specialized fat cells (called adipocytes) in which they are stored. (B) The fatty acid oxidation cycle. The cycle is catalyzed by a series of four enzymes in the mitochondrion. Each turn of the cycle shortens the fatty acid chain by two carbons (shown in *red*) and generates one molecule of acetyl CoA and one molecule each of NADH and FADH$_2$. (A, courtesy of Daniel S. Friend.)

Question 4–3 Many catabolic and metabolic reactions are based on reactions that are similar but work in opposite directions, such as the hydrolysis and condensation reactions described in Figure 3–32. This is true for fatty acid breakdown and fatty acid synthesis. From what you know about the mechanism of fatty acid breakdown outlined in Figure 4–9, would you expect the fatty acids found in cells to most commonly have an even or an odd number of carbon atoms?

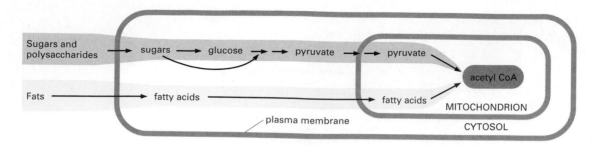

Figure 4–10 **The general pathways for the production of acetyl CoA from sugars and fats.** The mitochondrion in eucaryotic cells is the place where acetyl CoA is produced from both types of major food molecules. It is therefore the place where most of the cell's oxidation reactions occur and where most of its ATP is made.

cycle, also known as the *tricarboxylic acid cycle* or the *Krebs cycle.* The citric acid cycle accounts for about two-thirds of the total oxidation of carbon compounds in most cells, and its major end products are CO_2 and high-energy electrons in the form of NADH. The CO_2 is released as a waste product, while the high-energy electrons from NADH are passed to a membrane-bound electron-transport chain, eventually combining with O_2 to produce H_2O. Although the citric acid cycle itself does not use O_2, it requires O_2 in order to proceed because there is no other efficient way for the NADH to get rid of its electrons and thus regenerate the NAD^+ that is needed to keep the cycle going.

The citric acid cycle, which takes place inside mitochondria in eucaryotic cells, results in the complete oxidation of the carbon atoms of the acetyl groups in acetyl CoA, converting them into CO_2. The acetyl group is not oxidized directly. Instead, it is transferred from acetyl CoA to a larger, four-carbon molecule, *oxaloacetate,* to form the six-carbon tricarboxylic acid—*citric acid,* for which the subsequent cycle of reactions is named. The citric acid molecule is then gradually oxidized, and the energy of this oxidation is harnessed to produce energy-rich carrier molecules, in much the same manner as was described for glycolysis. The chain of eight reactions forms a cycle because at the end oxaloacetate is regenerated and enters a new turn of the cycle, as shown in outline in Figure 4–11.

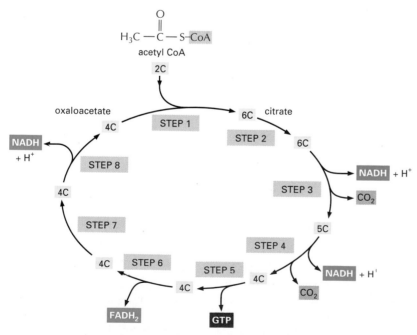

NET RESULT: ONE TURN OF THE CYCLE PRODUCES THREE NADH, ONE GTP, AND ONE FADH₂, AND RELEASES TWO MOLECULES OF CO₂

Figure 4–11 **Simple overview of the citric acid cycle.** The reaction of acetyl CoA with oxaloacetate starts the cycle by producing citrate (citric acid). In each turn of the cycle, two molecules of CO_2 are produced as waste products, plus three molecules of NADH, one molecule of GTP, and one molecule of $FADH_2$. The number of carbon atoms in each intermediate is shown in a *yellow box.*

As can be seen from Figure 4–11, we have thus far discussed only one of the three types of carrier molecules that are produced by the citric acid cycle, the NAD+-NADH pair. In addition to three molecules of NADH, each turn of the cycle also produces one molecule of **FADH₂** (reduced flavin adenine dinucleotide) from FAD and one molecule of the ribonucleotide **GTP** (guanosine triphosphate) from GDP. The structures of these two activated carrier molecules are illustrated in Figure 4–12. GTP is a close relative of ATP, and the transfer of its terminal phosphate group to ADP produces one ATP molecule in each cycle. Like NADH, FADH₂ is a carrier of high-energy electrons and hydrogen. As we discuss shortly, the energy that is stored in the readily transferred high-energy electrons of NADH and FADH₂ will be utilized subsequently for ATP production through the process of *oxidative phosphorylation,* the only step in the oxidative catabolism of foodstuffs that directly requires gaseous oxygen (O_2) from the atmosphere.

The complete citric acid cycle is presented in Panel 4–2 (pp. 122–123). The oxygen atoms required to make CO_2 from the acetyl groups entering the citric acid cycle are supplied not by molecular oxygen but by water. As illustrated in the panel, three molecules of water are split in each cycle, and the oxygen atoms of some of them are ultimately used to make CO_2.

In addition to pyruvate and fatty acids, some amino acids pass from the cytosol into mitochondria, where they are also converted into acetyl CoA or one of the other intermediates of the citric acid cycle (see Figure 4–2). Thus, in the eucaryotic cell, the mitochondrion is the center toward which all energy-yielding processes lead, whether they begin with sugars, fats, or proteins.

The citric acid cycle also functions as a starting point for important biosynthetic reactions by producing vital carbon-containing intermediates, such as *oxaloacetate* and *α-ketoglutarate*. These substances produced by catabolism are transferred back from the mitochondrion to the cytosol, where they serve in anabolic reactions as precursors for the synthesis of many essential molecules, such as amino acids.

Question 4–4 Looking at the chemistry detailed in Panel 4–2 (pp. 122–123), why do you suppose it is useful to link the acetyl group first to another carbon skeleton, oxaloacetate, before completely oxidizing both carbons to CO_2?

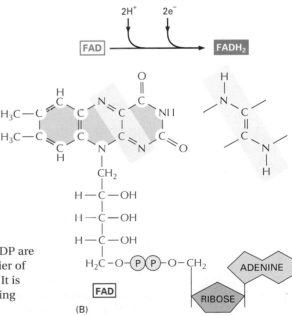

(A)

(B)

Figure 4–12 The structures of GTP and FADH₂. (A) GTP and GDP are close relatives of ATP and ADP, respectively. (B) FADH₂ is a carrier of hydrogens and high-energy electrons, like NADH and NADPH. It is shown here in its oxidized form (FAD) with the hydrogen-carrying atoms highlighted in *yellow.*

The Breakdown of Sugars and Fats

121

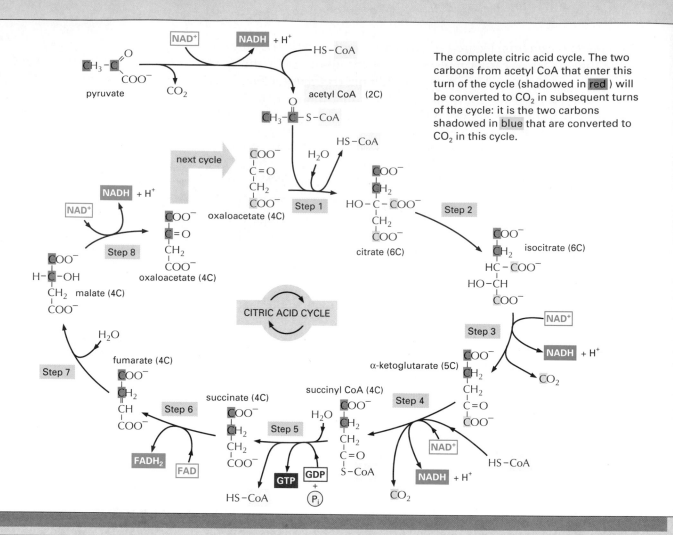

The complete citric acid cycle. The two carbons from acetyl CoA that enter this turn of the cycle (shadowed in red) will be converted to CO_2 in subsequent turns of the cycle: it is the two carbons shadowed in blue that are converted to CO_2 in this cycle.

Details of the eight steps are shown below. For each step, the part of the molecule that undergoes a change is shadowed in blue, and the name of the enzyme that catalyzes the reaction is in a yellow box.

Step 1 After the enzyme removes a proton from the CH_3 group on acetyl CoA, the negatively charged CH_2^- forms a bond to a carbonyl carbon of oxaloacetate. The subsequent loss by hydrolysis of the coenzyme A (CoA) drives the reaction strongly forward.

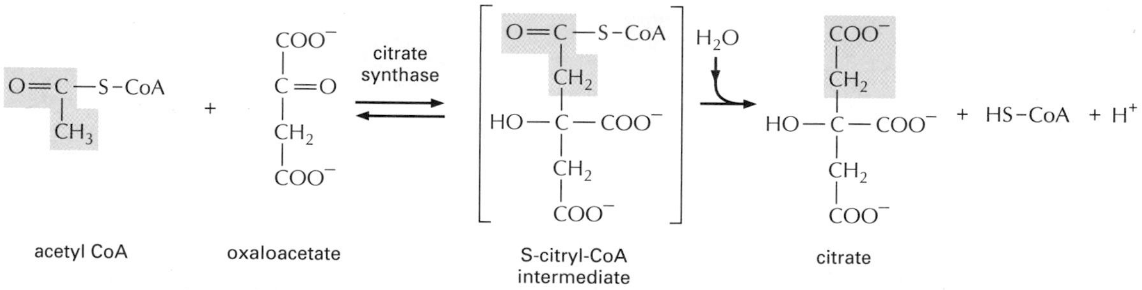

acetyl CoA   oxaloacetate   S-citryl-CoA intermediate   citrate

citrate synthase

Step 2 An isomerization reaction, in which water is first removed and then added back, moves the hydroxyl group from one carbon atom to its neighbor.

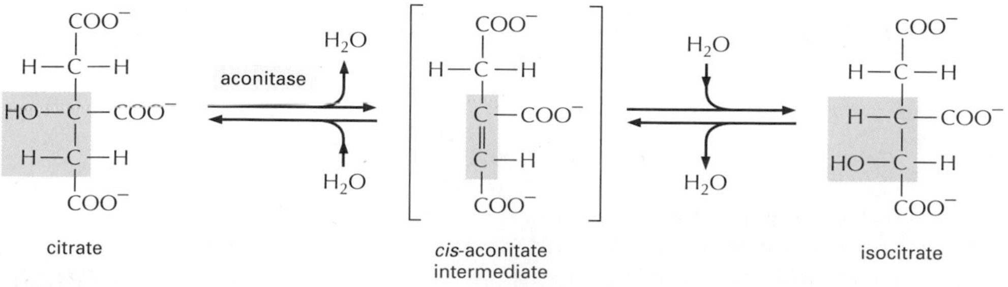

citrate   aconitase   cis-aconitate intermediate   isocitrate

Panel 4–2 The complete citric acid cycle.

Step 3 In the first of four oxidation steps in the cycle, the carbon carrying the hydroxyl group is converted to a carbonyl group. The immediate product is unstable, losing CO_2 while still bound to the enzyme.

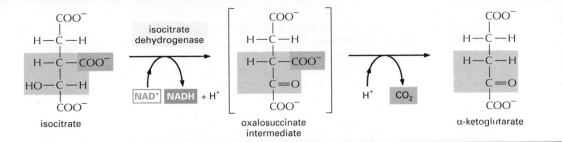

isocitrate

oxalosuccinate intermediate

α-ketoglutarate

Step 4 The *α-ketoglutarate dehydrogenase complex* closely resembles the large enzyme complex that converts pyruvate to acetyl CoA (*pyruvate dehydrogenase*). It likewise catalyzes an oxidation that produces NADH, CO_2, and a high-energy thioester bond to coenzyme A (CoA).

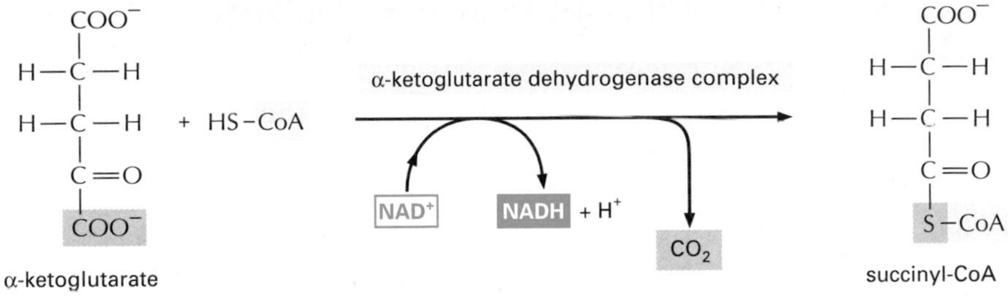

α-ketoglutarate

α-ketoglutarate dehydrogenase complex

succinyl-CoA

Step 5 A phosphate molecule from solution displaces the CoA, forming a high-energy phosphate linkage to succinate. This phosphate is then passed to GDP to form GTP. (In bacteria and plants, ATP is formed instead.)

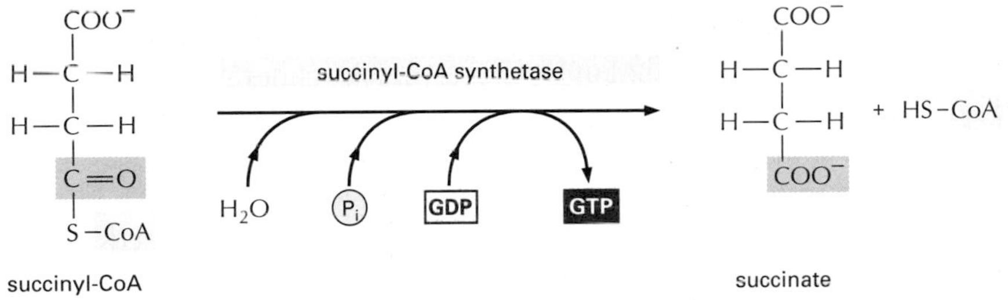

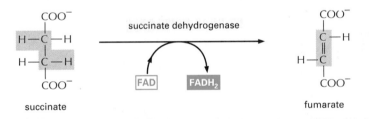

succinyl-CoA

succinyl-CoA synthetase

succinate

Step 6 In the third oxidation step in the cycle, FAD removes two hydrogen atoms from succinate.

succinate

succinate dehydrogenase

fumarate

Step 7 The addition of water to fumarate places a hydroxyl group next to a carbonyl carbon.

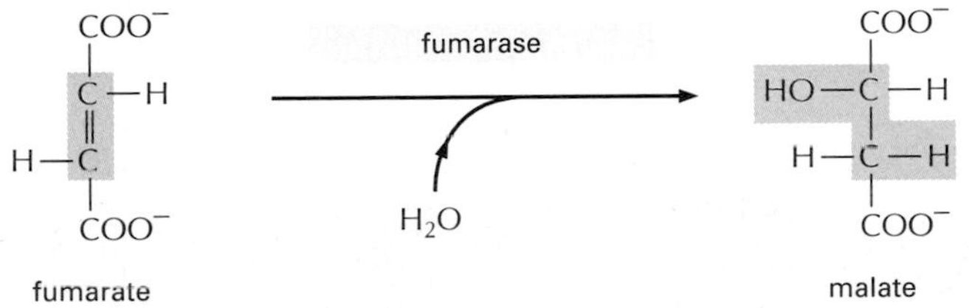

fumarate

fumarase

malate

Step 8 In the last of four oxidation steps in the cycle, the carbon carrying the hydroxyl group is converted to a carbonyl group, regenerating the oxaloacetate needed for step 1.

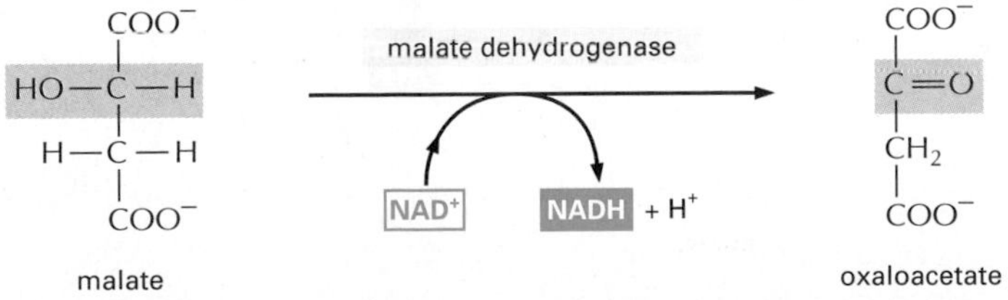

malate

malate dehydrogenase

oxaloacetate

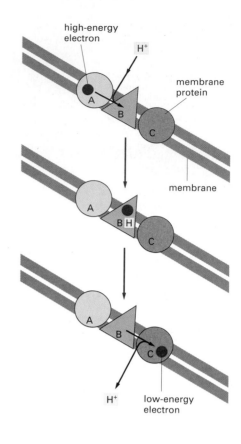

high-energy electron

H⁺

membrane protein

A

B

C

membrane

A

B H

C

A

B

C

H⁺   low-energy electron

Figure 4–13 The generation of an H⁺ gradient across a membrane by electron-transport reactions. A high-energy electron (derived, for example, from the oxidation of a metabolite) is passed sequentially by carriers A, B, and C to a lower energy state. In this diagram carrier B is arranged in the membrane in such a way that it takes up H⁺ from one side and releases it to the other as the electron passes. The result is an H⁺ gradient. This gradient represents a form of stored energy that is harnessed by other membrane proteins to drive the formation of ATP, as discussed in Chapter 13.

Electron Transport Drives the Synthesis of the Majority of the ATP in Most Cells

It is in the last step in the degradation of a food molecule that the major portion of its chemical energy is released. In this final process the electron carriers NADH and FADH$_2$ transfer the electrons that they have gained when oxidizing other molecules to the **electron-transport chain,** which is embedded in the inner membrane of the mitochondrion. As they pass along this long chain of specialized electron acceptor and donor molecules, the electrons fall to successively lower energy states. The energy they release in this process is used to drive H⁺ ions (protons) across the membrane—from the inner mitochondrial compartment to the outside (Figure 4–13). A gradient of H⁺ ions is thereby generated. This gradient serves as a source of energy, like a battery, that is tapped to drive a variety of energy-requiring reactions. The most prominent of these reactions is the generation of ATP by the phosphorylation of ADP.

At the end of this series of electron transfers, the electrons are passed to molecules of oxygen gas (O_2) that have diffused into the mitochondrion, which simultaneously combine with protons (H⁺) from the surrounding solution to produce molecules of water. The electrons have now reached their lowest energy levels, and therefore all the available energy has been extracted from the food molecule being oxidized. This process, termed *oxidative phosphorylation* (Figure 4–14), also occurs in the plasma membrane of bacteria. As one of the most remarkable achievements of cellular evolution, it will be a central topic of Chapter 13.

In total, the complete oxidation of a molecule of glucose to H_2O and CO_2 is used by the cell to produce about 30 molecules of ATP. In contrast, only 2 molecules of ATP are produced per molecule of glucose by glycolysis alone.

Question 4–5 What, if anything, is wrong with the following statement: "The oxygen consumed during the oxidation of glucose in animal cells is returned as part of CO_2 to the atmosphere." How could you support your answer experimentally?

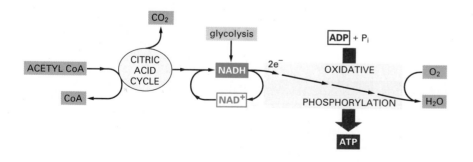

CO$_2$

glycolysis

ADP + P$_i$

ACETYL CoA → CITRIC ACID CYCLE → NADH → 2e⁻ → OXIDATIVE PHOSPHORYLATION → O$_2$

CoA ← ← NAD⁺ → H$_2$O

ATP

Figure 4–14 The final stages of oxidation of food molecules. Molecules of NADH and FADH$_2$ (not shown) produced by the citric acid cycle donate the electrons that are eventually used to reduce oxygen gas to water. A major portion of the energy released during an elaborate electron-transfer process in the mitochondrial inner membrane (or in the plasma membrane of bacteria) is harnessed to drive the synthesis of ATP.

Storing and Utilizing Food

Organisms Store Food Molecules in Special Reservoirs

All organisms need to restore their ATP pools constantly, if biological order is to be maintained in their cells. Yet animals have only periodic access to food, and plants need to survive overnight without sunlight, without the possibility of sugar production from photosynthesis.

To compensate for long periods of fasting, animals store food within their cells. Fatty acids are stored as fat droplets composed of water-insoluble triacylglycerols, largely in specialized fat cells (see Figure 4–9A). And sugar is stored as glucose subunits in the large branched polysaccharide **glycogen** (Figure 4–15), which is present as small granules in the cytoplasm of many cells, including liver and muscle. The synthesis and degradation of glycogen are rapidly regulated according to need. When more ATP is needed than can be generated from the food molecules taken in from the bloodstream, cells break down glycogen in a reaction that produces glucose 1-phosphate, which enters glycolysis.

Quantitatively, **fat** is a far more important storage form than glycogen, in part because the oxidation of a gram of fat releases about twice as much energy as the oxidation of a gram of glycogen. Moreover, glycogen differs from fat in binding a great deal of water, producing a sixfold difference in the actual mass of glycogen required to store the same

Figure 4–15 The storage of sugars and fats in animal and plant cells. (A) The structures of starch and glycogen. Both are storage polymers of the sugar glucose and differ only in the frequency of branch points (the region in *yellow* is shown enlarged below). There are many more branches in glycogen than in starch. (B) A thin section of a single chloroplast from a plant cell, showing the starch granules and lipid droplets that have accumulated as a result of the biosyntheses occurring there. (C) Fat droplets (stained red) beginning to accumulate in developing fat cells. (B, courtesy of K. Plaskitt; C, courtesy of Ronald M. Evans.)

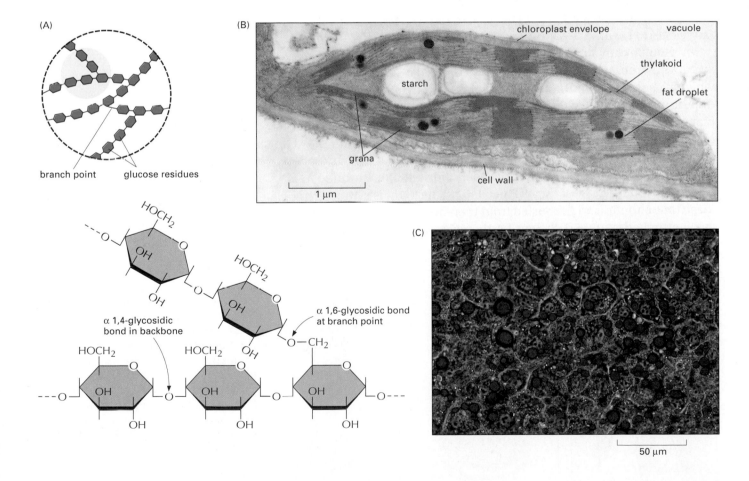

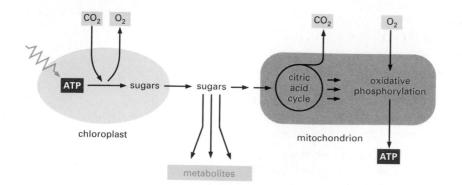

Figure 4–16 **In plants, the chloroplasts and mitochondria collaborate to supply cells with metabolites and ATP.**

amount of energy as fat. An average adult human stores enough glycogen for only about a day of normal activities but enough fat to last for nearly a month. If our main fuel reservoir had to be carried as glycogen instead of fat, body weight would need to be increased by an average of about 60 pounds.

Most of our fat is stored in adipose tissue, from which it is released into the bloodstream for other cells to utilize as needed. The need arises after a period of not eating; even a normal overnight fast results in the mobilization of fat, so that in the morning most of the acetyl CoA entering the citric acid cycle is derived from fatty acids rather than from glucose. After a meal, however, most of the acetyl CoA entering the citric acid cycle comes from glucose derived from food, and any excess glucose is used to replenish depleted glycogen stores or to synthesize fats. (While animal cells readily convert sugars to fats, they cannot convert fatty acids to sugars.)

Although plants produce NADPH and ATP by photosynthesis, this important process occurs in a specialized organelle, called a *chloroplast*, which is isolated from the rest of the plant cell by a membrane that is impermeable to both types of activated carrier molecules. Moreover, the plant contains many other cells—such as those in the roots—that lack chloroplasts and therefore cannot produce their own sugars or ATP. Therefore, for most of its ATP production, the plant relies on an export of sugars from its chloroplasts to the mitochondria that are located in all cells of the plant. Most of the ATP needed by the plant is synthesized in these mitochondria and exported from them to the rest of the plant cell, using exactly the same pathways for the oxidative breakdown of sugars that are utilized by nonphotosynthetic organisms (Figure 4–16).

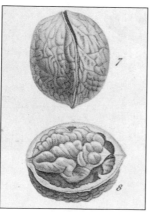
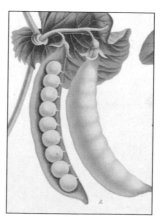

Figure 4–17 **Some plant seeds that serve as important foods for humans.** Corn, nuts, and peas all contain rich stores of starch and fat that provide the young plant embryo in the seed with energy and building blocks for biosynthesis. (Courtesy of the John Innes Foundation.)

During periods of excess photosynthetic capacity during the day, chloroplasts convert some of the sugars that they make into fats and into **starch,** a polymer of glucose analogous to the glycogen of animals. The fats in plants are triacylglycerols, just like the fats in animals, and differ only in the types of fatty acids that predominate. Fat and starch are both stored in the chloroplast as reservoirs to be mobilized as an energy source during periods of darkness (Figure 4–15B).

The embryos inside plant seeds must live on stored sources of energy for a prolonged period, until they germinate to produce leaves that can harvest the energy in sunlight. For this reason plant seeds often contain especially large amounts of fats and starch—which makes them a major food source for animals, including ourselves (Figure 4–17).

Question 4–6 After looking at the structures of sugars and fatty acids, give an intuitive explanation as to why oxidation of a sugar yields only about half the energy as the oxidation of an equivalent dry weight of a fatty acid.

Many Biosynthetic Pathways Begin with Glycolysis or the Citric Acid Cycle

Catabolism produces both energy for the cell and the building blocks from which many other molecules of the cell are made (see Figure 3–3). Thus far, we have emphasized energy production rather than the provision of the starting materials for biosynthesis. But many of the intermediates formed in glycolysis and the citric acid cycle are also siphoned off by other enzymes that use them to produce the amino acids, nucleotides, lipids, and other small organic molecules that the cell needs. Some idea of the complexity of this process can be gathered from Figure 4–18, which illustrates some of the branches from the central catabolic reactions that lead to biosyntheses.

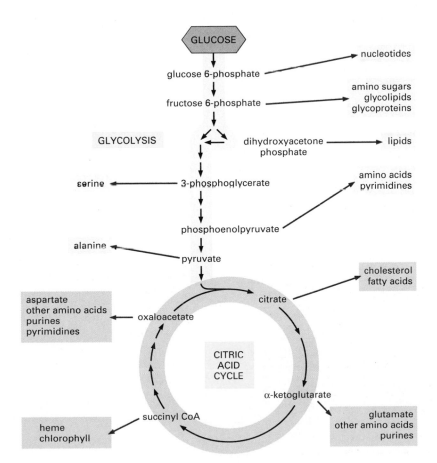

Figure 4–18 **Glycolysis and the citric acid cycle provide the precursors needed to synthesize many important biological molecules.** The amino acids, nucleotides, lipids, sugars, and other molecules—shown here as products—in turn serve as the precursors for the many macromolecules of the cell. Each *black arrow* in this diagram denotes a single enzyme-catalyzed reaction; the *red arrows* generally represent pathways with many steps that are required to produce the indicated products.

The existence of so many branching pathways in the cell requires that the choices at each branch be carefully regulated, as we discuss next.

Metabolism Is Organized and Regulated

One gets a sense of the intricacy of a cell as a chemical machine from Figure 4–19, which is a chart that represents only some of the enzymatic pathways in a cell. Highlighted in *red* on that chart are glycolysis and the citric acid cycle. It is obvious that our discussion of cell metabolism has dealt with only a tiny fraction of cellular chemistry.

All these reactions occur in a cell that is less than 0.1 mm in diameter, and each requires a different enzyme. As is clear from Figure 4–19, the same molecule can often be part of many different pathways. Pyruvate, for example, is a substrate for half a dozen or more different enzymes, each of which modifies it chemically in a different way. One enzyme converts pyruvate to acetyl CoA, another to oxaloacetate; a third enzyme changes pyruvate to the amino acid alanine, a fourth to

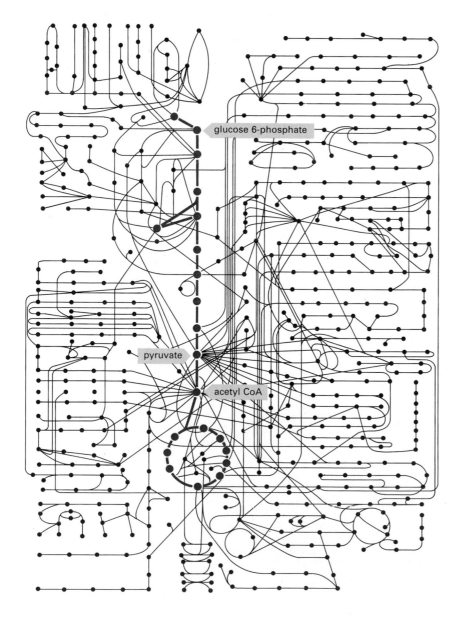

Figure 4–19 Glycolysis and the citric acid cycle are at the center of metabolism. Some 500 metabolic reactions of a typical cell are shown schematically with the reactions of glycolysis and the citric acid cycle in *red*. Other reactions either lead into these two central pathways—delivering small molecules to be catabolized with production of energy—or they lead outward and thereby supply carbon compounds for the purpose of biosynthesis.

lactate, and so on. All of these different pathways compete for the same pyruvate molecule, and similar competitions for thousands of other small molecules go on at the same time. One might think that the whole system would need to be so finely balanced that any minor upset, such as a temporary change in dietary intake, would be disastrous.

In fact, the metabolic balance of a cell is amazingly stable. Whenever the balance is perturbed, the cell reacts so as to restore the initial state. The cell can adapt and continue to function during starvation or disease. Mutations of many kinds can damage or even eliminate particular reaction pathways, and yet—provided that certain minimum requirements are met—the cell survives. It does so because an elaborate network of *control mechanisms* regulates and coordinates the rates of all of its reactions. These controls rest, ultimately, on the remarkable abilities of proteins to change their shape and their chemistry—and thereby be regulated—in response to changes in their immediate environment. The principles that underline how large molecules such as proteins are built and the chemistry behind their regulation will be our next concern.

> **Question 4–7** A cyclic reaction pathway requires that the starting material is regenerated and available at the end of each cycle. If compounds of the citric acid cycle are siphoned off as building blocks used in a variety of metabolic reactions, why does the citric acid cycle not quickly cease to exist?

Essential Concepts

- Glucose and other food molecules are broken down by controlled stepwise oxidation to provide chemical energy in the form of ATP and NADH.

- Three distinct stages in the breakdown of food molecules can be distinguished: glycolysis (which occurs in the cytosol), the citric acid cycle (in the mitochondrial matrix), and oxidative phosphorylation (on the inner mitochondrial membrane).

- The reactions of glycolysis degrade the six-carbon sugar glucose to two molecules of the three-carbon sugar pyruvate, producing small numbers of molecules of ATP and NADH.

- In the presence of oxygen, pyruvate is converted to acetyl CoA plus CO_2. The citric acid cycle then converts the acetyl group in acetyl CoA to CO_2 and H_2O. In eucaryotic cells these reactions occur in mitochondria. Much of the energy of oxidation released in these oxidation reactions is stored as high-energy electrons in the carriers NADH and $FADH_2$.

- The other major energy source in foods is fat. The fatty acids produced from fats are imported into mitochondria and oxidized to acetyl CoA molecules. These acetyl CoA molecules are then further oxidized through the citric acid cycle, just like the acetyl CoA derived from pyruvate.

- NADH and $FADH_2$ pass the electrons they are carrying to an electron-transport chain in the inner mitochondrial membrane, where a series of electron transfers is then used to drive the formation of ATP. Most of the energy captured during the breakdown of food molecules is harvested during this process of oxidative phosphorylation (described in Chapter 13).

- Cells store food molecules in special reservoirs. Glucose subunits are stored as glycogen in animals and as starch in plants; both animals and plants store food as fats. The food reservoirs produced by plants are major sources of food for animals, including ourselves.

- Molecules ingested as food are used not only as sources of metabolic energy but also as raw materials for biosynthesis. Thus many intermediates of glycolysis and the citric acid cycle are starting points for pathways that lead to the synthesis of proteins, nucleic acids, and other specialized molecules of the cell.

- The many thousands of different reactions carried out simultaneously by a cell are closely coordinated, enabling the cell to adapt and continue to function under a wide range of external conditions.

Key Terms	electron-transport chain	glycogen	oxidative phosphory-
acetyl CoA	FAD, FADH$_2$	glycolysis	lation
ADP, ATP	fat	glucose	pyruvate
citric acid cycle	fermentation	NAD$^+$, NADH	starch
	GDP, GTP		

Questions

Question 4–8   The oxidation of sugar molecules by the cell takes place according to the general reaction $C_6H_{12}O_6$ (glucose) $+ 6O_2 \rightarrow 6CO_2 + 6H_2O$ + energy. Which of the following statements are correct? Explain your answers.

A. All of the energy produced is in the form of heat.

B. None of the produced energy is in the form of heat.

C. The energy is produced by oxidation of carbon atoms.

D. The above reaction supplies the cell with essential water.

E. In cells the reaction takes place in more than one step.

F. Many steps in the oxidation of sugar molecules involve reaction with oxygen.

G. Some organisms carry out the reverse reaction.

H. Cells can grow in the absence of O_2 and without producing CO_2.

Question 4–9   An exceedingly sensitive instrument (yet to be devised) shows that one of the carbon atoms in Charles Darwin's last breath is resident in your bloodstream, where it forms part of a hemoglobin molecule. Suggest how this carbon atom might have traveled from Darwin to you, and list some of the molecules it could have entered en route.

Question 4–10   Yeast cells can grow both in the presence of molecular oxygen (aerobically) and in its absence (anaerobically). Under which of the two conditions would you expect the cells to grow better? Explain your answer.

Question 4–11   During movement, muscle cells require large amounts of ATP to fuel their contractile apparatus. These cells contain high levels of creatine phosphate (shown in Figure 4–7). Why is this a useful compound to store energy? Justify your answer with the information shown in Figure 4–7.

Question 4–12   Identical pathways that make up the complicated sequence of reactions of glycolysis, shown in Panel 4–1 (pp. 112–113), are found in most living cells, from bacteria to humans. One could envision, however, countless alternative chemical reaction mechanisms that would allow the oxidation of sugar molecules and that could, in principle, have evolved to take the place of glycolysis. Discuss this fact in the context of evolution.

Question 4–13   Assume that an animal cell is a cube that has a side length of 10 μm. The cell contains 10^9 ATP molecules that it uses up every minute. ATP is regenerated by oxidizing glucose molecules. After what amount of time will the cell have used up an amount of oxygen gas that is equal to its own volume? (Recall that one mole contains 6×10^{23} molecules. One mole of a gas has a volume of 22.4 liters.)

Question 4–14   Under the conditions existing in the cell, the free energies of the first few reactions in glycolysis (in Panel 4–1, pp. 112–113) are

step 1	$\Delta G = -8.0$ kcal/mole
step 2	$\Delta G = -0.6$ kcal/mole
step 3	$\Delta G = -5.3$ kcal/mole
step 4	$\Delta G = -0.3$ kcal/mole

Are these reactions energetically favorable? Using these values, draw to scale an energy diagram (A) for the overall reaction and (B) for the pathway composed of the four individual reactions.

Question 4–15   The chemistry of most metabolic reactions was deciphered by synthesizing metabolites containing atoms that are different isotopes from those naturally occurring. The products of reactions starting with isotopically labeled metabolites can be analyzed to determine precisely which atoms in the products are derived from which atoms in the starting material. The methods of detection exploit, for example, the fact that different isotopes have different masses that can be distinguished using biophysical techniques such as mass spectrometry. Moreover, some isotopes are radioactive and can therefore be readily recognized with Geiger counters or photographic film that becomes exposed by radiation.

A. Assume pyruvate containing radioactive ^{14}C in the carboxyl group is added to a cell extract that can support oxidative phosphorylation. What compound is produced that will contain the vast majority of the ^{14}C added?

B. Assume oxaloacetate containing radioactive ^{14}C in its keto group (refer to Panel 4–2, pp. 122–123) is added to the extract. Where would the ^{14}C atom be after precisely one turn of the cycle?

Question 4–16 In cells that can grow both aerobically and anaerobically, fermentation is inhibited in the presence of molecular oxygen. Suggest a reason for this observation.

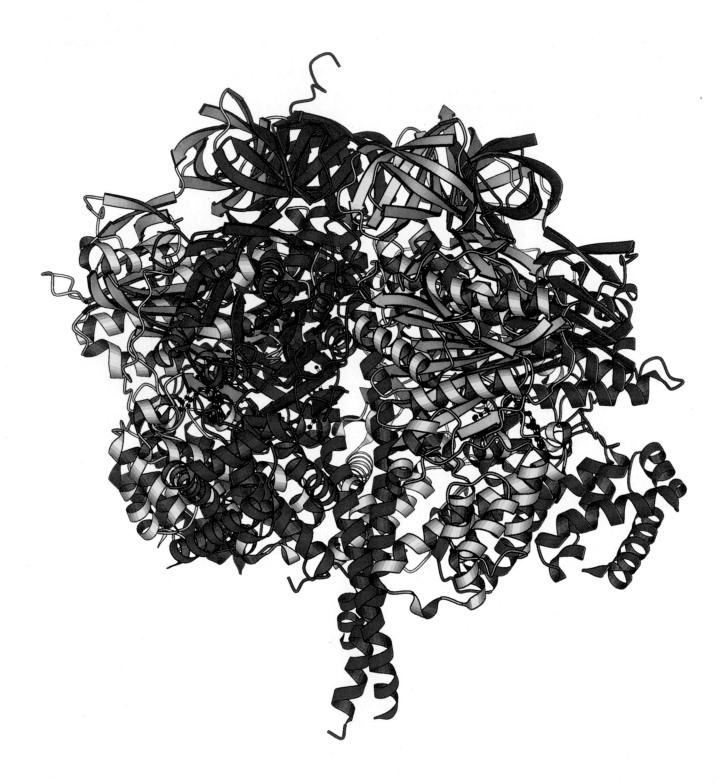

The main chemical energy currency of the cell is ATP, most of which is made in mitochondria by a remarkable protein machine called ATP synthase. Part of this machine, discussed in detail in Chapter 13, has a lollipop-like head made of multiple subunits. The head structure rotates at about 30 revolutions per second, producing 3–4 molecules of ATP at each revolution. The detailed three-dimensional atomic structure of this protein complex, shown here in simplified form, was determined by x-ray crystallography. (Courtesy of John Walker © Nature.)

5 Protein Structure and Function

Proteins constitute most of the dry mass of a cell. When we look at a cell through a microscope or analyze its electrical or biochemical activity we are, in essence, observing proteins. However, proteins are not only the building blocks from which cells are built; they also execute nearly all cell functions. For example, enzymes provide the intricate molecular surfaces in a cell that promote its many chemical reactions. Proteins embedded in the plasma membrane form channels and pumps that control the passage of small molecules into and out of the cell. Some proteins carry messages from one cell to another, while others act as signal integrators that relay sets of signals from the plasma membrane to the nucleus of individual cells. Yet others serve as tiny molecular machines with moving parts: *kinesin*, for example, propels organelles through the cytoplasm; *topoisomerase* can untangle knotted DNA molecules. Other specialized proteins act as antibodies, toxins, hormones, antifreeze molecules, elastic fibers, ropes, or sources of luminescence. Before we can hope to understand how genes work, how muscles contract, how nerves conduct electricity, how embryos develop, or how our bodies function, we must understand proteins. Some landmark discoveries that have contributed to our present understanding of proteins are listed in Table 5–1.

The multiplicity of functions performed by proteins (Panel 5–1, p. 135) arises from the huge number of different three-dimensional shapes they adopt: function follows structure. So we begin our description of these remarkable macromolecules by discussing their three-dimensional structures and the properties that these structures confer. In the second half of the chapter we look at how proteins work: how enzymes catalyze chemical reactions, how proteins act as molecular switches, and how proteins generate coherent movement. Methods for breaking open cells, for purifying proteins from them, and for determining protein structure are also described in this chapter in three separate panels, on pages 160–165.

Table 5–1 Historical Landmarks in Our Understanding of Proteins

1838 The name "protein" (from the Greek *proteios*, "primary") was suggested by **Berzelius** for the complex organic nitrogen-rich substance found in the cells of all animals and plants.

1819–1904 Most of the 20 common amino acids found in proteins were discovered.

1864 **Hoppe-Seyler** crystallized, and named, the protein hemoglobin.

1894 **Fischer** proposed a lock-and-key analogy for enzyme-substrate interactions.

1897 **Buchner and Buchner** showed that cell-free extracts of yeast can ferment sucrose to form carbon dioxide and ethanol, thereby laying the foundations of enzymology.

1926 **Sumner** crystallized urease in pure form, demonstrating that proteins could possess the catalytic activity of enzymes; **Svedberg** developed the first analytical ultracentrifuge and used it to estimate the correct molecular weight of hemoglobin.

1933 **Tiselius** introduced electrophoresis for separating proteins in solution.

1934 **Bernal and Crowfoot** presented the first detailed x-ray diffraction patterns of a protein, obtained from crystals of the enzyme pepsin.

1942 **Martin and Synge** developed chromatography, a technique now widely used to separate proteins.

1951 **Pauling and Corey** proposed the structure of a helical conformation of a chain of L-amino acids—the α helix—and the structure of the β sheet, both of which were later found in many proteins.

1955 **Sanger** completed the analysis of the amino acid sequence of insulin, the first protein to have its amino acid sequence determined.

1956 **Ingram** produced the first protein fingerprints, showing that the difference between sickle-cell hemoglobin and normal hemoglobin is due to a change in a single amino acid.

1960 **Kendrew** described the first detailed structure of a protein (sperm whale myoglobin) to a resolution of 0.2 nm, and **Perutz** proposed a lower-resolution structure for hemoglobin.

1963 **Monod, Jacob, and Changeux** recognized that many enzymes are regulated through allosteric changes in their conformation.

The Shape and Structure of Proteins

The structure and chemistry of each protein has been developed and fine-tuned over billions of years of evolutionary history. It is perhaps not surprising, therefore, that from a chemical point of view, proteins are by far the most structurally complex and functionally sophisticated molecules known. We start by considering how the location of each amino acid in the long string of amino acids that forms a protein determines its precise three-dimensional shape. An understanding of protein structure at the atomic level will then allow us to describe how the precise shape of the protein in turn determines its function.

The Shape of a Protein Is Specified by Its Amino Acid Sequence

Recall from Chapter 2 that there are 20 different types of amino acids in proteins, each with different chemical properties. A **protein** molecule is made from a long chain of these amino acids, each linked to its neighbor

ENZYME

function: Catalysis of covalent bond breakage or formation.

examples: Living cells contain thousands of different enzymes, each of which catalyzes (speeds up) one particular reaction. Examples include: *tryptophan synthetase*—used to make the amino acid tryptophan; *pepsin*—degrades dietary proteins in the stomach; *ribulose bisphosphate carboxylase*—helps to convert carbon dioxide into sugars in plants; *DNA polymerase*—makes DNA; *protein kinase*—adds a phosphate group to a protein molecule.

STRUCTURAL PROTEIN

function: Provides mechanical support to cells and tissues.

examples: Outside cells, *collagen* and *elastin* are common constituents of extracellular matrix and form fibers in tendons and ligaments. Inside cells, *tubulin* forms long, stiff microtubules and *actin* forms actin filaments that underlie and support the plasma membrane; α-*keratin* forms fibers that reinforce epithelial cells and is the major protein in hair and horn.

TRANSPORT PROTEIN

function: Carries small molecules or ions.

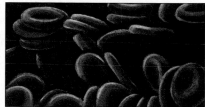

examples: In the bloodstream, *serum albumin* carries lipids, *hemoglobin* carries oxygen, and *transferrin* carries iron. Many proteins embedded in cell membranes transport ions or small molecules across the membrane. For example, the bacterial protein *bacteriorhodopsin* is a light-activated proton pump that transports H$^+$ ions out of the cell; the *glucose carrier* shuttles glucose into and out of liver cells; and a *Ca^{2+} pump* of muscle cells pumps the calcium ions needed to trigger muscle contraction into the endoplasmic reticulum, where they are stored.

MOTOR PROTEIN

function: Generates movement in cells and tissues.

examples: *Myosin* in skeletal muscle cells provides the motive force for animals to move; *kinesin* interacts with microtubules to move organelles around the cell; *dynein* enables eucaryotic cilia and flagella to beat.

STORAGE PROTEIN

function: Stores small molecules or ions.

examples: Iron is stored in the liver by binding to the small protein *ferritin*; *ovalbumin* in egg white is used as a source of amino acids for the developing bird embryo; *casein* in milk is a source of amino acids for baby animals.

SIGNALING PROTEIN

function: Carries signals from cell to cell.

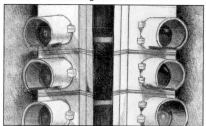

examples: Many of the hormones and growth factors that coordinate physiological function in animals are proteins; *insulin*, for example, is a small protein that controls glucose levels in the blood; *netrin* attracts growing nerve cells in a specific direction in a developing embryo; *nerve growth factor (NGF)* stimulates some types of nerve cells to grow axons; *epidermal growth factor (EGF)* stimulates the growth and division of epithelial cells.

RECEPTOR PROTEIN

function: Used by cells to detect signals and transmit them to the cell's response machinery.

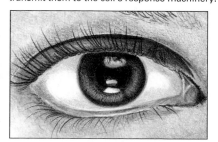

examples: *Rhodopsin* in the retina detects light; the *acetylcholine receptor* in the membrane of a muscle cell receives chemical signals released from a nerve ending; the *insulin receptor* allows a liver cell to respond to the hormone insulin by taking up glucose; the *adrenergic receptor* on heart muscle increases the rate of heartbeat when it binds adrenaline.

GENE REGULATORY PROTEIN

function: Binds to DNA to switch genes on or off.

examples: The *lactose repressor* in bacteria silences the gene for the enzymes that degrade the sugar lactose; many different *homeodomain proteins* act as genetic switches to control development in multicellular organisms, including humans.

SPECIAL PURPOSE PROTEIN

function: Highly variable.

examples: Organisms make many proteins with highly specialized properties. These molecules illustrate the amazing range of functions that proteins can perform. The *antifreeze proteins* of Arctic and Antarctic fishes protect their blood against freezing; *green fluorescent protein* from jellyfish emits a green light; *monellin*, a protein found in an African plant, has an intensely sweet taste; mussels and other marine organisms secrete *glue proteins* that attach them firmly to rocks, even when immersed in sea water.

Panel 5–1   A few examples of some general protein functions.

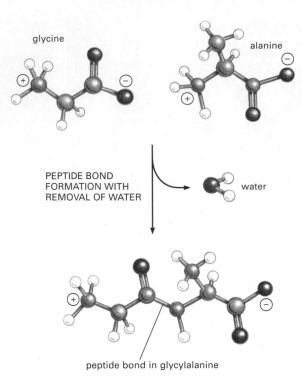

glycine

alanine

PEPTIDE BOND
FORMATION WITH
REMOVAL OF WATER

water

peptide bond in glycylalanine

Figure 5–1 **A peptide bond.** This covalent bond forms when the carbon atom from the carboxyl group of one amino acid shares electrons with the nitrogen atom (*blue*) from the amino group of a second amino acid. As indicated, a molecule of water is lost in this condensation reaction.

through a covalent peptide bond (Figure 5–1). The repeating sequence of atoms along the chain is referred to as the **polypeptide backbone.** Attached to this repetitive chain are the different amino acid **side chains**—those portions of the amino acids that are not involved in making a peptide bond and which give each amino acid its unique properties (Figure 5–2). Some are nonpolar and hydrophobic ("water hating"), some negatively or positively charged, some reactive and some unreactive, and so on. The structures and properties of the 20 amino acids in proteins were given in Panel 2–5 (pp. 62–63), and they are listed with their abbreviations in Figure 5–3. Each type of protein has a unique sequence of amino acids, exactly the same from one molecule to the next. Many thousands of different types of proteins are known, each with its own particular amino acid sequence.

Many of the covalent bonds in a long chain of amino acids allow free rotation of the atoms they join, so that the polypeptide backbone can in principle fold up in an enormous number of ways. Each folded chain will be constrained by many different sets of weak *noncovalent bonds*, formed both by atoms in the polypeptide backbone and by atoms in the amino acid side chains. These weak bonds are *hydrogen bonds, ionic bonds,* and *van der Waals attractions,* as described in Chapter 2 (see Panel 2–7, pp. 70–71). Individual noncovalent bonds are weak compared to covalent bonds, so that it takes many noncovalent bonds to hold two regions of a polypeptide chain together. The stability of each folded shape will therefore be affected by the combined strength of large numbers of such bonds (Figure 5–4).

A fourth weak force also plays a central part in determining the shape of a protein. As described in Chapter 2, hydrophobic molecules, including the nonpolar side chains of particular amino acids, tend to be forced together in an aqueous (watery) environment in order to minimize their disruptive effect on the hydrogen-bonded network of water

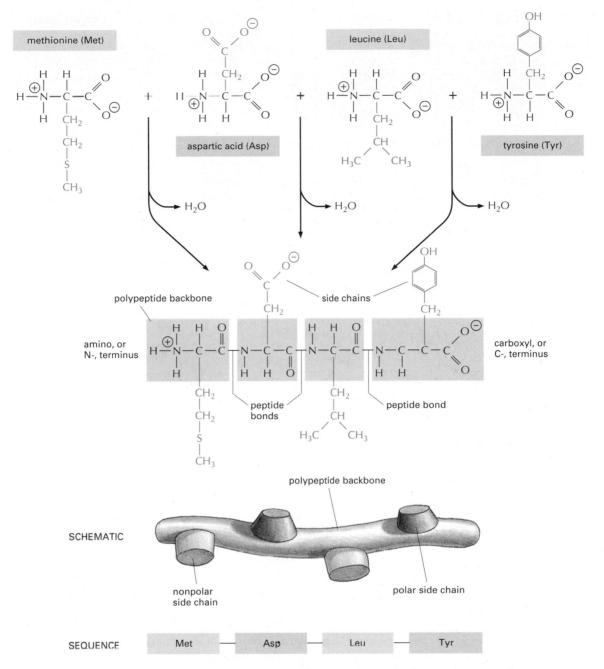

Figure 5–2 Proteins consist of a polypeptide backbone with attached side chains.

methionine (Met)

aspartic acid (Asp)

leucine (Leu)

tyrosine (Tyr)

polypeptide backbone

side chains

amino, or N-, terminus

carboxyl, or C-, terminus

peptide bonds

peptide bond

polypeptide backbone

SCHEMATIC

nonpolar side chain

polar side chain

SEQUENCE   Met   Asp   Leu   Tyr

molecules (see p. 48 and Panel 2–2, pp. 50–51). Therefore, an important factor governing the folding of any protein is the distribution of its polar and nonpolar amino acids. The nonpolar (hydrophobic) side chains in a protein—belonging to such amino acids as phenylalanine, leucine, valine, and tryptophan—tend to cluster in the interior of the molecule (just as hydrophobic oil droplets coalesce to form one large droplet). This enables them to avoid contact with the water that surrounds them inside a cell. In contrast, polar side chains—such as those belonging to arginine, glutamine, and histidine—tend to arrange themselves near the outside of the molecule, where they can form hydrogen bonds with water and with other polar molecules (Figure 5–5). When polar amino acids are buried within the protein, they are usually hydrogen-bonded to other polar amino acids or to the polypeptide backbone (Figure 5–6).

Each type of protein differs in its sequence and number of amino acids; therefore, it is the sequence of the chemically different side chains that makes each protein distinct. The two ends of a polypeptide chain are chemically different: the end carrying the free amino group (NH_3^+, also written NH_2) is the amino, or N-, terminus, and that carrying the free carboxyl group (COO^-, also written $COOH$) is the carboxyl, or C-, terminus. The amino acid sequence of a protein is always presented in the N to C direction, reading from left to right.

The Shape and Structure of Proteins

AMINO ACID			SIDE CHAIN	AMINO ACID			SIDE CHAIN
Aspartic acid	Asp	D	negative	Alanine	Ala	A	nonpolar
Glutamic acid	Glu	E	negative	Glycine	Gly	G	nonpolar
Arginine	Arg	R	positive	Valine	Val	V	nonpolar
Lysine	Lys	K	positive	Leucine	Leu	L	nonpolar
Histidine	His	H	positive	Isoleucine	Ile	I	nonpolar
Asparagine	Asn	N	uncharged polar	Proline	Pro	P	nonpolar
Glutamine	Gln	Q	uncharged polar	Phenylalanine	Phe	F	nonpolar
Serine	Ser	S	uncharged polar	Methionine	Met	M	nonpolar
Threonine	Thr	T	uncharged polar	Tryptophan	Trp	W	nonpolar
Tyrosine	Tyr	Y	uncharged polar	Cysteine	Cys	C	nonpolar

└────── POLAR AMINO ACIDS ──────┘ └────── NONPOLAR AMINO ACIDS ──────┘

Figure 5–3 **The 20 different amino acids found in proteins.** Both three-letter and one-letter abbreviations are listed. As shown, there are an equal number of polar and nonpolar side chains. For their atomic structures, see Panel 2–5 (pp. 62–63).

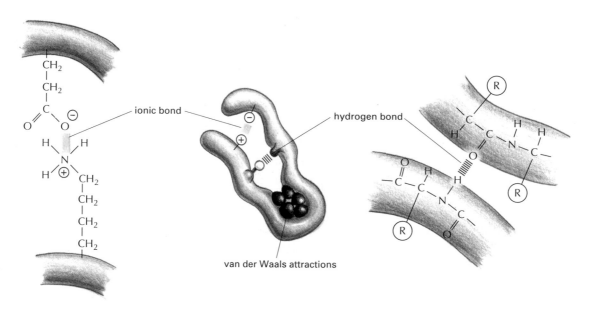

Figure 5–4 **Three types of noncovalent bonds that help proteins fold.** Although a single one of these bonds is quite weak, many of them often form together to create a strong bonding arrangement, as in the example shown. R is a general designation for a side chain.

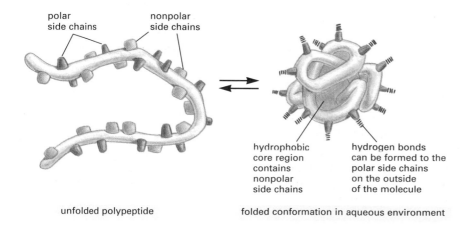

Figure 5–5 **How a protein folds into a compact conformation.** The polar amino acid side chains tend to gather on the outside of the protein, where they can interact with water; the nonpolar amino acid side chains are buried on the inside to form a highly packed hydrophobic core of atoms that are hidden from water. In this very schematic drawing, the protein contains only about 30 amino acids.

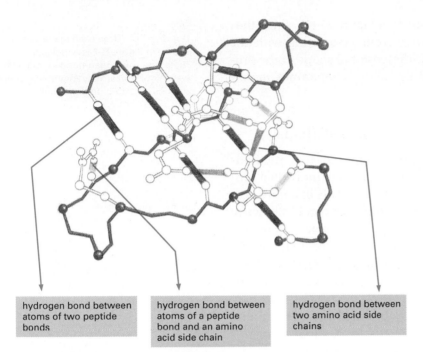

Figure 5–6 Hydrogen bonds in a protein molecule. Large numbers of hydrogen bonds form between adjacent regions of the folded polypeptide chain and help to stabilize its three-dimensional shape. The protein depicted is a portion of the enzyme lysozyme, and the hydrogen bonds between the three possible pairs of partners have been differently colored, as indicated. (After C.K. Matthews and K.E. van Holde, Biochemistry. Redwood City, CA: Benjamin/Cummings, 1996.)

hydrogen bond between atoms of two peptide bonds

hydrogen bond between atoms of a peptide bond and an amino acid side chain

hydrogen bond between two amino acid side chains

Proteins Fold into a Conformation of Lowest Energy

Each type of protein has a particular three-dimensional structure, which is determined by the order of the amino acids in its chain. The final folded structure, or **conformation,** adopted by any polypeptide chain is determined by energetic considerations, generally being that in which the free energy is minimized. Protein folding has been studied in the laboratory using highly purified proteins. A protein can be unfolded, or *denatured,* by treatment with certain solvents, which disrupt the noncovalent interactions holding the folded chain together. This converts the protein into a flexible polypeptide chain that has lost its natural shape. When the denaturing solvent is removed, the protein often refolds spontaneously, or *renatures,* into its original conformation (Figure 5–7). This indicates that all the information necessary to specify the three-dimensional shape of a protein is contained in its amino acid sequence.

Each protein normally folds up into a single stable conformation. However, this conformation will often change slightly when the protein interacts with other molecules in the cell. This change in shape is often crucial to the function of the protein, as we shall see later in this chapter.

Although a protein chain can fold into its correct conformation without outside help, protein folding in a living cell is generally assisted by special proteins called *molecular chaperones.* These are proteins that bind to partly folded chains and help to keep them folding along the most energetically favorable folding pathway. Chaperones are vital in

Figure 5–7 Refolding of a denatured protein. (A) This experiment demonstrates that the conformation of a protein is determined solely by its amino acid sequence. (B) The structure of urea, which is very soluble in water and unfolds proteins at high concentrations where there is about one urea molecule for every six water molecules.

(A)

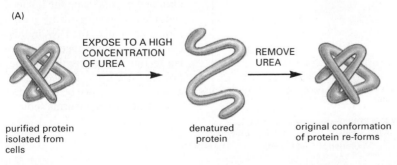

purified protein isolated from cells

EXPOSE TO A HIGH CONCENTRATION OF UREA

denatured protein

REMOVE UREA

original conformation of protein re-forms

(B)

the crowded conditions of the cytoplasm (see Figure 3–17), since they prevent newly synthesized protein chains from associating with the wrong partners. However, the final three-dimensional shape of the protein is still specified by its amino acid sequence: chaperones merely make the folding process more reliable.

Proteins Come in a Wide Variety of Complicated Shapes

As will be described in Panels 5–4, 5–5, and 5–6 (pp. 160–165), methods have been developed that allow cells to be broken open, individual proteins to be purified, and the precise order of amino acids in a pure protein to be determined. This order is known as the **amino acid sequence** (also called the protein sequence). For many years, protein sequencing was accomplished by analyzing the amino acids in the protein directly. The first protein to have its complete amino acid sequence determined was the hormone *insulin;* its sequence was published in 1955. The development of rapid methods for sequencing DNA (see Chapter 10) now makes it much easier to sequence a protein indirectly, by determining the order of the nucleotides in the DNA that encodes that protein— interpreting this as an amino acid sequence by applying the genetic code (see Chapter 7). Tens of thousands of proteins have already had their amino acid sequences determined this way. These proteins range in size from about 30 amino acids to more than 10,000 amino acids. However, the vast majority of proteins are between 50 and 2000 amino acids long (Figure 5–8).

All the information required for a polypeptide chain to fold is contained in its amino acid sequence. However, we have not yet learned how to "read" this information so as to predict from the amino acid sequence the detailed three-dimensional conformation of a protein— that is, the arrangement in space of all of its atoms. At present, the only way to discover the precise folding pattern of any protein is by experiment, using either the x-ray or nuclear magnetic resonance methods that will be described in Panel 5–6 (pp. 164–165). So far, more than 1000 proteins have been completely analyzed by these techniques. Each has a

Question 5–1 Urea used in the experiment shown in Figure 5–7 is a molecule that disrupts the hydrogen-bonded network of water molecules. Why might high concentrations of urea unfold proteins?

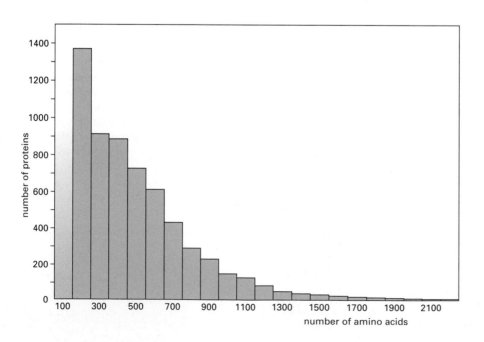

Figure 5–8 The sizes of proteins. The proteins in a cell are found in a wide range of sizes. A typical bacterial cell contains about 1000 different proteins, whereas a human cell is thought to contain about 10,000. Shown here is a plot of the proteins encoded by several chromosomes of the yeast *Saccharomyces cerevisiae,* showing the numbers of predicted proteins of different sizes. This information was obtained by analyzing the nucleotide sequence of the organism's genome; the sequence of all of the DNA in this small eucaryotic cell was recently determined. For technical reasons, the number of small proteins, especially those with 100 amino acids or less, cannot be determined accurately from the nucleotide sequence alone.

unique three-dimensional conformation so intricate and irregular that the structure of most proteins would require an entire chapter to describe in detail.

In order to show how proteins are constructed, we shall start with an unusually small protein and attempt to illustrate its structure clearly. Protein structures can be depicted in several ways, each emphasizing different features of the protein. Panel 5–2 (pp. 142–143) presents four different depictions of the phosphocarrier protein *HPr*, a transport protein that facilitates sugar transport into bacterial cells. Constructed from a string of 88 amino acids, the complete structure is displayed as a polypeptide backbone model in (A), as a ribbon model in (B), as a wire model that includes the amino acid side chains in (C), and as a space-filling model in (D). Each of the three horizontal rows shows the protein in a different orientation, and the image is colored in a way that allows the polypeptide chain to be followed from its N-terminus (*purple*) to its C-terminus (*red*).

Panel 5–2 makes it clear that a protein's conformation is amazingly complex, even for a protein as small as HPr. But the description of protein structures can be simplified by the recognition that several common structural motifs underlie these conformations, as we discuss next.

The α Helix and the β Sheet Are Common Folding Patterns

When the three-dimensional structures of many different protein molecules are compared, it becomes clear that, although the overall conformation of each protein is unique, two regular folding patterns are often found in parts of them. Both patterns were discovered about 50 years ago from studies of hair and silk. The first folding pattern to be discovered, called the **α helix,** was found in the protein *α-keratin,* which is abundant in skin and its derivatives—such as hair, nails, and horns. Within a year of the discovery of the α helix, a second folded structure, called a **β sheet,** was found in the protein *fibroin,* the major constituent of silk. These two patterns are particularly common because they result from hydrogen-bonding between the N–H and C=O groups in the polypeptide backbone, without involving the side chains of amino acids. Thus, they can be formed by many different amino acid sequences. In each case, the protein chain adopts a regular, repeating conformation. These two conformations, as well as the abbreviations that are used to denote them in ribbon models of proteins, are displayed in Figure 5–9.

The core of many proteins contains extensive regions of β sheet. As shown in Figure 5–10, these β sheets can form either from neighboring polypeptide chains that run in the same orientation (i.e., parallel chains) or from a polypeptide chain that folds back and forth upon itself, with each section of the chain running in the direction opposite to that of its immediate neighbors (i.e., antiparallel chains). Both types of β sheet produce a very rigid structure, held together by hydrogen bonds that connect the peptide bonds in neighboring chains (see Figure 5–9D).

An α helix is generated when a single polypeptide chain turns around itself to make a rigid cylinder. A hydrogen bond is made between every fourth peptide bond, linking the C=O of one peptide bond to the N–H of another (see Figure 5–9A). This gives rise to a regular helix with a complete turn every 3.6 amino acids. Note that the sample protein that was illustrated in Panel 5–2 contains both an α helix and a β sheet.

The Shape and Structure of Proteins

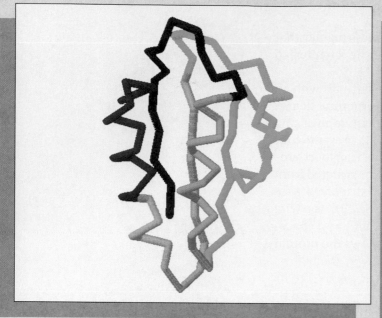

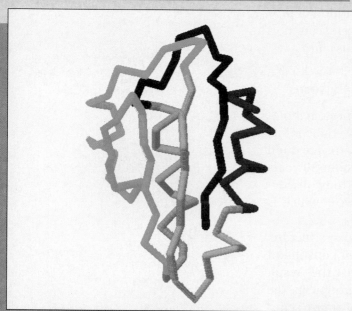
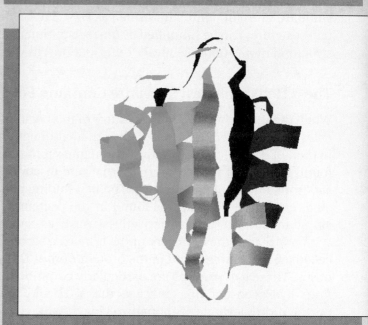
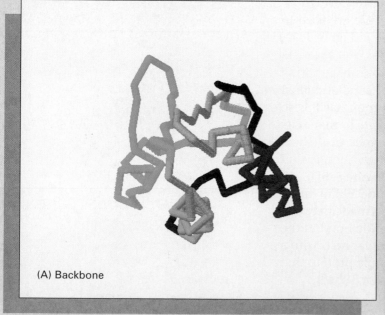
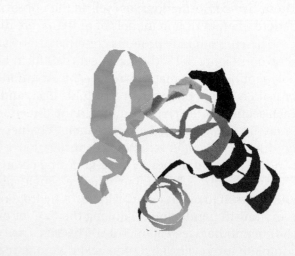

(A) Backbone

(B) Ribbon

Panel 5–2 **Four different ways of depicting a small protein.** (Images created using Ras Mol © Roger Sayle.)

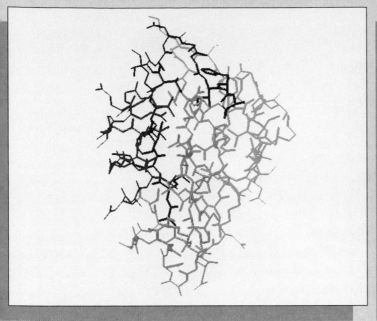

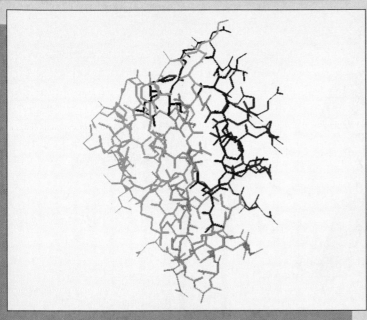

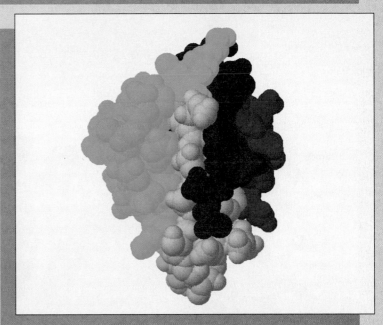

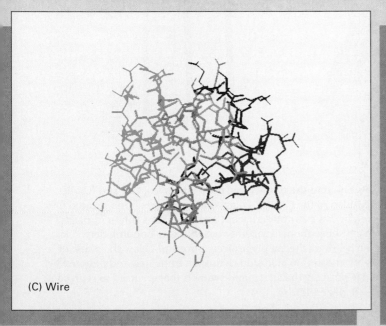

(C) Wire

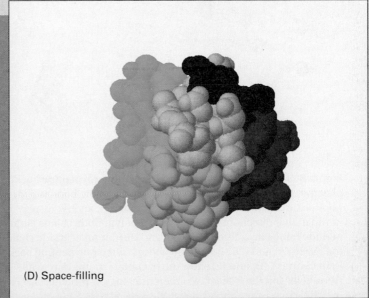

(D) Space-filling

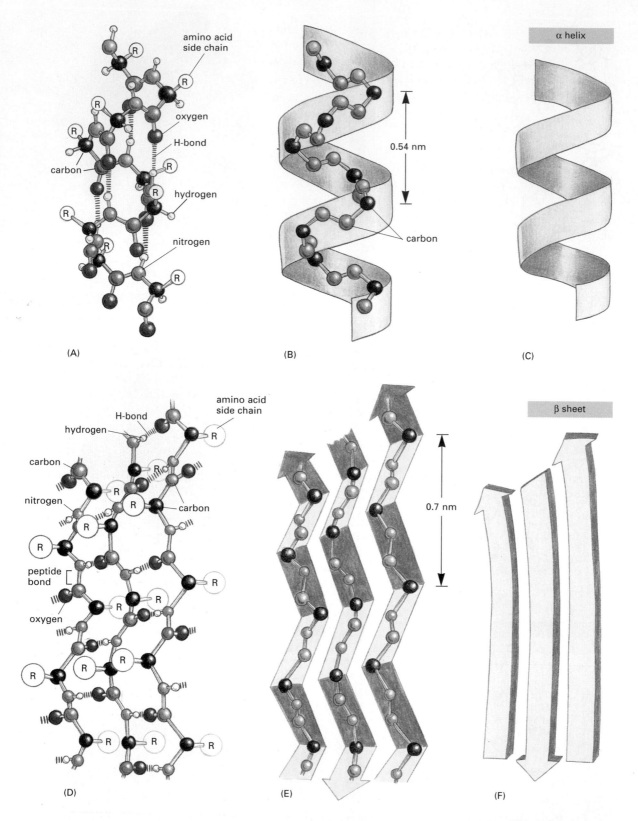

(A)

amino acid
side chain

oxygen

H-bond

carbon

R

hydrogen

nitrogen

(B)

0.54 nm

carbon

(C)

α helix

(D)

amino acid
side chain

H-bond

hydrogen

carbon

nitrogen

carbon

peptide
bond

oxygen

(E)

0.7 nm

(F)

β sheet

Figure 5–9 The regular conformation of the polypeptide backbone observed in the α helix and the β sheet. (A), (B), and (C) display the α helix, in which the N–H of every peptide bond is hydrogen-bonded to the C=O of a neighboring peptide bond located four peptide bonds away in the same chain. (D), (E), and (F) display a β sheet; in this example, adjacent peptide chains run in opposite (antiparallel) directions. The individual polypeptide chains (strands) in a β sheet are held together by hydrogen-bonding between peptide bonds in different strands, and the amino acid side chains in each strand alternately project above and below the plane of the sheet. (A) and (D) show all the atoms in the polypeptide backbone, but the amino acid side chains are truncated and denoted by R. In contrast, (B) and (E) show the backbone atoms only, while (C) and (F) display the shorthand symbols that are used to represent the α helix and the β sheet in ribbon drawings of proteins (see Panel 5–2B, page 142).

Short regions of α helix are especially abundant in the proteins located in cell membranes, such as transport proteins and receptors. We will see in Chapter 11 that those portions of a transmembrane protein that cross the lipid bilayer usually cross as an α helix that is composed largely of amino acids with nonpolar side chains. The polypeptide backbone, which is hydrophilic, is hydrogen-bonded to itself in the α helix and shielded from the hydrophobic lipid environment of the membrane by its protruding nonpolar side chains (see Figure 11–24).

Some pairs of α helices wrap around each other to form a particularly stable structure, known as a **coiled-coil.** This structure forms when the two α helices have most of their nonpolar (hydrophobic) side chains on one side, so that they can twist around each other with these side chains facing inward (Figure 5–11). Long rodlike coiled-coils form the structural framework for many elongated proteins. Examples are α-keratin, which forms the intracellular fibers that reinforce the outer layer of the skin and its appendages, and the myosin molecules responsible for muscle contraction (see Chapter 16).

Proteins Have Several Levels of Organization

Even a small protein molecule is built from thousands of atoms linked together by precisely oriented covalent and noncovalent bonds, and it is extremely difficult to visualize such a complicated structure without special tools. For this reason, various graphical and computer-based aids are used. A computer disk available from the publisher contains computer-generated images of selected proteins, designed to be displayed and rotated on the screen in a variety of formats.

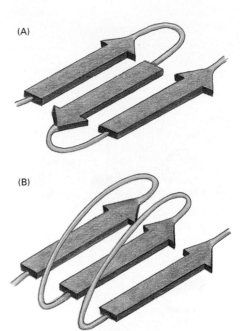

Figure 5–10 **Two types of β sheet structures.** (A) Antiparallel β sheet (see Figure 5–9D). (B) Parallel β sheet. Both of these structures are common in proteins.

Question 5–2   Remembering that the side chains projecting from each polypeptide backbone in a β sheet alternately point above and below the plane of the sheet (see Figure 5–9D), consider the following protein sequence: Leu-Lys-Val-Asp-Ile-Ser-Leu-Arg-Leu-Lys-Ile-Arg-Phe-Glu. Do you find anything remarkable about the arrangement of the amino acids in this sequence when incorporated into a β sheet? Can you make any predictions as to how the β sheet might be arranged in a protein? (Hint: consult the properties of the amino acids listed in Figure 5–3.)

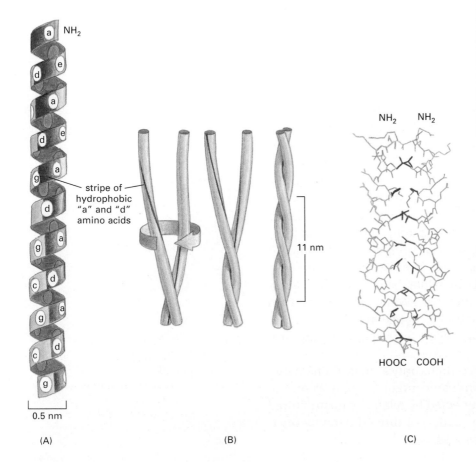

stripe of hydrophobic "a" and "d" amino acids

11 nm

NH₂   NH₂

HOOC   COOH

0.5 nm

(A)   (B)   (C)

Figure 5–11 **The structure of a coiled-coil.** In (A) a single α helix is shown, with successive amino acid side chains labeled in a sevenfold sequence "abcdefg" (from bottom to top). Amino acids "a" and "d" in such a sequence lie close together on the cylinder surface, forming a "stripe" (*shaded in red*) that winds slowly around the α helix. Proteins that form coiled-coils typically have nonpolar amino acids at positions "a" and "d." Consequently, as shown in (B), the two α helices can wrap around each other with the nonpolar side chains of one α helix interacting with the nonpolar side chains of the other, while the more hydrophilic amino acid side chains are left exposed to the aqueous environment. (C) The atomic structure of a coiled-coil determined by x-ray crystallography. The *red* side chains are nonpolar.

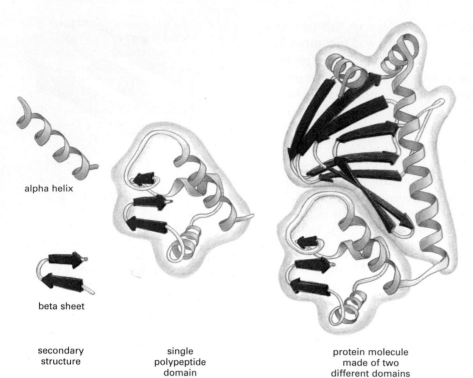

Figure 5–12 The relationship between protein domains and other structural elements. Elements of secondary structure such as α helices and β sheets pack together into stable globular elements called domains. A typical protein molecule is built from one or more domains, often linked through relatively unstructured regions of polypeptide chain. The ribbon diagram on the right is of the bacterial gene regulatory protein CAP, with one large (*blue*) and one small (*gray*) domain.

alpha helix

beta sheet

secondary
structure

single
polypeptide
domain

protein molecule
made of two
different domains

In considering the structure of a protein, it is helpful to distinguish various levels of organization. The amino acid sequence is the *primary structure* of the protein. Stretches of polypeptide chain that form α helices and β sheets constitute the protein's **secondary structure.** The three-dimensional conformation formed by a polypeptide chain is sometimes referred to as the protein's *tertiary structure,* and if a particular protein molecule is formed as a complex of more than one polypeptide chain, then the complete structure is designated as its *quaternary structure.*

Studies of the conformation, function, and evolution of proteins have also revealed the importance of a level of organization distinct from those just described. This is the **protein domain,** which is produced by any part of a polypeptide chain that can fold independently into a compact, stable structure. A domain usually contains between 50 and 350 amino acids, and it is the modular unit from which many larger proteins are constructed (Figure 5–12). The different domains of a protein are often associated with different functions. For example, the protein illustrated in Figure 5–12 is the *catabolite activator protein (CAP),* which is used by bacteria to turn genes on and off. The CAP protein has two domains: the small domain binds to DNA, while the large domain binds cyclic AMP, an intracellular signaling molecule. When the large domain binds cyclic AMP, it causes a conformational change in the protein that enables the small domain to bind to a specific DNA sequence and turn on adjacent genes.

A small protein molecule like HPr contains only a single domain (see Panel 5–2, pp. 142–143). Larger proteins can contain as many as several dozen domains, which are usually connected by relatively unstructured lengths of polypeptide chain. Ribbon models of three differently organized domains are presented in Figure 5–13.

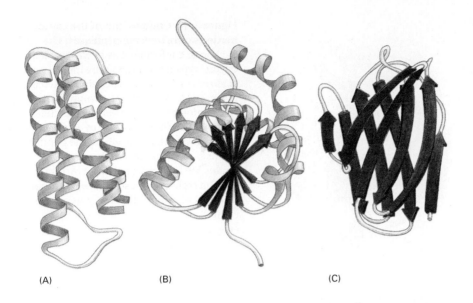

Figure 5–13 Ribbon models of three different protein domains. (A) Cytochrome b_{562}, a single-domain protein involved in electron transport in mitochondria. This protein is composed almost entirely of α helices. (B) The NAD-binding domain of the enzyme lactic dehydrogenase, which is composed of a mixture of α helices and β sheets. (C) The variable domain of an immunoglobulin (antibody) light chain, composed of a sandwich of two β sheets. In these examples, the α helices are shown in *green*, while strands organized as β sheets are denoted by *red arrows*. Note that the polypeptide chain generally traverses back and forth across the entire domain, making sharp turns only at the protein surface. The protruding *loop regions* (*yellow*) often form the binding sites for other molecules. (Drawings courtesy of Jane Richardson.)

Few of the Many Possible Polypeptide Chains Will Be Useful

In theory, a vast number of different polypeptide chains could be made. Since each of the 20 amino acids is chemically distinct and each can, in principle, occur at any position in a protein chain, there are $20 \times 20 \times 20 \times 20 = 160{,}000$ different possible polypeptide chains four amino acids long, or 20^n different possible polypeptide chains n amino acids long. For a typical protein length of about 300 amino acids, more than 10^{390} (20^{300}) different polypeptide chains could theoretically be made.

Only a very small fraction of this unimaginably large number of polypeptide chains would adopt a single stable three-dimensional conformation. The vast majority of individual protein molecules would have many different conformations of roughly equal stability, each conformation having different chemical properties. So why do virtually all proteins present in cells adopt unique and stable conformations? The answer is that a protein with many different conformations and variable properties would not be biologically useful. Such proteins would therefore have been eliminated by natural selection in the course of the long trial-and-error process that underlies cellular evolution (see Chapter 9).

Because of natural selection, not only is the amino acid sequence of a present-day protein such that a single conformation is extremely stable, but this conformation has the exact chemical properties that enable the protein to perform a particular catalytic or structural function in the cell. Proteins are so precisely built that the change of even a few atoms in one amino acid can sometimes disrupt the structure of a protein and thereby cause a catastrophic loss of function.

Proteins Can Be Classified into Families

Once a protein had evolved that folded up into a stable conformation with useful properties, its structure could be modified slightly during evolution to enable it to perform new functions. We know that this occurred quite often, because many present-day proteins can be grouped into **protein families,** with each family member having an amino acid sequence and a three-dimensional conformation that closely resembles all of the other family members.

Question 5–3 Random mutations only very rarely result in changes in a protein that improve its usefulness for the cell, yet useful mutations are selected in evolution. Because these changes are so rare, for each useful mutation there are innumerable mutations that lead to either no improvement or inactive proteins. Why then do cells not contain millions of different proteins that are of no use?

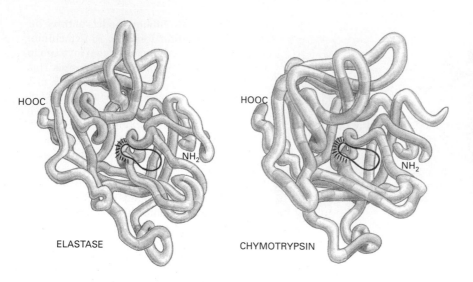

HOOC

NH₂

ELASTASE

HOOC

NH₂

CHYMOTRYPSIN

Figure 5–14 Comparison of the conformations of two serine proteases. The backbone conformations of elastase and chymotrypsin. Although only those amino acids in the polypeptide chain shaded in *green* are the same in the two proteins, the two conformations are very similar nearly everywhere. The active site of each enzyme is circled in *red;* this is where the peptide bonds of the proteins that serve as substrates are bound and cleaved by hydrolysis. The serine proteases derive their name from the amino acid serine, whose side chain is part of the active site of each enzyme and directly participates in the cleavage reaction.

Consider, for example, the *serine proteases,* a family of protein-cleaving (proteolytic) enzymes that includes the digestive enzymes chymotrypsin, trypsin, and elastase, and some of the proteases involved in blood clotting. When any two of these enzymes are compared, portions of their amino acid sequences are found to be nearly the same. The similarity of their three-dimensional conformations is even more striking: most of the detailed twists and turns in their polypeptide chains, which are several hundred amino acids long, are virtually identical (Figure 5–14). The various serine proteases nevertheless have distinct enzymatic activities, each cleaving different proteins or the peptide bonds between different types of amino acids. Each therefore carries out a distinct function in an organism.

Larger Protein Molecules Often Contain More Than One Polypeptide Chain

The same weak noncovalent bonds that enable a protein chain to fold into a specific conformation also allow proteins to bind to each other to produce larger structures in the cell. Any region of a protein's surface that interacts with another molecule through sets of noncovalent bonds is termed a *binding site.* A protein can contain binding sites for a variety of molecules, large and small. If a binding site recognizes the surface of a second protein, the tight binding of two folded polypeptide chains at this site can create a larger protein molecule with a precisely defined geometry. Each polypeptide chain in such a protein is called a protein **subunit.**

In the simplest case, two identical folded polypeptide chains will bind to each other in a "head-to-head" arrangement, forming a symmetrical complex of two protein subunits (called a *dimer*) that is held together by interactions between two identical binding sites. The CAP protein discussed previously is in fact a dimeric protein in the bacterial cell (Figure 5–15A), being formed from two identical copies of the protein subunit shown previously in Figure 5–12. Many other symmetrical protein complexes, formed from multiple copies of a single polypeptide chain, are commonly found in cells. The enzyme *neuraminidase,* for example, consists of a ring of four identical protein subunits (Figure 5–15B).

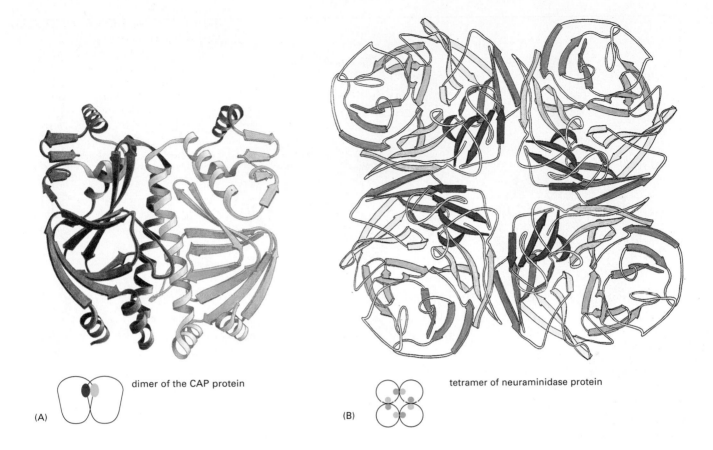

dimer of the CAP protein

(A)

tetramer of neuraminidase protein

(B)

Other proteins contain two or more different types of polypeptide chains. *Hemoglobin*, the protein that carries oxygen in red blood cells, is a particularly well-studied example (Figure 5–16). It contains two identical α globin subunits and two identical β globin subunits, symmetrically arranged. Such multisubunit proteins are numerous in cells and can be very large. Figure 5–17 presents a sampling of proteins whose exact structures are known, allowing the sizes and shapes of a few larger proteins to be compared to the relatively small proteins that we have thus far presented as models.

Proteins Can Assemble into Filaments, Sheets, or Spheres

Proteins can form even larger assemblies than those discussed so far. Most simply, a chain of identical protein molecules can be formed if the binding site on a protein is complementary to another region of the surface of the same protein. As discussed on page 152, since each protein molecule is bound to its neighbor in an identical way, the molecules will often be arranged in a *helix* that can be extended indefinitely (Figure 5–18). This type of arrangement can produce an extended protein filament. An actin filament, for example, is a long helical structure formed

Figure 5–15 **Many protein molecules contain multiple copies of a single protein subunit.** (A) A symmetrical dimer. The CAP protein exists as a complex of two identical polypeptide chains (see also Figure 5–12). (B) A symmetrical tetramer. The enzyme neuraminidase exists as a ring of four identical polypeptide chains. For both (A) and (B), a small schematic below the structure emphasizes how the repeated use of the same binding interaction forms the structure.

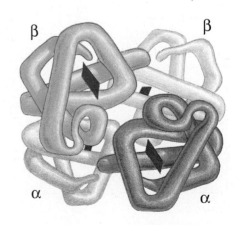

Figure 5–16 **A protein formed as a symmetrical assembly of two different subunits.** Hemoglobin is an abundant protein in red blood cells that contains two copies of α globin and two copies of β globin. Each of these four polypeptide chains contains a heme molecule (*red rectangle*) which is the site where oxygen (O_2) is bound. Thus, each molecule of hemoglobin in the blood carries four molecules of oxygen.

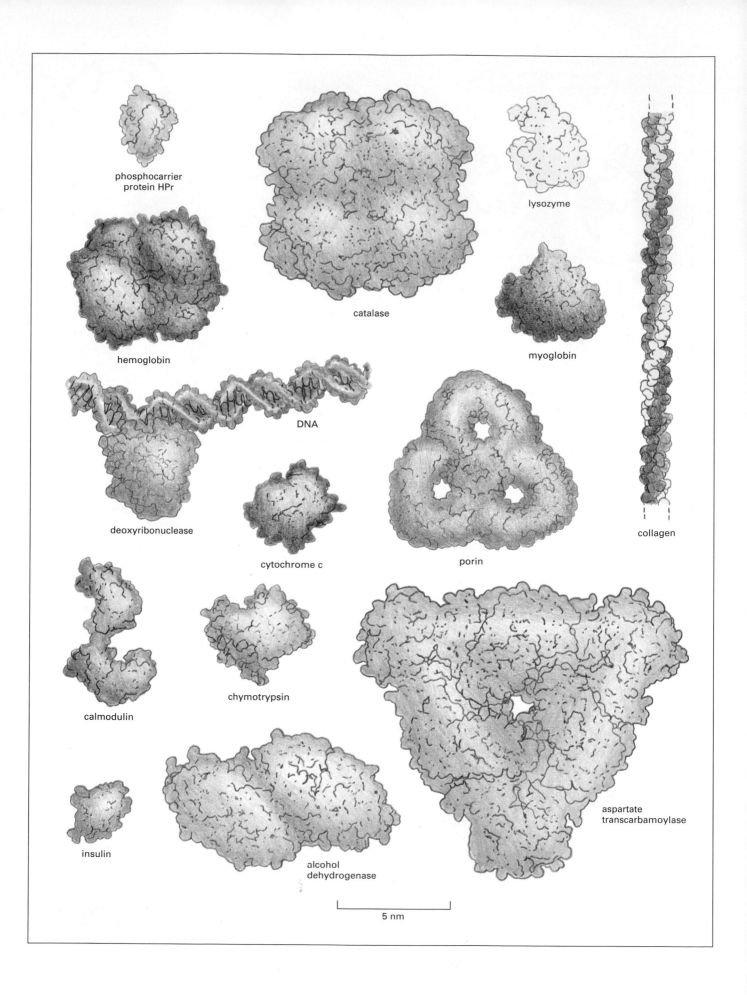

phosphocarrier
protein HPr

catalase

lysozyme

hemoglobin

myoglobin

DNA

deoxyribonuclease

cytochrome c

porin

collagen

calmodulin

chymotrypsin

insulin

alcohol
dehydrogenase

aspartate
transcarbamoylase

5 nm

Figure 5–17 (*opposite page*) **A collection of protein molecules, selected to show a range of sizes and shapes.** Each protein is shown as a space-filling model, represented at the same scale. In the top left corner is the phosphocarrier protein HPr, which was displayed in detail in Panel 5–2 (pp. 142–143). (After David S. Goodsell, Our Molecular Nature. New York: Springer-Verlag, 1996.)

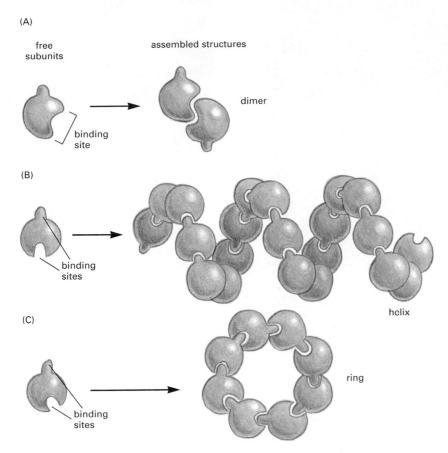

(A)

free subunits

assembled structures

binding site

dimer

(B)

binding sites

helix

(C)

binding sites

ring

Figure 5–18 **Protein assemblies.** (A) A protein with just one binding site can form a dimer with another identical protein. (B) Identical proteins with two different binding sites will often form a long helical filament. (C) If the two binding sites are disposed appropriately in relation to each other, the protein subunits will form a closed ring instead of a helix (see Figure 5–15B).

from many molecules of the protein actin (Figure 5–19). Actin is abundant in eucaryotic cells, where it forms one of the major filament systems of the cytoskeleton (Chapter 16). Other sets of proteins associate to form either extended sheets or tubes, as in the microtubules of the cell cytoskeleton, or cagelike spherical shells, as in the protein coats of virus particles.

Many large structures, such as viruses and ribosomes, are built from a mixture of one or more types of protein plus RNA or DNA molecules. All these structures can be isolated in pure form and dissociated into their constituent macromolecules. It is often possible to mix the isolated components back together and watch them reassemble spontaneously into the original structure. This demonstrates that all the information needed for assembly of the complicated structure is contained in the macromolecules themselves. Experiments of this type show that much of the structure of a cell is self-organizing: if the required proteins are produced, the appropriate structures will form.

actin helix

Figure 5–19 **An actin filament.** The helical array of protein molecules often extends for thousands of molecules, extending for micrometers in the cell.

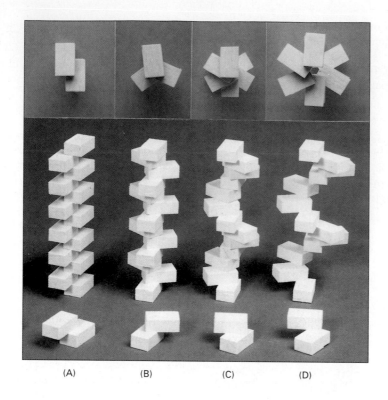

(A)   (B)   (C)   (D)

(E)

Figure 5–20 Some properties of a helix.
(A–D) A helix will form when a series of subunits bind to each other in a regular way. At the bottom, the interaction between two subunits is shown; behind them are the helices that result. These helices have two (A), three (B), and six (C and D) subunits per helical turn. At the top, the arrangement of subunits has been photographed from directly above the helix. Note that the helix in (D) has a wider path than that in (C), but the same number of subunits per turn. (E) A helix can be either right-handed or left-handed. As a reference, it is useful to remember that standard metal screws, which insert when turned clockwise, are right-handed. Note that a helix preserves the same handedness when it is turned upside down.

A Helix Is a Common Structural Motif in Biological Structures

As we have seen, biological structures are often formed by linking subunits that are very similar to each other—such as amino acids or protein molecules—into long, repetitive chains. If all the subunits are identical, the neighboring subunits in the chain will often fit together in only one way, adjusting their relative positions so as to minimize the free energy of the contact between them. In this case, each subunit will be positioned in exactly the same way in relation to its neighboring subunits, so that subunit 3 will fit onto subunit 2 in the same way that subunit 2 fits onto subunit 1, and so on. Because it is very rare for subunits to join up in a straight line, this arrangement will generally result in a **helix**—a regular structure that resembles a spiral staircase, as illustrated in Figure 5–20. Depending on the twist of the staircase, a helix is said to be either right-handed or left-handed (Figure 5–20E). Handedness is not affected by turning the helix upside down, but it is reversed if the helix is reflected in the mirror.

Helices occur commonly in biological structures, whether the subunits are small molecules that are covalently linked together (for example, amino acids in an α helix) or large protein molecules that are linked by noncovalent forces (for example, actin molecules in actin filaments). This is not surprising. A helix is an unexceptional structure, generated simply by placing many similar subunits next to each other, each in the same strictly repeated relationship to the one before.

Some Types of Proteins Have Elongated Fibrous Shapes

Most of the proteins we have discussed so far are **globular proteins,** in which the polypeptide chain folds up into a compact shape like a ball with an irregular surface. Most enzymes are globular proteins: even though many are large and complicated, with multiple subunits, most have an overall rounded shape (see Figure 5–17). In contrast, other proteins have roles in the cell which require that each protein molecule span a large distance. These proteins generally have a relatively simple, elongated three-dimensional structure and are commonly referred to as **fibrous proteins.**

One large class of intracellular fibrous proteins resembles α-keratin, which we met earlier. An α-keratin molecule is a dimer of two identical subunits, with the long α helices of each subunit forming a coiled-coil (see Figure 5–11). This class of proteins forms one component of the cell cytoskeleton. Their coiled-coil regions are capped at either end by globular domains containing binding sites. This allows the molecules in this class to assemble into ropelike *intermediate filaments* that create a structural scaffold for the cell's interior (see Chapter 16). Keratin filaments are extremely stable: long-lived structures such as hair, horn, and nails are composed mainly of this protein.

Fibrous proteins are especially abundant outside the cell, where they form the gel-like *extracellular matrix* that helps collections of cells to bind together to form tissues. These proteins are secreted by the cells into their surroundings, where they often assemble into sheets or long fibrils. *Collagen* is the most abundant of these fibrous proteins in animal tissues. The collagen molecule consists of three long polypeptide chains, each containing the nonpolar amino acid glycine at every third position. This regular structure allows the chains to wind around one another to generate a long regular triple helix (Figure 5–21A). Many collagen molecules then bind to one another side-by-side and end-to-end to create long overlapping arrays—thereby generating the extremely strong collagen fibrils that hold tissues together, as described in Chapter 19.

In complete contrast to collagen is another protein in the extracellular matrix, *elastin*. Elastin molecules are formed from relatively loose and unstructured polypeptide chains that are covalently cross-linked into a rubberlike elastic meshwork. The resulting elastic fibers enable skin and other tissues, such as arteries and lungs, to stretch and recoil

Figure 5–21 Collagen and elastin. (A) Collagen is a triple helix formed by three extended protein chains that wrap around one another. Many rodlike collagen molecules are cross-linked together in the extracellular space to form unextendable collagen fibrils (*top*) that have the tensile strength of steel. The striping on the collagen fibril is caused by the regular repeating arrangement of the collagen molecules within the fibril. (B) Elastin polypeptide chains are cross-linked together to form rubberlike, elastic fibers. Each elastin molecule uncoils into a more extended conformation when the fiber is stretched and will recoil spontaneously as soon as the stretching force is relaxed.

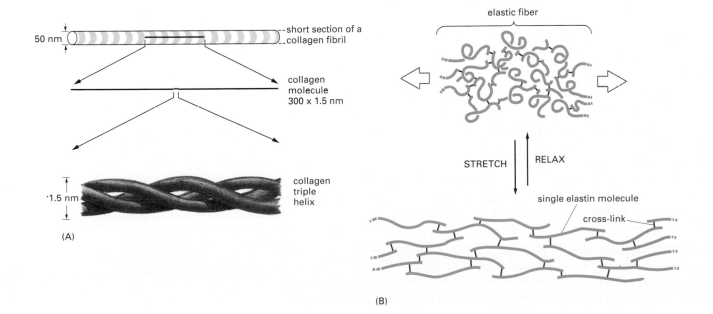

The Shape and Structure of Proteins

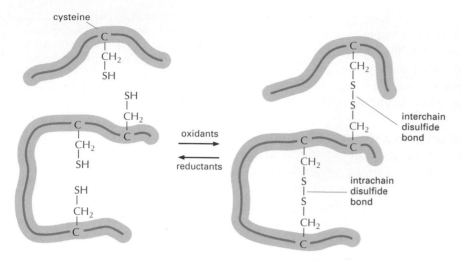

cysteine

oxidants →
← reductants

interchain disulfide bond

intrachain disulfide bond

Figure 5–22 **Disulfide bonds.** This diagram illustrates how covalent disulfide bonds form between adjacent cysteine side chains. As indicated, these cross-linkages can join either two parts of the same polypeptide chain or two different polypeptide chains. Since the energy required to break one covalent bond is much larger than the energy required to break even a whole set of noncovalent bonds (see Table 2–2, p. 43), a disulfide bond can have a major stabilizing effect on a protein.

without tearing. As illustrated in Figure 5–21B, the elasticity is due to the ability of the individual protein molecules to uncoil reversibly whenever they are stretched.

Extracellular Proteins Are Often Stabilized by Covalent Cross-Linkages

Many protein molecules are either attached to the outside of a cell's plasma membrane or secreted as part of the extracellular matrix. All such proteins are directly exposed to extracellular conditions. To help maintain their structures, the polypeptide chains in such proteins are often stabilized by covalent cross-linkages. These linkages can either tie two amino acids in the same protein together, or connect different polypeptide chains in a multisubunit protein. The most common cross-links in proteins are covalent sulfur–sulfur bonds. These **disulfide bonds** (also called *S–S bonds*) form as proteins are being exported from cells. Their formation is catalyzed in the endoplasmic reticulum by a special enzyme that links together two –SH groups of cysteine side chains that are adjacent in the folded protein (Figure 5–22). Disulfide bonds do not change the conformation of a protein, but instead act as atomic staples to reinforce its most favored conformation. For example, lysozyme—an enzyme in tears that dissolves bacterial cell walls—retains its antibacterial activity for a long time because it is stabilized by such cross-linkages.

   Disulfide bonds generally fail to form in the cell cytosol, where a high concentration of reducing agents converts such bonds back to cysteine –SH groups. Apparently, proteins do not require this type of reinforcement in the relatively mild environment inside the cell.

Question 5–4   Hair is composed largely of fibers of the protein keratin. Individual keratin fibers are covalently attached to one another ("cross-linked") by many disulfide bonds. If curly hair is treated with mild reducing agents that break a few of the cross-links, pulled straight and then oxidized again, it remains straight. Draw a diagram that illustrates the different stages of this chemical and mechanical process at the molecular level, focusing on the disulfide bonds. What do you think would happen if hair were treated with strong reducing agents that break all disulfide bonds?

How Proteins Work

We have seen that each type of protein consists of a precise sequence of amino acids, which allows it to fold up into a particular three-dimensional shape, or conformation. Because of their different amino acid sequences, proteins come in an enormous variety of different conformations—each with a unique surface topography of chemical groups. It is the conformation of a protein that endows it with a unique function (see Panel 5–1, p. 135). We now have to address the fundamental ques-

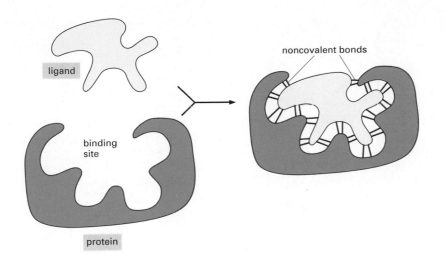

Figure 5–23 **The binding of a protein to another molecule is highly selective.** Many weak bonds are needed to enable a protein to bind tightly to a second molecule (a *ligand*). The ligand must therefore fit precisely into the protein's binding site, like a hand into a glove, so that a large number of noncovalent bonds can be formed between the protein and the ligand.

tion of how proteins accomplish their functions. In this part of the chapter, we shall explain how proteins bind to other selected molecules and how their activity depends on such binding. We shall see that the ability to bind other molecules allows proteins to act as catalysts, signal receptors, and tiny motors. The examples to be given here by no means exhaust the vast functional repertoire of proteins. However, the specialized functions of many of the proteins you will encounter elsewhere in this book are based on similar principles.

Proteins Bind to Other Molecules

The biological properties of a protein molecule depend on its physical interaction with other molecules. Thus, antibodies attach to viruses or bacteria as a signal to the body's defenses, the enzyme hexokinase binds glucose and ATP before catalyzing a reaction between them, actin molecules bind to each other to assemble into actin filaments, and so on. Indeed, all proteins stick, or *bind*, to other molecules. In some cases this binding is very tight; in others it is weak and short-lived. But the binding always shows great *specificity*, in the sense that each protein molecule can bind just one or a few molecules out of the many thousands of different molecules it encounters. Whether the substance that is bound by the protein is an ion, a small molecule, or a macromolecule, it is referred to as a **ligand** for that protein (from the Latin *ligare*, "to bind").

The ability of a protein to bind selectively and with high affinity to a ligand is due to the formation of a set of weak, noncovalent bonds—hydrogen bonds, ionic bonds, and van der Waals attractions—plus favorable hydrophobic interactions (see Panel 2–7, pp. 70–71). Each individual bond is weak, so that an effective interaction requires that many weak bonds be formed simultaneously. This is possible only if the surface contours of the ligand molecule fit very closely to the protein, matching it like a hand in a glove (Figure 5–23).

The region of a protein that associates with a ligand, known as its **binding site,** usually consists of a cavity in the protein surface formed by a particular arrangement of amino acids. These amino acids belong to widely separated regions of the polypeptide chain that are brought together when the protein folds (Figure 5–24). Other regions of the surface often provide binding sites for different ligands that regulate the protein's activity, as we shall see later. Yet other parts of the protein may

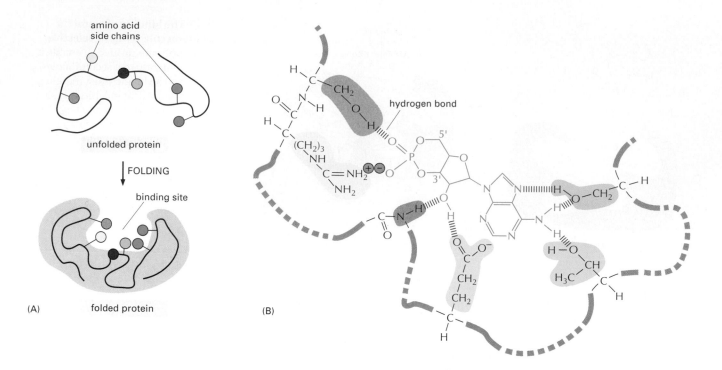

be required to place the protein in a particular location in the cell—for example, the hydrophobic α helix of a membrane-spanning protein that allows it to be inserted into the lipid bilayer of a cell membrane (see Chapter 11).

Although the atoms buried in the interior of the protein have no direct contact with the ligand, they provide an essential scaffold that gives the surface its contours and chemical properties. Even small changes to the amino acids in the interior of a protein molecule often change its three-dimensional shape and destroy the protein's ability to function.

We saw in Chapter 3 that molecules in the cell encounter each other rapidly because of their continual random movements due to thermal energy. When colliding molecules have poorly matching surfaces, few noncovalent bonds are formed and the two molecules dissociate as rapidly as they come together. This is what prevents incorrect and unwanted associations from forming between mismatched molecules. At the other extreme, when many noncovalent bonds are formed, the association can persist for a very long time (see Figure 2–28). Strong interactions occur in cells whenever a biological function requires that molecules remain tightly associated for a long time—for example, when a group of macromolecules come together to form a subcellular structure such as a ribosome.

The Binding Sites of Antibodies Are Especially Versatile

All proteins must bind to particular ligands to carry out their various functions. But this binding capacity seems to have been most highly developed for proteins in the antibody family, since our bodies have the capacity to produce an antibody molecule that will bind tightly to almost any other molecule.

Antibodies, or immunoglobulins, are proteins produced by the immune system in response to foreign molecules, such as those on the

Figure 5–24 The binding site of a protein. (A) The folding of the polypeptide chain typically creates a crevice or cavity on the protein surface. This crevice contains a set of amino acid side chains disposed in such a way that they can make noncovalent bonds only with certain ligands. (B) Close-up view of an actual binding site showing the hydrogen bonds and ionic interactions formed between a protein and its ligand (in this example, cyclic AMP is the bound ligand).

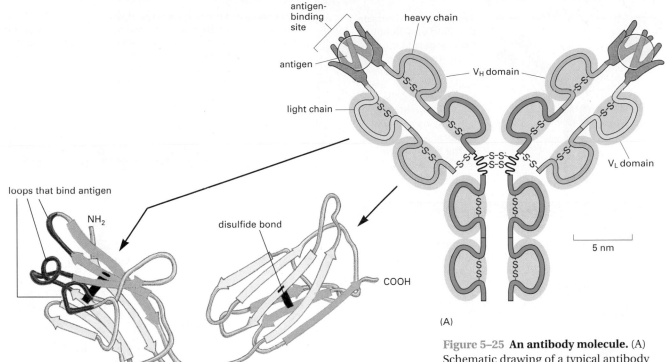

(A)

(B)

loops that bind antigen

NH2

disulfide bond

COOH

variable domain
of light chain (V_L)

constant domain
of light chain

antigen-
binding
site

heavy chain

antigen

V_H domain

light chain

V_L domain

5 nm

Figure 5–25 An antibody molecule. (A) Schematic drawing of a typical antibody molecule. As indicated, this protein is Y-shaped and has two identical binding sites for its antigen, one on either arm of the "Y." The protein is composed of four polypeptide chains (two identical heavy chains and two identical and smaller light chains) held together by disulfide bonds. Each chain is made up of several different domains, here shaded either *blue* or *gray*. The antigen-binding site is formed where a heavy chain variable domain (V_H) and a light chain variable domain (V_L) come close together. These are the domains that differ most in their sequence and structure in different antibodies. (B) Ribbon drawing of a light chain showing the parts of the V_L domain most closely involved in binding to the antigen in *red;* these contribute half of the fingerlike loops that fold around each of the antigen molecules in (A).

surface of an invading microorganism. Each antibody binds to a particular target molecule extremely tightly and thereby either inactivates the target directly or marks it for destruction. An antibody recognizes its target (called an **antigen**) with remarkable specificity, and because there are potentially billions of different antigens that a person might encounter, we have to be able to produce billions of different antibodies.

Antibodies are Y-shaped molecules with two identical binding sites that are each complementary to a small portion of the surface of the antigen molecule. A detailed examination of the antigen-binding sites of antibodies reveals that they are formed from several loops of polypeptide chain that protrude from the ends of a pair of closely juxtaposed protein domains (Figure 5–25). The amino acid sequence in these loops can be changed by mutation without altering the basic structure of the antibody. An enormous diversity of antigen-binding sites can be generated by changing only the length and amino acid sequence of the loops, and this is how the wide variety of different antibodies are formed.

Loops of this kind are ideal for grasping other molecules. They allow a large number of chemical groups to surround a ligand so that the protein can link to it with many weak bonds. For this reason, they are often used to form the ligand-binding sites in proteins. The properties of antibodies are summarized in Panel 5–3 (pp. 158–159).

Binding Strength Is Measured by the Equilibrium Constant

Different antibodies bind to their ligands with different strengths, and these strengths can be measured directly. For example, imagine a

THE ANTIBODY MOLECULE

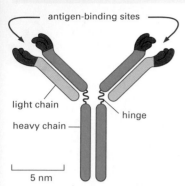

antigen-binding sites

light chain

heavy chain

hinge

5 nm

Antibodies are proteins that bind very tightly to their targets (antigens). They are produced in vertebrates as a defense against infection. Each antibody molecule is made of two identical light chains and two identical heavy chains, so the two antigen-binding sites are identical.

ANTIBODY SPECIFICITY

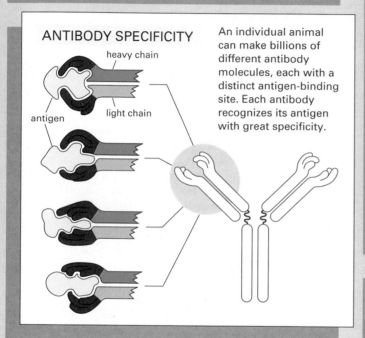

heavy chain

antigen

light chain

An individual animal can make billions of different antibody molecules, each with a distinct antigen-binding site. Each antibody recognizes its antigen with great specificity.

ANTIBODIES DEFEND US AGAINST INFECTION

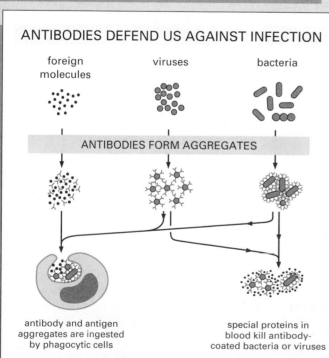

foreign molecules

viruses

bacteria

ANTIBODIES FORM AGGREGATES

antibody and antigen aggregates are ingested by phagocytic cells

special proteins in blood kill antibody-coated bacteria or viruses

B CELLS

Antibodies are made by a class of white blood cells, called B lymphocytes, or B cells. Each resting B cell carries a different membrane-bound antibody molecule on its surface that serves as a receptor for recognizing a specific antigen. When antigen binds to this receptor, the B cell is stimulated to divide and to secrete large amounts of the same antibody in a soluble form.

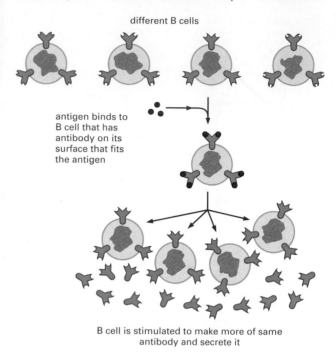

different B cells

antigen binds to B cell that has antibody on its surface that fits the antigen

B cell is stimulated to make more of same antibody and secrete it

RAISING ANTIBODIES IN ANIMALS

Antibodies can be made in the laboratory by injecting an animal (usually a mouse, rabbit, sheep, or goat) with antigen A.

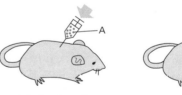

inject antigen A

take blood later

Repeated injections of the same antigen at intervals of several weeks stimulates specific B cells to secrete large amounts of anti-A antibodies into the bloodstream.

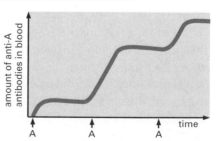

amount of anti-A antibodies in blood

A   A   A   time

Because many different B cells are stimulated by antigen A, the blood will contain a variety of anti-A antibodies, each of which binds A in a slightly different way.

ANTIBODIES USED TO PURIFY MOLECULES

IMMUNOPRECIPITATION

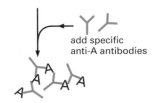

mixture of molecules

add specific
anti-A antibodies

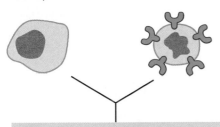

aggregate of A molecules plus
anti-A antibodies can be collected
by centrifugation

IMMUNOAFFINITY COLUMN CHROMATOGRAPHY

bead coated with
anti-A antibodies

column packed
with these beads

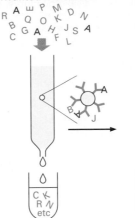

antigen A plus
unwanted molecules

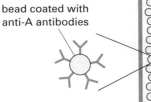

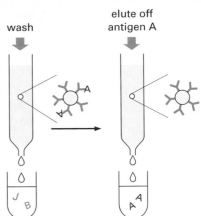

wash

elute off
antigen A

discard flow through   wash   pure antigen A

MONOCLONAL ANTIBODIES

Large quantities of a single type of antibody
molecule can be obtained by fusing a B cell
(taken from an animal injected with antigen A)
with a cell of a B cell tumor. The resulting
hybrid cell divides indefinitely and secretes
anti-A antibodies of a single type.

tumor cell from
cell culture divides
indefinitely but
does not make
antibody

B cell from animal
injected with antigen
A makes anti-A
antibody but does
not divide forever

**FUSE ANTIBODY-SECRETING
B CELL WITH TUMOR CELL**

hybrid cell
makes anti-A
antibody and
divides
indefinitely

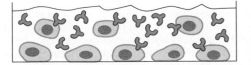

USING ANTIBODIES AS MOLECULAR TAGS

specific antibodies
against antigen A

couple to fluorescent dye,
colloidal gold particle, or
other special tag

labeled antibodies

MICROSCOPIC DETECTION

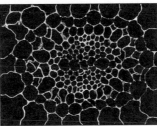

Fluorescent antibody binds to
antigen A in tissue and is
detected by fluorescence in a light
microscope. The antigen here is
pectin in plant cell walls.

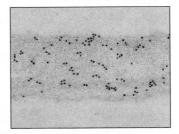

Gold-labeled antibody binds
to antigen A in tissue and is
detected in an electron microscope. The antigen is pectin
in a plant cell wall.

BIOCHEMICAL DETECTION

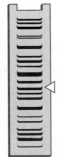

antigen A is
separated from
other molecules
by electrophoresis

incubation with the
labeled antibodies
that bind to antigen A
allows the position of the
antigen to be determined

labeled second antibody
(*blue*) binds to first
antibody (*green*)

antigen

Note: In all cases, the sensitivity can
be greatly increased by using multiple
layers of antibodies. This "sandwich"
method enables smaller numbers of
antigen molecules to be detected.

BREAKING CELLS AND TISSUES

The first step in the purification of most proteins is to disrupt tissues and cells in a controlled fashion.

Using gentle mechanical procedures, called homogenization, the plasma membranes of cells can be ruptured so that the cell contents are released. Four commonly used procedures are shown here.

The resulting thick soup (called a homogenate or an extract) contains large and small molecules from the cytosol, such as enzymes, ribosomes, and metabolites, as well as all the membrane-bounded organelles.

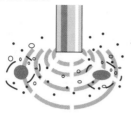

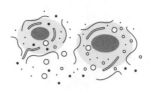

cell suspension or tissue

(1) break cells with high frequency sound

(2) use a mild detergent to make holes in the plasma membrane

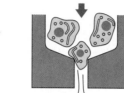

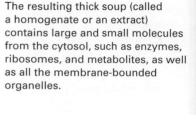

(3) force cells through a small hole using high pressure

(4) shear cells between a close-fitting rotating plunger and the thick walls of a glass vessel

When carefully applied, homogenization leaves most of the membrane-bounded organelles intact.

THE CENTRIFUGE

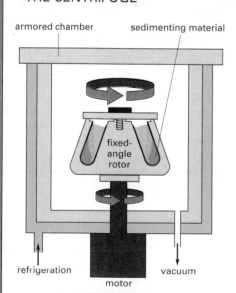

armored chamber

sedimenting material

fixed-angle rotor

refrigeration

motor

vacuum

swinging-arm rotor

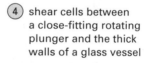

centrifugal force

tube

metal bucket

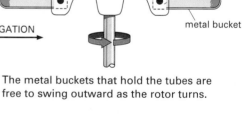

CENTRIFUGATION

Many cell fractionations are done in a second type of rotor, a swinging-arm rotor.

The metal buckets that hold the tubes are free to swing outward as the rotor turns.

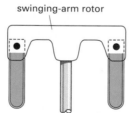

CELL HOMOGENATE before centrifugation

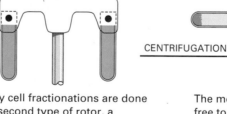

CENTRIFUGATION

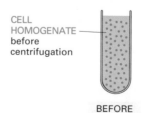

SUPERNATANT smaller and less dense components

PELLET larger and more dense components

BEFORE

AFTER

Centrifugation is the most widely used procedure to separate the homogenate into different parts, or fractions. The homogenate is placed in test tubes and rotated at high speed in a centrifuge (sometimes called an ultracentrifuge). Present-day ultracentrifuges rotate at speeds up to 100,000 revolutions per minute and produce enormous forces, as high as 600,000 times gravity.

At such speeds, centrifuge chambers must be refrigerated and evacuated so that friction does not heat up the homogenate. The centrifuge is surrounded by thick armor plating, since an unbalanced rotor can shatter with an explosive release of energy. A fixed-angle rotor can hold larger volumes than a swinging-arm rotor, but the pellet forms less evenly.

DIFFERENTIAL CENTRIFUGATION

Repeated centrifugation at progressively higher speeds will fractionate cell homogenates into their components.

Centrifugation separates cell components on the basic of size and density. The larger and denser components experience the greatest centrifugal force and move most rapidly. They sediment to form a pellet at the bottom of the tube, while smaller, less dense components remain in suspension above, called the supernatant.

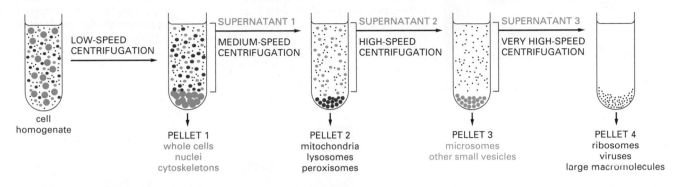

cell homogenate

LOW-SPEED CENTRIFUGATION

SUPERNATANT 1

MEDIUM-SPEED CENTRIFUGATION

SUPERNATANT 2

HIGH-SPEED CENTRIFUGATION

SUPERNATANT 3

VERY HIGH-SPEED CENTRIFUGATION

PELLET 1
whole cells
nuclei
cytoskeletons

PELLET 2
mitochondria
lysosomes
peroxisomes

PELLET 3
microsomes
other small vesicles

PELLET 4
ribosomes
viruses
large macromolecules

VELOCITY SEDIMENTATION

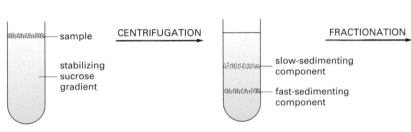

sample

stabilizing sucrose gradient

CENTRIFUGATION

slow-sedimenting component

fast-sedimenting component

FRACTIONATION

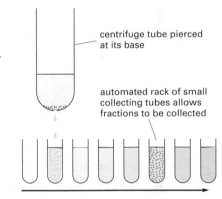

centrifuge tube pierced at its base

automated rack of small collecting tubes allows fractions to be collected

Subcellular components sediment at different speeds according to their size when carefully layered over a dilute salt solution. In order to stabilize the sedimenting components against convective mixing in the tube, the solution contains a continuous shallow gradient of sucrose that increases in concentration toward the bottom of the tube. This is typically 5–20% sucrose. When sedimented through such a dilute sucrose gradient, different cell components separate into distinct bands that can, after an appropriate time, be collected individually.

After an appropriate centrifugation time the bands may be collected, most simply by puncturing the plastic centrifuge tube and collecting drops from the bottom, as shown here.

EQUILIBRIUM SEDIMENTATION

The ultracentrifuge can also be used to separate cellular components on the basis of their buoyant density, independently of their size or shape. The sample is usually either layered on top of, or dispersed within, a steep density gradient that contains a very high concentration of sucrose or cesium chloride. Each subcellular component will move up or down when centrifuged until it reaches a position where its density matches its surroundings and then will move no further. A series of distinct bands will eventually be produced, with those nearest the bottom of the tube containing the components of highest buoyant density. The method is also called density gradient centrifugation.

the sample is distributed throughout the sucrose density gradient

at equilibrium, components have migrated to a region in the gradient that matches their own density

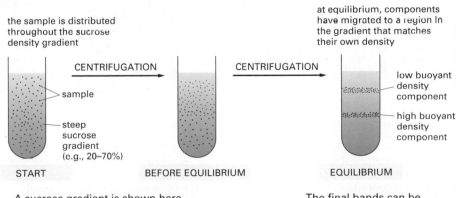

sample

steep sucrose gradient (e.g., 20–70%)

START

CENTRIFUGATION

BEFORE EQUILIBRIUM

CENTRIFUGATION

low buoyant density component

high buoyant density component

EQUILIBRIUM

A sucrose gradient is shown here, but denser gradients can be formed with cesium chloride that are particularly useful for separating the nucleic acids (DNA and RNA).

The final bands can be collected from the base of the tube, as shown above.

PROTEIN SEPARATION

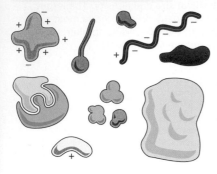

Proteins are very diverse. They differ by size, shape, charge, hydrophobicity, and their affinity for other molecules. All these properties can be exploited to separate them from one another so that they can be studied individually.

THREE KINDS OF CHROMATOGRAPHY

Many types of matrix are available for column chromatography, usually packed in the column in the form of small beads. A typical protein purification strategy might employ in turn each of the three kinds of matrix described below, with a final protein purification of up to 10,000-fold.

Purity can easily be assessed by gel electrophoresis (see *opposite page*).

COLUMN CHROMATOGRAPHY

Proteins are often fractionated by column chromatography. A mixture of proteins in solution is applied to the top of a cylindrical column filled with a permeable solid matrix immersed in solvent. A large amount of solvent is then pumped through the column. Because different proteins are retarded to different extents by their interaction with the matrix, they can be collected separately as they flow out from the bottom. According to the choice of matrix, proteins can be separated according to their charge, their hydrophobicity, their size, or their ability to bind to particular chemical groups (see *below*).

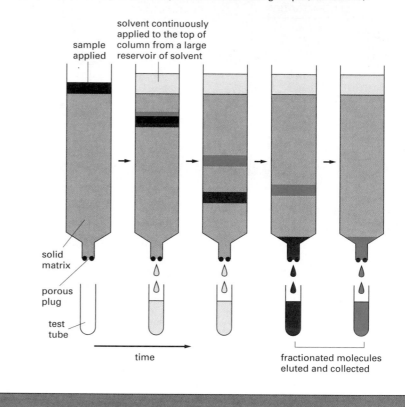

sample applied

solvent continuously applied to the top of column from a large reservoir of solvent

solid matrix

porous plug

test tube

time

fractionated molecules eluted and collected

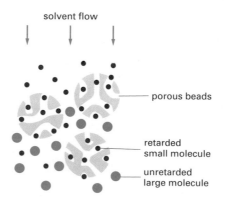

solvent flow

positively charged bead

bound negatively charged molecule

free positively charged molecule

(A) ION-EXCHANGE CHROMATOGRAPHY

Ion-exchange columns are packed with small beads that carry positive or negative charges that retard proteins of the opposite charge. The association between a protein and the matrix depends on the pH and ionic strength of the solution passing down the column. These can be varied in a controlled way to achieve an effective separation.

solvent flow

porous beads

retarded small molecule

unretarded large molecule

(B) GEL-FILTRATION CHROMATOGRAPHY

Gel-filtration columns separate proteins according to their size. The matrix consists of tiny porous beads. Protein molecules that are small enough to enter the holes in the beads are delayed and travel more slowly through the column. Proteins that cannot enter the beads are washed out of the column first. Such columns also allow an estimate of protein size.

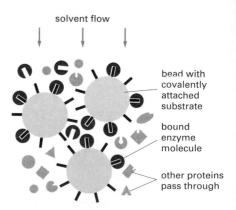

solvent flow

bead with covalently attached substrate

bound enzyme molecule

other proteins pass through

(C) AFFINITY CHROMATOGRAPHY

Affinity columns contain a matrix covalently coupled to a molecule that interacts specifically with the protein of interest (e.g., an antibody, or an enzyme substrate). Proteins that bind specifically to such a column can finally be released by a pH change or by concentrated salt solutions, and they emerge highly purified.

GEL ELECTROPHORESIS

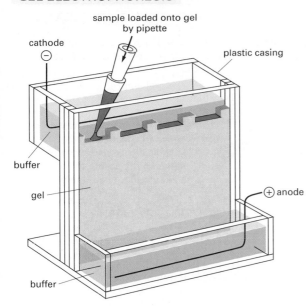

cathode ⊖

sample loaded onto gel by pipette

plastic casing

buffer

gel

⊕ anode

buffer

When an electric field is applied to a solution containing protein molecules, the molecules will migrate in a direction and at a speed that reflects their size and net charge. This forms the basis of the technique called electrophoresis.

the detergent sodium dodecyl sulfate (SDS) is used to solubilize proteins for SDS polyacrylamide-gel electrophoresis (see *below*)

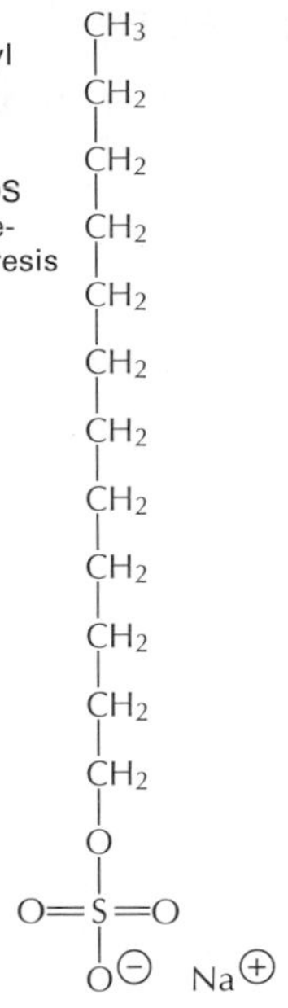

CH₃
CH₂
CH₂
CH₂
CH₂
CH₂
CH₂
CH₂
CH₂
CH₂
CH₂
CH₂
O
O=S=O
O⊖   Na⊕

SDS

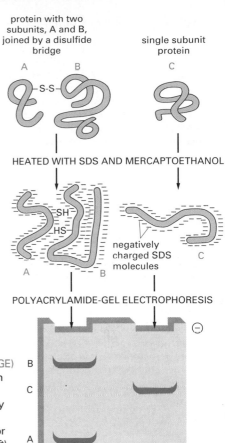

protein with two subunits, A and B, joined by a disulfide bridge

A   B

single subunit protein

C

HEATED WITH SDS AND MERCAPTOETHANOL

SH
HS

A   B

negatively charged SDS molecules

C

POLYACRYLAMIDE-GEL ELECTROPHORESIS

⊖

B
C
A

⊕

slab of polyacrylamide gel

SDS polyacrylamide-gel electrophoresis (SDS-PAGE) Individual polypeptide chains form a complex with negatively charged molecules of sodium dodecyl sulfate (SDS) and therefore migrate as a negatively charged SDS-protein complex through a slab of porous polyacrylamide gel. The apparatus used for this electrophoresis technique is shown above (*left*). A reducing agent (mercaptoethanol) is usually added to break any –S–S– linkages in or between proteins. Under these conditions, proteins migrate at a rate that reflects their molecular weight.

ISOELECTRIC FOCUSING

For any protein there is a characteristic pH, called the isoelectric point, at which the protein has no net charge and therefore will not move in an electric field. In isoelectric focusing, proteins are electrophoresed in a narrow tube of polyacrylamide gel in which a pH gradient is established by a mixture of special buffers. Each protein moves to a point in the gradient that corresponds to its isoelectric point and stays there.

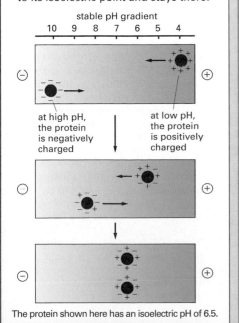

stable pH gradient

10 9 8 7 6 5 4

⊖   ⊕

at high pH, the protein is negatively charged

at low pH, the protein is positively charged

⊖   ⊕

⊖   ⊕

The protein shown here has an isoelectric pH of 6.5.

TWO-DIMENSIONAL POLYACRYLAMIDE-GEL ELECTROPHORESIS

Complex mixtures of proteins cannot be resolved well on one-dimensional gels, but two-dimensional gel electrophoresis, combining two different separation methods, can be used to resolve more than 1000 proteins in a two-dimensional protein map. In the first step, native proteins are separated in a narrow gel on the basis of their intrinsic charge using isoelectric focusing (see *left*). In the second step, this gel is placed on top of a gel slab, and the proteins are subjected to SDS-PAGE (see *above*) in a direction perpendicular to that used in the first step. Each protein migrates to form a discrete spot.

All the proteins in an *E. coli* bacterial cell are separated in this 2-D gel, in which each spot corresponds to a different polypeptide chain. They are separated according to their isoelectric point from left to right and to their molecular weight from top to bottom. (Courtesy of Patrick O'Farrell.)

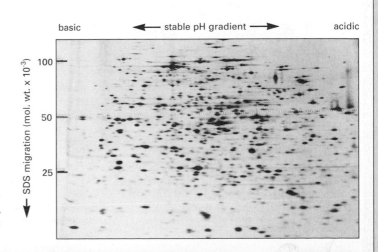

basic ← stable pH gradient → acidic

SDS migration (mol. wt. × 10⁻³)

100

50

25

THE AMINO ACID SEQUENCE OF PROTEINS

Before the 3-D structure of a protein can be determined, its amino acid sequence must be established. This is accomplished by repeating a series of chemical reactions (steps 1 and 2, *right*) that identify the amino-terminal amino acid after removing it. The reiterative nature of these reactions lends itself to automation, and machines called amino acid sequenators are commercially available for automatic determination of the amino acid sequence of peptides. Even with very small samples, for example, a single protein spot on a 2-D polyacrylamide gel (see Panel 5–5, pp. 162–163), the sequence of several dozen amino acids at the amino-terminal end can often be obtained overnight.

Peptides longer than about 50 amino acids, however, cannot be sequenced reliably. In addition, it is often found that the amino-terminal amino acid is chemically blocked, thus preventing the sequencing reactions from starting. To overcome these problems, the protein is first cleaved into smaller peptides.

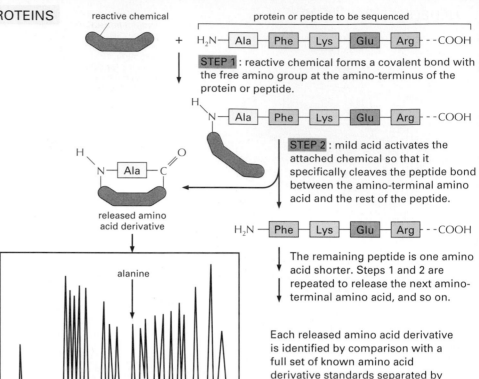

STEP 1: reactive chemical forms a covalent bond with the free amino group at the amino-terminus of the protein or peptide.

STEP 2: mild acid activates the attached chemical so that it specifically cleaves the peptide bond between the amino-terminal amino acid and the rest of the peptide.

The remaining peptide is one amino acid shorter. Steps 1 and 2 are repeated to release the next amino-terminal amino acid, and so on.

Each released amino acid derivative is identified by comparison with a full set of known amino acid derivative standards separated by column chromatography (see Panel 5–5). In this case, the amino-terminal amino acid is identified as alanine.

SELECTIVE PROTEIN CLEAVAGE

small protein subjected to selective cleavage

Ala — Phe — Lys — Glu — Arg — Gly — Met — Val — Ser — Asp

cleavage with trypsin   cleavage with cyanogen bromide

peptide   peptide

a. Ala — Phe — Lys   Val — Ser — Asp   d.

+   +

b. Glu — Arg   Ala — Phe — Lys — Glu — Arg — Gly — Met   e.

+

c. Gly — Met — Val — Ser — Asp

separation of resultant peptides by column chromatography

peptide b   peptide c
peptide a

peptide e
peptide d

The selective cleavage of a protein generates a distinctive set of peptide fragments. Proteolytic enzymes and chemical reagents are available that will cleave proteins between specific amino acid residues. The enzyme trypsin, for example, cuts on the carboxyl side of lysine or arginine residues, whereas the chemical cyanogen bromide cuts peptide bonds next to methionine. These reagents tend to produce a few relatively large peptides. These can be separated to create a peptide map that is diagnostic of the protein from which the peptides were generated. The sequences of overlapping peptides generated by different reagents can be used to laboriously piece together the sequence of proteins.

Knowing the sequence of as few as 20 amino acids of a protein is frequently enough to allow a DNA probe to be designed, so that the gene encoding the protein can be cloned (see Figure 10–17). Once the gene, or the corresponding cDNA, has been sequenced, the rest of the protein's amino acid sequence can be deduced by reference to the genetic code. Most proteins are sequenced in this indirect way.

X-RAY CRYSTALLOGRAPHY

A major technique that has been used to discover the three-dimensional structure of protein molecules, at atomic resolution, is x-ray crystallography.

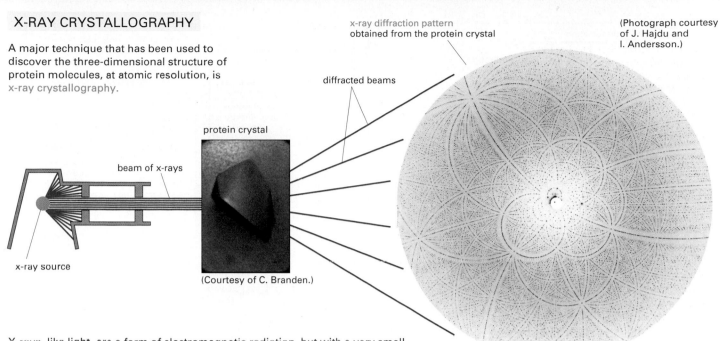

protein crystal

beam of x-rays

x-ray source

(Courtesy of C. Branden.)

diffracted beams

x-ray diffraction pattern obtained from the protein crystal

(Photograph courtesy of J. Hajdu and I. Andersson.)

X-rays, like light, are a form of electromagnetic radiation, but with a very small wavelength. If a narrow parallel beam of x-rays is directed at a well-ordered crystal of a pure protein, some of the beam will be scattered by the atoms in the crystal. The scattered waves will reinforce one another at certain points and will appear as a pattern of diffraction spots when the x-rays are recorded by a suitable detector.

The position and intensity of each spot in the x-ray diffraction pattern contains information about the position of the atoms in the protein crystal that gave rise to it. Computers can use this information to provide a three-dimensional electron density map of the protein molecule, which, together with the sequence of the protein, can be used to produce an atomic model. The complete atomic model is often hard to interpret, so simplified versions are derived that show the essential structural features (see Panel 5–2, pp. 142–143). The protein shown here is ribulose bisphosphate carboxylase, an enzyme that plays a central role in CO_2 fixation during photosynthesis (alpha helices are shown in *green,* and beta strands in *red*).

NMR SPECTROSCOPY

Nuclear magnetic resonance (NMR) spectroscopy has been used in the past to analyze the structure of small molecules and now is increasingly used to study the structure of small proteins or protein domains. The technique requires only a small volume of concentrated pure protein solution.

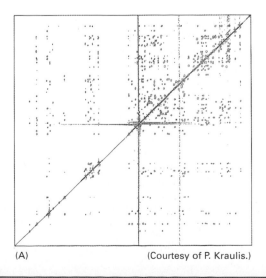

(A)   (Courtesy of P. Kraulis.)

(B)

The solution is placed in a strong magnetic field and subjected to radio frequency pulses of electromagnetic radiation. Signals from hydrogen nuclei in different amino acids can be identified that allow the distances between interacting pairs of hydrogen atoms to be measured. NMR gives information about the distances between the parts of a protein molecule, and by combining this with a knowledge of the amino acid sequence, it is possible to compute the 3-D structure of the protein. Only the structure of small proteins (20,000 daltons or less) can be determined by NMR spectroscopy.

In (A), a 2-D NMR spectrum derived from the carboxyl-terminal domain of the enzyme cellulase is shown. The spots represent interactions between neighboring H atoms. The resultant structures that satisfy the distance constraints equally well are shown in (B).

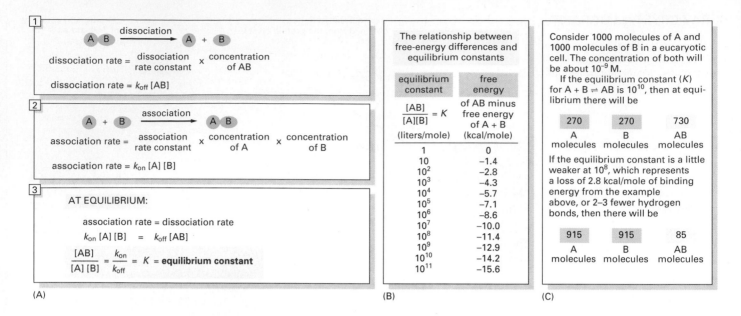

Figure 5–26 Relating binding energies to the equilibrium constant.
(A) The equilibrium between molecules A and B and the complex AB is maintained by a balance between the two opposing reactions shown in (1) and (2). Molecules A and B must collide in order to react, and the rate in reaction (2) is therefore proportional to the product of their individual concentrations [A] × [B], where the symbol "[ ]" indicates concentration. As shown in (3), the ratio of the rate constants for the association and the dissociation reactions is equal to the equilibrium constant (K) for the reaction. (B) The equilibrium constant in (3) is that for the reaction A + B ⇌ AB, and the larger its value, the stronger is the binding between A and B. For every 1.4 kcal/mole of free-energy difference, the equilibrium constant changes by a factor of 10 (see also Table 3–1, p. 92). (C) An example that illustrates the dramatic effect of the presence or absence of a few weak bonds in a biological context.

situation in which a population of identical antibody molecules suddenly encounters a population of ligands diffusing in the fluid surrounding them. At frequent intervals, one of the ligand molecules will bump into the binding site of an antibody and form an antibody–ligand complex. The population of antibody–ligand complexes will therefore increase, but not without limit: over time, a second process, in which individual complexes break apart because of thermally induced motion, will become increasingly important. Eventually, any population of antibody molecules and ligands will reach a steady state, or **equilibrium,** in which the number of binding (association) events per second is precisely equal to the number of "unbinding" (dissociation) events (see also Figure 3–20).

The concentrations of ligand, antibody, and antibody–ligand complex at equilibrium provide a convenient measure—termed the **equilibrium constant (K)**—of the strength of binding (Figure 5–26). The equilibrium constant becomes larger as the binding strength between the two molecules increases (due to an increased number of noncovalent bonds), reflecting an increase in the free-energy difference between the bound and free states (Figure 5–26B). Even a change of a few noncovalent bonds can have a striking effect on a binding interaction, as illustrated by the example in Figure 5–26C.

We have used the case of an antibody binding to its ligand to illustrate the effect of binding strength on the equilibrium state, but the same principles apply to any protein and its ligand. Many proteins are enzymes, which, as we shall now discuss, first bind to their ligands and then catalyze the breakage or formation of covalent bonds in these molecules.

Enzymes Are Powerful and Highly Specific Catalysts

For many proteins, binding to another molecule is all they do. An antibody molecule needs only to bind to its target bacterium or virus and its function is complete; an actin molecule need only associate with other actin molecules to form a filament. There are other proteins, however, for which ligand binding is simply a necessary first step in their function. This is the case for the large and very important class of proteins called **enzymes.** As described in Chapter 3, enzymes are remarkable molecules that determine all of the chemical transformations that occur in cells. They bind to one or more ligands, called **substrates,** and convert them into chemically modified *products,* doing this over and over again with amazing rapidity. Enzymes speed up reactions, often by a factor of a million or more, without themselves being changed—that is, they act as *catalysts* that permit cells to make or break covalent bonds at will. It is the catalysis of organized sets of chemical reactions by enzymes that creates and maintains the cell, making life possible.

Enzymes can be grouped into functional classes that carry out similar chemical reactions (Table 5–2). Each type of enzyme is highly specific, catalyzing only a single type of reaction. Thus, *hexokinase* adds a phosphate group to D-glucose but will ignore its optical isomer L-glucose; the blood-clotting enzyme *thrombin* cuts one type of blood protein between a particular arginine and its adjacent glycine and nowhere else, and so on. As discussed in detail in Chapters 3 and 4, enzymes work in teams, with the product of one enzyme becoming the substrate for the next. The result is an elaborate network of metabolic pathways that provides the cell with energy and generates the many large and small molecules that the cell needs (see Figure 3–2).

Lysozyme Illustrates How an Enzyme Works

To explain how enzymes catalyze chemical reactions, we shall use the example of an enzyme that acts as a natural antibiotic in egg white, saliva, tears, and other secretions. Lysozyme is an enzyme that catalyzes the cutting of polysaccharide chains in the cell walls of bacteria. Because the bacterial cell is under pressure due to osmotic forces, cutting even a small number of polysaccharide chains causes the cell wall to rupture and the cell to burst. Lysozyme is a relatively small and stable protein that can be isolated easily in large quantities. For these reasons, it has been intensively studied, and it was the first enzyme to have its structure worked out in atomic detail by x-ray crystallography (Panel 5–6, pp. 164–165).

The reaction catalyzed by lysozyme is a hydrolysis: a molecule of water is added to a single bond between two adjacent sugar groups in the polysaccharide chain, thereby causing the bond to break (see Figure 3–32). The reaction is energetically favorable because the free energy of

Question 5–5 Protein A binds to protein B to form a complex, AB. A cell contains an equilibrium mixture of protein A at a concentration of 1 μM, protein B at a concentration of 1 μM, and protein AB (produced when A binds to B) also at 1 μM.

A. Referring to Figure 5–26, calculate the equilibrium constant for the reaction A + B ⇌ AB.

B. What would the equilibrium constant be if A, B, and AB were each present in equilibrium at the much lower concentrations of 1 nM each?

C. How many extra hydrogen bonds would be needed to hold A to B at this lower concentration so that a similar portion of the molecules are found in the AB complex? (Remember that each hydrogen bond contributes about 1 kcal/mole.)

Question 5–6 Explain how an enzyme (such as hexokinase mentioned in the text) can distinguish substrates (here D-glucose) from their optical isomers (here L-glucose). (Hint: remembering that a carbon atom forms four single bonds that are tetrahedrally arranged and that the optical isomers are mirror images of each other around such a bond, draw the substrate as a simple tetrahedron with four different corners and then draw its mirror image. Using this drawing, indicate why only one compound might bind to a schematic active site of an enzyme.)

Table 5–2   Some Common Types of Enzymes

ENZYME	REACTION CATALYZED
HYDROLASES	General term for enzymes that catalyze a hydrolytic cleavage reaction.
NUCLEASES	Break down nucleic acids by hydrolyzing bonds between nucleotides.
PROTEASES	Break down proteins by hydrolyzing bonds between amino acids.
SYNTHASES	General name used for enzymes that synthesize molecules in anabolic reactions by condensing two smaller molecules together
ISOMERASES	Catalyze the rearrangement of bonds within a single molecule.
POLYMERASES	Catalyze polymerization reactions such as the synthesis of DNA and RNA.
KINASES	Catalyze the addition of phosphate groups to molecules. Protein kinases are an important group of kinases that attach phosphate groups to proteins.
PHOSPHATASES	Catalyze the hydrolytic removal of a phosphate group from a molecule.
OXIDO-REDUCTASES	General name for enzymes that catalyze reactions in which one molecule is oxidized while the other is reduced. Enzymes of this type are often called *oxidases, reductases,* and *dehydrogenases.*
ATPASES	Hydrolyze ATP. Many proteins with a wide range of roles have an energy-harnessing ATPase activity as part of their function, for example, motor proteins such as *myosin* and membrane transport proteins such as the *sodium-potassium pump.*

Enzyme names typically end in "-ase," with the exception of some enzymes, ~~such as~~ pepsin, trypsin, thrombin, lysozyme and so on, which were discovered and named before the convention became generally accepted at the end of the nineteenth century. The common name of an enzyme usually indicates the substrate and the nature of the reaction catalyzed. For example, citrate synthase catalyzes the synthesis of citrate by the addition of acetyl CoA to oxaloacetate.

the severed polysaccharide chain is lower than the free energy of the intact chain. However, the pure polysaccharide can sit for years in water without being hydrolyzed to any detectable degree. This is because there is an energy barrier to the reaction, as discussed in Chapter 3 (see Figure 3–12). In order for a colliding water molecule to break a bond linking two sugars, the polysaccharide molecule has to be distorted into a particular shape—the **transition state**—in which the atoms around the bond have an altered geometry and electron distribution. Because of this distortion, a large input of energy, called *activation energy,* must be supplied through random collisions before the reaction will take place. In aqueous solution at room temperature, the energy of collisions almost never exceeds the activation energy. Consequently, hydrolysis occurs extremely slowly, if at all.

This is where the enzyme comes in. Like all enzymes, lysozyme has a special binding site on its surface, termed an **active site,** that precisely fits the contours of its substrate molecule, and at which the catalysis of the chemical reaction occurs. The active site of lysozyme, since its substrate is a polymer, is a long groove that holds six linked sugars at the same time. As soon as the polysaccharide binds to form an enzyme–substrate complex, the enzyme cuts the polysaccharide by adding a water molecule across one of its sugar–sugar bonds. The product chains are then quickly released, freeing the enzyme for further cycles of reaction (Figure 5–27).

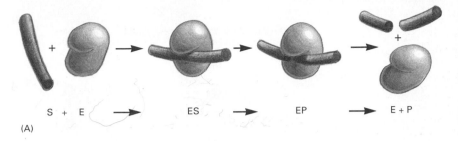

(A) S + E ⟶ ES ⟶ EP ⟶ E + P

(B)

Figure 5–27 **Outline of the reaction catalyzed by lysozyme.** (A) Schematic view of the enzyme lysozyme (denoted E), which catalyzes the cutting of a polysaccharide chain, which is its substrate (denoted S). The enzyme first binds to the chain to form an enzyme–substrate complex (ES) and then catalyzes the cleavage of a specific covalent bond in the backbone of the polysaccharide, forming an enzyme–product complex (EP) that rapidly dissociates. Release of the severed chain (the products P) leaves the enzyme free to act on another substrate molecule. (B) A space-filling model of the lysozyme molecule bound to a short length of polysaccharide chain prior to cleavage. (B, courtesy of Richard J. Feldmann.)

The chemistry that underlies the binding of lysozyme to its substrate is the same as that for antibody binding—the formation of multiple noncovalent bonds. However, lysozyme holds its polysaccharide substrate in a particular way, so that one of the two sugars involved in the bond to be broken is distorted from its normal, most stable conformation. The bond to be broken is also held close to two amino acids with acidic side chains: a glutamic acid and an aspartic acid within the active site.

Conditions are thereby created in the microenvironment of the lysozyme active site that greatly reduce the activation energy necessary for the hydrolysis to take place. Figure 5–28 shows the stages in this enzymatically catalyzed reaction. First, the enzyme stresses its bound substrate by bending some critical chemical bonds that will participate in the chemical reaction so that the shape of the substrate more closely resembles that of the high-energy transition state that forms halfway through the reaction. Second, a precisely positioned acidic side chain of the glutamic acid within the active site speeds up the hydrolysis by providing a high concentration of acidifying H+ ions, even though the solution surrounding the enzyme is at neutral pH. Third, the negatively charged aspartic acid further stabilizes the positively charged transition state. The overall chemical reaction, from the initial binding of the polysaccharide on the surface of the enzyme through the final release of the severed chains, therefore occurs many millions of times faster than it would in the absence of enzyme.

Similar mechanisms are used by other enzymes to lower the activation energy and speed up the reactions they catalyze. In reactions involving two or more reactants, the active site acts like a template or mold that brings the substrates together in the proper orientation for a reaction to occur between them (Figure 5–29A). As we saw for lysozyme, the active site of an enzyme contains precisely positioned atoms that speed up a reaction by using charged groups to alter the distribution of electrons in the substrates (Figure 5–29B). As we likewise saw, the binding to the enzyme will also change substrate shapes, bending bonds so as to drive a substrate toward a particular transition state (Figure 5–29C). Finally, many enzymes participate intimately in the reaction by briefly forming a covalent bond between the substrate and a side chain of the enzyme. Subsequent steps in the reaction restore the side chain to its original state, so that the enzyme remains unchanged after the reaction (see Figure 4–5).

V_{max} and K_M Measure Enzyme Performance

To catalyze a reaction, an enzyme must first bind its substrates. The substrates then undergo a reaction to form product molecules, which initially remain bound to the enzyme. Finally, the product is released and

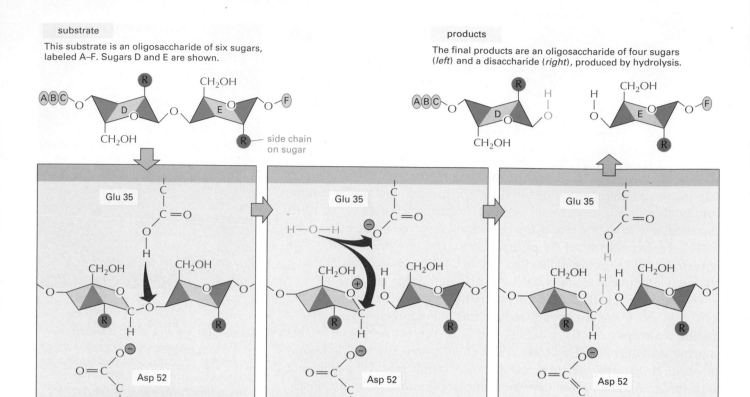

substrate

This substrate is an oligosaccharide of six sugars, labeled A–F. Sugars D and E are shown.

products

The final products are an oligosaccharide of four sugars (*left*) and a disaccharide (*right*), produced by hydrolysis.

In the enzyme–substrate complex (ES), the enzyme forces sugar D into a strained conformation, with Glu 35 positioned to serve as an acid that attacks the adjacent sugar–sugar bond by donating a proton (H⁺).

This is the unstable transition state, with a positive charge on sugar D. Both the strain on sugar D and the nearby negative charge on Asp 52 stabilize this intermediate, greatly lowering its energy on the enzyme surface.

The rapid addition of a water molecule (*green*) completes the hydrolysis and regenerates the protonated form of Glu 35, forming the enzyme–product complex (EP).

diffuses away, leaving the enzyme free to bind to another substrate molecule and catalyze another reaction (see Figure 5–27A). The rates of these different steps vary widely from one enzyme to another, and they can be measured by mixing purified enzymes and substrates together under carefully defined conditions.

If the concentration of the substrate is increased progressively from a very low value, the concentration of the enzyme–substrate complex, and therefore the rate at which product is formed, initially increases in a linear fashion in direct proportion to substrate concentration. However, as more and more enzyme molecules become occupied by substrate, this rate increase tapers off, until at a very high concentration of substrate it reaches a maximum value, termed V_{max}. At this point, the active sites of all enzyme molecules in the sample are fully occupied with sub-

Figure 5–28 Events at the active site of lysozyme. The top left and top right drawings depict the free substrate and the free products, respectively, whereas the other three drawings depict sequential events at the enzyme active site. Note the change in the conformation of sugar D in the enzyme–substrate complex; this is the sugar that acquires a positive charge in the unstable transition state.

(A) enzyme binds to two substrate molecules and orients them precisely to encourage a reaction to occur between them

(B) binding of substrate to enzyme rearranges electrons in the substrate, creating partial negative and positive charges that favor a reaction

(C) enzyme strains the bound substrate molecule, forcing it toward a transition state to favor a reaction

Figure 5–29 Some general features of enzyme catalysis. (A) Holding substrates together in a precise alignment. (B) Charge stabilization of reaction intermediates. (C) Altering bond angles in the substrate to increase the rate of a particular reaction.

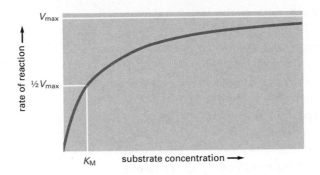

Figure 5–30 **Enzyme kinetics**. The rate of an enzyme reaction (V) increases as the substrate concentration increases until a maximum value (V_{max}) is reached. At this point all substrate-binding sites on the enzyme molecules are fully occupied, and the rate of reaction is limited by the rate of the catalytic process on the enzyme surface. For most enzymes the concentration of substrate at which the reaction rate is half-maximal (K_M) is a direct measure of how tightly the substrate is bound, with a large value of K_M corresponding to weak binding.

Question 5–7   The curve shown in Figure 5–30 is described by the simple equation:

$$\text{rate} = V_{max}\, [S/(S + K_M)]$$

This is called the Michaelis-Menten equation after the scientists who first used it to characterize enzymatic reactions. Can you convince yourself that the features qualitatively described in the text are accurately represented by this equation? In particular, how can the equation be simplified when the substrate concentration is in one of the following ranges: (A) the substrate concentration S is much smaller than the K_M, (B) the substrate concentration S equals the K_M, and (C) the substrate concentration S is much larger than the K_M?

strate, and the rate of product formation depends only on how rapidly the substrate molecule can be processed. For many enzymes, this **turnover number** is of the order of 1000 substrate molecules per second, although turnover numbers between 1 and 10,000 are known.

The concentration of substrate needed to make the enzyme work efficiently is often measured by a different parameter, termed the K_M, which is the concentration of substrate at which the enzyme works at half its maximum speed (0.5 V_{max}) (Figure 5–30). In general, a low value of K_M indicates that a substrate binds very tightly to the enzyme, and a large value corresponds to weak binding.

Tightly Bound Small Molecules Add Extra Functions to Proteins

Although we have emphasized the versatility of proteins as chains of amino acids that perform different functions, there are many instances in which the amino acids by themselves are not enough. Just as humans employ tools to enhance and extend the capabilities of their hands, so proteins often employ small nonprotein molecules to perform functions that would be difficult or impossible to do using amino acids alone. Thus the signal receptor protein *rhodopsin*—which is the purple light-sensitive pigment made by the rod cells in the retina—detects light by means of a small molecule, *retinal,* embedded in the protein (Figure 5–31A). Retinal changes its shape when it absorbs a photon of light, and this change is amplified by the protein to trigger a cascade of enzymatic reactions that eventually leads to an electrical signal being carried to the brain.

Another example of a protein that contains a nonprotein portion is hemoglobin (see Figure 5–16). A molecule of hemoglobin carries four

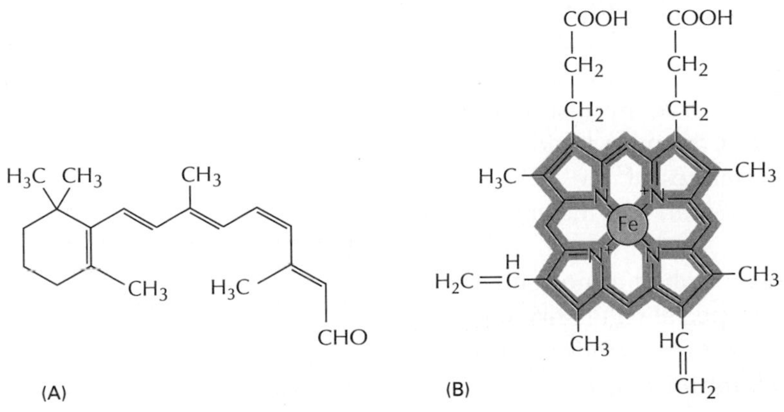

Figure 5–31 **Retinal and heme.** (A) The structure of retinal, the light-sensitive molecule attached to rhodopsin in our eyes. (B) The structure of a heme group, shown with the carbon-containing heme ring colored *red* and the iron atom at its center in *orange*. A heme group is tightly bound to each of the four polypeptide chains in hemoglobin, the oxygen-carrying protein whose structure was shown in Figure 5–16.

(A)

(B)

heme groups, ring-shaped molecules each with a single central iron atom (Figure 5–31B). Heme gives hemoglobin (and blood) its red color. By binding reversibly to oxygen gas through its iron atom, heme enables hemoglobin to pick up oxygen in the lungs and release it in the tissues.

Sometimes these small molecules are attached covalently and permanently to their protein, thereby becoming an integral part of the protein molecule itself. We will see in Chapter 11 that proteins are often anchored to cell membranes through covalently attached lipid molecules. And membrane proteins exposed on the surface of the cell, as well as proteins secreted outside the cell, are often modified by the covalent addition of sugars and oligosaccharides.

Enzymes frequently have a small molecule or metal atom tightly associated with their active site that assists with their catalytic function. *Carboxypeptidase*, for example, an enzyme that cuts polypeptide chains, carries a tightly bound zinc ion in its active site. During the cleavage of a peptide bond by carboxypeptidase, the zinc ion forms a transient bond with one of the substrate atoms, thereby assisting the hydrolysis reaction. In other enzymes, a small organic molecule serves a similar purpose. An example is *biotin*, which is found in enzymes that transfer a carboxylate group (–COO⁻) from one molecule to another (see Figure 3–31). Biotin participates in these reactions by forming a transient covalent bond to the –COO⁻ group to be transferred, being better suited for this function than any of the amino acids used to make proteins. Because it cannot be synthesized by humans, and must therefore be supplied in small quantities in the diet, biotin is a *vitamin*. Other vitamins are similarly needed to make small molecules that are essential components of our proteins; vitamin A, for example, is needed in the diet to make retinal, the light-sensitive part of rhodopsin.

The Catalytic Activities of Enzymes Are Regulated

A living cell contains thousands of enzymes, many of which operate at the same time and in the same small volume of the cytosol. By their catalytic action, the enzymes generate a complex web of metabolic pathways, each composed of chains of chemical reactions in which the product of one enzyme becomes the substrate of the next. In this maze of pathways there are many branch points where different enzymes compete for the same substrate. The system is so complex (see Figure 3–2) that elaborate controls are required to regulate when and how rapidly each reaction occurs.

Regulation occurs at many levels. At one level, the cell controls how many molecules of each enzyme it makes by regulating the expression of the gene that encodes that enzyme (discussed in Chapter 8). At another level, the cell controls enzymatic activities by confining sets of enzymes to particular subcellular compartments, enclosed by distinct membranes (discussed in Chapters 13 and 14). But the most rapid and general process that adjusts reaction rates operates at the level of the enzyme itself, through a change in the activity of an enzyme in response to other molecules that it encounters.

The most common type of control occurs when a molecule other than the substrates binds to an enzyme at a special regulatory site outside of the active site, thereby altering the rate at which the enzyme converts its substrates to products. In **feedback inhibition,** an enzyme act-

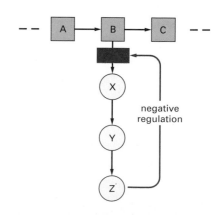

Figure 5–32 Feedback inhibition of a single biosynthetic pathway. The end product Z inhibits the first enzyme that is unique to its synthesis and thereby controls its own level in the cell. This is an example of negative regulation.

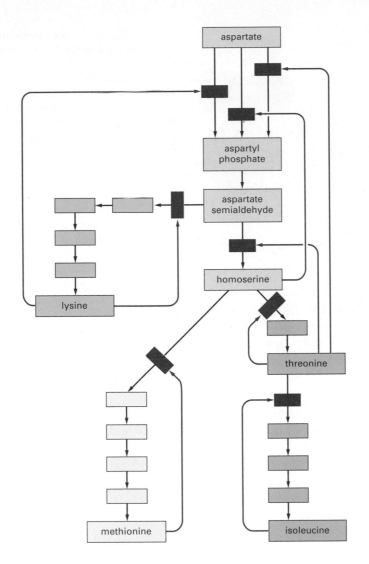

Figure 5–33 Multiple feedback inhibition. In this example, which shows the biosynthetic pathways for four different amino acids in bacteria, the *red arrows* indicate positions at which products feed back to inhibit enzymes. Each amino acid controls the first enzyme specific to its own synthesis, thereby controlling its own levels and avoiding a wasteful buildup of intermediates. The products can also separately inhibit the initial set of reactions common to all the syntheses; in this case, three different enzymes catalyze the initial reaction, each inhibited by a different product.

ing early in a reaction pathway is inhibited by a late product of that pathway. Thus, whenever large quantities of the final product begin to accumulate, the product binds to the first enzyme and slows down its catalytic action, thereby limiting further entry of substrates into that reaction pathway (Figure 5–32). Where pathways branch or intersect, there are usually multiple points of control by different final products, each of which works to regulate its own synthesis (Figure 5–33). Feedback inhibition can work almost instantaneously and is reversible.

Feedback inhibition is *negative regulation:* it prevents an enzyme from acting. Enzymes can also be subject to *positive regulation,* in which the enzyme's activity is stimulated by a regulatory molecule rather than being shut down. Positive regulation occurs when a product in one branch of the metabolic maze stimulates the activity of an enzyme in another pathway. As one example, the accumulation of ADP activates several enzymes involved in the oxidation of sugar molecules, thereby stimulating the cell to convert more ADP to ATP.

Allosteric Enzymes Have Two Binding Sites That Interact

A feature of feedback inhibition that was initially puzzling to those that discovered it was that the regulatory molecule often had a shape totally

Question 5–8 Consider the drawing in Figure 5–32. What will happen if, instead of the indicated feedback,

A. feedback inhibition from Z affects the step B → C only?

B. feedback inhibition from Z affects the step Y → Z only?

C. Z is a positive regulator of the step B → X?

D. Z is a positive regulator of the step B → C?

For each case, discuss how useful these regulatory schemes would be for a cell.

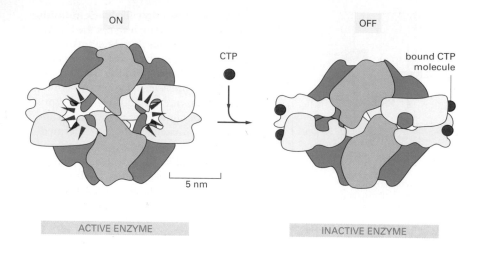

ON   OFF

CTP

bound CTP
molecule

|← 5 nm →|

ACTIVE ENZYME   INACTIVE ENZYME

Figure 5–34 A conformational change caused by feedback inhibition. An enzyme used in early studies of allosteric regulation was aspartate transcarbamoylase from *E. coli*. This large multisubunit enzyme (see Figure 5–17) catalyzes an important reaction that begins the synthesis of the pyrimidine ring of C, U, and T nucleotides. One of the final products of this pathway, cytosine triphosphate (CTP), binds to the enzyme to turn it off whenever CTP is plentiful. This diagram shows the conformational change that occurs when the enzyme is turned off by CTP binding.

different from the shape of the substrate of the enzyme. Indeed, when this form of regulation was discovered in the 1960s, it was termed *allostery* (from the Greek *allo*, "other," and *stere*, "solid" or "shape"). As more was learned about feedback inhibition, it was recognized that many enzymes must have at least two different binding sites on their surface—the active site that recognizes the substrates and a second site that recognizes a regulatory molecule. These two sites must "communicate" in a way that allows the catalytic events at the active site to be influenced by the binding of the regulatory molecule at its separate site on the protein's surface.

The interaction between separated sites on a protein molecule is now known to depend on a *conformational change* in the protein: binding at one of the sites causes a shift from one folded shape to a slightly different folded shape. Many enzymes have two conformations that differ in activity, each stabilized by the binding of different ligands. During feedback inhibition, for example, the binding of an inhibitor at one site on the protein causes the protein to shift to a conformation in which its active site—located elsewhere in the protein—is incapacitated (Figure 5–34).

Many—if not most—protein molecules are **allosteric:** they can adopt two or more slightly different conformations, and by a shift from one to another, their activity can be altered. This is true not only for enzymes but also for many other proteins—including receptors, structural proteins, and motor proteins. Because each conformation will have somewhat different contours on its surface, the binding site for one or more ligands can easily be altered. A particular ligand will favor the conformation that it binds to most strongly, and at high enough concentrations can "switch" the protein from one conformation to another (Figure 5–35).

A Conformational Change Can Be Driven by Protein Phosphorylation

We have just seen that many allosteric proteins are regulated by the binding of a small molecule. A second method that is commonly used by eucaryotic cells to regulate a protein's function is to add a phosphate group covalently to one of its amino acid side chains. Because each phosphate group carries two negative charges, the enzyme-catalyzed

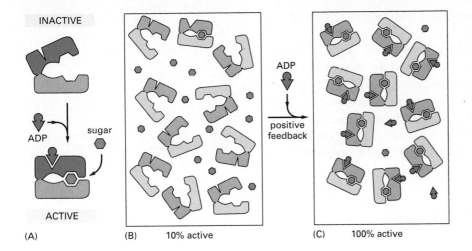

INACTIVE

ADP

sugar

ACTIVE

(A)

(B)   10% active

positive feedback

(C)   100% active

ADP

Figure 5–35 How the equilibrium between two conformations of a protein is affected by ligand binding. This schematic diagram shows a hypothetical enzyme in which an increase in the concentration of ADP molecules (*green wedges*) acts as an activator to increase the rate at which sugar molecules (*red hexagons*) are oxidized. (A) This hypothetical enzyme is allosterically regulated. It might, for example, catalyze a rate-limiting step in either glycolysis or the citric acid cycle, since when ADP accumulates it feeds back on such enzymes to accelerate sugar catabolism, thereby increasing the rate of production of ATP from ADP. (B) With no ADP present, only a small fraction of the molecules spontaneously adopt the active (closed) conformation; most are in the inactive (open) conformation. (C) Because ADP can bind only to the protein in its closed conformation, ADP addition lowers the energy of the closed conformation, causing nearly all of the enzyme molecules to switch to the active form.

addition of a phosphate group to a protein can cause a major conformational change by, for example, attracting a cluster of positively charged amino acid side chains. Removal of the phosphate group by a second enzyme returns the protein to its original conformation and restores its initial activity.

This reversible **protein phosphorylation** controls the activity of many different types of proteins in eucaryotic cells; more than a third of the 10,000 or so proteins in a typical mammalian cell are thought to be phosphorylated at any one time. The addition and removal of phosphate groups from specific proteins often occurs in response to signals that specify some change in a cell's state. For example, the complicated series of events that takes place as a eucaryotic cell divides is timed in this way (Chapter 18), and many of the signals generated by hormones and neurotransmitters are carried from the plasma membrane to the nucleus by a cascade of protein phosphorylation events (Chapter 15).

Protein phosphorylation involves the enzyme-catalyzed transfer of the terminal phosphate group of ATP to the hydroxyl group on a serine, threonine, or tyrosine side chain of the protein. This reaction is catalyzed by a **protein kinase,** whereas the reverse reaction of phosphate removal, or *dephosphorylation*, is catalyzed by a **protein phosphatase** (Figure 5–36). Cells contain hundreds of different protein kinases, each responsible for phosphorylating a different protein or set of proteins. There are also many different protein phosphatases; some of these are highly specific and remove phosphate groups from only one or a few proteins, whereas others act on a broad range of proteins. The state of phosphorylation of a protein at any moment in time, and thus its activ-

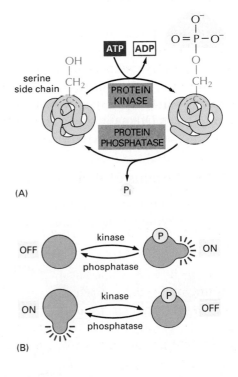

Figure 5–36 Protein phosphorylation. Many thousands of proteins in a typical eucaryotic cell are modified by the covalent addition of a phosphate group. (A) The general reaction, shown here, entails transfer of a phosphate group from ATP to an amino acid side chain of the target protein by a protein kinase. Removal of the phosphate group is catalyzed by a second enzyme, a protein phosphatase. In this example, the phosphate is added to a serine side chain; in other cases, the phosphate is instead linked to the –OH group of a threonine or a tyrosine in the protein. (B) The phosphorylation of a protein by a protein kinase can either increase or decrease the protein's activity, depending on the site of phosphorylation and the structure of the protein.

ity, will depend on the relative activities of the protein kinases and phosphatases that act on it.

For many proteins, a phosphate group is continually being added to a particular side chain and then removed. Phosphorylation cycles of this kind allow proteins to switch rapidly from one state to another, since the more rapidly the cycle is "turning," the faster the concentration of a phosphorylated protein can change in response to a sudden stimulus that changes its phosphorylation rate. The energy required to drive this cycle is derived from the free energy of hydrolysis of ATP, one molecule of which is consumed with each turn of the cycle.

GTP-binding Proteins Can Undergo Dramatic Conformational Changes

The loss of a phosphate group also drives a large conformational change in a major class of intracellular proteins called **GTP-binding proteins.** The activity of these proteins is controlled by binding a guanine nucleotide—either guanosine triphosphate (GTP) or guanosine diphosphate (GDP). In general, such proteins are in their active conformations with GTP bound; the protein itself then hydrolyzes this GTP to GDP, releasing a phosphate, and flips to an inactive conformation. As with protein phosphorylation, this process is reversible. The active conformation is regained by dissociation of the GDP, followed by the binding of a fresh molecule of GTP (Figure 5–37).

A large family of these GTP-binding proteins function as molecular switches in cells. The dissociation of GDP and its replacement by GTP, which turns the switch "on," is often stimulated in response to a signal received by the cell. The GTP-binding proteins often bind to other proteins to control enzyme activities, and their crucial role in intracellular signaling pathways will be discussed in Chapter 15. Here we shall look at their general mechanism of action through the bacterial elongation factor EF-Tu, a small GTP-binding protein that helps to load tRNA molecules onto ribosomes during protein synthesis.

Analysis of the three-dimensional structure of EF-Tu has revealed how an allosteric transition triggered by the gain or loss of a phosphate on the bound guanine nucleotide can cause a major shape change in a GTP-binding protein. Figure 5–38 shows how the loss of a single phosphate group, which initially causes only a tiny movement of 0.1 nm or so at the binding site, is magnified by the protein to create a movement 50 times larger. Dramatic shape changes of this type also underlie the very large movements created by the types of proteins that we consider next.

Motor Proteins Produce Large Movements in Cells

We have seen how conformational changes in proteins play a central part in enzyme regulation and cell signaling. We shall now discuss how

Figure 5–37 GTP-binding proteins form molecular switches. The activity of a GTP-binding protein generally requires the presence of a tightly bound GTP molecule (switch "on"). Hydrolysis of this GTP molecule produces GDP and inorganic phosphate (P_i), and it causes the protein to convert to a different, usually inactive, conformation (switch "off"). As shown here, resetting the switch requires that the tightly bound GDP dissociate, a slow step that is greatly accelerated by specific signals; once the GDP dissociates, a molecule of GTP is quickly rebound.

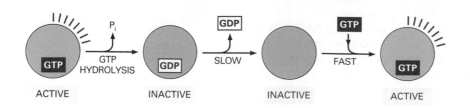

ACTIVE   INACTIVE   INACTIVE   ACTIVE

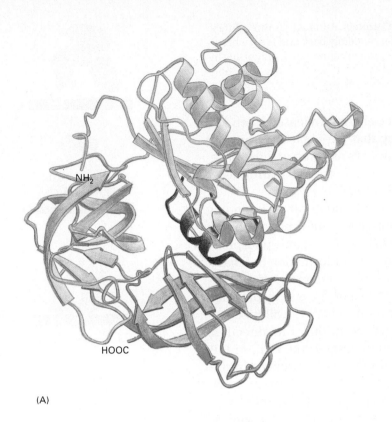

(A)

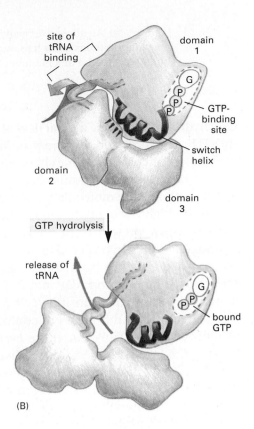

(B)

they enable proteins whose major function is to move other molecules, the **motor proteins,** to generate the forces responsible for muscle contraction and the dramatic movements of cells. Motor proteins also power smaller-scale intracellular movements: they help move chromosomes to opposite ends of the cell during mitosis (Chapter 17), move organelles along molecular tracks within the cell (Chapter 16), and move enzymes along a DNA strand during the synthesis of a new DNA molecule (Chapter 6). An understanding of how proteins can operate as molecules with moving parts is therefore essential for understanding the molecular basis of cell behavior.

How are shape changes in proteins used to generate orderly movements in cells? If, for example, a protein is required to walk along a narrow thread such as a DNA molecule, it can do this by undergoing a series of conformational changes—as illustrated in Figure 5–39. With nothing to drive these changes in an orderly sequence, however, they will be perfectly reversible and the protein will wander randomly back and forth along the thread. We can look at this situation in another way. Since the directional movement of a protein does net work, the laws of thermodynamics (see Chapter 3) demand that such movement utilize free energy from some other source (otherwise the protein could be used to make a perpetual motion machine). Therefore, without an input of energy, the protein molecule can only wander aimlessly.

How, then, can one make the series of conformational changes unidirectional? To force the entire cycle to proceed in one direction, it is enough to make any one of the changes in shape irreversible. For most proteins that are able to walk in one direction for long distances, this is achieved by coupling one of the conformational changes to the hydrolysis of an ATP molecule bound to the protein. The mechanism is similar to the one just discussed that drives allosteric shape changes by

Figure 5–38 A large conformational change is produced in response to nucleotide hydrolysis. (A) The elongation factor EF-Tu. (B) The hydrolysis of bound GTP in EF-Tu causes only a minute change in the position (equivalent to a few times the diameter of a hydrogen atom) of amino acids at the nucleotide-binding site. But this small change is magnified by conformational changes within the protein to produce a much larger movement. The hydrolysis of GTP releases an intramolecular bond, like a "latch" (*red dashes* in the upper diagram), which allows domains 2 and 3 to twist free and rotate by about 90° toward the viewer. This creates a major change in shape that releases the tRNA that was tightly held by the protein, as required to allow protein synthesis to proceed on the ribosome.

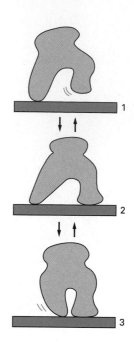

Figure 5–39 **An allosteric "walking" protein.** Although its three different conformations allow it to wander randomly back and forth while bound to a thread or a filament, the protein cannot move uniformly in a single direction.

GTP hydrolysis. Because a great deal of free energy is released when ATP (or GTP) is hydrolyzed, it is very unlikely that the nucleotide-binding protein will undergo a reverse shape change—as required to move backward—since this would require that it also reverse the ATP hydrolysis by adding a phosphate molecule to ADP to form ATP.

In the highly schematic model shown in Figure 5–40, ATP binding shifts a motor protein from conformation 1 to conformation 2. The bound ATP is then hydrolyzed to produce ADP and inorganic phosphate (P_i), causing a change from conformation 2 to conformation 3. Finally, the release of the bound ADP and P_i drives the protein back to conformation 1. Because the transition $2 \rightarrow 3$ is driven by the energy provided by ATP hydrolysis, this series of conformational changes will be effectively irreversible. Thus the entire cycle will go in only one direction, causing the protein molecule to walk continuously to the right in this example. Many motor proteins generate directional movement in this general way, including the muscle motor protein *myosin*—which "runs" along actin filaments to generate muscle contraction (Chapter 16)—and the *kinesin* protein involved in chromosome movements at mitosis (Chapter 17). Such movements can be rapid: some of the motor proteins involved in DNA replication propel themselves along a DNA strand at rates as high as 1000 nucleotides per second.

Proteins Often Form Large Complexes That Function as Protein Machines

As one progresses from small, single-domain proteins to large proteins formed from many domains, the functions that the proteins can perform become more elaborate. The most impressive tasks, however, are carried out by large protein assemblies formed from many protein molecules. Now that it is possible to reconstruct biological processes in cell-free systems in the laboratory, it is clear that each central process in a cell—such as DNA replication, protein synthesis, vesicle budding, or transmembrane signaling—is catalyzed by a highly coordinated, linked set of ten or more proteins. In most such **protein machines** the hydrolysis of bound nucleoside triphosphates (ATP or GTP) drives an ordered series of conformational changes in some of the individual protein subunits, enabling the ensemble of proteins to move coordinately. In this way, the appropriate enzymes can be moved directly into the positions where they are needed to carry out successive reactions in a series as, for example, in protein synthesis on a ribosome (Chapter 7), or in DNA replication—where a large multiprotein complex moves rapidly along the DNA. A simple mechanical analogy is illustrated in Figure 5–41.

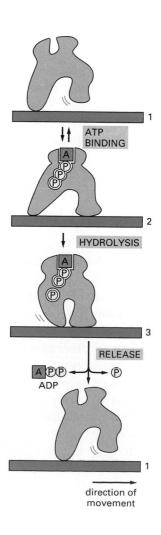

direction of movement

Figure 5–40 **An allosteric motor protein.** An orderly transition among three conformations is driven by the hydrolysis of a bound ATP molecule. Because one of these transitions is coupled to the hydrolysis of ATP, the entire cycle is essentially irreversible. By repeated cycles the protein moves continuously to the right along the thread.

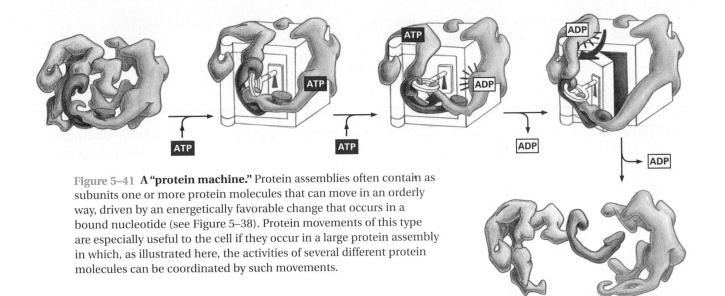

Figure 5–41 A "protein machine." Protein assemblies often contain as subunits one or more protein molecules that can move in an orderly way, driven by an energetically favorable change that occurs in a bound nucleotide (see Figure 5–38). Protein movements of this type are especially useful to the cell if they occur in a large protein assembly in which, as illustrated here, the activities of several different protein molecules can be coordinated by such movements.

Cells have evolved protein machines for the same reason that humans have invented mechanical and electronic machines: for accomplishing almost any task, manipulations that are spatially and temporally coordinated through linked processes are much more efficient than is the sequential use of individual tools.

> **Question 5–10** Explain why the enzymes in Figure 5–41 have a great advantage in opening the vault if they work as a protein complex, as opposed to working in an unlinked, sequential manner.

Essential Concepts

- Living cells contain an enormously diverse set of protein molecules, each made as a linear chain of amino acids covalently linked together.

- Each type of protein has a unique amino acid sequence that determines both its three-dimensional shape and its biological activity.

- The folded structure of a protein is stabilized by noncovalent interactions between different parts of the polypeptide chain.

- Hydrogen bonds between neighboring regions of the polypeptide backbone can give rise to regular folding patterns, known as α helices and β sheets.

- The structure of many proteins can be subdivided into smaller globular regions of compact three-dimensional structure, known as protein domains.

- The biological function of a protein depends on the detailed chemical properties of its surface and how it binds to other molecules, called ligands.

- Enzymes are proteins that first bind tightly to specific molecules, called substrates, and then catalyze the making and breaking of covalent bonds in these molecules.

- At the active site of an enzyme, the amino acid side chains of the folded protein are precisely positioned so that they favor the formation of the high-energy transition states that the substrates must pass through to react.

- The three-dimensional structure of many proteins has evolved so that the binding of a small ligand can induce a significant change in protein shape.

- Most enzymes are allosteric proteins that can exist in two conformations that differ in catalytic activity, and the enzyme can be turned on or off by ligands that bind to a distinct regulatory site to stabilize either the active or the inactive conformation.

- The activities of most enzymes within the cell are strictly regulated. One of the most common forms of regulation is feedback inhibition, in which an enzyme early in a metabolic pathway is inhibited by its binding to one of the pathway's end products.

- Many thousands of proteins in a typical eucaryotic cell are regulated either by cycles of phosphorylation and dephosphorylation, or by the binding and hydrolysis of GTP by a partner GTP-binding protein.

- The hydrolysis of ATP to ADP by motor proteins produces directed movements in the cell.

- Highly efficient protein machines are formed by assemblies of allosteric proteins in which conformational changes are coordinated to perform complex cellular functions.

Key Terms

active site	coiled-coil	helix	protein phosphatase
allosteric	conformation	K_M	protein phosphorylation
α helix	disulfide bond	ligand	secondary structure
amino acid sequence	enzyme	motor protein	side chain
antibody	equilibrium	polypeptide backbone	substrate
antigen	equilibrium constant	protein	subunit
β sheet	feedback inhibition	protein domain	transition state
binding site	fibrous protein	protein family	turnover number
	globular protein	protein kinase	V_{max}
	GTP-binding protein	protein machine	

Questions

Question 5–11 Which of the following statements are correct? Explain your answers.

A. The active site of an enzyme usually occupies only a small fraction of its surface.

B. Catalysis by some enzymes involves the formation of a covalent bond between an amino acid side chain and a substrate molecule.

C. A β sheet can contain up to five strands, but no more.

D. The specificity of an antibody molecule is contained exclusively in loops on the surface of the folded light chain domain.

E. Enzymes lower the activation energy barrier of a reaction, thereby changing the equilibrium point.

F. The possible linear arrangements of amino acids are so vast that new proteins almost never evolve by alteration of old ones.

G. The K_M of an enzyme is a measure of the maximal rate with which the enzyme can catalyze a reaction.

H. Allosteric enzymes have two or more binding sites.

I. Noncovalent bonds are too weak to influence the three-dimensional structure of macromolecules.

J. Affinity chromatography separates molecules according to their intrinsic charge.

K. Upon centrifugation of a cell homogenate, smaller organelles experience less friction and thereby sediment faster than larger ones.

Question 5–12 What common feature of α helices and β sheets makes them universal building blocks for proteins?

Question 5–13 Discuss the following statement: "Enzymes and heat are alike in that both can speed up reactions that—although thermodynamically feasible—do not occur at an appreciable rate because they require a high activation energy. Diseases that are treated by the careful application of heat, such as by ingestion of hot chicken soup, are therefore likely to be due to the insufficient function of an enzyme."

Question 5–14 Protein structure is determined only by a protein's amino acid sequence. Should a genetically engineered protein in which the order of all amino acids is reversed therefore have the same structure as the original protein?

Question 5–15 Consider the following protein sequence as an α helix: Leu-Lys-Arg-Ile-Val-Asp-Ile-Leu-Ser-Arg-Leu-Phe-Lys-Val. How many turns does this helix make? Do you find anything remarkable about the arrangement of the amino acids in this sequence when folded into an α helix? (Hint: consult the properties of the amino acids in Figure 5–3.)

Question 5–16   Simple enzyme reactions often conform to the equation

$$E + S \rightleftharpoons ES \rightarrow EP \rightleftharpoons E + P$$

where E, S, and P are enzyme, substrate, and product, respectively.

A. What does ES represent in this equation?

B. Why is the first step shown with bidirectional arrows and the second step as a unidirectional arrow?

C. Why does E appear at both ends of the equation?

D. One often finds that high concentrations of P inhibit the enzyme. Suggest why this might occur.

E. Compound X resembles S and binds to the active site of the enzyme but cannot undergo the reaction catalyzed by it. What effects would you expect the addition of X to the reaction to have? Compare the effects of X and of accumulation of P.

Question 5–17   Which of the following amino acids would you expect to find more often near the center of a folded globular protein? Which ones would you expect to find more often exposed to the outside? Explain your answers. Ser, Ser-P (a Ser residue that is phosphorylated), Leu, Lys, Gln, His, Phe, Val, Ile, Met, Cys-S-S-Cys (two Cys residues that are disulfide-bonded), and Glu. Where would you expect to find the most amino-terminal amino acid and the most carboxy-terminal amino acid?

Question 5–18   Assume you want to make and study fragments of a protein. Would you expect that any fragment of the polypeptide chain would fold the same way as the corresponding sequence folds in the intact protein? Consider the protein shown in Figure 5–12. Which fragments do you suppose are most likely to fold correctly?

Question 5–19   An enzyme isolated from a mutant bacterium grown at 20°C works when it is tested at 20°C but not at 37°C (37°C is the temperature of the gut, where this bacterium normally lives). Furthermore, once the enzyme has been exposed to the higher temperature, it no longer works at the lower one. Can you suggest what happens at the molecular level to this enzyme as the temperature increases?

Question 5–20   The rate of a simple enzyme reaction is given by the standard Michaelis-Menton equation

$$rate = V_{max} [S/(S + K_M)]$$

If the V_{max} of an enzyme is 100 μM/sec and the K_M is 1 mM, at what substrate concentration is the rate 50 μM/sec? Plot a graph of rate versus substrate concentration (S) for S = 0 to 10 mM. Convert this to a plot of 1/rate versus 1/S. Why is the latter plot a straight line?

Question 5–21   Select the correct options in the following and explain your choices. If S is much smaller than K_M, the active site of the enzyme is mostly *occupied/unoccupied*. If S is very much greater than K_M, the reaction rate is limited by the *enzyme/substrate* concentration.

Question 5–22

A. The reaction rates of the reaction S → P catalyzed by enzyme E were determined under conditions such that only very little product was formed. The following data were measured:

substrate concentration (μM)	reaction rate (μM/min)
0.08	0.15
0.12	0.21
0.54	0.7
1.23	1.1
1.82	1.3
2.72	1.5
4.94	1.7
10.00	1.8

Plot the above data as a graph. Use this graph to find the K_M and the V_{max} for this enzyme.

B. To determine these values more precisely, a trick is generally used in which the Michaelis-Menten equation is transformed so that it is possible to plot the data as a straight line. A simple rearrangement yields

$$\mathbf{1/rate} = (K_M/V_{max})\,(\mathbf{1/S}) + 1/V_{max}$$

that is, an equation of the form **y = ax + b**.

Calculate 1/rate and 1/S for the data given in the preceding question and then plot 1/rate versus 1/S as a new graph. Determine K_M and V_{max} from the intercept of the line with the axis where 1/S = 0, combined with the slope of the line. Do your results agree with the estimates made from the first graph of the raw data?

C. Note that it is stated above in part (A) of this question that only very little product was formed under the reaction conditions. Why is this important?

D. Assume the enzyme is regulated: upon phosphorylation, its K_M increases by a factor of 3 without changing its V_{max}. Is this an activation or inhibition? Plot the data you would expect for the phosphorylated enzyme in both the graph for (A) and the graph for (B).

Question 5–23 In cells, an enzyme catalyzes the reaction AB → A + B. It was isolated, however, as an enzyme that carries out the opposite reaction A + B → AB. Explain the apparent paradox.

Question 5–24 A motor protein moves along filaments in the cell. Why are the elements shown in the illustration not sufficient to provide unidirectionality to the movement? With reference to Figure 5–40, modify the illustration shown here to include other elements that are required to create a unidirectional motor and justify each modification you make to the illustration.

Question 5–25 Gel-filtration chromatography separates molecules according to size (see Panel 5–5, pp. 162–163). Smaller molecules diffuse faster in solution than larger ones, yet smaller molecules migrate slower through a gel-filtration column than larger ones. Explain this paradox. What should happen at very rapid flow rates?

6 DNA

Life depends on the ability of cells to store, retrieve, and translate the genetic instructions required to make and maintain a living organism. This *hereditary* information is passed on from a cell to its daughter cells at cell division, and from generation to generation of organisms through the reproductive cells. These instructions are stored within every living cell as the **genes**—the information-containing elements that determine the characteristics of a species as a whole and of the individuals within it.

Ever since the emergence of genetics as a science at the beginning of the twentieth century, scientists have asked what material genes could be made of. The genetic information in genes is copied and transmitted from cell to daughter cell millions of times during the life of a multicellular organism, and it survives the process essentially unchanged. What kind of molecule could be capable of such accurate and almost unlimited replication, and also be able to direct the development of an organism and the daily life of a cell? What kind of instructions does the genetic information contain? How are these instructions physically organized so that the enormous amount of information required for the development and maintenance of even the simplest organism can be contained within the tiny space of a cell?

The answers to some of these questions began to emerge in the 1940s, when it was discovered from studies in simple fungi that genetic information consists primarily of instructions for making proteins. Proteins are the macromolecules that carry out most cellular functions: they serve as building blocks for cellular structures (Chapter 5); they form the enzymes that catalyze all of the cell's chemical reactions (Chapter 3); they regulate gene expression (Chapter 8); and they enable cells to move (Chapter 16) and to communicate with each other (Chapter 15). The properties and functions of a cell are therefore almost entirely determined by the proteins it is able to make. With hindsight, it is hard to imagine what other instructions the genetic information could have encoded. The other crucial advance to be made in the 1940s was the identification of **deoxyribonucleic acid (DNA)** as the likely carrier of genetic information. But the mechanism whereby the hereditary information is copied for transmission from cell to cell, and how proteins are specified by the instructions in the DNA, remained mysterious until 1953, when the structure of DNA was determined by James Watson and Francis Crick. The structure immediately solved the problem of how DNA might be copied, or *replicated,* and provided the first clues about how a molecule of DNA might encode the instructions for making proteins. In this chapter we shall focus on the question of how the structure of DNA allows the faithful replication and maintenance of the genetic instructions, explaining only briefly how a DNA molecule can encode the information for specifying proteins, and leaving the details of this *genetic code* and the machinery for decoding it into proteins for the next chapter.

The amount of information stored in a cell's DNA is enormous. We shall see that each cell contains not only an elaborate machinery for accurately copying this information store, but also specialized enzymes for repairing DNA when it is damaged. Yet despite these systems for protecting the genetic instructions from copying errors and accidental damage, permanent changes, or *mutations,* sometimes do occur. Mutations in the DNA often affect the information it encodes. Occasionally, this can benefit the organism in which the mutation occurs: for example, mutations can make bacteria resistant to antibiotics that are used to kill them. Indeed, the accumulation of changes in DNA over millions of years provides the variety in genetic material that makes one species distinct from another. It also produces the smaller variations that underlie the differences between individuals of the same species that we can easily see in humans and other animals (Figure 6–1). The majority of mutations are, however, detrimental, and, in humans, are responsible for thousands of inherited diseases. Mutations that arise in the cells of the body throughout the lifetime of an individual may also cause disease, most notably the many types of cancer. Without the cellular systems that are continually monitoring and repairing damage to DNA, it is questionable whether life could exist at all.

Figure 6–1 **A family group photograph illustrating the passage of hereditary information from one generation to the next.** The three children resemble one another and their parents more closely than they resemble other humans because they inherit their particular genes from their parents. The cat shares many features with humans, but during the millions of years of evolution that have separated humans and cats, we both have accumulated many hereditary changes that now make us quite different species. The chicken is an even more distant relative.

The Structure and Function of DNA

By the early twentieth century, biologists had recognized that genes are carried on *chromosomes,* then known as threadlike structures in the nucleus of the eucaryotic cell that become visible as the cell begins to divide (Figure 6–2). Later, as biochemical analysis became possible, chromosomes were found to consist of DNA and proteins. Because DNA was perceived to be a relatively simple molecule chemically, it was at first assumed that genes had to be composed of protein, which is much more chemically diverse. We discuss in detail in Chapter 8 the role that proteins in fact play in chromosomes. In this chapter, we shall see how, despite its chemical simplicity, the structure and chemical properties of DNA make it ideally suited as the raw material for genes. The genes of every cell on earth are made of DNA, and insights into the relationship between DNA and genes have come from experiments carried out in a wide variety of organisms.

Genes Are Made of DNA

The first strong evidence that genes are made of DNA came in 1944, when it was shown that adding purified DNA to a bacterium changed its properties and that this change was faithfully passed on to subsequent generations of bacteria (Figure 6–3). Today, the fact that DNA is the genetic material is so fundamental to biological thought that it is difficult to realize the enormous intellectual gap that this discovery filled.

A DNA Molecule Consists of Two Complementary Chains of Nucleotides

Biologists in the 1940s had difficulty accepting DNA as the genetic material because of the apparent simplicity of its chemistry. DNA was thought of as simply a long polymer composed of only four types of subunits, which resemble one another chemically. Early in the 1950s, DNA was first examined by x-ray diffraction analysis, a technique for determining the three-dimensional atomic structure of a molecule (see Panel 5–5, pp. 162–163). The early x-ray diffraction results indicated that DNA was composed of two strands wound into a helix. The observation that DNA was double-stranded was of crucial significance. It provided one of the major clues that led, in 1953, to a correct model for the structure of DNA. Only when the Watson-Crick model of DNA structure was proposed did its potential for replication and information encoding become apparent.

A DNA molecule consists of two long polynucleotide chains composed of four types of nucleotide subunits. Each of these chains is known as a *DNA chain,* or a *DNA strand.* The two chains are held together by *hydrogen bonds* (see Panel 2–7, pp. 70–71) between the base portions of the nucleotides (Figure 6–4). As we saw in Chapter 2 (Panel

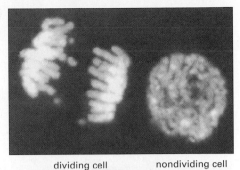
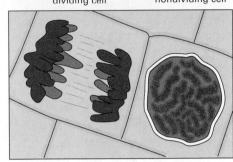

dividing cell   nondividing cell

Figure 6–2 DNA in cells. The top panel shows two adjacent plant cells photographed through a light microscope. The DNA has been stained with a fluorescent dye (DAPI) that binds to it. The DNA is present in chromosomes, which become visible as distinct structures in the light microscope only when they become compact structures in preparation for cell division, as shown on the left. The cell on the right, which is not dividing, contains the identical chromosomes, but they cannot be distinguished in the light microscope at this phase in the cell's life cycle, because they are in a much more extended conformation. The outlines of the two cells along with the state of their DNA are shown schematically in the lower panel. (Courtesy of Peter Shaw.)

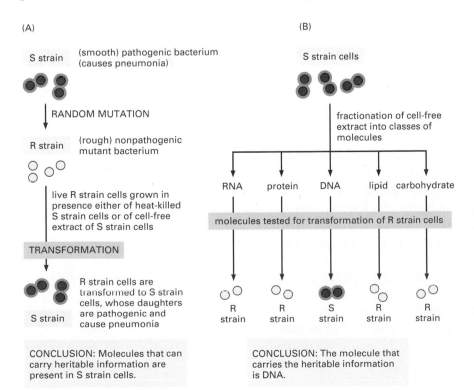

Figure 6–3 Experimental demonstration that DNA is the genetic material. Two closely related strains of the bacterium *Streptococcus pneumoniae* differ from each other both in their appearance under the microscope and in their pathogenicity. One strain appears smooth (S) and causes death when injected into mice, and the other appears rough (R) and is nonlethal. The experiment in (A) shows that a substance present in the S strain can change (or transform) the R strain into the S strain and that this change is inherited by subsequent generations of bacteria. The experiment in (B) identifies the substance as DNA.

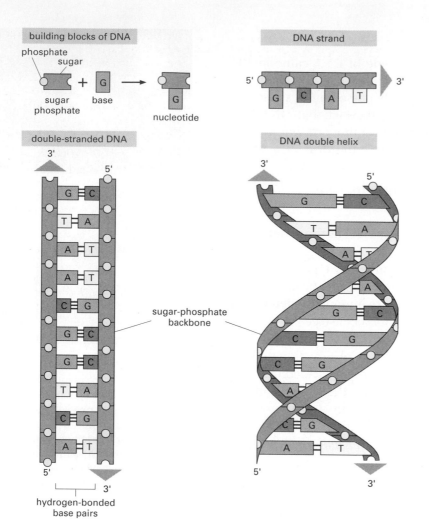

building blocks of DNA

DNA strand

double-stranded DNA

DNA double helix

sugar-phosphate backbone

hydrogen-bonded base pairs

Figure 6–4 DNA and its building blocks. DNA is made of four types of nucleotides, which are covalently linked together into polynucleotide chains with a sugar-phosphate backbone from which the bases (A, C, G, and T) extend. A DNA molecule is composed of two polynucleotide chains (DNA strands) held together by hydrogen bonds between the paired bases. The *arrows* on the DNA strands indicate the polarities of the two strands, which run antiparallel to each other in the DNA molecule. In the diagram at the bottom left of the figure, the DNA is shown straightened out; in reality, it is twisted into a double helix, as shown on the right.

2–6, pp. 66–67), nucleotides are composed of a five-carbon sugar to which are attached one or more phosphate groups and a nitrogen-containing base. In the case of the nucleotides in DNA, the sugar is deoxyribose attached to a single phosphate group (hence the name deoxyribonucleic acid), and the base may be either *adenine (A), cytosine (C), guanine (G),* or *thymine (T)*. The nucleotides are covalently linked together in a chain through the sugars and phosphates, which thus form a "backbone" of alternating sugar-phosphate-sugar-phosphate (Figure 6–4). Because it is only the base that differs in each of the four types of subunits, each polynucleotide chain in DNA can be thought of as a necklace (the backbone) strung with four types of beads (the four bases A, C, G, and T). These same symbols (A, C, G, and T) are also commonly used to denote the four different nucleotides, that is, the bases with their attached sugar and phosphate groups.

The way in which the nucleotide subunits are linked together gives a DNA strand a chemical polarity. If we imagine that each nucleotide has a knob (the phosphate) and a hole (see Figure 6–4), each completed chain, formed by interlocking knobs with holes, will have all of its subunits lined up in the same orientation. Moreover, the two ends of the chain will be easily distinguishable, as one has a hole and the other a knob. This polarity in a DNA chain is indicated by referring to one end as the 3′ end and the other as the 5′ end. This convention is based on the details of the chemical linkage between the nucleotide subunits.

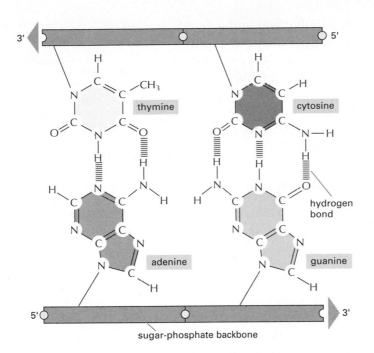

Figure 6–5 Complementary base pairs in the DNA double helix. The shapes and chemical structure of the bases allow hydrogen bonds to form efficiently only between A and T and between G and C, where atoms that are able to form hydrogen bonds (see Panel 2–1, pp. 46–47) can be brought close together without perturbing the double helix. Two hydrogen bonds form between A and T, while three form between G and C. The bases can pair in this way only if the two polynucleotide chains that contain them are antiparallel to each other.

The two polynucleotide chains in the DNA **double helix** are held together by hydrogen-bonding between the bases on the different strands. All the bases are therefore on the inside of the helix, with the sugar-phosphate backbones on the outside (see Figure 6–4). The bases do not pair at random, however: A always pairs with T, and G with C (Figure 6–5). In each case, a bulkier two-ring base (a purine, see Chapter 2) is paired with a single-ring base (a pyrimidine). This *complementary base-pairing* enables the **base pairs** to be packed in the energetically most favorable arrangement in the interior of the double helix. In this arrangement, each base pair is of similar width, thus holding the sugar-phosphate backbones an equal distance apart along the DNA molecule. In addition, the two sugar-phosphate backbones wind around each other to form a double helix (Figure 6–6).

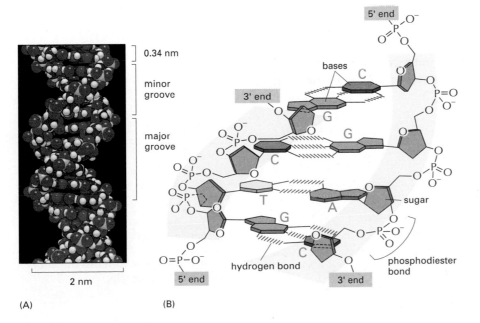

(A)

(B)

Figure 6–6 The DNA double helix. (A) Space-filling model of 1½ turns of the DNA double helix. The coiling of the two strands around each other creates two grooves in the double helix. As indicated in the figure, the wider groove is called the major groove, and the smaller, the minor groove. (B) A short section of the double helix viewed from its side. Four base pairs are shown. The nucleotides are linked together covalently by phosphodiester bonds through the 3′-hydroxyl (–OH) group of one sugar and the 5′-phosphate (P) of the next. Thus, each polynucleotide strand has a chemical polarity; that is, its two ends are chemically different. The 3′ end carries an unlinked –OH group attached to the 3′ position on the sugar ring; the 5′ end carries a free phosphate group attached to the 5′ position on the sugar ring.

The Structure and Function of DNA

The members of each base pair can fit together within the double helix only if the two strands of the helix are **antiparallel,** that is, only if the polarity of one strand is oriented opposite to that of the other strand (see Figure 6–4). A consequence of these base-pairing requirements is that each strand of a DNA molecule contains a sequence of nucleotides that is exactly **complementary** to the nucleotide sequence of its partner strand. This is of crucial importance for the copying of DNA, as we shall see in the next section of this chapter (pp. 189–198).

The Structure of DNA Provides a Mechanism for Heredity

Genes carry biological information that must be copied accurately and transmitted when a cell divides to form two daughter cells. Two central biological problems are posed by this idea: how can the information for specifying an organism be carried in chemical form, and how is it accurately copied? The discovery of the structure of the DNA double helix was a landmark in twentieth-century biology because it immediately suggested answers to these two questions, and thereby resolved at the molecular level the problem of heredity. We outline the answer to the first question before examining in more detail the answer to the second.

DNA encodes information in the order, or sequence, of the nucleotides along each strand. Each base—A, C, T, or G—can be considered as a letter in a four-letter alphabet that is used to spell out biological messages in the chemical structure of the DNA (Figure 6–7). Organisms differ from one another because their respective DNA molecules have different nucleotide sequences and, consequently, carry different biological messages. But how is the nucleotide alphabet used to make up messages, and what do they spell out?

It had already been established some time before the structure of DNA was determined that genes contain the instructions for producing proteins (Figure 6–8). The DNA messages, therefore, must somehow encode proteins. This immediately makes the problem easier to understand, because of the chemical character of proteins. As discussed in Chapter 5, the properties of a protein, which are responsible for its biological function, are determined by its three-dimensional structure, and the structure of a protein is determined in turn by the sequence of the amino acids of which it is composed. The linear sequence of nucleotides in a gene must therefore somehow spell out the linear sequence of amino acids in a protein. The exact correspondence between the four-letter nucleotide alphabet of DNA and the twenty-letter amino acid

(A) molecular biology is...

(B)

(C) • ▬ • • ▬ • •• • ▬ • ▬ • •

(D) 细胞生物学乐趣无穷

(E) TTCGAGCGACCTAACCTATAG

Figure 6–7 Examples of linear messages. The languages are, from top to bottom, English, a musical score, Morse code, Chinese, and DNA.

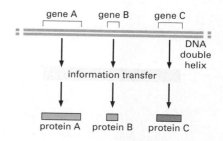

Figure 6–8 Each gene contains the information to make a protein.

Figure 6–9 Sequence of nucleotides in the human β-globin gene. This gene carries the information for the amino acid sequence of one of the two types of subunits of the hemoglobin molecule, which carries oxygen in the blood. A different gene, the α-globin gene, carries the information for the other type of hemoglobin subunit (a hemoglobin molecule has four subunits, two of each type). Only one of the two strands of the DNA double helix containing the β-globin gene is shown; the other strand has the exact complementary sequence. The sequence should be read left to right in successive lines down the page as if it were normal English text. The DNA sequences highlighted in color show the three regions of the gene that specify the amino sequence for the β-globin protein. We will see in Chapter 7 how the cell connects these three sequences together in order to synthesize a full-length β-globin protein.

```
CCCTGTGGAGCCACACCCTAGGGTTGGCCA
ATCTACTCCCAGGAGCAGGGAGGGCAGGAG
CCAGGGCTGGGCATAAAAGTCAGGGCAGAG
CCATCTATTGCTTACATTTGCTTCTGACAC
AACTGTGTTCACTAGCAACTCAAACAGACA
CCATGGTGCACCTGACTCCTGAGGAGAAGT
CTGCCGTTACTGCCCTGTGGGGCAAGGTGA
ACGTGGATGAAGTTGGTGGTGAGGCCCTGG
GCAGGTTGGTATCAAGGTTACAAGACAGGT
TTAAGGAGACCAATAGAAACTGGGCATGTG
GAGACAGAGAAGACTCTTGGGTTTCTGATA
GGCACTGACTCTCTCTGCCTATTGGTCTAT
TTTCCCACCCTTAGGCTGCTGGTGGTCTAC
CCTTGGACCCAGAGGTTCTTTGAGTCCTTT
GGGGATCTGTCCACTCCTGATGCTGTTATG
GGCAACCCTAAGGTGAAGGCTCATGGCAAG
AAAGTGCTCGGTGCCTTTAGTGATGGCCTG
GCTCACCTGGACAACCTCAAGGGCACCTTT
GCCACACTGAGTGAGCTGCACTGTGACAAG
CTGCACGTGGATCCTGAGAACTTCAGGGTG
AGTCTATGGGACCCTTGATGTTTTCTTTCC
CCTTCTTTTCTATGGTTAAGTTCATGTCAT
AGGAAGGGGAGAAGTAACAGGGTACAGTTT
AGAATGGGAAACAGACGAATGATTGCATCA
GTGTGGAAGTCTCAGGATCGTTTTAGTTTC
TTTTATTTGCTGTTCATAACAATTGTTTTC
TTTTGTTTAATTCTTGCTTTCTTTTTTTTT
CTTCTCCGCAATTTTTACTATTATACTTAA
TGCCTTAACATTGTGTATAACAAAAGGAAA
TATCTCTGAGATACATTAAGTAACTTAAAA
AAAAACTTTACACAGTCTGCCTAGTACATT
ACTATTTGGAATATATGTGTGCTTATTTGC
ATATTCATAATCTCCCTACTTTATTTTCTT
TTATTTTTAATTGATACATAATCATTATAC
ATATTTATGGGTTAAAGTGTAATGTTTTAA
TATGTGTACACATATTGACCAAATCAGGGT
AATTTTGCATTTGTAATTTTAAAAAATGCT
TTCTTCTTTTAATATACTTTTTTGTTTATC
TTATTTCTAATACTTTCCCTAATCTCTTTC
TTTCAGGGCAATAATGATACAATGTATCAT
GCCTCTTTGCACCATTCTAAAGAATAACAG
TGATAATTTCTGGGTTAAGGCAATAGCAAT
ATTTCTGCATATAAATATTTCTGCATATAA
ATTGTAACTGATGTAAGAGGTTTCATATTG
CTAATAGCAGCTACAATCCAGCTACCATTC
TGCTTTTATTTTATGGTTGGGATAAGGCTG
GATTATTCTGAGTCCAAGCTAGGCCCTTTT
GCTAATCATGTTCATACCTCTTATCTTCCT
CCCACAGCTCCTGGGCAACGTGCTGGTCTG
TGTGCTGGCCCATCACTTTGGCAAAGAATT
CACCCCACCAGTGCAGGCTGCCTATCAGAA
AGTGGTGGCTGGTGTGGCTAATGCCCTGGC
CCACAAGTATCACTAAGCTCGCTTTCTTGC
TGTCCAATTTCTATTAAAGGTTCCTTTGTT
CCCTAAGTCCAACTACTAAACTGGGGGATA
TTATGAAGGGCCTTGAGCATCTGGATTCTG
CCTAATAAAAAACATTTATTTTCATTGCAA
TGATGTATTTAAATTATTTCTGAATATTTT
ACTAAAAAGGGAATGTGGGAGGTCAGTGCA
TTTAAAACATAAAGAAATGATGAGCTGTTC
AAACCTTGGGAAAATACACTATATCTTAAA
CTCCATGAAAGAAGGTGAGGCTGCAACCAG
CTAATGCACATTGGCAACAGCCCCTGATGC
CTATGCCTTATTCATCCCTCAGAAAAGGAT
TCTTGTAGAGGCTTGATTTGCAGGTTAAAG
TTTTGCTATGCTGTATTTTACATTACTTAT
TGTTTTAGCTGTCCTCATGAATGTCTTTTC
```

alphabet of proteins is not obvious from the DNA structure, and it took over a decade from the discovery of the double helix before it was completely worked out. In Chapter 7, we describe this code and the mechanism by which a cell translates the nucleotide sequence of a gene into the amino acid sequence of a protein. Most genes are short stretches of DNA encoding a single protein, although not all of the DNA in genes is used to encode the proteins that they specify: much of the rest is concerned with determining when, and in what amounts, the protein encoded by each gene is made. We discuss in Chapter 8 these *regulatory regions* of genes and how they operate.

The complete set of information in an organism's DNA is called its **genome** (the term is also used to refer to the DNA that carries this information). The total amount of this information is staggering: a typical human cell contains a meter of DNA (3×10^9 nucleotides). Written out in the four-letter nucleotide alphabet, the nucleotide sequence of a small gene from humans occupies a quarter of a page of text (Figure 6–9), while the complete DNA sequence of the human genome would fill more than 1000 books the size of this one.

At each cell division, the cell must copy its genome in order to pass it to both daughter cells; the structure of DNA also revealed the principle that makes this copying possible.

DNA Replication

Because each strand of DNA contains a sequence of nucleotides that is exactly complementary to the nucleotide sequence of its partner strand, each strand can act as a **template,** or mold, for the synthesis of a new complementary strand (Figure 6–10). In other words, if we designate the

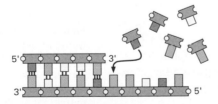

Figure 6–10 A DNA strand as a template. Preferential binding occurs between pairs of nucleotides (A with T, and G with C) that can form base pairs. This enables each strand to act as a template for forming its complementary strand.

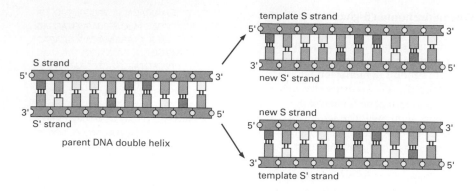

template S strand

new S' strand

new S strand

template S' strand

S strand
5'
3'

3'
5'
S' strand

parent DNA double helix

Figure 6–11 DNA acts as a template for its own duplication. As the nucleotide A will successfully pair only with T, and G with C, each strand of DNA can specify the sequence of nucleotides in its complementary strand. In this way, double-helical DNA can be copied precisely.

two DNA strands as S and S', strand S can serve as a template for making a new strand S', while strand S' can serve as a template for making a new strand S (Figure 6–11). Thus, the genetic information in DNA can be accurately copied by the beautifully simple process in which strand S separates from strand S', and each separated strand then serves as a template for the production of a new complementary partner strand that is identical to its former partner.

The ability of each strand of a DNA molecule to act as a template for producing a complementary strand enables a cell to copy, or *replicate*, its genes before passing them on to its descendants. But the task is awe-inspiring, as it can involve copying billions of nucleotide pairs every time a cell divides. The copying must be carried out with speed and accuracy: in about 8 hours, a dividing animal cell will copy the equivalent of 1000 books like this one and, on average, get no more than a single letter or two wrong. This feat is performed by a cluster of proteins that together form a "replication machine." **DNA replication** produces two complete double helices from the original DNA molecule, each new DNA helix identical (except for rare errors) in nucleotide sequence to the parental DNA double helix (see Figure 6–11). Because each parental strand serves as the template for one new strand, each of the daughter DNA double helices ends up with one of the original (old) strands plus one strand that is completely new (Figure 6–12).

DNA Synthesis Begins at Replication Origins

The DNA double helix is normally very stable: the two DNA strands are locked together firmly by the large numbers of hydrogen bonds between the bases on both strands (see Figure 6–4). As a result, only temperatures approaching those of boiling water supply sufficient thermal energy to separate these strands. In order to be used as a template, however, the double helix must first be opened up and the two strands separated to expose unpaired bases (see Figure 6–10). The process of DNA replication is begun by initiator proteins that bind to the DNA and pry the two strands apart, breaking the hydrogen bonds between the bases (Figure 6–13). Although the hydrogen bonds collectively make the DNA helix very stable, individually each hydrogen bond is weak (Chapter 2). Separating a short length of DNA does not therefore require a large energy input and can occur with the assistance of these proteins at normal temperatures.

The positions at which the DNA is first opened are called **replication origins** (see Figure 6–13), and they are marked by a particular sequence of nucleotides. In simple cells like those of bacteria or yeast, replication

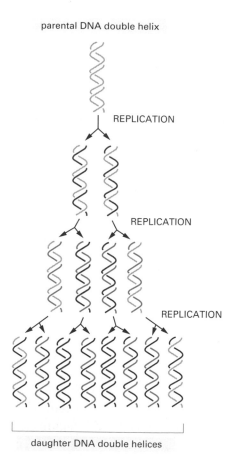

parental DNA double helix

REPLICATION

REPLICATION

REPLICATION

daughter DNA double helices

Figure 6–12 Replication of DNA. In each round of replication, each of the two strands of DNA is used as a template for the formation of a complementary DNA strand. The original strands, therefore, remain intact through many cell generations. DNA replication is "semiconservative" because each daughter DNA double helix is composed of one conserved strand and one newly synthesized strand.

origins span approximately 100 base pairs; they are composed of DNA sequences that attract the initiator proteins, as well as stretches of DNA that are especially easy to open. We saw in Figure 6–5 that an A-T base pair is held together by fewer hydrogen bonds than is a G-C base pair. Therefore, DNA rich in A-T base pairs is relatively easy to pull apart, and A-T-rich stretches of DNA are typically found at replication origins.

A bacterial genome, which is typically contained in a circular DNA molecule of several million nucleotide pairs, has a single origin of replication. The human genome, which is very much larger, has approximately 10,000 such origins. In humans, beginning DNA replication at many places at once allows a cell to replicate its DNA relatively quickly.

Once an initiator protein binds to DNA at the replication origin and locally opens up the double helix, it attracts a group of proteins that carry out DNA replication. This group cooperates as a *protein machine* with each member carrying out a specific function. We will introduce each of these proteins shortly, after we consider the overall appearance of DNA that is in the process of being replicated.

New DNA Synthesis Occurs at Replication Forks

DNA molecules caught in the act of being replicated can be seen in the electron microscope (Figure 6–14), where it is possible to see Y-shaped junctions in the DNA, called **replication forks.** At these forks, the replication machine is moving along the DNA, opening up the two strands of the double helix and using each strand as a template to make a new daughter strand. Two replication forks are formed starting from each replication origin, and they move away from the origin in both directions, unzipping the DNA as they go. DNA replication in bacterial and eucaryotic chromosomes is therefore termed *bidirectional.* The forks move very rapidly—at about 1000 nucleotide pairs per second in bacteria and 100 nucleotide pairs per second in humans.

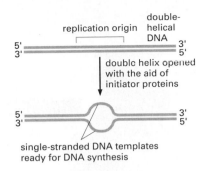

Figure 6–13 The opening of the DNA double helix at a replication origin. Replication initiator proteins recognize sequences of DNA at replication origins and locally pry apart the two strands of the double helix. The exposed single strands can then serve as templates for copying the DNA.

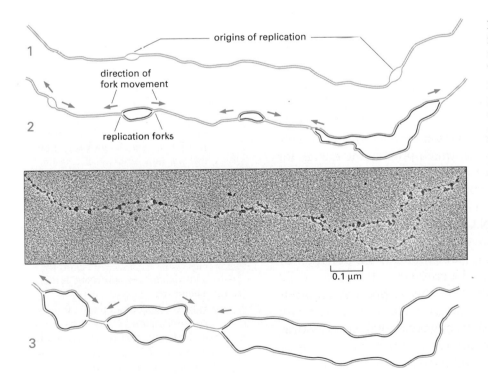

Figure 6–14 Replication forks move away in both directions from multiple replication origins in a eucaryotic chromosome. The electron micrograph shows DNA replicating in the early embryo of a fly. The particles visible along the DNA are nucleosomes, structures present in eucaryotic chromosomes in which the DNA is wrapped several times around a core of proteins (these structures will be described in Chapter 8). (1), (2), and (3) are drawings of the same portion of a DNA molecule as it might appear at successive stages of replication, drawn from electron micrographs. (2) is drawn from the electron micrograph shown here. The *yellow lines* represent the parental DNA strands; the *red lines* represent the newly synthesized DNA. (Electron micrograph courtesy of Victoria Foe.)

DNA Replication

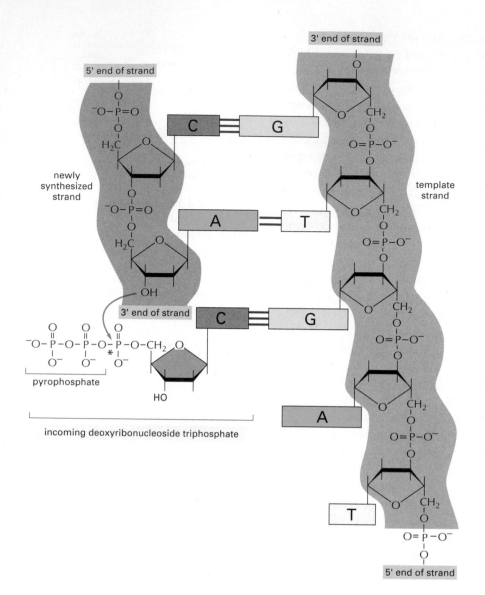

Figure 6–15 DNA synthesis. Addition of a deoxyribonucleotide to the 3′-hydroxyl end of a polynucleotide chain is the fundamental reaction by which DNA is synthesized; the new DNA chain is therefore synthesized in the 5′-to-3′ direction. Base-pairing between the incoming deoxyribonucleotide and the template strand guides the formation of a new strand of DNA that is complementary in nucleotide sequence to the template chain (see Figure 6–10). The enzyme DNA polymerase catalyzes the addition of nucleotides to the growing DNA strand. The nucleotides enter the reaction as nucleoside triphosphates. Breakage of a phosphoanhydride bond (indicated by the *asterisk*) in the incoming nucleoside triphosphate releases a large amount of free energy and thus provides the energy for the polymerization reaction.

At the heart of the replication machine is an enzyme called **DNA polymerase,** which synthesizes the new DNA using one of the old strands as a template. This enzyme catalyzes the addition of nucleotides to the 3′ end of a growing DNA strand by the formation of a phosphodiester bond between this end and the 5′-phosphate group of the incoming nucleotide (Figure 6–15). The nucleotides enter the reaction initially as energy-rich nucleoside triphosphates, which provide the energy for the polymerization reaction. The hydrolysis of one phosphoanhydride bond in the nucleoside triphosphate provides the energy for the condensation reaction that links the nucleotide monomer to the chain and releases pyrophosphate (PP_i). The DNA polymerase couples the release of this energy to the polymerization reaction. Pyrophosphate is further hydrolyzed to inorganic phosphate (P_i), which makes the polymerization reaction effectively irreversible (see Figure 3–35).

DNA polymerase does not dissociate from the DNA each time it adds a new nucleotide to the growing chain; rather, it stays associated with the DNA and moves along it stepwise for many cycles of the polymerization reaction. We will see, later in this chapter, how a special protein keeps the polymerase attached in this way.

Question 6–2 Look carefully at the micrograph in Figure 6–14.

A. Using the scale bar, estimate the lengths of the DNA strands between the replication forks. Numbering the replication forks sequentially from the left, how long will it take until forks #4 and #5, and forks #6 and #7, respectively, collide with each other? The distance between the bases in DNA is 0.34 nm, and eucaryotic replication forks move at about 100 nucleotides per second. For this question disregard the nucleosomes shown in Figure 6–14 and assume that the DNA is fully extended.

B. The fly genome is about 1.8×10^8 nucleotide pairs in size. How much of the total fly DNA is shown in the micrograph?

The Replication Fork Is Asymmetrical

The 5′-to-3′ DNA polymerization mechanism poses a problem at the replication fork. We saw in Figure 6–4 that the sugar-phosphate backbone of each strand of a DNA double helix has a unique chemical direction, or polarity, determined by the way that each sugar residue is linked to the next, and that the two strands in the double helix run in opposite orientations. As a consequence, at the replication fork, one new DNA strand is being made on a template running in one direction (3′-to-5′), whereas the other new strand is being made on a template running in the opposite direction (5′-to-3′) (Figure 6–16). The replication fork is therefore asymmetrical. Both the new DNA strands would appear to be growing in the same direction, that is, the direction in which the replication fork is moving. On the face of it this would suggest that one strand is being synthesized in the 3′-to-5′ direction and one is being synthesized in the 5′-to-3′ direction.

DNA polymerase, however, can catalyze the growth of the DNA chain in only one direction; it can add new subunits only to the 3′ end of the chain (see Figure 6–15). As a result, a new DNA chain can be synthesized only in a 5′-to-3′ direction by DNA polymerase. This easily can account for the synthesis of one of the two strands of DNA at the replication fork. One might have expected a second DNA polymerase to synthesize the other DNA strand—one that works by adding subunits to the 5′ end of a DNA chain. However, no such enzyme exists. Instead, the problem is solved by a "backstitching" maneuver. The DNA strand whose 5′ end must grow is made *discontinuously*, in successive separate small pieces, with the DNA polymerase working backward from the replication fork in the 5′-to-3′ direction for each piece. The pieces are later "stitched" together to form a continuous new strand (Figure 6–17). The DNA strand that is synthesized discontinuously in this way is called the **lagging strand;** the strand that is synthesized continuously is called the **leading strand.**

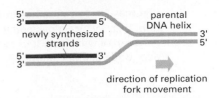

Figure 6–16 The polarity of DNA strands at a replication fork.

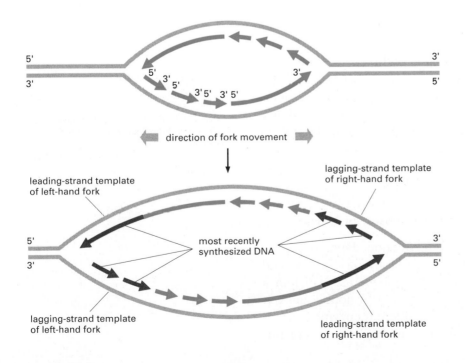

Figure 6–17 The asymmetric nature of DNA replication forks. Because both of the new strands are synthesized in the 5′-to-3′ direction, the DNA synthesized on the lagging strand must be made initially as a series of short DNA strands that are later joined together. The upper diagram shows two replication forks moving in opposite directions; the lower diagram shows the same forks a short time later. On the lagging strand, DNA polymerase "backstitches": it must synthesize short fragments (called Okazaki fragments) in the 5′-to-3′ direction, and then move in the opposite direction along the template strand (toward the fork) before it synthesizes the next fragment.

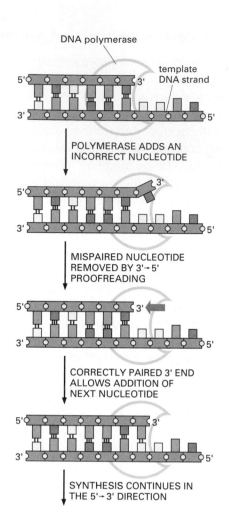

Figure 6–18 Proofreading by DNA polymerase during DNA replication. If an incorrect nucleotide is added to a growing strand, the DNA polymerase will cleave it from the strand and replace it with the correct nucleotide before continuing.

Although they differ in subtle details, the replication forks of all cells, procaryotic and eucaryotic, have leading and lagging strands. The common feature arises from the fact that all of the DNA polymerases used to replicate DNA polymerize in the 5′-to-3′ direction only. We shall look at events on the lagging strand in more detail later in this chapter; but first, we consider another feature of DNA polymerase that is common to all cells.

DNA Polymerase Is Self-correcting

DNA polymerase is so accurate that it makes only about one error in every 10^7 nucleotide pairs replicated. This error rate is much lower than can be accounted for simply by the accuracy of complementary base-pairing. Although A-T and C-G are by far the most stable base pairs, other less stable base pairs, for example G-T and C-A, can also be formed. Such incorrect base pairs are formed much less frequently than correct ones, but they occur often enough that they would kill the cell through an accumulation of mistakes in the DNA if they were allowed to remain. This catastrophe is avoided because DNA polymerase can correct its mistakes. As well as catalyzing the polymerization reaction, DNA polymerase has an error-correcting activity called **proofreading.** Before the enzyme adds a nucleotide to a growing DNA chain, it checks whether the previous nucleotide added is correctly base-paired to the template strand. If so, the polymerase adds the next nucleotide; if not, the polymerase removes the mispaired nucleotide by cutting the phosphodiester bond it has just made, releases the nucleotide, and tries again (Figure 6–18). Thus, DNA polymerase possesses both a 5′-to-3′ polymerization activity and a 3′-to-5′ nuclease (nucleic-acid-degrading) activity.

This proofreading mechanism explains why DNA polymerases synthesize DNA only in the 5′-to-3′ direction, despite the need this imposes for a cumbersome backstitching mechanism at the replication fork. As shown in Figure 6–19, a hypothetical DNA polymerase that synthesized in the 3′-to-5′ direction (and would thereby circumvent the need for backstitching) would be unable to proofread: if it removed an incorrectly paired nucleotide, the polymerase would create a chain end that is chemically dead, in the sense that it would no longer be able to elongate. Thus, in order for a DNA polymerase to function as a self-correcting enzyme that removes its own polymerization errors as it moves along the DNA, it must proceed only in the 5′-to-3′ direction.

Short Lengths of RNA Act as Primers for DNA Synthesis

We have seen that the accuracy of DNA replication depends on the requirement of the DNA polymerase for a correctly base-paired end before it can add nucleotides. But since the polymerase can join a nucleotide only to a base-paired nucleotide in a DNA double helix, it cannot start a completely new DNA strand. A different enzyme is needed to begin a new DNA strand, an enzyme that can begin a new polynu-

Question 6–3 DNA repair enzymes preferentially repair mismatched bases on the newly synthesized DNA strand, using the old DNA strand as a template. If mismatches were simply repaired without regard for which strand served as template, would this reduce replication errors? Explain your answer.

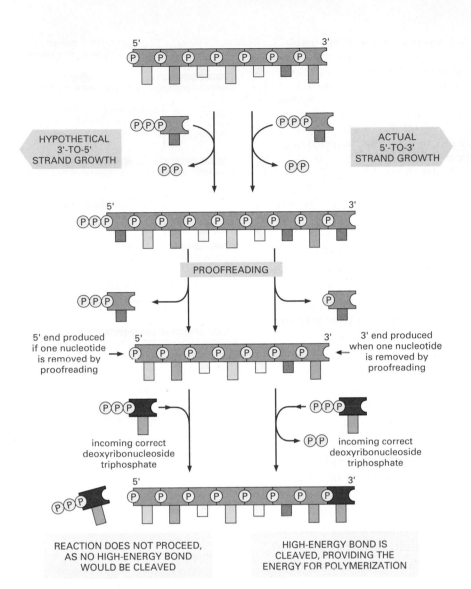

**Figure 6–19 Explanation for the 5′-to-3′
direction of DNA chain growth.** Growth in
the 5′-to-3′ direction, shown on the right,
allows the chain to continue to be elongated when a mistake in polymerization
has been removed by proofreading (see
Figure 6–18). In contrast, proofreading in
the hypothetical 3′-to-5′ polymerization
scheme, shown on the left, would block
further chain elongation. For convenience,
only one strand of the DNA double helix is
shown.

clcotide chain simply by joining two nucleotides together without the
need for a base-paired end. This enzyme does not, however, synthesize
DNA. It makes a short length of a closely related type of nucleic acid—
RNA (ribonucleic acid)—using the DNA strand as a template. This short
length of RNA, around 10 nucleotides long, is base-paired to the template strand and provides a base-paired 3′ end as a starting point for
DNA polymerase. It thus serves as a *primer* for DNA synthesis, and the
enzyme that synthesizes the RNA primer is known as *primase*. A strand
of RNA is very similar chemically to a single strand of DNA except that it
is made of ribonucleotide subunits, in which the sugar is ribose, not
deoxyribose; RNA also differs from DNA in that it contains the base
uracil (U) instead of thymine (T) (see Panel 2–6, pp. 55–67). However,
since U can form a base pair with A, the RNA primer is synthesized on
the DNA strand by complementary base-pairing in exactly the same way
as is DNA.

For the leading strand, an RNA primer is needed only to start replication at a replication origin; once a replication fork has been established, the DNA polymerase is continuously presented with a base-
paired 3′ end as it tracks along the template strand. But, on the lagging
strand, where DNA synthesis is discontinuous, new primers are needed

Figure 6–20 The synthesis of DNA fragments on the lagging strand. In eucaryotes the RNA primers are made at intervals of about 200 nucleotides on the lagging strand, and each RNA primer is 10 nucleotides long. In the bacterium *E. coli*, the primers and Okazaki fragments are about 5 and 1000 nucleotides long, respectively. Primers are erased by nucleases that recognize an RNA strand in an RNA/DNA helix and excise it; this leaves gaps that are filled in by a DNA repair polymerase that can proofread as it fills the gaps. The completed fragments are finally joined together by an enzyme called DNA ligase, which catalyzes the formation of a phosphodiester bond between the 3'-OH end of one fragment and the 5'-P end of the next, thus linking up the sugar-phosphate backbones.

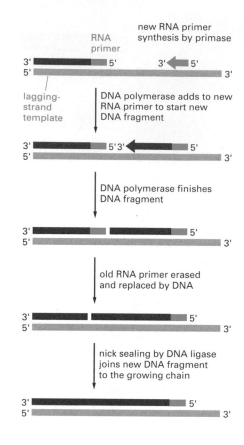

continually, as one can see from Figure 6–17. As the movement of the replication fork exposes a new stretch of unpaired bases, a new RNA primer is made at intervals along the lagging strand. DNA polymerase adds a deoxyribonucleotide to the 3' end of this primer to start a DNA strand, and it will continue to elongate this strand until it runs into the next RNA primer (Figure 6–20).

To produce a continuous new DNA strand from the many separate pieces of DNA made on the lagging strand (called *Okazaki fragments* after the biochemist who discovered them), three additional enzymes are needed. These act quickly to remove the RNA primer, replace it with DNA, and join the DNA fragments together: a *nuclease* breaks apart the RNA primer, a DNA polymerase called a *repair polymerase* replaces the RNA with DNA, and the enzyme *DNA ligase* joins the 5'-phosphate end of one new DNA fragment to the 3'-hydroxyl end of the next (see Figure 6–20). We will discuss these enzymes in more detail in the section on DNA repair later in this chapter.

Primase can begin new polynucleotide chains, but this is possible because it does not proofread its work. As a result, primers contain a high frequency of mistakes. But, since they are made of RNA instead of DNA, the primers have been marked out as "suspect copy" to be automatically removed and replaced by DNA. This DNA is put in by DNA repair polymerases, which, like the replicative polymerases, proofread as they synthesize. In this way, the cell's replication machinery is able to begin new DNA chains and, at the same time, ensure that all of the DNA is copied faithfully.

Proteins at a Replication Fork Cooperate to Form a Replication Machine

We saw earlier in this chapter that DNA replication requires a variety of proteins in addition to DNA polymerase. Here, we will discuss the additional proteins that, together with DNA polymerase and primase, form the protein machine that powers the replication fork forward and synthesizes new DNA behind it. (Although it would make good sense for the three proteins that replace RNA primers with DNA—nuclease, repair polymerase, and ligase—to also be a part of the replication machine, it is not yet known whether this is the case.)

At the head of the replication machine is a *helicase*, a protein that uses the energy of ATP hydrolysis to speed along DNA, opening the double helix as it moves (Figure 6–21). We saw earlier in this chapter that the DNA double helix must be opened to begin DNA replication, and it must

Question 6–4 Discuss the following statement: "Primase is a sloppy enzyme that makes many mistakes. Eventually, the RNA primers it makes are disposed of and replaced with DNA by a polymerase with higher fidelity. This is wasteful. It would be more energy efficient if a DNA polymerase made an accurate copy in the first place."

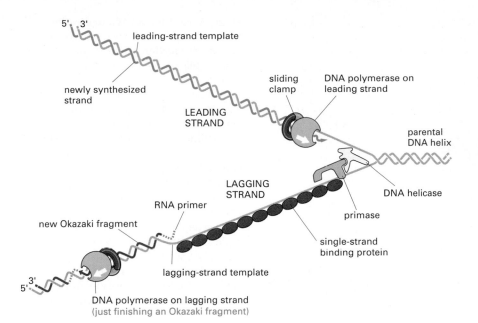

5' 3'
leading-strand template

newly synthesized
strand

LEADING
STRAND

sliding
clamp

DNA polymerase on
leading strand

parental
DNA helix

LAGGING
STRAND

RNA primer

new Okazaki fragment

DNA helicase

primase

single-strand
binding protein

lagging-strand template

5' 3'

DNA polymerase on lagging strand
(just finishing an Okazaki fragment)

Figure 6–21 Proteins that act at a replication fork. Two molecules of DNA polymerase are shown, one on the leading strand and one on the lagging strand. Both are held on to the DNA by a circular protein clamp that allows the polymerase to slide. DNA helicase uses the energy of ATP hydrolysis to propel itself forward and thereby separate the strands of the parental DNA double helix ahead of the polymerase. Single-stranded DNA-binding proteins maintain these separated strands as single-stranded DNA to provide access for the primase and polymerase. For simplicity, the figure shows the proteins working independently; in the cell they are held together in a large replication machine, a view of which is shown in Figure 6–22.

also be opened continuously as the replication fork progresses, in order to provide exposed templates for the polymerase. Another component of the replication machine—*single-strand binding protein*—clings to the single-stranded DNA exposed by the helicase and transiently prevents it from re-forming base pairs. Yet another protein, called a *sliding clamp*, keeps the DNA polymerase firmly attached to the DNA template; on the lagging strand, the sliding clamp releases the polymerase from the DNA each time an Okazaki fragment is completed. This clamp protein forms a ring around the DNA helix and binds polymerase, allowing it to slide along a template strand as it synthesizes new DNA (see Figure 6–21).

Most of the proteins involved in DNA replication are thought to be held together in a large multienzyme complex that moves as a unit along the DNA, enabling DNA to be synthesized on both strands in a coordinated manner. This complex can be likened to a tiny sewing machine composed of protein parts and powered by nucleoside triphosphate hydrolysis. Although the detailed structure of the replication machine is not known, some ideas as to its appearance have been proposed (Figure 6–22).

Our current understanding of DNA replication is more complete than that of many other areas of cell biology, yet many mysteries still remain. For example, it is not yet understood how the polymerase on the leading strand is connected with that on the lagging strand in order to allow replication to proceed synchronously on both strands. Moreover, although we know in some detail how DNA replication begins at replication origins in bacteria, our understanding of this process in humans is only just beginning.

Given the demands for accuracy and the lengths that cells go to to achieve it during DNA replication, it is not surprising, as we shall see in the following section, that cells have also evolved elaborate protein machines to scan the finished product, so as to correct mistakes (rare as they are) made during DNA replication and to repair DNA that becomes accidentally damaged by light, by chemicals in the cell, and by other agents.

Question 6–5 A gene encoding one of the proteins involved in DNA replication has been inactivated by a mutation in a cell. In the absence of this protein the cell attempts to replicate its DNA for the very last time. What DNA products would be generated in each case if the following protein were missing?

A. DNA polymerase

B. DNA ligase

C. Sliding clamp for DNA polymerase

D. Nuclease that removes RNA primers

E. DNA helicase

F. Primase

DNA Replication

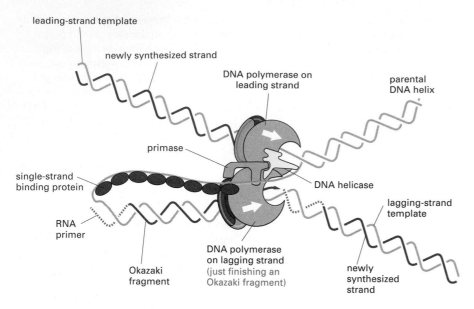

leading-strand template

newly synthesized strand

DNA polymerase on leading strand

parental DNA helix

primase

single-strand binding protein

DNA helicase

RNA primer

lagging-strand template

Okazaki fragment

DNA polymerase on lagging strand (just finishing an Okazaki fragment)

newly synthesized strand

Figure 6–22 A model of a replication fork. This diagram shows a current view of how the replication proteins are arranged at the replication fork when the fork is moving. The structure of Figure 6–21 has been altered by folding the DNA on the lagging strand to bring the lagging-strand DNA polymerase molecule in contact with the leading-strand DNA polymerase molecule. This folding process also brings the 3′ end of each completed Okazaki fragment close to the start site for the next Okazaki fragment (compare with Figure 6–21). Because the lagging-strand DNA polymerase molecule is held to the rest of the replication proteins, it can be reused to synthesize successive Okazaki fragments; in this diagram, it is about to let go of its completed DNA fragment and move to the RNA primer that will be synthesized nearby, as required to start the next DNA fragment on the lagging strand.

DNA Repair

The diversity of living organisms and their success in colonizing almost every part of the earth's surface depends on genetic changes accumulated gradually over millions of years, allowing organisms to adapt to changing conditions and to colonize new habitats. However, in the short term, and from the perspective of an individual organism, genetic change is almost always detrimental, especially in multicellular organisms, where a genetic change is more likely than not to upset an organism's extremely complex and finely tuned development and physiology. In order to survive and reproduce, individuals must be genetically stable. This stability is achieved not only through the extremely accurate mechanism for replicating DNA that we have just discussed, but also through mechanisms for correcting the rare copying mistakes made by the replication machinery and for repairing the accidental damage that is continually occurring to the DNA. Most of these changes in DNA are only temporary because they are immediately corrected by processes collectively called **DNA repair.**

Changes in DNA Are the Cause of Mutations

Only rarely do the cell's DNA replication and repair processes fail and allow a permanent change in the DNA. Such a permanent change is called a **mutation,** and it can have profound consequences. A mutation affecting just a single nucleotide pair can destroy an organism if the change occurs in a vital position in the DNA sequence. For example, humans use the protein hemoglobin to transport oxygen in the blood; the sequence of nucleotides that encodes the amino acid sequence of one of the two types of protein chains (the β-globin chain) of the hemoglobin molecule is shown in Figure 6–9. A permanent change in a single nucleotide in this sequence can cause cells to make a β-globin chain with an incorrect sequence of amino acids. Since the structure and activity of a protein depend on its amino acid sequence, a protein with an altered sequence may function poorly or not at all. For example, a muta-

Figure 6–23 **The mutation responsible for the disease sickle-cell anemia.** The complete nucleotide sequence of the β-globin gene is given in Figure 6–9. Only a small portion of the sequence near the beginning of the gene is shown in (A). The single nucleotide change (mutation) in the sickle-cell gene produces a β-globin that differs from normal β-globin only by a change from glutamic acid to valine at the sixth amino acid position. (The β-globin molecule contains a total of 146 amino acids.) Humans carry two copies of each gene (one inherited from each parent); a sickle-cell mutation in one of the two β-globin genes generally causes no harm to the individual, as it is compensated for by the normal gene. However, an individual who inherits two copies of the mutant β-globin gene displays the symptoms of sickle-cell anemia. Normal red blood cells are shown in (B), and those from an individual suffering from sickle-cell anemia in (C).

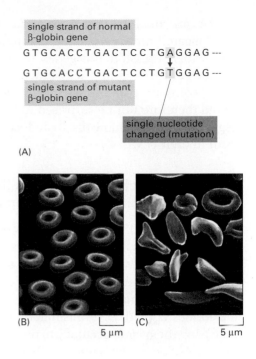

single strand of normal
β-globin gene

G T G C A C C T G A C T C C T G A G G A G ---

G T G C A C C T G A C T C C T G T G G A G ---

single strand of mutant
β-globin gene

single nucleotide
changed (mutation)

(A)

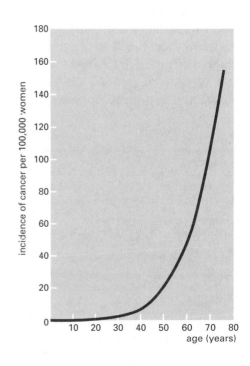

(B)　　5 μm　　(C)　　5 μm

tion affecting a single nucleotide pair is responsible for the disease *sickle-cell anemia* (Figure 6–23). The sickle-cell hemoglobin is less soluble than normal hemoglobin and forms fibrous precipitates, which lead to the characteristic sickle shape of affected red blood cells. Because these cells are more fragile and frequently break in the bloodstream, patients with the disease have a reduced number of red blood cells (Figure 6–23C), and this deficiency causes weakness, dizziness, and headaches, and can be life threatening.

The example of sickle-cell anemia, which is an inherited disease, illustrates the importance of protecting reproductive cells (*germ cells*) against mutation. A mutation in one of these will be passed on to all the cells in the body of the multicellular organism that develops from it, including the germ cells for production of the next generation. However, the many other cells in a multicellular organism (its *somatic cells*) must also be protected from genetic change to safeguard the health and well-being of the individual. Nucleotide changes that occur in somatic cells can give rise to variant cells, some of which grow in an uncontrolled fashion at the expense of the other cells in the organism. In the extreme case, the uncontrolled cell proliferation known as cancer results. This disorder, which is responsible for about 30% of the deaths that occur in Europe and North America, is due largely to a gradual accumulation of changes in the DNA sequences of somatic cells that is caused by random mutation (Figure 6–24). Therefore, a tenfold increase in the mutation frequency would cause a disastrous increase in the incidence of cancer by accelerating the rate at which somatic cell variants arise.

Figure 6–24 **Cancer incidence as a function of age.** The number of newly diagnosed cases of cancer of the colon in women in England and Wales in one year is plotted as a function of age at diagnosis. Since cells are continually experiencing accidental changes to their DNA that accumulate and are passed on to progeny cells, the chance that a cell will become cancerous increases greatly with age. The steep rise seen here in older women reveals that this cancer increases as the sixth power of age, suggesting that cancer arises only after a set of six or so random mutations have occurred in genes that regulate cell growth in the colon. (Data from C. Muir et al., Cancer Incidence in Five Continents, Vol. V. Lyon: International Agency for Research on Cancer, 1987.)

We can conclude that the high fidelity with which DNA sequences are replicated and maintained is important both for the reproductive cells, which transmit the genes to the next generation, and for the somatic cells, which normally function as carefully regulated members of the complex community of cells in a multicellular organism. We should therefore not be surprised to find that all cells have acquired an elegant set of mechanisms to reduce the number of mutations that occur in their DNA.

A DNA Mismatch Repair System Removes Replication Errors That Escape from the Replication Machine

In the first part of this chapter, we saw that the cell's replication machinery ensures against copying mistakes. Despite these safeguards, however, such mistakes do occur. The cell has a backup system—called *DNA mismatch repair*—which is dedicated to correcting these rare mistakes. The replication machine itself makes approximately one error per 10^7 nucleotides copied; DNA mismatch repair corrects 99% of these errors, increasing the overall accuracy to one mistake in 10^9 nucleotides copied. This level of accuracy is much higher than that generally encountered in the world around us (Table 6–1).

Whenever the replication machinery makes a copying mistake, it leaves a mispaired nucleotide (commonly called a mismatch) behind. If left uncorrected, the mismatch will result in a permanent mutation in the next round of DNA replication (Figure 6–25A). A complex of mismatch repair proteins recognizes these DNA mismatches, removes (excises) one of the two strands of DNA involved in the mismatch, and resynthesizes the missing strand (Figure 6–26). To be effective in correcting replication mistakes, this mismatch repair system must always excise only the newly synthesized DNA strand: excising the other strand (the old strand) would preserve the mistake instead of correcting it (Figure 6–25).

In eucaryotes, it is not yet known for certain how the mismatch repair machinery distinguishes the two DNA strands. However, there is evidence that newly replicated DNA strands—both leading and lagging—are preferentially nicked; it is these nicks (single-stranded breaks) that appear to provide the signal that directs the mismatch repair machinery to the appropriate strand (see Figure 6–26).

Table 6–1	Error Rates
U.S. Postal Service on-time delivery of local first-class mail	13 late deliveries per 100 parcels
Airline luggage system	1 lost bag per 200
A professional typist typing at 120 words per minute	1 mistake per 250 characters
Driving a car in the United States	1 death per 10^4 people per year
DNA replication (without mismatch repair)	1 mistake per 10^7 nucleotides copied
DNA replication (including mismatch repair)	1 mistake per 10^9 nucleotides copied

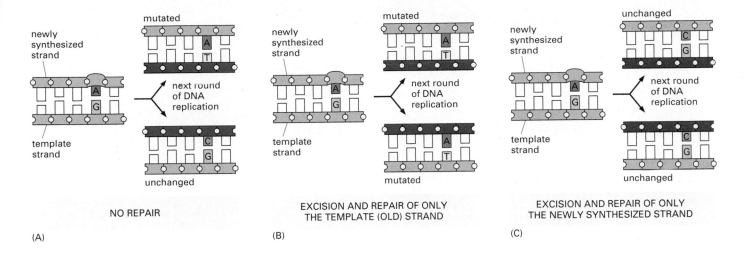

(A) NO REPAIR

(B) EXCISION AND REPAIR OF ONLY THE TEMPLATE (OLD) STRAND

(C) EXCISION AND REPAIR OF ONLY THE NEWLY SYNTHESIZED STRAND

The importance of mismatch repair in humans was recognized recently when it was discovered that an inherited predisposition to certain cancers (especially some types of colon cancer) is caused by a mutation in the gene responsible for producing one of the mismatch repair proteins. Humans inherit two copies of this gene (one from each parent), and individuals who inherit one damaged mismatch repair gene show no symptoms until the undamaged copy of the gene is accidentally mutated in a somatic cell. This gives rise to a clone of somatic cells that, because they are deficient in mismatch repair, accumulate mutations more rapidly than do normal cells. Since most cancers arise from cells that have accumulated multiple mutations (see Figure 6–24), a cell deficient in mismatch repair has a greatly enhanced chance of becoming cancerous. Thus, inheriting a damaged mismatch repair gene predisposes an individual to cancer.

Figure 6–25 DNA mismatches and their repair. (A) If uncorrected, the mismatch will lead to a permanent mutation in one of the two DNA molecules produced by the next round of DNA replication. (B) If the mismatch is "repaired" using the newly synthesized DNA strand as the template, both DNA molecules produced by the next round of DNA replication will contain a mutation. (C) If the mismatch is corrected using the original template (old) strand as the template, the possibility of a mutation is eliminated. The scheme shown in (C) is used by cells to repair mismatches, as shown in Figure 6–26.

DNA Is Continually Suffering Damage in Cells

Rare mistakes in DNA replication, as we have seen, can be corrected by the mismatch repair mechanism. There are also other ways in which the DNA can be damaged, and these require other mechanisms for their repair. Just like any other molecule in the cell, DNA is continually undergoing thermal collisions with other molecules. These often result in major chemical changes in the DNA. For example, during the time it takes to read this sentence, a total of about a trillion (10^{12}) purine bases (A and G) will be lost from the DNA of your cells by a spontaneous reac-

Figure 6–26 Mechanism of DNA mismatch repair in eucaryotes. A DNA mismatch distorts the geometry of the DNA double helix, and this distortion is recognized by the DNA mismatch repair proteins, which then remove the newly synthesized DNA. The gap in the newly synthesized DNA is replaced by a DNA polymerase that proofreads as it synthesizes and is sealed by DNA ligase. As shown in the figure, a nick in the DNA has been proposed as the signal that allows the mismatch repair proteins to distinguish the newly synthesized DNA (which contains the mistake) from the old DNA. Such nicks are known to occur in the lagging strands (see Figure 6–17) and are observed to also occur, although less frequently, in the leading strands. These nicks remain for only a short period after a replication fork passes (see Figure 6–20), so that mismatch repair must occur quickly.

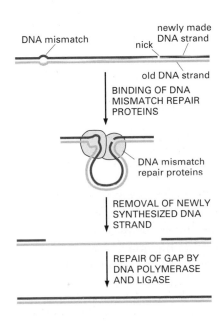

DNA mismatch   nick   newly made DNA strand

old DNA strand

BINDING OF DNA MISMATCH REPAIR PROTEINS

DNA mismatch repair proteins

REMOVAL OF NEWLY SYNTHESIZED DNA STRAND

REPAIR OF GAP BY DNA POLYMERASE AND LIGASE

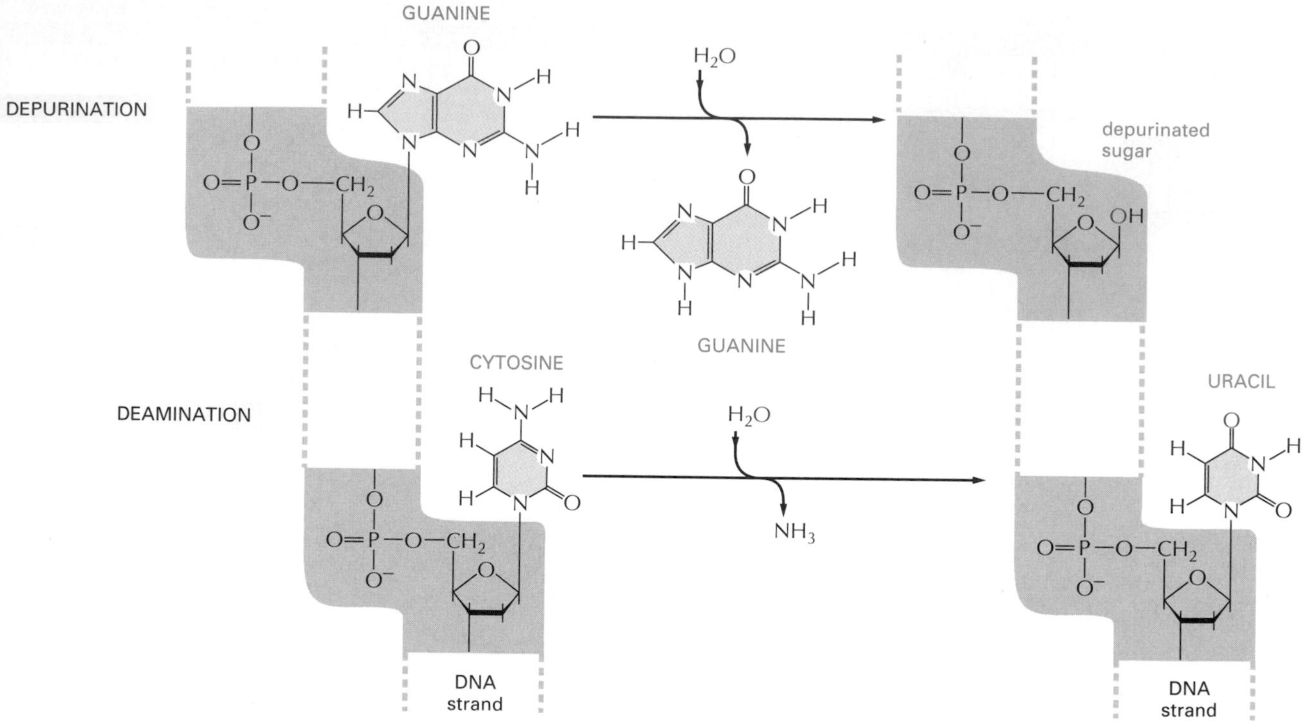

Figure 6–27 **Depurination and deamination.** These two reactions are the two most frequent spontaneous chemical reactions known to create serious DNA damage in cells. Depurination can release guanine (shown in the figure) as well as adenine from DNA. The major type of deamination reaction (shown in the figure) converts cytosine to an altered DNA base, uracil, but deamination occurs on other bases as well. These reactions take place on double-helical DNA; for convenience, only one strand is shown.

tion called *depurination* (Figure 6–27). Depurination does not break the phosphodiester backbone but, instead, gives rise to lesions that resemble missing teeth. Another major change is the spontaneous loss of an amino group (*deamination*) from cytosine in DNA to produce the base uracil (see Figure 6–27). Some chemically reactive by-products of metabolism also occasionally react with the bases in DNA, altering them in such a way that their base-pairing properties are changed. The ultraviolet radiation in sunlight is also damaging to DNA; it promotes covalent linkage between two adjacent pyrimidine bases, forming, for example, the thymine dimer shown in Figure 6–28.

These are only a few of many chemical changes that can occur in our DNA. If left unrepaired, many of them would lead either to the substitution of one nucleotide pair by another as a result of incorrect base-pairing during replication (Figure 6–29A) or to deletion of one or more nucleotide pairs in the daughter DNA strand after DNA replication (Figure 6–29B). Some types of DNA damage (thymine dimers, for example) often stall the DNA replication machinery at the site of the damage. All of these types of damage, if unrepaired, would have disastrous consequences for an organism.

Question 6–6 Discuss the following statement: "The DNA repair enzymes that correct defects introduced by deamination and depurination reactions must preferentially recognize such defects on newly synthesized DNA strands."

The Stability of Genes Depends on DNA Repair

Although thousands of random chemical changes are created every day in the DNA of a human cell by heat energy and metabolic accidents, the vast majority are eliminated by DNA repair. There are a variety of repair mechanisms, each catalyzed by a different set of enzymes. Nearly all these mechanisms depend on the existence of two copies of the genetic information, one in each strand of the DNA double helix: if the sequence in one strand is accidentally damaged, information is not lost irretrievably because a backup version of the altered strand remains in the complementary sequence of nucleotides in the other strand. Most damage

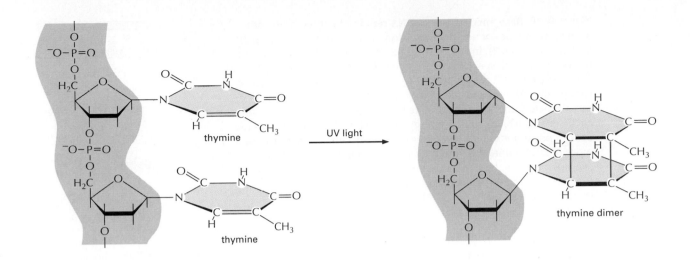

Figure 6–28 **DNA damage produced by ultraviolet radiation in sunlight.** Two adjacent thymine bases have become covalently attached to one another to form a thymine dimer. Skin cells that are exposed to sunlight are especially susceptible to this type of DNA damage.

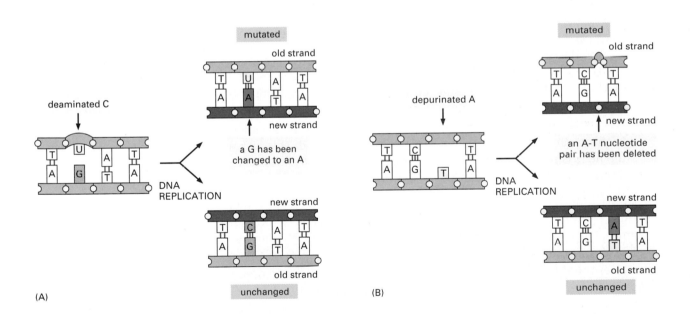

Figure 6–29 **How chemical modifications of nucleotides produce mutations.** (A) Deamination of cytosine, if uncorrected, results in the substitution of one base for another when the DNA is replicated. As shown in Figure 6–27, deamination of cytosine produces uracil. Uracil differs from cytosine in its base-pairing properties and preferentially base-pairs with adenine. The DNA replication machinery therefore inserts an adenine when it encounters a uracil on the template strand. (B) Depurination, if uncorrected, can lead to the loss of a nucleotide pair. When the replication machinery encounters a missing purine on the template strand, it can skip to the next complete nucleotide, thus producing a nucleotide deletion in the newly synthesized strand.

Figure 6–30 Basic mechanism of DNA repair. The three steps common to most types of repair are excision (step 1), resynthesis (step 2), and ligation (step 3). In step 1, the damage is cut out by one of a series of nucleases, each specialized for a type of DNA damage; in steps 2 and 3, the original DNA sequence is restored by a repair DNA polymerase, which fills in the gap created by the excision events. DNA ligase seals the nick left in the sugar-phosphate backbone of the repaired strand. Nick-sealing, which requires energy from ATP hydrolysis, remakes the broken phosphodiester bond between the adjacent nucleotides. Some types of DNA damage (deamination of cytosine [Figure 6–27], for example) involve the replacement of a single nucleotide, as shown in the figure. For the repair of other kinds of DNA damage, such as thymine dimers (see Figure 6–28), a longer stretch of 10 to 20 nucleotides is removed from the damaged strand.

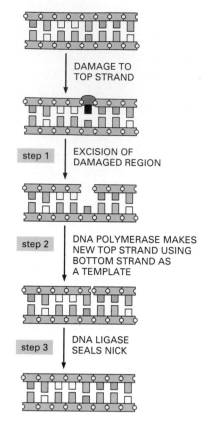

NET RESULT: REPAIRED DNA

creates structures that are never encountered in an undamaged DNA strand; thus the good strand is easily distinguished from the bad. The basic pathway for repairing damage to DNA is illustrated schematically in Figure 6–30. As indicated, it involves three steps:

1. The damaged DNA is recognized and removed by one of a variety of different nucleases, which cleave the covalent bonds that join the damaged nucleotides to the rest of the DNA molecule, leaving a small gap on one strand of the DNA double helix in this region.

2. A repair DNA polymerase binds to the 3′-hydroxyl end of the cut DNA strand. It then fills in the gap by making a complementary copy of the information stored in the undamaged strand. Although a different enzyme from the DNA polymerase that replicates DNA, a repair DNA polymerase synthesizes DNA strands in the same way. For example, it synthesizes chains in the 5′-to-3′ direction and has the same type of proofreading activity to ensure that the template strand is accurately copied. In many cells, this is the same enzyme that fills in the gap left after the RNA primers are removed in normal DNA replication (see Figure 6–20).

3. When the repair DNA polymerase has filled in the gap, a break remains in the sugar-phosphate backbone of the repaired strand. This nick in the helix is sealed by DNA ligase, the same enzyme that joins the lagging strand DNA fragments during DNA replication.

Steps 2 and 3 are nearly the same for most types of DNA repair, including mismatch repair. However, step 1 utilizes a series of different enzymes, each specialized for removing different types of DNA damage.

The importance of these repair processes is indicated by the large investment that cells make in DNA repair enzymes. Single-celled organisms such as yeasts contain more than 50 different proteins that function in DNA repair, and DNA repair pathways are likely to be even more complex in humans. The importance of these DNA repair processes is also evident from the consequences of their malfunction. Humans with the genetic disease *xeroderma pigmentosum*, for example, cannot repair thymine dimers (see Figure 6–28), because they have inherited a defective gene for one of the proteins involved in this repair process. Such individuals develop severe skin lesions, including skin cancer, because of the accumulation of thymine dimers in cells that are exposed to sunlight and the consequent mutations that arise in the cells that contain them.

WHALE GTGTGGTCTCGTGATCAAAGGCGAAAGGTGGCTCTAGAGAATCCC

HUMAN GTGTGGTCTCGCGATCAGAGGCGCAAGATGGCTCTAGAGAATCCC

Figure 6–31 Comparison of a section of the sex determination gene from two different animals. Although their body plans are strikingly different, humans and whales are built from the same proteins. Despite the length of time since humans and whales diverged, the nucleotide sequences of many of their genes are still closely similar. The sequences of a part of the gene encoding the protein that determines maleness in humans and whales are shown one above the other, and positions where the the two are identical are *shaded*.

The High Fidelity with Which DNA Is Maintained Means That Closely Related Species Have Proteins with Very Similar Sequences

We have seen in this chapter that DNA is replicated with remarkable fidelity and that accidental change, which could lead to permanent changes in DNA sequence, is efficiently repaired. As a consequence of these mechanisms to preserve DNA sequences, changes in the DNA accumulate remarkably slowly in the course of evolution. Of course, the rate of evolutionary change in the DNA of a species depends also on the effects of natural selection: DNA copying errors that have harmful consequences for the organism are eliminated from the population through the death or infertility of individuals carrying the misreplicated DNA. But the mechanisms of DNA replication and repair are so accurate that even where no such selection operates—at the many sites in the DNA where a change of nucleotide has no effect on the fitness of the organism—the genetic message is faithfully preserved over tens of millions of years. Thus humans and chimpanzees, after about five million years of divergent evolution, still have DNA sequences that are 98% identical. Even humans and whales, after ten or twenty times this period, still have chromosomes that are unmistakably similar in their DNA sequence and many proteins with amino acid sequences that are almost identical (Figure 6–31). Thus, in our genomes, we and our relatives receive a message from the distant past—a message that is longer and more detailed than any book. Thanks to the faithfulness of DNA replication and repair, a hundred million years have scarcely changed its essential content.

Question 6–7 Suppose a mutation affects an enzyme that is required to repair the damage to DNA caused by the loss of purine bases. This mutation causes the accumulation of about 5000 mutations in the DNA of each of your cells per day. As the average difference in DNA sequence between humans and chimpanzees is about 1%, how long will it take for you to turn into a monkey? What is wrong with this argument?

Essential Concepts

- Life depends on stable and compact storage of genetic information.

- Genetic information is carried by very long DNA molecules, encoded in the linear sequence of nucleotides A, T, G, and C.

- A molecule of DNA is in the form of a double helix composed of a pair of complementary strands of nucleotides held together by hydrogen bonds between G-C and A-T base pairs.

- Each strand of DNA has a chemical polarity due to the linkage of alternating sugars and phosphates in its backbone. The two strands of a DNA molecule run antiparallel—that is, in opposite orientations.

- Each of the two DNA strands can act as a template for the synthesis of the other strand. A DNA double helix thus carries the same information in each of its strands.

- A DNA molecule is duplicated (replicated) by the polymerization of new complementary strands onto each of the old strands of the DNA double helix. This process of DNA replication, in which two identical DNA molecules are formed from the original molecule, enables the genetic information to be copied and passed on from cell to daughter cell and from parent to offspring.

- As a DNA molecule replicates, its two strands are pulled apart to form one or more Y-shaped replication forks. The enzyme DNA polymerase, situated in the fork, lays down a new complementary DNA strand on each parental strand, thereby making two new double-helical molecules.

- DNA polymerase replicates a DNA template with remarkable fidelity, making less than one error in every 10^7 bases read. This is possible because the enzyme removes its own polymerization errors as it moves along the DNA (proofreading).

- Since DNA polymerase can synthesize new DNA in only one direction, only one of the strands in the replication fork, the leading strand, can be replicated in a continuous fashion. On the lagging strand DNA is synthesized by the polymerase in a discontinuous "backstitching" process, making short fragments of DNA that are later joined up by the enzyme DNA ligase to make a single continuous DNA strand.

- The proofreading feature of DNA polymerase makes it incapable of starting a new DNA chain. DNA synthesis is primed by an RNA polymerase, called primase, that makes short lengths of RNA, called primers, that are subsequently erased and replaced with DNA.

- DNA replication requires the cooperation of many proteins, which form a multienzyme replication machine, situated at the replication fork, that catalyzes DNA synthesis.

- Errors in the replication of DNA and chemical reactions that damage the nucleotides in DNA cause changes in the nucleotide sequence of DNA. If these changes were not efficiently corrected, they would give rise to mutations, many of which would be harmful to the organism. Genetic information can be stored stably in DNA sequences only because a variety of DNA repair enzymes continuously scan the DNA and correct replication mistakes and replace damaged nucleotides. DNA can be repaired easily because one strand can be corrected using the other strand as a template.

- The rare copying mistakes that slip through the DNA replication machinery are dealt with by the mismatch repair proteins, which monitor newly replicated DNA and repair copying mistakes. The overall accuracy of DNA replication, including mismatch repair, is one mistake per 10^9 nucleotides copied.

- DNA damage caused by chemical reactions and ultraviolet irradiation is corrected by a variety of enzymes that recognize damaged DNA and excise a short stretch of the DNA strand that contains it. The missing DNA is resynthesized by a repair DNA polymerase that uses the undamaged strand as a template. DNA ligase reseals the DNA to complete the repair process.

Key Terms

	DNA	gene	proofreading
	DNA polymerase	genome	replication fork
antiparallel	DNA repair	lagging strand	replication origin
base pair	DNA replication	leading strand	RNA
complementary	double helix	mutation	template

Questions

Question 6–8 Which of the following statements are correct? Explain your answers.

A. A DNA strand has a polarity because the bases contain hydrophilic amino groups.

B. The replication fork is asymmetrical because it contains two DNA polymerase molecules that are structurally distinct.

C. G-C base pairs are more stable than A-T base pairs.

D. Okazaki fragments are removed by an RNA nuclease.

E. The error rate of DNA replication is reduced both by proofreading of the DNA polymerase and by DNA repair enzymes.

F. In the absence of DNA repair, genes are unstable.

G. None of the aberrant bases formed by deamination occur naturally in DNA.

H. Cancer results from uncorrected mutations in somatic cells.

Question 6–9 An A-T base pair is stabilized by only two hydrogen bonds. Hydrogen-bonding schemes of very similar strengths can also be drawn between other base combinations, such as the A-C and the A-G pairs shown in Figure Q6–9. What would happen if these pairs formed during DNA replication and the inappropriate bases were incorporated? Discuss why this does not happen often. (Hint: see Figure 6–5.)

Figure Q6–9

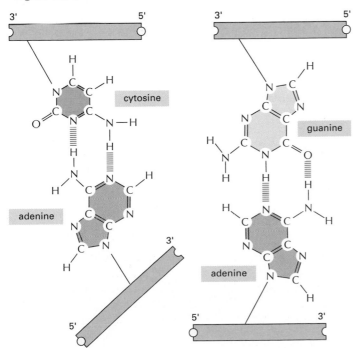

Question 6–10

A. A macromolecule isolated from an extraterrestrial source superficially resembles DNA but upon closer analyses reveals quite different base structures (Figure Q6–10) in place of A, T, G, and C. Look at these structures closely. Could these DNA-like molecules have been derived from a living organism that uses principles of genetic inheritance similar to those used by cells on earth? If so, what can you say about its properties?

Figure Q6–10

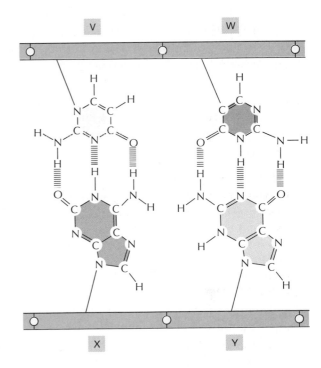

B. Simply judged by their potential for hydrogen-bonding, could any of these extraterrestrial bases replace terrestrial A, T, G, or C in terrestrial DNA? Explain your answers.

Question 6–11 The two strands of DNA double helix can be separated by heating. If you raised the temperature of a solution containing the following three DNA molecules, in what order do you suppose they would "melt"? Explain your answer.

A. 5'-GCGGGCCAGCCCGAGTGGGTAGCCCAGG-3'
   3'-CGCCCGGTCGGGCTCACCCATCGGGTCC-5'

B. 5'-ATTATAAAATATTTAGATACTATATTTACAA-3'
   3'-TAATATTTTATAAATCTATGATATAAATGTT-5'

C. 5'-AGAGCTAGATCGAT-3'
   3'-TCTCGATCTAGCTA-5'

Question 6–12 The total length of DNA in a human cell is about 1 meter, and the diameter of the double helix is about 2 nm. Nucleotides in a DNA double helix are stacked at an interval of 0.34 nm. If the DNA were enlarged so that its diameter equaled that of an electrical extension cord (5 mm), how long would the extension cord be from one end to the other (assuming that it is completely stretched out)? How close would the bases be to each other? How long would a gene of 1000 nucleotide pairs be?

Question 6–13   A compact disc (CD) stores about 4.8×10^9 bits of information in a 96 cm^2 area. This information is stored as a binary code—that is, every bit is either a 0 or a 1.

A. How many bits would it take to specify each nucleotide pair in a DNA sequence?

B. How many CDs would it take to store the information contained in the human genome?

Question 6–14   Being a born skeptic, you plan to confirm for yourself the results of a classic experiment originally performed in the 1960s by Meselson and Stahl from which they concluded that each daughter cell inherits one and only one strand of its mother's DNA. To do so, you "synchronize" (using established methods that need not concern us here) a culture of growing cells, so that virtually all cells in your flask begin and then complete DNA synthesis at the same time. Your cells are first grown in a normal growth medium and then, after one round of DNA synthesis, grown further in a specially concocted (and very expensive) growth medium that contains nutrients highly enriched in heavy isotopes of nitrogen and carbon (^{15}N and ^{13}C in place of the naturally abundant ^{14}N and ^{12}C). Cells growing on this medium use the heavy isotopes to build all of their macromolecules, including nucleotides and nucleic acids. You then isolate DNA from cells that have grown for a different number of generations in the heavy isotope medium and analyze the DNA for its density using a gradient centrifugation technique (see Panel 5–4, pp. 160–161). The more heavy isotopes have been built into the DNA, the heavier it appears in this analysis. Your data, plotting the amount of DNA isolated over its density, are shown in

Figure Q6–14. Are these results in agreement with your expectations? Explain the results.

Figure Q6–14

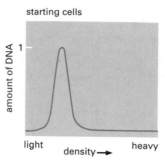

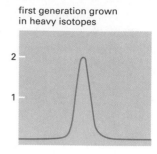

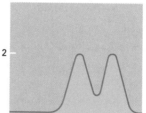

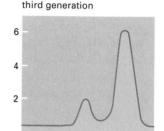

Question 6–15   The speed of DNA replication at a replication fork is about 100 nucleotides per second in human cells. What is the minimal number of origins of replication that a human cell must have in order to replicate its DNA once every 24 hours? Recall that a human cell contains two copies of the human genome, one inherited from the mother, the other from the father, each consisting of 3×10^9 nucleotide pairs.

Question 6–16   Look carefully at the structures of the compounds shown in Figure Q6–16. One or the other of the two compounds is added to a DNA replication reaction.

A. What would you expect if compound A were added in large excess over the concentration of the available deoxycytosine triphosphate (dCTP)?

Figure Q6–16

(A)

dideoxycytosine
triphosphate

(B)

dideoxycytosine
monophosphate

B. What would happen if it were added at 10% of the concentration of the available dCTP?

C. What effects would you expect if compound B were added under the same conditions?

Question 6–17   The genetic material of a hypothetical organism is structurally indistinguishable from DNA of normal cells. Surprisingly, analyses reveal that the DNA is synthesized from nucleoside triphosphates that contain free 5′-hydroxyl groups and triphosphate groups at the 3′ position. In what way must this organism's DNA polymerase differ from that of normal cells? Could it still proofread?

Question 6–18   Figure Q6–18 shows a snapshot of a replication fork in which the RNA primer has just been added to the lagging strand. Using this diagram as a guide, sketch the path of the DNA as the next Okazaki fragment is synthesized. Indicate the sliding clamp and the single-strand binding protein as appropriate.

Figure Q6–18

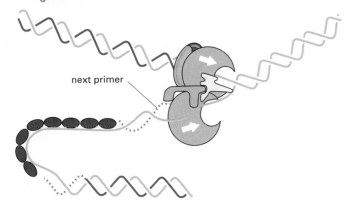

next primer

Question 6–19   Approximately how many high-energy bonds are used to replicate a bacterial chromosome? How much glucose (compared to its own weight of 10^{-12} g) does a bacterium need to consume to provide enough energy to copy its DNA once? The number of base pairs in the bacterial chromosome is 3×10^6. Oxidation of one glucose molecule yields about 30 high-energy phosphate bonds. The molecular weight of glucose is 180 g/mole. (Recall that there are 6×10^{23} molecules in a mole; see Chapter 2.)

Question 6–20   What, if anything, is wrong with the following statement: "Both reproductive cell DNA stability and somatic cell DNA stability are essential for the survival of a species." Explain your answer.

Question 6–21   A common type of error in DNA is produced by a spontaneous reaction termed "deamination" in which a nucleotide base loses an amino group (NH_2), which is replaced by a keto group (C=O) by the general reaction shown in Figure Q6–21. Write the structures of the bases A, G, C, T, and U and predict the products that will be produced by deamination. By looking at the products of this reaction—and remembering that, in the cell, these will need to be recognized and repaired—can you propose an explanation as to why DNA cannot contain uracil?

Figure Q6–21

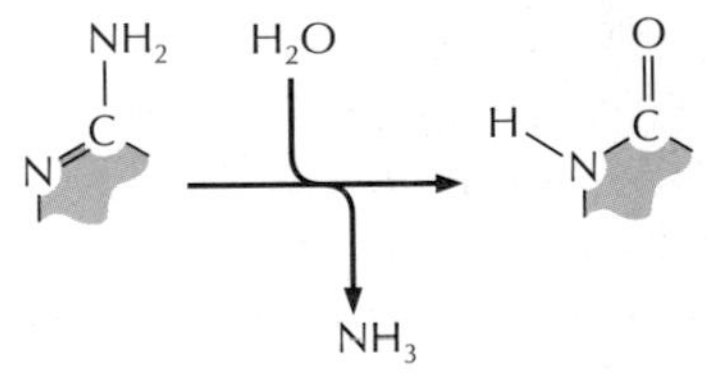

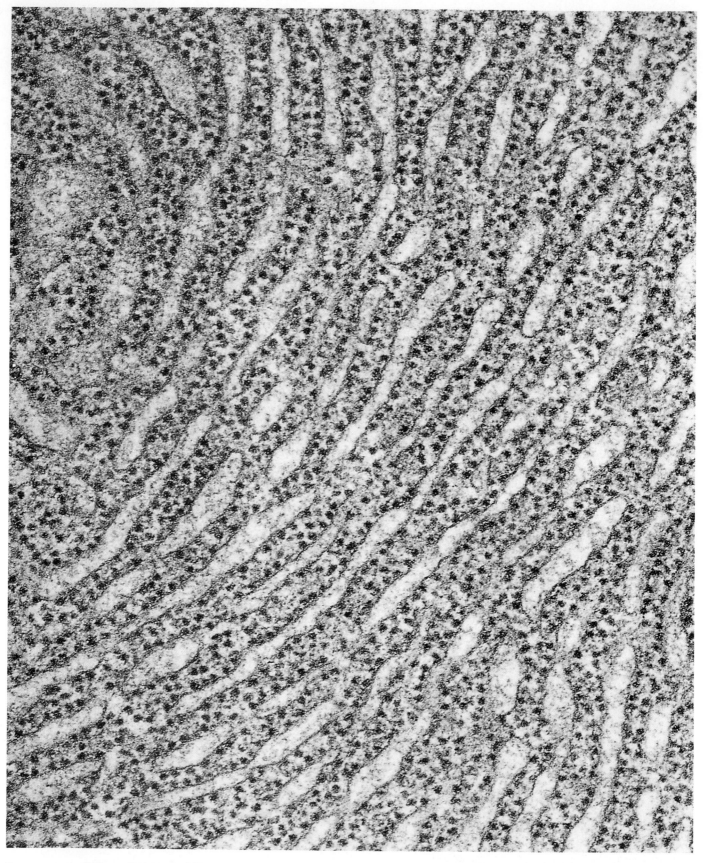

In a process called translation, the information in RNA is transferred to protein on large multisubunit molecular machines called ribosomes. In this electron micrograph of part of an eucaryotic cell, thousands of ribosomes can be seen associated with the membranes of the endoplasmic reticulum, where they are involved specifically in the production of proteins destined for secretion. (Courtesy of Daniel S. Friend.)

7

From DNA to Protein

Once the structure of DNA (deoxyribonucleic acid) was determined in the early 1950s, it became clear that the hereditary information in cells is encoded in DNA's sequence of nucleotides. We saw in Chapter 6 how this information can be passed on unchanged from a cell to its descendants through the process of DNA replication. But how does the cell decode and use the information? How do genetic instructions written in an alphabet of just four "letters"—the four different nucleotides in DNA—direct the formation of a bacterium, a fruit fly, or a human? We still have a lot to learn about how the information stored in an organism's genes produces even the simplest unicellular bacterium, let alone how it directs the development of complex multicellular organisms like ourselves. But the DNA code itself has been deciphered, and the language can be read.

Even before the DNA code was broken, it was known that the genetic information somehow directed the synthesis of proteins. Proteins are the principal constituents of cells and determine not only their structure but also their functions. In previous chapters, we have encountered some of the thousands of different kinds of proteins that cells can make. We have seen in Chapter 5 that the properties and function of a protein molecule are determined by the linear order—the *sequence* —of the different amino acid subunits in its polypeptide chain: each type of protein has its own unique amino acid sequence, and this sequence dictates how the chain will fold to give a molecule with a distinctive shape and chemistry. The genetic instructions carried by DNA must therefore specify the amino acid sequences of proteins. We shall see in this chapter exactly how this is done.

The DNA does not direct protein synthesis itself, but acts rather like a manager, delegating the various tasks required to a team of workers. When a particular protein is needed by the cell, the nucleotide sequence of the appropriate portion of the immensely long DNA molecule in a chromosome is first copied into another type of nucleic acid—*RNA (ribonucleic acid)*. It is these RNA copies of short segments of the DNA that are used as templates to direct the synthesis of the protein. Many thousands of these conversions from DNA to protein are occurring each second in every cell in our bodies. The flow of genetic information in cells is therefore from DNA to RNA to protein (Figure 7–1). All cells, from bacteria to humans, express their genetic information in this way—a principle so fundamental that it has been termed the *central dogma* of molecular biology.

The principal task of this chapter is to explain the mechanisms by which cells copy DNA into RNA (a process called *transcription*) and then use the information in RNA to make protein (a process called *translation*). In the final section of this chapter, we shall consider how the present scheme of information storage, transcription, and translation might have arisen from simpler systems in the earliest stages of cellular evolution.

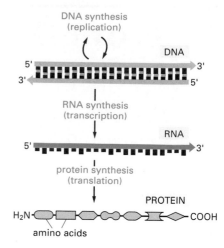

Figure 7–1 **From DNA to protein.** The flow of genetic information from DNA to RNA (transcription) and from RNA to protein (translation) occurs in all living cells.

From DNA to RNA

Transcription and translation are the means by which cells read out, or express, their genetic instructions—their *genes*. Many identical RNA copies can be made from the same gene and each RNA molecule can direct the synthesis of many identical protein molecules. Since there is usually only one copy of any particular gene in a cell, this successive amplification enables cells to synthesize the required amount of a protein much more rapidly than if the DNA itself were acting as the direct template for protein synthesis. Each gene can be transcribed and translated with a different efficiency, and this provides the cell with a way to make vast quantities of some proteins and tiny quantities of others (Figure 7–2). Moreover, as we shall see in Chapter 8, a cell can change (or regulate) the expression of each of its genes according to the needs of the moment. In this section, we shall begin with the production of RNA, the first step in gene expression.

Question 7–1   Consider the expression "central dogma," referring to the proposition that genetic information flows from DNA to RNA to protein. Is the word "dogma" appropriate in this scientific context?

Portions of DNA Sequence Are Transcribed into RNA

The first step a cell takes in reading out part of its genetic instructions is to copy the required portion of the nucleotide sequence of DNA—the gene—into a nucleotide sequence of RNA. The process is called **transcription** because the information, though copied into another chemi-

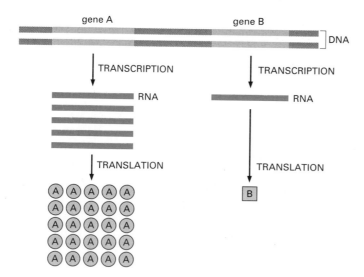

Figure 7–2 **Genes can be expressed with different efficiencies.** Gene A is transcribed and translated much more efficiently than is gene B. This allows the amount of protein A in the cell to be much higher than that of protein B.

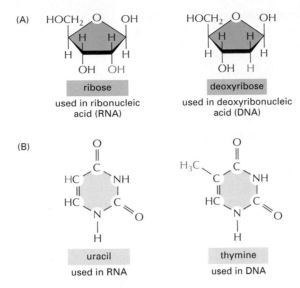

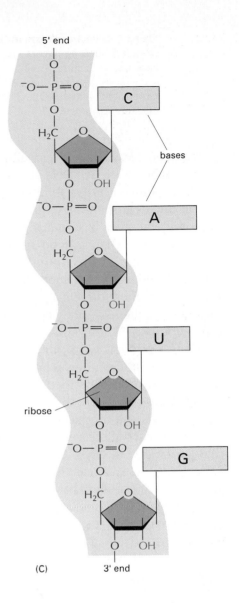

Figure 7–3 The chemical structure of RNA. (A) RNA contains the sugar ribose, which differs from deoxyribose, the sugar used in DNA, by the presence of an additional –OH group. (B) RNA contains the base uracil, which differs from thymine, the equivalent base in DNA, by the absence of a –CH₃ group. (C) A short length of RNA. The chemical linkage between nucleotides in RNA is the same as that in DNA.

cal form, is still written in essentially the same language—the language of nucleotides. Like DNA, RNA is a linear polymer made of four different types of nucleotide subunits linked together by phosphodiester bonds (Figure 7–3). It differs from DNA chemically in two respects: (1) the nucleotides in RNA are *ribonucleotides*—that is, they contain the sugar ribose (hence the name *ribo*nucleic acid) rather than deoxyribose; (2) although, like DNA, RNA contains the bases adenine (A), guanine (G), and cytosine (C), it contains uracil (U) instead of the thymine (T) in DNA. Since U, like T, can base-pair by hydrogen-bonding with A (Figure 7–4), the complementary base-pairing properties described for DNA in Chapter 6 apply also to RNA.

Despite these small chemical differences, DNA and RNA differ quite dramatically in overall structure. Whereas DNA always occurs in cells as a double-stranded helix, RNA is single-stranded; moreover, the RNA chain can fold up into a variety of shapes, just as a polypeptide chain folds up to form the final shape of a protein (Figure 7–5). As we shall see later in this chapter, the ability to fold into a complex three-dimensional shape allows RNA to carry out functions in cells in addition to conveying information between DNA and protein. Whereas DNA functions solely as an information store, there are various types of RNA, some having structural and even catalytic functions.

Transcription Produces RNA Complementary to One Strand of DNA

All of the RNA in a cell is made by transcription, a process that has certain similarities to DNA replication (Chapter 6). It begins with the

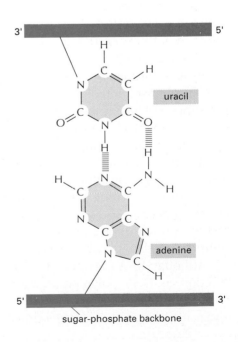

Figure 7–4 Uracil forms base pairs with adenine. Uracil has the same base-pairing properties as thymine. Thus, U-A base pairs closely resemble T-A base pairs (see Figure 6–5).

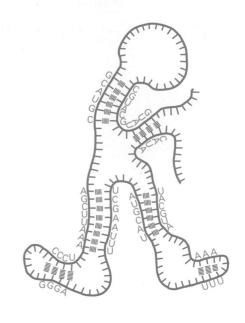

Figure 7–5 **RNA can form intramolecular base pairs.** RNA is single-stranded, but it often contains short stretches of nucleotides that can base-pair with complementary sequences found elsewhere on the same molecule. This allows an RNA molecule to fold into a three-dimensional structure that is determined by its sequence of nucleotides.

opening and unwinding of a small portion of the DNA double helix to expose the bases on each DNA strand. One of the two strands of the DNA double helix then acts as a template for the synthesis of RNA. As in DNA replication, the nucleotide sequence of the RNA chain is determined by the complementary base-pairing of incoming ribonucleotides on the DNA template. When a good match is made, the incoming ribonucleotide is covalently linked to the growing RNA chain in an enzymatically catalyzed reaction. The RNA chain produced by transcription—the *transcript*—is therefore elongated one nucleotide at a time and has a nucleotide sequence exactly complementary to the strand of DNA used as the template (Figure 7–6).

Transcription, however, differs from DNA replication in several crucial features. Unlike a newly formed DNA strand, the RNA strand does not remain hydrogen-bonded to the DNA template strand. Instead, just behind the region where the ribonucleotides are being added, the DNA helix re-forms and displaces the RNA chain. Thus, RNA molecules produced by transcription are single-stranded. As they are copied from only a limited region of DNA, RNA molecules are very much shorter than DNA molecules. Whereas the DNA molecule in a human chromosome can be up to 250 million nucleotide pairs long, most RNAs are no more than a few thousand nucleotides long, and many are very much shorter.

The enzymes that carry out transcription are called **RNA polymerases.** Like the DNA polymerase that catalyzes DNA replication (see Chapter 6), RNA polymerases catalyze the formation of the phosphodiester bonds that link the nucleotides together and form the sugar-phosphate backbone of the RNA chain. The RNA polymerase moves stepwise along the DNA, unwinding the DNA helix just ahead to expose a new region of the template strand for complementary base-pairing. In this way, the growing RNA chain is extended by one nucleotide at a time in the 5′-to-3′ direction (Figure 7–7) using nucleoside triphosphates (ATP, CTP, UTP, and GTP), the high-energy bonds of which provide the energy that drives the reaction forward (see Figure 6–15).

The almost immediate release of the RNA strand from the DNA as it is synthesized means that many RNA copies can be made from the same gene in a relatively short time, the synthesis of the next RNA usually being started before the first RNA is completed (Figure 7–8). A medium-sized gene (say 1500 nucleotide pairs) requires approximately 50 seconds for a molecule of RNA polymerase to transcribe it. There may be 15 polymerases speeding along the same stretch of DNA, hard on one another's heels, causing over a thousand transcripts to be synthesized in an hour from the one gene. For most genes, the amount of transcription is much less than this, however.

Although RNA polymerase catalyzes essentially the same chemical reaction as DNA polymerase, there are some important differences between the two enzymes. First, and most obvious, RNA polymerase catalyzes the linkage of ribonucleotides, not deoxyribonucleotides. Second,

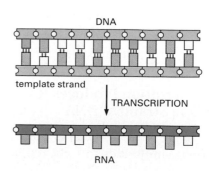

Figure 7–6 **Transcription produces an RNA complementary to one strand of DNA.**

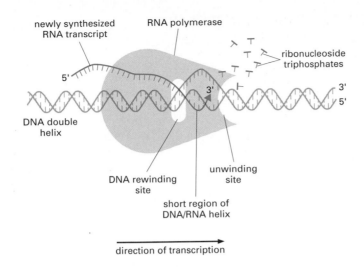

newly synthesized
RNA transcript

RNA polymerase

ribonucleoside
triphosphates

5′

3′

3′
5′

DNA double
helix

DNA rewinding
site

unwinding
site

short region of
DNA/RNA helix

direction of transcription

Figure 7–7 DNA is transcribed by the enzyme RNA polymerase. The RNA polymerase (*pale blue*) moves stepwise along the DNA, unwinding the DNA helix in front of it. As it progresses, the polymerase adds nucleotides (*small "T" shapes*) one by one to the RNA chain at the polymerization site. The polymerase rewinds the two DNA strands behind this site to displace the newly formed RNA. A short region of DNA/RNA helix is therefore formed only transiently, and the RNA transcript is a single-stranded complementary copy of one of the two DNA strands. The incoming nucleotides are in the form of ribonucleoside triphosphates (ATP, UTP, CTP, and GTP), whose hydrolysis provides the energy for the polymerization reaction (see Figure 6–15).

unlike the DNA polymerase involved in DNA replication, RNA polymerases do not possess nucleolytic *proofreading* activity; before adding the next nucleotide to the RNA chain, they do not check whether the previous nucleotide is correctly base-paired (see Figure 6–18). RNA polymerases can thus start an RNA chain without the need for a primer. This lack of proofreading reflects the fact that transcription does not need to be as accurate as DNA replication because RNA is not used as the permanent storage form of genetic information in cells. RNA polymerases make about one mistake for every 10^4 nucleotides copied into RNA (compared with an error rate for DNA polymerase of about one in 10^7 nucleotides).

Several Types of RNA Are Produced in Cells

The vast majority of genes carried in a cell's DNA specify the amino acid sequence of proteins, and the RNA molecules that are copied from these genes (and that ultimately direct the synthesis of proteins) are collectively called **messenger RNA (mRNA)**. The final product of other genes, however, is the RNA itself (Table 7–1). As we shall see in later sections of this chapter, these nonmessenger RNAs, like proteins, serve as structural and enzymatic components of cells and play key parts translating the genetic message into protein. Thus *ribosomal RNA (rRNA)* forms the core of the ribosomes, on which mRNA is translated into protein, and *transfer RNA (tRNA)* forms the adaptors that select amino acids and hold them in place on a ribosome for their incorporation into protein.

While DNA molecules are typically very long and carry the instructions for thousands of different proteins, an RNA molecule is very much shorter, as it carries the information from just one portion of the DNA. In eucaryotes, each mRNA typically carries information transcribed

Question 7–2 In the electron micrograph in Figure 7–8, are the RNA polymerase molecules moving from right to left or from left to right? Why are the RNA transcripts so much shorter than the length of the DNA that encodes them?

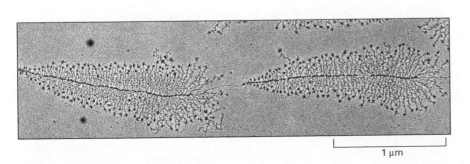

1 μm

Figure 7–8 Transcription of two genes as visualized in the electron microscope. The micrograph shows many molecules of RNA polymerase simultaneously transcribing each of two adjacent genes. Molecules of RNA polymerase are visible as a series of dots along the DNA with the transcripts (fine threads) attached to them. The RNA molecules (called rRNAs) transcribed from the genes shown in this example are not translated into protein but are instead used directly as components of ribosomes, the machines on which translation takes place. The particles at the 5′ end (the free end) of each rRNA transcript are believed to reflect the beginnings of ribosomal assembly. (Courtesy of Ulrich Scheer.)

Table 7–1	Types of RNA Produced in Cells
Type of RNA	**Function**
mRNAs	codes for proteins
rRNAs	forms part of the structure of the ribosome and participates in protein synthesis
tRNAs	used in protein synthesis as an adaptor between mRNA and amino acids
Small RNAs	used in pre-mRNA splicing, transport of proteins to the ER, and other cellular processes

from just one gene, coding for a single protein; in bacteria a set of adjacent genes is often transcribed as a single mRNA that therefore carries the information for several different proteins.

Signals in DNA Tell RNA Polymerase Where to Start and Finish

In order to begin transcription, RNA polymerase must be able to recognize the start of a gene and bind firmly to the DNA at this site. The way in which RNA polymerases recognize the transcription start site differs somewhat between procaryotes (bacteria) and eucaryotes. Because the situation in bacteria is simpler, we shall look at that first and defer discussing transcription initiation in eucaryotes to the next chapter. This is an important topic, because the initiation of transcription is the main point at which the cell regulates which proteins are to be produced and at what rate.

In bacteria, RNA polymerase molecules tend to stick weakly to the bacterial DNA when they make a random collision with it; the polymerase molecule then slides rapidly along the DNA. It latches tightly onto the DNA once it encounters a region called a **promoter,** which contains a sequence of nucleotides indicating the starting point for RNA synthesis. The polymerase protein can recognize this DNA sequence, even though the DNA is in its double-helical form, by making specific contacts with the portions of the bases that are exposed on the outside of the helix.

After the RNA polymerase makes contact with the promoter DNA and binds tightly, it opens up the double helix immediately in front of it to expose the nucleotides on a short stretch of DNA on each strand (Figure 7–9A). One of the two exposed DNA strands then acts as a template for complementary base-pairing with incoming ribonucleotides, two of which are joined together by the polymerase to begin the RNA chain. Chain elongation then continues until the enzyme encounters a second signal in the DNA, the terminator (or stop site), where the polymerase halts and releases both the DNA template and the newly made RNA chain (Figure 7–9B).

A subunit of bacterial polymerase, called *sigma* (σ) *factor,* is primarily responsible for recognizing the promoter sequence on DNA. Once the polymerase has latched onto the DNA at this site and has synthesized approximately 10 nucleotides of RNA, the sigma factor is released,

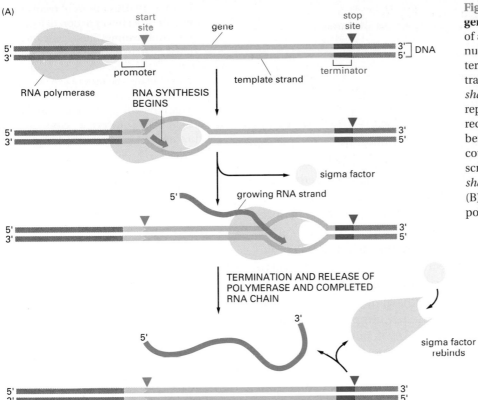

(A)

start site | gene | stop site

5' | 3' — RNA polymerase — promoter — template strand — terminator — DNA 5'

RNA polymerase

RNA SYNTHESIS BEGINS

sigma factor

growing RNA strand

TERMINATION AND RELEASE OF POLYMERASE AND COMPLETED RNA CHAIN

sigma factor rebinds

Figure 7–9 Transcription of a bacterial gene by RNA polymerase. (A) Production of an RNA molecule in bacteria. (B) The nucleotide sequences that signal to a bacterial RNA polymerase where to begin transcribing and where to stop. The *green-shaded* regions in the top diagram in (B) represent the DNA sequences that are required to create a promoter. The numbers represent the positions of nucleotides counting from the first nucleotide transcribed, which is designated +1. The *red-shaded* regions in the bottom diagram in (B) represent sequences that signal to RNA polymerase to terminate transcription.

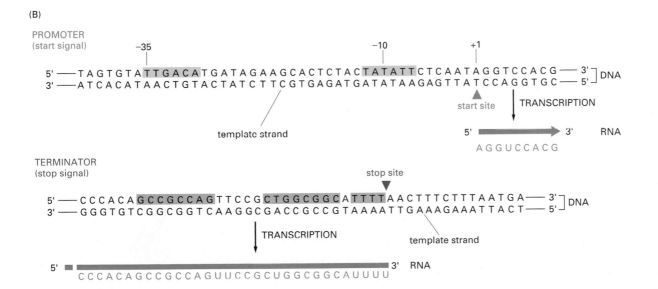

(B)

PROMOTER (start signal)

−35   −10   +1

5' — TAGTGTATTGACATGATAGAAGCACTCTACTATATTCTCAATAGGTCCACG — 3' DNA
3' — ATCACATAACTGTACTATCTTCGTGAGATGATATAAGAGTTATCCAGGTGC — 5'

template strand   start site   TRANSCRIPTION

5' ———————▶ 3'   RNA
AGGUCCACG

TERMINATOR (stop signal)

stop site

5' — CCCACAGCCGCCAGTTCCGCTGGCGGCATTTTAACTTTCTTTAATGA — 3' DNA
3' — GGGTGTCGGCGGTCAAGGCGACCGCCGTAAAATTGAAAGAAATTACT — 5'

TRANSCRIPTION

template strand

5' ■———————————————————————— 3'   RNA
CCCACAGCCGCCAGUUCCGCUGGCGGCAUUUU

enabling the polymerase to move forward and continue transcribing without it. After the polymerase is released at a terminator, it reassociates with a free sigma factor and searches for a promoter, where it can begin the process of transcription again.

Since DNA is double-stranded, two different RNA molecules could in principle be transcribed from any gene, using each of the two DNA strands as a template. However, the promoter is asymmetrical and binds the polymerase in only one orientation; thus, once properly positioned

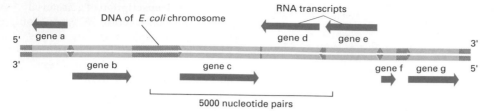

Figure 7–10 Directions of transcription along a short portion of a bacterial chromosome. Some genes are transcribed using one DNA strand as a template, while others are transcribed using the other DNA strand. The direction of transcription is determined by the promoter at the beginning of each gene (*green arrowheads*). Approximately 0.2% (10,000 base pairs) of the *E. coli* chromosome is depicted here. The genes transcribed from left to right use the bottom DNA strand as the template; those transcribed from right to left use the top strand as the template (see Figure 7–9).

on a promoter, the RNA polymerase has no option but to transcribe the appropriate DNA strand, since transcription can proceed only in the 5′ → 3′ direction. The direction of transcription with respect to the chromosome as a whole will vary from gene to gene (Figure 7–10).

The RNA polymerase's requirement for binding tightly to DNA before it can start transcription means that a portion of DNA can be transcribed only if it is preceded by a promoter sequence. This ensures that only those parts of a DNA molecule that contain a gene will be transcribed into RNA. In bacteria, genes tend to lie very close to one another in the DNA, with only very short lengths of nontranscribed DNA between them. But in plant and animal DNA, including that of humans, individual genes are widely dispersed, with stretches of DNA up to 100,000 nucleotide pairs long between one gene and the next. Unnecessary transcription of these spacer-DNA regions, which—as far as is known—encode no genetic instructions, would waste a cell's valuable resources.

Question 7–3   Could an RNA polymerase used for transcription be used as the polymerase that makes the RNA primer required for replication (discussed in Chapter 6)?

Eucaryotic RNAs Undergo Processing in the Nucleus

Although the templating principle by which DNA is transcribed into RNA is the same in all organisms, the way in which the RNA transcripts are handled before they can be used by the cell differs a great deal between bacteria and eucaryotes. Bacterial DNA lies directly exposed to the cytoplasm, which contains the *ribosomes* on which protein synthesis takes place. As mRNA molecules are transcribed, ribosomes immediately attach to the free 5′ end of the RNA transcript and protein synthesis starts. In eucaryotic cells, by contrast, DNA is enclosed within the *nucleus*. Transcription takes place in the nucleus, but protein synthesis takes place on ribosomes in the cytoplasm. So, before a eucaryotic mRNA can be translated, it must be transported out of the nucleus through small pores in the nuclear envelope (Figure 7–11). Before the RNA exits the nucleus, however, it goes through several different **RNA processing** steps. In eucaryotes, the RNA freshly produced by transcription, but not yet processed, is often called the **primary transcript.**

Depending on which type of RNA is being produced—mRNA or some other type—these transcripts are processed in various ways before leaving the nucleus. Two processing steps that occur only on primary transcripts destined to become mRNA molecules are *RNA capping* and *polyadenylation* (Figure 7–12):

1. RNA capping involves a modification of the 5′ end of the primary transcript, the end that is synthesized first during transcription. The 5′ end is capped by the addition of an atypical nucleotide—a guanine (G) nucleotide with a methyl group attached. Capping usually occurs just after the RNA polymerase has synthesized the 5′ end of the primary transcript and before it has completed transcribing the whole gene.

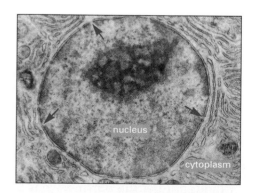

Figure 7–11 Section of a liver cell nucleus showing nuclear pores (*arrows*). mRNA molecules are made in the nucleus and have to move out into the cytoplasm via the pores in the nuclear envelope before they can be translated. (From D.W. Fawcett, A Textbook of Histology, 11th ed. Philadelphia: Saunders, 1986.)

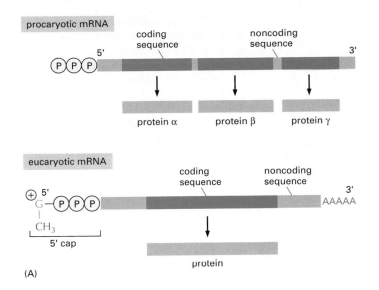

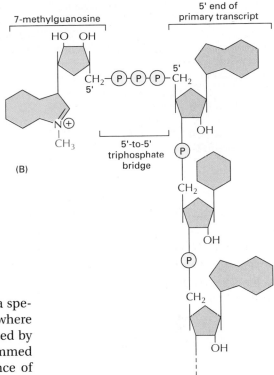

(B)

2. Polyadenylation provides most newly transcribed mRNAs with a special structure at their 3′, or tail, end. In contrast with bacteria, where the 3′ end of an mRNA is simply the end of the chain synthesized by the RNA polymerase, the 3′ ends of eucaryotic RNAs are first trimmed by an enzyme that cuts the RNA chain at a particular sequence of nucleotides and is then finished off by a second enzyme that adds a series of repeated adenine (A) nucleotides (a *poly(A) tail*) onto the cut end. The poly(A) tail is generally a few hundred nucleotides long.

These two modifications—capping and polyadenylation—are thought to increase the stability of the mRNA molecule and to aid its export from the nucleus to the cytoplasm. They are also used later by the protein-synthesis machinery as an indication that both ends of the mRNA are present and that the message is therefore complete.

Figure 7–12 A comparison of the structures of procaryotic and eucaryotic mRNA molecules. (A) The 5′ and 3′ ends of a bacterial mRNA are the unmodified ends of the chain synthesized by the RNA polymerase, which initiates and terminates transcription at those points, respectively. The corresponding ends of a eucaryotic mRNA are formed by adding a 5′ cap and by cleavage of the primary transcript and the addition of a poly(A) tail, respectively. The figure also illustrates another difference between the procaryotic and eucaryotic mRNAs: bacterial mRNAs can contain the instructions for several different proteins, whereas eucaryotic mRNAs nearly always contain the information for only a single protein. (B) The structure of the cap at the 5′ end of eucaryotic mRNA molecules. Note the unusual 5′-to-5′ linkage of the 7-methyl G to the remainder of the RNA. Many eucaryotic mRNAs carry an additional modification: the 2′-hydroxyl group on the second ribose sugar in the mRNA is methylated (not shown).

Eucaryotic Genes Are Interrupted by Noncoding Sequences

Most eucaryotic RNAs have to undergo a further processing step before they are functional. This step involves a far more radical modification of the primary RNA transcript than capping or polyadenylation, and it is the consequence of a surprising feature of eucaryotic gene arrangement. In the 1970s, cell biologists studying transcription in eucaryotic cells were puzzled by the behavior of the RNA in the nucleus, which seemed to be quite different from that of bacterial mRNAs. They found that nuclear RNAs that could be identified as prospective mRNAs by their G nucleotide caps and their poly(A) tails became progressively shorter while in the nucleus, although they retained both their caps and their tails. In all, only about 5% of the RNA initially transcribed in the nucleus ever reached the cytoplasm. This seemed not only strangely wasteful but also deeply mysterious: how could the middle part of an RNA molecule shrink?

The answer to this mystery came in 1977 with the unexpected discovery that the organization of eucaryotic genes is fundamentally different from that of bacterial genes. In bacteria, most proteins are encoded by an uninterrupted stretch of DNA sequence that is transcribed into an RNA that, without any further processing, can serve as an mRNA. Most eucaryotic genes, in contrast, have their coding sequences interrupted

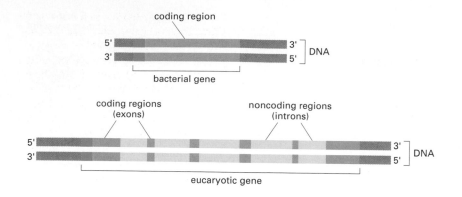

coding region

5'   3'] DNA
3'   5'

bacterial gene

coding regions   noncoding regions
(exons)   (introns)

5'   3'] DNA
3'   5'

eucaryotic gene

Figure 7–13 Comparison of a bacterial gene with a eucaryotic gene. The bacterial gene consists of a single stretch of uninterrupted nucleotide sequence that encodes the amino acid sequence of a protein. In contrast, the coding sequences of most eucaryotic genes (*exons*) are interrupted by noncoding sequences (*introns*). Promoters for transcription are indicated in *green*.

by noncoding sequences, called **introns** (Figure 7–13). The scattered pieces of coding sequence, called **exons**, are usually shorter than the introns, and the coding portion of a gene is often only a small fraction of the total length of the gene. Most introns range in length from about 80 nucleotides to 10,000 nucleotides, although even longer introns exist (Figure 7–14).

Introns Are Removed by RNA Splicing

To produce an mRNA in eucaryotic cells, the entire length of the gene, including introns as well as exons, is first transcribed into a long RNA molecule, the primary transcript. After capping and polyadenylation, but before the RNA leaves the nucleus, all of the intron sequences are removed and the exons joined together. The result is a much shorter RNA molecule, which now contains an uninterrupted coding sequence. When this step, called **RNA splicing,** has been completed, the RNA is a functional mRNA molecule that can now leave the nucleus and be translated into protein.

How does the cell determine which parts of the primary transcript to remove? Unlike the coding sequence of an exon, the exact nucleotide sequence of most of an intron seems to be unimportant. Although there is little resemblance between the nucleotide sequences of different introns, each intron contains a few short nucleotide sequences that act as cues for its removal. These sequences are found at or near each end of the intron and are the same or very similar in all introns (Figure 7–15).

Introns are removed from RNA by enzymes that, unlike most other enzymes, are composed of a complex of protein and RNA; these splicing enzymes are called **small nuclear ribonucleoprotein particles (snRNPs)**—pronounced "snurps." At each intron, a group of snRNPs assembles on the RNA, cuts out the intron, and rejoins the RNA chain—releasing the excised intron as a "lariat" (Figures 7–16 and 7–17). One role of the RNA in snRNPs is to recognize and—by using complementary base-pairing—to pair with the nucleotide sequences that mark the

Figure 7–14 Structure of two human genes showing the arrangement of exons and introns. (A) The nucleotide sequence of the β-globin gene, which encodes one of the subunits of the oxygen-carrying protein hemoglobin, was given in Figure 6–9. As indicated, it contains 3 exons. (B) The Factor VIII gene codes for a protein (Factor VIII) that functions in the blood-clotting pathway. Mutations in this large gene are responsible for the most prevalent form of hemophilia. As indicated, it contains 26 exons.

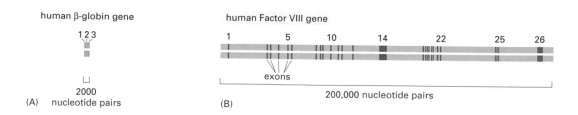

human β-globin gene   human Factor VIII gene

1 2 3   1   5   10   14   22   25   26

exons

2000   200,000 nucleotide pairs
(A) nucleotide pairs   (B)

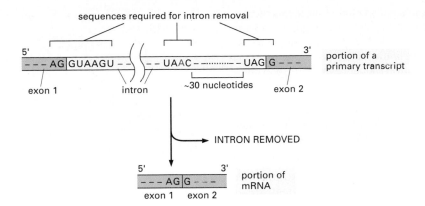

Figure 7–15 Nucleotide sequences that signal the beginning and the end of an intron. The three nucleotide sequences shown are required to remove an intron. The other positions in an intron can be occupied by any nucleotide. The special sequences are recognized by snRNPs, which cleave the RNA at the intron-exon borders and covalently link the exons together. The A highlighted in *red* forms the branch point of the lariat produced in the splicing reaction (Figure 7–16) and is typically located about 30 nucleotides from the 3′ end of the intron.

beginning and the branch point of each intron (see Figure 7–16). The snRNPs thereby bring the two ends of the intron together so that splicing can take place. Although the snRNPs are central to the splicing reaction, additional proteins are also required.

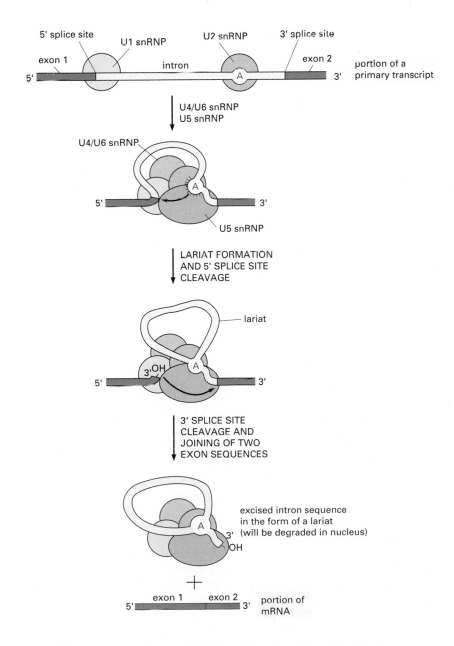

Figure 7–16 The RNA splicing mechanism. RNA splicing is catalyzed by an assembly of snRNPs (shown as *colored circles*) plus other proteins (not shown). One function of the complex of snRNPs is to bring the two ends of the intron together so that the reaction can take place. After the assembly of the snRNPs, a specific adenine nucleotide in the intron sequence (indicated in *red*) attacks the 5′ splice site and cuts the sugar-phosphate backbone of the RNA at this point. The cut 5′ end of the intron becomes covalently linked to the adenine nucleotide, forming a loop, or *lariat,* in the RNA molecule (see Figure 7–17). The free 3′-OH end of the first exon sequence then reacts with the beginning of the second exon sequence, cutting the intron at its 3′ end and, at the same time, joining the two exons together. The outcome of these splicing reactions is that the two exon sequences become joined into a continuous coding sequence, and the lariat containing the intron sequence is released and eventually degraded.

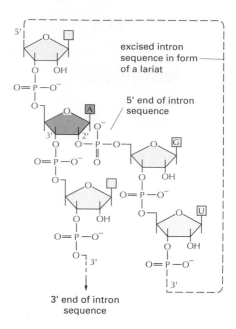

Figure 7–17 Structure of the branched RNA chain that forms during splicing. The nucleotide in *red* is the A highlighted in Figures 7–15 and 7–16. The branch is formed when the 5′ end of the intron sequence is covalently linked to the 2′-OH group of the A nucleotide. The branched chain remains in the final excised intron sequence and is responsible for its lariat form (see Figure 7–16).

The intron-and-exon type of gene arrangement seems at first wasteful, but it does have positive consequences. It is likely to have been profoundly important in the early evolutionary history of genes, where it is thought to have speeded up the emergence of new and useful proteins. The presence of numerous introns in DNA makes genetic recombination between exons of different genes more likely, as we shall explain in Chapter 9. This means that genes for new proteins could have evolved quite rapidly by the combination of parts of preexisting genes, resembling the assembly of a new type of machine from a kit of preexisting functional components. Indeed, many proteins in present-day cells resemble patchworks composed from a common set of protein pieces, called protein *domains* (see Chapter 5).

RNA splicing allows eucaryotes an additional advantage over procaryotes. The primary transcripts of many eucaryotic genes can be spliced in various ways to produce different mRNAs, depending on the cell type in which the gene is being expressed, or the stage of development of the organism. This allows different proteins to be produced from the same gene (Figure 7–18).

In conclusion, rather than being the wasteful process it seemed at first sight, RNA splicing enables eucaryotes to increase the already enormous coding potential of their genomes.

mRNA Molecules Are Eventually Degraded by the Cell

The length of time that an mRNA molecule persists in the cell affects the amount of protein produced from it, since the same mRNA molecule can be translated many times (see Figure 7–2). Each mRNA molecule is eventually degraded by the cell into nucleotides, but the lifetimes of mRNA molecules differ considerably—depending on the nucleotide sequence of the mRNA and the type of cell in which the mRNA is produced. Most mRNAs produced in bacteria are degraded rapidly, having a typical lifetime of about three minutes. The mRNAs in eucaryotic cells usually persist for longer. Some, such as that encoding β-globin, have

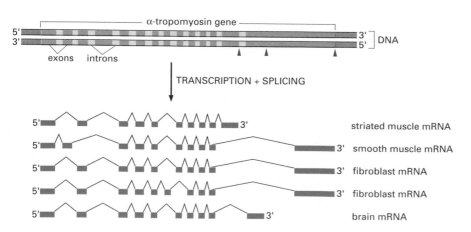

Figure 7–18 Alternative splicing of the α-tropomyosin gene from rat. α-Tropomyosin is a coiled-coil protein (see Figure 5–11) that regulates contraction in muscle cells. Its role in other cells is not well understood. Once the primary transcript is made, it can be spliced in different ways, as indicated in the figure, to produce distinct mRNAs that then give rise to variant proteins. Some of the splicing patterns are specific for certain types of cells. For example, the α-tropomyosin made in striated muscle is different from that made in smooth muscle. The arrowheads in the top part of the figure represent sites where cleavage and poly(A) addition can occur.

(A) EUCARYOTES

(B) PROCARYOTES

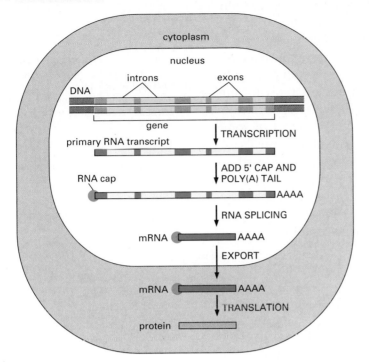

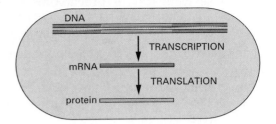

Figure 7–19 Summary of the steps leading from gene to protein. The final level of a protein in the cell depends on the efficiency of each step and on the rates of degradation of the RNA and protein molecules. (A) In eucaryotic cells, the initial RNA molecule produced by transcription (the primary transcript) contains both intron and exon sequences. Its two ends are modified, and the introns are removed by an enzymatically catalyzed RNA splicing reaction. The resulting mRNA is then transported from the nucleus to the cytoplasm, where it is translated into protein. Although these steps are depicted as occurring one at a time, in a sequence, in reality they often occur simultaneously. For example, the RNA cap is typically added and splicing typically begins before the primary transcript has been completed. (B) In procaryotes, the production of mRNA molecules is simpler. The 5′ end of an mRNA molecule is produced by the initiation of transcription by RNA polymerase, and the 3′ end is produced by the termination of transcription. Since procaryotic cells lack a nucleus, transcription and translation take place in a common compartment. In fact, translation of a bacterial mRNA often begins before its synthesis has been completed.

lifetimes of more than 10 hours, whereas other eucaryotic mRNAs have lifetimes of less than 30 minutes. These different lifetimes are in part signaled by nucleotide sequences that are present in the mRNA itself, most often the portion of RNA (called the 3′ untranslated region) that lies between the 3′ end of the coding sequence and the poly(A) tail. The different lifetimes of mRNA help the cell determine the level of each protein that it synthesizes. In general, proteins made at high levels, such as β-globin, are translated from mRNAs that have long lifetimes, whereas those proteins present at low levels, or those whose levels must change rapidly in response to signals, are typically synthesized from short-lived mRNAs.

The Earliest Cells May Have Had Introns in Their Genes

We have seen in this chapter that all cells, in order to use their genetic information, first transcribe it into RNA using the RNA polymerase enzyme and the principle of complementary base-pairing: this basic process of transcription is universal. But the way that the resulting transcript (called the primary transcript in eucaryotes) is handled by the cell differs considerably between eucaryotes and procaryotes (Figure 7–19). It may seem puzzling that although many aspects of mRNA production are similar between procaryotes and eucaryotes, some—especially RNA splicing—seem fundamentally different. Certainly, the discovery that eucaryotic genes contain introns came as a surprise to biologists, whose knowledge of gene structure up to that point derived from studies of procaryotes. How did this fundamental difference arise? We have seen in this chapter that RNA splicing does provide some advantages in evolutionary flexibility and in opportunities for control of gene expression. However, these advantages come with a cost: the cell has to maintain a larger genome. According to one school of thought, early cells—the common ancestors of procaryotes and eucaryotes—contained introns and these were lost in procaryotes during subsequent evolution. Procaryotes typically reproduce very rapidly, and the advantage of a smaller genome (allowing faster DNA replication) could have been one of the factors contributing to the gradual loss of introns in bacteria. Consistent with this idea, simple eucaryotes that reproduce rapidly (some yeasts, for example) have relatively few introns, and those introns that exist are usually much shorter than those of higher eucaryotes.

We have also seen in this section that the final product of some genes is an RNA molecule itself, such as those present in the snRNPs. We shall encounter more examples of RNA molecules that function as structural and catalytic components of cells. However, most genes in a cell produce RNA molecules that are used as intermediaries on the pathway to proteins. In the following section, we shall see how the cell converts the information carried in an mRNA molecule into a protein.

Question 7–4 Without looking back at the figures, draw a diagram depicting the essential steps of the splicing reaction. Indicate the intron-exon junctions in the primary transcript, the snRNPs, the excised intron, and the mature mRNA. During splicing, the same complex of snRNPs catalyzes both the cleavage of the RNA at the first intron-exon junction and the subsequent joining of the two exons. What might happen if the two reactions were catalyzed by separate enzymes that were not associated in a complex?

From RNA to Protein

By the end of the 1950s, biologists had agreed that the information encoded in DNA is copied first into RNA and then into protein. The debate then centered on the "coding problem": how is the information in a linear sequence of nucleotides in RNA translated into the linear sequence of a chemically quite different set of subunits—the amino acids in proteins? This fascinating question stimulated much excitement among scientists at the time. Here was a cryptogram set up by nature that, after more than three billion years of evolution, could finally be solved by one of the products of evolution—human beings! And indeed, not only was the code eventually cracked step by step, but the main features of the machinery by which cells make protein were pieced together.

An mRNA Sequence Is Decoded in Sets of Three Nucleotides

Once an mRNA has been produced, the information present in its nucleotide sequence can be used to synthesize a protein. Transcription as a means of information transfer is simple to understand, since DNA and RNA are chemically and structurally similar, and the DNA can act as a direct template for the synthesis of RNA by complementary base-pairing. As the term *transcription* signifies, it is as if a message written out by hand is being converted, say, into a typewritten text. The language itself and the form of the message do not change, and the symbols used are closely related.

Figure 7–20 The genetic code. The standard one-letter abbreviation for each amino acid is presented below its three-letter abbreviation (see Panel 2–5, pp. 62–63, for the full name of each amino acid and its structure). By convention, codons are always written with the 5′-terminal nucleotide to the left. Note that most amino acids are represented by more than one codon, and that there are some regularities in the set of codons that specifies each amino acid. Codons for the same amino acid tend to contain the same nucleotides at the first and second positions, and vary at the third position. Three codons do not specify any amino acid, but act as termination sites (stop codons), signaling the end of the protein-coding sequence. One codon—AUG—acts both as an initiation codon, signaling the start of a protein-coding message, and also as the codon that specifies methionine.

GCA GCC GCG GCU	AGA AGG CGA CGC CGG CGU	GAC GAU	AAC AAU	UGC UGU	GAA GAG	CAA CAG	GGA GGC GGG GGU	CAC CAU	AUA AUC AUU	UUA UUG CUA CUC CUG CUU	AAA AAG	AUG	UUC UUU	CCA CCC CCG CCU	AGC AGU UCA UCC UCG UCU	ACA ACC ACG ACU	UGG	UAC UAU	GUA GUC GUG GUU	UAA UAG UGA
Ala	Arg	Asp	Asn	Cys	Glu	Gln	Gly	His	Ile	Leu	Lys	Met	Phe	Pro	Ser	Thr	Trp	Tyr	Val	stop
A	R	D	N	C	E	Q	G	H	I	L	K	M	F	P	S	T	W	Y	V	

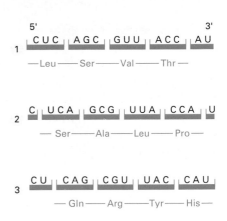

Figure 7–21 The three possible reading frames in protein synthesis. In the process of translating a nucleotide sequence (*blue*) into an amino acid sequence (*green*), the sequence of nucleotides in an mRNA molecule is read from the 5′ to the 3′ end in sequential sets of three nucleotides. In principle, therefore, the same RNA sequence can specify three completely different amino acid sequences, depending on the reading frame. In reality, however, only one of these reading frames encodes the actual message.

In contrast, the conversion of the information in RNA into protein represents a **translation** of the information into another language that uses quite different symbols. Since there are only four different nucleotides in mRNA and twenty different types of amino acids in a protein, this translation cannot be accounted for by a direct one-to-one correspondence between a nucleotide in RNA and an amino acid in protein. The rules by which the nucleotide sequence of a gene, through the medium of mRNA, is translated into the amino acid sequence of a protein are known as the **genetic code.** This code was deciphered in the early 1960s.

The sequence of nucleotides in the mRNA molecule is read consecutively in groups of three. Since RNA is a linear polymer of four different nucleotides, there are $4 \times 4 \times 4 = 64$ possible combinations of three nucleotides: AAA, AUA, AUG, and so on. However, only 20 different amino acids are commonly found in proteins. Either some nucleotide triplets are never used, or the code is redundant and some amino acids are specified by more than one triplet. The second possibility is, in fact, correct, as shown by the completely deciphered genetic code in Figure 7–20. Each group of three consecutive nucleotides in RNA is called a **codon,** and each specifies one amino acid.

This code is used universally in all present-day organisms. Although a few slight differences in the code have been found, these are chiefly in the DNA of mitochondria. Mitochondria have their own transcription and protein synthesis systems that operate quite independently from those of the rest of the cell (see Chapter 13), and they have been able to accommodate minor changes to the otherwise universal code.

In principle, an RNA sequence can be translated in any one of three different **reading frames,** depending on where the decoding process begins (Figure 7–21). However, only one of the three possible reading frames in an mRNA encodes the required protein. We shall see in a later section how a special punctuation signal at the beginning of each RNA message sets the correct reading frame.

Question 7–5 The genetic code was deciphered by experiments in which polynucleotides of repeating sequences were used as mRNAs to direct protein synthesis in cell-free extracts. In the test tube, artificial conditions were used that allow the ribosomes to start protein synthesis anywhere on an RNA molecule, i.e., without the need of a translation start codon that would be required in the living cell. When polynucleotides of repeating sequences—(1) UUUUUUUU..., (2) AUAUAUAU..., and (3) AUCAUCAUC...—were used as templates in such a cell-free extract, what types of polypeptides would you expect to be synthesized in each case? (You should consult the genetic code in Figure 7–20.)

tRNA Molecules Match Amino Acids to Codons in mRNA

The codons in an mRNA molecule do not directly recognize the amino acids they specify: the group of three nucleotides does not, for example, bind directly to the amino acid. Rather, the translation of mRNA into protein depends on adaptor molecules that can recognize and bind both to the codon and, at another site on their surface, to the amino acid. These adaptors consist of a set of small RNA molecules known as **transfer RNAs (tRNAs),** each about 80 nucleotides in length.

We saw earlier that an RNA molecule will generally fold up into a three-dimensional structure by forming base pairs between different

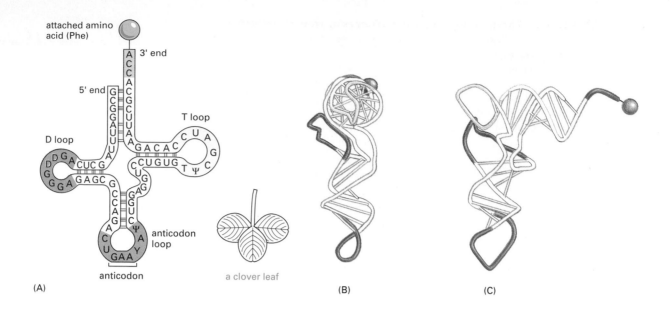

Figure 7–22 **A tRNA molecule.** In this series of diagrams, the same tRNA molecule—in this case a tRNA specific for the amino acid phenylalanine (Phe)—is depicted in various ways. (A) The cloverleaf structure, a convention used to show the complementary base-pairing (*red lines*) that creates the double-helical regions of the molecule. The anticodon is the sequence of three nucleotides that base-pairs with a codon in mRNA. The amino acid matching the codon/anticodon pair is attached at the 3′ end of the tRNA. tRNAs contain some unusual bases, which are produced by chemical modification after the tRNA has been synthesized. The bases denoted ψ (for pseudouridine) and D (for dihydrouridine) are derived from uracil. (B and C) Views of the actual L-shaped molecule, based on x-ray diffraction analysis. (D) The linear nucleotide sequence of the molecule, color-coded to match A, B, and C.

regions of the molecule (see Figure 7–5). If the base-paired regions are sufficiently extensive, they will form a double-helical structure, like that of double-stranded DNA. The tRNA molecules provide a striking example of this. Four short segments of the folded tRNA are double-helical, producing a molecule that looks like a cloverleaf when drawn schematically (Figure 7–22A). For example, a 5′-GCUC-3′ sequence in one part of a polynucleotide chain can form a relatively strong association with a 5′-GAGC-3′ sequence in another region of the same molecule. The cloverleaf undergoes further folding to form a compact L-shaped structure that is held together by additional hydrogen bonds between different regions of the molecule (Figure 7–22).

Two regions of unpaired nucleotides situated at either end of the L-shaped molecule are crucial to the function of tRNA in protein synthesis. One of these regions forms the **anticodon,** a set of three consecutive nucleotides that pairs with the complementary codon in an mRNA molecule. The other is a short single-stranded region at the 3′ end of the molecule; this is the site where the amino acid that matches the codon is attached to the tRNA.

We saw in the previous section that the genetic code is redundant; that is, several different codons can specify a single amino acid (see Figure 7–20). This redundancy implies either that there is more than one tRNA for many of the amino acids or that some tRNA molecules can base-pair with more than one codon. In fact, both situations occur. Some amino acids have more than one tRNA and some tRNAs are constructed so that they require accurate base-pairing only at the first two positions of the codon and can tolerate a mismatch (or *wobble*) at the third position. This wobble base-pairing explains why so many of the alternative codons for an amino acid differ only in their third nucleotide (see Figure 7–20). Wobble base-pairings make it possible to fit the 20 amino acids to their 61 codons with as few as 31 kinds of tRNA molecules. The exact number of different kinds of tRNAs, however, differs from one species to the next.

Specific Enzymes Couple tRNAs to the Correct Amino Acid

We have seen that in order to read the genetic code in DNA, cells make many different tRNAs. We now must consider how each tRNA molecule becomes linked to the one amino acid in 20 that is its appropriate partner. Recognition and attachment of the correct amino acid depends on enzymes called **aminoacyl-tRNA synthetases,** which covalently couple each amino acid to its appropriate set of tRNA molecules. There is a different synthetase enzyme for each amino acid (that is, 20 synthetases in all); one attaches glycine to all tRNAs that recognize codons for glycine, another attaches alanine to all tRNAs that recognize codons for alanine, and so on. Specific nucleotides in both the anticodon and the amino acid accepting arm allow the correct tRNA to be recognized by the synthetase enzyme. The synthetases are equal in importance to tRNA in the decoding process, because it is the combined action of synthetases and tRNAs that associates each codon in the mRNA molecule with its particular amino acid (Figure 7–23).

The synthetase-catalyzed reaction that attaches the amino acid to the 3′ end of the tRNA is one of many cellular reactions coupled to the energy-releasing hydrolysis of ATP (see Figure 3–27), and it produces a high-energy bond between the tRNA and the amino acid. The energy of this bond is used at a later stage in protein synthesis to covalently link the amino acid to the growing polypeptide chain.

The RNA Message Is Decoded on Ribosomes

The recognition of a codon by the anticodon on a tRNA molecule depends on the same type of complementary base-pairing as is used in DNA replication and transcription. However, accurate and rapid translation of mRNA into protein requires a large molecular machine, which

Figure 7–23 The genetic code is translated by means of two adaptors that act one after another. The first adaptor is the aminoacyl-tRNA synthetase, which couples a particular amino acid to its corresponding tRNA; the second adaptor is the tRNA molecule itself, whose *anticodon* forms base pairs with the appropriate *codon* on the mRNA. An error in either step will cause the wrong amino acid to be incorporated into a protein chain. In the sequence of events shown, the amino acid tryptophan (Trp) is selected by the codon UGG on the mRNA.

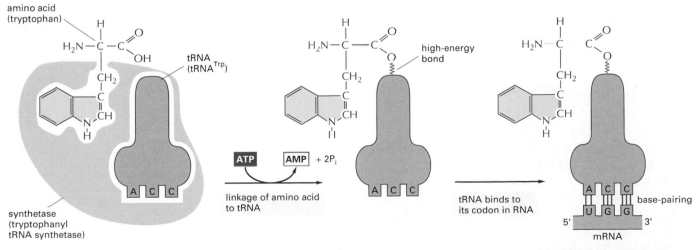

NET RESULT: AMINO ACID IS SELECTED BY ITS CODON

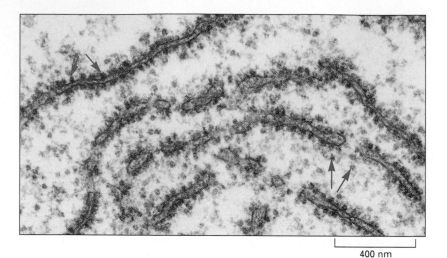

Figure 7–24 **Ribosomes in the cytoplasm of a eucaryotic cell.** This electron micrograph shows a thin section of a small region of cytoplasm. The ribosomes appear as black dots (*red arrows*). Some are free in the cytosol; others are attached to membranes of the endoplasmic reticulum. (Courtesy of Daniel S. Friend.)

400 nm

travels along the mRNA chain, capturing complementary tRNA molecules, holding them in position, and bonding together the amino acids that they carry so as to form a protein chain. This protein-manufacturing machine is the **ribosome**—a large complex made from more than 50 different proteins (the *ribosomal proteins*) and several RNA molecules called **ribosomal RNAs (rRNAs).** A typical living cell contains millions of ribosomes in its cytoplasm (Figure 7–24). In a eucaryote, the ribosomal subunits are made in the nucleus, by the association of newly tran-

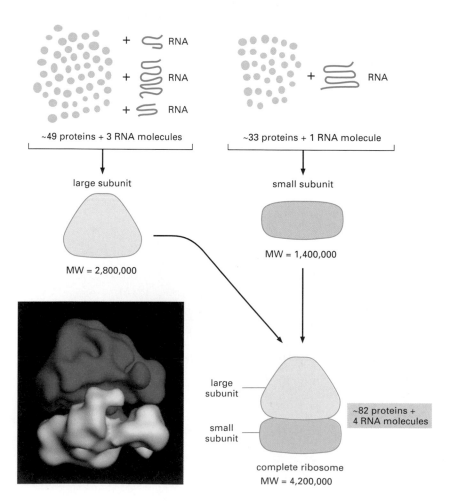

~49 proteins + 3 RNA molecules

~33 proteins + 1 RNA molecule

large subunit

small subunit

MW = 2,800,000

MW = 1,400,000

large subunit

small subunit

~82 proteins + 4 RNA molecules

complete ribosome
MW = 4,200,000

Figure 7–25 **The components of a eucaryotic ribosome.** The complete three-dimensional structure of the ribosome is not known yet, but a low-resolution model of the procaryotic ribosome, which has a similar structure to that in eucaryotes, has been determined by electron microscopy, and is shown at the lower left. The large subunit is shown in *blue*, and the small subunit in *yellow*. (Courtesy of Joachim Frank.)

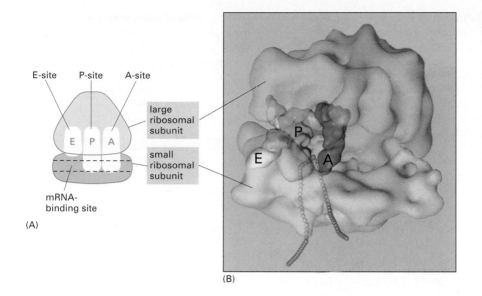

E-site   P-site   A-site

large
ribosomal
subunit

small
ribosomal
subunit

mRNA-
binding site

(A)

(B)

Figure 7–26 RNA-binding sites in the ribosome. Each ribosome has a binding site for mRNA and three binding sites for tRNA, the A-, P-, and E-sites (short for aminoacyl-tRNA, peptidyl-tRNA, and exit, respectively). (A) This representation of a ribosome, which will be used in subsequent figures, is highly schematic. (B) A model of the procaryote ribosome that depicts the arrangement of the mRNA (*orange beads*) and the positions of one tRNA molecule in the A-site of the ribosome (*pink*), one tRNA molecule in the P-site of the ribosome (*green*), and one tRNA molecule in the E-site of the ribosome (*yellow*). In this model the large ribosomal subunit is *light blue,* and the small subunit is *light green.* Although all three tRNA sites are shown occupied here, during the process of protein synthesis not more than two of these sites contain tRNA molecules at any one time (see Figure 7–27). (B, courtesy of Joachim Frank, Yanhong Li, and Rajendra Agarwal.)

scribed rRNAs with ribosomal proteins, which have been transported into the nucleus after their synthesis in the cytoplasm. The individual ribosomal subunits are then exported to the cytoplasm to take part in protein synthesis.

Eucaryotic and procaryotic ribosomes are very similar in design and function. Both are composed of one large and one small subunit that fit together to form a complete ribosome with a mass of several million daltons (Figure 7–25) (for comparison, an average-sized protein has a mass of 40,000 daltons). The small subunit matches the tRNAs to the codons of the mRNA, while the large subunit catalyzes the formation of the peptide bonds that link the amino acids together into a polypeptide chain (see Chapter 2). The two subunits come together on an mRNA molecule, usually near its beginning (5′ end), to begin the synthesis of a protein. The ribosome then moves along the mRNA, translating the nucleotide sequence into an amino acid sequence one codon at a time, using the tRNAs as adaptors to add each amino acid in the correct sequence to the end of the growing polypeptide chain. Finally, the two subunits of the ribosome separate when synthesis of the protein is finished. Ribosomes operate with remarkable efficiency: in one second, a single ribosome of a eucaryotic cell adds about 2 amino acids to a polypeptide chain, and the ribosomes of bacterial cells operate even faster, at a rate of about 20 amino acids per second.

How does the ribosome choreograph the movements required for translation? A ribosome contains four binding sites for RNA molecules: one is for the mRNA and three (called the A-site, the P-site, and the E-site) are for tRNAs. A tRNA molecule is held tightly at the A- and P-sites only if its anticodon forms base pairs (allowing for wobble) with a complementary codon on the mRNA molecule that is bound to the ribosome. The A- and P-sites are sufficiently close together that their two tRNA molecules are forced to form base pairs with adjacent codons on the mRNA molecule (Figure 7–26).

Once protein synthesis has been initiated, each new amino acid is added to the elongating chain in a cycle of reactions. We shall join the

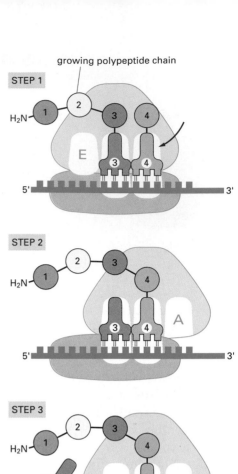

Figure 7–27 Translating an mRNA molecule. The three-step cycle shown is repeated over and over during the synthesis of a protein chain. An aminoacyl-tRNA molecule binds to the A-site on the ribosome in step 1, a new peptide bond is formed in step 2, and the small subunit moves a distance of three nucleotides along the mRNA chain in step 3, ejecting the spent tRNA molecule and "resetting" the ribosome so that the next aminoacyl-tRNA molecule can bind. As indicated, the mRNA is translated in the 5′-to-3′ direction, and the amino-terminal end of a protein is made first.

chain elongation process at a point where some amino acids have already been linked together and there is a tRNA joined to the growing polypeptide in the P-site on the ribosome (Figure 7–27, step 1). A tRNA carrying the next amino acid in the chain has bound to the vacant ribosomal A-site by forming base pairs with the codon in mRNA exposed at the A-site. In step 2, the carboxyl end of the polypeptide chain is uncoupled from the tRNA at the P-site (by breakage of the high-energy bond between the tRNA and its amino acid) and joined by a peptide bond to the free amino group of the amino acid linked to the tRNA at the A-site. This central reaction of protein synthesis is catalyzed by a *peptidyl transferase* enzyme activity, which is part of the ribosome. The catalytic part of the ribosome in this case appears not to be one of the proteins but rather one of the rRNAs in the large subunit. As shown in the figure, the peptidyl transferase reaction is thought to be accompanied by a shift of the small subunit, which holds onto the mRNA, relative to the large subunit. This shift moves the two tRNAs into the E- and P-sites of the large subunit. In step 3, the small subunit moves exactly 3 nucleotides along the mRNA molecule, bringing it back to its original position relative to the large subunit, and the tRNA occupying the E-site dissociates. The entire cycle of three steps is repeated each time that an amino acid is added to the polypeptide chain, with the chain growing from its amino to its carboxyl end until a stop codon is encountered.

Codons in mRNA Signal Where to Start and to Stop Protein Synthesis

The site where protein synthesis begins on the mRNA is crucial, since it sets the reading frame for the whole length of the message. An error of one nucleotide either way at this stage will cause every subsequent codon in the message to be misread, so that a nonfunctional protein with a garbled sequence of amino acids will result (see Figure 7–21). The initiation step is also of great importance in another respect, since it is the last point at which the cell can decide whether the mRNA is to be translated and the protein synthesized; the rate of initiation thus determines the rate at which the protein is synthesized.

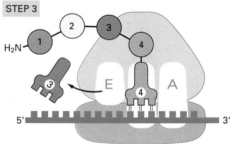

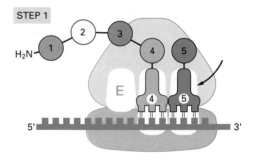

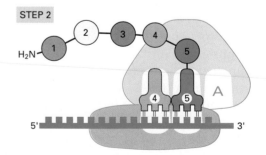

> **Question 7–6** The following sequence of nucleotides in a DNA strand was used as a template to synthesize an mRNA that was then translated into protein: 5′–T-T-A-A-C-G-G-C-T-T-T-T-T-T-C–3′. Predict the carboxy-terminal amino acid and the amino-terminal amino acid of the resulting polypeptide. Assume that the mRNA is translated without the need for a start codon.

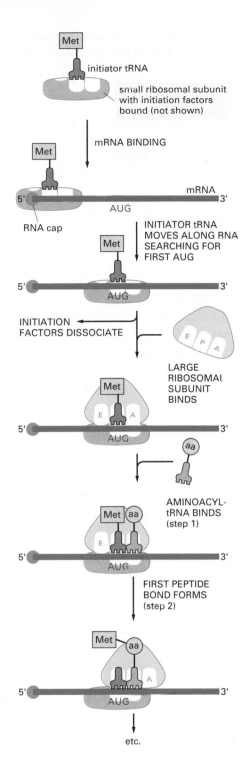

Figure 7–28 The initiation phase of protein synthesis in eucaryotes. Following initiation, the protein is elongated using the reactions outlined in Figure 7–27.

The translation of an mRNA begins with the codon AUG, and a special tRNA is required to initiate translation. This **initiator tRNA** always carries the amino acid methionine (in bacteria, a modified form of methionine—formylmethionine—is used) so that newly made proteins all have methionine as the first amino acid at their amino-terminal end, the end of a protein that is synthesized first. This methionine is usually removed later by a specific protease. The initiator tRNA is distinct from the tRNA that normally carries methionine.

In eucaryotes, the initiator tRNA (which is coupled to methionine) is first loaded into the small ribosomal subunit along with additional proteins called **initiation factors** (Figure 7–28). Of all the charged tRNAs in the cell, only the charged initiator tRNA is capable of tightly binding the small ribosome subunit. Next, the loaded ribosomal subunit binds to the 5′ end of an mRNA molecule, which is recognized in part by the cap that is present on eucaryotic mRNA (see Figure 7–12B). The small ribosomal subunit then moves forward (5′→ 3′) along the mRNA searching for the first AUG. When this AUG is encountered, several initiation factors dissociate from the small ribosomal subunit to make way for the large ribosomal subunit to assemble and complete the ribosome. Because the initiator tRNA is bound to the P-site, protein synthesis is ready to begin with the addition of the next tRNA coupled to its amino acid (see Figure 7–28).

The mechanism for selecting a start codon in bacteria is different. Bacterial mRNAs have no 5′ caps to tell the ribosome where to begin searching for the start of translation. Instead, they contain specific ribosome-binding sequences, up to six nucleotides long, that are located a few nucleotides upstream of the AUGs at which translation is to begin. Unlike a eucaryotic ribosome, a procaryotic ribosome can readily bind directly to a start codon that lies in the interior of an mRNA, as long as a ribosome-binding site precedes it by several nucleotides. As a result, procaryotic mRNAs are often polycistronic—that is, they encode several different proteins, each of which is translated from the same mRNA molecule (Figure 7–29). In contrast, a eucaryotic mRNA usually carries the information for a single protein.

The end of the protein-coding message is signaled by the presence of one of several codons (UAA, UAG, or UGA) called *stop codons* (see Figure 7–20). These are not recognized by a tRNA and do not specify an amino acid, but instead signal to the ribosome to stop translation. Proteins known as *release factors* bind to any stop codon that reaches the A-site on the ribosome, and this binding alters the activity of the peptidyl transferase in the ribosome, causing it to catalyze the addition of a

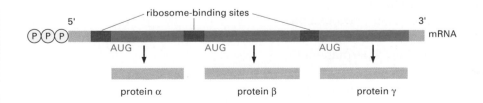

Figure 7–29 Structure of a typical procaryotic mRNA molecule. Unlike eucaryotic ribosomes, which recognize a 5′ cap, procaryotic ribosomes initiate transcription at ribosome-binding sites, which can be located in the interior of an mRNA molecule. This feature permits procaryotes to synthesize more than one type of protein from a single mRNA species.

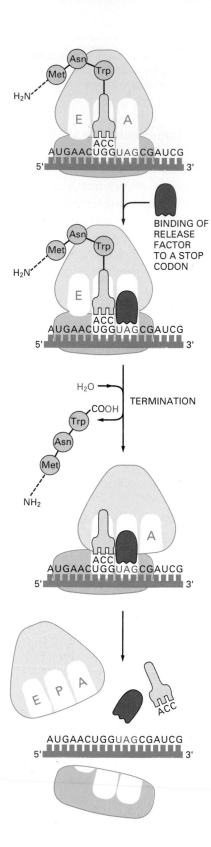

Figure 7–30 The final phase of protein synthesis. The binding of release factor to a stop codon terminates translation. The completed polypeptide is released, and the ribosome dissociates into its two separate subunits.

water molecule instead of an amino acid to the peptidyl-tRNA (Figure 7–30). This reaction frees the carboxyl end of the growing polypeptide chain from its attachment to a tRNA molecule, and since only this attachment normally holds the growing polypeptide to the ribosome, the completed protein chain is immediately released into the cytoplasm. The ribosome releases the mRNA and dissociates into its two separate subunits, which can assemble on another mRNA molecule to begin a new round of protein synthesis.

We saw in Chapter 5 that most proteins can spontaneously fold into their three-dimensional shape and that most do so as they are synthesized on the ribosome. Some proteins, however, require chaperones (see p. 139) to help them fold correctly. Such proteins are often met by the chaperones as they begin to be synthesized by the ribosome and are properly folded as they elongate.

Proteins Are Made on Polyribosomes

The synthesis of most protein molecules takes between 20 seconds and several minutes. But even during this very short period, multiple initiations usually take place on each mRNA molecule being translated. A new ribosome hops onto the 5' end of the mRNA molecule almost as soon as the preceding ribosome has translated enough of the nucleotide sequence to move out of the way. The mRNA molecules being translated are therefore usually found in the form of *polyribosomes* (also known as *polysomes*), large cytoplasmic assemblies made up of several ribosomes spaced as close as 80 nucleotides apart along a single mRNA molecule (Figure 7–31). These multiple initiations mean that many more protein molecules can be made in a given time than would be possible if each had to be completed before the next one could start.

Both bacteria and eucaryotes utilize polysomes, but bacteria can speed up the rate of protein synthesis even further. Because bacterial mRNA does not need to be processed and is also physically accessible to ribosomes while it is being made, ribosomes will attach to the free end of a bacterial mRNA molecule and start translating it even before the transcription of that RNA is complete, following closely behind the RNA polymerase as it moves along DNA.

Carefully Controlled Protein Breakdown Helps Regulate the Amount of Each Protein in a Cell

The number of copies of a protein in a cell depends, like the human population, not only on how quickly new individuals are made but also on how long they survive. So the breakdown of proteins into their constituent amino acids by cells is a way of regulating the amount of a particular protein present at a given time. Proteins vary enormously in their life span. Structural proteins that become part of a fairly permanent tissue such as bone or muscle may last for months or even years, whereas other proteins, such as metabolic enzymes and those that regulate the

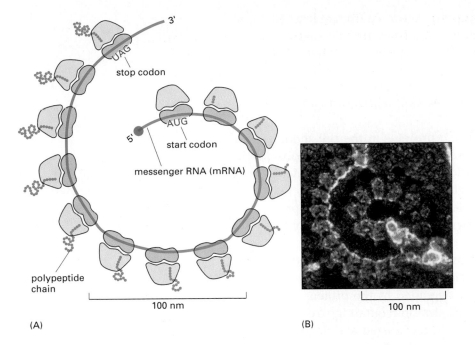

polypeptide chain

stop codon

UAG

3'

5'

AUG

start codon

messenger RNA (mRNA)

100 nm

(A)

100 nm

(B)

Figure 7–31 **A polyribosome.**
(A) Schematic drawing showing how a series of ribosomes can simultaneously translate the same mRNA molecule. (B) Electron micrograph of a polyribosome from a eucaryotic cell. (Courtesy of John Heuser.)

cycle of cell growth, mitosis, and division (discussed in Chapter 18) last only for days, hours, or even seconds. How does the cell control these lifetimes?

Cells have specialized pathways that enzymatically break proteins down into their constituent amino acids (a process termed *proteolysis*). The enzymes that degrade proteins, first to short peptides and finally to individual amino acids, are known collectively as **proteases.** Proteases act by cutting (hydrolyzing) the peptide bonds between amino acids (see Panel 2–5, pp. 62–63). One function of proteolytic pathways is to rapidly degrade those proteins whose lifetimes must be short. Another is to recognize and eliminate proteins that are damaged or misfolded. Although here we discuss how proteins are broken down in the cytosol, important protein degradation pathways in eucaryotic cells also operate in other compartments, such as lysosomes (Chapter 14).

Most proteins degraded in the cytosol of eucaryotic cells are broken down by large complexes of proteolytic enzymes, called **proteasomes.** A proteasome contains a central cylinder formed from proteases whose active sites are thought to face into an inner chamber. Each end of the

Figure 7–32 **Proteasomes degrade unwanted proteins in eucaryotic cells.**
(A) A large number of stained proteasome particles can be seen under the electron microscope. Many proteasomes are present throughout the cytoplasm, where they serve as trash cans for the cell's unwanted proteins. (B) A low-resolution three-dimensional structure of a single complete proteasome complex, derived by computer-based image processing of many such electron micrographs. (C) Diagram of the proteasome complex. The *red* portion of the protein to be degraded represents the ubiquitin tag. In the proteosome, the ubiquitin is removed intact so that it can be recycled. (Electron micrographs courtesy of Wolfgang Baumeister, from J.M. Peters et al., *J. Mol. Biol.* 234: 932–937, 1993.)

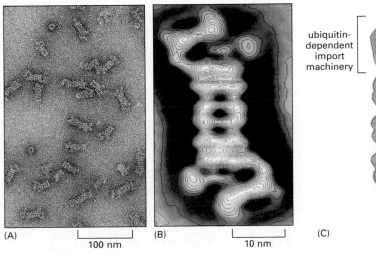

(A)

100 nm

(B)

10 nm

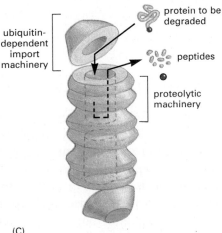

ubiquitin-dependent import machinery

protein to be degraded

peptides

proteolytic machinery

(C)

cylinder is stopped by a large protein complex formed from at least 10 different types of protein subunits (Figure 7–32). These protein stoppers are thought to bind the proteins destined for digestion and then feed them into the inner chamber of the cylinder; there the proteases degrade the proteins to short peptides, which are then released. This arrangement makes sense, as it keeps the proteases supplied with highly selected substrates.

How does the proteasome select which proteins in the cell should enter the cylinder and be degraded? Proteasomes act primarily on proteins that have been marked for destruction by the covalent attachment of a small protein called *ubiquitin*. Specialized enzymes tag selected proteins with one or more ubiquitin molecules; these ubiquinated proteins are then recognized and bound by the proteasome, probably by one of the proteins in the stopper.

Proteins that are meant to be short-lived are often distinguished by the presence of a short amino acid sequence that identifies the protein as one to be ubiquinated and ultimately degraded by the proteasome. Denatured or misfolded proteins, as well as proteins containing oxidized or otherwise abnormal amino acids, are also recognized and degraded by this ubiquitin-dependent proteolytic system. The enzymes that add ubiquitin to such proteins presumably recognize signals that become exposed on these proteins as a result of the misfolding or chemical damage—for example, amino acid sequences or conformational motifs that are normally buried and inaccessible.

There Are Many Steps Between DNA and Protein

We have seen so far in this chapter that many different types of chemical reactions are required to produce a protein from the information contained in a gene (Figure 7–33). The final level of a protein in a cell, therefore, depends upon the efficiency with which each of the many steps is carried out. We will see in the next chapter that cells have the ability to change the levels of most of their proteins according to their needs. In principle, all of the steps in Figure 7–33 can be regulated by the cell. However, as we shall see in the next chapter, the initiation of transcription is the most common point for a cell to regulate the expression of each of its genes.

Transcription and translation are universal processes that lie at the heart of life. However, when scientists came to consider how the flow of information from DNA to protein might have originated, they came to some unexpected conclusions.

RNA and the Origins of Life

To understand fully the processes occurring in present-day living cells, we need to consider how they arose in evolution. The most challenging and fundamental of all such problems is that of the expression of hereditary information, which today requires extraordinary complex machinery and proceeds from DNA to protein through an RNA intermediate. How did this machinery arise? The evolutionary arguments indicate that the sequence of events that led to formation of the first living cell probably *began* with RNA, and suggest that both DNA and protein molecules

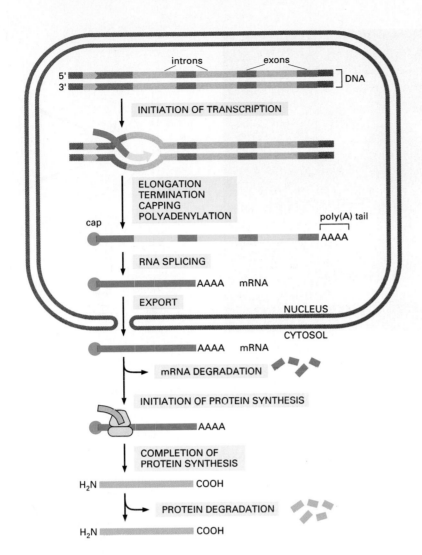

Figure 7–33 **The production of a protein by a eucaryotic cell.** The final level of each protein in a eucaryotic cell depends upon the efficiency of each step depicted. Although the steps are depicted as occurring one at a time, they often happen together in the cell. For example, an RNA molecule is often spliced before its transcription has been completed.

were later additions. RNA is well suited to act by itself, since it can both carry information, as DNA does, and catalyze specific chemical reactions and build macromolecular structures, as proteins do.

Simple Biological Molecules Can Form Under Prebiotic Conditions

Simple carbon-containing organic molecules are likely to have been produced spontaneously on the earth during the first billion years of its existence, over four billion years ago (Figure 7–34). The conditions that existed on the earth at this time to produce them are still a matter of dispute. Was the surface initially molten? Did the atmosphere contain ammonia or methane? There is, however, general agreement that the earth was a violent place, with volcanic eruptions, lightning, and torrential rains. There was little if any free oxygen and no layer of ozone to absorb the ultraviolet radiation from the sun. The radiation, by its photochemical action, may have helped to keep the atmosphere rich in reactive molecules. Today, if a mixture of the gases believed to be present in the early earth's atmosphere is heated with water and energized by electrical discharge in the laboratory (see Figure 7–34), representatives of most of the major classes of small organic molecules found in cells can be generated, including *amino acids, sugars,* and the *purines* and

(A)

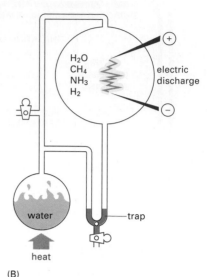

(B)

pyrimidines required to make *nucleotides.* Although such experiments cannot reproduce the early conditions on the earth exactly, they make it clear that the formation of organic molecules is surprisingly easy. And the developing earth had immense advantages over any human experimenter; it was very large and could produce an enormous spectrum of conditions. Above all, it had much more time—hundreds of millions of years. In such circumstances, it seems very likely that, at some time and place, many of the simple organic molecules found in present-day cells accumulated in high concentrations.

It is speculated that simple organic molecules such as nucleotides could then have become randomly linked together to form polymers. Despite the abundance of water on the early earth, which would have strongly favored hydrolysis over condensation reactions (the reactions that link nucleotides together [Chapters 2 and 3]), polynucleotides could have been synthesized in dry organic deposits that formed as shallow pools evaporated. One current theory holds that inorganic mineral catalysts might have helped these reactions to occur. Under laboratory conditions that may mimic such circumstances, polynucleotides of variable length and random nucleotide sequence are formed, in which the particular nucleotide added at any point depends on chance (Figure 7–35). Once a polymer has formed, however, it can in principle influence subsequent chemical reactions by acting as a catalyst itself.

Figure 7–34 The origin of organic molecules. (A) A lightning discharge. The massive discharge of electric energy in lightning strikes was one of the likely sources of energy for prebiotic synthesis of organic molecules on the early earth. (B) A typical experiment attempting to simulate conditions on the primitive earth. Water is heated in a closed apparatus containing methane (CH_4), ammonia (NH_3), and hydrogen (H_2)—all believed to have been plentiful in the atmosphere four billion years ago—and an electric discharge is passed through the vaporized mixture. Organic compounds accumulate in the U-tube trap. (A, courtesy of W. Faidley, Oxford Scientific Films.)

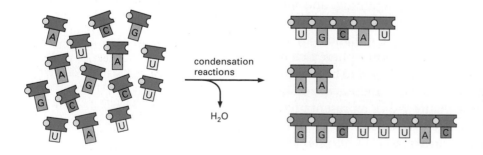

Figure 7–35 Formation of polynucleotides. Nucleotides can undergo spontaneous polymerization through condensation reactions (which involve the loss of water) (see Figures 2–26 and 3–33). The product is a mixture of polynucleotides that vary randomly in length and sequence. The polymerization of the four kinds of nucleotides present in RNA (represented by the letters A, U, G, and C) is depicted here.

The origin of life requires molecules possessing, if only to a small extent, one crucial property: the ability to catalyze reactions that lead, directly or indirectly, to production of more molecules like themselves. Catalysts with this special self-promoting property, once they had arisen by chance, would reproduce themselves and would therefore divert raw materials from the production of other substances. In this way one can envisage the gradual development of an increasingly complex chemical system of organic monomers and polymers that function together to generate more molecules of the same types, fueled by a supply of simple raw materials in the environment. Such an **autocatalytic** system would have some of the properties we think of as characteristic of living matter: the system would contain a far from random selection of interacting molecules; it would tend to reproduce itself; it would compete with other systems dependent on the same raw materials; and, if deprived of its raw materials or maintained at a temperature that upset the balance of reaction rates, it would decay toward chemical equilibrium and "die."

But what molecules could have had such autocatalytic properties? In present-day living cells the most versatile catalysts are polypeptides, which are able to adopt diverse three-dimensional forms that bristle with chemically reactive sites. But although polypeptides are versatile as catalysts, there is no known way in which a polypeptide can reproduce itself directly. Polynucleotides, however, can do all of these things, even if not quite so well as the modern-day specialists.

RNA Can Both Store Information and Catalyze Chemical Reactions

It is not difficult to imagine that polynucleotides could have been formed continually in the soup of molecules produced in prebiotic conditions. For life to evolve, however, individual polynucleotides would need to be able to replicate themselves. We have seen that complementary base-pairing allows a DNA strand to specify the sequence of a complementary polynucleotide, which, in turn, can specify the sequence of the original molecule, providing a way for the original polynucleotide to be replicated (Figure 7–36). RNA is quite capable of being used similarly as a template on which a complementary RNA can be synthesized, as indeed occurs today in some viruses, such as influenza virus, in which RNA is the primary genetic material (see Chapter 9).

But the synthesis of polynucleotides by such complementary templating mechanisms also requires catalysts to promote the polymerization reaction: without catalysts, polymer formation is slow, error-prone, and inefficient. Today, nucleotide polymerization is rapidly catalyzed by protein enzymes—such as the DNA and RNA polymerases. But how

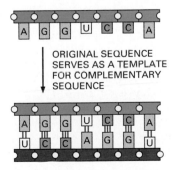

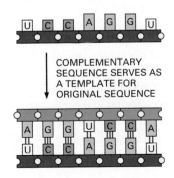

Figure 7–36 **Replication of an RNA molecule.** In the first step the original RNA molecule acts as a template to form an RNA molecule of complementary sequence. In the second step this complementary RNA molecule itself acts as a template, forming RNA molecules of the original sequence. Since each templating molecule can produce many copies of the complementary strand, these reactions can result in the "multiplication" of the original sequence.

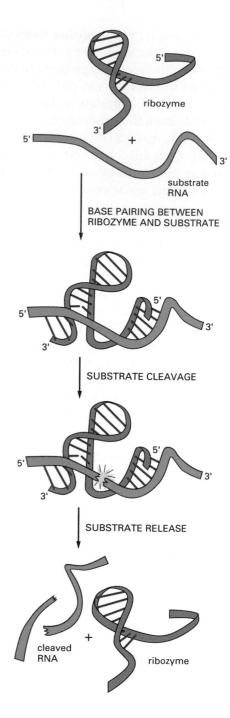

Figure 7–37 A ribozyme. This simple RNA molecule catalyzes the cleavage of a second RNA at a specific site. This ribozyme is found embedded in larger RNA genomes—called viroids—that infect plants. The cleavage, which occurs in nature at a distant location on the same RNA molecule that contains the ribozyme, is a step in the replication of the RNA genome. The catalysis requires a molecule of Mg (not shown), which is brought next to the site of cleavage on the substrate. (Adapted from T.R. Cech and O.C. Uhlenbeck, *Nature* 372:39–40, 1994.)

could it be catalyzed in the prebiotic soup before proteins with the appropriate catalytic specificity existed? The beginnings of an answer to this question were obtained in 1982, when it was discovered that RNA molecules themselves can act as catalysts. We have already seen in this chapter, for example, that a molecule of RNA is probably the catalyst for the peptidyl transferase reaction that takes place on the ribosome. The unique potential of RNA molecules to act both as information carriers and as catalysts is thought to have enabled them to play the central role in the origin of life.

In present-day cells, RNA is synthesized as a single-stranded molecule, and we have seen that complementary base-pairing can occur between nucleotides in the same chain (see Figure 7–5). This base-pairing (along with "nonconventional" hydrogen bonds) can cause each RNA molecule to fold up in a unique way that is determined by its nucleotide sequence. Such associations produce complex three-dimensional patterns of folding, where the molecule as a whole adopts a unique shape.

We have seen that a protein enzyme is able to catalyze a biochemical reaction because it has a surface with unique contours and chemical properties on which a given substrate can react (Chapter 5). In the same way, RNA molecules, with their unique folded shapes, can serve as enzymes (Figure 7–37), although the fact that they are constructed of only four different subunits limits their catalytic efficiency and the range of chemical reactions they can catalyze compared with proteins. Nonetheless, RNAs that catalyze quite a wide range of chemical reactions have now been found in nature or constructed in laboratories (Table 7–2). RNAs with catalytic properties are called **ribozymes.**

Table 7–2   Biochemical Reactions That Can Be Catalyzed by Ribozymes

Activity	Ribozymes
RNA cleavage, RNA ligation	self-splicing RNAs
DNA cleavage	self-splicing RNAs
Peptide bond formation in protein synthesis	ribosomal RNA (?)
DNA ligation	*in vitro* selected RNA
RNA splicing	RNAs of the spliceosome (?), self-splicing RNAs
RNA polymerization	*in vitro* selected RNA
RNA phosphorylation	*in vitro* selected RNA
RNA aminoacylation	*in vitro* selected RNA
RNA alkylation	*in vitro* selected RNA
Isomerization (C–C bond rotation)	*in vitro* selected RNA

Relatively few such catalytic RNAs exist in present-day cells; most catalytic functions in present-day cells have been taken over by proteins. But the processes in which catalytic RNAs still seem to have major roles include some of the most fundamental steps in the expression of genetic information—especially those steps where RNA molecules themselves are spliced or translated into protein.

RNA, therefore, has all the properties required of a molecule that could catalyze its own synthesis (Figure 7–38), and it is thought that such molecules were instrumental in the formation of early cells. Although self-replicating systems of RNA molecules have not been found in nature, scientists are confident that they can be constructed in the laboratory. While this demonstration would not prove that self-replicating RNA molecules were essential in the origin of life on earth, it would certainly suggest that such a scenario is possible.

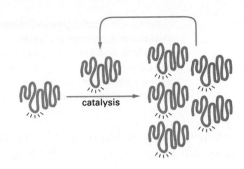

Figure 7–38 An RNA molecule that can catalyze its own synthesis. This hypothetical process would require the catalysis of both steps shown in Figure 7–36. The *red rays* represent the active site of this RNA enzyme.

RNA Is Thought to Predate DNA in Evolution

The first cells on earth would presumably have been much less complex and less efficient in reproducing themselves than even the simplest present-day cells, since catalysis by RNA molecules is less efficient than that by proteins. They would have consisted of little more than a simple membrane enclosing a set of self-replicating molecules and a few other components required to provide the materials and energy for their replication. If the evolutionary speculations about RNA outlined above are correct, these earliest cells would also have differed fundamentally from the cells we know today in having their hereditary information stored in RNA rather than DNA.

Evidence that RNA arose before DNA in evolution can be found in the chemical differences between them. Ribose (see Figure 7–3), like glucose and other simple carbohydrates, is readily formed from formaldehyde (HCHO), which is one of the principal products of experiments simulating conditions on the primitive earth. The sugar deoxyribose is harder to make, and in present-day cells it is produced from ribose in a reaction catalyzed by a protein enzyme, suggesting that ribose predates deoxyribose in cells. Presumably, DNA appeared on the scene later, but then proved more suited than RNA as a permanent repository of genetic information. In particular, the deoxyribose in its sugar-phosphate backbone makes chains of DNA chemically much more stable than chains of RNA, so that greater lengths of DNA can be maintained without breakage.

The other differences between RNA and DNA—the double-helical structure of DNA and the use of thymine rather than uracil—further enhance DNA stability by making the molecule easier to repair. We saw in the previous chapter that a damaged nucleotide on one strand of the double helix can be repaired using the other strand as a template. One of the most common unwanted chemical changes that happens to both RNA and DNA is the spontaneous loss of amino groups from its bases (deamination), which changes the nature of the base and thus its base-pairing abilities (see Figures 6–27 and 6–29). In order to protect the information carried by the molecule, it is vital that the cell should possess enzymes able to identify bases that have been altered by this process and to rectify the loss. Unfortunately, the product of deamination of the base C is, by chance, the base U, which already exists in RNA, so that such

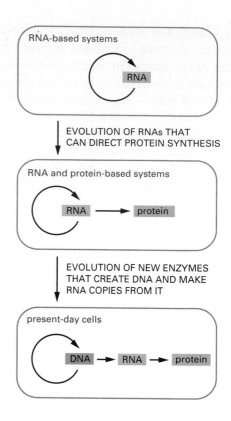

Figure 7–39 The hypothesis that RNA preceded DNA and proteins in evolution. In the earliest cells, RNA molecules combined genetic, structural, and catalytic functions. DNA is now the repository of genetic information, and proteins carry out almost all catalytic functions in cells. RNA now functions as a go-between in protein synthesis, while remaining a catalyst for a few crucial reactions.

changes would be impossible for repair enzymes to detect in an RNA molecule. However, in DNA, which has T rather than U, the U produced by the accidental degradation of C is easily detected and repaired.

Taken together, the evidence we have discussed points to the notion that RNA, having both genetic and catalytic properties, preceded DNA in evolution. This stage of cellular evolution has been termed the *RNA world.* As cells more closely resembling present-day cells appeared, it is believed that many of the functions originally performed by RNA were taken over by molecules more specifically fitted to the tasks required. Eventually DNA took over the primary genetic function and proteins became the major catalysts, while RNA remained primarily as the intermediary connecting the two (Figure 7–39). With the advent of DNA, cells were able to become more complex, for they could then carry and transmit more genetic information than could be stably maintained in an RNA molecule. Because of the greater chemical complexity of proteins and the variety of chemical reactions they can catalyze, the shift from RNA to proteins also provided a much richer source of structural components and enzymes. This enabled cells to evolve the great diversity of structure and function that we see in life today.

> **Question 7–7** Discuss the following: "During the evolution of life on earth, RNA has been demoted from its glorious position as the first self-replicating catalyst. Its role now is as a mere messenger in the information flow from DNA to protein."

Essential Concepts

- The flow of genetic information in all living cells is DNA → RNA → protein. The conversion of the genetic instructions in DNA into RNAs and proteins is termed gene expression.

- To express the genetic information carried in DNA, the nucleotide sequence of a gene is first transcribed into RNA. Transcription is catalyzed by the enzyme RNA polymerase. Nucleotide sequences in the DNA molecule indicate to the RNA polymerase where to start and stop transcribing.

- RNA differs in several respects from DNA. It contains the sugar ribose instead of deoxyribose and the base uracil (U) instead of thymine (T). RNAs in cells are synthesized as single-stranded molecules, which often fold up into precise three-dimensional shapes.

- Cells make several different functional types of RNA, including messenger RNA (mRNA), which carries the instructions for making proteins; ribosomal RNA (rRNA), which is a component of ribosomes; and transfer RNA (tRNA), which acts as an adaptor molecule in protein synthesis.

- In eucaryotic DNA, most genes are split into a number of smaller coding regions (exons) interspersed with noncoding regions (introns).

- When a eucaryotic gene is transcribed from DNA into RNA, both the exons and introns are copied to produce the primary RNA transcript.

- Introns are removed from primary RNA transcripts in the nucleus by the process of RNA splicing. In a reaction catalyzed by small ribonucleoprotein complexes known as snRNPs, the introns are excised from the primary transcript and the exons are joined together. The mRNA then moves to the cytoplasm.

- Translation of the nucleotide sequence of mRNA into a protein takes place in the cytoplasm on large ribonucleoprotein assemblies called ribosomes. These attach to the mRNA and move stepwise along the mRNA chain, translating the message into protein.

- The nucleotide sequence in mRNA is read in sets of three nucleotides (codons), each codon corresponding to one amino acid.

- The correspondence between amino acids and codons is specified by the genetic code. The possible combinations of the 4 different nucleotides in RNA give 64 different codons in the genetic code. Most amino acids are specified by more than one codon.

- tRNA acts as an adaptor molecule in protein synthesis. Enzymes called aminoacyl-tRNA synthetases link amino acids to their appropriate tRNAs. Each tRNA contains a sequence of three nucleotides, the anticodon, which matches a codon in mRNA by complementary base-pairing between codon and anticodon.

- Protein synthesis begins when a ribosome assembles at an initiation codon (AUG) in mRNA, a process that is regulated by proteins called initiation factors. The completed protein chain is released from the ribosome when a stop codon (UAA, UAG, or UGA) is reached.

- The stepwise linking of amino acids into a polypeptide chain is probably catalyzed by an rRNA molecule in the large ribosomal subunit.

- The degradation of proteins in the cell is carefully controlled. Some proteins are degraded in the cytosol by large protein complexes called proteasomes.

- From our knowledge of present-day organisms and the molecules they contain, it seems likely that living systems began with the evolution of RNA molecules that could catalyze their own replication.

- It has been proposed that, as cells evolved, the DNA double helix replaced RNA as a more stable molecule for storing increased amounts of genetic information, and proteins replaced RNAs as major catalytic and structural components.

- The flow of information in present-day living cells is DNA → RNA → protein, with RNA serving primarily as a go-between. Some important reactions, however, are still catalyzed by RNA; these are thought to provide a glimpse into the ancient, RNA-based world.

Key Terms

aminoacyl-tRNA synthetase	genetic code	protease	RNA processing
anticodon	initiation factor	proteasome	RNA splicing
autocatalytic	initiator tRNA	proteolysis	rRNA
codon	intron	reading frame	snRNP
exon	mRNA	ribosome	transcription
	primary transcript	ribozyme	translation
	promoter	RNA polymerase	tRNA

Questions

Question 7–8 Which of the following statements are correct? Explain your answers.

A. An individual ribosome can make only one type of protein.

B. All mRNAs fold into particular three-dimensional structures that are required for their translation.

C. The large and small subunits of an individual ribosome always stay together and never exchange partners.

D. Ribosomes are cytoplasmic organelles that are encapsulated by a single membrane.

E. Because the two strands of DNA are complementary, the mRNA of a given gene can be synthesized using either strand as a template.

F. An mRNA may contain the sequence ATTGACCC-CGGTCAA.

G. The amount of a protein present in the cell at a steady state depends on its rate of synthesis, its catalytic activity, and its rate of degradation.

Question 7–9 The Lacheinmal protein is a hypothetical protein that causes people to smile more often. It is inactive in many chronically unhappy people. The mRNA isolated from a number of different unhappy persons in the same family was found to lack an internal stretch of 173 nucleotides that are present in the Lacheinmal mRNA isolated from a control group of generally happy people. The DNA sequences of the Lacheinmal genes from the happy and unhappy families were determined and compared. They differed by just one nucleotide change—and no nucleotides were deleted. Moreover, the change was found in an intron. What can you say about the molecular basis of unhappiness in this family?

(Hints: consider the following two questions independently. [1] Can you hypothesize a molecular mechanism by which a single nucleotide change in a gene could cause the observed deletion in the mRNA? Note that it is an *internal* deletion. [2] What consequences for the Lacheinmal protein would result from removing a 173-ribonucleotide-long internal stretch from its mRNA? Assume that the 173 bases are deleted inside the coding region of the Lacheinmal mRNA.)

Question 7–10   Use the genetic code shown in Figure 7–20 to identify which of the following nucleotide sequences would code for the polypeptide sequence Arginine-Glycine-Aspartate.

1. 5′–AGA-GGA-GAU–3′
2. 5′–ACA-CCC-ACU–3′
3. 5′–GGG-AAA-UUU–3′
4. 5′–CGG-GGU-GAC–3′

Question 7–11   "The bonds that form between the anticodon of a tRNA molecule and the three nucleotides of a codon in mRNA are _____." Complete this sentence with each of the following options and explain why the resulting statements are correct or incorrect.

A. covalent bonds formed by GTP hydrolysis.

B. hydrogen bonds that form when the tRNA is at the A-site.

C. broken by the translocation of the ribosome along the mRNA.

Question 7–12   List the ordinary, dictionary definitions of the terms *replication, transcription,* and *translation* and by their side list the special meaning each term has when applied to the living cell.

Question 7–13   In an alien world, the genetic code is written in pairs of nucleotides. How many amino acids can it specify? In a different world a triplet code is used but the sequence of nucleotides is not important. It only matters which nucleotides are present. How many amino acids can this code specify? Would you expect to encounter any problems translating these codes?

Question 7–14   One remarkable regular feature of the genetic code is that amino acids with similar chemical properties often have similar codons. Thus, codons with U or C as the second nucleotide tend to specify hydrophobic amino acids. Can you suggest a possible explanation for this phenomenon in terms of the early evolution of the protein synthesis machinery?

Question 7–15   A mutation in DNA generates a UGA stop codon in the middle of the RNA coding for a particular protein. A second mutation in the cell leads to a single nucleotide change in a tRNA that allows the correct translation of the protein, that is, the second mutation "suppresses" the defect caused by the first. The altered tRNA translates the UGA as tryptophan. What nucleotide change has probably occurred in the mutant tRNA molecule? What consequences would the presence of such a mutant tRNA have for the translation of the normal genes in this cell?

Question 7–16   In a clever experiment performed in 1962, a cysteine already attached to its tRNA was chemically converted to an alanine. These "hybrid" tRNA molecules were then added to a cell-free translation system from which the normal cysteine-tRNAs had been removed. When the resulting protein was analyzed, it was found that alanine had been inserted at every point in the protein chain where cysteine was supposed to be. Discuss what this experiment tells you about the role of aminoacyl-tRNA synthetases during the normal translation of the genetic code.

Question 7–17   The charging of a tRNA with an amino acid can be represented by the following equation:

$$\text{amino acid} + \text{tRNA} + \text{ATP} \rightarrow \text{aminoacyl-tRNA} + \text{AMP} + \text{PP}_i$$

where PP_i is pyrophosphate (see Figure 3–34). In the aminoacyl-tRNA, the amino acid and tRNA are linked with a high-energy bond (discussed in Chapter 3); a large portion of the energy derived from the hydrolysis of ATP is thus stored in this bond and is available to drive peptide-bond formation at the later stages of protein synthesis. The free-energy change of the charging reaction shown in the equation is close to zero and therefore would not be expected to favor attachment of the amino acid to tRNA. Can you suggest a further step that could drive the reaction to completion?

Question 7–18

A. The average molecular weight of proteins in the cell is about 30,000 daltons. A few proteins, however, are very much larger. The largest known polypeptide chain made by any cell is a protein called titin (made by muscle cells), and it has a molecular weight of 3,000,000 daltons. Estimate how long it will take a muscle cell to translate an mRNA coding for titin (assume the average molecular weight of an amino acid to be 120, and a translation rate of two amino acids per second for eucaryotic cells).

B. Protein synthesis is very accurate: for every 10,000 amino acids joined together, only one mistake is made. What is the fraction of average-sized protein molecules and of titin molecules that are synthesized without any errors? (Hint: the probability P of obtaining an error-free protein is given by $P = (1 - E)^n$, where E is the error frequency and n the number of amino acids.)

C. The molecular weight of all eucaryotic ribosomal proteins combined is about 2.5×10^6 daltons. Would it be advantageous to synthesize them as a single protein?

D. Transcription occurs at a rate of about 30 nucleotides per second. Is it possible to calculate the time required to synthesize a titin mRNA from the information given here?

Question 7–19 Which of the following mutational changes would be predicted to harm an organism? Explain your answers.

a. insertion of a single nucleotide near the end of the coding sequence.

b. removal of a single nucleotide near the beginning of the coding sequence.

c. deletion of three consecutive nucleotides in the middle of the coding sequence.

d. deletion of four consecutive nucleotides in the middle of the coding sequence.

e. substitution of one nucleotide for another in the middle of the coding sequence.

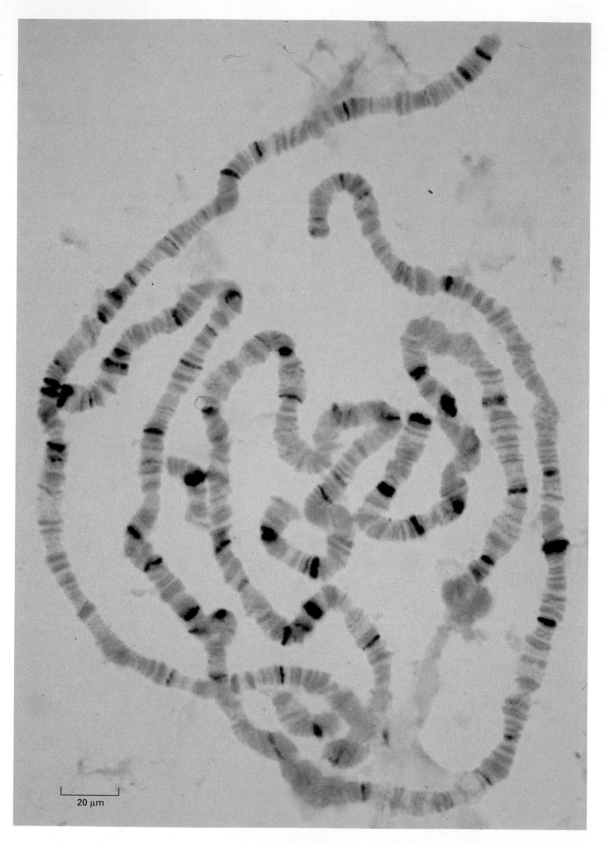

20 μm

In the salivary glands of the fly *Drosophila* are found giant cells with enormous chromosomes. Each of these chromosomes contains over a thousand identical DNA molecules, all aligned in register. This makes them very easy to see in the light microscope, where they display a characteristic and reproducible banding pattern. Such chromosomes have been widely used to study gene expression. In the micrograph shown here, the position of a protein that binds to the regulatory regions of over 60 genes, and controls their expression, is visualized directly as dark-staining bands by means of an antibody to the protein. (Courtesy of B. Zink and R. Paro, from R. Paro, *Trends Genet.* 6:416–421, 1990.)

8 Chromosomes and Gene Regulation

An organism's DNA encodes all of the RNA and protein molecules that are needed to make its cells. But a complete description of the DNA sequence of an organism—be it the few million nucleotides of a bacterium or the few billion nucleotides in each human cell—would no more enable us to reconstruct the organism than a list of English words in a dictionary would enable us to reconstruct a play by Shakespeare. We need to know how the elements in the DNA sequence—the genes—are used. Even the simplest single-celled bacterium can use its genes selectively, switching genes on and off so that it makes different metabolic enzymes depending on the food sources available.

In multicellular plants and animals, **gene expression** is under even more elaborate control. In the course of embryonic development, a fertilized egg cell gives rise to many cell types that differ dramatically in both structure and function. The differences between a mammalian neuron and a lymphocyte, for example, are so extreme that it is difficult to imagine that the two cells contain the same DNA (Figure 8–1). For this reason, and because cells in an adult organism rarely lose their distinctive characteristics, biologists originally suspected that genes might be selectively lost when a cell becomes specialized. We now know, however, that nearly all the cells of a multicellular organism contain the same genome. Cell differentiation is instead achieved by changes in gene expression.

Hundreds of different cell types carry out a range of specialized functions that depend upon genes that are only switched on in that cell type: for example, the β cells of the pancreas make the protein hormone insulin, while the α cells of the pancreas make the hormone glucagon; the lymphocytes of the immune system are the only cells in the body to make antibodies, while developing red blood cells are the only cells that make the oxygen-transport protein hemoglobin. The differences between a neuron, a lymphocyte, a pancreas cell, and a red blood cell depend upon the precise control of gene expression. In each case the cell is using only some of the genes in its total repertoire.

Large amounts of DNA are required to encode all the information needed to make just a single-celled bacterium, and far more DNA is

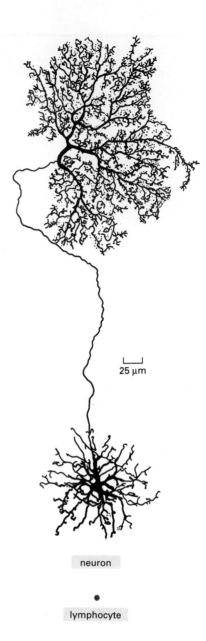

Figure 8–1 **A mammalian neuron and a lymphocyte.** The long branching processes of this neuron from the retina enable it to receive electrical signals from many cells and carry them to many neighboring cells. The lymphocyte is a white blood cell involved in the immune response to infection and moves freely through the body. Both of these cells contain the same genome, but they express different RNAs and proteins.

25 μm

neuron

lymphocyte

needed to encode the instructions for the development of a multicellular organism. This poses the problem of how the DNA can be folded and packaged so that it is compact enough to be managed easily within the cell nucleus. In eucaryotic cells, enormously long double-stranded DNA molecules are packaged by association with specialized proteins into *chromosomes* that fit readily inside the nucleus and can be apportioned correctly between the two daughter cells at each cell division. Packing has to be done in an orderly fashion so that the genes contained in the DNA molecules are available when they need to be transcribed or replicated.

We begin this chapter by describing how DNA is packaged into eucaryotic chromosomes and the changes in chromosome structure that occur at different phases of a cell's life cycle. Then, in the second half of the chapter, we shall discuss the main ways in which gene expression is controlled in bacterial and eucaryotic cells. Although some mechanisms of control apply to both sorts of cells, eucaryotic cells, through their more complex chromosomal structure, have ways of controlling gene expression that are not available to bacteria.

The Structure of Eucaryotic Chromosomes

The nucleus of a typical human cell is about 5–8 μm in diameter and contains about *2 meters* of DNA—equivalent to a tennis ball that contains 20 km (12 miles) of extremely fine thread. We will see in this section that the complex task of packaging DNA is accomplished by specialized proteins that bind to and fold the DNA, generating a series of coils and loops that provide increasingly higher levels of organization and prevent the DNA from becoming an unmanageable tangle. Amazingly, the DNA is compacted in a way that leaves it accessible to all of the enzymes and other proteins required for transcription, DNA replication, and DNA repair.

Eucaryotic DNA Is Packaged into Chromosomes

In eucaryotes, such as ourselves, the DNA in the nucleus is distributed among a set of different **chromosomes.** Each chromosome consists of a single, enormously long linear DNA molecule associated with proteins that fold and pack the fine DNA thread into a more compact structure. The complex of DNA and protein is called *chromatin* (from the Greek *chroma*, "colored," because of its staining properties). Many of the proteins associated with the DNA in chromosomes are involved in packaging the DNA, but chromosomes are also associated with other proteins involved in gene expression, DNA replication, and DNA repair.

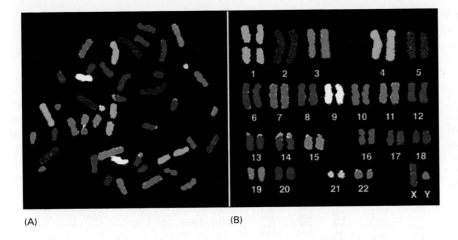

(A) (B)

Figure 8–2 **Human chromosomes.** The chromosomes, from a male, were isolated from a cell undergoing nuclear division (mitosis) and are therefore in a highly compact state. Each chromosome has been "painted" a different color to allow its unambiguous identification under the light microscope. Chromosome painting is carried out by exposing the chromosomes to a collection of human DNA molecules that have been coupled to a combination of fluorescent dyes. For example, DNA molecules derived from Chromosome 1 are labeled with one specific dye combination, those from Chromosome 2 with another, and so on. Because the labeled DNA can form base pairs, or hybridize, only to their chromosome of origin (see Chapter 10) each chromosome is differently labeled. For such experiments, the chromosomes are subjected to treatments that separate the double-helical DNA into individual strands, to enable base-pairing with the single-stranded labeled DNA while keeping the chromosome structure relatively intact. (A) The chromosomes as visualized as they originally spilled from the lysed cell. (B) The same chromosomes artificially lined up in order. This arrangement of the full chromosome set is called a karyotype. (From E. Schröck et al., *Science* 273:494–497, 1996.)

Bacteria typically have all of their genes on one circular DNA molecule, which is also associated with proteins that condense the DNA (different from those in eucaryotes) and is often called the bacterial "chromosome." It does not have the same structure as eucaryotic chromosomes, and less is known about how the bacterial DNA is packaged. Our discussion of chromosome structure in this chapter will therefore focus entirely on eucaryotic chromosomes.

Human cells, with the exception of the germ cells, each contain two copies of each chromosome, one inherited from the mother and one from the father; the maternal and paternal chromosomes of a pair are called *homologous chromosomes*. The only nonhomologous chromosome pairs are the sex chromosomes in males, where a *Y chromosome* is inherited from the father and an *X chromosome* from the mother. Figure 8–2 shows how *DNA hybridization* (which is described in Chapter 10) can be used to distinguish the human chromosomes by "painting" each one a different color. But the standard way of distinguishing one chromosome from another is to stain them with dyes that bind to certain types of DNA sequences. These dyes mainly distinguish DNA that is rich in A-T nucleotide pairs from DNA that is rich in G-C nucleotide pairs, and they produce a striking and reliable pattern of bands along each chromosome (Figure 8–3). The pattern of bands on each type of chromosome is unique, allowing each chromosome to be identified and numbered. If parts of chromosomes are lost, or switched between chromosomes, these changes can be detected by changes in the banding patterns. Cytogeneticists use alterations in banding patterns to detect chromosome abnormalities that are associated with some inherited defects or with certain types of cancer.

> **Question 8–1** In a DNA double helix, adjacent nucleotide pairs are 0.34 nm apart. Use Figure 8–3 to estimate the length of the DNA in human Chromosome 1 if it were unraveled and stretched out. If the actual length of Chromosome 1 at this stage of mitosis is approximately 10 μm, what is the degree of compaction of the DNA in this state?

Chromosomes Exist in Different States Throughout the Life of a Cell

Chromosomes are almost always depicted as they are shown in Figures 8–2 and 8–3; but these chromosomes are in fact in a highly condensed state that normally occurs only for a brief period when the cell is dividing. At all other times, the chromosomes are extended as long, thin, tangled threads of DNA in the nucleus and cannot be easily distinguished in the light microscope.

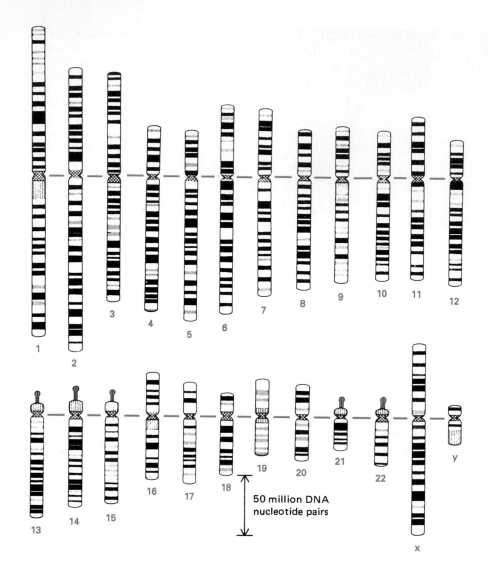

Figure 8–3 **The banding patterns of human chromosomes.** Chromosomes 1 through 22 are numbered in approximate order of size. A typical human somatic (that is, non-germ-line) cell contains two of each of these chromosomes plus two sex chromosomes—two X chromosomes in a female; one X and one Y chromosome in a male. The chromosomes used to make these maps were stained at an early stage in mitosis, when the chromosomes are compacted. The *horizontal line* represents the position of the centromere, which appears as a constriction on mitotic chromosomes; the *knobs* on Chromosomes 13, 14, 15, 21, and 22 indicate the positions of genes that code for the large ribosomal RNAs. These patterns are obtained by staining chromosomes with Giemsa stain, which produces dark bands in regions rich in A-T nucleotide pairs. (Adapted from U. Franke, *Cytogenet. Cell Genet.* 31:24-32, 1981.)

The state of condensation of the chromosomes thus varies according to the cell growth cycle, which goes through a series of stages, known collectively as the *cell cycle,* that will be discussed in detail in Chapter 18 and is briefly summarized in Figure 8–4. Only two of these stages need concern us in this chapter: *mitosis,* when the chromosomes have replicated and condensed and are distributed to the two daughter nuclei, and *interphase,* when the chromosomes are more extended. The highly condensed chromosomes in a dividing cell are known as **mitotic chromosomes;** we shall refer to chromosomes in their more usual extended state as *interphase chromosomes.*

Figure 8–4 **A simplified view of the eucaryotic cell cycle.** During interphase, the cell is actively transcribing its genes and synthesizing proteins. Still during interphase and before cell division, the DNA is replicated and the chromosomes are duplicated. Once DNA replication is complete, the cell can enter *M phase,* when mitosis occurs. Mitosis is the division of the nucleus. During this stage, the chromosomes condense, the nuclear envelope breaks down, and the mitotic spindle forms from microtubules and other proteins. The condensed chromosomes are captured by the mitotic spindle, and one complete set of chromosomes is pulled to each end of the cell. A nuclear envelope forms around each chromosome set, and in the final step of M phase, the cell divides to produce two daughter cells.

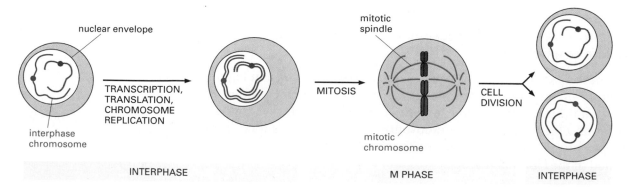

The condensed state of the mitotic chromosomes is important in allowing the duplicated chromosomes to be separated during cell division, and we discuss in Chapter 17 the process whereby they are captured and distributed to the two daughter cells by the *mitotic spindle* (Figure 8–4), which assembles from microtubules and other proteins in dividing cells. Specialized DNA sequences are required in each chromosome to allow it to engage with the spindle and ensure the delivery of a complete copy of the DNA to each of the two daughters of a dividing cell, and we describe these sequences in the next section before turning to the specialized proteins that package the DNA.

Specialized DNA Sequences Ensure That Chromosomes Replicate Efficiently

A chromosome operates as a distinct structural unit; in order for a copy to be passed on to each daughter cell at division, each chromosome must be able to replicate, and the newly replicated copies must subsequently be separated and partitioned correctly into the two daughter cells. These basic functions are controlled by three types of specialized DNA sequences found in all eucaryotic chromosomes (Figure 8–5). One type of nucleotide sequence acts as a DNA **replication origin,** at which duplication of the DNA begins (see Chapter 6). Most eucaryotic chromosomes contain many replication origins to ensure that the entire chromosome can be replicated rapidly. The presence of a second specialized DNA sequence, called a **centromere,** allows one copy of each duplicated chromosome to be pulled into each daughter cell when a cell divides. During mitosis, a protein complex called a *kinetochore* forms at the centromere and attaches the duplicated chromosomes to the spindle, allowing them to be pulled apart (see Chapter 17).

The third specialized DNA sequence, called a **telomere,** is found at each of the two ends of a chromosome. Telomeres contain repeated nucleotide sequences that enable the ends of chromosomes to be replicated. These specialized sequences are necessary because, as we saw in Chapter 6, DNA polymerases can synthesize DNA only in the $5' \to 3'$ direction. Thus synthesis of the lagging strand of a replication fork is in the form of discontinuous DNA fragments, each of which is primed with

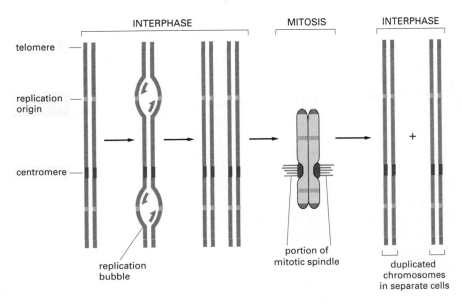

Figure 8–5 The three DNA sequence elements needed to produce a eucaryotic chromosome that can be replicated and then segregated at mitosis. Each chromosome has multiple origins of replication, one centromere, and two telomeres. Shown schematically is the sequence of events that a typical chromosome follows during the cell cycle. The DNA replicates in interphase. In M phase, the centromere attaches the duplicated chromosomes to the mitotic spindle so that one copy is distributed to each daughter cell during mitosis. The centromere also helps to hold the duplicated chromosomes together until they are ready to be moved apart.

The Structure of Eucaryotic Chromosomes

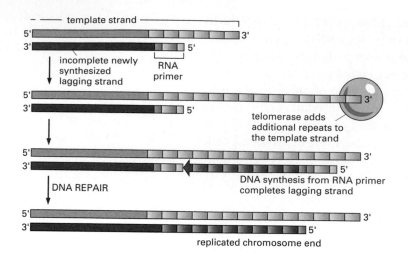

template strand

5' _____ 3'
3' _____ 5'

incomplete newly
synthesized   RNA
lagging strand   primer

5' _____ 3'
3' _____ 5'

telomerase adds
additional repeats to
the template strand

5' _____ 3'
3' _____ 5'

DNA synthesis from RNA primer
completes lagging strand

DNA REPAIR

5' _____ 3'
3' _____ 5'

replicated chromosome end

an RNA primer laid down by a separate enzyme (see Figure 6–20). There is no place to form an RNA primer to start this DNA synthesis at the very tip of a linear DNA molecule. Therefore, some DNA could easily be lost from the ends of a DNA molecule each time it is replicated.

Bacteria solve this "end-replication" problem by having circular DNA molecules as chromosomes. Eucaryotes solve it by having special nucleotide sequences at the ends of their chromosomes that attract an enzyme called *telomerase*. Telomerase adds multiple copies of the same telomere DNA sequence to the ends of the chromosomes, thereby producing a template that allows replication of the lagging strand to be completed (Figure 8–6). Telomeres also serve an additional function: the repeated telomere DNA sequences, together with the regions adjoining them, form structures that protect the DNA from attack by the DNA-digesting enzymes—called nucleases—that preferentially degrade the ends of DNA molecules in the cell.

Having described the three types of specialized DNA sequences that define a chromosome, we can return to the question of how the long DNA molecules are packaged by proteins so that they can be easily managed by cells.

Nucleosomes Are the Basic Units of Chromatin Structure

As previously mentioned, the complex of DNA and protein that forms a chromosome is called **chromatin.** We have seen that chromatin exists in different states at different phases of the cell's life cycle. In interphase nuclei it is in a more extended state, which allows access to the DNA for the enzymes and other proteins required for gene expression and DNA replication (Chapters 6 and 7). As the chromosomes prepare to enter mitosis, the chromatin thread coils up further, adopting a more and more compact structure, until the highly condensed mitotic chromosome has been formed (Figure 8–7). In the mitotic chromosomes, the DNA has already replicated and transcription has ceased. This ability of the eucaryotic cell to vary the packing of its DNA is also used during interphase as one mechanism for controlling gene expression, as we shall see shortly. Before we can hope to understand it, we will need to decipher exactly how the packing itself is achieved.

The first and most fundamental packing level of chromatin, the **nucleosome,** was discovered in 1974. When interphase nuclei are broken

Figure 8–6 Completion of DNA synthesis at the ends (telomeres) of chromosomes. On one of the DNA strands in a double-stranded DNA molecule, the synthesis of the new strand is discontinuous (see Chapter 6). The strand that is synthesized discontinuously is known as the lagging strand. (For simplicity, the other DNA strand being replicated is not shown in this diagram.) To synthesize the lagging strand, the machinery of DNA replication requires a length of template DNA extending beyond the DNA that is to be copied. In a linear DNA molecule, synthesis of the lagging strand thus stops short just before the end of the template. But the enzyme telomerase adds a series of repeats of a DNA sequence to the template strand, which allows the lagging strand to be completed by DNA polymerase, as shown. In humans, the nucleotide sequence of the repeat is GGGGTTA. The telomerase enzyme contains within it a short piece of RNA of complementary sequence to the DNA repeat sequence; this RNA acts as the template for the telomerase DNA synthesis.

Question 8–2   Describe the consequences that would arise if a eucaryotic chromosome . . .

A. contained only one origin of replication:

(i) at the exact center of the chromosome.

(ii) at one end of the chromosome.

B. lacked one or both telomeres.

C. had no centromere.

Assume that the chromosome is 150 million nucleotide pairs in length, a typical size for an animal chromosome, and that DNA replication in animal cells proceeds at about 100 nucleotides per second.

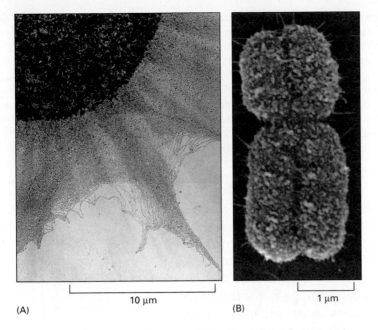

Figure 8–7 Comparison of extended interphase chromatin with the chromatin in a mitotic chromosome. (A) An electron micrograph showing an enormous tangle of chromatin spilling out of a lysed interphase nucleus. (B) A scanning electron micrograph of a mitotic chromosome, which is a duplicated chromosome in which the two new chromosomes are still linked together (see Figure 8–5). The constricted region indicates the position of the centromere. Note the difference in scales. (A, courtesy of Victoria Foe; B, courtesy of Terry D. Allen.)

(A) 10 μm

(B) 1 μm

open very gently and their contents examined under the electron microscope, most of the chromatin is in the form of a fiber with a diameter of about 30 nm (Figure 8–8A). If this chromatin is subjected to treatments that cause it to unfold partially, it can then be seen under the electron microscope as a series of "beads on a string" (Figure 8–8B). The string is DNA, and each bead is a "nucleosome core particle" that consists of DNA wound almost twice around a core of proteins.

The structure of nucleosomes was determined after first isolating them from the unfolded chromatin by digestion with particular nucleases that break down DNA by cutting between the nucleotides. After digestion for a short period, the exposed DNA between the nucleosome core particles, the *linker DNA,* is degraded. An individual nucleosome core particle consists of a complex of eight **histone** proteins—two molecules each of histones H2A, H2B, H3, and H4—and a double-stranded DNA of around 146 nucleotide pairs. The histone octamer forms a protein core around which the double-stranded DNA helix winds (Figure 8–9).

The term "nucleosome" refers to a nucleosome core particle plus an adjacent DNA linker. Formation of nucleosomes converts a DNA molecule into a chromatin thread approximately one-third of its initial length, and provides the first level of DNA packing.

(A)

(B)

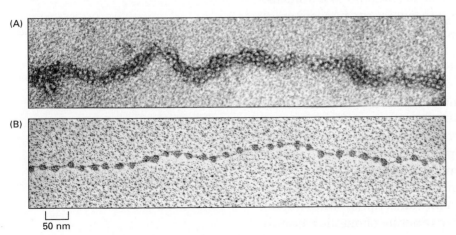

50 nm

Figure 8–8 Nucleosomes as seen in the electron microscope. (A) Chromatin isolated directly from an interphase nucleus appears in the electron microscope as a thread 30 nm thick. (B) This electron micrograph shows a length of chromatin that has been experimentally unpacked, or "decondensed," after isolation to show the nucleosomes. (A, courtesy of Barbara Hamkalo; B, courtesy of Victoria Foe.)

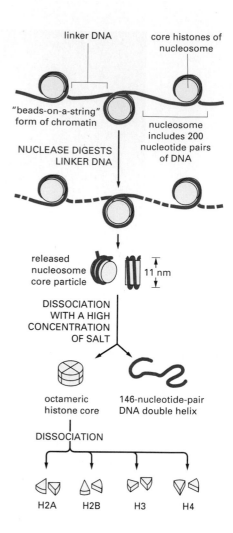

Figure 8–9 The nature of the nucleosome. A nucleosome contains a protein core made of eight histone molecules. As indicated, the nucleosome core particle is released from chromatin by digestion of the linker DNA with a nuclease, an enzyme that breaks down DNA. (The nuclease can degrade the exposed DNA but cannot attack the DNA wound around the nucleosome.) After dissociation of the isolated nucleosome into its protein core and DNA, the length of the DNA that was wound around the core can be determined. Its length of 146 nucleotide pairs is sufficient to wrap almost twice around the histone core.

Histones are small proteins with a high proportion of positively charged amino acids (lysine and arginine). These positive charges help the histones bind tightly to the negatively charged sugar-phosphate backbone of DNA, regardless of the precise nucleotide sequence. Nucleosomes are spaced along the DNA at intervals of around 200 nucleotide pairs (146 nucleotide pairs for DNA wound around the histone core and an average of about 50 nucleotide pairs for the linker between adjacent nucleosome cores). Histones are present in enormous quantities (about 60 million molecules of each type per cell), and their total mass in chromatin is about equal to that of the DNA itself.

The histones that form the nucleosome core are among the most highly conserved of all known eucaryotic proteins: there are only two differences between the amino acid sequences of histone H4 from peas and cows, for example. Recently, histones have been found in archaebacteria, a phylogenetic kingdom distinct from plants and animals (see Chapters 1 and 13). This extreme evolutionary conservation reflects the vital structural role of histones in forming chromatin.

Chromosomes Have Several Levels of DNA Packing

In the living cell, chromatin probably rarely adopts the extended beads-on-a-string form seen in Figure 8–8B. Instead, the nucleosomes are packed upon one another to generate a more compact structure, the *30-nm fiber* (see Figure 8–8A). Packing of nucleosomes into the 30-nm fiber depends on a fifth histone called histone H1, which is thought to pull the nucleosomes together into a regular repeating array. The structure that results is illustrated, as part of a larger schematic of the various levels of chromosome packing, in Figure 8–10.

We know that the 30-nm chromatin fiber can be compacted still further. We saw earlier in this chapter that during mitosis chromatin becomes so highly condensed that the chromosomes can be seen under the light microscope. How is the 30-nm fiber folded to produce mitotic chromosomes? The answer to this question is not yet known in detail, but it is thought that the 30-nm fiber is further organized into loops emanating from a central axis (Figures 8–10 and 8–11). Finally, this string of loops is thought to undergo at least one more level of packing to form the mitotic chromosome (see Figure 8–10).

The tight packing of chromatin that takes place in mitosis has an important consequence for the transcription of genes contained in the DNA: as the chromosomes become more compact, RNA synthesis ceases, in part because the DNA becomes so tightly packed that RNA polymerase and other proteins required for transcription cannot gain access to it. We shall now consider how the structure of mitotic chromo-

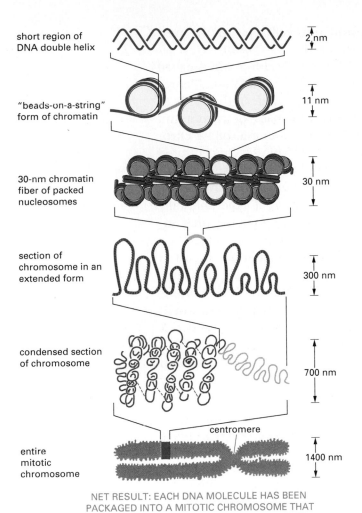

short region of DNA double helix — 2 nm

"beads-on-a-string" form of chromatin — 11 nm

30-nm chromatin fiber of packed nucleosomes — 30 nm

section of chromosome in an extended form — 300 nm

condensed section of chromosome — 700 nm

entire mitotic chromosome — centromere — 1400 nm

NET RESULT: EACH DNA MOLECULE HAS BEEN PACKAGED INTO A MITOTIC CHROMOSOME THAT IS 50,000x SHORTER THAN ITS EXTENDED LENGTH

Figure 8–10 Levels of chromatin packing. This schematic drawing shows some of the orders of chromatin packing thought to give rise to the highly condensed mitotic chromosome. The folding of naked DNA into nucleosomes is the best understood level of packing. The structures corresponding to the additional layers of chromosome packing are more speculative.

somes is related to the structure of interphase chromosomes, in which transcription can take place.

Interphase Chromosomes Contain Both Condensed and More Extended Forms of Chromatin

As the daughter cells complete their separation following mitosis, the nuclear membranes re-form and the mitotic chromosomes unfold into a more extended form—the interphase chromosomes (see Figure 8–4). However, the chromatin in an interphase chromosome is not in the same packing state throughout the chromosome. Some regions become more unfolded and extended than others, and it is likely that all of the different packing states shown in Figure 8–10 are found somewhere on an interphase chromosome. In general, regions of the chromosome that are being transcribed into RNA are more extended, while those that are transcriptionally quiet are more compact. Thus, the detailed structure of

Figure 8–11 Scanning electron micrograph of a region near one end of a typical mitotic chromosome. Each knoblike projection is believed to represent the tip of a separate loop of chromatin. The chromosome in this picture has duplicated, but the two new chromosomes (also called chromatids) are still held together (see Figure 8–7B). The ends of the two chromosomes can be easily distinguished in this micrograph. (From M.P. Marsden and U.K. Laemmli, *Cell* 17:849–858, 1989. © Cell Press.)

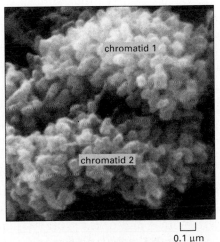

chromatid 1

chromatid 2

0.1 μm

an interphase chromosome differs from one cell type to the next depending on which genes are being expressed.

The most highly condensed form of interphase chromatin is called **heterochromatin** (from the Greek *heteros,* meaning "different," chromatin). It was first observed under the light microscope in the 1930s as discrete, strongly staining regions within the mass of chromatin. Heterochromatin typically makes up about 10% of an interphase chromosome, and in mammalian chromosomes, it is typically concentrated around the centromere region and at the ends of the chromosomes. Like mitotic chromatin, heterochromatin is transcriptionally inactive. As we will see in the next section, a normally active gene, if experimentally moved to a region of heterochromatin, will become inactive, presumably because it too becomes packed into heterochromatin.

Perhaps the most striking example of heterochromatin is found in the interphase X chromosomes of female mammals. Female cells contain two X chromosomes, while male cells contain one X and one Y. Presumably because a double dose of X-chromosome products would be lethal, female mammals permanently inactivate one of the two X chromosomes in each cell: at random, one or other of the two X chromosomes in each cell becomes highly condensed into heterochromatin early in embryonic development. Thereafter, in all of the many progeny of the cell, the condensed and inactive state of that X chromosome is inherited (Figure 8–12).

The rest of the interphase chromatin, which is in a variety of more extended states, is called *euchromatin* (from the Greek *eu,* meaning "true" or "normal," chromatin). In the typical differentiated eucaryotic cell, some 10% of this chromatin is in a state in which it is either actively being transcribed or is easily available for transcription; this is known as active chromatin and is the least condensed form of chromatin in the interphase chromosome.

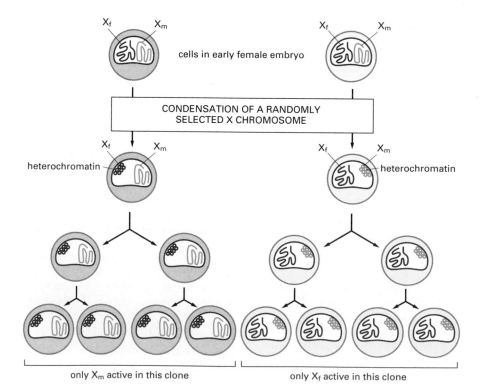

Figure 8–12 X-chromosome inactivation. Cells in the early female mammalian embryo contain two X chromosomes, one from the mother (X_m) and the other from the father (X_f). At an early stage of development, one of these two chromosomes in each cell becomes condensed into heterochromatin, apparently at random. At each cell division after this stage, the same chromosome becomes condensed in all the descendants of that original cell. In mice, X-chromosome inactivation occurs between the third and sixth days of development. In humans, too, X-inactivation occurs very early in development, before cells have been allocated to any particular developmental pathway. Thus the female ends up as a mosaic of cells bearing maternal or paternal inactivated X chromosomes. In most tissues and organs about half the cells will be of one type and the other half will be of the other.

It is not yet understood which of the levels of chromosome packing shown in Figure 8–10 best describes active chromatin. However, we know that it is folded into nucleosomes and that at least some of it may be a "loosened" form of the 30-nm fiber. The rest of the interphase chromosome is in a more condensed state. Although not packaged as tightly as heterochromatin nor permanently inactive, it is relatively difficult to transcribe.

It is interesting that the nucleosome itself does not obstruct the RNA polymerase once it has started transcribing. Even bacterial RNA polymerase, which does not normally encounter nucleosomes *in vivo*, can transcribe through them, suggesting that the nucleosome is constructed in such a way that it can be easily traversed. Similarly, DNA being replicated is also in the form of nucleosomes (you can see them in the electron micrograph of replicating DNA in Figure 6–14). We do not yet know exactly what happens in either case, but it is likely that the DNA partially detaches from the histone core of the nucleosome as a polymerase moves through. After the polymerase has passed, the nucleosome reassembles behind it.

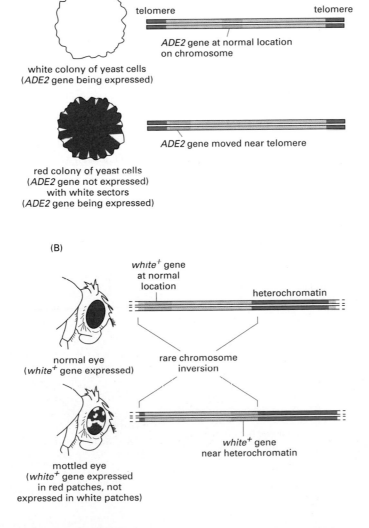

Figure 8–13 Position effects on gene expression. (A) The yeast *ADE2* gene at its normal chromosomal location is expressed in all cells. When moved near to the end of a yeast chromosome, which is thought to be folded into a particularly compact form of chromatin, the gene is no longer expressed in most cells of the population. *ADE2* encodes one of the enzymes of adenine biosynthesis, and the absence of the *ADE2* gene product leads to the accumulation of a red pigment. Therefore, a colony of cells that expresses *ADE2* is white, and one composed of cells where the *ADE2* gene is not expressed is red. The white sectors around the red colony represent cells where the *ADE2* gene has spontaneously become active. This may result from a heritable change in the packing state of chromatin near the *ADE2* gene in these cells.

(B) Position effects can also be observed for the *white* gene in the fruit fly *Drosophila*. The *white* gene controls eye pigment production and is named after the mutation that first identified it. Wild-type flies with a normal *white* gene (*white+*) have normal pigment production, which gives them red eyes, but if the *white* gene is mutated and inactivated, the mutant flies (*white−*) make no pigment and have white eyes. In flies in which a normal *white+* gene has been moved near a region of heterochromatin, the eyes are mottled, with both red and white patches. The white patches represent cells where the *white+* gene is silenced by the effects of the heterochromatin, and red patches represent cells that express the *white+* gene. The mottling occurs because the silencing of the *white+* gene by the heterochromatin is not complete. As for the yeast, the presence of large patches of red and white cells indicates that the state of transcriptional activity of the gene is inherited.

Figure image labels:
(A)
telomere   telomere
ADE2 gene at normal location on chromosome
white colony of yeast cells (*ADE2* gene being expressed)

ADE2 gene moved near telomere
red colony of yeast cells (*ADE2* gene not expressed) with white sectors (*ADE2* gene being expressed)

(B)
white+ gene at normal location
heterochromatin
normal eye (*white+* gene expressed)
rare chromosome inversion
white+ gene near heterochromatin
mottled eye (*white+* gene expressed in red patches, not expressed in white patches)

The Structure of Eucaryotic Chromosomes

Position Effects on Gene Expression Reveal Differences in Interphase Chromosome Packing

Some of the evidence that interphase chromosomes are a mixture of highly compact and less tightly packed chromatin comes from experiments in which the expression of a gene is altered when it is moved to a different location on a chromosome. Differences in expression of a gene that are dependent on its location in the genome are termed **position effects,** and they can be demonstrated in organisms such as brewer's yeast and fruit flies, which can be genetically manipulated in the laboratory (Figure 8–13). The properties of chromatin that give rise to position effects are still not well understood, although it is believed that the more compact forms of chromatin such as heterochromatin can spread along chromosomes, making nearby genes inaccessible to RNA polymerase and the additional proteins required for transcription.

Interphase Chromosomes Are Organized Within the Nucleus

Despite the fact that the chromosomes of cells in interphase are very much longer and finer than mitotic chromosomes, they are thought to be well-organized within the nucleus. The nucleus is delimited by a *nuclear envelope* formed by two concentric membranes. The nuclear envelope is supported by two networks of protein filaments (see Chapter 16): one, called the *nuclear lamina,* forms a thin layer underlying and supporting the inner nuclear membrane, while the other, less regularly organized, surrounds the outer nuclear membrane (Figure 8–14). The two membranes are punctured at intervals by nuclear pores, which actively transport selected molecules to and from the cytosol. For example, mRNAs are exported to the cytosol, whereas nuclear proteins such as histones are imported to the nucleus.

The interior of the nucleus is not a random jumble of its many DNA, RNA, and protein components. Each interphase chromosome probably occupies a particular region of the nucleus so that different chromosomes do not become entangled with each other. This organization is

Figure 8–14 The interphase nucleus.
(A) Electron micrograph of a thin section through the nucleus of a human fibroblast. The nucleus is surrounded by a nuclear envelope composed of a double membrane, perforated by nuclear pores. Inside the nucleus, the chromatin appears as a diffuse speckled mass, with heterochromatin (*dark staining*) located mainly around the periphery, immediately under the nuclear envelope. The large dark region is the nucleolus. (B) Diagrammatic view of a cross section of a typical cell nucleus. The nuclear envelope consists of two membranes, the outer one being continuous with the endoplasmic reticulum. Two networks of cytoskeletal filaments (*green*) provide mechanical support for the nuclear envelope; the one inside the nucleus forms a sheetlike *nuclear lamina* lining the internal face of the inner nuclear membrane. The nucleolus (*gray*) is the site of ribosomal RNA synthesis. (A, courtesy of E.G. Jordan and J. McGovern.)

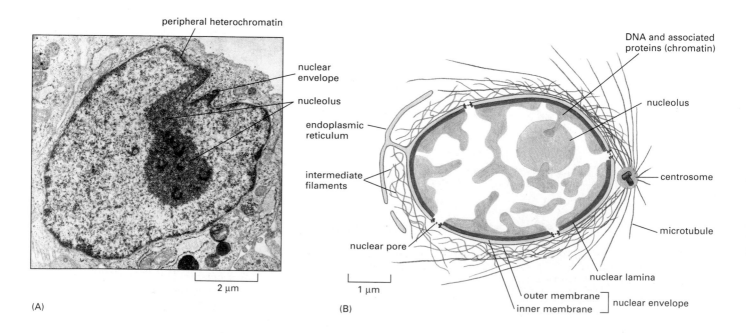

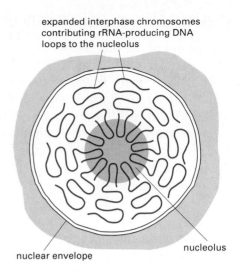

Figure 8–15 **The nucleolus.** This highly schematic view of a nucleolus in a human cell shows the contribution of loops of chromatin containing rRNA genes from 10 separate interphase chromosomes (see chromosome knobs in Figure 8–3).

thought to be achieved at least in part by attachment of parts of the chromosomes to sites on the nuclear envelope or the nuclear lamina.

The most obvious example of chromosome organization in the interphase nucleus is the **nucleolus,** which is the most prominent structure evident in the interphase nucleus under the light microscope. This is a region where the parts of the different chromosomes carrying genes for ribosomal RNA cluster together (Figure 8–15). The ribosomal RNA genes are transcribed and ribosomal subunits are assembled at the nucleolus, using ribosomal proteins imported from the cytoplasm. The partly assembled ribosomes are then exported to the cytosol through the nuclear pores.

Gene Regulation

Having glimpsed how the DNA in the nucleus of a eucaryotic cell is organized into chromosomes, we now turn to the question of how a cell specifies which of its many thousands of genes to express as proteins. This is an especially important problem for multicellular organisms because, as the organism develops, cell types such as muscle, nerve, and blood cells become different from one another, eventually leading to the wide variety of cell types seen in the adult organism (see Panel 1–3, pp. 32–33). This **differentiation** arises because cells make and accumulate different sets of RNA and protein molecules; that is, they express different genes.

As emphasized at the beginning of this chapter, cells generally change which genes are expressed without altering the nucleotide sequence of the DNA itself. How do we know this? If DNA were altered irreversibly during development, the chromosomes of a differentiated cell should be incapable of guiding the development of the whole organism. To test this possibility, a nucleus from a skin cell of an adult frog was injected into a frog egg whose own nucleus had been removed. In at least some cases the egg developed normally, indicating that the transplanted skin cell nucleus cannot have lost any important DNA sequences (Figure 8–16). A similar conclusion has been reached from experiments carried out in plants: for example, individual cells removed from a carrot can be shown to regenerate an entire adult carrot plant. These experiments prove that the DNA in specialized cell types still contains the entire set of instructions needed to form a whole organism. The cells of an organism therefore differ not because they contain different genes, but because they express them differently.

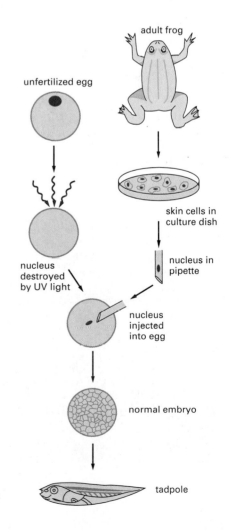

Figure 8–16 **The nucleus of a differentiated cell from the skin of an adult frog contains all the instructions, encoded in DNA, to control the formation of an entire tadpole.** A similar experiment was recently performed in sheep. (Modified from J.B. Gurdon, Gene Expression During Cell Differentiation. Oxford, UK: Oxford University Press, 1973.)

Cells Regulate the Expression of Their Genes

The extent of the differences in gene expression between different cell types may be roughly gauged by comparing the protein composition of cells in liver, heart, brain, and so on by the technique of two-dimensional gel electrophoresis (see Panel 5–5, pp. 162–163). Experiments of this kind reveal that many proteins are common to all the cells of a multicellular organism. These include the major structural proteins of the cytoskeleton and of chromosomes, the proteins essential to the endoplasmic reticulum and Golgi membranes, ribosomal proteins, and the enzymes that carry out glycolysis and other basic metabolic processes; these universal proteins are often called *housekeeping proteins,* and the genes that encode them, *housekeeping genes.* Each different cell type also produces specialized proteins that are responsible for the cell's distinctive properties. In mammals, for example, hemoglobin is made in reticulocytes, the cells that develop into red blood cells, but it cannot be detected in any other cell type.

Most proteins in a cell are in fact produced in such small numbers that they cannot be detected by the technique of gel electrophoresis. The mRNAs that encode such proteins can, however, be detected by more sensitive techniques (see Chapter 10) even in minute amounts. Estimates of the number of different mRNA sequences in cells suggest that a typical differentiated mammalian cell synthesizes perhaps 10,000 different proteins from a repertoire of about 60,000 genes. It is the expression of a different collection of genes in each cell type that causes the large variations seen in the size, shape, behavior, and function of differentiated cells.

How is the control of gene expression exercised? We saw in Chapter 7 that there are many steps in the pathway leading from DNA to protein, all of which can in principle be regulated. Thus, a cell can control the proteins it makes by: (1) controlling when and how often a given gene is transcribed, (2) controlling how the primary RNA transcript is spliced or otherwise processed, (3) selecting which mRNAs are translated by ribosomes, or (4) selectively activating or inactivating proteins after they have been made (Figure 8–17).

Although examples of regulation at each of the steps shown in Figure 8–17 are known, for most genes the control of transcription (step number 1) is paramount. This makes sense because only transcriptional control can ensure that no unnecessary intermediates are synthesized. For the remainder of this chapter, therefore, we concentrate on the DNA and protein components that determine which genes a cell transcribes into RNA.

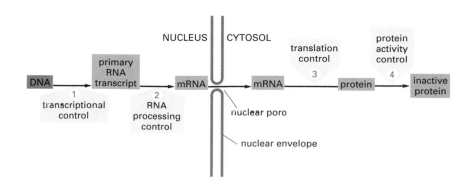

Figure 8–17 **Four steps at which eucaryotic gene expression can be controlled.** Examples of regulation at each of the steps are known, although for most genes the main site of control is step 1: transcription of a DNA sequence into RNA.

Transcription Is Controlled by Proteins Binding to Regulatory DNA Sequences

Control of transcription is usually exerted at the step at which transcription is initiated. In Chapter 7, we saw that the promoter region of a gene attracts the enzyme RNA polymerase and correctly orients it to begin its task of making an RNA copy of the gene. The promoters of both bacterial and eucaryotic genes include the *initiation site,* where transcription actually begins, and a sequence of approximately 50 nucleotides that extends "upstream" from the initiation site (if one likens the direction of transcription to the flow of a river). This region contains sites that are required for the RNA polymerase to bind to the promoter. In addition to the promoter, nearly all genes, whether bacterial or eucaryotic, have **regulatory DNA sequences** that are needed in order to switch the gene on or off. Whether a gene is expressed or not depends on a variety of factors, including the type of cell, its surroundings, its age, and extracellular signals.

Some regulatory DNA sequences are as short as 10 nucleotide pairs and act as simple gene switches that respond to a single signal. These simple switches predominate in bacteria. Other regulatory DNA sequences, especially those in eucaryotes, are very long (sometimes more than 10,000 base pairs) and act as molecular microprocessors, responding to a variety of signals that they integrate into an instruction that determines the rate at which transcription is initiated.

Regulatory DNA sequences do not work by themselves. To have any effect these sequences must be recognized by proteins called **gene regulatory proteins** that bind to the DNA. It is the combination of a DNA sequence and its associated protein molecules that acts as the switch to control transcription. Hundreds of regulatory DNA sequences have been identified, and each is recognized by one or more gene regulatory proteins.

Proteins that recognize a specific DNA sequence do so because the surface of the protein fits tightly against the special surface features of the double helix in that region. These will vary depending on the nucleotide sequence, and thus different proteins will recognize different nucleotide sequences. In most cases, the protein inserts into the major groove of the DNA helix (see Figure 6–6) and makes a series of molecu-

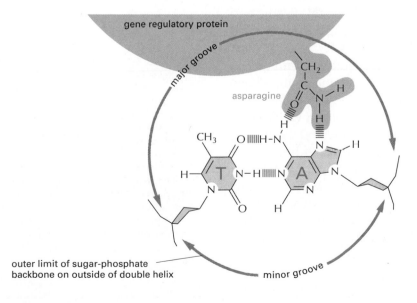

Figure 8–18 **The binding of a gene regulatory protein to DNA.** Only a single contact between the protein and one base pair in DNA is shown. Typically, the protein-DNA interface would consist of 10 to 20 such contacts, each involving a different amino acid and each contributing to the strength of the protein-DNA interaction.

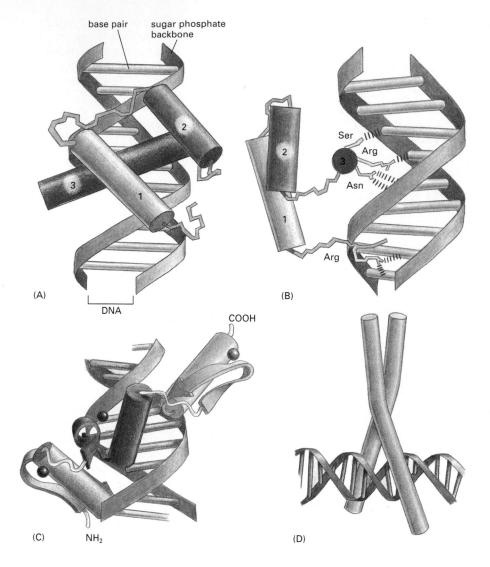

base pair   sugar phosphate backbone

(A)

DNA

(B)

COOH

(C)   NH₂

(D)

Figure 8–19 DNA-binding motifs in proteins. (A and B) Front and side views of the *homeodomain*—a structural motif in many DNA-binding proteins. It consists of three linked α helices, which are shown as *cylinders* in this figure. Most of the contacts with the DNA bases are made by helix number 3 (which is seen end-on in [B]). The asparagine (Asp) in this helix contacts an adenine in the manner shown in Figure 8–18. (C) The *zinc finger* motif is built from an α helix and a β sheet (the latter shown as a *twisted arrow*) held together by a molecule of zinc (indicated by a *sphere*). Zinc fingers are often found in clusters covalently joined together to allow the α helix of each finger to contact the DNA bases in the major groove. The illustration here shows a cluster of three zinc fingers. (D) A *leucine zipper* motif. This DNA-binding motif is formed by two α helices, each contributed by a different protein molecule. Leucine zipper proteins thus bind to DNA as dimers, gripping the double helix like a clothes pin on a clothes line.

These three motifs are found in gene regulatory proteins in virtually all eucaryotic organisms, where they are responsible for controlling the expression of thousands of different genes. Each motif makes many contacts with DNA. For simplicity, only the hydrogen-bond contacts are shown in (B), and none of the individual protein-DNA contacts are shown in (C) and (D). The names "homeodomain," "zinc finger," and "leucine zipper" all derive from historical terminology and are not meant to be accurate descriptions of these structures.

lar contacts with the base pairs. The protein forms hydrogen bonds, ionic bonds, and hydrophobic interactions with the edges of the bases, usually without disrupting the hydrogen bonds that hold the base pairs together (Figure 8–18). Although each individual contact is weak, the 20 or so contacts that are typically formed at the protein-DNA interface add together to ensure that the interaction is both highly specific and very strong; in fact protein-DNA interactions are among the tightest and most specific molecular interactions known in biology.

Although each example of protein-DNA recognition is unique in detail, many of the proteins responsible for gene regulation contain one of several particularly stable folding patterns. These fit into the major groove of the DNA double helix and form tight associations with a short stretch of DNA base pairs. The structures of some of these *DNA-binding motifs* are illustrated in Figure 8–19. In each example, an α helix of the protein contacts the major groove of DNA. Frequently, DNA-binding proteins bind in pairs (dimers) to the DNA helix. Dimerization roughly doubles the area of contact with the DNA, thereby increasing the strength and specificity of the protein-DNA interaction. Because two different proteins can be combined in pairs, dimerization also makes it possible for many different DNA sequences to be recognized by a limited number of proteins.

Question 8–4 Explain how DNA-binding proteins can make sequence-specific contacts to a double-stranded DNA molecule without breaking the hydrogen bonds that hold the bases together. Indicate how, by making such contacts, a protein can distinguish a T-A from a C-G pair. Give your answer in a similar form to Figure 8–18 and indicate what sorts of noncovalent bonds (hydrogen bonds, ionic bonds, or hydrophobic interactions; see Panel 2–1, pp. 46–47) would be made. There is no need to specify any particular amino acid on the protein. The structures of all the base pairs in DNA were given in Figure 6–5.

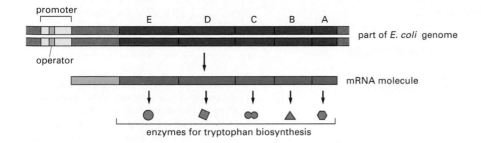

Figure 8–20
A cluster of bacterial genes transcribed from a single promoter. Each of these five genes encodes a different enzyme, all of which are needed to synthesize the amino acid tryptophan. They are transcribed as a single mRNA molecule, a feature that allows their expression to be coordinated. Clusters of genes transcribed as a single mRNA molecule are common in bacteria. Each such cluster is called an *operon*; expression of the tryptophan operon shown here is regulated at a regulatory sequence called the *operator*, situated within the promoter.

Repressors Turn Genes Off and Activators Turn Them On

The simplest and most completely understood examples of gene regulation occur in bacteria and their viruses. The genome of the bacterium *E. coli* consists of a single circular DNA molecule of about 4.6×10^6 nucleotide pairs. This DNA encodes about 4000 proteins, although only a fraction of these are being made at any one time. *E. coli* regulates the expression of many of its genes according to the food sources that are available in the environment. For example, five *E. coli* genes code for enzymes of the metabolic pathway that manufactures the amino acid tryptophan, and these genes are arranged in a cluster on the chromosome. They are transcribed from a single promoter as one long mRNA molecule from which the five proteins are translated (Figure 8–20). Genes arranged in this way and transcribed into a single mRNA are known as *operons*. Operons are common in bacteria but are not found in eucaryotes, where genes must be regulated individually. When tryptophan is present in the surroundings and enters the bacterial cell, these enzymes are no longer needed and their production is shut off. This situation arises, for example, when the bacterium is in the gut of a mammal that has just eaten a meal of protein.

We now understand in considerable detail how this switch functions. Within the promoter is a short DNA sequence (15 nucleotides in length) that is recognized by a gene regulatory protein. When the protein binds to this nucleotide sequence, termed the *operator*, it blocks access of RNA polymerase to the promoter; this prevents transcription of the operon and production of the tryptophan-producing enzymes. The gene regulatory protein is known as the tryptophan repressor, and it is regulated in an ingenious way: the repressor can bind to DNA only if it has also bound several molecules of the amino acid tryptophan (Figure 8–21).

Figure 8–21 Switching genes on and off with a repressor protein. If the level of tryptophan inside the cell is low, the tryptophan repressor protein does not bind tryptophan and thus cannot bind to its control region—the operator (*green*)—within the promoter (*yellow*). RNA polymerase can thus bind to the promoter and transcribe the five genes of the tryptophan operon (*left*). If the level of tryptophan is high, however, the repressor protein binds tryptophan, in which state it can bind to the operator, where it blocks the binding of RNA polymerase to the promoter (*right*). Whenever the level of intracellular tryptophan drops, the repressor releases its tryptophan and is released from the DNA, allowing the polymerase to again transcribe the operon (see also Figure 8–20).

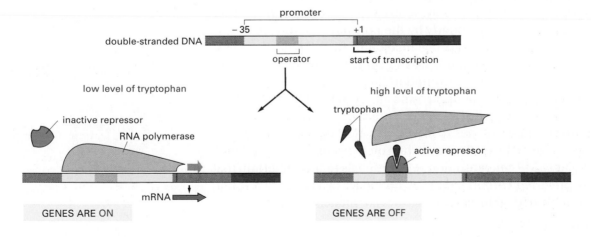

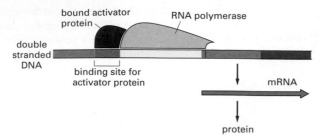

Figure 8–22 Controlling genes with an activator protein. An activator protein binds to a regulatory sequence on the DNA and then interacts with the RNA polymerase to help it initiate transcription. Without the activator, this promoter (*yellow*) fails to initiate transcription efficiently. In bacteria, the binding of the activator to DNA is often controlled by the interaction of a metabolite or other small molecule with the activator protein. For example, the bacterial activator protein CAP has to bind cyclic AMP (cAMP) before it can bind to DNA; thus CAP allows genes to be switched on in response to increases in intracellular cAMP concentration.

The tryptophan repressor is an allosteric protein: the binding of tryptophan causes a subtle change in its three-dimensional structure so that it can now bind to the operator DNA. When free tryptophan becomes low in the cell, the repressor no longer binds tryptophan and thus no longer binds to the DNA. The tryptophan operon is then transcribed. The repressor is thus a simple device that switches production of a set of biosynthetic enzymes on and off according to the availability of the end product of the pathway that the enzymes catalyze.

The bacterium can respond very rapidly to the rise in tryptophan level because the tryptophan repressor protein itself is always present in the cell. The gene that encodes it is continuously transcribed at a low level, so that a small amount of the repressor protein is always being made. Such unregulated gene expression is known as *constitutive* gene expression.

The tryptophan repressor, as its name suggests, is a **repressor** protein that switches genes off, or *represses* them. Other bacterial gene regulatory proteins operate in the opposite manner by switching genes on, or *activating* them. These **activator** proteins act on promoters that—in contrast to the promoter for the tryptophan operator—are only marginally functional in binding RNA polymerase on their own; they may, for example, be recognized only poorly by the polymerase. However, these poorly functioning promoters can be made fully functional by proteins that bind to a nearby site on the DNA and contact the RNA polymerase in a way that helps it initiate transcription (Figure 8–22).

Like a repressor, an activator protein's ability to bind to DNA is often affected by its interaction with a second molecule. For example, the bacterial activator protein CAP has to bind cyclic AMP before it can bind to DNA. Genes activated by CAP are switched on in response to an increase in intracellular cAMP concentration, which signals to the bacterium that glucose, its preferred carbon source, is no longer available; as a result, enzymes capable of degrading other sugars are made.

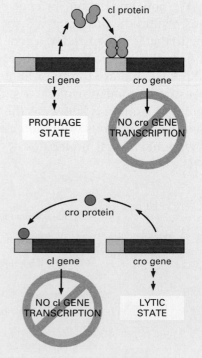

Initiation of Eucaryotic Gene Transcription Is a Complex Process

The genetic switches present in bacteria are telling examples of the economy and simplicity often observed in biology. In eucaryotes, however, a typical gene responds to many different signals, and its regulation is consequently more complex.

Initiation of transcription in eucaryotes differs in four important ways from that in bacteria.

1. The first difference lies in the RNA polymerases themselves. While bacteria contain a single type of RNA polymerase, eucaryotic cells have three—called *RNA polymerase I, RNA polymerase II,* and *RNA polymerase III*. These polymerases are responsible for transcribing different types of genes. RNA polymerases I and III transcribe the genes encoding transfer RNA, ribosomal RNA, and small RNAs that play a structural role in the cell (Table 8–1). RNA polymerase II transcribes the vast majority of eucaryotic genes, including all those that encode proteins, and our subsequent discussion will therefore focus on this enzyme.

2. A second difference is that while bacterial RNA polymerase is able to initiate transcription without the help of additional proteins, as we saw for the tryptophan operon (see Figure 8–21), eucaryotic RNA polymerases cannot. They require the help of a large set of proteins called *general transcription factors*, which must assemble at the promoter with the polymerase before the polymerase can begin transcription.

3. A third distinctive feature of gene regulation in eucaryotes is that gene regulatory proteins (repressors and activators) can influence the initiation of transcription even when they are bound to DNA thousands of nucleotide pairs away from the promoter. This feature allows a single promoter to be controlled by an almost unlimited number of regulatory sequences scattered along the DNA. In bacteria, in contrast, genes are often controlled by a single regulatory sequence that is typically located quite near the promoter (see Figures 8–21 and 8–22).

4. Last but not least, eucaryotic transcription initiation must take account of the packing of DNA into nucleosomes and more compact forms of chromatin structure.

We shall next consider each of the three latter features of eucaryotic gene regulation.

Table 8–1   **The Three RNA Polymerases in Eucaryotic Cells**

Type of Polymerase	Genes Transcribed
RNA polymerase I	most rRNA genes
RNA polymerase II	all protein-coding genes, plus some genes for small RNAs (e.g., those in spliceosomes)
RNA polymerase III	tRNA genes
	5S rRNA gene
	genes for some small structural RNAs

Eucaryotic RNA Polymerase Requires General Transcription Factors

The initial finding that, unlike bacterial RNA polymerase, purified eucaryotic RNA polymerase II could not on its own initiate transcription *in vitro* led to the discovery and purification of the **general transcription factors** required for this process. The general transcription factors are thought to position the RNA polymerase correctly at the promoter, to aid in pulling apart the two strands of DNA to allow transcription to begin, and to release RNA polymerase from the promoter once transcription begins.

The term "general" refers to the fact that these proteins assemble on all promoters transcribed by RNA polymerase II. In this they differ from the repressors and activators that we have just described in bacteria, which act only at particular genes or operons, and from the eucaryotic gene regulatory proteins we shall discuss below, which also act only at particular genes.

Figure 8–23 shows how the general transcription factors are thought to assemble at promoters utilized by RNA polymerase II. The assembly process starts with the binding of the general transcription factor TFIID to a short double-helical DNA sequence primarily composed of T and A nucleotides; for this reason, it is known as the TATA sequence, or **TATA box.** Upon binding to DNA, TFIID causes a dramatic local distortion in the DNA (Figure 8–24), which helps to serve as a landmark for the subsequent assembly of other proteins at the promoter. The TATA box is a key component of nearly all promoters utilized by RNA polymerase II, and it is typically located 25 nucleotides upstream from the transcription start site. Once the first general transcription factor is bound to this DNA site, then the other factors are assembled, along with RNA polymerase II, to form a complete *transcription initiation complex.*

After RNA polymerase II has been tethered to the promoter DNA in the transcription initiation complex, it must be released from the complex of transcription factors in order to begin its task of making an RNA

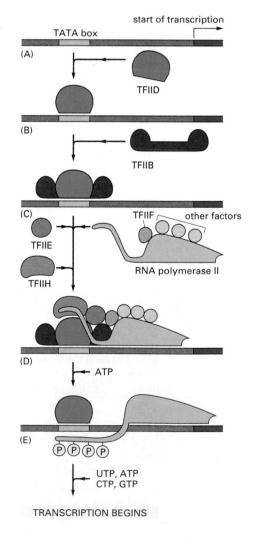

Figure 8–23 Initiation of transcription of a eucaryotic gene by RNA polymerase II. In order to begin transcription, RNA polymerase requires a number of general transcription factors (called TFIIA, TFIIB, and so on). (A) The promoter contains a DNA sequence called the TATA box, which is located 25 nucleotides away from the site where transcription is initiated. (B) The TATA box is recognized and bound by transcription factor TFIID, which then enables the adjacent binding of TFIIB (C). For simplicity the DNA distortion produced by the binding of TFIID (see Figure 8–24) is not shown. (D) The rest of the general transcription factors as well as the RNA polymerase itself assemble at the promoter. (E) TFIIH then uses ATP to phosphorylate RNA polymerase II, changing its conformation so that the polymerase is released from the complex and is able to start transcribing. As shown, the site of phosphorylation is a long polypeptide tail that extends from the polymerase molecule.

The exact order in which the general transcription factors assemble on the promoter is not known with certainty. One view holds that the general factors assemble off the DNA with the polymerase and that this whole assembly then binds to the DNA in a single step. The general transcription factors have been highly conserved in evolution; some of those from human cells can be replaced in biochemical experiments by the corresponding factors from simple yeasts.

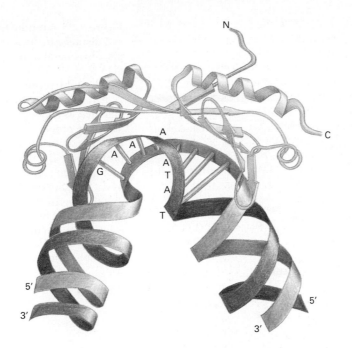

Figure 8–24 **Three-dimensional structure of TBP bound to a TATA box.** TBP (TATA-binding protein) is the subunit of the general transcription factor TFIID that is responsible for recognizing and binding to the TATA box sequence in the DNA (*red*). The unique DNA bending caused by TBP—two kinks in the double helix separated by partially unwound DNA—may serve as a landmark that helps attract the other general transcription factors. TBP is a single polypeptide chain that is folded into two very similar domains (*blue* and *green*). (Adapted from J.L. Kim et al., *Nature* 365:520–527, 1993.)

molecule. A key step in this release is addition of phosphate groups to the RNA polymerase, performed by the general transcription factor TFIIH, which contains a protein kinase enzyme as one of its subunits (see Figure 8–23E). This phosphorylation is thought to help the polymerase disengage from the cluster of transcription factors, allowing transcription to begin. The general transcription factors are then released so that they are available to initiate another round of transcription with a new RNA polymerase molecule.

Eucaryotic Gene Regulatory Proteins Control Gene Expression from a Distance

We have seen that bacteria utilize gene regulatory proteins (activators and repressors) to regulate the expression of their genes. Eucaryotic cells use the same basic strategy. Although the eucaryotic general transcription factors and RNA polymerase together can initiate transcription *in vitro* (see Figure 8–23), inside the cell these proteins on their own cannot efficiently initiate transcription. Nearly all eucaryotic promoters also require activator proteins to aid the assembly of the general transcription factors and RNA polymerase.

The DNA sites to which the eucaryotic gene activators bound were originally termed *enhancers,* since their presence "enhanced," or increased, the rate of transcription dramatically. It was surprising to biologists when, in 1979, it was discovered that these activator proteins could be bound thousands of nucleotide pairs away from the promoter. Moreover, eucaryotic activators could influence transcription of a gene when bound either upstream or downstream from it. How do enhancer sequences and the proteins bound to them function over these long distances? How do they communicate with the promoter?

Many models for "action at a distance" have been proposed, but the simplest of these seems to apply in most cases. The DNA between the enhancer and the promoter loops out to allow the activator proteins bound to the enhancer to come into contact either with RNA polymerase or with one of the general transcription factors bound to the pro-

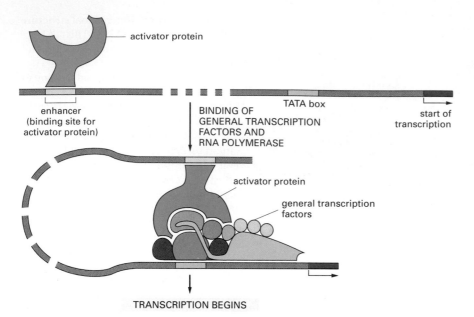

enhancer
(binding site for
activator protein)

BINDING OF
GENERAL TRANSCRIPTION
FACTORS AND
RNA POLYMERASE

TATA box

start of
transcription

activator protein

activator protein

general transcription
factors

TRANSCRIPTION BEGINS

Figure 8–25 A model for gene activation at a distance in eucaryotes. In this example, the general transcription factors and RNA polymerase do not assemble efficiently at the promoter on their own, and a gene regulatory protein bound at an enhancer is necessary to stimulate the assembly process. Looping of the DNA permits contact between the regulatory protein bound to the enhancer and the transcription complex bound to the promoter. The broken stretch of DNA signifies that the length of DNA between the enhancer and the start of transcription can be up to tens of thousands of nucleotide pairs.

moter (Figure 8–25). The DNA thus acts as a tether, causing a protein bound to an enhancer even thousands of nucleotide pairs away to interact with the complex of proteins bound to the promoter.

In eukaryotes, gene regulatory proteins bound to distant gene regulatory sequences can either increase or decrease the activity of RNA polymerase bound to the promoter. One of the ways they do this is by influencing the assembly of the transcription initiation complex. Activators will facilitate the assembly of the complex, whereas repressors will prevent or sabotage its correct assembly.

A comparison of bacterial and eukaryotic gene regulation is presented in Figure 8–26. In both cases, gene regulatory proteins can control gene expression by either aiding or preventing the assembly and activity of RNA polymerase at the promoter. Eukaryotic gene regulatory proteins bind to regulatory DNA sequences that are equivalent to the operators and activator-binding sequences in bacteria, but which are usually located at considerable distances from the promoter.

Packing of Promoter DNA into Nucleosomes Can Affect Initiation of Transcription

Initiation of transcription in eukaryotic cells must also take account of the packing of the DNA into chromatin. Can gene regulatory proteins, the general transcription factors, and RNA polymerase gain access to DNA that is packed into nucleosomes? How does the packing affect the initiation of transcription?

We saw earlier in the chapter that genes which are being expressed seem to be in an extended form of chromatin; however, the DNA is still packaged into nucleosomes. The presence of nucleosomes does not generally block the elongation stage of transcription, since the RNA polymerase can proceed through a nucleosome with only temporary disruption of nucleosome structure. Nucleosomes can, however, inhibit the initiation of transcription if they are positioned over a promoter, probably because they prevent the general transcription factors or RNA polymerase from assembling on the DNA.

Question 8–6 Some gene regulatory proteins (not covered in the text of this chapter) bind to DNA and cause the double helix to bend at a sharp angle. Such "bending proteins" can affect the initiation of transcription without contacting either the RNA polymerase, any of the general transcription factors, or any other gene regulatory proteins. Can you devise a plausible explanation for how these proteins might work to modulate transcription? Draw a diagram that illustrates your explanation.

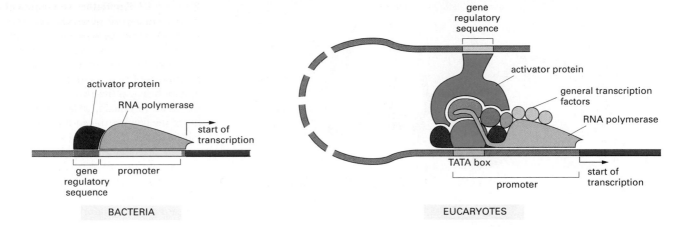

BACTERIA

EUCARYOTES

Since nucleosomes are placed along DNA at regular intervals with little apparent sequence specificity, they are likely to occur over promoter regions. Such nucleosomes are displaced when transcription of the gene is activated, although how this displacement occurs is not understood in detail. It is likely that the cell has specialized proteins, whose job it is to displace nucleosomes from promoters and clear the way for the general transcription factors to assemble. Another possibility is that, as a prelude to transcription initiation, the histones in the vicinity of a promoter are chemically modified, a step that would destabilize the affected nucleosomes.

Nucleosomes formed on regulatory DNA sequences might also interfere with gene expression by blocking the binding of gene regulatory proteins. This does not always seem to be the case, however. While there is evidence that some regulatory sequences are kept exposed in nucleosome-free regions of DNA, certain gene regulatory proteins seem able to bind to their regulatory DNA even when it is incorporated into a nucleosome, possibly destabilizing and partially disassembling the nucleosome in the process.

The interplay between chromatin structure and transcription initiation is only beginning to be understood in detail. The cell has several strategies for ensuring that transcription initiation can take place on DNA that is folded into nucleosomes. However, it is also clear that the more compact forms of chromatin (those found in mitotic chromosomes, inactive X chromosomes, and other regions of heterochromatin in interphase chromosomes) are resistant to transcription initiation. This is presumably because gene regulatory proteins, the general transcription factors, and RNA polymerase cannot gain access to the DNA when it is so tightly folded. However, the proteins responsible for the increased DNA compaction are only beginning to be discovered.

In the next sections, we discuss how, once packaged as active chromatin, a typical eucaryotic gene is regulated by combinations of activator and repressor proteins.

Eucaryotic Genes Are Regulated by Combinations of Proteins

Because eucaryotic gene regulatory proteins can control transcription when bound to DNA many nucleotide pairs away from the promoter, the DNA sequences that control the expression of a gene can be spread over long stretches of DNA. In animals and plants it is not unusual to

Figure 8–26 Summary of the principal differences in the mechanism of gene activation in bacteria and eucaryotes. In addition, eucaryotic DNA is packaged into nucleosomes (not shown).

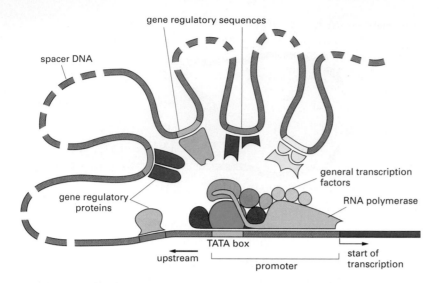

gene regulatory sequences

spacer DNA

gene regulatory proteins

general transcription factors

RNA polymerase

TATA box

upstream

promoter

start of transcription

find the regulatory sequences of a gene dotted over distances as great as 50,000 nucleotide pairs, although much of this DNA serves as "spacer" sequence and is not recognized by gene regulatory proteins.

So far in this chapter we have treated gene regulatory proteins as though each functioned individually to turn a gene on or off. While this idea holds for many bacterial activators and repressors, most eucaryotic gene regulatory proteins work as part of a "committee" of regulatory proteins, all of which are necessary to express the gene in the right cell, in response to the right conditions, at the right time, and at the required level.

The term **combinatorial control** refers to the way that groups of proteins work together to determine the expression of a single gene. As shown in Figure 8–27, many different proteins bind to regulatory sequences to influence whether transcription is initiated in eucaryotes. Most eucaryotic genes have control regions containing numerous sites for both positively acting and negatively acting gene regulatory proteins. The regulatory region that controls expression of the human β-globin gene is shown diagrammatically in Figure 8–28. The molecular mechanisms by which the effects of all of these proteins are added up to determine the final level of expression for a gene are only now beginning to be understood.

The Expression of Different Genes Can Be Coordinated by a Single Protein

Both bacteria and eucaryotes need to be able not only to switch genes on and off individually, but also to coordinate the expression of different genes. When a eucaryotic cell receives a signal to divide, for example, a number of hitherto unexpressed genes are turned on together to set in motion the events that lead eventually to cell division (see Chapter 15). One way bacteria coordinate the expression of a set of different genes is by having them clustered together in an operon under the control of a single promoter (see Figure 8–20). This is not the case in eucaryotes, in which each gene is regulated individually. So how do eucaryotes coordinate gene expression?

In particular, given that a eucaryotic cell uses a committee of regulatory proteins to control each of its genes, how can it rapidly and deci-

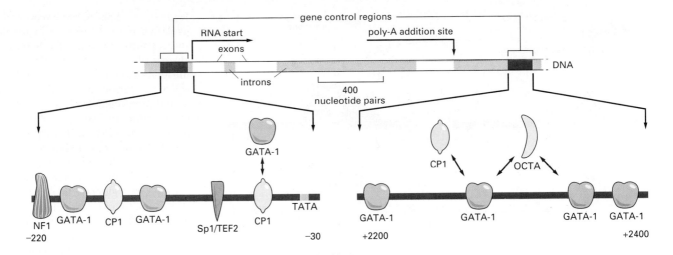

sively switch whole groups of genes on or off? The answer is that even though control of gene expression is combinatorial, the effect of a single gene regulatory protein can still be decisive in switching any particular gene on or off, simply by completing the combination needed to activate or repress that gene. This is like dialing in the final number of a combination lock: the lock will spring open if the other numbers have been previously entered. Just as the same number can complete the combination for different locks, the same protein can complete the combination for several different genes. If a number of different genes contain the regulatory site for the same gene regulatory protein, it can be used to regulate the expression of all of them.

An example of this in humans is the *glucocorticoid receptor protein*. This is a gene regulatory protein, but in order to bind to regulatory sites in DNA, it must first form a complex with a molecule of a glucocorticoid steroid hormone (e.g., cortisol, see Table 15–1, p. 485). This hormone is released in the body during times of starvation and intense physical activity, and, among its other activities, it stimulates cells in the liver to increase the production of glucose from amino acids and other small molecules. In response to glucocorticoid hormones, liver cells increase the expression of many different genes, including one coding for the enzyme tyrosine aminotransferase, which helps convert the amino acid tyrosine to glucose. These genes are all regulated by the binding of the hormone–glucocorticoid receptor complex to a regulatory site in the DNA of the gene. When the body has recovered and the hormone is no longer present, the expression of all of these genes drops to its normal level. In this way a single gene regulatory protein can control the expression of many different genes (Figure 8–29).

Combinatorial Control Can Create Different Cell Types

The ability to switch many different genes on or off using just one protein is not only useful in the day-to-day regulation of cell function. It is also one of the means by which eucaryotic cells differentiate into particular types of cells during embryonic development.

A striking example of the effect of a single gene regulatory protein on differentiation comes from studying the development of muscle cells. A mammalian skeletal muscle cell is a highly distinctive cell type. It is typically an extremely large cell that is formed by the fusion of many

Figure 8–28 Model for the control of the human β-globin gene. The diagram shows some of the gene regulatory proteins thought to control expression of this gene during red blood cell development. Some of the gene regulatory proteins shown, such as CP1, are found in many types of cells, while others, such as GATA-1, are present in only a few types of cells, including red blood cell precursors, and are therefore thought to contribute to the cell-type specificity of β-globin gene expression. As indicated by the bidirectional arrows, several of the binding sites for GATA-1 overlap those of other gene regulatory proteins; it is thought that occupancy of these sites by GATA-1 excludes binding of other proteins. (Adapted from B. Emerson, In Gene Expression: General and Cell-Type Specific (M. Karin, ed.), pp. 116–161. Boston: Birkhauser, 1993.)

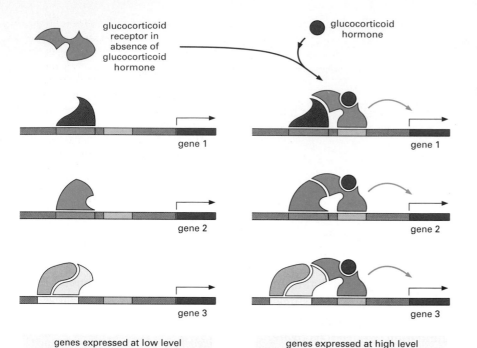

Figure 8–29 A single gene regulatory protein can coordinate the expression of several different genes. The action of the glucocorticoid receptor is illustrated. On the left is shown a series of genes, each of which has various gene activator proteins bound to its regulatory region. However, these bound proteins are not sufficient on their own to activate transcription efficiently. On the right is shown the effects of adding an additional gene regulatory protein—the glucocorticoid receptor in a complex with glucocorticoid hormone—that can bind to the regulatory region of each gene. The glucocorticoid receptor completes the combination of gene regulatory proteins required for efficient initiation of transcription, and the genes are now switched on as a set.

muscle precursor cells called *myoblasts* (and therefore contains many nuclei). The mature muscle cell is distinguished from other cells by the production of a large number of characteristic proteins, such as the actin and myosin that make up the contractile apparatus (see Chapter 16), as well as the receptor proteins and ion channel proteins in the cell membranes that make the muscle cell sensitive to nerve stimulation. Genes encoding these muscle-specific proteins are all switched on coordinately as the myoblasts begin to fuse. Studies of muscle cells differentiating in culture have identified key gene regulatory proteins, expressed only in potential muscle cells, that coordinate the gene expression and thus are crucial for muscle cell differentiation. These gene regulatory proteins activate the transcription of the genes that code for the muscle-specific proteins by binding to sites present in all their regulatory regions.

These key regulatory proteins can convert nonmuscle cells to muscle by activating the changes in gene expression typical of differentiating muscle cells. When the gene for one of these gene regulatory proteins, MyoD, is introduced into fibroblasts cultured from skin connective tissue, the fibroblasts start to behave like myoblasts and fuse to form musclelike cells. The dramatic effect of expressing MyoD in fibroblasts is shown in Figure 8–30. It appears that the fibroblasts, which are derived from the same broad class of embryonic cells as muscle cells, have already accumulated all the other necessary gene regulatory proteins required for the combinatorial control of the muscle-specific genes, and

Figure 8–30 The effect of expressing the MyoD protein in fibroblasts. As shown in this immunofluorescence micrograph, fibroblasts from the skin of a chick embryo have been converted to muscle cells by the experimentally induced expression of the *myoD* gene. The fibroblasts that have received the MyoD gene regulatory protein have fused to form elongated multinucleate musclelike cells, which are stained *green* with an antibody that detects a muscle-specific protein. Fibroblasts that did not receive the *myoD* gene are barely visible in the background. (Courtesy of Stephen Tapscott and Harold Weintraub.)

20 μm

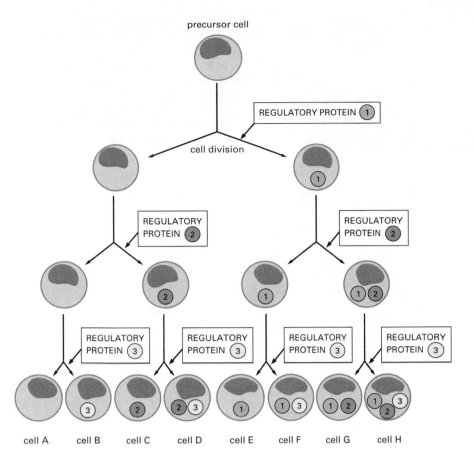

precursor cell

REGULATORY PROTEIN (1)

cell division

REGULATORY PROTEIN (2)

REGULATORY PROTEIN (2)

REGULATORY PROTEIN (3)

REGULATORY PROTEIN (3)

REGULATORY PROTEIN (3)

REGULATORY PROTEIN (3)

cell A   cell B   cell C   cell D   cell E   cell F   cell G   cell H

Figure 8–31 The importance of combinatorial gene control for generating different cell types. A hypothetical scheme illustrating how combinations of a few gene regulatory proteins can generate many different cell types during development. In this simple scheme a "decision" to make a new gene regulatory protein (shown as a *numbered circle*) is made after each cell division. Repetition of this simple rule enables eight cell types (A through H) to be created using only three different regulatory proteins. Each of these hypothetical cell types would then express different genes, as dictated by the combination of gene regulatory proteins that are present within it.

that addition of MyoD completes the unique combination that directs the cells to become muscle. Some other cell types fail to be converted to muscle by the addition of MyoD; these cells presumably have not accumulated the other required gene regulatory proteins during their developmental history.

How the accumulation of different gene regulatory proteins can lead to the generation of different cell types is illustrated schematically in Figure 8–31. This figure also illustrates how, thanks to the possibilities of combinatorial control and shared regulatory sequences, a limited set of gene regulatory proteins can control the expression of a much larger number of genes.

The conversion of one cell type (fibroblast) to another (muscle) by a single gene regulatory protein emphasizes one of the most important principles discussed in this chapter: the dramatic differences between cell types—such as size, shape, and function—are produced by differences in gene expression.

Stable Patterns of Gene Expression Can Be Transmitted to Daughter Cells

Although all cells, whether bacterial or eucaryotic, must be able to switch genes on and off, multicellular organisms require special gene-switching mechanisms for generating and maintaining their different types of cells. In particular, once a cell in a multicellular organism has become differentiated into a particular cell type, it will generally remain differentiated, and if it is able to divide, all its progeny cells will be of that same cell type. Some highly specialized cells never divide again once

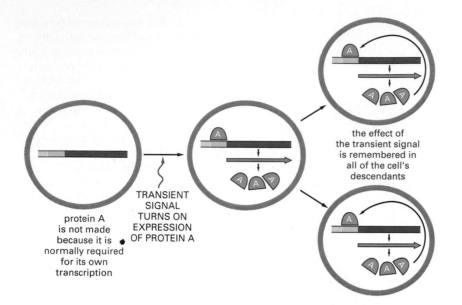

Figure 8–32 **Schematic diagram showing how a positive feedback loop can create cell memory.** Protein A is a gene regulatory protein that activates its own transcription. All of the descendants of the original cell will therefore "remember" that the progenitor cell had experienced a transient signal that had initiated the production of the protein.

protein A is not made because it is normally required for its own transcription

TRANSIENT SIGNAL TURNS ON EXPRESSION OF PROTEIN A

the effect of the transient signal is remembered in all of the cell's descendants

they have differentiated, for example, skeletal muscle cells and neurons. But there are many other differentiated cells, such as fibroblasts, smooth muscle cells, and liver cells (hepatocytes), that will divide many times in the life of an individual. All these cell types give rise only to cells like themselves when they divide: smooth muscle does not give rise to liver cells, nor liver cells to fibroblasts.

This means that the changes in gene expression that give rise to a differentiated cell must be remembered and passed on to its daughter cells through all subsequent cell divisions, unlike the temporary changes in gene expression that can occur in both bacterial and eucaryotic cells. For example, in the cells illustrated in Figure 8–31, the production of each gene regulatory protein, once begun, has to be perpetuated in the daughter cells of each cell division. How might this be accomplished?

Cells have several ways of ensuring that daughter cells "remember" what kind of cells they are supposed to be. One of the simplest is through a **positive feedback loop,** where a key gene regulatory protein activates transcription of its own gene in addition to that of other cell-type specific genes (Figure 8–32). For example, the MyoD gene regulatory protein discussed earlier functions in such a positive feedback loop. Another way of maintaining cell type is through the faithful propagation of a condensed chromatin structure from parent to daughter cell even though DNA replication intervenes. We saw an example of this in Figure 8–12,

Figure 8–33 **A general scheme that permits the direct inheritance of states of gene expression during DNA replication.** In this hypothetical model, portions of a cluster of chromosomal proteins cooperatively bound to the DNA are transferred directly from the parental DNA helix (*top left*) to both daughter helices. The inherited cluster then causes each of the daughter DNA helices to bind additional copies of the same proteins. Because the binding is cooperative and the protein is in short supply, DNA synthesized from an identical parental DNA helix that lacks the bound proteins (*top right*) will remain free of them.

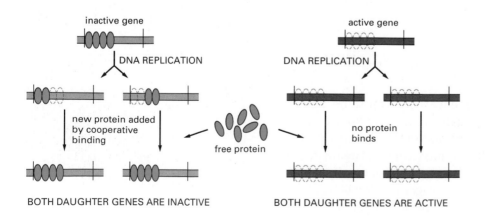

inactive gene

active gene

DNA REPLICATION

DNA REPLICATION

new protein added by cooperative binding

free protein

no protein binds

BOTH DAUGHTER GENES ARE INACTIVE

BOTH DAUGHTER GENES ARE ACTIVE

where the same X chromosome is inactived through many cell generations. The molecular mechanism through which the chromatin state is passed on is not understood in detail, but a general hypothesis is shown in Figure 8–33.

The Formation of an Entire Organ Can Be Triggered by a Single Gene Regulatory Protein

We have seen that even though combinatorial control is the norm for eucaryotic genes, a single gene regulatory protein, if it completes the appropriate combination, can be decisive in switching a whole set of genes on or off, and we have seen how this can convert one cell type into another. A dramatic extension of this principal comes from studies of eye development in *Drosophila*, mice, and humans. Here, a single gene regulatory protein (called Ey in flies and Pax-6 in vertebrates) is crucial for eye development. When expressed in the proper type of cell, Ey can trigger the formation of not just a single cell type but a whole organ (the eye), composed of different types of cells all properly organized in three-dimensional space.

The best evidence for the action of Ey comes from experiments in fruit flies in which the *ey* gene is artificially expressed early in development in cells that normally go on to form legs. This abnormal gene expression causes eyes to develop in the middle of the legs (Figure 8–34). The *Drosophila* eye is composed of thousands of cells, and how the Ey protein coordinates the specification of the different types of cells in the eye is not well understood. It seems likely, however, that Ey directly controls the expression of many other genes by binding to their regulatory regions. Some of the genes controlled by Ey are likely themselves to be gene regulatory proteins that, in turn, control the expresssion of other genes. So the action of just one regulatory protein can turn on a cascade of gene regulatory proteins whose actions result in forming an organized group of many different types of cells. One can begin to imagine how, by repeated applications of this principle, a complex organism is built up piece by piece.

Figure 8–34 Expression of the *Drosophila* *ey* gene in the precursor cells of the leg triggers the development of an eye on the leg. (A) Simplified diagrams showing the result when a fruit fly larvae contains either the normally expressed *ey* gene (*left*) or an *ey* gene that is additionally expressed artificially in cells that will give rise to legs (*right*). (B) Photograph of an abnormal leg that contains a misplaced eye. (B, courtesy of Walter Gehring.)

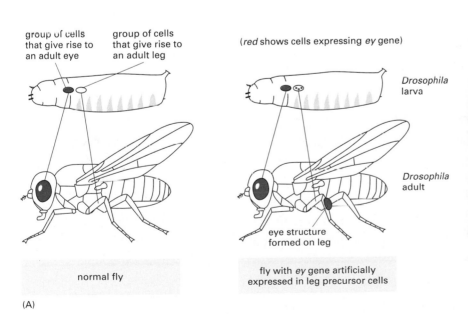

(A)

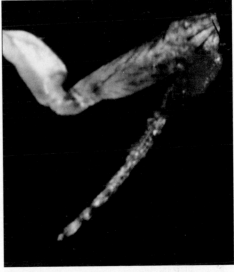

(B)

Essential Concepts

- The genetic material of a eucaryotic cell is contained within one or more chromosomes, each formed from a single, enormously long DNA molecule that contains many genes.

- The DNA in a chromosome contains, in addition to genes, many replication origins, one centromere, and two telomeres. These sequences ensure that the chromosome can be replicated efficiently and passed on to daughter cells.

- Chromosomes in eucaryotic cells consist of DNA tightly bound to a roughly equal mass of specialized proteins. These proteins fold up the DNA into a more compact form so that it can fit into a cell nucleus. The complex of DNA and protein in chromosomes is called chromatin.

- The proteins associated with DNA include the histones, which pack up DNA into a repeating array of DNA-protein particles called nucleosomes.

- Nucleosomes pack together, with the aid of histone H1 molecules, to form a 30-nm fiber. This fiber can be further coiled and folded.

- Some forms of chromatin are so highly compacted that the packaged genes are transcriptionally silent. This is the case for all genes in the chromosomes during nuclear division (mitosis) when the chromosomes become highly condensed.

- Specific regions of chromosomes, termed heterochromatin, are also condensed and inactive even in nondividing cells. Genes artificially moved into regions of heterochromatin are often silenced.

- A typical eucaryotic cell expresses only a fraction of its genes, and the distinct types of cells in multicellular organisms arise because different sets of genes are expressed as a cell differentiates.

- Although all of the steps involved in expressing a gene can in principle be regulated, for most genes the initiation of transcription is the most important point of control.

- The transcription of individual genes is switched on and off in cells by gene regulatory proteins. These act by binding to short stretches of DNA called regulatory DNA sequences.

- Although each gene regulatory protein has unique features, most bind to DNA using one of a small number of protein structure motifs. The precise amino acid sequence that is folded into the motif determines the particular DNA sequence that is recognized.

- RNA polymerase binds to the DNA and initiates transcription at a site called the promoter.

- In bacteria, regulatory proteins usually bind to regulatory DNA sequences close to the RNA-polymerase-binding site and either activate or repress transcription of the gene. In eucaryotes, these regulatory DNA sequences are often separated from the promoter by many thousands of nucleotide pairs.

- To initiate transcription, eucaryotic RNA polymerases require the assembly of a complex of general transcription factors at the promoter. Eucaryotic gene regulatory proteins are thought to act by affecting the assembly process, speeding it up (for activators) and slowing it down (for repressors).

- In eucaryotes, the expression of a gene is generally controlled by a combination of gene regulatory proteins.

- In multicellular plants and animals, the production of different gene regulatory proteins in different cell types ensures the expression of only those genes appropriate to that type of cell.

- A single gene regulatory protein, if expressed in the appropriate precursor cell, can trigger the formation of a specialized cell type or even an entire organ.

Key Terms

activator

centromere

chromatin

chromosome

combinatorial control

differentiation

gene expression

general transcription factors

gene regulatory protein

heterochromatin

histone

mitotic chromosome

nucleolus

nucleosome

position effect

positive feedback loop

regulatory DNA sequence

replication origin

repressor

TATA box

telomere

Questions

the very 3' end of the template for the lagging strand?

Question 8–7   Which of the following statements are correct? Explain your answers.

A. Each eucaryotic chromosome must contain the following DNA sequence elements: multiple origins of replication, two telomeres, and one centromere.

B. In bacteria the genes encoding ribosomal RNA, tRNA, and mRNA are transcribed by different RNA polymerases.

C. In bacteria, but not in eucaryotes, most mRNAs encode more than one protein.

D. Most DNA-binding proteins bind to the major groove of the double helix.

E. Of the four major control points in gene expression (transcription, RNA processing, translation, and control of a protein's activity), transcription initiation is used for the vast majority of gene regulation events.

F. The zinc atoms in DNA-binding proteins that contain zinc finger domains contribute to the binding specificity through sequence-specific interactions, which they form to the bases.

G. RNA polymerase stops transcribing at regions of DNA that are wrapped in nucleosomes.

H. Nucleosome core particles are 30 nm in diameter and, when lined up, form 30-nm filaments.

Question 8–8   Define the following terms and their relationships to one another.

A. interphase chromosome

B. mitotic chromosome

C. chromatin

D. heterochromatin

E. histones

F. nucleosome

Question 8–9

A. Explain why telomeres and telomerase are needed for replication of eucaryotic chromosomes but not for replication of a circular bacterial chromosome. Draw a diagram to illustrate your explanation.

B. Would you still need telomeres and telomerase to complete eucaryotic chromosome replication if DNA primase always laid down the RNA primer at the very 3' end of the template for the lagging strand?

Question 8–10   Carefully consider the result shown in Figure 8–13A. Each of the two colonies shown is a clump of approximately 100,000 yeast cells that has grown up from a single cell that is now somewhere in the middle of the colony. As explained in the figure caption, the lower colony contains mostly red cells, because the *ADE2* gene is inactivated when it is positioned near the telomere. Explain why the white sectors have formed near the rim of the colony. Based on the existence of these sectors, what can you conclude about the propagation of the transcriptional state of the *ADE2* gene from mother to daughter cells?

Question 8–11   Your task in the laboratory of Professor Quasimodo was to determine how far an enhancer (a binding site for an activator protein) could be moved from the promoter of the Straightspine gene and still activate transcription. You systematically varied the number of nucleotide pairs between these two sites and then determined the amount of transcription by measuring the production of Straightspine mRNA. At first glance, your data look confusing (Figure Q8–11). What would you have expected for the results of this experiment? Can you save your reputation and explain these results to Professor Quasimodo?

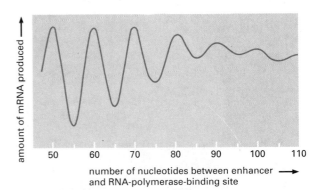

Figure Q8–11

y-axis: amount of mRNA produced

x-axis: number of nucleotides between enhancer and RNA-polymerase-binding site (50, 60, 70, 80, 90, 100, 110)

Question 8–12   Many gene regulatory proteins form dimers of identical or slightly different subunits on the DNA. Why is this advantageous? Describe three structural motifs that are often used to contact DNA. What are the particular features that suit them for this purpose?

Question 8–13 Bacterial cells can take up the amino acid tryptophan (Trp) from their surroundings, or, if there is an insufficient external supply, they can synthesize tryptophan from other small molecules. The Trp repressor is a bacterial gene regulatory protein that represses the transcription of genes that code for the enzymes required for the synthesis of tryptophan. Trp repressor binds to a site in the promoter of these genes only when molecules of tryptophan are bound to it (see Figure 8–21).

A. Why is this a useful property of the Trp repressor?

B. What would happen to the regulation of the tryptophan biosynthesis enzymes in cells that express a mutant form of Trp repressor that (i) cannot bind to DNA or (ii) binds to DNA even if no tryptophan is bound to it?

C. What would happen in scenarios (i) and (ii) if the cells, in addition, produced normal Trp repressor protein from a second, normal gene?

Question 8–14 The two electron micrographs in Figure Q8–14 show nuclei of two different cell types. Can you tell from these pictures which of the two cells is transcribing more of its genes? Explain how you arrived at your answer.

Figure Q8–14

Question 8–15 All differentiated cells in an organism contain the same genes. (One of the few exceptions to this rule are the cells of the mammalian immune system, where the formation of specialized cells is based on small rearrangements of the genome.) Describe an experiment that substantiates the first sentence of this question, and explain why it does.

Question 8–16 Figure 8–31 shows a simple scheme by which three gene regulatory proteins might be used during development to create eight different cell types. How many cell types could you create, using the same rules, with four different gene regulatory proteins? MyoD is a regulatory protein that by itself is sufficient to induce muscle-specific gene expression in fibroblasts. How does this observation fit the scheme in Figure 8–31?

Question 8–17 Discuss the following argument: "If the expression of every gene depends on a set of gene regulatory proteins, then the expression of these gene regulatory proteins must also depend on the expression of other gene regulatory proteins, and their expression must depend on the expression of still other gene regulatory proteins, and so on. Cells would therefore need an infinite number of genes, most of which would code for gene regulatory proteins." How does the cell get by without having to achieve the impossible?

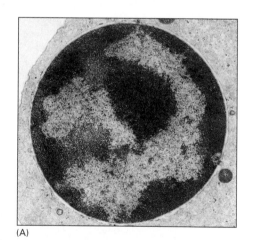

(A)

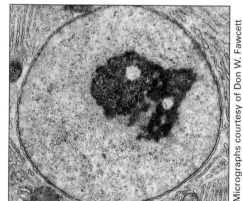

(B)

Micrographs courtesy of Don W. Fawcett

9 Genetic Variation

In the preceding three chapters we have seen how the information required to build and maintain a living organism is stored in the nucleotide sequence of its DNA and accurately replicated so that this information is inherited unchanged by a cell's progeny. However, DNA is not immutable. In any crowd of people, the individuals differ in a wide variety of heritable characteristics, such as eye color, skin color, hair color, height, build, and so on. Plainly their genomes do not contain exactly the same nucleotide sequences. Moreover, the vast diversity of life we see around us—plants, insects, mammals, for example—have arisen through changes in DNA that have accumulated since the first cells on earth arose some 3.5 billion years ago.

In its broadest sense the term *genetic variation* refers collectively to the diversity of genomes that give rise to different individuals on the planet. It includes the relatively small differences between genomes of members of the same species as well as the much greater difference between the genomes of different species. In this chapter, we shall consider how genomes change over evolutionary time. Evolution has favored systems where such genetic changes are not merely possible but inevitable, and the diversity of mechanisms that allow for genetic change suggest that it is advantageous for the maintenance of life on a changing earth.

The ultimate source of all genetic variation is changes in DNA (mutations) that alter its nucleotide sequence and thereby change its informational content. Some are small changes in nucleotide sequence such as those due to rare "mistakes" in DNA replication and repair (discussed in Chapter 6). There are, however, much more radical types of change, which result in large-scale alterations within a genome. Such changes include extensive rearrangements of nucleotide sequences, the duplication of individual genes or of even larger parts of the genome, deletion of parts of the genome, transfer of parts of one chromosome to another, and so on. These large-scale rearrangements are due to a variety of different processes including *DNA recombination* and the activities of viruses and *mobile genetic elements* that can move into and out of the DNA.

Mutation is not the only source of genetic variation. In sexually reproducing organisms, the genetic differences within individuals are largely the result of the reassortment of the gene pool of the species into new combinations in different individuals. Both mutation and gene reassortment will be discussed in this chapter as important sources of genetic change.

We begin by considering genetic variation in bacteria. Not only are bacteria among the most genetically simple and fastest growing of living cells, they are also among the most intensively studied. We are therefore closer to a full understanding of the life of a bacterium than we are to understanding any other type of organism. Despite their apparent genetic simplicity, bacteria possess almost all the mechanisms that give rise to the genetic variation that we consider in this chapter. The introduction to bacterial genetics here will also serve as an introduction to the key role of bacteria, their plasmids, and their viruses in the revolutionary techniques that make up *recombinant DNA technology*, which is the topic of Chapter 10.

In the second part of this chapter we look at some of the major ways in which eucaryotic genomes have diversified and evolved. Finally, we turn to *meiosis* and genetic recombination, which provide the main source of the genetic differences between individuals in sexually reproducing organisms.

Genetic Variation in Bacteria

Escherichia coli—*E. coli* for short—is a bacterium that lives in a symbiotic relationship with humans and other mammals. Billions of *E. coli* cells inhabit our large intestine (the colon), subsisting on nutrients that have escaped digestion or absorption in the stomach and small intestine. Living on these molecular crumbs, the bacteria synthesize amino acids and vitamins for their own use, some of which are absorbed into the bloodstream and are utilized by our bodies. For example, these bacteria are your chief source of vitamin K.

The widespread distribution of *E. coli*, the ease with which it grows on a wide variety of food sources, and its rapid rate of division led to its early notice by bacteriologists, and it was soon adopted as a standard organism for experimentation. Another advantage of studying bacteria such as *E. coli* is that they contain only one copy of their genome—that is, they are *haploid*. In contrast, *diploid* organisms like ourselves carry two copies of our genome in each cell (except the gametes, which are discussed later in this chapter). When a bacterial gene acquires a mutation, therefore, it often results in a detectable heritable change in the characteristics (the *phenotype*) of the bacterium. A mutation in a diploid organism, on the other hand, does not necessarily cause a change in phenotype, because each cell will also contain an unmutated copy of the same gene, which can often compensate for the effect of the mutated gene.

Many of the basic principles of molecular biology discussed in Chapters 6, 7, and 8 were discovered in *E. coli*, and today, after a century of investigation, we understand the workings of *E. coli* more completely than those of any other organism. It will serve as an excellent introduction to the fundamental mechanisms that generate genetic variability

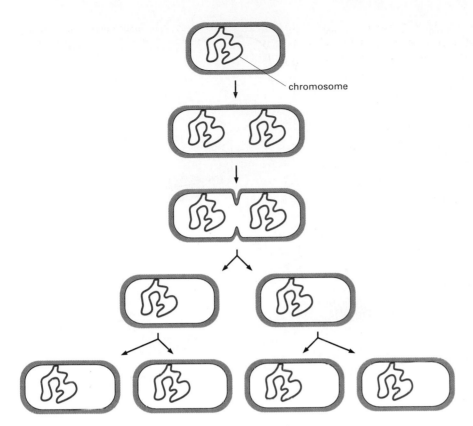

Figure 9–1 **Duplication of bacterial cells.**
The division of one bacterium into two
takes 20–25 minutes under ideal growth
conditions.

and show how environmental conditions exert selective forces that
operate on this variability to cause evolutionary change.

The Rapid Rate of Bacterial Division Means That Mutations Will Occur Over a Short Time Period

Bacteria like *E. coli* divide asexually by simple fission. Their DNA is replicated and segregated to the two ends of the growing bacterium; the bacterium then splits in two, producing two daughter cells, each containing
a genome identical to that of the parent cell, providing there have been
no mistakes in replication (Figure 9–1). In the laboratory, *E. coli* grows
rapidly in a liquid medium containing plentiful amino acids, sugars, and
salts. In such a medium, a population of *E. coli* cells will double in number every 20–25 minutes if the culture is aerated to provide oxygen and
kept at a temperature close to that of the human body (37°C). In less
than a day, a single *E. coli* can produce more descendants than there are
humans on earth. Individual bacteria are far too small to be seen with
the naked eye, but if grown on a solid surface, the descendants of a single cell will form a discrete colony, easily visible to the eye. Agar is an
inert polysaccharide obtained from seaweed that forms a solid gel. For
use as a culture medium for bacteria, it is melted, mixed with nutrient
broth, poured into a Petri dish, and allowed to solidify. A drop of a dilute
suspension of bacteria is then spread over the surface of the dish. If the
concentration of bacteria in the suspension is sufficiently low, individual
cells will be well separated from one another on the plate and each will
give rise to a round colony consisting of approximately 10^6 cells (Figure
9–2). Bacteria can then be picked from individual colonies and grown for
further study.

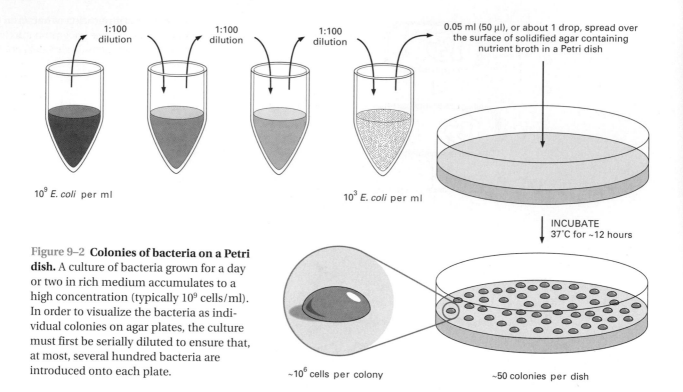

1:100 dilution   **1:100 dilution**   **1:100 dilution**

0.05 ml (50 μl), or about 1 drop, spread over the surface of solidified agar containing nutrient broth in a Petri dish

10^9 *E. coli* per ml

10^3 *E. coli* per ml

INCUBATE 37°C for ~12 hours

Figure 9–2 **Colonies of bacteria on a Petri dish.** A culture of bacteria grown for a day or two in rich medium accumulates to a high concentration (typically 10^9 cells/ml). In order to visualize the bacteria as individual colonies on agar plates, the culture must first be serially diluted to ensure that, at most, several hundred bacteria are introduced onto each plate.

~10^6 cells per colony

~50 colonies per dish

The *E. coli* genome is a circular DNA molecule consisting of 4–5 million nucleotide pairs that encode about 3000 different proteins. Although the replication of a cell's DNA is very accurate, it is not absolutely flawless. We saw in Chapter 6 that each time DNA is replicated, there is a very small probability (~$1/10^9$) that a given nucleotide will be miscopied and escape mismatch repair. Although this probability is very low, the rapid rate of division of *E. coli* means that very large populations of cells, in which many mistakes will have occurred, can be produced very rapidly by a human experimenter.

Mutations in Bacteria Can Be Selected by a Change in Environmental Conditions

To detect and isolate bacteria in which a mutation has occurred, the mutant bacteria must be identified and removed from the predominant population of unmutated bacteria. For example, the typical *E. coli* cell is killed by the antibiotic rifampicin. This drug binds tightly to RNA polymerase molecules inside the cell, and prevents them from transcribing DNA into RNA. This inhibition eventually blocks the synthesis of new proteins, and the bacterium dies. However, a large population of *E. coli* (for example, the 10^9 cells contained in a few milliliters of rapidly growing liquid culture) probably contains a few cells that are resistant to rifampicin. In such cells the gene encoding RNA polymerase has, during DNA replication, undergone a rare mutation so that it encodes an RNA polymerase that can still transcribe DNA into RNA but is now immune to attack by rifampicin. If rifampicin is added to a growing culture of *E. coli*, nearly all of the cells will be killed, but if there are a few rifampicin-resistant mutants already present in the culture, they will flourish, and within a few hours, they can take over the culture (Figure 9–3).

Question 9–1 In your microbiology lab course, you treated a bacterial culture with a chemical that causes mutations. In order to study individual mutant strains that arise from this treatment you need to prepare populations of bacteria, each of which originated from a single mutated cell. To do so, your instructor told you to plate 50 μl of your treated bacterial culture (which contains 1000 cells/ml) on agar plates (as shown in Figure 9–2) so that you can obtain bacterial colonies that arise from individual mutated cells. Next, cells from single colonies would be checked for the desired mutation.

The teaching assistant (an older student known widely for his famous shortcuts) suggests that you simply take 1 μl samples of your chemically treated bacterial culture and place each sample into a separate tube. These samples can then be individually diluted with fresh nutrient broth and incubated at 37°C. And voilà, you have many independent cultures, each arising from a single cell. Just skip the tedious plating!

Do you accept his advice—or does it reflect the reason he still has not graduated? Explain your answer.

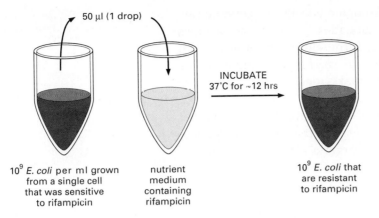

Figure 9–3 **Mutation and natural selection in a population of bacteria.** When a culture of bacteria is challenged with a new condition (the presence of the antibiotic rifampicin, for example), the rare mutants that are better able to survive (those resistant to rifampicin) will proliferate and take over the population.

This laboratory example neatly demonstrates the principles of genetic variation and natural selection in response to environmental conditions. First, rare DNA-replication errors that occur at random provide a source of genetic variability; a culture containing 10^9 E. coli cells has millions of mutant cells whose genomes differ from that of the ancestor cell, rifampicin-resistant mutants being only one of the many different variants present. (In general, the population will include thousands of variant forms of each gene.) Second, under a selective pressure (the presence of rifampicin in the above example), any rare individuals better able to survive (here, those that are rifampicin-resistant) will proliferate and can eventually take over the population.

Thus every large population of bacteria contains, in addition to the predominant genetic type, a pool of mutant bacteria that provides a generous supply of genetic variants. The large population of rifampicin-resistant E. coli that has now taken over the culture will also contain a pool of variants, and if conditions were to change again, yet another genetic type could have a selective advantage and become predominant. Thus, the fact that DNA replication is not perfectly accurate all the time allows a population of E. coli to quickly adapt genetically to changes in its surroundings.

In the absence of environmental change, mutants are unlikely to take over the population, since among those mutations that allow the bacterium to survive, many will have no functional effect on the phenotype and many will be slightly deleterious. For example, rifampicin-resistant mutants of E. coli are often less tolerant of temperature extremes than are wild-type E. coli, and so they would be gradually lost from a population if rifampicin were removed from the media.

Bacterial Cells Can Acquire Genes from Other Bacteria

Important as replication errors are to the genetic adaptability of E. coli, they are not the only way its genome can be altered. Although the intermixing of genes from different individuals by sexual reproduction has been recognized in plants and animals for a long time, only in the 1950s was it discovered that bacteria also possess mechanisms to intermix genes from different individuals in the population.

The transfer of genes from one bacterium to another can be demonstrated experimentally simply by mixing together bacteria with different genetic characteristics. For example, one laboratory strain of E. coli is

Question 9–2 The *lacY* gene of E. coli codes for a protein that transports the sugar lactose from outside the cell to inside, where it is broken down and used for fuel. An E. coli that has a disabling mutation in the *lacY* gene therefore cannot grow on medium in which lactose is the only source of fuel. If 10^9 *lacY* mutant E. coli are plated on agar that contains lactose as the only fuel, approximately 50 colonies are formed.

A. How are these colonies able to form?

B. If this experiment is repeated using a *lacY* mutant E. coli that has, in addition, a mutation that disables the mismatch repair system (see Chapter 6), 3000 colonies are formed. Can you suggest an explanation for this increase?

unable to manufacture the amino acid leucine because its genome carries a damaging mutation in the gene that codes for one of the enzymes required to synthesize this amino acid. It therefore cannot grow on a medium lacking leucine. Another strain of *E. coli* carries a mutation in a different gene that renders it unable to make methionine. It therefore cannot grow on a medium lacking methionine.

If these two strains of *E. coli* are mixed and allowed to grow together for a few hours and then transferred to a medium that lacks both leucine and methionine, many rapidly growing bacteria may be found in the new medium. These represent a new strain of *E. coli* that is able to synthesize both leucine and methionine, and can thus grow in medium lacking both amino acids. This new bacterial strain appears with a probability higher than could be accounted for by mutation due to replication errors. Instead, the genome of the new bacterial strain is composed of a normal gene for leucine synthesis from one strain and a normal gene for methionine synthesis from the other.

As we shall consider in more detail in the following sections, this genetic intermixing has occurred by the transfer of DNA from one bacterium to another, followed by the replacement of some of the recipient's genome with the corresponding genes from the incoming DNA. Replacement occurs by a process of DNA recombination, which can take place between any two DNA molecules of similar sequence. Gene transfer followed by recombination provides a powerful way for bacteria to adapt to changes in their surroundings by making the best use of the variation available within the population.

Bacterial Genes Can Be Transferred by a Process Called Bacterial Mating

Genes can be transferred from one bacterium to another by several different routes. One is by direct cell-to-cell transfer during **bacterial mating,** or *conjugation,* of two cells. Not all bacteria in a population can initiate mating and gene transfer. The ability to do so is conferred by genes contained in bacterial **plasmids:** small, circular, double-stranded DNA molecules that are separate from the larger bacterial chromosome. Plasmids contain their own replication origins that permit them to replicate independently of the bacterial chromosome. Plasmids come in many different types and exist in many different species of bacteria, but only a few are able to initiate mating and DNA transfer. A plasmid that commonly initiates conjugation in *E. coli* is the *F plasmid,* or fertility plasmid.

Mating can take place only between a bacterium containing an F plasmid and one that lacks it. When a bacterium carrying the F plasmid (the *donor*) encounters a bacterium lacking the plasmid (the *recipient*), a cytoplasmic bridge is formed between the two cells. The F plasmid DNA is replicated and transferred from the donor through the bridge to the recipient (Figures 9–4, 9–5, and 9–6). The bridge then breaks down, and the two bacteria, both of which now contain the F plasmid, can act as donors in subsequent encounters with recipient bacteria. The F plasmid is necessary for mating because it carries genes that encode some of the proteins required to make the bridge and to transfer the DNA.

The transfer of the F plasmid from one bacterium to another, while providing one of the simplest examples of gene transfer between

Figure 9–4 Electron micrograph showing two bacterial cells mating. The donor cell (*top*) is linked to the recipient cell by a sex pilus (see first panel of Figure 9–5). To make it more visible, the sex pilus has been labeled along its length by viruses that specfically adhere to it. (Courtesy of Charles C. Brinton Jr. and Judith Carnahan.)

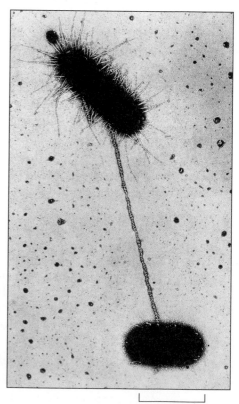

1 µm

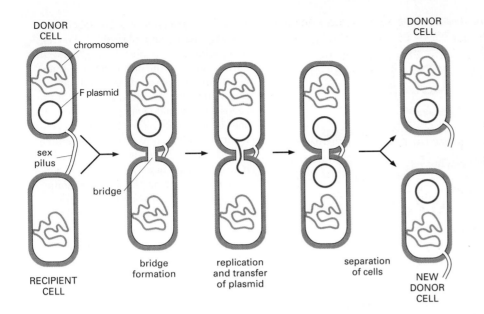

DONOR CELL
chromosome
F plasmid
sex pilus
bridge

RECIPIENT CELL

bridge formation

replication and transfer of plasmid

separation of cells

DONOR CELL

NEW DONOR CELL

Figure 9–5 Schematic diagram of bacterial mating. The fine appendage on the donor bacterium on the left is the sex pilus, which is used to attach the donor to the recipient. The DNA is transferred through a cytoplasmic bridge, which forms after attachment. After a copy of the plasmid has been transferred, the two bacteria separate. The bacterial genome and F plasmid both consist of double-stranded DNA. For convenience, however, each is depicted here as a single line. The unusual form of the replicating plasmid is called a rolling circle and is described in more detail in Figure 9–6.

individuals, is in itself limited in its usefulness as a means of creating genetic variability. Only the small number of genes residing on the plasmid are transferred, and most of these are concerned with mating. However, the F plasmid can also occasionally become an integral part of the large bacterial chromosome by a type of recombination called *integration*, which will be discussed later. In this state, the F plasmid is still able to initiate mating and transfer itself from one cell to another. But in doing so it can now also take much of the bacterial chromosome with it, since the two DNA molecules have become covalently joined (Figure 9–7). By this route virtually any gene from a donor bacterium can be transferred to a recipient.

Gene transfer enhances the genetic variability and therefore the adaptability of bacteria. One of the clearest examples has been brought about by the widespread use of antibiotics such as penicillin. Under the intense selective pressure applied by the use of these drugs, bacteria resistant to most common antibiotics soon arose. For example, roughly one-third of new clinical isolates of *Neisseria gonorrhoeae*, the bacterium that causes gonorrhea in humans, are resistant to penicillin. As a result, penicillin is no longer the first-line treatment for this disease. The analysis of antibiotic-resistant bacteria revealed that they carry plasmids (similar to the F plasmid) that encode proteins that inactivate the antibiotics. Because these plasmids can be transferred from one bacterium to the next by mating, they quickly spread, rendering large populations of bacteria resistant to these drugs. Many different plasmids have now been isolated from bacteria; although often a cause for serious medical

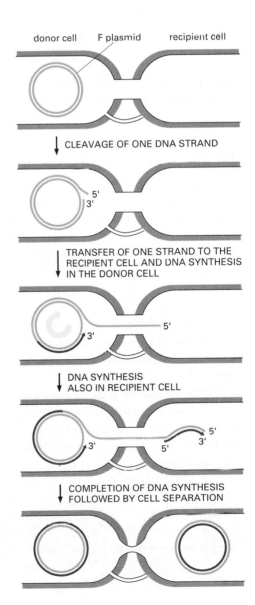

donor cell   F plasmid   recipient cell

CLEAVAGE OF ONE DNA STRAND

5'
3'

TRANSFER OF ONE STRAND TO THE RECIPIENT CELL AND DNA SYNTHESIS IN THE DONOR CELL

3'   5'

DNA SYNTHESIS ALSO IN RECIPIENT CELL

3'   5'   3'   5'

COMPLETION OF DNA SYNTHESIS FOLLOWED BY CELL SEPARATION

Figure 9–6 Synthesis and transfer of F plasmid DNA by rolling circle replication. The F plasmid replicates in a form known as a rolling circle. One strand of the plasmid DNA is cut and one end is transferred through the cytoplasmic bridge to the recipient cell. At the same time, DNA synthesis starts on the exposed plasmid DNA strand remaining in the donor bacterium, producing a new DNA strand (*red*). As these processes continue, the donor plasmid rotates (or "rolls") as indicated by the *yellow arrow*. As the single-stranded plasmid DNA strand enters the recipient cell, its complementary DNA strand is synthesized (*red*).

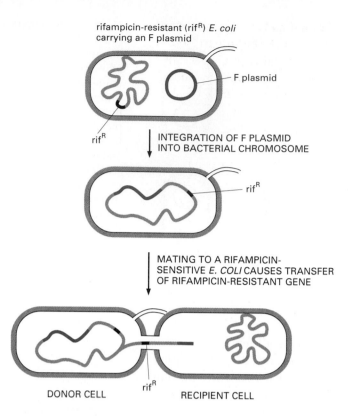

rifampicin-resistant (rif^R) *E. coli*
carrying an F plasmid

F plasmid

rif^R

INTEGRATION OF F PLASMID
INTO BACTERIAL CHROMOSOME

rif^R

MATING TO A RIFAMPICIN-
SENSITIVE *E. COLI* CAUSES TRANSFER
OF RIFAMPICIN-RESISTANT GENE

DONOR CELL rif^R RECIPIENT CELL

Figure 9–7 Transfer of bacterial chromosomal genes by an F plasmid. Following integration of the F plasmid DNA into the chromosome of the donor cell, the transfer of both the plasmid DNA and the attached bacterial chromosome DNA proceeds—using rolling circle replication that starts from a site in the plasmid DNA (see Figure 9–6). For convenience, the double-stranded DNA molecules are depicted here by a single line. Only rarely is a complete copy of the bacterial chromosome transferred. Mating is more commonly interrupted by breaking of the cytoplasmic bridge before transfer is complete. Only some bacterial genes are therefore transferred at each mating.

concern, these plasmids have also provided scientists with powerful research tools. For example, plasmids are used routinely in a large number of recombinant DNA applications, as discussed in Chapter 10.

Bacteria Can Take Up DNA from Their Surroundings

Although bacterial mating is probably a major route of gene transfer in *E. coli,* a different mechanism, called **transformation,** is the preferred route in some other species of bacteria, such as the soil bacterium *Bacillus subtilis.* These bacteria take up DNA molecules present in their surroundings as a result of the death and breakdown of other bacteria, and they pull the DNA through their cell membrane to the inside of the cell (Figure 9–8). The incoming DNA is often then incorporated into the genome by recombination. The term "transformation" originated from early observations of this phenomenon in which it appeared that one bacterial strain had become transformed into another. The transformation of one strain of bacterium with purified DNA derived from another strain provided one of the first proofs that DNA was indeed the genetic material (see Figure 6–3).

In a natural bacterial population, a source of DNA for transformation is provided by bacteria that have died and released their contents (including DNA) into the environment. One disadvantage of transfor-

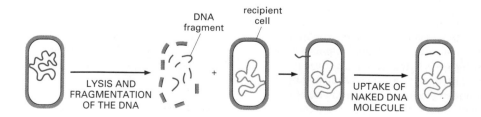

DNA
fragment

recipient
cell

LYSIS AND
FRAGMENTATION
OF THE DNA

UPTAKE OF
NAKED DNA
MOLECULE

Figure 9–8 Bacterial transformation. Some bacteria can take up small fragments of DNA derived from other bacterial cells that have died. Once inside the recipient cell, the donor DNA can become a part of the recipient genome through the process of homologous recombination, as shown in Figure 9–13.

mation as a means of genetic exchange is that naked DNA (that is, DNA without any of its normal chromosomal proteins attached) is exposed to the extracellular environment, which typically contains nucleases and other agents that damage DNA. In contrast, the DNA is protected during bacterial mating, as it never leaves the intracellular environment.

Bacterial transformation has had an enormous impact as a laboratory tool. A great advantage to the experimenter is that naked DNA from any source, not just the DNA of the same bacterial species, can be taken up by this route. *E. coli* does not normally use transformation as a major route of genetic exchange, but it can be coaxed in the laboratory to do so. As we shall see in Chapter 10, this allows DNA from complex organisms, such as humans, to be studied easily in the laboratory.

Homologous Recombination Can Take Place Between Two DNA Molecules of Similar Nucleotide Sequence

Once a new piece of DNA arrives inside a bacterium—either through bacterial mating or through transformation—several things can happen to it. If the incoming DNA is a plasmid, such as the F plasmid discussed earlier, it can replicate independently in its new host and be passed on to the progeny as the bacterium divides. If the incoming DNA cannot be replicated because it lacks an origin of DNA replication, it can be passed on to the bacterium's progeny only if it becomes part of the bacterial chromosome; otherwise, it will be lost rapidly from the population as the bacteria continue to divide. The most important route by which DNA becomes incorporated into a bacterial genome is called **homologous recombination.**

Homologous recombination occurs in all organisms and can take place between any two molecules of double-stranded DNA with regions that are similar in nucleotide sequence. Although the mechanism of homologous recombination is not completely understood, the following features, which are known to occur in bacteria, are probably common to all homologous recombination:

1. Two double-stranded DNA molecules that have regions of very similar (homologous) DNA sequence align so that their homologous sequences are in register. They can then "cross over": in a complex reaction, both strands of each double helix are broken and the broken ends are rejoined to the ends of the opposite DNA molecule to re-form two intact double helices, each made up of parts of the two different DNA molecules (Figure 9–9).

2. The site of exchange (that is, where a *red* double helix is joined to a *green* double helix in Figure 9–9) can occur anywhere in the homologous nucleotide sequences of the two participating DNA molecules.

3. No nucleotide sequences are altered at the site of exchange; the cleavage and rejoining events occur so precisely that not a single nucleotide is lost or gained.

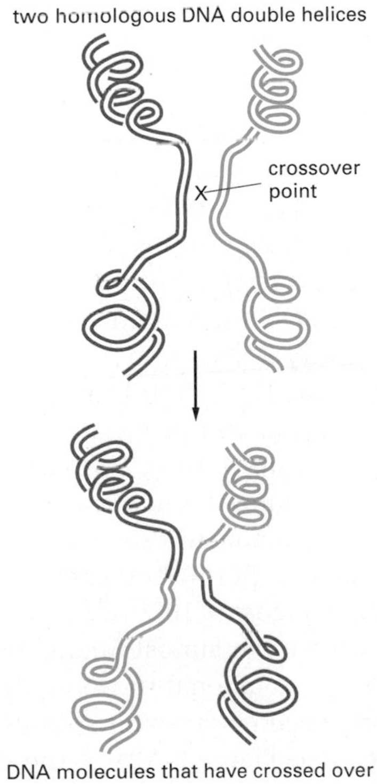

two homologous DNA double helices

crossover point

DNA molecules that have crossed over

Figure 9–9 **Homologous recombination.** The breaking and rejoining of two homologous DNA double helices creates two DNA molecules that have "crossed over." Although the two original DNA molecules must have similar nucleotide sequences in order to cross over, they do not have to be identical; thus a crossover can create DNA molecules of novel nucleotide sequence.

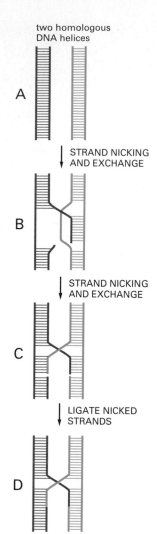

two homologous
DNA helices

A

STRAND NICKING
AND EXCHANGE

B

STRAND NICKING
AND EXCHANGE

C

LIGATE NICKED
STRANDS

D

two DNA molecules
joined by a cross-strand
exchange

Figure 9–10 (*left*) **The formation of a cross-strand exchange in homologous recombination.** (A) Two DNA molecules of similar sequence align. (B) A nick is made in one of the strands, and the nicked strand exchanges to pair with its complementary strand on the other DNA helix. (C) The displaced strand is nicked and exchanged so that two DNA strands from opposite DNA molecules cross over. (D) The nicks are sealed by a DNA ligase to form a structure in which there are two crossing strands and two noncrossing strands. There are several pathways that can lead from a single-strand exchange to a cross-strand exchange, but only one is shown here.

Figure 9–11 (*right*) **The rotation of a cross-strand exchange.** In step A, a cross-strand exchange has formed, as shown in Figure 9–10. Without a rotation (also called isomerization), cutting of the two crossing strands would terminate the exchange and genetic recombination would not occur. With rotation (steps B and C), cutting the two crossing strands creates two DNA molecules that have exchanged pieces of DNA (D and E).

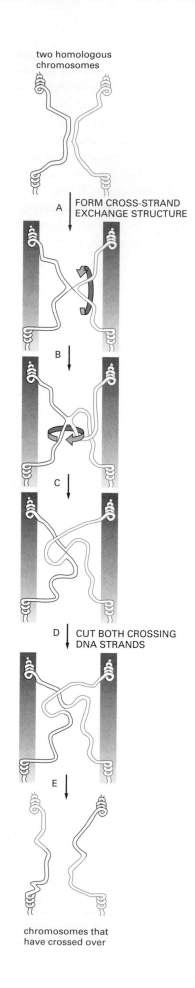

two homologous
chromosomes

A FORM CROSS-STRAND
   EXCHANGE STRUCTURE

B

C

D CUT BOTH CROSSING
   DNA STRANDS

E

chromosomes that
have crossed over

Homologous recombination is initiated by a nick in one of the DNA strands of one of the aligned double helices. The nicked strand unwinds from its double helix and invades the other DNA molecule, which unwinds locally so that the incoming strand can pair with its complementary strand (Figure 9–10B). This is the first stage of the crossover. The displaced DNA strand now breaks and crosses over to pair with its complementary DNA strand in the other DNA molecule (Figure 9–10C). The nicks in the DNA strands are sealed so that the two DNA molecules are now held together physically by a crossing-over of one of each of their strands (Figure 9–10D). This crucial intermediate in homologous recombination is known as a *cross-strand exchange*, or *Holliday junction*.

To regenerate two separate DNA molecules, the two crossing strands must be cut. But if they are cut while the structure is still in the form shown in Figure 9–10D, the two original DNA molecules would separate from each other almost unaltered. The structure can, however, undergo a series of rotational movements so that the two original noncrossing strands become crossing strands and vice versa (Figure 9–11, steps B and C, and Figure 9–12). If the crossing strands are cut after rotation

Figure 9–12 **Electron micrograph of a cross-strand exchange (Holliday junction).** This view of the molecule corresponds to the product of step B in Figure 9–11. (Courtesy of Huntington Potter and David Dressler.)

(Figure 9–11, step D), one section of each original DNA helix is joined to a section of the other DNA helix; in other words, the two DNA molecules have crossed over, and two molecules of novel DNA sequence have been produced (Figure 9–11, step E).

As shown in Figure 9–13A, two such exchanges can neatly replace a long stretch of bacterial DNA with a homologous but not identical DNA fragment from another source. Such a double exchange is a route by which a segment of DNA from one bacterium—whether transferred by mating (see Figure 9–7) or acquired by transformation (see Figure 9–8)—can be permanently incorporated into the genome of another.

The **integration** of the F plasmid into its host cell chromosome (see Figure 9–7) is also achieved by homologous recombination. The F plasmid contains a short stretch of DNA that is very similar in nucleotide sequence to stretches of the bacterial chromosome. Homologous recombination at one such site is sufficient for the integration of the circular F plasmid into the circular bacterial chromosome (see Figure 9–13B).

Cells utilize specialized proteins to facilitate homologous recombination; these proteins nick DNA, catalyze strand exchange, and cleave Holliday structures. Such recombination enzymes, although well characterized in bacteria, are only beginning to be understood in eucaryotic cells.

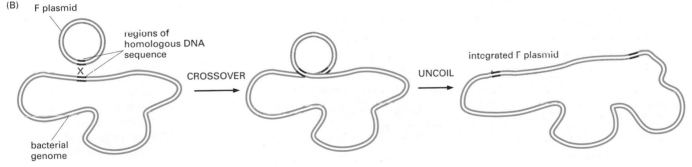

Figure 9–13 **Homologous recombination can lead either to an exchange of DNA between two molecules or to the combination of two circular DNA molecules into one.** (A) Replacement of one segment of bacterial DNA with a homologous segment by two crossovers, as in the incorporation of a portion of a fragmented genome taken up by transformation. The sites of crossovers are indicated by X's. (B) Integration of one circular DNA molecule into another by a single crossover, as in F plasmid integration into the main bacterial chromosome.

Figure 9–14 The life cycle of a hypothetical virus. The simple virus illustrated consists of a small double-stranded DNA molecule that encodes just a single type of viral coat protein. No known virus is quite this simple. In order to replicate, the viral genome must enter the cell. This is followed by replication of the viral DNA to form many copies. At the same time, some of the viral DNA is transcribed into RNA and translated to produce coat proteins. The viral genomes assemble spontaneously with the coat proteins to form new virus particles.

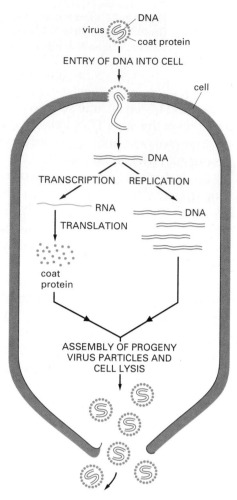

Genes Can Be Transferred Between Bacteria by Bacterial Viruses

A third route of gene transfer in bacteria is through viruses. Later in this chapter we shall discuss viruses that infect mammalian cells. Here we are concerned with viruses that attack bacterial cells—*bacteriophages*—which multiply inside the bacterial cell, commandeering the cell's biochemical machinery to make new copies of the viral genome and to synthesize the proteins of the viral coat (Figure 9–14). Virus reproduction is generally lethal to the infected cell; the cell bursts open (lyses) as a result of the infection, releasing the progeny viruses to infect neighboring cells. However, some bacteriophages can infect a bacterial cell without immediately multiplying to produce large numbers of virus particles. These bacteriophages instead enter a latent state, in which their genomes are present but inactive in the cell, and no progeny viruses are produced.

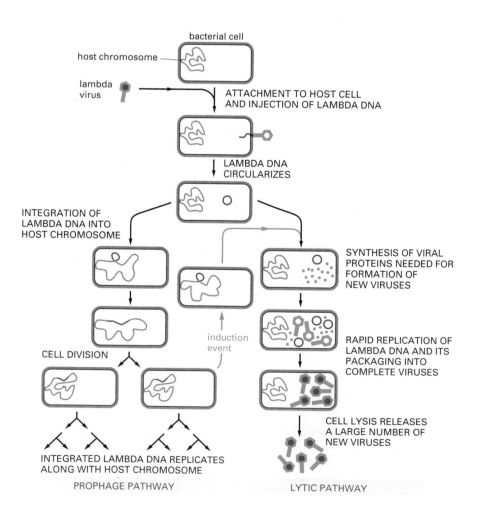

Figure 9–15 The life cycle of bacteriophage lambda. The linear double-stranded DNA lambda genome contains about 50,000 nucleotide pairs and encodes 50–60 different proteins. When the lambda DNA enters the cell the ends join to form a circular DNA molecule. The bacteriophage can multiply in *E. coli* by a lytic pathway, which destroys the cell, or it can enter a latent prophage state. Damage to a cell carrying a lambda prophage induces the prophage to exit from the host chromosome and shift to lytic growth (*green arrows*). The entrance and exit of the lambda DNA from the bacterial chromosome are site-specific recombination events.

The best-studied example of such a virus is bacteriophage lambda (λ), which has a small double-stranded DNA genome. When lambda infects an *E. coli* cell, it normally multiplies to produce several hundred new viruses that are released when the bacterial cell lyses; this is called a *lytic infection.* More rarely, however, the lambda genome becomes integrated into the circular host *E. coli* chromosome (Figure 9–15). The integrated lambda DNA is known as a *prophage,* and it lies dormant in the bacterial chromosome. The infected bacterium multiples normally, passing the lambda DNA on to all of its descendants. In this way the viral genome is replicated many times without destroying its host cell. An environmental insult, such as exposure to ultraviolet light, can induce the lambda genome to leave the host chromosome and begin a lytic phase of viral replication. Thus the integrated viral DNA need not perish with its damaged host cell but has a chance to escape as part of a virus particle.

The lambda genome has no extensive regions of similarity to the bacterial genome and so cannot integrate by homologous recombination. Instead it integrates by **site-specific recombination.** A specialized recombination enzyme encoded by the virus, called lambda *integrase*, recognizes a particular DNA sequence on the bacterial chromosome and a different sequence on the viral genome. This brings these two sequences together and catalyzes the breakage of both DNAs and their rejoining to each other to integrate the viral genome into the bacterial chromosome. The lambda DNA exits from the bacterial chromosome by carrying out this reaction in reverse.

On leaving a host chromosome, the lambda DNA will occasionally excise itself inaccurately and bring along a neighboring piece of host DNA in place of part of its own DNA. This host DNA will be packaged into a virus particle along with the lambda DNA (Figure 9–16). When such a modified lambda virus infects a new host, it introduces, along with the viral DNA, DNA derived from the previous host. This bacterial DNA can become part of the new host chromosome in one of several possible ways. If the incoming virus integrates into its new host chromosome, it will bring the passenger DNA along with it and both will become a constituent of the host chromosome. Alternatively, if the incoming virus does not destroy the host, the passenger bacterial DNA can become a permanent part of its host's genome by homologous recombination, by the pathway pictured in Figure 9–13A. This transfer process—called **transduction**—provides yet another way that genes can be transferred from one bacterium to another. As we shall see later, some of the viruses that infect animal cells can also transfer chromosomal genes by a transduction process.

Transposable Elements Create Genetic Diversity

Many bacterial genomes contain stretches of DNA called **transposable elements** (or transposons, for short), which can move from place to place within the chromosome by **transposition** and are an important source of genetic diversity. They range in length from several hundred to tens of thousands of base pairs (Figure 9–17). A typical laboratory strain of *E. coli* contains 10–20 different transposons per cell, many present in multiple copies per genome. Transposons move within the DNA of their host by means of special recombination enzymes (*transposases*)

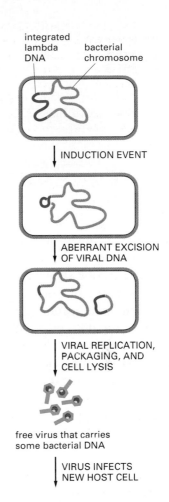

Figure 9–16 Transfer of DNA from one bacterium to another by transduction. Upon leaving the host chromosome, the lambda DNA sometimes excises itself inaccurately and packages some host DNA in place of part of the viral genome. When such viruses infect another cell, they can either lyse the cell or introduce the DNA from the original host without killing the new host. The newly introduced DNA can become a part of the infected cell's genome either through integration of the viral genome or through homologous recombination events.

Question 9–3 When it forms a prophage, bacteriophage lambda always integrates its genome into the same spot on the bacterial chromosome. How does this influence the range of bacterial genes that lambda can package into a virus (see Figure 9–16)?

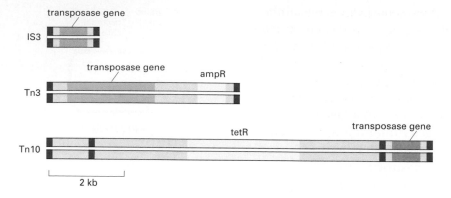

transposase gene

IS3

transposase gene

ampR

Tn3

tetR

transposase gene

Tn10

2 kb

Figure 9–17 Three of the many known types of transposons found in bacteria. Each of these transposons contains a gene that encodes a transposase, an enzyme that carries out some of the DNA breakage and joining reactions needed for the transposon to move. Each transposon also carries DNA sequences (indicated in *red*) that are recognized only by the transposase encoded by that element and are necessary for movement of the transposon. Some transposons carry, in addition, genes that encode enzymes that inactivate antibiotics such as ampicillin (ampR) and tetracycline (tetR). In fact, Tn10 is thought to have evolved by the chance landing of two short transposons on either side of a tetracyclin-resistance gene.

encoded by the transposable element (Figure 9–18), and one can consider the elements as tiny parasites hidden in the chromosomes of cells. However, these mobile elements provide a number of advantages to bacteria. For example, some of them contain transcriptional promoters, and when they land near a gene they can cause it to be expressed at a different level or under different cellular controls.

Transposons present in multiple copies also provide opportunities for bacterial genomes to become rearranged through homologous recombination between transposons. For example, the integration of the F plasmid into the *E. coli* genome (see Figure 9–13B) occurs through homologous recombination between one copy of a transposon DNA sequence located on the plasmid and another copy located in the genome. Finally, some transposons contain drug-resistance genes (see Figure 9–17). These transposons can jump from the bacterial genome into plasmids, which can subsequently be transferred to other bacteria by conjugation. It is thought that many of the drug-resistance plasmids now present in many clinically important bacteria arose in this way. Mobile genetic elements occur in eucaryotic genomes as well (in fact, they were originally discovered in maize in the 1950s), and they will be discussed again later in this chapter.

So far, we have seen that despite the remarkably high fidelity with which a bacterial cell copies its DNA and passes it on to the next generation, bacterial genomes are in a state of slow flux. Mutations accumulate, genes are transferred from one bacterium to another, viral genomes move in and out, and transposable elements change their positions.

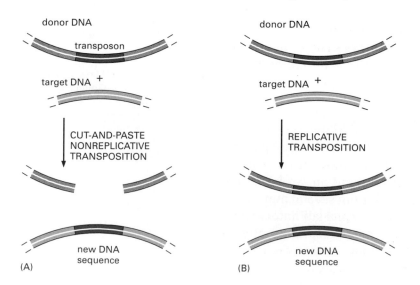

donor DNA

transposon

target DNA +

CUT-AND-PASTE
NONREPLICATIVE
TRANSPOSITION

new DNA
sequence

(A)

donor DNA

target DNA +

REPLICATIVE
TRANSPOSITION

new DNA
sequence

(B)

Figure 9–18 Two mechanisms by which bacterial transposons move. The donor and target DNAs shown can be part of the same molecule (the bacterial genome, for example) or reside on different molecules (a bacterial genome and a plasmid, for example). (A) In cut-and-paste nonreplicative transposition, the transposon is cut out of the donor DNA and inserted into the target DNA, leaving behind a broken donor DNA molecule. (The donor can be repaired in a variety of ways, but this sometimes results in deletions or rearrangements of the donor molecule.) (B) In the course of replicative transposition, the transposon is copied by DNA replication. The end products are a molecule that appears identical to the original donor and a target molecule that has a transposon inserted into it. In general, a particular type of transposon moves by only one of these mechanisms. However, the two mechanisms have many enzymatic similarities, and a few transposons can move by either mechanism.

Most alterations in the genome are harmful to the individual bacterium in which they occur, and these are quickly eliminated from the population by the death of the individuals. However, as bacteria migrate from one environment to another, new selective forces come into play, and these sources of genetic variation prove to be important for providing the raw material from which a better adapted individual can emerge as the predominant type. This principle has had especially important consequences in medicine, allowing disease-causing bacteria to become resistant to many of the antibiotics developed during the twentieth century.

It is clear that the ability to adapt genetically is a fundamental property of bacteria, and future strategies for combating infectious diseases must take this principle into account. In the next section of this chapter we shall see that eucaryotic cells share many of the same mechanisms of genetic diversity and that these mechanisms have had a profound effect on the evolution of all life on the planet.

Sources of Genetic Change in Eucaryotic Genomes

We have seen that procaryotes and eucaryotes share many genetic features, including the same genetic code, the same basic mechanism of transcription and translation, and many of the same basic strategies of DNA replication and repair. In this section, we will see how the organization of genetic information in a eucaryotic genome contributes to the genetic diversity of eucaryotic organisms.

As noted briefly in Chapter 7, bacterial genomes are highly streamlined in organization compared with those of eucaryotes. Genes are packed close together on bacterial chromosomes, with relatively little spacer DNA between them, and introns are extremely rare. In contrast, a typical mammalian genome contains enormous amounts of apparently nonessential DNA in the form of introns (discussed in Chapter 7) and much additional DNA apparently used as "spacer" between genes. Much of this spacer DNA consists of various types of repetitive DNA sequences with properties similar to those of the bacterial transposons discussed earlier. Although none of this DNA seems to encode anything useful to the organism, its presence has had far-reaching effects, as we shall see. Why have mammalian genomes retained this extra DNA? Bacteria and even simple unicellular eucaryotes such as yeast are generally under strong selective pressure to divide at the maximum rate permitted by the nutrients in their environment, and thus to minimize the amount of superfluous DNA in their genome, as DNA replication is costly in terms of energy and material resources. For larger cells, and especially cells in multicellular organisms, cell division occurs less frequently and nutrition is not usually a limiting factor. Therefore, the selective pressure to eliminate nonessential DNA sequences will not be so great. This is thought to be one of the reasons that such sequences have been allowed to accumulate in the genomes of higher eucaryotes.

Another feature of eucaryotic genomes that appears different from bacterial genomes is the large amount of *gene duplication* that has occurred over evolutionary time. This has resulted in the presence in eucaryotic genomes of many large families of highly related genes. In a mammalian haploid genome (that is, considering only one of the two sets of chromosomes present in an individual), there are multiple genes for most proteins, and each gene encodes a slightly different variant of the protein, as well as more distant relatives that can have a different function.

Finally, the cells of eucaryotic organisms (except the germ cells) are diploid; that is, except for the genes carried on the sex chromosomes, they carry two copies of each gene. These copies need not be identical. In principle, therefore, a human can carry two alternative forms (alleles) of each gene. For example, humans who carry one gene for normal β-globin and one gene for sickle-cell β-globin are more resistant to malaria than are humans who carry two normal β-globin alleles. Indeed, the "gene pool"—the entire collection of alleles of every gene present in a species as a whole—is vast for the human population.

Much of the history of genetic change is recorded in the genomes of present-day organisms and can be deciphered from a careful analysis of their DNA sequences. Hundreds of millions of nucleotides of DNA have now been sequenced from different organisms, and we are beginning to see how individual genes and whole genomes have evolved over the billions of years that have elapsed since the appearance of the first eucaryotic cell. The occasional changes that occur in present-day chromosomes provide additional clues to the mechanisms that have brought about evolutionary change in the past.

In this section we consider some of the mechanisms of genetic change in multicellular eucaryotes, whose large and complex genomes carry ample evidence of past events. In sexually reproducing organisms, like ourselves, the types of changes we will discuss occur in all diploid cells, including the somatic cells of the body. But they can be inherited and passed on to the next generation only if they occur in a diploid precursor of a germ cell.

Random DNA Duplications Create Families of Related Genes

As just mentioned, eucaryotic genomes contain many families of related genes. For example, different forms of the protein actin, which is a part of the contractile apparatus in muscle cells, as well as of the cytoskeleton (discussed in Chapter 16), are expressed in different types of muscle and nonmuscle cells; different opsins, proteins that detect light of different wavelengths, are expressed in different retinal cells; different collagen genes are expressed in the various types of connective tissue; and so on.

One of the best-studied *gene families* is the β-globin gene family. The human genome has a total of five β-globin genes. These genes encode the β subunits of the various hemoglobins produced at different times during embryonic, fetal, and adult life. These hemoglobins have slightly different oxygen-binding and -releasing characteristics, and each is especially well suited to the stage in development in which it is expressed. The expression of each of these genes is independently regulated so that each is expressed only at the appropriate time in development, and only in cells that give rise to red blood cells.

By comparing the nucleotide sequences of the β-globin genes and their arrangement in different organisms, the evolutionary history of this gene family has been reconstructed in great detail. This globin gene

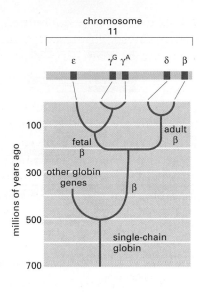

Figure 9–19 Evolution of the β-globin gene family in animals. An ancestral globin gene duplicated and gave rise to the β-globin family (shown here) as well as other globin genes (the α family). (A molecule of hemoglobin is formed from two α chains and two β chains.) The scheme shown was worked out from a comparison of β-globin genes from many different organisms. For example, the nucleotide sequences of the γ^G and γ^A genes are much more similar to each other than either of them is to the adult β gene.

family has been derived by repeated rounds of gene duplication, followed by mutation, starting from a single gene in an ancestor of the vertebrates (Figure 9–19). But how did such duplications occur? It is believed that they result from a rare recombination event between two homologous chromosomes that leads to the incorporation of an extra copy of the DNA containing the gene into one of the two chromosomes (Figure 9–20). Once a gene has been duplicated, homologous recombination can then generate extra copies or delete copies within the duplicated set by repetitions of the same mechanism.

Genes Encoding New Proteins Can Be Created by the Recombination of Exons

DNA duplication in evolution has also generated new genes by duplicating short DNA sequences that encode a protein domain. Many proteins in eucaryotes are composed of a series of repeating similar, discrete units of protein structure, or protein domains. Such proteins include the immunoglobulins (see Figure 5–25) and albumins, for example, as well as most fibrous proteins such as the collagens. These are encoded by genes that have evolved by repeated duplications of a single DNA sequence within the gene. In such genes, each individual protein domain is often encoded by a separate exon.

The evolution of new proteins in this way is thought to have been greatly facilitated by the organization of eucaryotic DNA coding

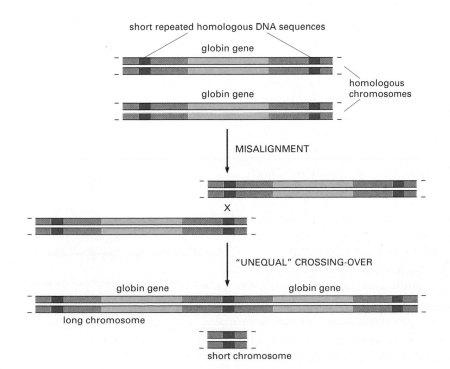

Figure 9–20 Gene duplication by unequal crossing-over. A pair of homologous chromosomes in a germ-line precursor cell undergoes a crossover (recombination) at a short sequence (*red*), which is repeated on each chromosome. In some cases, these sequences are believed to be transposons, which are present in many copies in the human genome. After the crossover, the long chromosome has two copies of the globin gene. Independent mutations in each of the genes, which could arise in subsequent generations, would lead to their diversification. (Since the short chromosome lacks the original globin gene, the individuals that inherit this chromosome would be expected to be lost from the population.)

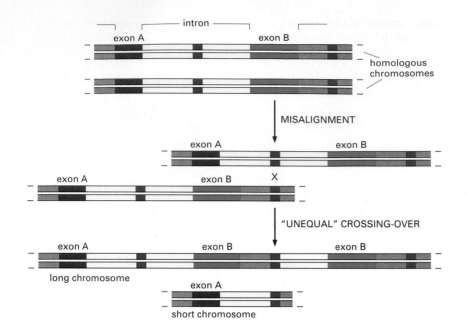

Figure 9–21 **Duplication of an exon by unequal crossing-over.** The general scheme is the same as that of Figure 9–20 except that an exon within a gene, rather than an entire gene, has been duplicated. The mRNA from the original gene contains two exons, A and B, whereas the long chromosome will produce an mRNA with three exons (A, B, and B). Since the duplicated exons are joined by an intron with its splicing sequences intact, the modified nucleotide sequence can be readily spliced after transcription to produce a functional mRNA.

sequences as a series of relatively short exons separated by long noncoding introns (see Figure 7–13). The duplications necessary to form a single gene with repeating sequences can occur by an unequal crossover anywhere in the long introns on either side of an exon encoding a protein domain (Figure 9–21). In contrast, without introns there would be very few sites in the original gene at which a recombinational exchange between homologous chromosomes could duplicate the domain without damaging it. Therefore, introns greatly increase the probability that DNA duplications will give rise to functional genes encoding functional proteins.

For the same reasons, the presence of introns greatly increases the probability that a chance recombination event can generate a functional hybrid gene by joining together two initially separate exons coding for quite different protein domains. The presumed results of such recombinations are seen in many present-day proteins, which are a patchwork of many different protein domains, many of which are found in other proteins as well (Figure 9–22). This evolutionary process is known as **exon shuffling.** It has been proposed that all the proteins encoded by the human genome (approximately 60,000) arose from the shuffling of only several thousand distinct exons, each encoding a protein domain of approximately 30–50 amino acids.

A Large Part of the DNA of Multicellular Eucaryotes Consists of Repeated, Noncoding Sequences

The genomes of multicellular eucaryotes contain not only introns but also a large amount of other seemingly nonessential DNA sequences. Most of these do not code for proteins and are not even transcribed.

In the human genome, about 70% of the total DNA is composed of so-called "unique" sequence DNA. This includes the DNA that encodes proteins and most types of RNA and also includes the intron DNA. The remaining 30%, however, contains nucleotide sequences that are repeated many times in the genome. Most of this repeated DNA does not encode proteins, and is of two types. About one-third is composed of

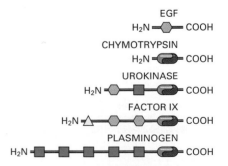

Figure 9–22 **Some results of exon shuffling.** Each type of symbol represents a different family of protein domain, and these have been joined together end-to-end, as shown, to create larger proteins, which are identified by name.

highly repeated short sequences that form serial arrays of DNA repeats known as *satellite DNA;* they have no known function and are mainly found clustered at the centromeres and at the ends of the chromosomes. The exact arrangement of satellite DNAs (e.g., number of different types and the number of repeats in each block) varies greatly between individuals of the same species.

The remainder of the repetitive sequence DNA is composed of more complex repeated sequences interspersed throughout the genome. In humans, most of this DNA derives from a few types of transposable DNA sequences similar to those found in bacterial genomes, which have multiplied to especially high numbers. Although all multicellular eucaryote genomes analyzed contain transposable elements, the DNA of humans and other primates is unusual in this respect. It contains a remarkably large number of copies of two of these transposable DNA sequences that seem to have nearly overrun our chromosomes, as we shall discuss next.

About 10% of the Human Genome Consists of Two Families of Transposable Sequences

Transposable DNA elements are probably present in all eucaryotic genomes, and many types of eucaryotic transposable elements can be distinguished. Some move from place to place within their host chromosomes using the cut-and-paste mechanism discussed earlier for bacterial transposons (see Figure 9–18A), and we shall not describe them further here. However, many eucaryotic transposons move not as DNA but via an RNA intermediate. These are called **retrotransposons** and are, as far as is known, unique to eucaryotes.

An example of a human retrotransposon is the *L1 transposable element,* or *LINE-1,* a highly repeated sequence that constitutes about 4% of the total mass of the human genome. The LINE-1 DNA sequence transposes by first being transcribed into RNA by cellular RNA polymerases. A DNA copy of this RNA is then made using the enzyme *reverse transcriptase,* an unusual DNA polymerase that can use RNA as a template. Reverse transcriptase is encoded by the LINE-1 element itself. The DNA copy can then reintegrate into another site in the genome (Figure 9–23). Even more abundant is the *Alu sequence,* which is unusually short (about 300 nucleotide pairs); it is present in about 500,000 copies in the haploid genome and constitutes about 5% of human DNA; thus it is present on average about once every 5000 nucleotide pairs. Only some of the Alu sequences in the genome can still be copied into RNA. This RNA

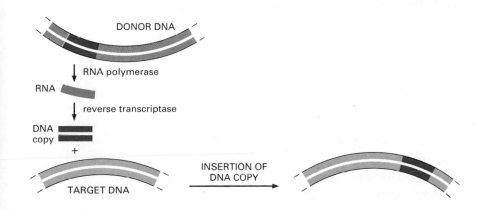

Figure 9–23 **The movement of a eucaryotic retrotransposon.** These transposable elements move by first duplicating themselves through an RNA intermediate. The DNA copy is then inserted into the target location. The target can be on the same or a different DNA molecule from the donor. Because of the similarity of this mechanism to the replication of a class of viruses known as retroviruses (see Figure 9–30), these transposable elements are called retrotransposons.

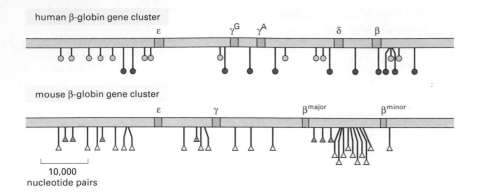

human β-globin gene cluster

mouse β-globin gene cluster

10,000
nucleotide pairs

Figure 9–24 Comparison of the β-globin gene cluster from the human and mouse genomes showing the location of transposable elements. This stretch of the human genome contains five functional β-globin-like genes (see Figure 9–19), while the comparable region from the mouse genome has only four. The positions of the human Alu sequences are indicated by *green circles*, and the human L1 elements by *red circles*. The mouse genome contains different transposable elements: the positions of B1 elements (which are related to the human Alu sequences) are indicated by *blue triangles*, and the positions of the mouse L1 elements (which are related to the human L1) are indicated by *yellow triangles*. Because the DNA sequences of the mouse and human transposable elements are distinct and because the positions of these transposable elements on the β-globin gene cluster are very different between human and mouse, it is believed that they accumulated in mammalian genomes relatively recently in evolutionary time. (Courtesy of Ross Hardison and Webb Miller.)

can, in rare circumstances, be copied back into DNA using reverse transcriptase encoded elsewhere in the gene.

Comparisons of the sequence and locations of the L1 and Alu-like elements in different mammals suggest that these sequences have multiplied to high copy numbers in primates relatively recently in evolutionary time (Figure 9–24). It is hard to imagine that these highly abundant sequences scattered throughout our genome have not had major effects on the expression of many neighboring genes. How many of our uniquely human qualities, for example, do we owe to these parasitic elements?

The Evolution of Genomes Has Been Accelerated by Transposable Elements

Transposable elements, especially those present in many copies in a genome, provide opportunities for genome rearrangements by serving as targets of homologous recombination. For example, the duplications that gave rise to the β-globin gene cluster are thought to have occurred by homologous recombination between Alu-like sequences (denoted in *red* in Figure 9–20). However, transposons also have more direct roles in the evolution of genomes. In addition to moving themselves, transposable elements occasionally move or rearrange neighboring DNA sequences of the host genome. For example, when two transposable elements that are recognized by the same transposase integrate into neighboring chromosomal sites, the DNA between them can become subject to transposition (see Figure 9–17 for a bacterial example). In eucaryotic genomes, this provides a particularly effective pathway for the movement of exons (another example of exon shuffling), generating new genes by creating novel arrangements of existing exons (Figure 9–25).

The insertion of a transposable element into the coding sequence of a gene or into its regulatory region is known to be a relatively frequent cause of spontaneous mutations in some organisms. Transposable elements can severely disrupt a coding sequence if they land directly in it, causing an insertion mutation that destroys the gene's capacity to encode a useful protein. For example, a number of the mutations in the Factor VIII gene that cause hemophilia result from insertion of transposable elements in the gene.

Most important for evolutionary pathways, an insertion of a transposable element in a regulatory region will often, by disrupting or adding short regulatory sequences of the sort described in Chapter 8 (see Figure 8–27), have a dramatic effect on gene expression (Figure

Question 9–4 Many transposable elements move within a genome by replicative mechanisms such as those shown in Figures 9–18A and 9–23 and so increase the number of copies in the genome at each transposition. Although individual transposition events are rare, many transposable elements are found in multiple copies in genomes. What do you suppose keeps the transposable elements from completely overrunning their hosts' genomes?

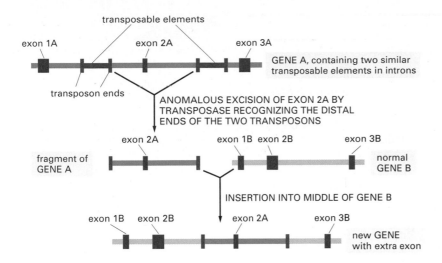

exon 1A   transposable elements   exon 2A   exon 3A

GENE A, containing two similar transposable elements in introns

transposon ends

ANOMALOUS EXCISION OF EXON 2A BY TRANSPOSASE RECOGNIZING THE DISTAL ENDS OF THE TWO TRANSPOSONS

exon 2A   exon 1B   exon 2B   exon 3B

fragment of GENE A   normal GENE B

INSERTION INTO MIDDLE OF GENE B

exon 1B   exon 2B   exon 2A   exon 3B

new GENE with extra exon

Figure 9–25 An example of exon shuffling that can be caused by transposable elements. When two transposable elements of the same type (*red* DNA) happen to insert near each other in a chromosome, the transposition mechanism may occasionally use the ends of two different elements (instead of the two ends of the same element) and thereby move the chromosomal DNA between them to a new chromosomal site. Since introns are very large relative to exons, the illustrated insertion of a new exon into a pre-existing intron is a frequent outcome of such a transposition event.

9–26). Transposable elements thus can be a significant source of mutation, and their presence makes the DNA sequences in chromosomes less stable than previously thought. They are certainly a source of genetic variation in the short term, and it is likely that they have been responsible for many important changes in genomes that have occurred over an evolutionary timescale.

As we discuss next, mobile genetic elements such as transposable elements and plasmids (which are also found in some eucaryotic cells) are thought to be the ancestors of viruses, which can thus be thought of as fully mobile genetic elements that can move independently between cells.

Viruses Are Fully Mobile Genetic Elements That Can Escape from Cells

Viruses were first noticed as disease-causing agents that, by virtue of their tiny size, passed through ultrafine filters that can hold back even the smallest cell. We now know that viruses are essentially genes enclosed by a protective coat. As we saw earlier in this chapter (see Figure 9–14), these genes must enter a cell and utilize the cell's machinery to express their genes as proteins and to replicate their chromosomes, so as to package themselves into newly made protective coats. Virus reproduction per se is often lethal to the cells in which it occurs; in many cases the infected cell breaks open (lyses) and thereby releases the

(A)

(B)

Figure 9–26 A change in the body plan of a fruit fly caused by a mutation. The fruit fly (*Drosophila melanogaster*) on the left is normal. In the fly on the right, the antennae have been transformed into legs. Although this particular change is not advantageous to the fly, it illustrates how a DNA rearrangement caused by a transposable element can produce a dramatic change in the organism. (A, courtesy of E.B. Lewis; B, courtesy of Matthew Scott.)

Table 9–1   Viruses That Cause Human Disease

Virus	Genome Type	Disease
Herpes simplex virus	double-stranded DNA	recurrent cold sores
Epstein-Barr virus (EBV)	double-stranded DNA	infectious mononucleosis
Varicella-zoster virus	double-stranded DNA	chicken pox and shingles
Smallpox virus	double-stranded DNA	smallpox
Hepatitis B virus	part single-, part double-stranded DNA	serum hepatitis
Human immuno-deficiency virus (HIV)	single-stranded RNA	acquired immunodeficiency syndrome (AIDS)
Influenza virus type A	single-stranded RNA	respiratory disease (flu)
Poliovirus	single-stranded RNA	infantile paralysis
Rhinovirus	single-stranded RNA	common cold
Hepatitis A virus	single-stranded RNA	infectious hepatitis
Hepatitis C virus	single-stranded RNA	non-A, non-B type hepatitis
Yellow fever virus	single-stranded RNA	yellow fever
Rabies virus	single-stranded RNA	rabies
Mumps virus	single-stranded RNA	mumps
Measles virus	single-stranded RNA	measles

progeny viruses and allows them access to nearby cells. Many of the medical symptoms of viral infection reflect the lytic effect of the virus. The cold sores formed by herpes simplex virus and the blisters caused by the chicken pox virus, for example, both reflect the localized killing of skin cells. Although the first viruses that were discovered attack mammalian cells, it is now recognized that many types of viruses exist. Some of these infect plant cells, and as we saw earlier in this chapter, others use bacterial cells as their hosts.

Viral genomes can be made of DNA or RNA and can be single-stranded or double-stranded (Table 9–1 and Figure 9–27). The amount of DNA or RNA that can be packaged inside a protein shell is limited, and is too small to encode the many different enzymes and other proteins that are required to replicate even the simplest virus. Viruses are therefore parasites that can reproduce themselves only inside a living cell, where they are able to hijack the cell's own biochemical machinery. Viral

Figure 9–27 Schematic drawings of several types of viral genomes. The smallest viruses contain only a few genes and can have an RNA or a DNA genome; the largest viruses contain hundreds of genes and have a double-stranded DNA genome. Some examples of these types of viruses are as follows: *single-stranded RNA*—tobacco mosaic virus, bacteriophage R17, poliovirus; *double-stranded RNA*—reovirus; *single-stranded DNA*—parvovirus; *single-stranded circular DNA*—M13 and φX174 bacteriophages; *double stranded circular DNA*—SV40 and polyoma-viruses; *double-stranded DNA*—T4 bacteriophage, herpesvirus; *double-stranded DNA with covalently linked terminal protein*—adenovirus; *double-stranded DNA with covalently sealed ends*—poxvirus. The peculiar ends (as well as the circular forms) present in some viral genomes overcome the difficulty of replicating the last few nucleotides at the end of a DNA chain (see Figure 8–6).

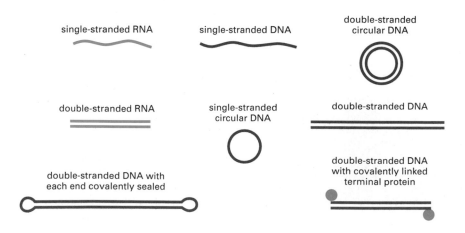

single-stranded RNA

single-stranded DNA

double-stranded circular DNA

double-stranded RNA

single-stranded circular DNA

double-stranded DNA

double-stranded DNA with each end covalently sealed

double-stranded DNA with covalently linked terminal protein

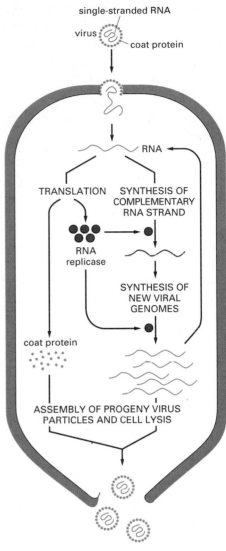

Figure 9–28 **Life cycle of a single-stranded RNA virus.** In the simplified case shown, the viral genome can be translated by the host ribosomes to produce viral proteins. *Red circles:* RNA-dependent RNA polymerase (RNA replicase); *green circles:* coat proteins. The complementary RNA strand is used as a template by the replicase to produce the new viral genomes. For comparison with the life cycle of a simplified DNA virus, see Figure 9–14.

genomes typically encode the viral coat proteins as well as proteins that attract host enzymes to replicate their genome at the viral replication origin. Viruses that have RNA genomes must, in addition, encode enzymes—called RNA replicases—that are needed to replicate their genomes (Figure 9–28).

The simplest viruses consist of a protein coat made up primarily of many copies of a single polypeptide chain surrounding a small genome composed of as few as three genes. More complex viruses have larger genomes of up to several hundred genes, surrounded by an elaborate shell composed of many different proteins (Figure 9–29).

Even the largest viruses depend heavily on their host cells for biosynthesis; no known virus makes its own ribosomes or generates the ATP needed for nucleic acid replication, for example. Clearly, cells must have evolved before viruses. The ancestors of the first viruses were probably small nucleic acid fragments that developed the ability to multiply independently of the chromosomes of their host cells. The nearest relatives of present-day viruses are probably plasmids and transposons. Unlike viruses, however, these elements do not make a protein coat and so cannot move independently from cell to cell.

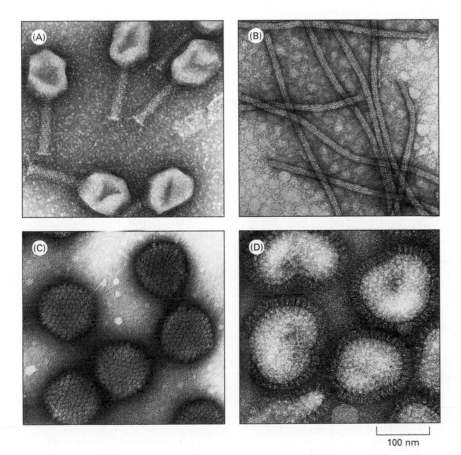

100 nm

Figure 9–29 **The coats of viruses.** These electron micrographs of virus particles are all shown at the same scale. (A) *T4,* a large DNA-containing virus that infects *E. coli* cells. The DNA is stored in the bacteriophage head and injected into the bacterium through the cylindrical tail. (B) *Potato virus X,* a plant virus that contains an RNA genome. (C) *Adenovirus,* a DNA-containing virus that can infect human cells. (D) *Influenza virus,* a large RNA-containing animal virus whose protein capsid is further enclosed in a lipid-bilayer-based envelope. The spikes protruding from the envelope are viral proteins embedded in the membrane bilayer (see Figure 9–30). (A, courtesy of James Paulson; B, courtesy of Graham Hills; C, courtesy of Mei Lie Wong; D, courtesy of R.C. Williams and H.W. Fisher.)

Sources of Genetic Change in Eucaryotic Genomes

If we accept the proposal outlined in Chapter 7 that the first cells would have had RNA-based information storage, the first virus may have formed from an RNA plasmid that had acquired a gene coding for a protein that could assemble into a protein coat. Protected by this coat, the viral RNA could now leave the cell and seek a new host.

Retroviruses Reverse the Normal Flow of Genetic Information

Although there are many similarities between bacterial and eucaryotic viruses, one important type of virus—the **retrovirus**—is found only in eucaryotic cells. In many respects, retroviruses resemble the retrotransposons we discussed earlier. A key feature found in both these genetic elements is a step where DNA is synthesized using RNA as a template (the term "retro" refers to this backward flow of the central dogma). The enzyme that carries out this step is *reverse transcriptase;* the retroviral genome (which is single-stranded RNA) encodes this enzyme, and a few molecules of the enzyme are packaged along with the RNA genome in each individual virus.

The life cycle of a retrovirus is shown in Figure 9–30. When the single-stranded RNA genome of the retrovirus enters a cell, the reverse transcriptase brought in with it makes a complementary DNA strand to form a DNA/RNA hybrid double helix. The RNA strand is removed, and the reverse transcriptase (which can use either DNA or RNA as a template) now synthesizes a complementary strand to produce a DNA double helix. This DNA is then integrated into a randomly selected site in the host genome by a virally encoded integrase enzyme. In this state, which resembles that of the lambda virus integrated into the bacterial genome (see Figure 9–15), the virus is latent: each time the host cell divides, it passes on a copy of the integrated viral genome to its progeny cells.

The next step in the replication of a retrovirus—which can take place long after its integration into the host genome—is the transcription of the integrated viral DNA by the host cell RNA polymerase, which can produce large numbers of single-stranded RNAs identical to the original infecting genome. These RNA molecules are then translated by the host cell machinery to produce the protein shell, the envelope proteins, and reverse transcriptase—all of which are assembled with the RNA genome into new virus particles.

The human immunodeficiency virus (HIV), which is the cause of AIDS, is a retrovirus. As with other retroviruses, the HIV genome can persist in the latent state as a DNA provirus embedded in the chromosomes of an infected cell. This ability of the virus to hide within host cells complicates any attempt to treat the infection with antiviral drugs. But because the HIV reverse transcriptase is not used by cells for any purpose of their own, one of the prime targets of drug development against AIDS is the viral reverse transcriptase (Figure 9–31).

Since some transposable elements that move through the genome via an RNA intermediate replicate in a manner similar to that of retroviral genomes (see Figure 9–23), it is thought that retroviruses are derived from such a retrotransposon that long ago acquired additional genes encoding coat proteins and other proteins required to make a virus particle. The RNA stage of its replicative cycle could then be packaged into a virus particle and leave the cell.

Question 9–5 The genomes of some single-stranded RNA viruses can serve as messenger RNA, and upon entering a cell, can be translated directly into protein by the host ribosomes (see Figure 9–28). The genomes of other RNA viruses, however, are single-stranded RNAs that are complementary to the mRNAs from which the viral proteins must be synthesized. These are known as "minus"-strand viruses. What modifications in Figure 9–28 must you make to explain the life cycles of this class of RNA viruses?

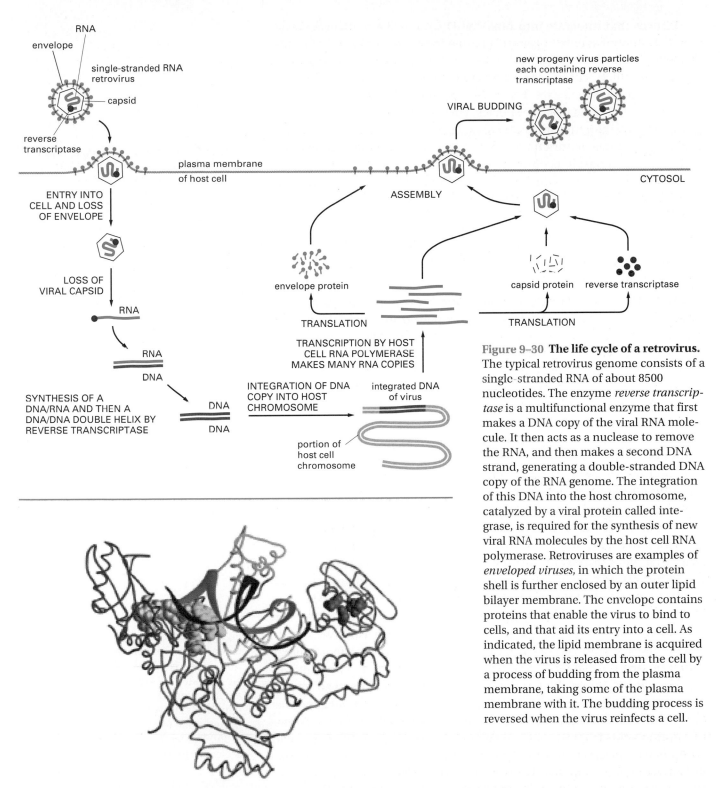

Figure 9–30 The life cycle of a retrovirus. The typical retrovirus genome consists of a single-stranded RNA of about 8500 nucleotides. The enzyme *reverse transcriptase* is a multifunctional enzyme that first makes a DNA copy of the viral RNA molecule. It then acts as a nuclease to remove the RNA, and then makes a second DNA strand, generating a double-stranded DNA copy of the RNA genome. The integration of this DNA into the host chromosome, catalyzed by a viral protein called integrase, is required for the synthesis of new viral RNA molecules by the host cell RNA polymerase. Retroviruses are examples of *enveloped viruses,* in which the protein shell is further enclosed by an outer lipid bilayer membrane. The envelope contains proteins that enable the virus to bind to cells, and that aid its entry into a cell. As indicated, the lipid membrane is acquired when the virus is released from the cell by a process of budding from the plasma membrane, taking some of the plasma membrane with it. The budding process is reversed when the virus reinfects a cell.

Figure 9–31 Structure of the HIV reverse transcriptase showing sites where inhibitory drugs bind. The enzyme has two subunits, one shown in *gray* and the other in several colors to indicate different structural domains within the subunit. The DNA that the enzyme is elongating is shown in *purple,* and the template strand (which can be either RNA or DNA) is shown in *dark brown*. The binding site for incoming deoxyribonucleoside triphosphates is shown in *gold;* this is the site where one class of inhibitors (including AZT and ddC) bind the enzyme, are added to the growing chain, and cause termination of the chain (see Figure 10–5). The binding site for a different class of reverse transcriptase inhibitors (which includes the drugs α-APA and nevirapine) is shown in *turquoise.* This class of inhibitors may act by subtly changing the conformation of the reverse transcriptase, rendering it unable to synthesize DNA efficiently. HIV reverse transcriptase also contains a domain that degrades RNA from the RNA/DNA duplex that is formed during the life cycle of the virus (see Figure 9–30). The active site for this enzymatic activity, also a target for drug development, is shown in *red*. (Courtesy of Edward Arnold.)

Viruses that integrate into host cell DNA are, like transposable elements, potential agents of genetic change for all the reasons mentioned earlier (see pp. 296–297). Like bacteriophage lambda, retroviruses can also acquire pieces of host cell DNA sequence and carry them (as RNA copies) to a new host (see Figure 9–16). Since many viruses are not restricted to infecting just one species, viruses also have, in principle, the ability to transfer genes from one species to another. There is evidence from the genomes of present-day organisms that this type of gene transfer has played a part in evolution.

In the following section we will see that the ability of retroviruses to acquire host genes provided crucial clues for scientists who were studying the causes of cancer.

Retroviruses That Have Picked Up Host Genes Can Make Cells Cancerous

The retroviruses are of interest not only for their unusual mode of replication, but for the central role they have played in proving that most cancers are caused by mutations in a small set of cellular genes (discussed in Chapter 19). Very few human cancers are caused by retrovirus infection, but these viruses are a prominent cause of cancers in some other animals.

Viruses that cause cancer are known as **tumor viruses;** the first was discovered more than 80 years ago in chickens, where it causes connective-tissue tumors known as *sarcomas*. The virus was called Rous sarcoma virus (after Peyton Rous, who discovered it), and it is a retrovirus. How can such a virus cause a cancer? When techniques for analyzing nucleic acid sequences became available, the Rous sarcoma virus turned out to have picked up a gene from its chicken host (Figure 9–32). This gene—called *src*—is unnecessary baggage from the point of view of the virus, but it has profound consequences for cells that are infected by the virus. It alone is responsible for the virus's ability to cause cancer.

The viral *src* gene (pronounced "sark") is an example of an **oncogene**—a gene that causes a normal cell to become cancerous. The normal *src* gene in the chicken genome encodes a protein kinase (see Chapter 5) that is involved in the control of cell division. However, the *src* gene carried by the virus is not quite identical to the normal cellular gene (the **proto-oncogene**), and the difference gives rise to its ability to cause a cancer. When the ancestral Rous sarcoma virus picked up the gene from the chicken genome, a small portion of the normal gene was altered, and in consequence the protein kinase it encodes differs by several amino acids from the normal enzyme in the chicken cell. This small difference is enough to release the enzyme from the normal cellular controls, making it hyperactive. Consequently, cells that express the viral *src* gene as a result of infection divide uncontrollably and eventually form a tumor.

Figure 9–32 The genome of the Rous sarcoma virus, a cancer-causing retrovirus that infects chickens. The genome is single-stranded RNA, and the genes it carries are indicated. The *gag* gene encodes a protein that is cleaved into several smaller proteins that form the capsid (see Figure 9–30). The *pol* gene encodes a protein that is cleaved to produce both the reverse transcriptase as well the integrase involved in integrating the viral genome (as double-stranded DNA) into the host genome. The *env* gene encodes the envelope proteins. *Src* is a modified cellular gene that the virus accidentally picked up from a host cell it once infected. The transport of host genes by retroviruses resembles in principle the transduction of bacterial genes by bacteriophages (see Figure 9–16).

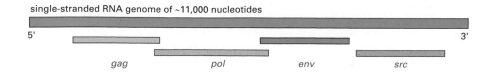

single-stranded RNA genome of ~11,000 nucleotides

5'   3'

gag   *pol*   *env*   *src*

Table 9–2   Some Oncogenes Originally Identified Through Their Presence in Retroviruses That Cause Cells to Become Cancerous

Oncogene	Proto-oncogene Function	Source of Virus	Virus-induced Tumor
abl	protein kinase (tyrosine)	mouse cat	pre-B-cell leukemia sarcoma
erb-B	epidermal growth factor (EGF) receptor	chicken	erythroleukemia; fibrosarcoma
fes	protein kinase (tyrosine)	cat/chicken	sarcoma
fms	macrophage colony-stimulating factor (M-CSF) receptor	cat	sarcoma
fos, jun	associate to form a gene regulatory protein	mouse chicken	osteosarcoma fibrosarcoma
myc	gene regulatory protein	chicken	sarcoma; myelocytoma carcinoma
raf	protein kinase (serine/threonine)	chicken/mouse	sarcoma
H-ras	GTP-binding protein	rat	sarcoma; erythroleukemia
rel	gene regulatory protein	turkey	rcticuloendotheliosis
sis	platelet-derived growth factor, B chain	monkey	sarcoma
src	protein kinase (tyrosine)	chicken	sarcoma

The *fos, jun, myc,* and *rel* proto-oncogenes encode proteins normally located in the nucleus, where they regulate gene transcription. The *erb-B* and *fms* proto-oncogenes encode proteins located on the outer membrane of cells that receive signals from outside the cell and transmit that information to the inside of the cell (see Chapter 15). *Abl, fes, ras, raf,* and *src* oncogenes encode proteins located in the cytoplasm that are components of signaling pathways (Chapter 15). The *sis* proto-oncogene encodes a secreted protein that signals other cells to divide.

Other cancer-causing retroviruses have picked up different cellular genes (Table 9–2). Each of these viral oncogenes has led to the discovery of the corresponding normal proto-oncogene, thus enabling many genes involved in the control of cell division to be identified. Once the crucial proto-oncogenes had been identified in human cells, it was found that changes in these genes as a result of mutation contributed to many types of cancer. As discussed in Chapter 19, human cancers tend to arise only after the accumulation of several mutations in a cell (see also Figure 6–24). We know now that many of these mutations damage proto-oncogenes, converting them to oncogenes.

Most proto-oncogenes code for proteins that regulate the social behavior of cells in the body (see Chapter 19). For example, differentiated cells often receive signals from other cells ordering them to stop dividing. However, mutations in some proto-oncogenes cause the receiving cell to become deaf to these instructions, helping to cause its uncontrolled proliferation. Many of our insights into human cancer have come from the study of oncogenes that were first discovered in retroviruses.

In this chapter so far, we have seen that changes in the genomes of cells provide a rich source of genetic diversity upon which natural selection can act. Mutations accumulate because errors in replication occur, transposable elements move around to different positions in the genome, and viruses move into and out of cells—sometimes moving bits of one genome to another cell. In the final section, we discuss how sexual reproduction has provided additional genetic diversity to life on our planet.

Sexual Reproduction and the Reassortment of Genes

Many organisms can reproduce without sex: bacteria and other unicellular organisms do so by simple cell division. Many plants also reproduce asexually, forming offshoots that later detach from the parent to make new independent plants. Even in the animal kingdom, some worms can be split into two halves, each of which regenerates its missing half. But while such asexual reproduction is simple and direct, it usually gives rise to offspring that are genetically identical to the parent organism. Sexual reproduction, on the other hand, involves the mixing of genomes from two individuals to produce offspring that are genetically distinct from one another and from both their parents. This mode of reproduction apparently has great advantages, as the vast majority of plants and animals have adopted it.

Sexual Reproduction Gives a Competitive Advantage to Organisms in an Unpredictably Variable Environment

The machinery of sexual reproduction is elaborate, and the resources spent on it are large. What benefits does it bring, and why did it evolve? Through the mixing of genes, sexually reproducing individuals beget unpredictably dissimilar offspring, whose patchwork genomes are at least as likely to represent a change for the worse as a change for the better. Why, then, should sexual individuals have a competitive advantage over individuals that breed true, by an asexual process? This problem continues to perplex population geneticists, but the general conclusion seems to be that the reshuffling of the genes in sexual reproduction helps a species survive in an unpredictably variable environment. If two parents produce many offspring with a wide variety of gene combinations, there is a better chance that at least one of them will have the combination of features necessary for survival.

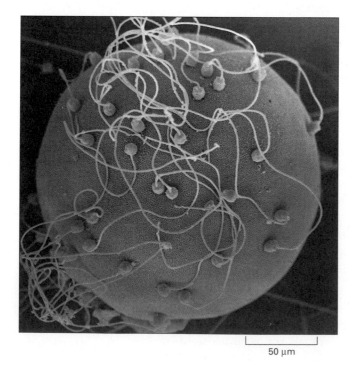

Figure 9–33 **Scanning electron micrograph of a clam egg with sperm bound to its surface.** (Courtesy of David Epel.)

50 μm

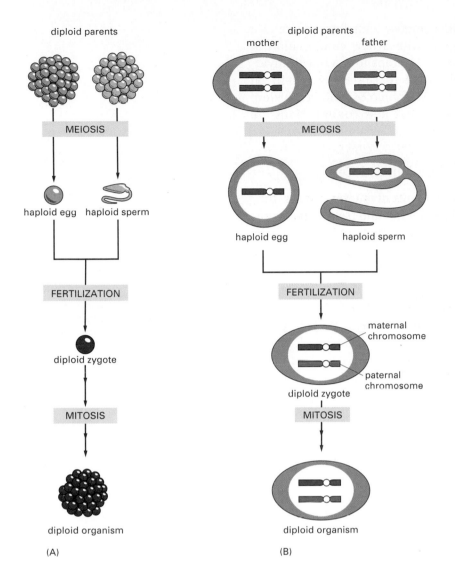

diploid parents

MEIOSIS

haploid egg   haploid sperm

FERTILIZATION

diploid zygote

MITOSIS

diploid organism

(A)

diploid parents
mother   father

MEIOSIS

haploid egg   haploid sperm

FERTILIZATION

maternal
chromosome

diploid zygote

paternal
chromosome

MITOSIS

diploid organism

(B)

Figure 9–34 The haploid-diploid cycle of sexual reproduction. (A) Cells in higher eucaryotic organisms proliferate in the diploid phase to form a multicellular organism; only the gametes (the egg and the sperm) are haploid. (B) Sexual reproduction parcels out chromosomes in new combinations. Humans, for example, have 23 chromosomes per gamete. For simplicity, only one chromosome is shown for each gamete.

Sexual Reproduction Involves Both Diploid and Haploid Cells

Sexual reproduction occurs in diploid organisms, in which each cell contains two sets of chromosomes, one inherited from each parent. Each diploid cell, therefore, carries two copies of each gene (with the exception of the genes carried on the sex chromosomes of males, which are present in only one copy), and as discussed earlier in the chapter, these two gene copies need not be identical. Moreover, for each gene, there are numerous varieties present in the gene pool of a species, and sexual reproduction ensures that new combinations of genes are continually tried out.

Unlike the other cells in a diploid organism, the specialized cells that actually carry out sexual reproduction—the **germ cells,** or *gametes*—are haploid. They each contain only one set of chromosomes. Typically, two types of gametes are produced. In animals one is large and nonmotile, and is referred to as the **egg;** the other is small and motile and is referred to as the **sperm** (Figure 9–33). These haploid germ cells are generated when a diploid cell undergoes the highly specialized process of cell division called *meiosis* (Figure 9–34). During meiosis the chromosomes of the double chromosome set are shared out, in fresh combinations, into single chromosome sets. The haploid gametes then fuse to make a diploid cell (the fertilized egg, or *zygote*) with a new combination of chromosomes. The zygote thus produced develops into a new individual with a set of chromosomes distinct from that of either parent. In this way, through cycles of diploidy, meiosis, haploidy, and cell fusion, old combinations of genes are broken up and new combinations are created.

Meiosis Generates Haploid Cells from Diploid Cells

As discussed in Chapter 17, when diploid cells divide by *mitosis*, they precisely duplicate their two sets of chromosomes and pass on two complete sets to each daughter cell. The process through which germ cells

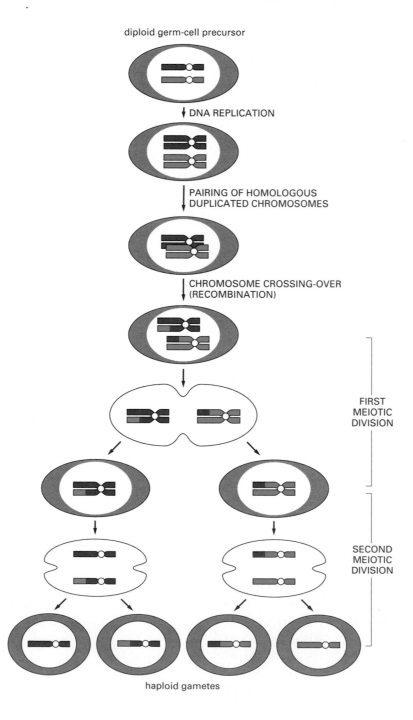

diploid germ-cell precursor

DNA REPLICATION

PAIRING OF HOMOLOGOUS
DUPLICATED CHROMOSOMES

CHROMOSOME CROSSING-OVER
(RECOMBINATION)

FIRST
MEIOTIC
DIVISION

SECOND
MEIOTIC
DIVISION

haploid gametes

Figure 9–35 Meiosis. Only one pair of homologous chromosomes is shown for simplicity. Following DNA replication in the original diploid cell, two cell divisions are required to produce the haploid gametes. Each diploid cell that enters meiosis (in animals only a specialized set of cells has this ability) therefore produces four haploid cells.

are formed is different, since only a single set of chromosomes is parceled out from the diploid starting cell to each germ cell. This process, called **meiosis,** involves a duplication of each chromosome followed by two successive cell divisions. The movement of chromosomes during meiosis resembles an elaborate dance, which we cover in detail at the end of Chapter 17. Here we only outline the process.

The specialized diploid cells in the ovaries or testis are the starting point for meiosis. Each contains two copies of each chromosome, one inherited from the father (the *paternal homologue*) and one from the mother (the *maternal homologue*). As outlined in Figure 9–35, the first step in meiosis, as in mitosis, is the duplication of the chromosomes of this diploid cell: their DNA is replicated, and the two daughter DNA molecules are packaged by proteins to form new chromosomes (see Chapter 8). As in mitosis, the new chromosomes remain attached to one another at first, like Siamese twins.

In the next phase of the process, each duplicated paternal chromosome pairs with its duplicated maternal homologue, an event unique to meiosis. Since the nucleotide sequences of the maternal and paternal homologues are identical in many places along their length, and they are held in close proximity, they can undergo homologous recombination while they are paired (see Figure 9–9). In the context of meiosis, such recombination events are called *chromosome crossovers,* or *chiasmata,* and they produce chromosomes that are hybrids of the maternal and paternal chromosomes (see Figure 9–35). Recombination during meiosis is a major source of genetic variation in sexually reproducing species, inasmuch as there is at least one crossover event per chromosome.

Two successive cell divisions now parcel out a complete set of single chromosomes to each gamete, as shown in Figure 9–35. Because of crossing-over, the chromosomes distributed to the gametes will include hybrids that contain DNA segments derived from both maternal and paternal chromosomes.

The sorting of chromosomes that takes place during meiosis is a remarkable feat of cellular bookkeeping: in humans, each meiosis requires that the starting cell keep track of 92 individual chromosomes (23 pairs, each of which has duplicated), handing out one complete set to each gamete. Not surprisingly, mistakes do occur, although rarely, in the distribution of chromosomes during meiosis. For example, Down syndrome, a human disease characterized by severe mental retardation, results from the distribution of an extra copy of Chromosome 21 to a gamete. If this gamete contributes to a fertilized egg, the resulting embryo will have three copies of Chromosome 21 and two copies of each of the remaining chromosomes. This chromosome imbalance interferes with the proper development of the embryo.

Meiosis Generates Enormous Genetic Variation

Identical twins, which develop from a single zygote, are genetically identical. Otherwise no two siblings are genetically the same. This is because, even before fertilization takes place, meiosis has produced two kinds of randomizing genetic reassortments.

First, the maternal and paternal chromosomes are shuffled and dealt out among the gametes during meiosis. Although the chromosomes are carefully parceled out so that each gamete receives one and

three pairs of homologous
chromosomes

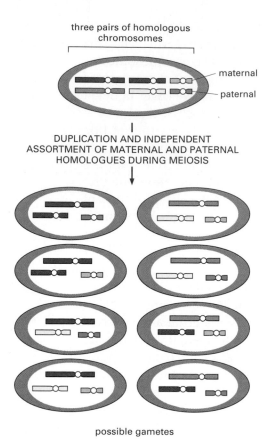

DUPLICATION AND INDEPENDENT
ASSORTMENT OF MATERNAL AND PATERNAL
HOMOLOGUES DURING MEIOSIS

possible gametes

Figure 9–36 Reassortment of chromosomes that occurs during meiosis. The independent assortment of the maternal and paternal homologues produces 2^n different haploid gametes for an organism with n chromosomes. Here $n = 3$, and there are 2^3, or 8, different possible gametes. For simplicity, chromosome crossing-over is not shown.

only one copy of each chromosome, each gamete receives a random mixture of paternal and maternal chromosomes (Figure 9–36). Thanks to this type of reassortment alone, an individual could in principle produce 2^n genetically different gametes, where n is the haploid number of chromosomes. Each human, for example, can in theory produce $2^{23} = 8.4 \times 10^6$ different gametes simply from the random assortment of chromosomes that takes place in meiosis.

The actual number of different gametes each person could produce is very much greater than this, because of the genetic reassortment caused by the homologous recombination that also takes place during meiosis. On average, between one and two crossovers occur on each pair of human chromosomes per meiosis. This process puts maternal and paternal genes that are initially on separate chromosomes onto the same chromosome, as illustrated in Figure 9–35. Since recombination occurs at more or less random sites along the length of a chromosome, each meiosis will produce gametes of different genetic makeup.

The reassortment of chromosomes in meiosis, together with the recombination of genes that arises from crossing-over, provides a nearly limitless source of genetic variation in the gametes produced by a single individual. Considering that each human is formed from the fusion of two such gametes, one from the father and one from the mother, the richness of human variation that we see around us is not at all surprising.

Question 9–7 Ignoring the effects of chromosome crossovers, an individual human can in principle produce $2^{23} = 8.4 \times 10^6$ genetically different gametes. How many of these possibilities can be realized in the average life of (a) a female? (b) a male?

Essential Concepts

- Their simplicity and rapid generation times have made bacteria—especially *E. coli*—among the best-understood organisms at the cellular level.

- Rare, spontaneous mistakes in copying DNA provide a population of bacteria with a constant source of genetic variants. When external conditions change, a variant that is better able to survive under the new conditions will proliferate and take over the population.

- Bacterial cells have three mechanisms through which they can acquire genes from other bacterial cells in the population: mating, transformation, and (with the help of viruses) transduction. This intermixing of the gene pool of the population is an important source of genetic variation.

- Homologous recombination is the process by which two double-stranded DNA molecules of similar nucleotide sequence can cross over to create DNA molecules of novel sequence.

- Transposable elements (transposons) are DNA sequences that can move from place to place in the genomes of their hosts. This movement creates change in the host genomes and provides another source of genetic variation.

- Many of the same mechanisms that provide genetic variation in bacteria also operate in eucaryotic genomes. However, eucaryotic genomes have a few additional features that have facilitated changes in their genomes during evolutionary time.

- In eucaryotes, unequal crossing-over can create gene duplications, which are thought to have led to the creation of families of related genes that are present in eucaryotic genomes. Unequal crossing-over is also one of the mechanisms believed to be responsible for duplicating exons to create proteins constructed of repeated short domains.

- Thirty percent of the human genome consists of DNA that is repeated many times in the genome. Approximately one-third of this repeated DNA (~10% of the total genome) consists of two trans-posable elements that have multiplied to especially high copy numbers in the genome.

- One type of transposable element found in eucaryotic genomes moves through an RNA intermediate and is called a retrotransposon.

- Transposable elements are thought to be responsible for many of the evolutionary changes that have led to present-day eucaryotic genomes. Transposition provides one of the mechanisms that creates "exon shuffling"—the joining together of different exons during evolution to produce novel proteins of "patchwork" construction.

- Viruses are little more than genes packaged in protective coats. They require host cells in order to reproduce themselves.

- Some viruses have RNA instead of DNA as their genomes. One group of RNA viruses—the retroviruses—must copy their RNA genomes into DNA in order to replicate.

- Retroviruses that carry oncogenes can cause the cells they infect to become cancerous.

- Sexual reproduction involves the cyclic alternation of diploid and haploid states: diploid cells divide by meiosis to form haploid gametes, and the haploid gametes from two individuals fuse at fertilization to form a new diploid cell.

- During meiosis, the maternal and paternal chromosomes of a diploid cell are parceled out to gametes so that each gamete receives one copy of each chromosome. Because the assortment of the two members of each chromosome pair occurs at random, many genetically different gametes can be produced from a single individual.

- Crossing-over enhances the genetic reassortment that occurs during meiosis by exchanging genes between homologous chromosomes.

- Sexual reproduction has probably been favored by evolution because the random mixing of genetic information improves the chances of producing at least some offspring that will survive in an unpredictably variable environment.

Key Terms

bacterial mating	homologous recombination	proto-oncogene	transduction
egg	integration	retrotransposon	transformation
exon shuffling	meiosis	retrovirus	transposable element
germ cell	oncogene	site-specific recombination	transposition
	plasmid	sperm	tumor virus
			virus

Questions

Question 9–8   Which of the following statements are correct? Explain your answers.

A. The egg and sperm cells of animals contain haploid genomes.

B. During meiosis chromosomes are allocated so that each daughter cell obtains one and only one copy of each of the different chromosomes.

C. Mutations that arise during meiosis are not transmitted to the next generation.

D. Cells can take up DNA molecules from their surroundings and, in some cases, express the genes encoded on them.

E. Viruses that integrate into the host chromosomes are replicated along with the host DNA. Each cell can therefore give rise to only a single mature virus particle.

F. Retroviruses contain reverse transcriptase inside the viral coat.

G. During crossing-over in meiosis, the regions that encode the same genes in the two chromosomes have to pair with each other.

H. Viral coats can be made of either DNA or RNA.

Question 9–9   Why is it advantageous for an organism to be diploid? Why is it a disadvantage to a geneticist who might wish to study the organism? Why is it advantageous for a diploid organism to produce haploid gametes in order to reproduce itself?

Question 9–10   It is striking that nearly all complex, present-day organisms have evolved through generations of sexual, rather than asexual, reproduction. Organisms that reproduce solely through asexual means, although plentiful on earth, seem to have remained relatively simple. Almost all of them are unicellular, for example. Why do you suppose this is?

Question 9–11   Discuss the following statement: "Viruses are parasites. They are harmful to the host organism, and therefore place it at an evolutionary disadvantage."

Question 9–12   Consider the following experiment: One million bacteria that have a damaging mutation in a gene required for the synthesis of the amino acid histidine are plated on agar that contains all necessary nutrients except histidine. As expected, no bacterial growth is observed after a day's incubation at 37°C. Now you place a tiny drop containing a sterilized extract prepared from moldy peanuts into the middle of the plate and return it to the incubator. To your sur-

prise, two days later you find a ring of colonies surrounding the drop (Figure Q9–12). If you take some bacteria from any of these growing colonies and put them onto a new plate (identical to the one described above but without the moldy peanut extract), they all grow into colonies. Can you propose an explanation for these results? Consider the following:

A. Why did some bacteria grow?

B. Why did those that grew form a ring around the drop, some distance away from it?

C. How many of the bacteria that had originally been plated in the ring region grew into colonies?

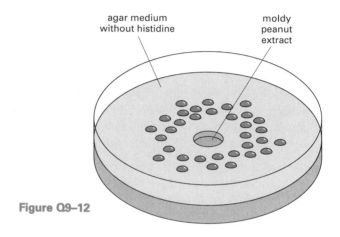

agar medium without histidine

moldy peanut extract

Figure Q9–12

Question 9–13   What are some of the differences between a plasmid and a virus?

Question 9–14   In this chapter it is argued that genetic variability is beneficial for a species because it enhances its ability to adapt to changing conditions. Why then does a cell go to great lengths to assure the fidelity of DNA replication?

Question 9–15   What would you expect to happen to mRNAs if animal cells contained high levels of reverse transcriptase in their cytoplasm?

Question 9–16   Discuss the following statement: "Viruses exist in the twilight zone of life: outside cells they are simply dead assemblies of molecules; inside cells, however, they are alive."

Question 9–17   Transposable DNA sequences, such as Alu sequences, are found in many copies in human DNA. In what ways could the presence of an Alu sequence affect a nearby gene?

Question 9–18   Assume that an mRNA in an animal cell is reverse transcribed into DNA (a very rare event that may be facilitated during a virus infection in which viral reverse transcriptase is produced). Assume the resulting double-stranded DNA becomes integrated at a position in a chromosome different from that of the original gene. How will this DNA sequence (referred to as a "pseudogene") differ from the original gene from which the mRNA was produced? Will the pseudogene be transcribed?

Question 9–19 Members of the highly repetitive Alu sequence family are found exclusively in the genomes of mammals. However, the progenitor of the Alu family is a small gene that codes for a structural RNA, called 7SL RNA, that is found in all eucaryotic cells. (It functions in directing proteins to their proper destinations in the cell, as discussed in Chapter 14.) This RNA folds up into a structure that is shown in Figure Q9–19, in which the *green* and *red* portions have similar sequences to the Alu repeat DNA. Can you propose a step-by-step mechanism to explain how the Alu DNA sequence might have arisen during evolution? For each of your proposed steps, state whether it is based on a reaction that is known to occur in present-day cells, or whether it requires a new mechanism that you must postulate.

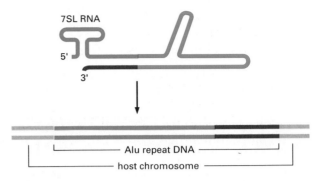

Figure Q9–19

Question 9–20 Assume a sexually reproducing animal has only four chromosomes, each of which can cross over at only three specific positions along the chromosome during meiosis. How many genetically different gametes can the animal produce?

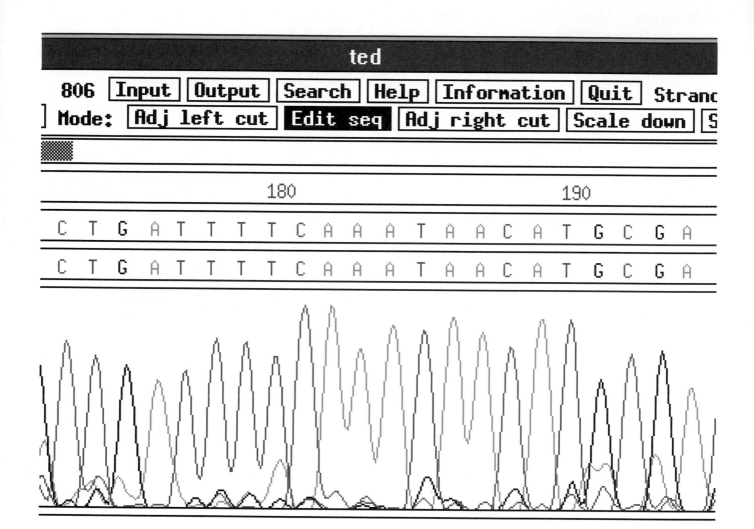

DNA technology has revolutionized the study of the cell and new, highly automated methods for sequencing DNA are now both rapid and reliable. This technology has made possible several major international genome-sequencing projects, aimed at elucidating the entire sequence of the DNA in the haploid genome of organisms from yeast to humans. Shown here is a tiny part of the data from an automated DNA-sequencing run as it appears on the computer screen. Each colored peak represents in turn the next nucleotide in the DNA sequence, and a clear stretch of nucleotide sequence can be read here between letters 173 and 195 from the start of the sequence. This particular example is taken from the international project to sequence the genome of the plant *Arabidopsis*. (Courtesy of George Murphy.)

10 DNA Technology

Humans have been experimenting with DNA, albeit without realizing it, for thousands of years. Modern garden rose varieties, for example, are the product of centuries of selective breeding between wild species of roses (Figure 10–1A). In a key development, four varieties of roses were brought by traders from China to Europe in the 1790s and subsequently crossed with the European varieties. The genomes of most modern roses are therefore made up of DNA derived from both European and Chinese species, an intermingling of genetic material that would probably never have been brought together without human intervention. Another visible result of a long series of DNA experiments is the large variety of dog breeds we see around us. The gray wolf—the ancestor of the modern dog—was first domesticated some 10,000–15,000 years ago. The different sizes, colors, shapes, and even behaviors of different breeds of dogs are the result of deliberate breeding experiments—with selection for desired traits—carried out since then (Figure 10–1B).

Although the selective breeding of crop plants and domestic animals has been going on since before recorded history, a technological revolution took place about 25 years ago that has dramatically increased the power of DNA experimentation. In the early 1970s, it became possible for the first time to isolate a given piece of DNA out of the many millions of nucleotide pairs in a typical chromosome. This in turn made it possible to create new DNA molecules in the test tube and to introduce them back into living organisms. These developments, called variously "recombinant DNA," "gene splicing," or "genetic engineering," make it

(A) (B)

Figure 10–1 **Early experiments with DNA.** (A) The oldest known depiction of a rose in Western art, from the Palace of Knossos in Crete, around 2000 BC. Modern roses are the result of centuries of breeding between such wild roses. (B) A poodle and a pug illustrate the range of dog breeds. All dogs, regardless of breed, belong to a single species. (B, courtesy of Heather Angel.)

possible to create chromosomes with combinations of genes that would never have formed naturally. The new techniques also made it possible to manipulate genes in ways that might conceivably happen in nature, but which could take thousands of years of chance events to occur.

Before the technological revolution of the 1970s, DNA was one of the most difficult molecules for the biochemist to analyze. The DNA molecules in chromosomes, even bacterial chromosomes, are enormously long (the chromosome of the *E. coli* bacterium, for example, contains 4.6 million nucleotides). Because DNA molecules are made up of only four types of nucleotides, with the biochemical analytical techniques then available they appeared chemically featureless with no landmarks to distinguish one part of the genome from another. The most important feature of a particular stretch of DNA—its sequence of nucleotides— could not be determined directly.

Today the situation has changed entirely. DNA has now become the easiest of the cellular macromolecules to analyze. It is now possible for a beginning student to cut out a region of DNA containing a specific gene from a genome, to produce a virtually unlimited number of exact copies of this DNA, and to determine its nucleotide sequence at a rate of thousands of nucleotides a day. By variations of these techniques, the isolated gene can be altered (or redesigned) in given ways in the laboratory and then transferred back into cells in culture in order to elucidate its function in the living cell. With more sophisticated techniques, the redesigned genes can be inserted into animals and plants, so that they become a functional and heritable part of the organism's genome.

In this chapter we survey the principal methods of this new field of **recombinant DNA technology.** These technical breakthroughs have had a dramatic impact on all aspects of cell biology by allowing cells and their macromolecules to be studied in ways not previously imagined. They have also made possible our present knowledge of the organization and evolutionary history of the complex genomes of eucaryotes (as discussed in Chapter 9) and led to the discovery of whole new classes of genes and proteins. They provide new means to determine the functions of proteins and of individual domains within proteins, revealing a host of unexpected relationships between them. And they give biologists an important set of tools for unraveling the mechanisms by which eucaryotic gene expression is regulated.

We now have the first complete nucleotide sequence of the genome of a eucaryotic organism (the single-celled brewer's yeast *Saccharomyces cerevisiae*), as well as the complete nucleotide sequences of several bacterial and archaebacterial genomes. The determination of the entire nucleotide sequence of our own genome is anticipated in about a decade. This latter achievement will provide us with the complete instructions necessary to produce a human, although exactly how these instructions are interpreted by cells to achieve this end will require many, many more decades of research.

Recombinant DNA technology has also had a profound influence on many aspects of human life outside of scientific research: it is used to detect the mutations in DNA that are responsible for inherited diseases; it is used in forensic science to identify or acquit possible suspects in a crime; it is used to produce an increasing number of pharmaceuticals, including insulin for diabetics and the blood-clotting protein Factor VIII for hemophiliacs. Even our laundry detergents, which contain heat-stable proteases that digest food spills and blood spots, make use of the products of DNA technology. Of all the discoveries described in this book, those of DNA technology are likely to have the greatest impact on our everyday lives.

In the first section of this chapter we discuss the basic techniques of DNA analysis. We then describe how DNA sequences can be isolated and produced in large numbers by the techniques of *DNA cloning* and the *polymerase chain reaction (PCR)*. In the final section of the chapter we look at the application of DNA technology to *genetic engineering*, the genetic manipulation of whole cells and organisms.

How DNA Molecules Are Analyzed

Ever since it was realized that genetic information was encoded in the sequence of nucleotides in DNA, scientists have wanted to be able to look directly at DNA and determine its nucleotide sequence in order to find out how genes work at the molecular level. Before the revolution in DNA technology, this was almost impossible. Enormous advances in understanding gene structure and regulation had been made by indirect genetic means in "model" organisms such as the *E. coli* bacterium and *Drosophila*, but the large and complex genomes of mammals were largely uncharted territory. At that time, the goal of isolating a single gene from a large chromosome seemed unattainable. Unlike a protein, a gene does not exist as a discrete entity in cells, but rather as a part of a much larger DNA molecule. Although DNA can be broken into small pieces by mechanical shear, the fragment containing a particular gene will still be only one among a hundred thousand or more DNA fragments that are obtained from a mammalian genome by these means. And in a sample containing many identical copies of the same large DNA molecule, each molecule would be broken up differently by shear, producing a confusing set of random fragments. How then, could any gene be isolated and purified?

The solution to this problem emerged with the discovery of a class of bacterial enzymes known as *restriction nucleases*. A nuclease catalyzes the hydrolysis of a phosphodiester bond in a nucleic acid. But these enzymes have a property that is distinct from other nucleases: they cut DNA only at particular sites, determined by a short sequence of nucleotides. Restriction nucleases can therefore be used to produce a set of specific DNA fragments from a genome. We start this section by describing how they work and how the DNA fragments produced by them can be separated from each other. We then explain how the order of nucleotides in a DNA fragment that has been isolated in this way is determined.

Restriction Nucleases Cut DNA Molecules at Specific Sites

Like most of the tools of DNA technology, restriction nucleases were first discovered by researchers studying a specialized biological problem that had gripped their interest. Certain bacteria, it was noticed, always degraded DNA from other bacteria that was experimentally introduced into them by the methods discussed in Chapter 9. A search for the cause of this degradation revealed a novel class of nucleases present inside the

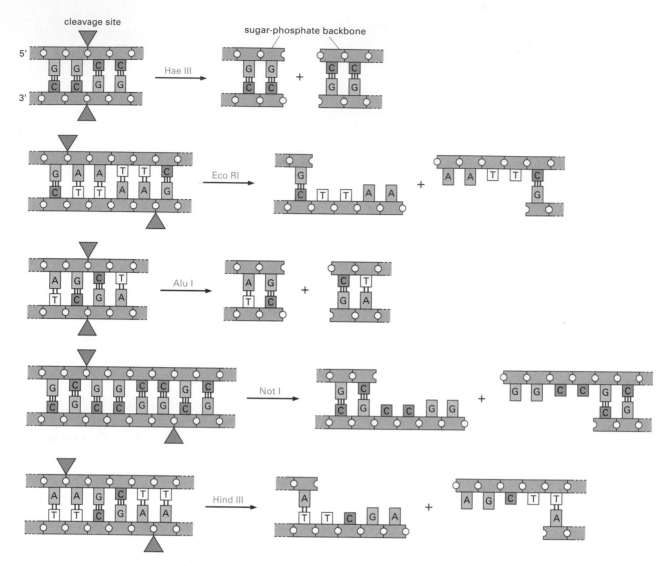

host bacterium. The most important feature of these nucleases is that they cleave DNA only at certain nucleotide sequences. The bacterium's own DNA is protected from cleavage by chemical modification of these same sequences. Because these enzymes restricted the transfer of DNA between certain strains of bacteria, the name **restriction nuclease** was given to them. Different bacterial species contain different restriction nucleases, each cutting at a different, specific sequence of nucleotides.

The restriction nucleases used in DNA technology come mainly from bacteria, and because their target sequences are short—generally 4–8 nucleotide pairs—they will occur, purely by chance, in any long DNA molecule. Thus they can be used to analyze DNA from any source. Restriction nucleases are now a hot commodity in DNA technology and are typically ordered through the mail; one supply catalog currently lists over 100 such enzymes, each able to cut a different DNA sequence. A few examples are shown in Figure 10–2.

Their ability to cleave DNA at specific sequences makes restriction nucleases crucial to all aspects of modern DNA technology. The reason for their usefulness is that a given restriction nuclease will always cut a given DNA molecule at the same sites. Thus, for a sample of DNA from, for example, human tissue, treatment with a given restriction nuclease will always produce the same set of DNA fragments.

Figure 10–2 The nucleotide sequences recognized and cut by five widely used restriction nucleases. As shown, the target sites at which these enzymes cut have a nucleotide sequence and length that depend on the enzyme. Target sequences are often palindromic (that is, the nucleotide sequence is symmetrical around a central point). In these examples, both strands of DNA are cut at specific points within the target sequence. Some enzymes, such as Hae III and Alu I, cut straight across the DNA double helix and leave two blunt-ended DNA molecules; for others, such as Eco RI, Not I, and Hind III, the cuts on each strand are staggered. Restriction nucleases are usually obtained from bacteria, and their names reflect their origins: for example, the enzyme Eco RI comes from *Escherichia coli.*

```
5'-AAGAATTGCGGAATTCGAGCTTAAGGGCCGCGCCGAAGCTTTAAA-3'
3'-TTCTTAACGCCTTAAGCTCGAATTCCCGGCGCGGCTTCGAAATTT-5'
```

Question 10–1 Which products are produced when the following piece of double-stranded DNA is digested with (A) Eco RI, (B) Alu I, (C) Not I, or (D) all three of these enzymes together.

The target sequences of restriction nucleases vary in the frequency with which they will occur in DNA. As shown previously in Figure 10–2, the enzyme Hae III cuts at a sequence of four nucleotide pairs; this sequence would be expected to occur purely by chance approximately once every 256 nucleotide pairs (1 in 4^4). By similar reasoning, the enzyme Not I, which has a target sequence of eight nucleotides, would be expected to cleave DNA on average once every 65,536 nucleotide pairs (1 in 4^8). The average size of the DNA fragments produced by different restriction nucleases can thus be very different. This makes it possible to cleave a long DNA molecule into the fragment sizes that are most suitable for each particular application.

Gel Electrophoresis Separates DNA Fragments of Different Sizes

After a large DNA molecule is cleaved into smaller pieces using a restriction nuclease, the DNA fragments need to be separated from each other.

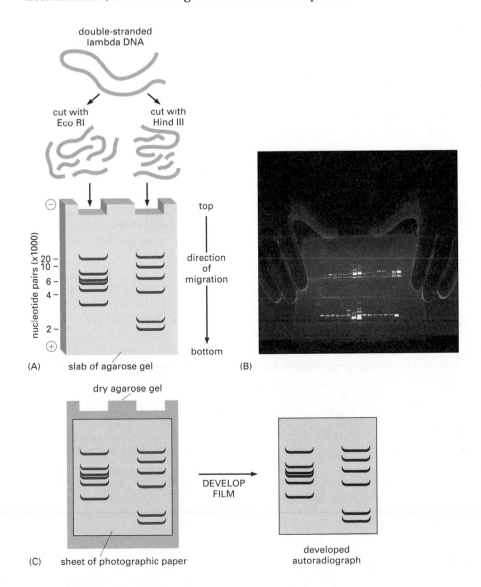

Figure 10–3 Separation and detection of DNA molecules by size using gel electrophoresis. (A) This schematic illustration compares the results of cutting the same DNA molecule (in this case the genome of the bacteriophage lambda; see Figure 9–15) with two different restriction nucleases—Eco RI (*left*) and Hind III (*right*). The fragments are then separated by gel electrophoresis. The mixture of DNA fragments obtained from treatment with the enzyme is placed at the top of a thin gel slab, and under the influence of an electric field, the fragments move through the gel toward the positive electrode. Larger fragments migrate more slowly than smaller fragments, and thus the fragments in the mixture become separated by size. For example, the two lowermost bands in the lane on the right correspond to the two smallest DNA fragments produced by Hind III digestion. To visualize the DNA bands, the gel has been soaked in a dye that binds to DNA and fluoresces brightly under ultraviolet light (B). (C) An alternative method for visualizing the DNA bands is autoradiography. Prior to cleavage with restriction enzymes, the DNA has been "labeled" with the radioisotope ^{32}P by substituting ^{32}P for some of the nonradioactive phosphorus atoms. This could be done, for example, by replicating the virus in the presence of ^{32}P. Since the β particles emitted from ^{32}P will expose photographic film, a sheet of film placed flat on top of the agarose gel will, when developed, show the position of all the DNA bands. (B, courtesy of J. C. Revy, Science Photo Library.)

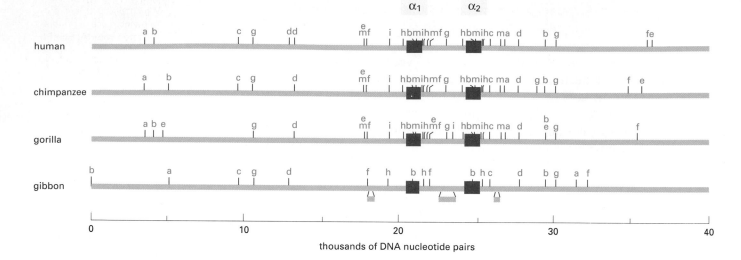

Figure 10–4 Comparison of the restriction maps of DNA containing the α-globin gene cluster from humans and other primates. The *thick green line* represents the chromosomal DNA containing the two α-globin genes (*red squares*) that are present in the primate genome (α globin is a subunit of hemoglobin, which contains two α-globin and two β-globin polypeptide chains). The *small letters* stand for sites cut by different restriction nucleases. The chimpanzee, the primate most closely related to humans, has the most similar restriction map to ours, whereas the gibbon, which is the most distantly related, has the most dissimilar map. The *small green bars* under the main line in the gibbon map represent positions where additional DNA is present in the gibbon genome compared with the other species. (Courtesy of Elizabeth Zimmer and Allan Wilson.)

This is usually accomplished using gel electrophoresis, which separates the fragments on the basis of their length. The mixture of DNA fragments is loaded at one end of a slab of agarose or polyacrylamide (the gel), which contains a microscopic network of pores. A voltage is then applied across the slab. Because DNA is negatively charged, the fragments migrate toward the positive electrode; the larger fragments migrate more slowly because their progress is more impeded by the agarose matrix. Over several hours, the DNA fragments become spread out across the gel according to size, forming a ladder of discrete bands, each composed of a collection of DNA molecules of identical length (Figure 10–3A). It is often a simple matter to isolate a particular DNA fragment: a small section of the gel containing the band is easily cut out using a scalpel or a razor blade.

DNA bands on agarose or polyacrylamide gels are invisible unless the DNA is labeled or stained in some way. One sensitive method of staining DNA is to expose it to a dye that fluoresces under ultraviolet light when it is bound to DNA (Figure 10–3B). An even more sensitive detection method involves incorporating a radioisotope into the DNA molecules before electrophoresis; ^{32}P is often used, as it can be incorporated into the phosphates of DNA and emits an energetic β particle that is easily detected by the technique of autoradiography (Figure 10–3C).

One of the earliest applications of restriction nuclease cleavage followed by the separation of individual DNA fragments was in the construction of *physical maps* of small regions of DNA. A physical map of DNA is one that characterizes a stretch of DNA by charting the position of various landmarks present along it. Restriction nuclease cleavage sites are one of the most useful types of landmarks. By comparing the sizes of the DNA fragments produced from a particular region of DNA after treatment with different combinations of restriction nucleases, a physical map of the region can be constructed showing the location of each cutting site. Such a map is known as a **restriction map.**

Since a restriction map depends on the nucleotide sequence of the corresponding DNA, it can reveal differences even between closely related DNAs (Figure 10–4). However, the ultimate physical map of a DNA is its complete nucleotide sequence, and in the next section we see how these nucleotide sequences can be determined.

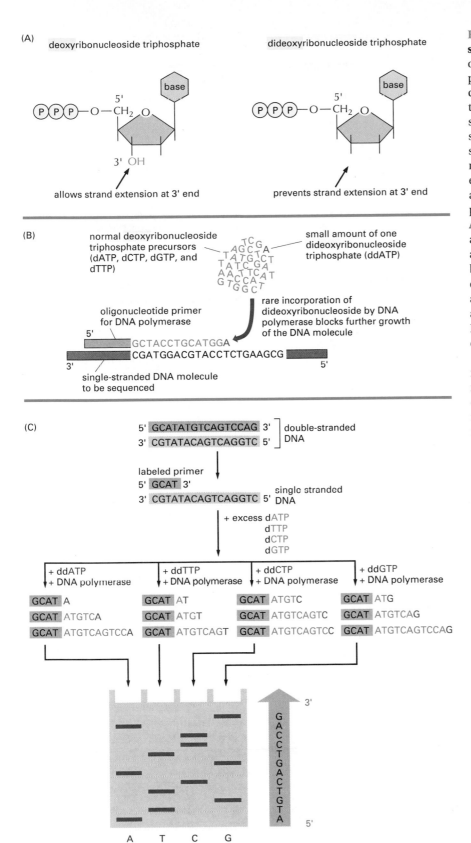

(A)

deoxyribonucleoside triphosphate

dideoxyribonucleoside triphosphate

5′

base

$\text{P}\ \text{P}\ \text{P}$—O—CH$_2$ O

3′ OH

allows strand extension at 3′ end

5′

base

$\text{P}\ \text{P}\ \text{P}$—O—CH$_2$ O

prevents strand extension at 3′ end

(B)

normal deoxyribonucleoside triphosphate precursors (dATP, dCTP, dGTP, and dTTP)

small amount of one dideoxyribonucleoside triphosphate (ddATP)

TCGA
AGCTGA
TATGTCT
TATCGAT
AACTTCAT
GTGGCT

rare incorporation of dideoxyribonucleoside by DNA polymerase blocks further growth of the DNA molecule

oligonucleotide primer for DNA polymerase

5′
GCTACCTGCATGGA

CGATGGACGTACCTCTGAAGCG

3′   5′

single-stranded DNA molecule to be sequenced

(C)

5′ GCATATGTCAGTCCAG 3′   ⎤ double-stranded
3′ CGTATACAGTCAGGTC 5′   ⎦ DNA

labeled primer

5′ GCAT 3′
3′ CGTATACAGTCAGGTC 5′   single stranded DNA

+ excess dATP
dTTP
dCTP
dGTP

+ ddATP   + ddTTP   + ddCTP   + ddGTP
+ DNA polymerase + DNA polymerase + DNA polymerase + DNA polymerase

GCAT A   GCAT AT   GCAT ATGTC   GCAT ATG
GCAT ATGTCA   GCAT ATGT   GCAT ATGTCAGTC   GCAT ATGTCAG
GCAT ATGTCAGTCCA GCAT ATGTCAGT   GCAT ATGTCAGTCC GCAT ATGTCAGTCCAG

3′

G
A
C
C
T
G
A
C
T
G
T
A

5′

A   T   C   G

Figure 10–5 The enzymatic method of sequencing DNA. (A) This method relies on the use of dideoxyribonucleoside triphosphates, derivatives of the normal deoxyribonucleoside triphosphates that lack the 3′ hydroxyl group. (B) Purified DNA is synthesized *in vitro* in a mixture that contains single-stranded molecules of the DNA to be sequenced (*gray*), the enzyme DNA polymerase, a short primer DNA (*orange*) to enable the polymerase to start replication, and the four deoxyribonucleoside triphosphates (dATP, dCTP, dGTP, dTTP: *green* A, C, G, and T). If a dideoxyribonucleoside analogue (*red*) of one of these nucleotides is also present in the nucleotide mixture, it becomes incorporated into a growing DNA chain. The chain now lacks a 3′ OH group, the addition of the next nucleotide is blocked, and the DNA chain terminates at that point. In the example illustrated, a small amount of dideoxyATP (ddATP, symbolized here as a *red* Λ) has been included in the nucleotide mixture. It competes with an excess of the normal deoxyATP (dATP, *green* A), so that ddATP is occasionally incorporated, at random, into a growing DNA strand. This reaction mixture will eventually produce a set of DNAs of different lengths complementary to the template DNA that is being sequenced and terminating at each of the different A's. (C) To determine the complete sequence of a DNA fragment, the double-stranded DNA is first separated into its single strands and one of the strands is used as the template for sequencing. Four different chain-terminating dideoxyribonucleoside triphosphates (ddATP, ddCTP, ddGTP, ddTTP, again shown in *red*) are used in four separate DNA synthesis reactions on copies of the same single-stranded DNA template (*gray*). Each reaction produces a set of DNA copies that terminate at different points in the sequence. The products of these four reactions are separated by electrophoresis in four parallel lanes of a polyacrylamide gel (labeled here A, T, C, and G). The newly synthesized fragments are detected by a label (either radioactive or fluorescent) that has been incorporated either into the primer or into one of the deoxyribonucleoside triphosphates used to extend the DNA chain. In each lane, the bands represent fragments that have terminated at a given nucleotide (e.g., A in the leftmost lane) but at different positions in the DNA. By reading off the bands in order, starting at the bottom of the gel and working across all lanes, the DNA sequence of the newly synthesized strand can be determined. The sequence is given in the *green arrow* to the right of the gel. This sequence is identical to that of the 5′ → 3′ strand (*green*) of the original double-stranded DNA.

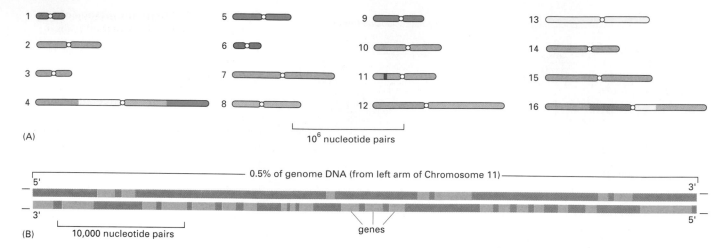

(A)

10^6 nucleotide pairs

0.5% of genome DNA (from left arm of Chromosome 11)

5'
3'

3'
5'

(B)

10,000 nucleotide pairs

genes

The Nucleotide Sequence of DNA Fragments Can Be Determined

In the late 1970s, methods were developed that allow the nucleotide sequence of any purified DNA fragment to be determined simply and quickly. Several schemes for sequencing DNA have been developed, but the most widely used employs DNA polymerase to make partial copies of the DNA fragment to be sequenced. These DNA replication reactions are carried out *in vitro* under conditions which ensure that the new DNA strands terminate when a given nucleotide (A, G, C, or T) is reached (Figure 10–5A and B). As illustrated in Figure 10–5C, this method produces a collection of different DNA copies that terminate at each position in the original DNA, and thus differ in length by a single nucleotide. These DNA copies can be separated on the basis of their length by gel electrophoresis, and the nucleotide sequence of the original DNA can be determined from the order of these DNAs in the gel, as explained in the figure.

The complete DNA sequences of tens of thousands of genes, several complete bacterial genomes, and the genome of the simple eucaryote *S. cerevisiae* (budding yeast) have now been determined (Figure 10–6). The complete DNA sequences of the fruit fly (*Drosophila melanogaster*) and nematode worm (*Caenorhabditis elegans*) genomes are anticipated before the year 2000. The volume of DNA sequence information contained in the computerized DNA databases is now so large (hundreds of millions of nucleotides) that highly sophisticated software is required to organize and analyze it.

As techniques for DNA sequencing have improved, scientists have begun to determine the nucleotide sequence of the entire human genome. This represents a total DNA sequence length of approximately 3×10^9 nucleotides. Because this sequence specifies all of the possible RNA and protein molecules that are used to construct the human body, its completion will provide us with a "dictionary of the human being," which will greatly expedite future studies of human cells and tissues.

Figure 10–6 Representation of the complete nucleotide sequence of the genome of *S. cerevisiae* (budding yeast). (A) The genome is distributed over 16 chromosomes, and its nucleotide sequence was determined by a cooperative effort involving scientists working in many different locations, as indicated (*gray* = Canada, *orange* = European Union [including scientists from many different European countries], *yellow* = United Kingdom, *blue* = Japan, *light green* = St. Louis, Missouri, *dark green* = Stanford, California). The constriction present on each chromosome represents the position of its centromere. The small region of Chromosome 11 highlighted in *red* is magnified in (B) to show the high density of genes characteristic of this species. As indicated, some genes are transcribed from the lower strand, while others are transcribed from the upper strand. There are about 6200 genes contained in the complete genome, which is 12,147,813 nucleotide pairs long.

Question 10–2 What are the consequences for a DNA sequencing reaction if the ratio of dideoxyribonucleoside triphosphates to deoxyribonucleoside triphosphates is increased? What happens if this ratio is decreased?

Nucleic Acid Hybridization

Once a gene has been isolated from a complete genome as a piece of DNA, we may want to know from which chromosome the gene came

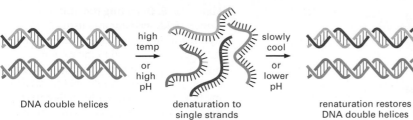

DNA double helices

high temp or high pH

denaturation to single strands (nucleotide pairs broken)

slowly cool or lower pH

renaturation restores DNA double helices (nucleotide pairs re-formed)

and where on the chromosome it is located. We may want to know in which cells of the organism the gene is transcribed. We may want to know whether other organisms have similar genes, or we might want to test a sample of human DNA for mutations in the gene suspected of causing an inherited disease. All of these questions can be answered in the laboratory by taking advantage of a fundamental property of DNA: through the formation of Watson-Crick base pairs, a strand of DNA can pair highly selectively with a second strand of *complementary* nucleotide sequence. The two strands of a DNA double helix are held together by relatively weak hydrogen bonds that can be broken by heating the DNA to around 90°C or subjecting it to extremes of pH. These treatments release the two strands from each other but do not break the covalent bonds that link nucleotides together within each strand. If this process is slowly reversed (that is, by slowly lowering the temperature to normal body temperature or by bringing the pH back to neutral), the complementary strands will readily re-form double helices. This process is called *renaturation*, or **hybridization,** and it results from a restoration of the complementary hydrogen bonds (Figure 10–7).

A similar hybridization reaction will occur between any two single-stranded nucleic acid chains (DNA/DNA, RNA/RNA, or RNA/DNA), provided they have complementary nucleotide sequences. The capacity of a single-stranded nucleic acid molecule to form a double helix only with a molecule complementary to it provides a powerful technique to detect specific nucleotide sequences in both DNA and RNA.

DNA Hybridization Facilitates the Prenatal Diagnosis of Genetic Diseases

In order to search by hybridization for a nucleotide sequence, one first needs a nucleic acid to search with, called a probe. A *DNA probe* is a short single-stranded DNA, an *oligonucleotide* typically 10–1000 nucleotides long, that is used in hybridization reactions to detect nucleic acid molecules containing a complementary sequence. At first, scientists were limited to using probes that could be obtained from natural sources. Today, because of advances in nucleotide chemistry, short lengths of DNA of any desired sequence can be synthesized non-enzymatically in the laboratory. Machines the size of a microwave oven can be programmed to string nucleotides together by *chemical synthesis* to produce single-stranded DNA chains of any sequence up to about 120 nucleotides long.

The following example illustrates one of the ways that DNA probes and hybridization can be used to detect possible carriers of genetic diseases. More than 3000 different human genetic diseases are caused by

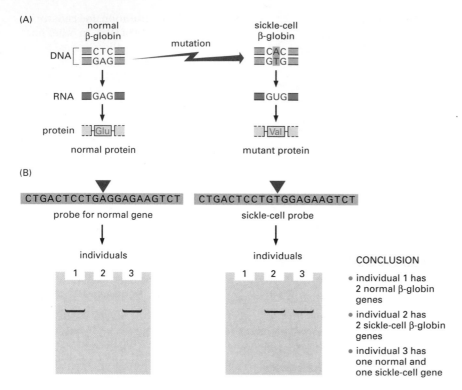

(A)

normal β-globin

DNA [CTC / GAG]

mutation →

sickle-cell β-globin

[CAC / GTG]

RNA GAG

GUG

protein Glu

Val

normal protein

mutant protein

(B)

CTGACTCCTGAGGAGAAGTCT

probe for normal gene

CTGACTCCTGTGGAGAAGTCT

sickle-cell probe

individuals

1 2 3

individuals

1 2 3

CONCLUSION

- individual 1 has 2 normal β-globin genes
- individual 2 has 2 sickle-cell β-globin genes
- individual 3 has one normal and one sickle-cell gene

Figure 10–8 Detecting the mutation that causes sickle-cell anemia using DNA hybridization. (A) Sickle-cell anemia is caused by a mutation in the gene for β-globin that results in a change of a single amino acid from glutamic acid to valine in the β-globin protein. Individuals carrying two defective β-globin genes inherit the disease; individuals carrying one mutant gene and one normal gene usually do not develop symptoms. (B) The mutant gene can be detected in fetal DNA by DNA hybridization. DNA samples from the fetus are first treated with restriction nucleases, and all the resulting DNA fragments (including those that contain the relevant portion of the β-globin gene) are electrophoresed through the gel. The gel is then treated with a DNA probe that detects only the restriction fragment that carries the β-globin gene. Two different synthetic DNA probes are used, one corresponding to the normal sequence and one corresponding to the mutant sequence. The probes are labeled either with radioactive isotopes or with a fluorescent dye incorporated when they are manufactured (see also Figure 10–9).

mutations in single genes. In most of these cases, the mutation is *recessive*—that is, it shows its effect only when an individual inherits two defective copies of the gene, one from each parent. For some of these diseases, it is now possible to identify early in a pregnancy fetuses that carry two copies of a defective gene; this information may be a factor in decisions relating to possible termination of the pregnancy.

Examining a single gene in the human genome requires searching through a total genome of over three billion nucleotides. However, the incredible specificity of DNA hybridization makes this possible. For the recessive genetic disease sickle-cell anemia, for example, the exact nucleotide change in the mutant gene is known; the sequence GAG is changed to GTG at a certain position in the DNA strand that codes for the β-globin chain of hemoglobin (Figure 10–8A). This causes a change in the amino acid encoded by that sequence from a glutamic acid to a valine; this small change is sufficient to alter the properties of the resulting hemoglobin molecules to produce the disease (see Figure 6–23). For prenatal diagnosis of sickle-cell anemia, DNA is extracted from fetal cells. Two DNA probes are used to test the fetal DNA—one corresponding to the normal gene sequence in the region of the mutation and the other corresponding to the mutant gene sequence. If the probes are short (about 20 nucleotides), they can be hybridized with DNA at a temperature at which only the perfectly matched helices will be stable. Using this technique, it is possible to distinguish whether DNA isolated from the fetus contains one, two, or no defective β-globin genes (Figure 10–8B). For example, a fetus carrying two copies of the mutant β-globin gene (which will result in the disease) can be recognized because its DNA will hybridize only with the probe that is complementary to the mutant DNA sequence. The laboratory procedure used to visualize the hybridization is called *Southern blotting,* as shown in Figure 10–9.

The same techniques can also be used to detect an individual's susceptibility to future disease. They can, for example, detect individuals

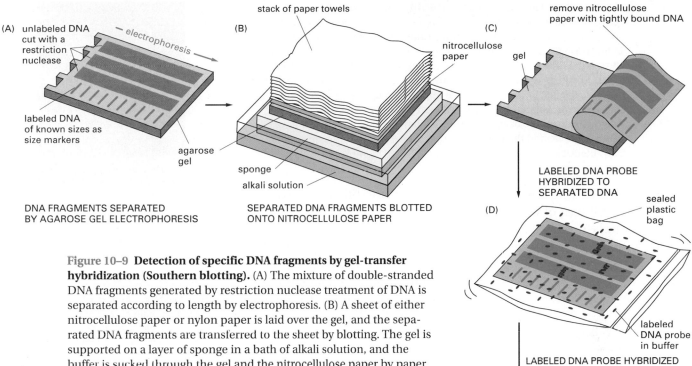

(A) unlabeled DNA cut with a restriction nuclease

– electrophoresis

labeled DNA of known sizes as size markers

agarose gel

DNA FRAGMENTS SEPARATED BY AGAROSE GEL ELECTROPHORESIS

(B) stack of paper towels

nitrocellulose paper

sponge
alkali solution

SEPARATED DNA FRAGMENTS BLOTTED ONTO NITROCELLULOSE PAPER

(C) remove nitrocellulose paper with tightly bound DNA

gel

LABELED DNA PROBE HYBRIDIZED TO SEPARATED DNA

(D) sealed plastic bag

labeled DNA probe in buffer

LABELED DNA PROBE HYBRIDIZED TO COMPLEMENTARY DNA BANDS VISUALIZED BY AUTORADIOGRAPHY

(E) positions of labeled markers

labeled bands

Figure 10–9 Detection of specific DNA fragments by gel-transfer hybridization (Southern blotting). (A) The mixture of double-stranded DNA fragments generated by restriction nuclease treatment of DNA is separated according to length by electrophoresis. (B) A sheet of either nitrocellulose paper or nylon paper is laid over the gel, and the separated DNA fragments are transferred to the sheet by blotting. The gel is supported on a layer of sponge in a bath of alkali solution, and the buffer is sucked through the gel and the nitrocellulose paper by paper towels stacked on top of the nitrocellulose. As the buffer is sucked through, it denatures the DNA and transfers the single-stranded fragments from the gel to the surface of the nitrocellulose sheet, where they adhere firmly. This transfer is necessary to keep the DNA firmly in place while the hybridization procedure (D) is carrried out. (C) The nitrocellulose sheet is carefully peeled off the gel. (D) The sheet containing the bound single-stranded DNA fragments is placed in a sealed plastic bag together with buffer containing a radioactively labeled DNA probe specific for the required DNA sequence. The sheet is exposed for a prolonged period to the probe under conditions favoring hybridization. (E) The sheet is removed from the bag and washed thoroughly, so that only probe molecules that have hybridized to the DNA on the paper remain attached. After autoradiography, the DNA that has hybridized to the labeled probe will show up as bands on the autoradiograph. An adaptation of this technique to detect specific sequences in RNA is called *Northern blotting*. In this case mRNA molecules are electrophoresed through the gel and the probe is usually a single-stranded DNA molecule.

who have inherited abnormal copies of a DNA mismatch repair gene. Because they cannot repair mistakes in DNA replication efficiently, such individuals (estimated at one in 200 North Americans) have a greatly increased risk of cancer, especially a certain type of colon cancer (see Chapter 6). They need to take protective measures, especially frequent checkups, to improve their prospects for remaining healthy.

In Situ Hybridization Locates Nucleic Acid Sequences in Cells or on Chromosomes

Nucleic acids, no less than other macromolecules, occupy precise positions in cells and tissues, and a great deal of potential information is lost when these molecules are extracted from cells. For this reason, techniques have been developed in which nucleic acid probes are used to

Question 10–3 DNA sequencing of your own two β-globin genes (one from each of your two Chromosome 11's) reveals a mutation in one of the genes. Given this information alone, how much should you worry about being a carrier of an inherited disease that could be passed on to your children? What other information would you like to have to assess your risk?

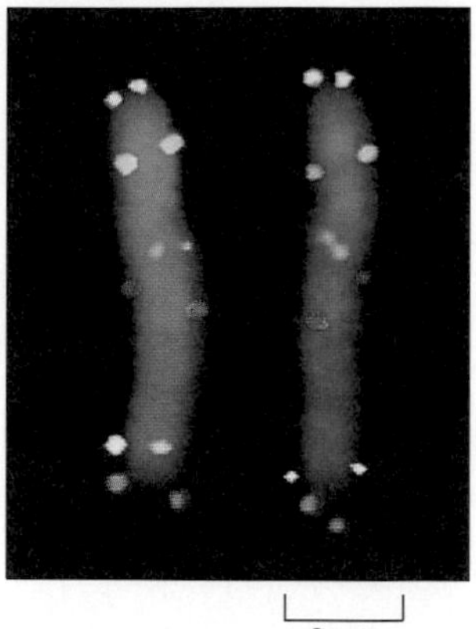

Figure 10–10 The use of *in situ* hybridization to locate genes on chromosomes. Here, six different DNA probes have been used to mark the location of their respective nucleotide sequences on human Chromosome 5 isolated in the metaphase stage of mitosis (see Figure 8–5, and Panel 17–1, pp. 554–555). The probes have been chemically labeled and are detected using fluorescent antibodies specific for the chemical label. Both the maternal and paternal copies of Chromosome 5 are shown, aligned side-by-side. Each probe produces two dots on each chromosome because chromosomes undergoing mitosis have already replicated their DNA and therefore each chromosome contains two identical DNA helices. (Courtesy of David C. Ward.)

2 µm

locate specific nucleic acid sequences while they are still in place within cells or are still part of chromosomes. This procedure is called *in situ* **hybridization** (from the Latin *in situ*, "in place"), and can be applied to detect either DNA sequences in chromosomes or RNA sequences in cells. For the former application, nucleic acid probes labeled with fluorescent dyes or radioactive isotopes are hybridized to whole chromosomes that have been exposed briefly to a very high pH to separate the two DNA strands. The chromosomal regions that bind the labeled probe can then be visualized (Figure 10–10).

In situ hybridization can also reveal the distribution of a particular RNA in cells in tissues (Figure 10–11). This technique in particular has led to great advances in our understanding of embryonic development, by making easily visible the many changes in patterns of gene expression that occur in different cells of the embryo.

DNA Cloning

We have seen that DNA molecules can be cleaved into shorter fragments using restriction nucleases and that these fragments can be separated from one another by gel electrophoresis. We also discussed how the sequence of nucleotides in a DNA fragment can be determined and how nucleic acid hybridization can be used to pick out a match with a DNA

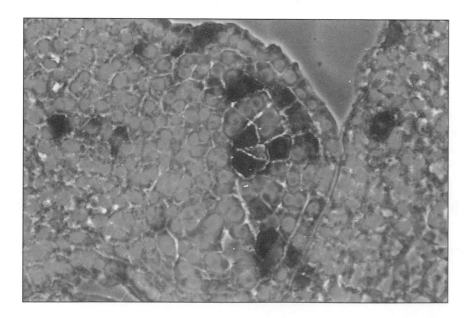

Figure 10–11 Cells expressing a particular mRNA can be visualized by *in situ* hybridization. This example shows a group of cells in a growing shoot tip of a snapdragon. Only a few of the cells (stained *dark blue*) are expressing an mRNA for a cyclin protein that triggers the cells to divide. The cells have been treated with a DNA probe to the cyclin mRNA. This probe was then specifically linked to an enzyme that produces a dark blue reaction product. The nuclei in other cells appear *light blue* because their DNA has been stained with the dye DAPI. (Courtesy of John Doonan.)

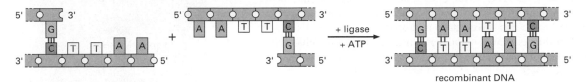

(A) JOINING TWO COMPLEMENTARY STAGGERED ENDS

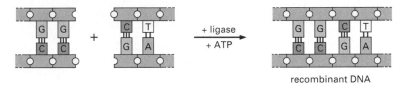

recombinant DNA

(B) JOINING TWO BLUNT ENDS

+ ligase
+ ATP

recombinant DNA

(C) JOINING A BLUNT END WITH A STAGGERED END

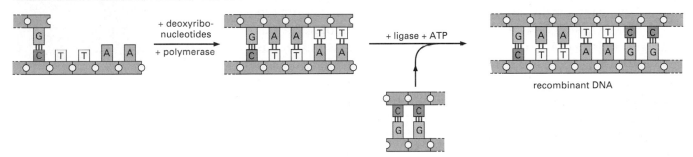

+ deoxyribo-
nucleotides

+ polymerase

+ ligase + ATP

recombinant DNA

probe of known sequence. In this section of the chapter, we shall see how these procedures are combined in order to clone an individual stretch of DNA. In cell biology, the term **DNA cloning** is used in two senses: in one it literally refers to the act of making many identical copies of a DNA molecule. However, it is also used to describe the isolation of a particular stretch of DNA (often a particular gene) from the rest of a cell's DNA, because this isolation is facilitated by making many identical copies of the DNA of interest.

DNA Ligase Joins DNA Fragments Together to Produce a Recombinant DNA Molecule

Modern DNA technology depends both on the ability to break long DNA molecules into conveniently sized fragments and on the ability to join these fragments back together in new combinations. As for most of the technological advances discussed in this chapter, the cell itself has provided the solution to this problem. As discussed in Chapter 6, the enzyme **DNA ligase** reseals the nicks in the DNA backbone that arise during DNA replication and DNA repair (see Figures 6–20 and 6–30). This enzyme has become one of the most important tools of recombinant DNA technology, as it allows scientists to join together any two DNA fragments (Figure 10–12). Since DNA has the same chemical structure in all organisms, this simple maneuver allows DNAs from any source to be joined together. In this way, isolated DNA fragments can be recombined in the test tube to produce DNA molecules not found in nature. Once two DNA molecules have been joined by ligase, the cell

Figure 10–12 The formation of recombinant DNA molecules *in vitro*. The enzyme DNA ligase can join any two DNA fragments together regardless of the source of the DNA. The ATP provides the energy necessary for the ligase to reseal the sugar-phosphate backbone of DNA. (A) The joining of two DNA fragments produced by the restriction nuclease Eco RI. Note that the staggered ends produced by this enzyme enable the ends of the two fragments to base-pair correctly with each other, greatly facilitating their rejoining. This ligation reaction also reconstructs the original restriction nuclease cutting site. (B) The joining of a DNA fragment produced by the restriction nuclease Hae III to one produced by Alu I. (C) The joining of DNA fragments produced by Eco RI and Hae III, respectively, using DNA polymerase to fill in the staggered cut produced by Eco RI. Each DNA fragment shown in the figure is oriented so that its 5′ ends are the left end of the upper strand and the right end of the lower strand, as indicated in (A).

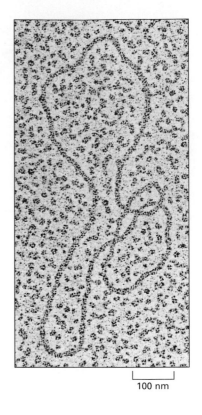

Figure 10–13 **Electron micrograph of a bacterial plasmid commonly used as a cloning vector.** This circular, double-stranded DNA molecule consists of several thousand nucleotide pairs. The staining required to make the DNA visible in the electron microscope makes the DNA appear much thicker than it actually is. (Courtesy of Brian Wells.)

100 nm

cannot tell that the two DNAs were originally separate and will treat the resulting DNA as a single molecule. If such a foreign stretch of DNA is appropriately introduced into the DNA of a host cell, it will be replicated and transcribed as if it were a normal part of the cell's DNA.

Bacterial Plasmids Can Be Used to Clone DNA

For many applications of DNA technology, it is necessary to clone (make identical copies) a stretch of DNA, often a gene. This can be accomplished in several ways. The simplest is to introduce the DNA to be copied into a rapidly dividing bacterium; each time the bacterium replicates its own DNA it also copies the introduced DNA. In order to maintain the foreign DNA in the bacterial cell, a bacterial plasmid (or alternatively a viral genome) is used as a carrier, or *vector*. A typical **plasmid** cloning vector is a relatively small circular DNA molecule of several thousand nucleotide pairs that can replicate inside a bacterium (Figure 10–13). A plasmid vector must contain a replication origin, which enables it to replicate in the bacterial cell independently of the bacterial chromosome. It also needs a cutting site for a restriction nuclease, so that the plasmid can be opened and a foreign DNA fragment can be inserted. The plasmid also usually contains a gene for some selectable property such as antibiotic resistance, which enables bacteria that take up the recombinant plasmid to be identified (see Chapter 9).

To insert the DNA to be cloned, the purified plasmid DNA is exposed to a restriction nuclease that cleaves it in just one place, and the DNA fragment to be cloned is covalently inserted into it using DNA ligase (Figure 10–14). This recombinant DNA molecule is then introduced into a bacterium (usually *E. coli*) by transformation (see Figure 9–8), and the

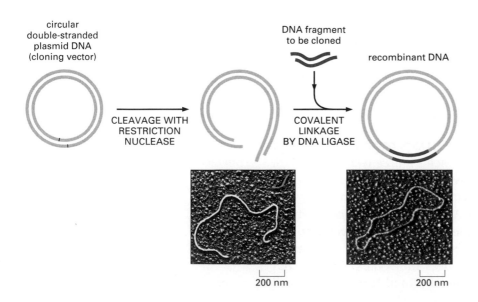

circular
double-stranded
plasmid DNA
(cloning vector)

DNA fragment
to be cloned

recombinant DNA

CLEAVAGE WITH
RESTRICTION
NUCLEASE

COVALENT
LINKAGE
BY DNA LIGASE

200 nm

200 nm

Figure 10–14 **The insertion of a DNA fragment into a bacterial plasmid using the enzyme DNA ligase.** The plasmid is cut open with a restriction nuclease (in this case one that produces staggered ends), and is mixed with the DNA fragment to be cloned (which has been prepared using the same restriction nuclease), DNA ligase, and ATP. The staggered ends base-pair, and the nicks in the DNA backbone are sealed by the DNA ligase to produce a complete recombinant DNA molecule. (Micrographs courtesy of Huntington Potter and David Dressler.)

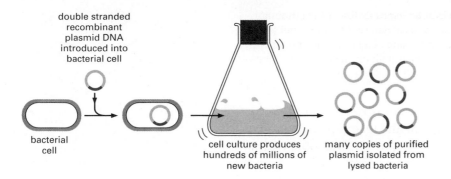

Figure 10–15 Cloning of a DNA fragment. In order to produce many copies of a particular fragment of DNA, it is first inserted into a plasmid vector, as shown in Figure 10–14. The resulting recombinant plasmid DNA is then introduced into a bacterium, where it can be replicated many millions of times as the bacterium multiplies.

double stranded recombinant plasmid DNA introduced into bacterial cell

bacterial cell

cell culture produces hundreds of millions of new bacteria

many copies of purified plasmid isolated from lysed bacteria

bacterium is allowed to grow in nutrient broth, where it doubles in number every 30 minutes. Each time it doubles, the number of copies of the recombinant DNA molecule also doubles, and after just a day, hundreds of millions of copies of the plasmid will have been produced. The bacteria are then lysed, and the much smaller plasmid DNA is purified away from the rest of the cell contents, including the large bacterial chromosome. The purified plasmid DNA will contain millions of copies of the original DNA fragment (Figure 10–15). This DNA fragment can be recovered by cutting it cleanly out of the plasmid DNA using the appropriate restriction enzyme and separating it from the plasmid DNA by gel electrophoresis (see Figure 10–3).

Human Genes Are Isolated by DNA Cloning

We have seen how any DNA fragment can be produced in large numbers. But how are these DNA fragments identified in the first place? In particular, how is an individual human gene (one of around 60,000 genes in the human genome) first isolated by cloning? As an example, we shall follow the steps by which the human gene for the blood-clotting protein *Factor VIII* was cloned. Although the exact methods by which human genes have been isolated and identified differ from one case to the next, this example illustrates many general features.

Defects in the gene for Factor VIII are the cause of the most common type of hemophilia—*hemophilia A*. This genetically determined disease has been recognized for over a thousand years and affects approximately one in 10,000 males. Sufferers from hemophilia A fail to produce fully active Factor VIII, and thus have repeated episodes of uncontrolled bleeding. Until recently, the standard treatment for this disease had been the injection of concentrated Factor VIII protein, pooled from many blood samples. Tragically, this treatment has exposed hemophiliacs to risk of infection by viruses, including HIV (the AIDS virus). The commercial production of pure Factor VIII using recombinant DNA technology offers a significant improvement in the treatment of this disease. This achievement required the cloning of the normal human gene that codes for Factor VIII and the piecing together of its coding sequence. This coding sequence was then used to produce large amounts of the protein, as described later in this chapter.

Dealing with the 3×10^9 nucleotide pairs of the complete human genome is a daunting task, and the first step in cloning a human gene is to break up the total genomic DNA into smaller, more manageable pieces to make it easier to work with. To do this, the total DNA extracted from a tissue sample or a culture of human cells is cut up into a set of

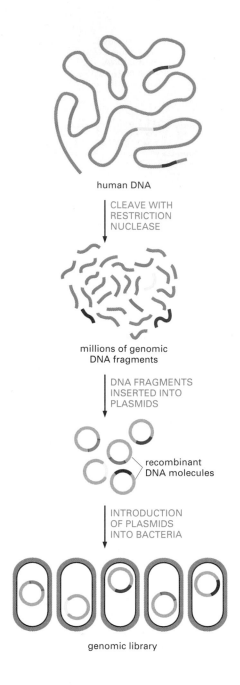

Figure 10–16 **Construction of a human genomic library.** A genomic library comprises a set of bacteria, each carrying a different small fragment of human DNA. For simplicity, cloning of just a few representative fragments (*colored*) is shown. In reality, all the *gray* DNA fragments will also be cloned.

human DNA

CLEAVE WITH
RESTRICTION
NUCLEASE

millions of genomic
DNA fragments

DNA FRAGMENTS
INSERTED INTO
PLASMIDS

recombinant
DNA molecules

INTRODUCTION
OF PLASMIDS
INTO BACTERIA

genomic library

DNA fragments by random mechanical shear or by restriction nuclease treatment, and each fragment is cloned using the techniques described previously. The collection of cloned DNA fragments thus obtained is known as a **DNA library.** In this case it is called a *genomic library,* as the DNA fragments are derived directly from the chromosomal DNA. As we see later, there are other types of DNA libraries. Genomic libraries of the DNAs of many different organisms now exist, and they are freely available to researchers. Their availability eliminates the need to go through the rather laborious process of constructing a library every time a scientist embarks on the isolation of a particular gene.

One general procedure for making a human genomic library is outlined in Figure 10–16. Human DNA is first cleaved with a restriction nuclease, which produces millions of different DNA fragments. The mixture of DNA fragments is inserted into plasmid vectors under conditions that favor the insertion of one DNA fragment for each plasmid molecule. These recombinant plasmids are mixed with a culture of *E. coli* at a concentration that ensures that no more than one plasmid molecule is taken up by each bacterium. The resulting bacterial culture represents the genomic library. If colonies derived from a single bacterium are isolated on Petri dishes (as described in Figure 9–2), each bacterial colony will represent a clone of one particular stretch of human DNA. A collection of several million colonies in this library should thus represent all of the human genome.

To find a particular gene, we now face a problem analogous to that of entering a real library, wishing to find a book, and realizing that there is no card catalogue or record of the millions of books in the library. How do we find a particular stretch of DNA (in our case the Factor VIII gene) in the vast human DNA library? The key is to exploit the hybridization properties of nucleic acids discussed earlier in this chapter. If we had a DNA probe for the Factor VIII gene, we could use it to find the matching clone in the library. But where does such a probe come from before the gene itself has been identified?

In this case, a small amount of the Factor VIII protein had been purified from human blood donors, and several short stretches of its amino acid sequence had been obtained by protein sequencing (see Panel 5–6, pp. 164–165). By applying the genetic code in reverse, the DNA

Figure 10–17 **An oligonucleotide probe for a gene based on a stretch of amino acid sequence from the protein.** Because of the degeneracy of the genetic code (that is, because each amino acid can be encoded by more than one DNA triplet; see Figure 7–20), a number of different nucleotide sequences are possible for each amino acid sequence. Although only one of these sequences will actually code for the protein in the genomic DNA, it is impossible to tell in advance which it is. Therefore, a mixture of the possible sequences—called a degenerate oligonucleotide probe—is used to search a genomic library for the gene. The example shown is from the human Factor VIII protein, and the degenerate probe shown was used in the original cloning of the gene in 1984.

protein sequence of a small portion
of the Factor VIII protein

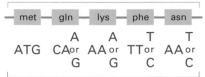

| met | gln | lys | phe | asn |

	A	A	T	T
ATG	CA or	AA or	TT or	AA or
	G	G	C	C

degenerate oligonucleotide probe
(a mixture of 16 different oligonucleotides)

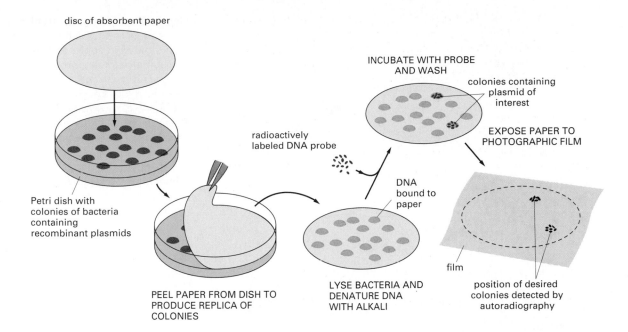

disc of absorbent paper

INCUBATE WITH PROBE AND WASH

colonies containing plasmid of interest

radioactively labeled DNA probe

EXPOSE PAPER TO PHOTOGRAPHIC FILM

Petri dish with colonies of bacteria containing recombinant plasmids

DNA bound to paper

film

PEEL PAPER FROM DISH TO PRODUCE REPLICA OF COLONIES

LYSE BACTERIA AND DENATURE DNA WITH ALKALI

position of desired colonies detected by autoradiography

sequences that code for these amino acid sequences can be deduced (Figure 10–17), and a suitable DNA probe can be prepared by chemical synthesis. Using this probe, those rare bacterial clones containing human DNA complementary to the probe can be identified by DNA hybridization (Figure 10–18).

By using the Factor VIII probe on a human genomic library, a single complementary clone was identified. However, the nucleotide sequence of this cloned DNA showed that it contained only a small portion of the Factor VIII gene. Subsequent analysis, after the complete gene had been pieced together, showed that the Factor VIII gene was 180,000 nucleotide pairs long and contained 25 introns (see Figure 7–14B), so it was hardly suprising that no single genomic clone contained the entire gene.

For many applications of DNA technology, it is advantageous to obtain a clone that contains only the coding sequence of a gene, that is, a clone that lacks the intron DNA. For example, in the case of the Factor VIII gene the complete genomic clone—introns and exons—is so large and unwieldly that it is necessary to analyze the gene in pieces. Moreover, if one wanted to deduce the complete amino acid sequence of the Factor VIII protein from the nucleotide sequence of its gene, it would be extremely inefficient to determine the nucleotide sequence of the gene isolated from a genomic library, because the vast majority of it corresponds to introns. It is, however, relatively simple to isolate a gene free of all its introns, as we see in the following section. For this purpose, a different type of library, called a *cDNA library*, is utilized.

cDNA Libraries Represent the mRNA Produced by a Particular Tissue

A human cDNA library is similar to the genomic library described earlier in that it also contains numerous clones containing many different human DNA sequences. But it differs in one important respect. The DNA that goes into a cDNA library is not genomic DNA (chromosomal DNA), but is instead DNA copied from the mRNAs present in a particular tissue or cell culture. To prepare a cDNA library, the total mRNA is extracted

Figure 10–18 A technique commonly used to detect a bacterial colony carrying a particular DNA clone. A replica of the arrangement of colonies on the Petri dish is made by pressing a piece of absorbent paper against the surface of the dish. This replica is treated with alkali (to lyse the cells and separate the plasmid DNA into single strands), and the paper is then hybridized to a highly radioactive DNA probe. Those bacterial colonies that have bound the probe are identified by autoradiography. Living cells containing the plasmid can then be isolated from the original dish.

Question 10–5 Each of the DNA probes shown in Figure 10–17 is only 15 nucleotides long. On average how many exact matches to this sequence are expected to be found in the haploid human genome (3×10^9 nucleotides)? How could you ascertain that a match corresponds to the Factor VIII gene?

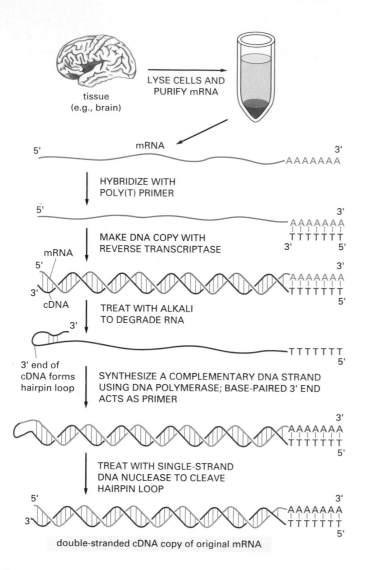

tissue
(e.g., brain)

LYSE CELLS AND
PURIFY mRNA

5'　　　　　　mRNA　　　　　　　　　3'
　　　　　　　　　　　　　　　　　　　AAAAAAA

HYBRIDIZE WITH
POLY(T) PRIMER

5'　　　　　　　　　　　　　　　　　3'
　　　　　　　　　　　　　　　　　AAAAAAA
　　　　　　　　　　　　　　　　　TTTTTTT
　　　　　　　　　　　　　　　3'　　　5'

mRNA
MAKE DNA COPY WITH
REVERSE TRANSCRIPTASE

5'　　　　　　　　　　　　　　　3'
3'　　　　　　　　　　　　　　AAAAAAA
　　　　　　　　　　　　　　TTTTTTT
cDNA　　　　　　　　　　　　5'

TREAT WITH ALKALI
TO DEGRADE RNA

3'
　　　　　　　　　　　　　TTTTTTT
　　　　　　　　　　　　　5'

3' end of
cDNA forms
hairpin loop

SYNTHESIZE A COMPLEMENTARY DNA STRAND
USING DNA POLYMERASE; BASE-PAIRED 3' END
ACTS AS PRIMER

　　　　　　　　　　　　　　　3'
　　　　　　　　　　　　　AAAAAAA
　　　　　　　　　　　　　TTTTTTT
　　　　　　　　　　　　　5'

TREAT WITH SINGLE-STRAND
DNA NUCLEASE TO CLEAVE
HAIRPIN LOOP

5'　　　　　　　　　　　　　　3'
3'　　　　　　　　　　　　AAAAAAA
　　　　　　　　　　　　TTTTTTT
　　　　　　　　　　　　5'

double-stranded cDNA copy of original mRNA

Figure 10–19 The synthesis of cDNA.
Total mRNA is extracted from a particular tissue, and DNA copies (cDNA) of the mRNA molecules are produced by the enzyme reverse transcriptase (see Figures 9–30 and 9–31). For simplicity, the copying of just one of these mRNAs into cDNA is illustrated here. A short oligonucleotide complementary to the poly(A) tail at the 3' end of the mRNA (discussed in Chapter 7) is first hybridized to the RNA to act as a primer for the reverse transcriptase, which then copies the RNA into a complementary DNA chain, thereby forming a DNA/RNA hybrid helix. Treating the DNA/RNA hybrid with alkali selectively degrades the RNA strand into its individual nucleotides. The remaining single-stranded cDNA is then copied into double-stranded cDNA by the enzyme DNA polymerase. The primer for this synthesis reaction is provided by the 3' end of the single-stranded DNA, which can fold back on itself and form a few chance complementary base pairs as shown.

from the cells, and DNA copies (**cDNA,** short for complementary DNA) of the mRNA molecules are produced by the enzyme reverse transcriptase (Figure 10–19). The cDNA molecules are then cloned, just like the genomic DNA fragments described earlier, to produce the cDNA library. Using such a cDNA library prepared from liver, the organ that normally produces Factor VIII, it was possible to isolate the complete coding sequence of the Factor VIII gene, devoid of introns. The Factor VIII cDNA was isolated from a cDNA library by using a portion of the genomic Factor VIII DNA as a probe and employing the procedure shown previously in Figure 10–18. We will see in the final part of this chapter how the coding sequence was used to produce purified human Factor VIII protein on a commercial scale.

There are several important differences between genomic DNA clones and cDNA clones, as illustrated in Figure 10–20. Genomic clones represent a random sample of all of the DNA sequences found in an organism's genome and, with very rare exceptions, will contain the same sequences regardless of the cell type from which the DNA came. Also, genomic clones from eucaryotes contain large amounts of repetitive DNA sequences, introns, gene regulatory regions, and spacer DNA, as well as sequences that code for proteins. By contrast, cDNA clones contain only coding sequences, and only those for genes that have been

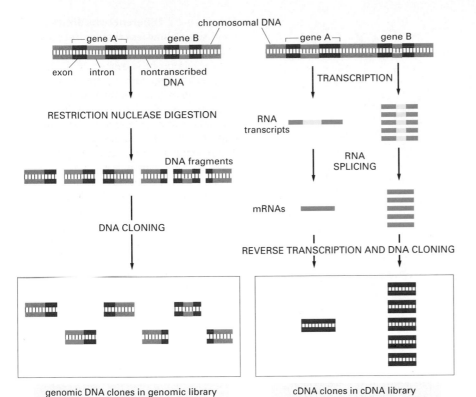

Figure 10–20 The differences between genomic DNA clones and cDNA clones derived from the same region of DNA. In this example, gene A is infrequently transcribed, whereas gene B is frequently transcribed, and both genes contain introns (*green*). In the genomic DNA library, both introns and the nontranscribed DNA (*pink*) are included in the clones, and most clones will contain only part of the coding sequence of a gene (*red*). In the cDNA clones the intron sequences (*yellow*) have been removed by RNA splicing during the formation of the mRNA (*blue*), and a continuous coding sequence is therefore present in each clone. Because gene B is transcribed more abundantly than is A in the cells from which the cDNA library was made, it will be represented much more often than A in the cDNA library. In contrast, A and B will in principle be represented equally in the genomic library.

transcribed into mRNA in the tissue from which the RNA came. As the cells of different tissues produce distinct sets of mRNA molecules, a different cDNA library will be obtained for each type of tissue. Patterns of gene expression change during development, and cDNA libraries are also constructed to reflect the genes expressed by cells at different stages in their development.

By far the most important advantage of cDNA clones is that they contain the uninterrupted coding sequence of the gene. Thus, if the aim of cloning the gene is either to deduce the amino acid sequence of the protein from the DNA or to produce the protein in bulk by expressing the cloned gene in a bacterial or yeast cell (neither of which can remove introns from mammalian RNA transcripts), it is essential to start with cDNA. Like genomic libraries, cDNA libraries of different tissues are research tools that are widely shared among investigators, and they are likewise available from commercial sources.

Hybridization Allows Even Distantly Related Genes to Be Identified

As described in Chapter 9, new genes arise during evolution by the duplication and divergence of existing genes and by the utilization of portions of old genes in new combinations. For this reason, most genes have a family of close relatives elsewhere in the genome, many of which are likely to have a related function. As we have seen, the effort required to isolate a gene from scratch is great, often requiring the prior determination of a protein sequence and then the manufacture of numerous DNA probes to discover one that works. However, once one member of a gene family has been isolated, other members of the family can often be isolated relatively easily by using sequences from the first gene as

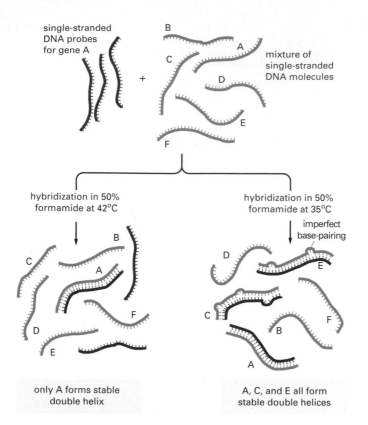

single-stranded
DNA probes
for gene A

mixture of
single-stranded
DNA molecules

hybridization in 50%
formamide at 42°C

hybridization in 50%
formamide at 35°C

imperfect
base-pairing

only A forms stable
double helix

A, C, and E all form
stable double helices

Figure 10–21 Different hybridization conditions allow less than perfect DNA matching. When only an identical match with a DNA probe is desired, the hybridization reaction is kept just a few degrees below the temperature at which a perfect DNA helix denatures in the solvent used (its *melting temperature*), so that all imperfect helices formed are unstable. When a DNA probe is being used to find DNAs that are related, but not identical in sequence, hybridization is carried out at a lower temperature. This allows even imperfectly paired double helices to form. Only the lower temperature hybridization conditions can be used to search libraries for genes (C and E in this example) that are nonidentical but related to gene A (see Figure 10–18).

DNA hybridization probes. Because the different members of a gene family do not usually have identical sequences, hybridization is carried out under conditions that allow even an imperfect match between the probe and its corresponding DNA to form a stable double helix (Figure 10–21). This approach also can allow a gene from one species (humans, for example) to be isolated using probes derived from the corresponding gene isolated from a more experimentally accessible species (a mouse or a fruit fly, for example).

The Polymerase Chain Reaction Amplifies Selected DNA Sequences

Although cloning via DNA libraries is the main route to gene isolation, a newer method invented about 10 years ago provides a quicker and less expensive alternative for cloning and many other applications of DNA technology. Known as the **polymerase chain reaction (PCR),** this new methodology can be carried out entirely *in vitro* without the use of cells. Using this technique, a given nucleotide sequence can be selectively and rapidly replicated in large amounts from any DNA that contains it. The polymerase chain reaction is now widely used, for example, in diagnostic tests for disease genes and to provide large amounts of any required gene from a small sample of human DNA.

PCR is based on the use of DNA polymerase to copy a DNA template in repeated rounds of replication. The polymerase is guided to the sequence to be copied by short primer oligonucleotides that are hybridized to the DNA template at the beginning and end of the desired DNA sequence. These primers are designed so that they provide a primer for DNA replication on each strand of the original double-stranded DNA.

(A)

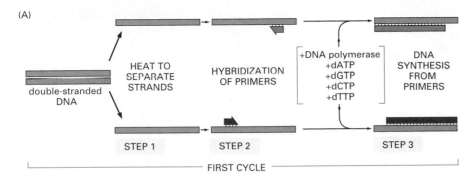

(B)

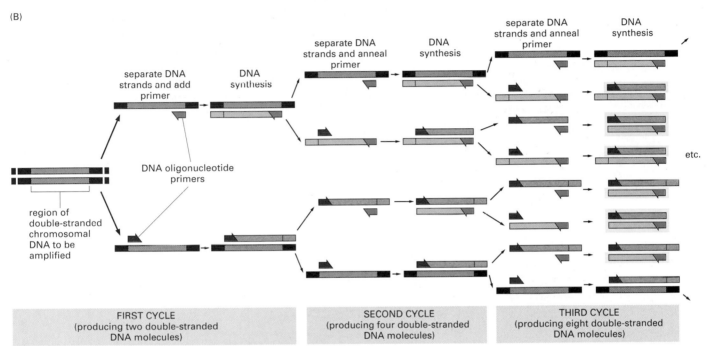

Figure 10–22 Amplification of DNA using the PCR technique. Knowledge of the DNA sequence to be amplified is used to design two synthetic DNA oligonucleotides, each complementary to the sequence on one strand of the DNA double helix at opposite ends of the region to be amplified. These oligonucleotides serve as primers for *in vitro* DNA synthesis, which is carried out by a DNA polymerase, and they determine the segment of the DNA that is amplified. (A) PCR starts with a double-stranded DNA, and each cycle of the reaction begins with a brief heat treatment to separate the two strands (step 1). After strand separation, cooling of the DNA in the presence of a large excess of the two primer DNA oligonucleotides allows these primers to hybridize to complementary sequences in the two DNA strands (step 2). This mixture is then incubated with DNA polymerase and the four deoxyribonucleoside triphosphates so that DNA is synthesized, starting from the two primers (step 3). The cycle is then begun again by a heat treatment to separate the newly synthesized DNA strands. The technique depends on the use of a special DNA polymerase isolated from a thermophilic bacterium; this polymerase is stable at much higher temperatures than eucaryotic DNA polymerases, so it is not denatured by the heat treatment shown in step 1. It therefore does not have to be added again after each cycle of reaction.

(B) As the procedure is carried out over and over again, the newly synthesized fragments serve as templates in their turn, and within a few cycles the predominant DNA is identical to the sequence bracketed by and including the two primers in the original template. In practice, 20–30 cycles are required for useful DNA amplification. Each cycle doubles the amount of DNA synthesized in the previous cycle. A single cycle takes only about 5 minutes, and automation of the whole procedure now enables cell-free cloning of a DNA fragment in a few hours, compared with the several days required for standard cloning procedures. Of the DNA put into the original reaction, only the sequence bracketed by the two primers is amplified because there are no primers attached anywhere else.

In the example illustrated in (B), three cycles of reaction produce 16 DNA chains, 8 of which (*boxed in yellow*) are the same length as and correspond exactly to one or the other strand of the original bracketed sequence shown at the far left; the other strands contain extra DNA downstream of the original sequence, which is replicated in the first few cycles. After three more cycles, 240 of the 256 DNA chains will correspond exactly to the original sequence, and after several more cycles, essentially all of the DNA strands will have this unique length.

333

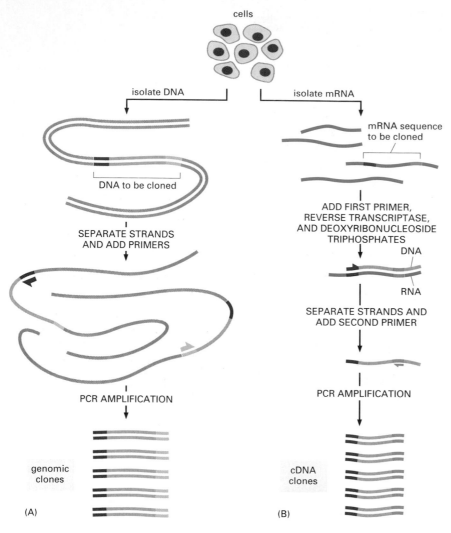

cells

isolate DNA · isolate mRNA

mRNA sequence to be cloned

DNA to be cloned

SEPARATE STRANDS AND ADD PRIMERS

ADD FIRST PRIMER, REVERSE TRANSCRIPTASE, AND DEOXYRIBONUCLEOSIDE TRIPHOSPHATES

DNA

RNA

SEPARATE STRANDS AND ADD SECOND PRIMER

PCR AMPLIFICATION

PCR AMPLIFICATION

genomic clones

cDNA clones

(A)

(B)

Figure 10–23 Use of PCR to obtain a genomic or cDNA clone. (A) To obtain a genomic clone using PCR, chromosomal DNA is first purified from cells. PCR primers that flank the stretch of DNA to be cloned are added, and many cycles of the PCR reaction are completed (see Figure 10–22). Since only the DNA between (and including) the primers is amplified, PCR provides a way to obtain selectively a short stretch of chromosomal DNA in an effectively pure form. (B) To use PCR to obtain a cDNA clone of a gene, mRNA is first purified from cells. The first primer is then added to the population of mRNAs, and reverse transcriptase is used to make a complementary DNA strand. The second primer is then added, and the single-stranded DNA molecule is amplified through many cycles of PCR, as shown in Figure 10–22.

Since the primers have to be chemically synthesized, PCR can be used only to clone DNA whose beginning and end sequences are known. Guided by these primers, DNA polymerase is then used to make many copies (billions are typical) of the sequence required. The method is outlined in Figure 10–22. PCR is extremely sensitive; it can detect a single copy of a DNA sequence in a sample by amplifying it so much that it becomes detectable by, for example, staining after separation by gel electrophoresis (see Figure 10–3B).

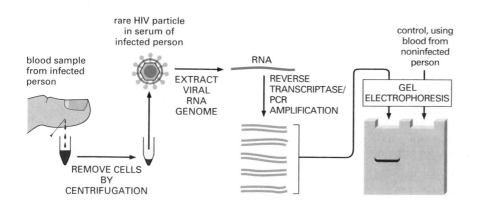

rare HIV particle in serum of infected person

blood sample from infected person

RNA

control, using blood from noninfected person

EXTRACT VIRAL RNA GENOME

REVERSE TRANSCRIPTASE/ PCR AMPLIFICATION

GEL ELECTROPHORESIS

REMOVE CELLS BY CENTRIFUGATION

Figure 10–24 Use of PCR to detect the presence of a viral genome in a sample of blood. The genome of the human immuno-deficiency virus (HIV), the cause of AIDS, is a single-stranded RNA molecule (see Figure 9–30). Because of its ability to enormously amplify the signal from every single molecule of nucleic acid, PCR is an extraordinarily sensitive method for detecting trace amounts of virus in a sample of blood or tissue without the need to purify the virus. In addition to HIV, many viruses that infect humans are now monitored in this way.

There are three especially useful applications of PCR. First, PCR can be used to clone directly a particular DNA fragment (a gene, for example) from a cell. The original template for the PCR reaction can be either DNA or RNA, so either a genomic DNA copy or a cDNA copy can be obtained by PCR (Figure 10–23).

Another application of PCR, which relies on its extraordinary sensitivity, is to detect viral infections at very early stages. Here, short sequences complementary to the viral genome are used as primers, and following many cycles of amplification, the presence or absence of even a single copy of a viral genome in a sample of blood can be ascertained (Figure 10–24). For many viral infections, PCR is the most sensitive method of detection; already it is replacing the use of antibodies against viral coat proteins to detect the presence of viruses in human samples.

Finally, PCR has great potential in forensic medicine. Its extreme sensitivity makes it possible to work with a very small sample—minute traces of blood and tissue that may contain the remnants of only a single cell—and still obtain a *DNA fingerprint* of the person from which it came. The genome of each human (other than identical twins) differs in DNA sequence from those of every other human; the DNA amplified by PCR using a particular primer pair is therefore quite likely to differ in sequence from one individual to another. Using a carefully selected set of primer pairs that cover the known highly variable parts of the human genome, PCR can generate a distinctive DNA fingerprint for each individual (Figure 10–25, next page).

Question 10–6

A. If the PCR reaction shown in Figure 10–22 is carried through an additional two rounds of amplification, how many of the DNA fragments labeled in *gray, green, red,* or outlined in *yellow* are produced? If many additional cycles are carried out, which fragments will predominate?

B. Assume you start with one double-stranded DNA molecule and amplify a 500–nucleotide pair sequence contained within it. Approximately how many cycles of PCR amplification will you need to produce 100 ng of this DNA? 100 ng is an amount that can be easily detected after staining with a fluorescent dye. (Hint: for this calculation, assume that each nucleotide has an average molecular weight of 330 g/mole.)

DNA Engineering

In this section we describe how extensions of the methods we have introduced so far in this chapter have revolutionized all other aspects of cell biology by providing new ways to study the functions of genes, RNA molecules, and proteins.

Completely Novel DNA Molecules Can Be Constructed

We have seen that recombinant DNA molecules are generally made by using DNA ligase to join together two DNA molecules (see Figure 10–12). For production of a DNA library, one of the two DNA molecules is the vector, derived from a bacterial plasmid or virus, while the other is either a fragment of a chromosome or a cDNA molecule (see Figure 10–16). Such a recombinant DNA molecule can in turn serve as a vector for the insertion and cloning of additional DNA, and by a repetition of this procedure, a cloned DNA can be generated that is different from any DNA that occurs naturally (Figure 10–26, p. 337). This new DNA can be made from the combination of naturally occurring DNA sequences, or its sequence can be entirely determined by the experimenter by using chemically synthesized DNA. By repeated DNA cloning steps, synthetic and naturally occurring DNAs can be joined together in various combinations to produce longer DNA molecules of any desired sequence. One of the most important uses of such custom-designed DNA sequences is for the high-level production of rare cellular proteins.

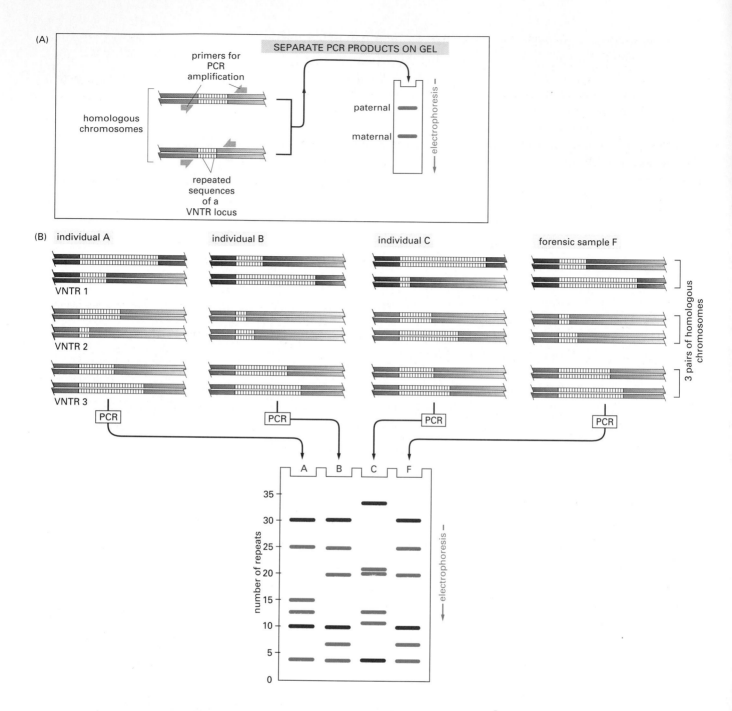

Figure 10–25 The use of PCR in forensic science. (A) The DNA sequences that create the variability used in this analysis contain runs of short, repeated sequences, such as GTGTGT . . . , which are found in various positions (loci) in the human genome. The number of repeats in each run is highly variable in the population, ranging from 4 to 40 in different individuals. A run of repeated nucleotides of this type is commonly referred to as a VNTR (*variable number of tandem repeat*) sequence. Because of the variability in these sequences, each individual will usually inherit a different variant of each VNTR locus from their mother and from their father; two unrelated individuals will therefore not usually contain the same pair of sequences. A PCR reaction using primers that bracket the locus produces a pair of bands of amplified DNA from each individual, one band representing the maternal variant and the other representing the paternal variant. The length of the amplified DNA, and thus its position after electrophoresis, will depend on the exact number of repeats at the locus. (B) The DNA bands obtained from a set of four different PCR reactions, each of which amplifies the DNA from a different VNTR locus. In the schematic example shown here, the same three VNTR loci are analyzed from three suspects (individuals A, B, and C), giving six bands for each person. You can see that although some individuals have several bands in common, the overall pattern is quite distinctive for each. The band pattern can serve as a "fingerprint" to identify an individual nearly uniquely. The fourth lane (F) contains the products of the same PCR reactions carried out on a forensic sample. The starting material for such a PCR reaction can be a single hair that was left at the scene of the crime. From the example, individuals A and C can be eliminated from enquiries, while B remains a clear suspect.

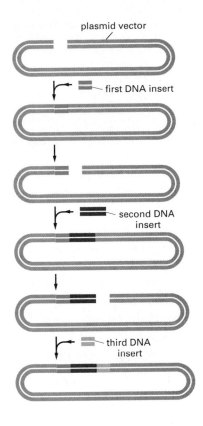

Figure 10–26 **Serial DNA cloning can be used to splice together a set of DNA fragments derived from different genes.** After each DNA insertion step, the recombinant plasmid is cloned to purify and amplify the new DNA (see Figure 10–15). The recombinant molecule is then cut once with a restriction nuclease, as indicated, and used as a cloning vector for the next DNA fragment.

Rare Cellular Proteins Can Be Made in Large Amounts Using Cloned DNA

Until recently, the only proteins that could be studied easily were those produced in relative abundance by cells. Starting with several hundred grams of cells, a major protein—one that constitutes 1% or more of the total cellular protein—can be purified by sequential chromatography steps (see Panel 5–5, pp. 162–163) to yield perhaps 0.1 g (100 mg) of pure protein. This amount is sufficient for conventional amino acid sequencing, for analysis of the protein's biological activity (for example, what biochemical reaction it might catalyze), and for the production of antibodies against the protein, which can then be used to detect it within the cell. Moreover, if suitable crystals of the protein can be grown (often a difficult task), its three-dimensional structure can be determined by x-ray crystallography (see Panel 5–6, pp. 164–165). The structure and function of many abundant proteins, including hemoglobin isolated from red blood cells, have been analyzed in this way.

But most of the thousands of different proteins in a eucaryotic cell, including many with crucially important functions, are present in very small amounts. For these it is extremely difficult, if not impossible, to obtain more than a few micrograms of pure material. One of the most important contributions of DNA cloning and genetic engineering to cell biology is that they make it possible to produce any protein, including the rare ones, in large amounts.

This is usually accomplished by using specially designed vectors known as *expression vectors*. Unlike the cloning vectors discussed earlier, these vectors also include appropriate gene regulatory and promoter DNA sequences necessary to enable an adjacent protein-coding DNA insert to be efficiently transcribed in cells. These vectors can thus direct the production of large amounts of mRNA, which can be translated into protein within the cell (Figure 10–27). Different expression vectors are designed for use in bacterial, yeast, insect, or mammalian cells, each containing the appropriate regulatory sequences for transcription and translation in these cells. The expression vector is replicated at each round of cell division, giving rise to a cell culture able to synthesize very large amounts of the protein of interest. Because the protein encoded by the expression vector is typically produced inside the cell, it must be purified away from the host cell proteins by cell lysis followed by

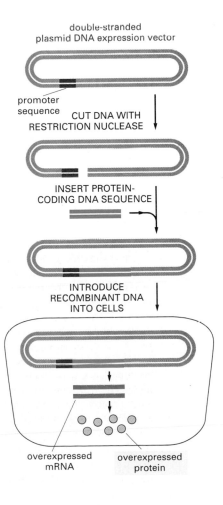

Figure 10–27 **Production of large amounts of a protein from a protein-coding DNA sequence cloned into an expression vector and introduced into cells.** A plasmid vector has been engineered to contain a highly active promoter, which causes unusually large amounts of mRNA to be produced from an adjacent protein-coding gene inserted into the plasmid vector. Depending on the characteristics of the cloning vector, the plasmid is introduced into bacterial, yeast, insect, or mammalian cells, where the inserted gene is efficiently transcribed and translated into protein.

DNA Engineering

337

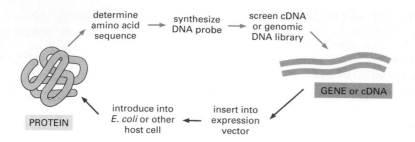

determine amino acid sequence → synthesize DNA probe → screen cDNA or genomic DNA library

GENE or cDNA

introduce into *E. coli* or other host cell ← insert into expression vector

PROTEIN

Figure 10–28 **Knowledge of the molecular biology of cells makes it possible to experimentally move from gene to protein and from protein to gene.** A small quantity of a purified protein is used to obtain a partial amino acid sequence. This provides sequence information that enables the corresponding gene to be cloned from a DNA library (see Figure 10–18). Once the gene has been cloned, its protein-coding sequence can be used to design a DNA that can then be used to produce large quantities of the protein from genetically engineered cells (see Figure 10–27).

chromatography (see Panel 5–4, pp. 160–161); but because it is so plentiful in the cell lysate (often comprising 1–10% of the total cell protein), purification is usually easy to accomplish in only a few steps.

These techniques have been used to produce many proteins of cell biological interest in large enough amounts for the detailed structural and functional studies that were previously possible only for a few (Figure 10–28). This technology is now also used to make large amounts of many medically useful proteins. The Factor VIII protein, for example, is now made commercially from cultures of genetically engineered mammalian cells and is thus free of viral contamination. Many other useful proteins, such as insulin, growth hormone, and viral coat proteins for use in vaccines, are also produced in this fashion.

RNAs Can Be Produced by Transcription *in Vitro*

We saw in the previous sections that recombinant DNA technology can be used to clone genes and to produce large quantities of even very rare proteins. Many of the discoveries discussed in this book were made using DNA and protein molecules produced in this way. However, many cell biological studies also require the use of purified RNA molecules. For example, studies of RNA splicing, protein synthesis, and RNA-based enzymes (see Chapter 7) are greatly facilitated by having available pure RNA molecules. Most RNAs are present only in tiny quantities in cells, and they are very difficult to purify away from the other cellular components, especially from the many thousands of other RNAs present in a cell. But DNA technology can be utilized to produce large amounts of any RNA molecule once its gene has been isolated. The most efficient method relies on the synthesis *in vitro* of the RNA of interest in the absence of any other RNA molecule (Figure 10–29).

Figure 10–29 **Large amounts of any RNA molecule can be prepared *in vitro* by transcription using a viral RNA polymerase.** The technique uses the unusually efficient RNA polymerases produced by certain bacterial viruses. To prepare a suitable DNA template for transcription, the DNA sequence encoding the RNA is cloned in a way that joins it to a second DNA segment that contains a start signal (promoter) recognized by the viral RNA polymerase. As indicated, the future 3′ end of the RNA is determined by cutting the DNA template at a unique site with a restriction nuclease. The purified DNA template is then mixed with a pure preparation of the RNA polymerase plus the four ribonucleoside triphosphates used in RNA synthesis (ATP, CTP, GTP, UTP). During many cycles of *in vitro* transcription, large amounts of the desired RNA are generated. The RNA can then be easily purified away from the DNA template and the RNA polymerase.

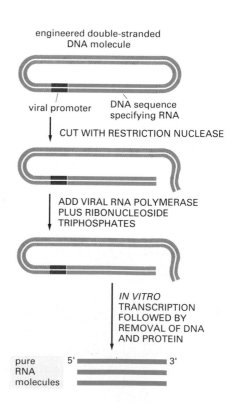

engineered double-stranded DNA molecule

viral promoter   DNA sequence specifying RNA

CUT WITH RESTRICTION NUCLEASE

ADD VIRAL RNA POLYMERASE PLUS RIBONUCLEOSIDE TRIPHOSPHATES

IN VITRO TRANSCRIPTION FOLLOWED BY REMOVAL OF DNA AND PROTEIN

pure RNA molecules   5′ ——— 3′

Mutant Organisms Best Reveal the Function of a Gene

So far in this chapter we have assumed that the function of a cloned gene is known. But suppose that you have cloned a gene that codes for a protein of unknown function. How do you discover what the protein does in the cell or in the organism as a whole? Now that genome-sequencing projects are rapidly identifying new genes from the DNA sequence alone, this has become a common question in cell biology. Neither the complete nucleotide sequence of a gene nor even the three-dimensional structure of the protein is sufficient to deduce a protein's function. Many proteins—such as those that have a structural role in the cell or normally form part of a large multienzyme complex—will have no obvious activity on their own. Even those that do have known biochemical activities (motor proteins and protein kinases, for example) could in principle participate in a number of different pathways in the cell, and it is not always clear from their biochemical activities how the proteins are actually used.

Genetics provides a powerful solution to this problem. Mutants that lack a particular protein may clearly reveal what the normal function or functions of the molecule are. Even more useful are mutants in which an abnormal protein is produced that functions only in a very narrow temperature range; such proteins can be inactivated by a small increase or decrease in temperature, and in such mutants the abnormality can thus be switched on and off simply by changing the temperature. Before the advent of gene cloning, the functions of most known genes were identified in this way, according to the cellular or physiological processes that were disrupted by a mutation. The genetic approach is most easily applicable to organisms that reproduce rapidly and can be easily mutated in the laboratory—such as bacteria, yeasts, nematode worms, and fruit flies.

Recombinant DNA technology has made possible a different type of genetic approach. Instead of starting with a randomly generated mutant and using it to identify a gene and its protein, one can start with a cloned gene and proceed to make mutations in it *in vitro*. Then, by reintroducing the altered gene back into the organism from which it originally came, one can produce a mutant organism in which the gene's function may be revealed. Using the techniques to be discussed shortly, the coding sequence of a cloned gene can be altered to change the functional properties of the protein product or even to eliminate it. Alternatively, the regulatory region of the gene can be altered so that the amount of protein made is altered, or so that the gene is expressed in a different cell type from normal or at a different time during development.

The ability to manipulate DNA *in vitro* makes it possible to introduce precise mutations, and genes can thus be altered in very subtle ways. It is often desirable, for example, to change the protein that the gene encodes by just one or a few amino acids. The use of the technique of **site-directed mutagenesis** to achieve this is outlined in Figure 10–30. By changing selected amino acids in this way, one can determine which parts of the polypeptide chain are crucial to such fundamental processes as protein folding, protein-ligand interactions, and enzymatic catalysis.

Question 10–7 After decades of work, Dr. Ricky M. isolated a small amount of attractase, an enzyme producing a powerful human pheromone, from hair samples of Hollywood celebrities. To produce attractase for his personal use, he obtained a complete genomic clone of the attractase gene, connected it to a strong bacterial promoter on an expression plasmid, and introduced the plasmid into *E. coli* cells. He was devastated to find that no attractase was produced in the cells. What is a likely explanation for the failure?

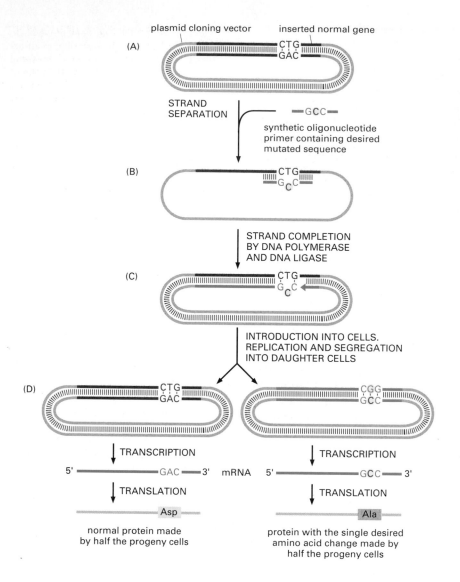

(A) plasmid cloning vector   inserted normal gene

CTG
GAC

STRAND
SEPARATION

—GCC—

synthetic oligonucleotide
primer containing desired
mutated sequence

(B)

CTG
G C C

STRAND COMPLETION
BY DNA POLYMERASE
AND DNA LIGASE

(C)

CTG
G C C

INTRODUCTION INTO CELLS.
REPLICATION AND SEGREGATION
INTO DAUGHTER CELLS

(D)

CTG
GAC

CGG
GCC

TRANSCRIPTION   TRANSCRIPTION

5′ —— GAC —— 3′   mRNA   5′ —— GCC —— 3′

TRANSLATION   TRANSLATION

Asp   Ala

normal protein made
by half the progeny cells

protein with the single desired
amino acid change made by
half the progeny cells

Figure 10–30 The use of a synthetic oligonucleotide to modify the protein-coding region of a gene by site-directed mutagenesis. (A) A recombinant plasmid containing a gene insert is separated into its two DNA strands. A synthetic oligo-nucleotide primer corresponding to part of the gene sequence but containing a single altered nucleotide at a predetermined point is added to the single-stranded DNA under conditions that permit less than perfect DNA hybridization (see Figure 10–21). (B) The primer hybridizes to the DNA, forming a single mismatched nucleotide pair. (C) The recombinant plasmid is made double stranded by *in vitro* DNA synthesis starting from the primer and sealing by DNA ligase. (D) The double-stranded DNA is introduced into a cell, where it is replicated. Replication of one strand produces a normal DNA molecule, but replication of the other (the strand that contains the primer) produces a DNA molecule carrying the desired mutation. Only half of the progeny cells will end up with a plasmid that contains the desired mutant gene; however, a progeny cell that contains the mutated gene can be identi-fied, separated from other cells, and cul-tured to produce a pure population of cells, all of which carry the mutated gene. Only one of the many changes that can be engineered in this way is shown here. With an oligonucleotide of the appropriate sequence, more than one amino acid sub-stitution can be made at a time, or one or more amino acids can be added or deleted.

Transgenic Animals Carry Engineered Genes

The ultimate test of the function of a gene mutated as described in Figure 10–30 is to insert it into the genome of an organism and see what effect it has. Organisms into which a new gene has been introduced, or those whose genomes have been altered in other ways using recombinant DNA techniques, are known as **transgenic organisms.**

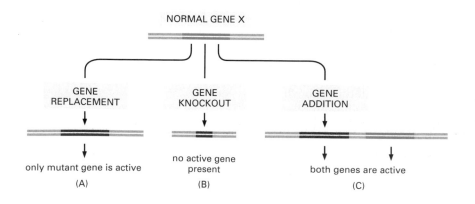

NORMAL GENE X

GENE
REPLACEMENT

GENE
KNOCKOUT

GENE
ADDITION

only mutant gene is active

(A)

no active gene
present

(B)

both genes are active

(C)

Figure 10–31 Gene replacement, gene knockout, and gene addition. A normal gene can be altered in several ways in a genetically engineered organism. (A) The normal gene can be completely replaced by a mutant copy of the gene, a process called gene replacement. This will provide information on the activity of the mutant gene, without interference from the nor-mal gene, and thus the effects of small and subtle mutations can be determined. (B) The normal gene can be inactivated com-pletely, for example, by making a large deletion in it; the gene is said to have suf-fered a knockout. This type of alteration is very widely used to obtain information on the possible function of the normal gene in the whole animal. (C) A mutant gene can simply be added to the genome. In some organisms this is the easiest type of genetic engineering to perform. Even this can still provide useful information when the introduced mutant gene overrides the function of the normal gene.

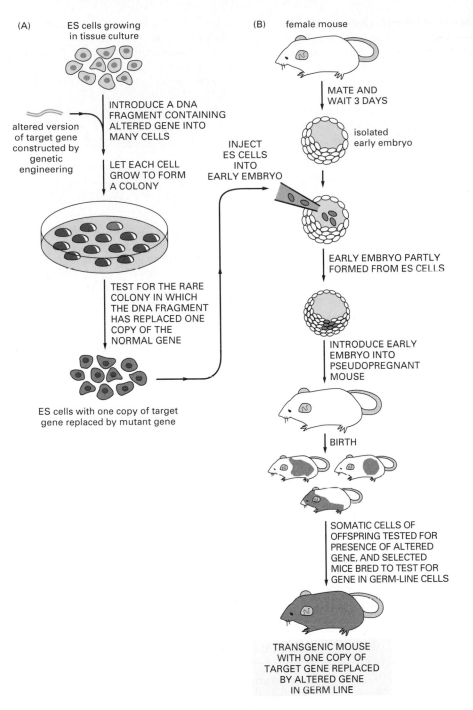

(A) ES cells growing in tissue culture

altered version of target gene constructed by genetic engineering

INTRODUCE A DNA FRAGMENT CONTAINING ALTERED GENE INTO MANY CELLS

LET EACH CELL GROW TO FORM A COLONY

TEST FOR THE RARE COLONY IN WHICH THE DNA FRAGMENT HAS REPLACED ONE COPY OF THE NORMAL GENE

ES cells with one copy of target gene replaced by mutant gene

(B) female mouse

MATE AND WAIT 3 DAYS

isolated early embryo

INJECT ES CELLS INTO EARLY EMBRYO

EARLY EMBRYO PARTLY FORMED FROM ES CELLS

INTRODUCE EARLY EMBRYO INTO PSEUDOPREGNANT MOUSE

BIRTH

SOMATIC CELLS OF OFFSPRING TESTED FOR PRESENCE OF ALTERED GENE, AND SELECTED MICE BRED TO TEST FOR GENE IN GERM-LINE CELLS

TRANSGENIC MOUSE WITH ONE COPY OF TARGET GENE REPLACED BY ALTERED GENE IN GERM LINE

Figure 10–32 Summary of the procedures used for making gene replacements in mice. In the first step (A), an altered version of the gene is introduced into cultured ES (embryonic stem) cells. A few rare ES cells will have their corresponding normal genes replaced by the altered gene through homologous recombination (see Figure 9–13A). Although often laborious, these rare cells can be identified and cultured to produce many descendants, each of which carries an altered gene in place of one of its two normal corresponding genes. In the next step of the procedure (B), these altered ES cells are injected into a very early mouse embryo; the cells are incorporated into the growing embryo, and a mouse produced by such an embryo will contain some somatic cells (indicated by *orange*) that carry the altered gene. Some of these mice will also contain germ-line cells that contain the altered gene. When bred with a normal mouse, some of the progeny of these mice will contain the altered gene in all of their cells. If two such mice are in turn bred, some of the progeny will contain two altered genes (one on each chromosome) in all of their cells. If the original gene alteration competely inactivates the function of the gene, these mice are known as "knockout" mice. It is often the case that mice missing genes that function during development die long before they reach adulthood (see Figure 10–33).

In order to study the function of a gene mutated *in vitro*, ideally one would like simply to replace the normal gene with the altered one so that the function of the mutant protein can be analyzed in the absence of the normal protein. Such gene replacement (Figure 10–31A) can be accomplished quite easily by homologous recombination between the introduced mutant DNA and the chromosomal DNA in some simple haploid organisms such as bacteria or yeasts (see Figure 9–13A). Homologous recombination with an introduced DNA that contains a deletion in the gene of interest can lead to a large deletion being made in the normal gene and its complete inactivation, or *knockout* (Figure 10–31B). A third possibility is that the mutant gene is added to the genome without any alteration being made to the normal gene in the process (Figure 10–31C).

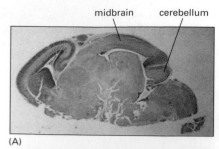

midbrain　　cerebellum

(A)

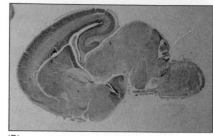

(B)

Figure 10–33 **A transgenic mouse in which both copies of the gene for the developmentally important growth factor *wnt-1* have been permanently inactivated (knocked out) by homologous recombination.** (A) Normal embryo; (B) mutant embryo. The mutant embryo lacks a cerebellum (*arrow*) and most of the midbrain and dies *in utero*. This experiment shows that the *wnt-1* gene, a member of a gene family that is important in the development of many different organisms, is required during early mammalian development for the formation of the cerebellum. (Photograph courtesy of Mario Capecchi.)

It is more difficult, but still possible, to achieve gene replacements in diploid organisms with large and complex genomes, such as mice (Figure 10–32). If the altered gene can be introduced into cells that will form part of the germ line, such transgenic animals are able to pass the altered gene on to at least some of their progeny as a permanent part of their genome. Technically, even the human germ line could now be altered in this way, although this is unlawful, even for therapeutic purposes, for fear of the unpredictable aberrations that might occur in such individuals. Similar techniques are, however, being explored to correct genetic defects in human somatic cells for use in somatic gene therapy. For example, some genetic diseases could be alleviated by the introduction of genetically corrected somatic cells into the tissue most affected by the disease. However, the alterations would not be passed on to descendants.

If the normal gene is replaced by a deletion or an otherwise inactive version of the gene (see Figure 10–31B), transgenic techniques make it possible to produce complex organisms that are missing certain gene products entirely. For example, many *knockout mice*—strains of mice that have a particular gene permanently inactivated—have now been produced. The study of mice with knockouts of genes that orchestrate development has led to considerable recent advances in identifying the precise roles of these genes in the complex process of mammalian development (Figure 10–33).

Essential Concepts

- Recombinant DNA technology has revolutionized the study of the cell, making it possible for researchers to pick out any gene at will from the thousands of genes in a cell and, after an amplification step, to determine the exact molecular structure of the gene.

- A crucial element in this technology is the ability to cut a large DNA molecule into a specific and reproducible set of DNA fragments using restriction nucleases, each of which cuts the DNA double helix only at a particular nucleotide sequence.

- DNA fragments can be separated from one another on the basis of size using gel electrophoresis.

- Techniques are now available for rapidly determining the nucleotide sequence of any isolated DNA fragment.

- The complete nucleotide sequence of the genome of several single-celled organisms (including several bacteria and a yeast) are now known. Those of more complex organisms (the nematode worm, *Drosophila*, several plants, and even humans) are anticipated in the next decade.

- Nucleic acid hybridization can detect any given DNA or RNA sequence in a mixture of nucleic acid fragments. This technique relies on the fact that a single strand of DNA or RNA will form a double helix only with another nucleic acid strand of the complementary nucleotide sequence.

- Single-stranded DNAs of known sequence and labeled with fluorescent dyes or radioisotopes are used as probes in hybridization reactions. Nucleic acid hybridization can be used to detect the precise location of genes in chromosomes, or RNAs in cells and tissues.

- DNA strands of any desired sequence up to about 120 nucleotides can be made by chemical (non-enzymatic) synthesis in the laboratory.

- DNA cloning techniques enable a DNA sequence to be selected from millions of other sequences and produced in unlimited amounts in pure form.

- DNA fragments can be joined together *in vitro* using DNA ligase to form recombinant DNA molecules not found in nature.

- The first step in a typical cloning procedure is to insert the DNA fragment to be cloned into a DNA molecule capable of replication, such as a plasmid or a viral genome. This recombinant DNA molecule is then introduced into a rapidly dividing host cell, usually a bacterium, so that the DNA is replicated at each cell division.

- A collection of cloned fragments of chromosomal DNA representing the complete genome of an organism is known as a genomic library. The library is often maintained as clones of bacteria, each clone carrying a different DNA fragment.

- cDNA libraries contain cloned DNA copies of the total mRNA of a particular cell type or tissue.

- Unlike genomic DNA clones, cloned cDNAs contain only protein-coding sequences; they lack introns, gene regulatory sequences, and promoters. They are thus most suitable for use when the cloned gene is to be expressed to make a protein.

- The polymerase chain reaction (PCR) is a powerful form of DNA amplification that is carried out *in vitro* using a purified DNA polymerase. PCR requires a prior knowledge of the sequence to be amplified, since two synthetic oligonucleotide primers must be synthesized that bracket the portion of DNA to be replicated.

- Cloned genes can be permanently inserted into the genome of a cell or an organism by the techniques of genetic engineering. Cloned DNA can be altered *in vitro* to create mutant genes to order and then reinserted into a cell or an organism to study gene function.

- Genetic engineering has far-reaching consequences. Bacteria, yeasts, and mammalian cells can be engineered to synthesize a particular protein from any organism in large quantities, thus making it possible to study otherwise rare or difficult-to-isolate proteins.

Key Terms

cDNA

DNA cloning

DNA hybridization

DNA library

DNA ligase

in situ hybridization

plasmid

polymerase chain reaction

recombinant DNA technology

restriction map

restriction nuclease

site-directed mutagenesis

transgenic organism

Questions

Question 10–8   Which of the following statements are correct? Explain your answers.

A. Restriction nucleases cut DNA at specific sites that are always located between genes.

B. DNA migrates toward the positive electrode during electrophoresis.

C. Clones isolated from cDNA libraries contain promoter sequences.

D. PCR utilizes a heat-stable DNA polymerase because for each amplification step, double-stranded DNA must be heat denatured.

E. Digestion of genomic DNA with Alu I, a restriction enzyme that recognizes a four-nucleotide sequence, produces fragments that are all exactly 256 nucleotides in length.

F. To make a cDNA library, both a DNA polymerase and a reverse transcriptase must be used.

G. DNA fingerprinting by PCR relies on the fact that different individuals have different numbers of repeats in VNTR regions in their genome.

H. It is possible for a coding region of a gene to be present in a genomic library prepared from a particular tissue, but not to be represented in a cDNA library prepared from the same tissue.

Question 10–9

A. Determine the sequence of the DNA that was used in the sequencing reaction shown in Figure Q10–9. The four lanes show the products of sequencing reactions that contained ddG (lane 1), ddA (lane 2), ddT (lane 3), and ddC (lane 4). The numbers to the right of the autoradiograph represent the positions of DNA fragments of 50 and 116 nucleotides.

B. The DNA was derived from the middle of a cDNA clone of a mammalian protein. What is the amino acid sequence of this portion of the protein?

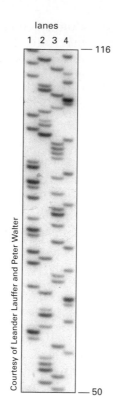

lanes
1 2 3 4
— 116

Courtesy of Leander Lauffer and Peter Walter

— 50

Figure Q10–9

Question 10–10

A. How many different DNA fragments would you expect to obtain if you cleaved human genomic DNA with Hae III? (Recall that there are 3×10^9 nucleotide pairs per haploid genome.) How many fragments would you expect with Eco RI, or with Not I?

B. Human genomic libraries are often made from fragments obtained by cleaving human DNA with Hae III in such a way that the DNA is only partially digested, that is, not all the possible Hae III sites have been cleaved. What is a possible reason for doing this?

Question 10–11 A molecule of double-stranded DNA was cleaved with three different restriction nucleases, and the resulting products were separated by electrophoresis (Figure Q10–11). DNA fragments of known sizes were electrophoresed on the same gel for use as size markers. The size of the DNA markers is given in kilobase pairs (kb), where 1 kb = a length of 1000 nucleotide pairs. Using the size markers as a guide, estimate the size of each restriction fragment obtained. From this information, deduce a map of the original DNA molecule that indicates the relative positions of all the restriction sites.

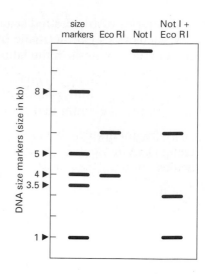

size markers Eco RI Not I Not I + Eco RI

DNA size markers (size in kb)

8 ▶
5 ▶
4 ▶
3.5 ▶

1 ▶

Figure Q10–11

Question 10–12 A mutation engineered *in vitro* as shown in Figure 10–30C introduces a mismatch into the DNA. Would you expect this mismatch to be recognized and repaired by DNA mismatch repair enzymes (see Figure 6–26) when the plasmid that contains the mismatch is introduced into cells? Explain your answer.

Question 10–13 You have isolated a small amount of a rare protein. You cleaved the protein into fragments using proteases, separated some of the fragments by chromatography, and determined their amino acid sequence (as described in Panel 5–6, pp. 164–165). Unfortunately, as it is often the case when only small amounts of protein are available to start with, you obtained only three short stretches of amino acid sequence from the protein:

1. Trp-Met-His-His-Lys
2. Leu-Ser-Arg-Leu-Arg
3. Tyr-Phe-Gly-Met-Gln

A. Using the genetic code (see Figure 7–20), design oligonucleotide probes that could be used to detect the gene in a cDNA library by hybridization. Which of the three sets of oligonucleotide probes would it be preferable to use first? Explain your answer.

B. You have also been able to determine that the Gln of your peptide #3 is the carboxyl-terminal (i.e., the final) amino acid of your protein. How would you go about designing oligonucleotide primers that could be used to amplify a portion of the gene from a cDNA library using PCR?

C. Suppose the PCR amplification in (B) yields a DNA that is precisely 300 nucleotides long. Upon determining the nucleotide sequence of this DNA, you find the sequence CTATCACGCTTTAGG approximately in its middle. What would you conclude from these observations?

Question 10–14 Discuss the following statement: "From the nucleotide sequence of a cDNA clone, the complete amino acid sequence of a protein can be deduced by applying the genetic code. Thus, protein biochemistry has become superfluous because there is nothing more that can be learned by studying the protein."

Question 10–15 Assume that a DNA sequencing reaction is carried out as shown in Figure 10–5, except that the four different dideoxyribonucleoside triphosphates are modified so that each contains a covalently attached dye of a different color (which does not interfere with its incorporation into the DNA chain). What would the products be if you added a mixture of all four of these labeled dideoxyribonucleoside triphosphates along with the four unlabeled deoxyribonucleoside triphosphates into a single sequencing reaction? What would the results look like if you electrophoresed these products in a single lane of a gel?

Question 10–16 As described in the answer to Question 10–10B, genomic DNA clones are often used to "walk" along a chromosome. In this approach, one cloned DNA is used to isolate other clones that contain overlapping DNA sequences (Figure Q10–16). Using this method, it is possible to build up a stretch of DNA sequence and thus identify new genes in near proximity to a previously cloned gene.

A. Would it be faster to use cDNA clones in this method, because they do not contain any intron sequences?

B. What would the consequences be if you encountered a repetitive DNA sequence, like the L1 transposon (see Figure 9–24), which is found in many copies and in many different places in the genome?

Question 10–17 One of the first organisms that was genetically engineered using modern DNA technology was a bacterium that normally lives on the surface of strawberry plants. This bacterium makes a protein, called ice-protein, that causes the efficient formation of ice crystals around it when the temperature drops to just below freezing. Thus, strawberries harboring this bacterium are particularly susceptible to frost damage because their cells get destroyed by the ice crystals. Consequently, strawberry farmers have a considerable interest in preventing this.

A genetically engineered version of this bacterium was constructed in which the ice-protein gene was knocked out. The mutant bacteria were then introduced in large numbers into strawberry fields, where they displaced the normal bacteria by competition for their ecological niche. This approach has been successful: strawberries bearing the mutant bacteria show a much reduced susceptibility to frost damage.

Nevertheless, at the time they were first carried out, the initial open-field trials triggered an intense debate because they represented the first release into the environment of an organism that had been genetically engineered using recombinant DNA techniques. Indeed, all preliminary experiments were carried out with extreme caution and in strict containment. The

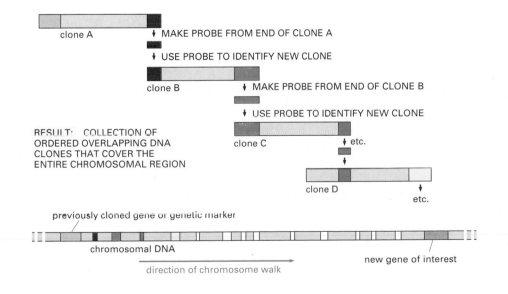

clone A   ↓ MAKE PROBE FROM END OF CLONE A

                 ↓ USE PROBE TO IDENTIFY NEW CLONE

clone B   ↓ MAKE PROBE FROM END OF CLONE B

                 ↓ USE PROBE TO IDENTIFY NEW CLONE

RESULT: COLLECTION OF
ORDERED OVERLAPPING DNA   clone C   ↓ etc.
CLONES THAT COVER THE   ↓
ENTIRE CHROMOSOMAL REGION

                                 clone D   ↓
                                                  etc.

previously cloned gene or genetic marker

chromosomal DNA

new gene of interest

Figure Q10–16   direction of chromosome walk

photograph gives an idea of the containment conditions during the initial applications of the bacteria to strawberry plants (Figure Q10–17).

Discuss some of the issues that arise from such applications of DNA technology. Do you think that bacteria lacking the ice-protein could be isolated without the use of modern DNA technology? Is it likely that such mutations have already occurred in nature? Would the use of a mutant bacterial strain isolated from nature be of lesser concern? Should we be concerned about the risks posed by the application of genetic engineering techniques in agriculture, medicine, and technology? Explain your answers.

Courtesy of John Bedbrook and DNA Plant Technology Corporation

Figure Q10–17

11 Membrane Structure

A living cell is a self-reproducing system of molecules held inside a container. The container is the **plasma membrane**—a fatty film so thin and transparent that it cannot be seen directly in the light microscope. It is simple in construction, being based on a sheet of lipid molecules about 5 nm—that is, about 50 atoms—thick. Its properties, however, are unlike those of any sheet of material that we are familiar with in the everyday world. Although it serves as a barrier to prevent the contents of the cell from escaping and mixing with the surrounding medium (Figure 11–1A), the plasma membrane does much more than that. Nutrients have to pass inward across it if the cell is to survive and grow, and waste products have to pass outward. Thus the membrane is penetrated by highly selective channels and pumps, formed from protein molecules, that allow specific substances to be imported while others are exported. Still other protein molecules in the membrane act as sensors to enable the cell to respond to changes in its environment. The mechanical properties of the membrane are equally remarkable. When the cell grows or changes shape, so does the membrane: it enlarges its area by addition of new membrane without ever losing its continuity, and it can deform without tearing (Figure 11–2). If it is pierced, it neither collapses like a balloon nor remains torn; instead, it quickly reseals.

The simplest bacteria have only a single membrane—the plasma membrane. Eucaryotic cells, however, contain in addition a profusion of *internal membranes* that enclose intracellular compartments. These other membranes are constructed on the same principles as the plasma membrane, and they too serve as selective barriers between spaces

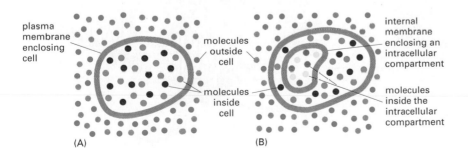

(A) (B)

plasma membrane enclosing cell

molecules outside cell

molecules inside cell

internal membrane enclosing an intracellular compartment

molecules inside the intracellular compartment

Figure 11–1 Cell membranes as barriers. Membranes serve as barriers between two compartments—either between the inside and the outside of the cell (A) or between two intracellular compartments (B). In either case the membrane prevents molecules on one side from mixing with those on the other.

347

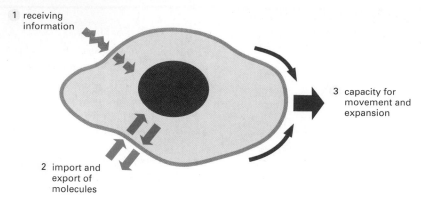

Figure 11–2 **Some functions of the plasma membrane.**

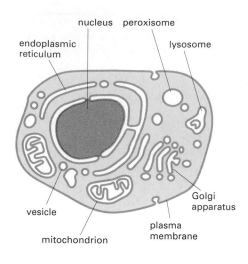

Figure 11–3 **Membranes form the many different compartments in a eucaryotic cell.** The membrane-bounded organelles in a typical animal cell are shown here. Note that the nucleus and mitochondria are each enclosed by two membranes.

containing distinct collections of molecules (Figure 11–1B). Thus the membranes of the endoplasmic reticulum, Golgi apparatus, mitochondria, and other membrane-bounded organelles (Figure 11–3) maintain the characteristic differences in composition and function between these organelles. These internal membranes act as more than just barriers: subtle differences between them, especially differences in the membrane proteins, are largely responsible for giving each organelle its distinctive character.

All cell membranes are composed of lipids and proteins and have a common general structure (Figure 11–4). The lipid component consists of many millions of lipid molecules arranged in two closely apposed sheets, forming a *lipid bilayer* (see Figure 11–4B). The lipid bilayer provides the basic structure and serves as a permeability barrier. The protein molecules mediate most of the other functions of the membrane and give different membranes their individual characteristics. In this chapter we first discuss membrane lipids and then membrane proteins.

The Lipid Bilayer

The **lipid bilayer** has been firmly established as the universal basis of cell-membrane structure. Its properties are responsible for the general properties of cell membranes. We begin this section by considering how

Figure 11–4 **Two views of a cell membrane.** (A) An electron micrograph of a plasma membrane (of a human red blood cell) seen in cross-section. (B) Schematic drawing showing a three-dimensional view of a cell membrane. (A, courtesy of Daniel S. Friend.)

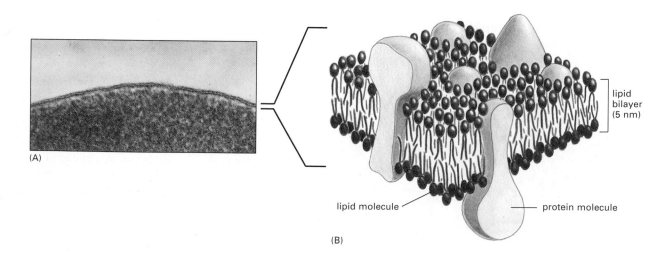

(A)

lipid molecule

protein molecule

lipid bilayer (5 nm)

(B)

the bilayer structure is a consequence of the way membrane lipid molecules behave in a watery (aqueous) environment.

Membrane Lipids Form Bilayers in Water

The lipids in cell membranes combine two very different properties in a single molecule: they have a hydrophilic ("water-loving") *head* and one or two hydrophobic ("water-hating") *hydrocarbon tails* (Figure 11–5). The most abundant membrane lipids are the **phospholipids,** in which the hydrophilic head group is linked to the rest of the molecule through a phosphate group. The most common type of phospholipid in most cell membranes is **phosphatidylcholine,** which has the small molecule choline attached to phosphate as its hydrophilic head and two long hydrocarbon chains as its hydrophobic tails (Figure 11–6).

Molecules with both hydrophilic and hydrophobic properties are termed **amphipathic.** This feature is also shared by other types of membrane lipids—*sterols* (such as cholesterol in animal cell membranes) and the *glycolipids,* which have sugars as their hydrophilic head (Figure 11–7)—and it plays a crucial part in driving the lipid molecules to assemble into bilayers.

As discussed in Chapter 2, hydrophilic molecules dissolve readily in water because they contain charged atoms or polar groups (that is, groups with an uneven distribution of positive and negative charges) that can form electrostatic bonds or hydrogen bonds with water molecules, which are themselves polar (Figure 11–8). Hydrophobic molecules, by contrast, are insoluble in water because all, or almost all, of their atoms are uncharged and nonpolar and therefore cannot form bonds with water molecules. Instead, they force adjacent water molecules to reorganize themselves into a cagelike structure around the

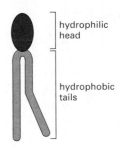

Figure 11–5 A schematic drawing of a typical membrane lipid molecule.

Figure 11–6 A phosphatidylcholine molecule. It is represented (A) schematically, (B) in formula, (C) as a space-filling model, and (D) as a symbol. This particular phospholipid is built from five parts: the hydrophilic head, *choline,* is linked via a *phosphate* to *glycerol,* which in turn is linked to two *hydrocarbon chains,* forming the hydrophobic tail. The two hydrocarbon chains originate as *fatty acids*— that is, hydrocarbon chains with a –COOH group at one end—which become attached to glycerol via their –COOH groups. A kink in one of the hydrocarbon chains occurs where there is a double bond between two carbon atoms; it is exaggerated in these drawings for emphasis. The *phosphatidyl-* part of the name of phospholipids refers to the phosphate–glycerol–fatty acid portion of the molecule.

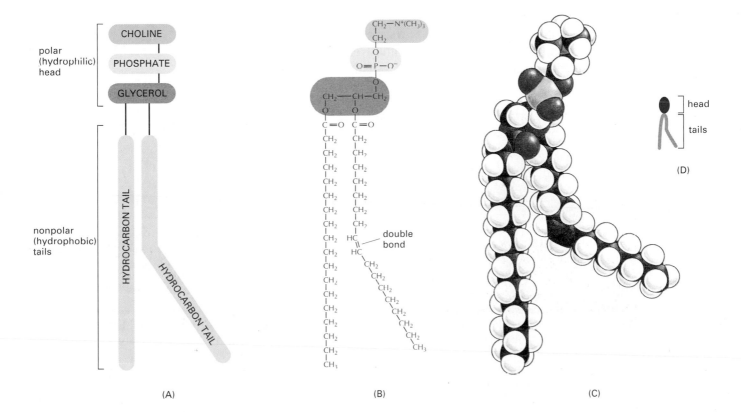

(A) (B) (C)

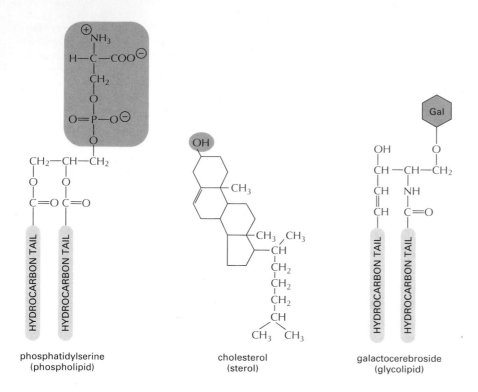

phosphatidylserine
(phospholipid)

cholesterol
(sterol)

galactocerebroside
(glycolipid)

hydrophobic molecule (Figure 11–9). Because the cagelike structure is more highly ordered than the surrounding water, its formation requires energy. The energy cost is minimized, however, if the hydrophobic molecules cluster together, so that the smallest possible number of water molecules are affected. Thus purely hydrophobic molecules, like the fats found in animal fat cells and the oils found in plant seeds (Figure 11–10), coalesce into a single large drop when dispersed in water.

Amphipathic molecules, such as phospholipids, are therefore subject to two conflicting forces: the hydrophilic head is attracted to water, while the hydrophobic tail shuns water and seeks to aggregate with other hydrophobic molecules. This conflict is beautifully resolved by the formation of a lipid bilayer—an arrangement that satisfies all parties

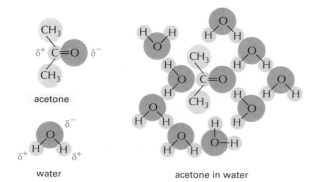

Figure 11–8 A hydrophilic molecule interacting with water molecules. Because acetone is polar, it can form favorable interactions with water molecules, which are also polar. Thus acetone readily dissolves in water. δ^- indicates a partial negative charge, and δ^+ indicates a partial positive charge. Polar atoms are shown in color (*pink* and *blue*); nonpolar groups are shown in *gray*.

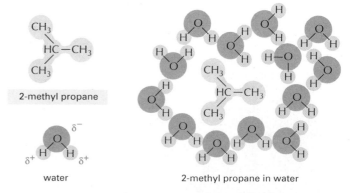

Figure 11–9 A hydrophobic molecule in water. Because the 2-methyl propane molecule is entirely hydrophobic, it cannot form favorable interactions with water and forces adjacent water molecules to reorganize into a cagelike structure around it.

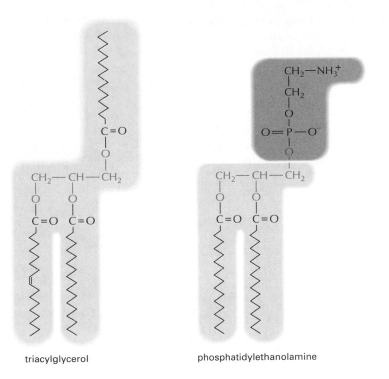

Figure 11–10 Comparison of a fat molecule (triacylglycerol) and a phospholipid molecule (phosphatidylethanolamine). The hydrophobic parts are shaded *gray*, and the hydrophilic parts are shaded *pink*. The fat molecule is entirely hydrophobic, whereas the phospholipid molecule is amphipathic. (The third hydrophobic tail of the triacylglycerol molecule is drawn here facing upward for comparison with the phospholipid, although normally it is depicted facing down.)

and is energetically most favorable. The hydrophilic heads face the water at each of the two surfaces of the sheet of molecules; the hydrophobic tails are all shielded from the water and lie next to one another in the interior of the sandwich (Figure 11–11).

The same forces that drive the amphipathic molecules to form a bilayer confer on that bilayer a self-sealing property. Any tear in the bilayer will create a free edge with water, and because this is energetically unfavorable, the molecules of the bilayer will spontaneously rearrange to eliminate the free edge. If the tear is small, this spontaneous rearrangement will lead to the repair of the bilayer, restoring a single continuous sheet. If the tear is large, the sheet may break up into separate vesicles. In either case, the overriding principle is that free edges are quickly eliminated.

Figure 11–11 A phospholipid bilayer seen in cross-section. (A) Computer simulation showing the phospholipid molecules (*red tails* and *yellow heads*) and the surrounding water molecules (*blue*). (B) Schematic drawing of a phospholipid bilayer in water. (A, adapted from *Science* 262:223–228 (1993), courtesy of R. Venable and R. Pastor.)

(A)

1 nm

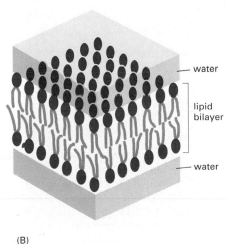

water

lipid bilayer

water

(B)

The Lipid Bilayer

The prohibition on free edges has a profound consequence: there is only one way a finite sheet can avoid having free edges, which is by form-ing a boundary around a closed space (Figure 11–12). Therefore, amphi-pathic molecules such as phospholipids necessarily assemble to form self-sealing containers defining closed compartments. This remarkable behavior, fundamental to the creation of a living cell, is in essence sim-ply a result of the property that each molecule is hydrophilic at one end and hydrophobic at the other.

The Lipid Bilayer Is a Two-dimensional Fluid

The aqueous environment inside and outside a cell prevents membrane lipids from escaping from the bilayer, but nothing stops these molecules from moving about and changing places with one another within the plane of the bilayer. The membrane therefore behaves as a two-dimen-sional fluid, which is crucial for membrane function. This property is distinct from *flexibility*, which is the ability of the membrane to bend. Membrane flexibility is also important, and it sets a lower limit of about 25 nm to the size of vesicle that cell membranes can form.

The fluidity of lipid bilayers can be studied using synthetic lipid bilayers, which are easily produced artificially by the spontaneous aggre-gation of amphipathic lipid molecules in water. Two types of synthetic lipid bilayers are commonly used in experiments. Closed spherical vesi-cles, called *liposomes*, form if pure phospholipids are added to water; they vary in size from about 25 nm to 1 mm in diameter (Figure 11–13). Alternatively, flat phospholipid bilayers can be formed across a hole in a partition between two aqueous compartments (Figure 11–14).

These simple artificial bilayers allow delicate measurements of the movements of the lipid molecules, revealing that some types of move-ment are rare while others are frequent and rapid. Thus, in synthetic lipid bilayers, phospholipid molecules very rarely flip from one mono-

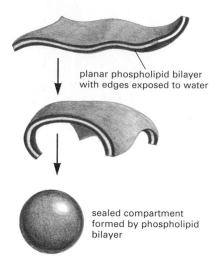

ENERGETICALLY UNFAVORED

planar phospholipid bilayer with edges exposed to water

sealed compartment formed by phospholipid bilayer

ENERGETICALLY FAVORED

Question 11–1 Water molecules are said "to arrange into a cagelike structure" around hydrophobic compounds (e.g., see Figure 11–9). This seems paradoxical because water molecules do not interact with the hydrophobic compound. So how could they "know" about its presence and change their behavior to interact differ-ently with one another? Discuss this argu-ment and in doing so develop a clear con-cept of what is meant by a "cagelike" structure. How does it compare to ice? Why would this cagelike structure be en-ergetically unfavorable?

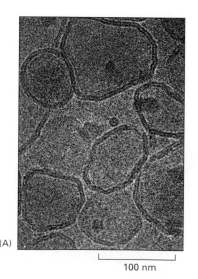

(A)

100 nm

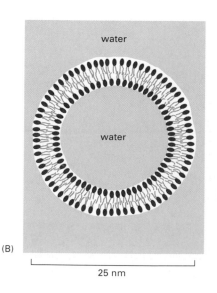

water

water

(B)

25 nm

Figure 11–13 Liposomes. (A) An electron micrograph of phospholipid vesicles (lipo-somes) showing the bilayer structure of the membrane. (B) A drawing of a small spherical liposome seen in cross-section. (A, courtesy of Jean Lepault.)

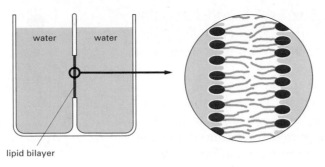

Figure 11–14 **A cross-sectional view of a synthetic phospholipid bilayer.** This planar bilayer forms across a small hole (about 1 mm in diameter) in a partition that separates two aqueous compartments. To make the bilayer across the hole, the hole is submerged in an aqueous solution and a phospholipid solution (in a nonaqueous solvent) is painted across the hole with a paintbrush.

layer (one half of the bilayer) to the other; without proteins to facilitate the process and under conditions similar to those in a cell, it is estimated that this event, called "flip-flop," occurs less than once a month for any individual lipid molecule. On the other hand, due to thermal motions, lipid molecules within a monolayer rotate very rapidly about their long axis and also constantly exchange places with their neighbors (Figure 11–15). This exchange leads to rapid diffusion in the plane of the membrane so that, for example, a lipid molecule in an artificial bilayer may diffuse a length equal to that of a large bacterial cell (~2 μm) in about one second. If the temperature is decreased, the drop in thermal energy decreases the rate of lipid movement, making the bilayer less fluid.

Similar results are obtained with isolated cell membranes and whole cells, indicating that the lipid bilayer of a cell membrane also behaves as a two-dimensional fluid in which the constituent lipid molecules are free to move within their own layer in any direction in the plane of the membrane; in cells, as in synthetic bilayers, individual phospholipid molecules are normally confined to their own monolayer and do not flip-flop spontaneously (see Figure 11–15).

The Fluidity of a Lipid Bilayer Depends on Its Composition

The degree of fluidity of a cell membrane—the ease with which its lipid molecules move about in the plane of the bilayer—is important for membrane function and has to be maintained within certain limits. Just how fluid a lipid bilayer is at a given temperature depends on its phospholipid composition and, especially, on the nature of the hydrocarbon tails: the closer and more regular the packing of the tails, the more viscous and less fluid the bilayer will be. There are two major properties of hydrocarbon tails that affect their packing in the bilayer—their length and their *unsaturation* (that is, the number of double bonds they contain). The hydrocarbon tails of the phospholipid molecules vary in length between 14 and 24 carbon atoms, 18–20 being most usual. A shorter chain length reduces the tendency of the hydrocarbon tails to interact with one another and therefore increases the fluidity of the bilayer. One of the two hydrocarbon tails of each phospholipid molecule usually has one or more double bonds between adjacent carbon atoms (see Figure 11–6). It therefore does not contain the maximum number of hydrogen atoms that could in principle be attached to its carbon backbone and is said to be **unsaturated** with respect to hydrogens. The second fatty acid tail usually has no double bonds, and because it has a full complement of hydrogen atoms, it is said to be **saturated.** Each double bond in an unsaturated tail creates a small kink in the hydrocarbon tail

Question 11–2 Five students in your class always sit together in the front row. This could be because (A) they really like each other or (B) nobody else in your class wants to sit next to them. Which explanation holds for the assembly of a lipid bilayer? Explain. Suppose that lipid molecules behaved in the other way. How would the properties of the lipid bilayer be different?

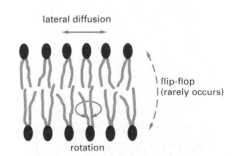

lateral diffusion

flip-flop (rarely occurs)

rotation

Figure 11–15 **Phospholipid mobility.** The drawing shows three types of movement possible for phospholipid molecules in a lipid bilayer.

(see Figure 11–6) that makes it more difficult for the tails to pack against one another. For this reason, lipid bilayers that contain a large proportion of unsaturated hydrocarbon tails are more fluid than those with lower proportions.

In bacterial and yeast cells, which have to adapt to varying temperatures, both the lengths and the unsaturation of the hydrocarbon tails in the bilayer are constantly adjusted to maintain the membrane at a relatively constant fluidity: at higher temperatures, for example, the cell makes membrane lipids with tails that are longer and that contain fewer double bonds. A related trick is used in the manufacture of margarine from vegetable oils. The fats produced by plants are generally unsaturated and therefore liquid at room temperatures, unlike animal fats such as butter or lard, which are generally saturated and therefore solid at room temperature. Margarine is made of hydrogenated vegetable oils, which have had their double bonds removed by adding hydrogen, so that they are more solid and butterlike at room temperature.

Membrane fluidity is important to a cell for many reasons. It enables membrane proteins to diffuse rapidly in the plane of the bilayer and to interact with one another, as is crucial, for example, in cell signaling (discussed in Chapter 15). It provides a simple means of distributing membrane lipids and proteins by diffusion from sites where they are inserted into the bilayer after their synthesis to other regions of the cell. It allows membranes to fuse with one another and mix their molecules, and it ensures that membrane molecules are distributed evenly between daughter cells when a cell divides. It is hard to imagine how a cell could live, grow, and reproduce if its membranes were not fluid.

In animal cells, membrane fluidity is modulated by the presence of the sterol **cholesterol,** which is absent in plants, yeast, and bacteria. These short, rigid molecules are present in especially large amounts in the plasma membrane, where they fill the spaces between neighboring phospholipid molecules that are caused by the kinks in their unsaturated hydrocarbon tails (Figure 11–16). In this way cholesterol stiffens the bilayer and makes it less fluid and less permeable.

The Lipid Bilayer Is Asymmetrical

Cell membranes are generally asymmetrical, presenting a very different face to the interior of the cell or organelle than to the exterior. The two halves of the bilayer often include strikingly different selections of phospholipids and glycolipids (Figure 11–17). Moreover, the proteins are embedded in the bilayer with a specific orientation, which is crucial for their function.

The lipid asymmetry begins at the point of manufacture. In cells, new phospholipid molecules are synthesized by membrane-bound enzymes that use as substrates fatty acids available in one half of the bilayer—that is, one monolayer—and release the newly made phospholipid into the same monolayer. To enable the membrane as a whole to grow, a proportion of the lipid molecules then have to be transferred to the opposite monolayer. This transfer is catalyzed by enzymes called *flippases* (Figure 11–18). It has been suggested that the flippases may transfer specific phospholipid molecules selectively, so that different types become concentrated in the two halves of the monolayer.

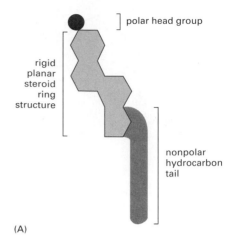

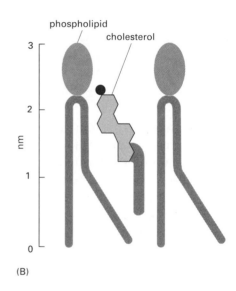

Figure 11–16 **The role of cholesterol in cell membranes.** (A) shows the structure of cholesterol. (B) shows how it fits into the gaps between phospholipid molecules in a lipid bilayer. The chemical formula of cholesterol is shown in Figure 11–7.

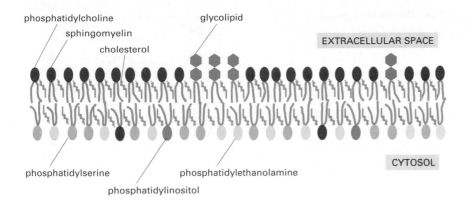

phosphatidylcholine
sphingomyelin
cholesterol
glycolipid

EXTRACELLULAR SPACE

CYTOSOL

phosphatidylserine
phosphatidylinositol
phosphatidylethanolamine

Figure 11–17 The asymmetrical distribution of phospholipids and glycolipids in a plasma membrane lipid bilayer. Five types of phospholipid molecules (labeled with *red* lettering) are shown in different colors. The glycolipids are drawn with *blue* hexagonal head groups to represent sugars. All the glycolipid molecules are in the external monolayer of the membrane, while cholesterol is distributed almost equally in both monolayers.

One-sided insertion and selective flippases are not the only ways of producing asymmetry in lipid bilayers, however. In particular, a different mechanism operates for glycolipids, which are the class of lipid molecules that show the most striking and consistent asymmetric distribution in animal cells. To explain their distribution it is necessary to look more carefully at the route by which new membrane is produced in eucaryotic cells.

Lipid Asymmetry Is Generated Inside the Cell

In eucaryotic cells nearly all new membrane synthesis occurs in one intracellular compartment—the *endoplasmic reticulum,* or *ER* (discussed in more detail in Chapter 14). The new membrane assembled there is exported to the other membranes of the cell by a cycle of vesicle budding and fusion: bits of membrane pinch off from the ER to form small vesicles, which then become incorporated into another membrane by fusing with it. Because the orientation of the bilayer relative to the cytosol is preserved during this vesicular transport process, all cell membranes, whether the external plasma membrane or an intracellular membrane surrounding an organelle, have distinct "inside" and "outside" faces: the inside, or *cytosolic,* face is adjacent to the cytosol, whereas the outside, or *noncytosolic,* face is exposed to either the cell exterior or the interior space of an organelle (Figure 11–19).

Glycolipids are located mainly in the plasma membrane, and they are found only in the noncytosolic half of the bilayer. Their sugar groups therefore are exposed on the exterior of the cell (see Figure 11–17), where they form part of the protective coat of carbohydrate that surrounds most animal cells. The glycolipid molecules acquire their sugar groups in the Golgi apparatus (discussed in Chapter 14). The enzymes that add the sugar groups are confined to the inside of the Golgi apparatus so that the sugars are added to lipid molecules in the noncytosolic half of the lipid bilayer. Once a glycolipid molecule has been created in this way, it remains trapped in this monolayer, as there are no flippases to transfer

lipid bilayer

flipped lipid molecule

FLIPPASE

new phospholipid synthesized

one side enlarged

both sides enlarged

Figure 11–18 The role of flippases in lipid bilayer synthesis. Although the newly synthesized phospholipid molecules are all added to one side of the bilayer, flippases transfer some of these to the opposite monolayer so that the entire bilayer expands.

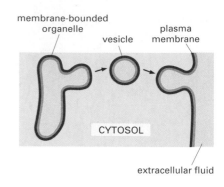

Figure 11–19 Budding and fusing of a membrane vesicle. A membrane vesicle is shown budding from a membrane-bounded organelle and fusing with the plasma membrane. Note that the orientation of the membrane is preserved during the process of vesicular budding and fusion, so that the cytosolic surface remains the cytosolic surface.

it to the cytosolic monolayer. Thus, when it is finally delivered to the plasma membrane, the glycolipid molecule faces away from the cytosol and displays its sugar on the exterior of the cell (see Figure 11–19).

Other lipid molecules show different types of asymmetric distributions, related to other functions. The *inositol phospholipids*, for example, are minor components of the plasma membrane, but they play a special role in relaying signals from the cell surface to the intracellular components that will respond to the signal (discussed in Chapter 15). They act only after the signal has been transmitted across the plasma membrane, and so they are concentrated in the cytosolic half of this lipid bilayer (see Figure 11–17).

Question 11–3 It seems paradoxical that a lipid bilayer can be fluid yet asymmetrical. Explain.

Lipid Bilayers Are Impermeable to Solutes and Ions

We saw earlier that a crucial function of any cell membrane is to act as a barrier that controls the passage of molecules across it. The hydrophobic interior of the lipid bilayer plays a major part here, as it creates a barrier to the passage of most hydrophilic molecules. These are as reluctant to enter a fatty environment as hydrophobic molecules are reluctant to enter water.

This barrier function of lipid bilayers can be demonstrated in synthetic bilayers of the kind illustrated in Figure 11–14. Given enough time, virtually any molecule will diffuse across such a bilayer. The *rate* at which it diffuses, however, varies enormously depending on the size of the molecule and its solubility properties. In general, the smaller the molecule and the more soluble it is in oil (that is, the more hydrophobic, or nonpolar, it is), the more rapidly it will diffuse across. Thus:

1. *Small nonpolar molecules,* such as molecular oxygen (O_2, molecular mass 32 daltons) and carbon dioxide (44 daltons), readily dissolve in lipid bilayers and therefore rapidly diffuse across them; indeed cells require this permeability to gases for the cellular respiration processes discussed in Chapter 13.

2. *Uncharged polar molecules* (molecules with an uneven distribution of electric charge) also diffuse rapidly across a bilayer, if they are small enough. Water (18 daltons) and ethanol (46 daltons), for example, cross fairly rapidly; glycerol (92 daltons) diffuses less rapidly, and glucose (180 daltons) diffuses hardly at all (Figure 11–20).

3. In contrast, lipid bilayers are highly impermeable to all *ions and charged molecules,* no matter how small. The molecules' charge and their strong electrical attraction to water molecules inhibit them from entering the hydrocarbon phase of the bilayer. Thus synthetic bilayers are a billion (10^9) times more permeable to water than they are to even such small ions as Na^+ or K^+.

Cell membranes therefore allow water and small nonpolar molecules to permeate by simple diffusion. But for cells to take up nutrients and release wastes, membranes must also allow the passage of many other molecules, such as ions, sugars, amino acids, nucleotides, and

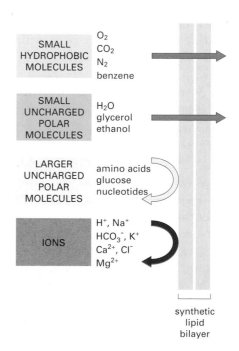

Figure 11–20 The relative permeability of a synthetic lipid bilayer to different classes of molecules. The smaller the molecule and, more important, the fewer its favorable interactions with water (that is, the less polar it is), the more rapidly the molecule diffuses across the bilayer. Note that many of the molecules that the cell uses as nutrients are too large and polar to pass through a pure lipid bilayer.

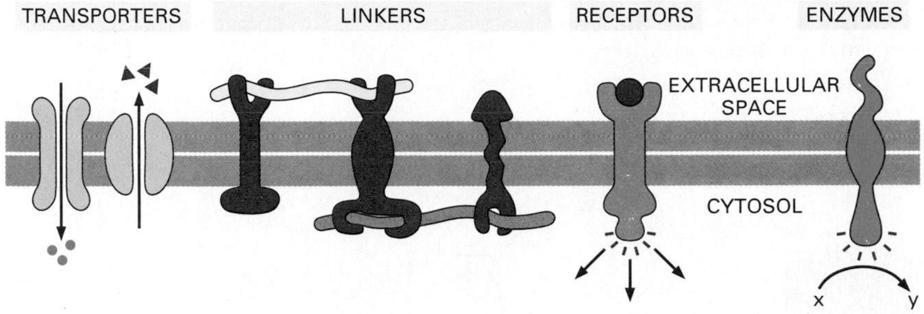

TRANSPORTERS   LINKERS   RECEPTORS   ENZYMES

EXTRACELLULAR SPACE

CYTOSOL

x   y

Figure 11–21 **Some functions of plasma membrane proteins.**

many cell metabolites. These molecules cross lipid bilayers far too slowly by simple diffusion, and specialized transport proteins are required to transfer them efficiently across cell membranes. These *membrane transport proteins* are the subject of Chapter 12. Before we can discuss them, however, we need to consider some general principles of how proteins are associated with the lipid bilayer in cell membranes.

Membrane Proteins

Although the lipid bilayer provides the basic structure of all cell membranes and serves as a permeability barrier, most membrane functions are carried out by **membrane proteins.** In animals, proteins constitute about 50% of the mass of most plasma membranes, the remainder being lipid plus relatively small amounts of carbohydrate. Because lipid molecules are much smaller than protein molecules, however, there are typically about 50 times more lipid molecules than protein molecules in a cell membrane (see Figure 11–4).

The proteins in membranes serve many functions besides that of transporting particular nutrients, metabolites, or ions across the lipid bilayer. Some anchor the membrane to macromolecules on either side. Others function as receptors that detect chemical signals in the cell's environment and relay them to the cell's interior, and still others work as enzymes to catalyze specific reactions (Figure 11–21; Table 11–1). Each type of cell membrane contains a different set of proteins, reflecting the particular functions of the particular membrane. In this section we discuss the structure of membrane proteins and illustrate the different ways that they can be associated with the lipid bilayer.

Table 11–1   **Some Examples of Plasma Membrane Proteins and Their Functions**

Functional Class	Protein Example	Specific Function
Transporters	Na⁺ pump	actively pumps Na⁺ out of cells and K⁺ in
Linkers	integrins	link intracellular actin filaments to extracellular matrix proteins
Receptors	platelet-derived growth factor (PDGF) receptor	binds extracellular PDGF and, as a consequence, generates intracellular signals that cause the cell to grow and divide
Enzymes	adenylate cyclase	catalyzes the production of intracellular cyclic AMP in response to extracellular signals

Membrane Proteins

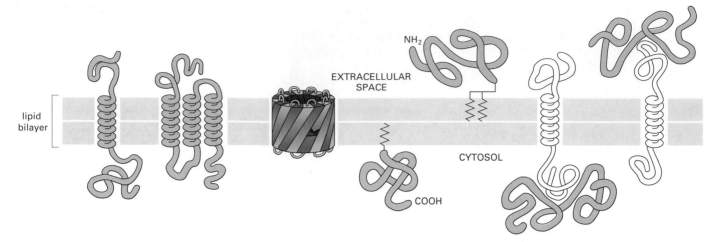

Membrane Proteins Associate with the Lipid Bilayer in Various Ways

There are three main ways in which proteins can be associated with the lipid bilayer of a cell membrane (Figure 11–22).

1. Many membrane proteins extend through the bilayer, with part of their mass on either side (Figure 11–22A). Like their lipid neighbors, these *transmembrane proteins* have both hydrophobic and hydrophilic regions. Their hydrophobic regions lie in the interior of the bilayer in contact with the hydrophobic tails of the lipid molecules. Their hydrophilic regions are exposed to the aqueous environment on either side of the membrane.

2. Other membrane proteins are located entirely outside the bilayer, being attached to the bilayer only by one or more covalently attached lipid groups (Figure 11–22B).

3. Yet other proteins are bound indirectly to one or the other face of the membrane only by interactions with other membrane proteins (Figure 11–22C).

All membrane proteins have a unique orientation in the membrane: a transmembrane protein, for example, always has the same region of the protein facing the cytosol. The orientation is a consequence of the way in which the protein is synthesized, as discussed in Chapter 14.

Proteins that are directly attached to membranes—whether transmembrane or lipid-linked—can be dissociated from membranes only by disrupting the lipid bilayer with detergents, as discussed later. Such proteins are known as *integral membrane proteins*. The remaining membrane proteins are known as *peripheral membrane proteins;* they can be released from the membrane by relatively gentle extraction procedures that interfere with protein-protein interactions but leave the lipid bilayer intact.

A Polypeptide Chain Usually Crosses the Bilayer as an α Helix

The portions of a transmembrane protein that extend outside the lipid bilayer are connected to each other by specialized membrane-spanning segments of the polypeptide chain. These segments, which run through

Figure 11–22 Ways in which membrane proteins associate with the lipid bilayer. (A) Transmembrane proteins can extend across the bilayer as a single α helix, as multiple α helices, or as a closed β sheet (a β barrel). (B) Other membrane proteins are attached to the bilayer solely by a covalent attachment to a lipid molecule (*red zigzag lines*). (C) Finally, many proteins are attached to the membrane only by relatively weak, noncovalent interactions with other membrane proteins.

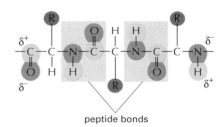

Figure 11–23 Peptide bonds. The peptide bonds (shaded in *gray*) that join adjacent amino acids together in a polypeptide chain are polar and therefore hydrophilic. δ^- indicates a partial negative charge, and δ^+ indicates a partial positive charge.

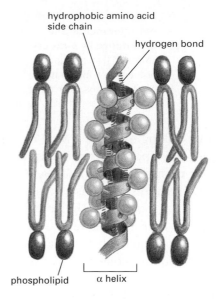

hydrophobic amino acid side chain

hydrogen bond

phospholipid

α helix

Figure 11–24 A segment of α helix crossing a lipid bilayer. The hydrophobic side chains of the amino acids forming the α helix contact the hydrophobic hydrocarbon tails of the phospholipid molecules, while the hydrophilic parts of the polypeptide backbone form hydrogen bonds with one another in the interior of the helix. About 20 amino acids are required to span a membrane in this way.

the hydrophobic environment of the interior of the lipid bilayer, are composed largely of amino acids with hydrophobic side chains. Because these side chains cannot form favorable interactions with water molecules, they prefer the lipid environment, where no water is present.

In contrast to the hydrophobic side chains, however, the peptide bonds that join the successive amino acids in a protein are normally polar, making the polypeptide backbone hydrophilic (Figure 11–23). Because water is absent from the bilayer, atoms forming the backbone are driven to form hydrogen bonds with one another. Hydrogen bonding is maximized if the polypeptide chain forms a regular α helix, and so the great majority of the membrane-spanning segments of polypeptide chains traverse the bilayer as α helices. In these membrane-spanning α helices, the hydrophobic amino acid side chains are exposed on the outside of the helix, where they contact the hydrophobic lipid tails, while parts of the polypeptide backbone form hydrogen bonds with one another on the inside of the helix (Figure 11–24).

In many transmembrane proteins the polypeptide chain crosses the membrane only once (see Figure 11–22A). Many of these proteins are receptors for extracellular signals: their extracellular part binds the signal molecule, while their cytoplasmic part signals to the cell's interior (see Figure 11–21).

Other transmembrane proteins form aqueous pores that allow water-soluble molecules to cross the membrane. Such pores cannot be formed by proteins with a single, uniformly hydrophobic transmembrane α helix. More complicated transmembrane proteins, in which the polypeptide chain crosses the bilayer a number of times, either as α helices or as a β barrel, are required for this task (see Figure 11–22A). In many of these proteins one or more of the transmembrane regions are formed from α helices that contain both hydrophobic and hydrophilic amino acid side chains. The hydrophobic side chains lie on one side of the helix, exposed to the lipids of the membrane. The hydrophilic side chains are concentrated on the other side, where they form part of the lining of a hydrophilic pore created by packing several helices side by side in a ring within the hydrophobic environment of the lipid bilayer (Figure 11–25). How such pores function in the selective transport of small water-soluble molecules across membranes is discussed in Chapter 12.

Although the α helix is by far the most common form in which a polypeptide chain crosses a lipid bilayer, the polypeptide chain of some transmembrane proteins crosses the lipid bilayer as a β sheet (discussed in Chapter 5), curved into a cylinder, forming an open-ended barrel called a *β barrel*. As expected, the amino acid side chains that face the inside of the barrel, and therefore line the aqueous channel, are mostly hydrophilic, while those on the outside of the barrel, which contact the hydrophobic core of the lipid bilayer, are exclusively hydrophobic. The

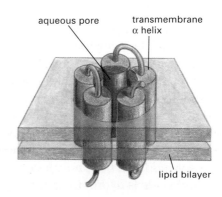

aqueous pore

transmembrane α helix

lipid bilayer

Figure 11–25 The formation of a transmembrane hydrophilic pore by multiple α helices. In this example, five transmembrane α helices form a water-filled channel across the lipid bilayer. The hydrophobic amino acid side chains (*green*) on one side of each helix contact the hydrophobic hydrocarbon tails, while the hydrophilic side chains (*red*) on the opposite side of the helices form a water-filled pore.

Membrane Proteins

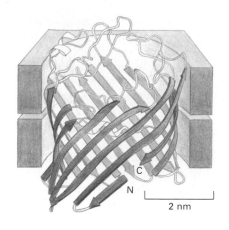

Figure 11–26 **The three-dimensional structure of a porin protein in the outer membrane of a bacterium (*Rhodobacter capsulatus*) as determined by x-ray crystallography.** The protein consists of a 16-stranded β sheet that is curved around on itself to form a transmembrane water-filled channel. Although not shown in the drawing, three porin proteins associate to form a trimer, which has three separate channels. (From S.W. Cowan, *Curr. Opin. Struct. Biol.* 3:501–507, 1993.)

most striking example is that of the *porin* proteins, which form large water-filled pores across the outer membrane of mitochondria and the outer membrane of some bacteria, where they allow the passage of nutrients and small ions, while preventing the entry of large antibiotics and toxin molecules. Unlike α helices, β barrels can form only wide channels, because there is a limit to how tightly the β sheet can be curved to form the barrel (Figure 11–26). In this respect a β sheet is less versatile than a collection of α helices.

Membrane Proteins Can Be Solubilized in Detergents and Purified

To understand a protein fully one needs to know its structure in detail, and for membrane proteins this presents special problems. Most biochemical procedures are designed for working with molecules dissolved in water or a simple solvent; membrane proteins, however, are designed to operate in an environment that is partly aqueous and partly fatty, and taking them out of this environment and purifying them while preserving their essential structure is no easy task.

To study an individual protein in detail, it is necessary to separate it from all other proteins. For most membrane proteins the first step in the separation process involves the solubilization of the membrane by agents that destroy the lipid bilayer by disrupting hydrophobic associations. The most useful of such agents are **detergents,** which are small, amphipathic, lipidlike molecules with both a hydrophilic and a hydrophobic region (Figure 11–27). Detergents differ from membrane phospholipids in that they have only a single hydrophobic tail and, consequently, behave in a significantly different way. Because of their single hydrophobic tail, detergent molecules are shaped more like cones than cylinders and in water tend to aggregate into small clusters called *micelles,* rather than forming a bilayer as do the phospholipids, which are more cylindrical. When mixed in great excess with membranes, the hydrophobic ends of detergent molecules bind to the membrane-spanning hydrophobic region of the transmembrane proteins, as well as to the hydrophobic tails of the phospholipid molecules, thereby separating the proteins from the phospholipids. Since the other end of the detergent molecule is hydrophilic, this binding tends to bring the membrane

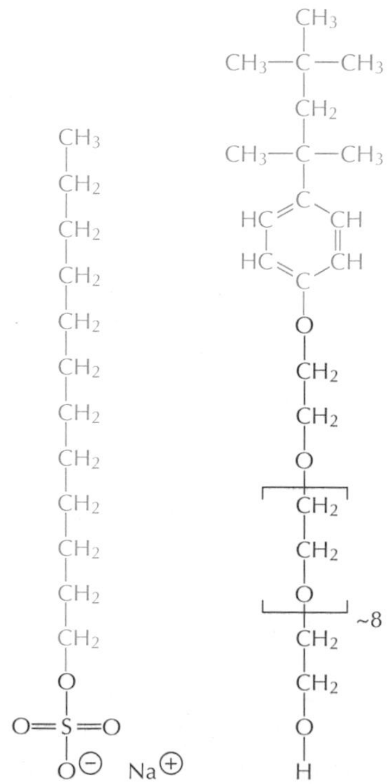

Question 11–4 Explain why the polypeptide chain of most transmembrane proteins crosses the lipid bilayer as an α helix or a β barrel.

Figure 11–27 **The structures of two commonly used detergents.** Sodium dodecyl sulfate (SDS) is a strong ionic detergent (that is, it has an ionized group at its hydrophilic end), and Triton X-100 is a mild non-ionic detergent (that is, it has a non-ionized but polar structure at its hydrophilic end). The hydrophobic portion of each detergent is shown in *green*, and the hydrophilic portion is shown in *red*. The bracketed portion of Triton X-100 is repeated about eight times. Strong ionic detergents like SDS not only displace lipid molecules from proteins but also unfold the proteins as well (see Panel 5–5, pp. 162–163).

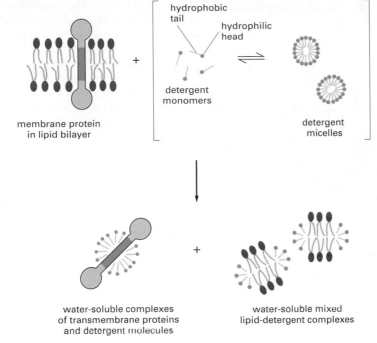

Figure 11–28 **Solubilizing membrane proteins with a mild detergent such as Triton X-100.** The detergent disrupts the lipid bilayer and brings the proteins into solution as protein-detergent complexes. The phospholipids in the membrane are also solubilized by the detergents. As illustrated, detergent molecules are cone-shaped in cross-section and in water tend to aggregate into clusters called micelles.

proteins into solution as protein-detergent complexes (Figure 11–28). At the same time, the detergent solubilizes the phospholipids. The protein-detergent complexes can then be separated from one another and from the lipid-detergent complexes by a technique such as SDS polyacrylamide gel electrophoresis (discussed in Chapter 5).

Question 11–5 Explain for the two detergents shown in Figure 11–27 why the *red* portions of the molecules are hydrophilic and the *green* portions hydrophobic. Draw a short stretch of a polypeptide chain made up of three amino acids with hydrophobic side chains (see Panel 2–5, pp. 62–63) and apply a similar color scheme.

The Complete Structure Is Known for Very Few Membrane Proteins

Much of what we know about the structure of membrane proteins has been learned by indirect means. The standard direct method for determining protein structure is x-ray crystallography (discussed in Chapter 5), but this requires ordered crystalline arrays of the molecule, and membrane proteins have proved hard to crystallize. Two notable exceptions, however, are *bacteriorhodopsin* and a *photosynthetic reaction center*, both of which are bacterial membrane proteins with important roles in the capture and utilization of energy from sunlight. The structures of these two proteins have revealed exactly how α helices cross the lipid bilayer, as well as how a set of different protein molecules can assemble to form functional complexes in a membrane.

The structure of **bacteriorhodopsin** is shown in Figure 11–29. This small protein (about 250 amino acids) is found in large amounts in the plasma membrane of an archaebacterium, *Halobacterium halobium*, that lives in salt marshes. Bacteriorhodopsin acts as a membrane transport protein that pumps H+ out of the bacterium. Pumping requires energy, and bacteriorhodopsin gets its energy directly from sunlight. Each bacteriorhodopsin molecule contains a single, light-absorbing, nonprotein molecule—called *retinal*—that gives the protein (and the bacterium) a deep purple color. This small hydrophobic molecule is covalently attached to one of bacteriorhodopsin's seven transmembrane α helices, and it lies in the plane of the lipid bilayer, entirely surrounded by the seven α helices (see Figure 11–29). When retinal absorbs a pho-

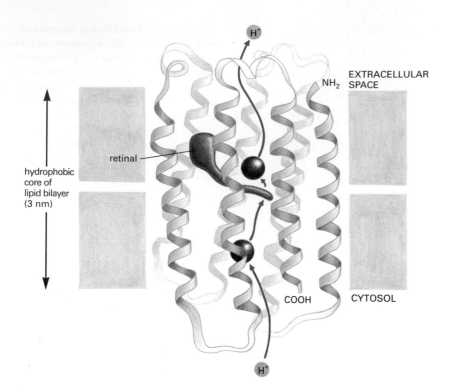

H+

EXTRACELLULAR
SPACE

NH₂

retinal

hydrophobic
core of
lipid bilayer
(3 nm)

COOH

CYTOSOL

H+

Figure 11–29 The three-dimensional structure of a bacteriorhodopsin molecule. The polypeptide chain crosses the lipid bilayer as seven α helices. The location of the retinal and the probable pathway taken by protons during the light-activated pumping cycle are shown; two polar amino acid side chains thought to be involved in the H⁺ transfer process are shown in *black*. Retinal is also used to detect light in our own eyes, where it is attached to a protein with a structure very similar to bacteriorhodopsin. (Adapted from R. Henderson et al., *J. Mol. Biol.* 213:899–929, 1990.)

ton of light, it changes its shape, and in doing so it causes the protein embedded in the lipid bilayer to undergo a series of small conformational changes. These changes result in the transfer of one H⁺ from the retinal to the outside of the bacterium: the H⁺ moves across the bilayer along a pathway of strategically placed polar amino acid side chains (see Figure 11–29). The retinal is then regenerated by taking up an H⁺ from the cytosol, returning the protein to its original conformation so that it can repeat the cycle. The overall outcome is the transfer of one H⁺ out of the bacterium, which lowers the H⁺ concentration inside the cell.

In the presence of sunlight, thousands of bacteriorhodopsin molecules pump H⁺ out of the cell, generating a concentration gradient of H⁺ across the bacterial membrane. This H⁺ gradient serves as an energy store, like water behind a dam. And just as the dammed water can be used to generate electricity when it flows downhill through a turbine, so the H⁺ gradient can be used to generate ATP when H⁺ flows back into the bacterium through a second membrane protein, called *ATP synthase*. The same type of ATP synthase generates much of the ATP in plant and animal cells, as discussed in Chapter 13.

The structure of a bacterial *photosynthetic reaction center* is shown in Figure 11–30. It is a large complex composed of four protein molecules. Three are transmembrane proteins; two of these (M and L) have multiple α helices passing through the lipid bilayer, while the other (H) has only one. The fourth protein (cytochrome) is associated with the outer surface of the membrane, bound to the transmembrane proteins. The entire protein complex serves as a protein machine, taking in light energy absorbed by chlorophyll molecules and producing high-energy electrons required for photosynthetic reactions (discussed in Chapter 13). Many membrane proteins are arranged in large complexes, and the structure of the photosynthetic reaction center is the best model we have for thousands of other membrane proteins whose structures are not known.

Question 11–6 Look at the structure of the photosynthetic reaction center in Figure 11–30. As you would expect, many α helices span the membrane. At the lower right-hand corner, however, there is a stretch of the polypeptide chain of the L subunit that forms a disordered loop within the hydrophobic core of the lipid bilayer. Does this invalidate the general rule that transmembrane proteins span the lipid bilayer as α helices or β sheets?

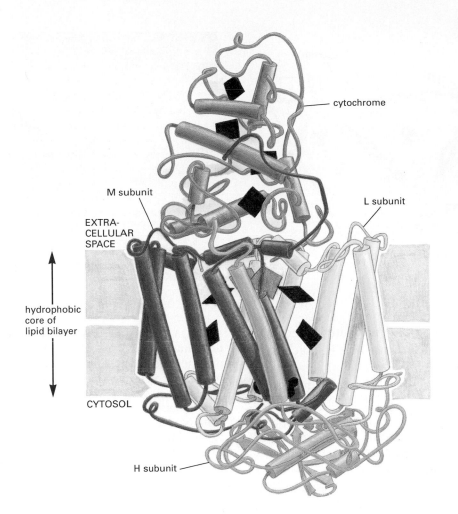

cytochrome

M subunit

EXTRA-
CELLULAR
SPACE

L subunit

hydrophobic
core of
lipid bilayer

CYTOSOL

H subunit

Figure 11–30 **The three-dimensional structure of the photosynthetic reaction center of the bacterium *Rhodopseudomonas viridis*.** The structure was determined by x-ray diffraction analysis of crystals of this transmembrane protein complex. The complex consists of four subunits—L, M, H, and a cytochrome. The L and M subunits form the core of the reaction center, and each contains five α helices that span the lipid bilayer. All the α helices are shown as cylinders. The locations of the various electron carrier groups, which are covalently bound to the protein subunits, are shown in *black*, except for the special pair of chlorophyll molecules that are excited by light, which are shown as *dark green rectangles* in the center of the drawing. Note that the cytochrome is bound to the outer surface of the membrane solely by its attachment to the transmembrane subunits (see Figure 11–22C). (Adapted from a drawing by J. Richardson based on data from J. Deisenhofer, O. Epp, K. Miki, R. Huber, and H. Michel, *Nature* 318:618–624, 1985.)

The Plasma Membrane Is Reinforced by the Cell Cortex

A cell membrane by itself is extremely thin and fragile. It would require nearly 10,000 cell membranes laid on top of one another to achieve the thickness of this paper. Most cell membranes are therefore strengthened and supported by a framework of proteins, attached to the membrane via transmembrane proteins. In particular, the shape of the cell and the

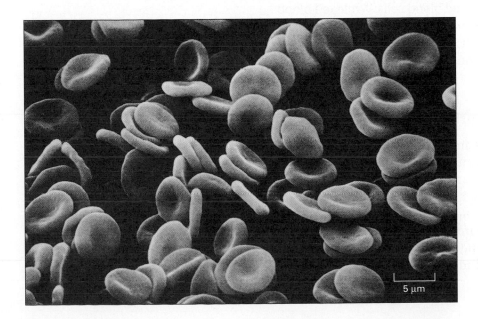

5 µm

Figure 11–31 **A scanning electron micrograph of human red blood cells.** They have a distinctive flattened shape and lack a nucleus and other intracellular organelles. (Courtesy of Bernadette Chailley.)

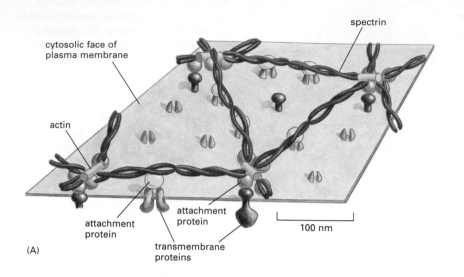

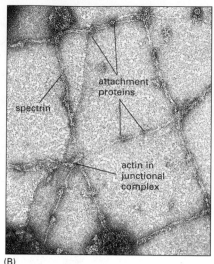

(A)

(B)

mechanical properties of the plasma membrane are determined by a meshwork of fibrous proteins, called the *cell cortex*, that is attached to the cytosolic surface of the membrane.

The cell cortex of human red blood cells is a relatively simple and regular structure and is by far the best understood. These cells are small and have a distinctive flattened shape (Figure 11–31). The main component of their cortex is the protein *spectrin,* a long, thin, flexible rod about 100 nm in length. It forms a meshwork that provides support for the plasma membrane and maintains cell shape. The spectrin meshwork is connected to the membrane through intracellular attachment proteins that link the spectrin to specific transmembrane proteins (Figure 11–32). The importance of this meshwork is seen in mice and humans that have genetic abnormalities in spectrin structure. These individuals are anemic: they have fewer red blood cells than normal, and the red cells they do have are spherical instead of flattened and are abnormally fragile.

Proteins similar to spectrin and to its associated attachment proteins are present in the cortex of most of our cells, but the cortex in these cells is much more complex than that in red blood cells. While red blood cells need their cortex mainly to provide mechanical strength as they are pumped through blood vessels, other cells also need their cortex to allow them to change their shape actively and to move, as we discuss in Chapter 16.

The Cell Surface Is Coated with Carbohydrate

We saw earlier that in eucaryotic cells many of the lipids in the outer layer of the plasma membrane have sugars covalently attached to them. The same is true for most of the proteins in the plasma membrane. The great majority of these proteins have short chains of sugars, called *oligosaccharides,* linked to them; they are called **glycoproteins.** Other membrane proteins have one or more long polysaccharide chains attached to them; they are called *proteoglycans.* All of the carbohydrate on the glycoproteins, proteoglycans, and glycolipids is located on one

Figure 11–32 The spectrin-based cell cortex of human red blood cells. (A) Spectrin molecules (together with a smaller number of actin molecules) form a meshwork that is linked to the plasma membrane by the binding of at least two types of attachment proteins (shown in *blue* and *yellow*) to two kinds of transmembrane proteins (shown in *brown* and *green*). The electron micrograph in (B) shows the spectrin meshwork on the cytoplasmic side of a red blood cell membrane. The meshwork has been stretched out to show the details of its structure; in the normal cell the meshwork shown would occupy only about one-tenth of this area. (B, courtesy of T. Byers and D. Branton, *Proc. Natl. Acad. Sci. USA* 82:6153–6157, 1985.)

Question 11–7 Look carefully at the transmembrane proteins shown in Figure 11–32A. What can you say about their mobility in the membrane?

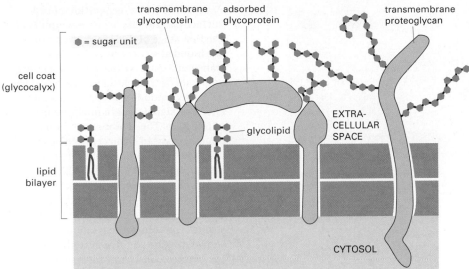

transmembrane glycoprotein · adsorbed glycoprotein · transmembrane proteoglycan

= sugar unit

cell coat (glycocalyx)

glycolipid

EXTRA-CELLULAR SPACE

lipid bilayer

CYTOSOL

Figure 11–33 Simplified diagram of the glycocalyx of a eucaryotic cell. The glycocalyx is made up of the oligosaccharide side chains attached to membrane glycolipids and glyco-proteins, and the polysaccharide chains on membrane proteoglycans. Glycoproteins and proteoglycans that have been secreted by the cell and then adsorbed back onto its surface can also contribute to the glycocalyx. Note that all the carbohydrate is on the extracellular (noncytosolic) surface of the plasma membrane.

side of the membrane, the noncytosolic side, where it forms a sugar coating called the **glycocalyx** (Figure 11–33).

The glycocalyx helps to protect the cell surface from mechanical and chemical damage. As the oligosaccharides and polysaccharides in the glycocalyx adsorb water, they give the cell a slimy surface. This helps mo-tile cells such as white blood cells to squeeze through narrow spaces, and it prevents blood cells from sticking to one another or to the walls of blood vessels.

Cell-surface carbohydrate does more than just protect and lubricate the cell, however. It has an important role in cell-cell recognition and adhesion. Just as many proteins will recognize and bind to a particular site on another protein, so some proteins (called *lectins*) are specialized to recognize particular oligosaccharide side chains and bind to them. The oligosaccharide side chains of glycoproteins and glycolipids, although short (typically fewer than 15 sugar units), are enormously diverse. Unlike polypeptide (protein) chains where the amino acids are all joined together linearly by identical peptide bonds (see Figure 11–23), sugars can be joined together in different ways and in varied sequences, often forming branched oligosaccharide chains (see Panel 2–3, pp. 56–57). Even three sugar groups can be put together in different com-binations of covalent linkages to form hundreds of different tri-saccharides.

In a multicellular organism the glycocalyx can thus serve as a kind of distinctive clothing, like a policeman's uniform, that is characteristic of cells specialized for a particular function and is recognized by other cells with which they must interact. Specific oligosaccharides in the glycoca-lyx are involved, for example, in the recognition of an egg by a sperm. They are also involved in inflammatory responses. In the early stages of a bacterial infection, for instance, the carbohydrate on the surface of white blood cells called *neutrophils* is recognized by a lectin on the cells lining the blood vessels at the site of infection. This recognition process causes the neutrophils to adhere to the blood vessels and then migrate from the bloodstream into the infected tissues, where they help to remove the bacteria (Figure 11–34).

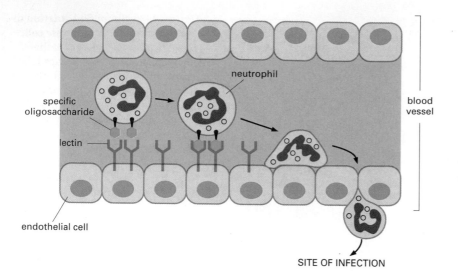

Figure 11–34 **The recognition of cell-surface carbohydrate on neutrophils is the first stage of their migration out of the blood at sites of infection.** Specialized transmembrane proteins (called lectins) are made by the cells lining the blood vessel (called endothelial cells) in response to chemical signals emanating from the site of infection. These proteins recognize particular groups of sugars carried by glycolipids and glycoproteins on the surface of neutrophils circulating in the blood. The neutrophils consequently stick to the blood vessel wall. This association is not very strong, but it leads to another, much stronger protein-protein interaction (not shown) that helps the neutrophil migrate out of the bloodstream between the endothelial cells into the tissue at the site of infection.

Cells Can Restrict the Movement of Membrane Proteins

Because a membrane is a two-dimensional fluid, many of its proteins, like its lipids, can move freely within the plane of the lipid bilayer. This can be neatly demonstrated by fusing a mouse cell with a human cell to form a double-size hybrid cell and then monitoring the distribution of mouse and human plasma membrane proteins. Although at first the mouse and human proteins are confined to their own halves of the newly formed hybrid cell, within half an hour or so the two sets of proteins become evenly mixed over the entire cell surface (Figure 11–35).

The picture of a membrane as a sea of lipid in which all proteins float freely is too simple, however. Cells have ways of confining particular plasma membrane proteins to localized areas within the bilayer, thereby creating functionally specialized regions, or **membrane domains,** on the cell or organelle surface. Some ways of restricting the movement of membrane proteins are summarized in Figure 11–36.

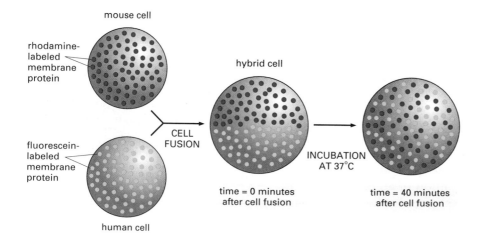

Figure 11–35 **Experiment demonstrating the mixing of plasma membrane proteins on mouse-human hybrid cells.** The mouse and human proteins are initially confined to their own halves of the newly formed hybrid cell plasma membrane, but they intermix within a short time. To reveal the proteins, two antibodies that bind to human and mouse proteins, respectively, are labeled with different fluorescent tags (rhodamine or fluorescein) and added to the cells. The two fluorescent antibodies can be distinguished in a fluorescence microscope because fluorescein is green, whereas rhodamine is red.

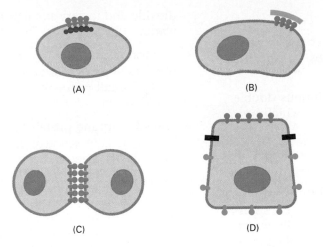

Figure 11–36 **Four ways of restricting the lateral mobility of plasma membrane proteins.** They can be tethered to the cell cortex inside the cell (A), to extracellular matrix molecules outside the cell (B), or to proteins on the surface of another cell (C). Finally, diffusion barriers (shown as *black* bars) can restrict proteins to a particular membrane domain (D).

The proteins can be linked to fixed structures outside the cell—for example, to molecules in the extracellular matrix (discussed in Chapter 19). Membrane proteins also can be anchored to relatively immobile structures inside the cell, especially to the cell cortex (see Figure 11–32). Finally, cells can create barriers that restrict particular membrane components to one membrane domain. In epithelial cells that line the gut, for example, it is important that transport proteins involved in the uptake of nutrients from the gut be confined to the *apical* surface of the cells (the surface that faces the gut contents) and that other proteins involved in the transport of solutes out of the epithelial cell into the tissues and bloodstream be confined to the *basal* and *lateral* surfaces (Figure 11–37). This asymmetric distribution of membrane proteins is maintained by a barrier formed along the line where the cell is sealed to adjacent epithelial cells by a so-called *tight junction*. At this site, specialized junctional proteins form a continuous belt around the cell where it contacts its neighbors, creating a seal between adjacent cell membranes. Membrane proteins cannot diffuse past the junction.

In the next chapter we examine the individual functions of the protein molecules that the cell takes such care to localize on its surface to transport molecules into and out of the cell.

Question 11–8 Describe the different methods that cells use to restrict proteins to specific regions of the plasma membrane. Is a membrane with many anchored proteins still fluid?

GUT LUMEN

protein A

apical plasma membrane

tight junction

lateral plasma membrane

protein B

basal plasma membrane

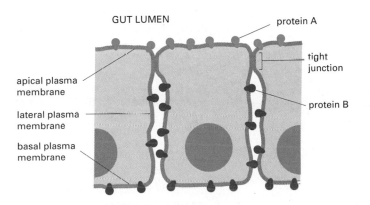

Figure 11–37 **Diagram of gut epithelium showing how a membrane protein is restricted to a particular domain of the plasma membrane of an epithelial cell.** Protein A (in the apical membrane) and protein B (in the basal and lateral membranes) can diffuse laterally in their own membrane domains but are prevented from entering the other domain by a specialized cell junction called a tight junction.

Essential Concepts

- Cell membranes enable a cell to create barriers that confine particular molecules to specific compartments.

- Cell membranes consist of a continuous double layer (a bilayer) of lipid molecules in which proteins are embedded.

- The lipid bilayer provides the basic structure and barrier function of all cell membranes.

- Membrane lipid molecules have both hydrophobic and hydrophilic regions. They assemble spontaneously into bilayers when placed in water, forming closed compartments that reseal if torn.

- There are three major classes of membrane lipid molecules: phospholipids, sterols, and glycolipids.

- The lipid bilayer is fluid, and individual lipid molecules are able to diffuse within their own monolayer; they do not, however, spontaneously flip from one monolayer to the other.

- The two layers of the lipid bilayer have a different lipid composition, reflecting the different functions of the two faces of a cell membrane.

- Cells adjust their membrane fluidity by modifying the lipid composition of their membranes.

- The lipid bilayer is impermeable to all ions and large polar molecules, but it is permeable to small nonpolar molecules such as oxygen and carbon dioxide and to very small polar molecules such as water.

- Membrane proteins are responsible for most of the functions of a membrane, such as the transport of small water-soluble molecules across the lipid bilayer.

- Transmembrane proteins extend across the lipid bilayer, usually as one or more α helices but sometimes as a β sheet in the form of a barrel.

- Other membrane proteins do not extend across the lipid bilayer but are attached to one or the other side of the membrane either by noncovalent association with other membrane proteins or by covalent attachment to lipids.

- Many of the proteins and some of the lipids exposed on the surface of cells have attached chains of sugars, which help protect and lubricate the cell surface and are involved in cell-cell recognition.

- Most cell membranes are supported by an attached framework of proteins. An example is the meshwork of fibrous proteins forming the cell cortex underneath the plasma membrane.

- Although many membrane proteins can diffuse rapidly in the plane of the membrane, cells have ways of confining proteins to specific membrane domains and of immobilizing particular proteins by attaching them to intracellular or extracellular macromolecules.

Key Terms

	detergent	lipid bilayer	phospholipid
amphipathic	glycocalyx	membrane domain	plasma membrane
bacteriorhodopsin	glycolipid	membrane protein	saturated
cholesterol	glycoprotein	phosphatidylcholine	unsaturated

Questions

Question 11–9   Which of the following statements are correct? Explain your answers.

A. Lipids in a lipid bilayer rotate rapidly around their long axis.

B. Lipids in a lipid bilayer rapidly exchange positions with one another in the plane of the membrane.

C. Lipids in a lipid bilayer do not flip-flop readily from one lipid monolayer to the other.

D. Hydrogen bonds that form between lipid head groups and water molecules are continually broken and re-formed.

E. Glycolipids move through different membrane-bounded compartments during their synthesis but remain restricted to one side of the lipid bilayer.

F. Margarine contains more saturated lipids than the vegetable oil from which it is made.

G. Some membrane proteins are enzymes.

H. The sugar coat that surrounds all cells is called a glycocalyx and makes cells more slippery.

Question 11–10 What is meant by the term "two-dimensional fluid"?

Question 11–11 The structure of a lipid bilayer is determined by the particular properties of its lipid molecules. What would happen if . . .

A. phospholipids had only one hydrocarbon chain instead of two?

B. the hydrocarbon chains were shorter than normal, say about 10 carbon atoms long?

C. all of the hydrocarbon chains were saturated?

D. all of the hydrocarbon chains were unsaturated?

E. the bilayer contained a mixture of two kinds of lipid molecules, one with two saturated hydrocarbon tails and the other with two unsaturated hydrocarbon tails?

F. each lipid molecule were covalently linked through the end carbon atom of one of its hydrocarbon chains to a lipid molecule in the opposite monolayer?

Question 11–12 What are the differences between a lipid molecule and a detergent molecule? How would the structure of a lipid molecule need to change to make it a detergent?

Question 11–13 List the following compounds in order of increasing membrane permeability: RNA, Ca^{2+}, glucose, ethanol, N_2, water.

Question 11–14

A. Lipid molecules exchange places with their lipid neighbors every 10^{-7} seconds. A lipid molecule diffuses from one end of a 2-μm-long bacterial cell to the other in about 1 second. Are these two numbers in agreement (assume that the diameter of a lipid head group is about 0.5 nm)? If not, can you think of a reason for the difference?

B. To get an appreciation for the great speed of molecular motions, assume that a lipid head group is about the size of a ping-pong ball (4 cm diameter) and that the floor of your living room (6 m × 6 m) is covered wall to wall with these balls. If two neighboring balls exchanged positions once every 10^{-7} seconds, what would their speed be in kilometers per hour? How long would it take for a ball to move from one end of the room to the other?

Question 11–15 Why does a red blood cell membrane need proteins?

Question 11–16 Consider a transmembrane protein that forms a hydrophilic pore across the plasma membrane of a eucaryotic cell, allowing Na^+ to enter the cell when it is activated upon binding a specific ligand on its extracellular side. It is made of five similar transmembrane subunits, each containing a membrane-spanning α helix with hydrophilic amino acid side chains on one surface of the helix and hydrophobic amino acid side chains on the opposite surface. Considering the function of the protein as an ion channel, propose a possible arrangement of the five membrane-spanning α helices in the membrane.

Question 11–17 In the membrane of a human red blood cell the ratio of the mass of protein (average molecular weight 50,000) to phospholipid (molecular weight 800) to cholesterol (molecular weight 386) is about 2:1:1. How many lipid molecules are there for every protein molecule?

Question 11–18 Draw a schematic diagram that shows a close-up view of two plasma membranes as they come together during cell fusion, as shown in Figure 11–35. Show the membrane proteins in both cells that were labeled from the outside by the binding of differently colored fluorescent antibody molecules. Indicate in your drawing the fates of these color tags as the cells fuse. Will they still be only on the outside of the hybrid cell (a) after cell fusion and (b) after the mixing of membrane proteins that occurs during the incubation at 37°C? How would the experimental outcome be different if the incubation was done at 0°C?

Question 11–19 Compare the hydrophobic forces that hold a membrane protein in the lipid bilayer to those that help proteins fold into a unique three-dimensional structure.

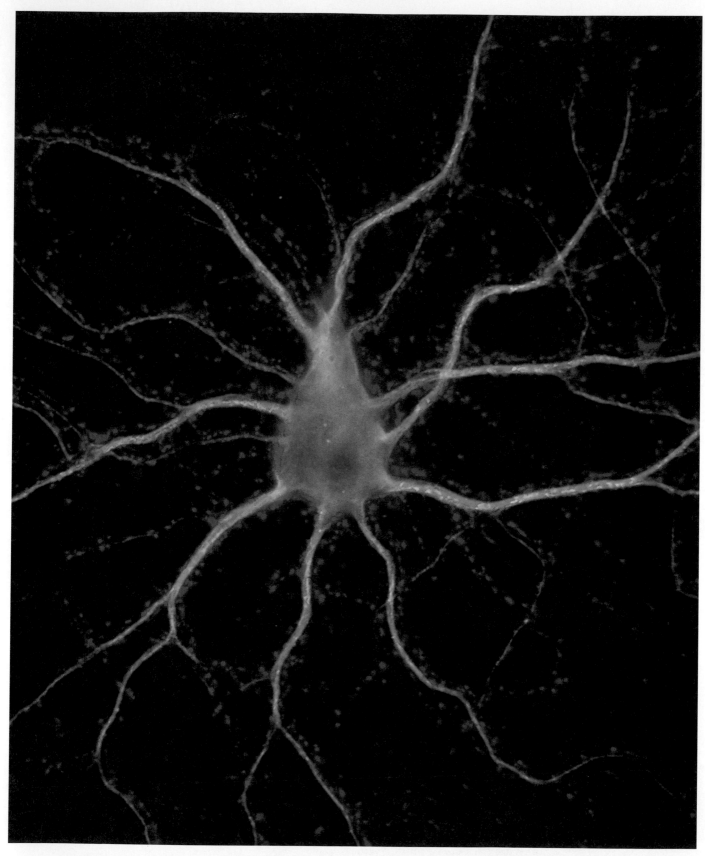

A rat nerve cell in culture. Its cell body and dendrites (*green*) are stained with a fluorescent antibody that recognizes a cytoskeletal protein. Thousands of axon terminals (*red*) from other nerve cells (not visible) make synapses on the cell's surface; they are stained with a fluorescent antibody that recognizes a protein in synaptic vesicles. Electrical signals are sent out along axons, relayed across synapses, and passed in along dendrites toward the nerve cell body. The signaling depends on movements of ions across the plasma membranes of the nerve cells. (Courtesy of Olaf Mundigl and Pietro de Camilli.)

12 Membrane Transport

Chapter 11 emphasized the role of cell membranes as barriers: the interior of the lipid bilayer is hydrophobic and blocks the passage of almost all water-soluble molecules. But cells live and grow by exchanging molecules with their environment, so that various water-soluble molecules must be able to cross the plasma membrane: nutrients such as sugars and amino acids have to be imported, waste products such as CO_2 have to be released, and intracellular concentrations of ions such as H^+, Na^+, K^+, and Ca^{2+} have to be adjusted. A few of these *solutes*, such as CO_2 and O_2, can simply diffuse across the lipid bilayer, but the vast majority cannot (see Figure 11–20). Instead, their transfer depends on **membrane transport proteins** that span the membrane and provide private passages across it for specific substances (Figure 12–1). Some passages, for example, are open to Na^+ but not K^+, others are open to K^+ but not Na^+, while yet others are open to glucose but not amino acids. The set of transport proteins in the plasma membrane, or in the membrane of an intracellular organelle, determines exactly what solutes can pass into and out of that cell or organelle. Each type of membrane therefore has its own characteristic set of transport proteins.

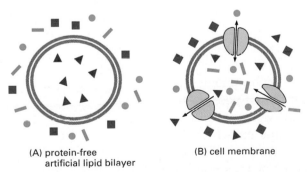

(A) protein-free artificial lipid bilayer

(B) cell membrane

Figure 12–1 Membrane transport proteins are responsible for transferring small water-soluble molecules across cell membranes. Whereas protein-free artificial lipid bilayers are impermeable to most water-soluble molecules (A), cell membranes are not (B). Note that each type of transport protein in a cell membrane transfers a particular type of molecule, causing a selective set of solutes to end up inside the membrane-bounded compartment.

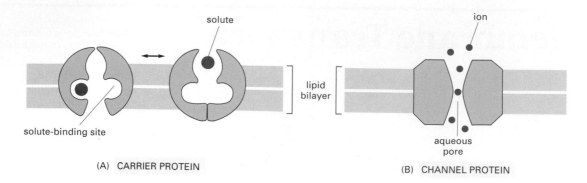

solute

solute-binding site

lipid bilayer

ion

aqueous pore

(A) CARRIER PROTEIN

(B) CHANNEL PROTEIN

Two major classes of membrane transport proteins can be distinguished. *Carrier proteins* bind a solute on one side of the membrane and deliver it to the other side through a change in the conformation of the carrier protein. The solutes transported in this way can be either small organic molecules or inorganic ions. *Channel proteins*, by contrast, form tiny hydrophilic pores in the membrane, through which solutes can pass by diffusion (Figure 12–2). Most channel proteins let through inorganic ions only and are therefore called *ion channels*. Cells can also selectively transfer macromolecules such as proteins across their membranes, but this requires more elaborate machinery, which is discussed in Chapter 14.

In the first section of this chapter we discuss carrier proteins and their functions in solute transport. In the second section we consider the behavior and functions of ion channels. Because ions are electrically charged, their movements can create powerful electric forces across the membrane. These forces enable nerve cells to carry out electrical signaling, as we discuss in the final section.

To provide a foundation for these discussions, we begin by considering the differences in intracellular ion composition between a cell and its environment. This will help make it clear why the transport of ions by both carrier proteins and ion channels is of such fundamental importance to cells.

The Ion Concentrations Inside a Cell Are Very Different from Those Outside

Ion transport across cell membranes is of central importance in biology. Cells maintain an internal ion composition that is very different from that in the fluid around them, and these differences are crucial for a cell's survival and function. Inorganic ions such as Na^+, K^+, Ca^{2+}, Cl^-, and H^+ (protons) are the most plentiful of all the solutes in a cell's environment, and their movements across cell membranes play an essential part in many cell processes. Animal cells, for example, pump Na^+ outward across their plasma membrane to keep the internal concentration of Na^+ low. This pumping helps balance the osmotic pressures on the two sides of the membrane: if the pumping fails, water flows into the cell by osmosis and causes it to swell and burst. Ion movements across cell membranes also play fundamental roles in the functioning of nerve cells, as we discuss later, and in the production of ATP by all cells, as we discuss in Chapter 13.

The ion concentrations inside and outside a typical mammalian cell are shown in Table 12–1. Na^+ is the most plentiful positively charged ion

Figure 12–2 A schematic view of the two classes of membrane transport proteins. A carrier protein undergoes a series of conformational changes to transfer small water-soluble molecules across the lipid bilayer. In contrast, a channel protein forms a hydrophilic pore across the bilayer through which specific inorganic ions can diffuse. As would be expected, channel proteins transport at a much greater rate than carrier proteins. Ion channels can exist in either an open or a closed conformation and transport only in the open conformation, which is shown. Their opening and closing is usually controlled by an external stimulus or by conditions within the cell.

Table 12–1 Comparison of Ion Concentrations Inside and Outside a Typical Mammalian Cell

Component	Intracellular Concentration (mM)	Extracellular Concentration (mM)
Cations		
Na^+	5–15	145
K^+	140	5
Mg^{2+}*	0.5	1–2
Ca^{2+}*	10^{-7}	1–2
H^+	7×10^{-5} ($10^{-7.2}$ M or pH 7.2)	4×10^{-5} ($10^{-7.4}$ M or pH 7.4)
Anions		
Cl^-	5–15	110
Fixed anions**	high	0

*The concentrations of Ca^{2+} and Mg^{2+} given are for the free ions in the cytosol. There is a total of about 20 mM Mg^{2+} and 1–2 mM Ca^{2+} in cells, but this is mostly bound to proteins and other substances and thus cannot leave the cell. Much of the total cell Ca^{2+} is stored within various organelles.
**The fixed anions are the negatively charged small and large organic molecules that are trapped inside the cell, being unable to cross the plasma membrane.

(cation) outside the cell, while K^+ is the most plentiful inside. If a cell is not to be torn apart by electrical forces, the quantity of positive charge inside the cell must be balanced by an almost exactly equal quantity of negative charge, and the same is true for the charge in the surrounding fluid. (Tiny excesses of positive or negative charge, concentrated in the neighborhood of the plasma membrane, are allowed and have important electrical effects, as we discuss later.)

The high concentration of Na^+ outside the cell is balanced chiefly by extracellular Cl^-. The high concentration of K^+ inside is balanced by a variety of negatively charged intracellular ions (anions). In fact, most intracellular constituents are negatively charged: in addition to Cl^-, cells contain inorganic ions such as bicarbonate (HCO_3^-) and phosphate (PO_4^{3-}), organic metabolites carrying negatively charged phosphate and carboxyl (COO^-) groups, and macromolecules such as proteins and nucleic acids that also carry numerous phosphate and carboxyl groups. The negatively charged organic molecules are sometimes called "fixed anions" because they are unable to escape from the cell by crossing the plasma membrane.

Carrier Proteins and Their Functions

Carrier proteins are required for the transport of almost all small organic molecules across cell membranes, with the exception of fat-soluble molecules and small uncharged molecules that can pass directly through the lipid bilayer by *simple diffusion*. Each carrier protein is highly selective, often transporting just one type of molecule. To guide and propel the complex traffic of small molecules into and out of the cell and between the cytosol and the different membrane-bounded organelles, each cell membrane contains a set of different carrier proteins appropriate to that particular membrane. Thus in the plasma membrane there are carriers to import nutrients such as sugars, amino

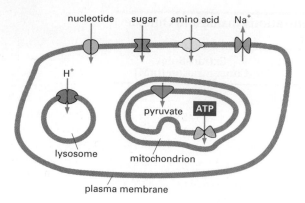

nucleotide   sugar   amino acid   Na⁺

H⁺

lysosome

pyruvate   ATP

mitochondrion

plasma membrane

Figure 12–3 **Some examples of sub-stances transported across cell membranes by carrier proteins.** Each cell membrane has its own characteristic set of carrier proteins.

acids, and nucleotides; in the inner membrane of mitochondria there are carriers for importing pyruvate and ADP and for exporting ATP; and so on (Figure 12–3).

As discussed in Chapter 11, the membrane transport proteins that have been studied in detail—both carriers and channels—have polypeptide chains that traverse the lipid bilayer multiple times—that is, they are multipass transmembrane proteins (see Figure 11–29). By crisscrossing back and forth across the bilayer, the polypeptide chain is thought to form a continuous protein-lined pathway that allows selected small hydrophilic molecules to cross the membrane without coming into direct contact with the hydrophobic interior of the lipid bilayer (Figure 12–4).

A basic difference between carrier proteins and channel proteins is the way they discriminate between solutes, transporting some solutes but not others. A channel protein discriminates mainly on the basis of size and electric charge: if the channel is open, molecules small enough and carrying the appropriate charge can slip through, as through an open, but narrow, trapdoor. A carrier protein acts more like a turnstile: it allows passage only to solute molecules that fit into a binding site on the

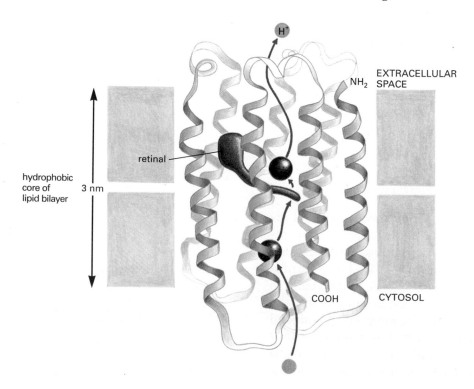

H⁺

NH₂   EXTRACELLULAR SPACE

retinal

hydrophobic core of lipid bilayer   3 nm

COOH   CYTOSOL

Figure 12–4 **Three-dimensional structure of a carrier protein.** The carrier is bacteriorhodopsin, which is discussed in Chapter 11 (see pp. 361–363). It is one of the few membrane transport proteins whose detailed three-dimensional structure is known. It pumps H⁺ (protons) out of the bacterium *Halobacterium halobium* in response to light, which is captured by the covalently attached, light-absorbing group (a chromophore) shown in *purple*. The polypeptide chain crosses the lipid bilayer as seven α helices. Note that the pathway taken by the protons (*red arrows*) enables them to avoid contact with the lipid bilayer. (Adapted from R. Henderson et al. *J. Mol. Biol.* 213:899-929, 1990.)

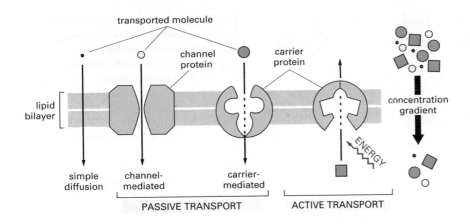

Figure 12–5 **Comparison of passive and active transport.** If uncharged solutes are small enough, they can move down their concentration gradients directly across the lipid bilayer itself by simple diffusion. Examples of such solutes are ethanol, carbon dioxide, and oxygen. Most solutes, however, can cross the membrane only if there is a membrane transport protein (a carrier protein or a channel protein) to transfer them. As indicated, passive transport, in the same direction as a concentration gradient, occurs spontaneously, whereas transport against a concentration gradient (active transport), requires an input of energy. Only carrier proteins can carry out active transport, but both carrier proteins and channel proteins can carry out passive transport.

protein, and it transfers these molecules across the membrane one at a time by changing its own conformation (see Figure 12–2A). Carrier proteins specifically bind their solutes in the same way that an enzyme binds its substrate, and it is this requirement for specific binding that makes the transport selective.

To understand fully how a carrier protein transfers solutes across a membrane, we would need to know its three-dimensional structure in detail, and as yet this information is available for only a very few membrane transport proteins. One is bacteriorhodopsin (see Figure 12–4), which functions as a light-activated H$^+$ pump, as discussed in Chapter 11. Nevertheless, the general principles underlying transport by carrier proteins are well understood.

Solutes Cross Membranes by Passive or Active Transport

An important question about any transport process is what causes it to go in one direction rather than another. Movements of molecules "downhill" from a region of high concentration to a region of low concentration occur spontaneously, provided a pathway exists. Such movements are called *passive*, because they need no other driving force. If, for example, a solute is present at a higher concentration outside the cell than inside and an appropriate channel or carrier protein is present in the plasma membrane, the solute will move spontaneously across the membrane into the cell by **passive transport** (sometimes called facilitated diffusion), without expenditure of energy by the transport protein.

To move a solute against its concentration gradient, however, a transport protein has to do work: it has to drive the "uphill" flow by coupling it to some other process that provides energy (as discussed in Chapter 3 for enzyme reactions). Transmembrane solute movement driven in this way is termed **active transport,** and it is carried out only by special types of carrier proteins that can harness some energy source to the transport process (Figure 12–5).

Electrical Forces as Well as Concentration Gradients Can Drive Passive Transport

A simple example of a carrier protein that mediates passive transport is the *glucose carrier* found in the plasma membrane of mammalian liver cells (and of many other types of cells). It consists of a protein chain that crosses the membrane at least 12 times. It is thought that the protein can

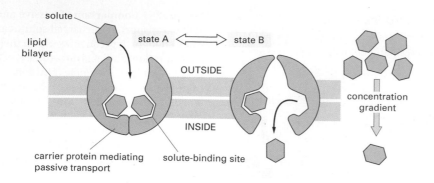

solute

lipid bilayer

state A ⟺ state B

OUTSIDE

INSIDE

concentration gradient

carrier protein mediating passive transport

solute-binding site

Figure 12–6 A hypothetical model showing how a conformational change in a carrier protein could mediate the passive transport of a solute like glucose. The carrier protein can exist in two conformational states; in state A the binding sites for the solute are exposed on the outside of the membrane; in state B the same sites are exposed on the other side of the membrane. The transition between the two states is proposed to occur randomly and independently of whether the solute is bound and to be completely reversible. If the concentration of the solute is higher on the outside of the membrane, it will be more often caught up in A → B transitions that carry it into the cell than in B → A transitions that carry it out, and there will be a net transport of the solute down its concentration gradient.

adopt at least two conformations and switches reversibly and randomly between them. In one conformation the carrier exposes binding sites for glucose to the exterior of the cell; in the other it exposes this site to the interior of the cell (Figure 12–6).

When glucose is plentiful outside the liver cell (after a meal), glucose molecules bind to the externally displayed binding sites; when the protein switches conformation, it carries these molecules inward and releases them into the cytosol, where the glucose concentration is low. Conversely, when blood sugar levels are low (when you are hungry), the hormone glucagon stimulates the liver cell to produce large amounts of glucose by the breakdown of glycogen. As a result, the glucose concentration is higher inside the cell than outside, and glucose binds to any internally displayed binding sites on the carrier protein; when the protein switches conformation in the opposite direction, the glucose is transported out of the cell. The flow of glucose can thus go either way, according to the direction of the glucose *concentration gradient* across the membrane—inward if glucose is more concentrated outside the cell than inside, and outward if the opposite is true. Transport proteins of this type, which permit a flux of solute but play no part in determining its direction, carry out passive transport. Although passive, the transport is highly selective: the binding sites in the glucose transporter bind only D-glucose and not, for example, its mirror image L-glucose, which the cell cannot use for glycolysis.

For glucose, which is an uncharged molecule, the direction of passive transport is determined simply by the concentration gradient. For electrically charged molecules, either small organic ions or inorganic ions, an additional force comes into play. For reasons we explain later, most cell membranes have a voltage across them, a difference in the electrical potential on each side of the membrane, which is referred to as the *membrane potential*. This exerts a force on any molecule that carries

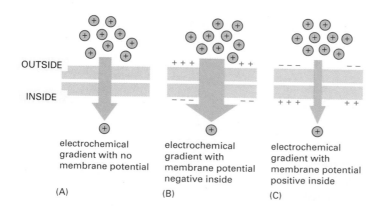

OUTSIDE

INSIDE

electrochemical gradient with no membrane potential

(A)

electrochemical gradient with membrane potential negative inside

(B)

electrochemical gradient with membrane potential positive inside

(C)

Figure 12–7 The two components of the electrochemical gradient. The net driving force (the electrochemical gradient) tending to move a charged solute (ion) across the membrane is the sum of the concentration gradient of the solute and the voltage across the membrane (the membrane potential). The width of the *green arrow* represents the magnitude of the electrochemical gradient for the same positively charged solute in three different situations. In (A), there is only a concentration gradient. In (B), the concentration gradient is supplemented by a membrane potential that increases the driving force. In (C), the membrane potential decreases the driving force that is caused by the concentration gradient.

an electric charge. The cytoplasmic side of the plasma membrane is usually at a negative potential relative to the outside, and this tends to pull positively charged solutes into the cell and drive negatively charged ones out. At the same time, a charged solute will also tend to move down its concentration gradient.

The net force driving a charged solute across the membrane is therefore a composite of two forces, one due to the concentration gradient and the other due to the voltage across the membrane. This net driving force is called the **electrochemical gradient** for the given solute. It is this gradient that determines the direction of passive transport across the membrane. For some ions, voltage and concentration gradient work in the same direction, creating a relatively steep electrochemical gradient (Figure 12–7B). This is the case with Na^+, for example, which is positively charged and at a higher concentration outside cells than inside. Na^+ therefore tends to enter cells if given a chance. If the voltage and concentration gradients have opposing effects, the resulting electrochemical gradient can be small (Figure 12–7C). This is the case for K^+, a positively charged ion that is present at a much higher concentration inside cells than outside. Because of opposing effects, K^+ has a small electrochemical gradient across the membrane, despite its large concentration gradient, and therefore there is little net movement of K^+ across the membrane.

Active Transport Moves Solutes Against Their Electrochemical Gradients

Cells cannot rely solely on passive transport. Active transport of solutes against their electrochemical gradient is essential to maintain the intracellular ionic composition of cells and to import solutes that are at a lower concentration outside the cell than inside. Cells carry out active transport in three main ways (Figure 12–8): (1) *Coupled transporters* couple the uphill transport of one solute across the membrane to the downhill transport of another. (2) *ATP-driven pumps* couple uphill transport to the hydrolysis of ATP. (3) *Light-driven pumps*, which are found mainly in bacterial cells, couple uphill transport to an input of energy from light, as mentioned earlier for bacteriorhodopsin.

Since a substance has to be carried uphill before it can flow downhill, the different forms of active transport are necessarily linked. Thus in the plasma membrane of an animal cell, an ATP-driven pump transports Na^+ out of the cell against its electrochemical gradient, and the Na^+ then

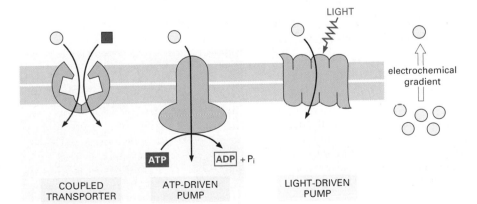

COUPLED TRANSPORTER   ATP-DRIVEN PUMP   LIGHT-DRIVEN PUMP

Figure 12–8 Three ways of driving active transport. The actively transported molecule is shown in *yellow*, and the energy source is shown in *red*.

Carrier Proteins and Their Functions

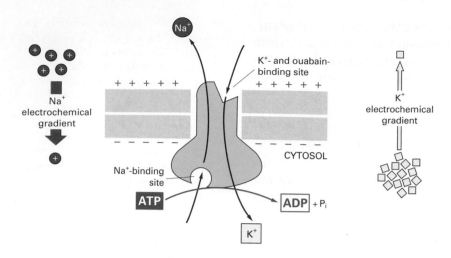

Figure 12–9 The Na⁺-K⁺ pump. This carrier protein uses the energy of ATP hydrolysis to pump Na⁺ out of the cell and K⁺ in, both against their electrochemical gradients. Ouabain is a drug that binds to the pump and inhibits its activity by preventing K⁺ binding.

flows back in, down its electrochemical gradient. Because it flows through Na⁺-coupled transporters, the influx of Na⁺ drives the active movement of many other substances into the cell against their electrochemical gradients. If the Na⁺ pump ceased operating, the Na⁺ gradient would soon run down, and transport through Na⁺-coupled transporters would come to a halt. The ATP-driven Na⁺ pump, therefore, has a central role in membrane transport in animal cells. In plant cells, fungi, and many bacteria, a similar role is played by ATP-driven pumps that create an electrochemical gradient of H⁺ ions (protons) by pumping H⁺ out of the cell, as we discuss later.

Animal Cells Use the Energy of ATP Hydrolysis to Pump Out Na⁺

The ATP-driven Na⁺ pump in animal cells hydrolyzes ATP to ADP to transport Na⁺ out of the cell and is, therefore, not only a carrier protein but also an enzyme—an ATPase. At the same time, it couples the outward transport of Na⁺ to an inward transport of K⁺. The pump is therefore known as the *Na⁺-K⁺ ATPase,* or the **Na⁺-K⁺ pump** (Figure 12–9).

This carrier protein plays a central part in the energy economy of animal cells, typically accounting for 30% or more of their total ATP consumption. Like a bilge pump in a leaky ship, it operates ceaselessly to expel the Na⁺ that is constantly entering through other carrier proteins and ion channels. In this way, it keeps the Na⁺ concentration in the cytosol about 10–30 times lower than in the extracellular fluid and the K⁺ concentration about 10–30 times higher (see Table 12–1, p. 373). Under normal conditions, the interior of most cells is at a negative electric potential compared to the exterior, so that positive ions tend to be pulled into the cell; thus the inward electrochemical driving force for Na⁺ is large, being the sum of a driving force due to the concentration gradient and a driving force in the same direction due to the voltage gradient (see Figure 12–7B).

Figure 12–10 Blyde River dam in South Africa. The water in the dam has potential energy, which can be used to drive energy-requiring processes. In the same way, an ion gradient across a membrane can be used to drive active processes in a cell, including the active transport of other molecules. (Courtesy of Paul Franklin © Oxford Scientific Films.)

The Na$^+$ outside the cell, on the uphill side of its electrochemical gradient, is like a large volume of water behind a high dam: it represents a very large store of energy (Figure 12–10). Even if one artificially halts the operation of the Na$^+$-K$^+$ pump with a toxin such as the plant glycoside *ouabain,* which binds to the pump and prevents K$^+$ binding (see Figure 12–9), the energy in this store is sufficient to sustain for many minutes the other transport processes that are driven by the downhill flow of Na$^+$.

For K$^+$ the situation is different. The electric force is the same as for Na$^+$, since this depends only on the charge carried by the ion. The concentration gradient, however, is in the opposite direction. The result, under normal conditions, is that the net driving force for movement of K$^+$ across the membrane is close to zero: the electric force pulling K$^+$ into the cell is almost exactly balanced against the concentration gradient tending to drive it out.

The Na$^+$-K$^+$ Pump Is Driven by the Transient Addition of a Phosphate Group

The Na$^+$-K$^+$ pump provides a beautiful illustration of how a protein couples one reaction to another, following the principles discussed in Chapter 3. The pump works in a cycle, as illustrated schematically in Figure 12–11. Na$^+$ binds to the pump at sites exposed intracellularly (stage 1), activating the ATPase activity. ATP is split, with the release of ADP and the transfer of a phosphate group into a high-energy linkage to

Figure 12–11 A schematic model of the pumping cycle of the Na$^+$-K$^+$ pump. The binding of Na$^+$ (1) and the subsequent phosphorylation by ATP of the cytosolic face of the pump (2) induce the protein to undergo a conformational change that transfers the Na$^+$ across the membrane and releases it on the outside (3). The high-energy linkage of the phosphate to the protein provides the energy to drive the conformational change. The binding of K$^+$ on the extracellular surface (4) and the subsequent dephosphorylation (5) return the protein to its original conformation, which transfers the K$^+$ across the membrane and releases it into the cytosol (6). These changes in conformation are analogous to the A $\rightleftharpoons$ B transitions shown in Figure 12–6 except that here the Na$^+$-dependent phosphorylation and the K$^+$-dependent dephosphorylation of the protein cause the conformational transitions to occur in an orderly manner, enabling the protein to do useful work. For simplicity, only one Na$^+$- and one K$^+$-binding site are shown. In the real pump in mammalian cells there are thought to be three Na$^+$- and two K$^+$-binding sites. The net result of one cycle of the pump is therefore to transport three Na$^+$ out of the cell and two K$^+$ in.

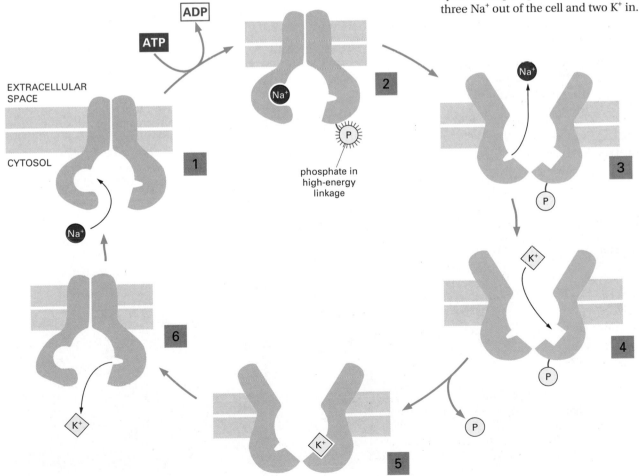

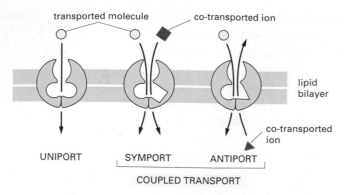

UNIPORT SYMPORT ANTIPORT

COUPLED TRANSPORT

Figure 12–12 **Three types of transport by carrier proteins**. Some carrier proteins transport a single solute across the membrane (uniports). In coupled transport, by contrast, the transfer of one solute depends on the simultaneous or sequential transfer of another solute, either in the same direction (symports) or in the opposite direction (antiports). Uniports, symports, and antiports are used for both passive and active transport.

the pump itself—that is, the pump phosphorylates itself (stage 2). Phosphorylation causes the pump to switch its conformation so as to release Na$^+$ at the exterior surface of the cell and, at the same time, to expose a binding site for K$^+$ at the same surface (stage 3). The binding of extracellular K$^+$ triggers the removal of the phosphate group (dephosphorylation) (stages 4 and 5), causing the pump to switch back to its original conformation, discharging the K$^+$ into the cell interior (stage 6). The whole cycle, which takes about 10 milliseconds, can then be repeated. Each step in the cycle depends on the one before, so that if any of the individual steps is prevented from occurring, all the functions of the pump are halted. This tight coupling ensures that the pump operates only when the appropriate ions are available to be transported, thereby avoiding useless ATP hydrolysis.

Animal Cells Use the Na$^+$ Gradient to Take Up Nutrients Actively

A gradient of any solute across a membrane, like the Na$^+$ gradient generated by the Na$^+$-K$^+$ pump, can be used to drive the active transport of a second molecule. The downhill movement of the first solute down its gradient provides the energy to drive the uphill transport of the second. The carrier proteins that do this are called **coupled transporters** (see Figure 12–8). They may couple the movement of one inorganic ion to

Figure 12–13 **One way in which a glucose pump could, in principle, be driven by a Na$^+$ gradient.** The pump oscillates randomly between two alternate states, A and B. In the A state the protein is open to the extracellular space; in the B state it is open to the cytosol. While Na$^+$ and glucose bind equally well to the protein in either state, they bind effectively only if both are present together: the binding of Na$^+$ induces a conformational change in the protein that greatly increases the protein's affinity for glucose and vice versa. Since the Na$^+$ concentration is much higher in the extracellular space than in the cytosol, glucose is more likely to bind to the pump in the A state; therefore, both Na$^+$ and glucose enter the cell (via an A → B transition) much more often than they leave it (via a B → A transition). The overall result is the net transport of both glucose and Na$^+$ into the cell. Note that, because the binding is cooperative, if one of the two solutes is missing, the other will fail to bind to the pump, and it will not be transported. An alternative way in which coupled transport may work is considered in Queston 12–3.

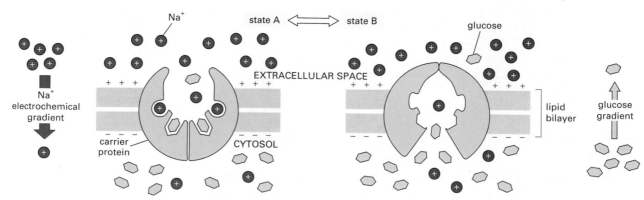

that of another, the movement of an inorganic ion to that of an organic molecule, or the movement of one organic molecule to that of another. If the transporter moves both solutes in the same direction across the membrane, it is called a *symport*. If it moves them in opposite directions, it is called an *antiport*. A carrier protein, like the passive glucose transporter mentioned earlier, that ferries only one type of solute across the membrane (and is therefore not a coupled transporter) is called a *uniport* (Figure 12–12).

Although we do not yet know the three-dimensional structure of any coupled transporter in detail, it is possible to envisage how the molecular mechanism of coupled transport might work, as outlined in Figure 12–13. This model is based on a simple modification of the mechanism proposed in Figure 12–6. Regardless of uncertainties about details, one principle of coupled transport is fundamental: if one of the co-transported solutes is absent, transport of its companion cannot occur.

In animal cells an especially important role is played by symports that use the inward flow of Na^+ down its steep electrochemical gradient to drive the import of other solutes into the cell. The epithelial cells that line the gut, for example, transfer glucose from the gut across the gut epithelium. If these cells had only the passive glucose transporters described earlier, they would release glucose into the gut after a sugar-free meal as freely as they take it up from the gut after a sugar-rich meal. But they also possess a glucose-Na^+ symport, with which they can take up glucose actively from the gut lumen even when the concentration of glucose is higher inside the cell than in the gut. If the gut epithelial cells had *only* this symport, however, they could never release glucose for use by the other cells of the body. These cells, therefore, have two types of glucose carriers. In the apical domain of the plasma membrane, which faces the lumen of the gut, they have glucose-Na^+ symports that take up glucose actively, creating a high glucose concentration in the cytosol. In the basal and lateral domains of the plasma membrane, they have passive glucose uniports that release the glucose down its concentration gradient for use by other tissues (Figure 12–14). The two types of glucose carriers are kept segregated in their proper domains of the plasma membrane by a diffusion barrier formed by a tight junction around the apex of the cell, which prevents mixing of membrane components between the apical and the basal and lateral domains, as discussed in Chapter 11 (see Figure 11–37).

Cells in the lining of the gut and in many other organs, such as the kidney, contain a variety of symports in their plasma membrane that are similarly driven by the electrochemical gradient of Na^+; each such carrier protein specifically imports a small group of related sugars or amino acids into the cell. Na^+-driven antiports are important too. An example is the *Na^+-H^+ exchanger* in the plasma membranes of many animal cells. This antiport uses the downhill influx of Na^+ to pump H^+ out of the cell and is one of the main devices that animal cells use to control the pH in their cytosol.

The Na^+-K^+ Pump Helps Maintain the Osmotic Balance of Animal Cells

The plasma membrane is permeable to water (see Figure 11–20), and if the total concentration of solutes is low on one side of it and high on the

> **Question 12–3** A transmembrane protein has the following properties: it has two binding sites, one for solute A and one for solute B. The protein can undergo a conformational change to switch between two states: either both binding sites are exposed exclusively on one side of the membrane or both binding sites are exposed exclusively on the other side of the membrane. The protein can switch between the two conformational states only if both binding sites are occupied or if both binding sites are empty, but cannot switch if only one binding site is occupied.
>
> A. What kind of protein do these properties define?
>
> B. Do you need to specify any additional properties to turn this protein into a symport that couples the movement of solute A up its concentration gradient to the movement of solute B down its electrochemical gradient?
>
> C. Write a set of rules that defines an antiport.

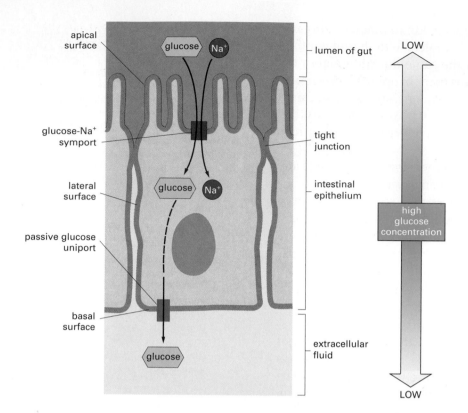

apical
surface

glucose   Na⁺

lumen of gut

LOW

glucose-Na⁺
symport

tight
junction

lateral
surface

glucose   Na⁺

intestinal
epithelium

high
glucose
concentration

passive glucose
uniport

basal
surface

glucose

extracellular
fluid

LOW

Figure 12–14 **Two types of glucose carriers enable gut epithelial cells to transfer glucose across the gut lining.** Glucose is actively transported into the cell by Na⁺-driven glucose symports at the apical surface, and it is released from the cell down its concentration gradient by passive glucose uniports at the basal and lateral surfaces. The two types of glucose carriers are kept segregated in the plasma membrane by the tight junction.

other, water will tend to move across it so as to make the solute concentrations equal. Such movement of water from a region of low solute concentration (high water concentration) to a region of high solute concentration (low water concentration) is called **osmosis.** The driving force for the water movement is equivalent to a difference in water pressure and is called the **osmotic pressure.** In the absence of any counteracting pressure, the osmotic movement of water into the cell will cause it to swell. Such effects are a severe problem for animal cells, which have no rigid external wall to prevent them from swelling. Placed in pure water, such cells will generally swell until they burst (Figure 12–15).

In the tissues of the animal body, cells are bathed by a fluid that is rich in solutes, especially Na⁺ and Cl⁻. This balances the concentration of organic and inorganic solutes confined inside the cell and so prevents osmotic disaster. But the osmotic balance is always in danger of being upset, as the external solutes are constantly leaking into the cell down their individual electrochemical gradients. The cells thus have to do continuous work, pumping out unwanted solutes to maintain the osmotic equilibrium (Figure 12–16A). This function is performed mainly by the Na⁺-K⁺ pump, which pumps out the Na⁺ that leaks in. At the same time,

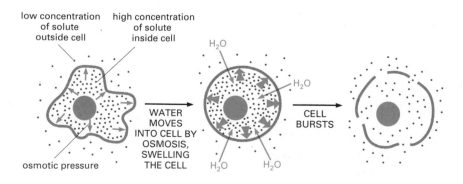

low concentration
of solute
outside cell

high concentration
of solute
inside cell

H₂O

WATER
MOVES
INTO CELL BY
OSMOSIS,
SWELLING
THE CELL

H₂O

CELL
BURSTS

osmotic pressure

H₂O   H₂O

Figure 12–15 **Osmosis.** If the concentration of solutes inside a cell is higher than that outside, water will move in by osmosis, causing the cell to swell. If the difference in solute concentration is great enough, the cell will burst.

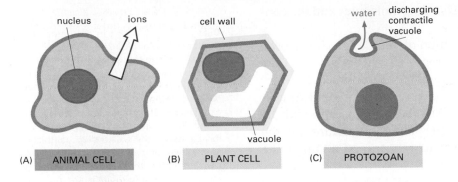

Figure 12–16 **Three ways in which cells avoid osmotic swelling.** The animal cell keeps the intracellular solute concentration low by pumping out ions (A). The plant cell is saved from swelling and bursting by its tough wall (B). The protozoan avoids swelling by periodically ejecting the water that moves into the cell (C).

by helping to maintain a membrane potential (as we explain later), the Na^+-K^+ pump also tends to prevent the entry of Cl^-, which is negatively charged. If the pump is halted with an inhibitor such as ouabain, or if the cell simply runs out of ATP to fuel the pump, Na^+ and Cl^- ions enter through carrier proteins and open ion channels, upsetting the osmotic balance so that the cell swells and eventually bursts.

Other cells cope with their osmotic problems in different ways. Plant cells are prevented from swelling and bursting by their tough cell walls and so can tolerate a large osmotic difference across their plasma membrane (Figure 12–16B). The cell wall exerts a counteracting pressure that tends to balance the osmotic pressure created by the solutes in the cell and thereby limits the movement of water into the cell. Osmosis, together with the active transport of ions into the cell, results in a *turgor pressure* that keeps plant cells distended with water, with their cell wall tense. Thus plant cells are like footballs whose leather outer case is held taut by the pressure in the pumped-up rubber bladder inside; the cell wall acts like the leather outer case, and the plasma membrane acts like the rubber bladder. The turgor pressure serves various functions. It holds plant stems rigid and leaves extended. It plays a part in regulating gas exchange through the stomata—the microscopic "mouths" in the surface of a leaf; the guard cells around the stomata regulate their own turgor pressure (by regulating movement of K^+ across the plasma membrane) to open and close the stomata.

In some protozoans living in fresh water, such as amoebae, the excess water that continually flows into the cell by osmosis is collected in contractile vacuoles that periodically discharge their contents to the exterior (Figure 12–16C). The cell first allows the vacuole to fill with a solution rich in solutes, which causes water to follow by osmosis. The cell then retrieves the solutes by actively pumping them back into the cytosol before emptying the vacuole to the exterior. But for most animal cells, the Na^+-K^+ pump is crucial for maintaining osmotic balance.

Intracellular Ca^{2+} Concentrations Are Kept Low by Ca^{2+} Pumps

Ca^{2+}, like Na^+, is also kept at a low concentration in the cytosol compared with its concentration in the extracellular fluid, but it is much less plentiful than Na^+, both inside and outside cells (see Table 12–1, p. 373). The movements of Ca^{2+} across cell membranes, however, are crucially important because Ca^{2+} can bind tightly to many other molecules in the cell, altering their activities. An influx of Ca^{2+} into the cytosol through Ca^{2+} channels, for example, is often used as a signal to trigger other intracellular events, such as the secretion of signaling molecules and the contraction of muscle cells.

The lower the background concentration of free Ca^{2+} in the cytosol, the more sensitive the cell is to an increase in cytosolic Ca^{2+}. Thus eucaryotic cells in general maintain very low concentrations of free Ca^{2+} in their cytosol (about 10^{-7} M) in the face of very much higher extracellular Ca^{2+} concentrations (typically 1–2 mM). This is achieved mainly by means of ATP-driven Ca^{2+} pumps in both the plasma membrane and the endoplasmic reticulum membrane, which actively pump Ca^{2+} out of the cytosol.

Like the Na^+-K^+ pump, the Ca^{2+} pump is an ATPase that is phosphorylated and dephosphorylated during its pumping cycle. It is thought to work in much the same way as depicted for the Na^+-K^+ pump in Figure 12–11 except that it returns to its original conformation without binding and transporting a second ion. These two ATP-driven pumps have similar amino acid sequences and structures, with about 10 membrane-spanning α helices in each subunit, and it is likely that they have a common evolutionary origin.

Question 12–4 A rise in the intracellular Ca^{2+} concentration causes muscle cells to contract. In addition to an ATP-driven Ca^{2+} pump, muscle cells that contract quickly and regularly, such as those of the heart, have an antiporter that exchanges Ca^{2+} for extracellular Na^+ across the plasma membrane. The majority of the Ca^{2+} ions that have entered the cell during contraction are rapidly pumped back out of the cell by this antiporter, thus allowing the cell to relax. Ouabain and digitalis are important drugs for treating patients with heart disease because they make the heart muscle contract more strongly. Both drugs function by partially inhibiting the Na^+-K^+ pump in the membrane of the heart muscle cell. Can you propose an explanation for the effects of the drugs in the patients? What will happen if too much of either drug is taken?

H^+ Gradients Are Used to Drive Membrane Transport in Plants, Fungi, and Bacteria

Plant cells, fungi (including yeasts), and bacteria do not have Na^+-K^+ pumps in their plasma membrane. Instead of an electrochemical gradient of Na^+, they mainly use an electrochemical gradient of H^+ to drive the transport of solutes into the cell. The gradient is created by H^+ pumps in the plasma membrane, which pump H^+ out of the cell, thus setting up an electrochemical proton gradient, with H^+ higher outside than inside; in the process, the H^+ pump also creates an acid pH in the medium surrounding the cell. The uptake of many sugars and amino acids into bacterial cells, for example, is driven by H^+ symports, which use the electrochemical gradient of H^+ across the plasma membrane in much the same way that animal cells use the electrochemical gradient of Na^+.

In some photosynthetic bacteria the H^+ gradient is created by the activity of light-driven H^+ pumps such as bacteriorhodopsin. In other bacteria the gradient is created by the activities of plasma membrane proteins that carry out the final stages of cellular respiration leading up to ATP synthesis, as discussed in Chapter 13. But plants, fungi, and many other bacteria set up their H^+ gradient by means of ATPases in their plasma membranes that use the energy of ATP hydrolysis to pump H^+ out of the cell; they resemble the Na^+-K^+ pumps and Ca^{2+} pumps in mammalian cells that we discussed earlier.

A different type of H^+ ATPase is found in the membranes of some intracellular organelles, such as the lysosomes of animal cells and the central vacuole of plant and fungal cells. Their function is to pump H^+ out of the cytosol into the organelle, which helps to keep the pH of the cytosol neutral and the pH of the interior of the organelle acidic. The acid environment in many organelles is crucial to their function, as we discuss in Chapter 14.

Some of the carrier proteins present in animal and plant cells that we have just discussed are shown in Figure 12–17. These and some other carrier proteins considered in this chapter are listed in Table 12–2 (p. 386).

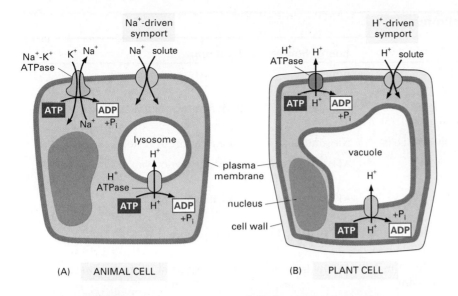

(A) ANIMAL CELL

(B) PLANT CELL

Figure 12–17 Some similarities and differences in carrier-mediated solute transport in animal and plant cells. Whereas an electrochemical gradient of Na^+, generated by the Na^+-K^+ pump (Na^+-K^+ ATPase), is often used to drive the active transport of solutes across the animal cell plasma membrane (A), an electrochemical gradient of H^+, usually set up by an H^+ ATPase, is often used for this purpose in plant cells (B), as well as in bacteria and fungi (including yeasts). The lysosomes in animal cells and the vacuoles in plant and fungal cells contain an H^+ ATPase in their membrane that pumps in H^+, helping to keep the internal environment of these organelles acidic. An electron micrograph showing the vacuole in plant cells in a young tobacco leaf (C). (C, courtesy of J. Burgess.)

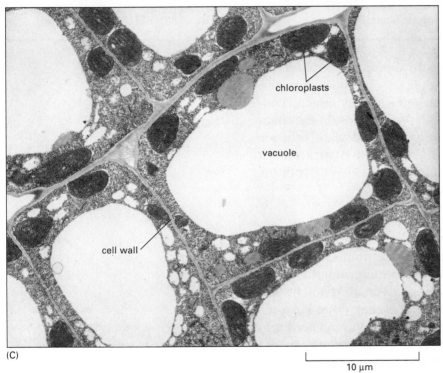

(C)

10 µm

We now turn to transport through ion channels and the generation of the membrane potential.

Ion Channels and the Membrane Potential

In principle, the simplest way to allow a small water-soluble molecule to cross from one side of a membrane to the other is to create a hydrophilic channel through which the molecule can pass. **Channel proteins** perform this function in cell membranes, forming transmembrane aqueous pores that allow the passive movement of small water-soluble molecules into or out of the cell or organelle.

A few channel proteins form relatively large pores: examples are the proteins that form *gap junctions* between two adjacent cells (see Figure

Table 12–2   Some Examples of Carrier Proteins

Carrier Protein	Location	Energy Source	Function
Glucose carrier	plasma membrane of most animal cells	none	passive import of glucose
Na$^+$-driven glucose pump	apical plasma membrane of kidney and intestinal cells	Na$^+$ gradient	active import of glucose
Na$^+$-H$^+$ exchanger	plasma membrane of animal cells	Na$^+$ gradient	active export of H$^+$ ions, pH regulation
Na$^+$-K$^+$ pump (Na$^+$-K$^+$ ATPase)	plasma membrane of most animal cells	ATP hydrolysis	active export of Na$^+$ and import of K$^+$
Ca^{2+} pump (Ca^{2+} ATPase)	plasma membrane of eucaryotic cells	ATP hydrolysis	active export of Ca^{2+}
H$^+$ pump (H$^+$ ATPase)	plasma membrane of plant cells, fungi, and some bacteria	ATP hydrolysis	active export of H$^+$ from cell
H$^+$ pump (H$^+$ ATPase)	membranes of lysosomes in animal cells and of vacuoles in plant and fungal cells	ATP hydrolysis	active export of H$^+$ from cytosol into vacuole
Bacteriorhodopsin	plasma membrane of some bacteria	light	active export of H$^+$ out of the cell

19–28) and the *porins* that form channels in the outer membrane of mitochondria and some bacteria (see Figure 11–26). But such large, permissive channels would lead to disastrous leaks if they directly connected the cytosol of a cell to the extracellular space. Most of the channel proteins in the plasma membrane of animal and plant cells are therefore quite different and have narrow, highly selective pores. Almost all of these proteins are **ion channels**, concerned exclusively with the transport of inorganic ions, mainly Na$^+$, K$^+$, Cl$^-$, and Ca^{2+}.

Ion Channels Are Ion Selective and Gated

Two important properties distinguish ion channels from simple aqueous pores. First, they show *ion selectivity*, permitting some inorganic ions to pass but not others. Ion selectivity depends on the diameter and shape of the ion channel and on the distribution of charged amino acids in its lining: the channel is narrow enough in places to force ions into contact with the wall of the channel so that only those of appropriate size and charge are able to pass. Narrow channels, for example, will not pass large ions, and channels with a negatively charged lining will deter negative ions from entering because of the mutual electrostatic repulsion between like charges. In this way channels have evolved that are selective for just one type of ion, such as Na$^+$ or Cl$^-$. Each ion in aqueous solution is surrounded by a small shell of water molecules, and it is thought that the ions have to shed most of their associated water molecules in order to pass, in single file, through the narrowest part of the channel. This step in the transport of an ion limits the maximum rate of passage through the channel. Thus, as ion concentrations are increased, the flow of ions through a channel at first increases proportionally but then levels off (saturates) at a maximum rate.

The second important distinction between simple aqueous pores and ion channels is that ion channels are not continuously open. Ion transport would be of no value to the cell if there were no means of controlling the flows and if all of the many thousands of ion channels in a

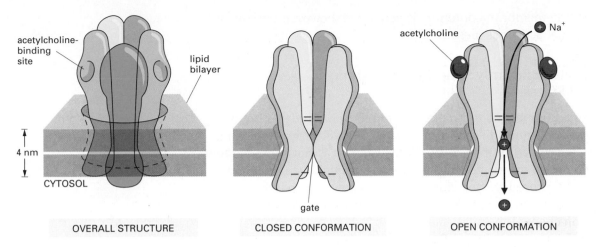

acetylcholine-binding site

lipid bilayer

4 nm

CYTOSOL

OVERALL STRUCTURE

gate

CLOSED CONFORMATION

acetylcholine

Na⁺

OPEN CONFORMATION

cell membrane were open all of the time. As we discuss later, most ion channels are *gated:* they can switch between an open and a closed state by a change in conformation, and this transition is regulated by conditions inside and outside the cell (Figure 12–18).

Ion channels have a large advantage over carrier proteins with respect to their maximum rate of transport. More than a million ions can pass through one channel each second, which is a rate 1000 times greater than the fastest rate of transport known for any carrier protein. On the other hand, channels cannot couple the ion flow to an energy source to carry out active transport. The function of most ion channels is simply to make the membrane transiently permeable to selected inorganic ions, mainly Na⁺, K⁺, Ca²⁺, or Cl⁻, allowing these to diffuse rapidly down their electrochemical gradients across the membrane when the channel gates are open.

Thanks to active transport by pumps and other carrier proteins, most ion concentrations are far from equilibrium across the membrane. When a channel opens, therefore, ions rush through it. The rush of ions amounts to a pulse of electric charge delivered either into the cell (as ions flow in) or out of the cell (as ions flow out). The ion flow changes the voltage across the membrane—the *membrane potential*—which alters the electrochemical driving forces for transmembrane movements of all the other ions. It also forces other ion channels, which are specifically sensitive to changes in the membrane potential, to open or close in a matter of milliseconds. The resulting flurry of electrical activity can spread rapidly from one region of the cell membrane to another, conveying an electrical signal, as we discuss later in the context of nerve cells. This type of electrical signaling is not restricted to animals; it also occurs in protozoans and plants: carnivorous plants like the Venus flytrap, for example, use electrical signaling to sense and trap insects (Figure 12–19).

The membrane potential is the basis of all electrical activity in cells, whether they are plant cells, animal cells, or protozoans. Before we

Figure 12–18 The structure of an ion channel. The ion channel shown is present in the plasma membrane of muscle cells and opens when the neurotransmitter acetylcholine, released by a nerve, binds to the channel (it is therefore an example of a transmitter-gated ion channel, as we discuss later). The ion channel is composed of five transmembrane protein subunits that combine to form an aqueous pore across the lipid bilayer. The pore is lined by five transmembrane α helices, one contributed by each subunit. Negatively charged amino acid side chains at either end of the pore ensure that only positively charged ions, mainly Na⁺ and K⁺, can pass. When the channel is in its closed conformation, the pore is occluded by hydrophobic amino acid side chains in the region called the *gate;* when acetylcholine binds, the protein undergoes a conformational change in which these side chains move apart and the gate opens, allowing Na⁺ and K⁺ to flow across the membrane, down their electrochemical gradients. As we discuss later, even with acetylcholine bound, the channel flickers randomly between the open and closed states; without acetylcholine bound, however, it rarely opens.

Figure 12–19 A Venus flytrap. The leaves snap shut in less than half a second when an insect moves on them. The response is triggered by touching any two of the three trigger hairs in succession in the center of each leaf. This opens ion channels and thereby sets off an electrical signal, which, by an unknown mechanism, leads to a rapid change in turgor pressure that closes the leaf. (Courtesy of J.S. Sira, Garden Picture Library.)

Ion Channels and the Membrane Potential

discuss how the membrane potential is generated, however, we look at how ion channel activity is measured.

Ion Channels Randomly Snap Between Open and Closed States

Electrical measurement is the main method used to study ion movements and ion channels in living cells. Amazingly, electrical recording techniques have been refined to the point where it is now possible to detect and measure the electric current flowing through a single channel molecule. The procedure for doing this is known as **patch-clamp recording,** and it has provided a direct and surprising picture of how individual ion channels behave.

In patch-clamp recording, a fine glass tube is used as a *microelectrode* to make electrical contact with the surface of the cell. The microelectrode is made by heating the glass tube and pulling it to create an extremely fine tip with a diameter of no more than a few micrometers; the tube is then filled with an aqueous conducting solution, and the tip is pressed against the cell surface. With gentle suction, a tight electrical seal is formed where the cell membrane contacts the mouth of the microelectrode (Figure 12–20A). If one wishes to expose the cytosolic face of the membrane, the patch of membrane held in the microelectrode is gently detached from the cell (Figure 12–20B). At the other, open end of the microelectrode, a metal wire is inserted. Current that enters

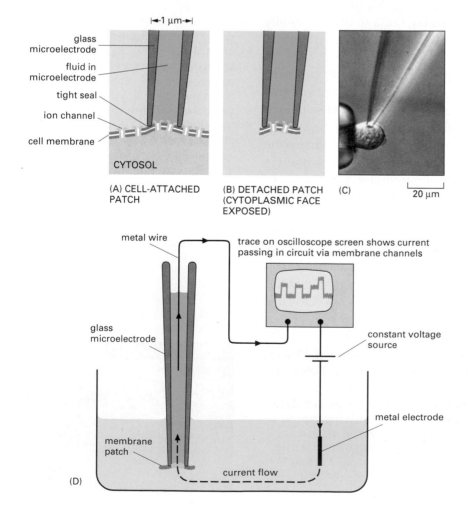

Figure 12–20 **The technique of patch-clamp recording.** Because of the extremely tight seal between the mouth of the microelectrode and the membrane, current can enter or leave the microelectrode only by passing through the channels in the patch of membrane covering its tip. The term "clamp" is used because an electronic device is employed to maintain, or "clamp," the membrane potential at a set value while recording the ionic current through individual channels. Recordings of the current through these channels can be made with the patch still attached to the rest of the cell, as in (A), or detached, as in (B). The advantage of the detached patch is that it is easy to alter the composition of the solution on either side of the membrane to test the effect of various solutes on channel behavior. The micrograph (C) shows a nerve cell from the eye held in a suction pipette (the tip of which is shown on the *left*) while a microelectrode (*upper right*) is being used for patch-clamp recording. (D) The circuitry for patch-clamp recording. (C, from T.D. Lamb, H.R. Matthews, and V. Torre, *J. Physiol.* 372:315–349, 1986.)

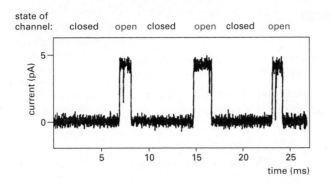

Figure 12–21 The current through a single ion channel, as recorded by the patch-clamp technique. The voltage (the membrane potential) across the isolated patch of membrane was held constant during the recording. In this example the membrane is from a muscle cell and contains a single channel protein, which is of the type shown in Figure 12–18 that is responsive to the neurotransmitter acetylcholine. This ion channel opens to allow passage of positive ions when acetylcholine binds to the exterior face of the channel, as is the case here. Even though acetylcholine is bound to the channel, however, the channel does not remain open all the time: instead, it flickers between open and closed states. In the time shown [about 25 milliseconds (ms)] the channel opened three times for a variable period each time. If acetylcholine were not bound to the channel, the channel would rarely open. (Courtesy of David Colquhoun.)

the microelectrode through ion channels in the small patch of membrane covering its tip passes via the wire, through measuring instruments, back into the bath of medium in which the cell or the detached patch is sitting (Figure 12–20D). Patch-clamp recording makes it possible to record from ion channels in all sorts of cell types—not only in large nerve cells, which are famous for their electrical activities, but also in cells such as yeasts that are too small for the electrical events in them to be detected by other means.

By varying the concentrations of ions in the medium on either side of the membrane patch, one can test which ions will go through the channels in it. With appropriate electronic circuitry, the voltage across the membrane patch—that is, the membrane potential—can also be set and held clamped at any chosen value (hence the term "patch-clamp"). This makes it possible to see how changes of membrane potential affect the opening and closing of the channels in the membrane.

With a sufficiently small area of membrane in the patch, sometimes only a single ion channel will be present. Modern electrical instruments are sensitive enough to reveal the ion flow through a single channel, detected as a minute electric current (of the order of 10^{-12} amps). These currents typically behave in a surprising way: even when conditions are held constant, the currents switch abruptly on and abruptly off again, as though an on/off switch were being randomly jiggled (Figure 12–21). This behavior implies that the channel has moving parts and is snapping to and fro between open and closed conformations (see Figure 12–18). As this behavior is seen even when conditions are constant, it presumably indicates that the channel protein is being knocked from one conformation to the other by the random thermal movements of the molecules in its environment. This is one of very few instances in which it is possible to follow the conformational change of a single protein molecule. The picture it conjures up, of jerky machinery subjected to a constant violent buffeting, is certain to apply also to other proteins with moving parts.

If ion channels randomly snap between open and closed conformations even when conditions on each side of the membrane are held constant, how can their state be regulated by conditions inside or outside the cell? The answer is that when the appropriate conditions change, the random behavior continues but with a greatly changed probability: if the altered conditions tend to open the channel, for example, the channel will now spend a much greater proportion of its time in the open conformation, although it will not remain open continuously (see Figure 12–21). When an ion channel is open, it is fully open, and when it is closed, it is fully closed.

Voltage-gated Ion Channels Respond to the Membrane Potential

More than a hundred types of ion channels have been discovered so far, and new ones are still being found. They differ from one another primarily with respect to (1) *ion selectivity*—the type of ions they allow to pass—and (2) *gating*—the conditions that influence their opening and closing.

For a **voltage-gated channel,** the probability of being open is controlled by the membrane potential (Figure 12–22A). For a **ligand-gated channel,** like the acetylcholine receptor shown in Figure 12–18, it is controlled by the binding of some molecule (the ligand) to the channel protein (Figure 12–22B, C). For a **stress-activated channel,** opening is controlled by a mechanical force applied to the channel (Figure 12–22D). The *auditory hair cells* in the ear are an important example of cells that depend on this type of channel. Sound vibrations pull stress-activated channels open, causing ions to flow into the hair cells; this sets up an electrical signal that is transmitted from the hair cell to the auditory nerve, which conveys the signal to the brain (Figure 12–23).

Voltage-gated ion channels play the major role in propagating electrical signals in nerve cells. They are present in many other cells too, including muscle cells, egg cells, protozoans, and even plant cells, where they enable electrical signals to travel from one part of the plant to another, as in the leaf-closing response of the mimosa tree (Figure 12–24). Voltage-gated ion channels have specialized charged protein domains called *voltage sensors* that are extremely sensitive to changes in the membrane potential: changes above a certain threshold value exert sufficient electrical force on these domains to encourage the channel to switch from its closed to its open conformation, or vice versa. A change in the membrane potential does not affect how widely the channel is open but alters the probability that it will be found in its open conformation. Thus in a large patch of membrane, containing many molecules of the channel protein, one might find that on average 10% of them are open at any instant when the membrane is held at one potential, while 90% are open when it is held at another potential.

To appreciate the function of voltage-gated ion channels in a living cell, we have to consider what controls the membrane potential. The

Question 12–5   Figure Q12–5 shows a recording from a patch-clamp experiment in which the electrical current passing across a patch of membrane is measured as a function of time. The membrane patch was plucked from the plasma membrane of a muscle cell by the technique shown in Figure 12–20 and contains molecules of the acetylcholine receptor, which is a ligand-gated cation channel that is opened by the binding of acetylcholine to the extracellular face of the channel protein. To obtain a recording, acetylcholine was added to the solution in the microelectrode. Describe what you can learn about the channels from this recording. How would the recording differ if acetylcholine were (a) omitted or (b) added to the solution outside the microelectrode only?

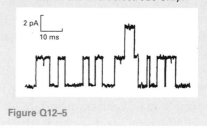

Figure Q12–5

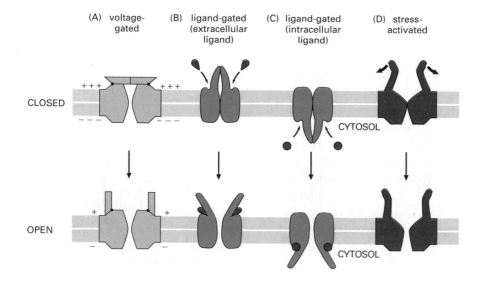

Figure 12–22 Gated ion channels. Depending on the type of ion channel, the gates open in response to a change in the voltage difference across the membrane (A), to the binding of a chemical ligand to the channel, outside (B) or inside the cell (C), or to mechanical stimulation (D).

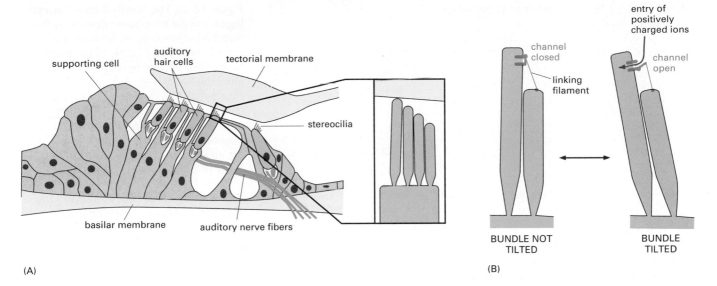

(A)

(B)

Figure 12–23 How stress-activated ion channels allow us to hear. (A) A section through the organ of Corti, which runs the length of the cochlea of the inner ear. Each auditory hair cell has a tuft of processes called stereocilia projecting from its upper surface. The hair cells are embedded in a sheet of supporting cells, which is sandwiched between the *basilar membrane* below and the *tectorial membrane* above. (These are not lipid bilayer membranes but sheets of extracellular matrix.) (B) Sound vibrations cause the basilar membrane to vibrate up and down, causing the stereocilia to tilt. Each stereocilium in the staggered array on each hair cell is attached to the next shorter stereocilium by a fine filament. The tilting stretches the filaments, which pull open stress-activated ion channels in the stereocilium membrane, allowing positively charged ions to enter from the surrounding fluid. The influx of ions activates the hair cells, which stimulate underlying nerve cells that convey the auditory signal to the brain. The hair-cell mechanism is astonishingly sensitive: the force required to open a single channel is estimated to be about 2×10^{-13} newtons, and the faintest sounds we can hear have been estimated to stretch the filaments by an average of about 0.04 nm, which is less than the diameter of a hydrogen ion.

simple answer is that ion channels themselves control it, and the opening and closing of these channels is what makes it change. This control loop, from ion channels → membrane potential → ion channels, is fundamental to all electrical signaling in cells. Having seen how the membrane potential can regulate ion channels, we now discuss how ion channels can control the membrane potential. In the last part of the chapter we consider how this control loop works as a whole.

The Membrane Potential Is Governed by Membrane Permeability to Specific Ions

All cells have an electrical potential difference, or **membrane potential,** across their plasma membranes. To understand how it arises, it is helpful to recall some basic principles of electricity.

(A)

(B)

(C)

Figure 12–24 Leaf-closing response in mimosa. (A) Resting leaf. (B and C) Successive responses to touch. A few seconds after the leaf is touched, the leaflets collapse. The response involves the opening of voltage-gated ion channels, generating an electric impulse. When the impulse reaches specialized hinge cells at the base of each leaflet, a rapid loss of water by these cells occurs, causing the leaflets to collapse suddenly and progressively down the leaf stalk. (Courtesy of G.I. Bernard. © Oxford Scientific Films.)

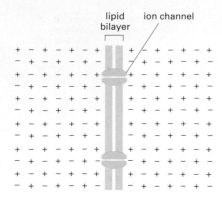

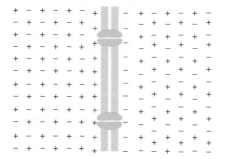

lipid bilayer   ion channel

exact balance of charges on each side of the membrane so that each positive ion is balanced by a negative counterion; membrane potential = 0

a few of the positive ions (*red*) cross the membrane from right to left, leaving their negative counterions (*red*) behind; this sets up a nonzero membrane potential

Figure 12–25 The distribution of ions giving rise to the membrane potential. The membrane potential results from a thin (<1 nm) layer of ions close to the membrane, held in place by their electrical attraction to their oppositely charged counterparts on the other side of the membrane. The number of ions that must move across the membrane to set up a membrane potential is a tiny fraction of the ions present. (6000 K^+ ions crossing 1 μm^2 of cell membrane are enough to shift the membrane potential by about 100 millivolts; the number of K^+ ions in 1 μm^3 of bulk cytoplasm is 70,000 times larger than this.)

While electricity in metals is carried by electrons, electricity in aqueous solutions is carried by ions, which are either positively (cations) or negatively (anions) charged. An ion flow across a cell membrane is detectable as an electric current, and an accumulation of ions, if not exactly balanced by an accumulation of oppositely charged ions, is detectable as an accumulation of electric charge, or a membrane potential (Figure 12–25).

To see how the membrane potential is generated and maintained, consider the ion movements into and out of a typical animal cell in an unstimulated "resting" state. The negative charges on the organic molecules confined within the cell are largely balanced by K^+, the predominant positive ion inside the cell. The high intracellular concentration of K^+ is in part created by the Na^+-K^+ pump, which actively pumps K^+ into the cell. This leads to a large concentration difference for K^+ across the plasma membrane, with the concentration of K^+ being much higher inside the cell than outside. The plasma membrane, however, also contains a set of K^+ channels known as *K^+ leak channels*. These channels randomly flicker between open and closed states no matter what the conditions are inside or outside the cell, and when they are open, they allow K^+ to move freely. In a resting cell, these are the main ion channels open in the plasma membrane, thus making the resting plasma membrane much more permeable to K^+ than to other ions.

K^+ will have a tendency to flow out of the cell through these channels down its steep concentration gradient. But any transfer of positive charge to the exterior leaves behind unbalanced negative charge within the cell, thereby creating an electrical field, or membrane potential, which will oppose any further movement of K^+ out of the cell. Within a millisecond or so, an equilibrium condition is established in which the membrane potential is just strong enough to counterbalance the tendency of K^+ to move out down its concentration gradient—that is, in which the electrochemical gradient for K^+ is zero, even though there is still a much higher concentration of K^+ inside the cell than out (Figure 12–26).

The *resting membrane potential* is the membrane potential in such steady-state conditions, in which the flow of positive and negative ions across the plasma membrane is precisely balanced, so that no further difference in charge accumulates across the membrane. The membrane

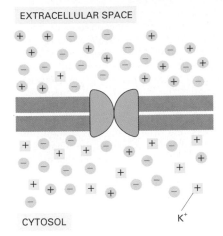

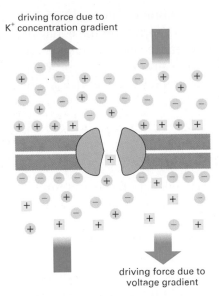

driving force due to
K⁺ concentration gradient

EXTRACELLULAR SPACE

CYTOSOL

K⁺

driving force due to
voltage gradient

(A) K⁺ channel closed, membrane potential = 0; more K⁺ inside the cell than outside, but zero *net* charge on each side (positive and negative charges balanced exactly)

(B) K⁺ channel open; K⁺ moves out, leaving negative ions behind, and this charge distribution creates a membrane potential that balances the tendency of K⁺ to move out

Figure 12–26 The role of K⁺ in the generation of the membrane potential across the plasma membrane. Starting from a hypothetical situation where the membrane potential is zero, K⁺ will tend to leave the cell moving down its concentration gradient through K⁺ leak channels. Assuming the membrane contains no open channels permeable to other ions, K⁺ ions will cross the membrane, but negative ions will be unable to follow them. The result will be an excess of positive charge on the outside of the membrane and of negative charge on the inside. This gives rise to a membrane potential that tends to drive K⁺ back in. At equilibrium, the effect of the K⁺ concentration gradient is exactly balanced by the effect of the membrane potential, and there is no net movement of K⁺.

potential is measured as a voltage difference across the membrane. In animal cells, the resting membrane potential varies between –20 and –200 millivolts (mV), depending on the organism and cell type. It is expressed as a negative value because the interior of the cell is negative with respect to the exterior, as the negative charges inside the cell are in slight excess over positive charges. The actual value of the resting membrane potential in animal cells is chiefly a reflection of the K⁺ concentration gradient across the plasma membrane because, at rest, this membrane is chiefly permeable to K⁺ and K⁺ is the main positive ion inside the cell. A simple formula called the **Nernst equation** (Figure 12–27) expresses the equilibrium quantitatively and makes it possible to calculate the theoretical resting membrane potential if the ratio of internal to external ion concentrations is known.

Suppose now that other channels permeable to some other ion—say Na⁺—are suddenly opened in the resting plasma membrane. Because Na⁺ is at a higher concentration outside the cell than inside, Na⁺ will move into the cell through these channels and the membrane potential will become less negative, maybe even reversing sign to become positive (so that the interior of the cell is positive with respect to the exterior). The membrane potential will shift toward a new value that is a compromise between the negative value that would correspond to equilibrium for K⁺ and the positive value that would correspond to equilibrium for Na⁺. Any change in the membrane's permeability to specific ions—that is, any change in the numbers of ion channels of different sorts that are open—thus causes a change in the membrane potential. The membrane potential, therefore, is determined by both the state of the ion channels and the ion concentrations in the cytosol and extracellular medium. Because the electrical processes at the plasma membrane occur very quickly compared to changes in the bulk ion concentrations, however, over the short term—milliseconds to seconds or minutes—it is the ion channels that are most important in controlling the membrane potential.

The force tending to drive an ion across a membrane is made up of two components: one due to the electrical membrane potential and one due to the concentration gradient. At equilibrium, the two forces are balanced and satisfy a simple mathematical relationship given by the

Nernst equation

$$V = 62 \log_{10} (C_o/C_i)$$

where V is the membrane potential in millivolts, and C_o and C_i are the outside and inside concentrations of the ion, respectively. This form of the equation assumes that the ion carries a single positive charge and that the temperature is 37°C.

Figure 12–27 The Nernst equation.

Ion Channels and the Membrane Potential

To see how the interplay between the membrane potential and ion channels is used for electrical signaling, we now turn from the behavior of ions and ion channels to the behavior of entire cells. We take nerve cells as our prime example, for they, more than any other cell type, have made a profession of electrical signaling and employ ion channels in the most sophisticated ways.

Ion Channels and Signaling in Nerve Cells

The fundamental task of a nerve cell, or **neuron,** is to receive, conduct, and transmit signals. Neurons carry signals inward from sense organs to the central nervous system, which consists of the brain and spinal cord. In the central nervous system, neurons signal from one to another, through networks of enormous complexity, to analyze, interpret, and respond to the signals coming in from the sense organs. From the central nervous system, neurons extend processes outward to convey signals for action to muscles and glands. To perform these functions, neurons are often extremely elongated: motor neurons in a human, carrying signals from the spinal cord to a muscle in the foot, for example, may be a meter long.

Every neuron consists of a *cell body* (containing the nucleus) with a number of long, thin processes radiating outward from it. Usually, there is one long **axon,** which conducts signals away from the cell body toward distant target cells, and several shorter, branching *dendrites,* which extend from the cell body like antennae and provide an enlarged surface area to receive signals from the axons of other neurons (Figure 12–28). The axon commonly divides at its far end into many branches, each of which ends in a **nerve terminal,** so that the neuron's message can be passed simultaneously to many target cells—either other neurons or muscle or gland cells. Likewise, the branching of the dendrites can be extensive, in some cases sufficient to receive as many as 100,000 inputs on a single neuron.

No matter what the meaning of the signal a neuron carries—whether it is visual information from the eye, a motor command to a muscle, or a step in signal analysis in the brain—the form of the signal is always the same: it consists of changes in the electrical potential across the neuron's plasma membrane.

Question 12–6 From the concentrations given in Table 12–1 (p. 373), calculate the equilibrium membrane potential of K$^+$ and Na$^+$ (assume the concentration of intracellular Na$^+$ is 10 mM). What membrane potential would you predict in a resting cell? Explain your answer. What would happen if a large number of Na$^+$ channels suddenly opened, making the membrane much more permeable to Na$^+$ than to K$^+$? (Note that because few ions need to move across the membrane to change the charge distribution across the membrane drastically, you should assume that the ion concentrations on either side of the membrane do not change significantly.) What would you predict would happen next if the Na$^+$ channels closed again?

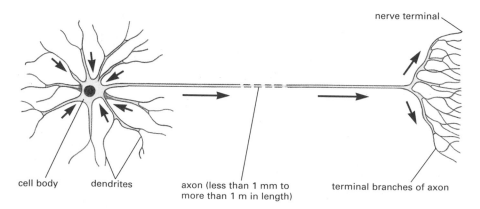

cell body   dendrites   axon (less than 1 mm to more than 1 m in length)   terminal branches of axon

Figure 12–28 Neurons. Schematic diagram of a typical neuron. The *red arrows* indicate the direction in which signals are conveyed. The single axon conducts signals away from the cell body, while the multiple dendrites receive signals from the axons of other neurons.

Action Potentials Provide for Rapid Long-Distance Communication

A neuron is stimulated by a signal—typically from another neuron—delivered to a localized site on its surface. This signal initiates a change in the membrane potential at that site. To transmit the signal onward, however, the change in membrane potential has to spread from this point, which is usually on a dendrite or the cell body, to the axon terminals, which relay the signal to the next cells in the pathway. Although a local change in membrane potential will spread passively along an axon or a dendrite to adjacent regions of the plasma membrane, it rapidly becomes weaker with increasing distance from the source. Over short distances this weakening is unimportant, but for long-distance communication such *passive spread* is inadequate. In the same way, a telephone signal can be transmitted without amplification the short distances through the wires in your hometown, but for transmission across an ocean by an undersea cable, the strength of the signal has to be boosted at intervals.

Neurons solve the problem by employing an active signaling mechanism: a local electrical stimulus of sufficient strength triggers an explosion of electrical activity in the plasma membrane that is propagated rapidly along the membrane of the axon, being sustained by automatic renewal all along the way. This traveling wave of electrical excitation, known as an **action potential,** or a *nerve impulse,* can carry a message, without the signal weakening, from one end of a neuron to the other at speeds of up to 100 meters per second.

All of the early research that established the mechanism of electrical signaling along nerve axons was done on the giant axon of the squid (Figure 12–29), which has such a large diameter that it is possible to record its electrical activity from an electrode inserted directly into it. From such studies it was deduced that action potentials are the direct consequence of the properties of voltage-gated ion channels (see Figure 12–22A) in the nerve cell membrane, as we now explain.

Figure 12–29 The squid *Loligo.* This animal has a large nerve cell with a giant axon that allows the squid to respond rapidly to threats in its environment. Long before patch-clamping allowed recordings from single channels in small cells, scientists were able to record action potentials in the squid giant axon and deduce the existence of ion channels in membranes. (Courtesy of Howard Hall © Oxford Scientific Films.)

Action Potentials Are Usually Mediated by Voltage-gated Na⁺ Channels

An action potential in a neuron is typically triggered by a sudden local *depolarization* of the plasma membrane—that is, by a shift in the membrane potential to a less negative value. We discuss later how such a depolarization is caused by the action of signaling molecules —*neurotransmitters*—released by another neuron. A stimulus that causes a sufficiently large depolarization to pass a certain threshold value promptly causes **voltage-gated Na⁺ channels** to open temporarily at that site, allowing a small amount of Na⁺ to enter the cell down its electrochemical gradient. The influx of positive charge depolarizes the membrane further (that is, it makes the membrane potential even less negative), thereby opening more voltage-gated Na⁺ channels, which admit more Na⁺ ions and cause still further depolarization. This process continues in a self-amplifying fashion until, within about a millisecond, the membrane potential in the local region of membrane has shifted from

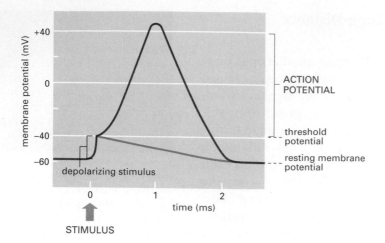

Figure 12–30 An action potential. The resting membrane potential in this neuron is –60 millivolts. The action potential is triggered when a stimulus depolarizes the plasma membrane by about 20 millivolts, making the membrane potential –40 millivolts, which is the threshold value in this cell for initiating an action potential. Once an action potential is triggered, the membrane rapidly depolarizes further: the membrane potential swings past zero and reaches +40 millivolts before it returns to its resting negative value, as the action potential terminates. The *green curve* shows how the membrane potential simply would have relaxed back to the resting value after the initial depolarizing stimulus if there had been no voltage-gated ion channels in the plasma membrane.

its resting value of about –60 mV to about +40 mV (Figure 12–30). This is close to the membrane potential at which the electrochemical driving force for movement of Na^+ across the membrane is zero—that is, at which the effects of the membrane potential and the concentration gradient for Na^+ are equal and opposite and Na^+ has no further tendency to enter or leave the cell. At this point the cell would get stuck with all of its voltage-gated Na^+ channels predominantly open if the channels continued indefinitely to respond to the altered membrane potential in the same way.

The cell is saved from this fate because the Na^+ channels have an automatic inactivating mechanism, which causes them to adopt rapidly (within a millisecond or so) a special inactive conformation, where the channel is unable to open again: even though the membrane is still depolarized, the Na^+ channels will remain in this *inactivated state* until a few milliseconds after the membrane potential returns to its initial negative value. A schematic illustration of these three distinct states of the voltage-gated Na^+ channel—*closed, open,* and *inactivated*—is shown in Figure 12–31. How they contribute to the rise and fall of the action potential is shown in Figure 12–32.

The membrane is also helped to return to its resting value by the opening of *voltage-gated K^+ channels*. These also open in response to depolarization of the membrane, but not so promptly as the Na^+ channels, and they then stay open as long as the membrane remains depolarized. As the action potential reaches its peak, K^+ ions (carrying positive charge) therefore start to flow out of the cell through these K^+ channels down their electrochemical gradient, unhindered by the negative membrane potential that restrains them in the resting cell. The rapid outflow

Figure 12–31 A voltage-gated Na^+ channel can adopt at least three conformations. The channel can flip from one conformation (state) to another, depending on the membrane potential. When the membrane is at rest (highly polarized), the closed conformation is the most stable. When the membrane is depolarized, however, the open conformation is more stable, and so the channel has a high probability of opening; but in the depolarized membrane the inactivated conformation is more stable still, and so, after a brief period spent in the open conformation, the channel becomes inactivated and cannot open. The *red arrows* indicate the sequence that follows a sudden depolarization, and the *black arrow* indicates the return to the original conformation after the membrane is repolarized.

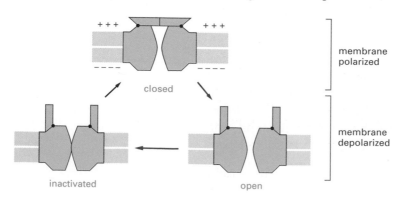

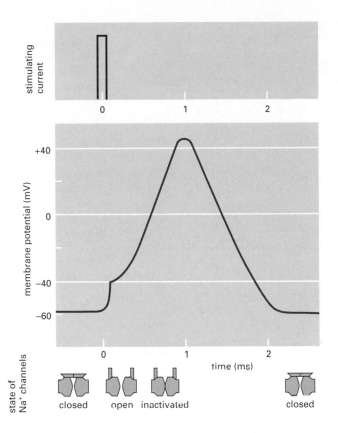

of K$^+$ through the voltage-gated K$^+$ channels brings the membrane back to its resting state more quickly than could be achieved by K$^+$ outflow through the K$^+$ leak channels alone.

The description just given of an action potential concerns only a small patch of plasma membrane. The self-amplifying depolarization of the patch, however, is sufficient to depolarize neighboring regions of membrane, which then go through the same self-amplifying cycle. In this way the action potential spreads outward as a traveling wave from the initial site of depolarization, eventually reaching the extremities of the axon (Figure 12–33).

Question 12–7 Explain in 100 words or less how an action potential is passed along an axon.

Voltage-gated Ca^{2+} Channels Convert Electrical Signals into Chemical Signals at Nerve Terminals

When an action potential reaches the ends of the axon—the *nerve terminals*—the signal must somehow be relayed to the *target cells* that the nerve terminals contact, which are usually neurons or muscle cells. The signal is transmitted at specialized sites of contact known as **synapses.** At most synapses the plasma membranes of the transmitting and receiving cells—the *presynaptic* and the *postsynaptic* cells, respectively—are separated from each other by a narrow *synaptic cleft* (typically 20 nm across), which the electrical signal cannot cross (Figure 12–34). For the message to be transmitted from one neuron to another, the electrical signal is converted into a chemical signal, in the form of a small signaling molecule known as a **neurotransmitter.**

Neurotransmitters are stored ready-made in the nerve terminals, packaged in membrane-bounded **synaptic vesicles** (see Figure 12–34).

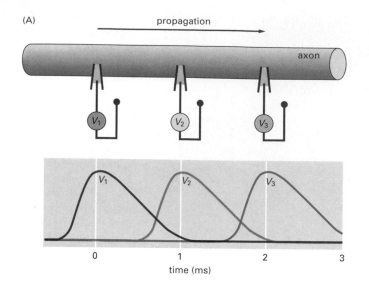

(A)

propagation

axon

V_1 V_2 V_3

V_1 V_2 V_3

0   1   2   3
time (ms)

Figure 12–33 The propagation of an action potential along an axon. (A) shows the voltages that would be recorded from a set of intracellular electrodes placed at intervals along the axon, whose width is greatly exaggerated in this schematic figure. Note that the action potential does not weaken as it travels. (B) shows the changes in the Na^+ channels and the current flows (*brown arrows*) that give rise to the traveling disturbance of the membrane potential. The region of the axon with a depolarized membrane is shaded in *pink*. An action potential can only travel away from the site of depolarization because Na^+-channel inactivation prevents the depolarization from spreading backward. (See also Figure 12–32.)

(B)

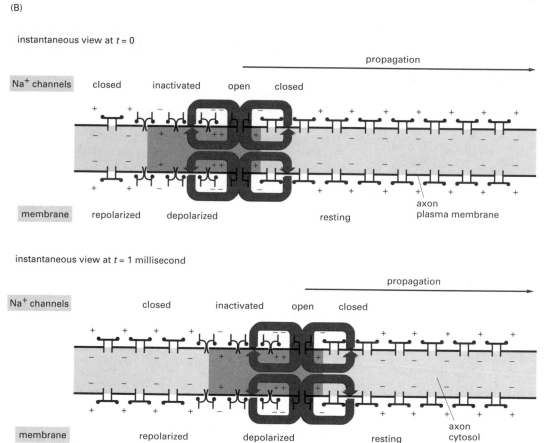

instantaneous view at $t = 0$

propagation

Na^+ channels   closed   inactivated   open   closed

membrane   repolarized   depolarized   resting

axon plasma membrane

instantaneous view at $t = 1$ millisecond

propagation

Na^+ channels   closed   inactivated   open   closed

membrane   repolarized   depolarized   resting

axon cytosol

They are released by exocytosis (discussed in Chapter 14) from the nerve ending when an action potential reaches the terminal. This link between the action potential and secretion involves the activation of yet another type of voltage-gated cation channel. The depolarization of the nerve terminal plasma membrane caused by the arrival of the action potential transiently opens *voltage-gated Ca^{2+} channels*, which are concentrated in the plasma membrane of the nerve terminal. Because the Ca^{2+} concentration outside the cell is more than 1000 times greater than the free Ca^{2+} concentration in the cytosol, Ca^{2+} rushes into the nerve terminal through the open channels. The resulting increase in Ca^{2+} concentration

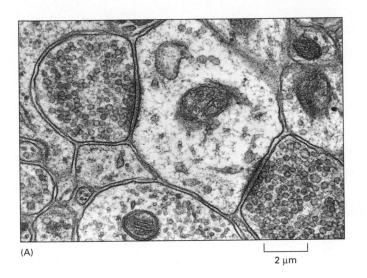

(A)

2 μm

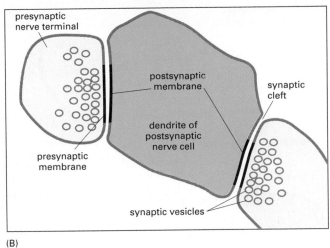

(B)

in the cytosol of the nerve terminal triggers the fusion of the synaptic vesicles with the plasma membrane, releasing the neurotransmitter into the synaptic cleft. Thanks to the voltage-gated Ca^{2+} channels, the electrical signal has now been converted into a chemical signal (Figure 12–35).

Transmitter-gated Channels in Target Cells Convert Chemical Signals Back into Electrical Signals

The released neurotransmitter rapidly diffuses across the synaptic cleft and binds to *neurotransmitter receptors* concentrated in the postsynaptic membrane on the target cell. The binding of neurotransmitter to its receptors results in a change in the membrane potential of the target cell, which can trigger the cell to fire an action potential. The neurotransmitter is rapidly removed, ensuring that when the presynaptic cell falls quiet, the postsynaptic cell will fall quiet also. The neurotransmitter is removed either by rapid breakdown by enzymes in the synaptic cleft or by reuptake—either into the nerve terminals that released it or into neighboring cells.

The neurotransmitter receptors can be of various types, some mediating relatively slow effects in the target cell, others mediating more rapid responses. Rapid responses—on a time scale of milliseconds—

Figure 12–34 Synapses. An electron micrograph (A) and drawing (B) of a cross section of two nerve terminals (*yellow*) forming synapses on a single nerve cell dendrite (*blue*) in the mammalian brain. Note that both the presynaptic and postsynaptic membranes are thickened at the synapse. (A, courtesy of Cedric Raine.)

RESTING NERVE TERMINAL

VOLTAGE-GATED Ca^{2+} CHANNEL (closed)

presynaptic nerve terminal

neurotransmitter

synaptic vesicle

synaptic cleft

neurotransmitter receptor

ACTIVATED NERVE TERMINAL

VOLTAGE-GATED Ca^{2+} CHANNEL (open)

nerve impulse (electrical signal)

Ca^{2+}

neurotransmitter released (chemical signal)

postsynaptic cell

Figure 12–35 Conversion of an electrical signal into a chemical signal at a nerve terminal. When an action potential reaches a nerve terminal, it opens voltage-gated Ca^{2+} channels in the plasma membrane, allowing Ca^{2+} to flow into the terminal. The increased Ca^{2+} in the nerve terminal stimulates the synaptic vesicles to fuse with the plasma membrane, releasing their neurotransmitter into the synaptic cleft.

Ion Channels and Signaling in Nerve Cells

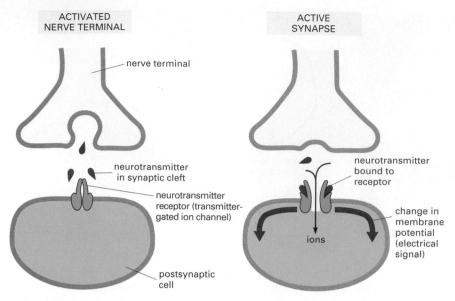

ACTIVATED NERVE TERMINAL

ACTIVE SYNAPSE

nerve terminal

neurotransmitter in synaptic cleft

neurotransmitter receptor (transmitter-gated ion channel)

postsynaptic cell

neurotransmitter bound to receptor

change in membrane potential (electrical signal)

ions

Figure 12–36 Conversion of a chemical signal into an electrical signal by transmitter-gated ion channels at a synapse. The released neurotransmitter binds to and opens the transmitter-gated ion channels in the plasma membrane of the postsynaptic cell. The resulting ion flows alter the membrane potential of the postsynaptic cell, thereby converting the chemical signal back into an electrical one.

Question 12–8 In the disease myasthenia gravis, the human body makes—by mistake—antibodies to its own acetylcholine receptor molecules. These antibodies bind to and inactivate acetylcholine receptors on the plasma membrane of muscle cells. The disease leads to a devastating progressive weakening of the patients. Early on, they may have difficulty opening their eyelids, for example, and, in an animal model of the disease, rabbits have difficulty holding their ears up. As the disease progresses, most muscles weaken, and patients have difficulty speaking and swallowing. Eventually, impaired breathing can cause death. Explain which step of muscle function is affected.

depend on receptors that are *transmitter-gated ion channels*. These are a subclass of ligand-gated ion channels (see Figure 12–22B), and their function is to convert the chemical signal carried by a neurotransmitter directly back into an electrical signal. The channels open transiently in response to the binding of the neurotransmitter, thus changing the permeability of the postsynaptic membrane to ions. This in turn causes a change in the membrane potential (Figure 12–36); if the change is big enough, it can trigger an action potential in the postsynaptic cell. A well-studied example of a transmitter-gated ion channel is found at the *neuromuscular junction*—the specialized type of synapse formed between a neuron and a muscle cell. In vertebrates the neurotransmitter here is *acetylcholine*, and the transmitter-gated ion channel is the *acetylcholine receptor* illustrated earlier (see Figure 12–18).

Neurons Receive Both Excitatory and Inhibitory Inputs

The response produced by a neurotransmitter at a synapse can be either excitatory or inhibitory. Some neurotransmitters (delivered by axon terminals of *excitatory neurons*) cause the postsynaptic cell to fire an action potential, whereas others (delivered by axon terminals of *inhibitory neurons*) prevent the postsynaptic cell from firing. The drug curare, which is used by surgeons to relax muscles during an operation, causes paralysis by blocking the delivery of excitatory signals at neuromuscular junctions, while the poison strychnine causes muscle spasms and death by blocking the delivery of inhibitory signals.

Excitatory and inhibitory neurotransmitters bind to different receptors, and it is the character of the receptor that makes the difference between excitation and inhibition. The chief receptors for excitatory neurotransmitters, mainly *acetylcholine* and *glutamate*, are ion channels that allow the passage of Na^+ and Ca^{2+}. When the neurotransmitter binds, the channels open to allow an influx mainly of Na^+, which depolarizes the plasma membrane toward the threshold potential required for triggering an action potential. Stimulation of these receptors thus tends to activate the postsynaptic cell. The receptors for inhibitory neurotransmitters, mainly *γ-aminobutyric acid (GABA)* and *glycine*, by contrast, are usually channels for Cl^-. When the neurotransmitter binds, the

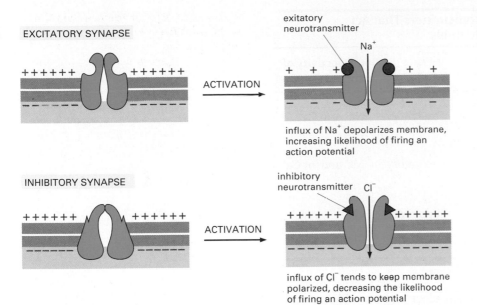

EXCITATORY SYNAPSE

exitatory
neurotransmitter

Na⁺

ACTIVATION

influx of Na⁺ depolarizes membrane,
increasing likelihood of firing an
action potential

INHIBITORY SYNAPSE

inhibitory
neurotransmitter Cl⁻

ACTIVATION

influx of Cl⁻ tends to keep membrane
polarized, decreasing the likelihood
of firing an action potential

Figure 12–37 Contrast between signaling at an excitatory and at an inhibitory synapse. Excitatory neurotransmitters activate ion channels that allow the passage of Na⁺ and Ca²⁺, whereas inhibitory neurotransmitters activate ion channels that allow the passage of Cl⁻.

channels open. Very little Cl⁻ enters the cell at this point because the driving force for movement of Cl⁻ across the membrane is close to zero at the resting membrane potential. If Na⁺ channels are also opened, however, Na⁺ will rush into the cell, causing the membrane potential to shift away from its resting value. This shift causes Cl⁻ to move into the cell, neutralizing the effect of the Na⁺ influx (Figure 12–37). In this way inhibitory neurotransmitters suppress the production of an action potential by making the target cell membrane harder to depolarize. The structures of these four neurotransmitters are shown in Table 12–3.

Most drugs used in the treatment of insomnia, anxiety, depression, and schizophrenia exert their effects at synapses in the brain, and many of them act by binding to transmitter-gated ion channels. The barbiturates and tranquilizers such as Valium, Halcion, and Temazepam, for example, bind to GABA-gated Cl⁻ channels. Their binding makes the channels easier to open by GABA, thus making the cell more sensitive to GABA's inhibitory action. By contrast, the antidepressant Prozac blocks the reuptake of an excitatory neurotransmitter, *serotonin*, increasing the amount of serotonin available at those synapses that use this transmitter. Why this should relieve depression is still a mystery.

The number of distinct types of neurotransmitter receptors is very large, although they fall into a small number of families. There are, for example, many subtypes of acetylcholine, glutamate, GABA, glycine, and serotonin receptors; they are usually located in different neurons and often differ only subtly in their properties. With such a large variety of receptors, it may be possible to design a new generation of psychoactive drugs that will act more selectively on specific sets of neurons to alleviate the mental illnesses that devastate so many people's lives. One percent of the human population, for example, has schizophrenia, and another 1 percent suffers from manic-depressive disease.

Synaptic Connections Enable You to Think, Act, and Remember

At a chemical synapse, the nerve terminal of the presynaptic cell converts an electrical signal into a chemical one, and the postsynaptic cell

Table 12–3 Some Common Neurotransmitters That Act on Ligand-gated Ion Channels

Neurotransmitter	Structure	Action	Ion Selectivity of Activated Channel
Acetylcholine	$H_3C-\overset{\overset{O}{\|\|}}{C}-O-CH_2-CH_2-\overset{\overset{CH_3}{\|}}{\underset{\underset{CH_3}{\|}}{N^+}}-CH_3$	excitatory	Na^+ and Ca^{2+}
Glutamate	$^-OOC-CH_2-CH_2-\overset{\overset{NH_3^+}{\|}}{CH}-COO^-$	excitatory	Na^+ and/or Ca^{2+}
γ-Aminobutyric acid (GABA)	$^-OOC-CH_2-CH_2-CH_2-NH_3^+$	inhibitory	Cl^-
Glycine	$^-OOC-CH_2-NH_3^+$	inhibitory	Cl^-

converts the chemical signal back into an electrical one. Interference with these processes, for good or ill, is of enormous practical importance to us. But why has evolution favored such an apparently inefficient way to pass on an electrical signal? It would seem more efficient to have a direct electrical connection between the pre- and postsynaptic cells, or do away with the synapse altogether and use a single continuous cell.

The value of chemical synapses becomes clear when we consider them in the context of a functioning nervous system—a huge network of neurons, interconnected by many branching pathways, performing complex computations, storing memories, and generating plans for action. To carry out these functions, neurons have to do more than merely generate and relay signals: they must also combine them, interpret them, and record them. Chemical synapses make these activities possible. A motor neuron in the spinal cord, for example, receives inputs

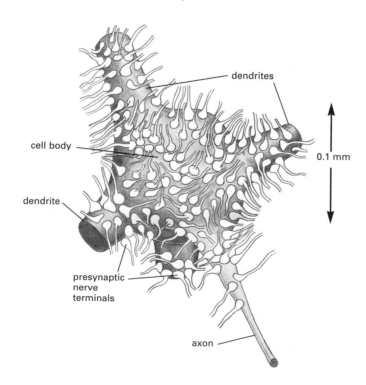

Figure 12–38 **Synapses on the cell body and dendrites of a motor neuron in the spinal cord.** Many thousands of nerve terminals synapse on the neuron, delivering signals from other parts of the animal to control the firing of action potentials along the neuron's axon.

dendrites

cell body

dendrite

presynaptic nerve terminals

axon

0.1 mm

Table 12–4   Some Examples of Ion Channels

Ion Channel	Typical Location	Function
K^+ leak channel	plasma membrane of most animal cells	maintenance of resting membrane potential
Voltage-gated Na^+ channel	plasma membrane of nerve cell axon	generation of action potentials
Voltage-gated K^+ channel	plasma membrane of nerve cell axon	return of membrane to resting potential after initiation of an action potential
Voltage-gated Ca^{2+} channel	plasma membrane of nerve terminal	stimulation of neuro-transmitter release
Acetylcholine receptor (acetylcholine-gated Na^+ and Ca^+ channel)	plasma membrane of muscle cell (at neuro-muscular junction)	excitatory synaptic signaling
GABA receptor (GABA-gated Cl^- channel)	plasma membrane of many neurons (at synapses)	inhibitory synaptic signaling
Stress-activated cation channel	auditory hair cell in inner ear	detection of sound vibrations

from hundreds or thousands of other neurons that make synapses on it (Figure 12–38). Some of these signals will tend to stimulate the neuron, while others will tend to inhibit it. The motor neuron has to combine all the information received and react either by firing action potentials along its axon to stimulate a muscle or by remaining quiet. This task of computing an appropriate output from the babble of inputs is achieved by a complicated interplay between different types of ion channels in the neuron's plasma membrane. Each of the hundreds of types of neurons in your brain has its own characteristic set of receptors and ion channels that enables the cell to respond in a particular way to a certain set of inputs and thus to perform its specialized task. Moreover, the ion channels and other components at a synapse can undergo lasting modifications according to the usage they have had, thereby preserving traces of past events. In this way, memories are stored. Ion channels, therefore, are at the heart of the machinery that enables you to act, think, feel, speak, and—most important of all—to read and remember everything in this book.

Some of the ion channels discussed in this chapter are summarized in Table 12–4.

Ion Channels and Signaling in Nerve Cells

Essential Concepts

- The lipid bilayer of cell membranes is highly impermeable to most water-soluble molecules and all ions. Transfer of nutrients, metabolites, and ions across the plasma membrane and internal cell membranes is carried out by membrane transport proteins.

- Cell membranes contain a variety of transport proteins, each of which is responsible for transferring a particular type of solute across the membrane. There are two classes of membrane transport proteins—carrier proteins and channel proteins.

- The electrochemical gradient represents the net driving force on an ion due to its concentration gradient and the electric field.

- In passive transport an uncharged solute moves spontaneously down its concentration gradient, and a charged solute (an ion) moves spontaneously down its electrochemical gradient. In active transport an uncharged solute or an ion is transported against its concentration or electrochemical gradient in an energy-requiring process.

- Carrier proteins bind specific solutes (inorganic ions, small organic molecules, or both) and transfer them across the lipid bilayer by undergoing conformational changes that expose the solute-binding site first on one side of the membrane and then on the other.

- Carrier proteins can act as pumps to transport a solute uphill against its electrochemical gradient, using energy provided by ATP hydrolysis, by a downhill flow of Na^+ or H^+ ions, or by light.

- The Na^+-K^+ pump in the plasma membrane of animal cells is an ATPase that actively transports Na^+ out of the cell and K^+ in, maintaining the steep Na^+ gradient across the plasma membrane that is used to drive other active transport processes and to convey electrical signals.

- Channel proteins form aqueous pores across the lipid bilayer through which solutes can diffuse. Whereas transport by carrier proteins can be active or passive, transport by channel proteins is always passive.

- Most channel proteins are selective ion channels that allow inorganic ions of appropriate size and charge to cross the membrane down their electrochemical gradients. Transport through ion channels is at least 1000 times faster than transport through any known carrier protein.

- Most ion channels are gated and open transiently in response to a specific stimulus, such as a change in membrane potential (voltage-gated channels) or the binding of a ligand (ligand-gated channels).

- Even when opened by its specific stimulus, ion channels do not remain continuously open: they flicker randomly between open and closed conformations. An activating stimulus increases the proportion of time the channel spends in the open state.

- The membrane potential is determined by the unequal distribution of electric charge on the two sides of the plasma membrane and is altered when ions flow through open channels. In most animal cells K^+-selective leak channels hold the resting membrane potential at a negative value, close to the value where the driving force for movement of K^+ across the membrane is close to zero.

- Neurons propagate signals in the form of action potentials, which can travel long distances along an axon without weakening. Action potentials are usually mediated by voltage-gated Na^+ channels that open in response to depolarization of the plasma membrane.

- Voltage-gated Ca^{2+} channels in nerve terminals couple electrical signals to transmitter release at synapses. Transmitter-gated ion channels convert these chemical signals back into electrical signals in the postsynaptic target cell.

- Excitatory neurotransmitters open transmitter-gated channels that are permeable to Na^+ and thereby depolarize the postsynaptic cell membrane toward the threshold potential for firing an action potential. Inhibitory neurotransmitters open transmitter-gated Cl^- channels and thereby suppress firing by keeping the postsynaptic cell membrane polarized.

Key Terms

action potential	electrochemical gradient	nerve terminal	stress-activated channel
active transport	ion channel	neuron	synapse
axon	ligand-gated channel	neurotransmitter	synaptic vesicle
carrier protein	membrane potential	osmosis	transmitter-gated ion channel
channel protein	membrane transport protein	osmotic pressure	voltage-gated channel
coupled transporter	Na^+-K^+ pump	passive transport	voltage-gated Na^+ channel
	Nernst equation	patch-clamp recording	

Questions

Question 12–9 Which of the following statements are correct? Explain your answers.

A. The plasma membrane is highly impermeable to all charged molecules.

B. Channel proteins must first bind to solute molecules before they can select those that they allow to pass.

C. Without a continual input of energy, cells will burst.

D. Carrier proteins allow solutes to cross a membrane at much faster rates than do channel proteins.

E. Certain H^+ pumps are fueled by light energy.

F. The plasma membrane of many animal cells contains open K^+ channels, yet the K^+ concentration in the cytosol is much higher than outside the cell.

G. A symport would function as an antiport if its orientation in the membrane were reversed (i.e., if the portion of the molecule normally exposed to the cytosol faced the outside of the cell instead).

H. The membrane potential of an axon temporarily becomes more negative when an action potential excites it.

Question 12–10 Name at least one similarity and at least one difference between the following: (It may help to review the definitions of the terms using the Glossary.)

A. symport and antiport

B. active transport and passive transport

C. membrane potential and electrochemical gradient

D. pump and carrier protein

E. axon and telephone wire

F. solute and ion

Question 12–11 Discuss the following statement: "The differences between a channel and a carrier protein are like the differences between a bridge and a ferry."

Question 12–12 The neurotransmitter acetylcholine is made in the cytosol and then transported into synaptic vesicles, where its concentration is more than 100-fold higher than in the cytosol. When synaptic vesicles are isolated from neurons, they can take up additional acetylcholine added to the solution in which they are suspended, but only when ATP is present. Na^+ ions are not required for acetylcholine uptake, but, curiously, raising the pH of the solution in which the synaptic vesicles are suspended increases the rate of acetylcholine uptake. Furthermore, transport is inhibited when drugs are added that make the membrane permeable to H^+ ions. Suggest a mechanism that is consistent with all of these observations.

Question 12–13 The resting membrane potential of a cell is about –70 mV, and the thickness of a lipid bilayer is about 4.5 nm. What is the strength of the electric field across the membrane in V/cm? What do you suppose would happen if you applied this voltage to two metal electrodes separated by a 1 cm air gap?

Question 12–14 Phospholipid bilayers form sealed spherical vesicles in water (see Chapter 11). Assume you have constructed lipid vesicles that contain Na^+-K^+ pumps as the sole membrane protein, and assume for the sake of simplicity that each pump transports one Na^+ one way and one K^+ the other way in each pumping cycle. All the Na^+-K^+ pumps have the portion of the molecule that normally faces the cytosol oriented toward the outside of the vesicles. With the help of Figure 12–11, determine what would happen if . . .

A. your vesicles were suspended in a solution containing both Na^+ and K^+ ions and had a solution with the same ionic composition inside them.

B. you add ATP to the suspension described in A.

C. you add ATP, but the solution—outside as well as inside the vesicles—contains only Na^+ ions and no K^+ ions.

D. half of the pump molecules embedded in the membrane of each vesicle were oriented the other way around so that the normally cytosolic portions of these molecules faced the inside of the vesicles. You then add ATP to the suspension.

E. you add ATP to the suspension described in A, but in addition to Na^+-K^+ pumps, the membrane of your vesicles also contains K^+ leak channels.

Question 12–15   Name the three ways in which an ion channel can be gated.

Question 12–16   One thousand Ca^{2+} channels open in the plasma membrane of a cell that is 1000 μm^3 in size and has a cytosolic Ca^{2+} concentration of 100 nM. For how long would the channels need to stay open in order for the cytosolic Ca^{2+} concentration to rise to 5 μM? There is virtually unlimited Ca^{2+} available in the outside medium (the extracellular Ca^{2+} concentration in which most animal cells live is a few millimolar), and each channel passes 10^6 calcium ions per second.

Question 12–17   Amino acids are taken up by animal cells using a symport in the plasma membrane. What is the most likely ion whose electrochemical gradient drives the import? Is ATP consumed in the process? If so, how?

Question 12–18   We shall see in Chapter 14 that an acidic pH inside endosomes, which are membrane-bounded intracellular organelles, is required for their function. Acidification is achieved by an H^+ pump in the endosomal membrane. The endosomal membrane also contains Cl^- channels. If the channels do not function properly (e.g., because of a mutation in the genes encoding the channel proteins), acidification is also impaired.

A. Can you explain how Cl^- channels might help acidification?

B. According to your explanation, would the Cl^- channels be absolutely required to lower the pH inside the endosome?

Question 12–19   Some bacterial cells can grow on either ethanol (CH_3CH_2OH) or acetate (CH_3COO^-) as their only carbon source. Dr. Schwips measured the rate at which the two compounds traverse the bacterial plasma membrane but, due to excessive inhalation of one of the compounds (which one?), failed to label his data accurately.

A. Plot the data from the table below.

B. Determine from your graph whether the data describing compound A correspond to the uptake of ethanol or acetate.

C. Determine the rates of transport for compounds A and B at 0.5 mM and 100 mM. (This part of the question requires that you are familiar with the principles of enzyme kinetics discussed in Chapter 5.)

Explain your answers.

Concentration of Carbon Source (mM)	Rate of Transport ($\mu mol/min$)	
	Compound A	Compound B
0.1	2.0	18
0.3	6.0	46
1.0	20	100
3.0	60	150
10.0	200	182

13 Energy Generation in Mitochondria and Chloroplasts

We can deduce from the geological record that there was no oxygen in the atmosphere at the time that life began on earth. One possible way for cells to grow and survive in such conditions would have been by breaking down, through some form of anaerobic fermentation, the organic molecules left by earlier geochemical processes. As discussed in Chapter 4, fermentation reactions occur in the cytosol of present-day cells; these reactions use the energy derived from the partial oxidation of energy-rich food molecules to form ATP, the chemical energy currency of cells. But very early in the history of life, a much more efficient method of energy generation and ATP synthesis appeared. This process is based on the transport of electrons along membranes. Billions of years later, it is so central to the survival of life on earth that we devote this entire chapter to it. As we shall see, this membrane-based mechanism is used by cells to acquire energy from a wide variety of sources: for example, it is central to the conversion of light energy into chemical bond energy in photosynthesis, and to the aerobic respiration that enables us to use oxygen to produce large amounts of ATP from food molecules.

The advent of this mechanism of energy conversion had a profound effect on the history of life on earth. The earth is 4.6 billion years old. Fossils formed 3.5 billion years ago suggest that photosynthetic bacteria that produced oxygen were already present at that time (Figure 13–1). The descendants of these cells have filled the atmosphere with oxygen gas (O_2), and they have crowded every corner and crevice of the land and the oceans with a wild menagerie of living forms.

Where do we come from and what are our relationships to other living things are questions that have fascinated humans since the beginning of recorded time. The progression of life on earth as we now understand it is briefly outlined in Figure 13–2. The story, worked out through a long chain of scientific investigation, is one of the most dramatic and exciting histories ever told. And we are not yet done. Each year, further discoveries in cell biology enable us to add more details through molecular detective work of dramatically increasing power. Absolutely central to life's progression was the provision of an abundant source of energy for cells. We shall now discuss the remarkable mechanism that made this all possible.

(A)

(B)

(C)

Figure 13–1 Microorganisms that carry out oxygen-producing photosynthesis changed the earth's atmosphere (A) Living stromatolites from a lagoon in western Australia. These structures are produced by large colonies of oxygen-producing photosynthetic cyanobacteria, which lay down successive layers of material, and are formed in specialized environments. (B) Cross section of a modern stromatolite, showing its layered structure. (C) Cross section through a fossil stromatolite in a rock 3.5 billion years old. Note the layered structure similar to that in (B). Fossil stromatolites are thought to have been formed by photosynthetic bacteria very similar to modern cyanobacteria. The activities of bacteria like these, which liberate O_2 gas as a waste product of photosynthesis, would have slowly changed the earth's atmosphere. (A, courtesy of Sally Birch. © Oxford Scientific Films. B and C, courtesy of S.M. Awramik, University of California/Biological Photo Service.)

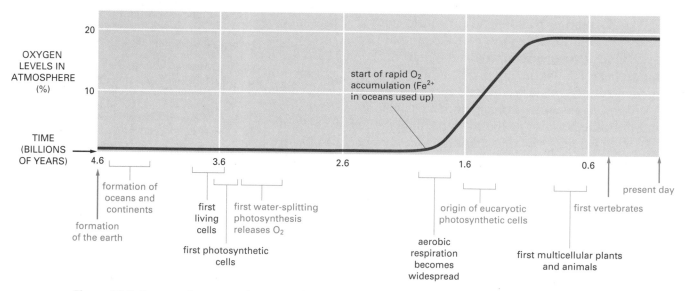

OXYGEN LEVELS IN ATMOSPHERE (%)

20

10

TIME (BILLIONS OF YEARS)

4.6   3.6   2.6   1.6   0.6

start of rapid O_2 accumulation (Fe^{2+} in oceans used up)

formation of oceans and continents

formation of the earth

first living cells

first photosynthetic cells

first water-splitting photosynthesis releases O_2

aerobic respiration becomes widespread

origin of eucaryotic photosynthetic cells

first multicellular plants and animals

first vertebrates

present day

Figure 13–2 Some major events that are believed to have occurred during the evolution of living organisms on earth. With the evolution of the membrane-based process of photosynthesis, organisms were no longer dependent on preformed organic chemicals. They could now make their own organic molecules from CO_2 gas. The delay of more than a billion years between the appearance of bacteria that split water and released O_2 during photosynthesis and the accumulation of high levels of O_2 in the atmosphere is thought to be due to the initial reaction of the oxygen with abundant ferrous iron (Fe^{2+}) dissolved in the early oceans. This iron would have removed oxygen from the atmosphere, and formed the enormous deposits of iron oxide found in some rocks of this age. Only when the iron was used up would oxygen have started to accumulate in the atmosphere. Membrane-based aerobic respiration presumably arose in response to the rising levels of oxygen in the atmosphere. As nonphotosynthetic oxygen-using organisms appeared, the concentration of oxygen in the atmosphere leveled out.

Cells Obtain Most of Their Energy by a Membrane-based Mechanism

The main chemical energy currency in cells is ATP (see Figure 3–25). In eucaryotic cells, small amounts of ATP are generated during glycolysis in the cytosol (as discussed in Chapter 4), but most ATP is produced by membrane-based processes in mitochondria (and also in chloroplasts in plant and algal cells). Very similar processes also occur in the cell membranes of many bacteria. The fundamental mechanism that is used to make all of this ATP arose very early in life's history, and it was so successful that its essential features have been retained in the long evolutionary journey from early procaryotes to modern cells. The process consists of two linked stages, both of which are carried out by protein complexes embedded in a membrane.

Stage 1. Electrons (derived from the oxidation of food molecules or from other sources discussed later) are transferred along a series of electron carriers—called an *electron-transport chain*—embedded in the membrane. These electron transfers release energy that is used to pump protons (H^+, derived from the water that is ubiquitous in cells) across the membrane and thus generate an electrochemical proton gradient. As discussed in Chapter 12, an ion gradient across a membrane is a form of stored energy; this energy can be harnessed to do useful work when the ions are allowed to flow back across the membrane down their electrochemical gradient.

Stage 2. H^+ flows back down its electrochemical gradient through a protein complex called *ATP synthase*, which catalyzes the energy-requiring synthesis of ATP from ADP and inorganic phosphate (P_i). This ubiquitous enzyme serves the role of a turbine, permitting the proton gradient to drive the production of ATP (Figure 13–3).

The linkage of electron transport, proton pumping, and ATP synthesis was called the *chemiosmotic hypothesis* when it was first proposed in the 1960s, because of the link between the chemical bond–forming reactions that synthesize ATP ("chemi") and the membrane-transport processes ("osmotic," from the Greek *osmos*, "to push"). It is now known as **chemiosmotic coupling.**

The basic chemiosmotic mechanism outlined in Figure 13–3 is used by the vast majority of living organisms, and a large variety of different

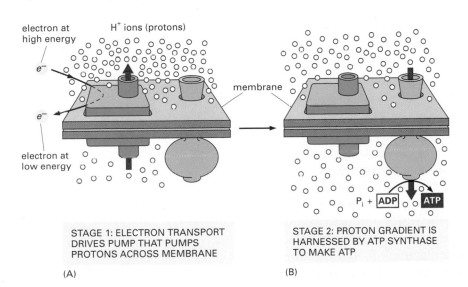

STAGE 1: ELECTRON TRANSPORT DRIVES PUMP THAT PUMPS PROTONS ACROSS MEMBRANE

(A)

STAGE 2: PROTON GRADIENT IS HARNESSED BY ATP SYNTHASE TO MAKE ATP

(B)

Figure 13–3 Harnessing energy for life. (A) The essential requirements for chemiosmosis are a membrane, in which are embedded a pump protein and an ATP synthase, and sources of high-energy electrons (e^-) and of protons (H^+). The pump harnesses the energy of electron transfer (details not shown here) to pump protons derived from water, creating a proton gradient across the membrane. (B) This gradient serves as an energy store that can be used to drive ATP synthesis by the ATP synthase. The *red arrow* shows the direction of proton movement at each stage.

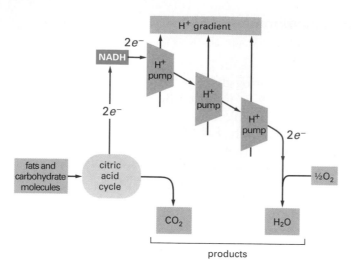

Figure 13–4 Proton pumping in mito-chondria. Only stage 1 of chemiosmotic coupling is shown. Inputs are *light green,* products are *blue,* and the path of electron flow is indicated by *red arrows.*

substances can serve as the source of the electrons that power proton pumping. For example, chemiosmotic coupling is the core reaction in the aerobic respiration that produces ATP in mitochondria and aerobic bacteria; here the electrons are derived ultimately from the oxidation of glucose or fatty acids (as discussed in Chapter 4), and molecular oxygen (O_2) acts as the final electron acceptor, producing water as a waste product (Figure 13–4). In chemiosmotic coupling in photosynthesis, the required electrons are derived from the action of light on the green pigment chlorophyll. And certain bacteria use inorganic substances such as hydrogen, iron, and sulfur as sources of the high-energy electrons that they need to make ATP.

Chemiosmotic coupling first evolved in bacteria. It is perhaps not surprising, therefore, that aerobic eucaryotic cells appear to have adopted the bacterial chemiosmotic mechanisms intact, first by engulfing aerobic bacteria to form mitochondria, and somewhat later—in the lineages leading to algae and plants—by engulfing cyanobacteria to form chloroplasts, as described in Chapter 1.

In this chapter we shall consider energy generation in both mitochondria and chloroplasts, emphasizing the common principles by which proton gradients are created and used in these organelles, as well as in bacteria. We start by describing the structure and function of mitochondria, looking in detail at the events that occur in the mitochondrial membrane to create the proton gradient and generate ATP. We next consider photosynthesis as it occurs in the chloroplasts of plant cells. Finally, the chapter ends with an examination of the life-style of some of our single-celled ancestors.

Question 13–1 Dinitrophenol (DNP) is a small molecule that renders membranes permeable to protons. In the 1940s, small amounts of this highly toxic compound were given to patients to induce weight loss. DNP was effective in producing weight loss, especially loss of fat reserves. Can you explain how it might cause such loss? As an unpleasant side reaction, however, patients had elevated temperature and sweated profusely during the treatment. Provide an explanation for these symptoms.

Mitochondria and Oxidative Phosphorylation

Mitochondria (singular, **mitochondrion**) are present in nearly all eucaryotic cells—in plants, animals, and in most eucaryotic microorganisms—and it is in these organelles that most of a cell's ATP is produced. As discussed in Chapter 4, mitochondria metabolize acetyl groups via the citric acid cycle, producing the waste product CO_2 and NADH—an activated carrier molecule that carries high-energy electrons. The NADH donates its high-energy electrons to the electron-transport chain in the mitochondrial membrane, and thus becomes oxidized to NAD$^+$. The electrons are quickly passed along the chain to molecular oxygen (O_2) to form water (H_2O). The energy released during passage of the electrons along the electron-transport chain is used to pump protons (see Figure 13–4), and the proton gradient in turn drives the synthesis of ATP to complete the chemiosmotic mechanism. This process involves both the consumption of O_2 and the synthesis of ATP through the addition of a phosphate group to ADP, and so is called **oxidative phosphorylation** (Figure 13–5).

The same events take place in aerobic bacteria, which do not possess mitochondria; here the plasma membrane carries out the chemiosmotic processes. But unlike a bacterial cell, which also has to carry out many other functions, the mitochondrion has become highly specialized for energy generation, as we see next.

A Mitochondrion Contains Two Membrane-bounded Compartments

Mitochondria are generally similar in size and shape to bacteria, although these attributes can vary depending on the cell type. They contain their own DNA and RNA, and a complete transcription and translation system including ribosomes, which allows them to synthesize some of their own proteins. Time-lapse movies of living cells reveal mitochondria as remarkably mobile organelles, constantly changing shape and position. Present in large numbers, they can form long moving chains in association with microtubules of the cytoskeleton (see Chapter 16). In other cells, they remain fixed in one cellular location to target ATP directly to a site of unusually high ATP consumption. In a heart muscle cell, for example, mitochondria are located close to the contractile apparatus, while in a sperm they are wrapped tightly around the motile flagellum (Figure 13–6).

Each mitochondrion is bounded by two highly specialized membranes—one wrapped around the other—that play a crucial part in its activities. The outer and inner mitochondrial membranes create two mitochondrial compartments: a large internal space called the **matrix** and the much narrower *intermembrane space* (Figure 13–7). If purified mitochondria are gently processed and fractionated into separate components by differential centrifugation (see Panel 5–4, pp. 160–161), the biochemical composition of each of the two membranes and of the spaces enclosed by them can be determined. Each contains a unique collection of proteins.

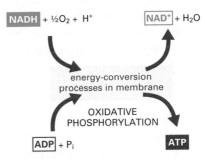

Figure 13–5 The major energy conversion catalyzed by the mitochondrion. In oxidative phosphorylation, the energy released by the oxidation of NADH to NAD$^+$ is harnessed, through energy-conversion processes in the membrane (electron transfer, proton pumping, and the flow of protons through ATP synthase), to the energy-requiring phosphorylation of ADP to form ATP. The high-energy electrons lost from NADH move along an electron-transport chain in the membrane and eventually combine with molecular oxygen and H$^+$ to form water. The net equation for this electron-transfer process is NADH + ½O_2 + H$^+$ → NAD$^+$ + H_2O, with two electrons passing from NADH to oxygen (see Figure 13–4).

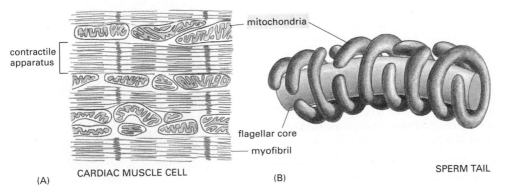

(A) CARDIAC MUSCLE CELL

(B) SPERM TAIL

Figure 13–6 Location of mitochondria near sites of high ATP utilization. (A) In a cardiac muscle cell, mitochondria are located close to the contractile apparatus, in which ATP hydrolysis provides the energy for contraction. (B) In a sperm, mitochondria are located in the tail, around the core of the motile flagellum that requires ATP for its movement.

The *outer membrane* contains many molecules of a transport protein called porin, which, as described in Chapter 11, forms wide aqueous channels through the lipid bilayer. As a result, the outer membrane is like a sieve that is permeable to all molecules of 5000 daltons or less, including small proteins. This makes the intermembrane space chemically equivalent to the cytosol with respect to the small molecules it contains. In contrast, the *inner membrane*, like other membranes in the cell, is impermeable to the passage of ions and most small molecules, except where a path is provided by membrane-transport proteins. The mitochondrial matrix therefore contains only molecules that can be selectively transported into the matrix across the inner membrane, and its contents are highly specialized.

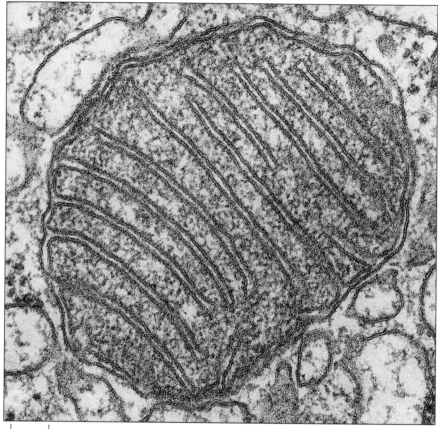

100 nm

Figure 13–7 The general organization of a mitochondrion. Each of the four regions in a mitochondrion contains a unique set of proteins that enables each compartment to perform its distinct functions. In liver mitochondria, an estimated 67% of the total mitochondrial protein is located in the matrix, 21% is located in the inner membrane, 6% in the outer membrane, and 6% in the intermembrane space. (Courtesy of Daniel S. Friend.)

Matrix. This large internal space contains a highly concentrated mixture of hundreds of enzymes, including those required for the oxidation of pyruvate and fatty acids and for the citric acid cycle. The matrix also contains several identical copies of the mitochondrial DNA genome, special mitochondrial ribosomes, tRNAs, and various enzymes required for expression of the mitochondrial genes.

Inner membrane. The inner membrane (*red*) is folded into numerous cristae, which greatly increases its total surface area. It contains proteins with three types of functions: (1) those that carry out the oxidation reactions of the electron-transport chain, (2) the ATP synthase that makes ATP in the matrix, and (3) transport proteins that allow the passage of metabolites into and out of the matrix. An electrochemical gradient of H$^+$, which drives the ATP synthase, is established across this membrane, and so it must be impermeable to ions and most small charged molecules.

Outer membrane. Because it contains a large channel-forming protein (called porin), the outer membrane is permeable to all molecules of 5000 daltons or less. Other proteins in this membrane include enzymes involved in mitochondrial lipid synthesis and enzymes that convert lipid substrates into forms that are subsequently metabolized in the matrix.

Intermembrane space. This space (*white*) contains several enzymes that use the ATP passing out of the matrix to phosphorylate other nucleotides.

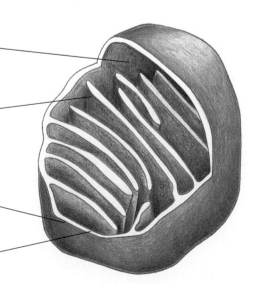

The inner mitochondrial membrane is the site of electron transport and proton pumping, and it contains the ATP synthase. Most of the proteins embedded in the inner mitochondrial membrane are components of the electron-transport chains required for oxidative phosphorylation. This membrane has a distinctive lipid composition and also contains a variety of transport proteins that allow the entry of selected small molecules, such as pyruvate and fatty acids, into the matrix.

The inner membrane is usually highly convoluted, forming a series of infoldings, known as *cristae,* that project into the matrix space to greatly increase the surface area of the inner membrane (see Figure 13–7). This provides a large surface on which ATP synthesis can take place; in a liver cell, for example, the inner mitochondrial membranes of all the mitochondria constitute about a third of the total membranes of the cell.

Question 13–2 Electron micrographs show that mitochondria in heart muscle have a much higher density of cristae than mitochondria in skin cells. Suggest an explanation for this observation.

High-Energy Electrons Are Generated via the Citric Acid Cycle

Mitochondria have been essential for the evolution of multicellular animals and plants. Without them, present-day eucaryotes would be dependent on the relatively inefficient process of glycolysis (described in Chapter 4) for all of their ATP production, and it seems unlikely that complex multicellular organisms could have been supported in this way. When glucose is converted to pyruvate by glycolysis, less than 10% of the total free energy potentially available from the glucose is released. In the mitochondria, the metabolism of sugars is completed, and the energy released is harnessed so efficiently that about 30 molecules of ATP are produced for each molecule of glucose oxidized. By contrast, only two molecules of ATP are produced per glucose molecule by glycolysis alone.

Mitochondria can use both pyruvate and fatty acids as fuel. Pyruvate comes from glucose and other sugars, and fatty acids come from fats. Both of these fuel molecules are transported across the inner mitochondrial membrane and then converted to the crucial metabolic intermediate *acetyl CoA* by enzymes located in the mitochondrial matrix (see p. 121). The acetyl groups in acetyl CoA are then oxidized in the matrix via the citric acid cycle. The cycle makes CO_2, which is released from the cell as a waste product, and it generates high-energy electrons, carried by the activated carrier molecules NADH and $FADH_2$ (Figure 13–8). These high-energy electrons are then transferred to the inner mitochondrial

Figure 13–8 How electrons are donated by NADH. In this drawing, the high-energy electrons are shown as two *red dots* on a *yellow* hydrogen atom. A hydride ion ($H:^-$, a hydrogen atom and an extra electron) is removed from NADH and is converted into a proton and two high-energy electrons: $H:^- \rightarrow H^+ + 2e^-$. Only the ring that carries the electrons in a high-energy linkage is shown; for the complete structure and the conversion of NAD^+ back to NADH, see the structure of the closely related NADPH in Figure 3–28. Electrons are also carried in a similar way by $FADH_2$, whose structure is shown in Figure 4–12B.

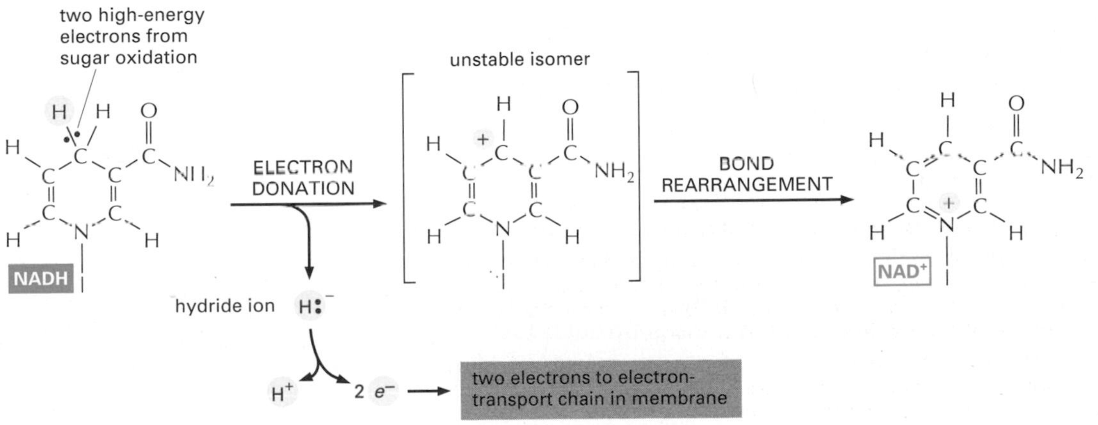

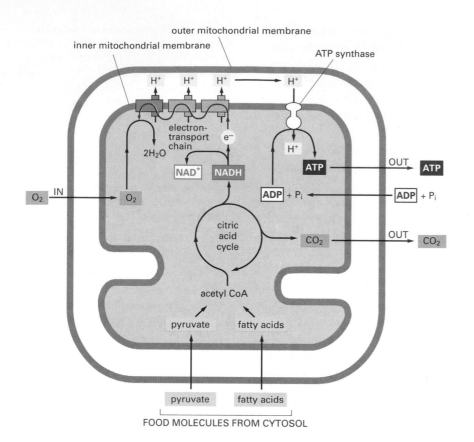

outer mitochondrial membrane

inner mitochondrial membrane

ATP synthase

H^+ H^+ H^+ → H^+

electron-transport chain

$2H_2O$

e^-

H^+

NAD^+ NADH

ATP

OUT ATP

ADP + P_i ← ADP + P_i

O_2 IN → O_2

citric acid cycle

CO_2 OUT CO_2

acetyl CoA

pyruvate fatty acids

pyruvate fatty acids

FOOD MOLECULES FROM CYTOSOL

Figure 13–9 A summary of energy-generating metabolism in mitochondria. Pyruvate and fatty acids enter the mitochondrion (*bottom*), are broken down to acetyl CoA, and are then metabolized by the citric acid cycle, which reduces NAD^+ to NADH (and FAD^+ to $FADH_2$, not shown). In the process of oxidative phosphorylation, high-energy electrons from NADH (and $FADH_2$) are then passed along the electron-transport chain in the inner membrane to oxygen (O_2). This electron transport generates a proton gradient across the inner membrane, which is used to drive the production of ATP by ATP synthase.

membrane, where they enter the electron-transport chain; the loss of electrons regenerates the NAD^+ and FAD^+ that is needed for continued oxidative metabolism. Electron transport along the chain now begins. The entire sequence of reactions is outlined in Figure 13–9.

Electrons Are Transferred Along a Chain of Proteins in the Inner Mitochondrial Membrane

The **electron-transport chain** that carries out oxidative phosphorylation is present in many copies in the inner mitochondrial membrane. Also known as the *respiratory chain*, it contains over 40 proteins, of which about 15 are directly involved in electron transport. Most of these proteins are embedded in the lipid bilayer and function only in an intact membrane, making them difficult to study. However, the components of the electron-transport chain, like other membrane proteins, can be solubilized using nonionic detergents (see Figure 11–28), purified, and then reconstituted in operational form in small membrane vesicles. Such studies reveal that most of the proteins involved in the mitochondrial electron-transport chain are grouped into three large *respiratory enzyme complexes,* each containing multiple individual proteins. Each complex includes transmembrane proteins that hold the entire protein complex firmly in the inner mitochondrial membrane.

The three respiratory enzyme complexes are (1) the *NADH dehydrogenase complex,* (2) the *cytochrome b-c$_1$ complex,* and (3) the *cytochrome oxidase complex.* Each contains metal ions and other chemical groups that form a pathway for the passage of electrons through the complex. The respiratory complexes are the sites of proton pumping, and each

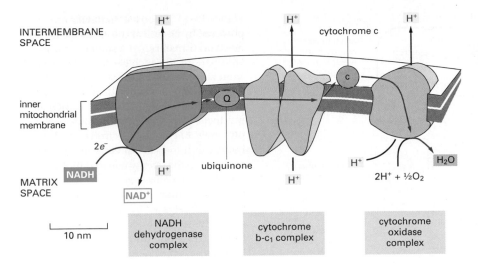

INTERMEMBRANE SPACE

inner mitochondrial membrane

$2e^-$

NADH

NAD$^+$

MATRIX SPACE

10 nm

H$^+$

H$^+$

H$^+$

cytochrome c

c

Q

ubiquinone

H$^+$

H$^+$

H$^+$

H$_2$O

$2H^+ + \frac{1}{2}O_2$

NADH dehydrogenase complex

cytochrome b-c$_1$ complex

cytochrome oxidase complex

Figure 13–10 The transfer of electrons through the three respiratory enzyme complexes in the inner mitochondrial membrane. The relative size and shape of each complex is indicated. During the transfer of electrons from NADH to oxygen (*red lines*), protons derived from water are pumped across the membrane by each of the respiratory enzyme complexes. The ubiquinone (Q) and cytochrome c (c) serve as mobile carriers that ferry electrons from one complex to the next. The structures and roles of these two molecules will be discussed later (see Figures 13–20 and 13–23).

can be thought of as a protein machine that pumps protons across the membrane as electrons are transferred through it.

Electron transport begins when a hydride ion (H:$^-$) is removed from NADH and is converted into a proton and two high-energy electrons: H:$^- \rightarrow$ H$^+$ + 2e^-, as previously explained in Figure 13–8. As illustrated in Figure 13–10, this reaction is catalyzed by the first of the respiratory enzyme complexes, the NADH dehydrogenase, which accepts the electrons. The electrons are then passed along the chain to each of the other enzyme complexes in turn, using mobile electron carriers that will be described later. The transfer of electrons along the chain is energetically favorable; the electrons start out at very high energy and lose energy at each step as they pass along the chain, eventually entering the cytochrome oxidase, where they combine with a molecule of O$_2$ to form water. This is the oxygen-requiring step of cellular respiration, and it uses the oxygen that we breathe.

Electron Transport Generates a Proton Gradient Across the Membrane

The proteins of the respiratory chain guide the electrons so that they move sequentially from one enzyme complex to another—with no short circuits that skip a complex. Each electron transfer is an oxidation-reduction reaction: as described in Chapter 3, the molecule or atom donating the electron becomes oxidized, while the receiving molecule or atom becomes reduced (see p. 84). Electrons will pass spontaneously from molecules that have a relatively low affinity for their available electrons, and thus lose them easily, to molecules with a higher electron affinity. For example, NADH with its high-energy electrons has a low electron affinity, so that its electrons are readily passed to the NADH dehydrogenase. The electrical batteries of our common experience are based on similar electron transfers between two chemical substances with different electron affinities.

In the absence of any means for harnessing the energy released by electron transfers, this energy is simply liberated as heat. But just as a battery can be connected to a device that pumps water (Figure 13–11), cells harness much of the energy of electron transfer by having the elec-

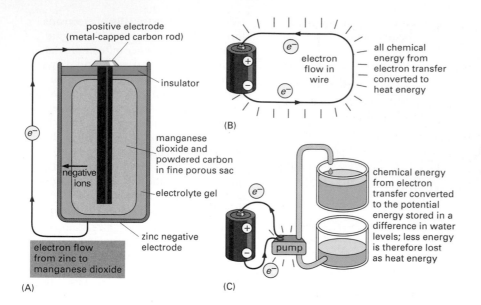

Figure 13–11 Electrical batteries are powered by chemical reactions based on electron transfers. (A) A standard flashlight battery, in which electrons are passed from zinc metal (Zn) to the manganese atom in manganese dioxide (MnO_2), forming Zn^{2+} and manganous oxide (MnO) as products. (The carbon in the battery simply serves to conduct electrons.) (B) If the battery terminals are directly connected to each other, the energy released by electron transfer is all converted into heat. (C) If the battery is connected to a pump, much of the energy released by electron transfers can be harnessed to do work instead (in this case to pump water). Cells can similarly harness the energy of electron transfer to a pumping mechanism, as illustrated in Figure 13–3A.

tron transfers take place within proteins that can pump protons (H⁺). Each of the respiratory enzyme complexes couples the energy released by electron transfer across it to an uptake of protons from water ($H_2O \rightarrow H^+ + OH^-$) in the mitochondrial matrix, followed by their release on the other side of the membrane into the intermembrane space. As a result, the energetically favorable flow of electrons along the electron-transport chain pumps protons across the membrane out of the matrix, creating an electrochemical proton gradient across the inner mitochondrial membrane (see Figure 13–10).

The active pumping of protons thus has two major consequences:

1. It generates a gradient of proton (H⁺) concentration (a gradient of pH) across the inner mitochondrial membrane, with the pH about one unit higher in the matrix (around pH 8) than in the intermembrane space, where the pH is generally close to 7. (Since H⁺, like any small molecule equilibrates freely across the outer membrane of the mitochondrion, the pH in the intermembrane space is the same as in the cytosol.) This is a tenfold drop in H⁺ concentration in the matrix compared with the intermembrane space.

2. It generates a membrane potential (discussed in Chapter 12) across the inner mitochondrial membrane, with the inside (the matrix side)

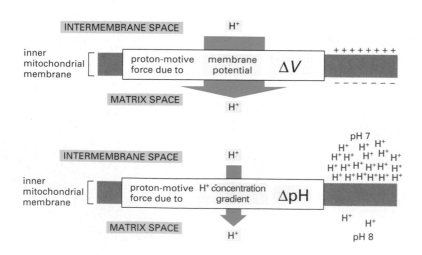

Figure 13–12 The two components of the electrochemical proton gradient across the inner mitochondrial membrane. The total electrochemical gradient of H⁺ across the inner mitochondrial membrane consists of a large force due to the membrane potential and a smaller force due to the H⁺ concentration gradient. Both forces combine to produce the total *proton-motive force* that drives H⁺ into the matrix space.

negative and the outside positive as a result of a net outflow of protons, which are positive ions.

As discussed in Chapter 12, the force driving the passive flow of ions such as Na^+ and K^+ across a membrane is proportional to the electrochemical gradient for that ion across the membrane. This in turn depends on the voltage across the membrane, which is measured as the membrane potential, and on the concentration gradient of the ion (see Figure 12–7). Since protons are positively charged, they will move more readily across a membrane if the membrane has an excess of negative electrical charges on the other side. In the case of the inner mitochondrial membrane, the pH gradient and the membrane potential work together to create a steep electrochemical proton gradient that makes it energetically very favorable for H^+ to flow back into the mitochondrial matrix. In all the energy-producing membranes that we will discuss in this chapter, the membrane potential adds to the driving force pulling H^+ back across the membrane; hence this potential increases the amount of energy stored in the proton gradient (Figure 13–12).

The Proton Gradient Drives ATP Synthesis

The electrochemical proton gradient across the inner mitochondrial membrane is used to drive ATP synthesis in the process of oxidative phosphorylation (Figure 13–13). The device that makes this possible is a large membrane-bound enzyme called **ATP synthase.** This enzyme creates a hydrophilic pathway across the inner mitochondrial membrane that allows protons to flow down their electrochemical gradient. As these ions thread their way through the ATP synthase, they are used to drive the energetically unfavorable reaction between ADP and P_i that makes ATP (see Figure 2–22). The ATP synthase is of ancient origin; the same enzyme occurs in the mitochondria of animal cells, the chloroplasts of plants and algae, and in the plasma membrane of bacteria.

The structure of ATP synthase is shown in Figure 13–14. It is a large multisubunit protein. A large enzymatic portion, shaped like a lollipop head, projects on the matrix side of the inner mitochondrial membrane and is attached through a thinner multisubunit "stalk" to a transmembrane carrier for protons. Although the detailed mechanism of action is not yet worked out, it is thought that the protons passing through the transmembrane carrier cause the stalk to spin rapidly within the head,

Question 13–3 When the drug dinitrophenol (DNP) is added to mitochondria, the inner membrane becomes permeable to protons (H^+). In contrast, when the drug nigericin is added to mitochondria, the inner membrane becomes permeable to K^+. (A) How will the electrochemical proton gradient change in response to DNP? (B) How will it change in response to nigericin?

Figure 13–13 The general mechanism of oxidative phosphorylation. (A) As a high-energy electron is passed along the electron-transport chain, some of the energy released is used to drive the three respiratory enzyme complexes that pump H^+ out of the matrix space. The resulting electrochemical proton gradient across the inner membrane drives H^+ back through the ATP synthase, a transmembrane protein complex that uses the energy of the H^+ flow to synthesize ATP from ADP and P_i in the matrix. (B) Electron micrograph of the inside surface of the inner mitochondrial membrane in a plant cell. Densely packed particles are seen, due to protruding portions of the ATP synthases and the respiratory enzyme complexes.

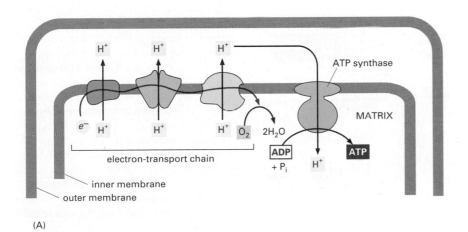

(A)

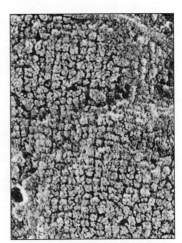

(B)

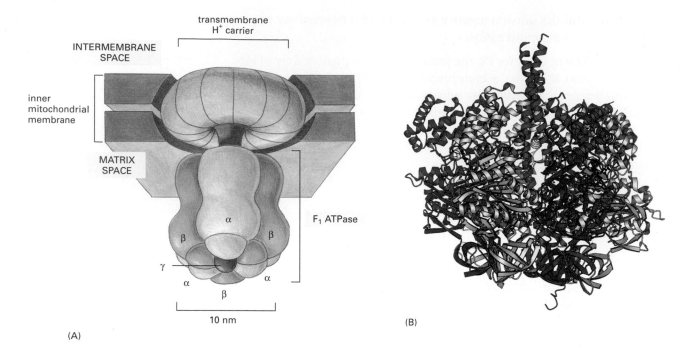

INTERMEMBRANE
SPACE

inner
mitochondrial
membrane

MATRIX
SPACE

α

β   β

γ

α   α

β

10 nm

F_1 ATPase

(A)

(B)

inducing the head to make ATP. The synthase is capable of producing more than 100 molecules of ATP per second, and about three protons need to pass through this marvelous device to make each molecule of ATP.

The ATP synthase is a reversible coupling device. It can either harness the flow of protons down their electrochemical gradient to make ATP (its normal role in mitochondria and the plasma membrane of bacteria growing aerobically) or use the energy of ATP hydrolysis to pump protons across a membrane, like the H⁺ pumps described in Chapter 12 (Figure 13–15). Whether the ATP synthase primarily makes or uses up ATP depends on the magnitude of the electrochemical proton gradient across the membrane in which it sits. In many bacteria that can grow either aerobically or anaerobically, the direction in which the ATP synthase works is routinely reversed when the bacterium runs out of O_2. At this point, the ATP synthase uses some of the ATP generated inside the

Figure 13–14 ATP synthase. (A) Schematic diagram. The enzyme is composed of a head portion, the F_1 ATPase, and a transmembrane H⁺ carrier, called F_0. They are both formed from multiple subunits. Greek letters denote the individual subunits of the F_1 ATPase. (B) Three-dimensional structure of the F_1 ATPase, as determined by x-ray crystallography. This part of the ATP synthase derives its name from its ability to carry out the reverse of the ATP synthesis reaction, namely the hydrolysis of ATP to ADP and P_i, when detached from the transmembrane portion. (B, courtesy of John Walker © Nature.)

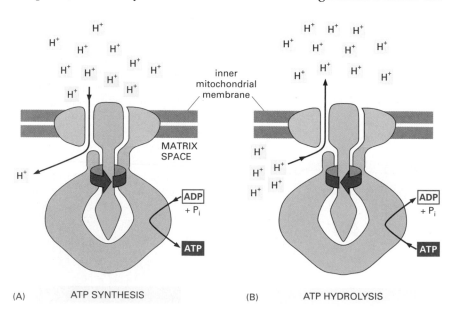

H⁺

inner
mitochondrial
membrane

MATRIX
SPACE

H⁺

ADP
+ P_i

ATP

H⁺

ADP
+ P_i

ATP

(A)   ATP SYNTHESIS

(B)   ATP HYDROLYSIS

Figure 13–15 The ATP synthase is a reversible coupling device that can convert the energy of the electrochemical proton gradient into chemical bond energy or vice versa. The ATP synthase can either synthesize ATP by harnessing the H⁺ gradient (A) or pump protons against their electrochemical gradient by hydrolyzing ATP (B). The direction of operation at any given instant depends on the net free-energy change (ΔG, discussed in Chapter 3) for the coupled processes of H⁺ translocation across the membrane and the synthesis of ATP from ADP and P_i. Thus, for example, if the electrochemical proton gradient falls below a certain level, the ΔG for the H⁺ transport into the matrix space will no longer be large enough to drive ATP production. Instead, ATP will be hydrolyzed by the ATP synthase to rebuild the gradient.

cell by glycolysis to pump protons out of the cell, creating the proton gradient that the bacterial cell needs to import its essential nutrients by active transport.

Coupled Transport Across the Inner Mitochondrial Membrane Is Driven by the Electrochemical Proton Gradient

The synthesis of ATP is not the only process driven by the electrochemical proton gradient. In mitochondria, many charged molecules, such as pyruvate, ADP, and P_i, are pumped into the matrix from the cytosol, while others, such as ATP, must be carried in the opposite direction. Many carrier proteins couple transport to the energetically favorable flow of H^+ into the mitochondrial matrix. Thus, pyruvate and inorganic phosphate (P_i) are cotransported inward with H^+ as the latter moves into the matrix. In contrast, ADP is cotransported with ATP in opposite directions by a single carrier protein. Since an ATP molecule has one more negative charge than ADP, each nucleotide exchange results in a total of one negative charge being moved out of the mitochondrion. The ADP-ATP cotransport is therefore driven by the charge difference across the membrane (Figure 13–16).

In eucaryotic cells, the proton gradient is consequently used to drive both the formation of ATP and the transport of certain metabolites across the inner mitochondrial membrane. In bacteria, the proton gradient across the bacterial plasma membrane serves all of these functions, but in addition, is itself an important source of directly usable energy: in motile bacteria, the gradient drives the rapid rotation of the bacterial flagellum, which propels the bacterium along (Figure 13–17).

Proton Gradients Produce Most of the Cell's ATP

As stated previously, glycolysis alone produces a net yield of 2 molecules of ATP for every molecule of glucose, which is the total energy yield for the fermentation processes that occur in the absence of O_2 (discussed in Chapter 4). In contrast, during oxidative phosphorylation each pair of electrons donated by the NADH produced in mitochondria is thought to

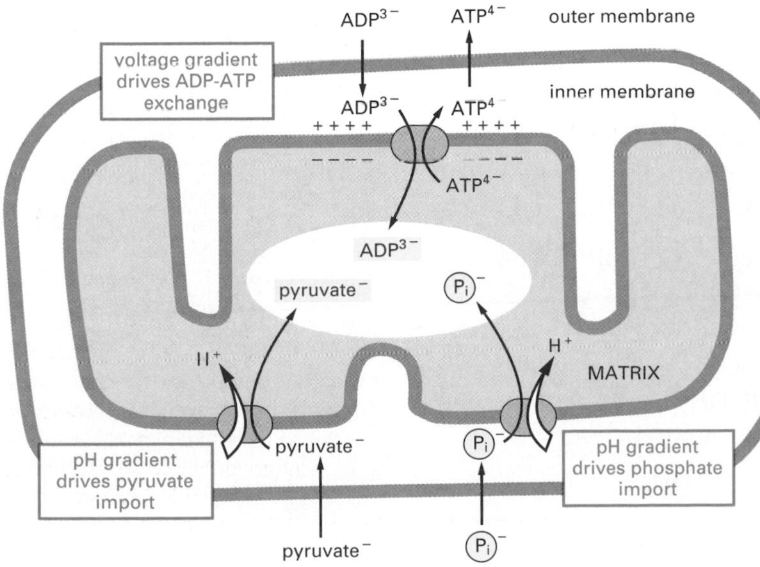

Question 13–4 The remarkable properties that allow ATP synthase to run in either direction allow the interconversion of energy stored in the H^+ gradient and energy stored in ATP in either direction. (A) If ATP synthase making ATP can be likened to a water-driven turbine producing electricity, what would be an appropriate analogy when it works in the opposite direction? (B) Under what conditions would one expect the ATP synthase to stall, running neither forward nor backward? (C) What determines the direction in which the ATP synthase operates?

Figure 13–16 Some of the active transport processes driven by the electrochemical proton gradient across the inner mitochondrial membrane. Pyruvate, inorganic phosphate, and ADP are moved into the matrix, while ATP is pumped out. The charge on each of the transported molecules is indicated for comparison with the membrane potential, which is negative inside, as shown. The outer membrane is freely permeable to all of these compounds. The active transport of molecules across membranes by carrier proteins is discussed in Chapter 12.

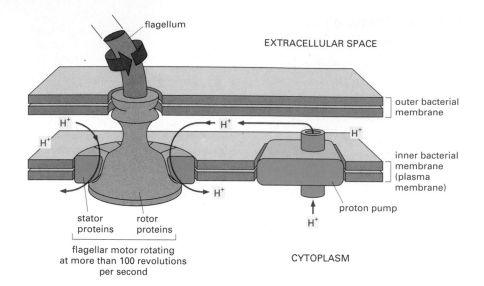

flagellum

EXTRACELLULAR SPACE

outer bacterial membrane

H^+

H^+

H^+

H^+

inner bacterial membrane (plasma membrane)

proton pump

H^+

stator proteins   rotor proteins

flagellar motor rotating at more than 100 revolutions per second

CYTOPLASM

Figure 13–17 The rotation of the bacterial flagellum is driven by H⁺ flow. The flagellum is attached to a series of protein rings (shown in *orange*), which are embedded in the outer and inner (plasma) membranes and rotate with the flagellum. The rotation is driven by a flow of protons through an outer ring of proteins (the *stator*), by mechanisms that are not yet understood.

provide energy for the formation of about 2.5 molecules of ATP, once one includes the energy needed for transporting this ATP to the cytosol. Oxidative phosphorylation also produces 1.5 ATP molecules per electron pair from $FADH_2$, or from the NADH molecules produced by glycolysis in the cytosol. From the product yields of glycolysis and the citric acid cycle summarized in Table 13–1, one can readily calculate that the complete oxidation of one molecule of glucose—starting with glycolysis and ending with oxidative phosphorylation—gives a net yield of about 30 ATPs. Thus, the vast majority of the ATP produced from the oxidation of glucose in an animal cell is produced by chemiosmotic mechanisms on the mitochondrial membrane. Oxidative phosphorylation in the mitochondrion also produces a large amount of ATP from the NADH and the $FADH_2$ that is derived from the oxidation of fats (Table 13–1; see also Figures 4–9 and 4–10).

Table 13–1	Summary of Product Yields from the Oxidation of Sugars and Fats

A. Net products from oxidation of one molecule of glucose

In cytosol (glycolysis)

glucose → 2 pyruvate + 2 NADH + 2 ATP

In mitochondrion (pyruvate dehydrogenase and citric acid cycle)

2 pyruvate → 2 acetyl CoA + 2 NADH

2 acetyl CoA → 6 NADH + 2 $FADH_2$ + 2 GTP

Net result in mitochondrion

2 pyruvate → 8 NADH + 2 $FADH_2$ + 2 GTP

B. Net products from oxidation of one molecule of palmitoyl CoA (activated form of palmitate, a fatty acid)

In mitochondrion (fatty acid oxidation and citric acid cycle)

palmitoyl CoA → 8 acetyl CoA + 7 NADH + 7 $FADH_2$

8 acetyl CoA → 24 NADH + 8 $FADH_2$

Net result in mitochondrion

palmitoyl CoA → 31 NADH + 15 $FADH_2$

Question 13–5

A. Calculate the number of ATP molecules produced per pair of electrons transferred from NADH to oxygen, if . . .

1. five protons are pumped across the inner mitochondrial membrane for each electron passed through the three respiratory enzyme complexes,

2. three protons must pass through the ATP synthase for each ATP molecule that it produces from ADP and inorganic phosphate inside the mitochondrion, and

3. one proton is used to produce the voltage gradient needed to transport each ATP molecule out of the mitochondrion to the cytosol where it is used.

B. Use the information in Table 13–1 together with your ATP yield per electron pair from (A) to calculate . . .

1. the number of ATP molecules produced per molecule of glucose as a result of the citric acid cycle alone. (Assume that the oxidation which results from the transfer of a pair of electrons from each $FADH_2$ molecule results in pumping of six protons, and that GTP and ATP are freely interconvertible according to the reaction GTP + ADP ⇌ GDP + ATP.)

2. the total number of ATP molecules produced by the complete oxidation of glucose. (Assume that electrons from each NADH molecule produced in the cytosol by glycolysis are also fed [indirectly] into the electron-transport chain; however, the yield of ATP is only 1.5 ATP molecules for each of these NADH molecules.)

The Rapid Conversion of ADP to ATP in Mitochondria Maintains a High ATP:ADP Ratio in Cells

Using the transport proteins in the inner mitochondrial membrane discussed earlier, the ADP molecules produced by ATP hydrolysis in the cytosol rapidly enter mitochondria for recharging, while the ATP molecules formed in the mitochondrial matrix by oxidative phosphorylation are rapidly pumped into the cytosol, where they are needed. A typical ATP molecule in the human body shuttles out of a mitochondrion and back into it (as ADP) for recharging more than once per minute, keeping the concentration of ATP in the cell about 10 times higher than that of ADP.

As discussed in Chapter 3, biosynthetic enzymes often drive energetically unfavorable reactions by coupling them to the energetically favorable hydrolysis of ATP (see Figure 3–24). The ATP pool is used to drive cellular processes in much the same way that a battery can be used to drive electric engines: if the activity of the mitochondria were halted, ATP levels would fall and the cell's battery would run down; eventually, energetically unfavorable reactions would no longer be driven and the cell would die. The poison cyanide, which blocks electron transport in the inner mitochondrial membrane, causes death in exactly this way.

Electron-Transport Chains and Proton Pumping

Having considered in general terms how a mitochondrion functions, we need to examine in more detail the mechanisms that underlie its membrane-based energy-conversion processes. In doing so, we will also be accomplishing a larger purpose. As emphasized at the beginning of this chapter, very similar energy-conversion devices are used by mitochondria, chloroplasts, and bacteria, and the basic principles that we shall discuss next therefore underlie the function of nearly all living things. To emphasize this central point, we shall end this chapter by describing how electron transport provides energy for one of the many types of bacteria that flourish in huge numbers, in total darkness and without oxygen, miles below the surface of the ocean.

For many years, the reason that electron-transport chains were embedded in membranes eluded the biochemists who were struggling to understand them. The puzzle was solved as soon as the fundamental role of transmembrane proton gradients in energy generation was proposed in the early 1960s. But the process of chemiosmotic coupling entails an interplay between chemical and electrical forces that is not easy to decipher at a molecular level; indeed, the idea was so novel that it was not widely accepted until many years later, after additional supporting evidence had accumulated from experiments designed as tests of the hypothesis.

Although investigators today are still unraveling many of the details of chemiosmotic coupling, the fundamentals are now clear. In this part of the chapter we shall look at some of the principles that underlie the electron-transport process and explain how it can generate a proton gradient.

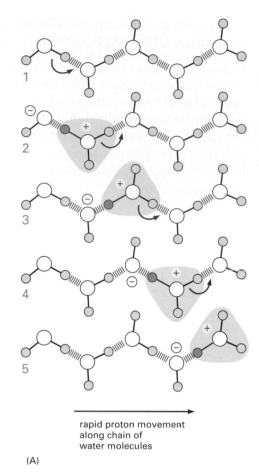

Figure 13–18 How protons behave in water. (A) Protons move very rapidly along a chain of hydrogen-bonded water molecules. In this diagram, proton jumps are indicated by *blue arrows,* and hydronium ions are indicated by *green shading.* As discussed in Chapter 2, naked protons rarely exist as such, and are instead associated with a water molecule in the form of a hydronium ion, H_3O^+. At a neutral pH (pH 7.0), the hydronium ions are present at a concentration of 10^{-7} M; however, for simplicity one usually refers to this as an H^+ concentration of 10^{-7} M (see Panel 2–2, pp. 50–51). (B) Electron transfer can cause the transfer of entire hydrogen atoms, because protons are readily accepted from or donated to water inside cells. In this example, A picks up an electron plus a proton when it is reduced, and B loses an electron plus a proton when it is oxidized.

Protons Are Readily Moved by the Transfer of Electrons

Although protons resemble other positive ions such as Na^+ and K^+ in their movement across membranes, in some respects they are unique. Hydrogen atoms are by far the most abundant type of atom in living organisms and are plentiful not only in all carbon-containing biological molecules but also in the water molecules that surround them. The protons in water are highly mobile, flickering through the hydrogen-bonded network of water molecules by rapidly dissociating from one water molecule in order to associate with its neighbor, as illustrated in Figure 13–18A. Thus, water, which is everywhere in cells, serves as a ready reservoir for donating and accepting protons.

Whenever a molecule is reduced by acquiring an electron, the electron (e^-) brings with it a negative charge. In many cases, this charge is rapidly neutralized by the addition of a proton (H^+) from water, so that the net effect of the reduction is to transfer an entire hydrogen atom, $H^+ + e^-$ (Figure 13–18B). Similarly, when a molecule is oxidized, the hydrogen atom can be readily dissociated into its constituent electron and proton, allowing the electron to be transferred separately to a molecule that accepts electrons, while the proton is passed to the water. Therefore, in a membrane in which electrons are being passed along an electron-transport chain, it is a relatively simple matter, in principle, to pump protons from one side of the membrane to another. All that is required is that the electron carrier be arranged in the membrane in a way that causes it to pick up a proton from one side of the membrane when it accepts an electron, while releasing the proton on the other side of the membrane as the electron is passed on to the next carrier molecule in the chain (Figure 13–19).

The Redox Potential Is a Measure of Electron Affinities

In biochemical reactions, any electrons removed from one molecule are always passed to another, so that whenever one molecule is oxidized, another is reduced. Like any other chemical reaction, the tendency of such oxidation-reduction, or **redox reactions,** to proceed spontaneously depends on the free-energy change (ΔG) for the electron transfer, which in turn depends on the relative affinities of the two molecules for electrons. (The role of free energy in chemical reactions is discussed in Chapter 3, p. 89.)

Because electron transfers provide most of the energy for living things, it is worth spending a little time to understand them. Many read-

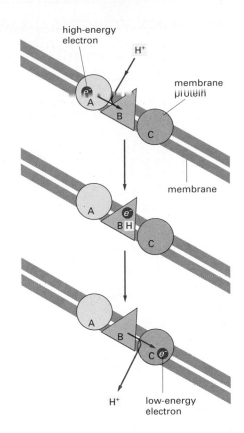

Figure 13–19 How protons can be pumped across membranes. As an electron passes along an electron-transport chain, it can bind and release a proton at each step. In this schematic diagram, electron carrier B picks up a proton (H⁺) from one side of the membrane when it accepts an electron (e^-) from carrier A; it releases the proton to the other side of the membrane when it donates its electron to carrier C.

ers are already familiar with acids and bases, which donate and accept protons (see Panel 2–2, pp. 50–51). Acids and bases exist in conjugate acid-base pairs, where the acid is readily converted into the base by the loss of a proton. For example, acetic acid (CH_3COOH) is converted into its conjugate base (CH_3COO^-) in the reaction:

$$CH_3COOH \rightleftharpoons CH_3COO^- + H^+$$

In exactly the same way, pairs of compounds such as NADH and NAD⁺ are called **redox pairs,** since NADH is converted to NAD⁺ by the loss of electrons in the reaction:

$$NADH \rightleftharpoons NAD^+ + H^+ + 2e^-$$

NADH is a strong electron donor: because its electrons are held in a high-energy linkage, the free-energy change for passing its electrons to many other molecules is favorable (see Figure 13–8). Conversely, it is difficult to form a high-energy linkage, and its pair, NAD⁺, is of necessity a weak electron acceptor.

The tendency to transfer electrons from any redox pair can be measured experimentally. All that is required is to form an electrical circuit that links a 1:1 (equimolar) mixture of the redox pair to a second redox pair that has been arbitrarily selected as a reference standard, so that the voltage difference can be measured between them (Panel 13–1, pp. 424). This voltage difference is defined as the **redox potential;** as defined, electrons will move spontaneously from a redox pair like NADH/NAD⁺ with a low redox potential (a low affinity for electrons) to a redox pair like O_2/H_2O with a high redox potential (a high affinity for electrons). Thus NADH is a good molecule to donate electrons to the respiratory chain, while O_2 is well suited to act as the "sink" for electrons at the end of the pathway. As explained in Panel 13–1, the difference in redox potential, $\Delta E_0'$, is a direct measure of the standard free-energy change ($\Delta G°$) for the transfer of an electron from one molecule to another.

Electron Transfers Release Large Amounts of Energy

As just discussed, those pairs of compounds that have the most negative redox potentials have the weakest affinity for electrons and therefore contain carriers with the strongest tendency to donate electrons. Conversely, those pairs that have the most positive redox potentials have the strongest affinity for electrons and therefore contain carriers with the strongest tendency to accept electrons. A 1:1 mixture of NADH and NAD⁺ has a redox potential of –320 mV, indicating that NADH has a strong tendency to donate electrons; a 1:1 mixture of H_2O and ½O_2 has a redox potential of +820 mV, indicating that O_2 has a strong tendency to accept electrons. The difference in redox potential is 1.14 volts (1140 mV), which means that the transfer of each electron from NADH to O_2 under these standard conditions is enormously favorable, where $\Delta G° = -26.2$ kcal/mole (–52.4 kcal/mole for the two electrons transferred per

HOW REDOX POTENTIALS ARE MEASURED

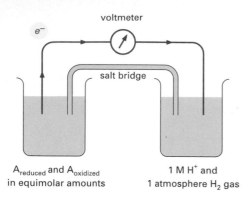

A_reduced and A_oxidized in equimolar amounts

1 M H^+ and 1 atmosphere H_2 gas

One beaker (*left*) contains substance A with an equimolar mixture of the reduced ($A_{reduced}$) and oxidized ($A_{oxidized}$) members of its redox pair. The other beaker contains the hydrogen reference standard ($2H^+ + 2e^- \rightleftharpoons H_2$), whose redox potential is arbitrarily assigned as zero by international agreement. (A salt bridge formed from a concentrated KCl solution allows the ions K^+ and Cl^- to move between the two beakers, as required to neutralize the charges in each beaker when electrons flow between them.) The metal wire (*red*) provides a resistance-free path for electrons, and a voltmeter then measures the redox potential of substance A. If electrons flow from $A_{reduced}$ to H^+, as indicated here, the redox pair formed by substance A is said to have a negative redox potential. If they instead flow from H_2 to $A_{oxidized}$, this redox pair is said to have a positive redox potential.

SOME STANDARD REDOX POTENTIALS AT pH 7

By convention, the redox potential for a redox pair is designated as E. For the standard state, with all reactants at a concentration of 1 M, including H^+, one can determine a standard redox potential, designated as E_0. Since biological reactions occur at pH 7, biologists use a different standard state in which $A_{reduced} = A_{oxidized}$ and $H^+ = 10^{-7}$ M. This standard redox potential is designated as E_0'. A few examples of special relevance to oxidative phosphorylation are given here.

redox reactions	redox potential E_0'
NADH $\rightleftharpoons$ NAD$^+$ + H$^+$ + 2e$^-$	–320 mV
reduced ubiquinone $\rightleftharpoons$ oxidized ubiquinone + 2H$^+$ + 2e$^-$	+30 mV
reduced cytochrome c $\rightleftharpoons$ oxidized cytochrome c + e$^-$	+230 mV
H$_2$O $\rightleftharpoons$ ½O$_2$ + 2H$^+$ + 2e$^-$	+820 mV

CALCULATION OF $\Delta G°$ FROM REDOX POTENTIALS

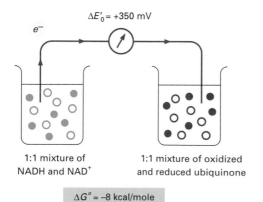

$\Delta E_0' = +350$ mV

1:1 mixture of NADH and NAD$^+$

1:1 mixture of oxidized and reduced ubiquinone

$\Delta G° = -8$ kcal/mole

$\Delta G° = -n(0.023) \Delta E_0'$ where n is the number of electrons transferred across a redox potential change of $\Delta E_0'$ millivolts (mV)

Example: The transfer of one electron from NADH to ubiquinone has a favorable $\Delta G°$ of –8.0 kcal/mole, whereas the transfer of one electron from ubiquinone to oxygen has an even more favorable $\Delta G°$ of –18.2 kcal/mole. The $\Delta G°$ value for the transfer of one electron from NADH to oxygen is the sum of these two values, –26.2 kcal/mole.

EFFECT OF CONCENTRATION CHANGES

As explained in Chapter 3 (see p. 89), the actual free-energy change for a reaction, ΔG, depends on the concentration of the reactants and generally will be different from the standard free-energy change, $\Delta G°$. The standard redox potentials are for a 1:1 mixture of the redox pair. For example, the standard redox potential of –320 mV is for a 1:1 mixture of NADH and NAD$^+$. But when there is an excess of NADH over NAD$^+$, electron transfer from NADH to an electron acceptor becomes more favorable. This is reflected by a more negative redox potential and a more negative ΔG for electron transfer.

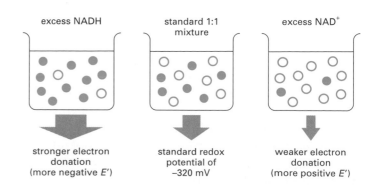

excess NADH

standard 1:1 mixture

excess NAD$^+$

stronger electron donation (more negative E')

standard redox potential of –320 mV

weaker electron donation (more positive E')

NADH molecule; see Panel 13–1). If we compare this free-energy change with that for the formation of the phosphoanhydride bonds in ATP ($\Delta G° = -7.3$ kcal/mole, see Figure 4–7), we see that more than enough energy is released by the oxidization of one NADH molecule to synthesize several molecules of ATP from ADP and P_i.

Living systems could certainly have evolved enzymes that would allow NADH to donate electrons directly to O_2 to make water in the reaction:

$$2H^+ + 2e^- + \tfrac{1}{2}O_2 \rightarrow H_2O$$

But because of the huge free-energy drop, this reaction would proceed with almost explosive force and nearly all of the energy would be released as heat. Cells do perform this reaction, but they make it proceed much more gradually by passing the high-energy electrons from NADH to O_2 via the many electron carriers in the electron-transport chain. Since each successive carrier in the chain holds its electrons more tightly, the energetically favorable reaction $2H^+ + 2e^- + \tfrac{1}{2}O_2 \rightarrow H_2O$ is made to occur in many small steps. This enables nearly half of the released energy to be stored, instead of being lost to the environment as heat.

Metals Tightly Bound to Proteins Form Versatile Electron Carriers

In the respiratory chain, the electron carriers are arranged in order of increasing redox potential, thus making possible the gradual release of the energy stored in NADH electrons. Within each of the three respiratory enzyme complexes, the electrons move mainly between metal atoms that are tightly bound to the proteins, and the electrons travel by skipping from one metal ion to the next.

In contrast, electrons are carried between the different respiratory complexes by molecules that diffuse along the lipid bilayer, picking up electrons from one complex and delivering them to another in an orderly sequence. *Ubiquinone*, a small hydrophobic molecule that dissolves in the lipid bilayer, is the only carrier that is not part of a protein. In the mitochondrial respiratory chain, ubiquinone picks up electrons from the NADH dehydrogenase complex and delivers them to the cytochrome b-c$_1$ complex (see Figure 13–10). As shown in Figure 13–20, a **quinone** (like ubiquinone) can pick up or donate either one or two electrons, and it picks up one H^+ from the surroundings with each electron that it carries. Its redox potential of +30 mV places ubiquinone about one-quarter of the way down the chain from NADH in terms of energy loss (Figure 13–21).

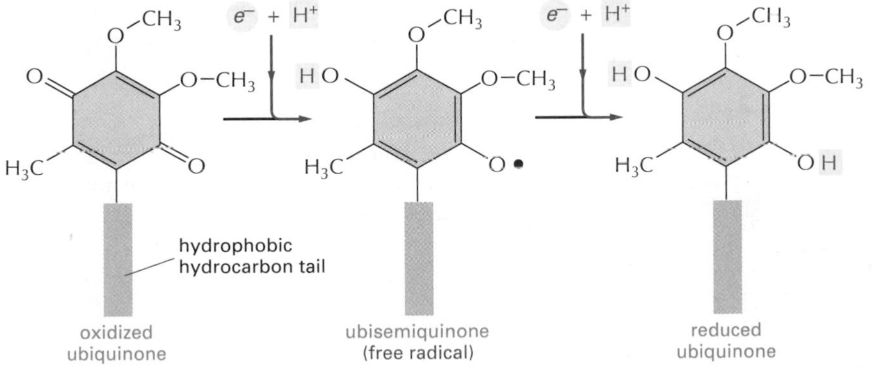

oxidized ubiquinone   ubisemiquinone (free radical)   reduced ubiquinone

hydrophobic hydrocarbon tail

Figure 13–20 Quinone electron carriers. Ubiquinone in the mitochondrial electron-transport chain picks up one H^+ from the aqueous environment for every electron it accepts, and it can carry either one or two electrons as part of a hydrogen atom (*yellow*). When reduced ubiquinone donates its electrons to the next carrier in the chain, these protons are released. The long hydrophobic tail confines ubiquinone to the membrane and consists of 6–10 five-carbon isoprene units, the number depending on the organism. The corresponding quinone electron carrier in the photosynthetic membranes of chloroplasts is plastoquinone, which is almost identical in structure. For simplicity, both ubiquinone and plastoquinone will be referred to as *quinone* and abbreviated as Q.

Electron-Transport Chains and Proton Pumping

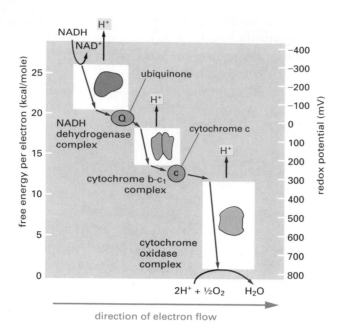

Figure 13–21 Redox potential increases along the mitochondrial electron-transport chain. For the symbols used here see Figure 13–10. Note that the big increases in redox potential occur across each of the three respiratory enzyme complexes, as required for each of them to pump protons.

All the rest of the electron carriers in the electron-transport chain are small molecules that are tightly bound to proteins. Between NADH and ubiquinone, inside the NADH dehydrogenase complex, the electrons are passed between a flavin group (see Figure 4–12 for structure) bound to the protein complex and a set of iron-sulfur centers of increasing redox potentials. **Iron-sulfur centers** have structures resembling those shown in Figure 13–22, and they carry one electron at a time. The final iron-sulfur center in the dehydrogenase donates its electrons to ubiquinone.

Iron-sulfur centers have relatively low affinities for electrons and thus would be less useful in the later part of the electron-transport chain, in the pathway from ubiquinone to O_2. In this part of the pathway, iron atoms in heme groups that are tightly bound to cytochrome proteins are commonly used as electron carriers, as in the cytochrome b-c$_1$ and cytochrome oxidase complexes. The **cytochromes** constitute a family of colored proteins (hence their name, from the Greek *chroma*, "color"); they each contain one or more heme groups whose iron atom changes from the ferric (FeIII) to the ferrous (FeII) state whenever it accepts an electron. As one would expect, the various cytochromes increase in redox potential as one progresses down the mitochondrial electron-transport chain toward oxygen. The structure of *cytochrome c,*

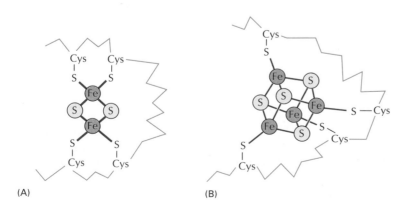

Figure 13–22 The structures of two types of iron-sulfur centers. (A) A center of the 2Fe2S type. (B) A center of the 4Fe4S type. Although they contain multiple iron atoms, each iron-sulfur center can carry only one electron at a time. There are many iron-sulfur centers in the mitochondrial electron-transport chain.

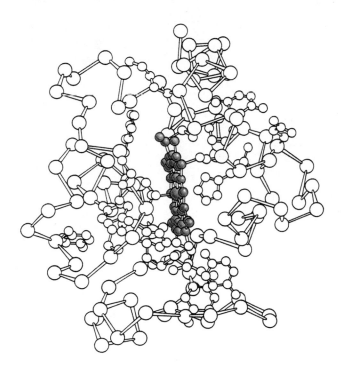

Figure 13–23 The three-dimensional structure of cytochrome c, an electron carrier in the electron-transport chain. This small protein contains just over 100 amino acids and is held loosely on the matrix face of the membrane by ionic interactions (see Figure 13–10). The iron atom (*orange*) in the bound heme (*blue*) can carry a single electron. The structure of the heme group in hemoglobin, which reversibly binds O_2 rather than an electron, was shown in Figure 5–31.

the small protein that shuttles electrons between the cytochrome b-c$_1$ complex and the cytochrome oxidase complex, is shown in Figure 13–23: its redox potential is +230 mV.

At the very end of the respiratory chain, just before oxygen, the electron carriers are those in the cytochrome oxidase complex. The carriers here are either iron atoms in heme groups or copper atoms that are tightly bound to the complex in specific ways that give them a high redox potential. Throughout the chain, the redox potential of each of the electron carriers has been fine-tuned to fit requirements by binding the carrier atom or molecule in a particular protein context, which alters its normal affinity for electrons.

Question 13–6   At many steps in the electron-transport chain Fe ions are used as part of heme or FeS clusters to bind the electrons in transit. Why do these functional groups that carry out the chemistry of electron transfer need to be bound to proteins? Provide several different reasons why this is necessary.

Protons Are Pumped Across the Membrane by the Three Respiratory Enzyme Complexes

Experiments have shown that H^+ is pumped across the membrane as electrons pass through each of the three respiratory enzyme complexes (see Figure 13–10). On average, nearly two H^+ are pumped per electron at each complex, or about five H^+ per electron in total (10 H^+ per NADH molecule oxidized). The three-dimensional structures of each of these complexes will be needed before we can hope to understand the details of the mechanisms that pump protons. The crystals of membrane proteins that are needed for x-ray diffraction studies (see Panel 5–5, pp. 162–163) are notoriously difficult to obtain. For this reason, complete structural information is so far available only for **cytochrome oxidase,** the final respiratory enzyme complex in the chain. We shall therefore discuss proton pumping in reference to this enzyme complex.

Cytochrome oxidase receives electrons from cytochrome c, thus oxidizing it (hence its name), and donates these electrons to oxygen. A schematic view of the reaction catalyzed by cytochrome oxidase is presented in Figure 13–24A, based on the complete three-dimensional

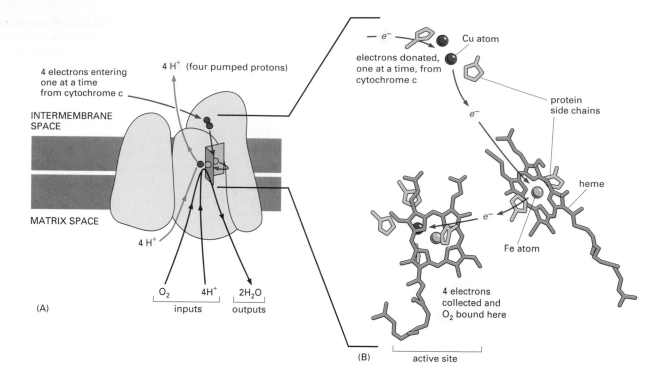

(A)

4 electrons entering
one at a time
from cytochrome c

4 H⁺ (four pumped protons)

INTERMEMBRANE
SPACE

MATRIX SPACE

4 H⁺

O₂   4H⁺   2H₂O

inputs   outputs

electrons donated,
one at a time, from
cytochrome c

Cu atom

protein
side chains

heme

Fe atom

4 electrons
collected and
O₂ bound here

(B)   active site

structure of this large transmembrane protein complex. In brief, four electrons from cytochrome c and four protons from the aqueous environment are added to each O_2 molecule in the reaction $4e^- + 4H^+ + O_2 \rightarrow 2H_2O$. In addition, four more protons are pumped across the membrane during electron transfer, building up the electrochemical proton gradient. Proton pumping is caused by allosteric changes in the conformation of the protein, which are driven by energy derived from electron transport.

At its active site, cytochrome oxidase contains a complex of a heme iron atom juxtaposed with a tightly bound copper atom (Figure 13–24B). It is here that nearly all of the oxygen we breathe is used, serving as the final repository for the electrons that NADH donated at the start of the electron-transport chain. Oxygen is useful for this purpose because of its very high affinity for electrons. But once O_2 picks up one electron, it forms the superoxide radical O_2^-; this radical is dangerously reactive and will avidly take up another three electrons wherever it can find them. One of the roles of cytochrome oxidase is to hold on tightly to its oxygen molecule until all four of the electrons needed to convert it to two H_2O molecules are in hand, thereby preventing a random attack on cellular molecules by superoxide radicals. The invention of cytochrome oxidase was therefore crucial to the evolution of cells that are able to use O_2 as an electron acceptor.

For a protein to pump protons actively across a membrane, the conformational change in the protein pump must be coupled to an energetically favorable reaction. For cytochrome oxidase, the energetically favorable reaction is the addition of electrons to O_2 at the enzyme's active site. The protons are thought to be carried at another site on the cytochrome oxidase by a histidine side chain that moves in response to the incoming electrons. This side chain is thought to bind H⁺ when facing the matrix and to release H⁺ when facing the intermembrane space (Figure 13–25).

Figure 13–24 The cytochrome oxidase complex. (A) The reactions catalyzed by cytochrome oxidase. The shapes and arrangement of the three subunits shown are based on the complete structure determined by x-ray diffraction. The inside of the membrane refers to the matrix space in mitochondria, and to the cytoplasm in aerobic bacteria. (B) Details of the active site in (A), showing the location of the protein-bound metal ions that carry electrons from cytochrome c to oxygen (O_2). As in (A), the two iron (Fe) atoms are *orange*, the three copper (Cu) atoms are *red*, and the two hemes that carry the iron atoms are *blue*. All of these metal ions are buried within the protein, and the electrons jump between them by a phenomenon called electron tunneling.

Question 13–7   Two different diffusible electron carriers, ubiquinone and cytochrome c, shuttle electrons between the three protein complexes of the electron-transport chain. Could, in principle, the same diffusible carrier be used for both steps? Explain your answer.

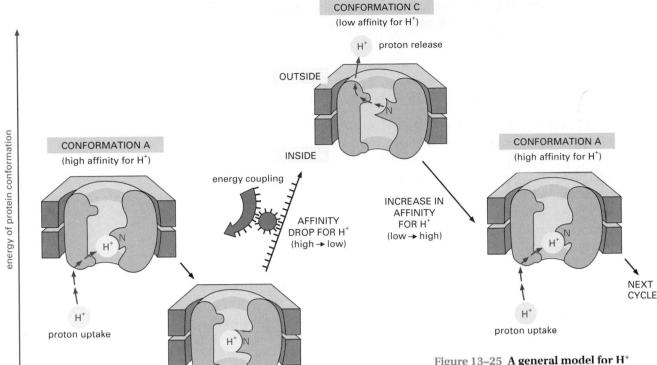

CONFORMATION C
(low affinity for H⁺)

H⁺ proton release

OUTSIDE

INSIDE

energy coupling

AFFINITY
DROP FOR H⁺
(high → low)

INCREASE IN
AFFINITY
FOR H⁺
(low → high)

CONFORMATION A
(high affinity for H⁺)

H⁺
proton uptake

CONFORMATION A
(high affinity for H⁺)

NEXT
CYCLE

H⁺
proton uptake

CONFORMATION B
(high affinity for H⁺)

energy of protein conformation

Figure 13–25 A general model for H⁺ pumping. This model for H⁺ pumping by a transmembrane protein is based on mechanisms that are thought to be utilized by both cytochrome oxidase and the light-driven bacterial proton pump, bacteriorhodopsin (see Figure 11–29). The protein is driven through a cycle of three conformations, denoted here as A, B, and C. As indicated by their vertical spacing, these protein conformations have different energies. In conformation A, the protein has a high affinity for H⁺, causing it to pick up an H⁺ on the inside of the membrane. In conformation C the protein has a low affinity for H⁺, causing it to release an H⁺ on the outside of the membrane. The transition from conformation B to conformation C that releases the H⁺ is energetically unfavorable, and it occurs only because it is driven by being allosterically coupled to an energetically favorable reaction occurring elsewhere on the protein (*blue arrow*). Because the other two conformational changes, A → B and C → A, lead to states of lower energy, they proceed spontaneously. The cycle A → B → C → A → B → C therefore goes only one way, causing H⁺ to be pumped from the inside (the matrix space in mitochondria) to the outside (the intermembrane space in mitochondria). For cytochrome oxidase, the energy for the transition B → C is provided by electron transport, whereas for bacteriorhodopsin this energy is provided by light. For yet other proton pumps, the energy is derived from ATP hydrolysis (discussed in Chapter 12).

Respiration Is Amazingly Efficient

The free-energy changes for burning fats and carbohydrates directly into CO_2 and H_2O can be compared to the total amount of energy generated and stored in the phosphate bonds of ATP during the corresponding biological oxidations. When this is done, it is seen that the efficiency with which oxidation energy is converted into ATP bond energy is often greater than 40%. This is considerably better than the efficiency of most nonbiological energy-conversion devices. If cells worked only with the efficiency of an electric motor or a gasoline engine (10–20%), an organism would have to eat voraciously in order to maintain itself. Moreover, since wasted energy is liberated as heat, large organisms would need more efficient mechanisms than they presently have for giving up heat to the environment.

Students sometimes wonder why the chemical interconversions in cells follow such complex pathways. The oxidation of sugars to CO_2 plus H_2O could certainly be accomplished more directly, eliminating the citric acid cycle and many of the steps in the respiratory chain. This would make respiration easier for students to learn, but it would have been a disaster for the cell. Oxidation produces huge amounts of free energy, which can be utilized efficiently only in small bits. Biological oxidative pathways involve many intermediates, each differing only slightly from its predecessor. The energy released is thereby parceled out into small packets that can be efficiently converted to high-energy bonds in useful molecules, such as ATP and NADH, by means of coupled reactions (see Figure 4–1).

Having seen how chemiosmotic coupling is used to generate ATP in mitochondria, we shall now look at how it harnesses light energy for the generation of ATP in chloroplasts.

Chloroplasts and Photosynthesis

Virtually all of the organic materials required by present-day living cells are produced by **photosynthesis**—the series of light-driven reactions that creates organic molecules from atmospheric carbon dioxide (CO_2). Organisms that carry out photosynthesis include plants, algae, and many types of photosynthetic bacteria. Plants, algae, and the most advanced photosynthetic bacteria, such as the cyanobacteria (see Figure 13–1A), use electrons from water and the energy of sunlight to convert atmospheric CO_2 into organic compounds. In the course of splitting water they liberate into the atmosphere vast quantities of oxygen gas. This oxygen is in turn required for cellular respiration—not only in animals but also in plants and many bacteria.

In plants, photosynthesis is carried out in a specialized intracellular organelle—the **chloroplast.** Chloroplasts perform photosynthesis during the daylight hours and thereby produce ATP and NADPH, which in turn are used to convert CO_2 into sugars inside the chloroplast. The sugars produced are exported to the surrounding cytosol, where they are used as fuel to make ATP and as starting materials for many of the other organic molecules that the plant cell needs. Sugar is also exported to all those cells in the plant that lack chloroplasts. As is the case for animal cells, most of the ATP present in the cytosol of plant cells is made by the oxidation of sugars and fats in mitochondria.

Chloroplasts Resemble Mitochondria but Have an Extra Compartment

Chloroplasts carry out their energy interconversions by means of proton gradients in much the same way that mitochondria do. Although larger (Figure 13–26A), they are organized on the same principles. They have a highly permeable outer membrane, a much less permeable inner membrane—in which membrane transport proteins are embedded—and a narrow intermembrane space in between. Together these membranes form the chloroplast envelope (Figure 13–26B). The inner membrane surrounds a large space called the **stroma,** which is analogous to the mitochondrial matrix and contains many metabolic enzymes. Like the mitochondrion, the chloroplast evolved from an engulfed bacterium (see Figure 1–14), and it still contains its own genome and genetic system. The stroma, therefore, like the mitochondrial matrix, also contains a special set of ribosomes, RNA, and DNA.

There is, however, an important difference between the organization of mitochondria and that of chloroplasts. The inner membrane of the chloroplast does not contain the electron-transport chains. Instead, the light-capturing systems, the electron-transport chains, and ATP synthase are all contained in the *thylakoid membrane*, a third membrane that forms a set of flattened disclike sacs, the *thylakoids* (Figure 13–26C). These are arranged in stacks, and the space inside each thylakoid is thought to be connected with that of other thylakoids, thereby defining

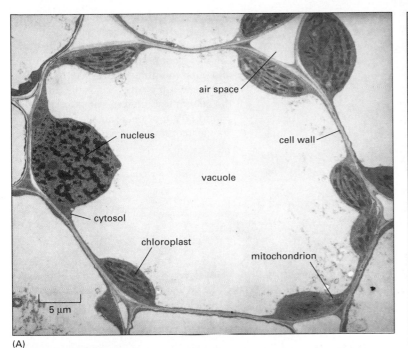

(A)

air space

nucleus

cell wall

vacuole

cytosol

chloroplast

mitochondrion

5 μm

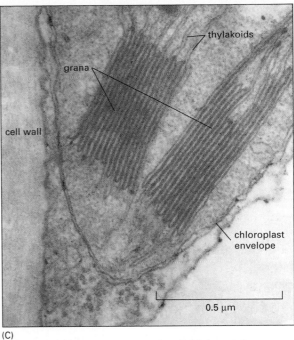

(C)

thylakoids

grana

cell wall

chloroplast
envelope

0.5 μm

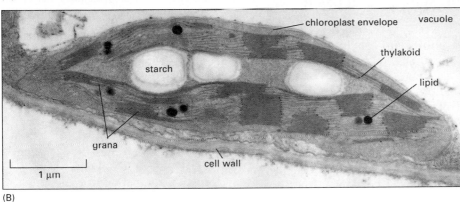

(B)

chloroplast envelope

vacuole

starch

thylakoid

lipid

grana

cell wall

1 μm

Figure 13–26 Electron micrographs of chloroplasts. (A) A wheat leaf cell in which a thin rim of cytoplasm containing nucleus, chloroplasts, and mitochondria surrounds a large vacuole. (B) A thin section of a single chloroplast, showing the chloroplast envelope, starch granules, and lipid (fat) droplets that have accumulated in the stroma as a result of the biosyntheses occurring there. (C) A high-magnification view of two *grana*, which is the name given to a stack of thylakoids. (Courtesy of K. Plaskitt.)

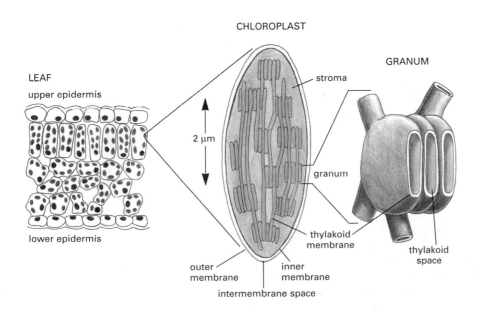

CHLOROPLAST

GRANUM

LEAF

upper epidermis

lower epidermis

stroma

2 μm

granum

thylakoid
membrane

outer
membrane

inner
membrane

intermembrane space

thylakoid
space

Figure 13–27 The chloroplast. This photosynthetic organelle contains three distinct membranes (the outer membrane, the inner membrane, and the thylakoid membrane) that define three separate internal compartments (the intermembrane space, the stroma, and the thylakoid space). The thylakoid membrane contains all of the energy-generating systems of the chloroplast, including its chlorophyll. In electron micrographs this membrane appears to be broken up into separate units that enclose individual flattened vesicles (see Figure 13–26C), but these are probably joined into a single, highly folded membrane in each chloroplast. As indicated, the individual thylakoids are interconnected, and they tend to stack to form grana.

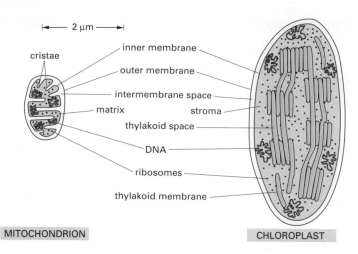

Figure 13–28 **Comparison of a mitochondrion and a chloroplast.** The chloroplast is generally much larger than a mitochondrion and contains, in addition to an inner and outer membrane, a thylakoid membrane enclosing a thylakoid space. Unlike the chloroplast inner membrane, the mitochondrial inner membrane is folded into cristae in order to increase its surface area.

a continuous third internal compartment that is separated from the stroma by the thylakoid membrane (Figure 13–27). The structural similarities and differences between mitochondria and chloroplasts are illustrated in Figure 13–28.

Chloroplasts Capture Energy from Sunlight and Use It to Fix Carbon

The many reactions that occur during photosynthesis in plants can be grouped into two broad categories.

1. In the *photosynthetic electron-transfer reactions* (also called the "light reactions") energy derived from sunlight energizes an electron in the green organic pigment *chlorophyll,* enabling the electron to move along an electron-transport chain in the thylakoid membrane in much the same way that an electron moves along the respiratory chain in mitochondria. The chlorophyll obtains its electrons from water (H_2O), producing O_2 as a by-product. During the electron-transport process, H^+ is pumped across the thylakoid membrane, and the resulting electrochemical proton gradient drives the synthesis of ATP in the stroma. As the final step in this series of reactions, high-energy electrons are loaded (together with H^+) onto $NADP^+$, converting it to NADPH. All of these reactions are confined to the chloroplast.

2. In the *carbon-fixation reactions* (also called the "dark reactions") the ATP and the NADPH produced by the photosynthetic electron-transfer reactions serve as the source of energy and reducing power, respectively, to drive the conversion of CO_2 to carbohydrate. The carbon-fixation reactions, which begin in the chloroplast stroma and continue in the cytosol, produce sucrose and many other organic molecules in the leaves of the plant. The sucrose is exported to other tissues as a source of both organic molecules and energy for growth.

Thus the formation of ATP, NADPH, and O_2 (which requires light energy directly) and the conversion of CO_2 to carbohydrate (which requires light energy only indirectly) are separate processes (Figure

Figure 13–29 **The reactions of photosynthesis in a chloroplast.** Water is oxidized and oxygen is released in the photosynthetic electron-transfer reactions, while carbon dioxide is assimilated (fixed) to produce sugars and a variety of other organic molecules in the carbon-fixation reactions

Question 13–8 Chloroplasts have a third internal compartment, the thylakoid space, bounded by the thylakoid membrane. This membrane contains the photosystems, reaction centers, electron-transport chain, and ATP synthase. In contrast, mitochondria use their inner membrane for electron transport and ATP synthesis. In both organelles, protons are pumped out of the largest internal compartment (the matrix in mitochondria and the stroma in chloroplasts). The thylakoid space is completely sealed off from the rest of the cell. Why does this arrangement allow a larger H^+ gradient in chloroplasts than can be achieved for mitochondria?

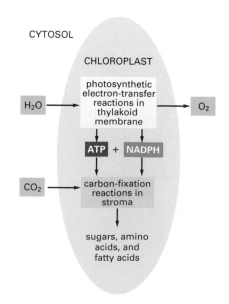

Figure 13–30 The structure of chlorophyll. A magnesium atom (*orange*) is held in the center of a porphyrin ring, which is structurally similar to the porphyrin ring that binds iron in heme. Light is absorbed by electrons within the bond network shown in *blue*, while the long hydrophobic tail helps to hold the chlorophyll in the thylakoid membrane.

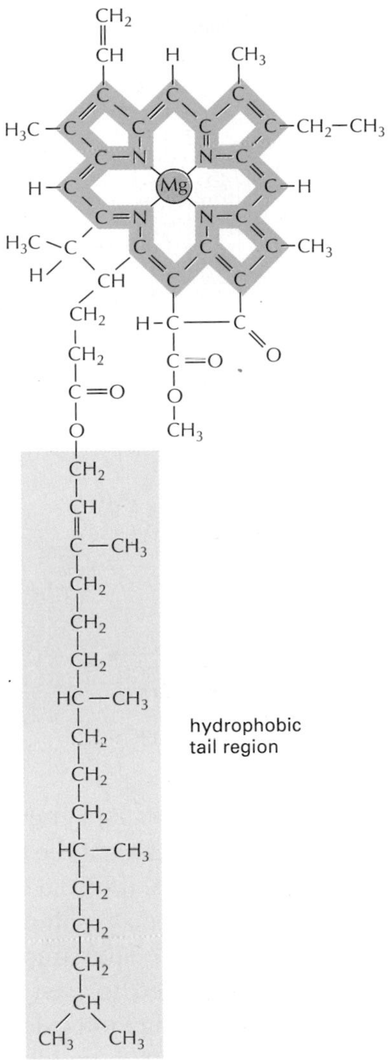

13–29), although elaborate feedback mechanisms interconnect the two. Several of the chloroplast enzymes required for carbon fixation, for example, are inactivated in the dark and reactivated by light-stimulated electron-transport processes.

Excited Chlorophyll Molecules Funnel Energy into a Reaction Center

Visible light is a form of electromagnetic radiation composed of many different wavelengths, ranging from violet (wavelength 400 nm) to deep red (700 nm). When we consider events at the level of a single molecule—such as the absorption of light by a molecule of chlorophyll, we have to picture light as being composed of discrete packets of energy called *photons*. Light of different colors is distinguished by photons of different energy, with longer wavelengths corresponding to lower energies. Thus photons of red light have a lower energy than photons of green light.

When sunlight is absorbed by a molecule of the green pigment **chlorophyll,** electrons in the molecule interact with photons of light and are raised to a higher energy level. The electrons in the extensive network of alternating single and double bonds in the chlorophyll molecule (Figure 13–30) absorb red light most strongly.

An isolated molecule of chlorophyll is incapable of converting the light it absorbs to a form of energy useful to living systems. It can do this only when it is associated with the appropriate proteins and embedded in a membrane. In plant thylakoid membranes and in the membranes of photosynthetic bacteria, the light-absorbing chlorophylls are held in large multiprotein complexes called **photosystems.** The *antenna* portion of a photosystem consists of hundreds of chlorophyll molecules that capture light energy in the form of excited (high-energy) electrons. These chlorophylls are arranged so that the energy of an excited electron can be passed from one chlorophyll molecule to another, funneling the energy into an adjacent protein complex in the membrane—the *reaction center*. There the energy is trapped and used to energize one electron in a *special pair* of chlorophyll molecules (Figure 13–31).

hydrophobic tail region

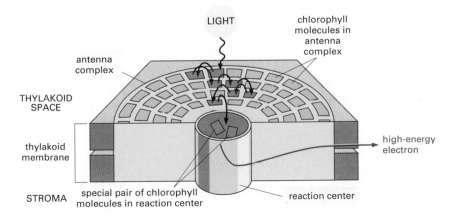

Figure 13–31 The reaction center and antenna in a photosystem. The antenna collects electrons that have been excited by light and funnels their energy to a special pair of chlorophyll molecules in the reaction center. The reaction center thereby acquires a high-energy electron that can be passed rapidly to the electron-transport chain in the thylakoid membrane.

Chloroplasts and Photosynthesis

433

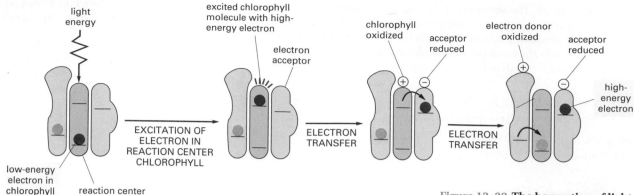

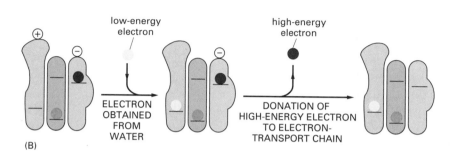

Figure 13–32 **The harvesting of light energy by a reaction center chlorophyll molecule.** (A) The initial events in a reaction center. A chlorophyll molecule of the special pair (*green*) is tightly held in a pigment-protein complex, positioned so that both a potential low-energy electron donor (*gray*) and a potential high-energy electron acceptor (*blue*) are available. As soon as the *red* electron in the chlorophyll is excited by light energy, it is passed to the electron acceptor and stabilized as a high-energy electron. The positively charged chlorophyll molecule quickly attracts the low-energy *orange* electron and returns to its resting state. These reactions require less than 10^{-6} seconds to complete. (B) The final production of a high-energy electron from a low-energy electron. In this process, which follows that in (A), the entire reaction center is restored to its resting state. In the process, a high-energy electron in the thylakoid membrane has been produced from a low-energy electron obtained from water.

The *reaction center* is a transmembrane complex of proteins and organic pigments that lies at the heart of photosynthesis. It is thought to have first evolved more than 3.5 billion years ago in primitive photosynthetic bacteria. The special pair of chlorophyll molecules in the reaction center acts as an irreversible trap for an excited electron, because these chlorophylls are poised to pass a high-energy electron to a precisely positioned neighboring molecule in the same protein complex. By moving the energized electron rapidly away from the chlorophylls, the reaction center transfers it to an environment where it is much more stable. The electron is thereby suitably positioned for subsequent photochemical reactions, which need more time to complete. In the process, the chlorophyll molecule in the reaction center loses an electron and becomes positively charged. As illustrated in Figure 13–32A, the chlorophyll rapidly regains an electron from an adjacent electron donor to return to its unexcited state. Then, in slower reactions, this electron donor is regenerated with an electron removed from water, and a high-energy electron is transferred to the electron-transport chain (Figure 13–32B).

Light Energy Drives the Synthesis of ATP and NADPH

As discussed in Chapter 3, biosynthesis is, in a sense, the opposite of oxidative breakdown. In order to make its cellular components, the cell needs not only energy in the form of ATP, but also reducing power in the form of the hydrogen carrier NADPH (see Figure 3–28). Since a primary function of photosynthesis is to synthesize organic molecules from CO_2, the process has a huge requirement for both ATP and reducing power. The need for reducing power is met by making NADPH from $NADP^+$, using the energy captured from sunlight to convert low-energy electrons in water into the high-energy electrons in NADPH.

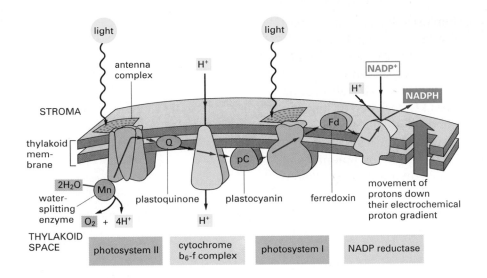

Figure 13–33 Electron flow during photosynthesis in the thylakoid membrane.
The mobile electron carriers in the chain are plastoquinone (which closely resembles the ubiquinone of mitochondria), plastocyanin (a small copper-containing protein), and ferredoxin (a small protein containing an iron-sulfur center). The cytochrome b_6-f complex closely resembles the cytochrome b-c_1 complex of mitochondria, and it is the sole site of active H^+-pumping in the chloroplast electron-transport chain. The H^+ released by water oxidation and the H^+ taken up during NADPH formation also contribute to generating the electrochemical proton gradient. The proton gradient drives an ATP synthase located in the same membrane (not shown here).

Photosynthesis in plants and cyanobacteria produces ATP and NADPH by a process requiring two photons of light. ATP is made after the first photon is absorbed, NADPH after the second. Two photosystems work in series. In outline, light energy is absorbed initially by one photosystem (for historical reasons confusingly called *photosystem II*), where it is used to produce a high-energy electron that is transferred via an electron-transport chain toward the second photosystem. While traveling down the electron-transport chain, the electron drives an H^+ pump in the thylakoid membrane and creates a proton gradient in the manner described previously for oxidative phosphorylation (Figure 13–33). An

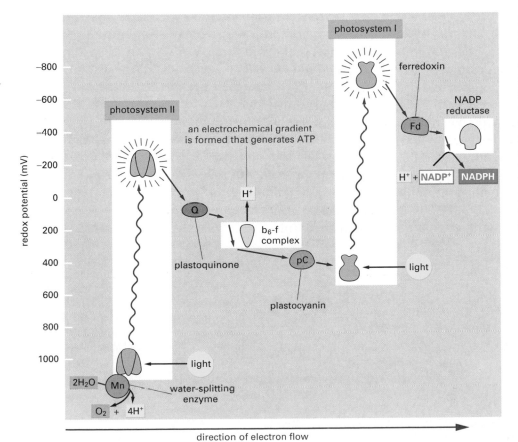

Figure 13–34 Changes in redox potential during photosynthesis.
The redox potential for each molecule is indicated by its position along the vertical axis. Photosystem II passes electrons from its excited chlorophyll through an electron-transport chain in the thylakoid membrane that leads to photosystem I. The net electron flow through the two photosystems linked in series is from water to $NADP^+$, and it produces NADPH as well as ATP. The ATP is synthesized by an ATP synthase (not shown) that harnesses the electrochemical proton gradient produced by electron transport.

Chloroplasts and Photosynthesis

ATP synthase in the thylakoid membrane then uses this proton gradient to drive the synthesis of ATP on the stromal side of the membrane.

Arriving at the second photosystem in the pathway (*photosystem I*), the electron fills a positively charged "hole" in the reaction center that has been created by an electron leaving this center due to the absorption of a second photon of light. Since photosystem I is designed to start at a higher energy level than photosystem II, it ends at a higher level and is thereby able to boost electrons to the very high energy level needed to make NADPH from NADP$^+$ (see Figure 13–33). The redox potentials of the components along this electron-transport chain are shown in Figure 13–34.

In the overall process described so far, an electron that has been removed from a chlorophyll molecule at the reaction center of photosystem II is donated to NADPH. This initial electron must be replaced to return the system to its unexcited state. The replacement electron comes from a low-energy electron donor, which, in plants and many photosynthetic bacteria, is water (see Figure 13–32B). The reaction center of photosystem II includes a water-splitting enzyme that holds the oxygen atoms of two water molecules bound to a cluster of manganese atoms in the protein (see Figures 13–33 and 13–34). This enzyme removes electrons one at a time from the water to fill the holes created by light in the chlorophyll molecules of the reaction center. As soon as four electrons have been removed from the two water molecules (which requires four photons of light), O_2 is released. It is this process, occurring over billions of years, that has generated all of the O_2 in the earth's atmosphere.

Carbon Fixation Is Catalyzed by Ribulose Bisphosphate Carboxylase

Having seen how the light reactions of photosynthesis generate ATP and NADPH, we must now consider how these compounds are used in the reactions of **carbon fixation.** When carbohydrates are oxidized to CO_2 and H_2O, a large amount of free energy is released. Clearly, the reverse reaction, in which CO_2 and H_2O combine to make carbohydrate, must therefore be energetically very unfavorable. In order to occur, it must be coupled to an energetically favorable reaction that drives it.

The central reaction of photosynthetic carbon fixation, in which an atom of inorganic carbon (as CO_2) is converted to organic carbon, is illustrated in Figure 13–35. CO_2 from the atmosphere combines with the five-carbon sugar derivative *ribulose 1,5-bisphosphate* plus water to give two molecules of the three-carbon compound 3-phosphoglycerate. This

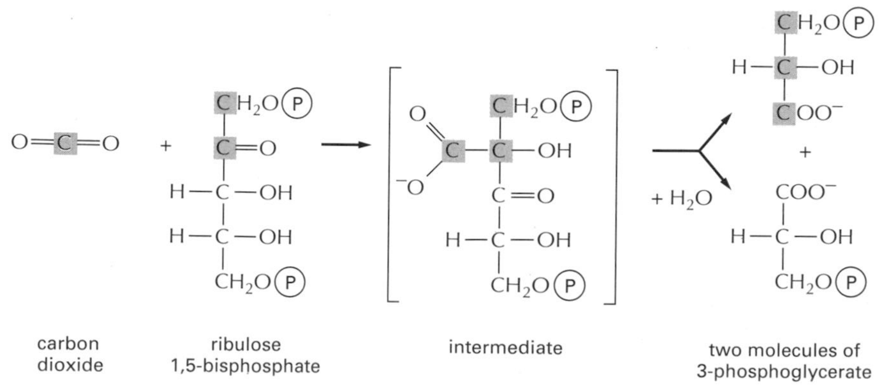

carbon dioxide   ribulose 1,5-bisphosphate   intermediate   two molecules of 3-phosphoglycerate

Figure 13–35 The initial reaction in carbon fixation. This reaction, in which carbon dioxide is converted into organic carbon, is catalyzed in the chloroplast stroma by the abundant enzyme ribulose bisphosphate carboxylase. The product is 3-phosphoglycerate.

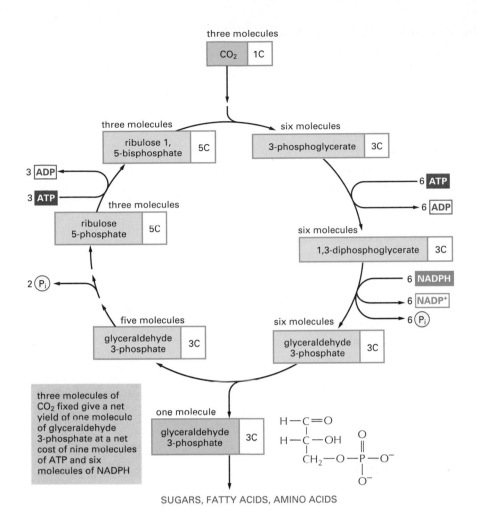

three molecules

CO_2 | 1C

three molecules

ribulose 1, 5-bisphosphate | 5C

six molecules

3-phosphoglycerate | 3C

3 ADP

3 ATP

6 ATP

6 ADP

three molecules

ribulose 5-phosphate | 5C

six molecules

1,3-diphosphoglycerate | 3C

6 NADPH

6 NADP⁺

6 P_i

2 P_i

five molecules

glyceraldehyde 3-phosphate | 3C

six molecules

glyceraldehyde 3-phosphate | 3C

three molecules of CO_2 fixed give a net yield of one molecule of glyceraldehyde 3-phosphate at a net cost of nine molecules of ATP and six molecules of NADPH

one molecule

glyceraldehyde 3-phosphate | 3C

$$H-C=O$$
$$H-C-OH$$
$$CH_2-O-P-O^-$$

SUGARS, FATTY ACIDS, AMINO ACIDS

Figure 13–36 The carbon-fixation cycle, which forms organic molecules from CO_2 and H_2O. The number of carbon atoms in each type of molecule is indicated in the *white box*. There are many intermediates between glyceraldehyde 3-phosphate and ribulose 5-phosphate, but they have been omitted here for clarity. The entry of water into the cycle is also not shown.

reaction, which was discovered in 1948, is catalyzed in the chloroplast stroma by a large enzyme called *ribulose bisphosphate carboxylase (rubisco)*. Since this enzyme works extremely sluggishly compared with most other enzymes (processing about three molecules of substrate per second compared with 1000 molecules per second for a typical enzyme), many enzyme molecules are needed. Ribulose bisphosphate carboxylase often represents more than 50% of the total chloroplast protein, and it is widely claimed to be the most abundant protein on earth.

The reaction in which CO_2 is initially fixed is energetically favorable, but only because it receives a continuous supply of the energy-rich compound ribulose 1,5-bisphosphate, to which each molecule of CO_2 is added (see Figure 13–35). The elaborate metabolic pathway by which this compound is regenerated requires both ATP and NADPH; it was worked out in one of the first successful applications of radioisotopes as tracers in biochemistry. This *carbon-fixation cycle* (or Calvin cycle) is outlined in Figure 13–36; it is a cyclic process, beginning and ending with ribulose 1,5-bisphosphate. However, for every three molecules of carbon dioxide that enter the cycle, one new molecule of *glyceraldehyde 3-phosphate* is produced—the three-carbon sugar that is the net product of the cycle. This sugar is then used for the synthesis of many other sugars and organic molecules.

In the carbon-fixation cycle, three molecules of ATP and two molecules of NADPH are consumed for each CO_2 molecule converted into carbohydrate. Thus both phosphate-bond energy (as ATP) and reducing

Chloroplasts and Photosynthesis

power (as NADPH) are required for the formation of sugar molecules from CO_2 and H_2O.

Carbon Fixation in Chloroplasts Generates Sucrose and Starch

Much of the glyceraldehyde 3-phosphate produced in chloroplasts is moved out of the chloroplast into the cytosol. Some of it enters the glycolytic pathway (see page 112), being converted to pyruvate that is then used to produce ATP by oxidative phosphorylation in plant cell mitochondria. The glyceraldehyde 3-phosphate is also converted into many other metabolites, including the disaccharide sucrose. *Sucrose* is the major form in which sugar is transported between plant cells: just as glucose is transported in the blood of animals, sucrose is exported from the leaves via the vascular bundle to provide carbohydrate to the rest of the plant.

The glyceraldehyde 3-phosphate that remains in the chloroplast is mainly converted to *starch* in the stroma. Like glycogen in animal cells, starch is a large polymer of glucose that serves as a carbohydrate reserve. The production of starch is regulated so that it is produced and stored as large grains in the chloroplast stroma during periods of excess photosynthetic capacity (see Figure 13–26B). At night, starch is broken down to sugars to help support the metabolic needs of the plant. Starch forms an important part of the diet of all animals that eat plants.

The Genetic Systems of Mitochondria and Chloroplasts Reflect Their Procaryotic Origin

The way that mitochondria divide, by growth followed by splitting in two (Figure 13–37), is evidence of their bacterial ancestry. As explained in Chapter 1, there is evidence that mitochondria arose more than a billion years ago, when an aerobic bacterium was endocytosed by a primitive anaerobic eucaryotic cell (see Figure 1–12). Since that time, most of the original bacterial genes have become transposed to the cell nucleus, leaving only relatively few genes inside the mitochondrion itself. In fact, animal mitochondria contain a uniquely simple genetic system: the human mitochondrial genome, for example, contains only 16,569 nucleotide pairs of DNA, containing 37 genes. The vast majority of mitochondrial proteins—including those needed to make the mitochondrial RNA polymerase, ribosomal proteins, and all of the enzymes of the citric acid cycle—are instead produced from nuclear genes, and these proteins must therefore be imported into the mitochondria from the cytosol, where they are made (discussed in Chapter 14).

The chloroplast also originated from an endocytosed bacterium, in this case a photosynthetic bacterium that derived its electrons from water, forming O_2 as a waste product (see Figure 1–14). As evidence of this history, the chloroplast contains many of its own genes, as well as a complete transcription and translation system that is required to produce proteins from these genes. Chloroplast genomes are considerably larger than mitochondrial genomes, and the genes are strikingly similar to the genes of the cyanobacteria from which they are thought to have been derived. Even so, like mitochondrial proteins, many chloroplast

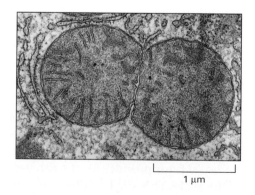

1 µm

Figure 13–37 Electron micrograph of a dividing mitochondrion. (Courtesy of Daniel S. Friend.)

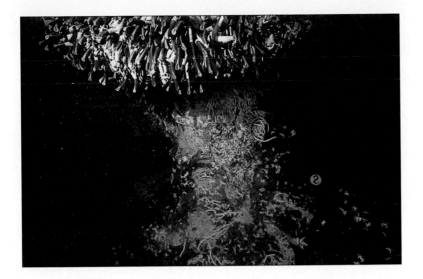

Figure 13–38 **Some organisms growing near a hydrothermal vent on the deep ocean floor.** These organisms were unknown before deep submersible vehicles reached these sites. (Courtesy of Woods Hole Oceanographic Institution.)

proteins are now encoded by nuclear genes and must be imported from the cytosol.

The earth is filled with millions of different types of organisms. Everywhere one looks there is life. Most of these organisms we have never seen—from the many strange life-forms that thrive in the hot water that streams up out of deep ocean hydrothermal vents (Figure 13–38), to the myriad single-celled microorganisms that thrive in the soil beneath our feet. The methods of recombinant DNA technology described in Chapter 10 now give us a powerful new lens to view this living world. Through this lens, we are discovering many new single-celled organisms that are helping us to make sense of the vast diversity of living things. We therefore end this chapter with a discussion of the nature of some of our single-celled ancestors, emphasizing their manner of obtaining energy.

Our Single-celled Ancestors

Our present understanding of the history of life on earth was briefly outlined in Figure 13–2. Microfossils have provided solid evidence for the presence of single-celled organisms 3.5 billion years ago that closely resemble present-day procaryotes. But fossils cannot tell us much about the chemistry of ancient cells, especially since we know that the superficial similarities of many procaryotes reveal very little about their vast biochemical diversity. The best clues we have for making sense of life's history are instead found inside present-day cells—in the form of the many molecules from which they are made. Like the fingerprints left at the scene of a crime, these molecules provide powerful evidence that is allowing us to trace the history of ancient events.

RNA Sequences Reveal Evolutionary History

The ribosomal RNAs (rRNAs) are found in all living cells. Comparisons of the nucleotide sequence of rRNAs in different organisms can be used to trace the evolutionary relatedness of organisms far back into time. The relatedness between organisms determined in this way is represented as

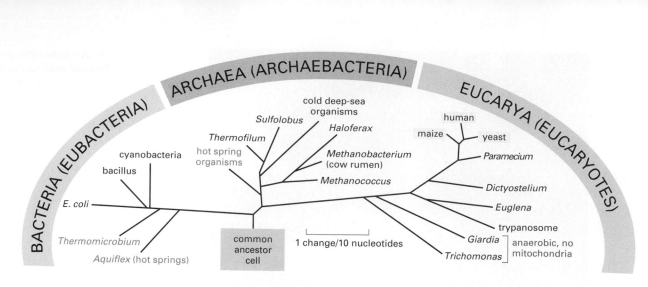

Figure 13–39 **Phylogenetic tree derived from ribosomal rRNA sequences.** This tree shows the relationships of a few selected organisms from the three proposed domains of living organisms: the Bacteria, the Archaea, and the Eucarya.

a *phylogenetic tree*, in which the total length of the line that connects any two organisms is proportional to the number of differences in the sequence of nucleotides in their two rRNAs—the more differences, the longer the line and the more distant the relationship. Figure 13–39 shows a phylogenetic tree constructed from rRNA sequences that indicates the positions of a few selected organisms in the whole domain of living things. The consistency of these trees with other biochemical data gives us confidence that the evolutionary relatedness of organisms can be reliably measured in this way.

The data now available from studies of molecular evolution has resulted in a major reappraisal of our view of the living world. On the basis of both cell biological data and molecular studies, the living world can be classified into three domains: two large and very diverse domains of procaryotes, the Archaea and the Bacteria, and a domain of eucaryotes, the Eucarya. We find that the organisms that we know best—the animals, plants, and fungi—occupy only one small corner of the universe of living things (see *yellow shading* in Figure 13–39).

Molecular studies also tell us that each of these three divisions of living organisms is very ancient—in particular, that the eucaryote lineage is much older than had previously been thought. The long line in the phylogenetic tree diagram that connects the Eucarya to a common ancestor with Archaea and Bacteria suggests that the eucaryotic lineage goes back more than 3 billion years. For the vast majority of this time, eucaryotes are thought to have been single-celled organisms that lived anaerobically, deriving all of their energy from fermentation pathways. In agreement with this idea, *Giardia* and *Trichomonas* lack mitochondria (see Figures 13–29 and 1–28). Only after they had acquired the bacterial symbionts that became mitochondria (and chloroplasts in the case of algae and plants) did eucaryotic cells embark on the amazing pathway of evolution that eventually led to complex multicellular organisms.

Ancient Cells Probably Arose in Hot Environments

The widespread use of recombinant DNA techniques to survey the organisms on the earth has led to the startling realization that the world is filled with many more single-celled organisms than we had ever suspected. Techniques such as DNA cloning and PCR analysis (see Chapter 10) can now be used to characterize all the different microorganisms that live in a particular habitat—for example, a sulfurous hot spring or a plot of soil. When compared with the standard microbiological techniques, which detect only those microbes that can survive and multiply in laboratory culture, the recombinant DNA studies tell us that fewer than one microbial species in a hundred could have been detected previously. The top six inches of soil alone is estimated to contain several tons of single-celled microorganisms per acre. It is therefore clear that a huge variety of new microbial species remain to be discovered, and that at present we have sampled only a tiny portion of the entire living world.

The species that are most closely related to the proposed common ancestor cell for the three domains in Figure 13–39 are the organisms in both Archaea and Bacteria that are denoted with *green lettering*. These organisms share the common feature of living at high temperatures (75°C to 95°C, close to the temperature of boiling water), suggesting that the common ancestor cell lived under very hot, anaerobic conditions.

The conditions today that most resemble those under which cells are thought to have lived 3.5–3.8 billion years ago may be those near deep-ocean hydrothermal vents. These vents represent places where the earth's molten mantle is breaking through the crust, expanding the width of the ocean floor (see Figure 13–38). We shall end this chapter with a brief discussion of one organism that lives in this environment. Its mode of existence gives us a hint of how early cells might have used electron transport to derive their energy and their carbon molecules from inorganic materials that were freely available on the hot early earth.

Methanococcus Lives in the Dark, Using Only Inorganic Materials as Food

The human genome, the collection of the DNA in all of our chromosomes, contains 3×10^9 nucleotide pairs. A large international project that aims to decipher this entire DNA sequence is well underway, with an anticipated completion date of 2005. But the technology developed for this project has made it possible to complete the much easier goal of sequencing the entire genome of many less complex "model organisms," with which our own DNA sequence can be compared. Thus far, a complete genome sequence is available for one eucaryote, the yeast *Saccharomyces cerevisiae*, whose genome of 12 million nucleotide pairs is 250 times smaller than ours (see Figure 10–6).

The yeast genome contains between 6000 and 7000 genes compared to our 60,000. Several procaryotic genomes have also been completely sequenced. These include the genome of the simplest cell known, the tiny bacterium *Mycoplasma genitalium*, with only about 500 genes, as well as the genome of *Methanococcus jannaschii*. *Methanococcus* contains about 1800 genes, and it was originally isolated from a hydrothermal vent more than a mile beneath the ocean surface.

Methanococcus is one of the Archaea that lives at high temperatures. It grows entirely on inorganic nutrients in the complete absence of light and gaseous oxygen, utilizing hydrogen gas (H_2), CO_2, and nitrogen gas (N_2) that bubble up from the vent. It reduces N_2 to ammonia (NH_3) by the addition of hydrogen, which makes it possible for the nitrogen to be incorporated into organic molecules. This process of **nitrogen fixation** requires a large amount of energy, as does the carbon-fixation process that the bacterium needs to convert CO_2 into sugars. Much of the energy required for both processes is derived from the transfer of

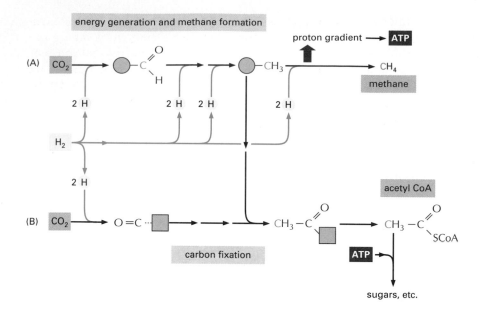

Figure 13–40 The pathways of energy generation and carbon fixation in *Methanococcus*. This deep-sea bacterium uses hydrogen gas (H_2) as the source of reducing power in both of the pathways shown. (A) Energy generation. The production of methane (CH_4) from CO_2 occurs in several stages. The initial reduction steps take place via enzyme-catalyzed reactions in the cytoplasm. In contrast, the final reduction step involves a membrane-based electron transfer that generates a proton gradient that drives ATP synthesis, while producing methane as a waste product. (B) Carbon fixation. The main pathway for the fixation of carbon dioxide results in the production of acetyl CoA. This is the source for the sugars, fatty acids, and nucleotides that the cell needs for biosynthesis. The *green* circles and squares in the diagram represent a series of coenzymes to which the indicated metabolic intermediates are bound.

electrons from H_2 to CO_2, with the release of large amounts of methane (CH_4) as a waste product (thus producing natural gas and giving the bacterium its name). Part of this electron transfer occurs in the bacterial membrane and results in protons (H^+) being pumped across it. The resulting electrochemical proton gradient drives an ATP synthase in the same membrane to make ATP (Figure 13–40A). Finding chemiosmotic coupling in an organism such as *Methanococcus* suggests that the storage of energy derived from electron transport in an H^+ gradient is an extremely ancient process.

The mechanism used by *Methanococcus* to fix carbon is completely different from the carbon-fixation pathway in plants, algae, and cyanobacteria discussed previously. As indicated in Figure 13–40B, in addition to being the source of the high-energy electrons for the membrane-based process that generates ATP, hydrogen gas (H_2) is used to reduce CO_2 to an enzyme-bound molecule of carbon monoxide (CO). The CO then reacts with a methyl group, produced as an intermediate in the process of methanogenesis, to form acetyl CoA. The acetyl CoA is then converted to sugars, amino acids, nucleotides, and many other small and large molecules through the familiar enzyme-catalyzed pathways that require ATP.

Methanococcus is by no means a simple organism; for example, among its 1800 genes are more than 60 that encode enzymes that function in the pathway from CO_2 to CH_4 alone. The earliest cells that contained proteins must have been much simpler; as a guess, they might have required less than a total of 100 genes. Many more genomes of diverse single-celled organisms are being sequenced, giving us ever more powerful tools for reconstructing the past. But major conceptual challenges remain. For example, we presently have no convincing way of describing how RNA-based cells, which are postulated to have existed very early in cellular evolution (see Figure 7–39), might have evolved to exploit the energy-yielding processes of fermentation or of membrane-based electron transport—the mechanisms that power all of the cells that we know about today.

Essential Concepts

- Mitochondria, chloroplasts, and many bacteria produce ATP by a membrane-based mechanism known as chemiosmotic coupling.

- Mitochondria produce most of an animal cell's ATP, using energy derived from oxidation of sugars and fatty acids.

- Mitochondria are bounded by two concentric membranes, the innermost of which encloses the mitochondrial matrix. The matrix space contains many enzymes, including those of the citric acid cycle. These enzymes produce large amounts of NADH and $FADH_2$ from the oxidation of acetyl CoA.

- In the inner mitochondrial membrane, high-energy electrons donated by NADH and $FADH_2$ pass along an electron-transport chain (the respiratory chain), eventually combining with molecular oxygen (O_2) in an energetically favorable reaction.

- Some of the energy released by electron transfers along the respiratory chain is harnessed to pump H^+ out of the matrix, thereby creating a transmembrane electrochemical proton (H^+) gradient. The proton pumping is carried out by three large respiratory enzyme complexes embedded in the membrane.

- The resulting electrochemical proton gradient across the inner mitochondrial membrane is then harnessed to make ATP by the flow of H^+ back into the matrix through ATP synthase, an enzyme located in the inner mitochondrial membrane.

- The electrochemical proton gradient also drives the active transport of metabolites into and out of the mitochondrion.

- In photosynthesis in chloroplasts and photosynthetic bacteria, high-energy electrons are generated when sunlight is absorbed by chlorophyll; this energy is captured by protein complexes known as photosystems, which are located in the thylakoid membranes of chloroplasts.

- Electron-transport chains associated with the photosystems transfer electrons from water to $NADP^+$ to form NADPH, with the concomitant production of an electrochemical proton gradient across the thylakoid membrane. Molecular oxygen (O_2) is generated as a by-product.

- As in mitochondria, the proton gradient across the thylakoid membrane is used by an ATP synthase embedded in the membrane to generate ATP.

- The ATP and the NADPH made by photosynthesis are used within the chloroplast to drive the carbon-fixation cycle in the chloroplast stroma, thereby producing carbohydrate from CO_2.

- This carbohydrate is exported to the cell cytosol, where it is metabolized to provide organic carbon, ATP (via mitochondria), and reducing power to the rest of the cell.

- Chemiosmotic coupling mechanisms are very widespread and of ancient origin. Both mitochondria and chloroplasts are thought to have evolved from bacteria that were endocytosed by primitive eucaryotic cells.

- Bacteria that live in environments very similar to those thought to have been present on the early earth also use chemiosmotic coupling to produce ATP.

Key Terms

ATP synthase	chloroplast	mitochondrion	quinone
carbon fixation	cytochrome	nitrogen fixation	redox pair
chemiosmotic coupling	cytochrome oxidase	oxidative phosphorylation	redox potential
chlorophyll	electron-transport chain	photosynthesis	redox reaction
	iron-sulfur center	photosystem	stroma
	matrix		

Questions

Question 13–11   Which of the following statements are correct? Explain your answers.

A. Many, but not all, electron-transfer reactions involve metal ions.

B. The electron-transport chain generates an electrical potential across the membrane because it moves electrons from the intermembrane space into the matrix.

C. The electrochemical proton gradient consists of two components: a pH difference and an electrical potential.

D. Ubiquinone and cytochrome c are both diffusible electron carriers.

E. Plants have chloroplasts and therefore can live without mitochondria.

F. Both chlorophyll and heme contain an extensive system of double bonds that allows them to absorb visible light.

G. The role of chlorophyll in photosynthesis is equivalent to that of heme in mitochondrial electron transport.

H. Most of the dry weight of a tree comes from the minerals that are taken up by the roots.

Question 13–12   A single proton moving down the electrochemical gradient into the mitochondrial matrix space liberates 4.6 kcal/mole of free energy. How many protons have to flow across the inner mitochondrial membrane to synthesize one molecule of ATP if the ΔG for ATP synthesis under intracellular conditions is between 11 and 13 kcal/mole? (ΔG is discussed in Chapter 3, p. 89.) Why is a range given for this latter value, and not a precise number? Under which conditions would the lower value apply?

Question 13–13   In the following statement, choose the correct one of the alternatives in italics and justify your answer. "If no O_2 is available, all components of the mitochondrial electron-transport chain will accumulate in their *reduced/oxidized* form. If O_2 is suddenly added again, the electron carriers in cytochrome oxidase will become *reduced/oxidized before/after* those in NADH dehydrogenase."

Question 13–14   Suppose that a large amount of ATP begins to be hydrolyzed by ATP-requiring reactions in the cytosol, causing the ATP:ADP ratio in the mitochondrial matrix to drop and to remain unusually low. How will this affect the electrochemical proton gradient across the inner mitochondrial membrane?

Question 13–15   Assume that the conversion of oxidized ubiquinone to reduced ubiquinone by NADH dehydrogenase occurs on the matrix side of the inner mitochondrial membrane and that its oxidation by cytochrome b-c₁ occurs on the intermembrane space side of the membrane (see Figure 13–10 and 13–20). What are the consequences of this arrangement for the generation of the H⁺ gradient across the membrane?

Question 13–16   If a voltage is applied to two platinum wires (electrodes) immersed in water, then water molecules become split into H_2 and O_2 gas. At the negative electrode, electrons are donated and H_2 gas is released; at the positive electrode, electrons are accepted and O_2 gas is produced. When photosynthetic bacteria and plant cells split water, they produce O_2, but no H_2. Why?

Question 13–17   In an insightful experiment performed in the 1960s, chloroplasts were first soaked in an acidic solution at pH 4, so that the stroma and thylakoid space became acidified (Figure Q13–17). They were then transferred to a basic solution (pH 8). This quickly increased the pH of the stroma to 8, while the thylakoid space temporarily remained at pH 4. A burst of ATP synthesis was observed, and the pH difference between the thylakoid and the stroma then disappeared.

A. Explain why these conditions lead to ATP synthesis.

B. Is light needed for the experiment to work?

C. What would happen if the solutions were switched so that the first incubation is in the pH 8 solution and the second one in the pH 4 solution?

D. Does the experiment support or question the chemiosmotic model?

Explain your answers.

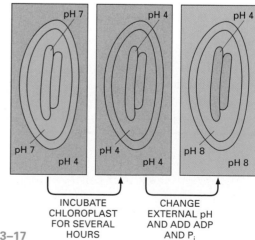

Figure Q13–17

INCUBATE CHLOROPLAST FOR SEVERAL HOURS

CHANGE EXTERNAL pH AND ADD ADP AND P$_i$

Question 13–18 As your first experiment in the laboratory, your advisor asks you to reconstitute purified bacteriorhodopsin, a light-driven H^+ pump from the plasma membrane of photosynthetic bacteria, and purified ATP synthase from ox heart mitochondria together into the same membrane vesicles—as shown in the Figure Q13–18. You are then to add ADP and P_i to the external medium and shine light into the suspension of vesicles.

A. What do you observe?

B. What do you observe if not all the detergent is removed, and the vesicle membrane therefore remains leaky to ions?

C. You tell a friend over dinner about your new experiments, and he questions the validity of an approach that utilizes components from so widely divergent, unrelated organisms: "Why would anybody want to mix vanilla pudding with brake fluid?" Defend your approach against his critique.

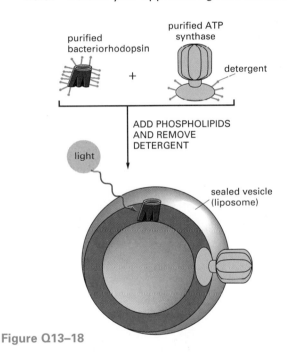

Figure Q13–18

Question 13–19 $FADH_2$ is produced in the citric acid cycle by a membrane-embedded enzyme complex, called succinate dehydrogenase, that contains bound FAD and carries out the reactions:

$$succinate + FAD \rightarrow fumarate + FADH_2$$

and

$$FADH_2 \rightarrow FAD + 2H^+ + 2e^-$$

The redox potential of $FADH_2$, however, is only –220 mV. Referring to Panel 13–1 (p. 424) and Figure 13–21, suggest a plausible mechanism by which its electrons could be fed into the electron-transport chain. Draw a diagram to illustrate your proposed mechanism.

Question 13–20 Some bacteria have become specialized to live in an environment of high pH (pH ~10). Do you suppose that these bacteria use a proton gradient across their plasma membrane to produce their ATP? (Hint: all cells must maintain their cytoplasm at a pH close to neutrality.)

Question 13–21 Figure Q13–21 summarizes the circuitry used by mitochondria and chloroplasts to interconvert different forms of energy. Is it accurate to say

A. that the products of chloroplasts are the substrates for mitochondria?

B. that the activation of electrons by the photosystems enables chloroplasts to drive electron transfer from H_2O to carbohydrate, which is opposite to the direction of electron transfer in the mitochondrion?

C. that the citric acid cycle is the reverse of the normal carbon-fixation cycle?

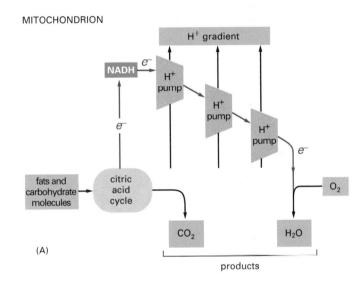

(A)

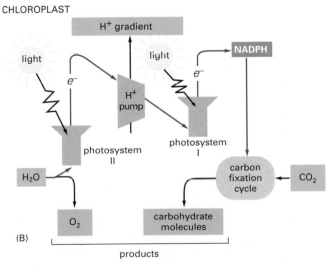

(B)

Figure Q13–21

Question 13–22   A manuscript has been submitted for publication to a prestigious scientific journal. In the paper the authors describe an experiment in which they have succeeded in trapping an individual ATP synthase molecule and then mechanically rotating its head by applying a force to it. The authors show that upon rotating the head of the ATP synthase, ATP is produced, in the absence of an H^+ gradient. What might this mean about the mechanism whereby ATP synthase functions? Should this manuscript be considered for publication in one of the best journals?

Question 13–23   You mix the following components in a solution. Assuming that the electrons must follow the path specified in Figure 13–10, in which experiments would you expect a net transfer of electrons to cytochrome c? Discuss why no electron transfer occurs in the other experiments.

A. reduced ubiquinone and oxidized cytochrome c

B. oxidized ubiquinone and oxidized cytochrome c

C. reduced ubiquinone and reduced cytochrome c

D. oxidized ubiquinone and reduced cytochrome c

E. reduced ubiquinone, oxidized cytochrome c, and cytochrome b-c_1 complex

F. oxidized ubiquinone, oxidized cytochrome c, and cytochrome b-c_1 complex

G. reduced ubiquinone, reduced cytochrome c, and cytochrome b-c_1 complex

H. oxidized ubiquinone, reduced cytochrome c, and cytochrome b-c_1 complex

14 Intracellular Compartments and Transport

At any one time, a typical eucaryotic cell is carrying out thousands of different chemical reactions, many of which are mutually incompatible. One series of reactions makes glucose, for example, while another breaks down glucose; some enzymes synthesize peptide bonds, whereas others hydrolyze them, and so on. Indeed, if the cells of an organ such as the liver are broken apart and their contents mixed together in a test tube, chemical chaos results, and the cells' enzymes and other proteins are quickly degraded by their own proteolytic enzymes. For the cell to operate effectively, the many different intracellular processes occurring simultaneously must somehow be segregated.

Cells have evolved several strategies for segregating and organizing their chemical reactions. One strategy used by both procaryotic and eucaryotic cells is to aggregate the different enzymes required to catalyze a particular sequence of reactions into a single large protein complex. As discussed in other chapters, such multiprotein complexes are used, for example, in the synthesis of DNA, RNA, and proteins. A second strategy, which is most highly developed in eucaryotic cells, is to confine different metabolic processes, and the proteins required to carry them out, within different membrane-bounded compartments. As discussed in Chapters 11 and 12, cell membranes provide selectively permeable barriers through which the transport of most molecules can be controlled. In this chapter we consider the compartmentalization strategy and some of its consequences.

In the first section we describe the principal membrane-bounded compartments, or *membrane-bounded organelles,* of eucaryotic cells and briefly consider their main functions and how they may have evolved. In the second section we discuss how the protein composition of the different compartments is set up and maintained. Each compartment contains a unique set of proteins, which have to be transferred selectively from the cytosol, where they are made, to the compartment in which they are used. This transfer process, called *protein sorting,* depends on signals built into the amino acid sequence of the proteins. In the third section we describe how certain of the membrane-bounded compartments of a eucaryotic cell communicate with one another by forming small membranous sacs, or *vesicles,* that pinch off from one compartment, move through the cytosol, and fuse with another compartment in a process called *vesicular transport.* In the last two sections we discuss how this constant vesicular traffic also provides the main routes for releasing proteins from the cell by the process of *exocytosis* and for taking them up by the process of *endocytosis.*

Membrane-bounded Organelles

Whereas a procaryotic cell consists of a single compartment, the **cytosol,** enclosed by the plasma membrane, a eucaryotic cell is elaborately subdivided by internal membranes. These membranes create enclosed compartments where sets of enzymes can operate without interference from reactions occurring in other compartments. When a cross section through a plant or an animal cell is examined in the electron microscope, numerous small membrane-bounded sacs, tubes, spheres, and irregularly shaped structures can be seen, often arranged without much apparent order (Figure 14–1). These structures are all distinct membrane-bounded organelles, or parts of such organelles, each of which contains a unique set of large and small molecules.

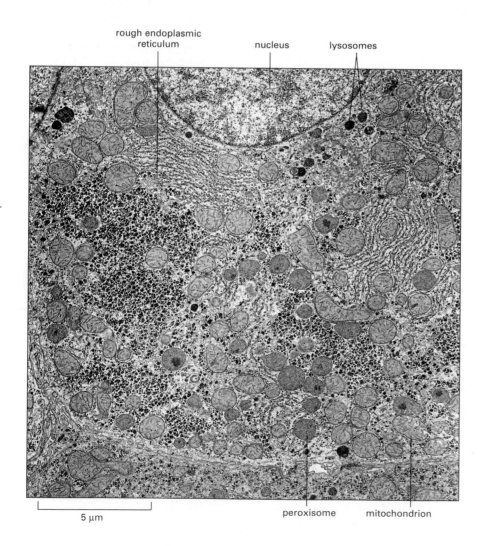

Figure 14–1 **Electron micrograph of part of a liver cell seen in cross-section.** Examples of many of the major membrane-bounded organelles can be seen. The small black granules seen between the membrane-bounded compartments are glycogen granules (glycosomes), which are aggregates of glycogen and the enzymes that control its synthesis and breakdown. (Courtesy of Daniel S. Friend.)

Eucaryotic Cells Contain a Basic Set of Membrane-bounded Organelles

The major **membrane-bounded organelles** of an animal cell are illustrated in Figure 14–2, and their functions are summarized in Table 14–1. They are surrounded by the cytosol, which is bounded by the plasma membrane. The *nucleus* is the most prominent organelle in all cells. It is surrounded by a double membrane, known as the *nuclear envelope,* and communicates with the cytosol via *nuclear pores* that perforate the envelope. The outer nuclear membrane is continuous with the membrane of the *endoplasmic reticulum (ER),* which is a continuous system of interconnected sacs and tubes of membrane, which often extends throughout most of the cell. The ER is the major site of new membrane synthesis in the cell. Large areas of it have ribosomes attached to the cytosolic surface and are therefore called *rough endoplasmic reticulum.* The ribosomes are actively engaged in synthesizing proteins that are delivered into the ER lumen or ER membrane. The *smooth ER* is scanty in most cells but is highly developed in others to perform particular functions: it is the site of steroid hormone synthesis in cells of the adrenal gland, for example, and the site where a variety of organic molecules, including alcohol, are detoxified in liver cells.

The *Golgi apparatus,* which is usually situated near the nucleus, receives proteins and lipids from the ER, modifies them, and then dispatches them to other destinations in the cell. *Lysosomes* are small sacs of digestive enzymes. They degrade worn-out organelles, as well as macromolecules and particles taken into the cell by endocytosis. On their way to lysosomes, endocytosed materials must first pass through a series of compartments called *endosomes,* which sort out some of the ingested molecules and recycle them back to the plasma membrane. *Peroxisomes* are small organelles bounded by a single membrane. They contain enzymes used in a variety of oxidative reactions that break down lipids and destroy toxic molecules. *Mitochondria* and (in plant cells) *chloroplasts* are each surrounded by a double membrane and are the

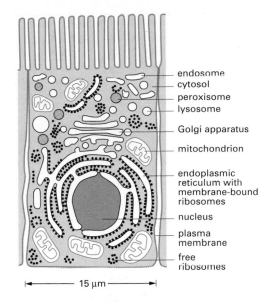

endosome
cytosol
peroxisome
lysosome
Golgi apparatus
mitochondrion
endoplasmic reticulum with membrane-bound ribosomes
nucleus
plasma membrane
free ribosomes

|← 15 μm →|

Figure 14–2 The major membrane-bounded organelles of an animal cell. This cell from the lining of the intestine contains the basic set of organelles found in most animal cells. The nucleus, endoplasmic reticulum (ER), Golgi apparatus, lysosomes, endosomes, mitochondria, and peroxisomes are distinct compartments separated from the cytosol (*gray*) by at least one selectively permeable membrane. Ribosomes are also shown, even though they are not enclosed by a membrane and are too small to be seen in a light microscope and therefore do not fit the original definition of an organelle. Some ribosomes are found free in the cytosol, while others are bound to the cytosolic surface of the ER.

Table 14–1	The Main Functions of the Membrane-bounded Compartments of a Eucaryotic Cell
Compartment	**Main Function**
Cytosol	contains many metabolic pathways (Chapters 3 and 4); protein synthesis (Chapter 7)
Nucleus	contains main genome (Chapter 8); DNA and RNA synthesis (Chapters 6 and 7)
Endoplasmic reticulum (ER)	synthesis of most lipids (Chapter 11); synthesis of proteins for distribution to many organelles and to the plasma membrane (this chapter)
Golgi apparatus	modification, sorting, and packaging of proteins and lipids for either secretion or delivery to another organelle (this chapter)
Lysosomes	intracellular degradation (this chapter)
Endosomes	sorting of endocytosed material (this chapter)
Mitochondria	ATP synthesis by oxidative phosphorylation (Chapter 13)
Chloroplasts (in plant cells)	ATP synthesis and carbon fixation by photosynthesis (Chapter 13)
Peroxisomes	oxidation of toxic molecules

Table 14–2	The Relative Volumes Occupied by the Major Membrane-bounded Organelles in a Liver Cell (Hepatocyte)	
Intracellular Compartment	Percent of Total Cell Volume	Approximate Number per Cell
Cytosol	54	1
Mitochondria	22	1700
Endoplasmic reticulum	12	1
Nucleus	6	1
Golgi apparatus	3	1
Peroxisomes	1	400
Lysosomes	1	300
Endosomes	1	200

sites of oxidative phosphorylation and photosynthesis, respectively (discussed in Chapter 13); they both contain membranes that are highly specialized for the production of ATP.

Many of the membrane-bounded organelles, including the ER, Golgi apparatus, mitochondria, and chloroplasts are held in their relative locations in the cell by attachment to the cytoskeleton, especially to microtubules. Cytoskeletal filaments provide tracks for moving the organelles around and for directing the traffic of vesicles between them. These movements are driven by motor proteins that use the energy of ATP hydrolysis to propel the organelles and vesicles along the filaments, as discussed in Chapter 16.

On average, the membrane-bounded organelles together occupy nearly half the volume of a eucaryotic cell (Table 14–2), and the total amount of membrane associated with them is enormous: in a typical mammalian cell, for example, the area of the endoplasmic reticulum membrane is 20–30 times greater than that of the plasma membrane. In terms of its area and mass, the plasma membrane is only a minor membrane in most eucaryotic cells.

Much can be learned about the composition and function of an organelle once it has been isolated from other cell structures. For the most part, organelles are far too small to be isolated by hand, but it is possible to separate one type of organelle from another by differential *centrifugation* (described in Panel 5–4, pp. 160–161). Once a purified sample of one type of organelle has been obtained, the organelle's proteins can be identified. In many cases the organelle itself can be incubated in a test tube under conditions that allow its functions to be studied. Isolated mitochondria, for example, can produce ATP from the oxidation of pyruvate to CO_2 and water, provided they are adequately supplied with ADP and O_2.

Membrane-bounded Organelles Evolved in Different Ways

In trying to understand the relationships between the different compartments of a modern eucaryotic cell, it is helpful to consider how they might have evolved. The compartments probably evolved in stages. The precursors of the first eucaryotic cells are thought to have been simple microorganisms, resembling bacteria, which had a plasma membrane but no internal membranes. The plasma membrane in such cells would

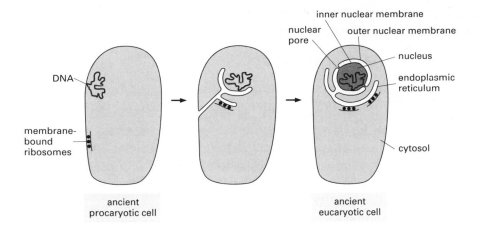

Figure 14–3 **A possible pathway for the evolution of the nuclear membranes and the ER.** In bacteria the single DNA molecule is typically attached to the plasma membrane. It is possible that in a very ancient procaryotic cell, the plasma membrane, with its attached DNA, could have invaginated and eventually formed a two-layered envelope of membrane completely surrounding the DNA. This envelope is presumed to have eventually pinched off completely from the plasma membrane, producing a nuclear compartment surrounded by a double membrane. This nuclear envelope is penetrated by channels called nuclear pores, which enable it to communicate directly with the cytosol. Other portions of the same membrane formed the ER, to which some of the ribosomes became attached. This hypothetical scheme would explain why the space between the inner and outer nuclear membranes is continuous with the lumen of the ER.

have provided all membrane-dependent functions, including ATP synthesis and lipid synthesis, as does the plasma membrane in most modern bacteria. Bacteria can get by with this arrangement because of their small size and thus their high surface-to-volume ratio; their plasma membrane is sufficient to sustain all the vital functions for which membranes are required. Present-day eucaryotic cells, however, have volumes 1000 to 10,000 times greater than that of a typical bacterium such as *E. coli*. Such a large cell has a small surface-to-volume ratio, and presumably could not survive with a plasma membrane as its only membrane. Thus the increase in size typical of eucaryotic cells probably could not have occurred without the development of internal membranes.

Membrane-bounded organelles are thought to have arisen in evolution in at least two ways. The nuclear membranes and the membranes of the ER, Golgi apparatus, endosomes, and lysosomes are believed to have originated by the invagination of the plasma membrane (Figure 14–3). These membranes, and the organelles they enclose, are all part of what is collectively called the *endomembrane system*. As we discuss later, the interiors of these organelles (with the exception of the nucleus) communicate extensively with one another and with the outside of the cell by means of small vesicles that bud off from one of these organelles and fuse with another. Consistent with this proposed evolutionary origin, the interiors of these organelles are treated by the cell in many ways as "extracellular," as we shall see. The hypothetical scheme shown in Figure 14–3 would also explain why the nucleus is surrounded by two membranes. Although membrane invagination is rare in present-day

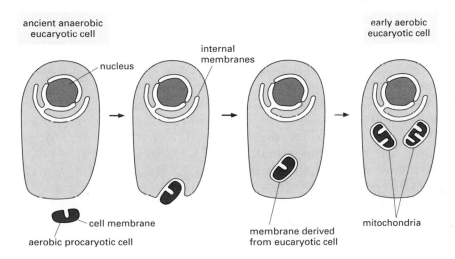

Figure 14–4 **The suspected evolutionary origin of mitochondria.** Mitochondria are thought to have originated when an aerobic procaryote was engulfed by a larger anaerobic eucaryotic cell. Chloroplasts are thought to have originated later in a similar way, when an aerobic eucaryotic cell engulfed a photosynthetic procaryote. This would explain why these organelles have two membranes and why they do not participate in the vesicular traffic that connects many other intracellular compartments.

Membrane-bounded Organelles

bacteria, it does occur in some photosynthetic bacteria in which the regions of the plasma membrane containing the photosynthetic apparatus are internalized, forming intracellular vesicles.

Mitochondria and chloroplasts are thought to have originated in a different way. They differ from all other organelles in that they possess their own small DNA genomes and can make some of their own proteins, as discussed in Chapter 13. The similarity of these genomes to those of bacteria and the close resemblance of some of their proteins to bacterial proteins strongly suggest that mitochondria and chloroplasts evolved from bacteria that were engulfed by primitive eucaryotic cells with which they initially lived in symbiosis. This would also explain why these organelles are enclosed by two membranes (Figure 14–4). As might be expected from their origins, mitochondria and chloroplasts remain isolated from the extensive vesicular traffic that connects the interiors of most of the other membrane-bounded organelles to one another and to the outside of the cell.

Having briefly reviewed the main membrane-bounded organelles of the eucaryotic cell, we turn now to the question of how each organelle acquires its unique set of proteins.

Question 14–1 As shown in the drawings in Figure 14–3, the lipid bilayer of the inner and outer nuclear membranes forms a continuous sheet, being joined around the nuclear pores. As membranes are two-dimensional fluids, this would imply that membrane proteins can diffuse freely between the two nuclear membranes. Yet each of these two nuclear membranes has a different protein composition, reflecting different functions. How could you reconcile this apparent paradox?

Protein Sorting

Before a eucaryotic cell reproduces by dividing in two, it has to duplicate its membrane-bounded organelles. A cell cannot make these organelles from scratch: it requires information in the organelles itself. Thus most of the organelles are formed from preexisting organelles, which grow and then divide. As cells grow during the cell-division cycle (discussed in Chapter 17), membrane-bounded organelles are enlarged by incorporation of new molecules; the organelles then divide and at cell division are distributed between the two daughter cells. The nuclear envelope, ER, and Golgi apparatus break up into small vesicles, which then coalesce again as the two daughter cells are formed (discussed in Chapter 17). Organelle growth requires a supply of new lipids to make more membrane and a supply of the appropriate proteins—both membrane proteins and soluble proteins that will occupy the interior of the organelle. Even in cells that are not dividing, proteins must be continuously and accurately delivered to organelles—some for eventual secretion from the cell and some to replace organelle proteins that have been degraded. Therefore the problem of how to make and maintain membrane-bounded organelles is largely one of how to direct newly made proteins to their correct organelle.

For some organelles, including the mitochondria, chloroplasts, peroxisomes, and the interior of the nucleus, proteins are delivered directly from the cytosol. For others, including the Golgi apparatus, lysosomes, endosomes, and the nuclear membranes, proteins and lipids are delivered indirectly via the ER, which is itself a major site of lipid and protein synthesis. Proteins enter the ER directly from the cytosol: some are retained there, but most are transported onward by vesicular transport to the Golgi apparatus and then onward to other organelles or the plasma membrane.

In this section we discuss the mechanisms by which proteins directly enter membrane-bounded organelles from the cytosol, leaving a detailed discussion of vesicular transport to later. Proteins made in the cytosol are dispatched to different locations in the cell according to specific address labels that they contain in their amino acid sequence. Once at the correct address, the protein enters the organelle.

Proteins Are Imported into Organelles by Three Mechanisms

The synthesis of virtually all proteins in the cell begins on ribosomes in the cytosol. The exceptions are the few mitochondrial and chloroplast proteins that are synthesized on ribosomes inside these organelles; most mitochondrial and chloroplast proteins, however, are made in the cytosol and subsequently imported. The fate of a protein molecule synthesized in the cytosol depends on its amino acid sequence, which can contain a *sorting signal* that directs the protein to the organelle in which it is required. Proteins that lack such signals remain as permanent residents in the cytosol; those that possess a sorting signal move from the cytosol to the appropriate organelle. Different sorting signals direct proteins into the nucleus, mitochondria, chloroplasts (in plants), peroxisomes, and the ER.

A general problem for a membrane-bounded organelle when it imports proteins from the cytosol or from another organelle is how to transport the protein across membranes that are normally impermeable to hydrophilic macromolecules. This is accomplished in different ways for different organelles, all requiring an input of energy.

1. Proteins moving from the cytosol into the nucleus are transported through the nuclear pores that penetrate the inner and outer nuclear membranes; the pores function as selective gates, which actively transport specific macromolecules but also allow free diffusion of smaller molecules (mechanism 1 in Figure 14–5).

2. Proteins moving from the cytosol into the ER, mitochondria, chloroplasts, or peroxisomes are transported across the organelle membrane by *protein translocators* located in the membrane; unlike transport through nuclear pores, the transported protein molecule usually must unfold in order to snake through the membrane (mechanism 2 in Figure 14–5). Bacteria have similar protein translocators in their plasma membrane.

3. Proteins moving from the ER onward and from one compartment of the endomembrane system to another are transported by a mechanism that is fundamentally different from the other two. They are ferried by *transport vesicles*, which become loaded with a cargo of proteins from the interior space, or *lumen*, of one compartment, as they pinch off from its membrane. The vesicles subsequently discharge their cargo into a second compartment by fusing with its membrane (mechanism 3 in Figure 14–5). In the process, membrane lipids and membrane proteins are also delivered from the first compartment to the second.

Signal Sequences Direct Proteins to the Correct Compartment

The typical sorting signal on proteins is a continuous stretch of amino acid sequence, typically 15–60 amino acids long. This **signal sequence** is often (but not always) removed from the finished protein once the

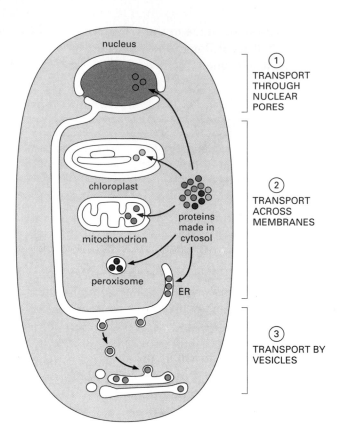

<figure>

nucleus

① TRANSPORT THROUGH NUCLEAR PORES

chloroplast

proteins made in cytosol

mitochondrion

② TRANSPORT ACROSS MEMBRANES

peroxisome

ER

③ TRANSPORT BY VESICLES

</figure>

Figure 14–5 Three main mechanisms by which membrane-bounded organelles import proteins. The protein remains folded during the transport steps in mechanisms 1 and 3 but usually has to be unfolded in mechanism 2. All of these processes require energy.

sorting decision has been executed. Some of the signal sequences used to specify different destinations in the cell are shown in Table 14–3.

Signal sequences are both necessary and sufficient to direct a protein to a particular organelle. This has been shown by experiments in which the sequence is either deleted or transferred from one protein to another by genetic engineering techniques (discussed in Chapter 10). Deleting a signal sequence from an ER protein, for example, converts it into a cytosolic protein, while placing an ER signal sequence at the beginning of a cytosolic protein redirects the protein to the ER (Figure 14–6). The signal sequences specifying the same destination can vary greatly even though they have the same function: physical properties,

Table 14–3	Some Typical Signal Sequences
Function of Signal	**Example of Signal Sequence**
Import into ER	^+H_3N-Met-Met-Ser-Phe-Val-Ser-Leu-Leu-Leu-Val-Gly-Ile-Leu-Phe-Trp-Ala-Thr-Glu-Ala-Glu-Gln-Leu-Thr-Lys-Cys-Glu-Val-Phe-Gln-
Retention in lumen of ER	-Lys-Asp-Glu-Leu-COO⁻
Import into mitochondria	^+H_3N-Met-Leu-Ser-Leu-Arg-Gln-Ser-Ile-Arg-Phe-Phe-Lys-Pro-Ala-Thr-Arg-Thr-Leu-Cys-Ser-Ser-Arg-Tyr-Leu-Leu-
Import into nucleus	-Pro-Pro-Lys-Lys-Lys-Arg-Lys-Val-
Import into peroxisomes	-Ser-Lys-Leu-

Positively charged amino acids are shown in *red,* and negatively charged amino acids in *green.* An extended block of hydrophobic amino acids is enclosed in a *yellow box.* ^+H_3N indicates the amino terminus of a protein; COO⁻ indicates the carboxyl terminus.

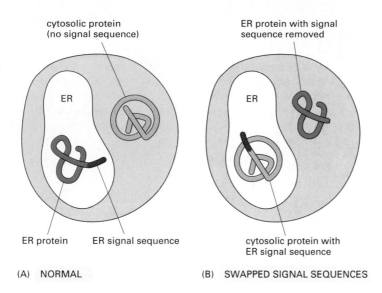

cytosolic protein
(no signal sequence)

ER protein with signal
sequence removed

ER

ER

ER protein   ER signal sequence

cytosolic protein with
ER signal sequence

(A) NORMAL

(B) SWAPPED SIGNAL SEQUENCES

Figure 14–6 **Role of signal sequences in protein sorting.** (A) Proteins destined for the ER possess an amino-terminal signal sequence that directs them to that organelle, whereas those destined to remain in the cytosol lack this sequence. (B) In the experiment illustrated, recombinant DNA techniques were used to attach an ER signal sequence to a cytosolic protein and to remove the signal sequence from an ER protein: in each case the altered protein ends up in an abnormal location in the cell, indicating that the ER signal sequence is both necessary and sufficient to direct a protein to the ER.

such as hydrophobicity or the placement of charged amino acids, often appear to be more important for their function than the exact amino acid sequence.

Proteins Enter the Nucleus Through Nuclear Pores

The **nuclear envelope** encloses the nuclear DNA and defines the nuclear compartment. It is formed from two concentric membranes. The inner nuclear membrane contains proteins that act as binding sites for the chromosomes (discussed in Chapter 8) and for the *nuclear lamina,* a finely woven meshwork of protein filaments that lines the inner face of this membrane and provides a rigid structural support for the nuclear envelope (discussed in Chapter 16). The composition of the outer nuclear membrane closely resembles the membrane of the ER, with which it is continuous (Figure 14–7).

The nuclear envelope in all eucaryotic cells is perforated by **nuclear pores,** which form the gates through which all molecules enter or leave the nucleus. Traffic occurs in both directions through the pores: newly made proteins destined for the nucleus enter from the cytosol; RNA molecules, which are synthesized in the nucleus, and ribosomal subunits, which are assembled in the nucleus, are exported. Messenger RNA molecules that are incompletely spliced are not exported from the nucleus, indicating that nuclear transport serves as a final quality-control step in mRNA synthesis and processing (discussed in Chapter 7).

A nuclear pore is a large elaborate structure composed of about 100 different proteins (Figure 14–8). Each pore contains one or more water-filled channels through which small water-soluble molecules can pass freely and nonselectively between the nucleus and the cytosol. Larger molecules (such as RNAs and proteins) and macromolecular complexes, however, cannot pass through the pores unless they carry an appropri-

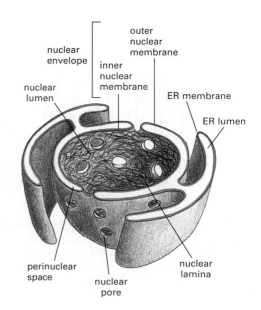

outer
nuclear
membrane

nuclear
envelope

inner
nuclear
membrane

nuclear
lumen

ER membrane

ER lumen

perinuclear
space

nuclear
pore

nuclear
lamina

Figure 14–7 **The nuclear envelope.** The double-membrane envelope is penetrated by nuclear pores. The outer membrane is continuous with the ER. The ribosomes that are normally bound to the cytosolic surface of the ER membrane and outer nuclear membrane are not shown.

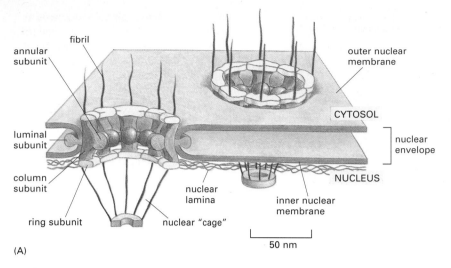

annular subunit
fibril
outer nuclear membrane
luminal subunit
CYTOSOL
nuclear envelope
column subunit
nuclear lamina
ring subunit
nuclear "cage"
inner nuclear membrane
NUCLEUS
50 nm

(A)

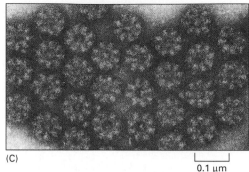

(C)

0.1 μm

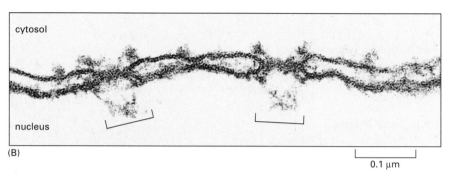

cytosol

nucleus

(B)

0.1 μm

Figure 14–8 The nuclear pore complex. (A) Drawing of a small region of the nuclear envelope showing two pore complexes. Each complex is composed of a large number of distinct protein subunits. Protein fibrils protrude from both sides of the complex; on the nuclear side they converge to form a cagelike structure. The spacing between the fibrils is wide enough so that the fibrils do not obstruct access to the pores. (B) Electron micrograph of a region of nuclear envelope showing a side view of two nuclear pores. (C) Electron micrograph showing a face-on view of nuclear pore complexes; the membranes have been extracted with detergent. (B, courtesy of Werner W. Franke; C, courtesy of Ron Milligan.)

ate sorting signal. The signal sequence (called a *nuclear localization signal*) that directs a protein from the cytosol into the nucleus typically consists of one or two short sequences containing several positively charged lysines or arginines (see Table 14–3, p. 454).

The initial interaction of a newly synthesized prospective nuclear protein with a nuclear pore requires the aid of other proteins in the

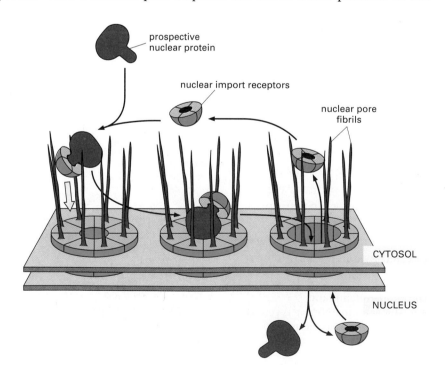

prospective nuclear protein

nuclear import receptors

nuclear pore fibrils

CYTOSOL

NUCLEUS

Figure 14–9 Schematic view of the mechanism of active transport through nuclear pores. First, specialized cytosolic proteins called nuclear import receptors bind to the prospective nuclear protein. The complex is guided to a nuclear pore by fibrils that extend from the pore into the cytosol. The binding of the nuclear protein to the pore opens the pore, and the nuclear protein, with its bound receptors, is actively transported into the nucleus. The receptors are then exported back through the pores into the cytosol for reuse.

cytosol. These cytosolic proteins, called *nuclear import receptors,* bind to the nuclear localization signal and help direct it to the pore by interacting with the nuclear pore fibrils (see Figure 14–9). The prospective nuclear protein is then actively transported into the nucleus by a process that uses the energy provided by GTP hydrolysis. A structure in the center of the nuclear pore functions like a close-fitting diaphragm: it opens just the right amount to allow the protein complex to pass through. The nuclear import receptors are then returned to the cytosol via the nuclear pore for reuse (Figure 14–9). The molecular basis of this gating mechanism and how it operates to pump macromolecules in both directions through the pores are unknown.

The nuclear pores transport proteins in their fully folded conformation and transfer ribosomal components as assembled particles. This distinguishes the transport mechanism from the mechanisms that transport proteins into other organelles. Proteins have to unfold during their transport across membranes in other organelles such as mitochondria, chloroplasts, and the ER, as we discuss next. Little is known about how proteins are transported into peroxisomes, and we shall not discuss these organelles further.

Question 14–2 Why do eucaryotic cells require a nucleus as a separate compartment when procaryotic cells can manage perfectly well without?

Proteins Unfold to Enter Mitochondria and Chloroplasts

Both mitochondria and chloroplasts are surrounded by inner and outer membranes, and both organelles specialize in the synthesis of ATP. Chloroplasts also contain a third membrane system, the thylakoid membrane (discussed in Chapter 13). Although both organelles contain their own genomes and make some of their own proteins, most mitochondrial and chloroplast proteins are encoded by genes in the nucleus and are imported from the cytosol. These proteins usually have a signal sequence at their amino terminus that allows them to enter either a mitochondrion or a chloroplast. The protein is translocated simultaneously across both the inner and outer membranes at specialized sites where the two membranes are in contact with each other. The protein is unfolded as it is translocated, and the signal sequence is cleaved off after translocation is completed (Figure 14–10). Chaperone proteins (discussed in Chapter 5) inside the organelles help to pull the protein across the two membranes and to refold the protein once it is inside. Subsequent transport to a particular site within the organelle, such as the inner or outer membrane or the thylakoid membrane, usually requires further sorting signals in the protein, which are often only

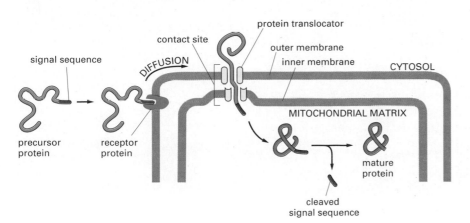

Figure 14–10 **Protein import into mitochondria.** The signal sequence is recognized by a receptor in the outer mitochondrial membrane. The complex of receptor and attached protein diffuses laterally in the membrane to a contact site, where the protein is translocated across both the outer and inner membranes by a protein translocator. The signal sequence is cleaved off by a signal peptidase inside the organelle. Proteins are imported into chloroplasts by a similar mechanism. The chaperone proteins that help to pull the protein across the membranes and help it to refold are not shown.

exposed after the first signal sequence is removed. The insertion of transmembrane proteins into the inner membrane, for example, is guided by signal sequences in the protein that start and stop the transfer process across the membrane, as we describe later for the insertion of transmembrane proteins in the ER membrane.

The growth and maintenance of mitochondria and chloroplasts requires not only the import of new proteins but also of new lipids into their membranes. Most of their membrane phospholipids are thought to be imported from the ER, which is the main site of lipid synthesis in the cell. Phospholipids are transported individually to these organelles by water-soluble lipid-carrying proteins that extract a phospholipid molecule from one membrane and deliver it into another.

Proteins Enter the Endoplasmic Reticulum While Being Synthesized

The **endoplasmic reticulum (ER)** is the most extensive membrane system in a eucaryotic cell (Figure 14–11A), and, unlike the organelles discussed so far, it serves as an entry point for proteins destined for other organelles, as well as for the ER itself. Proteins destined for the Golgi apparatus, endosomes, and lysosomes, as well as proteins destined for the cell surface, all first enter the ER from the cytosol. Once inside the ER or in the ER membrane, individual proteins will not reenter the cytosol during their onward journey. They will be ferried by transport vesicles from organelle to organelle and, in some cases, from organelle to the plasma membrane.

Two kinds of proteins are transferred from the cytosol to the ER: (1) water-soluble proteins are completely translocated across the ER membrane and are released into the ER lumen; (2) prospective transmembrane proteins are only partly translocated across the ER membrane and become embedded in it. The water-soluble proteins are destined either

Figure 14–11 Endoplasmic reticulum.
(A) Fluorescence micrograph of living plant cells showing the ER as a complex network of sheets and tubes. The cells have been genetically engineered so that they contain a fluorescent protein in their ER. The bright ellipses are chloroplasts. (B) An electron micrograph showing the rough ER in a cell from a dog's pancreas that makes and secretes large amounts of digestive enzymes. The cytosol is filled with closely packed sheets of ER studded with ribosomes. At the top left is a portion of the nucleus and its nuclear envelope; note that the outer nuclear membrane, which is continuous with the ER, is also studded with ribosomes. (A, courtesy of Jim Haseloff; B, courtesy of Lelio Orci.)

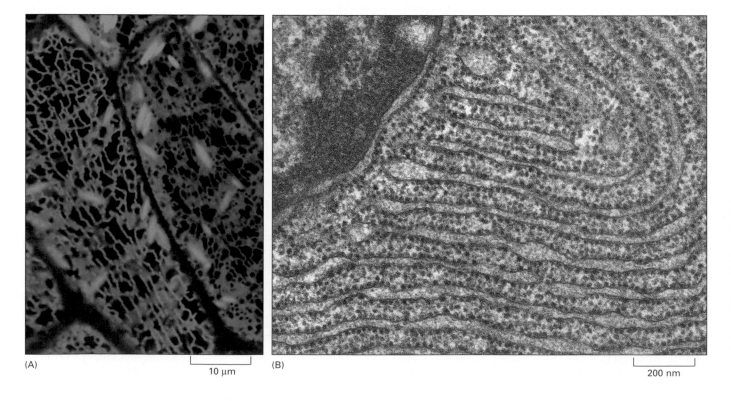

(A) 10 µm

(B) 200 nm

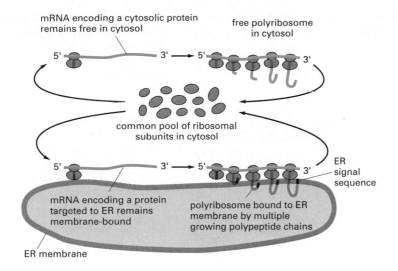

mRNA encoding a cytosolic protein remains free in cytosol

free polyribosome in cytosol

5' → 3'   5' → 3'

common pool of ribosomal subunits in cytosol

5' → 3'   5' → 3'

ER signal sequence

mRNA encoding a protein targeted to ER remains membrane-bound

polyribosome bound to ER membrane by multiple growing polypeptide chains

ER membrane

Figure 14–12 Free and membrane-bound ribosomes. A common pool of ribosomes is used to synthesize both the proteins that stay in the cytosol and those that are transported into membrane-bounded organelles, including the ER. It is the ER signal sequence on a growing polypeptide chain that directs the ribosome making the protein to the ER membrane. The mRNA molecule may remain bound to the ER as part of a polyribosome, while the ribosomes that move along it are recycled; at the end of each round of protein synthesis, the ribosomal subunits are released and rejoin the common pool in the cytosol.

for secretion (by release at the cell surface) or for the lumen of an organelle; the transmembrane proteins are destined to reside in either the ER membrane, the membrane of another organelle, or the plasma membrane. All of these proteins are initially directed to the ER by an *ER signal sequence,* a segment of eight or more hydrophobic amino acids (see Table 14–3, p. 454), which is also involved in the process of translocation across the membrane.

Unlike the proteins that enter the nucleus, mitochondria, chloroplasts, and peroxisomes, most of the proteins that enter the ER begin to be threaded across the ER membrane before the polypeptide chain is completely synthesized. This requires that the ribosome synthesizing the protein is attached to the ER membrane. These membrane-bound ribosomes coat the surface of the ER, creating regions termed **rough endoplasmic reticulum** because of the characteristic beaded appearance when viewed in an electron microscope (Figure 14–11B).

There are, therefore, two separate populations of ribosomes in the cytosol. *Membrane-bound ribosomes* are attached to the cytosolic side of the ER membrane (and outer nuclear membrane) and are making proteins that are being translocated into the ER. *Free ribosomes* are unattached to any membrane and are making all of the other proteins encoded by the nuclear DNA. Membrane-bound ribosomes and free ribosomes are structurally and functionally identical; they differ only in the proteins they are making at any given time. When a ribosome happens to be making a protein with an ER signal sequence, the signal sequence directs the ribosome to the ER membrane. As an mRNA molecule is translated, many ribosomes bind to it, forming a *polyribosome* (discussed in Chapter 7). In the case of an mRNA molecule encoding a protein with an ER signal sequence, the polyribosome becomes riveted to the ER membrane by the growing polypeptide chains, which have become inserted into the membrane (Figure 14–12).

Soluble Proteins Are Released into the ER Lumen

The ER signal sequence is guided to the ER membrane by at least two components: (1) a *signal-recognition particle (SRP)* that is present in the cytosol and binds to the ER signal sequence when it is exposed on the

Question 14–3 Explain how an mRNA molecule can remain attached to the ER membrane while individual ribosomes translating it are released and rejoin the cytosolic pool of ribosomes after each round of translation.

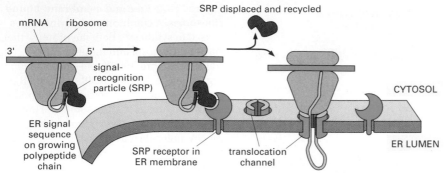

Figure 14–13 **How an ER signal sequence and an SRP direct a ribosome to the ER membrane.** The SRP binds to the exposed ER signal sequence and to the ribosome, thereby slowing protein synthesis by the ribosome. The SRP-ribosome complex then binds to an SRP receptor in the ER membrane. The SRP is then released, leaving the ribosome on the ER membrane. A protein translocation channel in the ER membrane then inserts the polypeptide chain into the membrane and starts to transfer it across the lipid bilayer.

ribosome, and (2) an *SRP receptor* that is embedded in the membrane of the ER. Binding of an SRP to a signal sequence causes protein synthesis by the ribosome to slow down, until the ribosome and its bound SRP bind to an SRP receptor. After binding to its receptor, the SRP is released and protein synthesis recommences, with the polypeptide now being threaded into the lumen of the ER through a *translocation channel* in the ER membrane (Figure 14–13). Thus the SRP and SRP receptor function as molecular matchmakers, connecting ribosomes that are synthesizing proteins containing ER signal sequences to available ER translocation channels.

After the ribosome-mRNA-SRP complex has bound to the ER membrane, the signal sequence, which for soluble proteins is almost always at the amino terminus of the protein, serves the additional function of opening the translocation channel; it remains bound to the channel while the rest of the protein chain is threaded through the membrane as a large loop. At some stage during translocation, the signal sequence is cleaved off by a signal peptidase located on the luminal side of the ER membrane; the signal peptide is then released from the translocation channel and rapidly degraded to amino acids. Once the carboxyl terminus of the protein has passed through the membrane, the protein is released into the ER lumen (Figure 14–14).

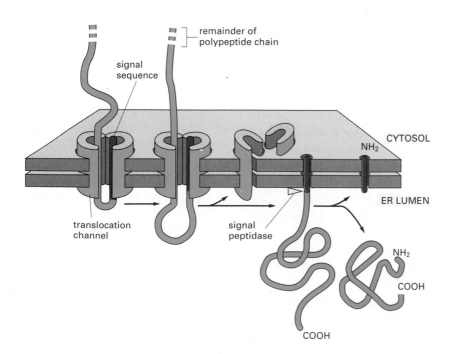

Figure 14–14 **The translocation of a soluble protein across the ER membrane into the lumen.** A protein translocation channel binds the signal sequence and actively transfers the rest of the polypeptide across the lipid bilayer as a loop. At some point during the translocation process, the translocation channel opens sideways and releases the signal sequence into the bilayer, where it is cleaved off by an enzyme (a signal peptidase). The translocated polypeptide is released as a soluble protein into the ER lumen. The membrane-bound ribosome is omitted from this and the following two figures for clarity.

Start and Stop Signals Determine the Arrangement of a Transmembrane Protein in the Lipid Bilayer

Not all proteins that enter the ER are released into the ER lumen. Some remain embedded in the ER membrane as transmembrane proteins. The translocation process for such proteins is more complicated than it is for soluble proteins, as some parts of the polypeptide chain must be translocated across the lipid bilayer while others are not.

In the simplest case, that of a transmembrane protein with a single membrane-spanning segment, the amino-terminal signal sequence initiates translocation, just as for a soluble protein. But the transfer process is halted by an additional sequence of hydrophobic amino acids, a *stop-transfer sequence*, further into the polypeptide chain. This second sequence is released sideways into the lipid bilayer from the translocation channel and forms an α-helical membrane-spanning segment that anchors the protein in the membrane. Simultaneously, the amino-terminal signal sequence is also released from the channel into the lipid bilayer and is cleaved off. As a result, the translocated protein ends up as a transmembrane protein inserted in the membrane with a defined orientation—the amino terminus on the luminal side of the lipid bilayer and the carboxyl terminus on the cytosolic side (Figure 14–15). Thus the ER translocation channel is gated in two directions: it must be able to open as a pore that spans from one side of the ER membrane to the other, and it must be able to open sideways to discharge transmembrane proteins into the lipid bilayer. As discussed in Chapter 11, once inserted into the membrane, a transmembrane protein does not change its orientation, which is retained throughout any subsequent vesicle budding and fusion events.

In some transmembrane proteins, an internal, rather than an amino-terminal, signal sequence is used to start transfer and is not removed. This arrangement occurs in some transmembrane proteins in which the polypeptide chain passes back and forth across the lipid bilayer. In these cases hydrophobic signal sequences are thought to work in pairs; an internal signal sequence (a *start-transfer sequence*) serves to

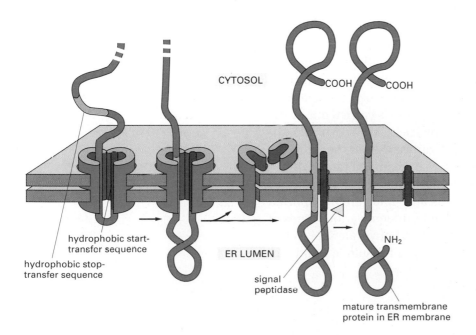

hydrophobic start-transfer sequence

hydrophobic stop-transfer sequence

CYTOSOL

COOH COOH

ER LUMEN

signal peptidase

NH₂

mature transmembrane protein in ER membrane

Figure 14–15 The integration of a transmembrane protein into the ER membrane. An amino-terminal ER signal sequence (*red*) initiates transfer as in Figure 14–14. In addition, the protein also contains a second hydrophobic sequence, a stop-transfer sequence (*orange*). When this enters the translocation channel, the channel discharges the protein sideways into the lipid bilayer, after which the amino-terminal signal sequence is cleaved off, leaving the transmembrane protein anchored in the membrane. Protein synthesis on the cytosolic side continues to completion.

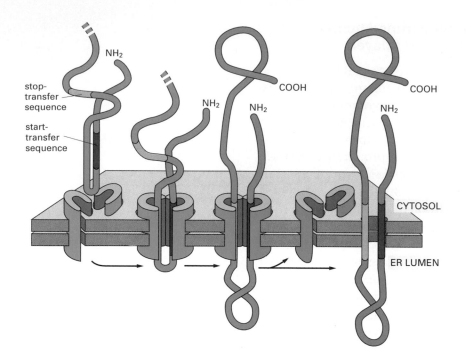

Figure 14–16 **The integration of a double-pass transmembrane protein into the ER membrane.** An internal ER signal sequence (*red*) acts as a start-transfer signal and initiates the transfer of the polypeptide chain. When a stop-transfer sequence (*orange*) enters the translocation channel, the channel discharges both sequences sideways into the lipid bilayer. Neither the start-transfer nor the stop-transfer sequence is cleaved off, and the entire polypeptide chain remains anchored in the membrane as a double-pass transmembrane protein. Proteins that span the membrane more times contain further pairs of stop and start sequences, and the same process is repeated for each pair.

initiate translocation, which continues until a stop-transfer sequence is reached; the two α-helical hydrophobic sequences are then released into the bilayer (Figure 14–16). In complex multipass proteins, in which many hydrophobic α helices span the bilayer, further pairs of stop and start sequences come into play: one sequence reinitiates translocation further down the polypeptide chain, and the other stops translocation and causes polypeptide release, and so on for subsequent starts and stops. Thus, multipass membrane proteins are stitched into the lipid bilayer as they are being synthesized by a mechanism resembling the workings of a sewing machine.

Having considered how proteins enter the ER lumen or become embedded in the ER membrane, we now discuss how they are carried onward by vesicular transport.

Vesicular Transport

Entry into the ER is usually only the first step on a pathway to another destination—which is, initially at least, the Golgi apparatus. Transport from the ER to the Golgi apparatus and from the Golgi apparatus to other compartments of the endomembrane system is carried out by the continual budding and fusion of **transport vesicles.** The transport pathways mediated by transport vesicles extend outward from the ER to the plasma membrane, and inward from the plasma membrane to lysosomes, and thus provide routes of communication between the interior of the cell and its surroundings. As proteins and lipids are transported outward along these pathways, many of them undergo various types of chemical modification, such as the addition of carbohydrate side chains (to both proteins and lipids) and the formation of disulfide bonds (in proteins) that stabilize protein structure.

Question 14–4

A. Predict the membrane orientation of a protein that is synthesized with an uncleaved, internal signal sequence (shown as the *red* start-transfer sequence in Figure 14–15) but does not contain a stop-transfer peptide.

B. Similarly, predict the membrane orientation of a protein that is synthesized with an amino-terminal, cleaved signal sequence followed by a stop-transfer sequence, followed by a start-transfer sequence.

C. What arrangement of signal sequences would enable the insertion of a multipass protein with an odd number of transmembrane segments?

Transport Vesicles Carry Soluble Proteins and Membrane Between Compartments

The vesicular traffic between membrane-bounded compartments of the endomembrane system is highly organized. A major outward *secretory pathway* leads from the biosynthesis of proteins on the ER membrane and their entry into the ER, through the Golgi apparatus, to the cell surface; at the Golgi apparatus a side branch leads off through endosomes to lysosomes. A major inward *endocytic pathway*, which is responsible for the ingestion and degradation of extracellular molecules, leads from the plasma membrane, through endosomes, to lysosomes (Figure 14–17).

To perform its function correctly, each transport vesicle that buds off from a compartment must take with it only the proteins appropriate to its destination and must fuse only with the appropriate target membrane. A vesicle carrying cargo from the Golgi apparatus to the plasma membrane, for example, must exclude proteins that are to stay in the Golgi apparatus, and it must fuse only with the plasma membrane and not with any other organelle. While participating in this constant flow of membrane components, each organelle must maintain its own distinct identity, that is, its own distinctive protein and lipid composition. All these recognition events depend on proteins associated with the transport vesicle membrane. As we shall see, different types of transport vesicles shuttle between the various organelles, each carrying a distinct set of molecules.

Vesicle Budding Is Driven by the Assembly of a Protein Coat

Vesicles that bud from membranes usually have a distinctive protein coat on their cytosolic surface and are therefore called **coated vesicles.** After budding is completed, the coat is lost, allowing the membrane of the vesicle to interact directly with the membrane to which it will fuse.

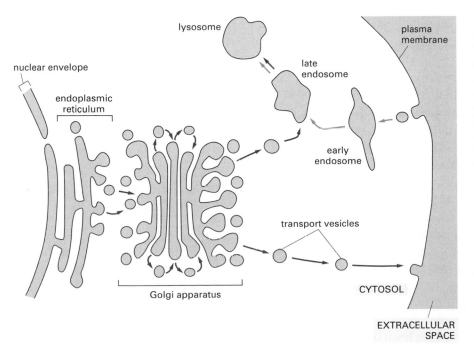

Figure 14–17 Vesicular traffic. The extracellular space and each of the membrane-bounded compartments (*shaded gray*) communicate with one another by means of transport vesicles, as shown. In the outward secretory pathway (*red arrows*) protein molecules are transported from the ER, through the Golgi apparatus, to the plasma membrane or (via late endosomes) to lysosomes. In the inward endocytic pathway (*green arrows*) extracellular molecules are ingested in vesicles derived from the plasma membrane and are delivered to early endosomes and then (via late endosomes) to lysosomes.

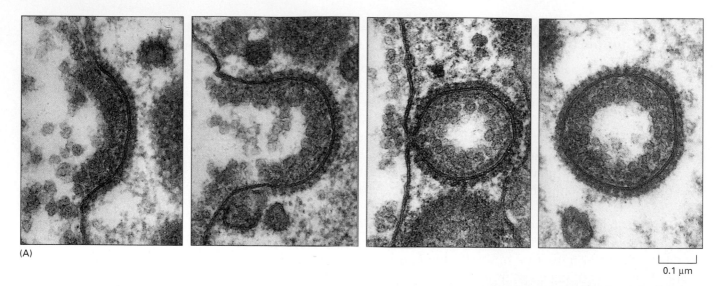

(A)

$\llcorner\underset{\text{0.1 μm}}{\quad\quad}\lrcorner$

Figure 14–18 Clathrin-coated pits and vesicles. (A) Electron micrographs showing the sequence of events in the formation of a clathrin-coated vesicle from a clathrin-coated pit. The clathrin-coated pits and vesicles shown here are unusually large and are being formed at the plasma membrane of a hen oocyte. They are involved in taking up particles made of lipid and protein into the oocyte to form yolk. (B) Electron micrograph showing numerous clathrin-coated pits and vesicles budding from the inner surface of the plasma membrane of cultured skin cells. (A, courtesy of M.M. Perry and A.B. Gilbert, *J. Cell Sci.* 39:257–272, 1979, by permission of The Company of Biologists; B, from J. Heuser, *J. Cell Biol.* 84:560–583, 1980, by copyright permission of the Rockefeller University Press.)

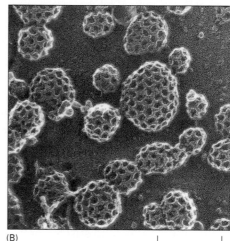

(B)

$\llcorner\underset{\text{0.2 μm}}{\quad\quad}\lrcorner$

There are several kinds of coated vesicles, each with a distinctive protein coat. The coat is thought to serve at least two functions: it shapes the membrane into a bud, and it helps to capture molecules for onward transport.

The best-studied vesicles are those that have coats made largely of the protein **clathrin** and are known as *clathrin-coated vesicles.* They bud from the Golgi apparatus on the outward secretory pathway and from the plasma membrane on the inward endocytic pathway. At the plasma membrane, for example, each vesicle starts off as a *clathrin-coated pit.* Clathrin molecules assemble into a basketlike network on the cytosolic surface of the membrane, and it is this assembly process that starts shaping the membrane into a vesicle (Figure 14–18). A small GTP-binding protein called *dynamin* assembles as a ring around the neck of each deeply invaginated coated pit. The dynamin then hydrolyzes its bound GTP, which is thought to cause the ring to constrict, thereby pinching off the vesicle from the membrane. Other kinds of transport vesicles, with different coat proteins, are also involved in vesicular transport. They form in a similar way and carry their own characteristic sets of molecules between the endoplasmic reticulum, the Golgi apparatus, and the plasma membrane. But how does a transport vesicle select its particular cargo? The mechanism is best understood for clathrin-coated vesicles.

Clathrin itself plays no part in capturing specific molecules for transport. This is the function of a second class of coat proteins in clathrin-coated vesicles, called *adaptins,* which both bind the coat to the vesicle membrane and help select cargo molecules for transport.

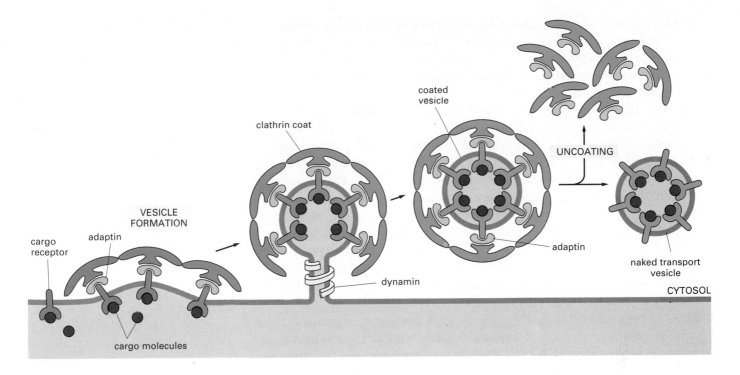

cargo receptor

adaptin

VESICLE FORMATION

cargo molecules

clathrin coat

dynamin

coated vesicle

UNCOATING

adaptin

naked transport vesicle

CYTOSOL

Molecules for onward transport carry specific *transport signals* that are recognized by *cargo receptors* in the compartment membrane. Adaptins help capture specific cargo molecules by trapping the cargo receptors that bind them. In this way a selected set of cargo molecules, bound to their specific receptors, is incorporated into the lumen of each newly formed clathrin-coated vesicle (Figure 14–19). There are at least two types of adaptins: those that bind cargo receptors in the plasma membrane are different from those that bind cargo receptors in the Golgi apparatus, reflecting the different cargo molecules transported in clathrin-coated vesicles from the plasma membrane and from the Golgi apparatus.

A different class of coated vesicles, called *COP-coated vesicles*, is involved in transporting molecules between the ER and the Golgi apparatus and from one part of the Golgi apparatus to another (Table 14–4).

The Specificity of Vesicle Docking Depends on SNAREs

After a transport vesicle buds off from a membrane, it must find its way to its correct destination to deliver its contents. If the distance is short,

Figure 14–19 Selective transport mediated by clathrin-coated vesicles. Cargo receptors, with their bound cargo molecules, are captured by adaptins, which also bind clathrin molecules to the cytosolic surface of the budding vesicle. Dynamin proteins assemble around the neck of budding vesicles; once assembled, they hydrolyze their bound GTP and pinch off the vesicle. After budding is complete, the coat proteins are removed and the naked vesicle can fuse with its target membrane. Functionally similar coat proteins are found in other types of coated vesicles.

Table 14–4	Some Types of Coated Vesicles		
Type of Coated Vesicle	**Coat Proteins**	**Origin**	**Destination**
Clathrin-coated	clathrin + adaptin 1	Golgi apparatus	lysosome (via endosomes)
Clathrin-coated	clathrin + adaptin 2	plasma membrane	endosomes
COP-coated	COP proteins	ER	Golgi apparatus
		Golgi cisterna	Golgi cisterna
		Golgi apparatus	ER

Question 14–5 The budding of clathrin-coated vesicles can be observed from eucaryotic plasma membrane fragments when adaptins, clathrin, and dynamin-GTP are added. What would you observe if you omitted (A) adaptins, (B) clathrin, or (C) dynamin? (D) What would you observe if the plasma membrane fragments were from a procaryotic cell?

as in moving from the ER to the Golgi apparatus, the vesicle moves by simple diffusion. If the distance is long, as in moving from the Golgi apparatus to the end of a long nerve cell process, the vesicle is actively transported by motor proteins that move along cytoskeletal fibers, as discussed in Chapter 16.

Once a transport vesicle has reached its target organelle it has to recognize it and dock with it. Only then can the vesicle membrane fuse with the target membrane and unload its cargo. The impressive specificity of vesicular transport suggests that all types of transport vesicles in the cell display on their surface molecular markers that identify the vesicle according to its origin and cargo. These markers must be recognized by complementary receptors on the appropriate target membrane, including the plasma membrane. Although the mechanism of this recognition process is not known for certain, it is thought to involve a family of related transmembrane proteins called **SNAREs:** SNAREs on the vesicle (called v-SNAREs) are recognized specifically by complementary SNAREs on the cytosolic surface of the target membrane (called t-SNAREs) (Figure 14–20). Each organelle and each type of transport vesicle is believed to carry a unique SNARE, and the interactions between complementary SNAREs ensure that transport vesicles fuse only with the correct membrane.

Once a transport vesicle has recognized its target membrane and docked there, the vesicle has to fuse with the membrane to deliver its cargo. Fusion not only delivers the contents of the vesicle into the interior of the target organelle, it also adds the vesicle membrane to the membrane of the organelle. Membrane fusion does not always follow immediately after docking, however, and must often wait to be triggered by a specific signal. Whereas docking requires only that the two membranes come close enough for proteins protruding from the two lipid bilayers to interact, fusion requires a much closer approach: the two lipid bilayers must come within 1.5 nm of each other so that they can fuse. For this close approach, water must be displaced from the hydrophilic surface of the membrane—a process that is energetically highly unfavorable. It is likely therefore that all membrane fusions in cells are catalyzed by specialized proteins that assemble at the fusion site to form a fusion complex that provides the means to cross this energy barrier (Figure 14–21). A number of cytosolic proteins that are required for vesicle fusion have been identified, but it is still not known how they act.

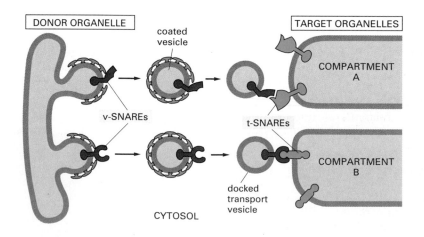

Figure 14–20 **A current model of transport vesicle docking.** Vesicles that bud from a membrane carry specific marker proteins called vesicle SNAREs (v-SNAREs) on their surface, which bind to complementary target SNAREs (t-SNAREs) on the target membrane. Many different complementary pairs of v-SNAREs and t-SNAREs are thought to play a crucial role in guiding transport vesicles to their appropriate target membranes.

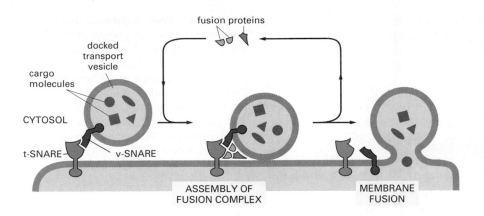

Figure 14–21 Transport vesicle fusion. Following the docking of a transport vesicle at its target membrane, a complex of membrane-fusion proteins assembles at the docking site and catalyzes the fusion of the vesicle with the target membrane. Fusion of the two membranes delivers the vesicle contents into the interior of the target organelle and adds the vesicle membrane to the target membrane.

Secretory Pathways

Vesicular traffic is not confined to the interior of the cell. It extends to and from the plasma membrane. Newly made proteins, lipids, and carbohydrates are delivered from the ER, via the Golgi apparatus, to the cell surface by transport vesicles that fuse with the plasma membrane in a process called **exocytosis.** Each molecule that travels along this route passes through a fixed sequence of membrane-bounded compartments and is often chemically modified en route.

In this section we follow the outward path of proteins as they travel from the ER, where they are made and modified, through the Golgi apparatus, where they are further modified and sorted, to the plasma membrane. As a protein passes from one compartment to another, it is monitored to check that it has folded properly and assembled with its appropriate partners, so that only correctly built proteins are released at the cell surface, while all of the others are degraded in the cell.

Most Proteins Are Covalently Modified in the ER

Most proteins that enter the ER are chemically modified there. Disulfide bonds are formed by the oxidation of pairs of cysteine side chains (discussed in Chapter 5), a reaction catalyzed by an enzyme that resides in the ER lumen. The disulfide bonds help to stabilize the structure of those proteins that may encounter changes in pH and degradative enzymes outside the cell—either after they are secreted or after they are incorporated into the plasma membrane. Disulfide bonds do not form in the cytosol because of the reducing environment there.

Many of the proteins that enter the ER lumen or ER membrane are converted to glycoproteins in the ER by the covalent attachment of short oligosaccharide side chains. This process of *glycosylation* is carried out by glycosylating enzymes found in the ER but not in the cytosol. Very few proteins in the cytosol are glycosylated, and those that are have only a single sugar residue attached to them. The oligosaccharides on proteins serve various functions, depending on the protein. They can protect the protein from degradation, retain it in the ER until it is properly folded, or help guide it to the appropriate organelle by serving as a transport signal for packaging the protein into appropriate transport vesicles (as in the case of lysosomal proteins discussed later). When displayed on the cell surface, oligosaccharides form part of the cell's glycocalyx (discussed in Chapter 11) and can function in the recognition of one cell by another.

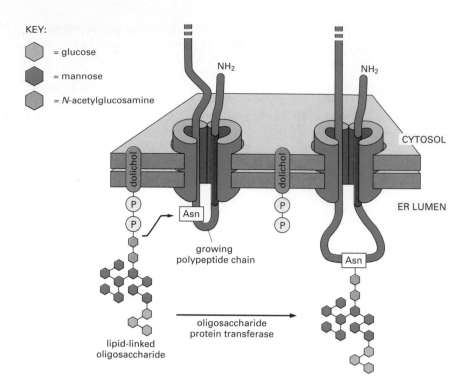

KEY:

= glucose

= mannose

= N-acetylglucosamine

CYTOSOL

ER LUMEN

NH₂

NH₂

dolichol

dolichol

P
P

P
P

Asn

growing
polypeptide chain

Asn

lipid-linked
oligosaccharide

oligosaccharide
protein transferase

Figure 14–22 Protein glycosylation in the ER. Almost as soon as a polypeptide chain enters the ER lumen, it is glycosylated by addition of oligosaccharide side chains to particular asparagines in the polypeptide. Each oligosaccharide chain is transferred as an intact unit to the asparagine from a lipid called dolichol. Asparagines that are glycosylated are always present in the tripeptide sequence asparagine, X, serine or threonine, where X can be any amino acid.

In the ER, individual sugars are not added one by one to the protein to create the oligosaccharide side chain. Instead, a preformed branched oligosaccharide, containing a total of 14 sugars, is attached en bloc to all proteins that carry the appropriate site for glycosylation. The oligosaccharide is originally attached to a specialized lipid, called *dolichol,* in the ER membrane and is then transferred to the amino (NH₂) group of an asparagine side chain on the protein immediately after the target asparagine emerges in the ER lumen during protein translocation (Figure 14–22). The addition takes place in a single enzymatic step catalyzed by a membrane-bound enzyme (an oligosaccharide protein transferase) that has its active site exposed on the luminal side of the ER membrane, which explains why cytosolic proteins are not glycosylated in this way. A simple sequence of three amino acids, of which the asparagine is one, defines which asparagine residues in a protein receive the oligosaccharide. Oligosaccharide side chains linked to an asparagine NH₂ group in a protein are said to be *N-linked* and are by far the most common type of linkage found on glycoproteins.

The addition of the 14-sugar oligosaccharide in the ER is only the first step in a series of further modifications before the mature glycoprotein emerges at the other end of the outward pathway. Despite their initial similarity, the *N*-linked oligosaccharides on mature glycoproteins are remarkably diverse. All of the diversity results from extensive modification of the original precursor structure shown in Figure 14–22. This *oligosaccharide processing* begins in the ER and continues in the Golgi apparatus.

Question 14–6 Why might it be advantageous to add a preassembled block of 14 sugar residues to a protein in the ER, rather than building the sugar chains step-by-step on the surface of the protein by the sequential addition of sugars by individual enzymes?

Exit from the ER Is Controlled to Ensure Protein Quality

Some proteins made in the ER are destined to function there. They are retained in the ER (and are returned to the ER when they escape to the Golgi apparatus) by a carboxyl-terminal four–amino acid sequence called an *ER retention signal* (see Table 14–3, p. 454), which is recognized

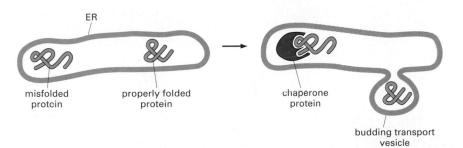

Figure 14–23 **Quality control in the ER.**
Misfolded proteins bind to chaperone proteins in the ER lumen and are thereby retained, while normally folded proteins are transported in transport vesicles to the Golgi apparatus. If the misfolded proteins fail to refold normally, they are transported into the cytosol where they are degraded.

by a membrane-bound receptor protein in the ER and Golgi apparatus. Most proteins that enter the ER, however, are destined for other locations; they are packaged into transport vesicles that bud from the ER and fuse with the Golgi apparatus. Exit from the ER, however, is highly selective. Proteins that fold up incorrectly, and dimeric or multimeric proteins that fail to assemble properly, are actively retained in the ER by binding to chaperone proteins that reside there. Interaction with chaperones holds the proteins in the ER until proper folding occurs; otherwise, the proteins are ultimately degraded (Figure 14–23). Antibody molecules, for example, are composed of four polypeptide chains (see Panel 5–3, pp. 158–159) that assemble into the complete antibody molecule in the ER. Partially assembled antibodies are retained in the ER until all four polypeptide chains have assembled; any antibody molecule that fails to assemble properly is ultimately degraded. In this way the ER controls the quality of the proteins that it exports to the Golgi apparatus.

Sometimes, however, this quality-control mechanism can be detrimental to the organism. The predominant mutation that causes the common genetic disease *cystic fibrosis,* for example, produces a plasma-membrane transport protein that is slightly misfolded; even though the mutant protein could function perfectly normally if it reached the plasma membrane, it is retained in the ER, with dire consequences. The devastating disease results not because the mutation inactivates an important protein but because the active protein is discarded by the cells before it is given an opportunity to function.

Proteins Are Further Modified and Sorted in the Golgi Apparatus

The **Golgi apparatus** is usually located near the cell nucleus, and in animal cells it is often close to the centrosome, or cell center. It consists of a collection of flattened, membrane-bounded sacs (*cisternae,* singular *cisterna*), which are piled like stacks of plates. Each stack contains three to twenty cisternae (Figure 14–24). The number of Golgi stacks per cell varies greatly depending on the cell type: some cells contain one large stack, while others contain hundreds of very small ones.

Each Golgi stack has two distinct faces: an entry, or *cis,* face and an exit, or *trans,* face. The *cis* face is adjacent to the ER, while the *trans* face points toward the plasma membrane. The outermost cisterna at each face is connected to a network of interconnected membranous tubes and vesicles (see Figure 14–24A). Soluble proteins and membrane enter the *cis Golgi network* via transport vesicles derived from the ER. The proteins travel through the cisternae in sequence by means of transport vesicles that bud from one cisterna and fuse with the next. Proteins exit from the *trans* **Golgi network** in transport vesicles destined for either

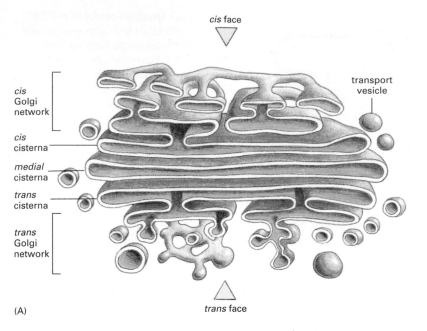

cis face

cis Golgi network

cis cisterna

medial cisterna

trans cisterna

trans Golgi network

transport vesicle

(A)

trans face

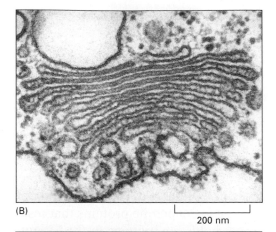

(B)

200 nm

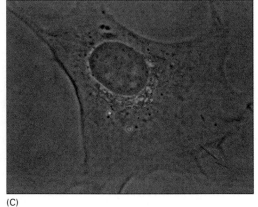

(C)

Figure 14–24 The Golgi apparatus. (A) Three-dimensional reconstruction of a Golgi stack. It was reconstructed from electron micrographs of the Golgi apparatus in a secretory animal cell. (B) Electron micrograph of a Golgi stack from a plant cell, where the Golgi apparatus is especially distinct. The Golgi apparatus is oriented as in (A). (C) The Golgi apparatus in a cultured fibroblast stained with a fluorescent antibody that labels the Golgi apparatus specifically. The *red arrow* indicates the direction of the cell's movement. Note that the Golgi apparatus is close to the nucleus and oriented toward the direction of movement. (A, redrawn from A. Rambourg and Y. Clermont, *Eur. J. Cell Biol.* 51:189–200, 1990; B, courtesy of George Palade; C, courtesy of John Henley and Mark McNiven.)

the cell surface or another compartment (see Figure 14–17). Both the *cis* and *trans* Golgi networks are thought to be important for protein sorting: proteins entering the *cis* Golgi network can either move onward through the Golgi stack or, if they contain an ER retention signal, be returned to the ER; proteins exiting the *trans* Golgi network are sorted according to whether they are destined for lysosomes or for the cell surface. We discuss some examples of sorting by the *trans* Golgi network later.

Many of the oligosaccharide groups that are added to proteins in the ER undergo further modifications in the Golgi apparatus. On some proteins, for example, complex oligosaccharide chains are created by a highly ordered process in which sugars are added and removed by a series of enzymes that act in a rigidly determined sequence as the protein passes through the Golgi stack. There is a clear correlation between the position of an enzyme in the chain of processing events and its localization in the Golgi stack: enzymes that act early are found in cisternae close to the *cis* face, while enzymes that act late are found in cisternae near the *trans* face.

Secretory Proteins Are Released from the Cell by Exocytosis

In all eucaryotic cells there is a steady stream of vesicles that bud from the *trans* Golgi network and fuse with the plasma membrane. This

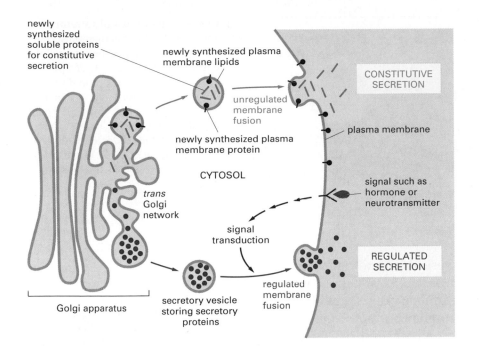

newly synthesized soluble proteins for constitutive secretion

newly synthesized plasma membrane lipids

unregulated membrane fusion

newly synthesized plasma membrane protein

CYTOSOL

trans Golgi network

CONSTITUTIVE SECRETION

plasma membrane

signal such as hormone or neurotransmitter

signal transduction

REGULATED SECRETION

regulated membrane fusion

Golgi apparatus

secretory vesicle storing secretory proteins

Figure 14–25 The regulated and constitutive pathways of exocytosis. The two pathways diverge in the *trans* Golgi network. Many soluble proteins are continually secreted from the cell by the constitutive secretory pathway, which operates in all cells. This pathway also supplies the plasma membrane with newly synthesized lipids and proteins. Specialized secretory cells also have a regulated exocytosis pathway, by which selected proteins in the *trans* Golgi network are diverted into secretory vesicles, where the proteins are concentrated and stored until an extracellular signal stimulates their secretion.

constitutive exocytosis pathway operates continually and supplies newly made lipids and proteins to the plasma membrane; it is the pathway for plasma membrane growth when cells enlarge before dividing. It also carries proteins to the cell surface to be released to the outside, a process called **secretion.** Some of the released proteins adhere to the cell surface and become peripheral proteins of the plasma membrane; some are incorporated into the extracellular matrix; still others diffuse into the extracellular fluid to nourish or to signal other cells.

In addition to the constitutive exocytosis pathway, which operates continually in all eucaryotic cells, there is a *regulated exocytosis pathway,* which operates only in cells that are specialized for secretion. Specialized *secretory cells* produce large quantities of particular products, such as hormones, mucus, or digestive enzymes, which are stored in **secretory vesicles.** These vesicles bud off from the *trans* Golgi network and accumulate near the plasma membrane. They fuse with the plasma membrane to release their contents to the outside only when the cell is stimulated by an extracellular signal (Figure 14–25). An increase in blood glucose, for example, signals cells in the pancreas to secrete the hormone insulin (Figure 14–26).

Proteins destined for secretory vesicles are sorted and packaged in the *trans* Golgi network. Proteins that travel by this pathway have special surface properties that cause them to aggregate with one another under the ionic conditions (acidic and high Ca^{2+}) that prevail in the *trans* Golgi network. The aggregated proteins are recognized by an unknown mechanism and packaged into secretory vesicles, which pinch off from the network. Proteins secreted by the constitutive pathway do not aggregate and are therefore carried automatically to the plasma membrane by the transport vesicles of the constitutive pathway. Selective aggregation has another function: it allows secretory proteins to be packaged into secretory vesicles at concentrations up to 200 times higher than the concentration of the unaggregated protein in the Golgi lumen. This enables secretory cells to release large amounts of the protein promptly when triggered to do so (see Figure 14–26).

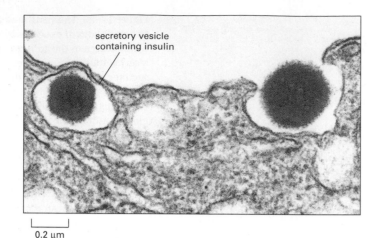

secretory vesicle
containing insulin

0.2 μm

Figure 14–26 Exocytosis of secretory vesicles. The electron micrograph shows the release of insulin into the extracellular space from a secretory vesicle of a pancreatic β cell. The insulin is stored in a highly concentrated form in each secretory vesicle and is released only when the cell is signaled to secrete by an increase in glucose levels in the blood. (Courtesy of Lelio Orci, from L. Orci, J.-D. Vassali, and A. Perrelet, *Sci. Am.* 256:85–94, 1988.)

When a secretory vesicle or transport vesicle fuses with the plasma membrane and discharges its contents by exocytosis, its membrane becomes part of the plasma membrane. Although this should greatly increase the surface area of the plasma membrane, it does so only transiently because membrane components are removed from other regions of the surface by endocytosis almost as fast as they are added by exocytosis. This removal returns both the lipids and the proteins of the vesicle membrane to the Golgi network, where they can be used again.

Question 14–7 What would you expect to happen in cells that secrete large amounts of protein through the regulated secretory pathway if the ionic conditions in the ER lumen could be changed to resemble those in the lumen of the *trans* Golgi network?

Endocytic Pathways

Eucaryotic cells are continually taking up fluid, as well as large and small molecules, by the process of endocytosis. Specialized cells are also able to internalize large particles and even other cells. The material to be ingested is progressively enclosed by a small portion of the plasma membrane, which first buds inward and then pinches off to form an intracellular *endocytic vesicle*. The ingested material is ultimately delivered to lysosomes, where it is digested. The metabolites generated by digestion are transferred directly out of the lysosome into the cytosol, where they can be used by the cell.

Two main types of endocytosis are distinguished on the basis of the size of the endocytic vesicles formed. *Pinocytosis* ("cellular drinking") involves the ingestion of fluid and molecules via small vesicles (<150 nm in diameter). *Phagocytosis* ("cellular eating") involves the ingestion of large particles, such as microorganisms and cell debris, via large vesicles called *phagosomes* (generally >250 nm in diameter). Whereas all eucaryotic cells are continually ingesting fluid and molecules by pinocytosis, large particles are ingested mainly by specialized *phagocytic cells*.

In this final section we trace the endocytic pathway from the plasma membrane to lysosomes. We start by considering the uptake of large particles by phagocytosis.

Specialized Phagocytic Cells Ingest Large Particles

The most dramatic form of endocytosis is **phagocytosis,** which was first observed over a hundred years ago. In protozoa, phagocytosis is a form of feeding: large particles such as bacteria are taken up into phagosomes; these then fuse with lysosomes, where the food particles are

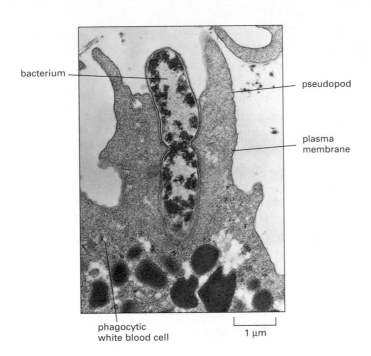

bacterium

pseudopod

plasma
membrane

phagocytic
white blood cell

1 μm

Figure 14–27 Phagocytosis of a bacterium by a white blood cell. Electron micrograph of a phagocytic white blood cell (a neutrophil) ingesting a bacterium, which is in the process of dividing. The white blood cell has extended surface projections called pseudopods, which progressively envelop the bacterium. (Courtesy of Dorothy F. Bainton.)

digested. Few cells in multicellular organisms are able to ingest large particles efficiently. In the animal gut, for example, large particles of food have to be broken down to individual molecules by extracellular enzymes before they can be taken up by the absorptive cells lining the gut.

Nevertheless, phagocytosis is important in most animals for purposes other than nutrition. It is most efficiently carried out by **phagocytic cells,** such as *macrophages,* which are widely distributed in tissues, and some white blood cells. Phagocytic cells defend us against infection by ingesting invading microorganisms. To be taken up by a macrophage or a white blood cell, particles must first bind to the phagocytic cell surface and activate one of a variety of surface receptors. Some of these receptors recognize antibodies, the proteins that protect us against infection by binding to the surface of microorganisms. Binding of antibody-coated bacteria to these receptors induces the phagocytic cell to extend sheetlike projections of the plasma membrane, called *pseudopods,* that engulf the bacterium (Figure 14–27) and fuse at their tips to form a *phagosome.*

Phagocytic cells also play an important part in scavenging dead and damaged cells and cellular debris. Macrophages, for example, ingest more than 10^{11} of our worn-out red blood cells each day (Figure 14–28).

Fluid and Macromolecules Are Taken Up by Pinocytosis

Eucaryotic cells continually ingest bits of their plasma membrane in the form of small pinocytic vesicles that are later returned to the cell surface. The rate at which plasma membrane is internalized by **pinocytosis** varies from cell type to cell type, but it is usually surprisingly large. A

Figure 14–28 Phagocytosis of red blood cells by a macrophage. Scanning electron micrograph of a macrophage phagocytosing two red blood cells. The *red arrows* point to the edges of fine sheets of membrane—pseudopods—that the macrophage is extending like collars to engulf the red cells. (Courtesy of Jean Paul Revel.)

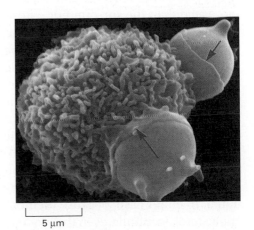

5 μm

macrophage, for example, ingests 25% of its own volume of fluid each hour. This means that it ingests 3% of its plasma membrane each minute, or 100% in about half an hour. Fibroblasts endocytose at a somewhat lower rate, whereas some phagocytic amoebae ingest their plasma membrane even more rapidly. Since a cell's total surface area and volume remain unchanged during this process, it is clear that as much membrane is being added to the cell surface by vesicle fusion (exocytosis) as is being removed by endocytosis.

Pinocytosis is mainly carried out by the clathrin-coated pits and vesicles that we discussed earlier (see Figures 14–18 and 14–19). After they pinch off from the plasma membrane, clathrin-coated vesicles rapidly shed their coat and fuse with an *endosome*. Extracellular fluid is trapped in a coated pit as it invaginates to form a coated vesicle, and so substances dissolved in the extracellular fluid are internalized and delivered to endosomes. This fluid intake is generally balanced by fluid loss during exocytosis.

Receptor-mediated Endocytosis Provides a Specific Route into Animal Cells

Pinocytosis, as just described, is indiscriminate. The endocytic vesicles simply trap any molecules that happen to be present in the extracellular fluid and carry them into the cell. In most animal cells, however, pinocytosis via clathrin-coated vesicles also provides an efficient pathway for taking up specific macromolecules from the extracellular fluid. The macromolecules bind to complementary receptors on the cell surface and enter the cell as receptor-macromolecule complexes in clathrin-coated vesicles. This process, called **receptor-mediated endocytosis,** provides a selective concentrating mechanism that increases the efficiency of internalization of particular macromolecules more than 1000-fold compared with ordinary pinocytosis, so that even minor components of the extracellular fluid can be taken up in large amounts without taking in a correspondingly large volume of extracellular fluid. An important example of receptor-mediated endocytosis is its use by animal cells to take up the cholesterol they need to make new membrane.

Cholesterol is extremely insoluble and is transported in the bloodstream bound to protein in the form of particles called *low-density lipoproteins,* or *LDL.* The LDL binds to receptors located on cell surfaces, and the receptor-LDL complexes are ingested by receptor-mediated endocytosis and delivered to **endosomes.** The interior of endosomes is more acid than the surrounding cytosol or the extracellular fluid, and in this acidic environment the LDL dissociates from its receptor: the receptors are returned in transport vesicles to the plasma membrane for reuse, while the LDL is delivered to lysosomes. In the lysosomes the LDL is broken down by hydrolytic enzymes; the cholesterol is released and escapes into the cytosol, where it is available for new membrane synthesis. The LDL receptors on the cell surface are continually internalized and recycled, whether they are occupied by LDL or not (Figure 14–29).

This pathway for cholesterol uptake is disrupted in individuals who inherit a defective gene encoding the LDL receptor protein. In some cases the receptors are missing; in others they are present but nonfunctional. In either case, because their cells are deficient in taking up LDL, cholesterol accumulates in the blood and predisposes the individuals to

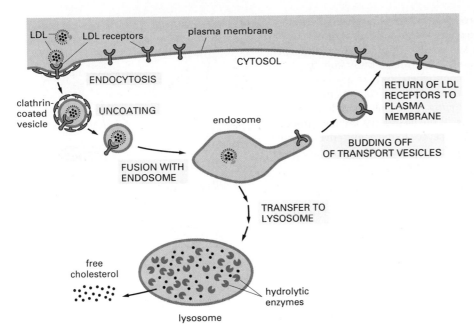

Figure 14–29 Receptor-mediated endocytosis of LDL. The LDL binds to receptors on the cell surface and is internalized in clathrin-coated vesicles. The vesicles lose their coat and then fuse with endosomes. In the acidic environment of the endosome, LDL dissociates from its receptors. Whereas the LDL ends up in lysosomes, where it is degraded to release free cholesterol, the LDL receptors are returned to the plasma membrane via transport vesicles to be used again. For simplicity, only one LDL receptor is shown entering the cell and returning to the plasma membrane. Whether it is occupied or not, an LDL receptor typically makes one round trip into the cell and back every 10 minutes, making a total of several hundred trips in its 20-hour life span.

develop atherosclerosis. Most die at an early age of heart attacks resulting from clogging of the arteries that supply the heart.

Receptor-mediated endocytosis is also used to take up many other essential metabolites, such as vitamin B_{12} and iron, that cells cannot take up by the processes of membrane transport discussed in Chapter 12. Vitamin B_{12} and iron are both required, for example, for the synthesis of hemoglobin, which is the major protein in red blood cells; they enter immature red blood cells as a complex with protein. Many cell-surface receptors that bind extracellular signaling molecules are also ingested by this pathway: some are recycled to the plasma membrane for reuse, while others are degraded in lysosomes. Unfortunately, receptor-mediated endocytosis can also be exploited by viruses: the influenza virus and HIV, which causes AIDS, gain entry into cells in this way.

Endocytosed Macromolecules Are Sorted in Endosomes

Because extracellular material taken up by pinocytosis is rapidly transferred to endosomes, it is possible to visualize the endosomal compartment by incubating living cells in fluid containing an electron-dense marker that will show up when viewed in an electron microscope. When loaded in this way, the endosomal compartment appears as a complex set of connected membrane tubes and larger vesicles. Two sets of endosomes can be distinguished in such loading experiments: the marker molecules appear first in *early endosomes,* just beneath the plasma membrane, and 5–15 minutes later in *late endosomes,* near the nucleus. The interior of the endosome compartment is kept acidic (pH 5–6) by an ATP-driven H^+ pump in the endosomal membrane that pumps H^+ into the endosome lumen from the cytosol.

The endosomal compartment acts as the main sorting station in the inward endocytic pathway, just as the *trans* Golgi network serves this function in the outward secretory pathway. The acidic environment of the endosome plays a crucial part in the sorting process by causing many receptors to release their bound cargo. The routes taken by receptors once they have entered an endosome differ according to the type of

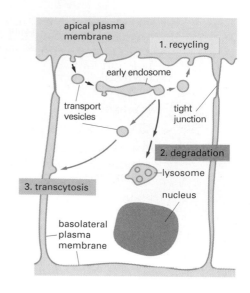

Figure 14–30 Possible fates for receptor proteins involved in receptor-mediated endocytosis. Three pathways from the endosomal compartment in an epithelial cell are shown. Receptors that are not specifically retrieved from early endosomes follow the pathway from the endosomal compartment to lysosomes, where they are degraded. Retrieved receptors are returned either to the same plasma membrane domain from which they came (*recycling*) or to a different domain of the plasma membrane (*transcytosis*). If the ligand that is endocytosed with its receptor stays bound to the receptor in the acidic environment of the endosome, it will follow the same pathway as the receptor; otherwise it will be delivered to lysosomes for degradation.

receptor: (1) most are returned to the same plasma membrane domain from which they came, as is the case for the LDL receptor discussed earlier; (2) some travel to lysosomes, where they are degraded; and (3) some proceed to a different domain of the plasma membrane, thereby transferring their bound cargo molecules from one extracellular space to another, a process called *transcytosis* (Figure 14–30).

Cargo molecules that remain bound to their receptors share the fate of their receptors. Those that dissociate from their receptors in the endosome are doomed to destruction in lysosomes along with most of the contents of the endosome lumen. It remains uncertain how molecules move from endosomes to lysosomes. One possibility is that they are carried in transport vesicles; another is that endosomes gradually convert into lysosomes.

Lysosomes Are the Principal Sites of Intracellular Digestion

Many extracellular particles and molecules ingested by cells end up in **lysosomes,** which are membranous sacs of hydrolytic enzymes that carry out the controlled intracellular digestion of both extracellular materials and worn-out organelles. They contain about 40 types of hydrolytic enzymes, including those that degrade proteins, nucleic acids, oligosaccharides, and phospholipids. All are optimally active in the acidic conditions (pH ~5) maintained within lysosomes. The membrane of the lysosome normally keeps these destructive enzymes out of the cytosol (whose pH is about 7.2), but the acid dependence of the enzymes protects the contents of the cytosol against damage even if some leakage should occur.

Like all other intracellular organelles, the lysosome not only contains a unique collection of enzymes but also has a unique surrounding membrane. The lysosomal membrane contains transport proteins that allow the final products of the digestion of macromolecules, such as amino acids, sugars, and nucleotides, to be transported to the cytosol, from where they can be either excreted or utilized by the cell. It also contains an ATP-driven H^+ pump, which, like that in the endosome membrane, pumps H^+ into the lysosome, thereby maintaining its contents at an acidic pH (Figure 14–31). Most of the lysosomal membrane proteins are unusually highly glycosylated; the sugars, which cover much of the protein surfaces facing the lumen, protect the proteins from digestion by the lysosomal proteases.

The specialized digestive enzymes and membrane proteins of the lysosome are synthesized in the ER and transported through the Golgi apparatus to the *trans* Golgi network. While in the ER and the *cis* Golgi

Question 14–8 Iron (Fe) is an essential trace metal that is needed by all cells. It is required, for example, for the synthesis of heme groups that are part of the active site of many enzymes involved in electron-transfer reactions; it is also required in hemoglobin, the main protein in red blood cells. Iron is taken up by cells by receptor-mediated endocytosis. The iron-uptake system has two components, a soluble protein called transferrin, which circulates in the bloodstream, and a transferrin receptor, a transmembrane protein that, like the LDL receptor in Figure 14–29, is continually endocytosed and recycled to the plasma membrane. Fe ions bind to transferrin at neutral pH but not at acidic pH. Transferrin binds to the transferrin receptor at neutral pH only when it has an Fe ion bound, but it binds to the receptor at acidic pH even in the absence of bound iron. From these properties, describe how iron is taken up and discuss the advantages of this elaborate scheme.

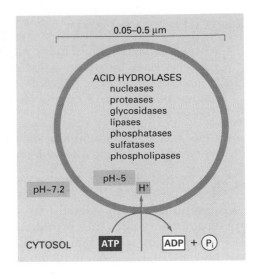

Figure 14–31 **A lysosome.** The acid hydrolases are hydrolytic enzymes that are active under acidic conditions. The lumen of the lysosome is maintained at an acidic pH by an H⁺ ATPase in the membrane that pumps H⁺ into the lumen.

network, the enzymes are tagged with a specific phosphorylated sugar group (mannose 6-phosphate) so that when they arrive in the *trans* Golgi network they can be recognized by an appropriate receptor, the mannose 6-phosphate receptor, and thereby be sorted and packaged into transport vesicles, which bud off and deliver their contents to lysosomes via late endosomes (see Figure 14–17).

Depending on their source, materials follow different paths to lysosomes. We have seen that extracellular particles are taken up into phagosomes, which fuse with lysosomes, and that extracellular fluid and macromolecules are taken up into smaller endocytic vesicles, which deliver their contents to lysosomes via endosomes. But cells have an additional pathway for supplying materials to lysosomes, which is used for degrading obsolete parts of the cell itself. In electron micrographs of liver cells, for example, one often sees lysosomes digesting mitochondria, as well as other organelles. The process seems to begin with the enclosure of the organelle by membrane derived from the ER, creating an *autophagosome,* which then fuses with lysosomes (Figure 14–32). It is not known what marks an organelle for such destruction.

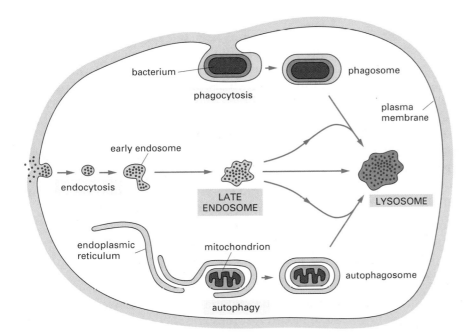

Figure 14–32 **Three pathways to degradation in lysosomes.** Each pathway leads to the intracellular digestion of materials derived from a different source. The compartments resulting from the three pathways can sometimes be distinguished morphologically—hence the terms "autophagolysosome," "phagolysosome," and so on. Such lysosomes differ, however, only because of the different materials they are digesting.

Essential Concepts

- Eucaryotic cells contain many membrane-bounded organelles, including a nucleus, endoplasmic reticulum (ER), Golgi apparatus, lysosomes, endosomes, mitochondria, chloroplasts (in plant cells), and peroxisomes.

- Cells make new membrane-bounded organelles by enlarging existing ones, which then divide.

- Most organelle proteins are made in the cytosol and transported into the organelle where they function. Sorting signals in the amino acid sequence guide the proteins to the correct organelle; proteins that function in the cytosol have no signals and remain where they are made.

- Nuclear proteins contain nuclear localization signals that help direct their active transport from the cytosol into the nucleus through nuclear pores, which penetrate the double-membrane nuclear envelope.

- Most mitochondrial and chloroplast proteins are made in the cytosol and are then actively transported into the organelles by protein translocators in their membranes.

- The ER is the membrane factory of the cell; it makes most of the cell's lipids and many of its proteins. The proteins are made by ribosomes bound to the surface of the rough ER.

- Ribosomes in the cytosol are directed to the ER if the protein they are making has an ER signal sequence, which is recognized by a signal-recognition particle (SRP) in the cytosol; the binding of the ribosome-SRP complex to a receptor on the ER membrane initiates the translocation process that threads the growing polypeptide across the ER membrane through a translocation channel.

- Soluble proteins destined for secretion or the lumen of an organelle pass completely into the ER lumen, while transmembrane proteins destined for the ER membrane or other cell membranes remain anchored in the lipid bilayer by one or more membrane-spanning α helices.

- In the ER lumen, proteins fold up, assemble with other proteins, form disulfide bonds, and become decorated with oligosaccharide chains.

- Exit from the ER is an important quality-control step; proteins that either fail to fold properly or fail to assemble with their normal partners are retained in the ER and are eventually degraded.

- Protein transport from the ER to the Golgi apparatus and from the Golgi apparatus to other destinations is mediated by transport vesicles that continually bud off from one membrane and fuse with another, a process called vesicular transport.

- Budding transport vesicles have distinctive coat proteins on their cytosolic surface; the assembly of the coat drives the budding process, and the coat proteins help incorporate receptors with their bound cargo molecules into the forming vesicle.

- Coated vesicles lose their protein coat soon after pinching off, enabling them to dock and then fuse with a particular target membrane; docking is thought to be mediated by v-SNAREs and t-SNAREs, while fusion is catalyzed by cytosolic proteins that assemble into a fusion complex at the docking site.

- The Golgi apparatus receives newly made proteins from the ER; it modifies their oligosaccharides, sorts the proteins, and dispatches them from the *trans* Golgi network to the plasma membrane, lysosomes, or secretory vesicles.

- In all eucaryotic cells, transport vesicles continually bud from the *trans* Golgi network and fuse with the plasma membrane, a process called constitutive exocytosis; the process delivers plasma membrane lipids and proteins to the cell surface and also releases molecules from the cell, a process called secretion.

- Specialized secretory cells also have a regulated exocytosis pathway, where molecules stored in secretory vesicles are released from the cell by exocytosis when the cell is signaled to secrete.

- Cells ingest fluid, molecules, and sometimes even particles, by endocytosis, in which regions of plasma membrane invaginate and pinch off to form endocytic vesicles.

- Much of the material that is endocytosed is delivered to endosomes and then to lysosomes, where it is degraded by hydrolytic enzymes; most of the components of the endocytic vesicle membrane, however, are recycled in transport vesicles back to the plasma membrane for reuse.

Key Terms

clathrin

coated vesicle

cytosol

endoplasmic reticulum

endosome

exocytosis

Golgi apparatus

lysosome

membrane-bounded organelle

nuclear envelope

nuclear pore

phagocytic cell

phagocytosis

pinocytosis

receptor-mediated endocytosis

rough endoplasmic reticulum

secretion

secretory vesicle

signal sequence

SNARE

trans Golgi network

transport vesicle

Questions

Question 14–9 Which of the following statements are correct? Explain your answers.

A. Ribosomes are cytoplasmic structures that, during protein synthesis, become linked by an mRNA molecule to form polyribosomes.

B. The amino acid sequence Leu-His-Arg-Leu-Asp-Ala-Gln-Ser-Lys-Leu-Ser-Ser is a signal sequence that directs proteins to the ER.

C. All transport vesicles in the cell must have a v-SNARE protein in their membrane.

D. Transport vesicles deliver proteins and lipids to the cell surface.

E. If the delivery of prospective lysosomal proteins from the *trans* Golgi network to the late endosomes were blocked, lysosomal proteins would be secreted by the constitutive secretion pathways shown in Figure 14–25.

F. Lysosomes digest only substances that have been taken up by cells by endocytosis.

G. *N*-linked sugar chains are found on glycoproteins that face the cell surface, as well as on glycoproteins that face the lumens of the ER, *trans* Golgi network, and mitochondria.

Question 14–10 How do you suppose that proteins with a nuclear export signal get into the nucleus?

Question 14–11 Influenza viruses are surrounded by a membrane that contains a fusion protein, which is activated by acidic pH. Upon activation, the protein causes the viral membrane to fuse with cell membranes. An old folk remedy against flu recommends that one should spend a night in a horse's stable. Odd as it may sound, there is a rational explanation for this advice. Air in stables contains ammonia (NH_3) generated by bacteria from the horses' urine. Sketch a diagram showing the pathway (in detail) by which flu virus enters cells and speculate how NH_3 may protect cells from virus infection. (Hint: NH_3 can neutralize acidic solutions by the reaction $NH_3 + H^+ \rightarrow NH_4^+$.)

Question 14–12 Consider the v-SNAREs that direct transport vesicles from the *trans* Golgi network to the plasma membrane. They, like all other v-SNAREs, are membrane proteins that are integrated into the membrane of the ER during their biosynthesis and are then transported by transport vesicles to their destination. Thus, transport vesicles budding from the ER contain at least two kinds of v-SNAREs—those that target the vesicles to the *cis* Golgi cisternae and those that are in transit to the *trans* Golgi network to be packaged in different transport vesicles destined for the plasma membrane. (A) Why might this be a problem? (B) Suggest possible ways in which the cell might solve it.

Question 14–13 A particular type of *Drosophila* mutant becomes paralyzed when the temperature is raised. The mutation affects the structure of dynamin, causing it to be inactivated at the higher temperature. Indeed, the function of dynamin was discovered by analyzing the defect in these mutant fruit flies. The complete paralysis at the elevated temperature suggests that synaptic transmission between nerve and muscle cells (discussed in Chapter 12) is blocked. Suggest why signal transmission at a synapse might require dynamin. On the basis of your hypothesis, what would you expect to see in electron micrographs of synapses of flies that were exposed to the elevated temperature?

Question 14–14 Edit the following statements, if required, to make them true: "Because nuclear localization sequences are not cleaved off by proteases following protein import into the nucleus, they can be reused to import nuclear proteins after mitosis, when cytosolic and nuclear proteins have become intermixed. This is in contrast to ER signal sequences, which are cleaved off by a signal peptidase once they reach the lumen of the ER. ER signal sequences cannot therefore be reused to import ER proteins after mitosis, when cytosolic and ER proteins have become intermixed; these ER proteins must therefore be degraded and resynthesized."

Question 14–15 Consider a protein that contains an ER signal sequence at its amino terminus and a nuclear localization sequence in its middle. What do you think the fate of this protein would be? Explain your answer.

Question 14–16 Compare and contrast protein import into the ER and into the nucleus. List at least two major differences in the mechanisms and speculate why the ER mechanism might not work for nuclear import and vice versa.

Question 14–17 During mitosis, the nuclear envelope breaks down and into small vesicles and intranuclear proteins completely intermix with cytosolic proteins. Is this consistent with the evolutionary scheme proposed in Figure 14–3?

Question 14–18 A protein that inhibits certain proteolytic enzymes (proteases) is normally secreted into the bloodstream by liver cells. This inhibitor protein, antitrypsin, is absent from the bloodstream of patients who carry a mutation that results in a single amino acid change in the protein. Antitrypsin deficiency causes a variety of severe problems, particularly in lung tissue, because of the uncontrolled activity of proteases. Surprisingly, when the mutant antitrypsin is synthesized in the laboratory, it is as active as the normal antitrypsin at inhibiting proteases. Why then does the mutation cause the disease? Think of more than one possibility and suggest ways in which you could distinguish between them.

Question 14–19 Dr. Outonalimb's claim to fame is her discovery of forgettin, a protein predominantly made by the pineal gland in human teenagers. The protein causes selective short-term unresponsiveness and memory loss when the auditory system receives statements like "Please take out the garbage!" Her hypothesis is that forgettin has a hydrophobic ER signal sequence at its carboxy terminus that is recognized by SRP and causes it to be translocated across the ER membrane by the mechanism shown in Figure 14–13. She predicts that the protein is secreted from pineal cells into the bloodstream, from where it exerts its devastating systemic effects. You are a member of the committee deciding whether she should receive a grant for further work on her hypothesis. Consider that grant reviews should be polite and constructive.

Question 14–20 Taking the evolutionary scheme in Figure 14–3 one step further, suggest how the Golgi apparatus could have evolved. Sketch a simple diagram to illustrate your ideas. For Golgi function to be useful, what else would have to have evolved?

Question 14–21 If membrane proteins are integrated into the ER membrane by means of the ER protein translocation channel (which is itself composed of membrane proteins), how do the first protein translocation channels become incorporated into the ER membrane?

Question 14–22 The sketch in Figure Q14–22 is a schematic drawing of the electron micrograph shown in the third panel of Figure 14–18A. Name the structures that are labeled in the sketch.

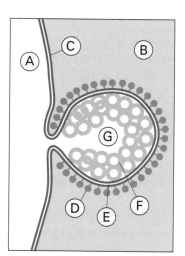

Figure Q14–22

15 Cell Communication

Individual cells, like multicellular organisms, need to sense and respond to their environment. A typical free-living cell, for example, must be able to sniff out nutrients, sense the difference between light and dark, and detect and avoid poisons and predators. If such a cell is to have a social life, it must be able to use its senses to communicate with other cells. When a yeast cell is ready to mate, for instance, it secretes a small protein called a mating factor, to which other yeast cells are sensitive: they detect the mating factor and respond by halting their progress through the cell cycle and putting out a protrusion toward the cell that emitted the signal (Figure 15–1).

In a multicellular organism, cells must coordinate their behavior in many different ways. As in any busy community, there is a constant hubbub of communication: neighbors carry on private conversations, public announcements are broadcast to the whole population, urgent messages are delivered from distant sites to individuals, alarms are rung when danger threatens. These social interactions are vital both for the survival of the community and for its creation. During development, cells in the embryo exchange signals to determine which specialized role each cell shall adopt, what position it shall occupy, and whether it shall live or die or divide; later, a huge variety of signals coordinate the animal's growth and its day-to-day physiology and behavior. In plants, too, cells are in constant communication with one another, responding to the conditions of light, dark, and temperature that guide the cycle of growth, flowering, and fruiting, and adjusting what happens in one part of the plant to what happens elsewhere.

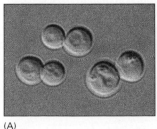
(A)
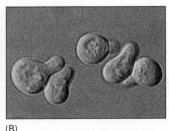
(B)

(C)

Figure 15–1 **Yeast cells responding to mating factor.** Cells of *Saccharomyces cerevisiae* are normally spherical, as shown in (A), but when exposed to mating factor produced by neighboring yeast cells they put out a protrusion toward the source of the factor, as shown in (B). Cells adopting this shape in response to the mating signal are called "shmoos" after Al Capp's 1940s cartoon character (C). (A and B, courtesy of Michael Snyder; C, copyright Capp Enterprises, Inc., all rights reserved.)

In this chapter we examine some of the most important means by which animal cells communicate and discuss how the cells interpret the signals they receive. We concentrate on animal cells, since the mechanisms of signal reception and interpretation in plant cells are far less well understood. We shall begin with an overview of the general principles of cell signaling and then consider two of the main systems used by animal cells to receive and interpret signals.

General Principles of Cell Signaling

To transmit a message from one person to another, information might first be written in ink on paper, then read aloud, then carried as electric impulses along a telephone wire, and finally expressed in the form of nerve impulses in the recipient's brain. At successive steps along this communication pathway, different forms of signals are used to represent the same information; the critical points in transmission occur where the information is converted from one form to another. This process of conversion is called **signal transduction** (Figure 15–2).

The signals that pass between cells are far simpler than human messages: typically, a particular type of molecule is produced by one cell—the *signaling cell*—and detected by another—the *target cell*—by means of a *receptor protein*, which recognizes and responds specifically to the signal molecule. The receptor protein performs the first step in a series of transduction processes at the receiving end of the signaling pathway, in the target cell, where the incoming extracellular signal is converted to intracellular signals that direct cell behavior. Most of this chapter will be concerned with signal reception and transduction. When cell biologists refer to **cell signaling,** it is these two aspects that they generally have in mind. First, however, we look briefly at the different types of signals that cells send to one another.

Signals Can Act Over Long or Short Range

Cells in multicellular organisms use hundreds of kinds of extracellular molecules to send signals to one another—proteins, peptides, amino acids, nucleotides, steroids, fatty acid derivatives, and even dissolved gases—but there are only a handful of basic styles of communication (Figure 15–3).

The most public style is to broadcast the signal over the whole body by secreting it into the bloodstream (of an animal) or the sap (of a plant). Signal molecules used in this way are called **hormones,** and in animals,

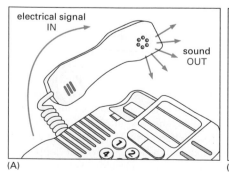

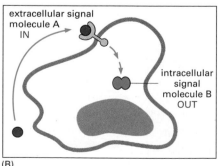

Figure 15–2 **Signal transduction.** In (A) a telephone receiver converts an electrical signal into a sound signal. In (B) a cell converts an extracellular signal (signal molecule A) into an intracellular signal (signal molecule B).

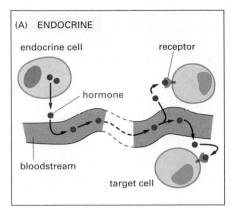

(A) ENDOCRINE

endocrine cell

receptor

hormone

bloodstream

target cell

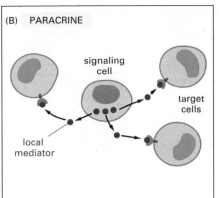

(B) PARACRINE

signaling cell

target cells

local mediator

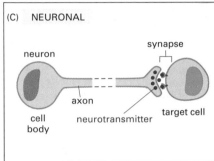

(C) NEURONAL

synapse

neuron

axon

cell body

neurotransmitter

target cell

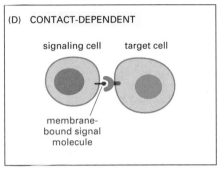

(D) CONTACT-DEPENDENT

signaling cell   target cell

membrane-bound signal molecule

Figure 15–3 Forms of cell signaling.
(A) Hormones produced in endocrine glands are secreted into the bloodstream and can be distributed widely throughout the body. (B) Paracrine signals are released by cells into the extracellular medium in their neighborhood and act locally. (C) Neuronal signals are transmitted along axons to remote target cells. (D) Contact-dependent signaling requires cells to be in direct membrane-to-membrane contact with each other. Many of the same types of signal molecules are used in endocrine, paracrine, and neuronal signaling. The crucial differences lie in the speed and selectivity with which the signals are delivered to their targets.

the cells that produce hormones are called *endocrine* cells (Figure 15–3A).

Less public is the process known as *paracrine* signaling. Here, the signal molecules diffuse locally through the extracellular medium, remaining in the neighborhood of the cell that secretes them: they act as **local mediators** (Figure 15–3B). Many of the signal molecules that regulate inflammation at sites of infection or cell proliferation in wound healing function in this way.

A third style of communication is **neuronal** signaling. As in hormonal signaling, messages are often delivered over long distances; in neuronal signaling, however, messages are delivered over private lines to individual cells very quickly (Figure 15–3C). As discussed in Chapter 12, the axon of a neuron terminates at specialized junctions (synapses) on target cells far away from the neuronal cell body. When activated by signals from the environment or from other nerve cells, the neuron sends electrical impulses along its axon at speeds of up to 100 meters per second. On reaching the axon terminal, the intracellular electrical signals are converted into an extracellular chemical form: each electrical impulse stimulates the terminal to secrete a pulse of a chemical signal called a **neurotransmitter.** Neurotransmitters diffuse across the narrow (<100 nm) gap between the axon-terminal membrane and the membrane of the target cell in less than a millisecond.

A fourth style of cell-cell communication—the most intimate and short-range of all—does not require the release of a secreted molecule. Instead, the cells make direct contact through signaling molecules in their plasma membranes. The message is delivered by the binding of a signal molecule anchored in the plasma membrane of the signaling cell to a receptor molecule embedded in the plasma membrane of the target cell (Figure 15–3D). While a neuronal signal is like a telephone call, such *contact-dependent signaling* is like a private face-to-face conversation.

General Principles of Cell Signaling

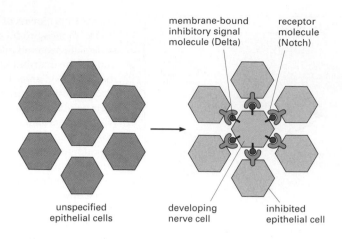

membrane-bound
inhibitory signal
molecule (Delta)

receptor
molecule
(Notch)

unspecified
epithelial cells

developing
nerve cell

inhibited
epithelial cell

Figure 15–4 **Contact-dependent signaling in the control of nerve-cell production.** The nervous system originates in the embryo from a sheet of epithelial cells. Isolated cells in this sheet begin to specialize as neurons, while their neighbors remain nonneuronal and maintain the epithelial structure of the sheet. Each future neuron delivers an inhibitory signal to the cells in contact with it, deterring them from specializing as neurons too. Both this signal molecule (called Delta) and the receptor molecule (called Notch) are transmembrane proteins. The same mechanism, mediated by essentially the same molecules, controls the detailed pattern of differentiated cell types in various other tissues, in both vertebrates and invertebrates. In mutants where the mechanism fails, some cell types (such as neurons) are produced in great excess at the expense of others.

Question 15–1 To remain a local response, paracrine signal molecules must be prevented from straying too far from their points of origin. Suggest different ways by which this could be accomplished. Explain your answers.

In embryonic development, for example, contact-dependent signaling has an important role in tissues where adjacent cells that are initially similar have to become specialized in different ways. Thus in the sheet of cells that gives rise to the nervous system, individual cells have to be singled out to differentiate as neurons while their neighbors remain nonneuronal. The signals that control this process are transmitted via cell–cell contacts: each future neuron inhibits its immediate neighbors from differentiating as neurons too (Figure 15–4).

Table 15–1 lists some examples of hormones, local mediators, neurotransmitters, and contact-dependent signal molecules. The action of some of these will be discussed in more detail later in this chapter.

Each Cell Responds to a Limited Set of Signals

A typical cell in a multicellular organism is exposed to hundreds of different signal molecules in its environment. These may be free in the extracellular fluid, or attached to the extracellular matrix, or bound to the surfaces of neighboring cells. The cell must respond selectively to this mixture of signals, disregarding some and reacting to others, according to the cell's specialized function.

Whether a cell reacts to a signal molecule depends first of all on whether it possesses a receptor for that signal. Without a receptor, the cell will be deaf to the signal and can make no response. By producing only a limited set of receptors out of the thousands possible, therefore, the cell restricts the range of signals that can affect it. But this limited range of signals can still be used to control the behavior of the cell in complex ways. The complexity is of two sorts.

First, one signal, binding to one type of receptor protein, can cause a multitude of effects in the target cell: shape, movement, metabolism, gene expression—all may be altered together. As we shall see, the signal from a cell-surface receptor is generally relayed into the cell interior through a set of other intracellular components that produce widespread effects. This intracellular relay system and the intracellular targets on which it acts vary from one type of specialized cell to another, so that different cells respond to the same signal in different ways. Thus, when a heart muscle cell is exposed to the neurotransmitter acetylcholine, it decreases the frequency of its contractions, but when a salivary gland is exposed to the same signal, it secretes components of saliva (Figure 15–5).

Table 15–1   Some Examples of Signal Molecules

Signal Molecule	Site of Origin	Chemical Nature	Some Actions
Hormones			
Adrenaline	adrenal gland	derivative of the amino acid tyrosine	increases blood pressure, heart rate, and metabolism
Cortisol	adrenal gland	steroid (derivative of cholesterol)	affects metabolism of proteins, carbohydrates, and lipids in most tissues
Estradiol	ovary	steroid (derivative of cholesterol)	induces and maintains secondary female sexual characteristics
Glucagon	α cells of pancreas	peptide	stimulates glucose synthesis, glycogen breakdown, and lipid breakdown in, e.g., liver and fat cells
Insulin	β cells of pancreas	protein	stimulates glucose uptake, protein synthesis, and lipid synthesis in, e.g., liver cells
Testosterone	testis	steroid (derivative of cholesterol)	induces and maintains secondary male sexual characteristics
Thyroid hormone (thyroxine)	thyroid gland	derivative of the amino acid tyrosine	stimulates metabolism of many cell types
Local Mediators			
Epidermal growth factor (EGF)	various cells	protein	stimulates epidermal and many other cell types to proliferate
Platelet-derived growth factor (PDGF)	various cells, including blood platelets	protein	stimulates many cell types to proliferate
Nerve growth factor (NGF)	various innervated tissues	protein	promotes survival of certain classes of neurons; promotes growth of their axons
Histamine	mast cells	derivative of the amino acid histidine	causes blood vessels to dilate and become leaky, helping to cause inflammation
Nitric oxide (NO)	nerve cells; endothelial cells lining blood vessels	dissolved gas	causes smooth muscle cells to relax; regulates nerve cell activity
Neurotransmitters			
Acetylcholine	nerve terminals	derivative of choline	excitatory neurotransmitter at many nerve-muscle synapses and in central nervous system
γ-Aminobutyric acid (GABA)	nerve terminals	derivative of the amino acid glutamic acid	inhibitory neurotransmitter in central nervous system
Contact-dependent Signaling Molecules			
Delta	prospective neurons; various other embryonic cell types	transmembrane protein	inhibits neighboring cells from becoming specialized in same way as signaling cell

The second kind of complexity arises because a typical cell possesses a whole collection of different receptors—a few dozen, maybe, rather than thousands, but still enough to make the cell simultaneously sensitive to many extracellular signals. These signals, by acting together, can evoke responses that are more than just the sum of the effects that each signal would evoke on its own. The intracellular relay system for the different signals interact, so that the presence of one signal modifies the responses to another. Thus one combination of signals may simply enable a cell to survive; another may drive it to differentiate in some specialized way; another may cause it to divide; and in the absence of any signals, the cell may be programmed to die (Figure 15–6). In this way, a relatively small number of signals can be used in different combinations to give subtle and complex control over cell behavior.

General Principles of Cell Signaling

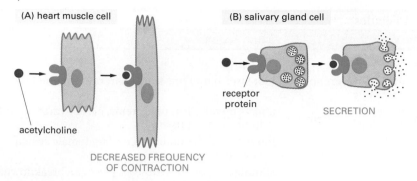

(A) heart muscle cell

acetylcholine

DECREASED FREQUENCY
OF CONTRACTION

(B) salivary gland cell

receptor
protein

SECRETION

(C) skeletal muscle cell

CONTRACTION

(D) acetylcholine

$$H_3C-\overset{\overset{\displaystyle O}{\|}}{C}-O-CH_2-CH_2-\overset{\overset{\displaystyle CH_3}{|}}{\underset{\underset{\displaystyle CH_3}{|}}{N^+}}-CH_3$$

Figure 15–5 **The same signal molecule can induce different responses in different target cells**. Different cell types are specialized to respond to the neurotransmitter acetylcholine in different ways. In (A) and (B), the signal molecule binds to similar receptor proteins, but these activate different responses in cells specialized for different functions. In (C) the cell produces a different type of receptor protein for the same signal. As we shall see, the different types of receptors generate quite different intracellular signals, and thus enable the different types of muscle cells to react differently to acetylcholine. (D) Chemical structure of acetylcholine.

Receptors Relay Signals via Intracellular Signaling Pathways

Before tracing in detail how any particular signal molecule controls cell behavior, it is helpful to take stock of some general principles.

Signal reception begins at the point where a signal originating outside the cell encounters a target molecule belonging to the cell itself. In virtually every case the target molecule is a **receptor protein,** and this is usually activated by just one type of signal. The receptor protein performs the primary transduction step: it receives the external signal, and it generates a new intracellular signal in response (see Figure 15–2B). As

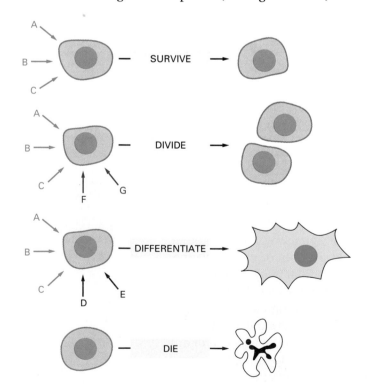

SURVIVE

DIVIDE

DIFFERENTIATE

DIE

Figure 15–6 **An animal cell depends on multiple extracellular signals.** Each cell type displays a set of receptor proteins that enables it to respond to a corresponding set of signal molecules produced by other cells. These signal molecules work in combinations to regulate the behavior of the cell. As shown here, many cells require multiple signals (*green arrows*) to survive, additional signals (*red arrows*) to divide, and still other signals (*black arrows*) to differentiate. If deprived of appropriate signals, most cells undergo a form of cell suicide known as programmed cell death, or apoptosis (discussed in Chapter 18).

a rule, this is only the first event in a subsequent chain of intracellular signal transduction processes. In these, the message is passed from one set of *intracellular signaling molecules* to another, each in turn provoking production of the next until, say, a metabolic enzyme is activated, the expression of a gene is switched on, or the cytoskeleton is kicked into action. This final outcome is the *response* of the cell.

These relay chains, or **signaling cascades,** of intracellular signaling molecules have several crucial functions (Figure 15–7):

1. They physically *transfer* the signal from the point at which it is received to the cell machinery that will make the response, which is often located in some other part of the cell.

2. They *transform* the signal into a molecular form that is able to stimulate that response.

3. In most cases, signaling cascades also *amplify* the signal received, making it stronger, so that a few extracellular signal molecules are enough to evoke a large response.

4. The signaling cascades can *distribute* the signal so as to influence several processes in parallel: at any step in the pathway, the signal can *diverge* and be relayed to a number of different intracellular targets, creating branches in the information flow diagram and evoking a complex response (see Figure 15–7).

5. Lastly, each step in this signaling cascade is open to interference by other factors, so that the transmission of the signal can be *modulated* according to conditions prevailing inside or outside the cell.

We shall consider first some of the simpler and more direct signaling pathways before going on to consider the longer cascades that relay signals from receptors at the cell membrane in animal cells.

Question 15–2 When a single photon of light is absorbed by a rhodopsin photoreceptor, it activates about 200 individual molecules of an intracellular signaling protein called transducin. Each molecule, in turn, binds to and activates an enzyme, phosphodiesterase, that hydrolyzes 4000 molecules of cyclic GMP per second. Cyclic GMP is a small molecule, similar to cyclic AMP, that in the cytosol of rod photoreceptor cells binds to the Na$^+$ channels in the plasma membrane and keeps them in their open conformation, as we discuss later (see Figure 15–26). What is the degree of signal amplification if each transducin molecule remains active for 100 milliseconds?

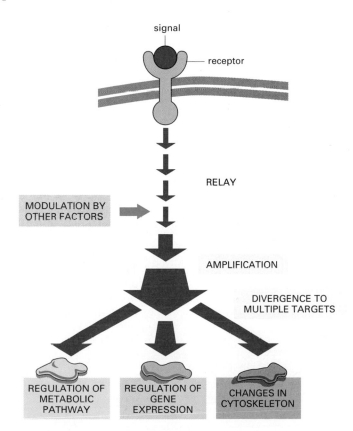

Figure 15–7 An intracellular signaling cascade. A receptor protein located on the cell surface transduces an extracellular signal into an intracellular signal, initiating a signaling cascade that relays the signal into the cell interior, amplifying and distributing it en route. Many of the steps in the cascade can be influenced (modulated) by other events in the cell.

Some Signal Molecules Can Cross the Plasma Membrane

Extracellular signal molecules in general fall into two classes, corresponding to two fundamentally different kinds of receptors. The first and largest class of signals consists of molecules that are too large or too hydrophilic to cross the plasma membrane of the target cell. The receptor proteins for these signal molecules therefore have to lie in the plasma membrane of the target cell and relay the message across the membrane (Figure 15–8A). The second, and smaller, class of signals consists of molecules that are sufficiently small and hydrophobic to diffuse across the plasma membrane. For these signal molecules, the receptors lie in the interior of the target cell and are generally either gene regulatory proteins (discussed in Chapter 8) or enzymes. These are activated when the signal molecule binds to them (Figure 15–8B).

The best-known hydrophobic signal molecules are the **steroid hormones** (including *cortisol, estradiol,* and *testosterone*) and the *thyroid hormones* such as *thyroxine* (Figure 15–9; see also Table 15–1, p. 485). All of these pass through the plasma membrane of the target cell and bind to receptor proteins located either in the cytosol or in the nucleus. The receptors for these hormones are gene regulatory proteins that are present in an inactive form in the unstimulated cell. When its hormone binds, the receptor protein undergoes a large conformational change that enables it to bind to its corresponding regulatory sequence in the DNA; it can then promote or inhibit the transcription of a selected set of genes (Figure 15–10). There are different receptor proteins for each type of hormone; each receptor acts at a different set of regulatory sites and thus regulates a different set of genes.

The essential role of the steroid hormone receptors is illustrated by the dramatic consequences of a lack of the receptor for testosterone in humans. The male sex hormone testosterone acts in the fetus and at puberty as a signal for development of male secondary sexual characteristics. Some very rare individuals are genetically male (that is, they have one X and one Y chromosome) but lack the testosterone receptor as a result of a mutation in the corresponding gene: they make the hormone, but their cells cannot respond to it. The consequence is that they develop, to all outward appearances, as females. This demonstrates the key role of the hormone receptor in the action of testosterone, and also shows that the receptor is required not just in one cell type to mediate

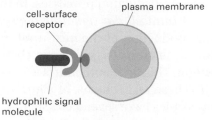

(A)   CELL-SURFACE RECEPTORS

cell-surface receptor

plasma membrane

hydrophilic signal molecule

(B)   INTRACELLULAR RECEPTORS

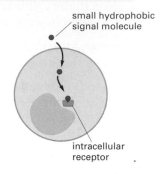

small hydrophobic signal molecule

intracellular receptor

Figure 15–8 Extracellular signal molecules bind either to cell-surface receptors or to intracellular receptors. (A) Most signal molecules are hydrophilic and are therefore unable to cross the plasma membrane directly; instead, they bind to cell-surface receptors, which in turn generate one or more signals inside the target cell. (B) Some small hydrophobic signal molecules, by contrast, diffuse across the plasma membrane and bind to receptors inside the target cell—either in the cytosol (as shown) or in the nucleus.

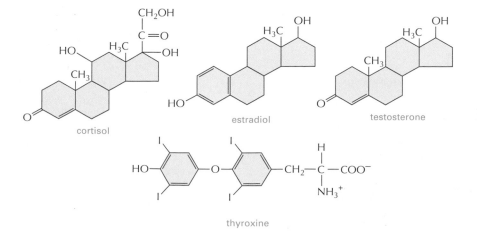

cortisol

estradiol

testosterone

thyroxine

Figure 15–9 Some small hydrophobic hormones whose receptors are intracellular gene regulatory proteins. The sites of origin and functions of these hormones are given in Table 15–1 (p. 485). The receptors for all of them are evolutionarily related, belonging to the *steroid-hormone receptor superfamily* of gene regulatory proteins.

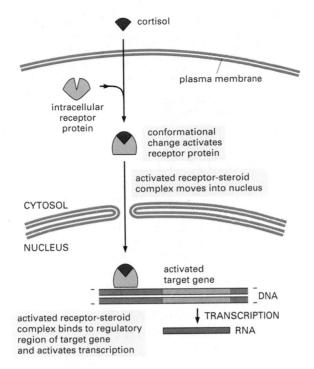

cortisol

plasma membrane

intracellular receptor protein

conformational change activates receptor protein

activated receptor-steroid complex moves into nucleus

CYTOSOL

NUCLEUS

activated target gene

activated receptor-steroid complex binds to regulatory region of target gene and activates transcription

DNA

TRANSCRIPTION

RNA

Figure 15–10 The steroid hormone cortisol acts by activating a gene regulatory protein. Cortisol diffuses directly across the plasma membrane and binds to its receptor protein, which is located in the cytosol. The complex of receptor and steroid is then transported into the nucleus via the nuclear pores. Steroid binding activates the receptor protein, which is then able to bind to specific regulatory sequences in the DNA and activate gene transcription, as discussed in Chapter 8. The receptors for cortisol and some other steroid hormones are located in the cytosol; those for the other signal molecules of this family are already located in the nucleus.

one effect, but in many cell types to produce the whole range of features that distinguish men from women.

Nitric Oxide Can Enter Cells to Activate Enzymes Directly

Responses involving changes of gene expression can take many minutes or even hours, as the genes must first be transcribed and the messenger RNA then translated into protein. Where a cell must respond to a signal within a few seconds or minutes, direct activation of an enzyme is a much quicker way to produce an effect. Some small signal molecules operate in this way. A remarkable example of a signal molecule that crosses the plasma membrane and activates an intracellular enzyme is **nitric oxide (NO)**. This dissolved gas is made from the amino acid arginine and operates as a local mediator in many tissues. Endothelial cells—the flattened cells that line every blood vessel—release NO in response to stimulation by nerve endings, and this NO causes smooth muscle cells in the vessel wall to relax, so that the vessel dilates and blood flows more freely through it (Figure 15–11). The effect of NO on blood vessels provides an explanation for the action of nitroglycerine, which has been used for almost 100 years to treat patients with angina (pain due to inadequate blood flow to heart muscle). The nitroglycerine is converted in the body to NO, which relaxes blood vessels, thereby

CH_3
$CH-CH_3$
CH_2
CH_2
CH_2
$CH-CH_3$
H_3C
H_3C
HO

cholesterol

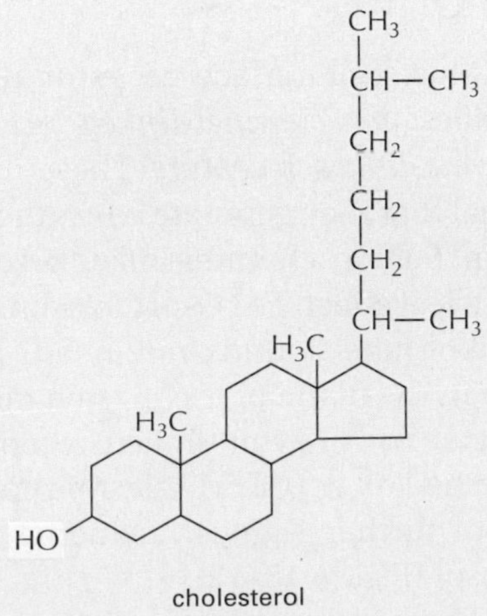

Figure Q15–3

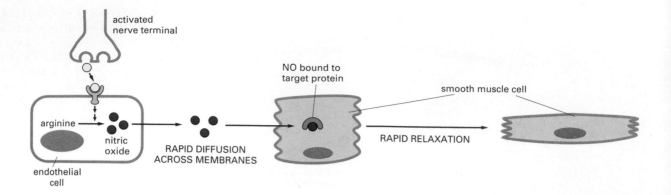

reducing the work load on the heart and, as a consequence, the oxygen requirement of the heart muscle. NO is also used by many nerve cells to signal neighboring cells: NO released by nerve terminals in the penis, for instance, causes the local blood-vessel dilation that is responsible for penile erection.

Dissolved NO diffuses readily out of the cell that makes it and into neighboring cells. It acts only locally because it has a short half-life (about 5–10 seconds) in the extracellular space before it is converted to nitrates and nitrites by reaction with oxygen and water. Once inside the target cell, the most common target of NO is the enzyme *guanylate cyclase*, which catalyzes the formation of *cyclic GMP* from the nucleotide GTP. Cyclic GMP is a small intracellular signaling molecule that then activates a short intracellular signaling pathway to evoke the cell's ultimate response. Cyclic GMP is very similar in structure and mechanism of action to *cyclic AMP*, a much more commonly used intracellular messenger molecule whose actions we discuss later.

Figure 15–11 The role of nitric oxide (NO) in smooth muscle relaxation in a blood-vessel wall. Acetylcholine released by nerve terminals in the blood-vessel wall activates endothelial cells lining the blood vessel to make and release NO. The NO diffuses out of the endothelial cells and into adjacent smooth muscle cells, causing the muscle cells to relax. Note that NO gas is highly toxic when inhaled and should not be confused with nitrous oxide (N_2O), also known as laughing gas.

There Are Three Main Classes of Cell-Surface Receptors

In contrast to the steroid and thyroid hormones, the vast majority of signal molecules are hydrophilic proteins, peptides, and other water-soluble molecules, which are unable to cross the plasma membrane of the target cell. Their receptor proteins therefore have to span the plasma membrane so that they can detect a signal on the outside and relay the message, in a new form, across the membrane into the interior of the cell.

Most cell-surface receptor proteins belong to one of three large families: *ion-channel-linked receptors*, *G-protein-linked receptors*, and *enzyme-linked receptors*. These differ in the nature of the intracellular signal that they generate when the extracellular signal molecule binds to them. For ion-channel-linked receptors, it is a flow of ions, producing an electrical effect. For G-protein-linked receptors, it is an activated form of a membrane-bound protein (a *G-protein subunit*) that is released to diffuse in the plane of the plasma membrane, initiating a cascade of other effects. For enzyme-linked receptors, it is an enzyme activity that is switched on at the cytoplasmic end of the receptor and generates a variety of further signals, including molecules that are released into the cytosol (Figure 15–12).

The number of different types of receptors in these three classes is even larger than the number of extracellular signals that act on them, since for many extracellular signal molecules there is more than one

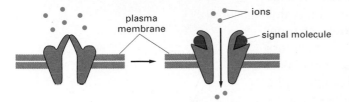

(A) ION-CHANNEL-LINKED RECEPTOR

plasma membrane

ions

signal molecule

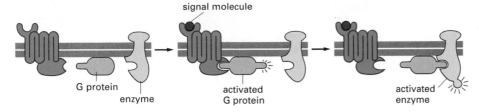

(B) G-PROTEIN-LINKED RECEPTOR

signal molecule

G protein

enzyme

activated G protein

activated enzyme

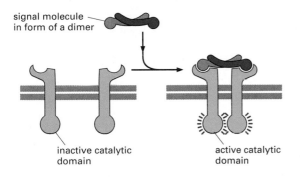

(C) ENZYME-LINKED RECEPTORS

signal molecule in form of a dimer

inactive catalytic domain

active catalytic domain

Figure 15–12 Three classes of cell-surface receptors. (A) An ion-channel-linked receptor opens (or closes, not shown) in response to binding of its ligand. (B) When a G-protein-linked receptor binds its ligand, the signal is passed first to a GTP-binding protein (a G protein) that associates with the receptor. This then leaves the receptor and activates a target enzyme (or ion channel, not shown) in the plasma membrane. For simplicity, the G protein is shown here as a single molecule; as we shall see, it is in fact a complex of three subunits that can dissociate. (C) An enzyme-linked receptor binds its extra-cellular ligand, switching on an enzyme activity at the other end of the receptor, on the opposite side of the membrane. In all three classes of receptors, binding of the signal molecule causes a change in the receptor, which relays the message onward without the need for the signal molecule itself to enter the cell.

type of receptor. The neurotransmitter acetylcholine, for example, acts on skeletal muscle cells via an ion-channel-linked receptor, whereas in heart muscle cells it acts through a G-protein-linked receptor (see Figure 15–5A and C). These two types of receptors generate different intra-cellular signals, and thus enable the two types of muscle cells to react to acetylcholine in different ways.

The multitude of different cell-surface receptors that the body requires for its own signaling purposes are also targets for many foreign substances that interfere with our physiology and sensations, from heroin and nicotine to tranquilizers and chili pepper. These substances either mimic the natural ligand for a receptor so as to occupy the normal ligand-binding site, or bind to the receptor at some other site, blocking or overstimulating the receptor's natural activity. Many drugs and poisons act in this way, and a large part of the pharmaceutical industry is devoted to the search for substances that will exert a precisely defined effect by binding to a specific type of cell-surface receptor.

Ion-Channel-linked Receptors Convert Chemical Signals into Electrical Ones

Ion-channel-linked receptors, also known as *transmitter-gated ion channels*, function in the simplest way, and they are discussed in Chapter 12. These are the receptors that serve for rapid transmission across synapses in the nervous system: they transduce a chemical signal, in the

Question 15–4 The signaling mechanisms used by a steroid-hormone receptor and by an ion-channel-linked receptor are both very simple and have very few components. Can they lead to an amplification of the initial signal, and, if so, how?

form of a pulse of neurotransmitter delivered to the outside of the target cell, directly into an electrical signal, in the form of a change in voltage across the plasma membrane. When the neurotransmitter binds, this type of receptor alters its conformation so as to open or close a channel for the flow of specific types of ions, such as Na^+, K^+, Ca^{2+}, or Cl^-, across the membrane (see Figure 15–12A). Driven by their electrochemical gradient, the ions rush into or out of the cell, creating a change in the membrane potential within a millisecond or so. This may trigger a nerve impulse, or alter the ability of other signals to do so. As we discuss later in this chapter, the opening of Ca^{2+} channels has special effects, as changes in the intracellular Ca^{2+} concentration can profoundly alter the activities of many enzymes.

Intracellular Signaling Cascades Act as a Series of Molecular Switches

Whereas ion-channel-linked receptors are a specialty of the nervous system, and of other electrically excitable cells such as muscle, G-protein-linked receptors and enzyme-linked receptors are used by practically every cell type of the body. Most of the remainder of this chapter will be concerned with them and with the signal transduction processes that they initiate.

Signals received via G-protein-linked or enzyme-linked receptors are transmitted to elaborate relay systems formed from cascades of intracellular signaling molecules. Apart from a few small molecules (such as cyclic GMP, cyclic AMP, and Ca^{2+}) these intracellular signaling molecules are proteins. Some serve as chemical transducers: in response to one type of chemical signal they generate another; others serve as messengers, receiving a signal in one part of the cell and moving to another to exert an effect.

Most of the key intracellular signaling proteins behave as **molecular switches:** receipt of a signal switches them from an inactive to an active state, and they then persist in that active state until some other process switches them off. The importance of the switching-off process is often not appreciated. If the signaling pathway is to recover after transmitting a signal and make itself ready to transmit another, every molecular switch must be reset to its original unstimulated state. Thus, at every step, for every activation mechanism there has to be an inactivation mechanism, and the two are equally important for the function of the system.

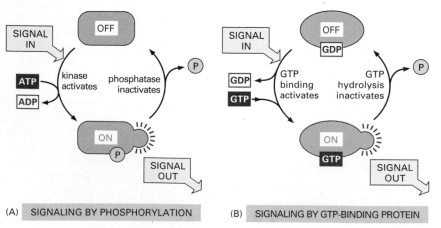

(A) SIGNALING BY PHOSPHORYLATION

(B) SIGNALING BY GTP-BINDING PROTEIN

Figure 15–13 **Intracellular signaling proteins that act as molecular switches.** In both cases an intracellular signaling protein is activated by the addition of a phosphate group and inactivated by the removal of the phosphate. In (A) the phosphate is added covalently to the protein by a protein kinase that transfers the terminal phosphate group from ATP to the signaling protein. The phosphate is removed by a protein phosphatase. In (B) a signaling protein is induced to exchange its bound GDP to GTP. Hydrolysis of the bound GTP to GDP switches the protein off.

Proteins that can act as molecular switches mostly fall into one of two main classes. The first and by far the largest class consists of proteins whose activity is turned on or off by phosphorylation, as discussed in Chapter 5 (see Figure 5–36). For these, the switch is thrown in one direction by a protein kinase, which adds a phosphate to the switch protein, and in the other direction by a protein phosphatase, which removes the phosphate from the switch protein (Figure 15–13A). Many of the switch proteins controlled by phosphorylation are themselves protein kinases, and these are often organized into *phosphorylation cascades:* one protein kinase, activated by phosphorylation, phosphorylates the next protein kinase in the sequence, and so on, transmitting the signal onward and, in the process, amplifying it, distributing it, and modulating it.

The other main class of switch proteins involved in signaling consists of GTP-binding proteins. These switch between an active and an inactive state according to whether they have GTP bound to them or GDP (Figure 15–13B); the mechanisms that control the switch on and the switch off will be described in the next section. GTP-binding proteins are important in several signaling pathways. One type, the G proteins, have a central role in signaling via G-protein-linked receptors, to which we now turn.

G-Protein-linked Receptors

G-protein-linked receptors form the largest family of cell-surface receptors, with hundreds of members already identified in mammalian cells. They mediate responses to an enormous diversity of extracellular signal molecules, including hormones, local mediators, and neurotransmitters. These signal molecules are as varied in structure as they are in function: they can be proteins, small peptides, or derivatives of amino acids or fatty acids, and for each one of them there is a different receptor or set of receptors.

Despite the diversity of the signal molecules that bind to them, all G-protein-linked receptors that have been analyzed have a similar structure, consisting of a single polypeptide chain that threads back and forth across the lipid bilayer seven times (Figure 15–14). This superfamily of *seven-pass transmembrane receptor proteins* includes rhodopsin (the light-activated photoreceptor protein in the vertebrate eye), as well as the olfactory (smell) receptors in the vertebrate nose. Its evolutionary origins are very ancient, since G-protein-linked receptors occur in very distantly related eucaryotic organisms, such as yeasts. There are even structurally similar membrane proteins in bacteria—such as the bacteriorhodopsin that functions as a light-driven H^+ pump (discussed in Chapter 11). In bacteria, however, such proteins do not act as G-protein-linked receptors. Bacteria do not possess G proteins, and the surface receptors by which they sense nutrients, for example, are coupled to different signal transduction systems.

Stimulation of G-Protein-linked Receptors Activates G-Protein Subunits

When an extracellular signaling molecule binds to a seven-pass transmembrane receptor, the receptor protein undergoes a conformational

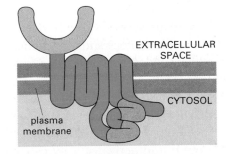

Figure 15–14 **Schematic drawing of a G-protein-linked receptor.** The cytoplasmic portions of the receptor that are mainly responsible for binding to the G protein are shown in *orange*. Receptors that bind protein signal molecules have a large extracellular ligand-binding domain formed by the part of the polypeptide chain shown in *light green*. Receptors for small signal molecules such as adrenaline, however, have small extracellular domains; the ligand-binding site is often formed deep within the plane of the membrane by parts of the transmembrane segments (not shown).

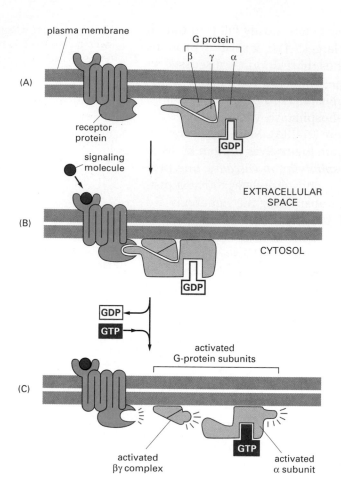

plasma membrane

G protein

β γ α

(A)

receptor
protein

signaling
molecule

(B)

EXTRACELLULAR
SPACE

GDP

CYTOSOL

GDP

GDP

GTP

activated
G-protein subunits

(C)

activated
βγ complex

GTP

activated
α subunit

Figure 15–15 G proteins disassemble into two signaling proteins when activated. (A) In the unstimulated state the receptor and the G protein are both inactive and are probably not in contact with each other. (B) Activation of the receptor by the extracellular signal molecule allows the G protein to associate with the receptor. (C) Binding to the activated receptor enables the α subunit of the G protein to exchange its GDP for GTP. This causes the G protein to break up into an activated α subunit and a βγ complex, which diffuse along the cytosolic surface of the plasma membrane until they encounter their target proteins, as shown in Figure 15–16. The receptor stays active while the external signal molecule is bound to it, and it can catalyze the activation of hundreds or thousands of molecules of G protein in this way.

change that alters the intracellular face of the receptor, enabling it to interact with a **G protein** located on the intracellular side of the plasma membrane. To explain the consequences we must first consider how G proteins are constructed and how they function.

There are several varieties of G proteins. Each is specific for a particular set of receptors and a particular set of downstream target proteins, as explained later. All of these G proteins, however, have a similar general structure and operate in a similar way. They are composed of three protein subunits—α, β, and γ. In the unstimulated state, the α subunit has GDP bound (Figure 15–15A) and the G protein is quiet. When an extracellular ligand binds to the receptor, the receptor binds to the G protein and activates it by causing the α subunit to eject its bound GDP and replace it with GTP. This causes the G protein to break up into a "switched-on" α subunit with GTP bound, and a detached βγ complex, giving rise to two separate molecules that can diffuse freely along the membrane (Figure 15–15B and C). The two activated parts of a G protein—the α subunit and the βγ complex—can both interact directly with targets located in the plasma membrane, which in turn may relay the signal to yet other destinations. The longer these targets have an α or a βγ subunit bound to them, the stronger and more prolonged the relayed signal will be.

The time for which the α and βγ subunits remain dissociated and available to act is limited by the behavior of the α subunit. The α subunit has an intrinsic GTP-hydrolyzing (*GTPase*) activity, and after a certain time hydrolyzes the bound GTP to GDP; the α subunit then reassociates

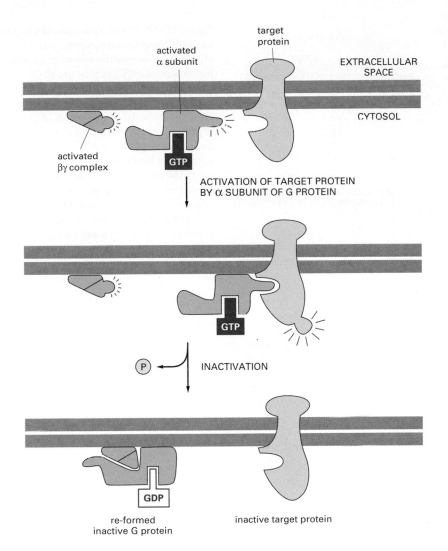

target protein

activated
α subunit

activated
βγ complex

GTP

EXTRACELLULAR
SPACE

CYTOSOL

ACTIVATION OF TARGET PROTEIN
BY α SUBUNIT OF G PROTEIN

GTP

(P) ← INACTIVATION

GDP

re-formed
inactive G protein

inactive target protein

Figure 15–16 The G-protein α subunit switches itself off by hydrolyzing its bound GTP. When an activated α subunit encounters and binds its target protein, it activates it (or in some cases inactivates it, not shown) for as long as the two proteins remain bound to each other. After a few seconds, the GTP on the α subunit is hydrolyzed to GDP by the α subunit's intrinsic GTPase activity. This inactivates the α subunit, which dissociates from its target protein and reassociates with a βγ complex to re-form an inactive G protein. The G protein is now ready to couple to another receptor, as in Figure 15–15B. Both the activated α subunit (as shown) and the free βγ complex can regulate target proteins.

with a βγ complex and the signal is shut off (Figure 15–16). This generally occurs a few seconds after the G protein has been activated.

This system allows us to reemphasize a general principle of cell signaling: the mechanisms that shut off a signal are as important as the mechanisms that turn it on. They offer as many opportunities for control, and as many dangers of mishap. The disease cholera provides an example. Cholera is caused by a bacterium that multiplies in the intestine, where it produces a protein called *cholera toxin*. This enters the cells lining the intestine and modifies the α subunit of a G protein so that it can no longer hydrolyze its bound GTP. The altered α subunit thus remains in the active state indefinitely, continuing to transmit a signal to its target proteins. In intestinal cells this results in a continual outflow of Na⁺ and water into the gut, resulting in catastrophic diarrhea and dehydration, often leading to death unless urgent steps are taken to replace the lost water and ions.

Some G Proteins Regulate Ion Channels

The target proteins for the G-protein subunits are either ion channels or membrane-bound enzymes. The different targets are affected by different types of G proteins (of which about 20 have so far been discovered in mammalian cells), and these various G proteins are themselves acti-

Question 15–5 G-protein-linked receptors activate G proteins by reducing the strength of GDP binding. This results in rapid dissociation of bound GDP, which is then replaced by GTP, which is present in the cytosol in much higher concentrations than GDP. What consequences would result from a mutation in the α subunit of a G protein that caused its affinity for GDP to be reduced without significantly changing its affinity for GTP? Compare the effects of this mutation with the effects of cholera toxin.

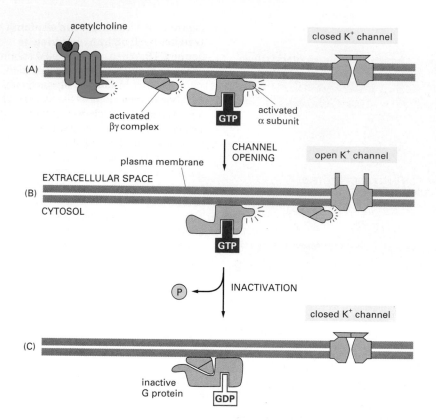

Figure 15–17 **A G protein couples recep-tor activation to the opening of a K⁺ channel in the plasma membrane of a heart muscle cell**. (A) Binding of the neurotransmitter acetylcholine to its G-protein-linked receptor on heart muscle cells results in the dissociation of the G protein into an activated βγ complex and an activated α subunit. (B) The activated βγ complex binds to and opens a K⁺ chan-nel in the heart cell plasma membrane. (C) Inactivation of the α subunit by hydro-lysis of bound GTP causes it to reassociate with the βγ complex to form an inactive G protein. The K⁺ channel closes.

vated by different classes of cell-surface receptors. In this way, binding of an extracellular signal molecule to a G-protein-linked receptor leads to effects on a particular subset of the possible target proteins, appropriate to that signal and that type of cell.

We look first at an example of G-protein regulation of ion channels. The heartbeat is controlled by two sets of nerve fibers; one set fires to speed the heart up, the other fires to slow it down. The nerve fibers that signal a slowdown in heartbeat do so by releasing acetylcholine, which binds to a G-protein-linked receptor on the heart muscle cells. When acetylcholine binds to the receptor, the G protein is activated—dissoci-ating into an α subunit and a βγ complex (Figure 15–17A). In this partic-ular example, the βγ complex is the active signaling component: it binds to the intracellular face of a K⁺ channel in the heart muscle cell plasma membrane, forcing the ion channel into an open conformation (Figure 15–17B). This alters the electrical properties of the heart muscle cell, causing it to contract less frequently. The action of the βγ complex is ended and the K⁺ channel recloses when the α subunit becomes inacti-vated by hydrolysis of its bound GTP and reassociates to form an inac-tive G protein (Figure 15–17C).

We shall now consider the interactions of G proteins with their other main targets—enzymes located in the plasma membrane.

Some G Proteins Activate Membrane-bound Enzymes

The interactions of G proteins with ion channels cause an immediate change in the state and behavior of the cell. Their interactions with enzyme targets have more complex consequences, leading to the pro-duction of further intracellular signaling molecules. The most frequent target enzymes for G proteins are *adenylate cyclase*, which is responsible

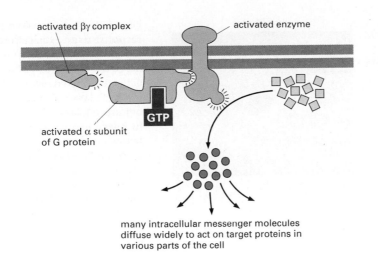

Figure 15–18 Enzymes activated by G proteins catalyze the synthesis of intracellular messenger molecules. As each activated enzyme generates many messenger molecules, the signal is greatly amplified at this step in the pathway. The signal is passed on by the messenger molecules binding to other target proteins in the cell and influencing their activity.

for production of the small signaling molecule *cyclic AMP*, and *phospholipase C*, which is responsible for production of the small signaling molecules *inositol trisphosphate* and *diacylglycerol*. These two enzymes are activated by different types of G proteins, coupling production of the small intracellular signaling molecules to different extracellular signals. The coupling may be either stimulatory (mediated by a stimulatory G protein) or inhibitory (mediated by an inhibitory G protein); for simplicity, we concentrate here on the cases where the activated G protein stimulates enzyme activity. The small intracellular signaling molecules in these pathways are often called **second messengers** (the "first messengers" being the extracellular signals); they are generated in large numbers when the membrane-bound enzyme is activated, and they then rapidly diffuse away from their source, broadcasting the signal throughout the cell (Figure 15–18).

We shall first consider the consequences of an increase in the concentration of cyclic AMP. This will take us along one of the signaling pathways leading from G-protein-linked receptors. We shall then discuss the actions of diacylglycerol and inositol trisphosphate, which take us along a different route.

The Cyclic AMP Pathway Can Activate Enzymes and Turn On Genes

Many extracellular signals acting via G-protein-linked receptors affect the activity of **adenylate cyclase** and thus alter the concentration of the messenger molecule **cyclic AMP** inside the cell. Most commonly, the activated G-protein α subunit switches on the adenylate cyclase, causing a dramatic and sudden increase in the synthesis of cyclic AMP from ATP (which is always present in the cell). A second enzyme, *cyclic AMP phosphodiesterase*, which is continuously active, rapidly breaks cyclic AMP down to ordinary AMP (Figure 15–19). Because cyclic AMP is so quickly broken down in the cell, its concentrations can change rapidly in response to extracellular signals, rising or falling tenfold in a matter of

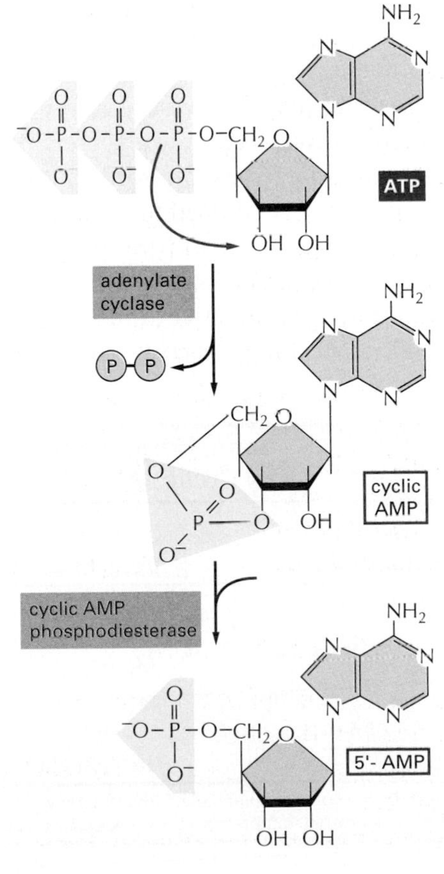

Figure 15–19 The synthesis and degradation of cyclic AMP (cAMP). Cyclic AMP is formed from ATP by a cyclization reaction that removes two phosphate groups from ATP and joins the "free" end of the remaining phosphate group to the sugar part of the ATP molecule. The degradation reaction breaks this second bond, forming AMP.

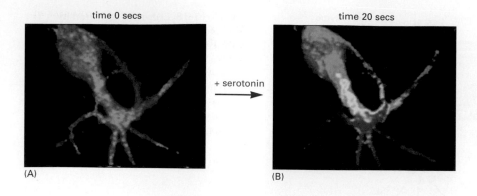

time 0 secs

time 20 secs

+ serotonin

(A)

(B)

Figure 15–20 **Increase in cyclic AMP in response to an extracellular signal.** A nerve cell in culture responds to the binding of the neurotransmitter serotonin to a G-protein-linked receptor by a rise in the intracellular level of cyclic AMP. The level of intracellular cyclic AMP was monitored by injecting into the cell a fluorescent protein whose fluorescence changes when it binds cyclic AMP. *Blue* indicates a low level of cyclic AMP, *yellow* an intermediate level, and *red* a high level. (A) In the resting cell, the cyclic AMP level is about 5×10^{-8} M. (B) Twenty seconds after adding serotonin to the culture medium, the intracellular level of cyclic AMP has increased to more than 10^{-6} M, an increase of more than twentyfold. (Courtesy of Roger Tsien.)

seconds (Figure 15–20). Cyclic AMP is a water-soluble molecule and so can carry the signal onward from the site on the membrane where it is synthesized to proteins located in the cytosol, the nucleus, or in other parts of the membrane.

Some examples of cell responses mediated by cyclic AMP are listed in Table 15–2. As the table shows, different target cells respond very differently to extracellular signals that change intracellular cyclic AMP levels. In many types of animal cells, activating cyclic AMP production is like stepping on the gas pedal in a car: it boosts the rate of consumption of metabolic fuel. When we are frightened or excited, for example, the adrenal gland releases the hormone adrenaline, which circulates in the bloodstream and binds to a class of G-protein-linked receptors (adrenergic receptors) that are present on many types of cells. The consequences vary from one cell type to another, but all help to prepare the body for sudden action. In skeletal muscle, for example, the consequence of adrenaline binding to its receptor is the G-protein-mediated stimulation of adenylate cyclase and a rise in the intracellular level of cyclic AMP, which causes the breakdown of glycogen (the polymerized storage form of glucose) to make more glucose available as fuel for anticipated muscular activity. Adrenaline also acts on fat cells, stimulating the breakdown of triacylglycerols (the storage form of fat) to fatty acids, an immediately usable form of cellular fuel (discussed in Chapter 4), which can be exported to other cells.

Cyclic AMP exerts these various effects mainly by activating the enzyme *cyclic-AMP-dependent protein kinase* (**A-kinase**): this kinase is normally held inactive in a complex with another protein; binding of cyclic AMP forces a conformational change that unleashes the active

Table 15–2   Some Hormone-induced Cell Responses Mediated by Cyclic AMP

Extracellular Signal Molecule*	Target Tissue	Major Response
Adrenaline	heart	increase in heart rate and force of contraction
Adrenaline	muscle	glycogen breakdown
Adrenaline, ACTH, glucagon	fat	fat breakdown
ACTH	adrenal gland	cortisol secretion

*Although all of the signal molecules listed here are hormones, some responses to local mediators and to neurotransmitters are also mediated by cyclic AMP.

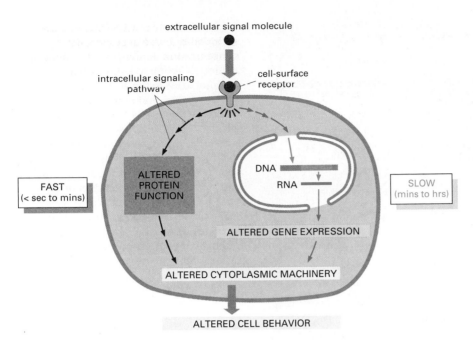

Figure 15–21 Fast and slow responses to extracellular signals. Extracellular signals regulate many aspects of cell behavior. Some types of altered cell behavior, such as increased cell growth and division, involve changes in gene expression and the synthesis of new proteins and therefore occur relatively slowly. Others, such as changes in cell movement, secretion, or metabolism, need not involve the nuclear machinery and therefore occur more quickly.

enzyme. The active protein kinase then catalyzes the phosphorylation of particular serines or threonines on certain intracellular proteins, thus altering their activity. In different cell types, different sets of target proteins are available to be phosphorylated, explaining why the effects of cyclic AMP vary with the target cell.

In some cases the effects are rapid, while in others they are slow (Figure 15–21). In skeletal muscle cells, for example, the activated A-kinase phosphorylates enzymes involved in glycogen metabolism, triggering the mechanism that breaks down glycogen to glucose. This response occurs within seconds. At the other extreme, there are effects of cAMP that take minutes or hours to develop. An important class of slow effects are those involving changes of gene expression. Thus in some cells, the A-kinase phosphorylates gene regulatory proteins that activate the transcription of selected genes. Figure 15–22 shows the extensive relay chain for such a pathway:

hormone → seven-pass transmembrane receptor → G protein →
adenylate cyclase → cyclic AMP → A-kinase →
gene regulatory protein → gene transcription

We have now followed one branch of a signaling pathway via G-protein-linked receptors and adenylate cyclase all the way from the plasma membrane to the cell nucleus. Now we must retrace our steps back to the plasma membrane, to examine the other enzyme-mediated signaling pathway that leads from G-protein-linked receptors—the pathway that begins with activation of the membrane-bound enzyme *phospholipase C*.

The Pathway Through Phospholipase C Results in a Rise in Intracellular Ca²⁺

More than 25 of the known G-protein-linked receptors exert their effects via a type of G protein that activates the membrane-bound enzyme **phospholipase C** instead of adenylate cyclase; a few examples are given in Table 15–3.

Question 15–6 Explain why cyclic AMP must be broken down rapidly in a cell to allow rapid signaling.

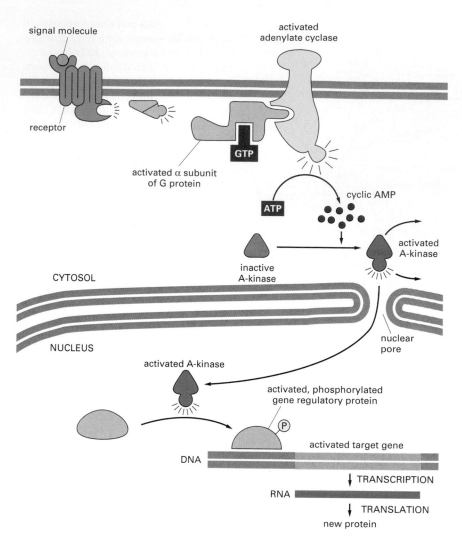

Figure 15–22 **Activation of gene transcription by a rise in cyclic AMP concentration**. Binding of a hormone or neurotransmitter to its G-protein-linked receptor leads to the activation of adenylate cyclase and a rise in cyclic AMP. Cyclic AMP activates A-kinase in the cytosol, which then moves into the nucleus and phosphorylates a gene regulatory protein. Once phosphorylated, the gene regulatory protein is able to stimulate the transcription of a whole set of target genes. This signaling pathway controls many processes in cells, ranging from hormone synthesis in endocrine cells to production of proteins required for long-term memory in the brain.

Phospholipase C acts on a membrane inositol phospholipid (a phospholipid with the sugar inositol attached) that is present in small quantities in the inner half of the plasma membrane lipid bilayer. Because of the involvement of this phospholipid, the signaling pathway begun by the action of phospholipase C is often known as the *inositol phospholipid pathway*. Phospholipase C generates two different messenger molecules. It cleaves off the hydrophilic sugar phosphate head of the phospholipid to produce the sugar phosphate *inositol 1,4,5-trisphosphate (IP$_3$)* while leaving the lipid *diacylglycerol (DAG)* tail embedded in the plasma membrane. Both molecules play a crucial part in signaling inside the cell, and we will consider them in turn.

Table 15–3   **Some Responses Mediated by Phospholipase C Activation**

Signal Molecule	Target Tissue	Major Response
Vasopressin (a protein hormone)	liver	glycogen breakdown
Acetylcholine	pancreas	secretes amylase (a digestive enzyme)
Acetylcholine	smooth muscle	contraction
Thrombin (a proteolytic enzyme)	blood platelets	aggregation

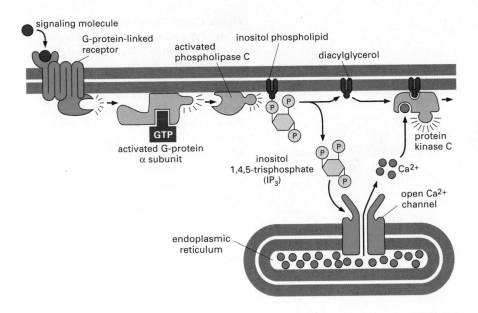

Figure 15–23 The signaling pathways activated by phospholipase C. Two intracellular messenger molecules are produced when a membrane inositol phospholipid is hydrolyzed by the activated phospholipase C. Inositol 1,4,5-trisphosphate (IP₃) diffuses through the cytosol and releases Ca²⁺ from the endoplasmic reticulum by binding to and opening Ca²⁺ channels in the endoplasmic reticulum membrane. The large electrochemical gradient for Ca²⁺ causes Ca²⁺ to rush out into the cytosol. Diacylglycerol remains in the plasma membrane and, together with Ca²⁺, helps activate the enzyme protein kinase C, which is recruited from the cytosol to the cytosolic face of the plasma membrane.

IP_3 leaves the plasma membrane and diffuses through the cytosol. When it reaches the endoplasmic reticulum, it binds to and opens Ca^{2+} channels in the endoplasmic reticulum membrane. Ca^{2+} stored inside the endoplasmic reticulum rushes out into the cytosol through these open channels (Figure 15–23), sharply raising the cytosolic concentration of free Ca^{2+}, which is normally kept very low.

The diacylglycerol remains attached to the plasma membrane, where it helps to activate a protein kinase. This one is called *protein kinase C* (**C-kinase**) because it also needs to bind Ca^{2+} to become active (see Figure 15–23). Once activated, C-kinase phosphorylates a set of intracellular proteins that varies depending on the cell type. The principles are the same as for A-kinase, although most of the target proteins are different.

A Ca²⁺ Signal Triggers Many Biological Processes

Ca^{2+} has such an important and widespread role as an intracellular messenger that we must digress to consider its functions more generally. A rise in the free Ca^{2+} level in the cytosol is the response to many different signals, not simply to those acting through G-protein-linked receptors. In egg cells, for example, a sudden rise in the intracellular Ca^{2+} concentration upon fertilization by a sperm triggers the start of embryonic development (Figure 15–24); in muscle cells Ca^{2+} triggers contraction; in many secretory cells, including nerve cells, it triggers secretion. Ca^{2+} carries out all these actions by binding to and influencing the activity of Ca^{2+}-sensitive proteins.

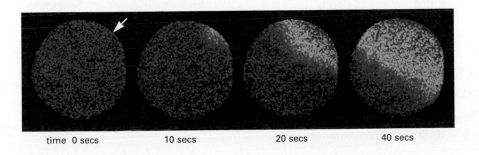

time 0 secs   10 secs   20 secs   40 secs

Figure 15–24 Fertilization of an egg by a sperm triggers an increase in cytosolic Ca²⁺. This starfish egg was injected with a Ca²⁺-sensitive fluorescent dye before it was fertilized. A wave of cytosolic Ca²⁺ (*red*), released from the endoplasmic reticulum, is seen to sweep across the egg from the site of sperm entry (*arrow*). This Ca²⁺ wave provokes a change in the egg cell membrane, preventing entry of other sperm, and it also initiates embryonic development. (Courtesy of Stephen A. Stricker.)

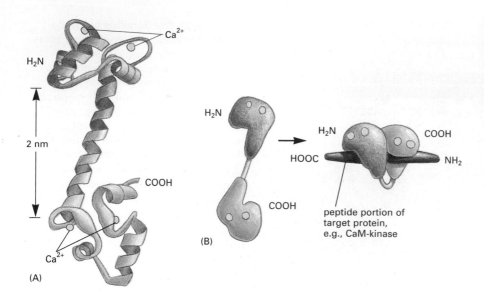

Figure 15–25 The structure of Ca²⁺-calmodulin based on x-ray diffraction and NMR studies. (A) The calmodulin molecule has a dumbbell shape, with two globular ends connected by a long, flexible α helix. Each end has two Ca²⁺-binding domains. (B) Simplified representation of the structure, showing the conformational changes in Ca²⁺-calmodulin that occur when it binds to a target protein. Note that the α helix has jack-knifed to surround the target protein. (A, based on x-ray crystallographic data from Y.S. Babu et al., *Nature* 315:37–40, 1985. © 1985 Macmillan Magazines Ltd.; B, based on x-ray crystallographic data from W.E. Meador, A.R. Means, and F.A. Quiocho, *Science* 257:1251–1255, 1992, and on NMR data from M. Ikura et al., *Science* 256:632–638, 1992. © 1992 the AAAS.)

The normal concentration of free Ca^{2+} in the cytosol is extremely low (10^{-7} M) compared to its concentration in the extracellular fluid and in the endoplasmic reticulum. These differences are maintained by membrane pumps that actively pump Ca^{2+} out of the cytosol, either across the plasma membrane or into the endoplasmic reticulum. As a result, there is a steep electrochemical gradient of Ca^{2+} across the endoplasmic reticulum membrane and across the plasma membrane (see Chapter 12). When a signal transiently opens Ca^{2+} channels in either of these membranes, Ca^{2+} rushes into the cytosol down its electrochemical gradient, triggering changes in Ca^{2+}-responsive proteins in the cytosol.

The inositol phospholipid signaling pathway, resulting in the release of Ca^{2+} into the cytosol from the endoplasmic reticulum, occurs in almost all eucaryotic cells, and affects a host of different target proteins, as well as the C-kinase mentioned earlier. The effects of Ca^{2+} on most proteins are indirect and are carried out through various transducer proteins known collectively as *Ca^{2+}-binding proteins*. The most widespread and common of these is the Ca^{2+}-responsive protein **calmodulin.** Calmodulin is present in the cytosol of all eucaryotic cells that have been examined, including those of plants, fungi, and protozoa. Calmodulin changes its conformational stability when it binds Ca^{2+}; this then enables it to bind to a wide range of target proteins in the cell and alter their activity (Figure 15–25). A particularly important class of targets are the **CaM-kinases.** These kinases are activated when calmodulin complexed with Ca^{2+} binds to them, and in their active state they influence other cell processes by phosphorylating selected proteins. In the mammalian brain, for example, a specific type of CaM-kinase is abundant at synapses, where it is thought to play a part in the creation of memory traces in response to the pulses of intracellular Ca^{2+} that occur during synaptic signaling. Mutant mice that lack this CaM-kinase show a marked inability to remember where things are.

Intracellular Signaling Cascades Can Achieve Astonishing Speed, Sensitivity, and Adaptability: Photoreceptors in the Eye

The steps in the signaling cascades associated with G-protein-linked receptors take a long time to describe, but they often take only seconds

Question 15–7 Why do you suppose cells have evolved intracellular Ca^{2+} stores even though extracellular Ca^{2+} is usually not limiting?

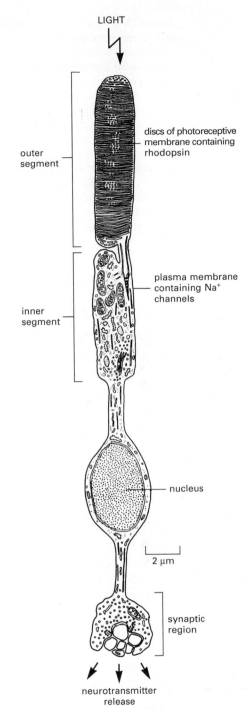

Figure 15–26 A rod photoreceptor cell from the retina. The light-absorbing molecules of rhodopsin are embedded in many pancake-shaped vesicles (discs) of membrane inside the outer segment of the cell; neurotransmitter is released from the opposite end of the cell to control firing of the retinal nerve cells that pass on the signal to the brain. When the rod cell is stimulated by light, a signal is relayed from the rhodopsin molecules in the discs, through the cytosol of the outer segment, to ion channels in the plasma membrane of the outer segment. The ion channels close in response to the signal, producing a change in the membrane potential of the rod cell. By mechanisms similar to those that control neurotransmitter release in ordinary nerve cells, the change in membrane potential alters the rate of neurotransmitter release from the synaptic region of the cell. (Adapted from T.L. Leutz, Cell Fine Structure. Philadelphia: Saunders, 1971.)

to execute. Consider how quickly a thrill can make your heart beat faster (adrenaline stimulates G-protein-linked receptors in heart muscle cells to accelerate the heartbeat), or how fast the smell of food can make you salivate (through G-protein-linked receptors for odors in your nose and G-protein-linked receptors for acetylcholine in salivary cells, stimulating secretion). Among the fastest of all responses mediated by a G-protein-linked receptor is that of the eye to bright light: it takes only 20 milliseconds for the most quickly responding photoreceptor cells of the retina (the cone photoreceptors) to produce their electrical response to a sudden flash of light.

This speed is achieved in spite of the necessity to relay the signal over several steps of an intracellular cascade. But photoreceptors also provide a beautiful illustration of the positive advantages of signaling cascades, especially of the spectacular amplification they can achieve and of their adaptability as detectors of signals of widely varying intensity. The quantitative details have been most thoroughly analyzed for the rod photoreceptor cells in the eye (Figure 15–26). Here, light impinges on *rhodopsin*, a G-protein-linked receptor. Light-activated rhodopsin activates a G protein called *transducin*. The activated α subunit of transducin then activates an intracellular signaling cascade that causes ion channels to close in the plasma membrane of the photoreceptor cell. This produces a change in the voltage across the cell membrane, with the ultimate consequence that a nerve impulse is sent to the brain.

The signal is repeatedly amplified as it is relayed along this pathway (Figure 15–27). When lighting conditions are dim (as on a moonless night) the amplification is enormous, and as few as a dozen photons absorbed in the entire retina will cause a perceptible signal to be delivered to the brain. In bright sunlight, when photons flood through each photoreceptor cell at a rate of billions per second, the signaling cascade *adapts*, stepping down the amplification by more than 10,000-fold so that the photoreceptor cells are not overwhelmed and can still register increases and decreases in the strong light. The adaptation depends on negative feedback: an intense response in the photoreceptor cell generates an intracellular signal (a change in Ca^{2+} concentration) that inhibits the enzymes responsible for signal amplification.

Adaptation also occurs in signaling pathways that respond to chemical signals, allowing a cell to remain sensitive to changes of signal intensity over a wide range of background levels of stimulation. Adaptation, in other words, makes the cell responsive both to messages that are whispered and to those that are shouted.

G-Protein-linked Receptors

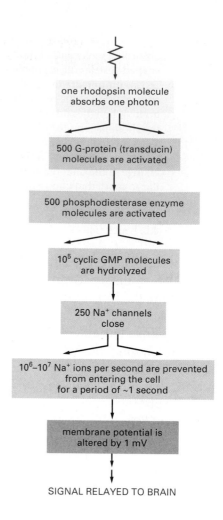

Figure 15–27 Amplification in the light-induced signaling cascade in rod photoreceptor cells. This figure shows the amplification that can be achieved when the photoreceptor is adapted for dim light. The intracellular signaling pathway from the G protein transducin uses different components from the pathways already described. In the absence of a light signal, the messenger molecule cyclic GMP (a molecule very similar to cyclic AMP) is continuously produced in the photoreceptor cell, and binds to Na⁺ ion channels in the photoreceptor cell plasma membrane, keeping them open. Activation of rhodopsin by light results in formation of activated transducin α subunits. These activate an enzyme called cyclic nucleotide phosphodiesterase, which breaks down cyclic GMP to GMP. The sharp fall in the intracellular level of cyclic GMP causes the bound cyclic GMP to dissociate from the channels, and they close. The *divergent arrows* indicate the steps at which amplification occurs.

Not only vision but also taste and smell depend on G-protein-linked receptors. It seems likely that this mechanism of signal reception, invented early in the evolution of the eucaryotes, has its origins in the basic and universal need of cells to sense and respond to their environment. We now turn to another class of cell-surface receptors that play a key part in controlling cell numbers, cell differentiation, and cell movement in multicellular animals.

Enzyme-linked Receptors

The enzyme-linked receptors, the third main class of cell-surface receptors (see Figure 15–12C), came to light through their role in responses to *growth factors*—extracellular signal proteins that regulate cell growth, proliferation, differentiation, and survival in animal tissues. Some examples are listed in Table 15–1 (p. 485). Most growth factors act as local mediators and are required only at very low concentrations (about 10^{-9} to 10^{-11} M). Responses to them are typically slow (of the order of hours) and require many intracellular transduction steps that eventually lead to changes in gene expression. Enzyme-linked receptors have since been found also to mediate direct, rapid effects on the cytoskeleton, controlling the way a cell moves and changes its shape. The extracellular signals for these latter responses are often not diffusible growth factors, but proteins attached to surfaces over which the cell is crawling. Disorders of cell proliferation, differentiation, survival, and migration are fundamental to cancer, and abnormalities of signaling via enzyme-linked receptors have a major role in the causation of this class of diseases.

Like G-protein-linked receptors, the enzyme-linked receptors are transmembrane proteins with their ligand-binding domains on the outer surface of the plasma membrane. Instead of the cytoplasmic domain of the receptor associating with a G protein, however, it acts as an enzyme, or forms a complex with another protein that acts as an enzyme. The largest class of enzyme-linked receptors are those with a cytoplasmic domain that functions as a tyrosine protein kinase, phosphorylating tyrosine side chains on selected intracellular proteins. Such receptors are called **receptor tyrosine kinases.** This category includes the great majority of growth factor receptors, on which we shall concentrate here.

Question 15–8 One important feature of any signaling cascade is its ability to turn off. Consider the cascade shown in Figure 15–27. Where would off switches be required? Which ones do you suppose are the most important?

Activated Receptor Tyrosine Kinases Assemble a Complex of Intracellular Signaling Proteins

To do its job as a signal transducer, an enzyme-linked receptor has to switch on the enzyme activity of its intracellular domain when an external signal molecule binds to its extracellular domain. Unlike the seven-pass G-protein-linked receptors, enzyme-linked receptor proteins usually have only one transmembrane segment, which is thought to span the lipid bilayer as a single α helix. There is, it seems, no way to transmit a conformational change through a single α helix, and so enzyme-linked receptors have a different mechanism for transducing the extracellular signal. Signal molecule binding causes two receptor molecules to come together in the membrane, forming a dimer. Contact between the two adjacent intracellular receptor tails activates their kinase function, with the result that each one phosphorylates the other.

This phosphorylation then triggers the assembly of an elaborate intracellular signaling complex on the receptor tails. The newly phosphorylated tyrosines serve as binding sites for a whole zoo of intracellular signaling proteins—perhaps as many as 10 or 20 different molecules—which themselves become activated upon binding. While it lasts, this signaling complex broadcasts the signal along several routes simultaneously to many destinations inside the cell, thus activating and coordinating the numerous biochemical changes that are required to trigger a complex response such as cell proliferation (Figure 15–28). To terminate the activation of the receptor, the cell contains *protein tyrosine phosphatases*, which remove the phosphates that were added in response to the extracellular signal. In many cases, activated receptors are also disposed of in a more brutal way: they are brought into the interior of the cell by endocytosis and then destroyed by digestion in lysosomes.

Different receptor tyrosine kinases recruit different collections of intracellular signaling proteins, producing different effects; but there are some components that seem to be very widely used. These include, for example, a phospholipase that functions in the same way as phospholipase C to activate the inositol phospholipid signaling pathway (see Figure 15–23). The main signaling pathway from receptor tyrosine kinases to the nucleus, however, takes another route, which has become well known for a sinister reason: mutations that activate this pathway

Figure 15–28 The activation of a receptor tyrosine kinase results in the formation of an intracellular signaling complex. Binding of a signal molecule to the extracellular domain of a receptor tyrosine kinase causes two receptor molecules to associate into a dimer. The signal molecule shown here is itself a dimer and thus can cross-link two receptor molecules. In other cases binding of the signal molecule changes the conformation of the receptor molecules in such a way that it dimerizes. Dimer formation brings the kinase domains on each intracellular receptor tail into contact with each other; this activates the kinases and enables them to phosphorylate each other on several tyrosine side chains. Each phosphorylated tyrosine serves as a specific binding site for a different intracellular signaling protein, which relays a signal into the cell's interior.

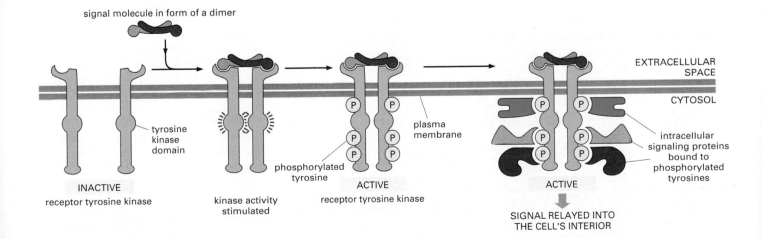

and thereby stimulate cell division inappropriately are a causative factor in many types of cancers. We shall conclude our discussion of receptor tyrosine kinases by following this pathway to the nucleus.

Receptor Tyrosine Kinases Activate the GTP-binding Protein Ras

Among the intracellular signaling proteins that bind to activated receptor tyrosine kinases are some that function as physical *adaptors,* which create an enlarged aggregate by coupling the receptor to other proteins, which in turn may bind to and activate yet other proteins, transmitting the signal onward. Through a linking complex of this type, activation of a receptor tyrosine kinase leads to activation of a small intracellular signaling protein called **Ras,** which is tethered to the cytoplasmic face of the plasma membrane. Virtually all receptor tyrosine kinases are thus coupled to Ras, from those that respond to platelet-derived growth factor (PDGF), mediating effects such as cell proliferation at a healing wound, to those that respond to nerve growth factor (NGF), preventing certain neurons from dying in the developing nervous system.

The Ras protein is a member of a large family of small, single-subunit GTP-binding proteins, often called the monomeric GTP-binding proteins to distinguish them from the trimeric G proteins that we encountered earlier in this chapter. Ras resembles the α subunit of a G protein and functions as a molecular switch in much the same way. It cycles between two distinct conformational states—active when GTP is bound and inactive when GDP is bound (see Figure 15–13B). Interaction with an activating protein causes Ras to exchange its GDP for GTP, thus switching to its activated state; after a delay Ras is switched off again by hydrolyzing its GTP to GDP. Figure 15–29 shows how Ras is coupled to an activated receptor tyrosine kinase by a short series of linking proteins, the last of which is a Ras-activating protein.

In its active state, Ras promotes the activation of a phosphorylation cascade in which a series of protein kinases phosphorylate and activate one another in sequence. This relay system carries the signal from the plasma membrane to the nucleus. The final protein kinase of the cascade phosphorylates certain gene regulatory proteins, altering their ability to regulate gene transcription. The result is a change in the

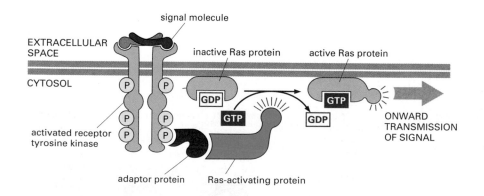

Figure 15–29 Activation of Ras by an activated receptor tyrosine kinase. An adaptor protein is docked on a particular phosphotyrosine on the activated receptor (the other signaling proteins shown bound to the receptor in Figure 15–28 are omitted for simplicity). The adaptor binds and activates a protein that functions as a Ras-activating protein. This in turn stimulates Ras to exchange its bound GDP for GTP. The activated Ras protein then stimulates the next step in the signaling pathway, which is shown in Figure 15–30.

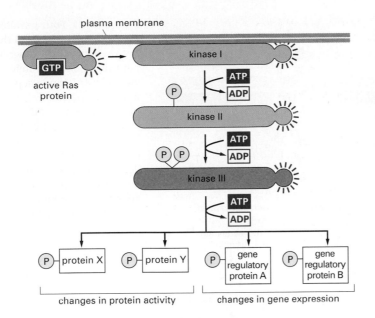

plasma membrane

active Ras protein

kinase I

ATP / ADP

kinase II

ATP / ADP

kinase III

ATP / ADP

protein X   protein Y   gene regulatory protein A   gene regulatory protein B

changes in protein activity   changes in gene expression

Figure 15–30 The Ras-activated phosphorylation cascade. A Ras protein activated by the process shown in Figure 15–29 triggers a phosphorylation cascade of three protein kinases, which amplify and distribute the signal. The final kinase in the cascade phosphorylates various downstream target proteins. These include other protein kinases and, most importantly, gene regulatory proteins that control gene expression. Changes in gene expression and protein activity result in complex changes in cell behavior such as cell proliferation and cell differentiation, typical outcomes of the Ras signaling pathway.

pattern of gene expression (Figure 15–30). This may stimulate cell proliferation, promote cell survival, or control cell differentiation: the precise outcome will depend on which other genes are active in the cell, and which other signals are acting on it.

If Ras is inhibited by an intracellular injection of Ras-inactivating antibodies, and the cell is exposed to a growth factor it would normally respond to, the normal cellular response does not occur. Experiments like this show that Ras is a vital component of the signaling pathway that elicits the normal cellular response. Conversely, if Ras activity is permanently switched on, as occurs with some hyperactive mutant forms of the Ras protein, the effect on the cell is the same as that of continuous stimulation of a receptor tyrosine kinase by a growth factor. The Ras protein was in fact first discovered in cancer cells, in which a mutation in the gene for Ras had led to the production of a hyperactive Ras. This mutant Ras protein stimulates the cells to divide even in the absence of growth factors, and the resulting uncontrolled cell proliferation contributes to the formation of a cancer.

About 30% of human cancers have mutations in the *ras* genes, and in many other cancers there are mutations in genes whose products lie in the same signaling pathway as Ras. Many of the genes that encode these intracellular signaling proteins were identified in the quest for cancer-promoting *oncogenes*, which are discussed in Chapter 18. The normal versions of the genes, which encode the normal signaling proteins, are often known as *proto-oncogenes*, since they are capable of being converted into oncogenes by mutation.

Cancer is a disease in which cells in the body behave in a selfish and antisocial way, destroying the harmony of the multicellular organism by proliferating when they should not and invading tissues that they should not enter. The common occurrence in cancer of mutations in genes for cell-signaling components reflects a familiar truth: the maintenance of a stable community depends above all on good communications.

Question 15–9 Would you expect to activate G-protein-linked receptors and receptor tyrosine kinases by exposing cells to antibodies raised to the respective proteins? (Hint: review Panel 5–3, on pp. 158–159, regarding the properties of antibody molecules.)

Enzyme-linked Receptors

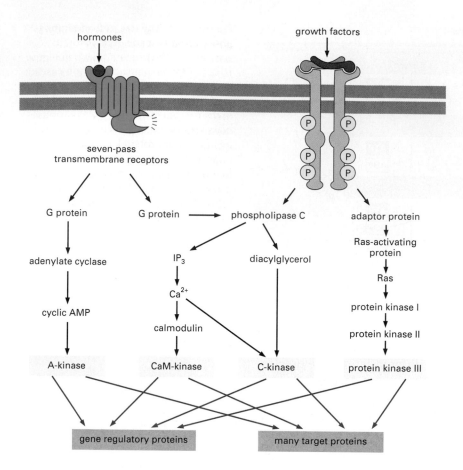

Figure 15–31 Four parallel intracellular signaling pathways and the connections between them. The diagram sketches the pathways from G-protein-linked receptors via cyclic AMP and via phospholipase C and from enzyme-linked receptors via phospholipase C and via Ras. The protein kinases in these pathways phosphorylate many proteins, including proteins belonging to the other pathways. The resulting dense network of regulatory interconnections is symbolized by the *red arrows* radiating from each kinase shaded in *yellow*, which phosphorylate some of the same target proteins.

Protein Kinase Networks Integrate Information to Control Complex Cell Behaviors

In this chapter we have outlined in succession several major pathways for signaling from the cell surface into the cell interior. Figure 15–31 compares the four pathways that we have discussed in most detail: the routes from G-protein-linked receptors via cyclic AMP and via phospholipase C, and the routes from enzyme-linked receptors via phospholipase C and via Ras. Each pathway is different from the others and yet confusingly similar in many of the principles of its operation. The different pathways share some of their components, and all of them, by eventually activating protein kinases, seem capable in principle of regulating practically any process in the cell.

In fact, the complexity is even greater than our account up to now has conveyed. First, there are a number of other pathways that we have neglected; second, the major pathways interact in ways that we have not described. These interactions are of many sorts, but the most extensive are those mediated by the protein kinases present in each of the pathways. These kinases phosphorylate, and hence regulate, components of the other signaling pathways as well as components of the pathway to which they themselves belong. Thus there is *cross-talk* between the different pathways (see Figure 15–31), and indeed between virtually all the control systems of the cell. To give an idea of the scale of the regulatory system, it is estimated that about 2% of our genes code for protein kinases, and that more than 1000 distinct types of protein kinases may be present in a single mammalian cell. How can we make sense of this

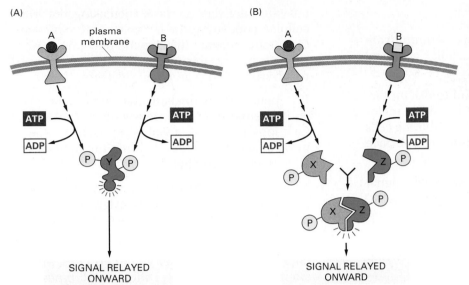

(A)

plasma membrane

A

ATP

ADP

P Y P

SIGNAL RELAYED ONWARD

(B)

A

B

ATP

ADP

ATP

ADP

P X

Z P

P X Z P

SIGNAL RELAYED ONWARD

Figure 15–32 Two possible simple mechanisms of signal integration. In (A), signals A and B activate different cascades of protein phosphorylations, each of which leads to the phosphorylation of protein Y but at different sites on the protein. Protein Y is activated only when both of these sites are phosphorylated, and therefore it is active only when signals A and B are simultaneously present. In (B), signals A and B lead to the phosphorylation of two proteins, X and Z, which then bind to each other to create the active protein XZ.

tangled web of control connections, and what is the function of such complexity?

Cells have to combine information from many different sources and use it to make the appropriate response—to live or die, to divide, to change shape, to secrete a chemical product, and so on. Through the cross-talk between signaling pathways, the cell is able to put two or more bits of information together and react to the combination. Thus some intracellular proteins can act as integrating devices, usually by having several potential phosphorylation sites, each of which can be phosphorylated by a different protein kinase. Information received from different sources can therefore converge on such proteins, which convert it to a single onward signal (Figure 15–32). The integrating proteins in turn can deliver a signal to many downstream targets. In this way, the intracellular signaling system may act like a network of nerve cells in the brain or of microprocessors in a computer, interpreting complex information and generating complex responses (Figure 15–33).

This chapter began by considering the signals that act on cells from outside, but it has led us deep into the cell's interior, into the workings of an elaborate intracellular control system. The components of this system convey messages between separate parts of the cell, combine and process information from different sources, serve as memory devices, and govern the timing of events. One important target of this control system is the cytoskeleton, which determines cell shape and is responsible for cell movements, as we discuss in the next chapter.

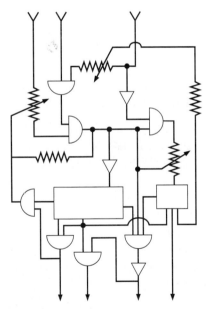

Figure 15–33 Network of integrating devices, as found in a computer.

Essential Concepts

- Cells in multicellular organisms communicate through a large variety of extracellular chemical signals.

- Hormones are carried in the blood to distant target cells, but most other extracellular signal molecules act over only a short range. Neighboring cells often communicate through direct cell-surface contacts.

- Cells are stimulated by an extracellular signal when it binds to and activates a receptor protein. Each receptor protein recognizes a particular signal molecule.

- Receptor proteins act as transducers, converting the signal from one physical form to another.

- Most extracellular signal molecules cannot pass through the cell membrane; they bind to receptor proteins located on the cell surface. These receptors transduce the signal across the plasma membrane.

- Small hydrophobic extracellular signal molecules such as steroid hormones and nitric oxide can diffuse directly across the plasma membrane; they activate intracellular receptor proteins, which are either gene regulatory proteins or enzymes.

- There are three main classes of cell-surface receptors: (1) ion-channel-linked receptors, (2) G-protein-linked receptors, and (3) enzyme-linked receptors.

- G-protein-linked receptors and enzyme-linked receptors respond to extracellular signals by initiating cascades of intracellular signaling reactions that alter the behavior of the cell.

- G-protein-linked receptors activate a class of trimeric GTP-binding proteins called G proteins, which act as molecular switches, transmitting the signal onward to the interior of the cell for a short period and then switching themselves off by hydrolyzing their bound GTP to GDP.

- Some G proteins directly regulate ion channels in the plasma membrane. Others activate the enzyme adenylate cyclase, increasing the intracellular concentration of cyclic AMP. Still other G proteins activate the enzyme phospholipase C, which generates the messenger molecules inositol trisphosphate (IP_3) and diacylglycerol.

- IP_3 opens ion channels in the membrane of the endoplasmic reticulum, releasing a flood of free Ca^{2+} ions into the cytosol. Ca^{2+} itself acts as an intracellular messenger, altering the activity of a wide range of proteins.

- A rise in cyclic AMP activates protein kinase A (A-kinase), while Ca^{2+} and diacylglycerol in combination activate protein kinase C (C-kinase).

- C-kinase and A-kinase phosphorylate selected target proteins on serines and threonines, thereby altering protein activity. Different cell types contain different sets of target proteins and are affected in different ways.

- In general, stimulation of G-protein-linked receptors produces rapid and reversible cell responses.

- Many enzyme-linked receptors have intracellular protein domains that function as enzymes; most are receptor tyrosine kinases, which are activated by growth factors and phosphorylate tyrosines on selected intracellular proteins.

- Activated receptor tyrosine kinases cause the assembly of an intracellular signaling complex on the intracellular tail of the receptor; a part of this complex serves to activate Ras, a small GTP-binding protein, which activates a cascade of protein kinases that relay the signal from the plasma membrane to the nucleus.

- Mutations that stimulate cell proliferation by making Ras hyperactive are a common feature of many cancers.

- The different intracellular signaling pathways interact, enabling cells to produce an appropriate response to a complex combination of signals. Some combinations of signals enable a cell to survive; other combinations of signals will cause it to proliferate; and in the absence of any signals, most cells will kill themselves.

Key Terms

adaptation	cell signaling	local mediator	receptor protein
adenylate cyclase	C-kinase	molecular switch	receptor tyrosine kinase
A-kinase	cyclic AMP	neuronal	second messenger
calmodulin	G protein	neurotransmitter	signaling cascade
CaM-kinase	G-protein-linked receptor	nitric oxide	signal transduction
	hormone	phospholipase C	steroid hormone
		Ras	

Questions

Question 15–10   Which of the following statements are correct? Explain your answers.

A. The signal molecule acetylcholine has different effects on different cell types in an animal and binds to different receptor molecules on different cell types.

B. After acetylcholine is secreted from cells it is long-lived, because it has to reach target cells all over the body.

C. Both the GTP-bound α subunits and nucleotide-free $\beta\gamma$ complexes, but not GDP-bound, fully assembled G proteins, activate other molecules downstream of G-protein-linked receptors.

D. IP_3 is produced directly from PIP_2, the inositol phospholipid from which it is derived, without incorporation of an additional phosphate group.

E. Calmodulin regulates the intracellular Ca^{2+} concentration.

F. Different signals originating from the plasma membrane can be integrated by cross-talk between different signaling pathways inside the cell.

G. *Ras* is an oncogene.

H. Tyrosine phosphorylation serves to build binding sites for other proteins to bind to receptor tyrosine kinases.

Question 15–11   The Ras protein functions as a molecular switch that is set to its on state by other proteins that cause it to bind GTP. A GTPase-activating protein resets the switch to the off state by inducing Ras to hydrolyze its bound GTP to GDP much more rapidly than it would without this encouragement. Thus Ras works like a light switch that one person turns on and another turns off. You are given a mutant cell that lacks the GTPase-activating protein. What abnormalities would you expect to find in the way that Ras activity responds to extracellular signals?

Question 15–12

A. Compare and contrast signaling by neurons to that carried out by endocrine cells, which secrete hormones.

B. Discuss the relative advantages of the two mechanisms.

Question 15–13   Two intracellular molecules, X and Y, are normally synthesized at a constant rate of 1000 molecules per second per cell. Molecule X is broken down slowly: each molecule of X survives on average

for 100 seconds. Molecule Y is broken down 10 times faster: each molecule of Y survives on average for 10 seconds.

A. Calculate how many molecules of X and Y the cell contains?

B. If the rates of synthesis of both X and Y are suddenly increased tenfold to 10,000 molecules per second per cell—without any change in their degradation rates—how many molecules of X and Y will there be after one second?

C. Which molecule would be preferred for rapid signaling?

Question 15–14   "One of the great kings of the past ruled an enormous kingdom that was more beautiful than anywhere else in the world. Every plant glistened as brilliantly as polished jade, and the softly rolling hills were as sleek as the waves of the summer sea. The wisdom of all of his decisions relied on a constant flow of information brought to him daily by messengers who told him about every detail of his kingdom so that he could take quick, appropriate actions whenever in need. Though despite all the abundance of beauty and efficiency, his people felt doomed to live under his rule, for he had an advisor who had studied cellular signal transduction and accordingly administered the king's Department of Information. The advisor had implemented the policy that all messengers shall be immediately beheaded whenever spotted by the Royal Guard, because for rapid signaling the lifetime of messengers ought to be short. Their plea 'Don't hurt me, I'm only the messenger!' was to no avail, and the people of the kingdom suffered terribly because of the rapid loss of their sons and daughters." Why is the analogy on which the king's advisor bases his policies inappropriate? Briefly discuss the features that set cell-signaling pathways apart from the human communication pathway described in the story.

Question 15–15   In a series of experiments, genes that code for mutant forms of a receptor tyrosine kinase are introduced into cells. The cells also express their own normal form of the receptor from their normal gene, although the mutant genes are constructed so that they are expressed at considerably higher levels than the normal gene. What would be the consequences of introducing a mutant gene that codes for a receptor tyrosine kinase (A) lacking its extracellular domain, or (B) lacking its intracellular domain?

Question 15–16   Discuss the following statement: "Membrane proteins that span the membrane many times can undergo a conformation change upon ligand binding that can be sensed on the other side of the membrane. Thus, individual protein molecules can transmit a signal across a membrane. In contrast, individual single-span membrane proteins cannot transmit a conformational change across the membrane but require oligomerization."

Question 15–17 What are the similarities and differences between the reactions that lead to the activation of G proteins and the reactions that lead to the activation of Ras?

Question 15–18 Why do you suppose cells use Ca^{2+} (which is kept by Ca^{2+} pumps at an intracellular concentration of 10^{-7} M) for intracellular signaling and not another ion such as Na^+ (which is kept by the Na^+ pump at an intracellular concentration of 10^{-3} M)?

Question 15–19 It seems counterintuitive that a cell, having a perfectly abundant supply of nutrients available, would commit suicide if not constantly stimulated by signals from other cells (see Figure 15–6). What do you suppose might be the purpose of such regulation?

Question 15–20 The contraction of the myosin/actin system in muscle cells is triggered by a rise in intracellular Ca^{2+}. Muscle cells have specialized Ca^{2+}-release channels—called ryanodine receptors because of their sensitivity to the drug ryanodine—that lie in the membrane of the sarcoplasmic reticulum, a specialized form of the endoplasmic reticulum. In contrast to the IP_3-gated Ca^{2+} channels in the endoplasmic reticulum shown in Figure 15–23, the ligand that opens ryanodine receptors is Ca^{2+} itself. Discuss the consequences of ryanodine channels for muscle cell contraction.

Question 15–21 Two protein kinases, K1 and K2, function sequentially in an intracellular signaling cascade. If either kinase contains a mutation that permanently inactivates its function, no response is seen in cells when an extracellular signal is received. A different mutation in K1 makes it permanently active, so that in cells containing that mutation a response is observed even in the absence of an extracellular signal. You characterize a double mutant cell that contains K2 with the inactivating mutation and K1 with the activating mutation. You observe that the response is seen even when no signal is received by these cells. In the normal signaling pathway, does K1 activate K2 or does K2 activate K1? Explain your answer.

16 Cytoskeleton

The ability of eucaryotic cells to organize the many components in their interior, to adopt a variety of shapes, and to carry out coordinated movements depends on the **cytoskeleton**—an intricate network of protein filaments that extends throughout the cytoplasm (Figure 16–1). This network of filaments helps to support the large volume of cytoplasm in a eucaryotic cell, a function that is particularly important in animal cells, which have no cell walls. Unlike our own bony skeleton, however, the cytoskeleton is a highly dynamic structure that is continuously reorganized as a cell changes shape, divides, and responds to its environment. The cytoskeleton is not only the "bones" of a cell but its "muscles" too, and it is directly responsible for large-scale movements such as the crawling of cells along a surface, contraction of muscle cells, and the changes in cell shape that take place as an embryo develops.

The interior of the cell itself is also in constant motion, and the cytoskeleton provides the machinery for intracellular movements such as the transport of organelles from one place to another, the segregation of chromosomes into the two daughter cells at mitosis, and the pinching apart of animal cells at cell division. The cytoskeleton seems to be a eucaryotic invention; it is absent from procaryotes, and it may have been a crucial factor in the evolution of large and structurally complex eucaryotic cells. The eucaryotic cell, like any factory making a complex product, has a highly organized interior in which specialized services are concentrated in different areas but linked by transport systems: the cytoskeleton both controls the location of the organelles that provide these services and provides the transport between them.

The cytoskeleton is built on a framework of three types of protein filaments: *intermediate filaments, microtubules,* and *actin filaments* (Figure 16–2). Each type of filament is formed from a different protein subunit. A family of fibrous proteins form intermediate filaments:

Figure 16–1 The cytoskeleton. A skin cell (fibroblast) in culture has been fixed and stained with Coomassie blue, a general stain for proteins. A variety of filamentous structures extend throughout the cell. The dark oval in the center is the nucleus. (Courtesy of Colin Smith.)

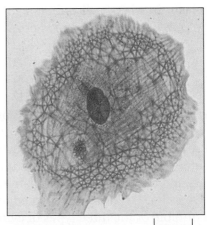

10 µm

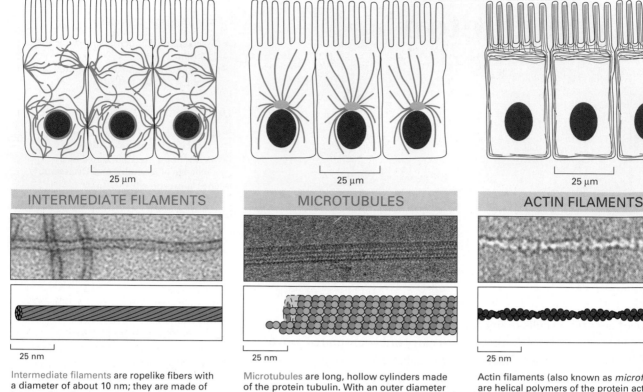

| INTERMEDIATE FILAMENTS | MICROTUBULES | ACTIN FILAMENTS |

25 nm (three times, under each panel)

Intermediate filaments are ropelike fibers with a diameter of about 10 nm; they are made of intermediate filament proteins, which constitute a large and heterogeneous family. One type of intermediate filament forms a meshwork called the nuclear lamina just beneath the inner nuclear membrane. Other types extend across the cytoplasm, giving cells mechanical strength and carrying the mechanical stresses in an epithelial tissue by spanning the cytoplasm from one cell-cell junction to another. (Micrograph courtesy of Roy Quinlan.)

Microtubules are long, hollow cylinders made of the protein tubulin. With an outer diameter of 25 nm, they are more rigid than actin filaments or intermediate filaments. Microtubules are long and straight and typically have one end attached to a single microtubule-organizing center called a *centrosome*, as shown here. (Micrograph courtesy of Richard Wade.)

Actin filaments (also known as *microfilaments*) are helical polymers of the protein actin. They appear as flexible structures, with a diameter of about 7 nm, that are organized into a variety of linear bundles, two-dimensional networks, and three-dimensional gels. Although actin filaments are dispersed throughout the cell, they are most highly concentrated in the *cortex*, just beneath the plasma membrane. (Micrograph courtesy of Roger Craig.)

tubulin is the subunit in microtubules, and *actin* is the subunit in actin filaments. In each case, thousands of subunits assemble into an unbranched thread of protein that sometimes extends across the entire cell.

In this chapter we consider the structure and function of the three types of protein filament networks in turn, beginning with the intermediate filaments that provide cells with mechanical strength. We shall also see how the actin cytoskeleton provides the motive force for a crawling fibroblast, and how cell appendages built from microtubules propel motile cells like protozoa and sperm.

Figure 16–2 The three types of protein filaments that form the cytoskeleton. The cells illustrated are epithelial cells lining the gut.

Intermediate Filaments

Intermediate filaments have great tensile strength, and their main function is to enable cells to withstand the mechanical stress that occurs when cells are stretched. They are called "intermediate" because their diameter (about 10 nm) is between that of the thin actin-containing filaments and the thicker myosin filaments of smooth muscle cells, where they were first discovered. Intermediate filaments are the toughest and most durable of the three types of cytoskeletal filaments: when cells are treated with concentrated salt solutions and nonionic detergents, the

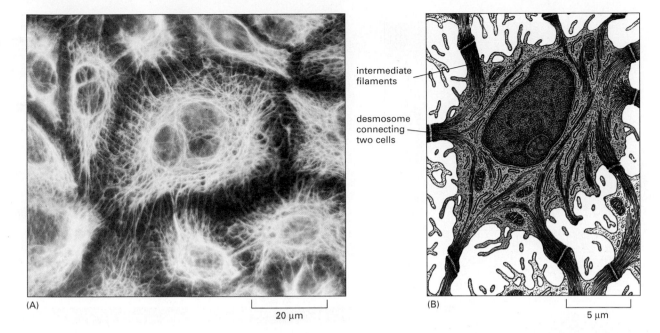

(A)

20 μm

(B)

intermediate
filaments

desmosome
connecting
two cells

5 μm

intermediate filaments remain while most of the rest of the cytoskeleton is destroyed.

Intermediate filaments are found in the cytoplasm of most, but not all, animal cells. They typically form a network throughout the cytoplasm, surrounding the nucleus and extending out to the cell periphery. There they are often anchored to the plasma membrane at cell-cell junctions, where the external face of the membrane is connected to that of another cell (Figure 16–3). Intermediate filaments are also found within the nucleus; a mesh of intermediate filaments, the *nuclear lamina*, underlies and strengthens the nuclear envelope in all eucaryotic cells.

Intermediate Filaments Are Strong and Durable

Intermediate filaments are like ropes with many long strands twisted together to give tensile strength. The strands of this rope—the subunits of intermediate filaments—are elongated fibrous proteins each composed of an amino-terminal globular head, a carboxyl-terminal globular tail, and a central elongated rod domain (Figure 16–4A). The rod domain consists of an extended α-helical region that enables pairs of intermediate filament proteins to form stable dimers by wrapping around each other in a coiled-coil configuration (Figure 16–4B), as described in Chapter 5. Two of these coiled-coil dimers then associate by noncovalent bonding to form a tetramer (Figure 16–4C), and the tetramers then bind to one another end-to-end and side-by-side, also by noncovalent bonding, to generate the final ropelike intermediate filament (Figure 16–4D and E).

The central rod domains of different intermediate filament proteins are all similar in size and amino acid sequence, so that when they pack together they always form filaments of similar diameter and internal structure. By contrast, the globular head and tail regions, which are exposed on the surface of the filament, are mainly concerned with its interactions with other components of the cytoplasm. They vary greatly in both size and amino acid sequence from one intermediate filament protein to another.

Figure 16–3 **Network of intermediate filaments.** (A) Immunofluorescence micrograph of a group of epidermal cells in culture stained to show the network of intermediate filaments in their cytoplasm. The filaments in each cell are indirectly connected to those of neighboring cells through specialized cell junctions called desmosomes (discussed in Chapter 19). (B) Drawing from an electron micrograph of a section of epidermis showing the bundles of intermediate filaments that traverse the cytoplasm and that are inserted at desmosomes. (A, courtesy of Michael Klymkowsky. B, from R. V. Krstić, Ultrastructure of the Mammalian Cell: An Atlas. Berlin: Springer, 1979.)

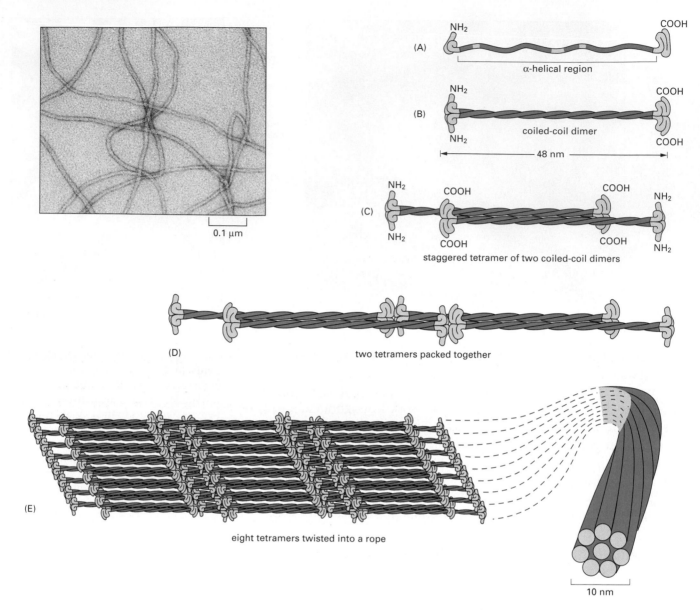

Intermediate Filaments Strengthen Cells Against Mechanical Stress

Intermediate filaments are particularly prominent in the cytoplasm of cells that are subject to mechanical stress. They are present in large numbers, for example, along the length of nerve cell axons, providing essential internal reinforcement to these extremely long and fine cell extensions. They are also abundant in muscle cells and in epithelial cells such as those of the skin, which are particularly subject to mechanical stress. In all these cells, intermediate filaments, by stretching and distributing the effect of locally applied forces, keep cells and their membranes from breaking in response to mechanical shear (Figure 16–5). A similar principle is used to manufacture reinforced concrete and other composite materials in which tension-bearing linear elements such as steel bars, glass, or carbon fibers are embedded in a space-filling matrix.

Intermediate filaments found in the cytoplasm can be grouped into three classes: (1) *keratin filaments* in epithelial cells; (2) *vimentin* and *vimentin-related filaments* in connective-tissue cells, muscle cells, and supporting cells of the nervous system (neuroglial cells); and (3) *neuro-*

Figure 16–4 The construction of an intermediate filament. The intermediate filament protein monomer shown in (A) consists of a central rod domain with globular regions at either end. Pairs of monomers associate to form a dimer (B), and two dimers then line up to form a staggered tetramer (C). Tetramers can pack together end-to-end as shown in (D) and assemble into a helical array (shown here spread out flat for clarity) that produces the final ropelike intermediate filament (E). An electron micrograph of the final 10-nm filament is shown at the upper left. (Electron micrograph courtesy of Roy Quinlan.)

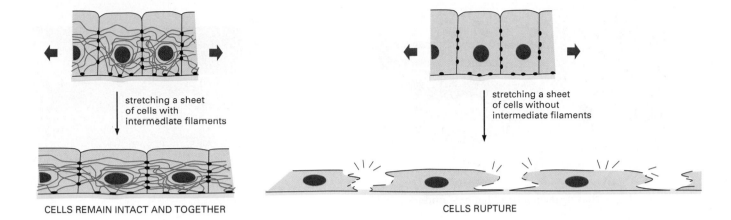

CELLS REMAIN INTACT AND TOGETHER   CELLS RUPTURE

filaments in nerve cells. Filaments of each class are formed by polymerization of their corresponding protein subunits (Figure 16–6).

The most diverse subunit family are the *keratins*. For example, different sets of keratins are found in different epithelia—one set in the lining of the gut, another in the epidermal layers of the skin. Specialized keratins occur in hair, feathers, and claws. In each case, the keratin filaments are formed from a mixture of different keratin subunits. Keratin filaments typically span the interiors of epithelial cells from one side of the cell to the other, and filaments in adjacent epithelial cells are indirectly connected through cell-cell junctions called *desmosomes* (see Figure 16–3), which are discussed in Chapter 19. The ends of the keratin filaments are anchored to the desmosomes, and they associate laterally with other cell components through their globular head and tail domains, which project from the surface of the assembled filament. This cabling of high tensile strength, formed by the filaments throughout the epithelial sheet, distributes the stress that occurs when the skin is stretched. The importance of this function is illustrated by the rare human genetic disease *epidermolysis bullosa simplex,* where mutations in the keratin genes interefere with the formation of keratin filaments in the epidermis. As a result, the skin is highly vulnerable to mechanical injury and even a gentle pressure can rupture its cells, causing the skin to blister.

While cytoplasmic intermediate filaments are like ropes, the intermediate filaments lining and strengthening the inside surface of the inner nuclear membrane are organized as a two-dimensional mesh (Figure 16–7). The intermediate filaments within this tough **nuclear lamina** are constructed from a class of intermediate filament proteins

Figure 16–5 Intermediate filaments strengthen animal cells. If a sheet of epithelial cells is stretched by external forces (due to the growth or movements of the surrounding tissues, for example), then the network of intermediate filaments and desmosomal junctions that extends through the sheet develops tension and limits the extent of stretching. If the junctions alone were present, then the same forces would cause a major deformation of the cells, even to the extent of causing their plasma membranes to rupture.

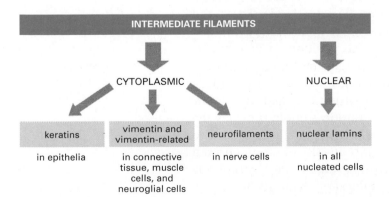

Figure 16–6 The major categories of intermediate filaments.

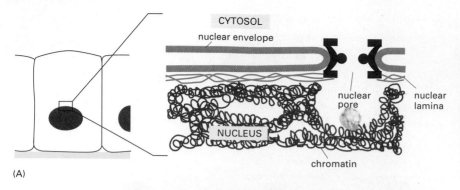

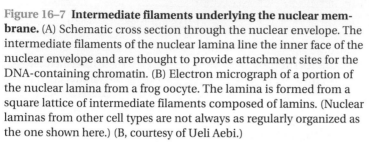

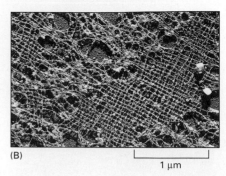

(A)

(B)

1 μm

Figure 16–7 Intermediate filaments underlying the nuclear membrane. (A) Schematic cross section through the nuclear envelope. The intermediate filaments of the nuclear lamina line the inner face of the nuclear envelope and are thought to provide attachment sites for the DNA-containing chromatin. (B) Electron micrograph of a portion of the nuclear lamina from a frog oocyte. The lamina is formed from a square lattice of intermediate filaments composed of lamins. (Nuclear laminas from other cell types are not always as regularly organized as the one shown here.) (B, courtesy of Ueli Aebi.)

called *lamins* (not to be confused with *laminin,* which is an extracellular matrix protein). In contrast to the very stable cytoplasmic intermediate filaments found in many cells, the intermediate filaments of the nuclear lamina disassemble and re-form at each cell division, when the nuclear envelope breaks down during mitosis and then re-forms in each daughter cell.

Disassembly and reassembly of the nuclear lamina is controlled by the phosphorylation and dephosphorylation (discussed in Chapter 5) of the lamins by protein kinases. When the lamins are phosphorylated, the consequent conformational change weakens the binding between the tetramers and causes the filament to fall apart. Dephosphorylation at the end of mitosis causes the lamins to reassemble (see Figure 17–14).

Question 16–1 Which of the following types of cells would you expect to contain a high density of intermediate filaments in their cytoplasm? Explain your answers.

A. *Amoeba proteus* (a free-living amoeba).

B. Skin epithelial cell.

C. Smooth muscle cell in the digestive tract.

D. *Escherichia coli.*

E. Nerve cell in the spinal cord.

F. Sperm cell.

G. Plant cell.

Microtubules

Microtubules have a crucial organizing role in all eucaryotic cells. They are long and relatively stiff hollow tubes of protein that can rapidly disassemble in one location and reassemble in another. In a typical animal cell, microtubules grow out from a small structure near the center of the cell, called the *centrosome.* Extending out toward the cell periphery, they create a system of tracks within the cell, along which vesicles, organelles, and other cell components can be moved (Figure 16–8A). These and other systems of cytoplasmic microtubules are the part of the cytoskeleton mainly responsible for determining the position of membrane-bounded organelles within the cell and guiding intracellular transport.

When a cell enters mitosis, the cytoplasmic microtubules disassemble and then reassemble into an intricate structure called the *mitotic spindle.* As described in Chapter 17, the mitotic spindle provides the machinery that will segregate the chromosomes equally into the two daughter cells just before a cell divides (Figure 16–8B). Microtubules can also form permanent structures, as exemplified by the rhythmically beating hairlike structures called *cilia* and *flagella* (Figure 16–8C). These extend from the surface of many eucaryotic cells, which use them either as a means of propulsion or to sweep fluid over the cell surface. The core of a eucaryotic cilium or flagellum is a highly organized and stable bundle of microtubules. (Bacterial flagella have an entirely different structure and act as propulsive structures by a different mechanism.)

In this section we first look at the structure and assembly of microtubules and then discuss their role in organizing the cytoplasm. Their organizing function depends on the association of microtubules with accessory proteins, especially the *motor proteins* that propel organelles along cytoskeletal tracks. Finally, we discuss the structure and function of cilia and flagella, where microtubules are permanently associated with motor proteins that power ciliary beating.

Microtubules Are Hollow Tubes with Structurally Distinct Ends

Microtubules are built from subunits—molecules of **tubulin**—each one of which is itself a dimer composed of two very similar globular proteins called *α-tubulin* and *β-tubulin,* bound tightly together by noncovalent bonding. The tubulin subunits stack together, again by noncovalent bonding, to form the wall of the hollow cylindrical microtubule. This appears as a cylinder made of 13 parallel *protofilaments,* each a linear chain of tubulin subunits with α- and β-tubulin alternating along its length (Figure 16–9). Each protofilament has a structural polarity, with α-tubulin exposed at one end and β-tubulin exposed at the other, and this polarity—the directional arrow embodied in the structure—is the same for all the protofilaments, giving a structural polarity to the microtubule as a whole. One end of the microtubule, thought to be the β-tubulin end, is called its *plus end,* and the other, the α-tubulin end, its *minus end.*

A microtubule grows from an initial ring of 13 tubulin molecules; tubulin dimers are added individually, gradually building up the structure of the hollow tube. *In vitro,* in a concentrated solution of pure tubulin, tubulin dimers will add to either end of a growing microtubule, although they add more rapidly to the plus end than the minus end (which is why the ends were originally named this way). The polarity of the microtubule—the fact that its structure has a definite direction, with the two ends being chemically different and behaving differently—is crucial, both for the assembly of microtubules and for their role once they are formed. If they had no polarity, they could not serve their function in defining a direction for intracellular transport, for example.

Microtubules Are Maintained by a Balance of Assembly and Disassembly

A living cell contains a mixture of microtubules and free tubulin subunits. In a typical fibroblast, for example, at any one time about half of the tubulin in the cell is in microtubules, while the remainder is free in the cytosol, forming a pool of subunits available for microtubule growth. This is quite unlike the situation with the more stable intermediate filaments, where the subunits are typically almost completely in the fully assembled form. The relative instability of microtubules allows them to undergo continual rapid remodeling, and this is crucial for microtubule

(A) INTERPHASE CELL

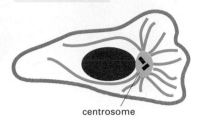
centrosome

(B) DIVIDING CELL

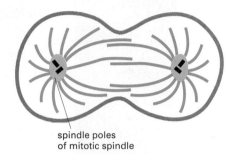
spindle poles of mitotic spindle

(C) CILIATED CELL

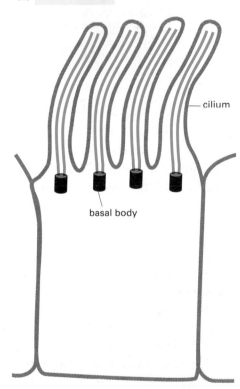
cilium

basal body

Figure 16–8 Three locations of microtubules in eucaryotic cells. Unlike intermediate filaments, microtubules (*dark green*) usually grow out of an organizing center such as (A) a centrosome, (B) a spindle pole, or (C) the basal body of a cilium.

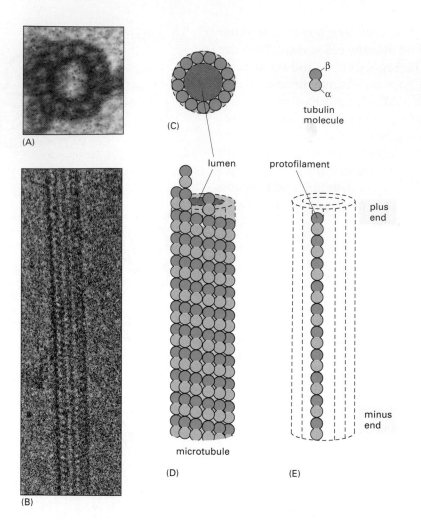

(A)

(C)

β
α

tubulin
molecule

lumen

protofilament

plus
end

minus
end

microtubule

(D)

(E)

(B)

Figure 16–9 **Structure of a microtubule.**
(A) Cross section of a microtubule with its
ring of 13 distinct subunits, each of which
corresponds to a separate tubulin dimer.
(B) Microtubule viewed lengthwise in an
electron microscope. (C and D) Schematic
diagrams of a microtubule showing how
tubulin molecules pack together in the
microtubule wall. (C) The 13 molecules in
cross-section. (D) A side view of a short
section of a microtubule with the tubulin
molecules aligned in rows, or protofilaments. (E) One tubulin molecule (an αβ
dimer) and one protofilament shown
schematically, together with their location
in the microtubule wall. Note that the
tubulin molecules are all arranged in the
protofilaments with the same orientation,
so that the microtubule has a definite
structural polarity. (A, courtesy of Richard
Linck; B, courtesy of Richard Wade.)

function, as demonstrated by the effect of drugs that prevent polymerization or depolymerization of tubulin.

Consider the mitotic spindle, the microtubule framework that
guides the chromosomes during mitosis (see Figure 16–8B). If a cell in
mitosis is exposed to the drug *colchicine*, which binds tightly to free
tubulin and prevents its polymerization into microtubules, the mitotic
spindle rapidly disappears and the cell stalls in the middle of mitosis,
unable to partition its chromosomes into two groups. This shows that
the mitotic spindle is normally maintained by a continuous balanced
addition and loss of tubulin subunits: when tubulin addition is blocked
by colchicine, tubulin loss continues until the spindle disappears.

The drug *taxol* has the opposite action at the molecular level. It
binds tightly to microtubules and prevents them from losing subunits.
Since new subunits can still be added, the microtubules can grow but
cannot shrink. However, despite the differences in molecular detail,
taxol has the same overall effect on the cell as colchicine, and it also
arrests dividing cells in mitosis. We learn from this that for the spindle to
function, microtubules must be able not only to assemble but also to
disassemble. The behavior of the spindle is discussed in more detail in
Chapter 17, when we consider mitosis.

The inactivation or destruction of the mitotic spindle eventually kills
dividing cells. Cancer cells, which are dividing more rapidly than most
other cells of the body, can therefore be killed preferentially by micro-

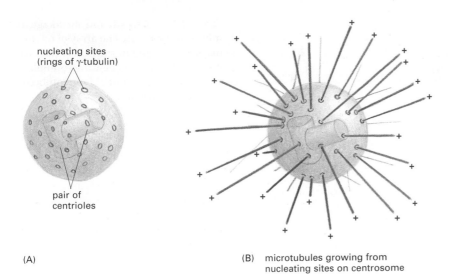

(A)

nucleating sites
(rings of γ-tubulin)

pair of
centrioles

(B) microtubules growing from
     nucleating sites on centrosome

Figure 16–10 **Polymerization of tubulin on a centrosome.** (A) Schematic drawing showing that a centrosome consists of an amorphous matrix of protein containing the γ-tubulin rings that nucleate microtubule growth. In animal cells, the centrosome contains a pair of centrioles, each made up of a cylindrical array of short microtubules. (B) A centrosome with attached microtubules. The minus end of each microtubule is embedded in the centrosome, having grown from a nucleating ring, whereas the plus end of each microtubule is free in the cytoplasm.

tubule-stabilizing and microtubule-destabilizing *antimitotic drugs.* Thus drugs derived from both colchicine and taxol are used in the clinical treatment of cancer.

The Centrosome Is the Major Microtubule-organizing Center in Animal Cells

Microtubules in cells are formed by outgrowth from specialized organizing centers, which control their number, their location, and their orientation in the cytoplasm. In animal cells, for example, the **centrosome,** which is typically present on one side of the cell nucleus when the cell is not in mitosis, organizes the array of microtubules that radiates outward from it through the cytoplasm (see Figure 16–8A). Centrosomes contain hundreds of ring-shaped structures formed of another type of tubulin, γ-tubulin, and each γ-tubulin ring serves as the starting point, or *nucleation site,* for the growth of one microtubule (Figure 16–10A). The αβ-tubulin dimers add to the γ-tubulin ring in a specific orientation, with the result that the minus end of each microtubule is embedded in the centrosome, and growth occurs only at the plus end—that is, the outward-facing end (Figure 16–10B).

The γ-tubulin rings in the centrosome should not be confused with the **centrioles,** curious structures each made of a cylindrical array of short microtubules embedded in the centrosome of most animal cells. The centrioles have no role in the nucleation of microtubules in the centrosome (the γ-tubulin rings alone are sufficient), and their function there remains something of a mystery, especially as plant cells lack them. Centrioles are, however, similar, if not identical, to the *basal bodies* that form the organizing centers for the microtubules in cilia and flagella (see Figure 16–8C), as we see later in this chapter.

Microtubules need nucleating sites such as those provided by the γ-tubulin rings in the centrosome, since it is much harder to start a new microtubule from scratch, by first assembling a ring of αβ-tubulin dimers, than to add such dimers to a preexisting microtubule structure. Purified free αβ-tubulin can polymerize spontaneously *in vitro* when at a high concentration, but in the living cell, the concentration of free αβ-tubulin is too low to drive the difficult first step of assembling the initial

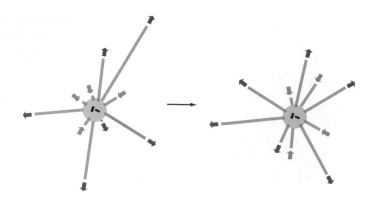

Figure 16–11 Growth and shrinkage in a microtubule array. The array of microtubules anchored in a centrosome is continually changing as new microtubules grow (*red arrows*) and old microtubules shrink (*blue arrows*).

ring of a new microtubule. By providing organizing centers containing nucleation sites, and keeping the concentration of free αβ-tubulin subunits low, cells can thus control where microtubules form.

Question 16–2 Why do you suppose it is much easier to add tubulin to existing microtubules than to start a new microtubule from scratch? Explain how γ-tubulin in the centrosome helps to overcome this hurdle.

Growing Microtubules Show Dynamic Instability

Once a microtubule has been nucleated, its plus end typically grows outward from the organizing center for many minutes by adding subunits. And then, unexpectedly, the microtubule suddenly undergoes a transition that causes it to shrink rapidly inward by losing subunits from its free end. It may shrink partially and then, no less suddenly, start growing again, or it may disappear completely, to be replaced by a new microtubule from the same γ-tubulin ring (Figure 16–11).

This remarkable behavior, known as **dynamic instability,** stems from the intrinsic capacity of tubulin molecules to hydrolyze GTP. Each free tubulin dimer contains one tightly bound GTP molecule that is hydrolyzed to GDP (still tightly bound) shortly after the subunit is added to a growing microtubule. The GTP-associated tubulin molecules pack efficiently together in the wall of the microtubule, whereas tubulin molecules carrying GDP have a different conformation and bind less strongly to each other.

Figure 16–12 How GTP hydrolysis is thought to control the growth of microtubules. Tubulin dimers carrying GTP (*red*) bind more tightly to one another than do tubulin dimers carrying GDP (*dark green*). Therefore, microtubules that have freshly added tubulin dimers with GTP bound tend to keep growing. From time to time, however, especially when microtubule growth is slow, the subunits in this "GTP cap" will hydrolyze their GTP to GDP before fresh subunits loaded with GTP have time to bind. The GTP cap is thereby lost; the GDP-carrying subunits are less tightly bound in the polymer and are readily released from the free end, so that the microtubule begins to shrink continuously.

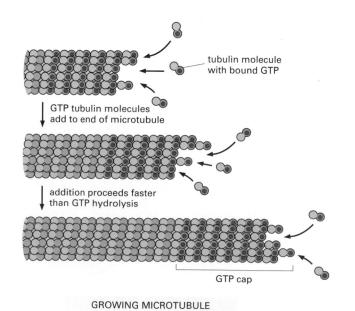

tubulin molecule with bound GTP

GTP tubulin molecules add to end of microtubule

addition proceeds faster than GTP hydrolysis

GTP cap

GROWING MICROTUBULE

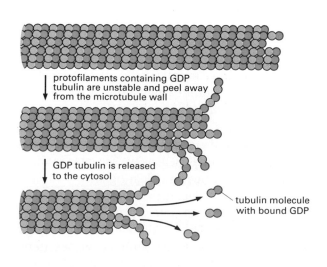

protofilaments containing GDP tubulin are unstable and peel away from the microtubule wall

GDP tubulin is released to the cytosol

tubulin molecule with bound GDP

SHRINKING MICROTUBULE

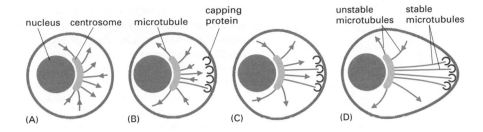

nucleus centrosome   microtubule   capping protein   unstable microtubules   stable microtubules

(A)   (B)   (C)   (D)

Figure 16–13 The selective stabilization of microtubules can polarize a cell. A newly formed microtubule will persist only if both its ends are protected from depolymerization. In cells, the minus ends of microtubules are generally protected by the organizing centers from which these filaments grow. The plus ends are initially free but can be stabilized by other proteins. Here, for example, a nonpolarized cell is depicted in (A) with new microtubules growing from and shrinking back to a centrosome in many directions randomly. Some of these microtubules happen by chance to encounter proteins (capping proteins) in a specific region of the cell cortex that can bind to and stabilize the free plus ends of microtubules (B). This selective stabilization will lead to a rapid reorientation of the microtubule arrays (C) and convert the cell to a strongly polarized form (D).

When polymerization is proceeding rapidly, tubulin molecules add to the end of the microtubule faster than the GTP they carry is hydrolyzed, resulting in the end of the microtubule being composed entirely of GTP-tubulin subunits. Because GTP-tubulin subunits bind strongly to one another, they form a cap at the end of the microtubule, known as a *GTP cap*, that prevents depolymerization. In this situation, since the microtubule can depolymerize only by loss of subunits from its free end, the growing microtubule will continue to grow. Because of the randomness of chemical processes, however, it will occasionally happen that tubulin at the free end of the microtubule hydrolyzes its GTP before the next tubulin has been added, so that the free ends of protofilaments are now composed of GDP-tubulin subunits. This tips the balance in favor of disassembly. Since the rest of the microtubule is composed of GDP-tubulin, once depolymerization has started, it will tend to continue, often at a catastrophic rate; the microtubule starts to shrink rapidly and continuously and may even disappear (Figure 16–12). The GDP-containing tubulin molecules that are freed as the microtubule depolymerizes join the pool of unpolymerized molecules in the cytosol. They then exchange their bound GDP for GTP, thereby becoming competent again to add to another microtubule that is in a growth phase.

What then is the function of this behavior? As a consequence of dynamic instability, the centrosome (or other organizing center) is continually shooting out new microtubules in an exploratory fashion in different directions and retracting them. A microtubule growing out from the centrosome can, however, be prevented from disassembling if its plus end is somehow permanently stabilized by attachment to another molecule or cell structure so as to prevent tubulin depolymerization. If stabilized by attachment to a structure in a more distant region of the cell, the microtubule will establish a relatively stable link between that structure and the centrosome (Figure 16–13). The centrosome can be compared to a fisherman casting a line: if there is no bite at the end of the line, the line is quickly withdrawn and a new cast is made; but if a fish bites, the line remains in place, tethering the fish to the fisherman. This simple strategy of random exploration and selective stabilization enables the centrosome and other nucleating centers to set up a highly organized system of microtubules linking selected parts of the cell. This system is used to position organelles relative to one another, as we now see.

Microtubules Organize the Interior of the Cell

Cells are able to modify the dynamic instability of their microtubules for particular purposes. As cells enter mitosis, for example, microtubules become initially more dynamic, switching between growing and shrink-

Question 16–3 Dynamic instability causes microtubules either to grow or to shrink rapidly. Consider an individual microtubule that is in its shrinking phase.

A. What must happen at the end of the microtubule in order for it to stop shrinking and to start growing?

B. How would a change in the tubulin concentration affect this switch?

C. What would happen if only GDP, but no GTP, were present in the solution?

D. What would happen if the solution contained an analogue of GTP that cannot be hydrolyzed?

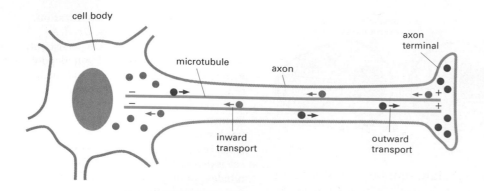

Figure 16–14 Transport along micro-tubules in a nerve cell axon. In nerve cells all the microtubules (from one to hundreds) in the axon point in the same direction, with their plus ends toward the axon terminal. The oriented microtubules serve as tracks for the directional transport of materials synthesized in the cell body but required at the axon terminal (such as membrane proteins required for growth). For an axon passing from your spinal cord to a muscle in your shoulder, say, the journey takes about two days. In addition to this *outward* traffic of material driven by one set of motor proteins (*red circles*) there is *inward* traffic in the reverse direction driven by another set of proteins (*blue circles*). The inward traffic carries materials ingested by the tip of the axon or produced by the breakdown of proteins and other molecules back toward the cell body.

ing much more frequently than cytoplasmic microtubules normally do. This enables them to disassemble rapidly and then reassemble into the mitotic spindle. On the other hand, when a cell has differentiated into a specialized cell type and taken on a definite fixed structure, the dynamic instability of its microtubules is often suppressed by proteins that bind to the ends of microtubules or along their length and stabilize them against disassembly. The stabilized microtubules then serve to maintain the organization of the cell.

Most differentiated animal cells are *polarized:* nerve cells, for example, put out an axon from one end of the cell and dendrites from the other; cells specialized for secretion have their Golgi apparatus positioned toward the site of secretion, and so on. The cell's polarity is a reflection of the polarized systems of microtubules in its interior, which help to position organelles in their required location within the cell and to guide the streams of traffic moving between one part of the cell and another. In the nerve cell, for example, all the microtubules in the axon point in the same direction, with their plus ends toward the axon terminal (Figure 16–14). Along these oriented tracks the cell is able to send cargoes of materials, such as membrane vesicles and proteins for secretion, that are made in the cell body but required far away at the end of the axon.

Some of these materials move at speeds in excess of 10 cm a day, which still means that it may take a week or more to travel to the end of a long axon in some animals. But movement along microtubules is immeasurably faster and more efficient than free diffusion. A protein molecule traveling by free diffusion would take years to reach the end of a long axon, if it arrived at all (see Question 16–10).

It is important to realize, also, that the microtubules in living cells do not act alone. Their functions, like those of other cytoskeletal filaments, depend on a large variety of accessory proteins that bind to microtubules and serve various functions. Some microtubule-associated proteins stabilize microtubules against disassembly, for example, while others link microtubules to other cell components, including the other types of cytoskeletal filaments. Thus all the components of the cytoskeleton can interact with each other and their functions can be coordinated. Microtubules also influence the distribution of membranes in a eucaryotic cell, especially by means of microtubule-associated motor proteins, which move along microtubules. As we discuss next, motor proteins use the energy of ATP hydrolysis to transport organelles, vesicles, and other cell materials along tracks in the cytoplasm provided by actin filaments and microtubules.

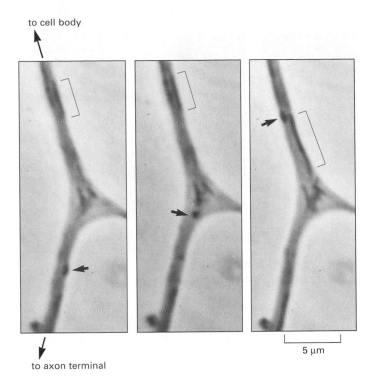

to cell body

to axon terminal

5 μm

Figure 16–15 Intracellular organelle movement. Three photographs of a living nerve cell axon, taken with less than a minute between the first and last frames. The *red arrow* highlights a large membrane-bounded organelle jerkily moving back toward the nerve cell body. The *blue bracket* indicates a mitochondrion that is slowly moving outward toward the axon terminal. (Courtesy of Peter J. Hollenbeck. Reproduced from *The Journal of Cell Biology* 121:307, 1993, by copyright permission of The Rockefeller University Press.)

Motor Proteins Drive Intracellular Transport

If a living cell is observed in a light microscope, its cytoplasm is seen to be in continual motion: mitochondria and the smaller membrane-bounded organelles and vesicles are moving in small jerky movements—that is, they move for a short period, stop, and then start again (Figure 16–15). This *saltatory movement* is much more sustained and directional than the continual, small Brownian movements caused by random thermal motions. Both microtubules and actin filaments are involved in saltatory and other directed intracellular movements in eucaryotic cells. In both cases the movements are generated by **motor proteins,** which bind to actin filaments or microtubules and use the energy derived from repeated cycles of ATP hydrolysis to travel steadily along the actin filament or the microtubule in a single direction (see Figure 5–40). At the same time, these motor proteins also attach other cell components, and thus transport this cargo along the filaments. Dozens of motor proteins have been identified. They differ in the type of filament they bind to, the direction in which they move along the filament, and in the cargo that they carry.

The motor proteins that move along cytoplasmic microtubules, such as those in the axon of a nerve cell, belong to two families: the **kinesins** generally move toward the plus end of a microtubule (away from the centrosome; outward in Figure 16–14), while the **dyneins** move toward the minus end (toward the centrosome; inward in Figure 16–14). Kinesins and dyneins both have two globular ATP-binding heads and a tail (Figure 16–16). The heads interact with microtubules in a stereospecific manner, so that kinesin, for example, will attach to a microtubule only if it is "pointing" in the correct direction. The tail of a motor protein generally binds stably to some cell component, such as a vesicle or an organelle, and thereby determines the type of cargo that the motor protein can transport (Figure 16–17). The globular heads of kinesin and

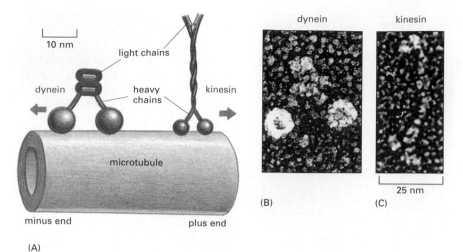

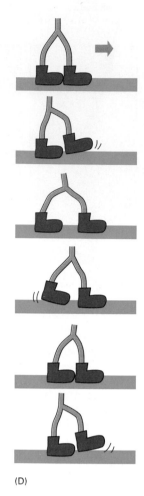

Figure 16–16 **Microtubule motor proteins.** (A) Kinesins and cytoplasmic dyneins are microtubule motor proteins that generally move in opposite directions along a microtubule. These proteins (drawn here to scale) are complexes composed of two identical heavy chains plus several smaller light chains. Each heavy chain forms a globular head that interacts with microtubules. (B) Freeze-etch electron micrograph of a molecule of cytoplasmic dynein. (C) Freeze-etch electron micrograph of a kinesin molecule. (D) Diagram of a motor protein showing ATP-dependent "walking" along a filament. (B and C, courtesy of John Heuser.)

dynein are enzymes with ATP-hydrolyzing (ATPase) activity. This reaction provides the energy for a cycle of conformational changes in the head that enable it to move along the microtubule by a cycle of binding, release, and rebinding to the microtubule (see Figure 5–40).

Organelles Move Along Microtubules

Microtubules and their associated motor proteins play an important part in positioning membrane-bounded organelles within a eucaryotic cell. In most animal cells, for example, the tubules of the endoplasmic reticulum reach almost to the edge of the cell, whereas the Golgi apparatus is located in the interior of the cell near the centrosome (Figure 16–18A). Both the endoplasmic reticulum and the Golgi apparatus depend on microtubules for their alignment and positioning. The membranes of the endoplasmic reticulum extend out from their points of connection with the nuclear envelope (see Figure 1–15), aligning with microtubules that extend from the centrosome out to the plasma membrane. As the cell develops and the endoplasmic reticulum grows, kinesins attached to the outside of the endoplasmic reticulum mem-

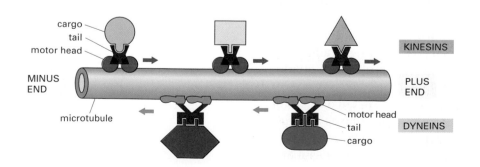

Figure 16–17 **The motor proteins that transport cargoes along microtubules.** Kinesins move toward the plus end of a microtubule, whereas dyneins move toward the minus end. Both types of microtubule motor proteins exist in many forms, each of which is thought to transport a different cargo. The tail of the motor protein determines what cargo the protein transports.

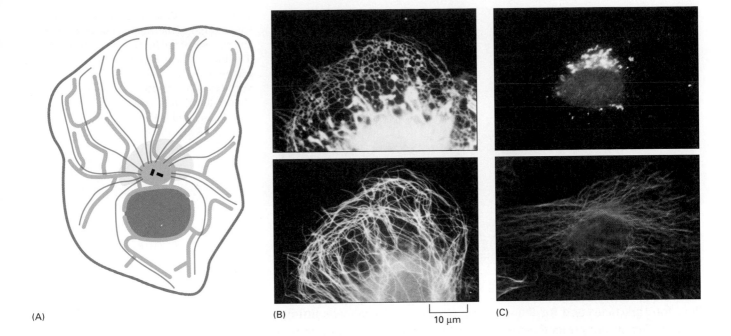

(A) (B) 10 µm (C)

brane pull it outward along microtubules, stretching it like a net. Dyneins pull the Golgi apparatus the other way along microtubules, inward toward the cell center. In this way the regional differences in internal membranes, on which their successful function depends, are created and maintained.

When cells are treated with a drug such as colchicine that causes microtubules to disassemble, both these organelles change their location dramatically. The endoplasmic reticulum, which has connections to the nuclear envelope, collapses to the center of the cell, while the Golgi apparatus, which is not attached to any other organelle, fragments into small vesicles, which disperse throughout the cytoplasm. When the drug is removed, the organelles return to their original positions, dragged by motor proteins moving along the re-formed microtubules. The normal positioning of these organelles is thought to be mediated by receptor proteins on their membranes that bind to motor proteins—to kinesins for the endoplasmic reticulum and to dyneins for the Golgi apparatus.

Figure 16–18 The placement of organelles by microtubules. (A) Schematic diagram of a cell showing the typical arrangement of microtubules (*dark green*), endoplasmic reticulum (*blue*), and Golgi apparatus (*yellow*). The nucleus is shown in *brown*, and the centrosome in *light green*. (B) Cell stained with antibodies to endoplasmic reticulum (*upper panel*) and to microtubules (*lower panel*). Motor proteins pull the endoplasmic reticulum out along microtubules. (C) Cell stained with antibodies to the Golgi apparatus (*upper panel*) and to microtubules (*lower panel*). In this case motor proteins move the Golgi apparatus inward to its position near the centrosome. (B, courtesy of Mark Terasaki, Lan Bo Chen, and Keigi Fujiwara; C, courtesy of Viki Allan and Thomas Kreis.)

Cilia and Flagella Contain Stable Microtubules Moved by Dynein

Earlier in this chapter we mentioned that many microtubules in cells are stabilized through their association with other proteins, and therefore no longer show dynamic instability. Stable microtubules are employed by cells as rigid girders to construct a variety of polarized structures, including the remarkable cilia and flagella, which enable eucaryotic cells to propel water over their surface. **Cilia** (singular, cilium) are hairlike structures about 0.25 µm in diameter that extend from the surface of many kinds of eucaryotic cells. A single cilium contains a core of stable microtubules, arranged in a bundle, which grow from a *basal body* in the cytoplasm, which is the organizing center for the cilium. The whole cilium is covered by plasma membrane (see Figure 16–8C). The primary function of cilia is to move fluid over the surface of a cell, or to propel single cells through a fluid. Some protozoa, for example, use cilia both to

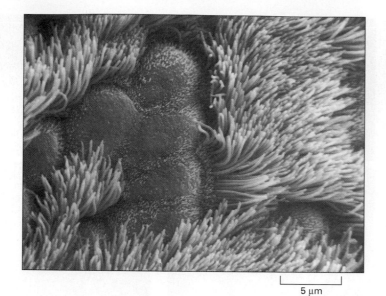

Figure 16–19 Cilia. Scanning electron micrograph of the ciliated epithelium on the surface of the human respiratory tract. The thick tufts of cilia on the ciliated cells are interspersed with the dome-shaped surfaces of nonciliated epithelial cells. (Reproduced from Tissues and Organs, R.G. Kessel and R.H. Karden, W.H. Freeman & Co, 1979.)

5 μm

collect food particles and for locomotion. On the epithelial cells lining the human respiratory tract (Figure 16–19), huge numbers of cilia (more than a billion per square centimeter) sweep layers of mucus with trapped dust particles and dead cells up toward the throat, to be swallowed and eventually eliminated from the body. Cilia on the cells of the oviduct wall create a current that helps to move eggs along the oviduct.

Each cilium moves with a whiplike motion, resembling the breast-stroke in swimming (Figure 16–20). The **flagella** (singular, **flagellum**) that propel sperm and many protozoa are much like cilia in their internal structure, but usually very much longer. They are designed to move the entire cell: instead of making whiplike movements, they propagate regular waves along their length that drive the cell through liquid (Figure 16–21).

The microtubules in cilia and flagella are slightly different from the cytoplasmic microtubules; they are arranged in a curious and distinctive pattern that was one of the most striking revelations of early electron microscopy. A cross section through a cilium shows nine doublet microtubules arranged in a ring around a pair of single microtubules (Figure 16–22A). This "9 + 2" array is characteristic of almost all forms of eucaryotic cilia and flagella, from those of protozoa to those found in humans.

The movement of a cilium or a flagellum is produced by the bending of its core as the microtubules slide against each other. The microtubules are associated with numerous proteins, which project at regular positions along the length of the microtubule bundle (Figure 16–22B). Some serve as cross-links to hold the bundle of microtubules together, whereas others generate the force that causes the cilium to bend.

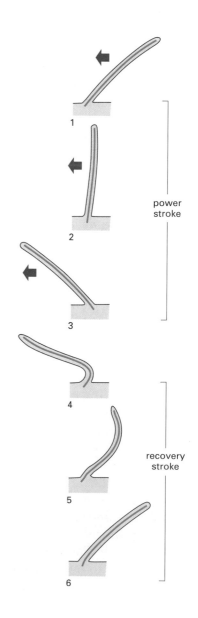

power stroke

recovery stroke

Figure 16–20 The beating of a cilium. Each cilium performs a repetitive cycle of movements consisting of a power stroke followed by a recovery stroke. In the fast power stroke, the cilium is fully extended and fluid is driven over the surface of the cell; in the slower recovery stroke the cilium curls back into position with minimal disturbance to the surrounding fluid. Each cycle typically requires 0.1–0.2 second and generates a force perpendicular to the axis of the cilium.

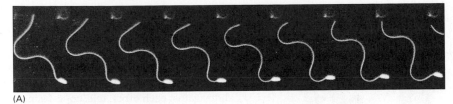

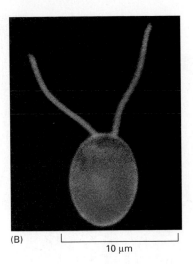

(A)

(B)

10 μm

Figure 16–21 **Flagella.** (A) The repetitive wavelike motion of a single flagellum on a tunicate sperm is seen in a series of images captured by stroboscopic illumination at 400 flashes per second. (B) The green alga *Chlamydomonas* swims by means of paired flagella that operate in a repetitive action reminiscent of a human breaststroke. (A, courtesy of Charles J. Brokaw; B, courtesy of Robert A. Bloodgood.)

The most important of these accessory proteins is the motor protein *ciliary dynein,* which generates the bending motion of the core. It closely resembles cytoplasmic dynein and functions in much the same way. It is attached by its tail to one microtubule, while its heads interact with an adjacent microtubule to generate a sliding force between the two microtubules. Because of the multiple links that hold the adjacent microtubule doublets together, what would be a simple parallel sliding movement between free microtubules is converted to a bending motion in the cilium (Figure 16–23).

Actin Filaments

Actin filaments are found in all eucaryotic cells and are essential for many of their movements, especially those involving the cell surface. Without actin filaments, for example, an animal cell could not crawl along a surface, engulf a large particle by phagocytosis, or divide in two. Like microtubules, many actin filaments are unstable, but they can also form stable structures in cells, such as the contractile apparatus of muscle. Actin filaments are associated with a large number of *actin-binding proteins* that enable the filaments to serve a variety of functions in cells. Depending on their association with different proteins, actin filaments can form stiff and relatively permanent structures such as the

Figure 16–22 **The arrangement of microtubules in a cilium or flagellum.** (A) Electron micrograph of a flagellum of *Chlamydomonas* shown in cross-section, illustrating the distinctive "9 + 2" arrangement of microtubules. (B) Diagram of the flagellum in cross-section. The nine outer microtubules (each a special paired structure) carry two rows of dynein molecules. The heads of these dyneins appear in this view like pairs of arms reaching toward the adjacent microtubule. In a living cilium, these dynein heads periodically make contact with the adjacent microtubule and move along it, thereby producing the force for ciliary beating. Various other links and projections shown are proteins that serve to hold the bundle of microtubules together and to convert the sliding motion produced by dyneins into bending, as illustrated in Figure 16–23. (A, courtesy of Lewis Tilney.)

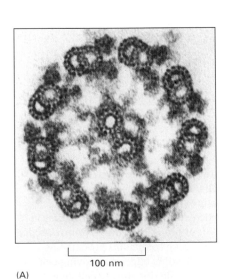

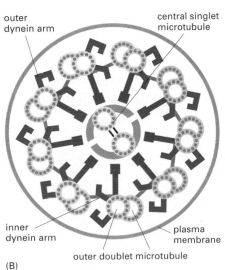

outer dynein arm

central singlet microtubule

inner dynein arm

outer doublet microtubule

plasma membrane

100 nm

(A)

(B)

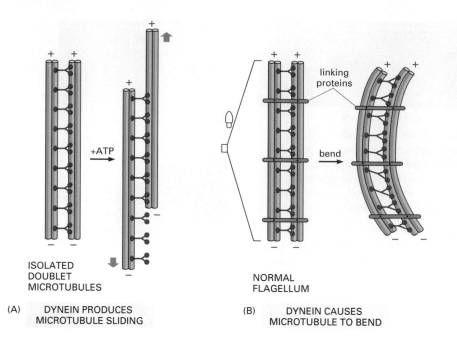

(A) DYNEIN PRODUCES MICROTUBULE SLIDING

(B) DYNEIN CAUSES MICROTUBULE TO BEND

Figure 16–23 **The movement of dynein causes the flagellum to bend.** (A) If the outer doublet microtubules and their associated dynein molecules are freed from other components of a sperm flagellum and then exposed to ATP, the doublets slide against each other, telescope-fashion, due to the repetitive action of their associated dyneins. (B) In an intact flagellum, however, the doublets are tied to each other by flexible protein links so that the action of the system produces bending rather than sliding.

micro-villi, on the brush-border cells lining the intestine (Figure 16–24A) or small *contractile bundles* in the cytoplasm that can contract and act like the "muscles" of a cell (Figure 16–24B), or temporary structures such as the protrusions formed at the leading edge of a crawling fibroblast (Figure 16–24C), or the *contractile ring* that pinches the cytoplasm in two when an animal cell divides (Figure 16–24D). Which of the many possible arrangements of actin filaments exist in a cell depends on the types of actin-binding proteins present. Even though actin filaments and microtubules are formed from unrelated types of proteins, we shall see that the principles according to which they assemble and disassemble, control cell structure, and bring about movement are strikingly similar.

Actin Filaments Are Thin and Flexible

Actin filaments appear in electron micrographs as threads about 7 nm in diameter. Each filament is a twisted chain of identical globular actin molecules, all of which "point" in the same direction along the axis of the chain. Like a microtubule, therefore, an actin filament has a structural polarity, with a plus end and a minus end (Figure 16–25).

Actin filaments are thinner, more flexible, and usually shorter than microtubules. There are many more individual actin filaments in a cell than microtubules; the total length of all the actin filaments in a cell is at least 30 times greater than the total length of all of the microtubules. Actin filaments rarely occur in isolation in the cell, being generally found

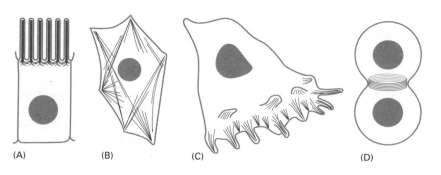

(A) (B) (C) (D)

Figure 16–24 **Bundles of actin filaments in cells.** The following actin-containing structures are shown here in *red.* (A) Microvilli. (B) Contractile bundles in the cytoplasm. (C) Sheetlike (*lamellipodia*) and fingerlike (*filopodia*) protrusions from the leading edge of a moving cell. (D) Contractile ring during cell division.

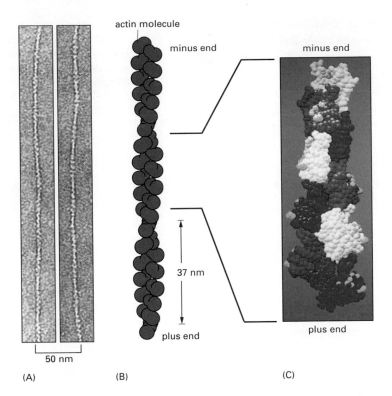

actin molecule

minus end

minus end

37 nm

plus end

plus end

50 nm

(A)   (B)   (C)

Figure 16–25 Actin filaments. (A) Electron micrographs of negatively stained actin filaments. (B) Arrangement of actin molecules in an actin filament. Each filament may be thought of as a two-stranded helix with a twist repeating every 37 nm. Strong interactions between the two strands prevent the strands from separating. (C) The identical subunits of an actin filament are depicted in different colors to emphasize the close interaction between each actin molecule and its four nearest neighbors. (A, courtesy of Roger Craig; C, from K.C. Holmes et al. *Nature* 347:44–49, 1990, copyright Macmillan Magazines Ltd.)

in cross-linked bundles and networks, which are much stronger than the individual filaments.

Actin and Tubulin Polymerize by Similar Mechanisms

Actin filaments can grow by addition of actin monomers at either end, but the rate of growth is faster at the plus end than at the minus end. A naked actin filament, like a microtubule without associated proteins, is inherently unstable, and it can disassemble from both ends. Each free actin monomer carries a tightly bound nucleoside triphosphate, in this case ATP, which is hydrolyzed to ADP soon after the incorporation of the actin monomer into the filament. As with the GTP of tubulin, hydrolysis of bound ATP to ADP in an actin filament reduces the strength of binding between monomers and decreases the stability of the polymer. Nucleotide hydrolysis thereby promotes depolymerization, helping the cell to disassemble filaments after they have formed (Figure 16–26).

As with microtubules, the ability to assemble and disassemble is required for many of the functions performed by actin filaments, such as their role in cell locomotion. Actin filament function can be perturbed experimentally by fungal toxins—both those that prevent actin polymerization, such as the *cytochalasins,* and those that stabilize actin fila-

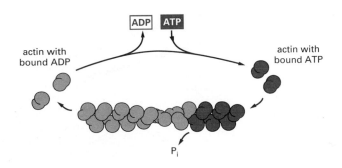

ADP   ATP

actin with bound ADP

actin with bound ATP

P_i

Figure 16–26 ATP hydrolysis during actin polymerization. Actin monomers in the cytosol carry ATP, which is hydrolyzed to ADP soon after assembly into a growing filament. The ADP molecules remain trapped within the actin filament, unable to exchange with ATP until the actin monomer that carries them dissociates from the filament to form a monomer once again.

ments against depolymerization, such as *phalloidin*. Addition of cytochalasin in low concentrations instantaneously freezes cell movements such as the crawling motion of a fibroblast. (Phalloidin has a similar effect but has to be injected into the cell because it does not readily cross the plasma membrane.) Thus, the function of actin filaments depends on a dynamic equilibrium between the actin filaments and the pool of actin monomers, with a typical filament persisting for only a few minutes after it forms.

Many Proteins Bind to Actin and Modify Its Properties

About 5% of the total protein in a typical animal cell is actin; about half of this is assembled into actin filaments, while the other half is free as actin monomers in the cytosol. The concentration of monomer is therefore high—much higher than the concentration required for purified actin monomers to polymerize *in vitro*. What then keeps the actin monomers in cells from polymerizing totally into filaments? The answer is that cells contain small proteins, such as *thymosin* and *profilin*, that bind to actin monomers in the cytosol, preventing them from adding to

Question 16–5 The formation of actin filaments in the cytosol is controlled by actin-binding proteins. Some actin-binding proteins significantly increase the rate at which formation of an actin filament is initiated. Suggest a mechanism by which they might do this.

Figure 16–27 Major classes of actin-binding proteins found in vertebrate cells. Actin is shown in *red*, and the actin-binding proteins are shown in *green*.

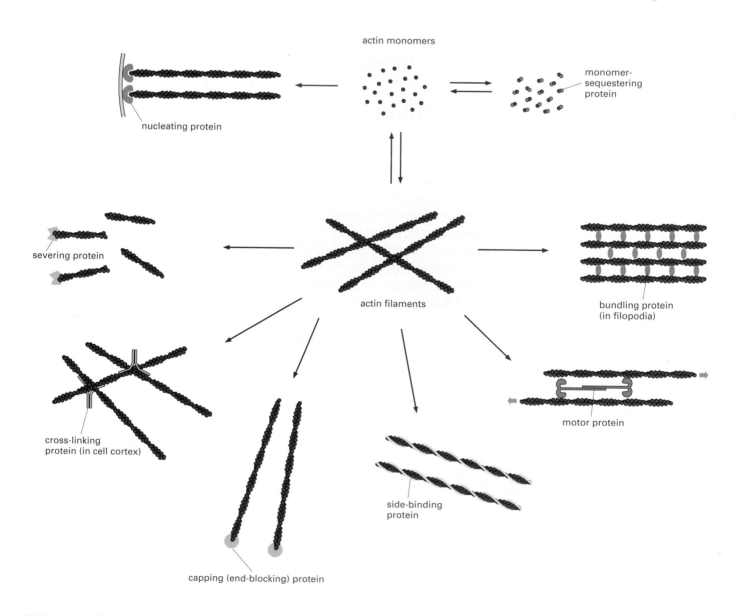

the ends of actin filaments. By keeping actin monomers in reserve until they are required, thymosin in particular plays a crucial role in regulating actin polymerization in cells.

There are a great many other actin-binding proteins in cells. Most of these bind to assembled actin filaments rather than to actin monomers, and control the behavior of the filaments (Figure 16–27). Actin-bundling proteins, for example, hold actin filaments together in parallel bundles in microvilli; cross-linking proteins hold actin filaments together in a gel-like meshwork within the cell cortex—the layer of cytoplasm just beneath the plasma membrane; filament-severing proteins, like *gelsolin*, fragment actin filaments into shorter lengths and thus can convert an actin gel to a more fluid state. Actin filaments can also be associated with motor proteins to form contractile bundles, as in muscle cells; they can also serve as tracks along which motor proteins transport organelles, a function that is especially conspicuous in plant cells.

In the remainder of this chapter, we consider some characteristic structures that actin filaments can form, and discuss how different types of actin-binding proteins are involved in their formation. We begin with the actin-rich *cell cortex* and its role in cell locomotion, and go on to consider the contractile apparatus of muscle cells as an example of a stable structure based on actin filaments.

An Actin-rich Cortex Underlies the Plasma Membrane of Most Eucaryotic Cells

Although actin is found throughout the cytoplasm of a eucaryotic cell, in most cells it is concentrated in a layer just beneath the plasma membrane. In this region, called the **cell cortex,** actin filaments are linked by actin-binding proteins into a meshwork that supports the outer surface of the cell and gives it mechanical strength. In red blood cells, as described in Chapter 11, a simple and regular network of fibrous proteins attached to the plasma membrane provides it with support necessary to maintain its simple discoid shape (see Figure 11–31). The cell cortex of other animal cells, however, is thicker and more complex and supports a far richer repertoire of shapes and movements. Although it also contains spectrin and ankyrin, it includes a dense network of actin filaments that project into the cytoplasm, where they become cross-linked into a three-dimensional meshwork. This cortical actin mesh governs the shape and mechanical properties of the plasma membrane and the cell surface. As we shall see, actin rearrangements within the cortex provide the molecular basis for changes in cell shape and for cell locomotion.

Cell Crawling Depends on Actin

Many cells move by crawling over surfaces, rather than by swimming by means of cilia or flagella. Carnivorous amoebae crawl continually, in search of food. White blood cells known as *neutrophils* migrate from the bloodstream into tissues when they "smell" small diffusing molecules released by bacteria, which the neutrophils eventually engulf and destroy. The advancing tip of a developing axon migrates in response to growth factors, following a path of substrate-bound and diffusible chemicals to its eventual synaptic target.

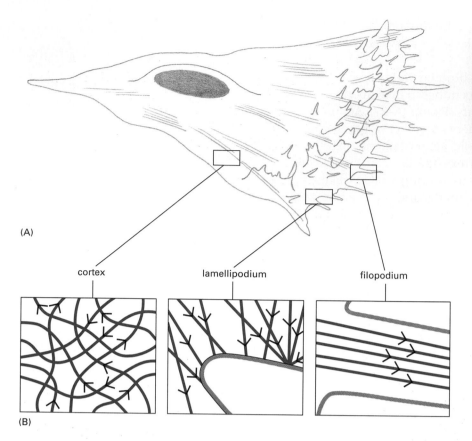

(A)

cortex   lamellipodium   filopodium

(B)

(C)

5 μm

Figure 16–28 Actin filaments in migrating animal cells. (A) Schematic drawing of a fibroblast showing flattened lamellipodia and fine filopodia projecting from its surface, especially in the regions of the leading edge. (B) Details of the arrangement of actin filaments in three regions of the fibroblast are shown, with *arrowheads* pointing toward the plus end of the filaments. (C) Scanning electron micrograph showing lamellipodia and filopodia at the leading edge of a human fibroblast migrating in culture. (C, courtesy of Julian Heath.)

The molecular mechanisms of these and other forms of cell crawling are difficult to dissect. They entail coordinated changes of many molecules in different regions of the cell, and no single, easily identifiable locomotory organelle, such as a flagellum, is responsible. In broad terms, however, three interrelated processes are known to be essential: (1) the cell pushes out protrusions at its "front," or leading edge; (2) these protrusions adhere to the surface over which the cell is crawling; and (3) the rest of the cell drags itself forward by traction on these anchorage points.

All three processes involve actin, but in different ways. The first step, the pushing forward of the cell surface, is driven by actin polymerization. The leading edge of a crawling fibroblast in culture regularly extends thin, sheetlike **lamellipodia** (Figure 16–28), which contain a dense meshwork of actin filaments, oriented so that most of the filaments have their plus ends close to the plasma membrane. Many cells also extend thin, stiff protrusions called **filopodia** (see Figure 16–28), both at the leading edge and elsewhere on their surface. These are about 0.1 μm wide and 5–10 μm long, and each contains a loose bundle of 10–20 actin filaments, again oriented with their plus ends pointing outward. The advancing tip (growth cone) of a developing nerve cell axon extends even longer filopodia, up to 50 μm long, which help it to probe its environment and find the correct path to its target (Figure 16–29A). Both lamellipodia and filopodia are exploratory, motile structures that form and retract with great speed. Both are thought to be generated by rapid local growth of actin filaments, which are nucleated at the plasma membrane and push out the membrane without tearing it as they elongate.

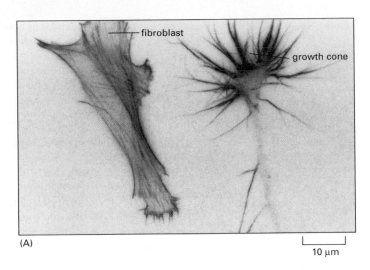
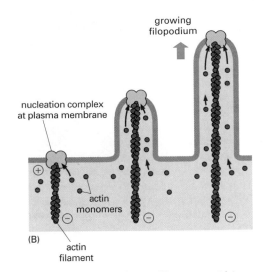

Figure 16–29 **Growth of filopodia.** (A) Two different cell types stained for actin. Fine bundles of actin filaments within filopodia can be seen at the leading edge of both the flattened fibroblast and the advancing tip (growth cone) of a nerve cell axon. (B) At the tip of each filopodium, a small aggregate of protein molecules on the inner face of the plasma membrane nucleates the assembly of actin monomers into a filament. Growth of the actin filament drives the plasma membrane outward to form a filopodium or other cell protrusion. For simplicity, only a single actin filament is shown, whereas most protrusions on the cell surface contain multiple actin filaments. (A, courtesy of Peter J. Hollenbeck.)

The plasma membrane at the leading edge of a cell seems to organize the actin filaments of lamellipodia and filopodia much as a centrosome organizes microtubules, by providing small aggregates of proteins that promote actin polymerization (Figure 16–29B). There is one crucial difference between actin filaments and microtubules, however. In the case of actin filaments, the organizing cluster of proteins is situated at the plus end of the filament, and actin monomers are added at this end to elongate the filament. Precisely how this nucleating complex of proteins operates at the molecular level is still a mystery.

When the lamellipodia and filopodia touch down on a favorable patch of surface, they stick: transmembrane proteins in their plasma membrane, known as *integrins*, adhere to molecules in the extracellular matrix or on the surface of another cell over which the moving cell is crawling. Meanwhile, on the intracellular face of the membrane of the crawling cell, integrins nucleate, or capture, actin filaments, thereby creating a robust anchorage for the system of actin filaments inside the crawling cell (Figure 16–30). To use this anchorage to drag its body

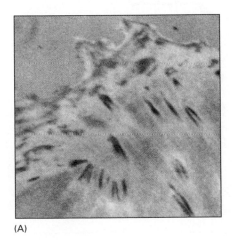
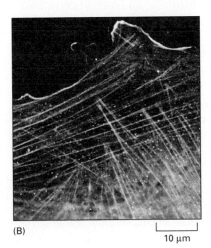

Figure 16–30 **The relation between external adhesions and intracellular actin filament bundles.** In (A), an optical trick has been used to reveal close contacts (dark patches) between a fibroblast and the glass slide on which it rests. (B) Staining the same cell (after fixation) with fluorescent antibodies to actin shows that most of the cell's actin filament bundles terminate at, or close to, these sites of contact. (Courtesy of Grenham Ireland.)

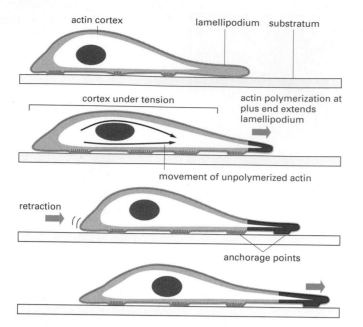

actin cortex   lamellipodium   substratum

cortex under tension   actin polymerization at plus end extends lamellipodium

movement of unpolymerized actin

retraction

anchorage points

Figure 16–31 One model of how the actin-rich cortex might move a cell forward. Actin polymerization at the leading edge of the cell pushes the plasma membrane forward and forms new regions of actin cortex, shown here in *red*. New points of anchorage are made between the actin filaments and the surface on which the cell is crawling (the substratum). Cortical tension then draws the body of the cell forward. As the rear of the cell detaches from the substratum and retracts, actin filaments in the retracting region depolymerize, and the actin molecules released move forward through the cytosol to sites of new polymerization. The same cycle is repeated over and over again, moving the cell forward in a stepwise fashion.

forward, the cell now makes use of internal contractions to exert a pulling force (Figure 16–31). These too depend on actin, but in a different way—through the interaction of actin filaments with motor proteins known as *myosins*. It is still not certain how this pulling force is produced: contraction of bundles of actin filaments in the cytoplasm, or contraction of the actin meshwork in the cell cortex, or both, may be responsible. The general principles of how myosin motor proteins interact with actin filaments to cause movement, however, is clear, as we now discuss.

Question 16–6 At the leading edge of a crawling cell, the plus ends of actin filaments are attached to the plasma membrane and actin monomers are added at these ends, pushing the membrane outward to form lamellipodia or filopodia. What do you suppose holds the filaments at their other ends to prevent them from just being pushed into the cell's interior?

Actin Associates with Myosin to Form Contractile Structures

All actin-dependent motor proteins belong to the **myosin** family. They bind and hydrolyze ATP, which provides the energy for their movement along actin filaments from the minus end of the filament toward the plus end. Myosin, along with actin, was first discovered in skeletal muscle, and much of what we know about the interaction of these two proteins was learned from studies of muscle. There are several different types of myosins in cells, of which the *myosin-I* and *myosin-II* subfamilies are most abundant (Figure 16–32). Myosin-I is found in all types of cells, and as it is simpler in structure and mechanism of action, we shall discuss it first.

Myosin-I molecules have only one head domain and a tail (see Figure 16–32A). The head domain interacts with actin filaments and has

Figure 16–32 Myosins. Two members of the myosin family of motor proteins are shown. (A) Myosin-I is a single molecule with one globular head and a tail that attaches to another molecule or organelle in the cell. In this way the attached molecule or organelle can be dragged along actin filament tracks by the motor activity of the myosin-I head. (B) Myosin-II is composed of a pair of identical myosin molecules and thus has two globular heads and a coiled-coil tail. (C) The tails of myosin-II can associate with one another to form a myosin filament in which the heads project outward along its length. The bare region in the middle of the filament consists of tails only.

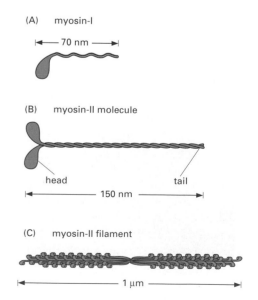

(A) myosin-I

|← 70 nm →|

(B) myosin-II molecule

head   tail

|← 150 nm →|

(C) myosin-II filament

|← 1 μm →|

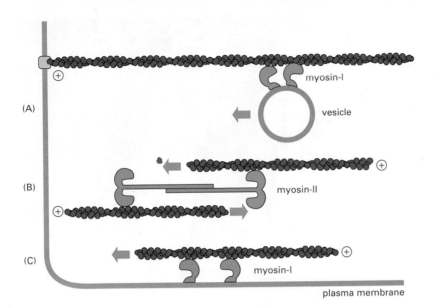

Figure 16–33 Some roles of myosin-I and myosin-II in a eucaryotic cell. The short tail of a myosin-I molecule contains sites that bind to various components of the cell, including membranes. This allows the head domain to move a vesicle relative to an actin filament (A), or an actin filament and the plasma membrane relative to each other (C). Small filaments composed of myosin-II molecules can slide actin filaments over each other, thus mediating local *shortening* of an actin filament bundle (B). Note that in every case illustrated here, the head group of the myosin walks toward the plus end of the actin filament it contacts.

ATP-hydrolyzing motor activity that enables it to move along the filament in a cycle of binding, detachment, and rebinding. The tail varies between the different types of myosin-I and determines what cell components will be dragged along by the motor. For example, the tail may bind to a particular type of membrane vesicle and propel it through the cell along actin filament tracks (Figure 16–33A), or it may bind to the plasma membrane and move it relative to cortical actin filaments, thus deforming, or pulling, the membrane into a different shape (Figure 16–33C).

Muscle myosin belongs to the myosin-II subfamily of myosins, all of which have two ATPase heads and a long, rodlike tail (see Figure 16–32B). Members of the myosin-II family are also abundant in nonmuscle cells. Myosin-II forms contractile structures with actin filaments; these are best known and most abundant in muscle, but they also occur in many other types of animal cells. Each myosin-II molecule is a dimer composed of a pair of identical myosin molecules held together by their tails; it has two globular ATPase heads at one end and a single coiled-coil tail at the other (see Figure 16–32B). Clusters of myosin-II molecules bind to each other through their coiled-coil tails, forming a *myosin filament* in which the heads project from the sides (see Figure 16–32C).

The myosin filament has a polarity like that of a double-headed arrow, with the two sets of heads pointing in opposite directions away from the center. One set of heads binds to actin filaments in one orientation and moves them one way; the other set of heads binds to other actin filaments in the opposite orientation and moves them in the opposite direction. The overall effect is to slide sets of oppositely oriented actin filaments past one another (Figure 16–33B). We can see how, therefore, if actin filaments and myosin filaments are organized together in a bundle, the bundle can generate a contractile force. This is seen most clearly in muscle contraction (discussed later), but it also occurs in the *contractile bundles* of actin filaments and myosin-II filaments (see Figure 16–24B) that assemble transiently in nonmuscle cells, and in the *contractile ring* that pinches a dividing cell in two by contracting and pulling inward on the plasma membrane (discussed in Chapter 17).

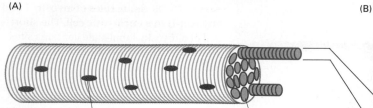

nucleus   myofibril

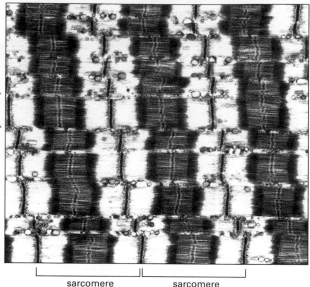

sarcomere   sarcomere

Figure 16–34 A skeletal muscle cell. (A) In an adult human these huge multinucleated cells (also called muscle fibers) are typically 50 μm in diameter, and they can be several centimeters long. They contain numerous myofibrils in which actin filaments and myosin filaments are arranged in a highly organized structure, which gives the myofibril a striped, or striated, appearance. (B) Low-magnification electron micrograph of a longitudinal section through a skeletal muscle cell of a rabbit showing the regular organization of sarcomeres, the contractile units of the myofibrils. (B, courtesy of Roger Craig.)

During Muscle Contraction Actin Filaments Slide Against Myosin Filaments

Muscle contraction is the most familiar and the best understood of animal cell movements. In vertebrates, running, walking, swimming, and flying all depend on the ability of *skeletal muscle* to contract strongly and move various bones. Involuntary movements such as heart pumping and gut peristalsis depend on *cardiac muscle* and *smooth muscle*, respectively, which are formed from muscle cells with a different structure from skeletal muscle, but which use actin and myosin in a similar way to contract.

The long fibers of skeletal muscle are huge single cells formed during development by the fusion of many separate smaller cells. The individual nuclei of the contributing cells are retained in the muscle fiber and lie just beneath the plasma membrane. The bulk of the cytoplasm is made up of **myofibrils,** the contractile elements of the muscle cell. These cylindrical structures are 1–2 μm in diameter and may be as long as the muscle cell itself (Figure 16–34A).

A myofibril consists of a chain of identical tiny contractile units, or **sarcomeres.** Each sarcomere is about 2.5 μm long, and in a chain, their regular pattern gives the vertebrate myofibril a striped, or striated, appearance (Figure 16–34B). Sarcomeres are highly organized assemblies of two types of filaments—actin filaments and filaments of muscle-specific myosin-II. Myosin filaments (the *thick filaments*) are centrally positioned in each sarcomere, whereas the thinner actin filaments (the *thin filaments*) extend inward from each end of the sarcomere (where they are anchored by their plus ends to a structure known as the *Z disc*) and overlap the ends of the myosin filaments (Figure 16–35).

Figure 16–35 Sarcomeres. (A) Detail of the skeletal muscle cell shown in the previous figure showing two myofibrils, with the extent of two sarcomeres marked. (B) Schematic diagram of a single sarcomere showing the origin of the light and dark bands seen in the microscope. Z discs at either end of the sarcomere are attachment points for actin filaments; the centrally located thick filaments are each composed of many myosin-II molecules. (A, courtesy of Roger Craig.)

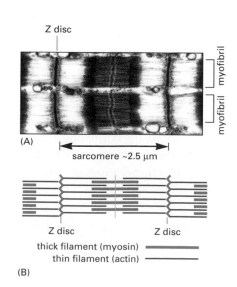

Z disc

myofibril

myofibril

(A)

sarcomere ~2.5 μm

Z disc   Z disc

thick filament (myosin)
thin filament (actin)

(B)

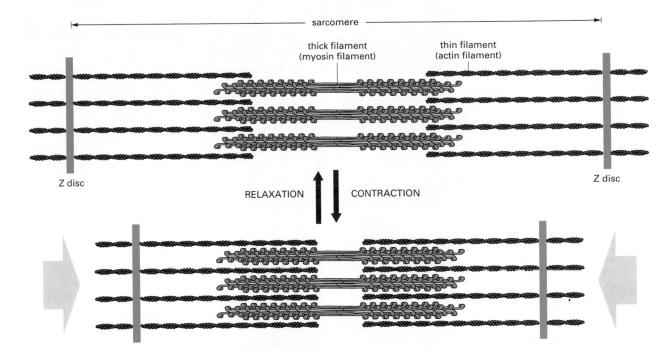

Figure 16–36 **The sliding filament model of muscle contraction.** (A) The myosin and actin filaments of a sarcomere overlap with the same relative polarity on either side of the midline. Recall that actin filaments are anchored by their plus ends to the Z disc and that myosin filaments are bipolar. (B) During contraction, the actin and myosin filaments slide past each other without shortening. The sliding motion is driven by the myosin heads walking toward the plus end of the adjacent actin filament.

The contraction of a muscle cell is caused by a simultaneous shortening of all the sarcomeres, which in turn is caused by the actin filaments sliding past the myosin filaments with no change in the length of either type of filament (Figure 16–36). The sliding motion is generated by myosin heads that project from the sides of the myosin filament and interact with adjacent actin filaments. When a muscle is stimulated to contract, the myosin heads start to walk along the actin filament in repeated cycles of attachment and detachment; their combined action pulls the actin and myosin filaments past each other, causing the sarcomere to contract. After a contraction is completed, the myosin heads lose contact with the actin filaments completely, and the muscle relaxes.

During each cycle of attachment and detachment, a myosin head binds and hydrolyzes one molecule of ATP. This is thought to cause a series of conformational changes in the myosin molecule that move the tip of the head by about 5 nanometers along the actin filament toward the plus end. This movement, repeated with each round of ATP hydrolysis, propels the myosin molecule unidirectionally along the actin filament (Figure 16–37). In so doing, the myosin heads pull against the actin filament, causing it to slide against the myosin filament. Each myosin filament has about 300 myosin heads; each myosin head can cycle about five times per second, sliding the myosin and actin filaments past one another at a speed of up to 15 μm per second. This speed is sufficient to take a sarcomere from a fully extended state (3 μm) to a fully contracted state (2 μm) in less than a tenth of a second. All of the sarcomeres of a muscle are coupled together and are triggered almost instantaneously by the system of signals described in the following section. Therefore, the entire muscle contracts extremely rapidly, usually within a tenth of a second.

Muscle Contraction Is Triggered by a Sudden Rise in Ca²⁺

The force-generating molecular interaction between myosin and actin filaments takes place only when the skeletal muscle receives a signal

Question 16–7 Compare the structure of intermediate filaments to that of the myosin-II filaments in skeletal muscle cells. What are the major similarities? What are the major differences? How do the differences in structure relate to their function?

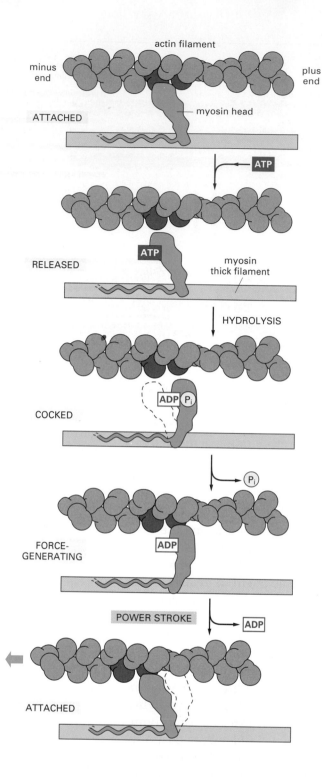

ATTACHED—At the start of the cycle shown in this figure, a myosin head lacking a bound nucleotide is locked tightly onto an actin filament in a *rigor* configuration (so named because it is responsible for *rigor mortis*, the rigidity of death). In an actively contracting muscle this state is very short-lived, being rapidly terminated by the binding of a molecule of ATP.

RELEASED—A molecule of ATP binds to the large cleft on the "back" of the head (that is, on the side farthest from the actin filament) and immediately causes a slight change in the conformation of the domains that make up the actin-binding site. This reduces the affinity of the head for actin and allows it to move along the filament. (The space drawn here between the head and actin emphasizes this change, although in reality the head probably remains very close to the actin.)

COCKED—The cleft closes like a clam shell around the ATP molecule, triggering a large shape change that causes the head to be displaced along the filament by a distance of about 5 nm. Hydrolysis of ATP occurs, but the ADP and P_i produced remain tightly bound to the protein.

FORCE-GENERATING—The weak binding of the myosin head to a new site on the actin filament causes release of the inorganic phosphate produced by ATP hydrolysis, concomitantly with the tight binding of the head to actin. This release triggers the power stroke—the force-generating change in shape during which the head regains its original conformation. In the course of the power stroke, the head loses its bound ADP, thereby returning to the start of a new cycle.

ATTACHED—At the end of the cycle, the myosin head is again locked tightly to the actin filament in a rigor configuration. Note that the head has moved to a new position on the actin filament.

Figure 16–37 The cycle of changes by which a myosin molecule walks along an actin filament.

from the nervous system. The signal from a nerve terminal triggers an action potential (discussed in Chapter 12) in the muscle cell plasma membrane. This electrical excitation spreads in a matter of milliseconds into a series of membranous tubes, called *transverse tubules*, that extend inward from the plasma membrane around each myofibril. The electrical signal is then relayed to the *sarcoplasmic reticulum*, an adjacent sheath of interconnected flattened vesicles that surrounds each myofibril like a net stocking (Figure 16–38A).

The sarcoplasmic reticulum is a specialized region of the endoplasmic reticulum in muscle cells. It contains a very high concentration of

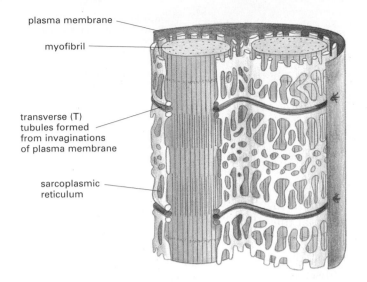

plasma membrane

myofibril

transverse (T) tubules formed from invaginations of plasma membrane

sarcoplasmic reticulum

(A)

Figure 16–38 T tubules and the sarcoplasmic reticulum. (A) Drawing of the two membrane systems that relay the signal to contract from the muscle cell plasma membrane to all of the myofibrils in the cell. (B) Schematic diagram showing how a Ca^{2+} release channel in the sarcoplasmic reticulum membrane is thought to be opened by a voltage-sensitive transmembrane protein in the adjacent T tubule.

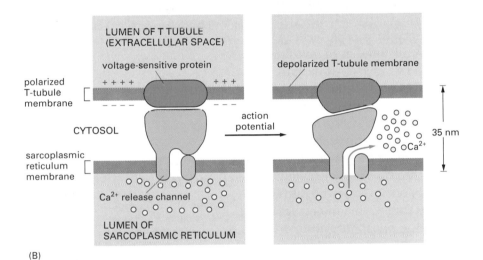

LUMEN OF T TUBULE (EXTRACELLULAR SPACE)

voltage-sensitive protein

depolarized T-tubule membrane

polarized T-tubule membrane

+ + + +   + + +

CYTOSOL

action potential

35 nm

sarcoplasmic reticulum membrane

Ca^{2+} release channel

Ca^{2+}

LUMEN OF SARCOPLASMIC RETICULUM

(B)

Ca^{2+}, and in response to the incoming electrical excitation, much of this Ca^{2+} is released into the cytosol through ion channels that open in the sarcoplasmic reticulum membrane in response to the change in voltage across the plasma membrane (Figure 16–38B). As discussed in Chapter 15, Ca^{2+} is widely used as an intracellular signal to relay a message from the exterior to the internal machinery of the cell. In the case of muscle, the Ca^{2+} interacts with a molecular switch made of specialized accessory proteins closely associated with the actin filaments (Figure 16–39A). One of these proteins is *tropomyosin*, a rigid rod-shaped molecule that binds in the groove of the actin helix, overlapping seven actin monomers, and prevents the myosin heads from associating with the actin filament. The other is *troponin*, a protein complex that includes a Ca^{2+}-sensitive protein (*troponin-C*), which is associated with the end of a tropomyosin molecule. When the level of Ca^{2+} rises in the cytosol, Ca^{2+} binds to troponin and induces a change in its shape. This in turn causes the tropomyosin molecules to shift their position slightly, allowing myosin heads to bind to the actin filament and initiating contraction (Figure 16–39B).

Actin Filaments

541

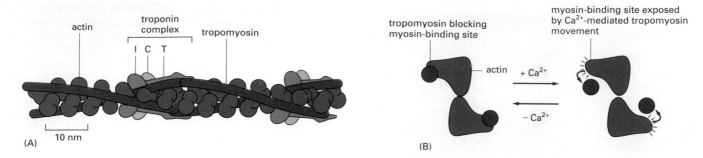

(A)   actin   troponin complex   tropomyosin   I C T

10 nm

(B)   tropomyosin blocking myosin-binding site   actin   + Ca²⁺   − Ca²⁺   myosin-binding site exposed by Ca²⁺-mediated tropomyosin movement

Because the signal from the plasma membrane is passed within milliseconds (via the transverse tubules and sarcoplasmic reticulum) to every sarcomere in the cell, all the myofibrils in the cell contract at the same time. The increase in Ca^{2+} in the cytosol ceases as soon as the nerve signal stops because the Ca^{2+} is rapidly pumped back into the sarcoplasmic reticulum by abundant Ca^{2+} pumps in its membrane (discussed in Chapter 12). As soon as Ca^{2+} concentrations have returned to their resting level, troponin and tropomyosin molecules move back to their original positions, where they block myosin binding and thus end contraction.

This highly specialized contractile machinery in muscle cells is thought to have evolved from the simpler contractile bundles of myosin and actin filaments found in all eucaryotic cells. The myosin-II in non-muscle cells is also activated by a rise in cytosolic Ca^{2+}, but the mechanism of activation is quite different. An increase in Ca^{2+} leads to the phosphorylation of myosin-II, which alters the myosin conformation and enables it to interact with actin. A similar activation mechanism operates in *smooth muscle*, which lies in the walls of the stomach, intestine, uterus, and arteries, and in many other structures in which slow and sustained contractions are needed. Contractions produced by this second mode are slower because time is needed for enzyme molecules to diffuse to the myosin heads and to carry out the phosphorylation or dephosphorylation. However, this mechanism has the advantage that it is less specialized and can be driven by a variety of incoming signals: thus smooth muscle, for example, is triggered to contract by adrenaline, serotonin, prostaglandins, and a variety of other extracellular signals.

The sarcomeres of skeletal muscle cells represent an extreme specialization of the basic components of the eucaryotic cell cytoskeleton. In Chapter 17 we examine one function of the cytoskeleton of such fundamental importance in all cells that it warrants a chapter to itself: the partitioning of the components of the cell into two duplicate sets in the process of cell division.

Figure 16–39 The control of skeletal muscle contraction by troponin. (A) A muscle thin filament showing the positions of tropomyosin and troponin along the actin filament. Each tropomyosin molecule has seven evenly spaced regions of homologous sequence, each of which is thought to bind to an actin monomer as shown. (B) A thin filament shown end-on, illustrating the slight movement of tropomyosin, itself caused by Ca^{2+} binding to troponin, that allows myosin heads to interact with actin.

Question 16–8

A. Recall from Figure 16–39 that troponin molecules are evenly spaced along an actin filament with one troponin found every seventh actin molecule. How do you suppose troponin molecules can be positioned this regularly? What does this tell you about the binding of troponin to actin filaments?

B. What do you suppose would happen if you mixed actin filaments with either (a) troponin alone, (b) tropomyosin alone, or (c) troponin plus tropomyosin, and then added myosin? Would the effects be dependent on Ca^{2+}?

Essential Concepts

- The cytoplasm of a eucaryotic cell is supported and spatially organized by a cytoskeleton of intermediate filaments, microtubules, and actin filaments.

- Intermediate filaments are stable, ropelike polymers of fibrous proteins that give cells mechanical strength. Some types underlie the nuclear membrane to form the nuclear lamina; others are distributed throughout the cytoplasm.

- Microtubules are stiff, hollow tubes formed by polymerization of tubulin dimer subunits. They are polarized structures with a slower-growing "minus" end and a faster-growing "plus" end.

- Microtubules are nucleated in, and grow out from, organizing centers such as the centrosome. The minus ends of the microtubules are embedded in the organizing center.

- Many of the microtubules in a cell are in a labile, dynamic state in which they alternate between a growing state and a shrinking state. These transitions, known as dynamic instability, are controlled by the hydrolysis of GTP bound to tubulin dimers.

- Each tubulin dimer has a tightly bound GTP molecule that is hydrolyzed to GDP after the tubulin assembles into a microtubule. GTP hydrolysis reduces the affinity of the subunit for its neighbors and decreases the stability of the polymer, causing it to disassemble.

- Microtubules can be stabilized by proteins that capture the plus end—a process that influences the position of microtubule arrays in a cell. Cells contain many microtubule-associated proteins that stabilize microtubules, bind them to other cell components, and harness them for specific functions.

- Kinesins and dyneins are motor proteins that use the energy of ATP hydrolysis to move unidirectionally along microtubules. They carry specific membrane vesicles and other cargoes and in this way maintain the spatial organization of the cytoplasm.

- Eucaryotic cilia and flagella contain a bundle of stable microtubules. Their beating is caused by bending of the microtubules, driven by a motor protein called ciliary dynein.

- Actin filaments are helical polymers of actin molecules. They are more flexible than microtubules and are often found in bundles or networks associated with the plasma membrane.

- Actin filaments are polarized structures with a fast- and a slow-growing end, and their assembly and disassembly are controlled by the hydrolysis of ATP tightly bound to each actin monomer.

- The varied forms and functions of actin filaments in cells depend on multiple actin-binding proteins. These control the polymerization of actin filaments, cross-link the filaments into loose networks or stiff bundles, attach them to membranes, or move them relative to one another.

- Myosins are motor proteins that use the energy of ATP hydrolysis to move along actin filaments; they can carry organelles along actin-filament tracks or cause adjacent actin filaments to slide past each other in contractile bundles.

- A network of actin filaments underneath the plasma membrane forms the cell cortex, which is responsible for the shape and movement of the cell surface, including the movements involved when a cell crawls along a surface.

- Muscle contraction depends on the sliding of actin filaments along myosin-II filaments, driven by the repetitive motion of the myosin heads.

- Contraction is initiated by a sudden rise in cytosolic Ca^{2+}, which delivers a signal via Ca^{2+}-binding proteins to the contractile apparatus.

Key Terms

actin filament	cilium	intermediate filament	myosin
cell cortex	cytoskeleton	kinesin	nuclear lamina
centriole	dynamic instability	lamellipodium	sarcomere
centrosome	dynein	microtubule	tubulin
	filopodium	motor protein	
	flagellum	myofibril	

Questions

Question 16–9 Which of the following statements are correct? Explain your answers.

A. Kinesin moves endoplasmic reticulum membranes along microtubules so that the network of ER tubules becomes stretched throughout the cell.

B. Without actin, cells can form a functional mitotic spindle and pull their chromosomes apart but cannot divide.

C. Lamellipodia and filopodia are "feelers" that a cell extends to find anchor points on the substratum that it will then crawl over.

D. GTP is hydrolyzed by tubulin to cause the bending of flagella.

E. Cells having an intermediate filament network that cannot be depolymerized would die.

F. The plus ends of microtubules grow faster because they have a larger GTP cap.

G. The transverse tubules in muscle cells are an extension of the plasma membrane, with which they are continuous, and likewise, the sarcoplasmic reticulum is an extension of the endoplasmic reticulum.

H. Activation of myosin movement on actin filaments is triggered by phosphorylation of troponin in some situations and by Ca^{2+} binding to troponin in others.

Question 16–10 The average time taken for a molecule or an organelle to diffuse a distance x cm is given by the formula:

$$t = x^2/2D$$

where t is the time in seconds and D is a constant for the molecule or particle called the diffusion coefficient. Using the above formula, calculate the time it would take for a small molecule, a protein, and a membrane vesicle to diffuse from one side to another of a cell 10 μm across. Typical diffusion coefficients in units of cm^2/sec are: small molecule 5×10^{-6}; protein molecule 5×10^{-7}; vesicle 5×10^{-8}. How long will a membrane vesicle take to reach the end of an axon 10 cm long by free diffusion?

Question 16–11 List differences between bacteria and animal cells that could have depended on the appearance during evolution of some or all of the components of the present eucaryotic cytoskeleton. Explain why a cytoskeleton might have been crucial for these differences to evolve.

Question 16–12 Examine the structure of an intermediate filament shown in Figure 16–4. Does the filament have a unique polarity—that is, could you distinguish one end from the other by chemical or other means? Explain your answer.

Question 16–13 There are no known motor proteins that move on intermediate filaments. Suggest an explanation for this.

Question 16–14 When cells enter mitosis, their existing array of cytoplasmic microtubules has to be rapidly broken down and replaced with the mitotic spindle that forms to pull the chromosomes into the daughter cells. The enzyme katanin, named after Japanese samurai swords and activated during the onset of mitosis, chops microtubules into short pieces. What do you suppose is the fate of the microtubule fragments created by katanin? Explain your answer.

Question 16–15 The drug taxol, extracted from the bark of yew trees, has the opposite effect of the drug colchicine, an alkaloid from autumn crocus. Taxol binds tightly to microtubules and stabilizes them; when added to cells, it causes much of the free tubulin to assemble into microtubules. In contrast, colchicine prevents microtubule formation. Taxol is just as pernicious to dividing cells as colchicine, and both are used as anticancer drugs. Based on your knowledge of microtubule dynamics, suggest why both drugs are toxic to dividing cells despite their opposite actions.

Question 16–16 A useful technique for studying microtubule motors is to attach them by their tails to a glass coverslip (which can be accomplished quite easily because the tails stick avidly to a clean glass surface) and then to allow microtubules to settle onto them. The microtubules may then be viewed in a light microscope as they are propelled over the surface of the coverslip by the heads of the motor proteins. Since the motor proteins attach at random orientations to the coverslip, however, how can they generate coordinated movement of individual microtubules rather than engaging in a tug-of-war? In which direction will microtubules crawl on a "bed" of kinesin molecules (i.e., plus end first, or minus end first)?

Question 16–17 A typical time course of polymerization of purified tubulin to form microtubules is shown in Figure Q16–17.

A. Explain the different parts of the curve (labeled A, B, and C). Draw a diagram that shows the behavior of tubulin molecules in each of the three phases.

B. How would the curve in the figure change if centrosomes were added at the outset?

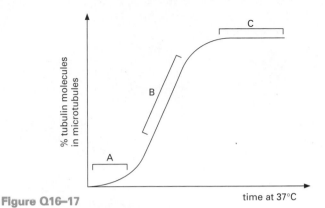

Figure Q16–17

Question 16–18   The electron micrograph shown in Figure Q16–18A was obtained from a population of microtubules that was growing rapidly. Figure Q16–18B was obtained from microtubules undergoing "catastrophic" shrinking. Comment on any differences between the two images and suggest likely explanations for the differences that you observe.

Figure Q16–18

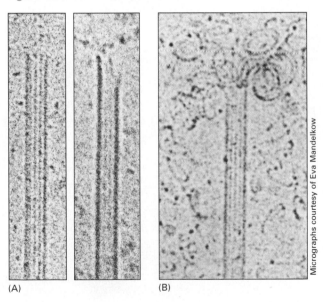

(A)   (B)

Micrographs courtesy of Eva Mandelkow

Question 16–19   The locomotion of fibroblasts in culture is immediately halted by the drug cytochalasin, whereas colchicine causes fibroblasts to cease to move directionally and to begin extending lamellipodia in seemingly random directions. Injection of fibroblasts with antibodies to vimentin has no discernible effect on their migration. What do these observations suggest to you about the involvement of the three different cytoskeletal filaments in fibroblast locomotion?

Question 16–20   Complete the following sentence accurately, explaining your reason for accepting or rejecting each of the four phrases. The role of calcium in muscle contraction is . . .

A. to detach myosin heads from actin.

B. to spread the action potential from the plasma membrane to the contractile machinery.

C. to bind to troponin, cause it to move tropomyosin, and thereby expose actin filaments to myosin heads.

D. to maintain the structure of the myosin filament.

Question 16–21   Which of the following changes takes place when a skeletal muscle contracts?

A. Z discs move farther apart.

B. Actin filaments contract.

C. Myosin filaments contract.

D. Sarcomeres become shorter.

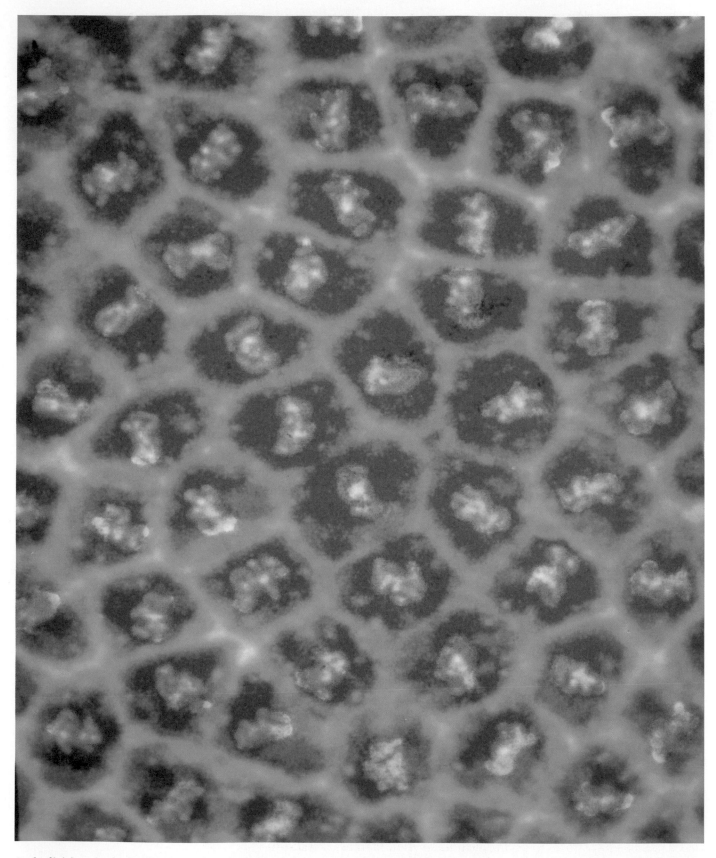

Early divisions in the fertilized egg of the fly *Drosophila* are synchronous and very rapid. The nuclei keep dividing in a common cytoplasm, without corresponding cell divisions. When about 6000 daughter nuclei have formed, they sit near the surface of the egg, and the plasma membrane then grows inward, enclosing each nucleus and forming a corresponding number of new cells. In this view of the egg surface the chromsosmes are stained *orange* and the actin cytoskeleton, associated with the membranes of the forming cells, is stained *green*. (Courtesy of William Sullivan.)

17 Cell Division

"Where a cell arises, there must be a previous cell, just as animals can only arise from animals and plants from plants." This *cell doctrine,* proposed by the German pathologist Rudolf Virchow in 1858, carried with it a profound message for the continuity of life. Cells are generated from cells, and the only way to make more cells is by division of those that already exist. All living organisms, from the unicellular bacterium through the multicellular mammal, are products of repeated rounds of cell growth and division extending back in time to the beginnings of life over three billion years ago.

A cell reproduces by carrying out an orderly sequence of events in which it duplicates its contents and then divides in two. This cycle of duplication and division, known as the **cell cycle,** is the essential mechanism by which all living things reproduce. In unicellular organisms, such as bacteria or yeasts, each cell division produces a complete new organism, while in multicellular organisms, many rounds of cell division are required to make a new individual from the single-celled egg. In many multicellular organisms, cell division occurs throughout life; even in the adult it is needed to replace cells that die. In adult humans, for example, whereas some cells, such as nerve cells and muscle cells, do not divide at all, many others do, although at very different rates: liver cells, for instance, divide once every year or so, whereas some of the epithelial cells lining the gut and many of the blood cell precursors in the bone marrow divide more than once every day. In fact, each of us must manufacture many millions of cells every second simply to survive: if all cell division is stopped—by exposure to a very large dose of x-rays, for instance—the individual will die within a few days.

The details of the cell cycle vary from organism to organism and at different times in an organism's life. Certain characteristics, however, are universal, as the cycle must comprise, at a minimum, a set of processes that a cell has to perform to accomplish its most fundamental task—to copy and pass on its genetic information to the next generation of cells. To produce two genetically identical daughter cells, the DNA in each chromosome must be faithfully replicated, and the replicated chromosomes must then be accurately separated into the two daughter cells so

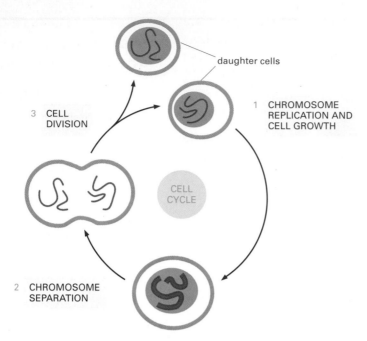

that each cell receives a copy of the entire genome (Figure 17–1). In most circumstances cells also double their mass in each cell cycle; otherwise they would get smaller each time they divide, as happens in some early embryos, as we discuss later.

Cell division is at its simplest and most rapid in bacteria, which do not have a nucleus and contain a single chromosome. In *Escherichia coli,* for example, the whole cell cycle can take as little as 20 minutes in favorable growth conditions. Its single circular chromosome, containing a single DNA molecule, is attached to the plasma membrane and remains attached while it replicates. The two new chromosomes then become separated as the cell grows. When the cell has approximately doubled in size, cell division occurs by simple fission—a process called *binary fission*—as new cell wall and plasma membrane is laid down between the two chromosomes to produce two separate daughter cells (Figure 17–2).

Cell division in a eucaryotic cell, even in single-celled eucaryotes such as yeasts and amoebae, is much more complex, partly because the main genetic information of the cell—its nuclear genome—is distributed between multiple chromosomes. (Recall that a eucaryotic cell also contains a small amount of DNA in its mitochondria and, if it is a plant cell, also in its chloroplasts.) The eucaryotic cell also contains an elaborate array of cytoplasmic organelles and cytoskeletal filaments, all of which must be duplicated and shared out equally between the two daughter cells. Thus complex cytoplasmic and nuclear processes have to be coordinated with one another during the eucaryotic cell cycle.

To explain how cells reproduce, we therefore have to consider three

Figure 17–2 **Cell division in bacteria.** The bacterial chromosome, containing a single circular DNA molecule, is attached to the plasma membrane and remains attached during and after replication. The two new DNA molecules become separated by cell growth, and the cell wall and plasma membrane grow inward between them, dividing the cell in two by simple fission. There are no organelles in a bacterial cell, so this method of division is all that is required to partition the genetic material and cytoplasmic contents equally between the daughter cells.

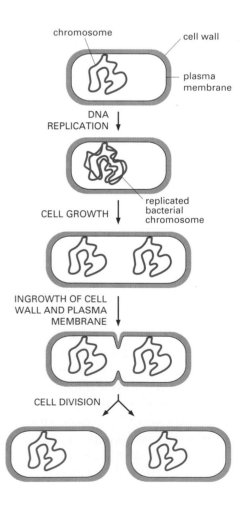

major questions: (1) How do cells duplicate their contents? (2) How do they partition the duplicated contents and split in two? (3) How do they coordinate all the machinery that is required for these two processes so as to ensure, for example, that division does not occur until duplication is completed? The first problem is discussed elsewhere. In Chapter 6 we discuss how the DNA is replicated, and in Chapters 7, 11, 14, and 16 we describe how the eucaryotic cell manufactures other components, such as proteins, membranes, organelles, and cytoskeletal filaments. In this chapter we tackle the second problem—the physical process of cell division. This is largely the job of the cytoskeleton, which pushes and pulls the duplicated components into two separate sets and divides the cell in two. In Chapter 18 we discuss the third and most difficult problem—how the eucaryotic cell coordinates the various steps of its reproductive cycle.

Most of this chapter is concerned with the events that occur during nuclear division (*mitosis*) and cytoplasmic division (*cytokinesis*) in a typical eucaryotic cell. In the final part of the chapter we briefly discuss the corresponding events that occur in *meiosis,* the specialized type of cell division by which haploid gametes—the sperm and eggs in animals—are produced from diploid precursor cells, as discussed in Chapter 9. To set the scene for our discussion of cell division, we begin by briefly outlining the events of a typical animal cell cycle.

Overview of the Cell Cycle

The duration of the cell cycle varies greatly from one cell type to another. A single-celled yeast can divide every 90–120 minutes in ideal conditions, while a mammalian liver cell divides on average less than once a year (Table 17–1). We focus here on the sequence of events in a fairly rapidly dividing mammalian cell, with a cell cycle time of about 24 hours.

The Eucaryotic Cell Cycle Is Divided into Four Phases

The eucaryotic cell cycle is traditionally divided into four stages, or *phases.* Seen in a microscope, the two most dramatic events are when the nucleus divides, a process called *mitosis,* and when the cell splits in two, a process called *cytokinesis.* These two processes together constitute the **M phase** of the cell cycle (Figure 17–3). In a typical mammalian cell the whole of M phase takes about an hour, which is only a small fraction of the total cell-cycle time.

The period between one M phase and the next is called **interphase.** Under the microscope it appears, deceptively, as an uneventful interlude

Table 17–1	**Some Eucaryotic Cell-Cycle Times**
Cell Type	**Cell-Cycle Times**
Early frog embryo cells	30 minutes
Yeast cells	1.5–3 hours
Intestinal epithelial cells	~12 hours
Mammalian fibroblasts in culture	~20 hours
Human liver cells	~1 year

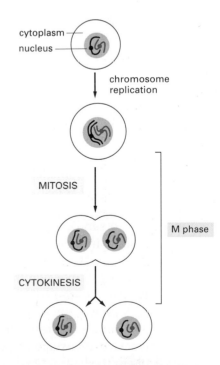

Figure 17–3 **The M phase of the cell cycle.** The M phase consists of nuclear division (mitosis) followed by cytoplasmic division (cytokinesis).

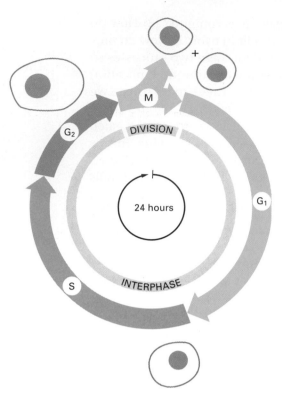

Figure 17–4 **The four successive phases of a standard eucaryotic cell cycle.** During interphase the cell grows continuously. During M phase growth stops, and the nucleus divides first, and then the cell divides. DNA replication is confined to the part of interphase known as S phase. G_1 phase is the gap between M phase and S phase; G_2 is the gap between S phase and M phase.

during which the cell simply grows in size. Interphase, however, is a very busy time for the cell, and it is divided into the remaining three phases of the cell cycle. During **S phase** (S = synthesis), the cell replicates its nuclear DNA, an essential prerequisite for cell division. S phase is flanked by two phases where the cell continues to grow. The **G_1 phase** (G = gap) is the interval between the completion of M phase and the beginning of S phase (DNA synthesis). The **G_2 phase** is the interval between the end of S phase and the beginning of M phase (Figure 17–4). As we discuss in Chapter 18, there are particular times in the G_1 and G_2 phases when the cell makes a decision whether to proceed to the next phase or pause to allow more time to prepare.

During all of interphase, a cell continues to transcribe genes, synthesize proteins, and grow in mass. Together, G_1 and G_2 phases provide additional time for the cell to grow and duplicate its cytoplasmic organelles: if interphase lasted only long enough for DNA replication, the cell would not have time to double its mass before it divided and would consequently get smaller with each division. Indeed, in some special circumstances that is just what happens. In some animal embryos, for example, the first few cell divisions after fertilization (called cleavage divisions) serve to subdivide a giant egg cell into many smaller cells as quickly as possible; in these cell cycles the cells do not grow before they divide, and the G_1 and G_2 phases are drastically shortened.

The first readily visible sign that a cell is about to enter M phase is the progressive *condensation* of its chromosomes, which were replicated earlier, during S phase (the two copies of each chromosome, however, remain tightly bound together). Chromosome condensation marks the end of G_2 phase. At this stage in the cell cycle the chromosomes first become visible in the light microscope as long threads, which gradually get shorter and thicker by the process of compaction, described in Chapter 8. This condensation makes the chromosomes less likely to get entangled and therefore physically easier to separate during mitosis.

Question 17–1 Cells from a growing population were stained with a dye that becomes fluorescent when it binds to DNA, so that the amount of fluorescence is directly proportional to the amount of DNA in each cell. To measure the amount of DNA in each cell, the cells were then passed through a fluorescence-activated cell sorter (FACS), an instrument that registers the level of fluorescence in individual cells. The number of cells with a given DNA content were plotted on a graph, as shown in Figure Q17–1. Indicate on the graph where you would expect to find cells that are in the following stages: G_1, S, G_2, and mitosis. Which is the longest phase of the cell cycle in this population of cells?

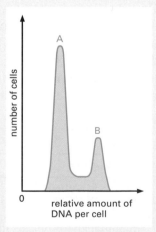

Figure Q17–1

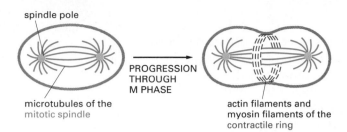

spindle pole

microtubules of the
mitotic spindle

PROGRESSION
THROUGH
M PHASE

actin filaments and
myosin filaments of the
contractile ring

Figure 17–5 Two transient cytoskeletal structures that mediate M phase in animal cells. The mitotic spindle assembles first to separate the chromosomes. Then the contractile ring assembles to divide the cell in two.

The Cytoskeleton Carries Out Both Mitosis and Cytokinesis

After the replicated chromosomes have condensed, two separate cytoskeletal structures are assembled in sequence to carry out the two mechanical processes that occur in M phase—nuclear division (mitosis) and cytoplasmic division (cytokinesis). Both structures are rapidly disassembled after they have performed their tasks.

To produce two genetically identical daughter cells, the eucaryotic cell has to perform the delicate job of separating the replicated chromosomes and allocating one copy of each chromosome to each daughter cell. This task is performed during mitosis and is carried out by the *mitotic spindle,* which is composed mainly of microtubules (discussed in Chapter 16) and starts to assemble late in G_2 phase. A microtubule-based mitotic spindle mediates mitosis in all eucaryotic cells, whether plant or animal or single-celled yeasts.

The second cytoskeletal structure that assembles in M phase is responsible for cytokinesis, but it is formed only in animal cells. It is called the *contractile ring* because it consists mainly of a contractile ring of actin filaments and myosin filaments (discussed in Chapter 16). It starts to assemble toward the end of mitosis, forming around the equator of the cell, just beneath the plasma membrane. As it contracts, it pulls the membrane inward, thereby dividing the cell in two (Figure 17–5). We discuss later how plant cells, which have a cell wall to contend with, divide their cytoplasm by a very different mechanism.

Some Organelles Fragment at Mitosis

The process of mitosis ensures that each daughter cell receives a full complement of chromosomes. But when a eucaryotic cell divides, each daughter cell must also inherit all of the other essential cell components, including the membrane-bounded organelles. Organelles like mitochondria and chloroplasts cannot assemble spontaneously from their individual components; they can arise only from the growth and division of the preexisting organelles. Likewise, cells cannot make a new endoplasmic reticulum or Golgi apparatus unless some part of it is already present, which can then be enlarged. How then are these various membrane-bounded organelles segregated when the cell divides? Organelles such as mitochondria and chloroplasts are usually present in large numbers and will be safely inherited if, on average, their numbers simply double once each cell cycle. Other organelles, such as the Golgi apparatus and the endoplasmic reticulum, break up into small fragments during mitosis, which increases the chance that they will be more or less evenly distributed between the daughter cells when a cell divides.

Question 17–2 The Golgi apparatus is partitioned into the daughter cells at cell division by a random distribution of fragments that are created at mitosis. Explain why random partitioning of chromosomes would not work.

Mitosis

Before nuclear division, or **mitosis,** begins, each chromosome has been replicated (as discussed in Chapter 8) and consists of two identical **chromatids** (called *sister chromatids*), which are joined together along their length by interactions between proteins on the surface of the two chromatids. During mitosis these proteins are cleaved, and the sister chromatids split apart to become independent daughter chromosomes, which are pulled to opposite poles of the cell by the mitotic spindle (Figure 17–6).

Mitosis proceeds as a continuous sequence of events, but it is traditionally divided into five stages: (1) during *prophase* the replicated chromosomes condense and the mitotic spindle begins to assemble outside the nucleus; (2) during *prometaphase* the nuclear envelope breaks down, allowing the spindle microtubules to contact the chromosomes and bind to them; (3) during *metaphase* the mitotic spindle gathers all of the chromosomes to the center (equator) of the spindle; (4) during *anaphase* the two sister chromatids in each replicated chromosome synchronously split apart, and the spindle draws them to opposite poles of the cell (see Figure 17–6); (5) during *telophase* a nuclear envelope reassembles around each of the two sets of separated chromosomes to form two nuclei. These five stages of mitosis, together with cytokinesis, are illustrated in Panel 17–1 (pp. 554–555). In this section on nuclear division we examine how the mitotic spindle assembles and how it works.

The Mitotic Spindle Starts to Assemble in Prophase

Most animal cells in interphase contain a cytoplasmic array of microtubules radiating out from the single centrosome, as discussed in Chapter 16. Toward the end of S phase the cell duplicates its centrosome to produce two daughter centrosomes, which initially remain together at one side of the nucleus. As **prophase** begins, the two daughter centrosomes separate and move to opposite poles of the cell, driven by centrosome-associated motor proteins that use the energy of ATP hydrolysis to move along microtubules. (The general mechanism of action of motor proteins is discussed in Chapter 16.) Each centrosome serves to organize its own array of microtubules, and the two sets of microtubules then interact to form the **mitotic spindle** (see Panel 17–1).

As discussed in Chapter 16, the microtubules that radiate from the centrosome in an interphase cell continuously polymerize and depolymerize by the addition and loss of the tubulin subunits that make up a microtubule: individual microtubules alternate between growing and shrinking—a process called *dynamic instability* (see Figure 16–11). The

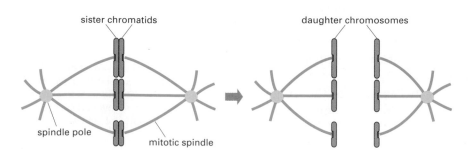

Figure 17–6 The separation of sister chromatids. Each pair of sister chromatids separates to become two daughter chromosomes, which are then pulled to opposite poles of the cell by the mitotic spindle.

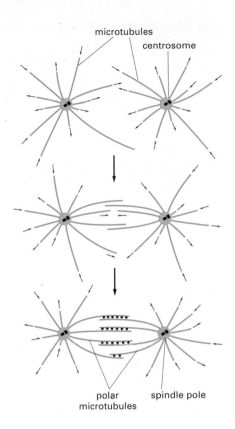

Figure 17–7 The formation of a bipolar mitotic spindle by the selective stabilization of interacting microtubules. New microtubules grow out in random directions from the two centrosomes. In animal cells a pair of centrioles is located at the center of each centrosome (see Figure 16–10). The two ends of a microtubule, called the plus and the minus ends, have different properties, and it is the minus end that is anchored in the centrosome (discussed in Chapter 16). The free plus ends are "dynamically unstable" and switch suddenly from uniform growth (*outward-pointing red arrows*) to rapid shrinkage (*inward-pointing red arrows*). When two microtubules from opposite centrosomes interact in an overlap zone, microtubule-associated proteins cross-link the microtubules together (*black dots*) in a way that stabilizes their plus ends by decreasing the probability of depolymerization.

microtubules that radiate from each of the separated daughter centrosomes at the start of mitosis display the same behavior, but they are even more dynamic, switching from polymerization to depolymerization at a rate that is 20 times faster than for interphase microtubules. Moreover, many more microtubules radiate from each centrosome, and they are, on average, much shorter. These differences are thought to reflect changes in the phosphorylation of particular centrosomal proteins that occur at the start of prophase.

The rapidly growing and shrinking microtubules extend in all directions from the two centrosomes, exploring the interior of the cell. During prophase, while the nuclear envelope is still intact, some of these microtubules become stabilized against disassembly to form the highly organized mitotic spindle: some of the microtubules growing from opposite centrosomes interact, binding the two sets of microtubules together to form the basic framework of the mitotic spindle, with its characteristic bipolar shape. These interacting microtubules are called *polar microtubules* (since they originate from the two poles of the spindle), and the two centrosomes that give rise to them are called **spindle poles** (Figure 17–7).

The next stage in mitosis involves the positioning of the replicated chromosomes on the spindle in such a way that, when the sister chromatids separate, they are drawn to opposite poles.

Chromosomes Attach to the Mitotic Spindle at Prometaphase

Prometaphase starts abruptly with the disassembly of the nuclear envelope, which breaks up into small membrane vesicles. As we discuss later, this process is triggered by the phosphorylation and consequent disassembly of the intermediate filament proteins of the nuclear lamina, the network of fibrous proteins that underlies and stabilizes the nuclear envelope (see Figure 14–8). The spindle microtubules, which have been lying in wait outside the nucleus, now gain access to the replicated chromosomes and bind to them (see Panel 17–1, pp. 554–555).

The spindle microtubules bind the chromosomes through specialized protein complexes called **kinetochores,** which are formed on the chromosomes during late prophase. As discussed in Chapter 8, each replicated chromosome consists of two sister chromatids joined along their length, and each chromatid is constricted at a region of specialized DNA sequence called the *centromere* (Figure 17–8; see also Figure 8–5).

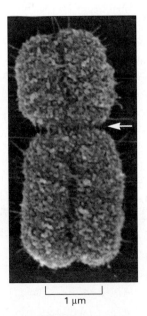

Figure 17–8 A replicated mitotic chromosome. Scanning electron micrograph of a human mitotic chromosome, consisting of two sister chromatids joined along their length. The constricted region (*arrow*) is the centromere region, where kinetochores assemble. (Courtesy of Terry D. Allen.)

Mitosis

CELL DIVISION AND THE CELL CYCLE

INTERPHASE

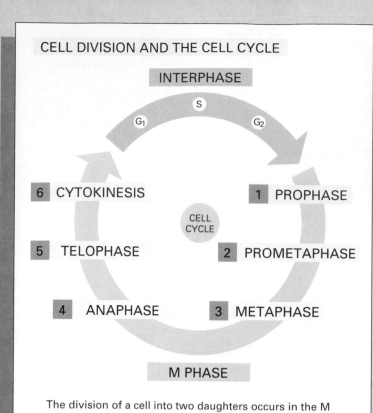

The division of a cell into two daughters occurs in the M phase of the cell cycle. M phase consists of nuclear division, or mitosis, and cytoplasmic division, or cytokinesis. In this figure M phase has been expanded for clarity. Mitosis is itself divided into five steps, and these, together with cytokinesis, are described in this panel.

INTERPHASE

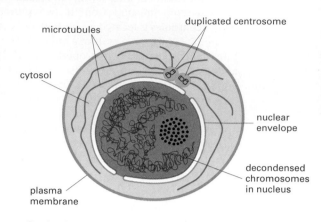

During interphase the cell increases in size. The DNA of the chromosomes is replicated, and the centrosome is duplicated.

The light micrographs shown in this panel are of a living cell from the lung epithelium of a newt. The same cell has been photographed at different times during its division into two daughter cells. (Courtesy of Conly L. Rieder.)

1 PROPHASE

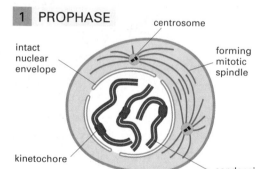

At prophase the replicated chromosomes, each consisting of two closely associated sister chromatids, condense. Outside the nucleus, the mitotic spindle assembles between the two centrosomes, which have replicated and moved apart.

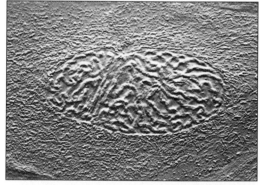

time = 0 min

2 PROMETAPHASE

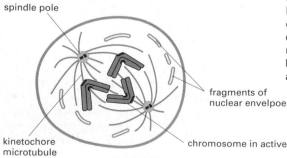

Prometaphase starts abruptly with the breakdown of the nuclear envelope. Chromosomes can now attach to spindle microtubules via their kinetochores, and undergo active movement.

time = 79 min

Panel 17–1   **The principal stages of M phase (mitosis and cytokinesis) in an animal cell.**

3 METAPHASE

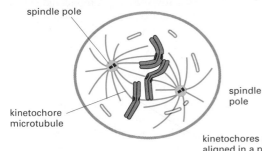

spindle pole

kinetochore
microtubule

spindle
pole

kinetochores of all chromosomes
aligned in a plane midway between
two spindle poles

At metaphase the chromosomes are aligned at the equator of the spindle, midway between the spindle poles. The paired kinetochore microtubules on each chromosome attach to opposite poles of the spindle.

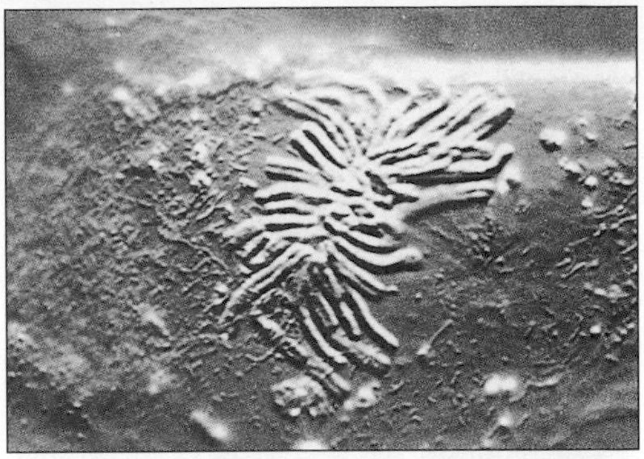

time = 250 min

4 ANAPHASE

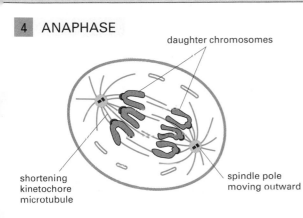

daughter chromosomes

shortening
kinetochore
microtubule

spindle pole
moving outward

At anaphase the paired chromatids synchronously separate to form two daughter chromosomes, and each is pulled slowly toward the spindle pole it faces. The kinetochore microtubules get shorter and the spindle poles also move apart, both contributing to chromosome separation.

time = 279 min

5 TELOPHASE

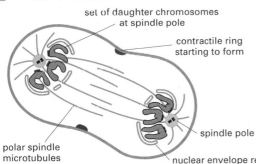

set of daughter chromosomes
at spindle pole

contractile ring
starting to form

polar spindle
microtubules

spindle pole

nuclear envelope reassembling
around individual chromosomes

During telophase the two sets of daughter chromosomes arrive at the poles of the spindle. A new nuclear envelope reassembles around each set, completing the formation of two nuclei and marking the end of mitosis. The division of the cytoplasm begins with the assembly of the contractile ring.

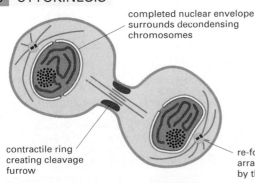

time = 315 min

6 CYTOKINESIS

completed nuclear envelope
surrounds decondensing
chromosomes

contractile ring
creating cleavage
furrow

re-formation of interphase
array of microtubules nucleated
by the centrosome

During cytokinesis of an animal cell the cytoplasm is divided in two by a contractile ring of actin and myosin, which pinches in the cell to create two daughters, each with one nucleus.

time = 362 min

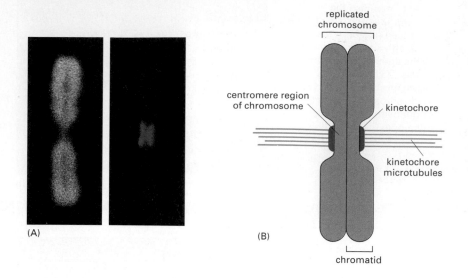

(A)

(B)

replicated
chromosome

centromere region
of chromosome

kinetochore

kinetochore
microtubules

chromatid

Figure 17–9 Kinetochores and kineto-chore microtubules. (A) A replicated chromosome at mitosis. The DNA is stained with a fluorescent dye (*left photo*), and the kinetochores are stained *red* with fluorescent antibodies that recognize the kinetochore proteins (*right photo*). These antibodies come from patients suffering from scleroderma (a disease that causes fibrosis of connective tissue in skin and other organs), who, for unknown reasons, produce antibodies against their own kinetochore proteins. (B) Schematic drawing of a mitotic chromosome showing its two sister chromatids attached to kinetochore microtubules via the kinetochores. Each kinetochore forms a plaque on the surface of the centromere. The plus ends of the microtubules bind to the kinetochores. (A, courtesy of B.R. Brinkley.)

Just before prometaphase, kinetochore proteins assemble into a large complex on each centromere. Each duplicated chromosome therefore has two kinetochores (one on each sister chromatid), which face in opposite directions. Kinetochore assembly depends entirely on the presence of the centromere DNA sequence: in the absence of this sequence, kinetochores fail to assemble and, consequently, the chromosomes fail to segregate properly during mitosis.

Once the nuclear envelope has broken down, a randomly probing microtubule encountering a kinetochore will bind to it, thereby capturing the chromosome. The microtubule is now called a *kinetochore microtubule,* and it links the chromosome to a spindle pole (Figure 17–9 and see Panel 17–1, pp. 554–555). As kinetochores on sister chromatids face in opposite directions, they tend to attach to microtubules from

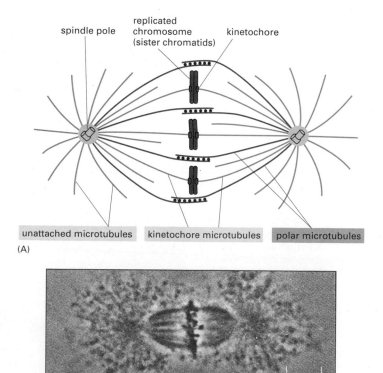

spindle pole

replicated
chromosome
(sister chromatids)

kinetochore

unattached microtubules   kinetochore microtubules   polar microtubules

(A)

(B)   10 μm

Figure 17–10 Three classes of microtubules in a mitotic spindle. (A) Schematic drawing of a spindle with chromosomes attached, showing the three types of spindle microtubules—unattached microtubules, kinetochore microtubules, and polar microtubules. In reality, the chromosomes are much larger than shown, and usually multiple microtubules are attached to each kinetochore. (B) Phase-contrast light micrograph of a mitotic spindle isolated at metaphase, clearly showing the chromosomes attached at the equator of the spindle. (B, courtesy of E.D. Salmon and R.R. Segall, from *J. Cell Biol.* 86:355–365, 1980, by copyright permission of the Rockefeller University Press.)

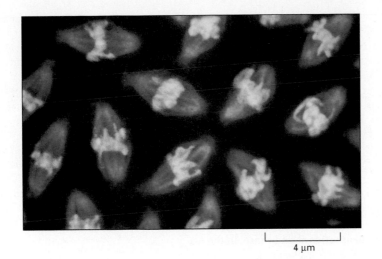

Figure 17–11 Multiple mitotic spindles at metaphase in a fruit fly (*Drosophila*) embryo. The microtubules are stained *red*, and the chromosomes are stained *green*. At this stage of *Drosophila* development, there are multiple nuclei in one large cytoplasmic compartment, and all of the nuclei divide synchronously. Although metaphase spindles are usually pictured in two dimensions, as they are here, when viewed in three dimensions the chromosomes are seen to be gathered at a plate-like region at the equator of the spindle—the so-called metaphase plate. (Courtesy of William Sullivan.)

4 µm

opposite poles of the spindle, so that each replicated chromosome becomes linked to the two spindle poles (Figure 17–10). The number of microtubules attached to each kinetochore varies between species: each human kinetochore binds 20–40 microtubules, for example, whereas a yeast kinetochore binds just one.

Chromosomes Line Up at the Spindle Equator at Metaphase

During prometaphase the chromosomes, now attached to the mitotic spindle, begin to move around, as if jerked first this way and then that. Eventually they align at the equator of the spindle, halfway between the two spindle poles, thereby forming the *metaphase plate*. This defines the beginning of **metaphase** (Figure 17–11). Although the forces that act to bring the chromosomes to the equator are not well understood, both the continual growth and shrinkage of microtubules and the action of microtubule motor proteins are thought to be involved. A continuous balanced addition and loss of tubulin subunits is required to maintain the spindle: when tubulin addition is blocked by the drug colchicine, tubulin loss continues until the spindle disappears.

The chromosomes gathered at the equator of the metaphase spindle oscillate back and forth, continually adjusting their positions, indicating that the tug-of-war between the microtubules attached to each pole of the spindle continues to operate after the chromosomes are all aligned. If one of the pair of kinetochore attachments is artificially severed with a laser beam during metaphase, the entire chromosome immediately moves toward the pole to which it remains attached. Similarly, if the attachment between sister chromatids is cut, the two chromatids separate and move toward opposite poles. These experiments show that the chromosomes at the metaphase plate are held there under considerable tension. Evidently, the forces that will ultimately pull the sister chromatids apart begin operating as soon as microtubules attach to the kinetochores.

Daughter Chromosomes Segregate at Anaphase

At the onset of **anaphase,** the connections between sister chromatids are cut by proteolytic enzymes, allowing each chromatid (now called a *daughter chromosome*) to be gradually pulled to the spindle pole to

Question 17–3 If fine glass needles are used to manipulate a chromosome inside a living cell during early M phase, it is possible to trick the kinetochores on the two sister chromatids into attaching to the same spindle pole. This arrangement is normally unstable, but the attachments can be stabilized if the needle is used to gently pull the chromosome so that both the microtubules that attach it to the pole are under tension. What does this suggest to you about the mechanism by which kinetochores normally become attached and stay attached to microtubules from opposite spindle poles during M phase? Is the finding consistent with the possibility that a kinetochore is programmed to attach to microtubules from a particular spindle pole? Explain your answers.

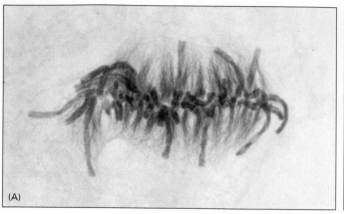

(A)

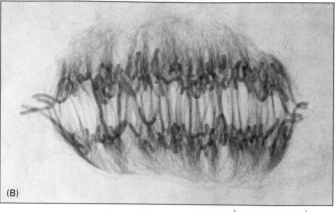

(B)

20 μm

Figure 17–12 Separation of sister chromatids at anaphase. In the transition from metaphase (A) to anaphase (B), sister chromatids separate and are pulled apart by spindle microtubules—as seen in these lily endosperm cells stained with gold-labeled antibodies to label the microtubules. Plant cells generally have less sharply defined spindle poles than animal cells, but they are present here at the top and bottom of each micrograph, although they cannot be seen. (Courtesy of Andrew Bajer.)

which it is attached (Figure 17–12). This movement distributes, or *segregates,* the two identical sets of chromosomes to opposite ends of the spindle (see Panel 17–1, pp. 554–555). All the newly separated chromosomes move at the same speed, typically about 1 μm per minute. The movement is the consequence of two independent processes mediated by different parts of the mitotic spindle. These processes are called *anaphase A* and *anaphase B,* and they occur more or less simultaneously. In anaphase A the kinetochore microtubules shorten by depolymerization, and the attached chromosomes move poleward. In anaphase B the spindle poles themselves move apart, further contributing to the segregation of the two groups of chromosomes.

Figure 17–13 The two processes that separate sister chromatids at anaphase. In *anaphase A* the daughter chromosomes are *pulled* toward opposite poles as the kinetochore microtubules depolymerize at the kinetochore. The force driving this movement is generated mainly at the kinetochore. In *anaphase B* the two spindle poles move apart as the result of two separate forces: (1) the elongation and sliding of the polar microtubules past one another *pushes* the two poles apart, and (2) outward forces exerted by outward-pointing microtubules at each spindle pole *pull* the poles away from each other, toward the cell cortex. All of these forces are thought to depend on the action of motor proteins associated with the microtubules.

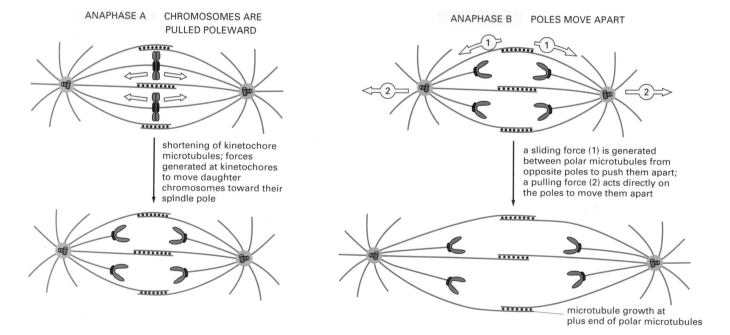

ANAPHASE A CHROMOSOMES ARE PULLED POLEWARD

shortening of kinetochore microtubules; forces generated at kinetochores to move daughter chromosomes toward their spindle pole

ANAPHASE B POLES MOVE APART

a sliding force (1) is generated between polar microtubules from opposite poles to push them apart; a pulling force (2) acts directly on the poles to move them apart

microtubule growth at plus end of polar microtubules

The driving force for the movements of anaphase A is thought to be provided mainly by the action of microtubule motor proteins operating at the kinetochore and partly by the loss of tubulin subunits that occurs mainly at the kinetochore end of the kinetochore microtubules. One possibility is that the motor proteins "walk" the kinetochore and its attached chromosome up the kinetochore microtubules, and the plus end of each microtubule depolymerizes as it becomes exposed.

The moving apart of the spindle poles in anaphase B is accompanied by the elongation of the polar microtubules, which polymerize at their plus (free) ends. The driving forces for this process are thought to be provided by two sets of motor proteins operating on the polar microtubules. One set acts on the long polar microtubules that form the spindle itself; these motor proteins slide the polar microtubules from opposite poles past one another at the equator of the spindle, pushing the spindle poles apart. The other set operates on the microtubules that extend from the spindle poles but point away from the chromosomes toward the cell cortex. These motor proteins are thought to be associated with the cell cortex and to pull each pole toward the adjacent cortex and away from the other pole (Figure 17–13).

The Nuclear Envelope Re-forms at Telophase

By the end of anaphase the daughter chromosomes have separated into two equal groups, one at each pole of the spindle. During **telophase,** the final stage of mitosis, a nuclear envelope reassembles around each group of chromosomes to form the two daughter nuclei. Vesicles of nuclear membrane first cluster around individual chromosomes and then fuse to re-form the nuclear envelope (see Panel 17–1, pp. 554–555). During this process the nuclear pores (discussed in Chapter 14) reassemble in the envelope, and the nuclear lamins, the intermediate filament protein subunits that were phosphorylated during prophase, are now dephosphorylated and reassociate to re-form the nuclear lamina (Figure 17–14). Once the nuclear envelope has re-formed, the pores

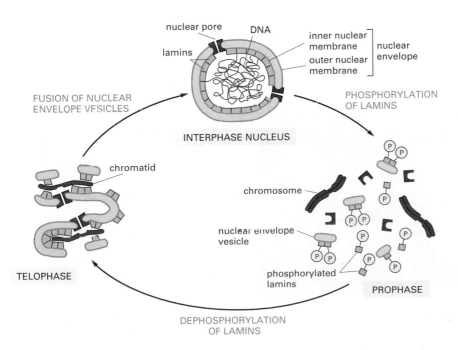

Figure 17–14 Breakdown and re-formation of the nuclear envelope during mitosis. The phosphorylation of the lamins at prophase helps trigger the disassembly of the nuclear lamina, which in turn causes the nuclear envelope to break up into vesicles. Dephosphorylation of the lamins at telophase helps reverse the process.

pump in nuclear proteins, the nucleus expands, the condensed mitotic chromosomes decondense into their interphase state, and, as a consequence of decondensation, gene transcription is able to resume. A new nucleus has been created, and mitosis is complete. All that remains is for the cell to complete its division into two.

Question 17–4 Consider the events that lead to the formation of the new nucleus at telophase. How do nuclear and cytosolic proteins become properly re-sorted so that the new nucleus contains nuclear proteins but not cytosolic proteins?

Cytokinesis

M phase involves more than the segregation of chromosomes and the formation of new nuclei. It is also the time during which other components of the cell—membranes, cytoskeleton, organelles, and soluble proteins—are distributed between the two daughter cells. This is achieved by **cytokinesis,** the process by which the cytoplasm is cleaved in two. Cytokinesis usually begins in anaphase but is not completed until after the two daughter nuclei have formed. Whereas mitosis involves a transient microtubule-based structure, the mitotic spindle, cytokinesis in animal cells involves a transient structure based on actin filaments, the *contractile ring* (see Figure 17–5). Both the plane of cleavage and the timing of cytokinesis, however, are determined by the mitotic spindle.

The Mitotic Spindle Determines the Plane of Cytoplasmic Cleavage

The first visible sign of cytokinesis in animal cells is a puckering and furrowing of the plasma membrane during anaphase (Figure 17–15). The furrowing invariably occurs perpendicular to the long axis of the mitotic spindle, ensuring that the *cleavage furrow* cuts between the two groups of daughter chromosomes so that the two daughter cells receive identical and complete sets of chromosomes. If, as soon as the furrow appears, the mitotic spindle is deliberately displaced (using a fine glass needle inserted into the cell), the starting furrow disappears and a new one develops at a site corresponding to the new spindle location and orientation. Once the furrowing process is well under way, however, cleavage proceeds even if the mitotic spindle is artificially sucked out of the cell or depolymerized using the drug colchicine. How the mitotic spindle dictates the position of the cleavage furrow remains a mystery.

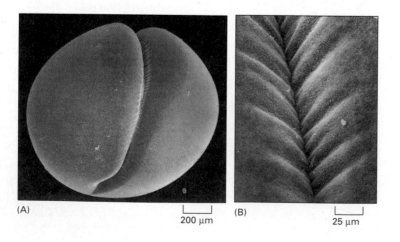

(A) 200 μm (B) 25 μm

Figure 17–15 Scanning electron micrographs of early cleavage in a fertilized frog egg. The furrowing of the cell membrane is caused by the activity of the *contractile ring* underneath it. The cleavage furrow is unusually well defined in this giant spherical cell. (A) Low-magnification view of egg surface. (B) Surface of the furrow at higher magnification. (From H.W. Beams and R.G. Kessel, *Am. Sci.* 36:279–290, 1976. Reprinted by permission of *American Scientist*, journal of Sigma Xi.)

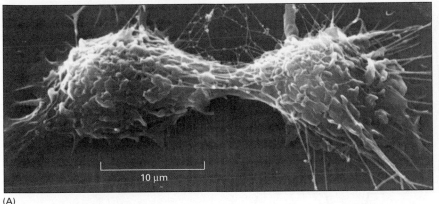

(A)

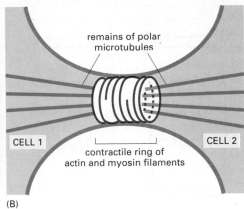

(B)

When the mitotic spindle is located centrally in the cell, the usual situation in most dividing cells, the cell divides symmetrically. The cleavage furrow forms around the equator of the parent cell, and the two daughter cells produced are of equal size and contain similar molecules. During embryonic development, however, there are many instances where cells divide asymmetrically. In these cases the mitotic spindle is positioned asymmetrically, and consequently, the furrow creates two cells that differ in size, that often differ in the molecules they contain, and that usually develop into different cell types.

Figure 17–16 **The contractile ring.** (A) Scanning electron micrograph of an animal cell in culture in the last stages of dividing. (B) Schematic diagram of the midregion of a similar cell showing the contractile ring beneath the plasma membrane and the remains of the two sets of polar microtubules. (A, courtesy of Guenter Albrecht-Buehler.)

The Contractile Ring of Animal Cells Is Made of Actin and Myosin

The contractile ring is composed mainly of an overlapping array of actin filaments and myosin filaments (Figure 17–16). It assembles at anaphase and is attached to membrane-associated proteins on the cytoplasmic face of the plasma membrane. Once assembled, it is capable of exerting a force strong enough to bend a fine glass needle inserted into the cell prior to cytokinesis. The force is generated by the sliding of the actin filaments against the myosin filaments, much as occurs during muscle contraction (see Figure 16–36). Unlike the contractile apparatus in muscle, however, the contractile ring is a transient structure: it assembles to carry out cytokinesis, gradually becomes smaller as cytokinesis progresses, and disassembles completely once the cell is cleaved in two.

Cell division in many animal cells is accompanied by large changes in cell shape and a decrease in the adherence of the cell to the extracellular matrix. These changes result from the reorganization of actin and

Figure 17–17 **Micrographs of a mouse fibroblast dividing in culture.** The same cell was photographed at successive times. Note how the cell rounds up as it enters mitosis; the two daughter cells then flatten out after cytokinesis is completed. (Courtesy of Guenter Albrecht-Buehler.)

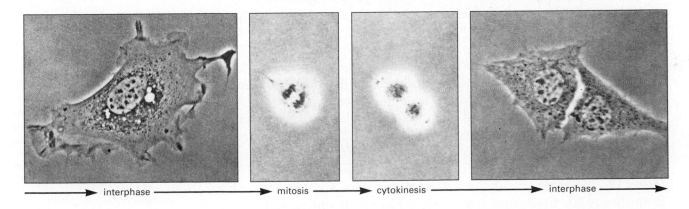

interphase   mitosis   cytokinesis   interphase

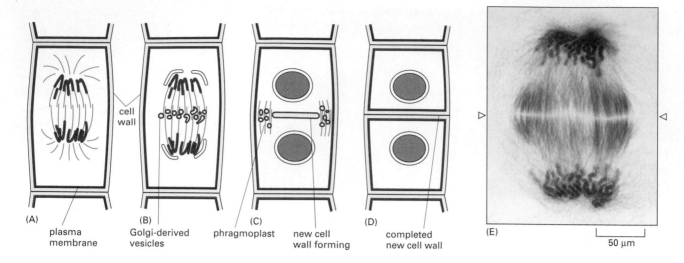

(A) plasma membrane

cell wall

(B) Golgi-derived vesicles

(C) phragmoplast

new cell wall forming

(D) completed new cell wall

(E)   50 μm

myosin filaments in the cell cortex, one aspect of which is the assembly of the contractile ring. Mammalian fibroblasts in culture, for example, are spread out flat during interphase as a result of the strong adhesive contacts they make with the surface they are growing on—called the substratum. As they enter M phase, however, the cells round up, at least partly because some of the plasma membrane proteins responsible for attaching the cells to the substratum—the *integrins* (discussed in Chapter 19)—become phosphorylated and thus weaken their grip. Once cytokinesis is complete, the daughter cells reestablish their contacts with the substratum and flatten out again (Figure 17–17). When cells divide in an animal tissue, this cycle of attachment and detachment presumably allows them to rearrange their contacts with neighboring cells and with the extracellular matrix, so that the new cells produced by cell division can be accommodated within the tissue.

Cytokinesis in Plant Cells Involves New Cell-Wall Formation

The mechanism of cytokinesis in higher plants is entirely different from that in animal cells, presumably because the plant cells are surrounded by a tough cell wall (discussed in Chapter 19). The two daughter cells are separated not by the action of a contractile ring at the cell surface but instead by the construction of a new wall inside the cell. The growing new cell wall is surrounded by a membrane, and it enlarges to divide the cytoplasm in two. The positioning of this new wall precisely determines the position of the two daughter cells relative to neighboring cells. Thus the planes of cell division, together with cell enlargement, determine the final form of the plant.

The new cell wall starts to assemble in the cytoplasm between the two sets of segregated chromosomes at the start of telophase. The assembly process is guided by a structure called the **phragmoplast,** which is formed by the remains of the polar microtubules at the equator of the old mitotic spindle. Small membrane-bounded vesicles, largely derived from the Golgi aparatus and filled with polysaccharides and glycoproteins required for the cell wall matrix, are transported along the microtubules to the equator of the phragmoplast. Here they fuse to form a disclike membrane-bounded structure, which expands outward by further vesicle fusion until it reaches the plasma membrane and orig-

Figure 17–18 Cytokinesis in a plant cell. At the beginning of telophase, after the chromosomes have segregated (A), a new cell wall starts to assemble inside the cell at the equator of the old spindle (B). The polar microtubules of the mitotic spindle remaining at telophase form the *phragmoplast* and guide the vesicles toward the center of the spindle. Here membrane-bounded vesicles, derived from the Golgi apparatus and filled with cell-wall material, fuse to form the growing new cell wall (C), which grows outward to reach the plasma membrane and original cell wall. The plasma membrane and the membrane surrounding the new cell wall (both shown in *red*) fuse, completely separating the two daughter cells (D). A light micrograph of a plant cell in telophase is shown in (E) at a stage corresponding to (B). The cell has been stained to show both the microtubules and the two sets of daughter chromosomes. The location of the growing new cell wall is indicated by the arrowheads. (E, courtesy of Andrew Bajer.)

Question 17–5 Draw a detailed view of the formation of the new cell wall that separates the two daughter cells when a plant cell divides. In particular, show where the membrane proteins of the Golgi-derived vesicles end up, indicating what happens to the part of a protein in the Golgi vesicle membrane that is exposed to the interior of the Golgi vesicle. (Refer to Chapter 11 if you need a reminder of membrane structure.)

inal cell wall and divides the cell in two (Figure 17–18). Later, cellulose microfibrils are laid down within the matrix to complete the construction of the new cell wall.

Meiosis

Meiosis was discovered in 1883, when it was observed that the fertilized egg of a particular worm contained four chromosomes, whereas the gametes of the worm (sperm in males and eggs in females) contained only two. This was the first time it was realized that gametes, cells specialized for sexual reproduction, are **haploid**, carrying only a single set of chromosomes and thus only a single copy of the organism's genetic information, while the other cells of the body, including the germ-line cells that give rise to the gametes, are **diploid**, carrying two sets of chromosomes, one derived from the mother and the other from the father (see Figure 9–34). This observation implied that sperm and eggs (Figure 17–19) must be formed by a special kind of cell division in which the number of chromosomes is precisely halved. This type of cell division is called **meiosis**, from the Greek, meaning "diminution," or "lessening."

In this final section we discuss the mechanics of meiosis, highlighting the differences between meiosis and mitosis. The genetic and biological significance of meiosis is considered in Chapter 9.

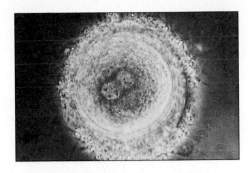

Figure 17–19 Light micrograph of a human egg with sperm bound to its surface. The egg is already fertilized and contains two nuclei (one from the egg and one from the sperm). Numerous unsuccessful sperm are still attached to the surface of the egg. (Courtesy of Peter Braude.)

Homologous Chromosomes Pair Off During Meiosis

With the exception of the chromosomes that determine sex (the *sex chromosomes*), a diploid nucleus contains two very similar versions of each chromosome, one from the male parent (paternal chromosome) and one from the female parent (maternal chromosome) (see Figure 8–2). The two versions of each chromosome, however, are not genetically identical, as they carry different variants of many of the genes (discussed in Chapter 9). They are thus called **homologous chromosomes**, or **homologues**, meaning similar but not identical. A diploid cell in a sexually reproducing organism therefore carries two similar sets of genetic information. In most cells the homologues maintain a completely separate existence as independent chromosomes.

In mitosis, as we saw earlier, each chromosome replicates and the replicated chromosomes line up in random order at the metaphase plate; the sister chromatids then separate from each other to become individual chromosomes, and the two daughter cells produced by cytokinesis inherit one copy of each paternal chromosome and one copy of each maternal chromosome. Thus both sets of genetic information are transmitted intact to both daughter cells, which are, therefore, diploid and genetically identical.

By contrast, when diploid cells divide by meiosis they form haploid gametes with only half the original number of chromosomes—only one chromosome of each type instead of a pair of homologues of each type. Thus each gamete acquires either the maternal copy or the paternal copy of a chromosome but not both. This reduction is required so that when two gametes (an egg and a sperm in animals) fuse at fertilization, the chromosome number is restored to the diploid number in the embryo. Since the assignment of maternal and paternal chromosomes

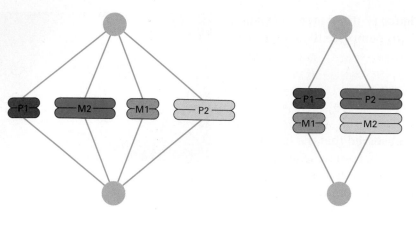

(A) Metaphase plate in mitosis: homologous chromosomes act independently

(B) Metaphase plate in meiosis: homologous chromosomes pair

Figure 17–20 Comparison of metaphase in mitosis and meiosis. In mitosis (A) the individual maternal (M) and paternal (P) chromosomes line up randomly at the metaphase plate, whereas in meiosis (B) the homologous maternal and paternal chromosomes pair off before lining up at the metaphase plate. In both cases the chromosomes have replicated before they become aligned. The spindle is shown in *green.*

to the gametes during meiosis is random, the original maternal and paternal chromosomes are reshuffled into different combinations.

The need to halve the number of chromosomes makes an extra demand on the cell-division machinery and leads to the first main difference between meiosis and mitosis. In a meiotic cell division the replicated homologous paternal and maternal chromosomes (including the two replicated sex chromosomes) pair up alongside each other before they line up on the spindle (Figure 17–20). As we shall see, this physical pairing of homologues enables the paternal and the maternal homologue to be segregated to different daughter cells. It is still not certain how the homologues (and the two sex chromosomes) recognize each other.

Meiosis Involves Two Cell Divisions Rather Than One

In principle, meiosis could occur by a simple modification of a normal mitotic cell division, such that DNA replication (S phase) is omitted. Unreplicated homologues could pair and then move to the metaphase plate; the separation of the homologues, followed by cytokinesis, would then produce two haploid cells directly. For unknown reasons, the actual meiotic process is more complicated and involves two cell divisions instead of one. Moreover, in some cells it can take very much longer than any mitosis: in a human oocyte, for example, meiosis can go on for up to 50 years.

Figure 17–21 shows the events that occur in the first meiotic cell division—*division I of meiosis.* Before the homologous maternal and paternal chromosomes pair, each one replicates to produce two sister

Figure 17–21 Division I of meiosis. For simplicity, only one pair of homologous chromosomes is shown, although all eucaryotic cells have more than one pair. After DNA replication each chromosome consists of two sister chromatids joined along their length. The pairing of replicated homologous chromosomes is unique to meiosis. As shown by the formation of chromosomes that are part *red* and part *black,* the chromosome pairing allows genetic recombination (crossing-over) to occur between homologous chromosomes, as explained in Chapter 9. After recombination has occurred, the meiotic spindle forms, the nuclear membrane breaks down, the paired homologues line up at the metaphase plate and are pulled apart by the spindle. The two daughter nuclei resulting from the first meiotic division each contain one of the duplicated homologues.

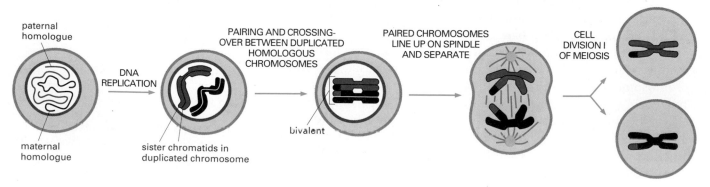

paternal homologue

DNA REPLICATION

PAIRING AND CROSSING-OVER BETWEEN DUPLICATED HOMOLOGOUS CHROMOSOMES

PAIRED CHROMOSOMES LINE UP ON SPINDLE AND SEPARATE

CELL DIVISION I OF MEIOSIS

maternal homologue

sister chromatids in duplicated chromosome

bivalent

chromatids, as in an ordinary mitotic cell division. It is only after DNA replication has been completed that the special features of meiosis come into play. Before the nuclear envelope breaks down, and before the chromosomes are assembled at the metaphase plate, each replicated chromosome pairs with its homologue, forming a structure called a *bivalent*, which contains four chromatids (Figure 17–22). It is this first prophase stage of meiosis that can last for years; the rest of meiosis occurs quickly. At the first meiotic cell division, the replicated homologues separate from each other and are segregated to separate daughter cells; the sister chromatids of each homologue, however, remain together through this division and behave as a single unit, as if chromosome duplication had not occurred (see Figure 17–21). They will not separate until the second meiotic cell division.

During the long prophase of the first meiotic division, the replicated homologous chromosomes in a bivalent are kept together and closely aligned by the *synaptonemal complex*. The complex consists of a long, ladderlike protein core, with the two replicated homologues aligned on opposite sides. The sister chromatids in each homologue are kept tightly packed together with their chromatin extending from the same side of the protein ladder in a series of loops.

As discussed in Chapter 9, the pairing of homologues in bivalents during the long prophase of the first meiotic division allows genetic recombination to occur, whereby a fragment of a maternal chromatid may be exchanged for the corresponding fragment of a homologous paternal chromatid. This process, called **crossing-over** (see Figure 17–21), is an important feature of meiosis. By scrambling the genetic constitution of each of the chromosomes in gametes, it helps to produce individuals with novel assortments of genes. At the end of the long prophase of the first meiotic division, the nuclear envelope breaks down, signaling the start of prometaphase.

The rest of the first meiotic division proceeds like a normal mitotic cell division. The bivalents become attached to the meiotic spindle and line up at the metaphase plate. At anaphase the two replicated homologues (each still consisting of two sister chromatids) separate and are pulled to opposite poles (see Figure 17–21). Because the sister chromatids are still closely joined and move as a single unit, when the cell divides, each daughter cell inherits two copies of either the maternal or the paternal homologue of each pair; these two copies are identical except where genetic recombination has occurred (see Figure 17–21). The two daughter cells of this division therefore each contain the same quantity of DNA as a diploid cell but differ from a true diploid cell in two ways: (1) both of the two DNA copies of each chromosome derive from only one of the two homologous chromosomes in the original cell (except where there has been genetic recombination), and (2) these two copies are inherited as closely associated sister chromatids, as if they were still a single chromosome (see Figure 17–21).

Figure 17–22 A bivalent. The maternal and paternal copies of chromosome 1 have each replicated and then paired to form a bivalent, which consists of four chromatids. When kinetochores assemble on the centromeres, those on the two sister chromatids of a replicated chromosome face in the same direction; thus spindle microtubules will attach the two replicated chromosomes to opposite poles of the spindle.

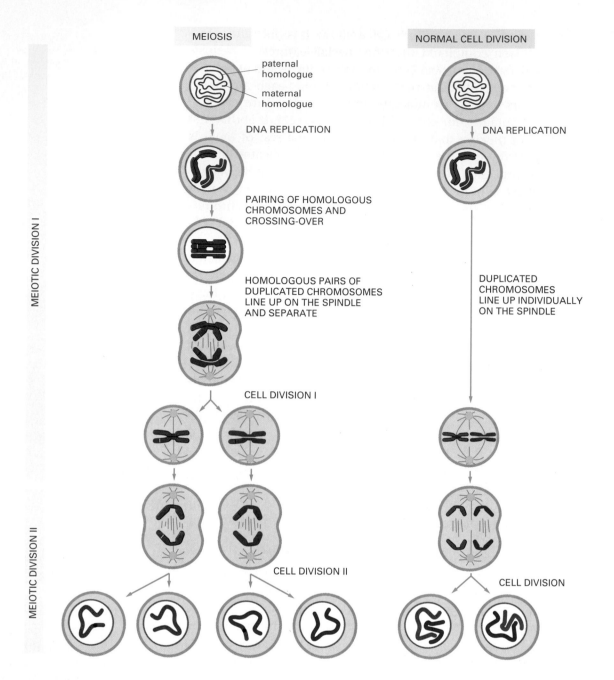

MEIOSIS

paternal homologue
maternal homologue

NORMAL CELL DIVISION

DNA REPLICATION

DNA REPLICATION

MEIOTIC DIVISION I

PAIRING OF HOMOLOGOUS CHROMOSOMES AND CROSSING-OVER

HOMOLOGOUS PAIRS OF DUPLICATED CHROMOSOMES LINE UP ON THE SPINDLE AND SEPARATE

DUPLICATED CHROMOSOMES LINE UP INDIVIDUALLY ON THE SPINDLE

CELL DIVISION I

MEIOTIC DIVISION II

CELL DIVISION II

CELL DIVISION

Formation of the actual gametes can now proceed through a second cell division, *division II of meiosis*, which occurs without further DNA replication and without any significant interphase period. A spindle forms, the chromosomes align at its equator, and the sister chromatids separate (as in normal mitosis) to produce daughter cells with a haploid DNA content. Meiosis thus consists of a single round of DNA replication followed by two cell divisions, so that *four haploid cells are produced from each diploid cell that enters meiosis*. Meiosis and mitosis are compared in Figure 17–23.

Occasionally, the meiotic process occurs abnormally and homologues fail to separate—a phenomenon known as **nondisjunction.** In this case some of the haploid cells that are produced lack a particular chromosome, while others have more than one copy of it. Such gametes form abnormal embryos, most of which die. Some survive, however. *Down syndrome* in humans, for example, is caused by an extra copy of

Figure 17–23 Comparison of meiosis and mitosis. As in Figure 17–21, only one pair of homologous chromosomes is shown. In meiosis, following DNA replication, two nuclear and cell divisions are required to produce the haploid gametes. Each diploid cell that enters meiosis therefore produces four haploid cells, whereas each diploid cell that divides by mitosis produces two diploid cells. The telophase stages and the prometaphase stages have been omitted for simplicity. Whereas mitosis and division II of meiosis usually occur within hours, division I of meiosis can last days, months, or even years, due to the long time spent in prophase.

Chromosome 21. This results from nondisjunction of a Chromosome 21 pair during meiosis, giving rise to a gamete that contains two copies of Chromosome 21 instead of one copy. When this abnormal gamete fuses with a normal gamete at fertilization, the resulting embryo contains three copies of Chromosome 21 instead of two.

Question 17–6 Why would it not be desirable for an organism to use the first steps of meiosis (up to and including meiotic cell division I) for the ordinary mitotic division of somatic cells?

Essential Concepts

- All cells are produced by the division of other cells. Bacteria divide by simple fission, whereas eucaryotic cells divide in a more complicated way.

- The eucaryotic cell cycle consists of several distinct phases. These include S phase, during which the nuclear DNA is replicated, and M phase, during which the nucleus divides (mitosis) and then the cytoplasm divides (cytokinesis).

- In most cells there is one gap phase (G_1) between M phase and S phase, and another (G_2) between S phase and M phase. These gaps give the cell more time to grow.

- The onset of M phase is signaled by the formation of a mitotic spindle made of microtubules, which segregates daughter chromosomes to opposite poles of the cell.

- Large membrane-bounded organelles such as the endoplasmic reticulum and Golgi apparatus break into many smaller fragments during M phase, ensuring an even distribution between the daughter cells.

- Before the mitotic spindle assembles at the start of M phase, the centrosome duplicates. The two daughter centrosomes separate and move to opposite sides of the nucleus to form the two poles of the spindle.

- Microtubules grow out from the centrosomes, and some of these interact with microtubules growing from the opposite pole, thereby becoming the polar microtubules of the spindle.

- When the nuclear envelope breaks down, the spindle microtubules invade the nuclear area. Some of these capture the replicated chromosomes by binding to protein complexes called kinetochores, associated with the centromere of each sister chromatid.

- Microtubules from opposite poles pull in opposite directions on each chromosome, bringing the chromosomes to the equator of the mitotic spindle.

- The daughter chromosomes are produced by the sudden separation of sister chromatids and are pulled by the spindle to opposite poles. The two poles also move apart, further separating the two sets of chromosomes.

- The movement of chromosomes by the spindle is driven both by microtubule motor proteins and by microtubule polymerization and depolymerization.

- A nuclear envelope re-forms around the two sets of segregated chromosomes to form two new nuclei, thereby completing mitosis.

- In animal cells, cell division is completed by a contractile ring of actin filaments and myosin filaments, which assembles midway between the spindle poles and contracts to divide the cytoplasm in two; in plant cells, by contrast, cell division is completed by the formation of a new cell wall inside the cell, which divides the cytoplasm in two.

- Gametes (sperm and eggs in animals) are formed by meiosis, a specialized form of cell division that produces haploid cells from diploid cells. Haploid cells are required by sexually reproducing organisms so that the chromosome number is not continually doubled at fertilization.

- Although most of the mechanical features of meiosis are similar to those of mitosis, the behavior of the chromosomes is different; meiosis produces four genetically dissimilar haploid cells by two consecutive cell divisions, whereas mitosis produces two genetically identical diploid cells by a single cell division.

Key Terms

anaphase

cell cycle

chromatid

crossing-over

cytokinesis

diploid

G_1 phase

G_2 phase

haploid

homologue

interphase

kinetochore

meiosis

metaphase

mitosis

mitotic spindle

M phase

nondisjunction

phragmoplast

prometaphase

prophase

S phase

spindle pole

telophase

Questions

Question 17–7 Which of the following statements are correct? Explain your answers.

A. Centrosomes are replicated independently of chromosomes.

B. The nuclear envelope becomes fragmented at mitosis. It is thus distributed between daughter cells like other membrane-bounded organelles such as the endoplasmic reticulum and the Golgi apparatus.

C. Two sister chromatids arise by replication of the DNA of the same chromosome and remain paired as they line up on the metaphase plate.

D. The total nuclear DNA content of a cell produced by meiosis is half that of a cell produced by mitosis.

E. Polar microtubules attach end to end and are therefore continuous from one spindle pole to the other.

F. Microtubule polymerization, depolymerization, and microtubule motor proteins are required for DNA replication.

G. Microtubules nucleate at the centromeres and then connect to the kinetochores, which are structures at the centrosome regions of chromosomes.

Question 17–8 Consider the following statement: "All present-day cells have arisen by an uninterrupted series of cell divisions extending back in time to the first cell division." Is it strictly true?

Question 17–9 Roughly how long would it take a single fertilized egg to make a cluster of cells weighing 70 kg by repeated divisions, if each cell weighs 1 nanogram just after cell division and each cell cycle takes 24 hours? Why does it take very much longer than this to make a 70-kg adult human?

Question 17–10 Which is the order in which the following events occur during cell division:

A. anaphase

B. metaphase

C. prometaphase

D. telophase

E. lunar phase

F. mitosis

G. prophase

Question 17–11 The shortest eucaryotic cell cycles of all—shorter even than those of many bacteria—occur in many early animal embryos. These divisions take place without any significant increase in the weight of the embryo. How can this be? Which phase of the cell cycle would you expect to be most reduced?

Question 17–12 Explain the relevance of each of the following for cell division in bacteria and for eucaryotic cell division.

A. microtubule assembly

B. DNA replication

C. meiosis

D. mitosis

E. cell-wall synthesis

F. cell growth

G. nuclear envelope reassembly

Question 17–13 The lifetime of a microtubule in a mammalian cell, between its formation by polymerization and its spontaneous disappearance by depolymerization, varies with the stage of the cell cycle. For an actively growing cell, the average lifetime is 5 minutes in interphase and 15 seconds in mitosis. If the average length of a microtubule in interphase is 20 μm, what will it be during mitosis, assuming that the rate of microtubule elongation due to the addition of tubulin subunits in the two phases are the same? If a typical centrosome in an interphase cell has 100 nucleation sites for microtubules, how many sites would you expect to find in centrosomes in a mitotic cell, assuming that the total number of tubulin molecules that are polymerized into microtubules is the same in both phases?

Question 17–14   What might cause the rare occurrence of two copies of the same chromosome ending up in the same daughter cell? What could be the consequences of this event occurring (a) in mitosis and (b) in meiosis?

Question 17–15   An antibody to myosin prevents the movement of myosin molecules along actin filaments (the interaction of actin and myosin are described in Chapter 16). How do you suppose the antibody exerts this effect? What might be the result of injecting this antibody into cells (a) on the movement of chromosomes at anaphase or (b) on cytokinesis. Explain your answers.

Question 17–16   Sketch the principal stages of mitosis, using Panel 17–1 (pp. 554–555) as a guide. Color one sister chromatid and follow it through mitosis and cytokinesis. What event commits this chromatid to a particular daughter cell? Once initially committed, can its fate be reversed? What may influence this commitment?

Question 17–17   Why do sister chromatids have to remain paired in division I of meiosis? Does the answer suggest a strategy for washing your socks?

Question 17–18   The polar movement of chromosomes during anaphase A is associated with microtubule shortening. In particular, microtubules depolymerize at the ends at which they are attached to the kinetochores. Sketch a model that explains how a microtubule can shorten and generate force, yet remain firmly attached to the chromosome.

This giant spherical egg cell of the frog *Xenopus* is just over 1 mm in diameter. When fertilized, it divides very rapidly, without intervening cell growth, to produce a young frog embryo with over 4000 small daughter cells, in about 7 hours. Studies of the rapid early embryonic cell cycles in *Xenopus* led to the identification of the central components of the cell-cycle control system. (Courtesy of Tony Mills.)

18 Cell-Cycle Control and Cell Death

Two types of machinery are required in the cell cycle: one to manufacture the new components of the growing cell, and another to haul the components into their correct places and to partition them appropriately when the cell divides in two, as described in Chapter 17. It is this *cell-cycle machinery* that first attracts attention, just as the working machines are what one notices first on visiting a factory. No less important, however, is the central control system that switches the machinery on and off at the correct times and coordinates all the activities that go to make the final product. In the cell, the control system that ensures correct progression through the cell cycle by regulating the cell-cycle machinery is the *cell-cycle control system,* which we discuss in the first part of this chapter.

As discussed in Chapter 17, the events of the cell cycle occur in a fixed sequence (see Panel 17–1, pp. 554–555). If we start with DNA replication (S phase), this is followed by a gap (G_2 phase), then by mitosis and cytokinesis (M phase), then another gap (G_1 phase), and then by DNA replication again, and so on (Figure 18–1). The cell-cycle control system has to activate the enzymes and other proteins responsible for carrying out each process at the appropriate time, and then it has to deactivate them once the process is completed. It must also ensure that each stage

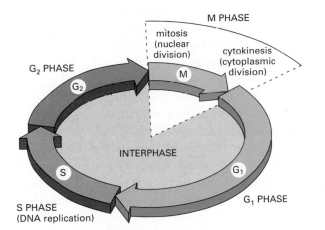

Figure 18–1 The phases of the cell cycle. Interphase includes all of the cell cycle except for M phase; it is a period of continuous cell growth, and it includes S phase, where the DNA is replicated. The nucleus and then the cytoplasm divides during M phase. G_1 phase is the gap between M phase and S phase; G_2 is the gap between S phase and M phase.

of the cycle is completed before the next one is begun: it has to make certain, for example, that DNA replication has been completed before mitosis begins, that mitosis has been completed before the cell splits in two, and that another round of DNA replication does not begin until the cell has passed through mitosis and grown to an appropriate size.

The control system must also take account of conditions outside the cell. Most important, in a multicellular organism the control system must be responsive to signals from other cells, such as those that stimulate cell division when more cells are needed. The cell-cycle control system therefore plays a central part in the regulation of cell numbers in the tissues of the body; when the system malfunctions, it can result in cancer. In the second half of this chapter we examine more closely how cell numbers are regulated. We shall see that this is not only a matter of controlling cell division; it also involves controlling whether cells survive or die by a process called *programmed cell death*.

The Cell-Cycle Control System

For many years cell biologists watched the puppet show of DNA synthesis, mitosis, and cytokinesis but had no idea of what lay behind the curtain controlling these events. The *cell-cycle control system* was simply a "black box" inside the cell. It was not even clear whether there was a separate control system, or whether the cell-cycle machinery somehow controlled itself. A breakthrough came with the identification of the key proteins of the control system and the realization that they are distinct from the components of the cell-cycle machinery—the enzymes and other proteins that perform the essential processes of DNA replication, chromosome segregation, and so on. We first consider some of the basic principles on which the cell-cycle control system operates, and then discuss the protein components of the control system and how they work together to trigger the different phases of the cycle.

A Central Control System Triggers the Major Processes of the Cell Cycle

The cell-cycle control system operates much like the control system of an automatic clothes-washing machine. The washing machine functions in a series of stages: it takes in water, mixes it with detergent, washes the clothes, rinses them, and spins them dry. These essential processes of the wash cycle are analogous to the essential processes of the cell cycle—DNA replication, mitosis, and so on. In both cases, a central controller triggers each process in a set sequence (Figure 18–2). The controller is itself regulated at certain critical points of the cycle by feedback from the processes that are being performed. In the washtub, sensors monitor the water levels, for example, and send signals back to the controller to prevent the next process from beginning before the previous one has finished. Without such feedback, an interruption or a delay in any of the processes could be disastrous.

The events of the cell cycle must also occur in a particular sequence, and this sequence must be preserved even if one of the steps takes

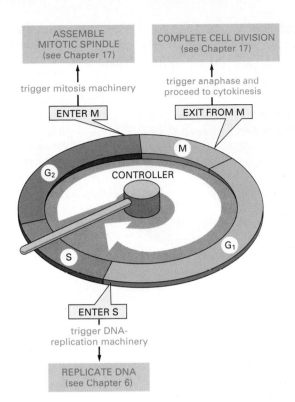

ASSEMBLE
MITOTIC SPINDLE
(see Chapter 17)

COMPLETE CELL DIVISION
(see Chapter 17)

trigger mitosis machinery

trigger anaphase and
proceed to cytokinesis

ENTER M

EXIT FROM M

M

G₂   CONTROLLER

G₁

S

ENTER S

trigger DNA-
replication machinery

REPLICATE DNA
(see Chapter 6)

Figure 18–2 The control of the cell cycle. The essential processes, such as DNA replication, mitosis, and cytokinesis, are triggered by a cell-cycle control system. By analogy with a washing machine, the cell-cycle control system is drawn as an indicator that rotates clockwise, triggering essential processes when it reaches specific points on the outer dial.

longer than usual. All of the nuclear DNA must be replicated before the nucleus begins to divide, which means that a complete S phase must precede M phase. If DNA synthesis is slowed down or stalled (because of DNA damage requiring repair, for example), mitosis and cell division must also be delayed. Similarly, it is crucial for most cells to double in size before dividing in two; otherwise, cells would get smaller with each division. The cell-cycle control system achieves all this by means of molecular brakes that can stop the cycle at various **checkpoints.** In this way the control system does not trigger the next step in the cycle before the previous one has been completed.

The control system in most cells has checkpoints for cell size, where the cell-cycle is halted until the cell has grown to an appropriate size. In G_1, a size checkpoint allows the system to halt and the cell to grow further, if necessary, before a new round of DNA replication is triggered. Cell growth depends on an adequate supply of nutrients and other factors in the extracellular environment, and the G_1 checkpoint also allows the cell to check that the environment is favorable for cell proliferation before committing itself to S phase. A second size checkpoint occurs in G_2, allowing the system to halt before it triggers mitosis. In addition, the G_2 checkpoint allows the cell to check that DNA replication is complete before proceeding to mitosis (Figure 18–3).

Checkpoints are important in another way: they are points in the cell cycle where the control system can be regulated by signals from other cells, such as growth factors and other extracellular signal molecules, which can either promote or inhibit cell proliferation. Later in this chapter we consider the factors that influence the decisions made at these checkpoints, but first we discuss the proteins that form the central control system.

Question 18–1 There is a point in G_1 in the cell cycle where a cell checks that it is of sufficient size before committing to the next step. A possible clue regarding the mechanism by which this checkpoint operates comes from cells that have an abnormal number of copies of their genome. For a given cell type, it seems to be a general rule that cell size is roughly proportional to the number of chromosome sets in the cell. With this observation in mind, suggest a mechanism by which a cell measures its own size in G_1.

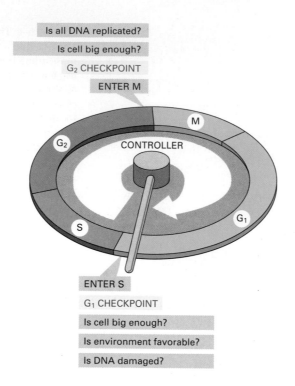

Is all DNA replicated?

Is cell big enough?

G₂ CHECKPOINT

ENTER M

M

G₂

CONTROLLER

S

G₁

ENTER S

G₁ CHECKPOINT

Is cell big enough?

Is environment favorable?

Is DNA damaged?

Figure 18–3 Two checkpoints in the cell cycle. Feedback from the intracellular events of the cell cycle, as well as signals from the cell's environment, determine whether the control system will pass through certain checkpoints. Two important checkpoints are shown here: the one in G_1, where the control system determines whether the cell proceeds to S phase, and the one in G_2, where the control system determines whether the cell proceeds to mitosis.

The Cell-Cycle Control System Is Based on Cyclically Activated Protein Kinases

The cell-cycle control system governs the cell-cycle machinery through the phosphorylation of key proteins that initiate or regulate DNA replication, mitosis, and cytokinesis. As discussed in Chapter 5, phosphorylation (and dephosphorylation) is one of the most common ways used by cells to alter the activity of a protein. The phosphorylation reactions that control the cell cycle are carried out by a specific set of protein kinases, enzymes that catalyze the transfer of a phosphate group from ATP to a particular amino acid side chain on the target protein. The effects of phosphorylation can be rapidly reversed by removal of the phosphate group (dephosphorylation), a reaction carried out by another set of enzymes, the protein phosphatases (discussed in Chapter 5).

The protein kinases of the cell-cycle control system are present in proliferating cells throughout the cell cycle. They are activated, however, only at appropriate times in the cycle, after which they quickly become deactivated again. Thus the activity of each of these kinases rises and falls in a cyclical fashion. Some of the protein kinases, for example, become active toward the end of G_1 phase and are responsible for driving the cell into S phase; another becomes active just before M phase and is responsible for driving the cell into mitosis.

Given that the protein kinases of the cell-cycle control system are present throughout the cycle, how is their enzyme activity switched on and off at the appropriate times? This is partly the responsibility of a second set of protein components of the control system—the *cyclins*. Cyclins have no enzymatic activity themselves, but they have to bind to the cell-cycle kinases before the kinases can become enzymatically active. The kinases of the cell-cycle control system are therefore known as *cyclin-dependent protein kinases*, or **Cdks** (Figure 18–4). Cyclins are so-called because, unlike the Cdks, their concentrations vary in a cycli-

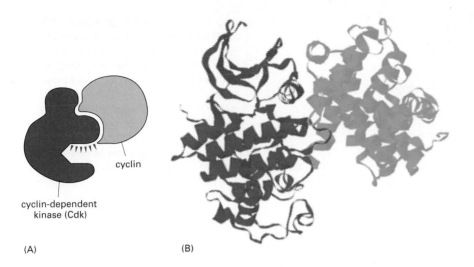

(A)   (B)

Figure 18–4 **The two key protein components of a cyclin-Cdk complex.** (A) The cyclin-dependent kinase (Cdk) is an enzyme that catalyzes protein phosphorylations, while the cyclin is a regulatory protein required for the Cdk to be enzymatically active. The active complex phosphorylates key proteins in the cell that are required to initiate a particular step in the cell cycle. (B) The three-dimensional structure of the cyclin-Cdk complex, called MPF, that controls the entry into M phase. (B, adapted from P.D. Jeffrey et al., *Nature* 376:313–320, 1995.)

cal fashion during the cell cycle. To illustrate how the cyclins help to activate the Cdks at the appropriate time, we shall focus on the cyclin-Cdk complex that is responsible for driving cells into mitosis.

MPF Is the Cyclin-Cdk Complex That Controls Entry into M Phase

The cyclin-Cdk complex that drives cells into M phase was first discovered through studies of cell division in frog eggs. The fertilized eggs of many animals are especially suitable for biochemical studies of the cell cycle because they are exceptionally large and divide rapidly. An egg of the frog *Xenopus*, for example, is just over 1 mm in diameter (Figure 18–5). After fertilization, it begins embryonic development with a series of rapid division cycles that serve to subdivide the egg into many smaller cells. These rapid cell cycles mainly consist of repeated S and M phases, with little or no G_1 or G_2 phases between them. There is no new gene transcription: all of the mRNAs, as well as most of the proteins, required for this early stage of embryonic development are already packed into the very large egg during its development as an oocyte in the ovary of the mother. In these early division cycles (called *cleavage divisions*), no cell growth occurs, and all the cells of the embryo divide in synchrony.

By taking frog eggs at a particular stage of the cell cycle, an extract can be prepared that is representative of that cell-cycle stage. The biological activity of such an extract can then be tested by injecting it into a *Xenopus* oocyte (the immature precursor of the unfertilized egg) and observing its effects on cell-cycle behavior. The *Xenopus* oocyte is a convenient test system for detecting an activity that drives cells into M phase, as it has completed DNA replication and is arrested just before M phase of the first meiotic division (discussed in Chapter 17). The oocyte is therefore at a stage in the cell cycle that is equivalent to the G_2 phase of a mitotic cell cycle.

In such experiments it was found that an extract from an M-phase egg instantly drives the oocyte into M phase, whereas cytoplasm from a cleaving egg at other phases of the cycle does not. When first discovered, the biochemical identity and mechanism of action of the factor responsible for this activity were unknown, and the activity was simply called *M-phase promoting factor,* or **MPF** (Figure 18–6). By testing cytoplasm

0.5 mm

Figure 18–5 **A mature *Xenopus* egg.** This giant cell is ready for fertilization. (Courtesy of Tony Mills.)

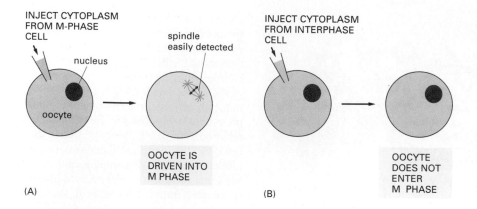

(A)

(B)

Figure 18–6 Experiment demonstrating MPF activity. (A) A *Xenopus* oocyte is injected with cytoplasm taken from a *Xenopus* egg in M phase. The cell extract drives the oocyte into M phase of the first meiotic division, causing the large nucleus to break down and a spindle to form. (B) When the cytoplasm is taken from a cleaving egg in interphase, it does not cause the oocyte to enter M phase. Thus the extract in (A) must contain some activity—M-phase promoting factor (MPF)—that triggers entry into M phase.

from different stages of the cell cycle, using the test shown in Figure 18–6, MPF activity was found to oscillate dramatically during the course of each cell cycle: it increased rapidly just before the start of mitosis and fell rapidly to zero toward the end of mitosis (Figure 18–7).

When MPF was finally purified, it was found to contain a single protein kinase, which is required for its activity. By phosphorylating key proteins, the kinase causes the chromosomes to condense, the nuclear envelope to break down, and the microtubules of the cytoskeleton to reorganize to form a mitotic spindle. The breakdown of the nuclear envelope, for example, is brought about by the disassembly of the nuclear lamina—the underlying shell of lamin filaments. As discussed in Chapter 17, the MPF kinase phosphorylates the lamin proteins, causing them to disassemble (see Figure 17–14). Similarly, the kinase phosphorylates microtubule-associated proteins, which alters the properties of the microtubules so that they form a mitotic spindle.

But the MPF kinase cannot act alone. As explained earlier, it has to have a specific cyclin bound to it in order to function. Cyclin binding is also thought to help guide the kinase to the proteins it will phosphorylate. The cyclins were first discovered in a different type of experiment, as we now describe.

Cyclin-dependent Protein Kinases Are Regulated by the Accumulation and Destruction of Cyclin

While intracellular injection experiments using frog eggs led to the discovery of MPF, biochemical experiments using cleaving clam eggs led to the discovery of cyclin. **Cyclin** was initially identified as a protein whose concentration rose gradually during interphase and then fell rapidly to zero as the cells went through M phase, repeating this performance in each cell cycle (Figure 18–8). As cyclin itself has no enzymatic activity, its role in cell-cyle control was initially obscure. The breakthrough occurred

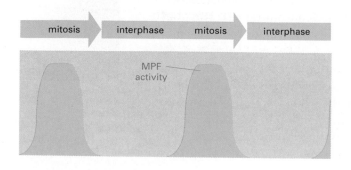

Figure 18–7 The oscillation of MPF activity during the cell cycle in *Xenopus* embryos. The activity assayed using the test outlined in Figure 18–6 rises rapidly just before the start of mitosis and falls rapidly to zero toward the end of mitosis. The same pattern of MPF activity is seen in all eucaryotic cells, from yeasts to humans.

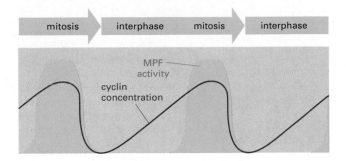

Figure 18–8 **The rise and fall in MPF activity and cyclin concentration during the cell cycle in a cleaving clam egg.** Whereas MPF activity rises and falls with each mitosis, cyclin concentration increases steadily during interphase, peaks at mitosis, and falls rapidly as mitosis ends. The concentration of the MPF kinase does not change during the course of the cell cycle; it is only its activity that changes.

when cyclin was found to be a component of MPF and to be required for the MPF kinase—the *mitotic Cdk*—to be active. Thus MPF is a protein complex containing two subunits—a regulatory subunit that is a cyclin and a catalytic subunit that is the mitotic Cdk. After the components of MPF were identified, other types of cyclins and Cdks were isolated whose concentrations or activities rise and fall at other stages in the cell cycle, as we discuss later.

While biochemists were identifying the proteins that regulate the cell cycles of frog and clam embryos, yeast geneticists were taking a different approach to dissecting the cell-cycle control system. Yeasts are unicellular fungi. Because they are relatively simple eucaryotes that multiply almost as fast as bacteria, they are favorite organisms for the genetic analysis of eucaryotic cell biology. Large numbers of yeast mutants that get stuck or misbehave at specific points in the cell cycle were isolated, and, using these mutants, many genes responsible for cell-cycle control were identified. Some of these genes turned out to encode cyclin and Cdk proteins, unmistakably similar—in both amino acid sequence and function—to their counterparts in frogs and clams. Similar genes were soon identified in human cells.

Many of the cell-cycle control genes have changed so little during evolution that the human version of the gene will function perfectly well in a yeast cell. For example, a yeast whose own copy of a particular *cdk* or *cyclin* gene is defective will fail to divide, but it will divide normally if a copy of the appropriate human gene is artificially introduced into the defective cell. Surely even Darwin would have been astonished at such clear evidence that humans and yeasts are cousins. Despite a billion years of divergent evolution, all eucaryotic cells (yeast, animal, or plant) use essentially the same molecules to control the events of their cell cycle.

The regulation of cyclin concentration plays an important part in timing the events of the cell cycle. Synthesis of the cyclin component of MPF, for example, starts immediately after cell division and continues steadily throughout interphase. The cyclin accumulates, so that its concentration rises gradually and helps time the onset of mitosis; its subsequent rapid decrease helps initiate the exit from mitosis (see Figure 18–8). The sudden fall in the cyclin concentration during mitosis is the result of the rapid destruction of the cyclin by the ubiquitin-dependent proteolytic system. Multiple molecules of ubiquitin are covalently attached to each of the cyclin molecules, targeting the cyclin for degradation in proteosomes (discussed in Chapter 7). This cyclin ubiquitination is an indirect result of the activation of the MPF kinase itself, as MPF activation initiates a process that—with a built-in delay—leads to the ubiquitination and degradation of the cyclin, turning the kinase off.

The Cell-Cycle Control System

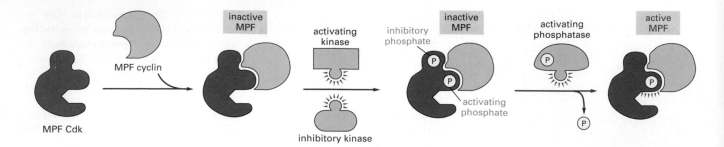

The Activity of Cdks Is Further Regulated by Their Phosphorylation and Dephosphorylation

Cyclin concentration increases gradually throughout interphase, whereas MPF kinase activity switches on abruptly at the end of interphase (see Figure 18–8). Thus, although cyclin is required for the activation of the MPF kinase, it cannot be the whole story. The kinase itself also has to be phosphorylated at one or more sites and dephosphorylated at others before it is enzymatically active. The removal of the inhibitory phosphate groups by a specific protein phosphatase is the step that activates the kinase at the end of interphase (Figure 18–9). Once activated, a cyclin-Cdk complex can activate more cyclin-Cdk complexes, as illustrated in Figure 18–10. This positive feedback produces the sudden, explosive increase in MPF kinase activity that drives the cell abruptly into M phase.

Different Cyclin-Cdk Complexes Trigger Different Steps in the Cell Cycle

There are many varieties of cyclin and, in most eucaryotes, many varieties of Cdk that are involved in cell-cycle control. Different cyclin-Cdk complexes trigger different steps of the cell cycle. Whereas the MPF cyclin-Cdk complex acts in G_2 to trigger entry into M phase, for example, a distinct set of other cyclins, called *S phase cyclins*, bind to Cdk molecules late in G_1 to trigger entry into S phase (Figure 18–11). Yet other cyclins, called *G_1 cyclins*, act earlier in G_1; they bind to Cdk molecules to help initiate the formation and activation of the S phase cyclin-Cdk complexes and thereby drive the cell toward S phase. We shall see later that the formation of G_1 cyclin-Cdk complexes in animal cells usually depends on extracellular growth factors that stimulate cells to divide.

The concentration of each type of cyclin rises and then falls sharply at a specific time in the cell cycle as a result of degradation by the ubiquitin pathway. The rise in concentration of each type of cyclin helps to activate its Cdk partner, while the rapid fall returns that Cdk to its inactive state (Figure 18–12). Thus the slow accumulation of a cyclin to a critical level is one way the cell-cycle control system measures the time intervals between one cell-cycle step and the next.

Figure 18–10 Activated MPF indirectly activates more MPF. Once activated, the MPF kinase phosphorylates, and thereby activates more activating phosphatase, which can now activate more MPF. Although not shown, activated MPF also inhibits the inhibitory kinase shown in Figure 18–9, and this also promotes the activation of MPF. In these ways activated MPF indirectly activates more MPF, so that the activation of MPF occurs explosively.

Figure 18–9 The activation of MPF. The MPF cyclin-Cdk complex is enzymatically inactive when first formed. Subsequently, the Cdk is phosphorylated at sites that are required for its activity and at other overriding sites that inhibit its activity. At this point the MPF remains inactive. As shown, it is finally activated by a phosphatase that removes the inactivating phosphate groups. It is still not clear how the timing of this activation is controlled.

Question 18–2 A small amount of cytoplasm isolated from a mitotic cell is injected into an unfertilized frog oocyte, causing the oocyte to enter M phase. A sample of its cytoplasm is then taken and injected into a second oocyte, causing this cell also to enter M phase. The process is repeated many times until essentially none of the original protein sample remains, and yet, cytoplasm taken from the last in the series of treated eggs is still able to trigger entry into M phase with undiminished efficiency. Explain this remarkable observation.

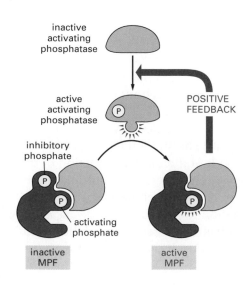

Chapter 18 : Cell-Cycle Control and Cell Death

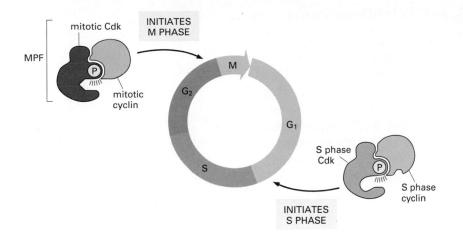

Figure 18–11 Two cyclin-Cdk complexes that act at different steps in the cell cycle. The S phase cyclin-Cdk complex shown drives the cell into S phase, whereas the mitotic cyclin-Cdk complex (MPF) drives the cell into M phase.

As is the case with MPK kinase, each of the different Cdks also has to be phosphorylated and dephosphorylated appropriately in order to act. Each cyclin-Cdk complex acts on a different set of target proteins in the cell. As a result, each type of complex triggers a different transition step in the cycle.

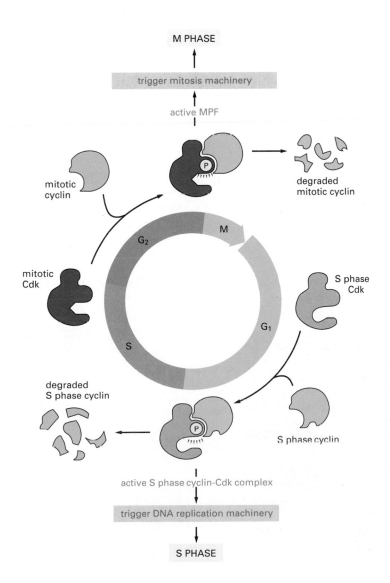

Figure 18–12 The regulation of Cdks by cyclin degradation. Only two types of cyclin-Cdk complexes are shown, one that triggers S phase and one that triggers M phase. In both cases the activation of the Cdk requires cyclin binding (as well as phosphorylation and dephosphorylation, as shown in Figure 18–9), and its inactivation depends on cyclin degradation.

The Cell Cycle Can Be Halted in G₁ by Cdk Inhibitor Proteins

We have seen that the cell-cycle control system triggers the events of the cycle in a specific order. It triggers mitosis, for example, only after all of the DNA is replicated, and it permits the cell to divide in two only after mitosis is completed. If one of the steps is delayed, the control system delays the activation of the next steps so that the sequence is maintained. This self-regulating property of the control system ensures, for example, that if DNA synthesis is halted for some reason during S phase, the cell will not proceed into M phase with its DNA only half replicated. As mentioned earlier, the control system accomplishes this feat through the action of molecular brakes that can stop the cell cycle at specific *checkpoints,* allowing the cell to monitor its internal state and its environment before continuing on in the cycle (see Figure 18–3).

For the most part the molecular mechanisms responsible for stopping cell-cycle progression at checkpoints are poorly understood. In some cases, however, specific **Cdk inhibitor proteins** come into play; these block the assembly or activity of one or more cyclin-Cdk complexes. One of the best understood checkpoints stops the cell cycle in G₁ if DNA is damaged, helping to ensure that a cell does not replicate damaged DNA. By an unknown mechanism, DNA damage causes an increase in both the concentration and activity of a gene regulatory protein called *p53*. When activated, the p53 protein stimulates the transcription of a gene encoding a Cdk inhibitor protein called *p21*. This increases the concentration of the p21 protein, which binds to the S phase cyclin-Cdk complexes responsible for driving the cell into S phase and blocks their action (Figure 18–13). The arrest of the cell cycle

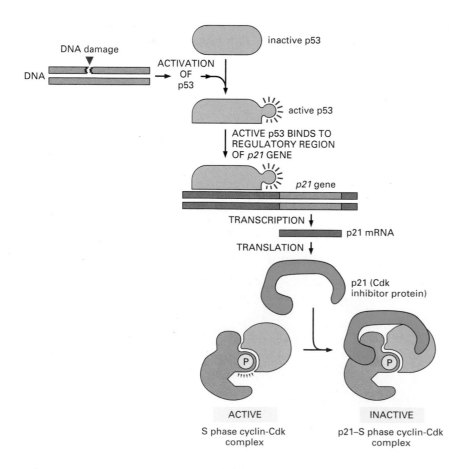

Figure 18–13 How p53 arrests the cell cycle in G₁. When DNA is damaged, the p53 protein increases (not shown), and the protein becomes activated. Activated p53 stimulates the transcription of the gene that encodes the Cdk inhibitor protein p21. The p21 protein binds to S phase cyclin-Cdk complexes and inactivates them, so that the cell cycle arrests in G₁. It is not known how DNA damage activates p53.

in G_1 allows the cell time to repair the damaged DNA before replicating it. If p53 is missing or defective, the unrestrained replication of damaged DNA leads to a high rate of mutation and the production of cells that tend to become cancerous. In fact, mutations in the *p53* gene that permit cells with damaged DNA to divide play an important part in the development of most human cancers, as we discuss later.

Cells Can Dismantle Their Control System and Withdraw from the Cell Cycle

The most radical decision that the cell-cycle control system has to implement is the decision for the cell to stop dividing altogether. This is a different matter from pausing in the middle of a cycle to cope with a temporary delay, and it has a special importance in multicellular organisms. In the human body, for example, nerve cells and skeletal muscle cells have to persist for a lifetime without dividing; they enter a modified G_1 state called G_0 (G zero), in which the cell-cycle control system is partly dismantled in that many of the Cdks and cyclins disappear. Some cell types, such as liver cells, normally divide only once every year or two, while certain epithelial cells in the gut divide more than twice a day in order to renew the lining of the gut continually. Most of our cells fall somewhere in between: they can divide if the need arises but normally do so infrequently.

It seems to be a general rule that mammalian cells will multiply (*proliferate*) only if they are stimulated to do so by signals from other cells. If deprived of such signals, the cell cycle arrests at a G_1 checkpoint and enters the G_0 state. Cells can remain in G_0 for days, weeks, or even years before dividing again. Most of the variation in cell-division rates in the adult body thus lies in the variation in the time cells spend in G_0 or in G_1; once past the G_1 checkpoint, a cell usually proceeds through the rest of the cell cycle quickly—typically within 12–24 hours in mammals.

The G_1 checkpoint is therefore sometimes called *Start*, because passing it represents a commitment to complete a full division cycle, although a better name might be Stop (Figure 18–14). Some of the main checkpoints in the cell cycle are summarized in Figure 18–15. The starting and stopping of cell proliferation are fundamentally important in controlling cell numbers and bodily proportions in a multicellular organism. But, as we discuss in the next section, controls on cell division are only half the story. On the other side of the balance sheet lie other equally important controls that determine whether a cell lives, or whether it dies by suicide.

Figure 18–14 Decisions at the G_1 checkpoint. The cell can commit to completing another cell cycle, pause transiently until conditions are right, or withdraw from the cell cycle altogether and enter G_0.

Question 18–3 Why do you suppose cells have evolved a special G_0 state to exit the cell cycle, rather than just stopping in a G_1 state at a G_1 checkpoint?

Figure 18–15 Summary of some cell-cycle checkpoints. The *red T's* represent potential checks on progress of the cell-cycle control system arising either from intracellular processes that are not completed or are deranged or from an unfavorable extracellular environment (highlighted in *yellow*). The checkpoint indicated in M phase ensures that all of the chromosomes are aligned on the mitotic spindle before progressing into anaphase, where the daughter chromosomes separate and move to opposite poles of the spindle (discussed in Chapter 17).

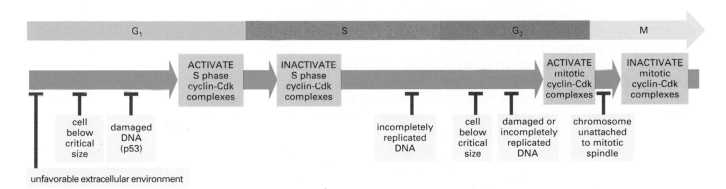

Control of Cell Numbers in Multicellular Organisms

A fertilized mouse egg and a fertilized human egg are similar in size, and yet an adult mouse is much smaller than an adult human. This difference in body size largely reflects a difference in cell numbers: an adult human is constructed from many more cells than an adult mouse. What is the difference in the control of cell behavior in humans and mice that generates such different numbers of cells? The same fundamental problem arises for each organ and tissue in an individual's body. What adjustment of cell behavior explains the length of an elephant's trunk or the size of its brain or its liver? These questions are largely unanswered, but it is at least possible to say what the ingredients of an answer must be. Cell proliferation, as well as cell survival and cell death, must be regulated by signals from other cells in the body (discussed in Chapter 15), combined with programs intrinsic to the individual cell.

In this section we first discuss the normal control of cell proliferation, cell survival, and cell death. We then consider how the faulty regulation of these processes can lead to cancer.

Cell Proliferation Depends on Signals from Other Cells

Unicellular organisms such as bacteria and yeasts tend to grow and divide as fast as they can, and their rate of proliferation depends largely on the availability of nutrients in the environment. The cells in a multicellular organism, by contrast, are specialized members of a highly organized community, and their proliferation must be controlled so that an individual cell divides only when another cell is required by the organism—either to allow growth or to replace cell loss. Thus, for an animal cell to proliferate, nutrients are not enough. It must also receive stimulatory chemical signals from other cells, usually its neighbors. These act to overcome intracellular braking mechanisms that tend to restrain cell growth and block progress through the cell cycle.

An important example of a brake that normally holds cell proliferation in check is the *Retinoblastoma (Rb) protein,* first identified through studies of a rare childhood eye tumor called retinoblastoma, in which the Rb protein is missing or defective. The Rb protein is abundant in the nucleus of all vertebrate cells. It binds to particular gene regulatory proteins, preventing them from stimulating the transcription of genes required for cell proliferation. Extracellular signals such as *growth factors* that stimulate cell proliferation lead to the activation of the G_1 cyclin-Cdk complexes mentioned earlier. These phosphorylate the Rb protein, altering its conformation so that it releases its bound gene regulatory proteins, which are then free to activate the genes required for cell proliferation to proceed (Figure 18–16).

The stimulatory signals that act to override the brakes on cell proliferation are mostly protein **growth factors.** These secreted signal proteins bind to cell-surface receptors, activating intracellular signaling pathways (discussed in Chapter 15) that stimulate cell growth and division. Most growth factors have been identified and characterized by their effects on cells growing in tissue culture (Figure 18–17). One of the first growth factors identified in this way was *platelet-derived growth factor,* or *PDGF,* whose effects are typical of many other growth factors.

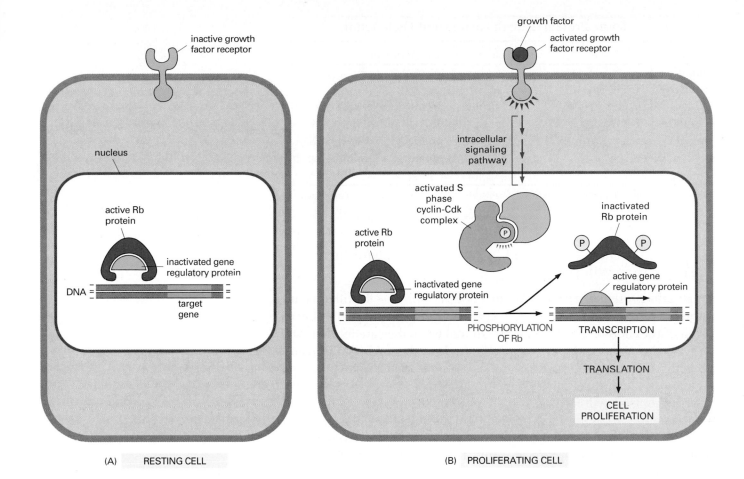

| (A) | RESTING CELL | (B) | PROLIFERATING CELL |

Figure 18–16 Simplified model of how growth factors stimulate cell proliferation. (A) In the absence of growth factors, dephosphorylated Rb protein holds specific gene regulatory proteins in an inactive state; these gene regulatory proteins are required to stimulate the transcription of target genes that encode proteins needed for cell proliferation. (B) Growth factors bind to cell-surface receptors and activate intracellular signaling pathways that lead to the formation and activation of the G_1 cyclin-Cdk complexes mentioned earlier. These complexes phosphorylate, and thereby inactivate, the Rb protein. The gene regulatory proteins are now free to activate the transcription of their target genes, leading to cell proliferation.

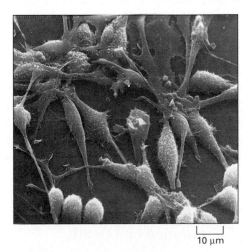

Figure 18–17 **Scanning electron micrograph of mammalian cells proliferating in culture.** The cells are rat fibroblasts, growing in the presence of calf serum, which contains growth factors that stimulate the cells to multiply. The spherical cells at the bottom of the micrograph have rounded up in preparation for cell division. (Courtesy of Guenter Albrecht-Buehler.)

When blood clots (in a wound, for example), blood platelets incorporated in the clot are triggered to release PDGF, which binds to receptor tyrosine kinases (discussed in Chapter 15) in surviving cells at the wound site, thereby stimulating them to proliferate and heal the wound. Similarly, if part of the liver is lost through surgery or acute injury, cells in the liver and elsewhere produce a protein called *hepatocyte growth factor*, which helps stimulate the surviving liver cells to proliferate.

Most animal cells require a specific combination of several growth factors to multiply in culture. Thus a relatively small number of growth factors can serve in different combinations to regulate selectively the proliferation of the many types of cells in an animal. Some protein growth factors and their actions are given in Table 18–1.

Table 18–1   Some Protein Growth Factors and Their Actions

Factor	Representative Actions*
Platelet-derived growth factor (PDGF)	stimulates proliferation of connective tissue cells
Epidermal growth factor (EGF)	stimulates proliferation of skin cells
Fibroblast growth factor (FGF)	stimulates proliferation of fibroblasts
Hepatocyte growth factor (HGF)	stimulates proliferation of liver cells
Erythropoietin	stimulates proliferation and differentiation of developing red blood cells

* Except for erythropoietin, all of the growth factors listed stimulate the proliferation of many additional cell types.

Animal Cells Have a Built-in Limitation on the Number of Times They Will Divide

Even in the presence of growth factors, normal animal cells will not continue dividing forever in culture. Cell types that maintain the ability to divide throughout the lifetime of the animal while they are in the intact body will usually stop dividing after a limited number of divisions in culture: fibroblasts taken from a human fetus, for example, will go through about 80 rounds of cell division before they stop, even if given ample nourishment, growth factors, and space to multiply. There is, however, a remarkable variation in the proliferation capacity of cells, depending on the age of the individual from whom the cells were derived. Fibroblasts taken from a 40-year-old adult, for example, stop after about 40 divisions.

This phenomenon is called **cell senescence** because of the correspondence with aging of the body as a whole, but its relationship to the aging of an organism is uncertain. Since fibroblasts from a mouse embryo halt their proliferation after only about 30 divisions in culture, it is possible that cell senescence may help determine body size. Perhaps mice are smaller than we are because their cells become insensitive to growth factor stimulation after fewer division cycles.

The mechanisms that halt the cell cycle in either developing or aging cells are largely mysterious, although the accumulation of Cdk inhibitor proteins and the loss of Cdks are likely to be involved. When we understand more about the molecular mechanisms that normally limit cell proliferation, it may become easier to establish the biological significance of phenomena such as cell senescence, and we may finally come to understand why we are bigger than mice.

> **Question 18–4** Besides the accumulation of Cdk inhibitors and the loss of Cdks discussed in the text, suggest any other plausible mechanisms by which cells could "know their age" and thereby limit the number of subsequent divisions.

Animal Cells Require Signals from Other Cells to Avoid Programmed Cell Death

Animal cells need signals from other cells not only to proliferate, but even to survive. If deprived of such **survival factors,** the cells activate an intracellular suicide program and die by a process called programmed cell death. This requirement for signals from other cells for survival helps to ensure that cells survive only when and where they are needed.

The amount of programmed cell death that occurs in both developing and adult tissues is astonishing. In the developing vertebrate nervous system, for example, more than half of the nerve cells normally die

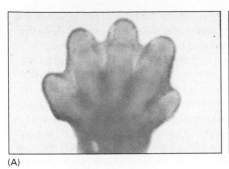 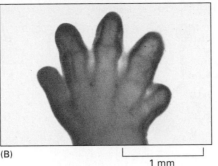

(A)   (B)

1 mm

Figure 18–18 Programmed cell death in the developing mouse paw sculpts the digits. (A) The paw shown has been stained with a dye that specifically labels (in *red*) cells that have undergone programmed cell death. These cell deaths eliminate the tissue between the developing digits, as seen in the paw shown one day later in (B). (From M.D. Jacobson et al., *J. Cell Biol.* 133:1041–1051, 1996.)

soon after they are formed. In a healthy adult human, billions of cells die in the bone marrow and intestine every hour. It seems remarkably wasteful for so many cells to die, especially as the vast majority are perfectly healthy at the time they kill themselves. What purposes does this massive cell death serve?

In some cases the answers are clear. Our hands and feet, for example, are sculpted by programmed cell death during embryonic development: they start out as spadelike structures, and the individual fingers and toes separate only as the cells between them die (Figure 18–18). In other cases, cells die when the structure they form is no longer needed. When a tadpole changes into a frog at metamorphosis, the cells in the tail die, and the tail, which is not needed in the frog, disappears (Figure 18–19). In still other cases cell death helps regulate cell numbers. In the developing nervous system, for example, cell death adjusts the number of nerve cells to match the number of target cells that require innervation. Nerve cells are produced in the embryo in excess and then compete for limiting amounts of survival factors secreted by the target cells they contact. Nerve cells that receive enough survival factor live, while the others die (Figure 18–20).

In adult tissues, cell death balances cell proliferation so as to keep the tissue from growing or shrinking. As mentioned earlier, if part of the liver is removed in an adult rat, liver cell proliferation increases to make up the loss, in part stimulated by an increased production of hepatocyte growth factor. Conversely, if a rat is treated with the drug phenobarbital—which stimulates liver cell division (and thereby liver enlargement) by an unknown mechanism—and then the phenobarbital treatment is stopped, cell death in the liver greatly increases until the liver has returned to its original size, usually within a week or so. Thus, the liver is kept at a constant size through the regulation of both the cell death rate and the cell birth rate.

Programmed Cell Death Is Mediated by an Intracellular Proteolytic Cascade

Cells that die as a result of acute injury typically swell and burst and spill their contents all over their neighbors (a process called *cell necrosis*),

Figure 18–19 Programmed cell death occurs during the metamorphosis of a tadpole into a frog. As a tadpole changes into a frog, the cells in the tadpole tail are induced to undergo programmed cell death, and as a consequence, the tail is lost. All of the changes that occur during metamorphosis, including the induction of programmed cell death in the tail, are stimulated by an increase in thyroid hormone in the blood.

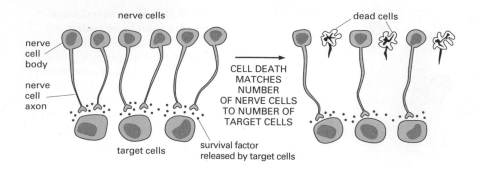

nerve cells

nerve cell body

nerve cell axon

target cells

survival factor released by target cells

CELL DEATH MATCHES NUMBER OF NERVE CELLS TO NUMBER OF TARGET CELLS

dead cells

Figure 18–20 Cell death helps match the number of developing nerve cells to the number of target cells they contact. More nerve cells are produced than can be supported by the limited amount of survival factor released by the target cells. Therefore some cells receive insufficient amounts of survival factor to keep their suicide program suppressed and, as a consequence, undergo programmed cell death. This strategy of overproduction followed by culling ensures that all target cells are contacted by nerve cells and that the "extra" nerve cells are automatically eliminated.

causing a potentially damaging inflammatory response. By contrast, a cell that undergoes **programmed cell death** (also called **apoptosis**) dies neatly, without damaging its neighbors. The cell shrinks and condenses. The cytoskeleton collapses, the nuclear envelope disassembles, and the nuclear DNA breaks up into fragments. Most important, the cell surface is altered, displaying properties that cause the dying cell to be phagocytosed immediately, either by its neighbors or by a macrophage (a specialized phagocytic cell, discussed in Chapter 14), before there is any leakage of its contents (Figure 18–21).

The machinery that is responsible for this kind of controlled cell suicide seems to be similar in all animal cells. It involves a family of proteases (enzymes that cut up other proteins), which are themselves activated by proteolytic cleavage in response to signals that induce programmed cell death. The activated suicide proteases cleave, and thereby activate, other members of the family, resulting in an amplifying proteolytic cascade. The activated proteases then cleave other key proteins in the cell, killing it quickly and neatly. One of the proteases, for example, cleaves the nuclear lamins, causing the irreversible breakdown of the nuclear lamina (Figure 18–22).

The suicide machinery is regulated by signals from other cells. Some act as killer signals, activating the suicide machinery of the cell. Thyroid

Figure 18–21 Cell death. Electron micrographs showing cells that have died by necrosis (A) or programmed cell death (B and C). The cells in (A) and (B) died in tissue culture, while the cell in (C) died in a developing tissue and has been engulfed by a neighboring cell. Note that the cell in (A) seems to have exploded, while those in (B) and (C) have condensed but seem relatively intact. The large vacuoles seen in the cytoplasm of the cell in (B) are a variable feature of programmed cell death. (Courtesy of Julia Burne.)

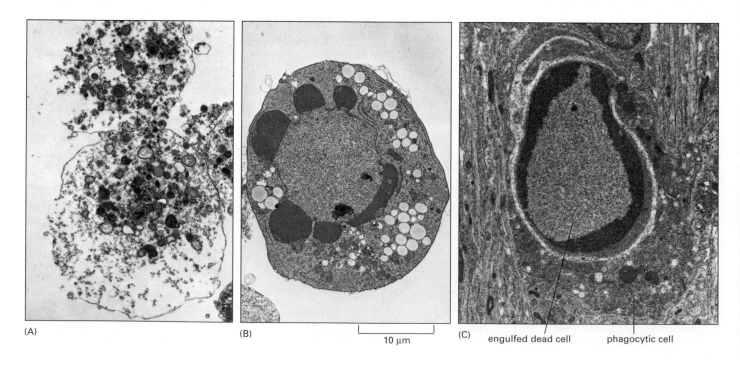

(A)

(B)

10 μm

(C) engulfed dead cell   phagocytic cell

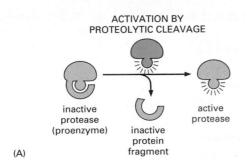

ACTIVATION BY
PROTEOLYTIC CLEAVAGE

inactive
protease
(proenzyme)

inactive
protein
fragment

active
protease

(A)

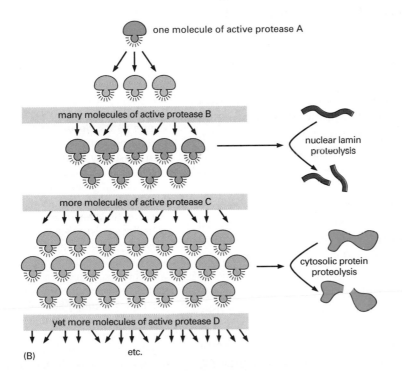

one molecule of active protease A

many molecules of active protease B

nuclear lamin
proteolysis

more molecules of active protease C

cytosolic protein
proteolysis

yet more molecules of active protease D

etc.

(B)

Figure 18–22 Programmed cell death is mediated by a proteolytic cascade. (A) Each suicide protease is made as an inactive proenzyme, which is itself activated by proteolytic cleavage by another member of the same protease family. (B) Each activated protease molecule can then cleave many proenzyme molecules, thereby activating them, and these can then activate even more protease molecules. In this way, an initial activation of a small number of protease molecules can lead, via an amplifying chain reaction (a cascade), to the explosive activation of a large number of protease molecules. Some of the activated proteases then break down a number of key proteins in the cell, such as nuclear lamins, leading to the controlled death of the cell.

hormone acts in this way in the tadpole tail at metamorphosis (see Figure 18–19). Others act as survival signals, suppressing the suicide machinery so as to keep the cell alive (see Figure 18–20).

For a multicellular organism, programmed cell death is a commonplace, normal, and generally benign event. It is the inappropriate proliferation and survival of cells that presents real dangers, as we now discuss.

Cancer Cells Disobey the Social Controls on Cell Proliferation and Survival

Cancers are the product of mutations that set cells free from the usual controls on cell proliferation and survival. A cell in the body mutates through a series of chance accidents and acquires the ability to proliferate without the normal restraints. Its progeny inherit the mutations and give rise to a tumor that can grow without limit.

As we discuss in the next chapter, faulty control of cell proliferation is not the only defect in cancer cells, but it is a central and essential feature. The mutations that make cancer cells defective in this respect affect two broad categories of genes: *proliferation genes*, which encode proteins that normally help to promote cell division, and *antiproliferation*

Question 18–5 Why do you think programmed cell death occurs by a different mechanism from the cell death that occurs in necrosis? What might be the consequences if programmed cell death were not achieved in so neat and orderly a fashion, in which the cell destroys itself from within and avoids leakage of the cell contents into the extracellular space?

genes, which encode proteins that normally help to apply the brakes that halt the cell cycle at the checkpoints discussed earlier.

A mutation in a proliferation gene that causes the protein produced by the gene to be overexpressed or hyperactive results in excessive cell multiplication. The mutant gene is then classified as an **oncogene** (that is, a cancer-promoting gene), while the normal gene is known as a *proto-oncogene.* Conversely, a mutation that inactivates an antiproliferation gene can release a cell from the normal restraints on cell multiplication and thereby also result in excessive cell proliferation. Thus the anti-proliferation genes present in normal cells are often referred to as **tumor-suppressor genes.** In a normal diploid cell there are two copies of each tumor-suppressor gene; both copies of the gene must typically be lost or inactivated to bring about the loss of proliferation control, as a single copy is usually enough for normal regulation of the cell cycle. In contrast, only one copy of a proto-oncogene needs to be mutated into an oncogene to bring about a similar effect.

Proto-oncogenes and tumor-suppressor genes are of many sorts, coding for different types of proteins. An oncogene can be created, for example, from any of the genes encoding proteins of the signaling pathways involved in the response of cells to growth factors (discussed in Chapter 15) through a mutation leading to the production of an abnormally active protein. The protein may be abnormally active because it is produced (1) by cells that normally do not make it, (2) in excessive amounts, or (3) in a form whose activity is uncontrolled. Such mutated genes can promote cancer by encouraging cells to proliferate, even in the absence of appropriate extracellular signals. One oncogene, for example, is produced from the gene for the growth factor PDGF by mutation of the control sequence governing its expression. Cells that carry this oncogene produce normal PDGF inappropriately, and if they also express PDGF receptors they will continuously stimulate themselves to proliferate. Other oncogenes encode hyperactive growth factor receptors, or intracellular signaling proteins (discussed in Chapter 15)

Figure 18–23 Normal cell proliferation compared to unrestrained cell proliferation caused by an oncogene. (A) In a normal resting cell the intracellular signaling proteins and genes that are normally activated by extracellular growth factors are inactive. (B) When the normal cell is stimulated by an extracellular growth factor, these signaling proteins and genes become active and the cell proliferates. (C) In this cancer cell, a mutation in a proto-oncogene that encodes an intracellular signaling protein that is normally activated by extracellular growth factors has created an oncogene. The oncogene encodes an altered form of the signaling protein that is active even in the absence of growth factor binding.

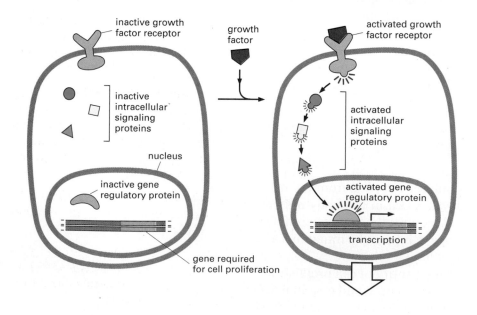

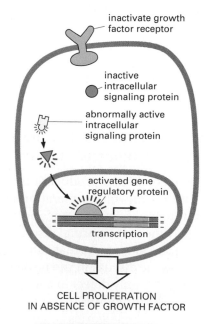

(A) NORMAL RESTING CELL

(B) NORMAL PROLIFERATING CELL

(C) PROLIFERATING CANCER CELL

that are hyperactive; cells expressing these oncogenes behave as though they are constantly receiving a signal to proliferate (Figure 18–23). Similarly, mutations affecting proteins that regulate cell survival and cell death can promote cancer by enabling the cell to survive in the absence of appropriate survival signals.

An example of a tumor-suppressor gene is the retinoblastoma (Rb) gene mentioned on page 582: inactivation of both copies of this gene removes an important brake on the cell cycle, drastically increasing the likelihood of cancer in certain susceptible tissues, especially the retina.

These examples show how the molecular lesions that underlie cancer are beginning to be understood in detail in terms of the mechanisms that regulate normal cell division, cell survival, and cell death. Rapid advances in understanding the molecular genetics of cancer promise to lead to greatly improved treatments of the disease, although this has not happened yet. For a full understanding of cancer, however, or indeed of almost any other human disease or bodily function, it is essential to go beyond molecular genetics and the individual cell and to examine more carefully how cells live together, interact, and cooperate as members of the large, highly organized cellular communities that constitute the tissues and organs of the body. This is the topic of the next and final chapter of the book.

Question 18–6 Explain why a mutation in only one of the two copies of a proto-oncogene in a diploid cell can promote the development of cancer, whereas both copies of a tumor-suppressor gene need to be mutated to do likewise.

Essential Concepts

- The cell-cycle control system coordinates the events of the cell cycle by cyclically switching on the appropriate parts of the cell-cycle machinery and then switching them off.

- The control system consists mainly of a set of protein complexes, each composed of a regulatory subunit called a cyclin and a catalytic subunit called a cyclin-dependent protein kinase (Cdk).

- The Cdks are cyclically activated by both cyclin binding and the phosphorylation of some amino acids and the dephosphorylation of others; when activated, Cdks phosphorylate key proteins in the cell.

- Cyclin concentrations rise and fall at specific times in the cell cycle, providing a timing mechanism; the rise results from steady synthesis, while the sudden fall results from rapid proteolysis.

- Different cyclin-Cdk complexes trigger different steps of the cell cycle: MPF, the mitotic cyclin-Cdk complex, drives the cell into mitosis; S phase cyclin-Cdk complexes drive the cell into S phase.

- The cell-cycle control system can halt the cycle at specific checkpoints to ensure that the next step in the cycle does not begin before the previous one has finished.

- The cell cycle can halt by at least two mechanisms: (1) Cdk inhibitor proteins can block the assembly or activity of one or more cyclin-Cdk complexes, or (2) components of the control system can stop being made.

- Animal cell numbers are regulated by a combination of intracellular programs and intercellular interactions (social controls) that control cell proliferation, cell survival, and cell death.

- Animal cells proliferate only if stimulated by growth factors, which activate intracellular signaling pathways to override the normal brakes that otherwise block cell-cycle progression; this mechanism ensures that a cell divides only when another cell is needed.

- Most normal animal cells have an internal mechanism of unknown nature that limits the number of times they can divide.

- Many normal cells die during the lifetime of an animal by activating an internal suicide program and killing themselves—a process called programmed cell death, or apoptosis.

- Programmed cell death depends on a family of proteolytic enzymes, which are themselves activated by a proteolytic cascade.

- Most animal cells require continuous signaling from other cells to avoid programmed cell death; this may be a mechanism to ensure that cells survive only when and where they are needed.

- Cancer results from an accumulation of mutations that activate proliferation-promoting genes (proto-oncogenes) and inactivate proliferation-suppressing genes (tumor-suppressor genes) in a single cell and its progeny, which therefore proliferate without restraint.

Key Terms

cancer

Cdk

Cdk inhibitor protein

cell senescence

checkpoint

cyclin

growth factor

MPF

oncogene

programmed cell death
    (apoptosis)

survival factor

tumor-suppressor gene

Questions

Question 18–7 Which of the following statements are correct? Explain your answers.

A. Cells do not pass from G_1 into M phase of the cell cycle unless there are sufficient nutrients to complete an entire cell cycle.

B. Programmed cell death is mediated by special intracellular proteases, one of which cleaves nuclear lamins.

C. Developing neurons compete for limited amounts of survival factors.

D. Some vertebrate cell-cycle control proteins function when expressed in yeast cells.

E. It is possible to study yeast mutants that are defective in cell-cycle control proteins, despite the fact that these proteins are essential for the cells to live.

F. Both the presence of a bound cyclin and its phosphorylation state determine whether a Cdk protein is enzymatically active.

G. When mutated, tumor-suppressor genes can turn into oncogenes, thus causing cancer.

Question 18–8 One of the functions of the mitotic Cdk (the MPF protein kinase) is to cause a precipitous drop in cyclin concentration halfway through M phase. Describe the consequences of this sudden decrease and suggest possible mechanisms by which it might occur.

Question 18–9 Compare the rules of cell proliferation in an animal to the rules that govern human behavior in society. What would happen to an animal if its cells behaved like people normally behave in our society? Could the rules that govern cell proliferation be applied to human societies?

Question 18–10 Figure 18–8 shows the rise of cyclin concentration and the rise of MPF activity in cells as they progress through the cell cycle. It is remarkable that the cyclin concentration rises slowly and steadily, whereas MPF activity increases suddenly. How do you think this difference arises?

Question 18–11 In his highly classified research laboratory Dr. Lawrence M. is charged with the task of developing a strain of dog-sized rats to be deployed behind enemy lines. In your opinion, which of the following strategies should Dr. M. pursue to increase the size of rats?

A. Delay cell senescence.

B. Block programmed cell death.

C. Block p53 function.

D. Overproduce survival factors.

E. Overproduce growth factors.

F. Obtain a taxi driver's license and switch careers.

Explain the likely consequences of each option.

Question 18–12 If PDGF is encoded by a proto-oncogene and can cause cancer when expressed inappropriately, why do cancers not arise at wounds when PDGF is released from platelets?

Question 18–13 One important biological effect of a large dose of ionizing radiation is to halt cell division.

A. How does this occur?

B. What happens if a cell has a mutation that prevents it from halting cell division after being irradiated?

C. What might be the effects of such a mutation if the cell is not irradiated?

D. An adult human who has reached maturity will die within a few days of receiving a radiation dose large enough to stop cell division. What does that tell you (other than that one should avoid large doses of radiation)?

Question 18–14 What do you suppose happens in mutant cells that . . .

A. cannot degrade cyclins?

B. always express high levels of p21?

C. cannot phosphorylate Rb?

Question 18–15 Many mutant yeasts have been isolated that are defective in the control of their cell cycle. They proliferate normally at low temperatures (30°C) but show abnormal patterns of cell growth and division when grown at a higher temperature (37°C). Two mutant strains (called "gee" and "wee") with defects at different sites in the same gene have very different responses to elevated temperature. Gee strain cells grow until they become enormous but no longer divide. Wee strain cells have very short cell cycles and divide when they are very much smaller than usual. Suggest a possible model to explain these observations, and suggest what the normal protein encoded by this gene might do.

Question 18–16 If cells are grown in a medium containing radioactive thymidine, the thymidine will be covalently incorporated into the cell's DNA during S phase. The radioactive DNA can be detected in the nuclei of individual cells by autoradiography (i.e., by placing a photographic film over the cells, radioactive cells will activate the film and show up as black dots when looked at in a microscope). Consider a simple experiment in which cells are radioactively labeled by this method for only a short amount of time (about 30 minutes). The radioactive thymidine medium is then replaced with one containing unlabeled thymidine,

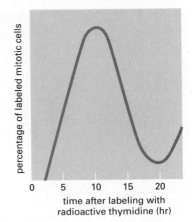

Figure Q18–16

and the cells are grown for some additional time. At different time points after replacement of the medium, cells are examined in a microscope. The fraction of cells in mitosis (that can be easily recognized because their chromosomes are condensed) that have radioactive DNA in their nuclei is then determined and plotted as a function of time after the labeling with radioactive thymidine (Figure Q18–16).

A. Would all cells (including cells at all phases of the cell cycle) contain radioactive DNA after the labeling procedure?

B. Note that initially there are no mitotic cells that contain radioactive DNA. Why is this?

C. Explain the rise and fall of the curve.

D. Estimate the length of the G_2 phase from these data.

Question 18–17 Liver cells proliferate both in patients with alcoholism and in patients with liver tumors. What are the differences in the mechanisms by which cell proliferation is induced in either disease?

Question 18–18 Look carefully at the electron micrographs in Figure 18–21. Describe the differences between the cell that died by necrosis and those that died by programmed cell death. How do the pictures confirm the differences between the two processes? Explain your answer.

Question 18–19 Could cancer be irradicated if tumor-suppressor genes could be expressed at abnormally high concentrations in all human cells? Explain your answer.

A tight cluster of developing flower buds in the plant *Arabidopsis* is seen here in the scanning electron microscope. At this stage each flower, with its petals, stamens, and carpels, is wrapped tightly in four interlocking sepals whose individual epidermal cells are clearly visible. These cells, of differing shapes and sizes, are attached tightly to their neighbors by their cell walls (or extracellular matrix) to form a coherent tissue. (Courtesy of Kim Findlay.)

19 Tissues

Cells are the building blocks of multicellular organisms. This seems a simple statement, but it raises deep problems. Cells are not like bricks: they are small and squishy. How can they be used to construct a giraffe or a giant redwood tree? Each cell is enclosed in a flimsy membrane less than a hundred-thousandth of a millimeter thick, and it depends on the integrity of this membrane for its survival. How then can cells be joined together robustly, with their membranes intact, to form a muscle that will lift an elephant's weight? Most mysterious of all, if cells are the building blocks, where is the builder and where are the architect's plans? How are all the different cell types in a plant or an animal produced, with each in its proper place in an elaborate pattern (Figure 19–1)?

Most of the cells in multicellular organisms are organized into cooperative assemblies called **tissues,** such as the nervous, muscle, epithelial, and connective tissues found in vertebrates (Figure 19–2). In this chapter we begin by discussing the architecture of tissues from a mechanical point of view. We shall see that tissues are composed not only of cells, with their internal framework of cytoskeletal filaments (Chapter 16), but also of **extracellular matrix,** which the cells secrete around themselves; it is this matrix that gives supportive tissues their strength. Cells can be bound together via the extracellular matrix, or directly by attachment to one another. We shall discuss the **cell junctions** that link cells together in the flexible, motile tissues of animals, transmitting forces from the cytoskeleton of one cell to that of the next, or from the cytoskeleton of a cell to the extracellular matrix.

But there is more to the organization of tissues than mechanics. Just as a building needs plumbing, telephone lines, and other fittings, so an animal tissue requires blood vessels, nerves, and other components formed from a variety of specialized cell types. All the tissue components have to be coordinated correctly, and many of them require continual maintenance and renewal. Cells die, and have to be replaced with new cells of the right type, in the right places, and in the right numbers. Disorders of tissue renewal are a major medical concern, and those due to the misbehavior of mutant cells are responsible for cancer.

The problem of how all the specialized cell types arise and how they become organized into the tissues of the body will bring us to one of the most ancient and fundamental questions in biology: how is a whole multicellular organism generated from a single fertilized egg? This problem of **development** is a complex one, but an answer can be given, and in the closing pages of the book we shall take a brief glimpse at its outline.

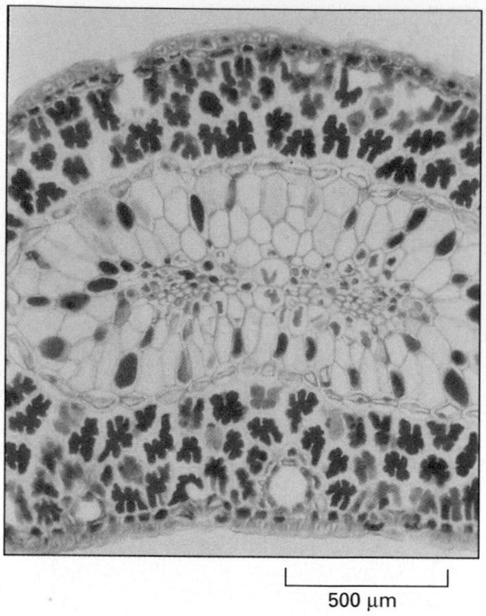

Extracellular Matrix and Connective Tissues

Plants and animals have evolved their multicellular organization independently, and their tissues are constructed on different principles. Animals are predators on other living things, and for this they must be strong and agile: they must possess tissues capable of rapid movement, and the cells that form those tissues must be able to generate and transmit forces and to change shape quickly. Plants, by contrast, are sedentary, their tissues are more or less rigid, and their cells, taken in isolation, are weak and fragile.

The strength of a plant tissue comes from the **cell walls,** formed like boxes, that enclose, protect, and constrain the shape of each of its cells (Figure 19–3). The cell wall is a type of extracellular matrix that the plant cell secretes around itself. The cell controls its composition: it can be thick and hard, as in wood, or thin and flexible, as in a leaf. But the principle of tissue construction is the same in either case: many tiny boxes are cemented together, with a delicate cell living inside each one. Indeed, as we noted in Chapter 1, it was this close-packed mass of microscopic chambers, seen in a slice of cork by Robert Hooke three centuries ago, that originally gave rise to the term "cell."

Animal tissues are more diverse. Like plant tissues, they consist of extracellular matrix as well as cells, but these components are organized in many different ways. In some tissues, such as bone or tendon, extracellular matrix is plentiful and mechanically all-important; in others, such as muscle or epidermis, extracellular matrix is scanty, and the cytoskeleton of the cells themselves carries the mechanical load. We begin with a brief discussion of plants, before moving on to animals.

Plant Cells Have Tough External Walls

A naked plant cell, artificially stripped of its wall, is a weak and vulnerable thing. With care, it can be kept alive in culture; but it is easily rup-

500 μm

Figure 19–1 Stained cross section through a leaf of a pine tree (a pine needle), showing the precisely organized pattern of different cell types.

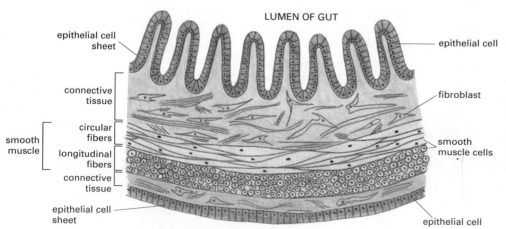

Figure 19–2 The organization of cells into tissues. Simplified drawing of a cross section through part of the wall of the intestine of a mammal. This long, tubelike organ is constructed from epithelial tissues (*red*), connective tissues (*green*), and muscle tissues (*yellow*). Each tissue is an organized assembly of cells held together by cell-cell adhesions, extracellular matrix, or both.

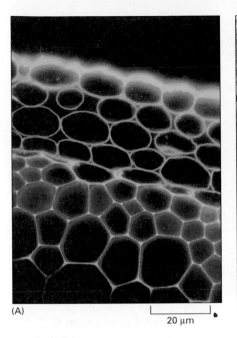

(A)

20 µm

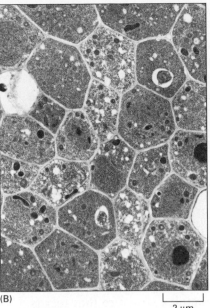

(B)

2 µm

Figure 19–3 **The plant cell wall.** (A) A cross section of part of the stem of the flowering plant *Arabidopsis* is shown, stained with fluorescent dyes that label two different cell wall components—cellulose in *blue,* and another polysaccharide (pectin) in *green.* The cells themselves are unstained and invisible in this preparation. (B) The cells and their walls are clearly seen in this electron micrograph of young cells in the root of the same plant. (Courtesy of Paul Linstead.)

tured, and even a small maladjustment of the osmotic strength of the culture medium can cause it to swell until it bursts. Its cytoskeleton lacks the tension-bearing intermediate filaments found in animal cells, and it has virtually no tensile strength. An external wall, therefore, is essential.

The plant cell wall has to be tough, but it does not necessarily have to be rigid. Osmotic swelling of the cell, limited by the resistance of the cell wall, can keep the chamber distended, and a mass of such swollen chambers cemented together forms a semirigid tissue (Figure 19–4). Such is the state of a crisp lettuce leaf. If water is lacking so that the cells shrink, the leaf wilts.

Most newly formed cells in a multicellular plant initially make relatively thin *primary cell walls* that are capable of slowly expanding to accommodate subsequent cell growth. The driving force for growth is the same that keeps the lettuce leaf crisp—a swelling pressure, called the

Figure 19–4 **Scanning electron micrograph of cells in a leaf.** Plump cells, swollen with water by osmotic forces, are stuck together via their walls to form a crisp leaf. (Courtesy of K. Findlay.)

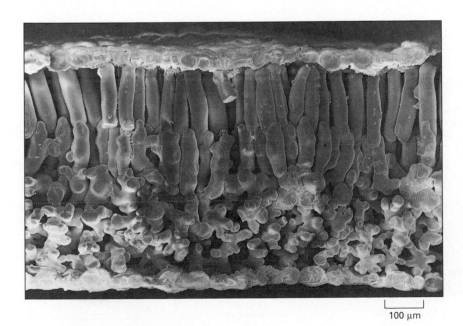

100 µm

Extracellular Matrix and Connective Tissues

turgor pressure, that develops due to osmotic imbalance between the interior of the cell and its surroundings (see Chapter 12). Once growth stops and the wall no longer needs to expand, a more rigid *secondary cell wall* is often produced, either by thickening of the primary wall or by deposition of new layers with a different composition underneath the old ones. When plant cells become specialized, they generally produce specially adapted types of walls: waxy, waterproof walls for the surface epidermal cells of a leaf; hard, thick, woody walls for the xylem cells of the stem; and so on, as shown in Panel 19–1 (pp. 598–599).

Cellulose Fibers Give the Plant Cell Wall Its Tensile Strength

Like all extracellular matrices, plant cell walls derive their tensile strength from long fibers oriented along the lines of stress. In higher plants the long fibers are generally made from the polysaccharide *cellulose,* the most abundant organic macromolecule on earth. The cellulose fibers are interwoven with other polysaccharides and some structural proteins, all bonded together to form a complex structure that resists compression as well as tension (Figure 19–5). In woody tissue, a highly cross-linked *lignin* network, composed of yet another class of molecules, is deposited within this matrix to make it more rigid and waterproof.

In order for a plant cell to grow or change its shape, the cell wall has to stretch or deform. Because the cellulose fibers resist stretching, their orientation governs the direction in which the growing cell enlarges: if, for example, they are arranged circumferentially as a corset, the cell will grow more readily in length than in girth (Figure 19–6). By the way that it lays down its wall, the plant cell consequently controls its own shaping and thus the direction of growth of the tissue to which it belongs.

Cellulose is produced in a radically different way from most other extracellular macromolecules. Instead of being made inside the cell and then exported by exocytosis (Chapter 14), it is synthesized on the outer surface of the cell by enzyme complexes embedded in the plasma membrane. These transport the sugar monomers across the membrane and

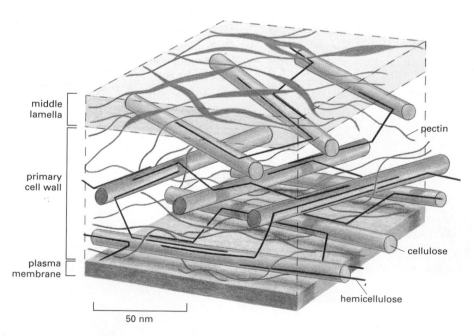

middle lamella

primary cell wall

plasma membrane

pectin

cellulose

hemicellulose

50 nm

Figure 19–5 **Scale model of a portion of a primary plant cell wall.** The *green bars* represent cellulose fibrils, providing tensile strength; other polysaccharides cross-link the cellulose fibrils (*red strands*) and fill the spaces between them (*blue strands*), providing resistance to compression. The middle lamella is the layer that cements one cell wall to another.

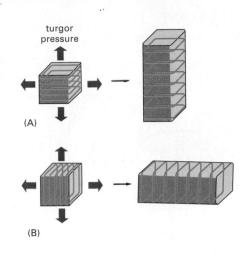

Figure 19–6 How the orientation of cellulose microfibrils within the cell wall influences the direction in which the cell elongates. The cells in (A) and (B) start off with identical shapes (shown here as *cubes*) but with different orientations of cellulose microfibrils in their walls. Although turgor pressure is uniform in all directions, each cell tends to elongate in a direction perpendicular to the orientation of the microfibrils, which have great tensile strength. The final shape of an organ, such as a shoot, is determined by the direction in which its cells expand.

incorporate them into a set of growing polymer chains at their points of membrane attachment. Each set of chains forms a cellulose microfibril. The enzyme complexes move in the membrane, spinning out new polymers and laying down a trail of oriented cellulose fibers behind them.

The paths followed by the enzyme complexes dictate the orientation in which cellulose is deposited in the cell wall; but what guides the enzyme complexes? Just underneath the plasma membrane, microtubules are aligned exactly with the cellulose fibers outside the cell. These microtubules are thought to serve as tracks to guide the movement of the enzyme complexes (Figure 19–7). In this curiously indirect way, the cytoskeleton controls the shaping of the plant cell and the

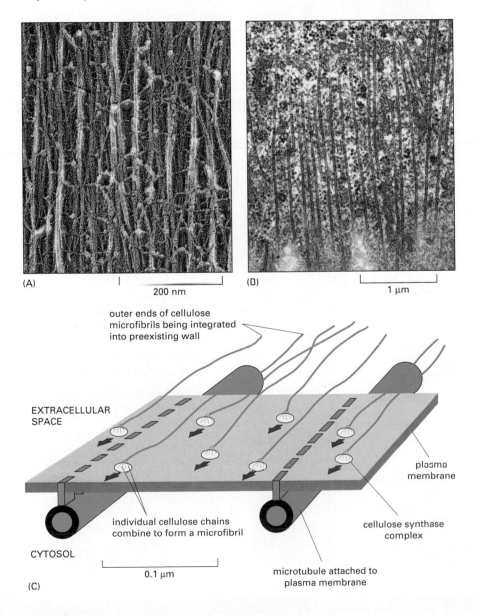

(A)
200 nm

(B)
1 μm

outer ends of cellulose microfibrils being integrated into preexisting wall

EXTRACELLULAR SPACE

plasma membrane

individual cellulose chains combine to form a microfibril

cellulose synthase complex

CYTOSOL

0.1 μm

microtubule attached to plasma membrane

(C)

Figure 19–7 Oriented deposition of cellulose in the plant cell wall. (A) Oriented cellulose fibrils in a plant cell wall, shown by electron microscopy. (B) Oriented microtubules just beneath a plant cell's plasma membrane. (C) One model of how the orientation of the newly deposited extracellular cellulose microfibrils might be determined by the orientation of the intracellular microtubules. The large *cellulose synthase* enzyme complexes are integral membrane proteins that continuously synthesize cellulose microfibrils on the outer face of the plasma membrane. The outer ends of the stiff microfibrils become integrated into the texture of the wall, and their elongation at the other end pushes the synthase complex along in the plane of the membrane. Because the cortical array of microtubules is attached to the plasma membrane in a way that confines the enzyme complex to defined membrane tracks, the microtubule orientation determines the direction in which the microfibrils are laid down. (A, courtesy of Brian Wells and Keith Roberts; B, courtesy of Brian Gunning.)

THE PLANT

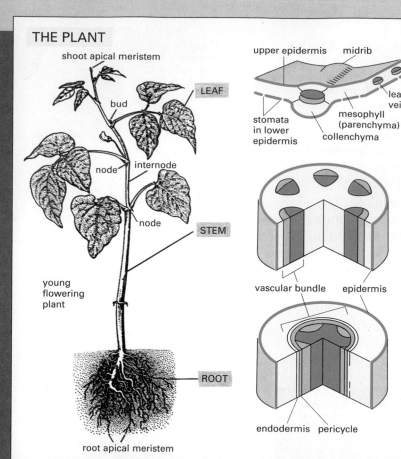

shoot apical meristem

bud

LEAF

node — internode

node

STEM

young flowering plant

ROOT

root apical meristem

upper epidermis — midrib

leaf vein

stomata in lower epidermis

mesophyll (parenchyma)

collenchyma

vascular bundle — epidermis

endodermis — pericycle

The young flowering plant shown on the *left* is constructed from three main types of organs: leaves, stems, and roots. Each plant organ in turn is made from three tissue systems: ground (▭), dermal (▬), and vascular (▭▬).

All three tissue systems derive ultimately from the cell proliferative activity of the shoot or root apical meristems, and each contains a relatively small number of specialized cell types. These three common tissue systems, and the cells that comprise them, are described in this panel.

THE THREE TISSUE SYSTEMS

Cell division, growth, and differentiation give rise to tissue systems with specialized functions.

DERMAL TISSUE: This is the plant's protective outer covering in contact with the environment. It facilitates water and ion uptake in roots and regulates gas exchange in leaves and stems.

VASCULAR TISSUE: Together the phloem and the xylem form a continuous vascular system throughout the plant. This tissue conducts water and solutes between organs and also provides mechanical support.

GROUND TISSUE: This packing and supportive tissue accounts for much of the bulk of the young plant. It also functions in food manufacture and storage.

GROUND TISSUE

The ground tissue system contains three main cell types called parenchyma, collenchyma, and sclerenchyma.

Parenchyma cells are found in all plant tissue systems. They are living cells, generally capable of further division, and have a thin primary cell wall. These cells have a variety of functions. The apical and lateral meristematic cells of shoots and roots provide the new cells required for growth. Food production and storage occur in the photosynthetic cells of the leaf and stem (called mesophyll cells); storage parenchyma cells form the bulk of most fruits and vegetables. Because of their proliferative capacity, parenchyma cells also serve as sources of new cells for wound healing and regeneration.

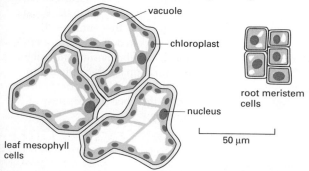

vacuole

chloroplast

root meristem cells

nucleus

leaf mesophyll cells

50 μm

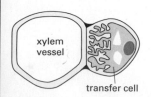

xylem vessel

transfer cell

A transfer cell, a specialized form of parenchyma cell, is readily identified by elaborate ingrowths of the primary cell wall. The increase in the area of the plasma membrane beneath these walls facilitates the rapid transport of solutes to and from cells of the vascular system.

Collenchyma are living cells similar to parenchyma cells except that they have much thicker cell walls and are usually elongated and packed into long ropelike fibers. They are capable of stretching and provide mechanical support in the ground tissue system of the elongating regions of the plant. Collenchyma cells are especially common in subepidermal regions of stems.

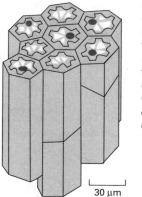

30 μm

typical locations of groups of supporting cells in a stem

sclerenchyma fibers

vascular bundle

collenchyma

Sclerenchyma, like collenchyma, have strengthening and supporting functions. However, they are usually dead cells with thick, lignified secondary cell walls that prevent them from stretching as the plant grows. Two common types are fibers, which often form long bundles, and sclereids, which are shorter branched cells found in seed coats and fruit.

fiber bundle

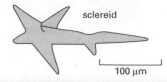

sclereid

10 μm

100 μm

Panel 19–1 **The cell types and tissues from which higher plants are constructed.**

DERMAL TISSUE

The epidermis is the primary outer protective covering of the plant body. Cells of the epidermis are also modified to form stomata and hairs of various kinds.

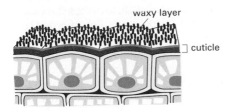

waxy layer

cuticle

The epidermis (usually one layer of cells deep) covers the entire stem, leaf, and root of the young plant. The cells are living, have thick primary cell walls, and are covered on their outer surface by a special cuticle with an outer waxy layer. The cells are tightly interlocked in different patterns.

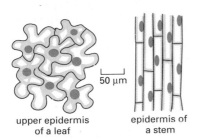

upper epidermis of a leaf

epidermis of a stem

50 μm

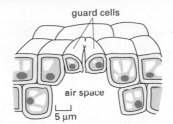

guard cells

air space

5 μm

Stomata are openings in the epidermis, mainly on the lower surface of the leaf, that regulate gas exchange in the plant. They are formed by two specialized epidermal cells called *guard cells,* which regulate the diameter of the pore. Stomata are distributed in a distinct species-specific pattern within each epidermis.

Vascular bundles

Roots usually have a single vascular bundle, but stems have several bundles. Each bundle is a complex of different types of cells providing pathways for fluid transport.

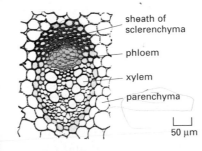

sheath of sclerenchyma

phloem

xylem

parenchyma

50 μm

a typical vascular bundle from the young stem of a buttercup

Hairs (or trichomes) are appendages derived from epidermal cells. They exist in a variety of forms and are commonly found in all plant parts. Hairs function in protection, absorption, and secretion; for example,

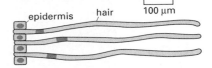

epidermis   hair   100 μm

young, single-celled hairs develop in the epidermis of the cotton seed. When these grow, the walls will be secondarily thickened with cellulose to form cotton fibers.

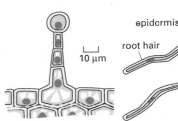

epidermis

root hair

10 μm

a multicellular secretory hair from a geranium leaf

Single-celled root hairs take up water and ions from the soil.

VASCULAR TISSUE

The phloem and the xylem together form a continuous vascular system throughout the plant. In young plants they are usually associated with a variety of other cell types in *vascular bundles.* Both phloem and xylem are complex tissues. Their conducting cells, or *elements,* are associated with parenchyma cells that maintain and exchange materials with the elements. Also, groups of collenchyma and sclerenchyma cells provide mechanical support.

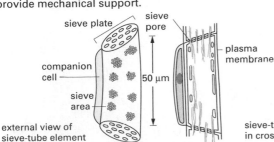

sieve plate

sieve pore

companion cell

sieve area

50 μm

plasma membrane

external view of sieve-tube element

sieve-tube element in cross-section

Phloem is involved in the transport of organic solutes in the plant. The main conducting elements are aligned to form tubes called *sieve tubes.* The sieve-tube elements at maturity are living cells, interconnected by perforations in their end walls formed from enlarged and modified plasmodesmata (sieve plates). These cells retain their plasma membrane, but they have lost their nuclei and much of their cytoplasm; they therefore rely on associated *companion cells* for their maintenance. These companion cells have the additional function of actively transporting soluble food molecules into and out of sieve-tube elements through porous sieve areas in the wall.

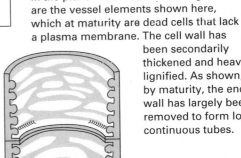

young vessel element in root tip

large, mature vessel element

Xylem carries water and dissolved ions in the plant. The main conducting cells are the vessel elements shown here, which at maturity are dead cells that lack a plasma membrane. The cell wall has been secondarily thickened and heavily lignified. As shown, by maturity, the end wall has largely been removed to form long, continuous tubes.

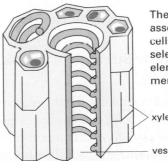

The vessel elements are closely associated with xylem parenchyma cells, which actively transport selected solutes into and out of the elements across their plasma membrane.

xylem parenchyma cells

vessel element

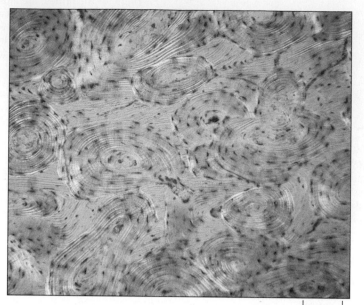

100 μm

Figure 19–8 **Cross section of bone.** The cells appear as small dark spidery objects, embedded in the bone matrix, which occupies most of the volume of the tissue and provides all its mechanical strength. The alternating light and dark bands are layers of matrix containing oriented collagen (made visible with the help of polarized light). Calcium phosphate crystals filling the interstices between the collagen fibrils make bone matrix resistant to compression as well as tension, like reinforced concrete.

modeling of the plant tissues. We shall see that animal cells use their cytoskeleton to control tissue architecture in a much more direct way.

Animal Connective Tissues Consist Largely of Extracellular Matrix

It is traditional to distinguish four major types of tissue in animals—connective tissues, epithelial tissues, nervous tissues, and muscle. But the basic architectural distinction is between connective tissues and the rest. In connective tissues, extracellular matrix is plentiful and carries the mechanical load. In other tissues, such as epithelia, extracellular matrix is scanty, and the cells are directly joined to one another and carry the mechanical load themselves. We discuss connective tissues first.

Animal connective tissues are enormously varied. They can be tough and flexible, like tendons or the dermis of the skin; hard and dense, like bone; resilient and shock-absorbing, like cartilage; or soft and transparent, like the jelly that fills the interior of the eye. In all these examples, the bulk of the tissue is occupied by extracellular matrix, and the cells that produce the matrix are scattered within it like raisins in a pudding (Figure 19–8). In all of these tissues, furthermore, the tensile strength—whether great or small—is provided not by a polysaccharide, as in plants, but by a fibrous protein—**collagen.** The various types of connective tissue owe their specific characters to the type of collagen that they contain, to its quantity, and, most importantly, to the other molecules that are interwoven with it in varying proportions.

Collagen Provides Tensile Strength in Animal Connective Tissues

Collagen is found in all multicellular animals, and it comes in many varieties. Mammals have about 20 different collagen genes, coding for the variant forms of collagen required in different tissues. Collagens are the chief proteins in bone, tendon, and skin (leather is pickled collagen),

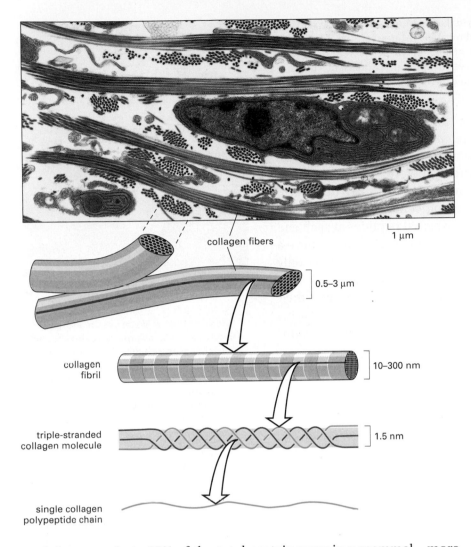

collagen fibers

0.5–3 μm

collagen
fibril
10–300 nm

triple-stranded
collagen molecule
1.5 nm

single collagen
polypeptide chain

1 μm

Figure 19–9 The organization and structure of collagen. The electron micrograph shows collagen fibrils in the connective tissue of embryonic chick skin. The fibrils are organized into bundles, some running in the plane of the section, others approximately at right angles to it. The cell in the photograph is a fibroblast, which secretes the collagen as well as other extracellular matrix components. The drawings show the molecular structure of the collagen fibrils. (Photograph from C. Ploetz, E.I. Zycband, and D.E. Birk, *J. Struct. Biol.* 106:73–81, 1991.)

and they constitute 25% of the total protein mass in a mammal—more than any other type of protein.

The characteristic feature of a typical collagen molecule is its long, stiff, triple-stranded helical structure, in which three collagen polypeptide chains are wound around one another in a ropelike superhelix (Figure 19–9). These molecules in turn assemble into ordered polymers called *collagen fibrils,* which are thin cables 10–300 nm in diameter and many micrometers long; these can pack together into still thicker *collagen fibers* (see Figure 19–9). Other collagen molecules decorate the surface of collagen fibrils and link the fibrils to one another and to other components in the extracellular matrix.

The connective-tissue cells that inhabit the matrix and manufacture it go by various names according to the tissue: in skin, tendon, and many other connective tissues they are called *fibroblasts* (Figure 19–10); in bone they are called *osteoblasts.* They make both the collagen and the

Figure 19–10 Scanning electron micrograph of fibroblasts in connective tissue. The tissue is from the cornea of a rat. The extracellular matrix surrounding the fibroblasts is composed largely of collagen fibrils. Other components that normally form a hydrated gel filling the spaces between the collagen fibrils have been removed by enzyme and acid treatment. (From T. Nishida et al. *Invest. Ophthalmol. Vis. Sci.* 29:1887–1890, 1988.)

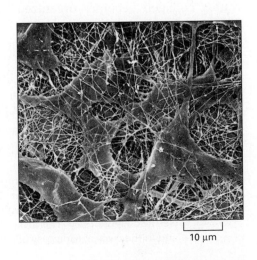

10 μm

Figure 19–11 Hyperextensible skin. James Morris, "the elastic skin man," from a photograph taken about 1890. Abnormally stretchable skin is part of a genetic syndrome that results from a fault in the assembly or cross-linking of collagen. In some individuals, this arises from a deficit of the collagenase that converts procollagen to collagen.

other organic components of the matrix. Almost all of these molecules are synthesized intracellularly and then released in the standard way, by exocytosis. Outside the cell they will assemble into huge, cohesive aggregates. If assembly were to occur prematurely, before secretion, the cell would become disastrously choked with its own products. In the case of collagen, the cells avoid this risk by secreting collagen molecules in a precursor form, called *procollagen,* with additional peptides at each end that obstruct assembly into collagen fibrils. An extracellular enzyme—a *collagenase*—cuts off these terminal domains to allow assembly only after the molecules have emerged into the extracellular space.

Some people have a genetic defect in the collagenase, so that their collagen fibrils do not assemble correctly. As a result, the skin and various other connective tissues have reduced tensile strength and are extraordinarily stretchable (Figure 19–11).

Cells Organize the Collagen That They Secrete

To do their job, collagen fibrils must be correctly aligned. In skin, for example, they are woven in a wickerwork pattern, or in alternating layers with different orientations so as to resist tensile stress in multiple directions (Figure 19–12). In tendons (which attach muscles to bone) they are aligned in parallel bundles along the major axis of tension.

The connective-tissue cells control this orientation, partly by depositing the collagen in an oriented fashion, partly by rearranging it subsequently. During development of the tissue, fibroblasts work on the collagen they have secreted, crawling over it and pulling on it—helping to compact it into sheets and draw it out into cables. This mechanical role of fibroblasts in shaping collagen matrices has been demonstrated dramatically in culture. When fibroblasts are mixed with a meshwork of randomly oriented collagen fibrils that form a gel in a culture dish, the fibroblasts tug on the meshwork, drawing in collagen from their surroundings and compacting it. If two small pieces of embryonic tissue containing fibroblasts are placed far apart on a collagen gel, the intervening collagen becomes organized into a compact band of aligned fibers that connect the two explants (Figure 19–13). The fibroblasts migrate out from the explants along the aligned collagen fibers. Thus the fibroblasts influence the alignment of the collagen fibers, and the collagen fibers in turn affect the distribution of the fibroblasts. Fibroblasts presumably play a similar role in generating long-range order in the extracellular matrix inside the body—in helping to create tendons, for example, and the tough, dense layers of connective tissue that ensheathe and bind together most organs.

Question 19–2 Mutations in the genes encoding collagens often have detrimental consequences resulting in severely crippling diseases. Particularly devastating are mutations that change glycine residues, which are required at every third position in the protein chain so that it can assemble into the characteristic triple-helical rod.

A. Would you expect that collagen mutations are detrimental if only one of the two copies of a collagen gene is defective?

B. A puzzling observation is that the change of a glycine residue into another amino acid is most detrimental if it occurs toward the amino terminus of the rod-forming domain. Suggest an explanation for this.

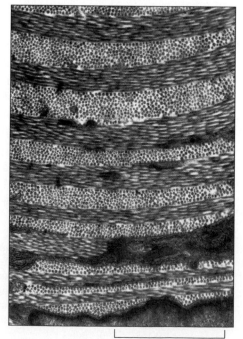

Figure 19–12 Electron micrograph of a cross section of tadpole skin. Note the plywoodlike arrangement of collagen fibrils, in which successive layers of fibrils are laid down nearly at right angles to each other (see also Figure 19–9). This arrangement is also found in mature bone and in the cornea. (Courtesy of Jerome Gross.)

5 µm

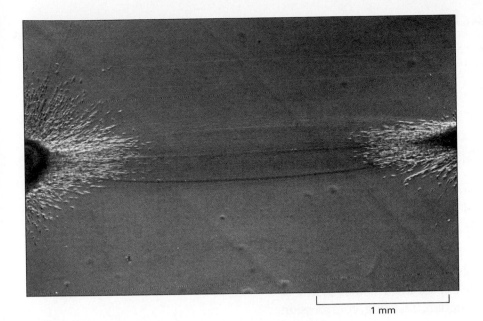

Figure 19–13 **The shaping of the collagen matrix by cells.** This micrograph shows a region between two pieces of embryonic chick heart (rich in fibroblasts as well as heart muscle cells) that have grown in culture on a collagen gel for four days. A dense tract of aligned collagen fibers has formed between the explants, presumably as a result of the fibroblasts in the explants tugging on the collagen. (From D. Stopak and A.K. Harris, *Dev. Biol.* 90:383–398, 1982.)

1 mm

Integrins Couple the Matrix Outside a Cell to the Cytoskeleton Inside It

If cells are to pull on the matrix and crawl over it, they must be able to attach to it. Cells do not attach well to bare collagen. Another extracellular matrix protein, *fibronectin*, provides a linkage: one part of the fibronectin molecule binds to collagen, while another part forms an attachment site for a cell (Figure 19–14A).

The cell binds to the specific site in the fibronectin by means of a receptor protein, called an *integrin*, that spans the cell's plasma membrane. While the extracellular domain of the integrin binds to fibronectin, the other end of the integrin molecule, in the cytosol, forms an attachment site for actin filaments. Thus, instead of being ripped out of the membrane when there is tension between the cell and the matrix, the integrin molecule transmits the stress from the matrix to the cytoskeleton (Figure 19–14C). Muscle cells couple their contractile apparatus in a similar way to the extracellular matrix at the junction between muscle and tendon, enabling them to exert large forces while remaining enveloped in a flimsy lipid bilayer membrane.

Figure 19–14 **The molecular linkage from extracellular matrix to cytoskeleton in an animal cell.** (A) Diagram and (B) electron micrograph of a fibronectin molecule. (C) The transmembrane linkage mediated by an integrin molecule. The integrin molecule transmits tension across the plasma membrane: it is anchored inside the cell to the cytoskeleton and externally via fibronectin to the extracellular matrix. The plasma membrane itself does not have to be strong. (B, from J. Engel et al., *J. Mol. Biol.* 150:97–120, 1981. Academic Press Inc. [London] Ltd.)

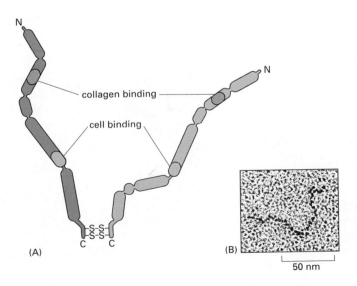

collagen binding

cell binding

S–S
S–S
C   C

(A)

(B)

50 nm

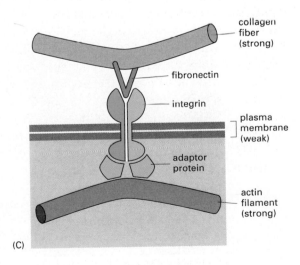

collagen fiber (strong)

fibronectin

integrin

plasma membrane (weak)

adaptor protein

actin filament (strong)

(C)

Extracellular Matrix and Connective Tissues

Figure 19–15 Hyaluronan, a relatively simple GAG. It consists of a single long chain of up to 25,000 repeated disaccharide units, each carrying negative charge. As in other GAGs, one of the sugar monomers in each disaccharide unit is an amino-sugar. Many GAGs have additional negatively charged side groups, especially sulfate groups.

glucuronic acid   *N*-acetylglucosamine

Gels of Polysaccharide and Protein Fill Spaces and Resist Compression

While collagen provides tensile strength to resist stretching, a completely different group of macromolecules in the extracellular matrix of animals provide the complementary function, resisting compression and serving as space-fillers. These are the *proteoglycans*, extracellular proteins linked to a special class of complex negatively charged polysaccharides, the **glycosaminoglycans (GAGs)** (Figure 19–15). Proteoglycans are extremely diverse in size, shape, and chemistry. Typically, many GAG chains are attached to a single core protein, which may in turn be linked at one end to another GAG, creating an enormous macromolecule resembling a bottlebrush, with a molecular weight in the millions (Figure 19–16).

In dense, compact connective tissues such as tendon and bone, the proportion of GAGs is small and the matrix consists almost entirely of collagen (or, in the case of bone, of collagen plus calcium phosphate crystals). At the other extreme, the jellylike substance in the interior of the eye consists almost entirely of one particular type of GAG, plus water,

Figure 19–16 A proteoglycan aggregate from cartilage. (A) Electron micrograph of an aggregate spread out on a flat surface. Many free subunits—themselves large proteoglycan molecules—are also seen. (B) Schematic drawing of the giant aggregate illustrated in (A), showing how it is built up from GAGs and proteins. The molecular weight of such a complex can be 10^8 or more, and it occupies a volume equivalent to that of a bacterium, which is about 2×10^{-12} cm^3. (A, courtesy of Lawrence Rosenberg.)

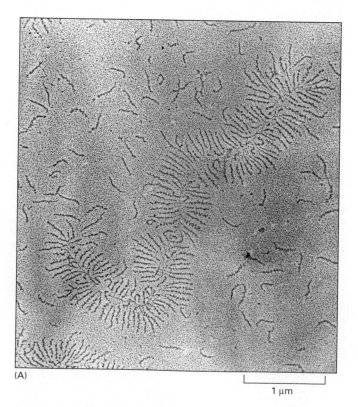

(A)

1 µm

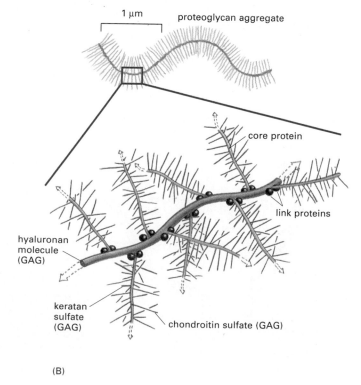

1 µm   proteoglycan aggregate

core protein

link proteins

hyaluronan molecule (GAG)

keratan sulfate (GAG)

chondroitin sulfate (GAG)

(B)

with only a little collagen. In general, GAGs are strongly hydrophilic and tend to adopt highly extended conformations, which occupy a huge volume relative to their mass (see Figure 19–16). They form gels even at very low concentrations, their multiple negative charges attracting a cloud of cations, such as Na^+, that are osmotically active, causing large amounts of water to be sucked into the matrix. This creates a swelling pressure that is balanced by tension in the collagen fibers that are interwoven with the proteoglycans. When the matrix is rich in collagen and large quantities of GAGs are trapped in its meshes, both the swelling pressure and the counterbalancing tension are enormous. Such a matrix is tough, resilient, and resistant to compression. The cartilage matrix that lines the knee joint, for example, has this character: it can support pressures of hundreds of kilograms per square centimeter.

Proteoglycans perform many sophisticated functions in addition to simply providing hydrated space around cells. They can forms gels of varying pore size and charge density that act as filters to regulate the passage of molecules through the extracellular medium. They can bind growth factors and other proteins that serve as signals for cells. They can block, encourage, or guide cell migration through the matrix. In all these ways, the matrix components influence the behavior of cells, often the same cells from which the matrix is produced—a reciprocal interaction that has important effects on cell differentiation. Much remains to be learned about how cells weave the tapestry of matrix molecules and how the chemical messages they leave in its fabric are organized and act.

Epithelial Sheets and Cell-Cell Junctions

There are more than 200 visibly different cell types in the body of a vertebrate. The majority of these are organized into epithelia—that is, they are joined together, side to side, so as to form multicellular sheets. In some cases the sheet is many cells thick, or *stratified*, as in the epidermal covering of the skin; in other cases it is only one cell thick, or *simple*, as in the lining of the gut. The cells may be tall and *columnar*, or *cuboidal*, or squat and *squamous* (Figure 19–17). They may be all alike, or a mixture of different types. They may simply act as a protective barrier, or they may have complex biochemical functions: they may secrete specialized products such as hormones, milk, or tears; they may serve to

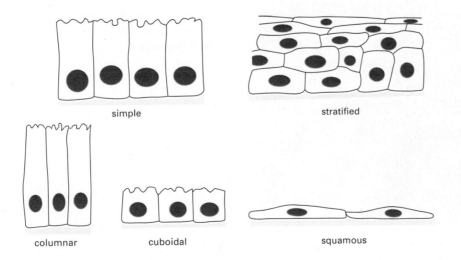

simple

stratified

columnar

cuboidal

squamous

Figure 19–17 **Different ways of packing cells together to form an epithelial sheet.**

absorb nutrients, as in the gut lining; or they may detect signals, like the photoreceptors of the eye or the auditory hair cells of the ear. Through these and many other variations, one can recognize a standard set of structural features that virtually all animal epithelia share. The epithelial arrangement of cells is so commonplace that one easily takes it for granted; yet it requires a collection of specialized devices, as we shall see, and these are common to a wide variety of different cell types.

Epithelia cover the external surface of the body and line all its internal cavities, and they must have been an early feature in the evolution of multicellular animals. Their importance is obvious. Cells joined together in an epithelial sheet create a barrier that has the same significance for the multicellular organism that the plasma membrane has for a single cell. It keeps some molecules in, and others out; it takes up nutrients and exports wastes; it contains receptors for environmental signals; and it protects the interior of the organism from invading microorganisms and fluid loss.

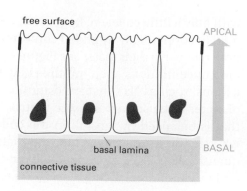

Figure 19–18 **The polarized organization of an epithelial sheet.**

Epithelial Sheets Are Polarized and Rest on a Basal Lamina

An epithelial sheet has two faces: the **apical** surface is free and exposed to the air or to a watery fluid; the **basal** surface rests on some other tissue—usually a connective tissue—to which it is attached (Figure 19–18). Supporting the basal surface of the epithelium there lies a thin tough sheet of extracellular matrix, called a **basal lamina** (Figure 19–19), composed of a specialized type of collagen (Type IV collagen) and various other molecules. These include a protein called *laminin*, which provides adhesive sites for integrin molecules in the plasma membrane of the epithelial cells, serving a linking role that resembles that of fibronectin in connective tissues.

The apical and basal faces of an epithelium are as a rule chemically different, reflecting a polarized internal organization of the individual

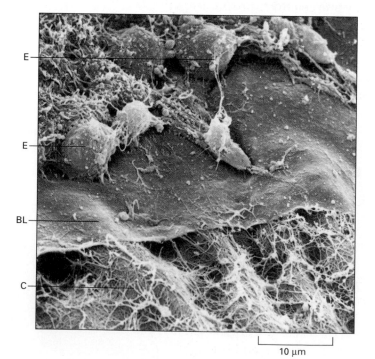

Figure 19–19 **Scanning electron micrograph of a basal lamina in the cornea of a chick embryo.** Some of the epithelial cells (E) have been removed to expose the upper surface of the matlike basal lamina woven from Type IV collagen (BL). A network of other collagen fibrils (C) in the underlying connective tissue interacts with the lower face of the lamina. (Courtesy of Robert Trelstad.)

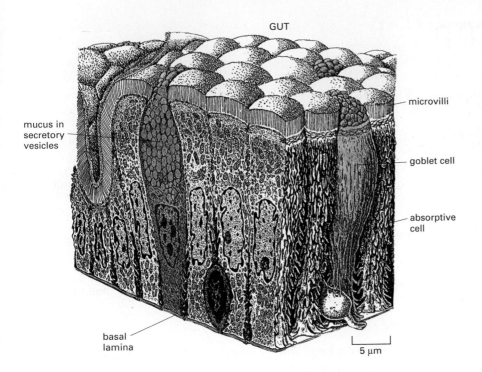

GUT

microvilli

mucus in
secretory
vesicles

goblet cell

absorptive
cell

basal
lamina

5 μm

Figure 19–20 Functionally polarized cell types in the lining of the gut. Absorptive cells, which take up nutrients from the gut, are mingled in the gut lining with goblet cells, which secrete mucus into the gut. The absorptive cells are often called *brush-border cells,* because of the brushlike mass of microvilli on their apical surface, serving to increase the area of membrane for transport of small molecules into the cell. The goblet cells owe their gobletlike shape to the mass of secretory vesicles that distends their apical region. (Adapted from R. Krstić, Human Microscopic Anatomy. Berlin: Springer, 1991.)

epithelial cells: each one has a top and a bottom, with different properties. This polarized organization is crucial for epithelial function. Consider, for example, the simple columnar epithelium that lines the small intestine of a mammal. This mainly consists of two cell types intermingled: *absorptive cells* to take up nutrients and *goblet cells* (so called from their shape) to secrete mucus that protects and lubricates the gut lining (Figure 19–20). Both cell types are polarized. The absorptive cells import food molecules through their apical surface from the gut lumen and export these molecules through their basal surface into the underlying tissues. To do this, they require different sets of membrane transport proteins in their apical and basal plasma membranes (see p. 381). The goblet cells also have to be polarized, but in a different way: they have to synthesize mucus and then discharge it from their apical ends only (see Figure 19–20): the Golgi apparatus, secretory vesicles, and cytoskeleton are all asymmetrically organized so as to bring this about. While many questions remain as to how this organization is maintained, it is clear that it depends on the junctions that the epithelial cells form with one another and with the basal lamina.

Tight Junctions Make an Epithelium Leak-proof and Separate Its Apical and Basal Surfaces

Epithelial cell junctions can be classified according to their function. Some provide a tight seal to prevent leakage of molecules across the epithelium through the gaps between its cells; some provide strong mechanical attachments; and some provide for a special type of intimate chemical communication. In most epithelia, all these types of junctions are present together (Figure 19–21).

The sealing function is served (in vertebrates) by **tight junctions.** These seal neighboring cells together so that water-soluble molecules cannot easily leak between the cells: if a tracer molecule is added to one side of an epithelial cell sheet, it will usually not pass beyond the tight

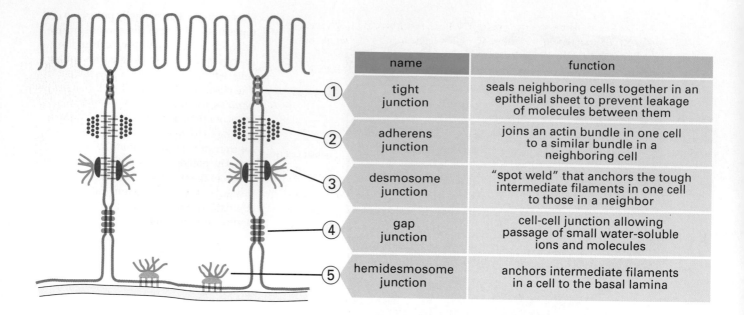

name	function
① tight junction	seals neighboring cells together in an epithelial sheet to prevent leakage of molecules between them
② adherens junction	joins an actin bundle in one cell to a similar bundle in a neighboring cell
③ desmosome junction	"spot weld" that anchors the tough intermediate filaments in one cell to those in a neighbor
④ gap junction	cell-cell junction allowing passage of small water-soluble ions and molecules
⑤ hemidesmosome junction	anchors intermediate filaments in a cell to the basal lamina

junction (Figure 19–22). Without tight junctions to prevent leakage, the pumping activities of absorptive cells such as those in the gut would be futile, and the composition of the medium on the two sides of the epithelium would become uniform. As we saw in Chapter 11, tight junctions also play a key part in maintaining the polarity of the individual epithelial cells: the tight-junction complex around the apical rim of each cell prevents diffusion of membrane proteins so as to keep the apical domain of the plasma membrane different from the basal (or basolateral) domain (see Figure 11–37).

Figure 19–21 Summary diagram of the main types of cell-cell junctions found in epithelia in animals. Tight junctions are peculiar to epithelia; the other types also occur, in modified forms, in various nonepithelial tissues.

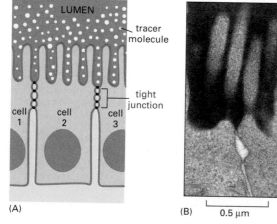

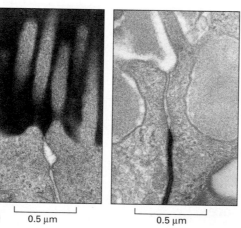

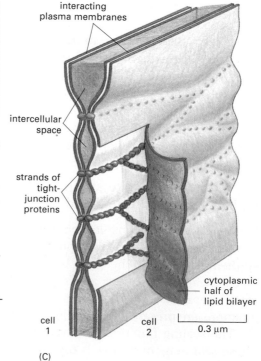

Figure 19–22 Tight junctions allow cell sheets to serve as barriers to solute diffusion. (A) Schematic drawing showing how a small extracellular tracer molecule added on one side of an epithelial cell sheet cannot traverse the tight junctions that seal adjacent cells together. (B) Electron micrographs of cells in an epithelium where a small, extracellular tracer molecule (*dark stain*) has been added to either the apical side (on the *left*) or the basolateral side (on the *right*); in both cases the tracer is stopped by the tight junction. (C) A model of the structure of a tight junction, showing how the cells are thought to be sealed together by proteins in the outer leaflet of the plasma membrane bilayer. (B, courtesy of Daniel Friend.)

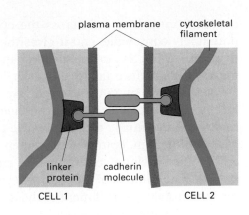

Figure 19–23 **Cadherin molecules mediate mechanical attachment of one cell to another.** Two similar cadherin molecules in the plasma membranes of adjacent cells bind to one another extracellularly; intracellularly they are attached, via linker proteins, to cytoskeletal filaments.

Cytoskeleton-linked Junctions Bind Epithelial Cells Robustly to One Another and to the Basal Lamina

The junctions that hold an epithelium together by forming mechanical attachments are of three main types. *Adherens junctions* and *desmosome junctions* bind one epithelial cell to another, while *hemidesmosomes* bind epithelial cells to the basal lamina. All of these junctions provide mechanical strength by the same strategy that we have already encountered in connective tissue (see Figure 19–14C): the molecule that forms the external adhesion spans the membrane and is linked inside the cell to strong cytoskeletal filaments. In this way, the cytoskeletal filaments are tied into a network that extends from cell to cell across the whole expanse of epithelial tissue.

Adherens junctions and desmosome junctions are both built around transmembrane proteins belonging to the same family, called **cadherins:** a cadherin molecule in the plasma membrane of one cell binds directly to an identical cadherin molecule in the plasma membrane of its neighbor (Figure 19–23). Such binding of like to like is called *homophilic*. In the case of cadherins, binding requires also that Ca^{2+} be present in the extracellular medium—hence the name.

At an **adherens junction,** each cadherin molecule is tethered inside its cell, via several linker proteins, to actin filaments. Often, the adherens junctions form a continuous *adhesion belt* around each of the interacting epithelial cells; this belt is located near the apical end of the cell, just below the tight junctions (Figure 19–24). Actin bundles are thus con-

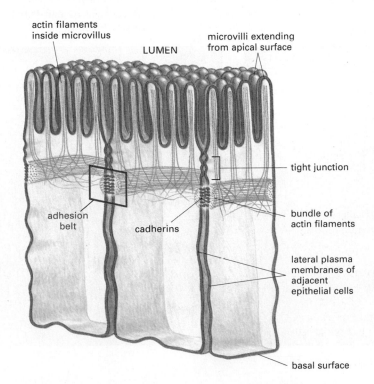

Figure 19–24 Adhesion belts in epithelial cells in the small intestine. A contractile bundle of actin filaments runs along the cytoplasmic surface of the plasma membrane near the apex of each cell, and these bundles of actin filaments in adjacent cells are linked to one another via cadherin molecules that span the cell membranes (see Figure 19–23).

nected from cell to cell across the epithelium. This actin network is potentially contractile, and it gives the epithelial sheet the capacity to develop tension and to change its shape in remarkable ways. By shrinking its apical surface along one axis, it can bend so as to roll itself up into a tube (Figure 19–25A). Alternatively, by shrinking its apical surface locally along both axes at once, the sheet can develop a cup-shaped concavity and eventually create a vesicle that may pinch off from the rest of the epithelium. Epithelial movements such as these are important in embryonic development, where they create structures such as the neural tube (the rudiment of the central nervous system) (Figure 19–25B) and the lens vesicle (the rudiment of the lens of the eye) (Figure 19–25C).

At a **desmosome junction,** by contrast, different members of the family of cadherin molecules are anchored inside each cell to intermediate filaments—specifically, to keratins, which are the type of interme-

Figure 19–25 The bending of an epithelial sheet to form a tube or a vesicle. (A) Diagram showing how contraction of apical bundles of actin filaments linked from cell to cell via adherens junctions causes the epithelial cells to narrow at their apex. According to whether the contraction is oriented along one axis or equal in all directions, the epithelium may be caused to roll up into a tube or to invaginate to form a vesicle. (B) Formation of the neural tube; the scanning electron micrograph shows a cross section through the trunk of a two-day chick embryo. Part of the epithelial sheet that covers the surface of the embryo has thickened, has rolled up into a tube by apical contraction, and is about to pinch off to become a separate internal structure. (C) Formation of the lens; a patch of surface epithelium overlying the embryonic rudiment of the retina of the eye has become concave and finally pinched off as separate vesicle— the lens vesicle—within the eye cup. (B, courtesy of Jean-Paul Revel; C, courtesy of K.W. Tosney.)

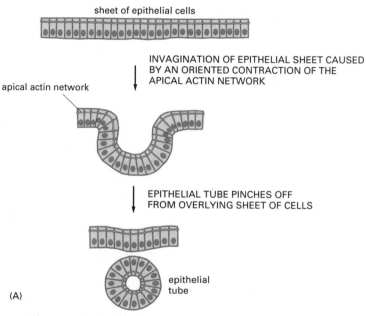

sheet of epithelial cells

INVAGINATION OF EPITHELIAL SHEET CAUSED BY AN ORIENTED CONTRACTION OF THE APICAL ACTIN NETWORK

apical actin network

EPITHELIAL TUBE PINCHES OFF FROM OVERLYING SHEET OF CELLS

epithelial tube

(A)

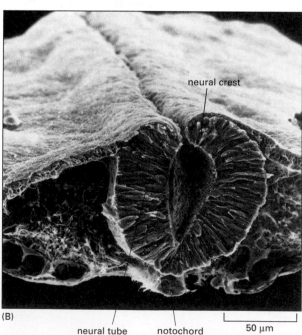

neural crest

(B)

neural tube   notochord   50 μm

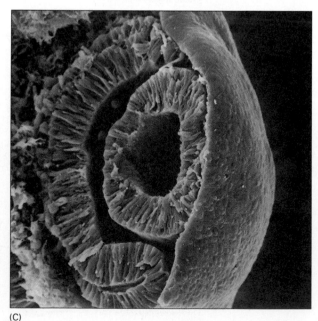

(C)

diate filament found in epithelia (see Chapter 16) (Figure 19–26). Thick bundles of ropelike keratin filaments, criss-crossing the cytoplasm and spot-welded via desmosome junctions to the bundles of keratin filaments in adjacent cells, confer great tensile strength and are particularly abundant in tough, exposed epithelia such as the epidermis of the skin.

Blisters are a painful reminder that it is not enough for epithelial cells to be firmly attached to one another: they must also be anchored to the underlying tissue. As we noted earlier, the anchorage is mediated by integrin proteins in the basal plasma membrane of the epithelial cells.

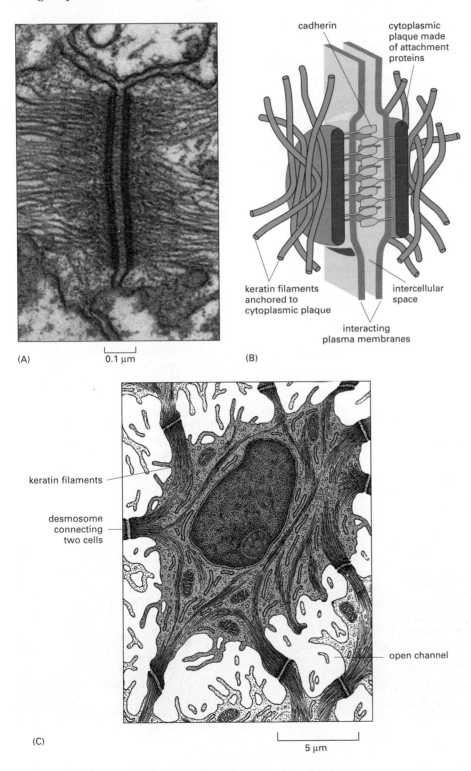

(A)   0.1 μm

(B)

cadherin

cytoplasmic plaque made of attachment proteins

keratin filaments anchored to cytoplasmic plaque

intercellular space

interacting plasma membranes

keratin filaments

desmosome connecting two cells

open channel

(C)   5 μm

Figure 19–26 **Desmosomes.** (A) An electron micrograph of a desmosome joining two cells in the epidermis of a newt, showing the attachment of keratin filaments. (B) Schematic drawing of a desmosome. On the cytoplasmic surface of each interacting plasma membrane is a dense plaque composed of a mixture of intracellular attachment proteins, to whose inner face keratin filaments are anchored. Proteins of the cadherin family are anchored in the outer face of each plaque and span the membrane to bind the two cells together. (C) Drawing from an electron micrograph of a section of the human epidermis, showing the bundles of keratin filaments that traverse the cytoplasm of one of the deep-lying cells and are inserted at the desmosome junctions that bind this cell (*red*) to its neighbors. Between adjacent cells in this deep layer of the epidermis there are also open channels that allow nutrients to diffuse freely through the metabolically active tissue. (A, from D.E. Kelly, *J. Cell Biol.* 28:51–59, 1966, by copyright permission of the Rockefeller University Press. C, from R.V. Krstić, Ultrastructure of the Mammalian Cell: An Atlas. Berlin: Springer, 1979.)

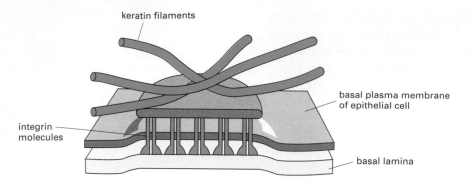

keratin filaments

integrin molecules

basal plasma membrane of epithelial cell

basal lamina

Figure 19–27 **Hemidesmosome junctions anchor the keratin filaments in an epithelial cell to the basal lamina.**

Externally, these integrins bind to laminin in the basal lamina; inside the cell, they are linked to keratin filaments, creating a structure that looks superficially like half a desmosome. These attachments of epithelial cells to the extracellular matrix beneath them are therefore called **hemidesmosomes** (Figure 19–27).

Gap Junctions Allow Ions and Small Molecules to Pass from Cell to Cell

The final type of epithelial cell-cell junction, found in virtually all epithelia and in many other types of tissue, serves a totally different purpose. It is called a **gap junction.** In the electron microscope it appears as a region where the membranes of two cells lie close together and exactly parallel, with a very narrow gap of 2–4 nm between them (Figure 19–28A). The gap is not empty, but is spanned by the protruding ends of many identical protein complexes that lie in the plasma membranes of the two apposed cells. These complexes, called *connexons*, form channels across the two plasma membranes and are aligned end-to-end so as to create narrow passageways that allow inorganic ions and small water-soluble molecules (up to a molecular mass of about 1000 daltons) to move directly from the cytoplasm of one cell to the cytoplasm of another (Figure 19–28B). This creates an electrical and a metabolic coupling between the cells. Gap junctions between heart muscle cells, for example, provide the electrical coupling that allows electrical waves of excitation to spread through the tissue, triggering coordinated contraction of the cells.

Curiously, plant tissues, though they lack all the other types of cell-cell junctions we have described earlier, have a functional counterpart of the gap junction. The cytoplasms of adjacent plant cells are connected

Question 19–4 Analogues of hemidesmosomes are the focal contact sites described in Chapter 16, which are also sites where the cell attaches to the extracellular matrix. These junctions are prevalent in fibroblasts but largely absent in epithelial cells. On the other hand, hemidesmosomes are prevalent in epithelial cells but absent in fibroblasts. In focal contact sites, intracellular connections are made to the actin cytoskeleton, whereas in hemidesmosomes connections are made to intermediate filaments. Why do you suppose these two different cell types attach differently to the extracellular matrix?

Figure 19–28 **Gap junctions.** (A) Thin-section electron micrograph of a gap junction between two cells in culture. (B) A model of a gap junction. The drawing shows the interacting plasma membranes of two adjacent cells. The apposed lipid bilayers (*red*) are penetrated by protein assemblies called *connexons* (*green*), each of which is thought to be formed by six identical protein subunits. Two connexons join across the intercellular gap to form an aqueous channel connecting the two cells. (From N.B. Gilula, in Cell Communication [R.P. Cox, ed.], pp. 1–29. New York: Wiley, 1974. Reprinted by permission of John Wiley & Sons, Inc.)

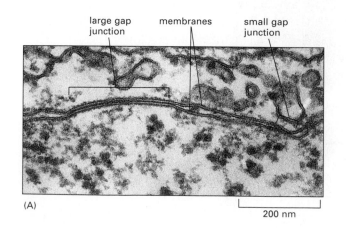

large gap junction   membranes   small gap junction

(A)

200 nm

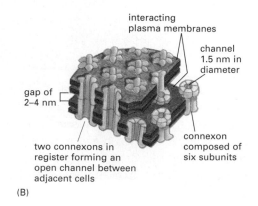

interacting plasma membranes

channel 1.5 nm in diameter

gap of 2–4 nm

two connexons in register forming an open channel between adjacent cells

connexon composed of six subunits

(B)

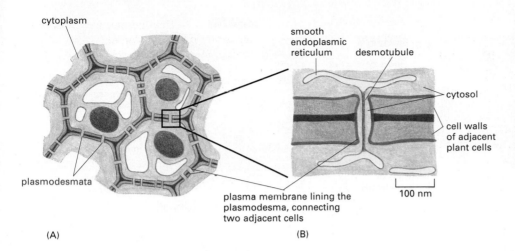

cytoplasm

smooth endoplasmic reticulum

desmotubule

cytosol

cell walls of adjacent plant cells

plasmodesmata

plasma membrane lining the plasmodesma, connecting two adjacent cells

100 nm

(A)

(B)

Figure 19–29 Plasmodesmata.
(A) The cytoplasmic channels of plasmodesmata pierce the plant cell wall and connect all cells in a plant together. (B) Each plasmodesma is lined with plasma membrane common to the two connected cells. It usually also contains a fine tubular structure, the desmotubule, derived from smooth endoplasmic reticulum. The plasmodesma usually allows only small molecules and ions to pass from cell to cell.

via minute communicating channels called *plasmodesmata*, which span the intervening cell wall (Figure 19–29). In contrast with gap-junctional channels, plasmodesmata are lined with plasma membrane, which is thus continuous from one plant cell to the next. In spite of their structural differences, plasmodesmata and gap junctions allow a similarly restricted range of ions and small molecules to pass. This suggests that neighboring cells in both plants and animals have a basic need to share these components while keeping their macromolecules segregated. It is still not clear why this should be.

Question 19–5 Gap junctions are dynamic structures that, like conventional ion channels, are gated: they can close by a reversible conformational change in response to changes in the cell. The permeability of gap junctions decreases within seconds, for example, when the intracellular Ca^{2+} is raised. Speculate why this form of regulation might be important for the health of a tissue.

Tissue Maintenance and Renewal, and Its Disruption by Cancer

Although the specialized tissues in our body differ in many ways, they all have certain basic requirements, usually provided for by a mixture of cell types, as illustrated for the skin in Figure 19–30. All tissues need mechanical strength, which is often supplied by a supporting bed or framework of connective tissue inhabited by fibroblasts. In this connective tissue, blood vessels lined with *endothelial cells* satisfy the need for oxygen, nutrients, and waste disposal. Likewise, most tissues are innervated by *nerve cell* axons, which are ensheathed by *Schwann cells. Macrophages* dispose of dying cells and other unwanted debris, and *lymphocytes* and other white blood cells combat infection. Most of these cell types originate outside the tissue and invade it, either early in the course of its development (endothelial cells, nerve cell axons, and Schwann cells) or continually during life (macrophages and other white blood cells). This complex supporting apparatus is required to maintain the principal specialized cells of the tissue: the contractile cells of the muscle, the secretory cells of the gland, or the blood-forming cells of the bone marrow, for example.

Almost every tissue is therefore an intricate mixture of many cell types that must remain different from one another while coexisting in the same environment. Moreover, in almost all adult tissues, cells are continually dying and being replaced; through all the hurly-burly of cell replacement and tissue renewal, the organization of the tissue must be preserved. Three main factors contribute to make this possible (Figure 19–31).

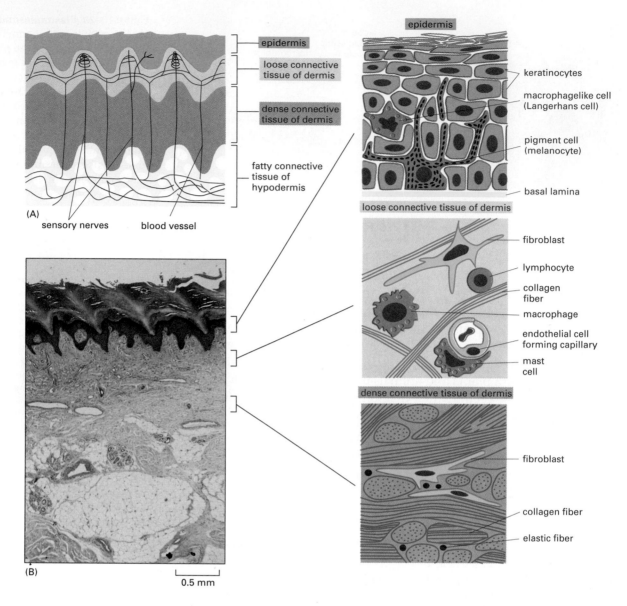

epidermis

epidermis

loose connective tissue of dermis

dense connective tissue of dermis

fatty connective tissue of hypodermis

(A)

sensory nerves   blood vessel

keratinocytes

macrophagelike cell (Langerhans cell)

pigment cell (melanocyte)

basal lamina

loose connective tissue of dermis

fibroblast

lymphocyte

collagen fiber

macrophage

endothelial cell forming capillary

mast cell

dense connective tissue of dermis

fibroblast

collagen fiber

elastic fiber

(B)

0.5 mm

1. *Cell communication:* each type of specialized cell continually senses its environment for signals, such as growth factors, from other cells, and adjusts its proliferation and properties accordingly; in fact, the very survival of most cells depends on such social signals (Chapter 15). These communications ensure that new cells are produced only when and where they are required.

2. *Selective cell-cell adhesion:* because different cell types have different cadherins and other adhesion molecules in their plasma membranes, they tend to stick selectively, by homophilic binding, to other cells of the same type. They may also form selective attachments to certain other cell types or to specific extracellular matrix components. The selectivity of adhesion prevents the different cell types in a tissue from becoming chaotically mixed.

3. *Cell memory:* as we saw in Chapter 8, specialized patterns of gene expression, evoked by signals that acted during embryonic development, are afterward stably maintained, so that cells autonomously preserve their distinctive character and pass it on to their progeny. The fibroblast divides to produce more fibroblasts, the endothelial cell divides to produce more endothelial cells. This principle, with

Figure 19–30 Mammalian skin.
(A) Schematic diagrams showing the cellular architecture of thick skin. (B) Photograph of a cross section through the sole of a human foot, stained with hematoxylin and eosin. The skin can be viewed as a large organ composed of two main tissues: epithelial tissue (the *epidermis*), which lies outermost, and connective tissue, which consists of the tough dermis (from which leather is made) and the underlying fatty *hypodermis*. Each tissue is composed of a variety of cell types. The dermis and hypodermis are richly supplied with blood vessels and nerves. Some nerve fibers extend also into the epidermis.

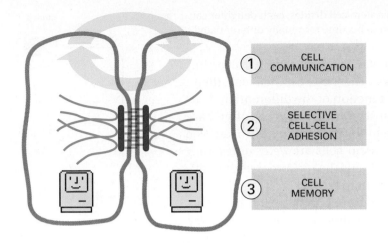

Figure 19–31 **Three key factors that maintain the cellular organization of tissues.**

elaborations that we shall explain later in this chapter, preserves the diversity of cell types in the tissue.

Different Tissues Are Renewed at Different Rates

Cells in tissues vary enormously in their rate and pattern of turnover. At one extreme are nerve cells, which last a lifetime without replacement. At the other extreme are the cells that line the intestine, which are completely replaced every few days. Between these extremes there is a spectrum of different rates and styles of cell replacement and tissue renewal. Bone, for example, has a turnover time of about ten years in humans, involving renewal of the matrix as well as of cells: old bone matrix is slowly eaten away by a set of cells called *osteoclasts,* akin to macrophages, while new matrix is deposited by another set of cells, *osteoblasts,* akin to fibroblasts. New red blood cells are continually generated in the bone marrow (from yet another class of cells) and released into the circulation, from which they are removed and destroyed after 120 days. In the skin, the outer layers of the epidermis are continually flaking off and being replaced from below, so that the epidermis is renewed with a turnover time of about two months. And so on.

Life depends on these renewal processes. A large dose of ionizing radiation, by blocking cell division, halts them: within a few days, the lining of the intestine, for example, becomes deprived of cells, leading to the devastating diarrhea and water loss characteristic of acute radiation sickness.

Question 19–6 Why does ionizing radiation stop cell division?

Stem Cells Generate a Continuous Supply of Terminally Differentiated Cells

Many of the differentiated cells that need continual replacement are themselves unable to divide. Red blood cells, surface epidermal cells, and the absorptive and goblet cells of the gut lining are all of this type. Referred to as *terminally differentiated,* they lie at the dead end of their developmental pathway.

More of such cells are generated from a stock of precursor cells, called **stem cells,** that are retained in the corresponding tissues along with the differentiated cells. Stem cells are not terminally differentiated and can divide without limit (or at least for the lifetime of the animal). When a stem cell divides, though, each daughter has a choice: either it

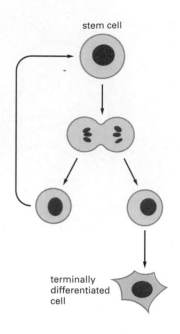

Figure 19–32 A stem cell. When a stem cell divides, each daughter can either remain a stem cell or go on to become terminally differentiated.

can remain a stem cell, or it can embark on a course leading irreversibly to terminal differentiation (Figure 19–32). The job of the stem cell, therefore, is not to carry out the specialized function of the differentiated cell but rather to produce cells that will. Consequently, stem cells often have a nondescript appearance, making them difficult to identify. However, although not terminally differentiated, stem cells are *determined:* they stably express sets of gene regulatory proteins that ensure that their differentiated progeny will be of the appropriate types.

The pattern of cell replacement varies from one stem-cell-based tissue to another. In the lining of the small intestine, for example, the absorptive and goblet cells together are arranged as a single-layered epithelium that covers the surfaces of the fingerlike *villi* that project into the lumen of the gut. This epithelium is continuous with the epithelium that lines the *crypts* that descend into the underlying connective tissue, and the stem cells lie near the bottoms of the crypts. Newborn absorptive and goblet cells generated from the stem cells are carried upward by a sliding movement in the plane of the epithelial sheet until they reach the exposed surfaces of the villi; at the tips of the villi the cells die and are shed into the lumen of the gut (Figure 19–33).

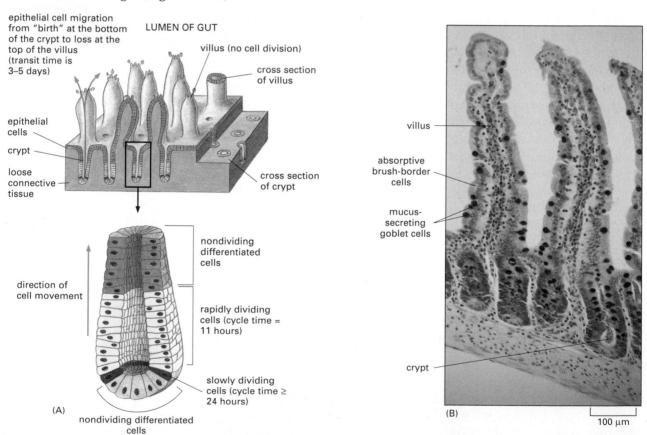

Figure 19–33 Renewal of the gut lining. (A) The pattern of cell turnover and the proliferation of stem cells in the epithelium that forms the lining of the small intestine. The nondividing differentiated cells at the base of the crypts also have a finite lifetime, terminated by programmed cell death, and are continually replaced by progeny of the stem cells. (B) Photograph of a section of part of the lining of the small intestine, showing the villi and crypts. Note how mucus-secreting goblet cells (stained *red*) are interspersed among the absorptive brush-border cells in the epithelium of the villi.

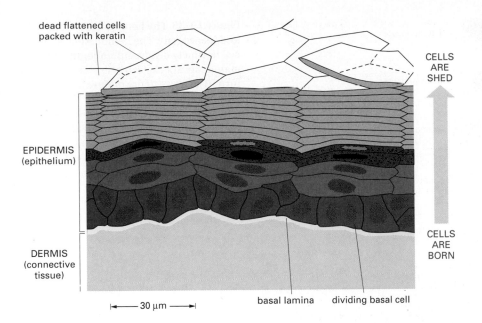

dead flattened cells
packed with keratin

CELLS
ARE
SHED

EPIDERMIS
(epithelium)

DERMIS
(connective
tissue)

CELLS
ARE
BORN

|← 30 µm →|

basal lamina   dividing basal cell

Figure 19–34 The pattern of cell replacement in the epidermis. New cells are born by division of stem cells in the basal layer, lose contact with the basal lamina, and move outward, differentiating as they go. Eventually, the cells undergo a special form of cell death: the nucleus disintegrates and the cell shrinks to the form of a flattened scale packed with keratin. The scale will ultimately be shed from the surface of the body.

A contrasting example is found in the epidermis. The epidermis is a many-layered epithelium, with stem cells in the basal layer, adherent to the basal lamina; the differentiating cells travel outward from their site of origin in a direction perpendicular to the plane of the cell sheet (Figure 19–34).

Often, a single type of stem cell gives rise to several types of differentiated progeny: the stem cells of the intestine, producing absorptive cells, goblet cells, and certain other cell types, are a case in point. The process of blood cell formation, or *hemopoiesis*, provides an extreme example of this phenomenon. All the different cell types in the blood—both the red blood cells that carry oxygen and the many types of white blood cells that fight infection (Figure 19–35)—ultimately derive from a shared *hemopoietic stem cell* that normally inhabits the bone marrow (Figure 19–36).

Because stem cells can proliferate as well as generate differentiated progeny, they allow for growth and tissue repair, as well as for normal maintenance. For example, by transfusing a few hemopoietic stem cells into a mouse whose own blood stem cells have been destroyed, it is possible to fully repopulate the animal with new blood cells and rescue it from death by anemia. A similar approach is the basis for a treatment of leukemia in humans through the use of bone marrow transplants.

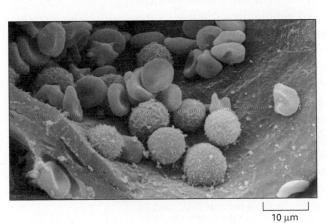

10 µm

Figure 19–35 Scanning electron micrograph of blood cells in a mammal. The larger, more spherical cells with a rough surface are white blood cells; the smaller, smoother, flattened cells are red blood cells. (From R.G. Kessel and R.H. Kardon, Tissues and Organs: A Text-Atlas of Scanning Electron Microscopy. San Francisco: Freeman, 1979. Copyright © 1979 W.H. Freeman and Company.)

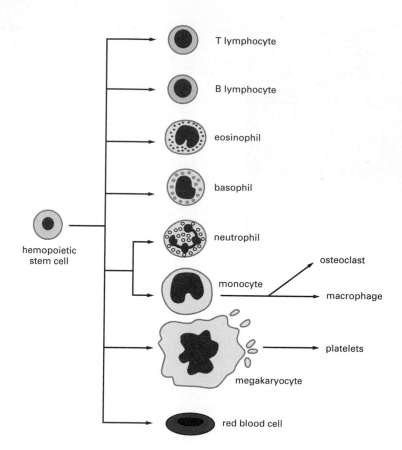

T lymphocyte

B lymphocyte

eosinophil

basophil

neutrophil

hemopoietic
stem cell

osteoclast

monocyte

macrophage

platelets

megakaryocyte

red blood cell

Figure 19–36 The hemopoietic stem cell and its progeny. The hemopoietic stem cell normally divides infrequently to generate more specialized cells that divide in turn to give rise to the mature blood cell types found in the circulation. The macrophages found in many tissues of the body and the osteoclasts that eat away bone matrix originate from the same source, as do a few other types of tissue cells not shown in this scheme.

Mutations in a Single Dividing Cell Can Cause It and Its Progeny to Violate the Normal Controls

Tissue renewal requires complex controls to coordinate the behavior of the individual cell with the needs of the organism as a whole. The cell must divide when new cells of its specific type are needed, and refrain from dividing when they are not; it must live as long as it is required to live, and die when it is required to die; it must maintain the appropriate specialized character; and it must occupy its proper place, and not stray into new territories.

Of course, in a large organism no significant harm is done if an occasional single cell misbehaves. But an insidious and potentially devastating breakdown of control occurs when a single cell suffers a *genetic* alteration that allows it to survive and divide when it should not, producing daughter cells that behave in the same bad way. The organization of the tissue, and eventually that of the body as a whole, may then become disrupted by a relentlessly expanding clone of abnormal cells. It is this catastrophe that happens in **cancer.**

Cancer cells are defined by two heritable properties: they and their progeny (1) reproduce in defiance of the normal constraints and (2) invade and colonize territories normally reserved for other cells. It is the combination of these features that creates the special danger. Cells that have the first property but not the second, so that they proliferate excessively but remain clustered together in a single mass, can form a *tumor,* but the tumor in this case is said to be *benign:* it can usually be removed cleanly and completely by surgery. A tumor is cancerous only if its cells have the ability to invade surrounding tissue, in which case it is said to

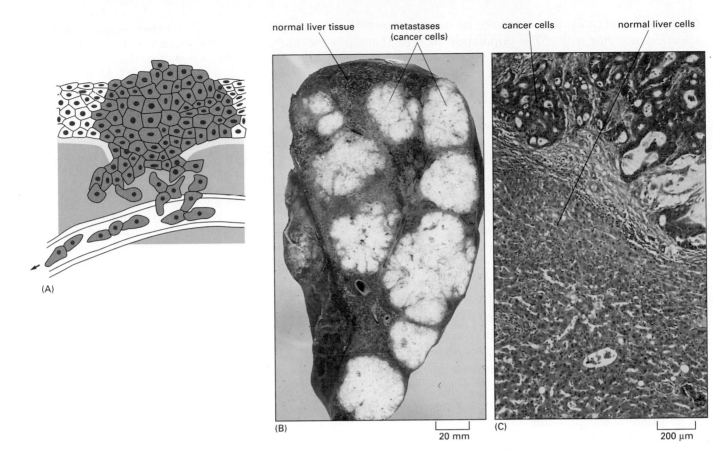

normal liver tissue   metastases (cancer cells)   cancer cells   normal liver cells

(A)

(B)   20 mm

(C)   200 μm

be *malignant.* Malignant tumor cells with this invasive property can break loose from the primary tumor, enter the bloodstream or lymphatic vessels, and form secondary tumors, or *metastases,* at other sites in the body (Figure 19–37). The more widely the cancer spreads, the harder it becomes to eradicate.

Cancer Is a Consequence of Mutation and Natural Selection Within the Population of Cells That Form the Body

Cancer is a genetic disease—a consequence of pathological changes in the information carried by DNA. It differs from other genetic diseases in that the mutations underlying cancer are mainly *somatic mutations*— those that occur in scattered individual cells of the mature body—as opposed to *germ-line mutations,* which are handed down via the germ cells from which the entire multicellular organism develops.

Some examples of the many and varied types of mutations involved in cancer have been discussed earlier in this book (pp. 302–303, 507, and 587–589). Dominant gain-of-function mutations may create **oncogenes** that promote development of cancer; for example, a mutation may result in a faulty receptor for a growth factor, such that the cell and its progeny behave as though they are receiving a signal to divide when in fact they are not. Conversely, recessive loss-of-function mutations may delete or inactivate **tumor-suppressor genes,** whose normal products are required to keep cell proliferation or invasiveness in check.

The key point, in either case, is that the mutations do not cripple the mutant cells. On the contrary, they give these cells a competitive advantage over their normal neighbors. It is this that leads to disaster for the

Figure 19–37 Metastasis. (A) To give rise to a colony, or metastasis, in a new site, the cells of a primary tumor in an epithelium must typically cross the basal lamina, migrate through connective tissue, and get into the bloodstream. (B) Metastases in a human liver, originating from a primary tumor in the colon. The metastases are the palely stained masses of cells. (C) Higher magnification, with a different stain to show the contrast between the normal cells and the tumor cells. (B and C, courtesy of Peter Isaacson.)

multicellular organism as a whole: in cancer, mutation and natural selection run riot within the population of cells that form the body, upsetting its regular structure.

Cancer Requires an Accumulation of Mutations

As we saw in Chapter 6, mutations are unavoidable. Even in an environment that is free of tobacco smoke, radioactivity, and all the other external mutagens that worry us, mutations will occur spontaneously at an estimated rate of about 10^{-6} mutations per gene per cell division—a value set by fundamental limitations on the accuracy of DNA replication and DNA repair. Something on the order of 10^{16} cell divisions take place in a human body in the course of a lifetime; thus every single gene in the genome is likely to have undergone mutation on about 10^{10} separate occasions in any individual human being. From this point of view, the problem of cancer seems to be not why it occurs, but why it occurs so infrequently.

The explanation, supported now by a great deal of evidence, is that it takes more than a single mutation to turn a normal cell into a cancer cell. Typically, at least five or six independent mutations must occur in the one cell to give it all the properties required. An epithelial stem cell in the skin or the lining of the gut, for example, must undergo changes that not only enable it to divide more frequently than it should, but also let its progeny escape being sloughed off in the normal way from the exposed surface of the epithelium, enable them to displace their normal neighbors, and let them attract a blood supply sufficient to nourish continued tumor growth. For the tumor to become invasive, further changes are needed to enable the cells to digest their way through the basal lamina of the epithelium into the underlying tissue and to settle and survive at new sites (Figure 19–38).

The mutations that give rise to a cancer do not all occur at once, but sequentially, usually over a period of many years. As an initial population of mutant cells grows, it slowly evolves: new chance mutations occur in the member cells, and some are favored by natural selection.

Figure 19–38 Tumor progression by accumulation of mutations. (A) A sequence of mutations that is thought to underlie many cases of cancer of the large intestine (colorectal cancer). The sequence of events shown here usually takes 10 to 20 years or more. It begins with loss of the *APC* (*adenomatous polyposis coli*) tumor-suppressor gene, originally identified through studies of rare individuals who have an inherited predisposition to colorectal cancer, developing the disease at an early age and at many independent sites in the large intestine. These people are born with only one functional copy of the gene, so that only one further mutation is required to abolish the gene function totally, as against the two somatic mutations that are required to achieve this in a genetically normal person. Loss of *APC* initiates growth of a benign tumor, called a polyp, which gives rise to a malignant tumor through further mutations. (B) Thousands of polyps are seen in the lining of the colon of a patient with an inherited *APC* mutation (one or two polyps might be seen in a genetically normal person). Through further mutations, some of these polyps will progress to become malignant cancers, if the tissue is not removed surgically. (C) Cross section of one such polyp; note the excessive quantities of deeply infolded epithelium. (B, courtesy of John Northover and the Imperial Cancer Research Fund; C, courtesy of Anne Campbell.)

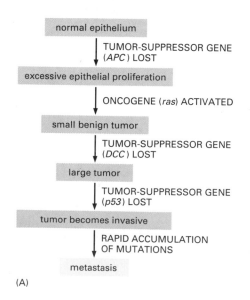

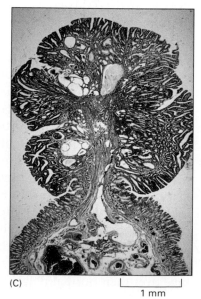

(A) (B) (C)

1 mm

Eventually cells emerge that have all the abnormalities required for full-blown cancer.

Cancer, therefore, is typically a disease of old age, because it takes a long time for an individual line of cells to accumulate a large number of somatic mutations. Occasionally, individuals are encountered who have inherited a germ-line mutation in a tumor-suppressor gene or an oncogene; for these people, the number of additional mutations required for cancer is less, and the disease occurs with higher frequency and on average at an earlier age. The families that carry such mutations are therefore unusually prone to cancer.

Question 19–8 About 10^{16} cell divisions take place in a human body during a lifetime, yet an adult human body consists of only 10^{12} cells. Why are these two numbers so different?

Development

Embryonic development is the antithesis of cancer. Cancer disrupts the orderly architecture of the body; development creates it. In cancer, cells misbehave as a result of chance mutations in their DNA sequence. In development, the DNA sequence is preserved unchanged, and cells alter their behavior through controlled changes in their pattern of gene expression. The underlying principles of cancer, once stated, seem obvious; but development, even now that we have begun to understand it, remains a profoundly astonishing process.

Development begins with a single cell—a fertilized egg. This divides repeatedly to give a clone of cells—more than 1,000,000,000,000 of them for a human being—all containing the same genome but specialized in different ways. This clone has a structure. It may take the form of a daisy or an oak tree, a sea urchin, a whale, or a mouse (Figure 19–39).

(A) 100 μm

(B)

(C) 50 μm

(D)

Figure 19–39 The genome of the egg determines the structure of the clone of cells that will develop from it. (A and B) A sea-urchin egg gives rise to a sea urchin; (C and D) a mouse egg gives rise to a mouse. (A, courtesy of David McClay; B, courtesy of M. Gibbs, Oxford Scientific Films; C, courtesy of Patricia Calarco, from G. Martin, *Science* 209:768–776, 1980; D, courtesy of O. Newman, Oxford Scientific Films.)

The structure is determined by the genome that the egg contains. The linear sequence of A, G, C, and T nucleotides in the DNA directs the production of a host of distinct cell types, each expressing different sets of genes, and arranged in a precise, intricate, three-dimensional pattern. The problem is to explain how this happens.

Programmed Cell Movements Create the Animal Body Plan

Although the final structure of an animal's body may be enormously complex, it is generated by a limited repertoire of cell activities. As we have seen in previous chapters, cells grow, divide, and die; they form mechanical attachments and generate forces for movement; they differentiate by switching on or off the production of specific sets of proteins; they produce molecular signals to influence neighboring cells, and they respond to signals that neighboring cells deliver to them. The genome, repeated identically in every cell, defines the rules according to which these various possible cell activities are called into play. Through its operation in each cell individually, it guides the whole intricate multicellular process of development by which an adult organism is generated from a fertilized egg.

The strategy is easy enough to state in the abstract, much harder to fathom in terms of concrete mechanisms and the functions of specific genes. The first step is to learn the anatomy of the process in the chosen organism—to follow the cells as they divide, move, and specialize to create the structure of the body. This can be a surprisingly difficult task, even for an embryo that is translucent and easily accessible. Figure 19–40 shows a series of stages in the development of one such embryo—that of a zebrafish. The egg is asymmetrical, with a large store of yolk in its lower part and clear cytoplasm containing the nucleus above. Fertilization triggers *cleavage* of this cell into many smaller cells, which then perform a complex series of movements. While some cells travel as free individuals, others become bound together by cell-cell junctions and move in concert as an epithelial sheet. A key early step in the development of the fish, as in that of virtually every sort of animal, is the creation of a gut through an intucking of cells from the exterior of the embryo into the interior—a process known as *gastrulation*. Gastrulation leads smoothly and rapidly into other cell movements; and with the magical slickness of a conjuring trick the rudiments of all the major features of the vertebrate body soon become distinguishable: a head and a tail, a back and a belly, eyes, ears, brain, musculature, heart, intestines.

Cells Switch On Different Sets of Genes According to Their Position and Their History

The movements that bring the body parts into being are foreshadowed by biochemical changes: cells in different regions of the embryo produce different sets of proteins, which control their adhesiveness, their motility, and their biochemical fate. At a time when most cells in the embryo look the same, well before they have visibly differentiated, it is possible to demonstrate complex changes in the patterns of expression of various genes that code for regulatory proteins (Figure 19–41). A central problem of development—some would say, *the* central problem of development—is to understand the mechanisms that generate these and the

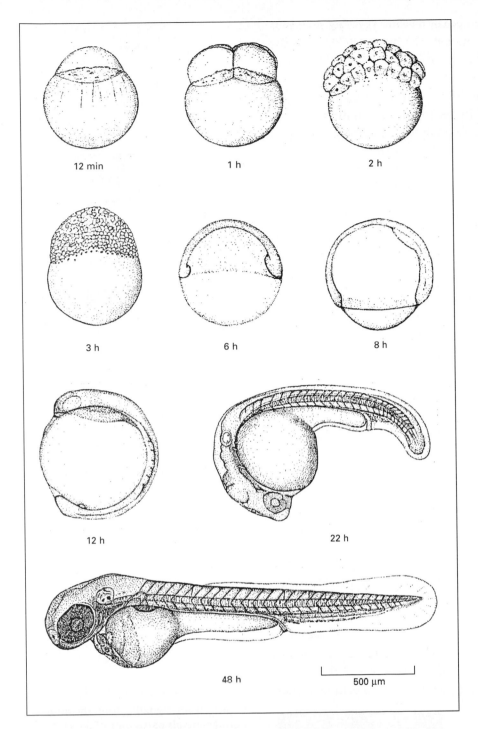

Figure 19–40 **Stages in the development of a zebrafish embryo.** Through complex cell movements and cell differentiation, the fertilized egg converts itself into a recognizable baby fish within 48 hours. (Adapted from C.B. Kimmel, et al., *Developmental Dynamics* 203:256–310, 1995.)

12 min

1 h

2 h

3 h

6 h

8 h

12 h

22 h

48 h

500 µm

even more intricate spatial patterns of gene expression that develop subsequently. How is each cell in the embryo caused to switch on the specific set of genes appropriate to its position?

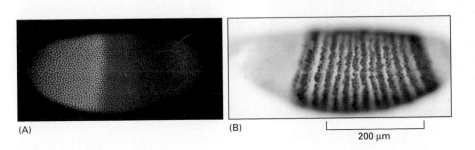

(A)

(B)

200 µm

Figure 19–41 **Gene expression patterns in an early *Drosophila* embryo.** The cells have not yet begun to differentiate overtly, but they already express different regulatory genes, according to their position in the embryo. (A) At 2.5 hours after fertilization, patterns of gene expression distinguish broad regions of the body; fluorescent antibodies have been used here to show the expression of two genes, *hunchback* (*green*) and *Krüppel* (*red*), coding for gene regulatory proteins. (B) At 3.5 hours after fertilization, the future repetitive pattern of body segments is foreshadowed in greater detail by the expression of numerous other genes; the two whose expression is shown here are called *even-skipped* (*gray* stain) and *fushi-tarazu* (*brown* stain). (A, courtesy of Jim Williams, Steve Paddock, Sean Carroll, and Howard Hughes Medical Institute; B, from P.A. Lawrence, The Making of a Fly. Oxford, UK: Blackwell, 1992.)

Figure 19–42 Refinement of pattern by repeated signaling. A series of signals causing changes of cell character can generate a complex pattern comprising many kinds of cells, starting from a simple pattern comprising only a few.

In many species, the origins of the body pattern can be traced back to asymmetries in the egg itself or in its initial environment. These beginning asymmetries result in differences between the cells produced by cleavage, and the early differences then give rise to others, in a continuing cascade. In general, cells act as sources of signaling molecules that influence other cells in their neighborhood. Neighbors at different distances from such a source are exposed to signals of differing intensity, evoking different responses. Some of the responding cells produce new signals in their turn. In this way, the original crude plan is modified and filled in with progressively finer and more intricate detail (Figure 19–42).

In this process of **pattern formation,** each cell is exposed to a succession of signals. The signals are transient, but the cell has a memory: **cell memory** preserves an internal trace of the signals to which the cell has been exposed—a historical record stamped on the genome in the form of persistent, stable states of activation or repression of particular genes (Chapter 8). The behavior of the cell is thus guided not simply by cues from its present environment but by an accumulation of remembered information. Thousands of examples of these principles of pattern formation could be given; we have space here for only one or two.

Diffusible Signals Can Provide Cells with Positional Information

During gastrulation in a vertebrate embryo, some of the cells that tuck into the interior become specialized and stick together to form a rodlike structure, the *notochord* (it can be seen in Figure 19–25B). This is the precursor of the vertebral column, and it defines the central axis of the body. Above the notochord lies the *neural plate,* a large patch of epithelial cells on the exterior of the embryo, which rolls up to form the neural tube (see Figure 19–25A). The notochord emits a signal that controls the regional specialization of cells within the neural plate and tube. Those that lie in the midline, closest to the notochord, are caused to differentiate into a structure called the *floorplate;* cells a little farther from the notochord, on either side of the floorplate, become specified as pre-

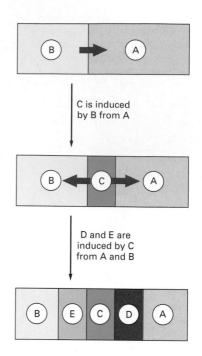

Question 19–9 During early development of many species, cells divide extremely rapidly and transcription is blocked so that no new mRNAs are produced. Transcription therefore cannot be regulated. Does this mean that the cells cannot become different from one another with regard to the sets of gene products that are active in them? Explain your answer.

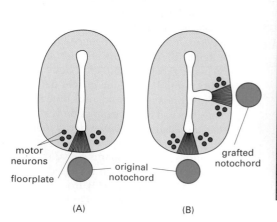

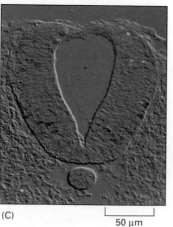

(A)   (B)   (C)

50 µm

Figure 19–43 Signaling from the notochord controls patterning of the embryonic central nervous system. (A) Diagram of a cross section through the normal structure: the cells closest to the notochord receive the strongest signal and differentiate into a floorplate, while cells a little farther away are caused to become motor neurons. (B) The effect of grafting a second notochord to one side of the normal location. (C) Cross section through the neural tube and notochord of a two-day chick embryo. Motor neurons have just begun to develop; their nuclei are stained *red* with an antibody. (C, courtesy of Julie Adam.)

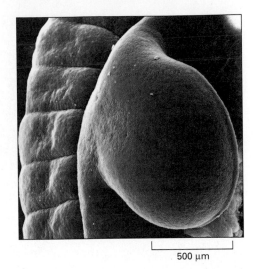

500 µm

cursors of motor neurons (Figure 19–43A). If a second notochord is grafted into a chick embryo to one side of the midline, a second floorplate flanked by motor neurons develops in the side wall of the central nervous system, next to the grafted cells (Figure 19–43B).

The signal from the notochord has been identified: it is a secreted signaling protein produced by a gene called *sonic hedgehog*. Mutations of this gene in mice and humans cause a loss of midline structures in the central nervous system: in extreme cases the embryo develops as a *cyclops*, with a single eye in the middle of its face and a rudimentary nose with a single nostril in the center of the forehead.

Given the cloned gene, one can make a probe to see which cells express it, using the technique of *in situ* hybridization (see pp. 323–324). This reveals *sonic hedgehog* mRNA not only in the notochord, but also, for example, in each of the limb buds—tongue-shaped protrusions from the flank of the vertebrate embryo that grow to form legs, arms, wings, or fins (Figure 19–44). Expression of *sonic hedgehog* here is limited to a small patch of tissue at the posterior side of each bud (the little-finger side). If an additional source of Sonic hedgehog protein is artificially placed on the opposite, anterior side of the limb bud (the thumb side), the limb develops with a mirror reduplication of its digits (Figure 19–45): the character of each digit that develops—its length, shape, and internal structure—is determined by its distance from the sources of Sonic hedgehog protein. Apparently, the concentration of the protein, or of some factor controlled by that protein, provides the cells in the limb bud with *positional information:* the cells respond by changes in the expression of genes that regulate the way they will interact with one another to form the bones, tendons, and skin of a particular type of digit.

The example of Sonic hedgehog shows how one and the same signaling protein is used repeatedly at many steps in development. Different groups of cells interpret the signal differently, according to their remembered developmental history: central nervous system cells

Question 19–10 You are studying the development of the unicorn, and you suspect that Sonic hedgehog may be the signal that triggers development of the horn in the middle of its forehead. You test this by injecting Sonic hedgehog protein into one side of the head: an additional horn develops there. Meanwhile, a colleague also working on unicorn embryology discovers a mutant strain of unicorns in which Sonic hedgehog is not expressed anywhere in the neighborhood of the developing horn, and yet the horn develops normally.. What conclusions could you draw about the role of Sonic hedgehog in inducing formation of a horn?

ANTERIOR

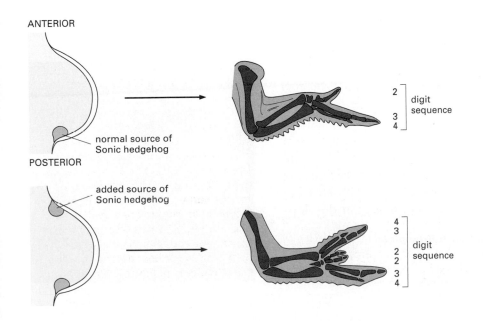

normal source of
Sonic hedgehog

POSTERIOR

added source of
Sonic hedgehog

2 ⎫
3 ⎬ digit
4 ⎭ sequence

4
3

2
2

3
4
⎫
⎬ digit
⎭ sequence

Figure 19–45 Sonic hedgehog signaling in limb development. *Sonic hedgehog* is normally expressed in a patch of cells on the posterior side of the limb bud; a second, additional source of Sonic hedgehog protein provided artificially on the anterior side of a chick wing bud causes a mirror reduplication of the pattern of digits.

Figure 19–46 **The fly** ***Drosophila melanogaster.*** (Courtesy of E.B. Lewis.)

respond to a strong signal by forming a floorplate, whereas limb bud cells respond by forming a little finger. Sonic hedgehog represents one of a handful of families of signaling proteins that together seem to be responsible for organizing the pattern of most of the body. Thanks to cell memory, a few basic signaling mechanisms apparently suffice to guide creation of an enormous variety of highly elaborate structures.

Studies in *Drosophila* Have Given a Key to Vertebrate Development

In *Drosophila* (Figure 19–46), mutants are relatively easy to make and to analyze. Through a heroic effort, involving the production and examination of hundreds of thousands of mutant fly embryos in many different laboratories, it has been possible to identify almost every gene in the fly that is capable of mutating to cause a visible developmental aberration. By cloning these developmental control genes and using their DNA sequences as probes to search for similar DNA sequences in the vertebrate genome, it has been possible to show that most—perhaps almost all—of the *Drosophila* developmental control genes have homologues serving similar functions in vertebrates. The comprehensive catalogue of these genes in the fly has thus provided a key to the problem of deciphering vertebrate development.

The gene family to which *sonic hedgehog* belongs—the *hedgehog* gene family—was in fact discovered by just this route, and it takes its name from the appearance of *Drosophila* embryos in which the gene is mutated. The microscopic spines (denticles) that normally appear in stripes on the belly surface of the larva cover this surface entirely in the mutant, reflecting a radical alteration in the organization of each of the segments that form the insect body (Figure 19–47). As in vertebrates, the protein operates repeatedly at different times and places in develop-

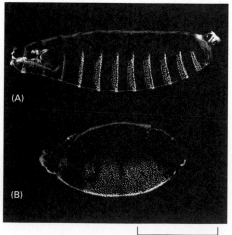

(A)

(B)

200 μm

Figure 19–47 **The *hedgehog* mutation in *Drosophila*.** (A) A normal young larva of *Drosophila*, showing the bands of microscopic spines (denticles) on its belly surface. (B) A *hedgehog* mutant: the spines cover the belly surface of the larva uniformly. (Courtesy of Marcel van den Heuvel and Antonio Jacinto.)

Figure 19–48 Effects of a mutation in a homeotic selector (Hox) gene—in this example, the *Antennapedia* gene. The antennae have been converted into legs; compare with the normal fly in Figure 19–46. (Courtesy of Matthew Scott.)

ment to provide cells with positional information. Some of the functional parallels seem absurdly close: for example, just as Sonic hedgehog protein controls the anteroposterior (thumb-to-little-finger) pattern of the wings of the chicken, so Hedgehog protein controls the anteroposterior pattern of the wings of the fly.

Similar Genes Are Used Throughout the Animal Kingdom to Give Cells an Internal Record of Their Position

The fundamental similarities between insects and vertebrates revealed by molecular genetics have astonished developmental biologists and have revolutionized their understanding of development in general. The parallels apply not only to signal molecules and their receptor proteins, but also to the larger and more complex set of components through which cells interpret, record, and react to the signals. Many of these components are gene regulatory proteins. The *Pax-6/ey* gene that acts as a master regulator of eye development (see p. 273) is a case in point. Perhaps the most illuminating example, however, is that of the *Drosophila* genes that carry each cell's internal record of which segment it belongs to along the head-to-tail axis of the body. Mutations in genes of this set—the so-called *homeotic selector genes,* or **Hox genes**—give rise to cells with mistaken information about their location: the mutant fly may sprout legs from its head in place of antennae (Figure 19–48), or an extra pair of wings from a body segment that should have none. Corresponding to the fly's single set of Hox genes, vertebrates have four sets, all homologous to those of the fly, and all similarly involved in specifying the differences between cells along the main head-to-tail body axis (Figure 19–49). Hox genes have now been found performing a similar function in the patterning of the main body axis not only in insects and vertebrates, but also in mollusks, worms, and hydroids—indeed, in virtually every class of animals. Like the signaling molecules that supply positional information, the Hox genes that record it in cell memory are also used in other parts of the body—in vertebrate limbs, for example, as well as along the main body axis.

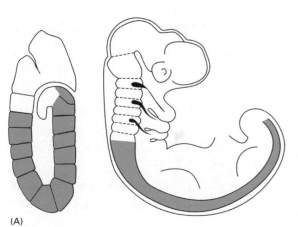

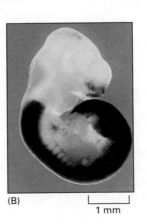

(A)

(B)

1 mm

Figure 19–49 Hox genes specify location along the main body axis in almost all animals. (A) Diagram of the expression pattern of the *Antennapedia* gene in a *Drosophila* embryo (*left*), compared with that of a corresponding Hox gene (*Hoxb-6*) in a mouse embryo (*right*). Expression of the *Antennapedia* gene or its vertebrate counterpart tells cells that they are located in the posterior rather than the anterior part of the embryo; in the mutant fly shown in Figure 19–48, it is expressed inappropriately in the cells that should form antennae (anterior structures), causing them to form legs (posterior structures) instead. (B) Photograph of the expression pattern of a Hox gene (*Hoxb-4*) in the mouse, showing how it demarcates posterior from anterior tissue. Different Hox genes have their boundaries of expression at different levels along the body axis. (B, courtesy of Robb Krumlauf.)

The recent progress of developmental biology has been rapid, like the climax in assembly of a jigsaw puzzle. Scattered fragments are beginning to fit together as parts of a coherent whole. Organisms that once were judged radically different now stand revealed as fundamentally similar. More and more, it seems that the miracle of development lies not in the complexity of the outcome, but in the wonderful economy of means by which it is achieved. Through a long evolutionary process, a relatively simple tool kit of cell-biological tricks has been repeatedly applied and adapted in endlessly inventive ways, guiding cell behavior so as to generate all the fantastic variety of multicellular animals that we find on earth today.

Essential Concepts

- Tissues are composed of cells plus extracellular matrix.

- In plants, each cell surrounds itself with extracellular matrix in the form of a cell wall made of cellulose and other polysaccharides.

- Naked plant cells are fragile but can exert an osmotic swelling pressure on the enveloping wall to keep the tissue turgid.

- Cellulose fibers in the plant cell wall confer tensile strength; other wall components give resistance to compression.

- The orientation in which cellulose is deposited controls the orientation of plant growth.

- Animal connective tissues provide mechanical support and consist of extracellular matrix with sparsely scattered cells.

- The organic components of the matrix are made by the connective-tissue cells embedded in it (called fibroblasts in most connective tissues).

- In extracellular matrix of animals, tensile strength is provided by a fibrous protein, collagen.

- Tension is transmitted from the cytoskeleton of the connective-tissue cell to the collagen fibers via a transmembrane protein, integrin, and an extracellular adaptor protein, fibronectin.

- Glycosaminoglycans (GAGs), complexed with proteins to form proteoglycans, act as space-fillers and provide resistance to compression.

- GAGs are negatively charged polysaccharides; the cloud of small positive ions that they attract draws in water by osmosis, creating a swelling pressure.

- Cells joined together in epithelial sheets line all external and internal surfaces of the animal body.

- In epithelial sheets, in contrast to connective tissues, tension is transmitted directly from cell to cell via cell-cell junctions.

- Proteins of the cadherin family span the epithelial cell membrane and bind to similar cadherins in the adjacent epithelial cell.

- At an adherens junction, the cadherins are linked intracellularly to actin filaments; at a desmosome junction, they are linked to keratin filaments.

- Actin bundles connected from cell to cell across an epithelium can contract, causing the epithelium to bend.

- Hemidesmosomes attach the basal face of an epithelial cell to the basal lamina.

- Tight junctions seal one epithelial cell to the next, creating a barrier to diffusion across the epithelium.

- Gap junctions form channels allowing passage of small molecules and ions from cell to cell; plasmodesmata in plants have the same function, but have a different structure.

- Most tissues in vertebrates are complex mixtures of cell types that are subject to continual turnover.

- New terminally differentiated cells are generated from stem cells.

- Tissue organization is actively maintained through cell communication, selective cell-cell adhesion, and cell memory.

- Cancer cells selfishly violate the constraints that normally maintain tissue organization: they proliferate when they should not, and they invade regions that they should keep out of.

- Cancers arise from the accumulation of many somatic mutations in a single cell lineage.

- In embryonic development, all the cells inherit the same genome from the fertilized egg and become different through controlled changes of gene expression.

- The genome, repeated in each cell, defines the rules of cell behavior that generate the multicellular pattern of the body.

- Cells act as sources of signaling molecules that provide neighboring cells with positional information to guide cell specialization.

- Cells have a memory: transient signals leave lasting effects on gene expression and cell behavior.

- The pattern of specialized cells is refined progressively by means of signals acting in succession.

- The same few basic signaling mechanisms are used repeatedly in different contexts.

- Cells can interpret the same signal differently according to their genome and their developmental history.

- The cells of vertebrate and invertebrate animals use essentially similar sets of genes to send, detect, interpret, and record the signals that guide development.

Key Terms

adherens junction	cancer	development	oncogene
apical	cell junction	extracellular matrix	pattern formation
basal	cell memory	gap junction	stem cell
basal lamina	cell wall	glycosaminoglycan	tight junction
cadherin	collagen	hemidesmosome	tissue
	desmosome junction	Hox gene	tumor-suppressor gene

Questions

Question 19–11 Which of the following statements are correct? Explain your answers.

A. Gap junctions connect the cytoskeleton of one cell to that of a neighboring cell or to the extracellular matrix.

B. A wilted plant leaf can be likened to a deflated bicycle tire.

C. Because of their rigid structure, proteoglycans can withstand a large amount of compressive force.

D. The basal lamina is a specialized layer of extracellular matrix to which sheets of epithelial cells are attached.

E. Skin cells are continually shed and are renewed every few weeks; for a permanent tattoo, it is therefore necessary to deposit pigment below the epidermis.

F. Although stem cells are not differentiated, they are determined and therefore give rise only to specific cell types.

Question 19–12 Which of the following substances would you expect to spread from one cell to the next through (a) gap junctions and (b) plasmodesmata: glutamic acid, mRNA, cyclic AMP, Sonic hedgehog protein, Ca^{2+}, G proteins, and plasma membrane phospholipids?

Question 19–13 Discuss the following statement: "If plant cells contained intermediate filaments to provide the cells with tensile strength, their cell walls would be dispensable."

Question 19–14 Through the exchange of small metabolites and ions, gap junctions provide metabolic and electrical coupling between cells. Why, then, do you suppose that neurons communicate primarily through synapses rather than through gap junctions?

Question 19–15 Gelatin is primarily composed of collagen, which is responsible for the remarkable tensile strength of connective tissue. It is the basic ingredient of jello; yet, as you probably experienced many times yourself while consuming the strawberry-flavored variety, jello has virtually no tensile strength. Why?

Question 19–16 The text on page 622 states: "The structure [of an organism] is determined by the genome that the egg contains." What is the evidence on which this statement is based? Indeed, a friend challenges you and suggests that you replace the DNA of a stork's egg with human DNA to see if a human baby results. How would you answer him?

Question 19–17 There are two essentially different mechanisms (Figure Q19–17) by which a diffusible signal can be passed from one cell to others to affect their development according to their relative position: (1) the signaling molecule diffuses from cell A that secretes it, and neighboring cells B and C determine their position by measuring its concentration (higher closer to the source for cell B and lower at a distance for cell C); alternately (2) the signal is relayed from one cell to the next. According to the second scenario, cell B senses the signal molecule and responds by putting out a signal of its own, which is then recognized by cell C. Suggest a way in which one could distinguish experimentally between these two possibilities.

mechanism 1

cell A   cell B   cell C

mechanism 2

cell A   cell B   cell C

Figure Q19–17

Question 19–18 Leukemias—that is, cancers arising through mutations that cause excessive production of white blood cells—have an earlier average age of onset than other cancers. Propose an explanation for why this might be the case.

Question 19–19 Carefully consider the graph in Figure Q19–19, showing the number of cases of colon cancer diagnosed per 100,000 women per year as a function of age. Why is this graph so steep and curved, if mutations occur with a similar frequency throughout a person's life span?

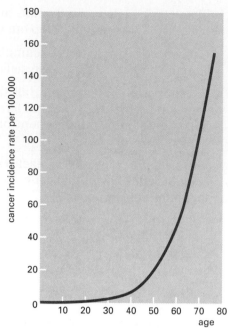

Figure Q19–19

Question 19–20 Heavy smokers or industrial workers exposed for a limited time to a chemical carcinogen that induces mutations in DNA do not usually begin to develop cancers characteristic of their habit or occupation until 10, 20, or even more years after the exposure. Suggest an explanation for this long delay.

Question 19–21 High levels of the female sex hormone estrogen increase the incidence of some forms of cancer. Thus some early types of contraceptive pills containing high concentrations of estrogen were eventually withdrawn from use because this was found to increase the risk of cancer of the lining of the uterus. Male transsexuals who use estrogen preparations to give themselves a female appearance have an increased risk of breast cancer. High levels of androgens (male sex hormones) increase the risk of some other forms of cancer, such as cancer of the prostate. Can one infer that estrogens and androgens are mutagenic?

Question 19–22 Do cells seem simpler or more complicated than you supposed before you read this book? Explain your reasons.

Glossary

A-kinase (cyclic-AMP-dependent protein kinase)
Enzyme that phosphorylates target proteins in response to a rise in intracellular cyclic AMP.

acetyl group
Chemical group derived from acetic acid.

acetyl CoA (acetyl coenzyme A)
Small water-soluble molecule that carries acetyl groups in cells. Contains an acetyl group linked to coenzyme A (CoA) by an easily hydrolyzable thioester bond.

acid
Substance that dissociates in water to release protons (H^+ ions) thereby producing a low pH; these protons associate with H_2O, generating hydronium (H_3O^+) ions.

actin filament
Protein filament, about 7 nm thick, formed from a chain of globular actin molecules. A major constituent of the cytoskeleton of all eucaryotic cells and especially abundant in muscle cells.

action potential
Rapid, transient, self-propagating electrical signal in the plasma membrane of a cell such as a neuron or muscle cell. A nerve impulse.

activated carrier
Small molecule carrying a chemical group in a high-energy linkage, serving as a donor of energy or of the chemical group in many different chemical reactions. Examples include ATP, acetyl CoA, and NADH.

activation energy
Extra energy that a molecule must acquire in order to surmount an energy barrier so as to undergo a particular chemical reaction.

activator
In bacteria, a protein that binds to a specific region of DNA to permit transcription of an adjacent gene.

active site
Region of an enzyme surface to which a substrate molecule binds before it undergoes a catalyzed reaction.

active transport
Movement of a molecule across a membrane driven by ATP hydrolysis or other form of metabolic energy.

acyl group
Functional group derived from a carboxylic acid. (R represents an alkyl group, such as methyl.)

adaptation
Adjustment of sensitivity of a cell or an organism following repeated stimulation. Allows a response even when there is a high background level of stimulation.

adenylate cyclase
Membrane-bound enzyme that catalyzes the formation of cyclic AMP from ATP. An important component of some intracellular signaling pathways.

adherens junction
Cell junction in which the cytoplasmic face of the junctional membrane is attached to actin filaments.

ADP (adenosine 5′-diphosphate)
Nucleotide that is produced by hydrolysis of the terminal phosphate of ATP. (*See* Figure 3–25.)

alcohol
Organic compound containing a hydroxyl group (–OH) bound to a saturated carbon atom—for example, ethyl alcohol.

aldehyde
Reactive organic compound that contains the –CH=O group—for example, glyceraldehyde.

alkaline—*see* **basic**

alkane
Compound made of carbon and hydrogen atoms that has only single covalent bonds. An example is ethane.

alkene
Hydrocarbon with one or more carbon-carbon double bonds. An example is ethylene.

alkyl group
General term for a group of covalently linked carbon and hydrogen atoms such as a methyl (–CH$_3$) or ethyl (–CH$_2$CH$_3$) group.

allosteric protein
Protein that exists in two or more conformations depending on the binding of a molecule (a ligand) at a site other than the catalytic site. Allosteric proteins composed of multiple subunits often display a cooperative response to ligand binding.

alpha helix (α helix)
Common structural motif of proteins in which a linear sequence of amino acids folds into a right-handed helix stabilized by internal hydrogen-bonding between backbone atoms.

amide
Molecule containing a carbonyl group linked to an amine.

amino acid
Organic molecule containing both an amino group and a carboxyl group. α amino acids (those in which the amino and carboxyl groups are linked to the same carbon atom) serve as the building blocks of proteins. (*See* Panel 2–5, p. 62.)

amino group (–NH₂)
Weakly basic functional group, derived from ammonia (NH₃). In aqueous solution an amino group can accept a proton and carry a positive charge.

amino terminus—*see* **N-terminus**

amoeba (plural **amoebae**)
General term for free-living single-celled carnivorous organisms that crawl: a subdivision of protozoa. *Amoeba proteus* is a species of giant freshwater amoeba widely used in studies of cell crawling.

AMP (adenosine 5′ monophosphate)
One of the four nucleotides in RNA. AMP is produced by the energetically favorable hydrolysis of ATP. (*See* Figure 3–35.)

amphipathic
Having both hydrophobic and hydrophilic regions, as in a phospholipid or a detergent molecule.

anabolic
Pertaining to a biochemical reaction or reaction pathway in which large molecules are made from smaller ones. Biosynthetic.

anaerobic
(Of a cell, organism, or metabolic process) functioning in the absence of air or, more precisely, in the absence of molecular oxygen.

anaphase
Stage of mitosis during which the two sets of chromosomes separate and move away from each other. Composed of anaphase A (chromosomes move toward the two spindle poles) and anaphase B (spindle poles move apart).

anion
Negatively charged ion, such as Cl⁻ or CH₃COO⁻.

antibody (immunoglobulin)
Protein produced by B lymphocytes in response to a foreign molecule or an invading organism. Binds tightly to the foreign molecule or cell, thereby inactivating it or marking it for destruction.

anticodon
Sequence of three nucleotides in a transfer RNA molecule that is complementary to the three-nucleotide codon on a messenger RNA molecule; the anticodon is matched to a specific amino acid covalently attached to the transfer RNA molecule.

antigen
Molecule that is recognized by and binds specifically to an antibody molecule; so called because the antigen generates (provokes) the immune response that produces the antibodies.

antiparallel
Arranged in parallel, but with opposite orientations; the two strands of a DNA double helix are an example.

antiport
Membrane carrier protein that transports two different ions or small molecules across a membrane in opposite directions, either simultaneously or in sequence.

apical
Situated at the tip of a cell, a structure, or an organ. The apical surface of an epithelial cell is the exposed free surface (here shown in *red*) opposite to the basal surface.

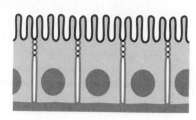

apoptosis—*see* **programmed cell death**

atom
The smallest particle of an element that still retains its distinctive chemical properties.

atomic number
The number of protons in the nucleus of an atom of an element.

atomic weight
Mass of an atom of an isotope expressed in daltons.

ATP (adenosine 5′-triphosphate)
Nucleoside triphosphate composed of adenine, ribose, and three phosphate groups; the principal carrier of chemical energy in cells. The terminal phosphate groups are highly reactive in the sense that their hydrolysis, or transfer to another molecule, takes place with release of a large amount of free energy. (*See* Figure 2–21.)

ATP synthase
Membrane-associated enzyme complex that catalyzes the formation of ATP during oxidative phosphorylation and photosynthesis. Found in mitochondria, chloroplasts and bacteria.

Avogadro's number
The number of daltons (molecular mass units) in one gram; equal to the number of molecules in M grams of a substance whose molecular weight is M daltons. It has a value of 6.02×10^{23}.

axon
Long thin nerve cell process capable of rapidly conducting nerve impulses over long distances so as to deliver signals to other cells.

bacteriorhodopsin
Light-absorbing purple-colored protein found in the plasma membrane of a salt-loving bacterium, *Halobacterium halobium*; it pumps protons out of the cell in response to light.

bacterium (plural **bacteria**)
General name for a procaryotic cell. The bacteria fall into two evolutionary ancient groups: the Eubacteria (also called simply Bacteria) and the Archaebacteria (also called Archaea).

basal
Situated near the base. The basal surface of a cell is opposite the apical surface. (*See* **basal lamina**.)

basal body—*see* **centriole**

basal lamina
Thin mat of extra-cellular matrix that separates epithelial sheets, and many types of cells such as muscle cells or fat cells, from connective tissue. Sometimes called a basement membrane.

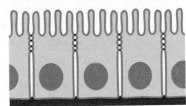

base
Molecule that accepts a proton in solution. Also used to refer to the purines and pyrimidines in DNA and RNA.

base pair
Two nucleotides in an RNA or a DNA molecule that are paired by hydrogen bonds—for example, G with C, and A with T or U.

basic
Having the properties of a base. Alkaline.

beta sheet
Folding pattern found in many proteins in which neighboring regions of the polypeptide chain associate with each other through hydrogen bonds to give a rigid, flat structure.

binding site
Region on the surface of a protein—typically a cavity or groove—that is complementary in shape to another molecule (the ligand) and therefore can bind it through the formation of multiple weak (noncovalent) bonds.

biochemistry
The study of the chemical compounds and reactions that occur in living organisms.

biosynthetic
Pertaining to the processes by which living cells make organic molecules.

bond—*see* **chemical bond**

bond length
The distance between two atoms linked by a specific chemical bond (usually a covalent bond).

bond energy
The strength of the chemical linkage between two atoms, measured by the energy in kilocalories/mole needed to break it.

C-kinase
Protein kinase activated by Ca^{2+} and diacylglycerol.

C-terminus (carboxyl terminus)
That end of a polypeptide chain which carries an unattached carboxylic acid group.

calmodulin (CaM)
Small Ca^{2+}-binding protein that modifies the activity of many enzymes and other proteins in response to changes of Ca^{2+} concentration.

calorie
Unit of heat. One calorie (lowercase "c") is the amount of heat needed to raise the temperature of 1 gram of water by 1°C. One kilocalorie (1000 calories) is sometimes called a Calorie (uppercase "C").

cancer
Disease arising from mutant cells that escape normal controls on cell division and invade and colonize the tissues of the body.

carbohydrate
General term for sugars and related compounds with the general formula $(CH_2O)_n$.

carbon fixation
Process by which green plants incorporate carbon atoms from atmospheric carbon dioxide into sugars. The second stage of photosynthesis.

carbonyl group (–C=O)
Chemical group consisting of a carbon atom linked to an oxygen atom by a double bond.

carboxyl group (–COOH)
Chemical group consisting of a carbon atom linked both to an oxygen atom by a double bond and to a hydroxyl group. Molecules containing a carboxyl group are weak acids (carboxylic acids).

carboxyl terminus—*see* **C-terminus**

carrier protein
Membrane transport protein that binds to a solute and transports it across the membrane by undergoing a series of conformational changes.

cascade—*see* **signaling cascade**

catabolic
(Of a biochemical reaction or reaction pathway) involving degradation of larger molecules to smaller molecules with production of useful energy.

catabolism
The system of enzyme-catalyzed reactions in a cell by which complex molecules are degraded to simpler ones with release of energy. Intermediates in these reactions are sometimes called catabolites.

catalyst
Substance that accelerates a chemical reaction without itself undergoing a change. Enzymes are catalysts made of protein.

cation
Positively charged ion, such as Na^+ or $CH_3NH_3^+$. (Pronounced "cat-ion.")

Cdk (cyclin-dependent kinase)
Protein kinase that has to be complexed with a cyclin protein in order to act. Different Cdk-cyclin complexes trigger different steps in the cell-division cycle by phosphorylating specific target proteins.

cell
The basic unit from which living organisms are made, consisting of an aqueous solution of organic molecules enclosed by a membrane. All cells arise from existing cells, usually by a process of division into two.

cell body
Main part of a nerve cell that contains the nucleus. The other parts are axons and dendrites.

cell cortex
Specialized layer of cytoplasm on the inner face of the plasma membrane. In animal cells it is an actin-rich layer responsible for cell-surface movements.

cell cycle
Reproductive cycle of the cell: the orderly sequence of events by which a cell duplicates its contents and divides into two.

cell division
Separation of a cell into two daughter cells. In eucaryotic cells it entails division of the nucleus (mitosis) closely followed by division of the cytoplasm (cytokinesis).

cell junction
Specialized region of connection between two cells or between a cell and the extracellular matrix.

cell line
Clone of cells of plant or animal origin capable of dividing indefinitely in culture.

cell locomotion (cell migration)
Active movement of a cell from one location to another. Particularly the migration of a cell over a surface.

cell senescence
The normal aging of cells of a higher animal whereby, after an allotted number of divisions, either in the body or in culture, they cease to divide and eventually die.

cell signaling
Communication between cells by extracellular chemical signals; especially, the molecular mechanisms by which cells detect and respond to these signals.

cellulose
Structural polysaccharide consisting of long chains of covalently linked glucose units. It provides tensile strength in plant cell walls.

cell wall
Mechanically strong fibrous layer deposited by a cell outside its plasma membrane. Prominent in most plants, bacteria, algae, and fungi but not present in most animal cells.

central dogma
The principle that genetic information flows from DNA to RNA to protein.

centriole
Short cylindrical array of microtubules, usually found (in animal cells) in a paired arrangement at the center of a centrosome. Similar structures are found at the base of cilia and flagella, where they are called basal bodies.

centromere
Constricted region of a mitotic chromosome that holds sister chromatids together; also the site on the DNA

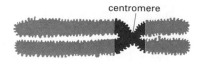

centromere

where the kinetochore forms as an attachment site for microtubules from the mitotic spindle.

centrosome (cell center)
Centrally located organelle of animal cells that is the primary microtubule-organizing center and is duplicated to form the spindle poles during mitosis. In most animal cells it contains a pair of centrioles.

channel protein
Protein that forms a narrow hydrophilic pore across a membrane, allowing ions or small molecules to move passively from one side to the other.

checkpoint
Point in the eucaryotic cell-division cycle where progress through the cycle can be halted until conditions are suitable for the cell to proceed to the next stage.

chemical bond
Linkage between two atoms that holds them together in a chemical compound. Types found in living cells include ionic bonds, covalent bonds, polar bonds, and hydrogen bonds.

chemical group
Set of covalently linked atoms, such as a hydroxyl group ($-OH$) or an amino group ($-NH_2$), that occurs in many different molecules and has a distinctive chemical character.

chemiosmotic coupling
Mechanism in which a gradient of hydrogen ions (a pH gradient) across a membrane is used to drive an energy-requiring process, such as ATP production or the transport of a molecule across a membrane.

Chlamydomonas
Unicellular green alga with two flagella.

chlorophyll
Light-absorbing pigment that plays a central part in photosynthesis.

chloroplast
Specialized organelle in algae and plants that contains chlorophyll and in which photosynthesis takes place.

cholesterol
Lipid molecule with a characteristic four-ringed steroid structure that is an important component of the plasma membranes of animal cells. (*See* Figure 11–7.)

chromatid
One of the two identical copies of a chromosome that are formed by DNA replication but are still joined at the centromere; in mitosis, the chromatids detach from one another to become separate daughter chromosomes.

chromatin
Complex of DNA, histones, and nonhistone proteins found in the nucleus of a eucaryotic cell. The material of which chromosomes are made.

chromosome
Long threadlike structure composed of DNA and associated proteins that carries part or all of the genetic information of an organism. Especially evident in plant and animal cells undergoing mitosis or meiosis. (*See also* **interphase chromosome; mitotic chromosome**.)

ciliate

Type of single-celled eucaryotic organism (protozoan) characterized by numerous cilia on its surface. The cilia are used for swimming, feeding, or the capture of prey.

cilium (plural cilia)

Hairlike extension on the surface of a cell with a core bundle of microtubules and capable of performing repeated beating movements. Cilia in large numbers drive the movement of fluid over epithelial sheets, as in the lungs.

cis

On the near side of; for example, the *cis* Golgi network is that part of the Golgi apparatus closest to the endoplasmic reticulum.

citric acid cycle (TCA, or tricarboxylic acid cycle)

Central metabolic pathway in all aerobic organisms that oxidizes acetyl groups derived from food molecules to CO_2. In eucaryotic cells these reactions are located in the mitochondrial matrix.

cloning

Making many identical copies of a cell or a DNA molecule.

cloning vector—*see* **vector**

codon

Sequence of three nucleotides in a DNA or messenger RNA molecule that represents the instruction for incorporation of a specific amino acid into a polypeptide chain.

coenzyme A (CoA)

Small molecule used in the enzymatic transfer of acyl groups in the cell. (*See also* **acetyl CoA** and Figure 3–30.)

coiled-coil

Especially stable rodlike protein structure formed by two alpha helices coiled around each other.

collagen

Fibrous protein rich in glycine and proline that is a major component of the extracellular matrix in animal tissues. Exists in many forms: type I, the most common, is found in skin, tendon, and bone; type II is found in cartilage; type IV is present in basal laminae; and so on.

combinatorial

Describes any process that is governed by a specific combination of factors (rather than by any one single factor), with different combinations having different effects.

combinatorial control

(Of the expression of a gene) control dependent on the presence or absence of a specific combination of regulatory proteins.

complementary

(Of two nucleic acid sequences) capable of precisely base-pairing with one another by matching of G with C and A with T or U, like the two strands of a DNA double helix.

complementary DNA (cDNA)

DNA molecule made as a copy of mRNA and therefore lacking the introns that are present in genomic DNA. Used to determine the amino acid sequence of a protein by DNA sequencing or to make the protein in large quantities by cloning followed by expression.

complex—*see* **molecular complex**

condensation reaction

Type of chemical reaction in which two organic molecules become linked to each other by a covalent bond with concomitant removal of a molecule of water. Also called a dehydration reaction.

conformation

Spatial arrangement of the atoms of a molecule. The precise shape of a protein or other macromolecule in three dimensions.

coupled reaction

Linked pair of chemical reactions in which free energy released by one serves to drive the other.

coupled transport

Membrane transport process in which the transfer of one molecule depends on the simultaneous or sequential transfer of a second molecule.

covalent bond

Stable chemical link between two atoms produced by sharing one or more pairs of electrons.

crossing-over

Process whereby two homologous chromosomes break at corresponding sites and rejoin to produce two recombined chromosomes.

cyclic AMP (cAMP)

Nucleotide generated from ATP in response to hormonal stimulation of cell surface receptors. cAMP acts as a signaling molecule by activating A-kinase; it is hydrolyzed to AMP by a phosphodiesterase.

cyclic-AMP-dependent kinase—*see* **A-kinase**

cyclin

Protein that periodically rises and falls in concentration in step with the eucaryotic cell cycle. Cyclins activate specific protein kinases (*see* **Cdk**) and thereby help control progression from one stage of the cell cycle to the next.

cyclin-dependent kinase—*see* **Cdk**

cytochrome

Colored, heme-containing protein that transfers electrons during cellular respiration and photosynthesis.

cytokinesis

Division of the cytoplasm of a plant or animal cell into two, as distinct from the division of its nucleus (which is mitosis).

cytoplasm

Contents of a cell that are contained within the plasma membrane but, in the case of eucaryotic cells, outside the nucleus.

cytoskeleton

System of protein filaments in the cytoplasm of a eucaryotic cell that gives the cell a polarized shape and the capacity for directed movement. Its most abundant components are actin filaments, microtubules, and intermediate filaments.

cytosol
Aqueous solution of large and small molecules that fills the main compartment of the cytoplasm. Excludes membrane-bounded organelles such as endoplasmic reticulum and mitochondria.

dalton
Unit of molecular mass. Defined as one-twelfth the mass of an atom of carbon 12 (1.66×10^{-24} g); approximately equal to the mass of a hydrogen atom.

dehydration reaction—*see* **condensation reaction**

denature
To cause a dramatic change in conformation of a protein or nucleic acid by heating or by chemical treatments. Usually results in the loss of biological function.

dendrite
Extension of a nerve cell, typically branched and relatively short, that receives stimuli from other nerve cells.

deoxyribonucleic acid—*see* **DNA**

desmosome
Specialized cell-cell junction, usually formed between two epithelial cells, mediated by cadherin molecules and characterized by dense plaques of protein into which intermediate filaments in the two adjoining cells insert.

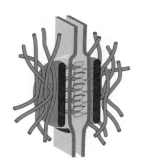

detergent
Soapy compound with a hydrophobic tail and a hydrophilic head; widely used by biochemists to solubilize membrane proteins and other hydrophobic molecules.

development
Succession of changes through which a fertilized egg gives rise to a multicellular plant or animal.

differentiation
Process in which a cell undergoes a change to a distinctively specialized character.

diffusion
The spread of molecules and small particles from one location to another through random, thermally driven movements.

digestion
The enzymatic breakdown of large molecules ingested as food. The small molecules produced then enter the cytosol of cells and are metabolized.

dimer
A structure composed of two equivalent subunits. The term "heterodimer" is sometimes used when the two subunits are not identical.

diploid
Containing two sets of homologous chromosomes and hence two copies of each gene or genetic locus.

disaccharide
Carbohydrate molecule, such as sucrose, consisting of two covalently joined monosaccharide units.

disulfide bond (S–S bond)
Covalent linkage formed between two sulfhydryl groups on cysteines. Common way to join two proteins or to link together different parts of the same protein in the extracellular space.

DNA (deoxyribonucleic acid)
Double-stranded polynucleotide formed from two separate chains of deoxyribonucleotide units; serves as the carrier of genetic information.

DNA cloning—*see* **cloning**

DNA library
Collection of cloned DNA molecules, usually representing either an entire genome (genomic library) or copies of the mRNA extracted from a cell or tissue sample (cDNA library).

DNA ligase—*see* **ligase**

DNA polymerase—*see* **polymerase**

DNA repair
Collective term for the enzymatic processes that correct deleterious changes affecting the continuity or sequence of a DNA molecule.

DNA replication
The process by which a copy of a DNA molecule is made.

DNA transcription—*see* **transcription**

domain
Individual region of a larger structure. A protein domain is a compact and stably folded region of polypeptide. A membrane domain is a region of bilayer with a characteristic lipid and protein composition.

double bond
A type of chemical linkage between two atoms, formed by sharing four electrons.

double helix
The typical conformation of a DNA molecule in which two strands are wound around each other with base-pairing between the strands.

Drosophila melanogaster
Species of small fly, commonly called a fruit fly, much used in genetic studies of development.

dynamic instability
The property shown by microtubules of growing and shrinking repeatedly through the addition and loss of tubulin subunits from their exposed ends.

dynein
Member of a family of large motor proteins that undergo ATP-dependent movement along microtubules. Dynein is responsible for the bending of cilia.

egg
The female germ cell, usually large, nonmotile, and having abundant cytoplasm.

electrochemical gradient
Driving force that causes an ion to move across a membrane. Caused by differences in ion concentration and in electrical potential on either side of the membrane.

electron
Fundamental subatomic particle with a unit negative charge (e^-).

electron acceptor
Atom or molecule that takes up electrons readily; upon gaining an electron, it is said to become reduced.

electron carrier
Molecule such as cytochrome c that transfers an electron from a donor molecule to an acceptor molecule.

electron donor
Molecule that easily gives up an electron; in the process, it is said to become oxidized.

electron transport
Movement of electrons from a higher to a lower energy level along a series of electron carrier molecules (termed an *electron-transport chain*) as in oxidative phosphorylation and photosynthesis.

element
Substance that cannot be broken down to any other chemical form; composed of a single type of atom.

endocytosis
Uptake of material into a cell by an invagination of the plasma membrane leading to internalization in a membrane-bounded vesicle. (*See also* **pinocytosis** and **phagocytosis**.)

endoplasmic reticulum (ER)
Labyrinthine, membrane-bounded compartment in the cytoplasm of eucaryotic cells, where lipids are secreted and membrane-bound proteins are made.

enhancer
Regulatory DNA sequence to which gene regulatory proteins bind, influencing the rate of transcription of a structural gene that can be many thousands of base pairs away.

entropy
Thermodynamic quantity that measures the degree of disorder in a system; the higher the entropy, the more the disorder.

enzyme
A protein that catalyzes a specific chemical reaction.

epithelium
Sheet of cells covering or lining an external surface or cavity.

equilibrium
In a chemical context, a state in which forward and reverse reactions proceed at exactly equal rates so that no net chemical change occurs.

equilibrium constant (K)
Number that characterizes the equilibrium state for a reversible chemical reaction; given by the ratio of forward and reverse rate constants of the reaction. (*See* Table 3–1, p. 92.)

Escherichia coli (E. coli)
Rodlike bacterium normally found in the colon of humans and other mammals and widely used in biomedical research.

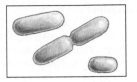

eucaryote (eukaryote)
Living organism composed of one or more cells with a distinct nucleus and cytoplasm. Includes plants, animals, fungi, and protozoa; excludes bacteria (procaryotes).

evolution
Gradual change in a population of living organisms, taking place through mutation and natural selection over many generations; the process by which new species of living organisms are generated.

exocytosis
Process by which most secreted molecules are exported from a eucaryotic cell. The molecules are packaged in membrane-bounded vesicles that fuse with the plasma membrane, releasing their contents to the outside.

exon
Segment of a eucaryotic gene that is transcribed into RNA and codes for the amino acid sequence of part of a protein. (*See also* **intron**.)

exon shuffling
Evolutionary process by which new genes are formed by linking together combinations of initially separate exons encoding different protein domains.

expression—*see* **gene expression**

extracellular matrix
Complex network of polysaccharides (such as glycosaminoglycans or cellulose) and proteins (such as collagen) secreted by cells. A structural component of tissues that also influences their development and physiology.

FADH$_2$ (reduced flavin adenine dinucleotide)
Major electron carrier in metabolism; produced by reduction of FAD during the oxidation of catabolites such as succinate.

fat
Lipid used by living cells to store metabolic energy. Mainly composed of triacylglycerols. (*See* Panel 2–4, pp. 58–59.)

fatty acid
Compound such as palmitic acid that has a carboxylic acid attached to a long hydrocarbon  chain. Used as a major source of energy during metabolism and as a starting point for the synthesis of phospholipids. (*See* Panel 2–4, p. 58.)

feedback inhibition
A form of metabolic control in which the end product of a chain of enzymatic reactions reduces the activity of an enzyme early in the pathway.

fermentation
The breakdown of organic molecules without the involvement of molecular oxygen. Oxidation is less complete than in aerobic processes and yields less energy.

fibroblast
Common cell type, found in connective tissue, that secretes an extracellular matrix rich in collagen and other macromolecules. Migrates and proliferates readily in wound repair and in tissue culture.

fibrous protein
A protein with an elongated shape; typically one such as collagen or intermediate filament protein that is able to associate into long filamentous structures.

filopodium (plural **filopodia**)
Long, thin, actin-containing extension on the surface of an animal cell. Sometimes has an exploratory function, as in the growth cone of a developing nerve cell.

flagellum (plural **flagella**)
Long, whiplike protrusion that drives a cell through a fluid medium by its beating. Eucaryotic flagella are longer versions of cilia; bacterial flagella are completely different, being smaller and simpler in construction.

free energy (G)
Energy that can be extracted from a system to do useful work, such as driving a chemical reaction.

free-energy change (ΔG)
"Delta G": the difference in free energy between reactant and product molecules in a chemical reaction. A large negative value of ΔG indicates that the reaction has a strong tendency to occur spontaneously.

gamete
Cell type in a diploid organism that carries only one set of chromosomes and is specialized for sexual reproduction. A sperm or an egg.

gap junction
Communicating cell-cell junction that allows ions and small molecules to pass from the cytoplasm of one cell to the cytoplasm of the next.

gene
Region of DNA that controls a discrete hereditary characteristic of an organism, usually corresponding to a single protein or RNA.

gene expression
The process by which a gene exerts its effect on a cell or an organism, usually by directing the synthesis of an RNA molecule that can be translated into a protein with a characteristic activity.

gene regulatory protein
General name for any protein that binds to a specific DNA sequence to alter the expression of a gene.

genetic code
Set of rules specifying the correspondence between nucleotide triplets (codons) in DNA or RNA and amino acids in proteins.

genetic engineering—*see* **recombinant DNA technology**

general transcription factor
Any of the proteins whose assembly around the TATA box is required for the initiation of transcription of most eucaryotic genes.

genome
The total genetic information carried by a cell or an organism (or the DNA molecules that carry this information).

genotype
The specific set of genes carried by an individual cell or organism.

germ cell
Cell belonging to the specialized cell lineage that gives rise to gametes (eggs or sperm) in a multicellular animal or plant.

globular protein
Any protein with an approximately rounded shape. Most enzymes are globular.

glucose
Six-carbon sugar that plays a major role in the metabolism of living cells. Stored in polymeric form as glycogen in animal cells and as starch in plant cells. (*See* Panel 2–3, pp. 56–57.)

glycocalyx
Coat of polysaccharides, including the polysaccharide portions of proteoglycans and oligosaccharides attached to protein or lipid molecules, on the outer surface of a cell.

glycogen
Polysaccharide composed exclusively of glucose units used to store energy in animal cells. Large granules of glycogen are especially abundant in liver and muscle cells.

glycolipid
Membrane lipid molecule with a short carbohydrate chain attached to a hydrophobic tail.

glycolysis
Ubiquitous metabolic pathway in the cytosol in which sugars are incompletely degraded and ATP is produced. (From Greek, for "sugar splitting.")

glycoprotein
Any protein with one or more covalently linked oligosaccharide chains. Includes most secreted proteins and most proteins exposed on the outer surface of the plasma membrane.

Golgi apparatus
Membrane-bounded organelle in eucaryotic cells, where the proteins and lipids made in the endoplasmic reticulum are modified and sorted. (Named after its discoverer, Camillo Golgi.)

G protein
One of a large family of GTP-binding proteins that are important intermediaries in signaling pathways. Usually activated by the binding of a hormone or other ligand to a membrane receptor.

G₁ phase
Gap 1 phase of the eucaryotic cell cycle, between the end of cytokinesis and the start of DNA synthesis.

G₂ phase

Gap 2 phase of the eucaryotic cell cycle, between the end of DNA synthesis and the beginning of mitosis.

group—*see* **chemical group**

growth factor

Extracellular polypeptide signal molecule that stimulates a cell to grow or proliferate. Examples are epidermal growth factor (EGF) and platelet-derived growth factor (PDGF).

GTP (guanosine 5′-triphosphate)

Major nucleoside triphosphate used in the synthesis of RNA and in some energy-transfer reactions. Has a special role in microtubule assembly, protein synthesis, and cell signaling.

GTP-binding protein

An allosteric protein whose conformation is determined by its association with either GTP or GDP. Includes many proteins involved in cell signaling, such as G proteins.

haploid

Possessing only one set of chromosomes, as in a sperm cell or a bacterium. (*See also* **diploid**.)

helix

An elongated structure that twists in regular corkscrew fashion around a central axis.

α helix—*see* **alpha helix**

heterochromatin

Region of a chromosome that remains unusually condensed and transcriptionally inactive during interphase.

high-energy bond

Covalent bond whose hydrolysis releases an unusually large amount of free energy under the conditions existing in a cell. Examples include the phosphodiester bonds in ATP and the thioester linkage in acetyl CoA.

histone

One of a group of basic proteins, rich in arginine and lysine, that are associated with DNA in chromosomes.

homologous

Similar by virtue of a common evolutionary origin. Homologous genes or proteins generally show similarities in their sequence.

homologous chromosome

One of the two copies of a particular chromosome in a diploid cell, one from the father and the other from the mother.

homologue

(1) A homologous chromosome. (2) A macromolecule that has a close evolutionary relationship to another.

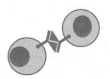

homophilic

(Of an adhesion molecule) binding to other molecules of the same kind.

hormone

A chemical signal produced by one set of cells in a multicellular organism and transported via the circulation to remote target tissues on which it exerts specific effects.

hybridization

Experimental process in which two complementary nucleic acid strands are allowed to bind to each other selectively; a powerful technique for detecting specific nucleotide sequences.

hydrogen bond

A weak chemical bond between an electronegative atom such as nitrogen or oxygen and a hydrogen atom bound to another electronegative atom.

hydrogen ion (H⁺)

A proton in aqueous solution—the basis of acidity. Such protons readily combine with water molecules to form H_3O^+ (*see also* **hydronium ion**), so that hydrogen ions in a strict sense are a rarity.

hydrolysis (adjective hydrolytic)

Cleavage of a covalent bond with accompanying addition of water, –H being added to one product of the cleavage and –OH to the other.

hydronium ion (H₃O⁺)

The ion created by addition of a proton to a water molecule.

hydrophilic

Charged or polar molecule or part of a molecule that forms enough hydrogen bonds to water to dissolve readily in water. (From Greek, for "water loving.")

hydrophobic (lipophilic)

Nonpolar molecule or part of a molecule that cannot form favorable bonding interactions with water molecules and therefore does not dissolve in water. (From Greek, for "water hating.")

hydroxyl group (–OH)

Chemical group consisting of a hydrogen atom linked to an oxygen, as in an alcohol.

hypertonic

(Of a solution bathing a cell) containing a sufficiently high concentration of solutes to cause water to move out of the cell by osmosis. (From Greek, *hyper*, over.)

hypotonic

(Of a solution bathing a cell) containing a sufficiently low concentration of solutes to cause water to move into the cell by osmosis. (From Greek *hypo*, under.)

initiation factor

Protein that promotes the proper association of ribosomes with mRNA and is required for the initiation of protein synthesis.

inositol

Sugar molecule with six hydroxyl groups that forms the sugar component of inositol phospholipids.

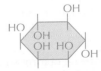

inositol phospholipid (phosphoinositide)

Minor lipid components of plasma membranes containing phosphorylated inositol derivatives; important in signal transduction in eucaryotic cells.

in situ hybridization
Technique in which a single-stranded RNA or DNA probe is used to locate a gene or an mRNA molecule in a cell or tissue.

integration
Process by which one DNA molecule recombines with and becomes physically part of another DNA molecule. This is the way in which, for example, the genome of a retrovirus becomes incorporated into a host genome.

intermediate filament
Fibrous protein filament (about 10 nm in diameter) that forms ropelike bundles in animal cells. Often provides tensile strength to withstand tension applied to the cell from outside.

interphase chromosome
A chromosome in the decondensed (opened-up) state adopted during interphase of the cell cycle, when the DNA is accessible for transcription. (*See also* **chromosome; mitotic chromosome.**)

internal membrane
Eucaryotic cell membrane other than the plasma membrane. The membranes of the endoplasmic reticulum and the Golgi apparatus are examples.

intron
Region of a eucaryotic gene that does not code for protein but is transcribed into an RNA molecule and then excised by RNA splicing to produce mRNA. (So called because it intervenes between the segments of RNA that code for protein.)

in vitro
Term used by biochemists to describe a process taking place in an isolated cell-free extract. Also used by cell biologists to refer to cells growing in culture. (Latin for "in glass.")

in vivo
In an intact cell or organism. (Latin for "in life.")

ion
Atom or molecule carrying an electrical charge, either positive or negative.

ion channel
Transmembrane protein that forms a water-filled channel across the lipid bilayer through which specific inorganic ions can diffuse down their electrochemical gradients.

ionic bond
Attractive force that holds together two ions, one positive and the other negative.

iron-sulfur center
Grouping of atoms within a type of protein, consisting of iron atoms linked to sulfur atoms and cysteine side chains.

isoform
One of a set of forms of the same protein that differ somewhat in their amino acid sequence. They can be produced by different genes or by alternative splicing of RNA transcripts from the same gene.

isomer (stereoisomer)
One of a set of substances that contain the same atoms and have the same molecular formula (such as $C_6H_{12}O_6$)

but differ in the spatial arrangement of these atoms. Optical isomers differ only by being mirror images of each other.

isotope
One of a set of forms of an atom that have the same chemistry but differ in atomic weight. May be either stable or radioactive.

K_M
The concentration of substrate at which an enzyme works at half its maximum rate. Large values of K_M usually indicate that the enzyme binds to its substrate with low affinity.

kilocalorie (kcal)
Unit of heat equal to 1000 calories. Often used to express the energy content of food or molecules: bond strengths, for example, are measured in kcal/mole. An alternative unit in wide use is the kilojoule; one kilocalorie equals 4.2 kilojoules.

kilojoule (kJ)
A unit of energy in the meter-kilogram-second system, equal to 0.239 kilocalories. A mass of 2 metric tons, moving at a speed of 1 meter per second, has a kinetic energy of exactly 1 kilojoule.

kinase
Enzyme that transfers a phosphate group from ATP (or from another nucleoside triphosphate) onto another molecule. (*See also* **protein kinase.**)

kinesin
One of the types of motor proteins that use the energy of ATP hydrolysis to move along microtubules.

kinetochore
Complex protein-containing structure on a mitotic chromosome to which microtubules attach. The kinetochore forms on the part of the chromosome known as the centromere.

lagging strand
One of the two newly made strands of DNA found at a replication fork. The lagging strand is made in discontinuous segments that are later joined covalently.

lamellipodium (plural lamellipodia)
Dynamic sheetlike protrusion of the plasma membrane of an animal cell, especially one migrating over a surface.

leading strand
One of the two newly made strands of DNA found at a replication fork. The leading strand is made by continuous synthesis in the 5′-to-3′ direction.

ligand
Molecule such as a hormone or a neurotransmitter that binds to a specific site on a protein.

ligand-gated channel
Ion-channel protein that opens when it binds a small molecule such as a neurotransmitter.

ligase
Enzyme that joins two segments of DNA or RNA together end-to-end.

lipid
Organic molecule, generally containing hydrocarbon chain(s), that is insoluble in water but dissolves readily in

nonpolar organic solvents. One class, the phospholipids, forms the structural basis of biological membranes.

lipid bilayer
Thin bimolecular sheet of mainly phospholipid molecules that forms the structural basis for all cell membranes. The two layers of phospholipid molecules are 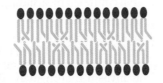 packed with their hydrophobic tails pointing inward and their hydrophilic heads outward, exposed to water.

lipophilic—*see* **hydrophobic**

lumen
Cavity enclosed by an epithelial sheet (in a tissue) or by a membrane (in a cell)—for example, the lumen of the endoplasmic reticulum. (From Latin, *lumen*, light or opening.)

lymphocyte
White blood cell that mediates the specific immune response to a foreign molecule (an antigen). B lymphocytes (B cells) produce antibody molecules; T lymphocytes (T cells) recognize and respond to foreign molecules displayed on cell surfaces and help to regulate the behavior of B lymphocytes.

lysosome
Intracellular membrane-bounded organelle containing digestive enzymes. The interior of a lysosome is strongly acidic, and its enzymes are active at an acid pH.

M phase
Period of the eucaryotic cell cycle during which the chromosomes are condensed and the nucleus and cytoplasm divide.

M-phase-promoting factor—*see* **MPF**

macromolecule
Polymer molecule, such as a protein, nucleic acid, or polysaccharide, with a molecular mass greater than a few thousand daltons. (From Greek, *makros*, big.)

macrophage
Cell found in animal tissues that is specialized for the uptake of particulate material by phagocytosis; derived from a type of white blood cell. (From Greek, *makros*, big, *phagein*, to eat.)

meiosis
Special type of cell division by which eggs and sperm cells are made, involving reduction from a diploid (double) to a haploid (single) chromosome set. Two successive nuclear divisions with only one round of DNA replication generate four haploid daughter cells from the initial diploid cell. (From Greek, *meiosis*, diminution.)

membrane
Thin sheet of lipid molecules and associated proteins that encloses all cells and forms the boundaries of many eucaryotic organelles.

membrane domain—*see* **domain**

membrane potential
Voltage difference across a membrane due to a slight excess of positive ions on one side and of negative ions on the other. A typical membrane potential for an animal cell plasma membrane is –60 mV (inside negative), measured relative to the surrounding fluid.

membrane protein
A protein associated with a lipid bilayer. Integral membrane proteins are embedded in, and generally span, the bilayer; peripheral membrane proteins are attached to its surface.

membrane transport protein
Any protein embedded in a membrane that enables ions or small molecules to pass from one side of the membrane to the other.

messenger RNA (mRNA)
RNA molecule that specifies the amino acid sequence of a protein. Produced by RNA splicing (in eucaryotes) from a larger RNA molecule made by RNA polymerase as a complementary copy of DNA. It is translated into protein in a process catalyzed by ribosomes.

metabolic pathway
Sequence of enzymatic reactions in which the product of one reaction is the substrate of the next.

metabolism
The sum total of the chemical reactions that take place in a living cell or multicellular organism.

metaphase
Stage of mitosis at which chromosomes are firmly attached to the mitotic spindle at its equator but have not yet segregated toward opposite poles.

methyl group (–CH$_3$)
Hydrophobic chemical group derived from methane (CH$_4$).

micro-
Prefix denoting 10^{-6}.

micrograph
Picture taken through a microscope. A light micrograph is taken through a light microscope, an electron micrograph, through an electron microscope.

micrometer (μm or micron)
Unit of length equal to 10^{-6} meter or 10^{-4} centimeter.

microtubule
Long, stiff, cylindrical intracellular structure, 20 nanometers 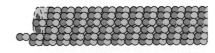 in diameter, composed of the protein tubulin; one of the major components of the cytoskeleton, used by eucaryotic cells to regulate their shape and control their movements.

milli-
Prefix denoting 10^{-3}.

mitochondrion (plural mitochondria)
Membrane-bounded organelle, about the size of a bacterium, that carries out oxidative phosphorylation and produces most of the ATP in eucaryotic cells.

mitosis
Division of the nucleus of a eucaryotic cell, involving condensation of the DNA into visible chromosomes. (From Greek, *mitos,* a thread, referring to the threadlike appearance of the condensed chromosomes.)

mitotic chromosome
A chromosome in the condensed state adopted during the mitotic phase of the cell cycle. Mitotic chromosomes, in contrast with interphase chromosomes, are individually visible in the light microscope as dense threadlike or sausagelike bodies. (*See also* **chromosome; interphase chromosome.**)

mitotic spindle
Array of microtubules and associated molecules that forms between the two poles of a eucaryotic cell during mitosis and serves to move the duplicated chromosomes apart.

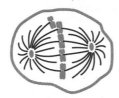

mole
M grams of a substance, where M is its relative molecular mass (molecular weight). One mole of any substance consists of 6.02×10^{23} molecules.

molecular complex
An assembly of molecules, usually macromolecules, held together by noncovalent bonds and performing a specific function, such as DNA replication or the synthesis of phospholipids.

molecular recognition
Binding of one molecule selectively to another through matching of complementary features. For example, an enzyme recognizes its substrate, and an antibody recognizes its antigen.

molecular specificity
Selective affinity of one molecule for another that permits the two to bind or react even in the presence of many unrelated molecular species.

molecular weight
Mass of a molecule expressed in daltons.

molecule
Group of atoms joined together by covalent bonds.

monomer
Small molecule that can be linked to others of the same type to form a larger molecule. (*See also* **polymer.**)

motor protein
Protein such as myosin or kinesin that uses energy derived from ATP hydrolysis to propel itself along a protein filament.

MPF (M-phase-promoting factor)
Protein complex containing cyclin and a protein kinase that triggers a cell to enter M phase. (Originally called maturation-promoting factor.)

mutation
A heritable alteration in the nucleotide sequence of a chromosome.

myofibril
Long, highly organized contractile bundle of actin, myosin, and other proteins in the cytoplasm of a muscle cell; contracts by a sliding-filament mechanism.

myosin
Motor protein that uses ATP to drive movements along actin filaments. Myosin II is a large protein that forms the thick filaments of skeletal muscle. Smaller myosins, such as myosin I, are widely distributed, and responsible for many actin-based movements inside cells.

N-terminus (amino terminus)
The end of a polypeptide chain that carries a free α-amino group.

NAD⁺ (nicotine adenine dinucleotide)
Molecule that participates in an oxidation reaction by accepting a hydride ion (H^-) from a donor molecule thereby producing NADH, which serves as an activated carrier of electrons; important in the energy-producing breakdown of sugars and fats. (*See* Figure 3–28.)

NADPH (nicotine adenine dinucleotide phosphate)
A carrier molecule closely related to NADH used as an electron donor in biosynthetic pathways.

Na⁺-K⁺ pump (Na⁺-K⁺ ATPase, sodium pump)
Transmembrane carrier protein found in the plasma membrane of most animal cells that pumps Na^+ out of and K^+ into the cell, using the energy derived from ATP hydrolysis.

nanometer (nm)
Unit of length commonly used to measure molecules and cell organelles. $1 \text{ nm} = 10^{-3} \, \mu m = 10^{-9} \, m$.

Nernst equation
Quantitative expression that relates the equilibrium ratio of concentrations of an ion on either side of a permeable membrane to the voltage difference across the membrane. (*See* Figure 12–27.)

nerve cell—*see* **neuron**

nerve terminal
Ending of an axon from which signals are relayed to a target cell or cells, usually via a synapse.

neuron (nerve cell)
Cell with long processes (axon and dendrites) specialized to receive, conduct, and transmit signals in the nervous system.

neurotransmitter
Small signaling molecule secreted by a nerve cell at a chemical synapse to signal to the postsynaptic cell. Examples include acetylcholine, glutamate, GABA, and glycine.

neutron
Fundamental subatomic particle found in the atomic nucleus; it has a mass similar to that of a proton, but no electric charge.

nitric oxide (NO)
Small highly diffusible molecule widely used for signaling between cells; it is able to diffuse across cell membranes without the help of membrane transport proteins.

nitrogen fixation
Incorporation of nitrogen from the atmosphere into nitrogen-containing organic molecules; performed by soil bacteria and by cyanobacteria.

NO—*see* **nitric oxide**

noncovalent bond
Chemical bond in which, in contrast to a covalent bond, no electrons are shared. Noncovalent bonds are relatively weak, but they can sum together to produce strong, highly specific interactions between molecules.

nondisjunction
An event that occurs occasionally during meiosis in which a pair of homologous chromosomes fails to separate so that the resulting germ cell has either too many or too few chromosomes.

nonpolar molecule
Molecule that lacks any local accumulation of positive or negative charge. Nonpolar molecules are generally insoluble in water.

nuclear envelope
Envelope surrounding the nucleus in a eucaryotic cell. It consists of two lipid bilayer membranes—an outer and an inner—and is perforated by nuclear pores.

nuclear lamina
Fibrous layer on the inner surface of the inner nuclear membrane, consisting of a mesh of intermediate filaments made from nuclear lamins.

nuclear pore
Channel through the nuclear envelope that allows selected molecules to move between the nucleus and cytoplasm.

nucleic acid
RNA or DNA; consists of a chain of nucleotides joined together by phosphodiester bonds.

nucleolus
Structure in the nucleus where ribosomal RNA is transcribed and ribosomal subunits are assembled.

nucleoside
Compound composed of a purine or pyrimidine base linked to either a ribose or a deoxyribose sugar. (*See* Panel 2–6, p. 67.)

nucleosome

Structural, beadlike unit of a eucaryotic chromosome composed of a short length of DNA wrapped around a core of histone proteins; the fundamental subunit of chromatin.

nucleotide
Nucleoside with one or more phosphate groups joined in ester linkages to the sugar part. DNA and RNA are polymers of nucleotides.

nucleus
(1) In a eucaryotic cell, the major organelle, containing DNA organized into chromosomes. (2) In an atom, the massive central body composed of neutrons and protons.

oligo-
Prefix denoting an object, such as a short polymer, consisting of a small number of subunits. An oligomer (a short polymer) may be made of amino acids (oligopeptide), sugars (oligosaccharide), or nucleotides (oligonucleotide). (From Greek, *oligos,* few or little.)

oncogene
Gene that makes a cell cancerous. Typically a mutant form of a normal gene (proto-oncogene) involved in the control of cell growth or division.

organic chemistry
The branch of chemistry concerned with compounds made of carbon. Includes essentially all of the molecules from which living cells are made, apart from water.

organelle
A discrete structure or subcompartment of a eucaryotic cell specialized to carry out a particular function; especially a cellular substructure that is visible in the light microscope. Examples include mitochondria and the Golgi apparatus.

osmosis
Movement of water molecules across a semipermeable membrane driven by a difference in concentration of solute on either side. The membrane must be permeable to water but not to the solute molecules.

osmotic pressure
Water pressure that must be exerted on a semipermeable membrane to counteract the tendency of water to flow across the membrane by osmosis.

oxidation
Loss of electron density from an atom, as occurs upon addition of oxygen to a molecule or upon removal of hydrogen. The opposite of *reduction.* (*See* Figure 3–11.)

oxidative phosphorylation
Process in bacteria and mitochondria in which ATP formation is driven by the transfer of electrons from food molecules to molecular oxygen. Involves the intermediate generation of a pH gradient across a membrane, which drives ATP synthesis by a chemiosmotic coupling.

palindromic sequence
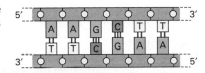
Nucleotide sequence that is identical to its complementary strand when each is read in the same $(5' \rightarrow 3')$ chemical direction.

passive transport
The movement of a small molecule or ion across a membrane driven by a difference in concentration or electrical potential.

patch-clamp recording
Technique in which the tip of a small glass electrode is sealed onto a patch of cell membrane, thereby making it possible to record the flow of current through individual ion channels in the patch.

PCR—*see* **polymerase chain reaction**

peptide bond
Chemical bond between the carbonyl group of one amino acid and the amino group of a second amino acid—a special form of amide linkage. (*See* Panel 2–5, p. 62.)

peroxisome
Small membrane-bounded organelle that uses molecular oxygen to oxidize organic molecules. Contains some enzymes that produce hydrogen peroxide and others that degrade hydrogen peroxide (H_2O_2).

phagocytic cell
A cell such as a macrophage or neutrophil that is specialized to take up particles and microorganisms by phagocytosis.

phagocytosis
The process by which particulate material is engulfed by a cell. Prominent in predatory protozoa, such as *Amoeba proteus*, and in certain specialized cells of multicellular animals such as macrophages. (From Greek, *phagein*, to eat.)

phenotype
The observable character of a cell or an organism.

phosphatidylcholine
A common phospholipid abundant in most cell membranes. (*See* Figure 11–6.)

phosphodiester bond
A covalent chemical bond formed when two hydroxyl groups are linked in ester linkage to the same phosphate group, as in the case of adjacent nucleotides in RNA or DNA. (*See* Panel 2–6, p. 67.)

phospholipase C
Enzyme associated with the plasma membrane that performs a crucial step in inositol phospholipid signaling pathways: it cleaves inositol phospholipid molecules to release diacylglycerol (DAG) and inositol trisphosphate (IP_3).

phospholipid
Type of lipid molecule used to make biological membranes. Generally composed of two fatty acids linked through glycerol phosphate to one of a variety of polar groups.

phosphorylate
To add a covalently linked phosphate group to a small molecule or protein. In the cell, phosphorylations are catalyzed by enzymes (kinases) and most often use ATP as the source of the phosphate group.

photosynthesis
The process by which plants and some bacteria use the energy of sunlight to drive the synthesis of organic molecules from carbon dioxide and water.

photosystem
Large multiprotein complex containing chlorophyll that captures light energy.

phragmoplast
Structure made of microtubules and membrane vesicles that forms in the equatorial region of a dividing plant cell.

pH scale
Common measure of the acidity of a solution: "p" refers to power of 10, "H" to hydrogen. Defined as the negative logarithm of the hydrogen ion concentration in moles per liter (M). Thus an acidic solution with pH 3 will contain 10^{-3} M hydrogen ions.

pinocytosis
Uptake of fluid by a cell by means of endocytosis. (From Greek, *pinein*, to drink.)

plasma membrane
Membrane that surrounds a living cell.

plasmid
Small circular DNA molecule that replicates independently of the genome. Used extensively as a vector for DNA cloning.

plasmodesma (plural plasmodesmata)
Cell-cell junction in plants in which a channel of cytoplasm lined by membrane connects two adjacent cells through a small pore in their cell walls.

polar
Refers to a molecule, or a covalent bond in a molecule, in which bonding electrons are attracted more strongly to some atoms than to others, thereby creating an uneven (or polarized) distribution of electric charge.

polymer
Large and usually linear molecule made by linking together multiple identical or similar units (monomers) in a repetitive fashion.

polymerase
General term for an enzyme that catalyzes addition of subunits to a polymer. DNA polymerase, for example, makes DNA, while RNA polymerase makes RNA.

polymerase chain reaction (PCR)
Technique for amplifying specific regions of DNA by multiple cycles of DNA polymerization, each followed by a brief heat treatment to separate complementary strands.

polynucleotide
A molecular chain of nucleotides chemically bonded by a series of phosphodiester linkages. RNA or DNA.

polypeptide
Linear polymer composed of multiple amino acids. Proteins are large polypeptides, and the two names can be used interchangeably.

polypeptide backbone
The linear chain of atoms containing repeating peptide bonds that runs through a protein molecule and to which amino acid side chains are attached.

polysaccharide
Linear or branched polymer composed of sugars. Examples are glycogen, hyaluronic acid, and cellulose.

position effect
Refers to differences in the expression of a gene dependent on its location in the genome.

primary transcript—*see* **transcription**

procaryote (prokaryote)
Type of living cell distinguished by the absence of a distinct nucleus. Bacterium.

programmed cell death (apoptosis)
Normal benign process of cell suicide, in which the cell shrinks, dissolves its contents, and activates phagocytosis by neighboring cells.

prometaphase
Stage of mitosis that precedes metaphase.

promoter
Nucleotide sequence in DNA to which RNA polymerase binds to begin transcription.

proofreading
The process by which DNA polymerase corrects its own errors as it moves along DNA.

prophase
First stage of mitosis during which the chromosomes are condensed but not yet attached to a mitotic spindle. Also a superficially similar stage in meiosis.

protease (proteinase, proteolytic enzyme)
Enzyme such as trypsin that degrades proteins by hydrolyzing some of their peptide bonds.

proteasome
Large protein complex in the cytosol that is responsible for degrading cytosolic proteins that have been marked for destruction by ubiquitination or by some other means.

protein
Linear polymer of amino acids linked together in a specific sequence by peptide bonds.

protein domain—*see* **domain**

protein kinase
Enzyme that transfers a phosphate group from ATP to a specific amino acid of a target protein. Cells contain hundreds of different protein kinases, phosphorylating different sets of target proteins.

protein phosphatase (phosphoprotein phosphatase)
Enzyme that removes, by hydrolysis, a phosphate group from a protein, usually with high specificity.

proteolysis
Degradation of a protein by means of a protease.

proton
Positively charged subatomic particle found in atomic nuclei; the nucleus of a hydrogen atom. Also exists as an independent chemical species as a positive hydrogen ion (H^+).

proto-oncogene—*see* **oncogene**

protozoan
A free-living, nonphotosynthetic, single-celled, motile eucaryotic organism. Most protozoa live, like *Paramecium* or *Amoeba,* by feeding on other organisms, as predators or as parasites.

pseudopodium (plural pseudopodia)
Large cell-surface protrusion formed by amoeboid cells as they crawl. More generally, any dynamic actin-rich extension of the surface of an animal cell. (From Greek, for "false foot.")

pump
Transmembrane protein that drives the active transport of ions and small molecules across the lipid bilayer.

purine
One of the two categories of nitrogen-containing ring compounds found in DNA and RNA. Examples are adenine and guanine. (*See* Panel 2–6, p. 66.)

pyrimidine
One of the two categories of nitrogen-containing ring compounds found in DNA and RNA. Examples are cytosine, thymine, and uracil. (*See* Panel 2–6, p. 66.)

reading frame
The choice of how to subdivide a nucleotide sequence into successive triplets to be read as codons for translation into protein. A nucleotide sequence can be read in any one of three reading frames depending on the starting point.

receptor-mediated endocytosis
Mechanism of selective uptake by animal cells in which a macromolecule binds to a receptor in the plasma membrane and enters the cell in clathrin-coated vesicles.

receptor protein
Protein that detects a stimulus, usually a change in concentration of a specific signal molecule, and thereupon initiates a response in the cell. Many receptor proteins are located in the plasma membrane, and interact with hormones, neurotransmitters and other molecules in the external medium.

recognition—*see* **molecular recognition**

recombinant DNA technology (genetic engineering)
The collection of techniques by which DNA segments from different sources are combined to make new DNA. Recombinant DNAs are widely used in the cloning of genes, in the genetic modification of organisms, and in molecular biology generally.

recombination
Process in which chromosomes or DNA molecules are broken and the fragments are rejoined in new combinations. Can occur in the living cell—for example, through crossing-over during meiosis—or in the test tube using purified DNA and enzymes that break and ligate DNA strands.

redox potential
A measure of the tendency of a given system to donate electrons (act as a reducing agent) or to accept electrons (act as an oxidizing agent).

redox reaction
A reaction in which electrons are transferred from one chemical species to another. An oxidation-reduction reaction.

reduction
Addition of electrons to an atom, as occurs during the addition of hydrogen to a molecule or the removal of oxygen from it. The opposite of oxidation. (*See* Figure 3–11.)

replication fork
Y-shaped region of a replicating DNA molecule at which the two parent strands separate and the two new daughter strands are synthesized.

replication origin
Site on a bacterial, viral, or eucaryotic chromosome at which DNA replication begins.

repressor
A protein that binds to a specific region of DNA to prevent transcription of an adjacent gene.

respiration
General term for any process in a cell in which the uptake of O_2 molecules is coupled to the production of CO_2.

restriction enzyme (restriction nuclease)
Nuclease that recognizes a specific short sequence of nucleotides in DNA and cleaves the DNA wherever this sequence occurs; different restriction enzymes recognize different nucleotide sequences. Extensively used in recombinant DNA technology.

retrovirus
RNA-containing virus that reproduces by entering a host cell, synthesizing a DNA copy of its own RNA genome by reverse transcription, and integrating this DNA into the genome of the host cell. The integrated viral DNA then directs synthesis of many molecules of fresh viral RNA to make new virus particles.

reverse transcriptase
Enzyme, present in retroviruses, that makes a double-stranded DNA copy from a single-stranded RNA template molecule.

ribosomal RNA (rRNA)
Any one of a number of specific RNA molecules that form part of the structure of a ribosome and participate in the synthesis of proteins.

ribosome
Particle composed of ribosomal RNAs and ribosomal proteins that associates with messenger RNA and catalyzes the synthesis of protein.

ribozyme
An RNA molecule possessing catalytic properties.

RNA (ribonucleic acid)
Polymer formed from covalently linked ribonucleotide units.

RNA polymerase
Enzyme that catalyzes the synthesis of an RNA molecule on a DNA template from nucleoside triphosphate precursors.

RNA primer
Short length of RNA made on the lagging strand during DNA replication and subsequently removed.

RNA splicing
Process in which intron sequences are excised from RNA molecules in the nucleus during formation of messenger RNA.

rough endoplasmic reticulum (RER)
Region of the endoplasmic reticulum associated with ribosomes; involved in the synthesis of secreted and membrane-bound proteins.

S phase
Period of a eucaryotic cell cycle in which DNA is synthesized.

saccharide
Used as a suffix to denote a sugar, as in disaccharide (made of two sugars) or polysaccharide (made of many sugars).

sarcomere
Repeating unit of a myofibril in a muscle cell, about 2.5 µm long, composed of an array of overlapping thick (myosin) and thin (actin) filaments.

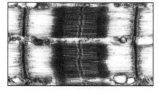

saturated
(Of an organic molecule) containing C–C but no C=C or C≡C bonds.

second messenger
Small intracellular molecule formed in or released into the cytosol in response to an extracellular signal. The second messenger relays the signal to the interior of the cell. Examples include cAMP, IP_3, and Ca^{2+}.

secretion
Production and release of a substance from a cell.

secretory vesicle
Membrane-bounded organelle in which molecules destined for secretion are stored prior to release. Sometimes called secretory granule because darkly staining contents make the organelle appear as a small dense object.

sequence
The linear order of subunits in a polymer chain—for example, of amino acids in a protein or of nucleotides in DNA. In general the sequence of a macromolecule specifies its precise biological function.

sex chromosome
Chromosome that may be present or absent, or present in a variable number of copies, according to the sex of the individual. In mammals, the X and Y chromosomes.

side chain
In an amino acid, the portion of the molecule not involved in making peptide bonds and which gives each amino acid its unique properties.

signaling cascade
Sequence of linked protein reactions, often including phosphorylation and dephosphorylation, acting as a relay chain to transmit a signal within a cell.

signal sequence
Amino acid sequence that directs a protein to a specific location in the cell, such as the nucleus, the mitochondria, or the endoplasmic reticulum.

signal transduction
Conversion of a signal from one physical or chemical form to another. In cell biology, the process by which a cell produces a response to an extracellular signal.

smooth endoplasmic reticulum (SER)
Region of the endo-plasmic reticulum not associated with ribo-somes; involved in the synthesis of lipids.

SNARE
One of a family of membrane proteins responsible for the selective fusion of vesicles with appropriate target membranes.

sodium pump—*see* **Na⁺-K⁺ pump**

solute
Any molecule that is dissolved in a liquid. The liquid is the *solvent*.

specificity—*see* **molecular specificity**

sperm
The male gamete, usually small and highly motile, and produced in large numbers.

spindle pole
One of the pair of centrosomes in a cell undergoing mitosis. Microtubules radiating from these centrosomes form the mitotic spindle.

starch
Polysaccharide composed exclusively of glucose units and used as an energy store in plant cells.

steroid hormone
Lipophilic hormone molecule related to cholesterol. Examples include estrogen and testosterone.

stress-activated channel
Membrane protein that allows the selective entry of specific ions into a cell and is opened by mechanical force.

stroma
(1) The connective tissue in which a glandular or other epithelium is embedded. (2) The large interior space of a chloroplast, containing enzymes that incorporate CO_2 into sugars for photosynthesis.

substrate
The substance on which an enzyme acts.

substratum
Solid surface to which a cell adheres.

subunit
A chemical group or molecule that forms part of a larger molecule; a monomer. Many proteins, for example, are complexes built from multiple polypeptide chains held together by noncovalent bonds; each constituent polypeptide chain is a protein subunit.

sugar
A substance made of carbon, hydrogen, and oxygen with the general formula $(CH_2O)_n$. A carbohydrate or saccharide. The "sugar" of everyday usage is a specific sweet-tasting disaccharide produced by beet or sugar cane.

sulfhydryl group (–SH, thiol)
Chemical group containing sulfer and hydrogen found in the amino acid cysteine and other molecules. Two sulfhydryls can join to produce a disulfide bond.

survival factor
Extracellular signaling molecule that must be present in order to prevent programmed cell death.

symbiosis
Intimate association between two organisms of different species from which both derive a long-term selective advantage.

symport
Form of co-transport in which a membrane carrier protein transports two solute species across the membrane in the same direction.

synapse
Specialized junction between a nerve cell and another cell (nerve cell, muscle cell, or gland cell) across which the nerve cell transmits its signal. In most synapses the signal is carried by a neurotransmitter, which is secreted by the nerve cell terminal as a result of a nerve impulse and diffuses to the target cell.

synaptic vesicle
Small membrane sac filled with neurotransmitter inside a nerve terminal at a synapse. Synaptic vesicles release their contents by exocytosis when a nerve impulse (an action potential) reaches the nerve terminal.

TATA box
Characteristic sequence rich in T's and A's found in the promoter region of many eucaryotic genes and serving to specify the site where transcription is initiated.

telomere
Structure at the end of a chromosome, associated with a characteristic DNA sequence that is replicated in a special way. (From Greek, *telos*, end, *meros*, part.)

telophase
Final stage of mitosis in which the two sets of separated chromosomes decondense and become enclosed by nuclear envelopes.

template
A molecular structure that serves as a pattern for the production of other molecules. Thus a specific sequence of nucleotides in DNA can act as a template to direct the synthesis of a new strand of DNA with a complementary sequence.

tight junction
Cell-cell junction that seals adjacent epithelial cells together, preventing dissolved molecules in the extracellular medium from passing from one side of the epithelial sheet to the other.

tissue
Organized assembly of specialized cells that forms a distinct part of a plant or an animal.

thioester bond
High-energy bond formed by a condensation reaction between an acid (acyl) group and a thiol group (–SH); seen, for example, in acetyl CoA and in many enzyme-substrate complexes.

trans
Beyond, or on the other side.

transcription

Copying of one strand of DNA into a complementary RNA sequence by the enzyme RNA polymerase.

transcription factor

Term loosely applied to any protein required to initiate or regulate transcription in eucaryotes. Includes gene regulatory proteins as well as the general transcription factors.

transduction

(In genetics) virus-mediated transfer of host DNA from one cell to another. (*See also* **signal transduction**.)

transfer RNA (tRNA)

Small RNA molecule used in protein synthesis as an interface (adaptor) between a specific codon or set of codons, mRNA, and specific amino acids. Each type of tRNA molecule is covalently linked to a particular amino acid and recognizes a particular codon or set of codons by basepairing.

transformation

Process by which cells take up DNA molecules from their surroundings and then express genes in that DNA.

transgenic

(Of a plant or an animal) possessing, in a stably incorporated form, one or more genes derived from another cell or organism and able to pass these genes on to successive generations.

trans Golgi network (TGN)

That part of the Golgi apparatus that is farthest from the endoplasmic reticulum. Proteins and lipids destined for lysosomes, secretory vesicles, or the cell surface originate from the TGN.

transition state

Chemical structure that forms transiently in the course of a reaction and has the highest free energy of any reaction intermediate.

translation

Process by which the sequence of nucleotides in a messenger RNA molecule directs the incorporation of amino acids into protein; occurs on a ribosome.

transport vesicle

Membrane vesicle that carries proteins from one intracellular compartment to another, for example, from the ER to the Golgi apparatus.

transposon (transposable element)

Segment of DNA that can move from one location to another in a chromosome or from one chromosome to another in the same cell. An important source of genetic variation in most organisms.

triacylglcerol

Glycerol ester of fatty acids; the main constituent of fat droplets in animal tissues (where the fatty acids are saturated) and of vegetable oil (where the fatty acids are mainly unsaturated).

tubulin

Protein from which microtubules are made.

tumor-suppressor gene

A gene that functions in a normal tissue cell to restrain cell proliferation or invasive behavior. Loss or inactivation of both copies of such a gene from a diploid cell leads to a loss of control and helps to convert the cell into a cancer cell.

unsaturated

(Of an organic molecule) containing one or more double or triple carbon-carbon bonds.

V_{max}

The maximum rate of an enzymatic reaction, attained immediately after addition of the substrate at saturating levels.

van der Waals force

Weak attractive force due to fluctuating electrical charges that comes into play between two atoms that are 0.3–0.4 nm apart. At a shorter distance, strong repulsive forces begin to operate.

vector

(In molecular genetics) genetic element, usually a bacteriophage or plasmid, that can incorporate a fragment of foreign DNA and carry it into a recipient cell. Widely used in gene cloning.

vesicle

Small, membrane-bounded, spherical organelle in the cytoplasm of a eucaryotic cell.

virus

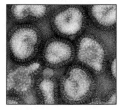

Infectious particle consisting of nucleic acid (RNA or DNA) enclosed in a protein coat; replicates by parasitizing the reproductive machinery of a host cell, from which new virus particles are released to infect other cells. Often the cause of disease.

voltage-gated channel

Membrane protein that selectively allows ions such as Na$^+$ to cross a membrane and which is opened by changes in membrane potential. Found mainly in electrically excitable cells such as nerve cells and muscle cells.

wild type

Normal, nonmutant form of a species resulting from breeding under natural conditions.

X chromosome

One of the two sex chromosomes in mammals. The cells of males contain one X and one Y chromosome; those of females contain two X chromosomes.

Y chromosome

One of the two sex chromosomes of mammals, peculiar to males. (*See also* **X chromosome**.)

yeast

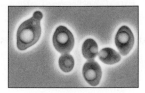

Common term for several families of unicellular fungi. Includes species used for brewing beer and making bread, as well as species that cause disease.

Answers to Questions

Chapter 1

Answer 1–1 See Figure A1–1.

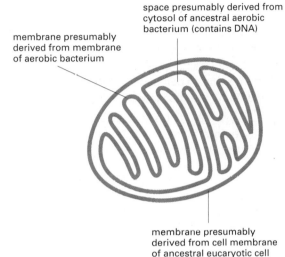

membrane presumably derived from membrane of aerobic bacterium

space presumably derived from cytosol of ancestral aerobic bacterium (contains DNA)

membrane presumably derived from cell membrane of ancestral eucaryotic cell

Figure A1–1

Answer 1–2 By engulfing substances, such as food particles, eucaryotic cells can sequester them and feed on them selfishly. Bacteria, in contrast, have no way of capturing lumps of food; they can export substances that help break down food substances in the environment, but the products of this labor must then be shared with other cells in the same neighborhood.

Answer 1–3 Light microscopy is much easier to use and requires much simpler instruments. Objects that are 1 μm in size can easily be resolved; the lower limit of resolution is 0.2 μm, which is a theoretical limit imposed by the wavelength of visible light. Visible light is nondestructive and passes readily through water, making it possible to observe living cells. Electron microscopy, on the other hand, is much more complicated, both in the preparation of the sample (which needs to be extremely thinly sliced, stained with electron-dense heavy metal, and completely dehydrated) and in the nature of the instrument. Cells cannot be observed while alive. The resolution of electron microscopy is much higher, however, and objects as small as 10 nm can easily be resolved. To see any structural detail, microtubules, mitochondria, and bacteria would need to be analyzed by electron microscopy. It is possible, however, to stain them with specific dyes and then determine their location by light microscopy.

Answer 1–4 Most random changes to the shoe design would result in defects that would make the shoes less useful: shoes with multiple heels, with no soles, or with awkward sizes would obviously not sell and would therefore be selected against by market forces. Other changes would be neutral, such as minor variations in color or in size. A minority of changes, however, might result in more desirable shoes: deep scratches in a previously flat sole, for example, might create shoes that would perform better in wet conditions; the loss of high heels might produce shoes that are far more comfortable to wear. The example illustrates that random changes can lead to meaningful improvements if the number of trials is large enough and selective pressures are imposed.

Answer 1–5 6×10^{39} ($= 6 \times 10^{27}$ g $/ 10^{-12}$ g) bacteria would have the same mass as the earth. And $6 \times 10^{39} = 2^{t/20}$, according to the equation describing exponential growth. Solving this equation for t results in $t = 2642$ min (or 44 hours). This represents only 132 generation times (!), whereas 5×10^{14} generation times have passed during the last 3.5 billion years. Obviously, the total mass of bacteria on this planet is nowhere close to the mass of the earth. This illustrates that exponential growth can occur only for very few generations, i.e., for minuscule periods of time compared to evolution. In any realistic scenario, food supplies become very quickly limiting.

This simple calculation shows us that the ability to grow and divide quickly when food is ample is only one factor in the survival of a species. Food is generally scarce, and those cells that are most adaptable to changing environments and that have acquired more sophisticated means to allow them to utilize different food sources often have a major competitive advantage.

Answer 1–6 The two different evolutionary relationships can be summarized as

    ancestor of all cells → *Giardia* → eucaryotes

    ancestor of all cells → eucaryotes → *Giardia*

To make the distinction it would therefore be important to know how closely related *Giardia* and procaryotes are. This could be done, for example, by looking at the level of similarity of their proteins or nucleic acids. Such approaches show that *Giardia* is almost as distant in the evolutionary tree from other eucaryotes as it is from procaryotes, thereby supporting the first model.

Answer 1–7 Because the basic workings of cells are so similar, a great deal has been learned from studying model systems. Brewer's yeast is a good model system because yeast cells are much simpler than human cancer cells. We can grow yeast inexpensively and in vast quantities, and we can manipulate yeast cells genetically and biochemically much more easily than human cells. This allows us to decipher the ground rules governing how cells divide and grow. Cancer cells divide when they should not (and therefore give rise to tumors), and a basic understanding of how cell division is controlled is therefore directly relevant to the cancer problem. Indeed, the National Cancer Institute, the American Cancer Society, and many other institutions that are devoted to finding a cure for cancer strongly support basic research on various aspects of cell division in different model systems, such as yeast.

Answer 1–8 Check your answers using the Glossary and Panel 1–2 (p. 18).

Answer 1–9

   A. False. The hereditary information is encoded in the cell's DNA, which in turn encodes its proteins.

   B. True. Bacteria do not have a nucleus.

   C. False. Plants are composed of eucaryotic cells that contain chloroplasts as cytoplasmic organelles. The chloroplasts are thought to be evolutionarily derived from procaryotic cells.

   D. True. The number of chromosomes varies from one organism to another, but is constant in all cells of the same organism.

   E. False. The cytosol is the cytoplasm excluding all organelles.

   F. True. The nuclear envelope is a double membrane, and mitochondria are surrounded by both an inner and an outer membrane.

   G. False. Protozoans are single-cell organisms and therefore do not have different tissues. They have a complex structure, however, that has highly specialized parts.

   H. Sort of true. Peroxisomes and lysosomes contain enzymes that catalyze the breakdown of substances produced in the cytosol or taken up by the cell. One can argue, however, that many of these substances are degraded to generate food molecules, and as such are certainly not "unwanted."

Answer 1–10 One average brain cell weighs 10^{-8} g (= 1000 g / 10^{11}). Because 1 g of water occupies 1 ml = 1 cm^3 (= 10^{-6} m^3), the volume of one cell is 10^{-14} m^3 (= 10^{-8} g × 10^{-6} m^3/g). Taking the cubic root yields a side length of 2.1 × 10^{-5} m, or 21 μm (10^6 μm = 1 m) for each cell.

The page of the book has a surface of 0.057 m^2 (= 21 cm × 27.5 cm), and each cell has a footprint of 441 × 10^{-12} m^2 (2.1 × 10^{-5} m × 2.1 × 10^{-5} m). Therefore, 129 × 10^6 (= 0.057 m^2 / 441 × 10^{-12} m^2) cells fit on this page when spread out as a single layer. Thus, 10^{11} cells would occupy 775 pages (= 10^{11} / 129 × 10^6).

Answer 1–11 In this plant cell, A is the nucleus, B is a vacuole, C is the cell wall, and D is a chloroplast. The scale bar is about 10 μm, the width of the nucleus.

Answer 1–12 The three major filaments are actin filaments, intermediate filaments, and microtubules. Actin filaments are involved in rapid cell movement, such as muscle contraction; intermediate filaments provide mechanical stability; and microtubules function as "railroad tracks" for intracellular movements, and are responsible for the separation of chromosomes during cell division. Other functions of all these filaments are discussed in Chapter 16.

Answer 1–13 It takes only 20 hours, i.e., less than a day, before mutant cells become more abundant in the culture. Using the equation provided in the question, we see that the number of the original ("wild-type") bacterial cells at time t minutes after the mutation occurred is $10^6 × 2^{t/20}$. The number of mutant cells at time t is $1 × 2^{t/15}$. To find out when the mutant cells "overtake" the wild-type cells, we simply have to make these two numbers equal to each other (i.e., $10^6 × 2^{t/20} = 2^{t/15}$). Taking the logarithm to base 10 of both sides of this equation and solving it for t, results in $t = 1200$ min (or 20 hours). At this time, the culture contains 2 × 10^{24} cells ($10^6 × 2^{60} + 1 × 2^{80}$). Incidentally, 2 × 10^{24} bacterial cells, each weighing 10^{-12} g, would weigh 2 × 10^{12} g (= 2 × 10^9 kg, or two million tons!). This can only have been a thought experiment.

Answer 1–14 Bacteria continually acquire mutations in their DNA. In the population of cells that are exposed to the poison, one or a few may acquire a mutation that makes them resistant to the action of the drug. Antibiotics that are poisonous to bacteria because they bind to certain bacterial proteins, for example, would not work if the proteins have a slightly changed surface so that binding occurs more weakly or not at all. These mutant bacteria would continue dividing rapidly when their cousins are slowed down. The antibiotic-resistant bacteria would soon become the predominant species in the culture.

Answer 1–15 $10^{13} = 2^{(t/1)}$. Therefore, it would take only 43 days [$t = 13/\log(2)$]. This explains why some cancers can progress extremely rapidly. Many cancer cells drink much more slowly, however, or die because they do not have sufficient blood supply, and the actual progression of cancer is therefore usually slower.

Answer 1–16 Living cells evolved from nonliving matter, but grow and replicate. Like the material they originated from, they are governed by the laws of physics, thermodynamics, and chemistry. Thus, for example, they cannot create energy *de novo* or build ordered structures without the expenditure of energy. We can understand virtually all cellular events, such as metabolism, catalysis, membrane assembly, and DNA replication, as complicated chemical reactions that can be experimentally reproduced, manipulated, and studied in test tubes.

Despite this fundamental reduction, a living cell is more than the sum of its parts. We cannot randomly mix proteins, nucleic acids, and other chemicals together in a test tube, for example, and make a cell. The cell functions by virtue of its organized structure, and this is a product of its evolutionary history. Cells always come from preexisting cells, and the division of a mother cell passes both chemical constituents and structures to its daughters. The plasma membrane, for example, never has to form *de novo,* but always by expansion of a preexisting membrane; there will always be a ribosome, in part made up of proteins whose function it is to make more proteins including those that build more ribosomes.

Answer 1–17 Eucaryotic cells have acquired sophistication at the expense of rapid cell growth. Because of their more complex structure, they can specialize: different cells in an organism can take on dedicated functions and cooperate with one another. With this increased sophistication, multicellular organisms are able to exploit food sources that are inaccessible to single-cell organisms. A plant, for example, can reach the soil with its roots to take up water and nutrients and at the same time harvest light energy and CO_2 from the air through its leaves. Single-cell organisms like bacteria, on the other hand, are specialized for rapid growth and are able to adapt rapidly by mutation to changing environmental conditions. The ability to reproduce and change rapidly would be limited if the bacteria were organized into multicellular structures.

Answer 1–18 The volume and the surface area are 5.24×10^{-19} m^3 and $3.14 \, 10^{-12}$ m^2 for the bacterial cell, and 1.77×10^{-15} m^3 and 7.07×10^{-10} m^2 for the animal cell, respectively. From these numbers, the surface-to-volume ratios are 6×10^6 m^{-1} and 4×10^5 m^{-1}, respectively. In other words, although the animal cell has a 3375-fold larger volume, its membrane surface is increased only 225-fold. If internal membranes are included in the calculation, however, the surface-to-volume ratios of both cells are about equal. Thus, because of their internal membranes, eucaryotic cells can grow bigger and still maintain a sufficiently large membrane area that—as we shall discuss in more detail in later chapters—is required for many essential functions. We have already encountered the mitochondrial inner membrane as an example: it is highly invaginated (i.e., folded internally), to increase its

membrane area (see Figure 1–11). This membrane is very important in the production of ATP, the energy carrier that drives most cellular reactions.

Answer 1–19 There are many lines of evidence for a common ancestor. Analyses of modern-day living cells show an amazing degree of similarity in the basic components that make up the inner workings of otherwise vastly different cells. Many metabolic pathways, for example, are conserved from one cell to another, and the compounds that make up nucleic acids and proteins are the same in all living cells, even though it is easy to imagine that a different choice of compounds (e.g., amino acids with different side chains) would have worked just as well. Similarly, it is not uncommon to find that important proteins have a closely similar detailed structure in procaryotic and eucaryotic cells. Theoretically, there would be many different ways to build proteins that could perform the same functions. The evidence overwhelmingly shows that most important processes were "invented" only once and then became fine-tuned during evolution to suit the particular needs of specialized cells.

It seems highly unlikely, however, that the first cell survived to become the primordial founder cell of today's living world. As evolution is not a directed process with a purposeful progression, it is more likely that there were a vast number of unsuccessful trials cells that replicated for a while and then became extinct because they could not adapt to changes in the environment or could not survive in competition with other types of cells. We can therefore speculate that the primordial ancestor cell was a "lucky" cell that ended up in a relatively stable environment in which it had a chance to replicate and evolve.

Chapter 2

Answer 2–1

A. The atomic number is 6; the atomic weight is 12 (= 6 protons + 6 neutrons).

B. The number of electrons is six (= the number of protons).

C. The first shell can accommodate two and the second shell eight electrons. Carbon therefore needs four additional electrons (or would have to give up four electrons) to obtain a full outermost shell. Carbon is most stable when it shares four additional electrons with other atoms (including other carbon atoms) by forming four covalent bonds.

D. Carbon 14 has two additional neutrons in its nucleus. Because the chemical properties of an atom are determined by its electrons, carbon 14 is chemically identical to carbon 12.

Answer 2–2 The statement is correct. Both ionic and covalent bonds are based on the same principles: electrons can be shared equally between two interacting atoms, forming a nonpolar covalent bond; electrons can be shared unequally between two interacting atoms, forming a polar covalent bond; or electrons can be completely lost from one atom and gained by the other, forming an ionic bond. There are bonds of every conceivable intermediate state, and for borderline cases it becomes arbitrary whether a bond is described as a very polar covalent bond or an ionic bond.

Answer 2–3 The statement is correct. The hydrogen-oxygen bond in water molecules is polar, such that the oxygen atom carries a more negative charge than the hydrogen atoms. These partial negative charges are attracted to the positively charged sodium ions, but are repelled from the negatively charged chloride ions.

Answer 2–4

A. Hydronium (H_3O^+) ions result from water dissociating into protons and hydroxyl ions, each proton binding to a water molecule to form a hydronium ion ($2H_2O \rightarrow H_2O + H^+ + OH^- \rightarrow H_3O^+ + OH^-$). At neutral pH, i.e., in the absence of an acid providing more H_3O^+ ions or a base providing more OH^- ions, the concentrations of H_3O^+ ions and OH^- ions are equal. We know that at neutrality the pH = 7.0, and therefore, the H^+ concentration is 10^{-7} M. The H^+ concentration equals the H_3O^+ concentration.

B. To calculate the ratio of H_3O^+ ions to H_2O molecules, we need to know the concentration of water molecules. The molecular weight of water is 18 (i.e., 18 g/mole) and 1 liter of water weighs 1 kg. Therefore, the concentration of water is 55.6 M (= 1000 [g/l]/[18 g/mole]), and the ratio of H_3O^+ ions to H_2O molecules is 1.8×10^{-9} (= $10^{-7}/55.6$), i.e., only two water molecules in a billion are dissociated at neutral pH.

Answer 2–5 No apologies (sorry). Note that the small hydrogen atoms are those that are linked to an oxygen atom, whereas the ones linked to a carbon atom are larger. This reflects the polarity of the respective bonds; the H–C bond is nonpolar, whereas the H–O bond is polar. Oxygen more strongly draws the shared electrons away from the hydrogen, resulting in a smaller radius of the electron cloud of the hydrogen atom.

Answer 2–6 The synthesis of a macromolecule with a unique structure requires that in each position only one stereoisomer is used. Changing one amino acid from its L- to its D-form would result in a different protein. Thus, if for each amino acid a random mixture of the D- and L-forms were used to build a protein, its amino acid sequence could not specify a single structure, but many different structures (2^N different structures would be formed, where N is the number of amino acids in the protein).

Why L-amino acids were selected in evolution as the exclusive building blocks of proteins is a mystery; we could easily imagine a cell in which certain (or even all) amino acids were used in the D-forms to build proteins, as long as these particular stereoisomers were used exclusively.

Answer 2–7 The term "polarity" can refer to two different principles. In one meaning it refers to directional asymmetry, for example, in linear polymers such as polypeptides, which have an N-terminus and a C-terminus, or nucleic acids, which have a 3′ and a 5′ end. Because bonds form only between the amino and the carboxyl groups of the amino acids in a polypeptide, and between the 3′ and the 5′ ends of nucleotides, nucleic acids and polypeptides always have two different ends, which gives the chain a defined polarity.

In the other meaning, polarity refers to a separation of electric charge in a bond or molecule. This kind of polarity allows hydrogen-bonding, and because the water solubility, or hydrophilicity, of a molecule depends upon its being polar in this sense, the term "polar" is sometimes used to indicate water solubility.

Answer 2–8 A major advantage of condensation reactions is that they are readily reversible by hydrolysis (and water is readily available in the cell). This allows cells to break down their macromolecules (or macromolecules of other organisms that were ingested as food) and to recover the subunits intact so that they can be "recycled," i.e., used to build new macromolecules.

Answer 2–9 Many of the functions that macromolecules perform rely on their ability to associate and dissociate readily. This allows cells, for example, to remodel their interior when they move or divide, and to transport components from one organelle to another. Some macromolecules, however, are indeed joined to each other by covalent bonds. This occurs primarily in situations where extreme structural stability is required, such as in the cell walls of many bacteria, fungi, and plants, and in the extracellular matrix that provides structural support for most animal cells.

Answer 2–10

A. True. All nuclei are made of positively charged protons and uncharged neutrons; the only exception is the hydrogen nucleus, which consists of only one proton.

B. False. Atoms are electrically neutral. The number of positively charged protons is always balanced by an equal number of negatively charged electrons.

C. True—but only for the cell nucleus (see Chapter 1), and not for the atomic nucleus discussed in this chapter.

D. False. Elements can have different isotopes, which differ only in their number of neutrons.

E. True. In certain isotopes the number of neutrons destabilizes the nucleus, which decomposes in a process called radioactive decay.

F. True. Examples include granules of glycogen, a polymer of glucose, found in liver cells; and fat droplets, made of aggregated triacylglycerols, found in fat cells.

G. True. Individually, these bonds are weak and readily broken by thermal motion, but because interactions between two macromolecules involve a large number of such bonds, the overall binding can be quite strong, and because hydrogen bonds form only between correctly positioned groups on the interacting macromolecules, they are very specific.

Answer 2–11

A. One cellulose molecule has a molecular weight of $n \times (12 \text{ [C]} + 2 \times 1 \text{ [H]} + 16 \text{ [O]})$. We do not know n, but we can determine the ratio with which the individual elements contribute to the weight of cellulose. The contribution of carbon atoms is 40% [$= 12/(12 + 2 + 16) \times 100\%$]. Therefore, 2 g (40% of 5 g) of carbon atoms are contained in the cellulose that makes up this page. The atomic weight of carbon is 12 g/mole, and there are 6×10^{23} atoms or molecules in a mole. Therefore, 10^{23} carbon atoms [$= (2g/12 \text{ [g/mole]}) \times 6 \times 10^{23}$ (molecules/mole)] make up this page.

B. The volume of the page is $4 \times 10^{-6} \text{ m}^3$ ($= 21 \text{ cm} \times 27.5 \text{ cm} \times 0.07 \text{ mm}$), which equals a cube with a side length of 1.6 cm ($= \sqrt[3]{4 \times 10^{-6} \text{ m}^3}$). Since we know from part A that the page contains 10^{23} carbon atoms, geometry tells us that there are about 4.6×10^7 carbon atoms ($= \sqrt[3]{10^{23}}$) lined up along each side of this cube. Therefore, about 200,000 carbon atoms ($= 4.6 \times 10^7 \times 0.07 \times 10^{-3} \text{ m} / 1.6 \times 10^{-2} \text{ m}$) span the thickness of the page.

C. If tightly stacked, 350,000 carbon atoms with a 0.2 nm diameter would span the 0.07 mm thickness of the page. The 1.7-fold difference in the two calculations reflects (1) that carbon is not the only atom in cellulose and (2) that paper is not an atomic lattice of precisely arranged molecules (as a diamond would be for precisely arranged carbon atoms), but a random meshwork of fibers containing many voids.

Answer 2–12

A. The occupancies of the three innermost electron levels are 2, 8, 8.

B.

hydrogen	gain 1 or lose 1
helium	already has full level
oxygen	gain 2
carbon	gain 4 or lose 4
sodium	lose 1
chlorine	gain 1

C. Helium with its fully occupied electron level is chemically unreactive. Sodium and chlorine on the other hand are extremely reactive and readily form stable Na^+ and Cl^- ions that form ionic bonds. Oxygen has to gain two electrons, whereas hydrogen can go either way, i.e., gain or lose an electron. In a covalent bond between oxygen and hydrogen, such as in water or in hydroxyl groups, electrons are pulled toward the oxygen atom. The bond is therefore polar with a relative negative charge at the oxygen atom and a relative positive charge at the hydrogen atom. An oxygen-carbon bond is similarly polar. In contrast, in a hydrogen-carbon bond both interacting atoms share the electrons more equally and the bond is nonpolar.

Answer 2–13 A sulfur atom is much larger than an oxygen atom, and because of its larger size, the outermost electrons are not as strongly attracted to the nucleus of the sulfur atom as they are in an oxygen atom. Consequently, the hydrogen-sulfur bond is much less polar than the hydrogen-oxygen bond. Because of the reduced polarity, the sulfur in a H_2S molecule is not strongly attracted to the hydrogen atoms in an adjacent H_2S molecule, and hydrogen bonds do not form.

Answer 2–14

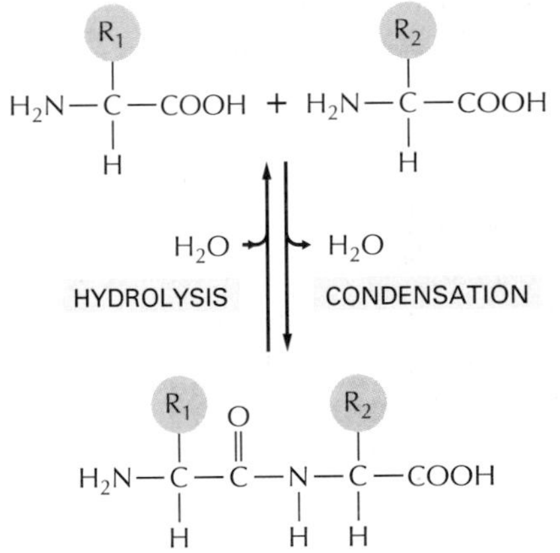

Figure A2–14

where R_1 and R_2 are amino acid side chains.

Answer 2–15

A. False. The properties of a protein depend on both the amino acids it contains and the order in which they are linked together. The diversity of proteins is due to the almost unlimited number of ways in which 20 different amino acids can be combined in a linear sequence.

B. False. Lipids assemble into bilayers by noncovalent forces. A membrane is therefore not a macromolecule.

C. True. The backbone of nucleic acids is made up of alternating ribose (or deoxyribose in DNA) and phosphate groups. Ribose and deoxyribose are sugars.

D. True. About half of the 20 naturally occurring amino acids have hydrophobic side chains. In folded proteins, many of these side chains face toward the inside of the folded-up proteins, because they are repelled from water.

E. True. Hydrophobic hydrocarbon tails contain only nonpolar bonds. Thus, they cannot participate in hydrogen-bonding and are repelled from water. We consider the underlying principles in more detail in Chapter 11.

F. False. RNA contains the four listed bases, but DNA contains T instead of U. T and U are very much alike, however, and differ only by a single methyl group.

Answer 2–16

A. (a) 400 (= 20^2); (b) 8000 (= 20^3); (c) 160,000 (= 20^4).

B. A protein with a molecular weight of 4800 is made of about 40 amino acids; thus there are 1.1×10^{52} (= 20^{40}) different ways to make such a protein. Each individual protein molecule weighs 8×10^{-21} g (= $4800/6 \times 10^{23}$); thus a mixture of one molecule each weighs 9×10^{31} g (= 8×10^{-21} g $\times 1.1 \times 10^{52}$), which is 15,000 times the total weight of the planet earth, weighing 6×10^{24} kg. You would need a quite large container, indeed.

C. Given that most cellular proteins are even larger than the one used in this example, it is clear that only a minuscule fraction of the total possible amino acid sequences is used in living cells.

Answer 2–17 Because all living cells are made up of chemicals and because all chemical reactions (whether in living cells or in test tubes) follow the same rules, an understanding of the basic chemical principles is fundamentally important to the understanding of biology. In the course of this book, we will frequently refer back to these principles, on which all of the more complicated pathways and reactions that occur in cells are based.

Answer 2–18

A. Hydrogen bonds require specific groups to interact; one is always a hydrogen atom linked in a polar bond to an oxygen or a nitrogen, and the other is usually a nitrogen or an oxygen atom. Van der Waals attractions are weaker and occur between any two atoms that are in close enough proximity. Both hydrogen bonds and van der Waals attractions are short-range interactions that come into play only when two molecules are already close. Both types of bonds can therefore be thought of as means of "fine-tuning" an interaction, i.e., helping position two molecules correctly with respect to each other once they have been brought together by diffusion.

B. Van der Waals attractions would form in all three examples. Hydrogen bonds would form in (c) only.

Answer 2–19 Noncovalent interactions form between the subunits of macromolecules—e.g., the side chains of amino acids in a polypeptide chain—and cause the polypeptide chain to assume a unique shape. These interactions include hydrogen bonds, ionic interactions, van der Waals interactions, and hydrophobic interactions. Because these interactions are weak, they can be broken with relative ease; thus, most macromolecules can be unfolded by heating, which increases thermal motion.

Answer 2–20 Amphipathic molecules have both a hydrophilic and a hydrophobic end. Their hydrophilic ends can hydrogen-bond to water, but their hydrophobic ends are repelled from water because they interfere with the water structure. Consequently, the hydrophobic ends of amphipathic molecules tend to be exposed to air at air-water interfaces, or will cluster together. (See Figure A2–20.)

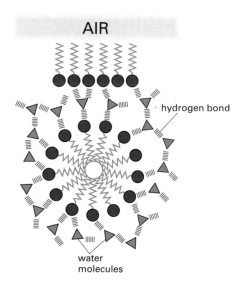

Figure A2–20

Answer 2–21

A,B. (A) and (B) are both correct formulas of the amino acid phenylalanine. In formula (B) phenylalanine is shown in the ionized form that exists in water solution, where the basic amino group is protonated and the acidic carboxylic group is deprotonated.

C. Incorrect. This structure of a peptide bond is missing a hydrogen atom bound to the nitrogen.

D. Incorrect. This formula of an adenine base features one double bond too many, creating a five-valent carbon atom and a four-valent nitrogen atom.

E. Incorrect. In this formula of a nucleoside triphosphate there should be two additional oxygen atoms, one between each of the phosphorus atoms.

F. This is the correct formula of ethanol.

G. Incorrect. Water does not hydrogen-bond to hydrogens bonded to carbon. The lack of the capacity to hydrogen-bond makes hydrocarbon chains hydrophobic, i.e., water-hating.

H. Incorrect. Na and Cl form an ionic bond, Na^+Cl^-, but a covalent bond is drawn.

I. Incorrect. The oxygen atom attracts electrons more than the carbon atom; the polarity of the two bonds should therefore be reversed.

J. This structure of glucose is correct.

K. Almost correct. It is more accurate to show that only one hydrogen is lost from the $-NH_2$ group and the $-OH$ group is lost from the $-COOH$ group.

Chapter 3

Answer 3–1 The equation represents the "bottom line" of a large set of individual reactions that are catalyzed by many individual enzymes. Because sugars are more complicated molecules than CO_2 and H_2O, the reaction generates a more ordered state inside the cell. As demanded by the second law of thermodynamics, heat is generated at many steps along the pathway of the reactions that are summarized in this equation.

Answer 3–2 Oxidation is defined as removal of electrons. Therefore, (A) is an oxidation, and (B) is a reduction. The red carbon atom in (C) remains largely unchanged; the neighboring black carbon atom, however, loses a hydrogen atom (i.e., an electron and a proton) and hence becomes oxidized. The red carbon atom in (D) becomes oxidized because it loses a hydrogen atom, whereas the carbon atom in (E) becomes reduced because it gains a hydrogen atom.

Answer 3–3 The reaction rates might be limited by (1) the concentration of the substrate, i.e., how often a molecule of CO_2 collides with the active site on the enzyme; (2) how many of these collisions are energetic enough to lead to a reaction; and (3) how fast the enzyme can release the products of the reaction and therefore be free to bind the next substrate. The diagram in Figure A3–3 shows that by lowering the activation energy barrier, more molecules have sufficient energy to undergo the reaction. The area under the curve from point A to infinite energy or from point B to infinite energy indicates the total number of molecules that will react without or with the enzyme, respectively. Although not drawn to scale, the ratio of these two areas should be 10^7.

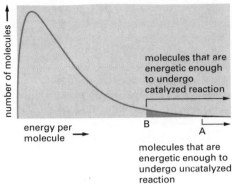

Figure A3–3

Answer 3–4

A. Both states of the coin, H and T, have an equal probability. There is therefore no driving force, i.e., no energy difference, that would favor H turning to T or vice versa. Therefore, $\Delta G° = 0$ for this reaction. However, a reaction proceeds if H and T coins are not present in the box in equal numbers. Now the concentration difference between H and T creates a driving force, a $\Delta G \neq 0$, for the reaction until it reaches equilibrium, i.e., until there are equal numbers of H and T.

B. The amount of shaking corresponds to the temperature as it results in the "thermal" motion of the coins. The activation energy of the reaction is the energy that needs to be expended to flip the coin, i.e., to stand it on its rim from where it can fall back facing either side up. Jigglase would speed up the flipping by lowering the energy required for this; it could, for example, be a magnet that is suspended above the box and helps lift the coins. Jigglase would not affect where the equilibrium lies (at an equal number of H and T), but would speed up the process of reaching the equilibrium, because in the presence of jigglase more coins would flip back and forth.

Answer 3–5 See Figure A3–5 (*next page*). Note that $\Delta G°_{X \to Y}$ is positive, whereas $\Delta G°_{Y \to Z}$ and $\Delta G°_{X \to Z}$ are negative. The graph also shows that $\Delta G°_{X \to Z} = \Delta G°_{X \to Y} + \Delta G°_{Y \to Z}$. We do not know from the information given in Figure 3–12 how high the activation energy barriers are; they are therefore drawn to an arbitrary height (*solid lines*). They would be lowered by enzymes that catalyze these reactions, thereby speeding up the reaction rates (*dotted lines*).

Answer 3–6

A. The rocks in Figure 3–24B provide the energy to lift the bucket of water. ATP is driving the reaction; thus ATP corresponds to the rocks on top of the cliff. The broken debris corresponds to ADP and Ⓟ, the products of ATP after it has released its energy and performed its work. In the reaction,

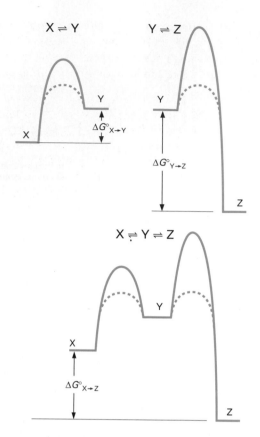

Figure A3-5

ATP hydrolysis is coupled to the conversion of X to Y. X, therefore, is the starting material, the bucket on the ground, which is converted to Y, the bucket at its highest point.

B. (a) The rock hitting the ground would be the futile hydrolysis of ATP. In the absence of an enzyme that uses the energy of ATP hydrolysis to drive an otherwise unfavorable reaction, the energy stored in the phosphoanhydride bond would be lost as heat. (b) The energy stored in Y could be used to drive another reaction. If Y represented the activated form of an amino acid X, for example, it could undergo a condensation reaction to form a peptide bond during protein synthesis.

Answer 3–7   The free energy ΔG derived from ATP hydrolysis depends on both $\Delta G°$ *and* the concentrations of substrate and products. In this case,

$$\Delta G = -12 \text{ kcal/mole} = -7.3 \text{ kcal/mole} + 0.616 \ln \frac{[\text{ADP}] \times [\text{P}]}{[\text{ATP}]}$$

ΔG is smaller than $\Delta G°$, because the ATP concentration in cells is high (in the millimolar range) and the ADP concentration is low (in the 10 μM range). The concentration term of this equation is therefore smaller than 1 and its logarithm is a negative number.

$\Delta G°$ is a constant for the reaction that will not vary with reaction conditions. ΔG, in contrast, depends on the concentrations of ATP, ADP, and phosphate, which may be somewhat different between cells.

Answer 3–8   Reactions B, C, D, and E all require coupling to other, energetically favorable reactions. In each case, higher-order structures are formed that are more complicated and have higher-energy bonds than the starting materials. In contrast, reaction A is a catabolic reaction that leads to compounds in a lower energy state and will occur spontaneously.

Answer 3–9

A. Nearly true, but strictly speaking, false. Because enzymes enhance the rate but do not change the equilibrium point of a reaction, a reaction will always occur in the absence of the enzyme, though often at a minuscule rate. Moreover, competing reactions may use up the substrate more quickly, thus further impeding the desired reaction. Thus, in practical terms, without an enzyme, some reactions may never occur to an appreciable extent.

B. False. High-energy electrons are more easily transferred, i.e., more loosely bound to the donor molecule. This does not mean that they move any faster.

C. True. Hydrolysis of an ATP molecule to form AMP also produces a pyrophosphate (PP_i) molecule, which in turn is hydrolyzed into two phosphate molecules. This second reaction releases almost the same amount of energy as the initial hydrolysis of ATP, thereby approximately doubling the energy yield.

D. True; oxidation is the removal of electrons, which reduces the diameter of the carbon atom.

E. True. ATP, for example, can donate both energy and/or a phosphate group.

F. False. Living cells have selected a particular kind of chemistry in which most oxidations are energy-releasing events; under different conditions, however, such as in a hydrogen-containing atmosphere, reductions would be energy-releasing events.

G. False. All cells, including those of cold- and warm-blooded animals, radiate comparable amounts of heat as a consequence of their metabolic reactions. For bacterial cells, for example, this becomes apparent when a compost pile heats up.

H. False. The equilibrium constant of the reaction X ⇌ Y remains unchanged. If Y is removed by a second reaction, more X is converted to Y so that the ratio of X to Y remains constant.

Answer 3–10   The free-energy difference ($\Delta G°$) between B and A due to three hydrogen bonds is –3 kcal/mole. (Note that the free energy of B is lower than that of A, because energy would need to be expended to break the bonds to convert B to A. The value for $\Delta G°$ for the transition A → B is therefore negative.) The equilibrium constant for the reaction is therefore about 100 (from Table 3–1, p. 92), i.e., there are 100 times more molecules of B than of A at equilibrium. An additional three hydrogen bonds would increase $\Delta G°$ to –6 kcal/mole and increase the equilibrium constant by another 100-fold to 10^4. Thus, relatively small differences in energy can have a major effect on equilibria.

Answer 3–11   The statement is correct. The criterion for whether a reaction proceeds spontaneously is ΔG, not $\Delta G°$, and takes the concentrations of the reacting components into account. A reaction with a negative $\Delta G°$, for example, would not proceed spontaneously under conditions where there is already an excess of products, i.e., more than at equilibrium. Conversely, a reaction with a positive $\Delta G°$ might spontaneously go forward under conditions where there is a huge excess of substrate.

Answer 3–12

A. The energy available in 57 ATP molecules (= 686/12) corresponds to the energy released by the complete oxidation of glucose to CO_2 and H_2O.

B. The overall efficiency of ATP production would be about 53%, calculated as the ratio of actually produced ATP molecules (= 30) divided by the number of ATP molecules that could be obtained if all the energy stored in a glucose molecule could be harvested as chemical energy in ATP (= 57).

C. During the oxidation of 1 mole of glucose, 322 kcal (the remaining 47% of the available 686 kcal in one mole of glucose, which are not stored as chemical energy in ATP) would be released as heat. This amount of energy would heat your body by 4.3°C (= 357 kcal/75 kg). This is a significant amount of heat, considering that 4°C of elevated temperature would be a quite incapacitating fever and that 1 mole (180 g) of glucose is no more than two cups of sugar.

D. If the energy yield were only 20%, then instead of 53% in the example above, 80% of the available energy would be released as heat and would need to be dissipated by your body. The heat production would be more than 1.5-fold higher, and your body would certainly overheat.

E. The chemical formula of ATP is $C_{10}H_{12}O_{13}N_5P_3$ and its molecular weight is therefore 503 g/mole. Your resting body therefore hydrolyzes about 80 moles (= 40 kg/0.503 kg/mole) of ATP in 24 hours (this corresponds to about 1000 kcal of liberated chemical energy). Because every mole of glucose yields 30 moles of ATP, this amount of energy could be produced by oxidation of 480 g glucose (= 180 g/mole × 80 moles/30).

Answer 3–13   This scientist is definitely a fake. The 57 ATP molecules would store 684 kcal (= 57 × 12 kcal) of chemical energy, which implies that the efficiency of

ATP production from glucose would have been an impossible 99.7%. This would leave virtually no energy to be released as heat, and this release is required according to the laws of thermodynamics.

Answer 3–14

A. The mutant organism would probably be safe to eat. ATP hydrolysis can provide approximately –12 kcal/mole of energy. This amount of energy shifts the equilibrium point of a reaction by an enormous factor: about 10^8-fold (from Table 3–1, p. 92, we see that –5.7 kcal/mole corresponds to an equilibrium constant of 10^4; thus, –12 kcal/mole corresponds to about 10^8. Note that for coupled reactions energies are additive, while equilibrium constants are multiplied). Therefore, if the energy of ATP hydrolysis cannot be utilized by the enzyme, 10^8-fold less poison is made. This example illustrates that coupling a reaction to the hydrolysis of an activated carrier molecule can shift the equilibrium point drastically.

B. It would be risky to consume this mutant organism. Slowing down the reaction rate would not affect its equilibrium point, and if the reaction were allowed to proceed for a long enough time, the mutant organism would likely be loaded with poison. Perhaps the equilibrium of the reaction would not be reached, but it would not be advisable to take any chances.

Answer 3–15

A. From Table 3–1 (p. 92) we know that a free-energy difference of 4.3 kcal/mole corresponds to an equilibrium constant of 10^{-3}, i.e., [A*]/[A] = 10^{-3}. The concentration of A* is therefore 1000-fold lower than that of A.

B. The ratio of A to A* would be unchanged. Lowering the activation energy barrier would accelerate the rate of the reaction, i.e., it would allow more molecules in a given time period to convert from A → A* and from A* → A, but would not affect the equilibrium point.

Answer 3–16   Enzyme A is beneficial. It allows the interconversion of two energy carrier molecules, both of which are required as the triphosphate form for many metabolic reactions. Any ADP that is formed is quickly converted to ATP, and thus the cell maintains a high ATP-ADP ratio. Because of enzyme A, called nucleotide phosphokinase, some of the ATP is used to keep the GTP-GDP ratio similarly high.

Enzyme B would be highly detrimental to the cell. Cells use NAD^+ as an electron acceptor in catabolic reactions and must maintain high concentrations of this form of the carrier so as to break down glucose to make ATP. In contrast, NADPH is used as an electron donor in biosynthetic reactions and is kept at high levels in the cells so as to allow the synthesis of nucleotides, fatty acids, and other essential molecules. Since enzyme B would deplete the cell's reserves of both NAD⁺ and NADPH, it would reduce the rates of both catabolic *and* biosynthetic reactions.

Chapter 4

Answer 4–1 In order to keep glycolysis going, cells need to regenerate NAD⁺ from NADH. There is no efficient way to do this without fermentation. In the absence of regenerated NAD⁺, step 6 of glycolysis (the oxidation of glyceraldehyde 3-phosphate to 1,3-bis-phosphoglycerate, Panel 4–1, pp. 112–113) could not occur and the product glyceraldehyde 3-phosphate would accumulate. The same thing would happen in cells unable to make either pyruvate or ethanol: neither would be able to regenerate NAD⁺, and so glycolysis would be blocked at the same step.

Answer 4–2 Arsenate instead of phosphate becomes attached in step 6 of glycolysis to form 1-arseno-3-phosphoglycerate (Figure A4–2). Because of its sensitivity to hydrolysis in water, the high-energy bond is destroyed before the molecule that contains it can diffuse to reach the next enzyme. The product of the hydrolysis, 3-phosphoglycerate, is the same product normally formed in step 7 by the action of phosphoglycerate kinase. But because hydrolysis occurs nonenzymatically, the energy liberated by breaking the high-energy bond cannot be captured to generate ATP. In Figure 4–6 therefore the reaction corresponding to the downward-pointing arrow would still occur, but the wheel that provides the coupling to ATP synthesis is missing. Arsenate wastes metabolic energy by uncoupling many phosphotransfer reactions by the same mechanism, and this is why it is so poisonous.

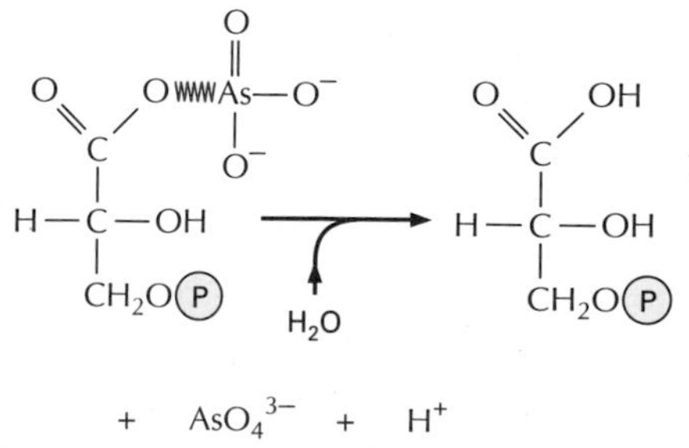

Figure A4–2

Answer 4–3 The oxidation of fatty acids breaks the carbon chain into two-carbon units (acetyl groups) that become attached to CoA. Conversely, during biogenesis fatty acids are constructed by linking together acetyl groups. Most fatty acids therefore have an even number of carbon atoms.

Answer 4–4 Because the function of the citric acid cycle is to harvest the energy released during the oxidation, it is advantageous to break the overall reaction into as many steps as possible (see Figure 4–1). Using a two-carbon compound, the available chemistry would be much more limited, and it would be impossible to generate as many intermediates. Note, for example, that many of the reactions of the cycle oxidize carbon atoms that have been through more than one cycle and have therefore not come directly from the most recently added acetyl group.

Answer 4–5 It is true that oxygen is returned as part of CO_2 to the atmosphere. The CO_2 released from the cells, however, does not contain those specific oxygen atoms that were consumed as part of the oxidative phosphorylation reaction and converted into water (see Figure 4–14). One can show this directly by incubating living cells in an atmosphere that contains molecular oxygen that contains a different isotope, ¹⁸O, instead of the naturally abundant isotope ¹⁶O. In such an experiment one finds that all the CO_2 released from cells contains only ¹⁶O. Therefore, the oxygen atoms in the released CO_2 molecules do not come directly from the atmosphere but from organic molecules that the cell has first made and then oxidized as fuel.

Answer 4–6 The carbon atoms in sugar molecules are already partially oxidized, in contrast to all but the very first carbon atoms in the acyl chains of fatty acids. Consequently, two carbon atoms from glucose are lost as CO_2 during the conversion of pyruvate to acetyl CoA, and only four of the six carbon atoms of the sugar molecule are recovered and can enter the citric acid cycle, where most of the energy is captured. In contrast, all carbon atoms of a fatty acid are converted into acetyl CoA.

Answer 4–7 The cycle continues because intermediates are replenished as necessary by reactions leading to the citric acid cycle (instead of away from it). One of

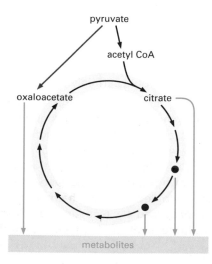

Figure A4–7

the most important reactions of this kind is the conversion of pyruvate to oxaloacetate by the enzyme pyruvate carboxylase:

pyruvate + CO_2 + ATP + H_2O → oxaloacetate + ADP + P_i + $2H^+$

This is one of the many examples of how metabolic pathways are elegantly balanced and work together to maintain appropriate concentrations of all metabolites required by the cell (see Figure A4–7, *previous page*).

Answer 4–8

A. False. If this were the case, then the reaction would be useless for the cell. No chemical energy would be harvested in a useful form (e.g., ATP) to be used for metabolic processes. (Cells would be nice and warm, though!)

B. False. No energy conversion process can be 100% efficient. Recall that the entropy in the universe always has to increase, and for most reactions this is accomplished by releasing heat.

C. True. The carbon atoms in glucose are in a reduced state compared to those in CO_2, in which they are fully oxidized.

D. False. The reaction does indeed produce some water, but water is so abundant in the biosphere that this is no more than "a drop in the bucket."

E. True. If it occurred in only one step, then all the energy would be released at once and it would be impossible to harness it efficiently to drive other reactions, such as the synthesis of ATP.

F. False. Molecular oxygen (O_2) is used only in the very last step of the reaction (see Figure 4–14).

G. True. Plants convert CO_2 into sugars by harvesting the energy of light in photosynthesis. O_2 is produced in the process and released by plant cells.

H. Almost true. Anaerobically growing cells use glycolysis to oxidize sugars to pyruvate. Animal cells convert pyruvate to lactate, and no CO_2 is produced; yeast cells, however, convert pyruvate to ethanol and CO_2. It is this CO_2 gas, released from yeast cells during fermentation, that makes bread dough rise and that carbonates beer and champagne.

Answer 4–9 Darwin exhaled the carbon atom, which therefore must be the carbon atom of a CO_2 molecule. After spending some time in the atmosphere, the CO_2 molecule must have entered a plant cell, where it became "fixed" by photosynthesis and converted into part of a sugar molecule. While it is certain that these early steps must have happened this way, there are many different paths from there that the carbon atom could have taken. The sugar could have been broken down by the plant cell into pyruvate or acetyl CoA, for

example, which then could have entered biosynthetic reactions to build an amino acid. The amino acid might have been incorporated into a plant protein, maybe an enzyme or a protein that builds the cell wall. You might have eaten the delicious leaves of the plant in your salad, and digested the protein in your gut to produce amino acids again. After circulating in your bloodstream, the amino acid might have been taken up by a developing red blood cell to make its own protein, such as the hemoglobin in question.

If we wish, of course, we can make our food chain scenario more complicated. The plant, for example, might have been eaten by an animal that in turn was consumed by you during lunch break. Moreover, because Darwin died more than 100 years ago, the carbon atom could have traveled such a route many times. In each round, however, it would have started again as fully oxidized CO_2 gas and entered the living world following its reduction during photosynthesis.

Answer 4–10 Yeast cells grow much better aerobically. Under anaerobic conditions they cannot perform oxidative phosphorylation and therefore have to produce all their ATP by glycolysis, which is less efficient. Whereas one glucose molecule yields a net gain of two ATP molecules by glycolysis, the additional use of the citric acid cycle and oxidative phosphorylation boosts the energy yield up to about 30 ATP molecules.

Answer 4–11 The amount of free energy stored in the phosphate bond in creatine phosphate is larger than that of the anhydride bonds in ATP. Hydrolysis of creatine phosphate can therefore be directly coupled to the production of ATP.

creatine phosphate + ADP → creatine + ATP

The $\Delta G°$ for this reaction is –3 kcal/mole, indicating that it proceeds rapidly to the right, as written.

Answer 4–12 The extreme conservation of glycolysis is evidence that all present cells are derived from a single founder cell as discussed in Chapter 1. The elegant reactions of glycolysis would therefore have evolved only once, and then they would have been inherited as cells evolved. The later invention of oxidative phosphorylation allowed follow-up reactions to capture 15 times more energy than is possible by glycolysis alone. This remarkable efficiency is close to the theoretical limit and hence virtually eliminates the opportunity for further improvements. Thus the generation of alternative pathways would result in no obvious growth advantage that would have been selected in evolution.

Answer 4–13 As discussed in the text and in Questions 4–8 and 4–10, 30 ATP molecules are produced from each glucose molecule that is oxidized according to the reaction $C_6H_{12}O_6$ (glucose) $+ 6O_2 \rightarrow$ $6CO_2 + 6H_2O$ + energy. Thus, one O_2 molecule is consumed for every five ATP molecules produced. The cell therefore consumes 2×10^8 O_2 molecules/min, which corresponds to the consumption of 3.3×10^{-16} moles (= $2 \times 10^8 / 6 \times 10^{23}$) or 7.4×10^{-15}/liters (= $3.3 \times 10^{-16} \times 22.4$) each minute. The volume of the cell is 10^{-15} m^3 [= $(10^{-5})^3$], which is 10^{-12}/liter. The cell therefore consumes about 0.7% (= $100 \times 7 \times 10^{-15}/10^{-12}$) of its volume of O_2 gas every minute, or its own volume of O_2 gas in 2 hours and 15 minutes.

Answer 4–14   The reactions each have negative ΔG values and are therefore energetically favorable (see Figure A4–14 for energy diagrams).

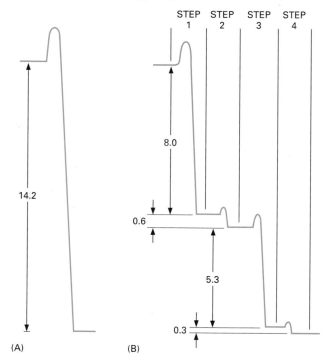

Figure A4–14

Answer 4–15

A. Pyruvate is converted to acetyl CoA, and the labeled ^{14}C atom is released as $^{14}CO_2$ gas (see Figure 4–8B).

B. By following the ^{14}C-labeled atom through every reaction in the cycle, shown in Panel 4–2 (pp. 122–123), you find that the added ^{14}C label would

COO⁻   COO⁻
| |
^{14}C=O   C=O
| |
CH₂ $^{14}CH_2$
| |
COO⁻   COO⁻

radioactive   radioactive
oxaloacetate   oxaloacetate
added to   isolated after
the extract   one turn of
                     citric acid cycle

Figure A4–15

be quantitatively recovered in oxaloacetate. The analysis also reveals, however, that it is no longer in the keto group but in the methylene group of oxaloacetate (Figure A4–15).

Answer 4–16   In the presence of molecular oxygen, oxidative phosphorylation converts most of the cellular NADH to NAD^+. Since fermentation requires NADH, it is severely inhibited by the availability of oxygen gas.

Chapter 5

Answer 5–1   Urea is a very small molecule that functions both as an efficient hydrogen-bond donor (through its –NH_2 groups) and as an efficient hydrogen-bond acceptor (through its –C=O group). As such, it can squeeze between hydrogen bonds that stabilize protein molecules and thus destabilize protein structures. In addition, nonpolar side chains are held together in the interior of folded proteins because they disrupt the structure of water if they are exposed. At high concentrations of urea, the hydrogen-bonded network of water molecules becomes disrupted so that these hydrophobic forces are significantly diminished. Proteins unfold in urea as a consequence of its effect on these two forces.

Answer 5–2   The amino acid sequence consists of an alternation of nonpolar and charged or polar amino acids. The resulting strand in a β sheet would therefore be polar on one side and hydrophobic on the other. Such a strand would probably be surrounded on either side by similar strands that together form a β sheet with a hydrophobic and a polar face. In a protein, such a β sheet (called "amphipathic" from the Greek *amphi*, "of both kinds," and *pathos*, "passion," because of its two surfaces with such different properties) would be positioned so that the hydrophobic side would face the protein's interior and the polar side would be on the surface, exposed to the water outside.

Answer 5–3   Mutations that are beneficial to an organism are selected in evolution because they confer a growth or survival advantage to the organism. Examples are the better utilization of a food source, enhanced resistance to environmental insults, such as heat or concentrated salt, or an enhanced ability to attract a mate for sexual reproduction. In contrast, the accumulation of useless proteins is detrimental to organisms. Useless mutant proteins waste the metabolic energy required to make them. If such mutant proteins were made in excess, the synthesis of normal proteins would suffer because the capacity of the cell is limited. In more severe cases, mutant proteins interfere with the normal workings of the cell; a mutant enzyme that still binds an activated carrier molecule but does not catalyze its reaction, for example, may compete for a limited amount of this carrier and therefore inhibit normal processes. Natural selection therefore provides a strong driving force that leads to the loss of both useless and harmful proteins.

Answer 5–4 Strong reducing agents that broke all of the S–S bonds would cause all of the keratin filaments to separate. Hair therefore disintegrates. Indeed, strong reducing agents are used commercially in hair removal creams sold by your local pharmacist. However, mild reducing agents are used in treatments that either straighten or curl hair, the latter requiring hair curlers. (See Figure A5–4.)

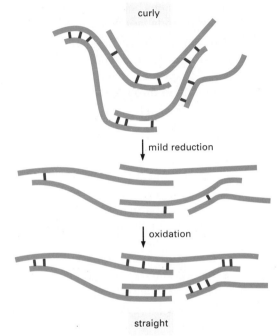

Figure A5–4

Answer 5–5

A. The equilibrium constant is defined as $K = [AB]/([A] \times [B])$. The square brackets indicate the concentration. Thus, if A, B, and AB are each 1 μM (10^{-6} M), K will be 10^6 M^{-1} [$= 10^{-6} / (10^{-6} \times 10^{-6})$].

B. Similarly, if A, B, and AB are each 1 μM (10^{-9}M), then K will be 10^9 M^{-1}.

C. This example illustrates that interacting proteins that are present in cells in lower concentrations need to bind to each other with high affinities so that a significant fraction of the molecules is bound at equilibrium. In this particular case, lowering the concentration by three orders of magnitude (from μM to nM) requires a change in the equilibrium constant by three orders of magnitude to maintain the AB protein complex (corresponding to –4.3 kcal of free energy, Figure 5–26). This corresponds to about 4–5 extra hydrogen bonds.

Answer 5–6

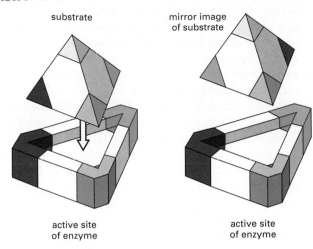

Figure A5–6

Answer 5–7

A. When S $\ll K_M$, the term (S + K_M) approaches K_M. Therefore, the equation is simplified to rate = V_{max}/K_M (S), that is, the rate is proportional to S.

B. When S = K_M, the term [S/(S + K_M)] equals ½. Therefore, the reaction rate is half of the maximal rate V_{max}.

C. If S $\gg K_M$, the term (S + K_M) approaches S. Therefore, [S/(S + K_M)] equals 1 and the reaction occurs at its maximal rate V_{max}.

Answer 5–8

A. Feedback inhibition from Z to the reaction B → C would increase the flux through the B → X → Y → Z pathway, because the conversion of B to C is inhibited. Thus, the more Z there is, the more production of Z would be stimulated. This is likely to result in an uncontrolled "runaway" amplification of this pathway.

B. Feedback inhibition from Z affecting Y → Z controls the production of Z. In this scheme, however, X and Y are still made at normal rates, even though both of these intermediates are no longer needed at this level. This pathway is therefore less efficient than the one shown in Figure 5–32.

C. If Z is a positive regulator of the step B → X, then the more Z there is, the more B would be converted to X and therefore shunted into the pathway producing more Z. This would result in a runaway amplification similar to that described for (A).

D. If Z is a positive regulator of the step B → C, then accumulation of Z leads to a redirection of the pathway to make more C. This is a second possible way, in addition to that shown in the figure, to balance the distribution of compounds into the two branches of the pathway.

Answer 5–9 Both nucleotide binding and phosphorylation can induce allosteric changes in proteins. These can have a multitude of consequences, such as altered enzyme activity, drastic shape changes, and changes in affinity for other proteins or small molecules. Both mechanisms are quite versatile. An advantage of nucleotide binding is the fast rate with which a small nucleotide can diffuse to the protein; the shape changes that accompany the function of motor proteins, for example, require quick nucleotide replenishment. If the different conformational states of a motor protein were controlled by phosphorylation, for example, a protein kinase would either need to diffuse into position at each step, a much slower process, or be associated permanently with each motor. One advantage of phosphorylation is that it requires only a single amino acid residue on the protein's surface, rather than a specific binding site. Phosphates can therefore be added to many different side chains on the same protein (as long as protein kinases with the proper specificities exist), thereby vastly increasing the complexity of regulation that can be achieved for a single protein.

Answer 5–10 In working as a complex, all three proteins contribute to the specificity (by binding the safe and key directly), help position one another correctly, and provide the mechanical bracing that allows them to perform a task that they could not perform individually (the key is grasped by two subunits, for example). Moreover, their functions are generally coordinated in time (for example, the binding of ATP to one subunit is likely to require that ATP has already been hydrolyzed to ADP by another).

Answer 5–11

A. True. Only a few amino acid side chains contribute to the active site. The rest of the protein is required to maintain the polypeptide chain in the correct position, provide additional binding sites for regulatory purposes, and localize the protein in the cell.

B. True. Some enzymes form covalent intermediates with their substrates (see Figure 4–5); however, in all cases the enzyme is restored to its original structure after the reaction.

C. False. β sheets can, in principle, contain any number of strands because the two strands that form the rims of the sheet are available for hydrogen-bonding to other strands. (β sheets in known proteins contain from 2 to 16 strands.)

D. False. It is true that the specificity of an antibody molecule is exclusively contained in loops on its surface; however, these loops are contributed by both the folded light and heavy chain domains (see Figure 5–25).

E. False. Enzymes do not change the equilibrium point for the reaction they catalyze.

F. False. The possible linear arrangements of amino acids that lead to a stably folded protein domain are so few that most new proteins evolve by alteration of old ones.

G. False. The K_M of an enzyme is a measure of its affinity for the substrate and not of the rate at which it catalyzes the reaction.

H. True. Allosteric enzymes generally bind one or more molecules that function as regulators at sites that are distinct from the active site.

I. False. Noncovalent bonds are a major contributor to the three-dimensional structure of macromolecules.

J. False. Affinity chromatography separates specific macromolecules because of their interactions with specific ligands, not because of their charge.

K. False. The larger an organelle is, the more centrifugal force it experiences and the faster it sediments, despite an increased frictional resistance from the fluid through which it moves.

Answer 5–12 In an α helix and in the central strands of a β sheet, all peptide bonds are engaged in hydrogen bonds. This gives considerable stability to these secondary structure elements, and it allows them to form from many different amino acid sequences.

Answer 5–13 Because enzymes are catalysts, enzyme reactions have to be thermodynamically feasible; the enzyme only lowers the activation energy barrier that otherwise slows the rate with which the reaction occurs. Heat, in contrast, confers more kinetic energy to the substrates so that a higher fraction of them can surmount the activation energy barrier. Many substrates, however, have many different ways in which they could react, and all of these potential pathways will be enhanced by heat, whereas an enzyme confers selectivity and facilitates only one particular pathway that, in evolution, was selected to be useful for the cell. Heat, therefore, cannot substitute for enzyme function, and chicken soup must exert its beneficial effects by other mechanisms that remain to be discovered.

Answer 5–14 No. It would not have the same or even a similar structure, because the peptide bond has a polarity. Looking at two sequential amino acids in a polypeptide chain, the amino acid that is closer to the amino-terminal end contributes the carboxyl group and the other amino acid contributes the amino group to the peptide bond that links the two. Changing their order would put the side chains into a different position with respect to the peptide backbone and therefore change their chemical environment.

Answer 5–15 As it takes 3.6 amino acid residues to complete a turn of an α helix, this sequence of 14 amino acids would make close to 4 full turns. It is remarkable because its polar and hydrophobic amino acids are spaced so that all polar residues are on one side of an α helix and all the hydrophobic residues are on the other. It is therefore likely that such an amphipathic α helix is exposed on the protein surface with its hydrophobic side facing the protein's interior. In addition, two such helices might wrap around each other as shown in Figure 5–11.

Answer 5–16

A. "ES" represents the enzyme–substrate complex.

B. Enzyme and substrate are in equilibrium between their free and bound states; once bound to the enzyme, a substrate molecule may either dissociate again (hence the bidirectional arrows) or be converted to product. As substrate is converted to product (with the concomitant release of free energy), however, a reaction often proceeds strongly in the forward direction, as indicated by the unidirectional arrow.

C. The enzyme is a catalyst and is therefore liberated in an unchanged form after the reaction; thus, "E" appears at both ends of the equation.

D. Often the products of a reaction resemble the substrates sufficiently that they can also bind to the enzyme. Any enzyme molecules that are bound to product (i.e., the "EP" complex) are unavailable for catalysis; excess P therefore inhibits the reaction by lowering the concentration of free E.

E. Compound X is an inhibitor of the reaction and works similarly by forming an EX complex. However, since P has to be made before it can inhibit the reaction, it takes longer to act than X, which is present from the beginning of the reaction.

Answer 5–17 The polar amino acids Ser, Ser-P, Lys, Gln, His, and Glu are more likely to be found on a protein's surface, and the hydrophobic amino acids Leu, Phe, Val, Ile, and Met are more likely to be found in its interior. The oxidation of two cysteine residues to form a disulfide bond eliminates their potential to form hydrogen bonds and therefore makes them even more hydrophobic. Disulfide bonds are usually found in the interior of proteins. Irrespective of the nature of their side chains, the most amino-terminal amino acid and the most carboxyl-terminal amino acid each contain a charged group, the amino and carboxyl groups that mark the ends of the polypeptide chain, and hence are usually found on the protein's surface.

Answer 5–18 Many secondary structure elements are not stable in isolation but require the presence of additional parts of the polypeptide chain. Hydrophobic regions that would normally be hidden in the inside of a folded domain would be exposed on the outside, and because such regions are energetically disfavored in water solution, the fragments tend to aggregate nonspecifically. Such fragments therefore would not have a defined structure, and they would be inactive for ligand binding even if they contained all of the amino acids that would normally contribute to the ligand-binding site. A protein *domain,* in contrast, is considered a folding unit, and fragments of a polypeptide chain that correspond to intact domains are often able to fold correctly. Thus, separated protein domains often retain their activities, such as ligand binding, if the binding site is contained entirely within this domain. Thus the most likely place in which the polypeptide chain of the protein in Figure 5–12 could be severed to give rise to stable fragments is at the boundary between the two domains (i.e., at the loop between the two α helices at the bottom right of the structure shown).

Answer 5–19 The heat inactivation of the enzyme suggests that the mutation causes the enzyme to have a less stable structure. For example, a hydrogen bond that is normally formed between two amino acid side chains might no longer be formed because the mutation replaces one of these amino acids with a different one that cannot participate in the bond. Lacking such a bond that normally helps to keep the polypeptide chain folded properly, the protein unfolds at a temperature at which it normally would be stable. Polypeptide chains that are denatured when the temperature is raised often aggregate, and they rarely refold into active proteins when the temperature is decreased.

Answer 5–20 The substrate concentration is 1 mM. This value can be obtained by substituting values in the equation, but it is simpler to note that the desired rate (50 μM/sec) is exactly half of the maximum rate,

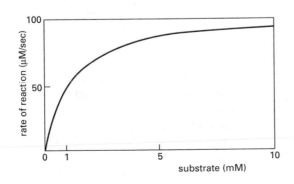

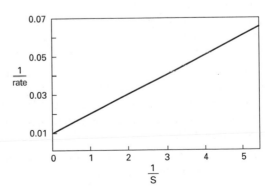

Figure A5–20

V_{max}, where the substrate is typically equal to the K_M. The two plots requested are shown in Figure A5–20 (*previous page*). A plot of 1/rate versus 1/S is a straight line because rearranging the standard equation yields the equation listed in Question 5–22B.

Answer 5–21   If S is very much smaller than K_M, the active site of the enzyme is mostly *unoccupied*. If S is very much greater than K_M, the reaction rate is limited by the *enzyme* concentration (because most of the catalytic sites are fully occupied).

Answer 5–22

A and B. The data in the boxes have been used to plot the *red* curve and *red* line in Figure A5–22, *below*. From the plotted data, the K_M is 1 µM and the V_{max} is 2 µM/min. Note that the data are much easier to interpret in the linear plot, because the curve in (A) approaches but never reaches V_{max}.

C. It is important that only a small quantity of product is made, because otherwise the rate of reaction would decrease as the substrate was depleted and product accumulated. Thus the measured rates would be lower than they should be.

D. If the K_M increases, then the concentration of substrate needed to give a half-maximal rate is increased. As more substrate is needed to produce the same rate, the enzyme-catalyzed reaction has been inhibited by the phosphorylation. The expected data plots for the phosphorylated enzyme are the green curve and the green line in Figure A5–22, *below*.

Answer 5–23   All reactions are reversible. If the compound AB can dissociate to produce A and B, then it must also be possible for A and B to associate to form AB. Which of the two reactions predominates depends on the equilibrium constant and the concentration of A, B, and AB (as discussed in Figure 5–26). Presumably, when this enzyme was isolated its activity was detected by supplying A and B in relatively large amounts and measuring the amount of AB generated. We can suppose, however, that in the cell there is a large concentration of AB so that under normal conditions the enzyme actually catalyzes AB →

A + B. (This question is based on an actual example in which an enzyme was isolated and named according to the reaction in one direction, but was later shown to catalyze the reverse reaction in living cells.)

Answer 5–24   The motor protein in the illustration can move just as easily to the left as to the right and so will not move steadily in one direction. However, if just one of the steps is coupled to ATP hydrolysis (for example, by making detachment of one foot dependent on binding of ATP and coupling the reattachment to hydrolysis of the bound ATP), then the protein will show unidirectional movement that requires the continued consumption of ATP. Note that, in principle, it does not matter which step is coupled to ATP hydrolysis (Figure A5–24).

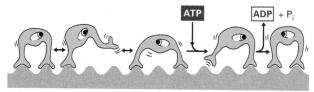

Figure A5–24

Answer 5–25   The slower migration of small molecules through a gel-filtration column results because smaller molecules have access to many more spaces in the porous beads that are packed into the column than do larger molecules. However, it is important to give the smaller molecules sufficient time to diffuse into the spaces inside the beads. At very rapid flow rates, all molecules will move rapidly around the beads, so that large and small molecules will now tend to exit together from the column.

Chapter 6

Answer 6–1

A. The complementary strand reads 5′-TGATTGTG-GACAAAAATCC-3′. Paired DNA strands have opposite polarity, and the convention is to write a single-stranded DNA sequence in the 5′-to-3′ direction.

Figure A5–22

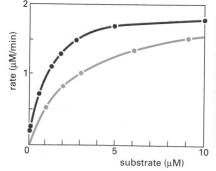

S (µM)	$\frac{1}{S} \left[\frac{1}{\mu M}\right]$	rate (µM/min)	$\frac{1}{rate} \left[\frac{min}{\mu M}\right]$
0.08	12.50	0.15	6.7
0.12	8.30	0.21	4.8
0.54	1.85	0.70	1.4
1.23	0.81	1.1	0.91
1.82	0.56	1.3	0.77
2.72	0.37	1.5	0.67
4.94	0.20	1.7	0.59
10.00	0.10	1.8	0.56

DATA FOR A AND B

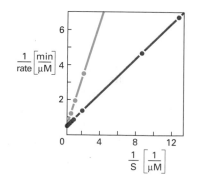

B. The DNA is made of four nucleotides (100% = 13% A + x% T + y% G + z% C). Because A pairs with T, the two nucleotides are represented in equimolar proportions in DNA. Therefore, the bacterial DNA in question contains 13% thymidine. This leaves 74% [= 100% − (13% + 13%)] for G and C, which also form base pairs and hence are equimolar. Thus $y = z = 74/2 = 37$.

C. A single-stranded DNA molecule that is N nucleotides long can have any one of 4^N possible sequences, but the number of possible double-stranded DNA molecules is more difficult to calculate. Many of the 4^N single-stranded sequences will be the complement of another possible sequence in the list; for example, 5′-AGTCC-3′ and 5′-GGACT-3′ form the same double-stranded DNA molecule and therefore count as a single double-stranded possibility. If N is an odd number, then every single-stranded sequence will complement another sequence in the list so that the number of double-stranded sequences will be 0.5×4^N. If N is an even number, then there will be slightly more than this, since some sequences will be self-complementary (such as 5′-ACTAGT-3′) and the actual value can be calculated to be $0.5 \times 4^N + 0.5 \times 4^{N/2}$.

D. To specify a unique sequence which is N nucleotides long, 4^N has to be larger than 3×10^6. Thus, $4^N > 3 \times 10^6$, solved for N, gives $N > \ln(3 \times 10^6)/\ln(4) = 10.7$. Thus, on average a sequence of only 11 nucleotides in length is unique in the genome. Performing the same calculation for the genome size of an animal cell yields a minimal stretch of 16 nucleotides. This shows that a relatively short sequence can mark a unique position in the genome and is sufficient, for example, to serve as an identity tag for one specific gene.

Answer 6–2

A. The distance between replication forks #4 and #5 is about 280 nm, corresponding to 824 nucleotides (= 280/0.34). These two replication forks would collide in about 8 seconds. Forks #7 and #8 move away from each other and would therefore never collide.

B. The total length of DNA shown in the electron micrograph is about 1.5 μm, corresponding to 4400 nucleotides. This is only about 0.002% [= $(4400/1.8 \times 10^8) \times 100\%$] of the total DNA in a fly cell.

Answer 6–3 If the old strand were "repaired" using the new strand that contains a replication error as the template, then the error would become a permanent mutation in the genome. The old information would be erased in the process. Therefore, if repair enzymes did not distinguish between the two strands, there would be only a 50% chance that any given replication error would be corrected.

Answer 6–4 While the process may seem wasteful, it is not possible to proofread during primer formation. To start a new primer on a piece of single-stranded DNA, one nucleotide needs to be put in place and then linked to a second and then to a third and so on. Even if these first nucleotides were perfectly matched to the template strand, such short oligonucleotides bind only with very low affinity and it would consequently be difficult to distinguish the correct from incorrect bases by a hypothetical proofreading activity. The task of the primase is therefore to "just get anything down that binds reasonably well and don't worry about accuracy." Later these sequences are removed and replaced by DNA polymerase, which uses the accurate, correctly proofread newly synthesized DNA as its primer. The latter enzyme has the advantage—which primase does not have—of putting the new nucleotide onto an already existing strand. The newly added nucleotide is thus firmly held in place, and the accuracy of its base-pairing to the next nucleotide on the template strand can be checked. Therefore, as DNA polymerase fills the gap, it can proofread the new DNA strand that it makes.

Answer 6–5

A. Without DNA polymerase, no replication can take place at all. RNA primers will be laid down at the origin of replication.

B. DNA ligase links the DNA fragments that are produced on the lagging strand. In the absence of ligase, the newly replicated DNA strands will remain as fragments, but no nucleotides will be missing.

C. Without the sliding clamp, the DNA polymerase will frequently fall off the DNA template. In principle, it can rebind and continue, but the continual falling off and rebinding will be time-consuming and greatly slow down the rate of DNA replication.

D. In the absence of RNA excision enzymes, the RNA fragments will remain covalently attached to the newly replicated DNA fragments. No ligation will take place, because the ligase will not link DNA to RNA. The lagging strand will therefore consist of fragments composed of both RNA and DNA.

E. Without DNA helicase, the DNA polymerase will stall because it cannot separate the strands of the template DNA ahead of it. Little or no new DNA will be synthesized.

F. In the absence of primase, RNA primers cannot begin either on the leading or the lagging strand. DNA replication therefore cannot begin.

Answer 6–6 DNA defects introduced by deamination and depurination reactions occur spontaneously. They are not the result of replication and are therefore equally likely to occur on either strand. If DNA repair enzymes recognized such defects only on newly synthesized DNA strands, half of the defects would go uncorrected. The statement is therefore incorrect.

Answer 6–7 The argument is severely flawed. You cannot transform one species into another simply by introducing 1% random changes into the DNA. It is exceedingly unlikely that the 5000 mutations that would accumulate every day in the absence of DNA repair would be in the very positions where human and monkey DNA sequences are different. It is also very likely that at such a high mutation frequency many essential genes would be inactivated, leading to cell death. Furthermore, your body is made up of about 10^{13} cells. For you to turn into a monkey, not just one but many of these cells would need to be changed. And, even then, many of these changes would have to occur during development to effect changes in your body plan (like the inclusion of a tail).

Answer 6–8

A. False. The polarity of a DNA strand commonly refers to the orientation of its sugar-phosphate backbone.

B. False. Identical DNA polymerase molecules catalyze DNA synthesis on the leading and lagging strands. The replication fork is asymmetrical, because the lagging strand is synthesized in pieces that are then "stitched" together.

C. True. G-C base pairs are held together by three hydrogen bonds, whereas A-T base pairs are held together by only two.

D. False. The RNA primers are removed by RNA nuclease; the Okazaki fragments are the pieces of newly synthesized DNA that are eventually joined to form the new lagging strand.

E. True. DNA polymerase has an error rate of one in 107, which is amazingly low because of its proofreading activity, and 99% of its errors are corrected by DNA repair enzymes, bringing the final error rate to one in 10^9.

F. True. Mutations would accumulate rapidly, destroying the genes.

G. True. If an aberrant base occurred naturally in DNA, it might be recognized as a mismatch by DNA repair enzymes, but the enzymes could not tell on which strand the error was introduced. They would therefore have only a 50% chance of fixing the right strand.

H. True. Usually, multiple mutations of specific types need to accumulate before a cell turns into a cancer cell. A mutation in a gene that codes for a DNA repair enzyme can make a cell more liable to accumulate further mutations, thereby accelerating the onset of cancer.

Answer 6–9 If the wrong bases were frequently incorporated during DNA replication, genetic information

could not be inherited accurately. Life, as we know it, could not exist. Although the bases can form hydrogen-bonded pairs as indicated, these do not fit into the structure of the double helix. Thus, the angle with which the A residue is attached to the sugar-phosphate backbone is vastly different in the A-C pair, and the spacing between the two sugar-phosphate strands is considerably increased in the A-G pair, where two large purine rings interact. Consequently, it is energetically unfavorable to incorporate the wrong bases into the DNA chain, and such errors occur only very rarely.

Answer 6–10

A. The bases V, W, X, and Y can form a DNA-like double-helical molecule with virtually identical properties to those of bona fide DNA. V would always pair with X, and W with Y. Therefore, the macromolecules could be derived from a living organism using the same principles of replication of its genome. In principle, different bases, such as V, W, X, and Y, could have been selected during evolution as building blocks of DNA on earth. (Similarly, there are many more conceivable amino acid side chains than the set of 20 selected in evolution that make up all proteins.)

B. None of the bases V, W, X, or Y can replace A, T, G, or C. To preserve the distance between the two sugar-phosphate strands in the double helix, a pyrimidine always has to pair with a purine (see, e.g., Question 6–1). Thus, the eight possible combinations are V-A, V-G, W-A, W-G, X-C, X-T, Y-C, and Y-T. Because of the positions of hydrogen-bond acceptors and hydrogen-bond donor groups, however, no stable base pairs would form in any of these combinations, as shown for the pairing of V-A in Figure A6–10, where only a single hydrogen bond could form.

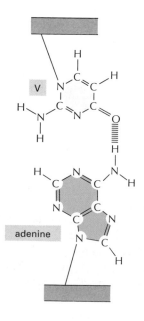

Figure A6–10

Answer 6–11 As the strands are held together by hydrogen bonds between the bases, the stability of the helix is largely dependent on the number of hydrogen bonds that can be formed. Thus two parameters determine the stability: the number of nucleotide pairs and the number of hydrogen bonds that each nucleotide pair contributes. As shown in Figure 6–5, an A-T pair contributes two hydrogen bonds, whereas a G-C pair contributes three hydrogen bonds. Therefore, helix C (containing a total of 34 hydrogen bonds) melts at the lowest temperature, helix B (containing a total of 65 hydrogen bonds) melts next, and helix A (containing a total of 78 hydrogen bonds) is the most stable, largely owing to its high GC content. Indeed, the DNA of organisms that grow in extreme temperature environments, such as certain bacteria that grow in geothermal vents, has an unusually high GC content.

Answer 6–12 The DNA would be enlarged by a factor of 2.5×10^6 ($= 5 \times 10^{-3}/2 \times 10^{-9}$ m). Thus the extension cord would be 2500 km long. This is approximately the distance from London to Istanbul, San Francisco to Kansas City, Tokyo to the southern tip of Taiwan, and Melbourne to Cairns. Adjacent nucleotides would be about 0.85 mm apart (which is only about the thickness of a stack of 12 pages of this book). A gene that is 1000 nucleotide pairs long would be about 85 cm in length.

Answer 6–13

A. It takes two bits to specify each nucleotide pair (for example, 00, 01, 10, and 11 would be the binary codes for the four different nucleotides, each paired with its appropriate partner).

B. The entire human genome (3×10^9 nucleotide pairs) could be stored on two CDs ($3 \times 10^9 \times 2$ bits/4.8×10^9 bits).

Answer 6–14 DNA isolated from your starting cells grown under normal conditions has a light density, as you would expect. After one generation of growth in a medium containing heavy isotopes, the DNA has uniformly shifted to medium density: synthesis of the new DNA from heavy nucleotides (e.g., the *red* strand shown in Figure 6–12 that is made during the first round of replication) results in hybrid DNA molecules that contain one light, maternal strand and one heavy, newly synthesized strand. After another round of replication in a medium containing heavy isotopes, two forms of DNA appear in about equal proportions: one form is again a hybrid of a light and a heavy strand (the *orange/red* DNA in Figure 6–12) and has medium density, while the other form is composed of two heavy strands (the *red/green* DNA in Figure 6–12) and has a heavy density. During the subsequent rounds of replication, more heavy DNA is formed, and the proportion of medium density DNA diminishes. Your results are therefore in complete agreement with the hypothesis that you set out to test.

Answer 6–15 With a single origin of replication that launches two DNA polymerases in opposite directions on the DNA each moving at 100 nucleotides per second, the number of nucleotides replicated in 24 hours will be 1.73×10^7 ($= 2 \times 100 \times 24 \times 60 \times 60$). To replicate all the 6×10^9 nucleotides of DNA in the cell in this time, therefore, will require at least 348 ($= 6 \times 10^9/1.73 \times 10^7$) origins of replication. The estimated 10,000 origins of replication in the human genome are therefore more than enough to satisfy this minimal requirement.

Answer 6–16 Compound A is dideoxycytosine triphosphate (ddCTP), identical to dCTP except that it lacks the 3′-hydroxyl group on the sugar ring. ddCTP is recognized by DNA polymerase as dCTP and becomes incorporated into DNA; because it lacks the crucial 3′-hydroxyl group, however, its addition to a growing DNA strand creates a dead end to which no further nucleotides can be added. Thus, if ddCTP is added in large excess, strands will be synthesized until the first G (the nucleotide complementary to C) is encountered on the template strand. ddCTP will then be incorporated instead of C, and the extension of this strand will be terminated. If ddCTP is added at about 10% of the concentration of the available dCTP, there is a 1 in 10 chance of its being incorporated whenever a G is encountered on the template strand. Thus a population of DNA fragments will be synthesized, and from their lengths one can deduce where the G residues are located on the template strand. This experiment forms the basis of methods used to determine the sequence of nucleotides in a stretch of DNA (discussed in Chapter 10).

The same chemical phenomenon is exploited by a drug, Cortisiepin, that is now commonly used in HIV-infected patients to treat AIDS. Cortisiepin is converted in cells to ddCTP and thus blocks DNA synthesis and replication of the virus. Since this treatment is not selective for viral DNA only but also affects replication and repair of all cellular DNA, its side effects are severe.

Compound B is dideoxycytosine monophosphate (ddCMP), which lacks the 5′-triphosphate group as well as the 3′-hydroxyl group of the sugar ring. It therefore cannot provide the energy that drives the polymerization reaction of nucleotides into DNA and therefore will not be incorporated into the replicating DNA. The compound is therefore expected not to affect DNA replication.

Answer 6–17 To use the energy of hydrolysis of the 3′-triphosphate group for polymerization, strand growth would need to occur in the opposite, that is, the 3′-to-5′, direction. Proofreading could then occur by a 5′-to-3′ nuclease activity. This scenario describing the hypothetical organism would be the same as that shown in Figure 6–19, except that all phosphate and triphosphate groups would be on the right sides of the DNA and nucleotide structures as drawn.

Answer 6–18 See Figure A6–18.

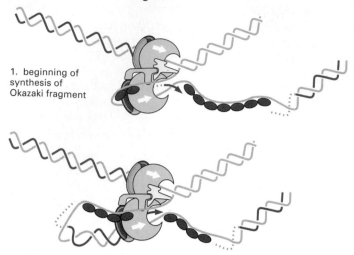

1. beginning of synthesis of Okazaki fragment

2. midpoint of synthesis of Okazaki fragment

Figure A6–18

Answer 6–19 Both strands of the bacterial chromosome contain 6×10^6 nucleotides. During the polymerization of nucleotide triphosphates into DNA, two phosphoanhydride bonds are broken for each nucleotide added: the nucleotide triphosphate is hydrolyzed to produce the nucleotide monophosphate added to the growing DNA strand, and the released pyrophosphate is hydrolyzed to phosphate. Therefore, 1.2×10^7 high-energy bonds are hydrolyzed during each round of bacterial DNA replication. This requires 4×10^5 ($= 1.2 \times 10^7/30$) glucose molecules, which weigh 1.2×10^{-16} g ($= 4 \times 10^5$ molecules $\times$ 180 g/mole/6×10^{23} mole-cules/mole), which is 0.01% of the total weight of the cell.

Answer 6–20 The statement is correct. If the DNA in somatic cells is not sufficiently stable (that is, if it accumulates mutations too rapidly), the organism dies (of cancer, for example), and because this may often happen before the organism can reproduce, the species will die out. If the DNA in reproductive cells is not sufficiently stable, many mutations will accumulate and be passed on to future generations, so that the species will not be maintained.

Answer 6–21 As shown in Figure A6–21 (*top right*), thymine and uracil contain no amino groups and therefore cannot be deaminated. Deamination of adenine and guanine produces purine rings that are not found in nucleic acids. In contrast, deamination of cytosine produces uracil. Therefore, if uracil were a naturally occurring base in DNA, repair enzymes could not distinguish whether a uracil is the appropriate base at that position or an error that is the result of spontaneous deamination. This dilemma is not encountered, however, because thymine is used in DNA. Therefore, if a uracil base is found in DNA, it can be automatically recognized as an erroneous base that is excised and replaced by cytosine.

Figure A6–21

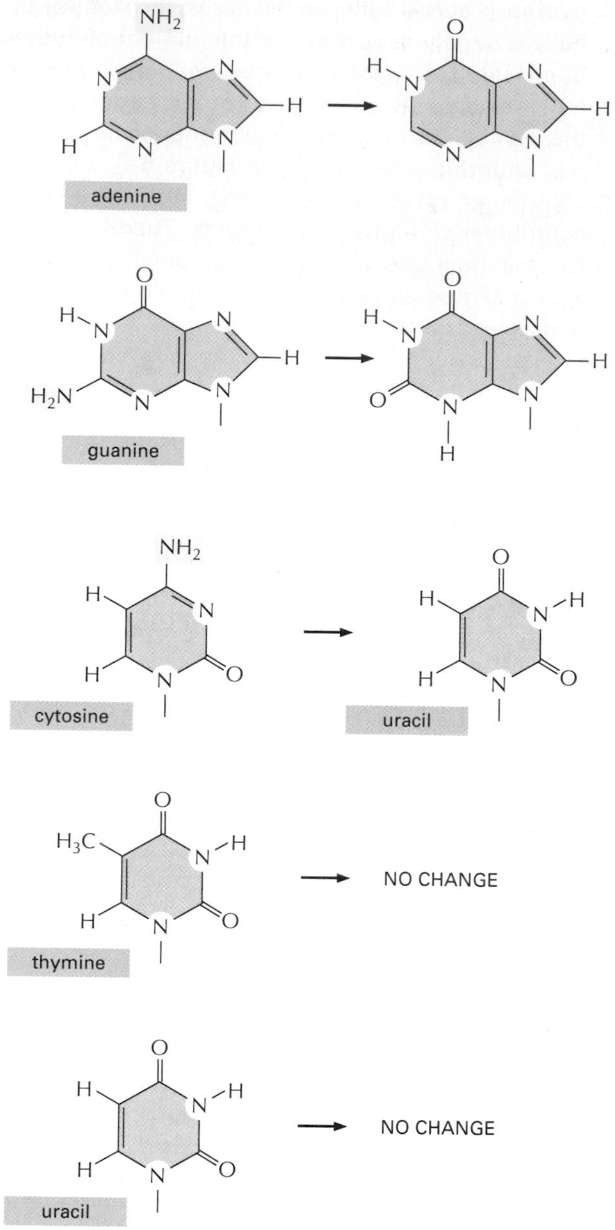

adenine

guanine

cytosine → uracil

thymine → NO CHANGE

uracil → NO CHANGE

Chapter 7

Answer 7–1 The answer is best given in a reflection of Francis Crick himself, who coined the term in the mid-1950s: "I called this idea the central dogma for two reasons, I suspect. I had already used the obvious word hypothesis in the sequence hypothesis, which proposes that genetic information is encoded in the sequence of the DNA bases, and in addition I wanted to suggest that this new assumption was more central and more powerful. . . . As it turned out, the use of the word dogma caused more trouble than it was worth. Many years later Jacques Monod pointed out to me that I did not appear to understand the correct use of the word dogma, which is a belief that cannot be doubted. I did appreciate this in a vague sort of way but since I thought that all religious beliefs were without serious foundation, I used the word in the way I myself thought about it, not as the world does, and simply applied it to a grand hypothesis that, however plausible, had little direct experimental support at the time." (Francis Crick, *What Mad Pursuit,* p. 109.)

Answer 7–2 Actually, the RNA polymerases are not moving at all, because they have been fixed and coated with metal to prepare the sample that can be viewed in the electron microscope. However, before they were fixed, they were moving from left to right, as indicated by the gradual lengthening of the RNA transcripts.

The RNA transcripts are shorter because they begin to fold up (i.e., to acquire a three-dimensional structure) as they are synthesized (see, e.g., Figure 7–5), whereas the DNA is an extended double helix.

Answer 7–3 At a first glance, the catalytic activities of an RNA polymerase used for transcription could replace the primase adequately. Upon further reflection, however, there are some serious problems: (1) The RNA polymerase used to make primer would need to initiate every few hundred bases, which is much more frequent than promoters are spaced on the DNA. Initiation would therefore need to occur in a promoter-independent fashion or many more promoters would have to be present in the DNA, both of which would be problematic for the control of transcription. (2) Similarly, the RNA primers used in replication are much shorter than mRNAs. The RNA polymerase would therefore need to terminate much more frequently than during transcription. Termination would need to occur spontaneously, i.e., without requiring a terminator sequence in the DNA, or many more terminators would need to be present. Again, both of these scenarios would be problematic for the control of transcription.

While it might be possible to overcome this problem if special control proteins became attached to RNA polymerase during replication, the problem has been solved during evolution by using separate enzymes with specialized properties. Some small DNA viruses, however, do utilize the host RNA polymerase to make primers for their replication.

Answer 7–4 As shown in Figure 7–16, your illustration should include the following steps: (1) snRNP binding to the precursor RNA, (2) lariat formation and 5′ splice site cleavage, and (3) 3′ splice site cleavage, joining of the two exon sequences, and release of the spliced mRNA. For the reaction it is important that the snRNP complex holds on to the 5′ exon after the first cleavage reaction, so that it can be joined to the 3′ exon in the next step. If the reactions were performed by separate enzymes, exons from one primary transcript could become mixed and join with exons from another.

Answer 7–5

1. UUUUUUUU... codes for Phe-Phe-Phe-..., i.e., a polymer of phenylalanine residues

2. AUAUAUAU... codes for Ile-Tyr-Ile-Tyr-..., i.e., a polymer of alternating isoleucine and tyrosine residues. Because the start point of the ribosome on the RNA is random, however, the polymer is a mixture of molecules that begin with an Ile residue and molecules that begin with a Tyr residue.

3. AUCAUCAUC... codes for a mixture of three different polymers. Because the start point of the ribosome on the RNA is random, ribosomes start translation in each of the three possible reading frames: AUC-AUC-AUC-... coding for Ile-Ile-Ile-..., UCA-UCA-UCA-... coding for Ser-Ser-Ser-..., and CAU-CAU-CAU-... coding for His-His-His-....

Answer 7–6 The mRNA will have a 5′-3′ polarity opposite that of the DNA strand which serves as a template. Thus, the mRNA sequence will read 5′-GAAAAAAGC-CGUUAA-3′. UAA specifies a stop codon. The carboxyl-terminal amino acid preceding the stop codon is therefore coded for by CGU and is an arginine residue. The amino-terminal amino acid coded for by GAA is a glutamic acid residue. Note that the convention in describing the sequence of a gene is to give the sequence of the DNA strand that is *not* used as a template for RNA synthesis; this sequence is the same as that of the RNA transcript, with T written in place of U.

Answer 7–7 The first statement is factually correct: RNA is thought to have been the first self-replicating catalyst and in modern cells is no longer self-replicating. We can debate, however, whether this represents a "demotion." The role of RNA now is more than that of messenger alone: it serves as primers for DNA replication (discussed in Chapter 6) and catalyzes some of the most fundamental processes in cells.

Answer 7–8

A. False. All ribosomes are equivalent and can make any protein that is specified by the particular mRNA that they are translating. After translation, ribosomes are released from the mRNA and can then start translating a new mRNA.

B. False. mRNAs are translated as linear polymers; there is no requirement that they have any particular folded structure. In fact, such structures that are formed by mRNA can inhibit translation because the ribosome has to unfold the mRNA in order to read the message it contains.

C. False. Ribosomal subunits exchange partners after each round of translation. After a ribosome is released from an mRNA, its two subunits dissociate and enter a pool of free small and large subunits from which new ribosomes are formed upon translation of a new mRNA.

D. False. Ribosomes are cytoplasmic organelles, but they are not individually enclosed in a membrane.

E. False. The position of the promoter determines the direction in which transcription proceeds and which DNA strand is used as the template. Transcription in the opposite direction would produce an mRNA with a completely different (and probably meaningless) sequence.

F. False. RNA contains uracil but not thymine.

G. False. The level of a protein depends on its rate of synthesis and degradation but not on its catalytic activity.

Answer 7–9 Because the deletion in the Lacheinmal mRNA is internal, it is likely that the deletion arises from a splicing defect. The simplest interpretation is that the Lacheinmal gene contains a 173-nucleotide-long exon (labeled "E2" in Figure A7–9), and that this exon is lost during the processing of the mutant precursor mRNA. This could occur, for example, if the mutation changed the 3′ splice site in the preceding intron ("I1") so that it was no longer recognized by the splicing machinery (a change in the UAG sequence shown in Figure 7–15 could do this). The snRNP would search for the next available 3′ splice site, which is found at the 3′ end of the next intron ("I2"), and the splicing reaction would therefore remove E2 together with I1 and I2, resulting in a shortened mRNA. The mRNA is then translated into a defective protein, resulting in the Lacheinmal deficiency.

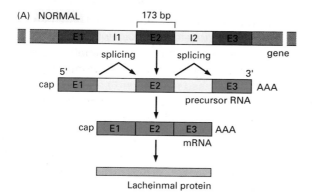

(A) NORMAL

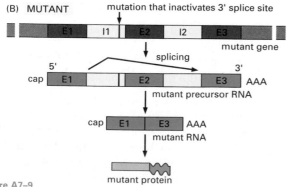

(B) MUTANT

Figure A7–9

Because 173 nucleotides do not amount to an integral number of codons, the lack of this exon in the mRNA will shift the reading frame at the splice junction. Therefore, the Lacheinmal protein would be made correctly only through exon E1. As the ribosome begins translating sequences in exon E3, it will be in a different reading frame and therefore will produce a protein sequence that is unrelated to the Lacheinmal sequence normally encoded by exon E3. Most likely, the ribosome will soon encounter a stop codon, which in RNA sequences that do not code for protein would be expected to occur on average about once in every 21 codons (there are 3 stop codons in the 64 codons of the genetic code).

Answer 7–10 Sequence 1 and sequence 4 both code for the peptide Arg-Gly-Asp. Because the genetic code is redundant, different nucleotide sequences can encode the same amino acid sequence.

Answer 7–11

A. Incorrect. The bonds are not covalent, and their formation does not require input of energy.

B. Correct. The aminoacyl-tRNA enters the ribosome at the A-site.

C. Correct. The ribosome moves along the mRNA, and the tRNAs that have donated their amino acid to the growing polypeptide chain are released from the ribosome and the mRNA.

Answer 7–12 *Replication:* Dictionary definition: the creation of an exact copy; molecular biology definition: the act of duplicating DNA. *Transcription:* Dictionary definition: the act of writing out a copy, especially from one physical form to another; molecular biology definition: the act of copying the information stored in DNA into RNA. *Translation:* Dictionary definition: the act of putting words into a different language; molecular biology definition: the act of polymerizing amino acids into a defined linear sequence from the information provided by the linear sequence of nucleotides in mRNA. (Note that "translation" is also used in a quite different sense, both in ordinary language and in scientific contexts, to mean a movement from one place to another.)

Answer 7–13 A code of two nucleotides could specify 16 different amino acids (= 4^2), and a triplet code in which the position of the nucleotides is not important could specify 20 different amino acids (= 4 possibilities of 3 of the same bases + 12 possibilities of 2 bases the same and one different + 4 possibilities of 3 different bases). In both cases, these maximal amino acid numbers would need to be reduced by at least one, because of the need to specify translation stop codons. It is relatively easy to envision how a doublet code could be translated by a mechanism similar to that used in our world by providing tRNAs with only two relevant bases in the anticodon loop. It is more difficult to envision how the nucleotide composition of a stretch of three nucleotides could be translated without regard to their order, because base-pairing can then no longer be used: an AUG, for example, will not base-pair with the same anticodon as a UGA.

Answer 7–14 In present-day cells, there is some wobble in the matching of codons to anticodons: in a number of cases, the same tRNA can pair with several codons that differ slightly in their nucleotide sequence. It seems likely that in the early world, without such highly evolved ribosomes as we have now to help in the pairing process, the converse may also have been true: several different tRNAs, with slightly different anticodons, may have been able to bind to the same codon. This would have played havoc with the translation of the genetic message into protein, unless the amino acids carried by all of the tRNAs capable of binding to the same codon were chemically similar. Natural selection would thus have ensured that tRNAs with similar anticodons carried chemically similar amino acids.

Perhaps in the early world, before modern aminoacyl-tRNA synthetases had evolved, there was also some "wobble" in the matching of tRNAs with appropriate amino acids: the same tRNA might have been liable to become coupled to any of a number of amino acids that were chemically similar. One can imagine the evolution of the genetic code by refinement of a matching process that was originally imprecise and gave only a blurred relationship between sets of roughly similar codons and sets of roughly similar amino acids.

Answer 7–15 The codon for Trp is 5′-UGG-3′. Thus, a normal Trp-tRNA contains the sequence 5′-CCA-3′ in its anticodon loop. If this tRNA contains a mutation so that its anticodon is changed to UCA, it will recognize a UGA codon and lead to the incorporation of a tryptophan residue instead of causing translation to stop. Many other protein-encoding sequences, however, contain UGA codons as their natural stop sites, and these stops would also be affected by the mutant tRNA. All these proteins would therefore be made with additional amino acids at the carboxyl-terminal end. The additional lengths would depend on the number of codons before the ribosomes encounter a non-UGA stop codon in the mRNA in the reading frame in which the protein is translated.

Answer 7–16 This experiment beautifully demonstrates that the ribosome does not check the amino acid that is attached to a tRNA. Once an amino acid has been coupled to a tRNA, the ribosome will "blindly" incorporate that amino acid into the position according to the match between the codon and anticodon. We can therefore conclude that a significant part of the correct reading of the genetic code, i.e., the matching of a codon with the correct amino acid, is performed by the synthetase enzymes that correctly match tRNAs and amino acids.

Answer 7–17 One effective way of driving a reaction to completion is to remove one of the products, so that the reverse reaction cannot occur. ATP contains *two* high-energy bonds that link the three phosphate groups. In the reaction shown, PP_i is released, consisting of two phosphate groups linked by one of these high-energy bonds. Thus, PP_i can be hydrolyzed with a considerable gain of free energy, and thereby be efficiently removed. This happens rapidly in cells, and reactions that produce and further hydrolyze PP_i are therefore virtually irreversible (see Chapter 3).

Answer 7–18

A. A titin molecule is made of 25,000 amino acids. It therefore takes about 3.5 hours to synthesize a single molecule of titin in muscle cells.

B. Because of its large size, the probability of making a titin molecule without any mistakes is only 0.08 [$= (1 - 10^{-4})^{25,000}$], i.e., only 8 in 100 titin molecules synthesized are free of mistakes. In contrast, over 97% of newly synthesized proteins of average size are made correctly.

C. The error rate limits the sizes of proteins that can be synthesized accurately. As for titin, a large portion (87%) of the newly synthesized hypothetical giant ribosomal protein would be expected to contain at least one mistake. It is more advantageous to make ribosomal proteins individually, because in this way only a small proportion of each type of protein will be defective, and these few bad molecules can be individually eliminated to ensure that there are no defects in the ribosome as a whole.

D. To calculate the time it takes to transcribe a titin mRNA, you would need to know the size of its gene, which is likely to contain many introns. Transcription of the exons alone requires about 42 minutes. Because introns can be quite large, the time required to transcribe the entire gene is likely to be considerably longer.

Answer 7–19 Mutations of the type described in (b) and (d) are often the most harmful. In both cases, the reading frame would be changed, and because this frame shift occurs near the beginning or in the middle of the coding sequence, much of the protein will contain a nonsensical and/or truncated sequence of amino acids. In contrast, a reading-frame shift that occurs toward the end of the coding sequence [as described in scenario (a)], will result in a largely correct protein that may be functional. Deletion of three consecutive nucleotides [scenario (c)] leads to the deletion of an amino acid, but does not alter the reading frame. The deleted amino acid may or may not be important for the folding or activity of the protein; in many cases such mutations are silent, i.e., they have no or only minor consequences for the organism. Substitution of one nucleotide for another, as in (e), is often completely harmless, not even causing a change in the amino acid sequence; in other cases it may change one amino acid in the protein sequence; at worst, it may create a new stop codon, giving rise to a truncated protein.

Chapter 8

Answer 8–1 The scale bar in Figure 8–3 is in millions of nucleotide pairs. Using this to estimate the amount of DNA packaged into Chromosome 1 we obtain approximately 256 million nucleotide pairs. This would give a total length for the DNA of 8.7 cm ($256 \times 10^6 \times 0.34$ nm; 1 nm = $1/10^9$ m) and a compaction of 8.7 cm/10 μm = 8700-fold.

Answer 8–2

A. If the single origin of replication were located exactly in the center of the chromosome, it would take more than eight days to replicate the DNA [= 75×10^6 nucleotides/(100 nucleotides/sec)]. The rate of replication would therefore severely limit the rate of cell division. (If the origin were located off-center, the time required for replication would be even longer.)

B. A chromosome end that is not "capped" with a telomere would lose nucleotides during each round of DNA replication and would gradually shrink. Eventually, essential genes would be lost, leading to cell death.

C. Without a centromere that attaches them to the mitotic spindle, the two new chromosomes that result from replication cannot be partitioned accurately between the two daughter cells. Therefore many daughter cells would die, because they failed to receive a full set of chromosomes.

Answer 8–3 Men have only one copy of the X chromosome; a defective gene carried on it therefore has no backup copy. Women, on the other hand, have two copies of the X chromosome, one inherited from each parent, so a defective copy of the gene on one X chromosome can generally be compensated for by a normal copy on the other chromosome. This is the case with regard to the gene that causes color blindness. However, during female development, transcription from one X chromosome is shut down because it is compacted into heterochromatin (Figure 8–12). This occurs at random to one or the other of the two X chromosomes, and therefore some cells of the woman will express the defective mutant copy of the gene, whereas others will express the normal copy. This results in a retina in which on average only every other cone cell has functional color photoreceptors, and women carrying the mutant gene on one X chromosome therefore see color images with reduced resolution.

A woman who is color blind must have two defective copies of this gene, one inherited from each parent. Her father must therefore carry the mutation on his X chromosome; because this is his only copy of the gene, he would be color blind. Her mother could carry the defective gene on either or both of her X chromosomes. Her mother could therefore either be color blind (defective genes on both X chromosomes) or have color vision but reduced resolution as described above. Several different types of inherited color blindness are found in the human population; this question applies to only one type.

Answer 8–4 Contacts can form between the protein and the edges of the base pairs that are exposed in the major groove of the DNA. The contacts that can form are shown in Figure A8–4. The bonds responsible for sequence-specific contacts are hydrogen bonds and a hydrophobic interaction that can occur with the methyl group on the pyrimidine ring of T. Note that the arrangement of hydrogen-bond donors and hydrogen-

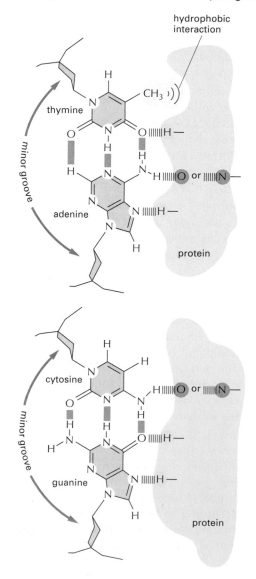

Figure A8–4

bond acceptors of a T-A pair is different from that of a C-G pair. Similarly, the arrangement of hydrogen-bond donors and hydrogen-bond acceptors of A-T and G-C pairs would be different from one another and from the two pairs shown in the figure. In addition to the contacts shown in the figure, ionic interactions between the positively charged amino acid side chains of the protein and the negatively charged phosphate groups in the DNA backbone usually stabilize DNA-protein interactions.

Answer 8–5

A. UV light throws the switch from the prophage to the lytic state: when cl protein is destroyed, cro is made and turns off the production of new cl. The virus starts to produce coat proteins, and new virus particles are made.

B. When the UV light is switched off, the virus remains in the lytic state. Thus, cl and cro form a gene regulatory switch that "memorizes" its previous setting.

C. This switch makes sense in the viral life cycle: UV light is prone to damage the bacterial DNA (see Figure 6–28), thereby rendering the bacterium an unreliable host for the virus. A prophage virus will therefore switch to the lytic state and leave an irradiated cell in search for new host cells to infect.

Answer 8–6   Bending proteins can help to bring distant DNA regions together that normally would contact each other only inefficiently (Figure A8–6). Such proteins are found in both procaryotes and eucaryotes and are involved in many examples of transcriptional regulation.

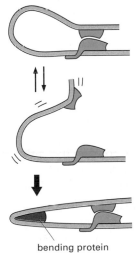

bending protein

Figure A8–6

Answer 8–7

A. True (see also Question 8–2, p. 250).

B. False. Bacteria have only one RNA polymerase that transcribes all genes; in contrast, eucaryotic

cells have three different polymerases, each dedicated to one of the three classes of genes.

C. True. Procaryotic mRNAs are often transcripts of entire operons. Ribosomes can initiate translation at internal AUG start sites of these "polycistronic" mRNAs (see Figure 7–29).

D. True. The major groove of double-stranded DNA is sufficiently wide to allow a protein surface, such as one face of an α helix, access to the base pairs.

E. True. It is advantageous to exert control at the earliest possible point in a pathway. This conserves metabolic energy because unnecessary products are not made in the first place.

F. False. The zinc atoms in zinc finger domains are required for the correct folding of the protein domain; they are internal to these domains and do not contact the DNA.

G. False. RNA polymerase readily transcribes through nucleosomes.

H. False. Nucleosome core proteins are approximately 11 nm in diameter. A model for the way they are packed to form a 30-nm-diameter filament is shown in Figure 8–10.

Answer 8–8   The definitions of the terms can be found in the Glossary. DNA assembles with specialized proteins to form *chromatin.* At a first level of packing, *histones* form the core of *nucleosomes.* In a nucleosome, the DNA is wrapped twice around this core. Between nuclear divisions, that is, in interphase, the *chromatin* of the *interphase chromosomes* is in a relatively extended form and is dispersed in the nucleus, although some regions of it, the *heterochromatin,* remain densely packed and are transcriptionally inactive. During nuclear division, that is, in mitosis, replicated chromosomes become condensed into *mitotic chromosomes,* which are transcriptionally inactive, and which are designed to be distributed between the daughter cells.

Answer 8–9

A. Without telomeres and telomerase, chromosome ends would shrink during each round of replication, because there would be no 3' –OH group to prime the DNA synthesis that fills in after removal of the primer of the last DNA fragment laid down on the lagging strand. Because bacterial chromosomes have no ends, the problem does not arise; there will always be a 3' –OH group available to prime the DNA polymerase that replaces the RNA primer with DNA (Figure A8–9, *next page*). Telomeres and telomerase prevent the shrinking of chromosomes because the telomeres extend the 3' end of a DNA strand with DNA repeats that are added directly by telomerase without the need for a template (see Figure 8–6). A few of these

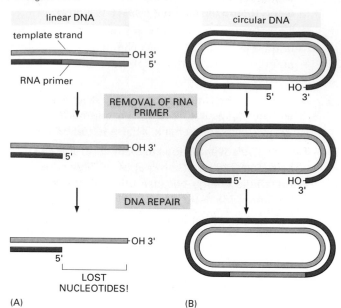

linear DNA

circular DNA

template strand

RNA primer

REMOVAL OF RNA PRIMER

DNA REPAIR

LOST NUCLEOTIDES!

(A) (B)

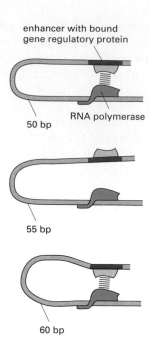

enhancer with bound gene regulatory protein

RNA polymerase

50 bp

55 bp

60 bp

repeats are lost during each round of replication, but they are replenished.

B. As shown in Figure A8–9, telomeres and telomerase are still needed even if the last fragment of the lagging strand were initiated by primase at the very 3′ end of chromosomal DNA, inasmuch as an RNA primer must still be used.

Answer 8–10   Colonies are clumps of cells that originate from a single founder cell and grow outward as the cells divide more and more times. In the lower colony of Figure 8–13A, the *ADE2* gene is inactivated when placed near a telomere, but apparently it can become spontaneously activated in a few cells, which then turn white. Once spontaneously activated in a cell, the *ADE2* gene continues to be active in the descendants of that cell, resulting in clumps of white cells (the white sectors) in the colony. This result shows both that the inactivation of a gene positioned close to a telomere can be reversed and that this change is passed on to further generations (see Figure 8–33). This change in *ADE2* expression probably results from a spontaneous decondensation of the chromatin structure around the gene.

Answer 8–11   From our knowledge of enhancers one would expect its function to be relatively independent of its distance from the RNA-polymerase-binding site—possibly weakening as it is moved farther and farther away. The surprising feature of the data (which have been adapted from an actual experiment) is the periodicity: the enhancer is maximally active at certain distances from the RNA-binding site (50, 60, or 70 nucleotides), but almost inactive at intermediate distances (55 or 65 nucleotides). The periodicity of 10 suggests that the mystery can be explained by considering

the structure of double-helical DNA, which has very close to 10 base pairs per turn. Thus placing an enhancer on the side of the DNA opposite to that of the promoter (Figure A8–11) would make it more difficult to interact with the proteins bound at the promoter. At longer distances, there is more DNA to absorb the twist, and the effect diminishes.

Answer 8–12   Two advantages of dimeric DNA-binding proteins are (i) that the binding affinity can be very high because the number of potential contacts with DNA is double that possible with a monomer and (ii) that several different subunits can be combined in many different combinations to increase the number of DNA-binding specificities that are available to cells. Three of the most frequently used protein domains involved in DNA binding are leucine zippers, homeodomains, and zinc fingers. Each provides a particularly stable fold in the polypeptide chain that positions an α helix appropriately on the protein's surface so that it can insert into the major groove of the DNA helix and contact the sides of the base pairs (see Figure 8–19).

Answer 8–13

A. If sufficient tryptophan is present in the cells, Trp repressor will block the synthesis of enzymes that would make more tryptophan. Likewise, if cells are starved for tryptophan, the unoccupied repressor would not bind to the DNA, and the enzymes that synthesize tryptophan would be induced. This simple and elegant form of feedback inhibition (see Chapter 5) allows cells to adjust the rate of tryprophan synthesis to their needs.

B. Transcription of the genes encoding the tryptophan biosynthetic enzymes would no longer be regulated by the absence or presence of trypto-phan; the enzymes would be permanently on in scenario (i) and permanently off in scenario (ii).

C. In scenario (i), the normal tryptophan repressor molecules would completely restore the regulation of the tryptophan biosynthesis enzymes. In contrast, expression of the normal protein would have no effect in scenario (ii), because the Trp-repressor-binding sites on the DNA would remain permanently occupied by the mutant protein.

Answer 8–14   In the electron micrographs, one can detect chromatin regions of two different densities; the densely stained regions correspond to heterochromatin, while less condensed chromatin is more lightly stained. The chromatin in nucleus A is mostly in the form of condensed, transcriptionally inactive heterochromatin, whereas most of the chromatin in nucleus B is decondensed and therefore potentially active for transcription. Nucleus A is from a reticulocyte, a red blood cell precursor, which is largely devoted to making a single protein, hemoglobin. Nucleus B is from a lymphocyte, which is active in transcribing many different genes.

Answer 8–15   The experiment is one that shows that a single differentiated cell taken from a specialized tissue can re-create a whole organism. This proves that the cell must contain all the information required to produce a whole organism, including all of its specialized cell types. See Figure 8–16.

Answer 8–16   You could create 16 different cell types with 4 different gene regulatory proteins (all the 8 cell types shown in Figure 8–31, plus another 8 created by adding an additional gene regulatory protein). MyoD by itself is sufficient only to induce muscle-specific gene expression in certain cell types, such as some kinds of fibroblasts. The action of MyoD is therefore consistent with the model shown in Figure 8–31: if muscle cells were specified, for example, by the combination of gene regulatory proteins 1, 3, and MyoD, then the addition of MyoD will convert only two of the cell types of Figure 8–31 (cells F and H) to muscle.

Answer 8–17   Many gene regulatory proteins are always being made in the cell; that is, their expression is constitutive and the activity of the protein is controlled by signals from inside or outside the cell (e.g., the availability of nutrients, as for the Trp repressor, or by hormones, as for the glucocorticoid receptor), thereby adjusting the transcriptional program to the physiological needs of the cell. Moreover, a given gene regulatory protein usually regulates the expression of many different genes. Gene regulatory proteins are often used in various combinations and can affect each other's activity, thereby further increasing the possible regulatory repertoire of gene expression with a limited set of proteins. Nevertheless, the cell devotes a large fraction of its genome to the control of

transcription: an estimated 10% of all genes in eucaryotic cells code for gene regulatory proteins.

Chapter 9

Answer 9–1   The result of the suggested shortcut would be undesirable. The 1 μl samples of the culture would contain a statistical distribution of cells: some would have one cell, as desired, but others would have none, two, or even more cells. After growing cultures from these samples you will therefore not know whether the cells in them really arose from a single cell. (The distribution of cells in the samples is described by the Poisson distribution, which indicates that 37% of the initial 1 μl samples would contain no cells, 37% would contain one cell, 18% two cells, and 8% three or more cells.)

Answer 9–2

A. The colonies are formed by bacteria that have acquired a new mutation that reverses the effect of the original mutation in *lacY*. The new mutation is often one that changes the nucleotide sequence of the mutant gene back to that of the wild-type sequence.

B. *E. coli* cells that lack a functional mismatch repair system accumulate random mutations at an increased frequency, because errors that occur in DNA replication will not be corrected. A small fraction of those random mutations will correct the *lacY* mutant gene; because the random mutation rate of the *E. coli strain* deficient in mismatch repair is about 100 times higher than that of a repair-competent strain, more colonies able to use lactose will appear on the agar plate.

Answer 9–3   Bacteriophage lambda will primarily package bacterial genes that are in proximity to its integration site. Other bacterial viruses (such as P1), however, can package any bacterial gene.

Answer 9–4   Each time another copy of a transposable element is inserted into the chromosome, the change can be either neutral, beneficial, or detrimental for the organism. Since individuals that accumulate detrimental insertions would be selected against, the proliferation of transposable elements is controlled by natural selection. If a transposable element arose that proliferated uncontrollably, it is unlikely that a viable host organism could be maintained. For this reason, most transposons have evolved to transpose only rarely. Many transposons, for example, synthesize only infrequent bursts of very small amounts of the transposase that is required for their movement.

Answer 9–5   Minus-strand RNA viruses must synthesize a complementary RNA strand using their genome as a template before any viral proteins can be synthesized. Since the enzyme that carries this out—an RNA replicase—is encoded by the virus, the virus particle must carry a few molecules of the replicase protein inside its coat. Upon infection, both the genome and the replicase are released into the cell.

Answer 9–6 The AIDS virus (the human immunodeficiency virus, HIV) is a retrovirus, and thus synthesizes DNA from an RNA template using reverse transcriptase. This leads to frequent mutation of the viral genome. In fact, AIDS patients often harbor many different variants of HIV that are genetically distinct from the original virus that infected them. This poses great problems in treating the infection: drugs that block essential viral enzymes work only temporarily, because new virus strains that are resistant to these drugs arise rapidly by mutation.

RNA replicases (RNA polymerases that synthesize RNA using RNA as a template, see Figure 9–28) do not proofread either. Thus RNA viruses that replicate RNA genomes directly also mutate frequently. In such a virus, this tends to produce changes in the coat proteins that cause the mutated virus to appear "new" to the immune system; the virus is therefore not suppressed by immunity that has arisen to the previous version. This is part of the explanation for the new strains of the influenza (flu) virus and the common cold virus that regularly appear.

Answer 9–7 A typical human female produces less than 1000 mature eggs in her lifetime (12 per year over about 40 years); this is less than a tenth of a percent of the possible gametes excluding the effects of meiotic crossing-over. A typical human male produces billions of sperm during a lifetime, so in principle, all the possible chromosome combinations are each sampled many times.

Answer 9–8

A. True.

B. True.

C. False. Mutations that occur during meiosis can be propagated, unless they give rise to nonviable gametes.

D. True. Cells can take up DNA molecules, and if the genes contained in the DNA contain promoters and ribosome initiation sites that are recognized by the recipient cell, they are expressed. Many bacteria take up DNA very efficiently, but most eucaryotic cells need to be coaxed in the laboratory into doing it.

E. False. Cells can produce many copies of the virus from the integrated viral genome.

F. True. Because the single-stranded RNA genome of a retrovirus cannot be directly translated into protein, retroviruses require a premade reverse transcriptase in order to produce a DNA copy of their genome once they have entered a cell. This copy

is then integrated into the host genome and transcribed by RNA polymerase to produce mRNA that can be translated into protein. As cells do not generally contain reverse transcriptase that can efficiently work on the viral genome, the virus brings its own enzyme into the cell.

G. True. If noncorresponding regions of chromosomes could recombine, large-scale genomic rearrangements would occur that would almost always be detrimental to the organism. However, a few rare "unequal crossovers" have provided strong advantages to the organism, and records of these rare beneficial examples are found in present-day genomes (see Figures 9–19, 9–20, and 9–21).

H. False. Viral coats are made of proteins or—in the case of enveloped viruses—of an internal protein coat surrounded by a lipid bilayer membrane containing additional proteins. The DNA or RNA genome is protected inside this viral coat.

Answer 9–9 Diploid organisms contain two copies of most of their genes in each cell, providing "backup copies" for genes that may become damaged by mutation. This complicates genetic studies. Many mutations in diploids do not cause any detectable defect in cellular processes because a normal copy of the gene is also expressed in the same cell and can often compensate for the mutated gene. If a diploid organism produced diploid gametes, its number of chromosomes would double in each generation. In addition, it has been suggested that haploid gametes can allow a diploid organism to eliminate many detrimental mutations that might not be apparent in diploid cells, because impaired gametes are likely to be less successful in fertilization. Gametes are usually made in numbers greatly exceeding what might seem to be required, and must pass competitive "fitness tests." Millions of mammalian sperm cells, for example, compete in a fierce race to reach and penetrate a single egg cell.

Answer 9–10 Evolution depends to a large extent on competition among individuals, each carrying slight variants of the gene set. Suppose that two cells in a population each undergo a beneficial mutation that affects a different gene. In a strictly asexual species, each of these cells will give rise to a population of mutant cells. Cells of the two populations will compete with each other until one or the other predominates: only one of the mutations will therefore spread through the population, while the other will eventually be lost. But if one of the original mutant cells has evolved a mechanism that allows it to incorporate genes from other cells, such a cell may also acquire the second beneficial mutation. This creates a cell that carries both beneficial mutations. Such a cell will be the most successful of all, and its success will also ensure the propagation of the trait that confers the ability to incorporate genes from other cells. This rudimentary sexual capability will thus be favored by natural selection.

Answer 9–11 With their ability to transfer nucleic acid sequences from cell to cell, viruses have almost certainly played an important part in the evolution of the organisms they infect. Many pick up small pieces of host chromosome at random and carry them to different cells or organisms. Therefore, viruses can speed up evolution by promoting the mixing of gene pools. While often harmful to the individual organism, they are probably beneficial to a species as a whole.

Answer 9–12

A. A component of the peanut extract must cause a change in some of the bacteria that allows them to grow in the absence of histidine. Because this change is heritable (recall that the cells isolated from the ring of colonies grow in the absence of histidine when placed on fresh plates), the change must be a mutation that reverses the effect of the original mutation that made the cells require histidine. Your moldy peanut extract must therefore contain a mutagen—a compound that produces mutations by damaging DNA. So don't eat moldy peanuts!

B. A ring of colonies forms because the mutagen contained in the extract diffuses, forming a concentration gradient that is highest at the center of the plate where you originally placed the drop of extract and diminishes toward the plate's rim. Cells too close to the drop receive a very high concentration of the mutagen, which seriously damages their DNA, resulting in cell death. At just the right distance from the drop, the mutation rate is high but not high enough to kill all of the cells. At distances farther from the center of the plate, the mutation rate is so low that the required mutation is produced too rarely to be detectable.

C. The thirty-three colonies each arose from a single mutated cell. The mutagen can cause mutations all over the bacterial chromosome, but only a very few of these will reverse the effects of the original mutation. Therefore, only a very small fraction of the bacteria that were originally plated in the ring region of the plate become mutated in this way and grow into colonies.

The mutagen involved here is aflatoxin, which is one of the most powerful mutagens known. It is produced by a particular mold that grows on peanuts. Aflatoxin is a cause of human stomach cancer in countries where this mold is prevalent and peanuts are a significant part of the diet. Since most human cancers derive from mutations accumulated in the cells of the body (see Figure 6–24), it makes sense that ingestion of a mutagen would increase the likelihood of cancer. The sort of experiment described in this question is used as one of the standard tests (the Ames test) to determine whether substances are mutagenic and therefore might cause cancer in humans. Components of cigarette smoke test positive in this assay, for example.

Answer 9–13 The main difference is that viruses encode coat proteins. Whereas plasmids can exist only as naked DNA, viral genomes are surrounded and protected by the protein coat. Viruses can thus survive outside cells. Coat proteins also enable viruses to attach to the outside of host cells and facilitate the entry of their genomes into these cells.

Answer 9–14 The answer lies in the need for the cell to maintain a balance between stability and change. If the mutation rate were too high, a species would eventually die out because all its individuals would accumulate too many mutations in genes essential for survival. For a species to be successful—in evolutionary terms—it is important for individual members to have good genetic memory, that is, fidelity in DNA replication, but also to introduce occasional variations. If the change leads to an improvement, it will persist by selection; if it proves disastrous, the individual organism that was the unfortunate subject of nature's experiment will die—not the entire population.

Answer 9–15 mRNAs could be converted back into DNA sequences. These DNA sequences would differ from the DNA sequences of genes because they would contain mutations, since reverse transcriptase does not copy as accurately as do DNA polymerases; and they would lack introns, since these are removed by splicing before the mRNA enters the cytoplasm (see Chapter 7).

These DNA sequences could undergo homologous recombination events similar to that shown in Figure 9–13A and replace the chromosomal DNA sequences. This process would eventually introduce many mutations into the genome, putting the species at risk of extinction.

Answer 9–16 Viruses cannot exist as free-living organisms: they have no metabolism, do not communicate with other viruses, and cannot reproduce themselves. They thus have none of the attributes that one normally associates with life. In fact, they can even be crystallized. Inside cells, they redirect normal cellular biosynthetic activities to the task of making more copies of themselves, but a virus cannot reproduce without exploiting a host cell. Thus the only aspect of "living" that viruses display is their capacity to direct their own reproduction once inside a cell.

Answer 9–17 Transposable elements could provide opportunities for homologous recombination events, thereby causing genomic rearrangements. They could insert into genes, possibly obliterating splice sites, thereby changing the gene structure. They could also insert into the regulatory region of a gene, where insertion between an enhancer and a transcription start site could block the function of the enhancer and therefore reduce the level of expression of a gene. In addition, the transposable element could itself contain an enhancer and thereby change the time and place in the organism where the gene is expressed.

Answer 9–18 The pseudogene will lack introns because it was derived from mRNA that had already been spliced. Since it will probably also lack a promoter, it is unlikely that it will be transcribed. Many such inactive pseudogenes have been discovered in mammalian genomes; they are thought to be harmless relics of some accidental reverse transcription event in the history of the species. Pseudogenes are often incomplete, because reverse transcriptase may not have copied the complete mRNA.

Answer 9–19 Alu sequences lack the middle portion of 7SL RNA, the *blue* portion in Figure Q9–19 (p. 311). We can therefore postulate that a cleavage and rejoining event (perhaps an aberrant splicing reaction) took place that left a shortened RNA intermediate, as shown in Figure A9–19. Reverse transcription of this RNA intermediate would produce a single-stranded DNA molecule. This DNA molecule, after removal of the RNA template, could then serve as a template for a reverse transcriptase to synthesize the other strand to give a double-stranded DNA molecule (see Figures 9–23 and 9–30). This DNA fragment could then have integrated into chromosomal DNA. When transcribed and converted into DNA, the transposable element could insert additional copies of itself elsewhere into the host genome. All the reactions described here are conceptually feasible and use mechanisms that are known to occur in present-day cells.

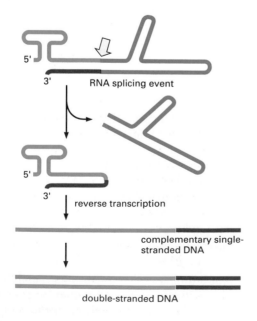

Figure A9–19

5'
3' RNA splicing event

reverse transcription

complementary single-stranded DNA

double-stranded DNA

Answer 9–20 Without chromosome crossovers, the animal could produce $2^4 = 16$ genetically different gametes. If the effects of the restricted crossing-over are included (each chromosome has 2^3 possible combinations) the total number of possible gametes is $2^7 = 208$. If, as in real life, the sites of chromosome crossing-over were not restricted (that is, crossovers between homologous chromosomes could occur anywhere along the lengths of chromosomes), the number of possible gametes would be much higher.

Chapter 10

Answer 10–1

A. Digestion with Eco RI produces two products:

```
5'-AAGAATTGCGG      AATTCGAGCTTAAGGGCCGCGCCGAAGCTTTAAA-3'
3'-TTCTTAACGCCTTAA      GCTCGAATTCCCGGCGCGGCTTCGAAATTT-5'
```

B. Digestion with Alu I produces three products:

```
5'-AAGAATTGCGGAATTCGAG    CTTAAGGGCCGCGCCGAAG    CTTTAAA-3'
3'-TTCTTAACGCCTTAAGCTC    GAATTCCCGGCGCGGCTTC    GAAATTT-5'
```

C. The sequence contains no Not I cleavage site.

D. Digestion with all three enzymes therefore produces:

```
5'-AAGAATTGCGG    AATTCGAG    CTTAAGGGCCGCGCCGAAG    CTTTAAA-3'
3'-TTCTTAACGCCTTAA    GCTC    GAATTCCCGGCGCGGCTTC    GAAATTT-5'
```

Answer 10–2 If the ratio of dideoxyribonucleoside triphosphates to deoxyribonucleoside triphosphates is increased, DNA polymerization is terminated more frequently and thus shorter DNA strands are produced. Such conditions are favorable for determining short nucleotide sequences, that is, the sequences that are close to the DNA primer used in the reaction. In contrast, decreasing the ratio of dideoxyribonucleoside triphosphates to deoxyribonucleoside triphosphates allows one to determine nucleotide sequences more distant from the primer.

Answer 10–3 The presence of mutation in a gene does not necessarily mean that the protein expressed from it is defective. For example, the mutation could change one codon into another that still specifies the same amino acid, and so does not change the amino acid sequence of the protein. Or, the mutation may cause a change from one amino acid to another in the protein, but in a position that is not important for the folding or function of the protein. In assessing the likelihood that such a mutation might cause a defective protein, information on the known β-globin mutations that are found in humans is essential. You would therefore want to know the precise nucleotide change in your mutant gene, and whether this change has any known or predictable consequences for the function of the encoded protein. If your mate has two normal copies of the globin gene, none of your children would manifest disease arising from defective hemoglobin (thalassemia); however, on average, 50% of your children would be carriers of your one defective gene.

Answer 10–4 Although several explanations are possible, the simplest is that the DNA probe has hybridized predominantly with its corresponding mRNA, which when expressed is typically present in many more copies than is the gene. The strongly hybridizing cells probably express the gene at high levels and therefore have high levels of the mRNA.

Answer 10–5 There are approximately 10^9 different sequences that can be formed by 15 nucleotides (4^{15} = 1,073,741,824). By chance, therefore, any sequence specified by 15 nucleotides is expected to be found on average three times in the human genome. As the probe shown in Figure 10–17 is a mixture of 16 different oligonucleotides, each 15 nucleotides long, there are predicted to be approximately 48 (= 3 × 16) matching sites in the genome, of which only one corresponds to a sequence in the Factor VIII gene. To ascertain that a sequence identified with the probe indeed contains the Factor VIII gene, one usually obtains an additional stretch of protein sequence and repeats the search with a second oligonucleotide probe. If these probes hybridize near one another on the genome, it is a strong indication that one really has isolated the Factor VIII gene.

Answer 10–6

A. After an additional round of amplification there will be 2 *gray,* 4 *green,* 4 *red,* and 22 *yellow* outline fragments; after a second additional round there will be 2 *gray,* 5 *green,* 5 *red,* and 52 *yellow* outline fragments. Thus the DNA fragments outlined in *yellow* increase exponentially and will eventually overrun the other reaction products. Their length is precisely determined by the DNA sequence that spans the distance between the two primers used in the amplification.

B. The mass of one DNA molecule 500 nucleotide pairs long is 5.5×10^{-19} g [= 2 x 500 × 330 (g/mole)/ 6×10^{23} (molecules/mole)]. Ignoring the complexities of the first few steps of the amplification reaction (which produce longer products that eventually make an insignificant contribution to the total DNA amplified), this amount of product approximately doubles for every amplification step. Therefore, 100×10^{-9} g = 2^N x 5.5×10^{-19} g, where N is the number of amplification steps of the reaction. Solving this equation for $N = \log(1.81 \times 10^{11})/ \log(2)$ gives $N = 37.4$. Thus, only about 40 cycles of PCR amplification are sufficient to amplify DNA from a single molecule to a quantity that can be readily handled and analyzed biochemically. This whole procedure takes only about 4 hours in the laboratory.

Answer 10–7 Like the vast majority of mammalian genes, it is likely that the attractase gene contains introns. Bacteria do not have the splicing machinery required to remove introns, and therefore the correct protein cannot be expressed from the gene. For expression of most mammalian genes in bacterial cells, a cDNA version of the gene must be used.

Answer 10–8

A. False. Restriction sites are found throughout the DNA, that is, within, as well as between, genes.

B. True. DNA bears a negative charge at each phosphate, giving DNA an overall negative charge.

C. False. Clones isolated from cDNA libraries never contain promoter sequences. These sequences are not transcribed and are therefore not part of the mRNAs that are used as the templates to make cDNAs.

D. True. Each polymerization reaction produces double-stranded DNA that must, at each cycle, be denatured to allow new primers to hybridize so that the DNA strand can be copied again.

E. False. Digestion of genomic DNA with restriction nucleases that recognize four-nucleotide sequences produces fragments that are *on average* 256 nucleotides long. However, the actual lengths of the fragments produced will vary considerably on both sides of the average.

F. True. Reverse transcriptase is first needed to copy the mRNA into single-stranded DNA, and DNA polymerase is then required to make the second DNA strand.

G. True. Using a sufficient number of VNTRs, individuals can be uniquely "fingerprinted" (see Figure 10–25).

H. True. If cells of the tissue do not transcribe the gene of interest, it will not be represented in a cDNA library prepared from this tissue. However, it will be represented in a genomic library prepared from the same tissue.

Answer 10–9

A. The DNA sequence, from its 5′ end to its 3′ end, is read starting from the bottom of the gel, where the smallest DNA fragments migrate. Each band results from the incorporation of the appropriate dideoxyribonucleoside triphosphate, and as expected, there are no two bands that have the same mobility. This allows one to determine the DNA sequence by reading off the bands in strict order, proceeding upward from the bottom of the gel, and assigning the correct nucleotide according to which lane the band is in.

The nucleotide sequence of the top strand (Figure A10–9A, *next page*) was obtained from Figure Q10–9, and the bottom strand was deduced from the complementary base-pairing rules.

(A) 5'-TATAAACTGGACAACCAGTTCGAGCTGGTGTTCGTGGTCGGTTTTCAGAAGATCCTAACGCTGACG-3'
     3'-ATATTTGACCTGTTGGTCAAGCTCGACCACAAGCACCAGCCAAAAGTCTTCTAGGATTGCGACTGC-5'

(B) 5' top strand of DNA   3'
     TATAAACTGGACAACCAGTTCGAGCTGGTG TTCGTGGTCGGTTTTCAGAAGATCCTAACGCTGACG
   1 LeuLysLeuGluAsnGlnPheGlnLeuVal PheValValGlyPheGlnLysIleLeuThrLeuThr
   2 IleAsnTrpThrThrSerSerSerTrpCy sSerTrpSerValPheArgArgSer   Arg
   3   ThrGlyGlnProValArgAlaGlyV alArgGlyArgPheSerGluAspProAsnAlaAsp

B. The DNA sequence can then be translated into an amino acid sequence using the genetic code. However, there are two possible strands of DNA that could be transcribed into RNA, and translated into protein and three possible reading frames, for each strand. Thus there are six amino acid sequences that can in principle be encoded by this stretch of DNA. Of the three reading frames possible from the top strand, only one is not interrupted by a stop codon (Figure A10–9B).

From the bottom strand, two of the three reading frames also have stop codons (not shown). The third frame gives the following sequence:

SerAlaLeuGlySerSerGluAsnArgProArgThrProAlaArgThrGlyCysProValIle

It is not possible from the information given to tell which of the two "open reading frames" corresponds to the actual protein encoded by this stretch of DNA.

Answer 10–10

A. Cleavage of human genomic DNA with Hae III would generate about 11×10^6 different fragments [= $3 \times 10^9 / 4^4$], with Eco RI about 730,000 different fragments [= $3 \times 10^9 / 4^6$], and with Not I about 46,000 different fragments [= $3 \times 10^9 / 4^8$]. There will also be some additional fragments generated because the maternal and paternal chromosomes are very similar but not identical in DNA sequence.

B. A set of overlapping DNA fragments will be generated. Libraries constructed from sets of overlapping fragments of a genome are valuable because they can be used to order cloned sequences in relation to their original order on the chromosomes, and thus obtain the DNA sequence of a long stretch of DNA. Sequences from the end of one cloned DNA are used as DNA hybridization probes to find other clones in the library that contain those sequences and that therefore might overlap. By repeating this procedure, a long stretch of continuous DNA sequence can be gradually built up (see Question 10–16, p. 345). This laborious technique of "chromosome walking" has been used to build up the sequence of areas of chromosomes known to contain important genes, but where no prior information on the nucleotide sequence exists. By careful inspection of the nucleotide sequences of the areas obtained by chromosome walking, the positions and sequences of candidate genes can be identified.

Answer 10–11 By comparison with the positions of the size markers, we find that Eco RI treatment gives two fragments of 4 kb and 6 kb; Not I treatment gives one fragment of 10 kb; and treatment with Eco RI + Not I gives three fragments of 6 kb, 3 kb, and 1 kb. This gives a total length of 10 kb calculated as the sum of the fragments in each lane. Thus the original DNA molecule must be 10 kb (10,000 nucleotide pairs) long. Since treatment with Not I gives a fragment 10 kb long it could be that the original DNA is a linear molecule with no cutting site for Not I. But we can rule that out by the results of the Eco RI + Not I digestion. We know that Eco RI cleavage alone produces two fragments of 6 kb and 4 kb, and in the double digest this 4 kb fragment is further cleaved by Not I into a 3 kb and a 1 kb fragment. The DNA therefore contains a Not I cleavage site, and thus it must be circular, as only a single fragment of 10 kb is produced when it is cut with Not I alone. Arranging the cutting sites on a circular DNA to give the appropriate sizes of fragments produces the map illustrated in Figure A10–11.

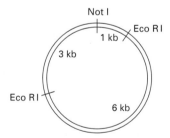

Figure A10–11

Answer 10–12 If the repair enzymes act on the plasmid before it is replicated, the plasmid will indeed be repaired in cells. However, the repair enzymes cannot distinguish which strand of the DNA contains the mutation and which one contains the normal nucleotide. Therefore, in half of the cells that have been transformed with the mismatched plasmid, a normal gene is restored, whereas in the other half of the cells the normal strand is converted to match the mutated strand and the mutation is thus propagated. Cells containing a plasmid with the desired mutation can be identified, for example, by hybridization with a single-stranded DNA probe that allows one to distinguish between the normal and mutant genes.

Answer 10–13

A. The genetic code is degenerate, and there is more than one possible codon for each amino acid, with the exception of tryptophan and methionine. Therefore, to detect a nucleotide sequence known only from the amino acid sequence of the protein it encodes, many oligonucleotides must be made and pooled in order to ensure that the mixture will contain one oligonucleotide that exactly matches the DNA sequence of the gene. For the three peptide sequences given in this question, the following oligonucleotide probes need to be made (alternative bases at the same position are given in parentheses):

```
Peptide 1: 5'-TGGATGCA(C,T)CA(C,T)AA(A,G)-3'
```

Because of the three twofold degeneracies, you would need eight (= 2^3) different oligonucleotide sequences in the mixture.

```
Peptide 2: 5'-(T,C)T(G,A,T,C)(A,T)(G,C)(G,A,T,C)(A,C)G-
           (G,A,T,C)(T,C)T(G,A,T,C)(A,C)G(G,A,T,C)-3'
```

The oligonucleotide mixture representing peptide sequence #2 is much more complicated. Leu, Ser, and Arg are each encoded by six different codons; you would therefore need to synthesize a mixture of 7776 (= 6^5) different oligonucleotides. This could not be done, however, simply by using more than one different nucleotide in any one position because the different bases in each codon are not independent. (Ser, for example, has A or T as the first base of the codon, G or C as the second base, and G, A, T, or C as the third base; when the first base is A, however, the second base is always G and the third base can be only T or C.)

```
Peptide 3: 5'-TA(C,T)TT(C,T)GG(G,A,T,C)ATGCA(A,G)-3'
```

Because of three twofold and one fourfold degeneracies, you would need 32 (= $2^3 \times 4$) different sequences in the mixture.

You would presumably first use probe #1 to screen your library by hybridization. Because there are only eight DNA sequences, the ratio of the one correct sequence to the incorrect ones is highest, giving you a strong signal when screening your library and thus a high likelihood that the desired sequence is among ones that you isolate from the library (see Question 10–5, p. 329). Probe #2 is practically useless, because only 1/7776th of the DNA in the mixture would perfectly hybridize to your gene of interest. You could use probe #3 to analyze your isolated sequences again by hybridization. Those sequences that hybridize to

probes #1 and #3 are very likely to contain the gene of interest.

B. Knowing that peptide sequence #3 contains the last amino acid of the protein is valuable information because it tells you that the other two peptide sequences must precede it, that is, they must be located farther toward the amino-terminal end of the protein. Knowing this order is important, because DNA primers can be extended by DNA polymerases only from their 3' ends; thus, the 3' ends of two primers need to "face" each other during a PCR amplification reaction (see Figure 10–22). A PCR primer based on peptide sequence #3 must therefore be the complementary sequence of probe #3 (so that its 3' end corresponds to the first nucleotide of the sequence complementary to the Trp codon):

```
5'-(TC)TGCAT(G,A,T,C)CC(G,A)AA(G,A)TA-3'
```

Probe #1 could be your choice for the second primer. Probe #2, again because of its high degeneracy, would be a much less suitable choice.

C. The ends of the final amplification product are derived from the primers, which are each 15 nucleotides long. Therefore, a 270-nucleotide segment of the cDNA of the gene has been amplified. This will encode 90 amino acids; adding the amino acids encoded by the primers gives you a protein-coding sequence of 100 amino acids. This is unlikely to represent the whole protein because we do not know the location of peptide #1 in the sequence of the complete protein. To your satisfaction, however, you note that CTATCACGCTT-TAGG encodes peptide sequence #2. This information therefore confirms that your PCR product indeed encodes a fragment of the protein you originally isolated.

Answer 10–14 Protein biochemistry is still very important, mostly because it provides the link between the amino acid sequence (which can be deduced from DNA sequences) and the functional properties of the protein. We are, for example, still not able to predict the folding of a polypeptide chain from its amino acid sequence, so that in many cases information regarding the function of the protein, such as its catalytic activity, cannot be deduced but must instead be obtained experimentally by analyzing the properties of proteins biochemically. Furthermore, the structural information that can be deduced from DNA sequences is necessarily incomplete. We cannot obtain information, for example, about modifications of protein side chains (such as phosphorylation), proteolytic processing, the presence of tightly bound coenzymes, or the association of the protein with other polypeptide chains.

Answer 10–15 The products will comprise a large number of different single-stranded DNA molecules, one for each nucleotide in the sequence. However, these products will be of four colors, depending on which of the four dideoxyribonucleotides terminated the polymerization reaction of that DNA chain. Separation by gel electrophoresis will generate a ladder of bands, each one nucleotide apart, where the color of each band indicates the nucleotide of the template at that position in the sequence. The method described here forms the basis for the DNA sequencing strategy used in most automated DNA sequencing machines (Figure A10–15).

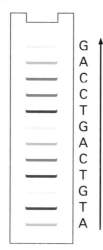

Figure A10–15

Answer 10–16

A. cDNA clones could not be used because there is no overlap between two cDNA clones even if they are derived from genes adjacent to each other in the chromosome.

B. Such repetitive DNA sequences can confuse chromosome walks, because the walk would appear to branch off in many different directions at once. The general strategy for avoiding these problems is to use genomic clones that are sufficiently long to overspan the repetitive DNA sequence.

Answer 10–17 Mutant bacteria that do not produce ice-protein have probably arisen many times in nature. However, bacteria that produce ice-protein have a slight growth advantage over bacteria that do not, so it would be difficult to find such mutants in the wild (and it would probably be difficult for the ice-protein-free mutant bacteria to survive in the long run facing the competition of their natural counterparts). Genetic engineering, using genes deliberately mutated *in vitro*, simply makes these mutants much easier to obtain. The consequences, both advantageous and disadvantageous, of using a genetically engineered organism for a practical application, therefore, are nearly indistinguishable from those that would follow the use of a natural mutant. Indeed, bacterial and yeast strains

have been selected for centuries for desirable genetic traits that make them suitable for industrial-scale applications such as cheese and wine production. The possibilities of genetic engineering are endless, however, and as with any technology, there is a finite risk that unforeseen consequences will arise. Recombinant DNA experimentation, therefore, is regulated, and the risks of individual projects are carefully assessed by review panels before permissions are granted. The state of our knowledge is sufficiently advanced that the consequences of some changes, such as the disruption of a bacterial gene in the example above, can be predicted with reasonable certainty. Other applications, such as germ-line gene therapy to correct human disease, may have far more complex outcomes, and it will take many more years of research and ethical debate to determine whether such treatments will eventually be used.

Chapter 11

Answer 11–1 Water is a liquid, and thus hydrogen bonds between water molecules are not static; they are continually formed and broken again by thermal motion. When a water molecule happens to be next to a hydrophobic molecule, it is more restricted in motion and has fewer neighbors with which it can interact, because it cannot form any hydrogen bonds in the direction of the hydrophobic molecule. It will therefore form hydrogen bonds to the more limited number of water molecules in its proximity. Bonding to fewer partners results in a more ordered water structure, which represents the cagelike structure in Figure 11–9. This structure has been likened to ice, although it is a more transient, less organized, and less extensive network than even a tiny ice crystal. The formation of any ordered structure decreases the entropy of the system (see Chapter 3), which is energetically unfavorable.

Answer 11–2 (B) is the correct analogy for lipid bilayer assembly because exclusion from water rather than attractive forces between the lipid molecules is involved. If the lipid molecules formed bonds with one another, the bilayer would be less fluid, and might even become rigid, depending on the strength of the interaction.

Answer 11–3 The fluidity of the bilayer is strictly confined to one plane: lipid molecules can diffuse laterally but do not readily flip from one monolayer to the other. Specific types of lipid molecules inserted into one monolayer therefore remain in it unless they are actively transferred by an enzyme—a flippase.

Answer 11–4 In both an α helix and a β barrel the polar peptide bonds of the polypeptide backbone can be completely shielded from the hydrophobic environment of the lipid bilayer by the hydrophobic amino acid side chains. Internal hydrogen bonds between the peptide bonds stabilize the α helix and β barrel.

Answer 11–5 The sulfate group in SDS is charged and therefore hydrophilic. The OH group and the C–O–C groups in Triton X-100 are polar; they can form hydrogen bonds with water and are therefore hydrophilic. In contrast, the *green* portions of the molecules are either hydrocarbon chains or aromatic rings, neither of which have polar groups that could hydrogen-bond with water molecules; they are therefore hydrophobic. (See Figure A11–5.)

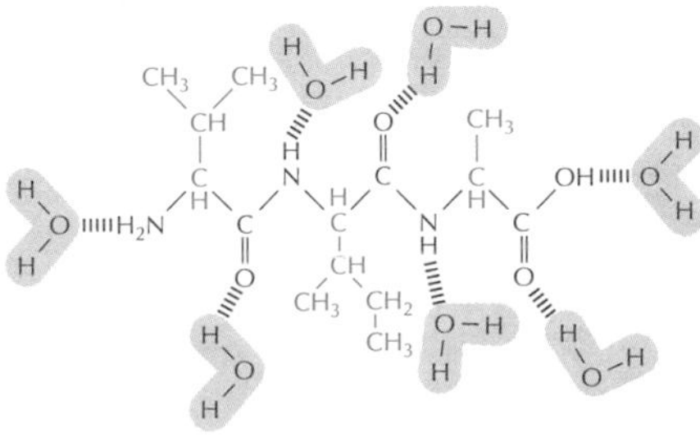

valine   isoleucine   alanine

Figure A11–5

Answer 11–6 Alpha helices in proteins are often used to span lipid bilayers. These structures are well suited for this purpose because they expose hydrophobic amino acid side chains to the hydrophobic interior of the lipid bilayer but sequester the polar peptide bonds of the polypeptide backbone away from the hydrophobic phase (see Figures 11–23 through 11–26). There are, however, other, less regular ways to fold up a polypeptide chain to achieve the same result, as seen in the small loop in the photosynthetic reaction center. This illustrates the importance of determining three-dimensional structures, which to date are known for only a few membrane proteins.

Answer 11–7 Some of the molecules of the two different transmembrane proteins are anchored to the spectrin filaments of the cell cortex. These molecules are not free to rotate or diffuse within the plane of the membrane. There is an excess of transmembrane proteins over the available attachment sites in the cortex, however, so that some of the transmembrane protein molecules are not anchored but rotate and diffuse freely within the plane of the membrane. Indeed, measurements of protein mobility show that there are two different populations of each transmembrane protein, corresponding to those proteins that are anchored and those that are not.

Answer 11–8 The different ways in which membrane proteins can be restricted to different regions of the membrane are summarized in Figure 11–36. The mobility of the membrane proteins is drastically reduced if they are bound to other proteins such as those of the cytoskeleton or the extracellular matrix.

The mobility of proteins is not affected if the proteins are not bound to other proteins but are confined to membrane domains by barriers, such as tight junctions. The fluidity of the lipid bilayer is not significantly affected by the anchoring of membrane proteins; the sea of lipid molecules flows around anchored membrane proteins like the water around the posts of a pier.

Answer 11–9 All of the statements are correct.

A., B., C., D. The lipid bilayer is fluid because the lipid molecules in the bilayer can undergo these motions.

E. Glycolipids are mostly restricted to the monolayer of membranes that faces away from the cytosol. Some special glycolipids, such as phosphatidylinositol (discussed in Chapter 15), are found specifically in the cytosolic monolayer.

F. The reduction of double bonds (by hydrogenation) allows lipid molecules to pack more tightly against one another and therefore increases the viscosity—that is, it turns oil into margarine.

G. Examples are enzymes involved in signaling (discussed in Chapter 15).

H. Polysaccharides are the main constituents of mucus and slime; the glycocalyx, which is made up of polysaccharides and oligosaccharides, is a very important lubricant—for example, for cells that line blood vessels or circulate in the bloodstream.

Answer 11–10 In a two-dimensional fluid the molecules are free to move only in one plane; the molecules in a normal fluid, in contrast, can move in three dimensions.

Answer 11–11

A. You would have a detergent. The diameter of the lipid head would be much larger than that of the hydrocarbon tail, so that the shape of the molecule would be a cone rather than a cylinder and the molecules would aggregate to form micelles rather than bilayers.

B. Lipid bilayers formed would be much more fluid. The bilayers would also be less stable, as the shorter hydrocarbon tails would be less hydrophobic, so the forces that drive the formation of the bilayer would be reduced.

C. The lipid bilayers formed would be much less fluid. Whereas a normal lipid bilayer has the viscosity of olive oil, a bilayer made of the same lipids but with saturated hydrocarbon tails would have the consistency of bacon fat.

D. The lipid bilayers formed would be much more fluid. Also, because the lipids would pack together less well, there would be more gaps and the bilayer would be more permeable to small water-soluble molecules.

E. If we assume that the lipid molecules are completely intermixed, the fluidity of the membrane would be unchanged. In such bilayers, however, the saturated lipid molecules would tend to aggregate with one another because they can pack so much more tightly and would therefore form patches of much-reduced fluidity. The bilayer would not, therefore, have uniform properties over its surface. Because normally one saturated and one unsaturated hydrocarbon tail are linked to the same hydrophilic head in lipid molecules, such segregation does not occur in cell membranes.

F. The lipid bilayers formed would have virtually unchanged properties. Each lipid molecule would now span the entire membrane, with one of its two head groups exposed at each surface. Such lipid molecules are found in the membranes of thermophilic bacteria, which can live at temperatures approaching boiling water. Their bilayers do not come apart at elevated temperatures, as usual bilayers do, because the original two monolayers are now covalently linked into a single membrane.

Answer 11–12 Lipid molecules are approximately cylindrical in shape. Detergent molecules, by contrast, are conical or wedge shaped. A lipid molecule with only one hydrocarbon tail, for example, would be a detergent. To make a lipid molecule into a detergent, you would have to make its hydrophilic head larger or remove one of its tails so that it could form a micelle. Detergent molecules also usually have shorter hydrocarbon tails than lipid molecules. This makes them slightly water soluble, so that detergent molecules leave and reenter micelles frequently in aqueous solution. Because of this, some monomeric detergent molecules are always present in aqueous solution and therefore can enter lipid bilayers to solubilize membrane proteins (see Figure 11–28).

Answer 11–13 The permeabilities are: N_2 (small and nonpolar) > ethanol (small and slightly polar) > water (small and polar) > glucose (large and polar) > Ca^{2+} (small and charged) > RNA (very large and charged).

Answer 11–14 When lined up, there are about 4000 lipid molecules (each 0.5 nm wide) between a lipid molecule at one end of the bacterial cell and one at the other end. Thus, if one of these molecules started to move toward the other, exchanging places with a neighboring molecule every 10^{-7} sec, it would take only 4×10^{-4} sec (= 4000×10^{-7} sec) to reach the other end. In reality, however, the lipid molecule would move in a random path rather than in a defined direction, so it would take considerably longer (1 sec) to reach the other end. If a 4-cm ping-pong ball exchanged places with a neighbor every 10^{-7} sec, it would travel at a speed of 1,440,000 km/h (= 4 cm/10^{-7} sec). If its movement were only in one direction, it would reach the other wall in 1.5×10^{-5} sec; in a random walk it would take considerably longer (~2 msec).

Answer 11–15 Membrane proteins anchor the lipid bilayer to the cytoskeleton, which strengthens the plasma membrane so that it can withstand the forces on it when the red blood cell is pumped through small blood vessels. Membrane proteins also transport nutrients and ions across the plasma membrane.

Answer 11–16 The hydrophilic faces of the five membrane-spanning α helices, each contributed by a different subunit, are thought to come together to form a pore across the lipid bilayer that is lined with the hydrophilic amino acid side chains. Ions can pass through this hydrophilic pore. The hydrophobic side chains interact with the hydrophobic lipid tails in the bilayer.

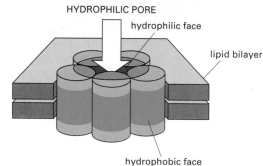

Figure A11–16

Answer 11–17 There are about 100 lipid molecules (i.e., phospholipid + cholesterol) for every protein molecule in the membrane [= ($2 \times 50,000$)/($800 + 386$)]. A similar protein:lipid ratio is seen in many cell membranes.

Answer 11–18 Membrane fusion does not alter the orientation of the membrane proteins with their attached color tags: the portion of each transmembrane protein that is exposed to the cytosol always remains exposed to cytosol, and the portion exposed to the outside always remains exposed to the outside. At 0°C the fluidity of the membrane is reduced and the mixing of the membrane proteins is significantly slowed.

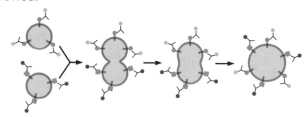

Figure A11–18

Answer 11–19 The exposure of hydrophobic amino acid side chains to water is energetically unfavorable. There are two ways that such side chains can be sequestered from water to achieve an energetically more favorable state. First, they can form transmembrane segments that span a lipid bilayer. This requires about 20 of them to be located sequentially in a polypeptide chain. Second, the hydrophobic amino acid side chains can be sequestered in the interior of the folded polypeptide chain. This is one of the major forces that lock the polypeptide chain into a unique three-dimensional structure. In either case the hydrophobic forces in the lipid bilayer or in the interior of a protein are based on the same principles.

Chapter 12

Answer 12–1

A. Transport mediated by a carrier protein can be described by a strictly analogous equation:

$$(1)\ CP + S \rightleftharpoons CPS \rightarrow CP + S^*$$

where S is the solute, S* is the solute on the other side of the membrane (i.e., although it is still the same molecule, it is now located in a different environment), and CP is the carrier protein.

B. This equation is useful because it describes a binding step, followed by a delivery step. The mathematical treatment of this equation would be very similar to that described for enzymes (see Figure 5–27A, p. 169); thus, carrier proteins are characterized by a K_M value that describes their affinity for a solute and a V_{max} value that describes their maximal rate of transfer. To be more accurate, one could include the conformational change of the carrier protein in the reaction scheme:

$$(2a)\ CP + S \rightleftharpoons CPS \rightleftharpoons CP^*S^* \rightarrow CP^* + S^*$$

$$(2b)\ CP \rightleftharpoons CP^*$$

where CP* is the carrier protein after the conformational change that exposes its solute-binding site on the other side of the membrane. This account requires a second equation (2b) that allows the carrier protein to return to its starting conformation.

C. The equations do not describe the behavior of channels, because solutes passing through channels do not bind to them in the way that a substrate binds to an enzyme.

Answer 12–2 By analogy to the Na⁺-K⁺ pump shown in Figure 12–11, ATP might be hydrolyzed and donate a phosphate group to the carrier protein when—and only when—it has the solute bound on the "inside" face of the membrane (step 1 → 2). The attachment of the phosphate would trigger an immediate conformational change (step 2 → 3), thereby capturing the solute and exposing it to the "outside." The phosphate would be removed from the protein when—and only when—the solute has dissociated, and the now empty, nonphosphorylated carrier protein would switch back to the starting position (step 3 → 4) (Figure A12–2, *below*).

Answer 12–3

A. The properties define a symport.

B. No additional properties need to be specified. The important feature that provides the coupling of the two solutes is that the protein cannot switch its conformation if only one of the two solutes is bound. Solute B, which is driving the transport of solute A, is in excess on the side of the membrane from which transport initiates and occupies its binding site most of the time. In this state, the carrier protein, prevented from switching its conformation, waits until once in a while a solute A molecule binds. With both binding sites occupied, the carrier protein switches conformation. Now exposed to the other side of the membrane, the binding site for solute B is mostly empty because there is little of it in the solution on this side of the membrane. Although the binding site for A is now more frequently occupied, the carrier can switch back only after solute A is unloaded as well.

C. An antiport could be similarly constructed as a transmembrane protein with the following properties: it has two binding sites, one for solute A and one for solute B. The protein can undergo a conformational change to switch between two states: either both binding sites are exposed exclusively on one side of the membrane or both binding sites are exposed exclusively on the other side of the membrane. The protein can switch between the

Figure A12–2

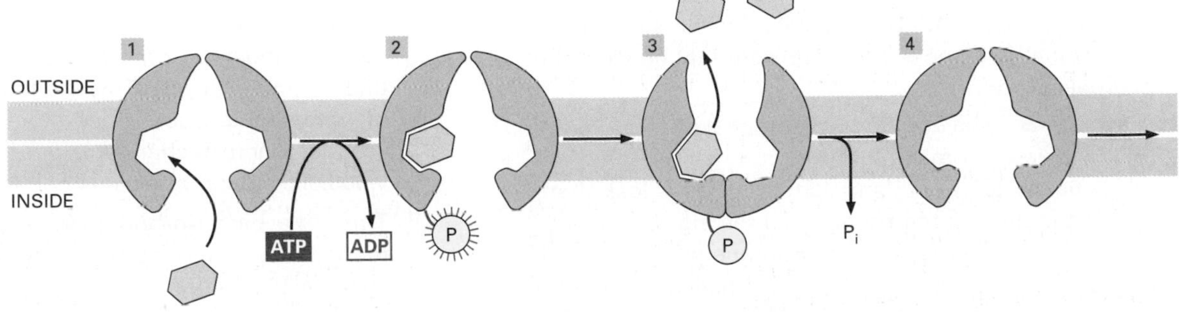

two conformational states only if one binding site is occupied, but cannot switch if both binding sites are occupied or if both binding sites are empty.

Note that these rules provide an alternative model to that shown in Figure 12–13. Thus there are two possible ways to couple the transport of two solutes: (1) provide cooperative solute-binding sites and allow the pump to switch between the two states randomly as shown in Figure 12–13, or (2) allow independent binding of both solutes and make the switch between the two states conditional on the occupancy of the binding sites. As the structure of a coupled transporter has not yet been determined, we do not know which of the two mechanisms such pumps use.

Answer 12–4 If the Na^+-K^+ pump is not working at full capacity because it is partially inhibited by ouabain or digitalis, it generates an electrochemical gradient of Na^+ that is less steep than that in untreated cells. Consequently, the Ca^{2+}-Na^+ antiporter works less efficiently, and Ca^{2+} is removed from the cell more slowly. When the next cycle of muscle contraction begins, there is still an elevated level of Ca^{2+} left in the cytosol. The entry of the same number of Ca^{2+} ions into the cell leads therefore to a higher Ca^{2+} concentration than in untreated cells, which in turn leads to a stronger and longer lasting contraction. Because the Na^+-K^+ pump fulfills essential functions in all animal cells, both to maintain osmotic balance and to generate the Na^+ gradient used to power many transporters, the drugs are deadly poisons at higher concentrations.

Answer 12–5 Each of the rectangular peaks corresponds to the opening of a single channel that allows a small current to pass. You note from the recording that the channels present in the patch of membrane open and close frequently. Each channel remains open for a very short, somewhat variable time, averaging about 10 milliseconds. When open, the channels allow a small current with a unique amplitude (4 pA; one picoampere = 10^{-12} A) to pass. In one instance, the current doubles, indicating that two channels in the same membrane patch opened simultaneously.

If acetylcholine is omitted or added to the solution outside the pipette, you would measure only the baseline current. Acetylcholine must bind to the extracellular portion of the acetylcholine receptor molecules to allow the channel to open, and in the membrane patch shown in Figure 12–20, the cytoplasmic side of the membrane is exposed to the solution outside the microelectrode.

Answer 12–6 The equilibrium potential of K^+ is –90 mV [= 62 mV $\log_{10}$ (5 mM/140 mM)] and that of Na^+ is +72 mV [= 62 mV $\log_{10}$ (145 mM/10 mM)]. The K^+ leak channels in the plasma membrane of a resting cell allow K^+ to come to equilibrium; the membrane potential of the cell is therefore close to –90 mV. When Na^+ channels

open, Na^+ rushes in, and, as a result, the membrane potential reverses its polarity to a value nearer to +72 mV, the equilibrium value for Na^+. Upon closure of the Na^+ channels, the K^+ leak channels allow K^+, now no longer at equilibrium, to exit the cell until the membrane potential is restored to the equilibrium value for K^+, about –90 mV.

Answer 12–7 When the resting membrane potential of an axon drops below a threshold value, voltage-gated Na^+ channels in the immediate neighborhood open and allow an influx of Na^+. This depolarizes the membrane further, causing more distant voltage-gated Na^+ channels to open as well. This creates a wave of depolarization that spreads rapidly along the axon, called the action potential. Because Na^+ channels become inactivated soon after they open, the flow of K^+ through voltage-gated K^+ channels and K^+ leak channels is able to restore the original resting membrane potential rapidly after the action potential has passed. (96 words)

Answer 12–8 If the number of functional acetylcholine receptors is reduced by the antibodies, the neurotransmitter (acetylcholine) that is released from the nerve terminals cannot (or can only weakly) stimulate the muscle to contract.

Answer 12–9

A. False. The plasma membrane contains proteins that confer selective permeability to many charged molecules. In contrast, a pure lipid bilayer lacking proteins is highly impermeable to all charged molecules.

B. False. Channel proteins do not bind the solute that passes through them. Selectivity of a channel protein is achieved by the size of the internal pore and by charged regions at the entrance of the pore that attract or repel ions of the appropriate charge.

C. True for animal cells. Cells contain a concentrated solution of many molecules that will cause the osmotic influx of water. Unless ions are constantly pumped out to maintain an osmotic balance, cells will eventually burst. False for plant, yeast, and bacterial cells. Although water will tend to enter them by osmosis, these cells are surrounded by a tough cell wall that prevents the plasma membrane from rupturing.

D. False. Carrier proteins are slower. They have enzymelike properties, i.e., they bind solutes and need to undergo conformational changes during their functional cycle. This limits the maximal rate of transport to about 1000 solute molecules per second, whereas channel proteins can pass up to 1,000,000 solute molecules per second.

E. True. The bacteriorhodopsin of some photosynthetic bacteria moves H^+, using energy captured from visible light.

F. True. Most animal cells contain K^+ leak channels in their plasma membrane that are predominantly open. The K^+ concentration inside the cell still remains higher than outside, because the membrane potential is negative and therefore inhibits the positively charged K^+ from leaking out. K^+ is also continually pumped into the cell by the Na^+-K^+ pump.

G. False. A symport binds two different solutes on the same side of the membrane. Turning it around would not change it into an antiport, which must also bind to different solutes, but on opposing sides of the membrane.

H. False. The peak of an action potential corresponds to a transient shift of the membrane potential from a negative to a positive value. The influx of Na^+ causes the membrane potential first to move toward zero and then to reverse, rendering the cell positively charged on its inside. Eventually, the resting potential is restored by an efflux of K^+ through voltage-gated K^+ channels and K^+ leak channels.

Answer 12–10

A. Both couple the movement of two different solutes across the membrane. Symports transport both solutes in the same direction, whereas antiports transport the solutes in opposite directions.

B. Both are mediated by membrane transport proteins. Passive transport of a solute occurs downhill, in the direction of its concentration or electrochemical gradient, whereas active transport occurs uphill and therefore needs an energy source. Active transport can be mediated by carrier proteins but not by channel proteins, whereas passive transport can be mediated by either.

C. Both terms describe energy changes involved in moving an ion from one side of a membrane to the other. The membrane potential refers to the electrical energy change; the electrochemical gradient is a composite of this electrical energy change and the chemical energy change associated with moving between a region of high concentration and a region of low concentration. The membrane potential is defined independently of the choice of ion, whereas an electrochemical gradient depends on the concentration gradient of the particular ionic solute and is therefore a solute-specific parameter.

D. A pump is a specialized carrier protein that uses energy to transport a solute uphill against an electrochemical gradient.

E. Both transmit signals by electrical activity. Wires are made of copper, axons are not. The signal passing down an axon does not diminish in strength because it is self-amplifying, whereas the signal in a wire decreases over distance (by leakage of current across the insulating sheath).

F. Both affect the osmotic pressure of the cell. An ion is a solute that bears a charge.

Answer 12–11 A bridge allows vehicles to pass over a river in a steady stream; the entrance can be designed to exclude, for example, oversized trucks, and it can be intermittently closed to traffic by a gate. By analogy, channels allow ions to flow in a gated stream across the membrane, imposing size and charge restrictions.

A ferry, in contrast, loads vehicles on one riverbank and then, after movement of the ferry itself, unloads on the other side of the river. This process is slower. During loading, particular vehicles could be selected from the waiting line because they fit particularly well on the car deck. By analogy, carrier proteins bind solutes on one side of the membrane and then, after a conformational movement, release them on the other side. Specific binding leads to the selection of the molecules to be transported. As in the case of coupled transport, sometimes you have to wait until the ferry is full before you can go.

Answer 12–12 Acetylcholine is being transported into the vesicles by an H^+-acetylcholine antiporter in the vesicle membrane. The H^+ gradient that drives the uptake is generated by an ATP-driven H^+ pump in the vesicle membrane, which pumps H^+ into the vesicle (hence the dependence of the reaction on ATP). Raising the pH of the solution surrounding the vesicles increases the H^+ gradient: at an elevated pH there are fewer H^+ ions in the solution outside the vesicles while the number inside remains the same. This explains the observed enhanced rate of uptake.

Answer 12–13 The voltage gradient across the membrane is about 150,000 V/cm. This extremely powerful electric field is close to the limit at which insulating materials—such as the lipid bilayer—break down and cease to act as insulators. The large field corresponds to the large amount of energy that can be stored in electrical gradients across the membrane, as well as to the extreme electrical forces that proteins can experience in a membrane. A voltage of 150,000 V would instantly discharge in an arc across a 1 cm wide gap (that is, air would be an insufficient insulator for this strength of field).

Answer 12–14

A. Nothing. You require ATP to drive the Na^+-K^+ pump.

B. The ATP becomes hydrolyzed, and Na^+ is pumped into the vesicles, generating a concentration gradient of Na^+ across the membrane. At the same time, K^+ is pumped out of the vesicles, generating a concentration gradient of K$^+$ of opposite polarity. When all the K$^+$ had been pumped out of the vesicle or the ATP ran out, the pump would stop.

C. The Na$^+$-K$^+$ pump would go through states 1, 2, and 3 in Figure 12–11. Because all reaction steps must occur strictly sequentially, however, dephosphorylation and the conformation switch cannot occur in the absence of K$^+$. The Na$^+$-K$^+$ pump will therefore become stuck in the phosphorylated state, waiting indefinitely for a potassium ion. The number of sodium ions transported would be minuscule, because each pump molecule would have functioned only a single time.

Similar experiments, leaving out individual ions and analyzing the consequences, were used to determine the sequence of steps by which the Na$^+$-K$^+$ pump works.

D. ATP would become hydrolyzed and Na$^+$ and K$^+$ would be pumped across the membrane as described in scenario A. However, the pump molecules that sit in the membrane in the reverse orientation would be completely inactive (i.e., they would not—as one might have erroneously assumed—pump ions in the opposite direction), because ATP would not have access to phosphorylate the appropriate site on these molecules. This site is normally exposed to the cytosol. ATP is highly charged and cannot cross membranes without the help of specific carrier proteins.

E. ATP becomes hydrolyzed and Na$^+$ and K$^+$ are pumped across the membrane, as described in scenario A. K$^+$, however, immediately flows back into the vesicles through the K$^+$ leak channels. K$^+$ moves down the K$^+$ concentration gradient formed by the action of the Na$^+$-K$^+$ pump. With each K$^+$ that moves into the vesicle through the leak channel, a positive charge is moved across the membrane, building a membrane potential that is positive on the inside of the vesicles. Eventually, K$^+$ will stop flowing through the leak channels when the membrane potential balances the concentration gradient. The scenario described here is a slight oversimplification: the Na$^+$-K$^+$ pump in mammalian cells actually moves three sodium ions out of cells for each two potassium ions that it pumps into the cell, thereby driving an electric current across the membrane and making a small additional contribution to the resting membrane potential (which therefore corresponds only approximately to a state of equilibrium for K$^+$ moving via K$^+$ leak channels.

Answer 12–15 Ion channels can be ligand-gated, voltage-gated, or mechanically gated.

Answer 12–16 The cell has a volume of 10^{-12} liters (= 10^{-15} m^3) and thus contains 6×10^4 calcium ions (= 6

$\times 10^{23}$ molecules/mole $\times 100 \times 10^{-9}$ moles/liter $\times 10^{-12}$ liters). Therefore, to raise the intracellular Ca^{2+} concentration fiftyfold, another 2,940,000 calcium ions have to enter the cell (note that at 5 μM concentration there are 3×10^6 ions in the cell, of which 60,000 are already present before the channels are opened). Because each of the 1000 channels allows 10^6 ions to pass per second, each channel has to stay open for only 3 milliseconds.

Answer 12–17 Animal cells drive most transport processes across the plasma membrane with the electrochemical gradient of Na$^+$. ATP is needed to fuel the Na$^+$-K$^+$ pump to maintain the Na$^+$ gradient.

Answer 12–18

A. If H$^+$ is pumped across the membrane into the endosomes, an electrochemical gradient of H$^+$ results—composed of both an electrical potential and a concentration gradient, with the interior of the vesicle positive. Both of these components add to the energy that is stored in the gradient and that must be supplied to generate it. The electrochemical gradient will therefore limit the transfer of more H$^+$. If, however, the membrane also contains Cl$^-$ channels, the negatively charged Cl$^-$ will flow into the endosomes and diminish the electrical potential. It therefore becomes energetically less expensive to pump more H$^+$ across the membrane, and the interior of the endosomes can become more acidic.

B. Yes. As explained in (A), some acidification would still occur in their absence.

Answer 12–19

A. See Figure A12–19(A) (next page).

B. The transport rates of compound A are proportional to its concentration, indicating that compound A can diffuse through membranes on its own. Compound A is likely to be ethanol, because it is a small and relatively nonpolar molecule that can diffuse readily through the lipid bilayer.

In contrast, the transport rates of compound B saturate at high concentrations, indicating that compound B is transported across the membrane by a transport protein. Transport rates cannot increase beyond a maximal rate at which this protein can function. Compound B is likely to be acetate, because it is a charged molecule that could not cross the membrane without the help of a membrane transport protein.

C. For ethanol, we measured a linear relationship between concentration and transport rate. Thus, at 0.5 mM the transport rate would be 10 μmol/min, and at 100 mM the transport rate would be 2 μmol/min.

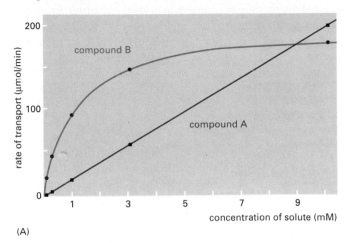

(A)

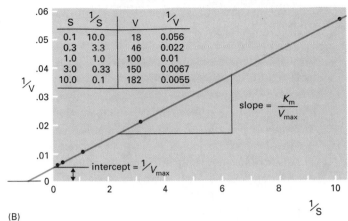

(B)

For the carrier protein–mediated transport of acetate, the relationship between concentration, S, and transport rate can be described by the Michaelis-Menten equation, which describes simple enzyme reactions:

(1) transport rate = $V_{max} \times S / [K_M + S]$

Recall from Chapter 5 (see Question 5–7, p. 171) that to determine the V_{max} and K_M, a trick is used in which the Michaelis-Menten equation is transformed so that it is possible to plot the data as a straight line. A simple transformation yields

(2) **1/rate** = (K_M/V_{max}) **(1/S)** + $1/V_{max}$
(i.e., an equation of the form **y** = a**x** + b)

Calculation of 1/rate and 1/S for the given data and plotting them in a new graph as in Figure A12–19(B) (*above right*) gives a straight line. The K_M (= 1.0 mM) and V_{max} (= 200 µmol/min) are determined from the intercept of the line with the y axis $(1/V_{max})$ and from its slope (K_M/V_{max}). Knowing the values for K_M and V_{max} allows you to calculate the transport rates for 0.5 mM and 100 mM acetate using equation (1). The results are 67 µmol/min and 198 µmol/min, respectively.

Chapter 13

Answer 13–1 By making membranes permeable to protons, DNP collapses—or at very small concentrations diminishes—the proton gradient across the inner mitochondrial membrane. Cells continue to oxidize food molecules to feed electrons into the electron-transport chain, but H^+ ions pumped across the membrane flow back into the mitochondria in a futile cycle. Their energy cannot be tapped to drive ATP synthesis, and hence is released as heat. Patients who have been given small doses of DNP lose weight because their fat reserves are more rapidly used to feed the electron-transport chain, since this whole process merely "wastes" energy.

A similar mechanism of heat production is used by a specialized tissue composed of brown fat cells, which is abundant in newborn humans and in hibernating animals. These cells are packed with mitochondria that leak part of their H^+ gradient futilely back across the membrane for the sole purpose of warming up the organism. These cells are brown because they are packed with mitochondria, which contain high concentrations of pigmented proteins, such as cytochromes.

Answer 13–2 The inner mitochondrial membrane is the site of oxidative phosphorylation, and it produces most of the cell's ATP. Cristae are portions of the mitochondrial inner membrane that are folded inward. Mitochondria that have a higher density of cristae have a larger area of inner membrane and therefore a greater capacity to carry out oxidative phosphorylation. Heart muscle expends a lot of energy during its continuous contractions, whereas skin cells have a lesser energy demand. An increased density of cristae therefore increases the ATP-production capacity of the heart muscle cell. This is a remarkable example of how cells adjust the abundance of their individual components according to need.

Answer 13–3

A. The DNP collapses the electrochemical proton gradient completely. H^+ ions that are pumped to one side of the membrane flow back freely, and therefore no energy can be stored across the membrane.

B. An electrochemical gradient is made up of two components: a concentration gradient and an electrical potential. If the membrane is made permeable to K^+ with nigericin, K^+ will be driven into the matrix by the electrical potential of the inner membrane (minus inside, positive outside). The influx of positively charged K^+ will abolish the membrane's electrical potential. In contrast, the concentration component of the H^+ gradient (the pH difference) is unaffected by nigericin. Therefore, only part of the driving force that makes it energetically favorable for H^+ ions to flow back into the matrix is lost.

Answer 13–4

A. Such a turbine running in reverse is an electrically driven water pump, which is analogous to what the ATP synthase becomes when it uses the energy of ATP hydrolysis to pump protons against their electrochemical gradient across the inner mitochondrial membrane.

B. The ATP synthase should stall when the energy that it can draw from the proton gradient is just equal to the free-energy change required to make ATP; at this equilibrium point there will be neither net ATP synthesis nor net ATP consumption.

C. As the cell uses up ATP, the ATP:ADP ratio in the matrix falls below the equilibrium point just described, and ATP synthase uses the energy stored in the proton gradient to synthesize ATP in order to restore the original ATP:ADP ratio. Conversely, when the electrochemical proton gradient drops below that at the equilibrium point, ATP synthase uses ATP in the matrix to restore this gradient.

Answer 13–5

A. An electron pair causes 10 H^+ to be pumped across the membrane when passing from NADH to O_2 through the three respiratory complexes. Four H^+ are needed to make each ATP: three for synthesis from ADP and one for ATP export to the cytosol. Therefore, 2.5 ATP molecules are synthesized from each NADH molecule.

B. Twenty ATP molecules are produced per molecule of glucose as the result of the citric acid cycle, as compared to 30 ATP molecules that are produced in total from the complete oxidation of one molecule of glucose:

Process	Direct Product	Final ATP
Glycolysis	2 NADH (cytosolic)	3
	2 ATP	2
Pyruvate oxidation	2 NADH (mitochondrial matrix)	5
Acetyl-CoA oxidation	6 NADH (mitochondrial matrix)	15
(two per glucose)	2 $FADH_2$	3
	2 GTP	2
Total yield per molecule of glucose		30

Answer 13–6   One can describe four essential roles for the proteins in the process. First, the chemical environment provided by a protein's amino acid side

chains sets the redox potential of each Fe ion such that electrons can be passed in a defined order from one component to the next, giving up their energy in small steps and becoming more firmly bound as they proceed. Second, the proteins position the Fe ions so that the electrons can move efficiently between them. Third, the proteins prevent electrons from skipping an intermediate step; thus, as we have learned for other enzymes (discussed in Chapter 5), they channel the electron flow along a defined path. Fourth, the proteins couple the movement of the electrons down their energy ladder to the pumping of protons across the membrane, thereby harnessing the energy that is released and storing it in a proton gradient that is then used for ATP production.

Answer 13–7   It would not be productive to use the same carrier in two steps. If ubiquinone, for example, could transfer electrons directly to the cytochrome oxidase, the cytochrome b-c_1 complex would often be skipped when electrons are collected from NADH dehydrogenase. Given the large difference in redox potential between ubiquinone and cytochrome oxidase, a large amount of energy would be released as heat and thus be wasted. Electron transfer directly between NADH dehydrogenase and cytochrome c would similarly allow the cytochrome b-c_1 complex to be bypassed.

Answer 13–8   Protons pumped across the inner mitochondrial membrane into the intermembrane space equilibrate with the cytosol, which functions as a huge H^+ sink. Both the mitochondrial matrix and the cytosol house many metabolic reactions that require a pH around neutrality. The H^+ concentration difference, ΔpH, that can be achieved between mitochondrial matrix and cytosol is therefore relatively small (one pH unit). Much of the energy stored in the mitochondrial electrochemical proton gradient is instead due to the electrical potential of the membrane (see Figure 13–12).

In contrast, chloroplasts have a smaller, dedicated compartment into which H^+ ions are pumped. Much higher concentration differences can be achieved (up to a thousandfold, or 3 pH units), and much of the energy stored in the thylakoid H^+ gradient is due to the H^+ concentration difference between the thylakoid space and the stroma.

Answer 13–9   All statements are correct.

A. This is a necessary condition. If it were not true, electrons could not be removed from water and the reaction that splits water molecules ($H_2O \rightarrow 2H^+ + \frac{1}{2}O_2 + 2e^-$) would not occur.

B. This transfer allows the energy of the photon to be harnessed as energy that can be utilized in chemical conversions.

C. It can be argued that this is one of the most important obstacles that had to be overcome during the evolution of photosynthesis: partially reduced oxygen molecules, such as the superoxide radical O_2^-, are dangerously reactive and will attack and destroy almost any biologically active molecule. These intermediates, therefore, have to remain tightly bound to the metals in the active site of the enzyme until all four electrons have been removed from two water molecules. This requires the sequential capture of four photons by the same reaction center.

Answer 13–10

A. Photosynthesis produces sugars, most importantly sucrose, that are transported from the photosynthetic cells through the sap to root cells. There, the sugars are oxidized by glycolysis in the root cell cytoplasm and by oxidative phosphorylation in the root cell mitochondria to produce ATP, as well as being used as the building blocks for many other metabolites.

B. Mitochondria are required even during daylight hours in chloroplast-containing cells to supply the cell with ATP derived by oxidative phosphorylation. Glyceraldehyde 3-phosphate made by photosynthesis in chloroplasts moves to the cytosol and is eventually used as a source of energy to drive ATP production in mitochondria.

Answer 13–11

A. True. NAD$^+$ and quinones are examples of compounds that do not have metal ions but can participate in electron transfer.

B. False. The potential is due to protons (H$^+$) that are pumped across the membrane from the matrix to the intermembrane space. Electrons remain bound to electron carriers in the inner mitochondrial membrane.

C. True. Both components add to the driving force that makes it energetically favorable for H$^+$ to flow back into the matrix.

D. True. Both move rapidly in the plane of the membrane.

E. False. Not only do plants need mitochondria to make ATP in cells that do not have chloroplasts, such as root cells, but mitochondria make most of the cytosolic ATP in all plant cells.

F. True. Chlorophyll's physiological function requires it to absorb light; heme just happens to be a colored compound from which blood derives its red color.

G. False. Chlorophyll absorbs light and transfers energy in the form of an energized electron, whereas the iron in heme is a simple electron carrier.

H. False. Most of the dry weight of a tree comes from carbon derived from the CO_2 that has been fixed during photosynthesis.

Answer 13–12   It takes three protons. The precise value of the ΔG for ATP synthesis depends on the concentrations of ATP, ADP, and P$_i$ (as described in Chapter 3). The higher the ratio of the concentration of ATP to ADP, the more energy it takes to make additional ATP. The lower value of 11 kcal/mole therefore applies to conditions where cells have expended a lot of energy and have therefore decreased the normal ATP : ADP ratio.

Answer 13–13   If no O_2 is available, all components of the mitochondrial electron-transport chain will accumulate in their *reduced* form. This is the case because electrons derived from NADH enter the chain but can not be transferred to O_2. The electron-transport chain therefore stalls with all of its components in the reduced form. If O_2 is suddenly added again, the electron carriers in cytochrome oxidase will become *oxidized before* those in NADH dehydrogenase. This is true because, after O_2 addition, cytochrome oxidase will donate its electrons directly to O_2, thereby becoming oxidized. A wave of increasing oxidation then passes backward with time from cytochrome oxidase through the components of the electron-transport chain, as each component regains the opportunity to pass on its electrons to downstream components.

Answer 13–14   An unusually large number of protons will flow into the matrix, producing ATP at a higher rate, until equilibrium is reached—that is, until the proton flux is so great that the energy available in the proton gradient drops to the energy required to make ATP. At a lower ATP : ADP ratio, the ΔG required for ATP synthesis is smaller. Therefore, the proton gradient will drop to a new constant level in these cells with a lower ATP : ADP ratio.

Answer 13–15   As oxidized ubiquinone becomes reduced, it picks up two electrons but also two protons from water (Figure 13–20). Upon oxidation, these protons are released. If reduction occurs on one side of the membrane and oxidation at the other side, a proton is pumped across the membrane for each electron transported. Electron transport by ubiquinone thereby contributes directly to the generation of the H$^+$ gradient.

Answer 13–16   Photosynthetic bacteria and plant cells use the electrons derived in the reaction $2H_2O \rightarrow 4e^- + 4H^+ + O_2$ to reduce NADP$^+$ to NADPH, which is then used to produce useful metabolites. If the electrons were used instead to produce H_2 in addition to O_2, the cells would lose any benefit they derive from carrying out the reaction, because the electrons could not be used in metabolically useful reactions.

Answer 13–17

A. The switch in solutions creates a pH gradient across the thylakoid membrane. The flow of H$^+$ ions down its electrochemical potential drives ATP synthase, which converts ADP to ATP.

B. No light is needed, because the H⁺ gradient is established artificially without a need for the light-driven electron-transport chain.

C. Nothing. The H⁺ gradient would be in the wrong direction; ATP synthase would not work.

D. The experiment provided early supporting evidence for the chemiosmotic model by showing that an H⁺ gradient alone is sufficient to drive ATP synthesis.

Answer 13–18

A. When the vesicles are exposed to light, H⁺ ions (derived from H_2O) pumped into the vesicles by the bacteriorhodopsin flow back out through the ATP synthase, causing ATP to be made in the solution surrounding the vesicles in response to light.

B. If the vesicles are leaky, no H⁺ gradient can form and thus ATP synthase cannot work.

C. Using components from widely divergent organisms can be a very powerful experimental tool. Because the two proteins come from such different sources, it is very unlikely that they form a direct functional interaction. The experiment therefore strongly suggests that electron transport and ATP synthesis are separate events. This approach is therefore a valid one.

Answer 13–19   The redox potential of $FADH_2$ is too low to transfer electrons to the NADH dehydrogenase complex, but high enough to transfer electrons to ubiquinone (Figure 13–21). Therefore, electrons from $FADH_2$ can enter the electron-transport chain only at this step (Figure A13–19). Because the NADH dehydrogenase complex is bypassed, fewer H⁺ ions are pumped across the membrane and less ATP is made. This example shows the versatility of the electron-transport chain. The ability to use vastly different sources of electrons from the environment to feed

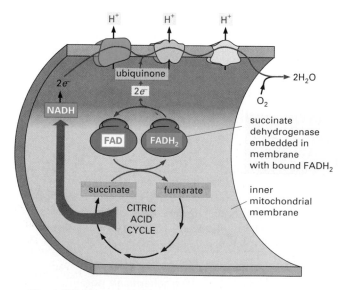

Figure A13–19

electron transport is thought to have been an essential feature in the early evolution of life.

Answer 13–20   If these bacteria used a proton gradient to make their ATP in an fashion analogous to other bacteria (that is, fewer protons inside than outside), they would need to raise their cytoplasmic pH even higher than that of their environment (pH 10). Cells with a cytoplasmic pH greater than 10 would not be viable. These bacteria, therefore, must use gradients of ions other than H⁺, such as Na⁺ gradients, in the chemiosmotic coupling between electron transport and an ATP synthase.

Answer 13–21   Statements A and B are accurate. Statement C is incorrect, because the chemical reactions that are carried out in each cycle are completely different, even though the net effect is the same as that expected for simple reversal.

Answer 13–22   This experiment would suggest a two-step model for ATP synthase function. According to this model, the flow of protons through the base of the synthase drives rotation of the head, which in turn causes ATP synthesis. In their experiment, the authors have succeeded in uncoupling these two steps. If rotating the head mechanically is sufficient to produce ATP in the absence of any applied proton gradient, the ATP synthase is a protein machine that indeed functions like a "molecular turbine." This would be a very exciting experiment, indeed, because it would directly demonstrate the relationship between mechanical movement and enzymatic activity. There is no doubt that it should be published and that it would become a "classic."

Answer 13–23   Only under condition (E) is electron transfer observed, with cytochrome c becoming reduced. A portion of the electron-transport chain has been reconstituted in this mixture, so that electrons can flow in the energetically favored direction from reduced ubiquinone to the cytochrome b-c₁ complex to cytochrome c. Although energetically favorable, the transfer in (A) cannot occur spontaneously in the absence of the cytochrome b-c₁ complex to catalyze this reaction. No electron flow occurs in the other experiments, whether the cytochrome b-c₁ complex is present or not: in experiments (B) and (F) both ubiquinone and cytochrome c are oxidized; in experiments (C) and (G) both are reduced; and in experiments (D) and (H) electron flow is energetically disfavored because reduced cytochrome c has a lower free energy than oxidized ubiquinone.

Chapter 14

Answer 14–1   Although the nuclear envelope forms one continuous membrane, it has specialized regions, containing special proteins and having a characteristic appearance. One such specialized region is the inner nuclear membrane. Membrane proteins can indeed diffuse between the inner and outer nuclear membranes, at the connections formed around the nuclear pores. Those with particular functions in the inner membrane, however, are usually anchored there by their interaction with other components such as chromosomes and the nuclear lamina, a protein meshwork underlying the inner nuclear membrane that helps give structural integrity to the nuclear envelope.

Answer 14–2 Eucaryotic gene expression is more complicated than procaryotic gene expression. In particular, procaryotic cells do not have introns that interrupt the coding sequences of their genes, so that an mRNA can be translated immediately after it is transcribed, without a need for further processing (discussed in Chapter 7). In fact, in procaryotic cells ribosomes start translating most mRNAs before transcription is finished. This would have disastrous consequences in eucaryotic cells, because most RNA transcripts have to be spliced before they can be translated. The nuclear envelope separates the transcription and translation processes in space and time. A primary RNA transcript is held in the nucleus until it is properly processed to form an mRNA, and only then is it allowed to leave the nucleus so that ribosomes can translate it.

Answer 14–3 An mRNA molecule is attached to the ER membrane by the ribosomes translating it. This ribosome population, however, is not static; the mRNA is continuously moved through the ribosome. Those ribosomes that have finished translation dissociate from the 3′ end of the mRNA and from the ER membrane, but the mRNA itself remains bound by other ribosomes, newly recruited from the cytosolic pool, that have attached to the 5′ end of the mRNA and are still translating the mRNA. Depending on its length, there are about 10–20 ribosomes attached to each membrane-bound mRNA molecule.

Answer 14–4

A. The internal signal sequence functions as a membrane anchor, as shown in Figure 14–15. Because there is no stop-transfer sequence, however, the carboxyl-terminal end of the protein continues to be translocated into the ER lumen. The resulting protein therefore has its amino-terminal domain in the cytosol, followed by a single transmembrane segment, and a carboxyl-terminal domain in the ER lumen (Figure A14–4A).

B. The amino-terminal signal sequence initiates translocation of the amino-terminal domain of the protein until translocation is stopped by the stop-transfer sequence. A cytosolic domain is synthesized until the start-transfer sequence initiates translocation again. The situation now resembles that described in (A), and the carboxyl-terminal

domain of the protein is translocated into the lumen of the ER. The resulting protein therefore spans the membrane twice. Both its amino-terminal and carboxyl-terminal domains are in the ER lumen, and a loop domain between the two transmembrane regions is exposed in the cytosol (Figure A14–4B).

C. It would need a cleaved signal sequence, followed by an internal stop-transfer sequence, followed by pairs of start- and stop-transfer sequences (Figure A14–4[C]).

These examples demonstrate that complex protein topologies can be achieved by simple variations and combinations of the two basic mechanisms shown in Figures 14–15 and 14–16.

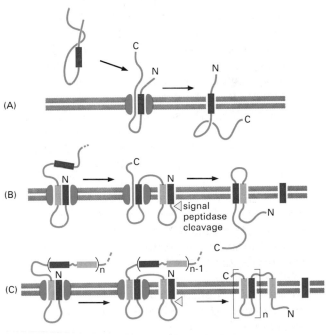

Figure A14–4

Answer 14–5

A. Clathrin coats cannot assemble in the absence of adaptins that link the clathrin to the membrane. At high clathrin concentrations and under the appropriate ionic conditions, clathrin cages assemble in solution, but they are empty shells, lacking other proteins, and they contain no membrane. This shows that the information to form clathrin baskets is contained in the clathrin molecules themselves, which are therefore able to self-assemble.

B. Without clathrin, adaptins still bind to receptors in the membrane, but no clathrin coat can form and thus no clathrin-coated pits or vesicles are produced.

C. Deeply invaginated clathrin-coated pits form on the membrane, but they do not pinch off to form closed vesicles.

D. Procaryotic cells do not perform endocytosis. A procaryotic cell therefore does not contain any receptors with appropriate cytosolic tails that could mediate adaptin binding. Therefore, no clathrin can bind and no clathrin coats can assemble.

Answer 14–6 The preassembled sugar chain allows for better quality control. The assembled oligosaccharide chains can be checked for accuracy before they are added to the protein; if a mistake were made in adding sugars individually to the protein, the whole protein would have to be discarded. Since far more energy is used in building a protein than in building a short oligosaccharide chain, this is a much more economical strategy. Also, once a sugar tree is added to a protein, it is more difficult for enzymes to modify its branches, compared to modifying them on the free sugar tree. This difficulty becomes apparent as the protein moves to the cell surface: although sugar chains are continually modified by enzymes in various compartments of the secretory pathway, these modifications are often incomplete and result in considerable heterogeneity of the glycoproteins that leave the cell. This heterogeneity is largely due to the restricted access that the enzymes have to the sugar trees attached to the surface of proteins. The heterogeneity also explains why glycoproteins are more difficult to study and purify than nonglycosylated proteins.

Answer 14–7 Aggregates of the secretory proteins would form in the ER, just as they do in the *trans* Golgi network. As the aggregation is specific for secretory proteins, ER proteins would be excluded from the aggregates. The aggregates eventually would be degraded.

Answer 14–8 Transferrin without Fe bound does not interact with its receptor and circulates in the bloodstream until it catches an Fe ion. Once iron is bound, the iron-transferrin complex can bind to the transferrin receptor on the surface of a cell and be endocytosed. Under the acidic conditions of the endosome, the transferrin releases its iron, but the transferrin remains bound to the transferrin receptor, which is recycled back to the cell surface, where it encounters the neutral pH environment of the blood. The neutral pH causes the receptor to release the transferrin into the circulation, where it can pick up another Fe ion to repeat the cycle. The iron released in the endosome, like the LDL in Figure 14–29, moves on to lysosomes, from where it is transported into the cytosol.

The system allows cells to take up iron efficiently even though the concentration of iron in the blood is extremely low. The iron bound to transferrin is concentrated at the cell surface by binding to transferrin receptors; it becomes further concentrated in clathrin-coated pits, which collect the transferrin receptors. In this way, transferrin cycles between the blood and endosomes, delivering the iron that cells need to grow.

Answer 14–9

A. True.

B. False. The signal sequences that direct proteins to the ER contain a core of eight or more hydrophobic amino acids. The sequence shown here contains many hydrophilic amino acid side chains, including the charged amino acids His, Arg, Asp, and Lys, and the uncharged hydrophilic amino acids Gln and Ser.

C. True. Otherwise they could not dock at the correct target membrane or recruit a fusion complex to a docking site.

D. True.

E. True. Lysosomal proteins are selected in the *trans* Golgi network and packaged into transport vesicles that deliver them to the late endosome. If not selected, they would enter by default into transport vesicles that move constitutively to the cell surface.

F. False. Lysosomes also digest internal organelles by autophagocytosis.

G. False. Mitochondria do not participate in vesicular transport, and therefore *N*-linked glycoproteins, which are exclusively assembled in the ER, cannot be transported to mitochondria.

Answer 14–10 They must contain a nuclear localization signal as well. Proteins with nuclear export signals shuttle between the nucleus and the cytosol. An example is the A1 protein, which binds to mRNAs in the nucleus and guides them through the nuclear pores. Once in the cytosol, a nuclear localization signal ensures that the A1 protein is reimported so that it can participate in the export of further mRNAs.

Answer 14–11 Influenza virus enters cells by endocytosis and is delivered to endosomes, where it encounters an acidic pH that activates its fusion protein. The viral membrane then fuses with the membrane of the endosome, releasing the viral genome into the cytosol (Figure A14–11, *next page, top*). NH_3 is a small molecule that readily penetrates membranes. Thus it can enter all intracellular compartments, including endosomes, by diffusion. Once in a compartment that has an acidic pH, NH_3 binds H^+ to form NH_4^+, which is a charged ion and therefore cannot cross the membrane by diffusion. NH_4^+ ions therefore accumulate in acidic compartments, raising their pH. When the pH of the endosome is raised, viruses are still endocytosed, but because the viral fusion protein cannot be activated, the virus cannot enter the cytosol. Remember this the next time you have the flu and have access to a stable.

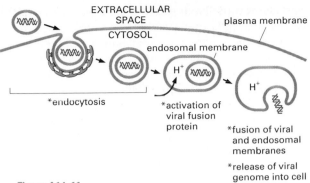

EXTRACELLULAR SPACE

plasma membrane

CYTOSOL

endosomal membrane

*endocytosis

*activation of viral fusion protein

*fusion of viral and endosomal membranes

*release of viral genome into cell

Figure A14–11

Answer 14–12

A. The problem is that vesicles having two different kinds of v-SNAREs in their membrane could dock on either of two different membranes.

B. The answer to this puzzle is presently not known, but we can predict that cells must have ways of turning the docking ability of SNAREs on and off. This may be achieved through other proteins that are, for example, copackaged in the ER with SNAREs into transport vesicles and facilitate the interactions of the correct v-SNARE with the t-SNARE in the *cis* Golgi network.

Answer 14–13 Synaptic transmission involves the release of neurotransmitters by exocytosis. During this event, the membrane of the synaptic vesicles fuse with the plasma membrane of the nerve terminals. To make new synaptic vesicles, membrane must be retrieved from the plasma membrane by endocytosis. This endocytosis step is blocked if dynamin is defective, as the protein seems to be required to pinch off the clathrin-coated endocytic vesicles. The clue to deciphering the role of dynamin came from electron micrographs of synapses of the mutant flies (Figure A14–13). Note that there are many flasklike invaginations of the plasma membrane, representing deeply invaginated clathrin-coated pits that cannot pinch off. The collars visible around the necks of these invaginations are made of mutant dynamin.

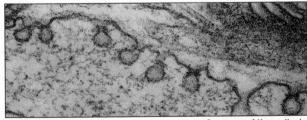

Courtesy of Kazuo Ikeda

Figure A14–13

Answer 14–14 The first two sentences are correct. The third is not. It should read: "Because the contents of the lumen of the ER or any other compartment in the secretory or endocytic pathways never mix with the cytosol, proteins that enter these pathways will never

need to be imported again." When the nuclear envelope and the ER break down in mitosis, they form vesicles whose contents remain separated from the cytosol by the vesicle membrane.

Answer 14–15 The protein is translocated into the ER. Its ER signal sequence is recognized as soon as it emerges from the ribosome. The ribosome then becomes bound to the ER membrane, and the growing polypeptide chain is transferred through the ER translocation channel. The nuclear localization sequence is therefore never exposed to the cytosol. It will never encounter nuclear import receptors, and the protein will not enter the nucleus.

Answer 14–16 (1) Proteins are imported into the nucleus after they have been synthesized, folded, and, if appropriate, assembled into complexes. In contrast, unfolded polypeptide chains are translocated into the ER as they are being made by the ribosomes. Ribosomes are assembled in the nucleus yet function in the cytosol, and the enzyme complexes that catalyze RNA transcription and splicing are assembled in the cytosol yet function in the nucleus. Thus both ribosomes and these enzyme complexes need to be transported through the nuclear pores intact. (2) Nuclear pores are gates, which are always open to small molecules; in contrast, translocation channels in the ER membrane are normally closed and open only after the ribosome attaches to the membrane and the translocating polypeptide chain seals the channel from the cytosol. It is important that the ER membrane remains impermeable to small molecules during the translocation process, as the ER is a major store for Ca^{2+} in the cell, and Ca^{2+} release into the cytosol must be tightly controlled (discussed in Chapter 15). (3) Nuclear localization signals are not cleaved off after protein import into the nucleus; in contrast, ER signal peptides are usually cleaved off. Nuclear localization signals are needed to repeatedly reimport nuclear proteins after they have been released into the cytosol during mitosis, when the nuclear envelope breaks down.

Answer 14–17 The transient intermixing of nuclear and cytosolic contents during mitosis supports the idea that the nuclear interior and the cytosol are indeed evolutionarily related. In fact, one can consider the nucleus as a subcompartment of the cytosol that has become surrounded by the nuclear envelope, with access only through the nuclear pores.

Answer 14–18 The actual explanation is that the single amino acid change causes the protein to misfold slightly so that, although it is still active as a protease inhibitor, it is prevented by chaperone proteins in the ER from exiting this organelle. It therefore accumulates in the ER lumen and is eventually degraded. Alternative interpretations might have been: (1) the mutation affects the stability of the protein in the bloodstream so that it is degraded much faster in the blood than the normal protein, or (2) the mutation inactivates the ER signal sequence and prevents the protein from entering the ER. (3) Another explanation could have been that the mutation altered the sequence to create an ER retention signal, which would have retained the mutant protein in the ER. One could distinguish between these possibilities by using fluorescent-tagged antibodies against the protein to follow its transport in the cells (see Panel 5–3, pp. 158–159).

Answer 14–19 Critique: "Dr. Outonalimb proposes to study the biosynthesis of forgettin, a protein of significant interest. The main hypothesis on which this proposal is based, however, requires further support. In particular it is questionable whether forgettin is indeed a secreted protein, as proposed. ER signal sequences are normally found at the amino terminus. Carboxyl-terminal hydrophobic sequences will be exposed outside the ribosome only after protein synthesis has already terminated and can therefore not be recognized by SRP during translation. It is therefore unlikely that forgettin will be translocated by an SRP-dependent mechanism, and it may therefore remain in the cytosol. Dr. Outonalimb should take these considerations into account when submitting a revised application."

Answer 14–20 The Golgi apparatus may have evolved from specialized patches of ER membrane. These regions of the ER might have pinched off, forming a new compartment (Figure A14–20), which still communicates with the ER by vesicular transport. For the newly evolved Golgi compartment to be useful, transport vesicles would also have to have evolved.

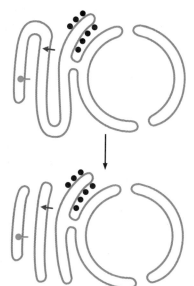

Figure A14–20

Answer 14–21 This is a chicken-and-egg question. In fact, the situation never arises in present-day cells,

although it must have posed a considerable problem for the first cells that evolved. New cell membranes are made by expansion of existing membranes, and the ER is never made *de novo*. There will always be an existing piece of ER with translocation channels to integrate new translocation channels. Inheritance is therefore not limited to the propagation of the genome; a cell's organelles must also be passed from generation to generation. In fact, the ER translocation channels can be traced back to structurally related translocation channels in the procaryotic plasma membrane.

Answer 14–22

  A. extracelluar space

  B. cytosol

  C. plasma membrane

  D. clathrin coat

  E. membrane of deeply invaginated clathrin-coated pit

  F. captured cargo particles

  G. lumen of deeply invaginated clathrin-coated pit

Chapter 15

Answer 15–1 Most paracrine signaling molecules are very short-lived and are rapidly degraded after they are released from cells. In addition, some molecules become attached to the extracellular matrix and are thus prevented from diffusing too far, or are released into a confined space, such as the synaptic cleft between a nerve and a muscle cell, from which diffusion into the surrounding space is restricted.

Answer 15–2 Each photon causes the hydrolysis of 80,000 cyclic GMP molecules, i.e., the signal is amplified 80,000-fold (= $200 \times 4000 \times 0.1$).

Answer 15–3 Polar groups are hydrophilic, and cholesterol, having only one –OH group, would be too hydrophobic to be an effective hormone. A lipid that is virtually insoluble in water could not move readily as a messenger from one cell to another via the extracellular fluid.

Answer 15–4 In the case of the steroid receptor, a one-to-one complex of steroid and receptor binds to DNA to activate transcription; there is thus no amplification between ligand binding and transcription activation. Amplification occurs later, because transcription of a gene gives rise to many mRNAs that are each translated to give many protein molecules (discussed in Chapter 7). For the ion-channel-linked receptors, a single ion channel will let through thousands of ions in the time it remains open; this serves as the amplification step in this type of signaling system.

Answer 15–5   The mutant G protein would be almost continuously activated as GDP would spontaneously dissociate, allowing GTP to bind even in the absence of an activated G-protein-linked receptor. The consequences for the cell would therefore be similar to those of cholera toxin, which modifies the α subunit so that it cannot hydrolyze GTP to shut itself off. In contrast to the cholera toxin case, however, the mutant G protein would not stay permanently activated: it would switch itself off normally but then instantly become activated again as the GDP dissociates and GTP rebinds.

Answer 15–6   Rapid breakdown keeps the cyclic AMP levels low. The lower the cAMP levels are, the larger the increase achieved upon activation of adenylate cyclase, which makes new cyclic AMP. If you have $100 in the bank, and you deposit another $100, you have doubled your wealth; if you have only $10 to start with and you deposit $100, you have increased your wealth tenfold, a much larger proportional increase resulting from the same deposit.

Answer 15–7   Recall that the plasma membrane constitutes a rather small area compared to the total membrane surfaces in a cell (discussed in Chapter 14). The endoplasmic reticulum is generally much more abundant and spans the entire volume of the cell as a vast network of membrane tubes and sheets. This allows homogeneous Ca^{2+} release throughout the cell, which is important because the fast clearing of Ca^{2+} ions from the cytosol by Ca^{2+} pumps prevents Ca^{2+} from diffusing any significant distance in the cytosol.

Answer 15–8   Each reaction involved in the amplification scheme must be turned off to reset the signaling pathway to a resting level. Each of these off switches is equally important.

Answer 15–9   Because each antibody has two antigen-binding sites, binding to the receptors can induce receptor clustering on the cell surface. This is likely to activate receptor tyrosine kinases, which are activated by self-phosphorylation after individual kinase domains of the receptors have been brought into proximity. The activation of G-protein-linked receptors is more complicated, because the ligand has to induce a particular conformational change. Only very special antibodies mimic receptor ligands sufficiently well to induce a conformational change that activates the receptor.

Answer 15–10

A. True. Acetylcholine, for example, decreases the beating of heart muscle cells by binding to a G-protein-linked receptor and stimulates the contraction of skeletal muscle cells by binding to a different acetylcholine receptor, which is a ligand-gated ion channel.

B. False. Acetylcholine is short-lived and exerts its effects locally. Indeed, the consequences of pro-longing its lifetime are disastrous. Compounds that inhibit the enzyme acetylcholine esterase, which normally breaks down acetylcholine at a nerve-muscle synapse, are extremely toxic: an example is the nerve gas sarin, used in chemical warfare.

C. True. Nucleotide-free βγ complexes can activate ion channels, and GTP-bound a subunits can activate enzymes. The GDP-bound form of trimeric G proteins is the inactive state.

D. True. PIP_2 contains three phosphate groups, one of which links the sugar to the diacylglycerol lipid. IP_3 is generated by a simple hydrolysis reaction.

E. False. Calmodulin senses but does not regulate the intracellular Ca^{2+} levels.

F. True.

G. False. Ras is a proto-oncogene. It becomes an oncogene, i.e., promotes the development of cancer, if it harbors mutations that keep it in an active state all the time.

H. True. See Figure 15–28.

Answer 15–11

1. You would expect a high background level of Ras activity because Ras cannot be turned off efficiently.

2. As some Ras molecules are already GTP-bound, Ras activity in response to an extracellular signal would be greater than normal, but would be liable to saturate when all Ras molecules are converted to the GTP-bound form.

3. The response to a signal would be much less rapid, because the signal-dependent increase in GTP-bound Ras would occur over an elevated background of preexisting GTP-bound Ras.

Answer 15–12

A. Both types of signaling can occur long range: neurons can send action potentials along very long axons (think of the neck of a giraffe, for example), and hormones are passed through the bloodstream throughout the organism. Because at a synapse neurons secrete large amounts of neurotransmitters into a small, well-defined space between two cells, the concentrations are high; neurotransmitter receptors, therefore, need to bind to neurotransmitters with only low affinity. Hormones, in contrast, are diluted vastly in the bloodstream, where they circulate at often minuscule concentrations; hormone receptors, therefore, generally bind their hormone with extremely high affinities.

B. Whereas neuronal signaling is a private affair, one neuron talking to a select group of target cells through specific synaptic connections, hormonal signaling is a public announcement, with target cells sensing the hormone levels in the blood. Neuronal signaling is very fast, limited only by the speed of propagation of the action potential and the workings of the synapse, whereas hormonal signaling is slower, limited by blood flow and diffusion over larger distances.

Answer 15–13

A. There are 100,000 molecules of X and 10,000 molecules of Y in the cell (= rate of synthesis × average lifetime).

B. After one second, the concentration of X will have increased by 10,000 molecules. The concentration of X, therefore, one second after its synthesis is increased, is about 110,000 molecules per cell—which is a 10% increase over the concentration of X before the boost of its synthesis. The concentration of Y will also increase by 10,000 molecules, which, in contrast to X, represents a full twofold increase in its concentration (for simplicity, we can neglect the breakdown in this estimation because X and Y are relatively stable during the one-second stimulation).

C. Because of its larger proportional increase, Y is the preferred signal molecule. This calculation illustrates the surprising but important principle that the time it takes to switch a signal on is determined by the lifetime of the signal molecule.

Answer 15–14 The information transmitted by a cell-signaling pathway is contained in the *concentration* of the messenger, be it a small molecule or a phosphorylated protein. Therefore, to allow detection of a change, the original messenger is destroyed. The more unstable the messenger is, the faster the system can respond to changes. Human communication relies on messages that are delivered only once, and that are generally not interpreted by their abundance, but by their *content*. So do not kill the messengers; they can be used more than once.

Answer 15–15

A. The mutant receptor tyrosine kinase is inactive, and its presence has no consequences for the function of the normal receptor kinase; lacking its extracellular ligand-binding domain, the mutant receptor cannot be activated and thus is inert.

B. This mutant receptor is also inactive, but its presence will block signaling by the normal receptors. When ligand binds to either receptor, it will induce their dimerization. Two normal receptors have to come together to activate each other by phosphorylation. In the presence of an excess of mutant receptor, however, normal receptors will usually form mixed dimers, in which their intracellular domain cannot be activated because their partner is mutant (Figure A15–15, *top, next column*).

Figure A15–15

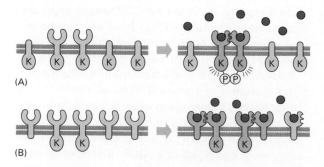

(A)

(B)

Answer 15–16 The statement is correct. Upon ligand binding, transmembrane helices of multispanning receptors, like the G-protein-linked receptors, shift and rearrange with respect to one another (Figure A15–16). This conformational change is sensed on one side of the membrane because of a change in the arrangement of the cytoplasmic loops. A single transmembrane segment is not sufficient to transmit a signal across the membrane directly; no rearrangements in the membrane are possible upon ligand binding. Upon ligand binding, single-span receptors such as receptor tyrosine kinases tend to dimerize, thereby bringing the intracellular enzyme domains into proximity so that they can activate each other.

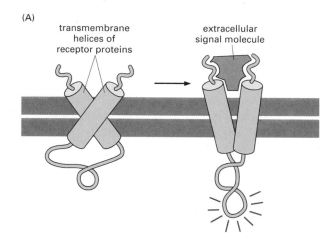

(A)

transmembrane helices of receptor proteins

extracellular signal molecule

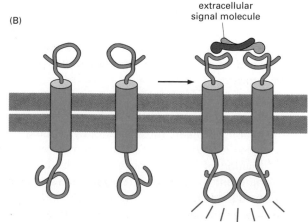

(B)

extracellular signal molecule

Figure A15–16

enzyme domain of receptors

Answer 15–17 Activation in both cases depends on proteins that catalyze GDP/GTP exchange on the G protein or Ras protein. Whereas the G-protein-linked receptors perform this function directly for G proteins, enzyme-linked receptors assemble multiple adaptor proteins into a signaling complex when the receptors are activated by phosphorylation and one of these recruits a Ras-activating protein that fulfills this function for Ras.

Answer 15–18 Because the intracellular concentration of Ca^{2+} is so low, an influx of relatively few Ca^{2+} ions leads to large changes in its cytosolic concentration. Thus, a tenfold increase in Ca^{2+} can be achieved by raising the concentration of Ca^{2+} into the micromolar range, which would require much fewer ions than would be required to change significantly the concentration of a more abundant ion such as Na^+. In muscle, a greater than tenfold change can be achieved in microseconds by releasing Ca^{2+} from the intracellular stores of the sarcoplasmic reticulum, a task that would be difficult to accomplish if changes in the millimolar range were required.

Answer 15–19 In a multicellular organism such as an animal, it is important that cells survive only when and where they are needed. Having cells depend on signals from other cells may be a simple way of ensuring this. A misplaced cell, for example, would probably fail to get the survival signals it needs (as its neighbors would be inappropriate) and would therefore kill itself. This strategy can also help regulate cell numbers: if cell type A depends on a survival signal from cell type B, the number of B cells could control the number of A cells by making a limited amount of the survival signal, so that only a certain number of A cells could survive. There is indeed evidence that such a mechanism does operate to help regulate cell numbers—both in developing and adult tissues.

Answer 15–20 Ca^{2+}-activated Ca^{2+} channels create a positive feedback loop: the more Ca^{2+} that is released, the more channels open. The Ca^{2+} signal in the cytosol is therefore propagated explosively throughout the entire muscle cell, thereby ensuring that all myosin/actin filaments contract almost synchronously.

Answer 15–21 K2 activates K1. If K1 is permanently activated, a response is observed independent of the status of K2. If the order were reversed, K1 would need to activate K2, which cannot occur because in our example K2 contains an inactivating mutation.

Chapter 16

Answer 16–1 Cells that migrate rapidly from one place to another, such as amoebae (A) and sperm cells (F), do not in general need intermediate filaments in their cytoplasm, since they do not develop or sustain

large tensile forces. Plant cells (G) are pushed and pulled by the forces of wind and water, but they resist these forces by means of their rigid cell walls, rather than by their cytoskeleton. Epithelial cells (B), smooth muscle cells (C), and the long axons of nerve cells (E) are all rich in cytoplasmic intermediate filaments, which prevent them from rupturing as they are stretched and compressed by the movements of their surrounding tissues.

All of the above eucaryotic cells possess at least intermediate filaments in their nuclear lamina. Bacteria, such as *Escherichia coli* (D), have none whatsoever.

Answer 16–2 Two tubulin dimers have a lower affinity for each other (because of a more limited number of interaction sites) than a tubulin dimer has for the end of a microtubule (where there are multiple possible interaction sites, both end-to-end of tubulin dimers adding to a protofilament and side-to-side of the tubulin dimers interacting with tubulin subunits in adjacent protofilaments forming the ringlike cross section). Thus, to initiate a microtubule from scratch, enough tubulin dimers have to come together and remain bound to one another for long enough for other tubulin molecules to add to them. Only when a number of tubulin dimers have already assembled will the binding of the next subunit be favored. The formation of these initial "nucleating sites" is therefore rare and will not occur spontaneously at cellular tubulin concentrations.

Centrosomes contain preassembled rings of γ-tubulin (in which the γ-tubulin subunits are held together in much tighter side-to-side interactions than αβ-tubulin can form) to which αβ-tubulin dimers can bind. The binding conditions of αβ-tubulin dimers resemble those of adding to the end of an assembled microtubule. The γ-tubulin rings in the centrosome can therefore be thought of as permanently preassembled nucleation sites.

Answer 16–3

A. The microtubule is shrinking because it has lost its GTP cap, i.e., the tubulin subunits at its end are all in their GDP-bound form. GTP-loaded tubulin subunits from solution will still add to this end, but they will be short-lived—either because they hydrolyze their GTP or because they fall off as the microtubule rim around them disassembles. If, however, enough GTP-loaded subunits are added quickly enough to cover up the GDP-containing tubulin subunits at the microtubule end, then a new GTP cap can form and regrowth is favored.

B. The rate of addition of GTP-tubulin will be greater at higher tubulin concentrations. The frequency with which shrinking microtubules switch to the growing mode will therefore increase with increasing tubulin concentration. The consequence of this regulation is that the system is self-balancing: the more microtubules shrink (resulting in a higher concentration of free tubulin), the more frequently microtubules will start to grow again. Conversely, the more microtubules grow, the lower the concentration of free tubulin will become and the rate of GTP-tubulin addition will slow down; at some point GTP hydrolysis will catch up with new GTP-tubulin addition, the GTP cap will be destroyed, and the microtubule will switch to the shrinking mode.

C. If only GDP were present, microtubules would continue to shrink and eventually disappear, because tubulin dimers with GDP have very low affinity for each other and will not add stably to microtubules.

D. If GTP is present but cannot be hydrolyzed, microtubules will continue to grow until all free tubulin subunits have been used up.

Answer 16–4 If all the dynein arms were equally active, there could be no significant relative motion of one microtubule to the other as required for bending (think of a circle of nine weight lifters, each trying to lift his neighbor off the ground: if they all succeeded, the group would levitate!). Thus, a few ciliary dynein molecules must be activated selectively on one side of the cilium. As they move their neighboring microtubules toward the tip of the cilium, the cilium bends away from the side containing the activated dyneins.

Answer 16–5 Any actin-binding protein that stabilizes complexes of two or more actin monomers without blocking the ends required for filament growth will facilitate the initiation of a new filament (nucleation).

Answer 16–6 Cells contain actin-binding proteins that bundle and cross-link actin filaments (see Figure 16–27). The filaments extending from lamellipodia and filopodia become firmly connected to the filamentous meshwork of the cell cortex, thus providing the mechanical anchorage required for the growing rod-like filaments to deform the cell membrane.

Answer 16–7 Both filaments are composed of subunits in the form of protein dimers that are held together by coiled-coil interactions. Moreover, in both cases, the dimers polymerize through their coiled-coil domains into filaments. Whereas intermediate filament dimers assemble head-to-head, however, and thereby create a filament that has no polarity, all myosin molecules in the same half of the myosin filament are oriented with their heads pointing in the same direction. This polarity is necessary for them to be able to develop a contractile force in muscle.

Answer 16–8

A. Successive actin molecules in an actin filament are identical in position and conformation. After a first protein (such as troponin) had bound to the actin filament, there would be no way a second protein

could recognize every seventh monomer in a naked actin filament. Tropomyosin, however, binds along the length of an actin filament, spanning precisely seven monomers, and thus provides a molecular "ruler" that measures the length of seven actin monomers. Troponin becomes localized by binding to the end of a tropomyosin molecule.

B. Calcium ions influence force generation in the actin/myosin system only if both troponin (to bind the calcium ions) and tropomyosin (to transmit the information that troponin has bound calcium to the actin filament) are present. (a) Troponin cannot bind to actin without tropomyosin. The actin filament would be permanently exposed to the myosin, and the system would be continuously active, independent of whether calcium ions were present or not (a muscle cell would therefore be continuously contracted with no possibility of regulation). (b) Tropomyosin would bind to actin and block binding of myosin completely; the system would be permanently inactive, no matter whether calcium ions were present, as tropomyosin is not affected by calcium. (c) The system will contract in response to calcium ions.

Answer 16–9

A. True. A continual outward movement of ER is required; in the absence of microtubules, the ER collapses toward the center of the cell.

B. True. Actin is needed to make the contractile ring that causes the physical cleavage between the two daughter cells, whereas the mitotic spindle that partitions the chromosomes is composed of microtubules.

C. True. Both extensions are associated with transmembrane proteins that protrude from the plasma membrane and enable the cell to form new anchor points on the substratum.

D. False. To cause bending, ATP is hydrolyzed by the dynein motor proteins that are attached to the outer microtubules in the flagellum.

E. False. Cells could not divide without rearranging their intermediate filaments, but many terminally differentiated and long-lived cells, such as nerve cells, have stable intermediate filaments that are not known to depolymerize.

F. False. The rate of growth is independent of the size of the GTP cap. The plus and minus ends have different growth rates because they have physically distinct binding sites for the incoming tubulin subunits; the rate of addition of tubulin subunits differs at the two ends.

G. True. Both are nice examples of how the same membrane can have regions that are highly specialized for a particular function.

H. False. Myosin movement is activated by phosphorylation of the myosin light chains, or by calcium binding to troponin.

Answer 16–10   The average time taken for a small molecule (such as ATP) to diffuse a distance of 10 μm is given by the calculation:

$$(10^{-3})^2 / (2 \times 5 \times 10^{-6}) = 0.1 \text{ sec}$$

Similarly, a protein takes 1 second and a vesicle 10 seconds on average to travel 10 μm. A vesicle would require on average 10^9 seconds, or more than 30 years, to diffuse to the end of a 10 cm axon. This calculation makes it clear why kinesin and other motor proteins evolved to carry molecules and organelles along microtubules.

Answer 16–11   (1) Animal cells are much larger, diversely shaped, and do not have a cell wall. Cytoskeletal elements are required to provide mechanical strength and shape in the absence of a cell wall. (2) Animal cells, and all other eucaryotic cells, have a nucleus that is shaped and held in place in the cell by intermediate filaments; the nuclear lamins attached to the inner nuclear membrane support and shape the nuclear membrane, and a meshwork of intermediate filaments surrounds the nucleus and spans the cytosol. (3) Animal cells can move by a process that requires a change in cell shape. Actin filaments and myosin motor proteins are required for these activities. (4) Animal cells have a much larger genome than bacteria; this genome is fragmented into many chromosomes. For cell division, chromosomes need to be accurately distributed to the daughter cells, requiring the function of the microtubules that form the mitotic spindle. (5) Animal cells have internal organelles. Their localization in the cell is dependent on motor proteins that move them along microtubules. A remarkable example is the long-distance travel of membrane-bounded vesicles (organelles) along microtubules in an axon that can be up to 1 m (~3 feet) long in the case of the nerve cells that extend from your spinal cord to your feet.

Answer 16–12   The ends of an intermediate filament are indistinguishable from each other, because the filaments are built by assembly of symmetrical tetramers made from two coiled-coil dimers. In contrast to microtubules and actin filaments, intermediate filaments therefore have no polarity.

Answer 16–13   Intermediate filaments have no polarity; their ends are chemically indistinguishable. It would therefore be difficult to envision how a hypothetical motor protein that bound to the middle of the filament could sense a defined direction. Such a motor protein would be equally likely to attach to the filament facing one end or the other.

Answer 16–14   Katanin breaks microtubules along their length, and at positions remote from their GTP caps. The fragments that form therefore contain GDP-tubulin at their exposed ends and rapidly depolymerize. Katanin thus provides a very quick means of destroying existing microtubules.

Answer 16–15   Cell division depends on the ability of microtubules both to polymerize and to depolymerize. This is most obvious when one considers that the formation of the mitotic spindle requires the prior depolymerization of other cellular microtubules to free up the tubulin required to build the spindle. This rearrangement is not possible in taxol-treated cells, whereas in colchicine-treated cells, division is blocked because a spindle cannot be assembled. On a more subtle but no less important level, both drugs block the dynamic instability of microtubules and would therefore interfere with the workings of the mitotic spindle, even if one could be properly assembled.

Answer 16–16   Motor proteins are unidirectional in their action; kinesin always moves toward the plus end of a microtubule and dynein toward the minus end. Thus if kinesin molecules are attached to glass, only those individual motors that have the correct orientation in relation to the microtubule that settles on them can attach to the microtubule and exert force on it to propel it forward. Since kinesin moves toward the plus end of the microtubule, the microtubule will always crawl minus-end first over the coverslip.

Answer 16–17

A. Phase A corresponds to a lag phase, during which tubulin molecules assemble to form nucleation centers (Figure A16–17[A], *next page, top right*). Nucleation is followed by a rapid rise (phase B) to a plateau value as tubulin dimers add to the ends of the elongating microtubules (Figure A16–17[B]). At phase C, equilibrium is reached with some microtubules in the population growing, whereas others are rapidly shrinking (Figure A16–17[C]). The concentration of free tubulin is constant at this point, because polymerization and depolymerization are balanced (see also Question 16–3, p. 523).

B. The addition of centrosomes introduces nucleation sites that eliminate the lag phase A as shown by the *red* curve in Figure A16–17 (*next page, top left*). The rate of microtubule growth (i.e., the slope of the curve in the elongation phase B) and equilibrium level of free tubulin remain unchanged, because the presence of centrosomes does not affect the rates of polymerization and depolymerization.

Answer 16–18   The ends of the shrinking microtubule are visibly frayed, and the individual protofilaments appear to come apart and curl as the end depolymerizes. This micrograph therefore suggests that the GTP cap (which is lost from shrinking microtubules) holds the protofilaments properly aligned with each other,

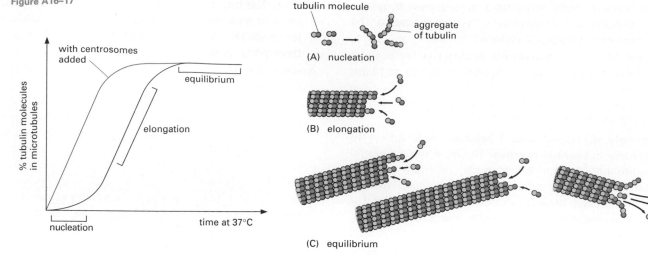

(A) nucleation

(B) elongation

(C) equilibrium

perhaps by strengthening the side-to-side interactions between αβ-tubulin subunits when they are in their GTP-bound form.

Answer 16–19 Cytochalasin interferes with actin filament formation, and its effect on the cell demonstrates the importance of actin to cell locomotion. The experiment with colchicine shows that microtubules are required to give a cell a polarity that then determines which end becomes the leading edge (see Figure 16–13). In the absence of microtubules, cells still go through the motions normally associated with cell movement, such as the extension of lamellipodia, but in the absence of cell polarity these are futile exercises because they happen indiscriminately in all directions.

Antibodies bind tightly to the antigen (in this case vimentin) to which they were raised (see Panel 5–3, pp. 158–159). When bound, an antibody can interfere with the function of the antigen by preventing it from interacting properly with other cell components. The antibody injection experiment therefore suggests that intermediate filaments are not required for the maintenance of cell polarity or for the motile machinery.

Answer 16–20 Either (B) or (C) would complete the sentence correctly. The direct result of the action potential in the plasma membrane is the release of Ca^{2+} into the cytosol from the sarcoplasmic reticulum; muscle cells are triggered to contract by this rapid rise in cytosolic Ca^{2+}. Calcium ions at high concentrations bind to troponin, which in turn causes tropomyosin to move to expose myosin-binding sites on the actin filaments. (A) and (D) would be wrong because Ca^{2+} has no effect on the detachment of the myosin head from actin, which is the result of ATP hydrolysis. Nor does it have any role in maintaining the structure of the myosin filament.

Answer 16–21 Only (D) is correct. Upon contraction, the Z discs move closer together, and neither actin nor myosin filaments contract. The answer to this question

will become clear if you reexamine Figures 16–35 and 16–36.

Chapter 17

Answer 17–1 Cells in peak B contain twice as much DNA as those in the peak A, indicating that they contain replicated DNA while the cells in peak A contain unreplicated DNA. Peak A therefore contains cells that are in G_1, and peak B contains cells that are in G_2 and mitosis. Cells in S phase have begun but not finished DNA synthesis; they therefore have various intermediate amounts of DNA and are found in the region between the two peaks. Most cells are in G_1, indicating that it is the longest phase of the cell cycle (see Figure 17–4).

Answer 17–2 In a eucaryotic organism the genetic information that the organism needs to survive and reproduce is distributed between multiple chromosomes. It is crucial therefore that each daughter cell receives a copy of each chromosome when a cell divides; if a daughter cell receives too few or too many chromosomes, the effects are usually deleterious or even lethal. Only two copies of each chromosome are produced by chromosome replication in mitosis. If the cell were to randomly distribute the chromosomes when it divided, it would be very unlikely that each daughter cell would receive precisely one copy of each chromosome. In contrast, the Golgi apparatus fragments into many thousands of vesicles that are all alike, and by random distribution it is very likely that each daughter cell will receive an approximately equal number of them.

Answer 17–3 The experiment shows that kinetochores are not preassigned to one or the other spindle pole; microtubules attach to the kinetochores that they are able to reach. To remain attached, tension has to be exerted. Tension is normally achieved by the opposing pulling forces from both spindle poles. The requirement for such tension ensures that if two sister kinetochores ever become attached to the same spindle pole, so that tension is not generated, one or both of the connections would break, and microtubules from the opposing spindle pole would have another chance to attach properly.

Answer 17–4 Recall from Figure 17–14 that the new nuclear envelope reassembles on the surface of the chromosomes. The close apposition of the envelope to the chromosomes prevents cytosolic proteins from being trapped between the chromosomes and the envelope. Nuclear proteins are then selectively imported through the nuclear pores, causing the nucleus to expand while maintaining its characteristic protein composition.

Answer 17–5 The membranes of the Golgi vesicles fuse to form the new plasma membranes of the two daughter cells. The interiors of the vesicles, which are filled with cell-wall material, become the new cell-wall matrix separating the two daughter cells. Proteins in the membranes of the Golgi vesicles thus become plasma membrane proteins. Those parts of the proteins that were exposed to the lumen of the Golgi vesicle will end up exposed to the new cell wall (Figure A17–5).

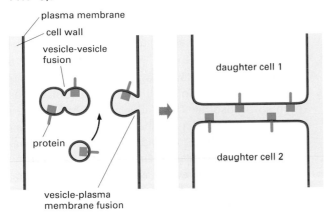

Figure A17–5

Answer 17–6 Although each daughter cell ends up with a diploid amount of DNA after the first meiotic division, each cell has effectively only a haploid set of chromosomes (albeit in two copies), representing only one or other homologue of each type of chromosome (although some mixing will have occurred during crossing-over). Since the maternal and paternal chromosomes of a pair will carry different versions of many of the genes (discussed in Chapter 9), these daughter cells will not be genetically identical. In contrast, somatic cells dividing by mitosis inherit a diploid set of chromosomes, and all daughter cells are genetically identical. The role of gametes produced by meiosis is to mix and reassort gene pools during sexual

reproduction, and thus it is a definite advantage for each of them to have a slightly different genetic constitution. The role of somatic cells on the other hand is to build an organism that contains the same genes in all its cells.

Answer 17–7

A. True. Centrosomes replicate as the first step of mitosis. The mechanism by which one centrosome gives rise to two and only two centrosomes is unknown.

B. True. In fact, the nuclear envelope is continuous with the ER membrane (discussed in Chapter 14), and some of the fragments therefore contain membrane derived from both origins.

C. True. Sister chromatids separate only at the start of anaphase.

D. True. The gametes formed by meiosis are haploid.

E. False. The ends of polar microtubules overlap and attach to one another via proteins that bridge between the microtubules.

F. False. Microtubules play no role in DNA replication.

G. False. In the correct statement "centromere" and "centrosome" must be switched.

Answer 17–8 Since all cells arise by division of another cell, this statement is correct, assuming that "first cell division" refers to the division of the successful founder cell from which all life as we know it has derived. There were probably many other unsuccessful attempts to start the chain of life.

Answer 17–9 Since the cell population is growing exponentially, doubling its weight at every cell division, the weight of the cell cluster after N number of cell divisions is $2^N \times 10^{-9}$ g. Therefore, 70 kg (70×10^3 g) $= 2^N \times 10^{-9}$ g, or $2^N = 7 \times 10^{13}$. Taking the logarithm on both sides allows you to solve the equation for N. Therefore, $N = \ln(7 \times 10^{13}) / \ln 2 = 46$—i.e., it would take only 46 days if cells grew exponentially. Cell division in animals is tightly controlled, however, and most cells in the human body stop dividing when they become highly specialized. The example demonstrates that exponential cell growth occurs only for very brief periods, even during embryonic development.

Answer 17–10 The order is G, C, B, A, D. Together, these six steps are referred to as mitosis (F). No step in mitosis is influenced by the phases of the moon (E).

Answer 17–11 Many egg cells are big and contain stores of enough cellular components to last for many cell divisions. The daughter cells that form during the first cell divisions after the egg is fertilized are progressively smaller in size and thus can be formed without a need for new protein or RNA synthesis. Whereas normally dividing cells would wait in G_1 and G_2 until their size has doubled, both G_1 and G_2 are virtually absent in the first few cell divisions of egg cells. As G_1 is usually longer than G_2, G_1 is the most drastically reduced.

Answer 17–12

A. Microtubules are used to form the mitotic spindle in all eucaryotic cells. Bacteria do not have microtubules, and their replicated chromosomes are assigned to the daughter cells by attachment to the plasma membrane.

B. Both bacterial and eucaryotic cells must replicate their DNA before they can divide.

C. Bacteria do not undergo meiosis. Meiosis is required in eucaryotes to produce haploid gametes from diploid cells; since bacteria have only one chromosome, they cannot undergo anything resembling meiosis.

D. Mitosis is the process of nuclear division in eucaryotic cells. Bacteria do not have a nucleus and divide by a process called binary fission.

E. Synthesis of new cell wall is required for bacterial cells to divide and is part of the mechanism for separating the replicated bacterial chromosomes; plant cells also need to synthesize new cell wall to undergo cytokinesis.

F. Both bacteria and eucaryotic cells usually grow in size before they divide.

G. In eucaryotic cells the nuclear envelope breaks down at the beginning of mitosis and is reassembled around the segregated chromosomes after mitosis is complete.

Answer 17–13   If the growth rate of microtubules is the same in mitotic and in interphase cells, their length is proportional to their lifetime. Thus, the average length of microtubules in mitosis is 1 mm (= 20 μm × 15 s/300 s). If microtubules are on average 20 times shorter, but in total contain the same number of tubulin monomers, then there must be 20 times as many microtubules, or 2000 nucleation sites per centrosome in mitosis.

Answer 17–14   Two copies of the same chromosome can end up in the same daughter cell if one of the microtubule connections breaks before sister chromatids are separated. Alternatively, microtubules from the same spindle pole could attach to both kinetochores of the chromosome. As the consequence of this severe and rare error, one daughter cell would contain only one copy of all the genes carried on that chromosome and the other daughter cell would contain three copies. The changed gene dosage, leading to correspondingly changed amounts of the mRNAs and proteins produced, is in many cases detrimental to the cell. If the mistake happens during meiosis, in the process of gamete formation, it will be propagated in all cells of the organism. A severe form of mental retardation called Down syndrome, for example, is due to the presence of three copies of Chromosome 21 in all of the nucleated cells in the body.

Answer 17–15   Antibodies bind tightly to the antigen (in this case myosin) to which they were raised. When bound, an antibody can interfere with the function of the antigen by preventing it from interacting properly with other cell components. (a) The movement of chromosomes at anaphase depends on microtubules and does not depend on actin or myosin. Injection of an antimyosin antibody into a cell will therefore have no effect on chromosome movement during mitosis. (b) Cytokinesis, on the other hand, depends on the assembly and contraction of a ring of actin and myosin filaments, which forms the cleavage furrow that splits the cell in two. Injection of the antimyosin antibodies will therefore block cytokinesis.

Answer 17–16   The sister chromatid becomes committed when a microtubule from one of the spindle poles attaches to the kinetochore of the chromatid. Microtubule attachment is still reversible until a second microtubule from the other spindle pole attaches to the kinetochore of its partner sister chromatid so that the duplicated chromosome is now put under mechanical tension by pulling forces from both poles. The tension ensures that both microtubules remain attached to the chromosome. The position of the chromosome in the cell at the time the nuclear envelope breaks down will influence which spindle pole the chromatid will be pulled to, as a kinetochore is most likely to become attached to the spindle pole toward which it is facing.

Answer 17–17   Meiosis begins with DNA replication, producing a tetraploid cell containing four copies of each chromosome. These four copies have to be distributed equally during the two sequential meiotic divisions into four haploid cells. Sister chromatids remain paired so that (1) the cells resulting from the first division receive two complete sets of chromosomes and (2) the chromosomes can be evenly distributed again in the second meiotic division. If the sister chromatids did not remain paired, it would not be possible in the second division to distinguish which chromatids belong together, and it would therefore be difficult to ensure that precisely one copy of each chromatid is pulled into each daughter cell. Keeping two sister chromatids paired in the first meiotic division is therefore an easy way to keep track of which chromatids belong together.

This biological principle suggests that you might consider clamping your socks together in matching pairs before putting them into the laundry. In this way the cumbersome process of sorting them out afterward—and the seemingly inevitable mistakes that occur during that process—could be avoided.

Answer 17–18   Two alternative models of how the kinetochore may generate a poleward force on its

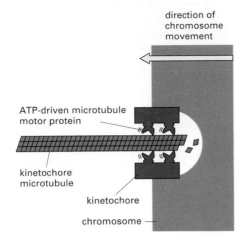

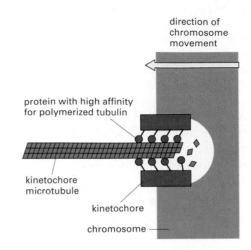

direction of
chromosome
movement

ATP-driven microtubule
motor protein

kinetochore
microtubule

kinetochore

chromosome

(A) ATP-driven chromosome movement drives
microtubule disassembly

protein with high affinity
for polymerized tubulin

kinetochore
microtubule

kinetochore

chromosome

(B) microtubule disassembly drives
chromosome movement

chromosome during anaphase A are shown in Figure A17–18. In (A) microtubule motor proteins are part of the kinetochore and use the energy of ATP hydrolysis to pull the chromosome along its bound microtubules. In (B) chromosome movement is driven by microtubule disassembly: as tubulin subunits dissociate, the kinetochore is obliged to slide poleward in order to maintain its binding to the walls of the microtubule.

Chapter 18

Answer 18–1   The precise mechanism of the G_1 cell-size checkpoint is not yet known, but one can imagine in a general way how it might work. The amount of DNA in a G_1 cell is invariant; thus its concentration relative to that of cytosolic components steadily decreases as the cell gets bigger in G_1. The observation that the size of each particular type of cell is roughly proportional to its total amount of DNA suggests that the control mechanism depends on comparing the quantity of cytoplasmic components against the quantity of DNA, with the cell cycle being delayed until a threshold ratio of one or more cytoplasmic components to DNA mass is attained by the biosyntheses responsible for cell growth.

Answer 18–2   The frog oocytes must contain an inactive form of MPF (as inactive cyclin-Cdk complexes). Upon injection into a new oocyte, a small amount of active MPF thus activates a larger pool of inactive MPF by regulation of the enzymes that cause phosphorylation and dephosphorylation of the inactive MPF at the appropriate sites (see Figure 18–10). An extract of the second oocyte, now in M phase itself, therefore contains as much active MPF as the original cytoplasmic extract, and so on.

Answer 18–3   For multicellular organisms, the control of cell division is extremely important. Individual cells must not proliferate unless it is to the benefit of the whole organism. The G_0 state offers protection from aberrant activation of cell division, because the cell-cycle control system is largely dismantled. If a cell just paused in G_1, on the other hand, it would still contain all the cell-cycle machinery and might still be induced to divide. It would also have to remake the "decision" not to divide almost continuously. To reenter the cell cycle from G_0, a cell has to resynthesize the components that have disappeared.

Answer 18–4   Any mechanism explaining cell senescence must involve a change in cells that occurs cumulatively over many cell generations. This could, for example, be the accumulation of mutations, possibly in some specific genes that are not being repaired. Another possibility is that the telomeres at the ends of chromosomes could become progressively shorter with each round of DNA replication if the activity of telomerase (the enzyme that replicates the telomeres) becomes impaired with age. While many such correlations have been observed, the actual causes of cell senescence still remain a mystery.

Answer 18–5   As programmed cell death occurs on a large scale in both developing and adult tissues, it must not trigger alarm reactions that are normally associated with cell injury. Tissue injury, for example, leads to the release of PDGF, which induces the proliferation of surrounding cells so that the wound heals, and other signals are released that can cause a destructive inflammatory reaction. Moreover, the release of intracellular contents could elicit an immune response against molecules that are normally not encountered by the immune system. In normal development, such reactions would be self-defeating if they occurred in response to cell death.

Answer 18–6 Proto-oncogenes tend to encode proteins that stimulate cell proliferation; mutation of one proto-oncogene to produce an oncogene that stimulates cell proliferation in an uncontrolled way is therefore sufficient to promote the development of cancer. In contrast, tumor suppressor genes tend to inhibit cell proliferation, and only one good copy is normally sufficient to hold proliferation in check. Therefore both copies of a tumor suppressor gene have to be inactivated to promote the development of cancer.

Answer 18–7

A. False. There is no G_1 to M phase transition. The statement is correct, however, for the G_1 to S phase transition, where cells commit themselves to a division cycle.

B. True. Programmed cell death is an active process carried out by special proteases dedicated to this function.

C. True. This mechanism is thought to adjust the number of neurons to the number of specific target cells to which the neurons connect.

D. True. An amazing evolutionary conservation!

E. True. Such studies employ so-called conditional mutations, which lead to the production of proteins that usually are stable and functional at one temperature, but unstable or inactive at another temperature. Cells can be grown at the temperature at which the mutant protein functions normally and then can be shifted to a temperature at which the protein's function is lost.

F. True. Association of a Cdk protein with a cyclin is required for its activity (hence its name <u>c</u>yclin-<u>d</u>ependent <u>k</u>inase). Furthermore, phosphorylation at specific sites and dephosphorylation at other sites on the Cdk protein are required for the cyclin-Cdk complex to be active.

G. False. When mutated, proto-oncogenes can become oncogenes, which produce proteins that stimulate uncontrolled cell proliferation. Tumor suppressor genes, on the other hand, normally block cell proliferation, and therefore mutations that inactivate them promote cell proliferation.

Answer 18–8 Loss of cyclin leads to inactivation of the mitotic Cdk. As the result, its target proteins become dephosphorylated by phosphatases, and the cells exit mitosis—they disassemble the mitotic spindle, reassemble the nuclear envelope, decondense their chromosomes, and so on. Cyclin is degraded by ubiquitin-dependent destruction in proteosomes, and the activation of the mitotic Cdk most likely causes the ubiquitination of the cyclin, but with a substantial delay. As discussed in Chapter 5, ubiquitination tags proteins for degradation in proteasomes.

Answer 18–9 Cells in an animal must behave for the good of the organism as a whole—to a much greater extent than people generally act for the good of society as a whole. In the context of an organism, unsocial behavior would lead to a loss of organization and to cancer. Many of the rules that cells have to obey would be unacceptable in a human society. Most people, for example, would be reluctant to kill themselves for the good of society, yet cells do it all the time.

Answer 18–10 Cyclin accumulates gradually, as it is steadily synthesized. As it accumulates, it will tend to form complexes with the Cdk molecules that are present. After a certain threshold level has been reached, a sufficient amount of MPF complex has been formed that it is activated by the appropriate kinases and phosphatases that phosphorylate and dephosphorylate it. Once activated, MPF acts to enhance the activity of the activating phosphatase; this positive feedback leads to the explosive activation of MPF. Thus cyclin accumulation acts like a slow-burning fuse, which eventually helps trigger the explosive self-activation of MPF. The precipitous destruction of cyclin terminates MPF activity, and a new round of cyclin accumulation begins.

Answer 18–11 The most likely approach to success (if that is what the goal should be called) is plan E, which should result in an increase in cell numbers. The problem is, of course, that cell numbers of each tissue must be increased similarly to maintain balanced proportions in the organism, yet different cells respond to different growth factors. As shown in Figure A18–11, however, the approach has indeed met with limited success: a mouse producing very large quantities of growth hormone (*left*) grows to almost twice the size of a normal mouse (*right*). To achieve this twofold change in size, however, growth hormone was massively overproduced (about fiftyfold).

The other approaches have conceptual problems:

A. Delaying cell senescence is more likely to lead to an animal with an extended lifespan—possibly a desirable goal in itself, but as we do not yet know much about the molecular mechanisms that govern cell senescence, it is not obvious how to approach the problem experimentally.

Figure A18–11   Courtesy of Ralph Brinster

B. and D. Blocking programmed cell death (or over-producing survival factors that inhibit programmed cell death) might lead to defects in development, as development requires the selective death of cells. It is unlikely that a viable animal would be obtained.

C. Blocking p53 function would eliminate an important checkpoint of the cell cycle that detects DNA damage and stops the cycle so the cell can repair the damage; removing p53 would increase mutation rates and lead to cancer. Indeed, mice without p53 usually develop normally, but die of cancer at a young age.

F. Given the circumstances, switching careers might not be a bad option.

Answer 18–12 The on-demand, limited release of PDGF at a wound site triggers cell division of neighboring cells for a limited amount of time, until the PDGF is degraded. This is different from the continuous release of PDGF from mutant cells, where PDGF is made in an uncontrolled way at high levels. Moreover, the mutant cells that make PDGF often inappropriately express their own PDGF receptor, so that they can stimulate their own proliferation, thereby promoting the development of cancer.

Answer 18–13

A. Radiation leads to DNA damage, which activates a feedback control (mediated by p53 and p21; see Figure 18–13) that arrests the cell cycle until the DNA has been repaired.

B. The cell will replicate damaged DNA and thereby introduce mutations to the daughter cells when the cell divides.

C. The cell will be able to divide normally, but it will be prone to mutations, because some DNA damage always occurs as the result of natural irradiation caused, for example, by cosmic rays. The checkpoint mediated by p53 is mainly required as a safeguard against the devastating effects of DNA damage, but not for the natural progression of the cell cycle in undamaged cells.

D. Cell division is an ongoing process that does not cease upon reaching maturity. Blood cells, epithelial cells in the skin or lining the gut, and the cells of the immune system, for example, are being constantly produced by cell division to meet the body's needs; your body produces about 1011 new red blood cells alone each day.

Answer 18–14 All three types of mutant cells would be unable to divide. The cells . . .

A. would enter mitosis but would not be able to exit mitosis.

B. would arrest permanently in G_1 because its G_1 cyclin-Cdk complexes would be inactivated.

C. would not be able to activate the transcription of genes required for cell division because the required gene regulatory proteins would be sequestered by unphosphorylated Rb.

Answer 18–15 Normally, cells divide only when they have grown to a certain size. This size control is clearly defective in the two mutant strains. In the case of gee cells, cell size increases without ever triggering cell division, suggesting that the mutant cell-cycle control protein has lost its ability to monitor cell size. It might, for example, now permanently inhibit MPF. In wee cells, on the other hand, the mutant control protein triggers cell division prematurely, before cells have grown to the appropriate size. This could be a control protein, for example, that no longer inhibits MPF, so that MPF becomes active prematurely. In fact, there is a yeast cell-cycle control protein called Wee1, which is a kinase that phosphorylates MPF on a site that causes its inactivation; yeast cells with a mutation in the *wee-1* gene have a short cell cycle and are small.

Answer 18–16

A. Only the cells that were in the S phase of their cell cycle (i.e., those cells making DNA) during the 30-minute labeling period contain any radioactive DNA.

B. Initially, mitotic cells contain no radioactive DNA because these cells were not engaged in DNA synthesis during the labeling period. Indeed it takes about two hours before the first labeled mitotic cells appear.

C. The initial rise of the curve corresponds to cells that were just finishing DNA replication when the radioactive thymidine was added. The curve rises as more labeled cells enter mitosis; the peak corresponds to those cells that just started S phase when the radioactive thymidine was added. The labeled cells then exit mitosis, being replaced by unlabeled mitotic cells again that were not yet in S phase during the labeling period. After 20 hours the curve starts rising again, because the labeled cells enter their second round of mitosis.

D. The intial two-hour lag before any labeled mitotic cells appear corresponds to the G_2 phase, which is the time between the end of S phase and the beginning of mitosis. The first labeled cells seen in mitosis were those that were just finishing S phase (DNA synthesis) when the radioactive thymidine was added.

Answer 18–17 In alcoholism, liver cells proliferate because the organ is overburdened and becomes damaged by the large amounts of alcohol that have to be metabolized. This need for more liver cells activates the control mechanisms that normally regulate proliferation. Unless badly damaged, the liver will usually shrink back to a normal size after the patient stops drinking excessively. In a liver tumor, in contrast, mutations abolish normal cell proliferation control, and as a result, cells divide and keep on dividing in an uncontrolled manner.

Answer 18–18 The plasma membrane of the cell that died by necrosis in Figure 18–21A is ruptured; a clear break is visible, for example, at a position corresponding to the 11 o'clock mark on a watch. The cell's contents, mostly membranous and cytoskeletal debris, are seen spilling into the surroundings through these breaks. The cytosol stains lightly, as most soluble components have been lost before the cell was fixed. In contrast, the cell that underwent programmed cell death in Figure 18–21B is surrounded by an intact membrane, and its cytosol is densely stained, indicating a normal concentration of cellular components. The cell's interior is remarkably different from a normal cell, however. Particularly characteristic are the large "blobs" that extrude from the nucleus, probably as the result of the breakdown of the nuclear lamina. The cytosol also contains many large, round membrane-bounded vesicles of unknown origin that are not normally seen in healthy cells. The pictures visually confirm the notion that necrosis involves cell lysis, whereas cells undergoing programmed cell death remain relatively intact until they are digested inside a normal cell.

Answer 18–19 Tumor supressor genes function normally as antiproliferation genes—that is, they encode proteins that stop the cell cycle. During normal cell division, these proteins must be turned off. If the proteins were overexpressed in all cells, it is likely that the machinery that keeps these proteins turned off would be overwhelmed, and cell division would stop. Thus, this cure for cancer might be successful, but the patient would be dead.

Chapter 19

Answer 19–1 The horizontal orientation of the microtubules will be associated with a horizontal orientation of cellulose fibers deposited in the cell walls. The growth of the cells will therefore be in a vertical direction, expanding the distance between the cellulose fibers without stretching these fibers. In this way, the stem will rapidly elongate; in a typical natural environment, this will hasten emergence from darkness into light.

Answer 19–2

A. As three collagen chains have to come together to form the triple helix, a defective molecule will impair assembly, even if normal collagen chains are present at the same time. Collagen mutations are therefore dominant, i.e., they have a deleteri-

ous effect even in the presence of a normal copy of the gene.

B. The different severity of the mutations results from a polarity in the assembly process. Collagen monomers assemble into the triple-helical rod starting from their amino-terminal ends. A mutation in an "early" glycine therefore allows only short rods to form, whereas a mutation farther downstream allows for longer, more normal rods.

Answer 19–3 The remarkable ability to swell and thus occupy a large volume of space depends on the negative charges. These attract a cloud of positive ions, chiefly Na^+, which by osmosis draw in large amounts of water, thus giving proteoglycans their unique properties. Uncharged polysaccharides such as cellulose, starch, and glycogen, by contrast, are easily compacted into fibers or granules.

Answer 19–4 Focal contact sites are common in connective tissue and in cell culture where cell crawling is observed. The forces driving crawling movement are generated by the actin cytoskeleton. In mature epithelium, focal contact sites are presumably less prominent because the cells are largely fixed in place and have no need to crawl over the basal lamina.

Answer 19–5 Suppose a cell is damaged so that its plasma membrane becomes leaky. Ions present in high concentration in the extracellular fluid, such as Na^+ and Ca^{2+}, then rush into the cell, and valuable metabolites leak out. If the cell were to remain connected to its healthy neighbors, these too would suffer from the damage. But the influx of Ca^{2+} into the sick cell causes its gap junctions to close immediately, effectively isolating the cell and preventing damage from spreading in this way.

Answer 19–6 Ionizing (high-energy) radiation tears through matter, knocking electrons out of their orbits and breaking chemical bonds. In particular, it creates breaks and other damage in DNA, and thus causes cells to arrest in the cell cycle (see Chapter 18). If the damage is so severe that it cannot be repaired, cells become permanently arrested and undergo apoptosis, i.e., they activate a suicide program.

Answer 19–7 Cells in the gut epithelium are exposed to a quite hostile environment, containing digestive enzymes and many other substances that vary drastically from day to day depending on the food intake of the organism. The epithelial cells also form a first line of defense against potentially hazardous compounds and mutagens that are ubiquitous in our environment. The rapid turnover protects the organism from harmful consequences, as wounded and sick cells are discarded. If an epithelial cell started to divide inappropriately as the result of a mutation, for example, it and its unwanted progeny would most often simply be discarded by natural disposal from the tip of a villus: even though such mutations must occur often, they rarely give rise to a cancer.

A neuron, on the other hand, lives in a very protected environment, insulated from the outside world. Its function depends on a complex system of connections with other neurons—a system that is created during development and is not easy to reconstruct if the neuron dies subsequently.

Answer 19–8 Every cell division generates one additional cell; so if the cells were never lost or discarded from the body, the number of cells in the body should equal the number of divisions plus one. The number of divisions is 10,000-fold greater than the number of cells because, in the course of a lifetime, 10,000 cells are discarded and replaced for every cell that is retained in the body.

Answer 19–9 Cells become different even in the absence of new transcription. Components that are asymmetrically distributed in the egg are inherited by the daughter cells in different amounts, making the two daughters different. Moreover, such components can, even in the absence of transcription, regulate gene expression by selectively modulating the translation of different mRNAs or by turning the activity of proteins on and off—e.g., by phosphorylation.

Answer 19–10 The first experiment shows that Sonic hedgehog is able to induce a horn, but it does not prove that it performs this role in normal development. The second experiment shows that Sonic hedgehog is not required for formation of a horn, but there could be at least two reasons for this: it might play no part in this process in normal development; or it might act in parallel with other signals that have similar actions and are sufficient to induce a horn even in the absence of Sonic hedgehog. This is in fact a common phenomenon: many developmental processes are controlled by multiple signals acting together in a redundant fashion.

Answer 19–11

A. False. Gap junctions are not connected to the cytoskeleton; their role is to provide cell-to-cell communication by allowing small molecules to pass from cell to cell.

B. True. Upon wilting, the turgor pressure in the plant cell is reduced, and consequently, the cell walls, having tensile but little compressive strength like a rubber tire, no longer provide rigidity.

C. False. Proteoglycans can withstand a large amount of compressive force but do not have a rigid structure. Their space-filling properties result from their tendency to absorb large amounts of water.

D. True.

E. True.

F. True. Stem cells stably express control genes that ensure that their daughter cells will be of the appropriate differentiated cell types.

Answer 19–12 Small cytosolic molecules, such as glutamic acid, cyclic AMP, and Ca^{2+} ions, pass readily through both gap junctions and plasmodesmata, whereas large cytosolic macromolecules, such as mRNA and G proteins, are excluded. Sonic hedgehog is a secreted protein and therefore does not ever encounter either type of junction. Plasma membrane phospholipids diffuse in the plane of the membrane through plasmodesmata because the plasma membranes of adjacent cells are continuous through these junctions. This traffic is not possible through gap junctions, because the membranes of the connected cells remain separate.

Answer 19–13 Plants are exposed to extreme changes in the environment, which often are accompanied by huge fluctuations in the osmotic properties of their surroundings. An intermediate filament network as we know it from animal cells would not be able to provide full osmotic support for cells: the sparse rivetlike attachment points would not be able to prevent the membrane from bursting in response to a huge osmotic pressure applied from the inside of the cell.

Answer 19–14 Action potentials can be passed from cell to cell through gap junctions. Indeed, heart muscle cells are connected this way, which ensures that they contract synchronously when stimulated. This mechanism of passing the signal from cell to cell is rather limited, however. As we have seen in Chapter 12, synapses are far more sophisticated and allow signals to be modulated and to be integrated with other signals received by the cell. Thus gap junctions are like simple soldered joints between electrical components, while synapses are like complex relay devices, enabling systems of neurons to perform computations.

Answer 19–15 To make jello, gelatin is boiled in water, which denatures the collagen fibers. Upon cooling, the disordered fibers form a tangled mess that solidifies into a gel. This gel actually resembles the collagen as it is initially secreted by fibroblasts, i.e., before the fibers become aligned and cross-bridged.

Answer 19–16 The evidence that DNA is the blueprint that specifies all the structural characteristics of an organism is based on observations that small changes in the DNA by mutation result in changes in the organism. While DNA provides the plans that specify structure, these plans need to be executed during development. This requires a suitable environment (a human baby would not fit into a stork's egg shell), suitable nourishment, suitable tools (such as the appropriate gene regulatory proteins required for early development), suitable spatial organization (such as the asymmetries in the egg cell required to allow for appropriate cell differentiation during the early cell divisions), and so on. Thus inheritance is not restricted to the passing on of the organism's DNA, because development requires appropriate conditions to be set up by the parent. Nevertheless, when all these conditions are met, the plans that are archived in the genome will determine the structure of the organism to be built.

Answer 19–17 One way to distinguish between mechanisms 1 and 2 is to modify cell B so that it cannot respond to a signal from A; for example, a mutation can be created in B to deprive it of the receptor for the signal molecule from A. This will block transmission of the signal to C if it depends on mechanism 2, but not if it depends on mechanism 1.

Answer 19–18 White blood cells circulate in the bloodstream and migrate into and out of the tissues in performance of their normal function of defending the body against infection: they are naturally invasive. Once mutations have occurred to upset the normal controls on production of these cells, there is no need for additional mutations to enable the cells to spread through the body. Thus the number of mutations that have to be accumulated in order to give rise to leukemia is less than for other types of cancer.

Answer 19–19 The shape of the curve reflects the need for multiple mutations to accumulate in a cell before a cancer results. If a single mutation were sufficient, the graph would be a straight horizontal line: the likelihood of occurrence of a particular mutation, and therefore of cancer, would be the same at any age. If two specific mutations were required, the graph would be a straight line sloping upward from the origin: the second mutation has an equal chance of occurring at any time, but will tip the cell into cancerous behavior only if the first mutation has already occurred in the same lineage; and the likelihood that the first mutation has already occurred will be proportional to the age of

the individual. The steeply curved graph shown in the figure goes up approximately as the fifth power of the age, indicating that five prior mutations have to be accumulated before a final, sixth mutation can turn the cell into a cancer cell.

Answer 19–20 During exposure to the carcinogen, mutations are induced, but the number of relevant mutations in any one cell is usually not enough to convert it directly into a cancer cell. Over the years, the cells that have become predisposed to cancer through the induced mutations accumulate progressively more mutations. Eventually, one of them will turn into a cancer cell. The long delay between exposure and cancer has made it extremely difficult to hold cigarette manufacturers or producers of industrial carcinogens legally responsible for the damage that is caused by their products.

Answer 19–21 By definition, a carcinogen is any substance that promotes the occurrence of one or more types of cancer. The sex hormones can therefore be classified as naturally occurring carcinogens. Although most carcinogens act by directly causing mutations, carcinogenic effects are also often exerted in other ways. The sex hormones increase both the rate of cell division and the numbers of cells in hormone-sensitive organs such as breast, uterus, and prostate. The first effect increases the mutation rate per cell, because mutations, regardless of environmental factors, are spontaneously generated in the course of DNA replication and chromosome segregation; the second effect increases the number of cells at risk. In these and possibly other ways, the hormones can favor development of cancer, even though they do not directly cause mutations.

Answer 19–22 We would be interested to know your answer to this question. E-mail (ecb@garland.com) or fax (212-308-9399) your answer to us; we may use the most thought-provoking answers in the next edition of this book.

Index

Page numbers in **boldface** refer to a major text discussion of the entry; page numbers with an F refer to a figure, with an FF to figures that follow consecutively; page numbers with a T refer to a table.

Page numbers in **boldface** refer to a major text discussion of the entry; F refers to a figure, FF to figures that follow consecutively; T refers to a table.

I-3

Page numbers in **boldface** refer to a major text discussion of the entry; F refers to a figure, FF to figures that follow consecutively; T refers to a table.

---

recombinant, *see* Recombinant DNA technology

Dolichol, 59F, 468, 468F

Double bonds, 45, 46F

in fatty acids, 55, 58F

Down's syndrome, 307, 566–567

Drosophila, as model eucaryote, 29–30, 29F

Drosophila development, **626–628,** 626FF

of eye, and Ey, 273, 273F

gene expression patterns, early embryo, 622–623, 623F

Hox genes, 627, 627F

as model for vertebrate, 626–627, 627F

sonic hedgehog gene family, 626–627, 626F

Drugs

antimitotic, 520–521

HIV, 300, 301F

psychoactive, 401

Dynamic instability, of microtubules, 522–523, 522FF

Dynamin, clathrin-coated pits, 464, 465F

Dyneins, 135F

ciliary, 529, 529FF

cytoplasmic, 525–526, 526F

E

E. coli, 23F

see also Bacteria

advantages for study, 278

division and growth rate, 279, 279F, 548, 548F

gene regulation in, 261–262, 261FF

genetic intermixing, demonstration *in vitro*, 281–282

genome, 280

mating, F plasmid and, 282–283, 282FF

as model organism, **25**

mutations

from replication errors, 280–281, 281F

in vitro selection, 280–281, 281F

transposons in, 289

E-site, of ribosome, 229–230, 229FF

Earth

age, 407

origin of organic molecules, 235–236, 236F

Eco RI (restriction nuclease), 316FF

EF-Tu, 176, 177F

EGF, *see* Epidermal growth factor

Egg, 33F, 304FF, 305, 563F

see also Cleavage; Fertilization

in cell-cycle studies, MPF and cyclin discovery, 575–577, 575FF

cleavage division, 575

developmental body pattern and, 624, 624F

Elastase, 148F

Elastin, 32F, 135F, 153–154, 153F

Electrical signaling

neurons, 394–400, 394FF

protozoans and plants, 387, 387F

Electrochemical gradient

calcium ions, 373T, 384, 502

defined, 377

Nernst equation, 393, 393F

proton, inner mitochondrial membrane, 416–417, 416F

sodium and potassium ions, 372–373, 373T, 377, 378, 379

transport across membranes and, 376–377, 376F

Electron carriers

cytochrome oxidase complex, 427–428, 428F

cytochromes, 426–427, 427F

heme group, 427, 427F

iron-sulfur centers, 426, 426F

NADH and NADPH as, 98–100, 99F

in photosynthetic electron-transfer chain, 433–436, 435F

quinone as, 425, 425F

respiratory chain complexes, 425–427, 425FF

Electron microscope

resolution, 3F, 5F

scanning, 5F, 8

transmission, 5F, 8

Electron microscopy

immunogold, 5F

preparation, 6

in study of cells, 19T

Electron shell

chemical bond formation and, 40–42, 41F

defined, 40

stability of atom and, 40–42, 41F

Electron transfers

in cytochrome oxidase complex, 427–428, 428F

electrical batteries, 415, 416F

energy released, 423, 425

hydrogen atom transfer and, 422, 422F

in *Methanococcus*, 441–442, 442F

in oxidation and reduction, 84–85, 84F

proton pump, coupling to, 415–416, 415F

in respiratory chain, 415, 415F

by metals in enzyme complexes, 425–427, 425FF

Electron-transport chains

in chloroplasts, 432, 433–434, 433FF, **435–436**

components, 414–415, 415F

electron transfer along, 415, 415F

outline, in chemiosmotic coupling, 409–410, 410F

proton gradient generation by, 124, 124F

proton pumping, 415–416, 415F

detailed mechanisms, **422–429,** 422FF

schematic, 422, 423F

redox potentials, 423, 424F, 425, 426F

Electrons

in activated carriers, 98–100, 99F, 100T

defined, in atoms, 38, 38F, 40F

harvesting light energy by chlorophyll, 434, 434F

interaction of atoms and, 39–42, 41FF

in ionic bonds, **42–43,** 42F, 43T

Electrophoresis, types, 163F

Elements

defined, 37

electron shells and reactivity, 40–42, 41F

in living organism, 39, 39FF

in nonliving world, 39, 39F

Elongation factor, 176, 177F

Endocrine glands

in cell signaling, 483, 483F, 485T

defined, 33F

Endocytic pathways, **472–477,** 473FF

see also Endocytosis

to lysosomes, summary, 477, 477F

outline, 463, 463F

Endocytosis

see also Endocytic pathways

general, 472

macromolecule sorting in endosomes, 475–476, 476F

phagocytosis, *see* Phagocytosis

pinocytosis, *see* Pinocytosis

receptor-mediated, **474–476,** 475FF

Endomembrane system

defined, 451

protein movement in, 453, 454F

protein synthesis in ER for, 458

vesicular traffic, summary, 463, 463F

Endoplasmic reticulum (ER), 7F, 13–15, 14FF, 18F

in cell division, 551

coated transport vesicles for and from, 465T

evolution, 451, 451F

membrane assembly and export, 355

placement in cell, microtubules and, 526–527, 527F

protein import

signal recognition mechanism, **459–460,** 460F

signal sequence, 454T

translocation, **460,** 460F

protein retention, 468–469, 469F

signal sequence, 454T

storage
    in animal cells, 125–126, 125F
    in plants, 126–127, 126F
Forensic medicine, PCR in, 335, 336F
Free energy
    see also ΔG; $\Delta G°$
    in chemical reactions, 85
    defined, 89, 90F
Formylmethionine, 231
Frog
    see also *Xenopus*
    development from adult DNA, 257, 257F
Fructose, formula, structure, 56F
Fructose 1,6-bisphosphate, in glycolysis, 111F, 112F
Fructose 6-phosphate
    in anabolism, 127F
    in glycolysis, 112F
Fungi
    H⁺ gradients and H⁺ ATPases, 384, 385F
    S. cerevisiae as model, 26F, 27
Fushi-tarazu gene expression, 623F

G

G, see Free energy; ΔG
ΔG
    of ATP hydrolysis to ADP and AMP, 104
    concentration of reactants and, 89, 91FF, 92–93
    defined, 89, 90F
    energetically favorable and unfavorable reactions, 89, 89FF
    negative and positive, 89, 90F
    redox reactions and, 422
$\Delta G°$
    defined, some values, 91F, 92
    equilibrium constant and, 91F, 92–93, 92T, 92FF, 166, 166F
    of glyceraldehyde 3-phosphate oxidation, 115
    of hydrolysis for some bonds, 91F
    from redox potentials, 423, 424F
    sequential reactions and, 93–94, 94F
    of some phosphate bonds, 117F, 118
G-protein-linked receptors, **493–504,** 493FF, 498T, 500T
    activation by ligand, 493–494, 494F
    cardiac cells, 496, 496F
    general system, 490, 491F
    photoreceptor cells, signaling pathway, 503, 504F
    rhodopsin, 503, 504F
    speed of signaling cascades, 502–503
    structure, superfamily of, 493, 493F
G proteins
    see also GTP-binding proteins; Ras protein; Transducin

inactivation, **494–495,** 495F
in selectivity of target cell response, 495–496
stimulatory or inhibitory, 497
subunits, and activation, **494,** 494F
targets
    adenylate cyclase, 496–497, 496F, **497–499,** 497FF, 498T
    ion channels, **495–496,** 496F
    phospholipase C, 497, 497F, **499–501,** 500T, 501F
G_0 phase, 581, 581F
G_1 checkpoint, 573, 574F
    arrest by p21, 580–581, 580F
    start, 581
G_1 phase, overview, 550, 550F
G_2 checkpoint, 573, 574F
G_2 phase, overview, 550, 550F
GABA, 400–401, 401F, 402T, 405T
GABA receptor, 400–401, 401F, 402T
GAGs (glycosaminoglycans), 604–605, 604F
Galactose, 56F
Gametes, *see* Germ cells
Gap junctions, 385, 608F, 612, 612F
Gastrulation, 622, 622F
GATA-1 (gene regulatory protein), 269F
Gel electrophoresis
    in DNA analysis, 317–318, 317F
    types, 163F
Gel-filtration chromatography, 162F
Gelsolin, 532F, 533
Gene
    addition, transgenic organisms, 340FF, 341
    bacterial and eucaryotic, compared, 219–220, 220F
    chromosomes and, 10
    defined, 21, 183
    in differentiated cells, 257, 257F
    DNA as, 185, 185F
    duplication, *see* Gene duplication
    early theories and studies, 184–185, 185F
    evolution, role of introns, 222
    exons, defined, 220, 220F
    families
        evolution of by gene duplication, 292–293, 293F
        isolation by cloning, 331–332, 331F
    function, mutants reveal, 339, 340F
    housekeeping, 258
    introns
        defined, 219–220, 220F
        in evolution, 223–224
    isolation
        by cloning, 327–329, 328FF
        of related, 331–332, 331F
        using cDNA, 329–330, 330FF

knockout, 340FF, 341–342
location on chromosomes, *in situ* hybridization, 324, 324F
protein, coding for, 188–189, 188F
regulation, *see* Gene regulation; Gene regulation, eucaryotes; Gene regulatory proteins
replacement, transgenic organisms, 340FF, 341–342
size (β-globin), 189, 189F
stability and DNA repair, 202, 204
transfer, *see* Gene transfer, bacteria; Gene transfer, eucaryotes
transgenic organisms, 340FF, 341–342
Gene duplication, 291–292, **292–294,** 293FF
    of exons, 294, 294F
    gene families, evolution of, 292–293, 293F
    protein domains, evolution of, 293–294, 294F
Gene expression
    see also Transcription; Translation
    alternative RNA splicing and, 222, 222F
    cell memory of, 271–273, 272F
    central dogma, from DNA to protein, 212, 212F
    chromatin packing and, 253–255, 254F
    constitutive, 262
    coordination of different, 261, 261F
        eucaryotes, 268–269, 270F
    DNA to protein, steps in eucaryotes, 234, 235F
    efficiency, differences among genes, 212, 212F
    pattern in early development, 622–624, 623F
    position effects, 255F, 256
    in tissue maintenance, 614–615, 615F
    transposons and, 296–297, 297F
Gene regulation, **257–273,** 257FF
    by activator, bacteria, 262, 262F, 265–266, 266FF
    cAMP and, 499, 500F
    bacteria, 261–262, 261FF
    gene regulatory proteins, **259–260,** 259FF
    operons, 261, 261F
    points of control in expression, 258, 258F
    Ras and, 507F
    regulatory DNA sequences, 259–260, 259F, 263, 265–266, 266FF
    by repressor proteins, bacteria, 261–262, 261FF
    transcriptional control, **259–273,** 259FF
    transposons, 290
Gene regulation, eucaryotes
    by activator, 265–266, 266FF
    chromatin structure and, 266–267
    combinatorial control, **268–271,** 268FF

Membrane transport proteins
see also Carrier proteins; Channel proteins; Ion channels; and individual transport proteins
carriers and channels, summarized, 372F, 374–375
defined, 371, 371F
examples, 135F
Memory, Ca²⁺ and, 502
Mental functions, synapses and integrations and, 401–403, 402F
Mesophyll cells, 598F
Messenger RNA, see mRNA
Metabolic pathways
$\Delta G°$ and sequential reactions, 93–94, 94F
enzyme catalysis and, 86, 87F
regulation
levels of, 172
need for, 128–129, 128F
negative and positive, 172–173, 172FF
stability of balance in, 129
Metabolic processes, segregation strategies, 447
Metabolism
efficiency of oxidative, 429
general scheme, 78, 79F
Metals
electron shells of, 42
in electron-transfer chain, 425–427, 425FF
Metamorphosis, programmed cell death, 585, 585F
Metaphase, 552, 552F, 555F, **557**, 557F
Metaphase plate, 557
division I meiosis, 564F, 565
homologous chromosomes and mitosis and meiosis compared, 563–564, 564F
Metastasis, 619, 619F
Methanococcus, genome, energy pathways, metabolism, 439F, 441–442, 442F
Methionine, 63F, 137F
Methyl group, 46F, 52
in activated carriers, 100T, 102
7-Methylguanosine, as mRNA cap, 218, 219F
Mice, transgenic, 342, 342F
Micelles, of detergent, 360–361, 361F
Microelectrodes, 388, 388F
Micrometers, defined, 3, 3F
Microscopy, see Electron microscopy; Light microscopy
Microsomes, isolation techniques, 160FF
Microtubule-associated proteins, 524
Microtubule-organizing center, centrosome as, 521–522, 521F
Microtubules, 16–17, 16FF, 18F, **518–529,** 519FF
actin filaments, compared, 530–531

cell polarity and, 523–524, 523FF
in cilia and flagella, 528–529, 529FF
colchicine and taxol on, 520
functions, summary, 518–519, 519F
growth
dynamic instability and, 522–523, 522F
from microtubule-organizing centers, 521–522, 521F
GTP cap, 522F, 523
kinetochore, 554F, 556–557, 556F
anaphase A, 558–559, 558F
locations in cell, 518–519, 519F
maintenance of, by assembly and disassembly, 519–520
in mitotic spindle, 551, 551F
classes, 556F
formation, **552–553,** 553F
motor proteins and intracellular transport, 525–526, 526F
MPF and, 576
nucleus and, diagram, 256F
organelle placement in cell and, 526–527, 527F
organizing centers for, 518, 519F
other cytoskeleton filaments and, 524
overview, 513–514, 514F
plant cell elongation and, 597, 597F, 600
polar, 553F, 556, 556F
anaphase B, 558, 558F, 559
plant cell division, 562, 562F
selective stabilization, 523–524, 523F
structure, polarity, 519, 520F
tubulin, 519, 520F
Microvilli, 32F, 530, 530F
Mimosa, leaf-closing response, 391F
Mitochondria, 7F, 449–450, 448FF, 449T, 450T
see also ATP synthase; Electron-transport chains; Oxidative phosphorylation; Proton pump
acetyl CoA production in, 118–119, 118FF
in cell division, 551
chloroplasts, compared, 430, 432, 432F
cristae, 412F, 413
energy-generating metabolism
details, **414–430,** 415FF
summary, 413–414, 414F
in eucaryotic evolution, 413
evolutionary origin, 23, 26F, 438, 452
function, general, 10, 12
genome and protein synthesis, 438
growth, 438
isolation techniques, 160FF
lipid and phospholipid import, 458
location in cells, 411, 411F
matrix, 411–413, 412F
membranes, inner and outer, 411–412, 412F

origin, 11F
overview, **10–12,** 10FF, 18F
oxidative phosphorylation, see Oxidative phosphorylation
in plants, role, 126, 126F
porin, 360, 360F
protein import
mechanism, 453, **457–458,** 457F
signal sequence, 454T
proteins, nuclear origin, 438
structure, organization, composition, 411–413, 412F
Mitosis, **552–560,** 552FF
see also individual stages
chromatid separation, **552–553,** 552F, 554FF, **556–557**
chromosome
assortment, paternal and maternal, 563, 564F
segregation, 555F, **557–559,** 558F
state during, 248–249, 248F
cytoskeletal structures in, 551, 551F
defined, 549
meiosis compared, 566F
mitotic spindle, see Mitotic spindle
nuclear envelope during, 518, 553, 554FF, 559–560, 559F
organelles during, 551
stages, overview, 552, 554FF
Mitotic chromosomes
see also Chromatin; Chromosome
banding pattern and karyotype, 247FF
defined, 248
kinetochore assembly, 553, 553F, **556,** 556F
level of chromatin packing, 252–253, 253F
at metaphase plate, **557,** 557F
segregation in anaphase, 555F, **557–559,** 558F
separation of sister chromatids, 249, 552–553, 552F, 554FF, **556–557**
Mitotic spindle, 249, 518, 519F
colchicine effect on, 520
cytokinesis timing, 560
cytoplasmic cleavage plane, 560–561
formation, **552–553,** 553F
in metaphase, **557,** 557F
MPF and, 576
overview, 551, 551F
plant cell division, 562, 562F
Molar solutions, defined, 39, 39F
Mole, defined, 39, 39F
Molecular evolution
measurement, rRNA sequences for, 439–440, 440F
phylogenetic tree, 440, 440F
Molecular switches
general principles, 492, 492F

GTP-binding proteins as, 176, 176F, 492F, 493

protein kinases and phosphorylation cascades, summary, 492F, 493

Molecular weight, defined, 38

Molecules, defined, 37, 43

Monellin, 135F

Monomers, defined, 65

Monosaccharides

see also Sugars

defined, 53, 56F

Motor neurons

embryonic development, 624–625, 624F

synapses and signal integration, 402–403, 402F

Motor proteins

allosteric "walking," 177, 178F

ATP hydrolysis, driven by, 177–178, 178F

examples, 135F

general, 525

intracellular transport by, **525–526,** 526F

of microtubules

in anaphase chromosome segregation, 558–559, 558F

cytoplasmic, 525–526, 526F

eucaryotic cilia and flagella, 529, 529FF

in muscle contraction, 539, 539FF

myosin, *see* Myosin

some examples, 177

Mouse

development, Hox genes, 627, 627F

β-globin gene cluster transposons, 296, 296F

as model organism, 29

MPF, **575–578,** 575FF

activation, 578, 578F

activity during cell cycle, 576, 576F

discovery, 575–576, 575FF

regulation by cyclin, 578, 579F

structure, 575F

Multicellular organisms

cell communication, need for, 481

cell division in, 547

cell number, control of, **582–589,** 583FF, 584T

cell proliferation, **582–583,** 583F, 584T

cell types, 31, 32FF

development from adult DNA, 257, 257F

epithelia, importance of, 606

genome constancy in all cells, 245, 246F, 257, 257F

Muscle cells

see also Cardiac muscle; Skeletal muscle; Smooth muscle

cardiac

acetylcholine effects, 486F, 496, 496F

G-protein activation, 496, 496F

gap junctions and electrical coupling, 612, 612F

differentiation, control, 269–271, 270FF

skeletal

acetylcholine effects, 486F

structure, 538, 538F

smooth, NO effects, 489–490, 490F

types, 33F

Muscle contraction

ATP hydrolysis, 539, 540F

control

skeletal muscle, **539–542,** 541FF

smooth, 542

troponin and tropomyosin, **541–542,** 542F

sliding filament mechanism, **538–539,** 539FF

speeds, 539

trigger, by Ca^{2+}, **539–542,** 541FF

Mutagenesis

genetically engineered, 339, 340F

to show protein function, 339

transgenic organisms, 340–342, 340FF

Mutations

in cancer, 199, 199F, 587–589, 588F, **619–620**

multiple, 620–621, 620F

from chemical modifications of nucleotides, 202, 203F

consequences, 184, 198–199, 199F

from DNA repair, lack of, 200, 201F

in evolution of cells, 21–22

in germ cells, 199–200, 199F, 619

Hox genes, 627, 627F

rate per gene and human genome, 620

selection for in culture, 280–281, 281F

in sickle-cell anemia, 198–199, 199F

in somatic cells, 199–200, 199F, 619

sonic hedgehog genes, 626–627, 626F

transposons and, 296–297, 297F

Myoblasts, 269–270, 270F

MyoD, 270–271, 270FF

Myofibrils

muscle contraction model, 539, 539FF

structure, 538, 538F

Myoglobin, 150F

Myosin, 135F, **536–537,** 536FF

in cell crawling, 536

coiled-coil, 145

as motor protein, 178

of muscle, 536FF, 537

phosphorylation, control by, 542

Myosin filaments, **536–537,** 536FF

contractile ring, 555F, **561,** 561F

in M phase, 551, 551F

in muscle contraction model, 539, 539FF

in sarcomeres, 538, 538F

N

Na$^+$-K$^+$ ATPase, *see* Na$^+$-K$^+$ pump

Na$^+$-K$^+$ pump, 357T, **378–380,** 378FF, 385F, 386T

membrane potential and, 392

pumping cycle, model, 379–380, 379F

uses

for coupled transport, 380–381, 380F

for osmotic balance, 381–383, 382F

in transport, 377–378

NaCl, *see* Sodium chloride

NADH dehydrogenase complex, 414, 415F

electron carriers within, 426, 426F

NADH/NAD$^+$

as activated carrier, 98, 100, 100T

in citric acid cycle, 119–121, 120F

detail, 122FF

electron donation, structural formulas, 413, 413F, 415

electron-transport chain and, 124, 124F

energy from oxidation to O$_2$, 423, 425

in fermentation, 114, 115F

in glycolysis

detail, 113F

outline, 110–111, 111F, 114

steps 6 and 7, 114–118, 116FF

steps 6 and 7, energy changes, 117F, 118

in metabolism, general scheme, 109–110, 109F

in pyruvate oxidation, to acetyl CoA, 118, 118F

as redox pair, 423, 424F, 425

NADP reductase, 435F

as activated carrier, 98–100, 99F

in carbon-fixation cycle, 437–438, 437F

in photosynthesis, 435–436, 435F

structures, 99F

Nanometers, defined, 3F, 6

Natural selection

bacteria in culture, 280–281, 281F

cancer and, 619–620

of proteins of stable conformation, 147

Nematode (*Caenorhabditis elegans*), 29, 29F

Nernst equation, 393, 393F

Nerve cells, 613, 614F

action potential, *see* Action potentials

development

cell death in, 585, 586F

contact-dependent signaling, 484, 484F

Page numbers in **boldface** refer to a major text discussion of the entry; F refers to a figure, FF to figures that follow consecutively; T refers to a table.

Page numbers in **boldface** refer to a major text discussion of the entry; F refers to a figure, FF to figures that follow consecutively; T refers to a table.

Page numbers in **boldface** refer to a major text discussion of the entry; F refers to a figure, FF to figures that follow consecutively; T refers to a table.

I-29